Applications of Nanofluid Transportation and Heat Transfer Simulation

Mohsen Sheikholeslami
Babol Noshirvani University of Technology, Iran

A volume in the Advances in Chemical and Materials Engineering (ACME) Book Series

Published in the United States of America by
IGI Global
Engineering Science Reference (an imprint of IGI Global)
701 E. Chocolate Avenue
Hershey PA, USA 17033
Tel: 717-533-8845
Fax: 717-533-8661
E-mail: cust@igi-global.com
Web site: http://www.igi-global.com

Library of Congress Cataloging-in-Publication Data

Names: Sheikholeslami, Mohsen, 1988- author.
Title: Applications of nanofluid transportation and heat transfer simulation / by Mohsen Sheikholeslami.
Description: Hershey, PA : Engineering Science Reference, [2019]
Identifiers: LCCN 2018031663| ISBN 9781522575955 (h/c) | ISBN 9781522575962 (eISBN)
Subjects: LCSH: Nanofluids--Simulation methods. | Heat--Transmission.
Classification: LCC TJ853.4.M53 S54 2019 | DDC 620.1/064--dc23 LC record available at https://lccn.loc.gov/2018031663

This book is published in the IGI Global book series Advances in Chemical and Materials Engineering (ACME) (ISSN: 2327-5448; eISSN: 2327-5456)

British Cataloguing in Publication Data
A Cataloguing in Publication record for this book is available from the British Library.

Advances in Chemical and Materials Engineering (ACME) Book Series

J. Paulo Davim
University of Aveiro, Portugal

ISSN:2327-5448
EISSN:2327-5456

Mission

The cross disciplinary approach of chemical and materials engineering is rapidly growing as it applies to the study of educational, scientific and industrial research activities by solving complex chemical problems using computational techniques and statistical methods.

The **Advances in Chemical and Materials Engineering (ACME) Book Series** provides research on the recent advances throughout computational and statistical methods of analysis and modeling. This series brings together collaboration between chemists, engineers, statisticians, and computer scientists and offers a wealth of knowledge and useful tools to academics, practitioners, and professionals through high quality publications.

Coverage

- Wear of Materials
- Multifuncional and Smart Materials
- Chemical Engineering
- Fracture Mechanics
- Thermo-Chemical Treatments
- Computational methods
- Electrochemical and Corrosion
- Biomaterials
- Sustainable Materials
- Coatings and surface treatments

IGI Global is currently accepting manuscripts for publication within this series. To submit a proposal for a volume in this series, please contact our Acquisition Editors at acquisitions@igi-global.com or visit: https://www.igi-global.com/publish/.

The Advances in Chemical and Materials Engineering (ACME) Book Series (ISSN 2327-5448) is published by IGI Global, 701 E. Chocolate Avenue, Hershey, PA 17033-1240, USA, www.igi-global.com. This series is composed of titles available for purchase individually; each title is edited to be contextually exclusive from any other title within the series. For pricing and ordering information please visit https://www.igi-global.com/book-series/advances-chemical-materials-engineering/73687. Postmaster: Send all address changes to above address.

Titles in this Series

For a list of additional titles in this series, please visit: www.igi-global.com/book-series

The Geometry of Higher-Dimensional Polytopes
Gennadiy Vladimirovich Zhizhin (Russian Academy of Sciences, Russia)
Engineering Science Reference • copyright 2019 • 286pp • H/C (ISBN: 9781522569688) • US $195.00 (our price)

Composites and Advanced Materials for Industrial Applications
K. Kumar (Birla Institute of Technology, India) and J. Paulo Davim (University of Aveiro, Portugal)
Engineering Science Reference • copyright 2018 • 402pp • H/C (ISBN: 9781522552161) • US $225.00 (our price)

Emerging Synthesis Techniques for Luminescent Materials
Ratnesh Tiwari (Bhilai Institute of Technology, India) Vikas Dubey (Bhilai Institute of Technology, India) and Sanjay J. Dhoble (Rashtrasant Tukadoji Maharaj Nagpur University, India)
Engineering Science Reference • copyright 2018 • 505pp • H/C (ISBN: 9781522551706) • US $215.00 (our price)

Handbook of Research on Ergonomics and Product Design
Juan Luis Hernández Arellano (Autonomous University of Ciudad Juárez, Mexico) Aide Aracely Maldonado Macías (Autonomous University of Ciudad Juárez, Mexico) Juan Alberto Castillo Martínez (University of Rosario, Colombia) and Porfirio Peinado Coronado (Autonomous University of Ciudad Juárez, Mexico)
Engineering Science Reference • copyright 2018 • 446pp • H/C (ISBN: 9781522552345) • US $235.00 (our price)

Mechanical Properties of Natural Fiber Reinforced Polymers Emerging Research and Opportunities
Sarah S. Gebai (International University of Beirut, Lebanon) Ali M. Hallal (International University of Beirut, Lebanon) and Mohammad S. Hammoud (International University of Beirut, Lebanon)
Engineering Science Reference • copyright 2018 • 228pp • H/C (ISBN: 9781522548379) • US $165.00 (our price)

Advanced Solid Catalysts for Renewable Energy Production
Sergio González-Cortés (Oxford University, UK) and Freddy Emilio Imbert (Univsersidad de Los Andes, Venezuela)
Engineering Science Reference • copyright 2018 • 520pp • H/C (ISBN: 9781522539032) • US $225.00 (our price)

Production, Properties, and Applications of High Temperature Coatings
Amir Hossein Pakseresht (University of Tehran, Iran & Materials and Energy Research Center, Iran)
Engineering Science Reference • copyright 2018 • 557pp • H/C (ISBN: 9781522541943) • US $235.00 (our price)

Energetic Materials Research, Applications, and New Technologies
Rene Francisco Boschi Goncalves (Federal University of Para, Brazil) José Atilio Fritz Fidel Rocco (Aeronautics Institute of Technology, Brazil) and Koshun Iha (Aeronautics Institute of Technology, Brazil)
Engineering Science Reference • copyright 2018 • 367pp • H/C (ISBN: 9781522529033) • US $225.00 (our price)

IGI Global
DISSEMINATOR OF KNOWLEDGE

701 East Chocolate Avenue, Hershey, PA 17033, USA
Tel: 717-533-8845 x100 • Fax: 717-533-8661
E-Mail: cust@igi-global.com • www.igi-global.com

Table of Contents

Preface

In this book, I provide readers the various applications of nanofluid flow and heat transfer. Different numerical and analytical methods have been employed to find the solution of governing equations.

The first chapter of this book deals with the necessary fundamentals of nanotechnology. Nanofluids are fluids containing the solid nanometer size particle dispersion. Two main methods were introduced namely single-phase and two-phase modeling. In first method the combination of nanoparticle and base fluid is considered as a single-phase mixture with steady properties and the second method the nanoparticle properties and behaviors are considered separately from the base fluid properties and behaviors. Moreover, nanofluid flow and heat transfer can be studied in presence of thermal radiation, electric field, magnetic field, and porous media.

In the second chapter, influence of uniform magnetic field on nanofluid flow and heat transfer is presented. Natural convection under the influence of a magnetic field has great importance in many industrial applications such as crystal growth, metal casting and liquid metal cooling blankets for fusion reactors. Existence of magnetic field has a noticeable effect on heat transfer reduction under natural convection while in many engineering applications increasing heat transfer from solid surfaces is a goal. At this circumstance, the use of nanofluids with higher thermal conductivity can be considered as a promising solution.

The third chapter deals with effect of non-uniform magnetic field on magnetic nanofluid behavior. Magnetic nanofluid (Ferrofluid) is a magnetic colloidal suspension consisting of base liquid and magnetic nanoparticles with a size range of 5–15 nm in diameter coated with a surfactant layer. The effect of magnetic field on fluids is worth investigating due to its numerous applications in wide range of fields. The study of interaction of the magnetic field or the electromagnetic field with fluids have been documented e.g. among nuclear fusion, chemical engineering, medicine and transformer cooling. The goal of nanofluid is to achieve the highest possible thermal properties at the smallest possible concentrations by uniform dispersion and stable suspension of nano particles in host fluids. In chapter, influence of external magnetic field on ferrofluid flow and heat transfer is investigated. Both effects of Ferrohydrodynamic (FHD) and Magnetohydrodynamic (MHD) have been taken in to account. So, effects of Lorentz and Kelvin forces on hydrothermal behavior are examined.

Application of nanofluid for saving thermal energy in energy storage systems is reported in Chapter 4. Latent Heat Thermal Energy Storage Systems (LHTESS) which work based on energy storage and retrieval during solid-liquid phase change is used to establish balance between energy supply and demand. LHTESS stores and retrieves thermal energy during solid-liquid phase change, while in SHTESS phase change doesn't occur during the energy storage and retrieval process. LHTESS has a lot of advantages in comparison to SHTESS, The most important one is storing a large amount of energy during phase change process, which makes the energy storage density in LHTESS much higher than SHTESS. Because of this property, LHTESS have a wide application in different cases, such as solar Air Dryer, HVAC Systems, Electronic Chip Cooling and engine heat recovery. The main restriction for these systems is thermal conductivity weakness of common PCMs. In this chapter, the method of adding nanoparticles to pure PCM and making Nano-Enhanced Phase Change Material (NEPCM) and using fin with suitable array are presented to accelerate solidification process. The numerical approach which is used in this chapter is Standard Galerkin Finite Element Method.

Behavior of nanofluid in a porous media is studied in Chapter 5. The study of convective heat transfer in fluid-saturated porous media has many important applications in technology geothermal energy recovery such as oil recovery, food processing, fiber and granular insulation, porous burner and heater, combustion of low-calorific fuels to diesel engines and design of packed bed reactors. Also the flow in porous tubes or channels has been under considerable attention in recent years because of its various applications in biomedical engineering, transpiration cooling boundary layer control and gaseous diffusion. Nanofluids are produced by dispersing the nanometer-scale solid particles into base liquids with low thermal conductivity such as water, ethylene glycol (EG), oils. In this chapter, nanofluid hydrothermal behavior in porous media has been investigated.

In Chapter 6, various shapes of nanoparticles have been used to find the best shape for heat transfer enhancement. Shape of nanoparticle can changes the thermal conductivity of nanofluid. So, effect of shape factor on nanofluid flow and heat transfer has been reported in this chapter. Governing equations are presented in vorticity stream function formulation. Control volume based finite element method (CVFEM) is utilized to obtain the results. Results indicate that Platelet shape has the highest rate of heat transfer.

Chapter 7 gives a complete account of electric field effects on nanofluid forced convection heat transfer. In this chapter, effect of electric field on forced convection heat transfer of nanofluid is presented. The governing equation are derived and presented in vorticity stream function formulation. Control Volume based Finite Element Method (CVFEM) is employed to solve the final equations. Results indicate that the flow style is depends on supplied voltage and this effect is more sensible for low Reynolds number.

Non-Darcy model for porous media is introduced in Chapter 8. Both natural and forced convection heat transfer can be analyzed with this model. The governing equations in forms of vorticity stream function are derived and then they are solved via Control Volume based Finite Element Method (CVFEM). Effect of Darcy number on nanofluid flow and heat transfer is examined.

Effect of Magnetic field dependent (MFD) viscosity on nanofluid treatment is shown in Chapter 9. In this chapter, effect of magnetic field dependent (MFD) viscosity on free convection heat transfer of nanofluid in an enclosure is investigated. Single phase nanofluid model is utilized considering Brownian motion. Control Volume based Finite Element Method is applied to simulate this problem. The effects of viscosity parameter, Hartmann number and Rayleigh number on hydrothermal behavior have been examined.

Chapter 10 deals with effect of melting surface on nanofluid convective heat transfer. In this chapter, effect of melting surface heat transfer on Magnetohydrodynamic nanofluid free convection is analyzed by means of Control Volume based Finite Element Method (CVFEM). KKL model is taken in to account to obtain viscosity and thermal conductivity of CuO-water nanofluid. Roles of melting parameter, nanofluid volume fraction, Hartmann and Rayleigh numbers are illustrated.

Several examples exist in this book which helps the reader to understand all scientific topics. The user (bachelor's, master's and PhD students, university teachers and even in research centers in different fields) can encounter such systems in confidently. In the different chapters of the book, not only are the basic ideas of the methods broadly discussed, but also applied examples are practically solved by the proposed methodology.

Mohsen Sheikholeslami
Babol Noshirvani University of Technology, Iran

Chapter 1
Nanotechnology as Effective Passive Technique for Heat Transfer Augmentation

ABSTRACT

Nanofluids are fluids containing the solid nanometer-sized particle dispersion. Two main methods are introduced in this chapter, namely single-phase and two-phase modeling. In first method, the combination of nanoparticle and base fluid is considered as a single-phase mixture with steady properties, and in the second method, the nanoparticle properties and behaviors are considered separately from the base fluid properties and behaviors. Moreover, nanofluid flow and heat transfer can be studied in the presence of thermal radiation, electric field, magnetic field, and porous media. In this chapter, a definition of nanofluid and its applications have been presented.

1. INTRODUCTION

Nanofluids are produced by dispersing the nanometer-scale solid particles into base liquids with low thermal conductivity such as water, ethylene glycol (EG), oils, etc. Control of heat transfer in many energy systems is crucial due to the increase in energy prices. In recent years, nanofluids technology is proposed and studied by some researchers experimentally or numerically to control heat transfer in a process. The nanofluid can be applied to engineering problems, such as heat exchangers, cooling of electronic equipment and chemical processes. There are two ways for simulation of nanofluid: single phase and two phase. In first method, researchers assumed that nanofluids treated as the common pure fluid and conventional equations of mass, momentum and energy are used and the only effect of nanofluid is its thermal conductivity and viscosity which are obtained from the theoretical models or experimental data. These researchers assumed that nanoparticles are in thermal equilibrium and there aren't any slip velocities between the nanoparticles and fluid molecules, thus they have a uniform mixture of nanoparticles. In second method, researchers assumed that there are slip velocities between nanoparticles and fluid molecules. So the volume fraction of nanofluids may not be uniform anymore and there would be

DOI: 10.4018/978-1-5225-7595-5.ch001

a variable concentration of nanoparticles in a mixture. There are several numerical and semi analytical methods which have been used by several authors in order to simulate nanofluid flow and heat transfer.

1.1. Definition of Nanofluid

Low thermal conductivity of conventional heat transfer fluids such as water, oil, and ethylene glycol mixture is a serious limitation in improving the performance and compactness of many engineering equipment such as heat exchangers and electronic devices. To overcome this disadvantage, there is strong motivation to develop advanced heat transfer fluids with substantially higher conductivity. An innovative way of improving the thermal conductivities of fluids is to suspend small solid particles in the fluid. Various types of powders such as metallic, non-metallic and polymeric particles can be added into fluids to form slurries. The thermal conductivities of fluids with suspended particles are expected to be higher than that of common fluids. Nanofluids are a new kind of heat transfer fluid containing a small quantity of nanosized particles (usually less than 100 nm) that are uniformly and stably suspended in a liquid. The dispersion of a small amount of solid nanoparticles in conventional fluids changes their thermal conductivity remarkably. Compared to the existing techniques for enhancing heat transfer, the nanofluids show a superior potential for increasing heat transfer rates in a variety of cases (Khanafer, Vafai, & Lightstone, 2003).

1.2. Model Description

In the literature, convective heat transfer with nanofluids can be modeled using mainly the two-phase or single approach. In the two-phase approach, the velocity between the fluid and particles might not be zero (Li, 2000) due to several factors such as gravity, friction between the fluid and solid particles, Brownian forces, Brownian diffusion, sedimentation and dispersion. In the second approach, the nanoparticles can be easily fluidized and therefore, one may assume that the motion slip between the phases, if any would be considered negligible (Xuan & Roetzel, 2000). The latter approach is simpler and more computationally efficient.

1.3. Conservation Equations

1.3.1. Single Phase Model

Although nanofluids are solid–liquid mixtures, the approach conventionally used in most studies of natural convection handles the nanofluid as a single-phase (homogenous) fluid. In fact, due to the extreme size and low concentration of the suspended nanoparticles, the particles are assumed to move with same velocity as the fluid. Also, by considering the local thermal equilibrium, the solid particle–liquid mixture may then be approximately considered to behave as a conventional single-phase fluid with properties that are to be evaluated as functions of those of the constituents. The governing equations for a homogenous analysis of natural convection are continuity, momentum, and energy equations with their density, specific heat, thermal conductivity, and viscosity modified for nanofluid application. The specific governing equations for various studied enclosures are not shown here and they can be found in different references (Abu-Nada & Oztop, 2009). It should be mentioned that sometime this assumption is not correct. For example, Ding and Wen (2005) this assumption may not always remain true for a nanofluid. They investigated the particle migration in a nanofluid for a pipe flow and stated that at Peclet numbers exceeding 10 the particle distribution is significantly non-uniform. Nevertheless, many studies have performed the numerical simulation using single-phase assumption and reported acceptable results for the heat transfer and hydrodynamic properties of the flow.

1.3.2. Two-Phase Model

Several authors have tried to establish convective transport models for nanofluids (Akbarinia & Laur, 2009). Nanofluid is a two-phase mixture in which the solid phase consists of nano-sized particles. In view of the nanoscale size of the particles, it may be questionable whether the theory of conventional two-phase flow can be applied in describing the flow characteristics of nanofluid. On the other hand, several factors such as gravity, friction between the fluid and solid particles and Brownian forces, the phenomena of Brownian diffusion, sedimentation, and dispersion may affect a nanofluid flow. Consequently, the slip velocity between the fluid and particles cannot be neglected for simulating nanofluid flows. Since the two phase approach considers the movement between the solid and fluid molecule, it may have better prediction in nanofluid study. To fully describe and predict the flow and behavior of complex flows, different multiphase theories have been proposed and used. The large number of published articles concerning multiphase flows typically employed the Mixture Theory to predict the behavior of nanofluids (Mirmasoumi & Behzademehr, 2008). A comprehensive survey of convective transport in nanofluids was made by Buongiorno (2006), using a model in which Brownian motion and thermophoresis are accounted for. Buongiorno developed a two-component four-equation non-homogeneous equilibrium model for mass, momentum, and heat transfer in nanofluids. The nanofluid is treated as a two-component mixture (base fluidþnanoparticles) with the following assumptions: No chemical reactions; Negligible external forces; Dilute mixture ($\phi = 1$); Negligible viscous dissipation; Negligible radiative heat transfer; Nanoparticle and base fluid locally in thermal equilibrium. Invoking the above assumptions, the following equations represent the mathematical formulation of the non- homogenous single phase model for the governing equations as formulated by Buongiorno (2006):

1.3.2.1. Continuity Equation

$$\nabla . v = 0 \tag{1}$$

where v is the velocity

1.3.2.2. Nanoparticle Continuity Equation

$$\frac{\partial \phi}{\partial t} + v.\nabla \phi = \nabla . \left(D_B \nabla \phi + D_T \frac{\nabla T}{T} \right) \tag{2}$$

Here ϕ is nanoparticle volume fraction, D_B is the Brownian diffusion coefficient given by the Einstein–Stokes's equation:

$$D_B = \frac{k_B T}{3\pi\mu d_p} \tag{3}$$

where μ is the viscosity of the fluid, d_p is the nanoparticle diameter, $k_B = 1.385 \times 10^{-23}$ is Boltzmann constant and D_T is the thermophoretic diffusion coefficient, which is defined as

$$D_T = \left(\frac{\mu}{\rho}\right)\left(0.26\frac{k}{k+k_p}\right) \tag{4}$$

In Equation (4), k and k_p are the thermal conductivity of the fluid and particle materials, respectively.

1.3.2.3. Momentum Equation

$$v.\nabla v = -\frac{1}{\rho_{nf}}\nabla p + \nabla.\tau + g \tag{5}$$

where

$$\tau = -\mu_{nf}\left(\nabla v + \left(\nabla v\right)^t\right) \tag{6}$$

where the superscript 't' indicates the transpose of ∇v. Also p is pressure.

1.3.2.4. Energy Equation

$$v.\nabla T = \nabla\left(\alpha_{nf}\nabla T\right) + \frac{\rho_p c_p}{\rho_{nf} c_{nf}}\left(D_B \nabla\phi.\nabla T + D_T\frac{\nabla T.\nabla T}{T}\right) \tag{7}$$

where ϕ and T are nanoparticle concentration and temperature of nanofluid, respectively.

This nanofluid model can be characterized as a 'two-fluid' (nanoparticles + base fluid), four-equation (mass, momentum, energy), non-homogeneous (nanoparticle/fluid slip velocity allowed) equilibrium (nanoparticle/fluid temperature differences not allowed) model. Note that the conservation equations are strongly coupled. That is, v depends on ϕ via viscosity; ϕ depends on T mostly because of thermophoresis; T depends on ϕ via thermal conductivity and also via the Brownian and thermophoretic terms in the energy equation: ϕ and T obviously depends on v because of the convection terms in the nanoparticle continuity and energy equations, respectively.

In a numerical study by Behzadmehr et al. (2007) for the first time a two-phase mixture model were implemented to investigate the behavior of Cu–water nanofluid in a tube and the results were also compared with previous works using a single-phase approach. The authors claimed that the simulation done by assuming that basefluid and particles behave separately possessed results that are more precise compared to the previous computational modeling. They implemented the mixture theory for their work. It was suggested that the continuity, momentum and energy equations be written for a mixture of fluid and a solid phase. Some assumptions were also stated for the model such as a strong coupling between two phases and the fluid being closely followed by the particles with each phase owning a different velocity leading to a term called slip velocity of nanoparticles as in Equation (8):

$$V_{pf} = V_P - V_f = \frac{\rho_p d_p^2}{18\mu_f f_{drag}} \frac{\left(\rho_p - \rho_m\right)}{\rho_p} a, a = g - \left(V_m.\nabla\right) V_m, f_{drag} = \begin{cases} 1 + 0.15\,\mathrm{Re}_p^{0.687} \\ 0.0183\,\mathrm{Re}_p \end{cases} \tag{8}$$

The conservation equations (continuity, momentum and energy respectively) will be written for the mixture as follows:

$$\nabla.\left(\rho_m V_m\right) = 0 \tag{9}$$

$$\nabla.\left(\rho_m V_m V_m\right) = -\nabla P_m + \nabla.\left[\tau - \tau_t\right] + \rho_m g + \nabla.\left(\sum_{k=1}^{n} \phi_k \rho_k V_{dr,k} V_{dr,k}\right) \tag{10}$$

$$\nabla.\left(\phi_p V_k \left(\rho_k h_k + p\right)\right) = \nabla.\left(k_{eff} \nabla T - C_p \rho_m vt\right) \tag{11}$$

where $V_{dr,p}$ is the particle draft velocity that is related to the slip velocity and is defined as:

$$V_{dr,p} = V_P - V_f = V_{pf} - \sum_{k=1}^{n} \frac{\phi_k \rho_k}{\rho_m} V_{fk} \tag{12}$$

1.4. Physical Properties of the Nanofluids for Single-Phase Model

Base nanofluid properties have been published over the past few years in literature. However, only recently have some data on temperature-dependent properties been provided, even though they are only for nanofluid effective thermal conductivity and effective absolute viscosity.

1.4.1. Density

In the absence of experimental data for nanofluid densities, constant-value temperature independent values, based on nanoparticle volume fraction, are used:

$$\rho_{nf} = \rho_f(1-\phi) + \rho_p\phi \tag{13}$$

1.4.2. Specific Heat Capacity

It has been suggested that the effective specific heat can be calculated using the following equation as reported in Pak and Cho, (1998) as

$$\left(C_p\right)_{nf} = \left(C_p\right)_f(1-\phi) + \left(C_p\right)_p\phi \tag{14}$$

Other authors suggest an alternative approach based on heat capacity concept (Eastman et al., 1999):

$$\left(\rho C_p\right)_{nf} = \left(\rho C_p\right)_f(1-\phi) + \left(\rho C_p\right)_p\phi \tag{15}$$

These two formulations may of course lead to different results for specific heat. Due to the lack of experimental data, both formulations are considered equivalent in estimating nanofluid specific heat capacity (Palm et al., 2006).

1.4.3. Thermal Expansion Coefficient

Thermal expansion coefficient of nanofluid can be obtained as follows (Khanafer, Vafi, & Lightstone, 2003):

$$\left(\rho\beta\right)_{nf} = \left(\rho\beta\right)_f(1-\phi) + \left(\rho\beta\right)_p\phi \tag{16}$$

1.4.4. The Electrical Conductivity

The effective electrical conductivity of nanofluid was presented by Maxwell (1904) as below:

$$\sigma_{nf}/\sigma_f = 1 + 3\left(\sigma_P/\sigma_f - 1\right)\phi / \left(\left(\sigma_P/\sigma_f + 2\right) - \left(\sigma_P/\sigma_f - 1\right)\phi\right) \tag{17}$$

1.4.5. Dynamic Viscosity

Various models have been suggested to model the viscosity of a nanofluid mixture that take into account the percentage of nanoparticles suspended in the base fluid. The classic Brinkman model (1952) seems to be a proper one which has been extensively used in the studies on numerical simulation concerning nanofluids. Equation (1) shows the relation between the nanofluid viscosity, basefluid viscosity as well as the nanoparticle concentration in this model.

$$\mu_{nf} = \mu_f / \left(1-\phi\right)^{2.5} \tag{18}$$

However, in some recent computational studies, other models have been selected to be used in the numerical process, like the work done by Abu-neda and Chamkha (2010) to investigate the convection of CuO–ethylene glycol–water nanofluid in a an enclosure where Namburu correlation for viscosity (Namburu et al., 2007) were applied:

$$\log\left(\mu_{nf}\right) = Ae^{-BT} \tag{19}$$

where

$$\begin{aligned} A &= 1.837\phi^2 - 29.643\phi + 165.65 \\ B &= 4\times10^{-6}\phi^2 - 0.001\phi + 0.0186 \end{aligned} \tag{20}$$

In their study, the results where compared to that of viscosity modeled by Brinkman. It was outlined that as far as a value for normalized average Nusselt number for the fluid is concerned, for various values of Rayleigh number, Brinkman model owns a prediction of higher value compared to that for Namburu model showing the notable role of viscosity model used in the calculations. The authors also state that a combination of different models might as well be implemented that will show different dependence on volume concentration as well as the geometry aspect ratio yet along with the limitation that the models include only the ones mentioned in the study. Other studies have also shown that different models might lead to different results, like that by a number of suggested relations for viscosity models used in numerical studies are also presented in table 1.

1.4.6. Thermal Conductivity

Different nanofluid models based on a combination of the different formulas for the thermal conductivity adopted in the studies of natural convection are summarized in Table 2. Also Table 3 demonstrates values of thermo physical properties for different materials used as suspended particles in nanofluids.

2. SIMULATION OF NANOFLUID FLOW AND HEAT TRANSFER

Several semi analytical and numerical methods have been applied successfully in order to simulate nanofluid flow and heat transfer. In following sections we presented these works.

2.1. Semi Analytical Methods

Forced convective heat transfer to Sisko nanofluid past a stretching cylinder in the presence of variable thermal conductivity was presented by Khan and Malik (2016). They used Homotopy Analysis Method (HAM) to solve the governing equations. They found that the curvature parameter assisted the temperature as well as concentration profiles. Momentum and heat transfer characteristics from heated spheroids in water based nanofluids has been investigated by Sasmal and Nirmalkar (2016). They showed that smaller the nanoparticles size better in heat transfer at low Reynolds number and volume fraction. Hayat

Table 1. Different models for viscosity of nanofluids used in simulation

Model	Equation
Einstein model (1956)	$\mu_{nf} = \left(2.5\phi + 1\right)\mu_f,\ \phi < 0.05$
Pak and Cho's Correlation (1998)	$\mu_{nf} = \mu_f\left(1 + 39.11\phi + 533.9\phi^2\right)$
Jang et al. model (2007)	$\mu_{nf} = \left(2.5\phi + 1\right)\mu_f\left[1 + \eta\left(d_p / H\right)^{-2\varepsilon}\phi^{2/3}\left(\varepsilon + 1\right)\right]$
Koo and Kleinstreuer (2004)	$\mu_{nf} = 5\times10^4\beta\phi\rho_f\sqrt{\frac{k_B T}{d_p\rho_p}}f\left(T,\phi\right), \begin{cases}\beta = 0.0137\left(100\phi\right)^{-0.8229} & for\,\phi < 1\% \\ \beta = 0.0011\left(100\phi\right)^{-0.7272} & for\,\phi > 1\%\end{cases}$
Maiga model (2005)	$\mu_{nf} = \mu_f\left(1 + 7.3\phi + 123\phi^2\right)$
Brownian model (Orozco, 2005)	$\mu_{nf} = \mu_f\left(1 + 2.5\phi + 6.17\phi^2\right)$
Nguyen model (2003)	$\mu_{nf} = \mu_f\left(1 + 0.025\phi + 0.015\phi^2\right)$
Masoumi et al. (2009)	$\mu_{nf} = \mu_f + \rho_p V_B d_p^2 / \left(72C\delta\right)$
Gherasim et al.(2009)	$\mu_{nf} = \mu_f 0.904 e^{14.8\phi}$

et al. (2016) studied the effects of homogeneous–heterogeneous reactions in flow of magnetite-Fe3O4 nanoparticles by a rotating disk. They showed that the axial, radial and azimuthal velocity profiles are decreasing function of Hartman number. Sheikholeslami et al. (2013a) utilized Least Square and Galerkin Methods to investigate MHD nanofluid flow in a Semi-Porous Channel. They indicate that Velocity boundary layer thickness decrease with increase of Reynolds number and it increases as Hartmann number increases. Sheikholeslami et al. (2013b) studied the squeezing unsteady nanofluid flow using Adomian Decomposition Method (ADM). They showed that Nusselt number increases with increase of nanoparticle volume fraction and Eckert number. Sheikholeslami and Ganji (2013c) applied Homotopy perturbation method (HPM) to analysis heat transfer of Cu-water nanofluid flow between parallel plates. They indicated that Nusselt number has direct relationship with nanoparticle volume fraction, the squeeze number and Eckert number when two plates are separated. Application of ADM for nanofluid Jeffery-Hamel flow with high magnetic field has been presented by Sheikholeslami et al. (2012a). They proved that in greater angles or Reynolds numbers high Hartmann number are needed to reduction of backflow.

Flow and Heat Transfer of Cu-Water Nanofluid between a Stretching Sheet and a Porous Surface in a Rotating System was studied by Sheikholeslami et al. (2012b). They showed that for both suction and injection, the heat transfer rate at the surface increases with increasing in nanoparticle volume fraction, Reynolds number and injection/suction parameter and it decreases with power of rotation parameter.

Table 2. Different models for thermal conductivity of nanofluids used in simulation

Model	Equation
Koo and Kleinstreuer (2004)	$k_{nf} = k_f \left[\frac{k_p + 2k_f - 2\phi\left(k_f - k_p\right)}{k_p + 2k_f + \phi\left(k_f - k_p\right)} \right] + 5\times10^4 \beta\phi\rho_f \left(C_p\right)_f \sqrt{\frac{k_B T}{d_p \rho_p}} f\left(T, \phi\right)$
Jang and Choi model (2004)	$\frac{k_{nf}}{k_f} = \left(1-\phi\right) + Bk_p\phi + 18\times10^6 \frac{3d_f}{d_p} k_f \mathrm{Re}_{dp}^2 \mathrm{Pr}\,\phi$
Bruggeman model (Wang et al., 2003)	$k_{nf} = 0.25k_f \left(3\phi - 1\right) k_p / k_f + \left[3\left(1-\phi\right) - 1\right] + \sqrt{\Delta_B}$ $\Delta_B = \left[\left(3\phi - 1\right) k_p / k_f + \left(3\left(1-\phi\right) - 1\right)\right]^2 + 8k_p / k_f$
Chon et al. model (2005)	$k_{nf} = k_f \left[1 + 64.7\phi^{0.7640} \left(d_f / d_p\right)^{0.369} \left(k_f / k_p\right)^{0.7476} \mathrm{Pr}_T^{0.9955} \mathrm{Re}^{1.2321}\right]$
Charuyakorn et al. (1991)	$\frac{k_{nf}}{k_f} = \left[\frac{k_p + 2k_f - 2\phi\left(k_f - k_p\right)}{k_p + 2k_f + \phi\left(k_f - k_p\right)} \right] \left(1 + b\phi Pe_p^m\right)$
Staionary model (Eastman et al., 2000)	$k_{nf} = k_f \left[1 + k_p \phi d_f / \left(k_f \left(1-\phi\right) d_p\right)\right]$
Yu and Choi (2003)	$\frac{k_{nf}}{k_f} = \left[\frac{k_p + 2k_f - 2\phi\left(k_f - k_p\right)\left(1+\eta\right)^3}{k_p + 2k_f + \phi\left(k_f - k_p\right)\left(1+\eta\right)^3} \right]$
Patel et al. (2005)	$\frac{k_{nf}}{k_f} = 1 + \frac{k_p d_f \phi}{k_f d_p \left(1-\phi\right)} \left[1 + c\frac{2k_B T d_p}{\pi \alpha_f \mu_f d_p^2}\right]$
Mintsa et al. (2009)	$k_{nf} = k_f \left(1.72\phi + 1.0\right)$

Table 3. The thermo physical properties of the nanofluid

	$\rho(kg / m^3)$	$C_p(j / kgk)$	$k(W / m.k)$	$\beta(K^{-1})$	$\sigma\left(\Omega \cdot m\right)^{-1}$
Pure water	997.1	4179	0.613	21×10^{-5}	0.05
Copper(Cu)	8933	385	401	1.67×10^{-5}	5.96×10^{7}
Silver(Ag)	10 500	235	429	1.89×10^{-5}	3.60×10^{7}
Alumina(Al_2O_3)	3970	765	40	0.85×10^{-5}	1×10^{-10}
Titanium Oxide(TiO_2)	4250	686.2	8.9538	0.9×10^{-5}	1×10^{-12}

Sheikholeslami et al. (2014) used HAM to describe nanofluid flow over a permeable stretching wall in a porous medium. They found that increase in the nanoparticle volume fraction will decrease momentum boundary layer thickness and entropy generation rate while this increases the thermal boundary layer thickness. Sheikholeslami and Ganji (2014) utilized Galerkin Optimal Homotopy Asymptotic Method for investigate magnetohydrodynamic nanofluid flow in a permeable channel. They showed that velocity boundary layer thickness decreases with increase of Reynolds number and nanoparticle volume fraction and it increases as Hartmann number increases. Sheikholeslami et al. (2015) presented an application of HPM for simulation of two phase unsteady nanofluid flow and heat transfer between parallel plates in presence of time dependent magnetic field. Nanofluid flow and heat transfer between parallel plates considering Brownian motion has been investigated by Sheikholeslami and Ganji (2015). They used Differential Transformation Method (DTM) to solve the governing equations. They showed that skin friction coefficient increases with increase of the squeeze number and Hartmann number. Sheikholeslami et al. (2016) studied the steady nanofluid flow between parallel plates. They indicated that Nusselt number augments with increase of viscosity parameters but it is decreases with augment of Magnetic parameter, thermophoretic parameter and Brownian parameter. DTM has been applied by Domairry et al. (2012) to solve the problem of free convection heat transfer of non-Newtonian nanofluid between two vertical flat plates. They showed that as the nanoparticle volume fraction increases, the momentum boundary layer thickness increases. Table 4 shows the summary of the semi analytical method studies on nanofluid.

2.2. Runge-Kutta Method

Slip effects on unsteady stagnation point flow of a nanofluid over a stretching sheet has been investigated by Malvandi et al. (2014). They showed that Cu-water nanofluids exhibits a better thermal performance among the other considered nanofluids. Malvandi (2015) investigated the unsteady flow of a nanofluid in the stagnation point region of a time-dependent rotating sphere. Ashorynejad et al. (2013) studied nanofluid flow and heat transfer due to a stretching cylinder in the presence of magnetic field. They showed that choosing copper (for small of magnetic parameter) and alumina (for large values of magnetic parameter) leads to the highest cooling performance for this problem. Heated permeable stretching surface in a porous medium was studied by Sheikholeslami and Ganji (2014a). Three dimensional nanofluid flow, heat and mass transfer in a rotating system has been presented by Sheikholeslami and Ganji (2014b). They showed that Nusselt number has direct relationship with Reynolds number while it has reverse relationship with Rotation parameter, Magnetic parameter.

Sheikholeslami et al. (2014) studied the nanofluid flow and heat transfer in a rotating system in the presence of a magnetic field. Sheikholeslami and Ganji (2014c) studied two phase modeling of nanofluid in a rotating system with permeable sheet. Unsteady nanofluid flow and heat transfer in presence of magnetic field considering thermal radiation has been investigated by Sheikholeslami and Ganji (2015). Sheikholeslami et al. (2014) studied MHD nanofluid flow and heat transfer considering viscous dissipation. They showed that the magnitude of the skin friction coefficient is an increasing function of the magnetic parameter, rotation parameter and Reynolds number and it is a decreasing function of the nanoparticle volume fraction. Sheikholeslami et al. (2015) studied the effect of thermal radiation on magnetohydrodynamics nanofluid flow and heat transfer by means of two phase model. Sheikholeslami (2014) used KKL model for simulating nanofluid flow and heat transfer in a permeable channel. Effect of uniform suction on nanofluid flow and heat transfer over a cylinder has been studied by Sheikholeslami (2015). Sheikholeslami and Abelman (2015) studied two phase simulation of nanofluid flow and heat

Table 4. Summary of the semi analytical method studies on nanofluid

Authors	Gometry of study	Method	Type of nanofluid	Properties	Remarks
Sheikholeslami et al. (2013a)	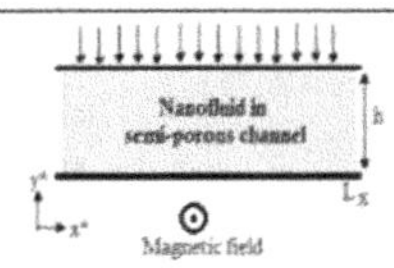	Least Square and Galerkin Methods	Cu-water	$1 \le \mathrm{Re} \le 10$ $0 \le \phi \le 0.06$ $0 \le Ha \le 10$ $\mathrm{Pr} = 6.8$	Velocity boundary layer thickness decrease with increase of Reynolds number and it increases as Hartmann number increases
Sheikholeslami et al. (2013b)	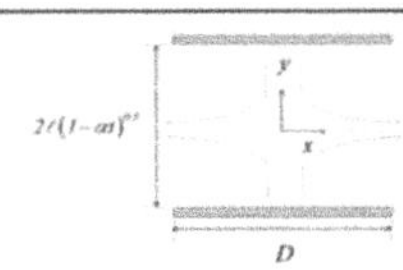	Adomian Decomposition Method	Cu-water Ag-water Al_2O_3-water TiO_2-water	$-1 \le S \le 1$ $0 \le \phi \le 0.06$ $0.01 \le Ec \le 2$ $\mathrm{Pr} = 6.2$	Nusselt number increases with increase of nanoparticle volume fraction and Eckert number but it decreases with increase of the squeeze number
Sheikholeslami and Ganji (2013c)	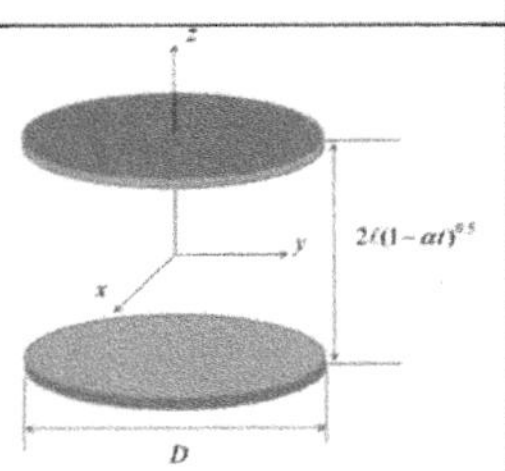	Homotopy perturbation method	Cu-water	$-1 \le S \le 1$ $0 \le \phi \le 0.06$ $0 \le Ec \le 1$ $0 \le \delta \le 1$ $\mathrm{Pr} = 6.2$	Nusselt number has direct relationship with nanoparticle volume fraction, δ, the squeeze number and Eckert number when two plates are separated but it has reverse relationship with the squeeze number when two plates are squeezed
Sheikholeslami et al. (2012a)	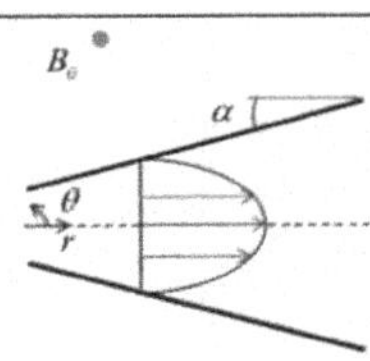	Adomian decomposition method	Cu-water	$100 \le \mathrm{Re} \le 300$ $2.5° \le \alpha \le 5°$ $0 \le \phi \le 0.06$ $0 \le Ha \le 2000$ $\mathrm{Pr} = 6.2$	In greater angles or Reynolds numbers high Hartmann number are needed to reduction of backflow
Sheikholeslami et al. (2012b)	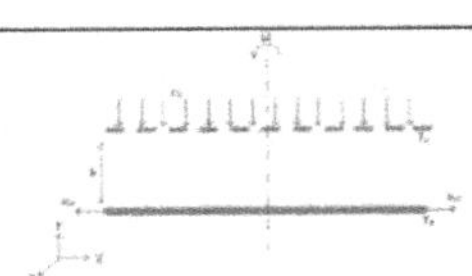	Homotopy Analysis Method	Cu-water	$5 \le \mathrm{R} \le 20$ $0.5 \le Kr \le 6$ $-1 \le \lambda \le 1$ $0 \le \phi \le 0.2$ $0 \le Ha \le 2000$ $\mathrm{Pr} = 6.2$	For both suction and injection, the heat transfer rate at the surface increases with increasing in nanoparticle volume fraction, Reynolds number and injection/suction parameter and it decreases with power of rotation parameter

continued on following page

Table 4. Continued

Sheikholeslami et al. (2014)	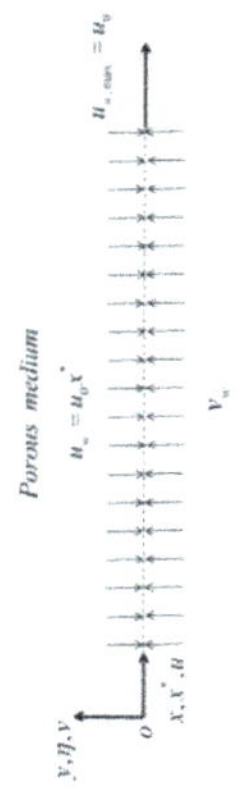	Homotopy Analysis Method	Cu-water Ag-water Al_2O_3-water TiO_2-water	$0.1 \le Re \le 1.5$ $-0.1 \le f_w \le 1$ $0 \le n \le 5$ $0 \le \phi \le 0.2$ $0 \le Ha \le 2000$ $Pr = 6.2$	Increase in the nanoparticle volume fraction will decrease momentum boundary layer thickness and entropy generation rate while this increases the thermal boundary layer thickness. Such effects are found to be more noticeable in the Ag-water solution than in the other solutions.
Sheikholeslami and Ganji (2014)	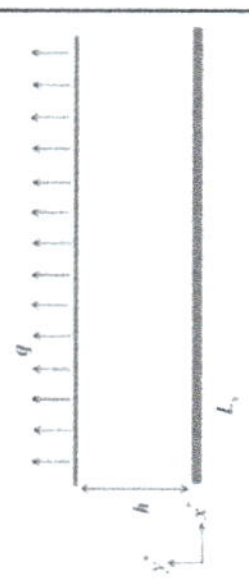	Galerkin Optimal Homotopy Asymptotic Method	Cu-water	$0 \le Ha \le 20$ $1 \le Re \le 20$ $0 \le \phi \le 0.06$ $Pr = 6.2$	Velocity boundary layer thickness decreases with increase of Reynolds number and nanoparticle volume fraction and it increases as Hartmann number increases.
Sheikholeslami et al. (2015)	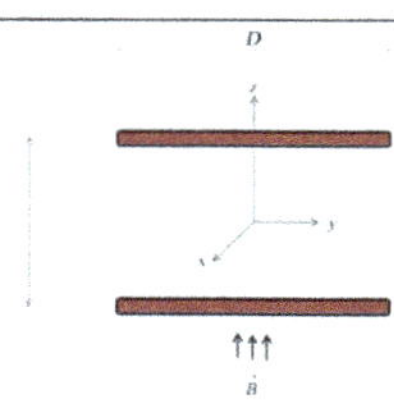	Homotopy perturbation method	Al_2O_3-water	$0 \le Ha \le 12$ $0 \le S \le 12$ $0.01 \le Ec \le 0.4$ $0.5 \le Sc \le 12$ $Pr = 10$	Nusselt number is an increasing function of Hartmann number, Eckert number and Schmidt number but it is a decreasing function of squeeze number
Sheikholeslami and Ganji (2015)	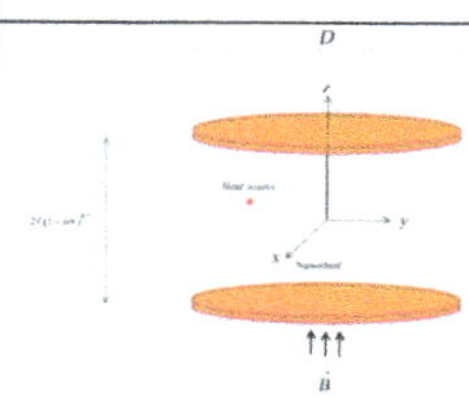	Differential Transformation Method	CuO-water, Al_2O_3-water	$0 \le S \le 10$ $0 \le Ha \le 20$ $-1 \le Hs \le -10$ $0.01 \le Ec \le 0.4$ $0.5 \le Sc \le 12$ $Pr = 10$	Skin friction coefficient increases with increase of the squeeze number and Hartmann number but it decreases with increase of nanofluid volume fraction. Nusselt number increases with augment of nanoparticle volume fraction, Hartmann number while it decreases with increase of the squeeze number

continued on following page

Table 4. Continued

Sheikholeslami et al. (2016)	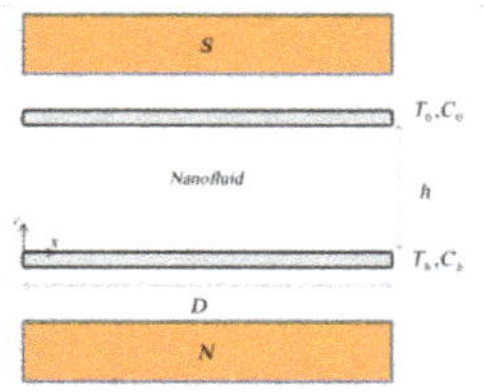	Differential Transformation Method	Al_2O_3-water	$0 \le S \le 10$ $0 \le Ha \le 20$ $-1 \le Hs \le -10$ $0.01 \le Ec \le 0.4$ $0.5 \le Sc \le 12$ $Pr = 10$	Nusselt number augments with increase of viscosity parameters but it is decreases with augment of Magnetic parameter, thermophoretic parameter and Brownian parameter
Domairry et al. (2012)	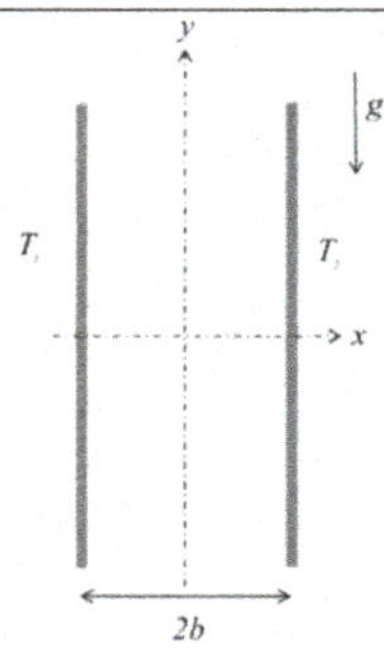	Differential Transformation Method	Cu-water	$0 \le Ha \le 20$ $0 \le \delta \le 100$ $0 \le \phi \le 0.1$ $0 \le Ec \le 6$ $Pr = 6.2$	As the nanoparticle volume fraction increases, the momentum boundary layer thickness increases, whereas the thermal boundary layer thickness decreases

transfer in an annulus in the presence of an axial magnetic field. Nanofluid spraying on an inclined rotating disk for cooling process has been investigated by Sheikholeslami et al. (2015). Sheikholeslami et al. (2015) investigated nanofluid flow and heat transfer over a stretching porous cylinder considering thermal radiation. They showed that Skin friction coefficient increases with increase of Reynolds number and suction parameter but it decreases with increase of nanoparticle volume fraction. Table 5 shows the summary of the Runge-Kutta method studies on nanofluid. Chamkha and Aly (2011) have studied the boundary layer flow of a nanofluid past a vertical flat plate. They have considered the Brownian motion and the thermophoresis effect. They have transformed the governing equations to a non-similar form and used numerical techniques to solve the same. They have reported that the local skin-friction coefficient increased as either of the suction, injection parameter, thermophoresis parameter, Lewis number, or heat generation or absorption parameter increased, while it decreased as either of the buoyancy ratio, Brownian motion parameter, or the magnetic field parameter increased.

2.3. Finite Difference Method

Chamkha and Rashad (2014) have studied the flow of a nanofluid around a non-isothermal wedge. They have considered the Brownian movement and the thermophoresis effects. They have concluded that the local skin-friction coefficient, local Nusselt number, and the local Sherwood number reduced as either of the magnetic parameter or the pressure gradient parameter was increased. The presence of the Brownian motion and the thermophoresis effects caused the local Nusselt number to decrease and the Sherwood number to increase. Sheremet and Pop (2014) used Buongiorno's mathematical model for conjugate natural convection in a square porous cavity filled with nanofluid. They showed that high thermophoresis parameter, low Brownian motion parameter, low Lewis and Rayleigh numbers and high thermal conductivity ratio reflect essential non-homogeneous distribution of the nanoparticles inside

Table 5. Summary of the Runge-Kutta method studies on nanofluid

Authors	Gometry of study	Method	Type of nanofluid	Properties	Remarks
Malvandi et al. (2014)	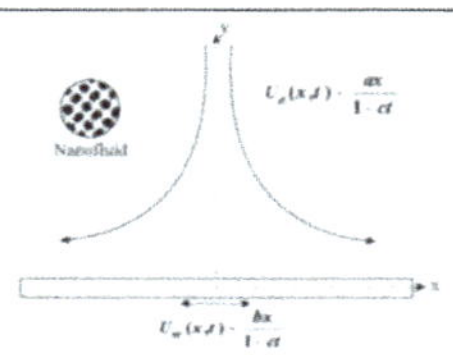	Runge-Kutta-Fehlberg method	Cu-water, Al_2O_3-water and TiO_2-water nanofluids	$1 \le d_p \le 100nm$ $0 \le \phi \le 0.2$ $\Pr = 6.2$	Cu-water nanofluids exhibits a better thermal performance among the other considered nanofluids. Slip velocity at the walls increase the heat transfer rate.
Malvandi (2015)	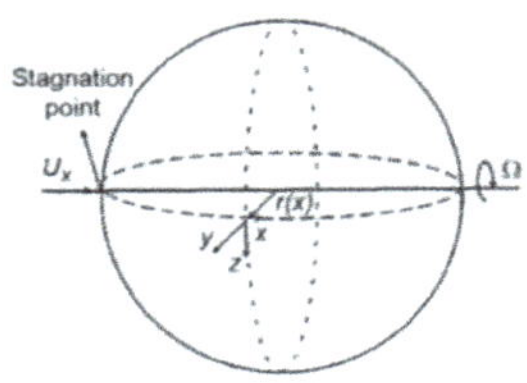	Runge-Kutta-Fehlberg method	Non-homogeneous mixtures	$1 \le d_p \le 100nm$ $0 \le \phi \le 0.2$	Increasing the thermophoresis is found to decrease heat transfer and concentration rates. This decrease suppresses for higher thermophoresis number. In addition, it was observed that unlike the heat transfer rate, a rise in Brownian motion, leads to an increase in concentration rate
Ashorynejad et al. (2013)	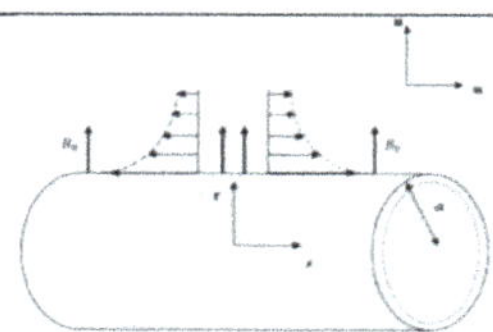	Fourth order Runge–Kutta method	Cu-water Ag-water Al_2O_3-water TiO_2--water	$1 \le \mathrm{Re} \le 5$ $0 \le M \le 10$ $0 \le \phi \le 0.2$ $\Pr = 6.2$	Choosing copper (for small of magnetic parameter) and alumina (for large values of magnetic parameter) leads to the highest cooling performance for this problem
Sheikholeslami and Ganji (2014a)	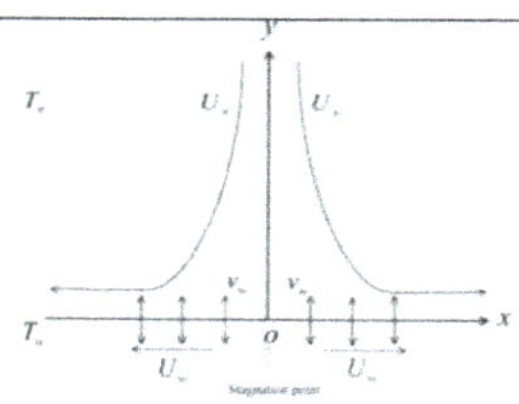	Fourth order Runge–Kutta method	Cu-water Ag-water Al_2O_3-water TiO_2-water Cu- Ethylene glycol	$1 \le \mathrm{Re} \le 5$ $-0.5 \le \gamma \le 1$ $0 \le \varepsilon \le 1$ $0 \le \phi \le 0.2$ $\Pr = 6.2$	Choosing Titanium oxide as the nanoparticle and Ethylene glycol as base fluid proved to have the highest cooling performance for this problem.
Sheikholeslami and Ganji (2014b)	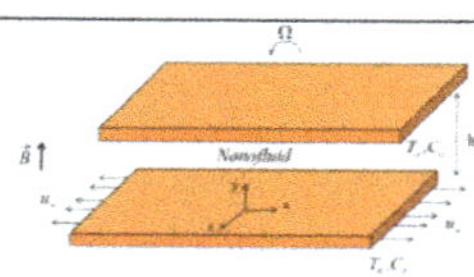	Fourth order Runge–Kutta method	Al_2O_3-water	$1 \le \mathrm{R} \le 10$ $0 \le M \le 12$ $-0.5 \le \gamma \le 1$ $K_r = 0.5, Nt = 0.1$ $Nb = 0.1, Sc = 0.1$ $\Pr = 10$	Nusselt number has direct relationship with Reynolds number while it has reverse relationship with Rotation parameter, Magnetic parameter, Schmidt number, Thermophoretic parameter and Brownian parameter

continued on following page

Table 5. Continued

Sheikholeslami et al. (2014)	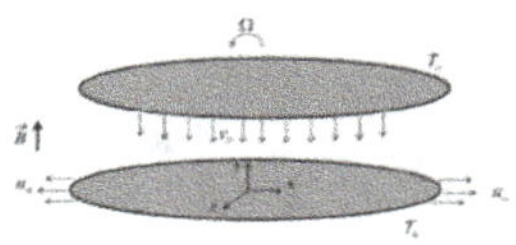	Fourth order Runge–Kutta method	Cu-water Ag-water Al_2O_3-water TiO_2--water	$5 \le R \le 20$ $0 \le M \le 50$ $0.5 \le Kr \le 6$ $0 \le \phi \le 0.2$ $-0.5 \le \lambda \le 1$ $Pr = 6.2$	The highest values are obtained when titanium oxide used as nanoparticle. Also it can be found that Nusselt number decreases with increase of magnetic parameter due to presence of Lorentz forces.
Sheikholeslami and Ganji (2014c)	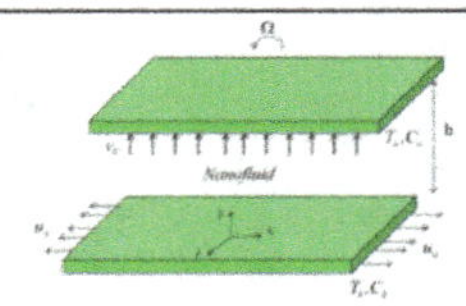	Fourth order Runge–Kutta method	Al_2O_3-water	$1 \le R \le 10$ $0.5 \le Kr \le 6$ $0 \le \lambda \le 3$ $Nt = Nb = 0.1$ $Pr = 10$	Nusselt number has direct relationship with Reynolds number and injection parameter while it has reverse relationship with Rotation parameter, Schmidt number, Thermophoretic parameter and Brownian parameter
Sheikholeslami and Ganji (2015)	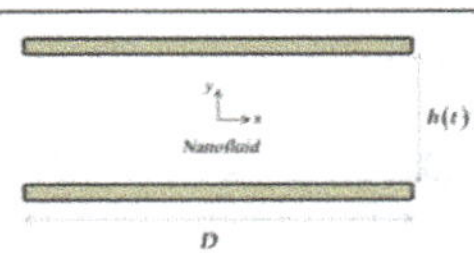	Fourth order Runge–Kutta method	Al_2O_3-water	$0.5 \le S \le 12$ $0.01 \le Ec \le 0.4$ $0.01 \le Nt \le 0.4$ $0.5 \le Sc \le 20$ $0 \le Rd \le 16$ $Pr = 10$	Concentration boundary layer thickness increases with increase of Radiation parameter
Sheikholeslami et al. (2014)	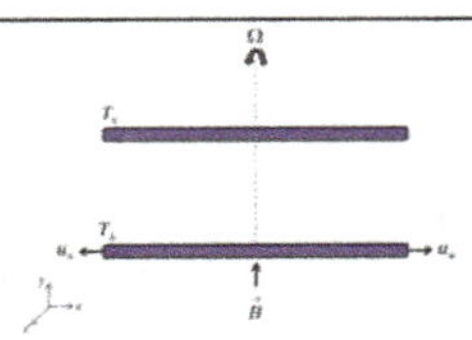	Fourth order Runge–Kutta method	CuO-water, Al_2O_3-water	$0.1 \le R \le 2$ $0 \le M \le 9$ $0 \le \phi \le 0.04$ $1 \le Kr \le 6$ $0.01 \le Ec \le 0.04$ $Pr = 10$	The magnitude of the skin friction coefficient is an increasing function of the magnetic parameter, rotation parameter and Reynolds number and it is a decreasing function of the nanoparticle volume fraction
Sheikholeslami et al. (2015)	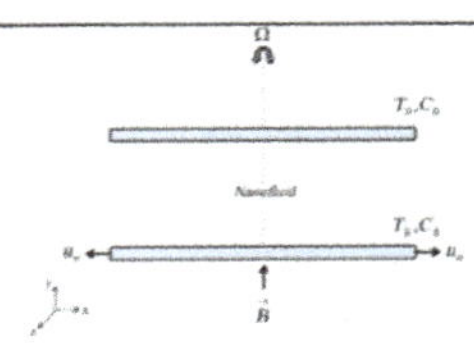	Fourth order Runge–Kutta method	Al_2O_3-water	$0.1 \le R \le 20$ $0 \le Rd \le 12$ $0.01 \le Sc \le 6$ $0 \le M \le 16$ $0 \le \phi \le 0.04$ $0.1 \le Kr \le 12$ $Pr = 10$	Nusselt number has direct relationship with radiation parameter and Reynolds number while it has reverse relationship with other active parameters. Also it can be found that concentration boundary layer thickness decreases with the increase of radiation parameter

continued on following page

Table 5. Continued

Sheikholeslami (2014)	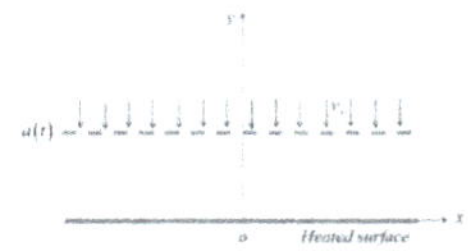	Fourth order Runge–Kutta method	Al_2O_3-water	$-1 \le \alpha \le 2$ $1 \le R \le 4$ $0 \le \phi \le 0.04$ $0 \le m \le 3$ $Pr = 6.8$	Heat transfer enhancement has direct relationship with Reynolds number when power law index is equals to zero but opposite trend is observed for other values of power law index.
Sheikholeslami (2015)	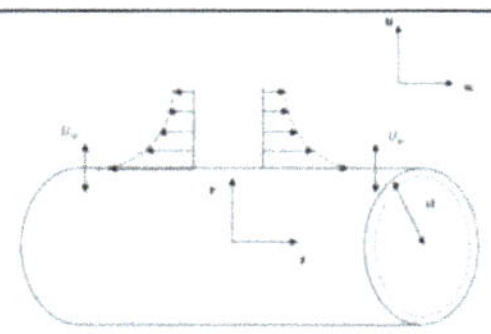	Fourth order Runge–Kutta method	CuO-water, Al_2O_3-water	$0.1 \le Re \le 2.7$ $0 \le \gamma \le 2$ $0 \le \phi \le 0.04$ $Pr = 6.2$	Adding nanoparticle into the base fluid of this problem is capable to change the flow pattern. It is found that Nusselt number is an increasing function of nanoparticle volume fraction, suction parameter, Reynolds number
Sheikholeslami and Abelman (2015)	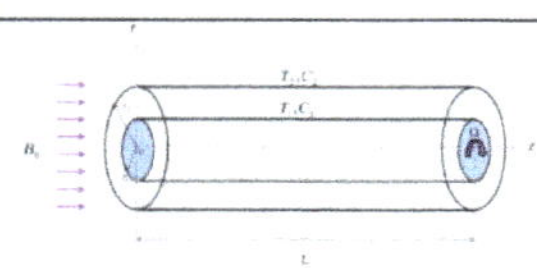	Fourth order Runge–Kutta method	Al_2O_3-water	$1 \le R \le 4$ $1 \le M \le 4$ $0.01 \le Ec \le 0.1$ $0.01 \le Nt \le 0.4$ $0.01 \le Nt \le 0.6$ $0.001 \le Ec \le 0.02$ $Pr = 10$	Nusselt number has a direct relationship with the aspect ratio and Hartmann number but it has a reverse relationship with the Reynolds number, Schmidt number, Brownian parameter, thermophoresis parameter and Eckert number.
Sheikholeslami et al. (2015)	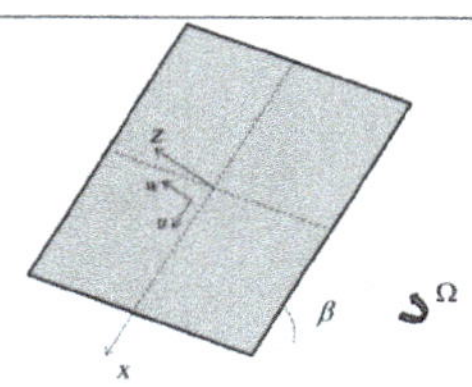	Fourth order Runge–Kutta method	Al_2O_3-water	$0.2 \le \delta \le 1$ $0.1 \le Nt = Nb \le 0.4$ $0.1 \le Sc \le 8$ $Pr = 10$	Nusselt number is an increasing function of each active parameter. Latent heat has direct relationship with Schmidt number, Brownian parameter and thermophoretic parameter but it has reverse relationship with Normalized thickness.
Sheikholeslami et al. (2015)	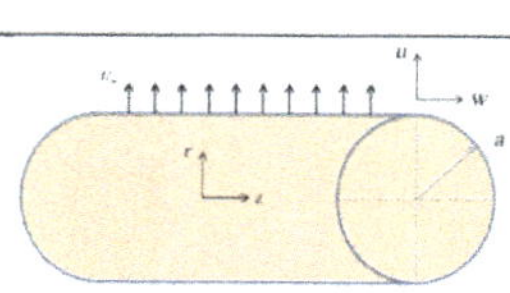	Fourth order Runge–Kutta method	Cu-water	$0.5 \le \gamma \le 2$ $0.5 \le Rd \le 2$ $0 \le \phi \le 0.04$ $Pr = 6.8$	Skin friction coefficient increases with increase of Reynolds number and suction parameter but it decreases with increase of nanoparticle volume fraction

the porous cavity. Sheremet et al. (2015) studied the three-dimensional natural convection in a porous enclosure filled with a nanofluid using Buongiorno's mathematical model. Sheremet et al. (2015) investigated the effect of thermal stratification on free convection in a square porous cavity filled with a nanofluid using Tiwari and Das' nanofluid model.

Ghalambaz et al. (2015) studied the free convection heat transfer in a porous cavity filled with a nanofluid using Tiwari and Das' nanofluid model. Double-diffusive mixed convection in a porous open cavity filled with a nanofluid using Buongiorno's model has been studied by Sheremet et al. (2015). Sheremet and Pop (2015) studied nanofluid free convection in a triangular porous cavity porous. Natural convection in a horizontal cylindrical annulus filled with a porous medium saturated by a nanofluid has been investigated by Sheremet and Pop (2015). Magnetic field effect on the unsteady natural convection in a wavy-walled cavity filled with a nanofluid has been studied by Sheremet et al. (2016). Khan et al. (2015) studied the three-dimensional flow of nanofluid induced by an exponentially stretching sheet. They showed that the existence of interesting Sparrow-Gregg-type hills for temperature distribution corresponding to some range of parametric values. Nanofluid flow with multimedia physical features for conjugate mixed convection and radiation has been studied by Hsiao (2014). Table 6 shows the summary of the Finite difference method studies on nanofluid.

2.4. Finite Volume Method

Garoosi and Hoseininejad (2016) investigated the natural and mixed convection heat transfer between differentially heated cylinders in an adiabatic enclosure filled with nanofluid. Garoosi et al. (2014) applied Buongiorno model for mixed convection of the nanofluid in heat exchangers. Two-phase mixture modeling of mixed convection of nanofluids in a square cavity with internal and external heating has been studied by Garoosi et al. (2015). Teamah et al. (2012) studied the augmentation of natural convective heat transfer in square cavity by utilizing nanofluids in the presence of magnetic field. They showed that weak magnetic field; the addition of nanoparticles is necessary to enhance the heat transfer but for strong magnetic field there is no need for nanoparticles because the heat transfer will decrease. Santra et al. (2008) studied the heat transfer augmentation in a differentially heated square cavity using copper-water nanofluid. Das and Ohal (2009) investigated Natural convection heat transfer augmentation in a partially heated and partially cooled square cavity utilizing nanofluids. Oztop et al. (2011) analyzed the non- isothermal temperature distribution on natural convection in nanofluid filled enclosures. They showed that an enhancement in heat transfer rate was registered for the whole range of Rayleigh numbers. Table 7 shows the summary of the Finite volume method studies on nanofluid.

2.5. Finite Element Method

MHD mixed convection of nanofluid filled partially heated triangular enclosure with a rotating adiabatic has been investigated by Selimefendigil and Oztop (2014). They showed that local and average heat transfer and total entropy generation enhance as the solid volume fraction of nanoparticle and angular rotational speed of the cylinder increases and Hartmann number decreases. Heat transfer enhancements around 30% are achieved for the highest volume fraction compared to base fluid. Selimefendigil and Oztop (2015) studied the natural convection and entropy generation of nanofluid filled cavity having different shaped obstacles under the influence of magnetic field and internal heat generation. Selimefendigil and Oztop (2014) studied pulsating nanofluids jet impingement cooling of a heated horizontal surface.

Table 6. Summary of the Finite difference method studies on nanofluid

Authors	Gometry of study	Method	Type of nanofluid	Properties	Remarks
Sheremet and Pop (2014)	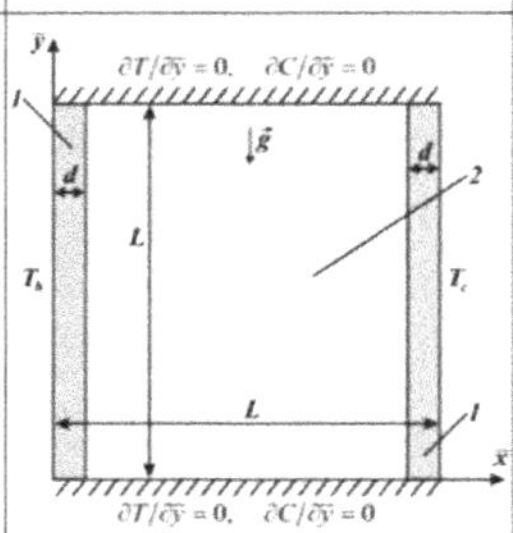	Finite difference method	Porous cavity with water based nanofluid	Ra = 100–500; Le = 1–10; Nb = 0.1–0.4; Nt = 0.1–0.4; Nr = 0.1–0.4; K_r = 0.1–10.0; D = 0.1–0.3.	High thermophoresis parameter, low Brownian motion parameter, low Lewis and Rayleigh numbers and high thermal conductivity ratio reflect essential non-homogeneous distribution of the nanoparticles inside the porous cavity.
Sheremet et al. (2015)	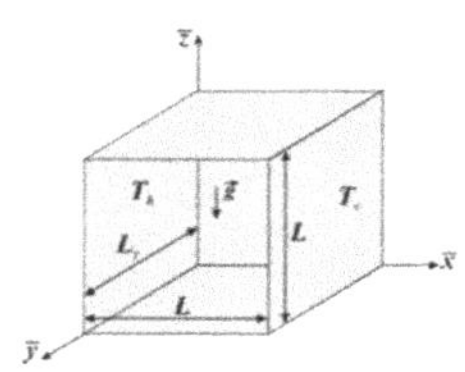	Finite difference method	Porous cavity with water based nanofluid	Ra = 30–500, Le = 1–10, Nr = 0.1–0.4, Nb = 0.1–0.4, Nt = 0.1–0.4, aspect ratio (A = 0.2–5.0).	Average Nusselt number at the hot surface is an increasing function of the Rayleigh number and Brownian motion parameter, and a decreasing function of the Lewis number, buoyancy-ratio and thermophoresis parameters. The average Sherwood number at the hot surface is an increasing function of the Rayleigh number and thermophoresis parameter, and a decreasing function of the Lewis number, buoyancy-ratio and Brownian motion parameters.
Sheremet et al. (2015)	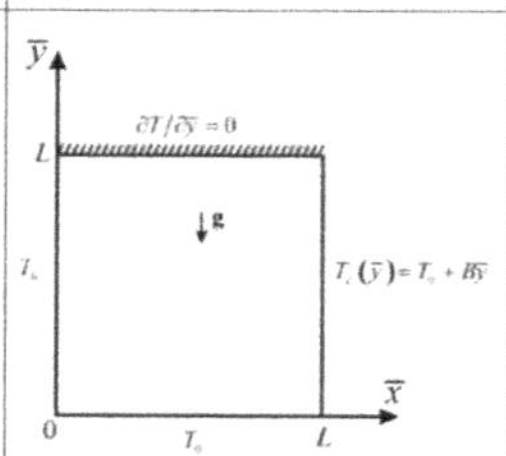	Finite difference method	Cu-water	Ra = 10–500, φ = 0.0–0.05, porosity of porous medium (ε = 0.1–0.8), thermal stratification parameter (b = 0–1.0) solid matrix of the porous medium (aluminum foam and glass balls)	An increase in the thermal stratification parameter leads to an essential reduction of the convective core sizes, displacement of this core close to the lower left corner and rotation of this vortex along the horizontal axis. More intensive attenuation of convective flow with ε and b occurs for low thermal conductivity material for the solid matrix of the porous medium.
Ghalambaz et al. (2015)	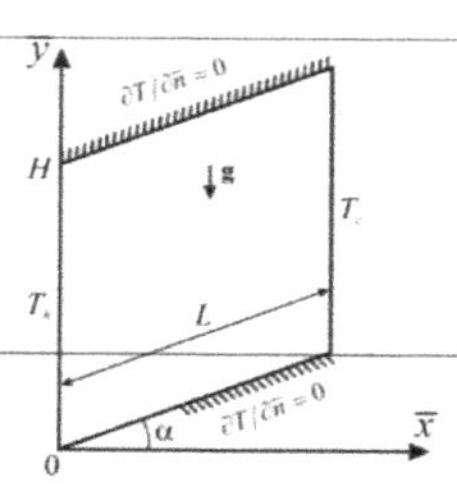	Finite difference method	Cu-water	-60° < α < 60°, aspect ratio 0.1 < A < 10, 10<Ra<1000,	Presence of nanoparticles deteriorates the heat transfer in all studied cases. The decrease of the porosity increases the porous matrix thermal conductivity while the decrease

continued on following page

Table 6. Continued

				$0 < \varphi < 0.1$, $\varepsilon = 0.3, 0.5, 0.7$, $k_s = 2k_{bf}$	of the inclination angle and of the aspect ratio would boost the deterioration of heat transfer.
Sheremet et al. (2016)	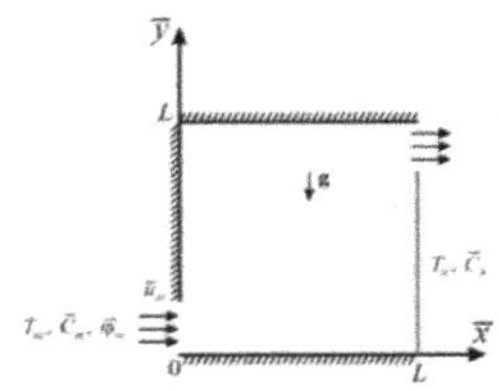	Finite difference method	Porous cavity with water based nanofluid	$Re = 10–100$, $Pr = 0.2$, $Ra = 50–500$, usual Lewis number ($Le = 1–50$), Dufour-contaminant Lewis number ($Ld = 1–50$), nanofluid Lewis number ($Ln = 1–50$), $Nr = 0.1–0.4$, regular double-diffusive buoyancy ratio ($Nc = 0.1–0.4$), $Nb = 0.1–0.4$, $Nt = 0.1–0.4$, modified Dufour parameter ($Nd = 0.1–0.4$)	Average Nusselt number at hot vertical wall is an increasing function of the Rayleigh and Reynolds numbers, and a decreasing function of the usual Lewis number. While the average Sherwood number at this vertical wall is an increasing function of the usual Lewis. Effects of the Rayleigh and Reynolds numbers on $\lvert\overline{Sh}\rvert$ and the thermophoresis parameter on the average Nusselt and Sherwood numbers are non-monotonic. It has been shown that in the present porous problem the Richardson number does not define the prevailing of the forced or natural convection modes.
Sheremet and Pop (2015a)	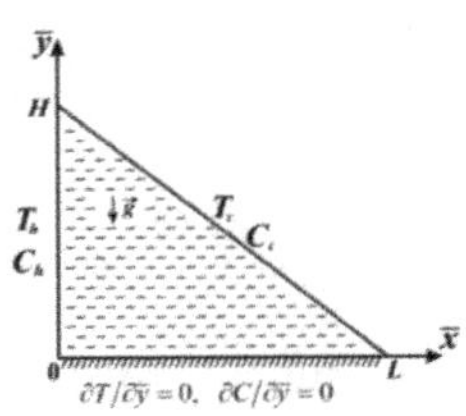	Finite difference method	Porous cavity with water based nanofluid	$Ra = 100–500$; $Le = 1–10$; $Nb = 0.1–0.4$; $Nt = 0.1–0.4$; $Nr = 0.1–0.4$; $A = H/L = 1.0$.	Average Nusselt number is an increasing function of *Ra*, *Le*, and a decreasing function of *Nr*, *Nb*, *Nt*. At the same time the average Sherwood number is an increasing function of *Ra*, *Le*, *Nb*, *Nt* and a decreasing function of *Nr*.
Sheremet and Pop (2015b)	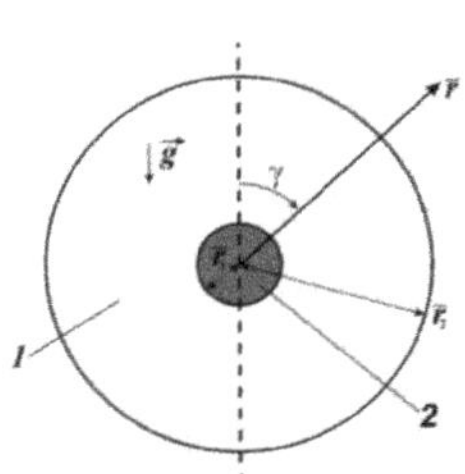	Finite difference method	Cu-water	$Ra = 200–700$, $\varepsilon = 0.1–0.8$, $\varphi = 0.0–0.05$, the annulus radius ratio ($R = 1.5, 2.0, 2.5$), solid matrix of porous medium (aluminum foam and glass balls)	Average Nusselt number is an increasing function of the Rayleigh number and annulus radius ratio and a decreasing function of the porosity of porous medium regardless of the solid matrix material. A decrease in the thermal conductivity of the solid matrix material leads to an attenuation of the convective heat transfer inside the annulus for high values of the porosity and solid

continued on following page

Table 6. Continued

					volume fraction of nanoparticles.
Sheremet et al. (2016)	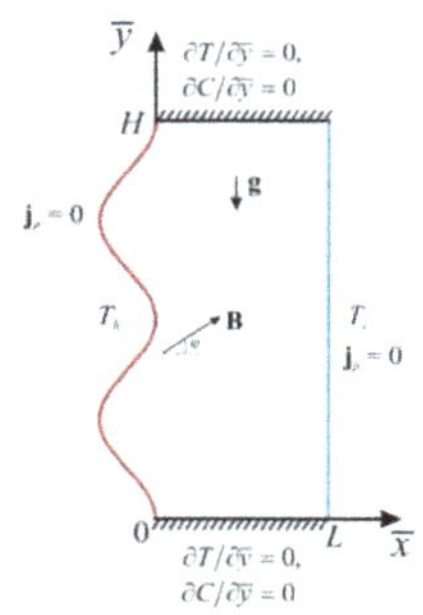	Finite difference method	Water based nanofluid	$Ra = 10^5$, $Le = 10$, $Pr = 6.26$, $Nr = 0.1$, $Nb = 0.1$, $Nt = 0.1$, $A = 1$, $Ha = 0$–100, undulation number ($\kappa = 1$–3), wavy contraction ratio ($b = 0.1$–0.3), $\varphi = 0$–π, $\tau = 0$–0.13.	An increase in the Hartmann number leads to an attenuation of convective flow and heat transfer and a formation of a double-core convective cell for high values of *Ha*. An increase in the wavy contraction ratio leads to an increase in the wave amplitude and an attenuation of the convective flow with more intensive heating of the wavy troughs.
Khan et al. (2015)	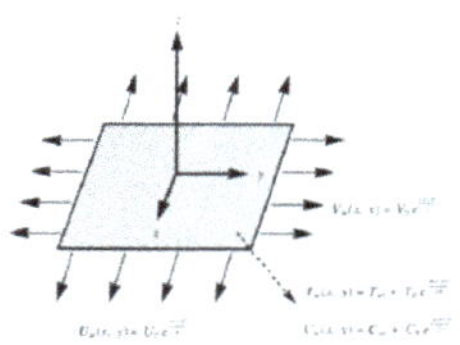	Finite difference scheme known as Keller-box method	Water based nanofluid	$0 \le \lambda \le 1$ $0.1 \le Nb \le 0.9$ $0.3 \le Nt \le 0.5$ $5 \le \Pr \le 10$ $-1 \le A \le 3$	The existence of interesting Sparrow-Gregg-type hills for temperature distribution corresponding to some range of parametric values
Kai-Long Hsiao (2014)	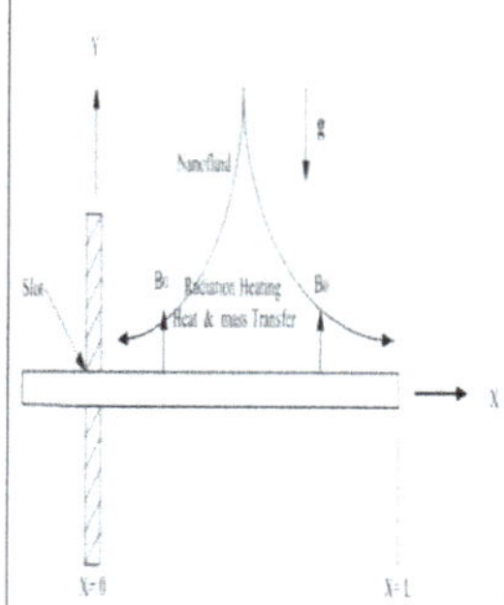	An improved Finite-Difference with Box method	Cu-water	K=400 (W/mK): $\rho = 8940(\text{kg}/\text{m}^3)$ $C_p = 385(\text{J}/\text{kgK})$ $10^1 \le \text{Re} \le 10^6$ $0 \le \Pr \le 1000$	The results show that dimensionless heat transfer effects increase with increasing values of Pr, R0, Nb or G but decrease with increasing M, Nt or Ec. Those, parameters Pr, R0, Nb or G are important factors in this study for increasing the heat transfer effects. The fourth, for mass transfer that the values Sc and Nb can be obtained a good mass diffusion effect, but the parameter Nt has not been processed this kind of function.

They showed that the combined effect of pulsation and inclusion of nanoparticles is not favorable for the stagnation point heat transfer enhancement for some combinations of Reynolds number and nanoparticle volume fraction. Selimefendigil and Oztop (2014) studied MHD mixed convection in a nanofluid filled lid driven square enclosure with a rotating cylinder. Selimefendigil and Oztop (2015) investigated numerical investigation and reduced order model of mixed convection at a backward facing step with a

Table 7. Summary of the Finite volume method studies on nanofluid

Authors	Gometry of study	Method	Type of nanofluid	Properties	Remarks
Garoosi and Hoseininejad (2016)	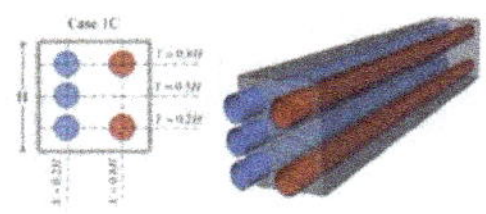	Finite volume method	Cu-water TiO_2-water Al_2O_3-water	$10^4 \le Ra \le 10^7$ $0 \le \phi \le 0.05$ $25 \le d_p \le 145$ $Pr = 6.8$	Changing the location of the heat source/sink from bottom–top to top–bottom configuration the heat transfer rate decreases significantly.
Garoosi et al. (2014)	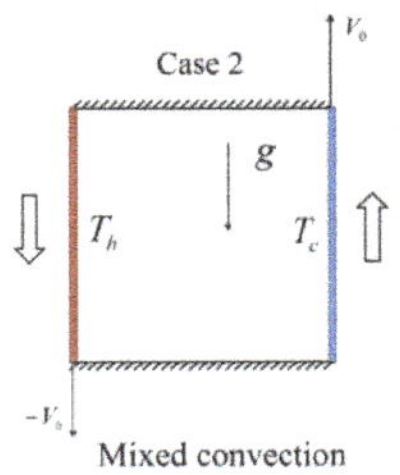	Finite volume method	Al_2O_3-water	$10^4 \le Ra \le 10^7$ $0.01 \le Ri \le 1000$ $10^2 \le Gr \le 10^4$ $0 \le \phi \le 0.05$ $25 \le d_p \le 145$	At low Rayleigh numbers and high Richardson numbers, the particle distribution is fairly non-uniform while at high Ra and low Ri values particle distribution remains almost uniform for free and mixed convection cases, respectively
Garoosi et al. (2015)	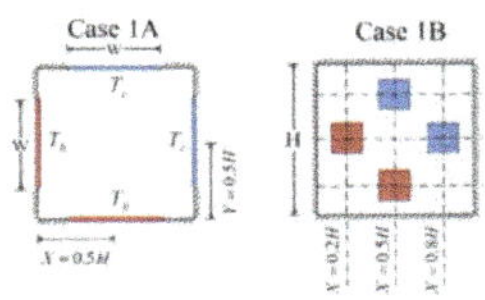	Finite volume method	Cu-water TiO_2-water Al_2O_3-water	$10^4 \le Ra \le 10^7$ $0.01 \le Ri \le 1000$ $0 \le \phi \le 0.05$ $25 \le d_p \le 145$	Thermophoretic effects are negligible for nanoparticles with high thermal conductivity. As a result, in such conditions use of homogeneous and single-phase models is valid at any Ra and Ri
Teamah et al. (2012)	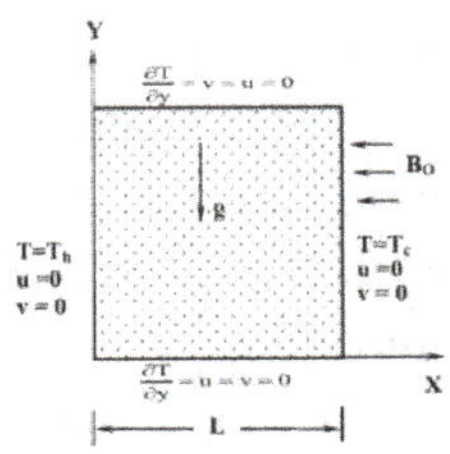	Finite volume method	Water based nanofluid	$10^4 \le Ra \le 10^7$ $0 \le \phi \le 0.05$ $0 \le Ha \le 60$	Weak magnetic field; the addition of nanoparticles is necessary to enhance the heat transfer but for strong magnetic field there is no need for nanoparticles because the heat transfer will decrease.

continued on following page

Table 7. Continued

Santra et al. (2008)	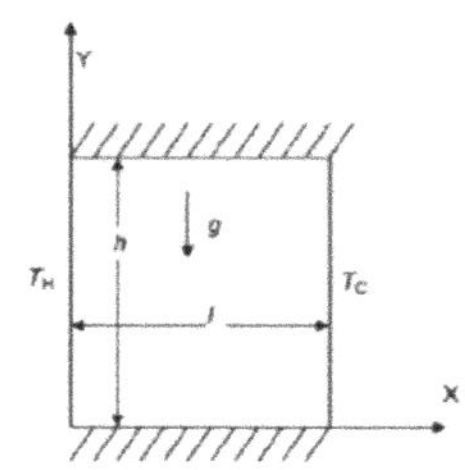	Finite volume method	Cu-water	$10^4 \leq Ra \leq 10^7$ $0 \leq \phi \leq 0.05$ $A = 1$	The heat transfer rate decreases with increase in solid volume fraction for a particular Rayleigh number. However, it increases with Rayleigh number for a particular solid volume fraction.
Das and Ohal (2009)	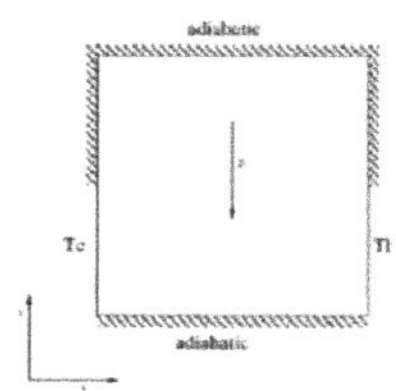	Finite volume method	Cu-water	$10^4 \leq Gr \leq 10^7$ $0 \leq \phi \leq 0.02$ $A = 1$	The rate of heat transfer increases with an increase in the nanoparticles volume fraction
Oztop et al. (2011)	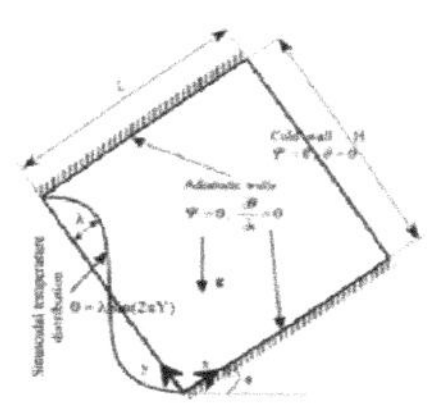	Finite volume method	Cu-water TiO_2-water Al_2O_3-water	$10^4 \leq Gr \leq 10^7$ $0 \leq \phi \leq 0.01$ $0 \leq \varphi \leq 90^\circ$ $A = 1$	An enhancement in heat transfer rate was registered for the whole range of Rayleigh numbers

rotating cylinder subjected to nanofluid. Effect of nanoparticle shape on mixed convection due to rotating cylinder in an internally heated and flexible walled cavity filled with SiO2–water nanofluids has been investigated by Selimefendigil et al. (2016). They indicated that Nussetl number enhances with external Rayleigh number and nanoparticle volume fraction while the opposite behavior is seen as the value of internal Rayleigh number and flexibility of the wall increases. Conjugate natural convection in a cavity with a conductive partition and filled with different nanofluids on different sides of the partition has been studied by Selimefendigil and Oztop (2016). They proved that as the value of the Grashof number, thermal conductivity ratio (Kr) and nanoparticle volume fraction increase, average Nusselt number increase. Table 8 shows the summary of the Finite element method studies on nanofluid.

2.6. Control Volume Based Finite Element Method

Heatline analysis has been used by Sheikholeslami et al. (2013) to investigate two phase simulation of nanofluid flow and heat transfer. They found that Nusselt number decreases as buoyancy ratio number increases until it reaches a minimum value and then starts increasing. As Lewis number increases, this minimum value occurs at higher buoyancy ratio number. Natural convection heat transfer in a cavity with sinusoidal wall filled with CuO-water nanofluid in presence of magnetic field has been studied by

Table 8. Summary of the Finite element method studies on nanofluid

Authors	Gometry of study	Method	Type of nanofluid	Properties	Remarks
Selimefendigil and Oztop (2014)	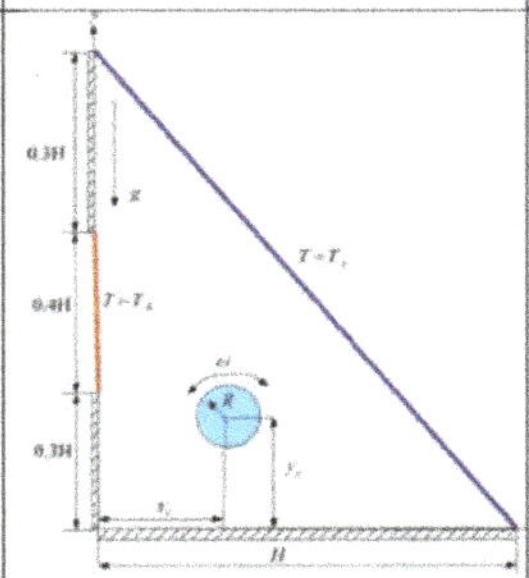	Finite element method	Cu-water	$10^4 \le Gr \le 10^6$ $0 \le \phi \le 0.05$ $0 \le Ha \le 50$ $-20 \le \omega \le 20$ $\Pr = 7.1$	Local and average heat transfer and total entropy generation enhance as the solid volume fraction of nanoparticle and angular rotational speed of the cylinder increases and Hartmann number decreases. Heat transfer enhancements around 30% are achieved for the highest volume fraction compared to base fluid.
Selimefendigil and Oztop (2015)	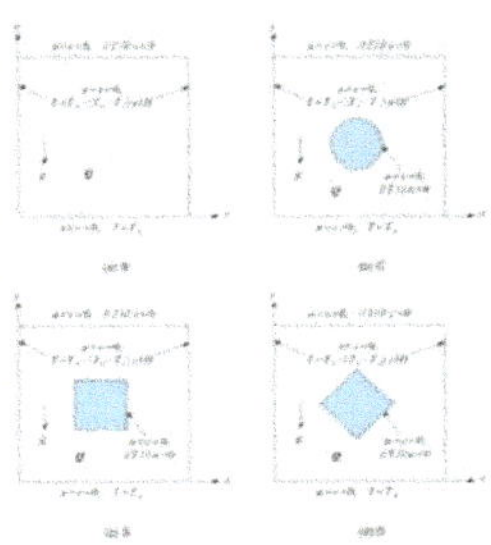	Finite element method	Cu-water	$10^4 \le Ra_E \le 10^6$ $10^4 \le Ra_I \le 10^6$ $0 \le \phi \le 0.05$ $0 \le Ha \le 50$ $\Pr = 6.9$	As the value of the external Rayleigh number decreases, internal Rayleigh number and Hartmann number increase, average heat transfer enhance. Average Nusselt number enhances by about 40%-60% at the highest solid volume fraction when compared to base fluid with different obstacles installed within the cavity.
Selimefendigil and Oztop (2014)	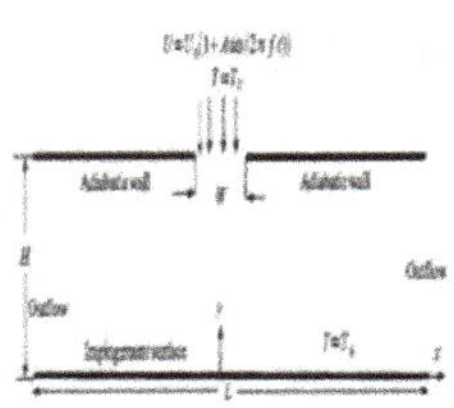	Finite volume method	Al_2O_3-water	$100 \le \mathrm{Re} \le 400$ $1 \le f \le 10$ $0 \le \phi \le 0.06$ $\Pr = 7.1$	The combined effect of pulsation and inclusion of nanoparticles is not favorable for the stagnation point heat transfer enhancement for some combinations of Reynolds number and nanoparticle volume fraction.
Selimefendigil and Oztop (2014)	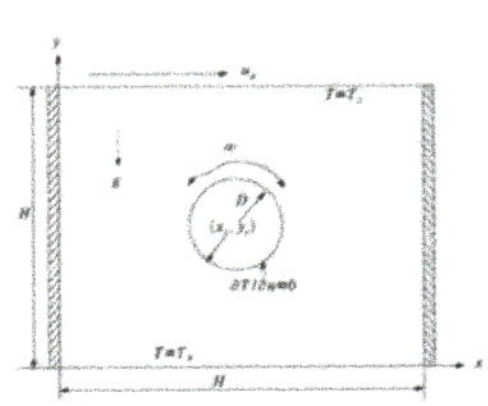	Finite element method	Cu-water	$0.001 \le Ri \le 10$ $0 \le Ha \le 50$ $0 \le \phi \le 0.05$ $-10 \le \Omega \le 10$ $\Pr = 6.2$	Average heat transfer enhances with Richardson number, nanoparticle volume fraction and cylinder rotation while magnetic field reduces the convection. The combined effect of magnetic field and cylinder rotation acts in a way to enhance local heat transfer for some locations along the heated wall.

continued on following page

Table 8. Continued

Selimefendigil and Oztop (2015)	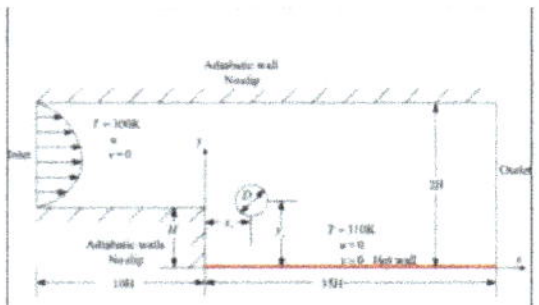	Finite element method	Al_2O_3-water	$50 \le Re \le 200$ $Gr = 10^4$ $0 \le \phi \le 0.05$ $-4 \le \Omega \le 4$ $Pr = 6.2$	Cylinder rotation affects the flow and thermal patterns behind the step. The addition of the nanoparticles enhances the averaged heat transfer along the bottom wall downstream of the step. There is almost a linear relation between heat transfer enhancement and nanoparticle volume fraction.
Selimefendigil et al. (2016)	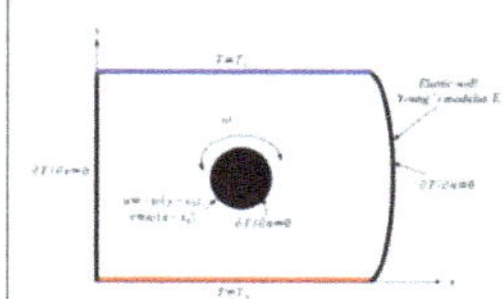	Finite element method	SiO2-water	$10^3 \le Ra_E \le 5\times10^5$ $10^4 \le Ra_I \le 10^6$ $5\times10^2 \le E \le 10^6$ $0 \le \phi \le 0.05$ $-2000 \le \Omega \le 2000$	Average Nussetl number enhances with external Rayleigh number and nanoparticle volume fraction while the opposite behavior is seen as the value of internal Rayleigh number and flexibility of the wall increases. Among different nanoparticle types cylindrical ones show the best performance in terms of heat transfer enhancement.
Selimefendigil and Oztop (2016)	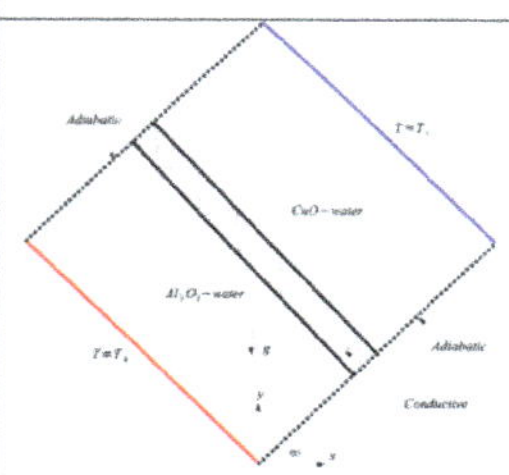	Finite element method	Al_2O_3-water and CuO-water	$10^3 \le Gr \le 10^6$ $0^o \le \theta \le 270^o$ $0.01 \le Kr \le 10$ $0 \le \phi \le 0.04$	As the value of the Grashof number, thermal conductivity ratio (Kr) and nanoparticle volume fraction increase, average Nusselt number increase. When nanoparticles with low thermal conductivity on the right cavity is added it is more effective for the heat transfer enhancement compared to adding nanoparticles with high thermal conductivity.

Sheikholeslami et al. (2014). Effects of a magnetic field on natural convection in different enclosures filled with naofluids have been examined by Sheikholeslami et al. (2013; 2014; 2012). Soleimani et al. (2012) studied the natural convection heat transfer in a nanofluid filled semi-annulus enclosure. They found that there is an optimum angle of turn in which the average Nusselt number is maximum for each Rayleigh number. Moreover, the angle of turn has an important effect on the streamlines, isotherms and maximum or minimum values of local Nusselt number. Effects of MHD on Cu-water nanofluid flow and heat transfer has been studied by Sheikholeslami et al. (2014). Constant temperature and heat flux boundary condition for Al2O3-water nanofluid filled enclosure have been examined by Sheikholeslami

Table 9. Summary of the control volume based finite element method studies on nanofluid

Authors	Gometry of study	Method	Type of nanofluid	Properties	Remarks
Sheikholeslami et al. (2013)	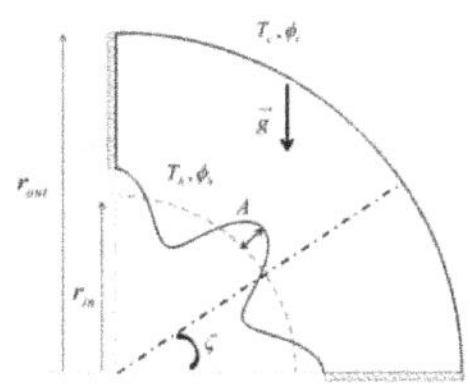	Control Volume based Finite Element Method	Al_2O_3-water	$10^3 \le Ra \le 10^5$ $0.1 \le Nr \le 4$ $1 \le Le \le 8$ $Nb = Nt = 0.5$ $\Pr = 10$	Average Nusselt number decreases as buoyancy ratio number increases until it reaches a minimum value and then starts increasing. As Lewis number increases, this minimum value occurs at higher buoyancy ratio number.
Sheikholeslami et al. (2014)	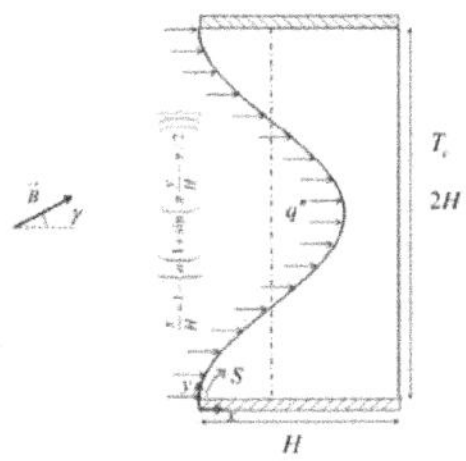	Control Volume based Finite Element Method	CuO-water	$10^3 \le Ra \le 10^5$ $0.1 \le a \le 0.3$ $0 \le Ha \le 100$ $0 \le \phi \le 0.04$ $\Pr = 6.2$	Nusselt number is an increasing function of nanoparticles volume fraction, dimensionless amplitude of the sinusoidal wall and Rayleigh number while it is a decreasing function of Hartmann number
Sheikholeslami et al. (2013)	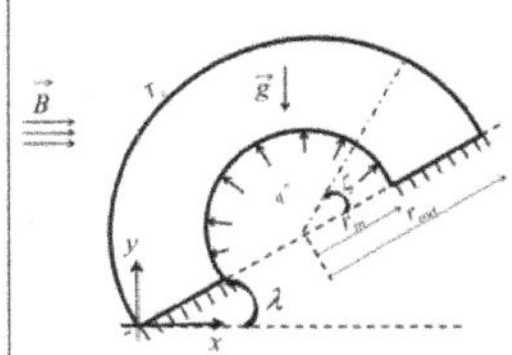	Control Volume based Finite Element Method	Cu-water	$10^3 \le Ra \le 10^5$ $0^\circ \le \lambda \le 90^\circ$ $0 \le \phi \le 0.06$ $0 \le Ha \le 100$ $\Pr = 6.2$	Hartmann number and the inclination angle of the enclosure can be control parameters at different Rayleigh number. In presence of magnetic field velocity field retarded and hence convection and Nusselt number decreases
Sheikholeslami et al. (2014)	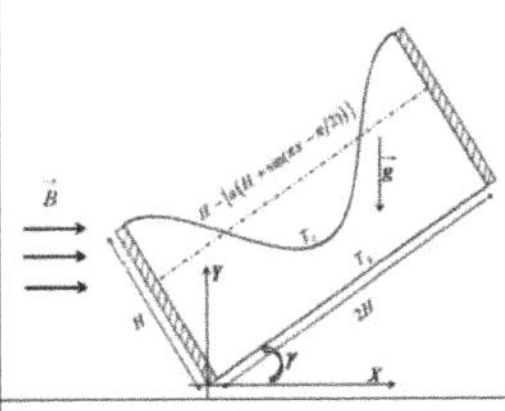	Control Volume based Finite Element Method	Cu-water	$10^3 \le Ra \le 10^5$ $0^\circ \le \gamma \le 90^\circ$ $0 \le Ha \le 100$ $0 \le \phi \le 0.06$ $\Pr = 6.2$	At $Ra = 10^3$, maximum value of enhancement for low Hartmann number is obtained at $\gamma = 0^\circ$, but for higher values of Hartmann number, maximum values of E occurs at $\gamma = 90^\circ$
Sheikholeslami et al. (2012)	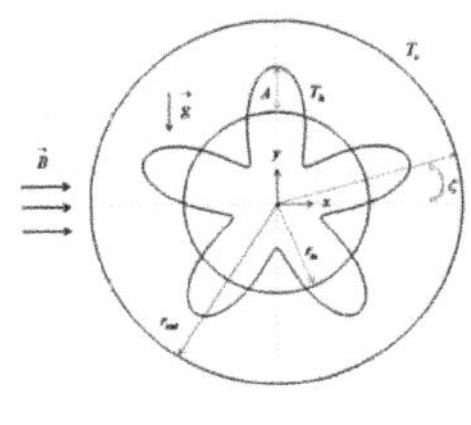	Control Volume based Finite Element Method	Cu-water	$10^3 \le Ra \le 10^5$ $3 \le N \le 6$ $0 \le Ha \le 100$ $0 \le \phi \le 0.06$ $A = 0.5, \Pr = 6.2$	Nusselt number is an increasing function of nanoparticle volume fraction, the number of undulations and Rayleigh numbers while it is a decreasing function of Hartmann number.

continued on following page

Table 9. Continued

Soleimani et al. (2012)	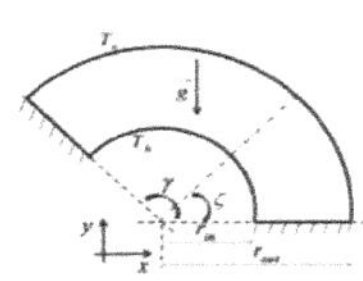	Control Volume based Finite Element Method	Cu-water	$10^3 \le Ra \le 10^5$ $45^\circ \le \gamma \le 180^\circ$ $0 \le \phi \le 0.06$ $Pr = 6.2$	There is an optimum angle of turn in which the average Nusselt number is maximum for each Rayleigh number. Moreover, the angle of turn has an important effect on the streamlines, isotherms and maximum or minimum values of local Nusselt number
Sheikholeslami et al. (2014)	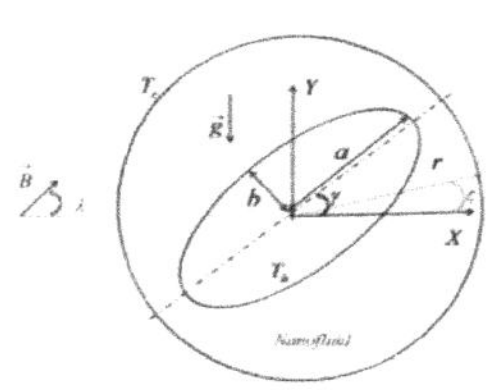	Control Volume based Finite Element Method	Cu-water	$10^3 \le Ra \le 10^5$ $0^\circ \le \gamma \le 90^\circ$ $0 \le \phi \le 0.06$ $\varepsilon = 0.9, a = 0.8L$ $Pr = 6.2$	Increasing Rayleigh number leads to decrease heat transfer enhancement while opposite trend is observed with augment of Hartmann number
Sheikholeslami et al. (2014)	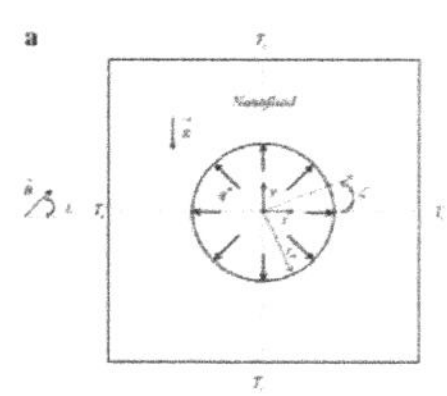	Control Volume based Finite Element Method	Al_2O_3-water	$10^3 \le Ra \le 10^5$ $0^\circ \le Ha \le 100$ $0 \le \phi \le 0.04$ $0.2 \le r_{in} / L \le 0.4$ $Pr = 6.2$	Domination of conduction mechanism causes heat transfer enhancement to increase. So enhancement in heat transfer increases with increase of Hartmann number and aspect ratio while it decreases with augment of Rayleigh number
Sheikholeslami et al. (2014)	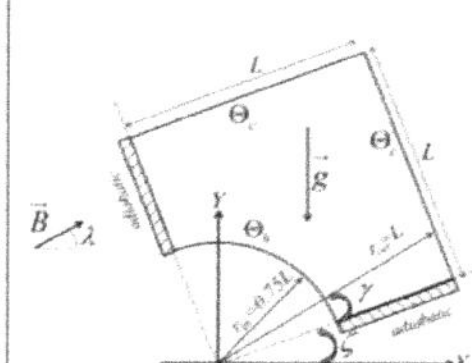	Control Volume based Finite Element Method	Al_2O_3-water	$10^3 \le Ra \le 10^5$ $-90^\circ \le \gamma \le 0^\circ$ $0 \le \phi \le 0.04$ $0 \le Ha \le 100$ $Pr = 6.2$	The heat transfer between cold and hot regions of the enclosure cannot be well understood by using isotherm patterns so heatline visualization technique is used to find the direction and intensity of heat transfer in a domain.
Sheikholeslami et al. (2014)		Control Volume based Finite Element Method	Al_2O_3-water	$30^\circ \le \gamma \le 90^\circ$ $0.1 \le Nr \le 4$ $2 \le Le \le 8$ $Ra = 10^5$ $Nb = Nt = 0.5$ $Pr = 10$	Lewis number has no significant effect on Nusselt number at low values of buoyancy ratio number
Sheikholeslami et al. (2014)	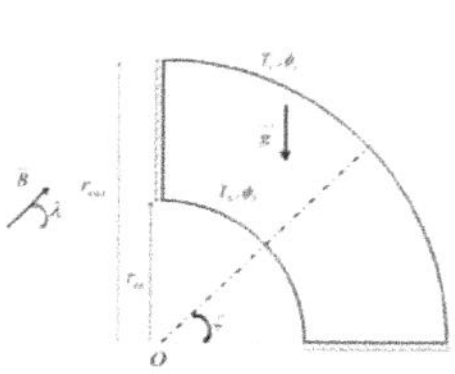	Control Volume based Finite Element Method	Al_2O_3-water	$10^3 \le Ra \le 10^5$ $0 \le Ha \le 100$ $30^\circ \le \gamma \le 90^\circ$ $0.1 \le Nr \le 4$ $2 \le Le \le 8$ $Nb = Nt = 0.5$ $Pr = 10$	As buoyancy ratio number increases the effects of other active parameters are more pronounced

continued on following page

Table 9. Continued

Sheikholeslami et al. (2014)	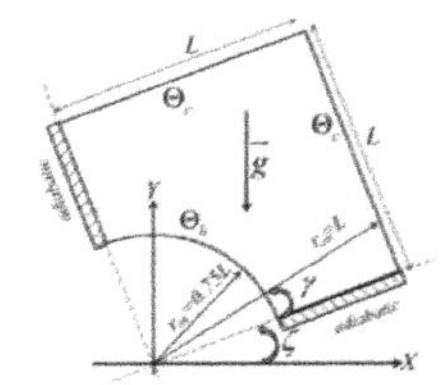	Control Volume based Finite Element Method	Cu-water	$10^3 \le Ra \le 10^5$ $-45^\circ \le \zeta \le 45^\circ$ $0 \le \phi \le 0.06$ $Pr = 6.2$	The results show that for $Ra = 10^4$ the maximum and minimum average Nusselt number are corresponding to $\zeta = -45^\circ$ and 45° respectively, whereas opposite trend is observed for $Ra = 10^5$.
Sheikholeslami and Ganji (2014)	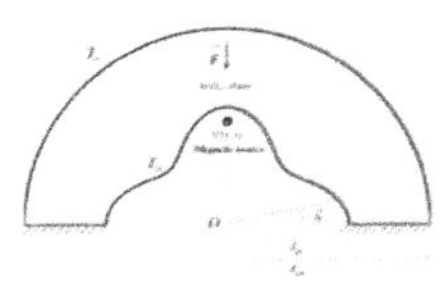	Control Volume based Finite Element Method	Fe_3O_4-water	$10^3 \le Ra \le 10^5$ $0 \le Ha \le 5$ $0 \le Mn_F \le 500$ $0 \le \phi \le 0.04$ $Pr = 6.8$ $Ec = 10^{-5}$	Nusselt number increases with augment of Rayleigh number and nanoparticle volume fraction but it decreases with increase of Hartmann number. Magnetic number has different effect on Nusselt number corresponding to Rayleigh number
Sheikholeslami et al. (2015)	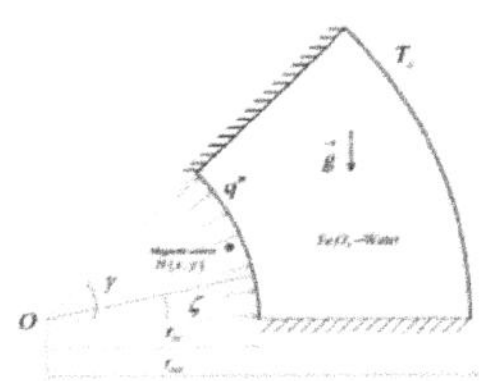	Control Volume based Finite Element Method	Fe_3O_4-water	$10^3 \le Ra \le 10^5$ $0 \le Ha \le 10$ $0 \le Mn_F \le 100$ $0 \le Nr \le 0.027$ $0 \le \phi \le 0.04$ $Pr = 6.8$ $Ec = 10^{-5}$	Nusselt number is an increasing function of Rayleigh number, nanoparticle volume fraction, Magnetic number while it is a decreasing function of with Hartmann number and Radiation parameter.
Sheikholeslami Kandelousi (2014)	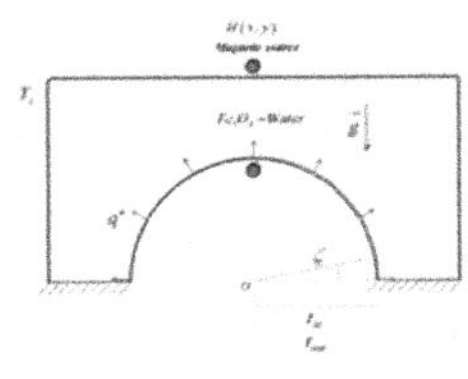	Control Volume based Finite Element Method	Fe_3O_4- water	$10^3 \le Ra \le 10^5$ $0 \le Mn_F \le 100$ $0 \le Ha \le 10$ $0 \le \phi \le 0.04$ $Pr = 6.2$	Enhancement in heat transfer decreases with increase of Rayleigh number and Magnetic number but it increases with increase of Hartmann number.
Sheikholeslami et al. (2014)	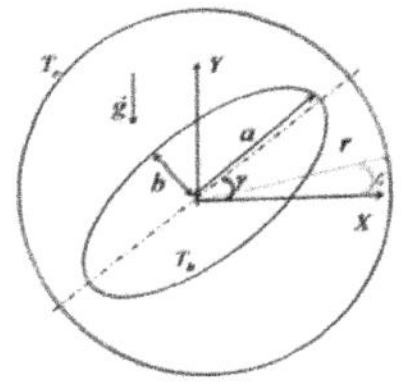	Control Volume based Finite Element Method	CuO-water	$10^3 \le Ra \le 10^5$ $0^\circ \le \gamma \le 90^\circ$ $0 \le \phi \le 0.04$ $Pr = 6.2$	Increasing Rayleigh number leads to decrease in ratio of heat transfer enhancement. For high Rayleigh number the minimum heat transfer enhancement ratio occurs at $\gamma = 90°$.

continued on following page

Table 9. Continued

Sheikholeslami and Rashidi (2015)	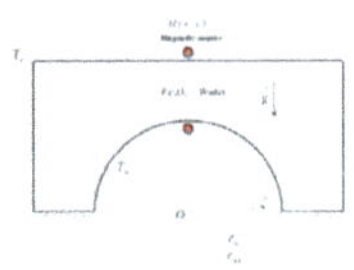	Control Volume based Finite Element Method	Fe_3O_4- water	$10^3 \le Ra \le 10^5$ $0 \le Mn_F \le 100$ $0 \le Ha \le 10$ $0 \le \phi \le 0.04$ $Pr = 6.2$	Nusselt number is an increasing function of Magnetic number, Rayleigh number and nanoparticle volume fraction while it is a decreasing function of Hartmann number
Sheikholeslami et al. (2015)		Control Volume based Finite Element Method	Fe_3O_4- water	$10 \le Re \le 10^3$ $0 \le Ha \le 20$ $0 \le \phi \le 0.04$ $Pr = 6.2$	Nusselt number has direct relationship with Reynolds number, nanoparticle volume fraction while it has reverse relationship with Hartmann number
Sheikholeslami and Ellahi (2015)	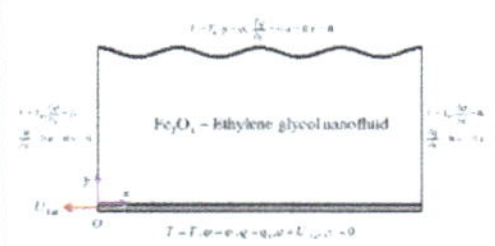	Control Volume based Finite Element Method	Fe_3O_4- Ethylene glycol	$3000 \le Re \le 6000$ $0 \le \Delta\varphi \le 10$ $0 \le \phi \le 0.04$ $Pr = 149.54$	Heat transfer rises with augment of supplied voltage and Reynolds number
Sheikholeslami et al. (2015)	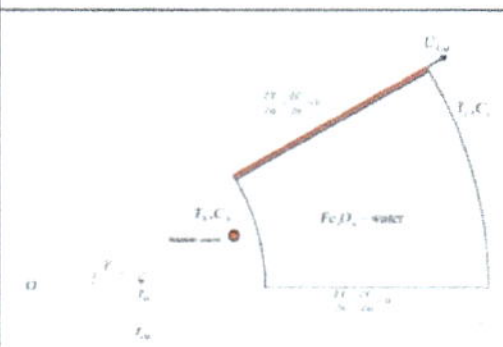	Control Volume based Finite Element Method	Fe_3O_4- water	$10 \le Re \le 500$ $2 \le Le \le 8$ $0 \le Ha \le 20$	Nusselt number has direct relationship with Reynolds number while it has reverse relationship with Hartmann number and Lewis number
Sheikholeslami et al. (2016)	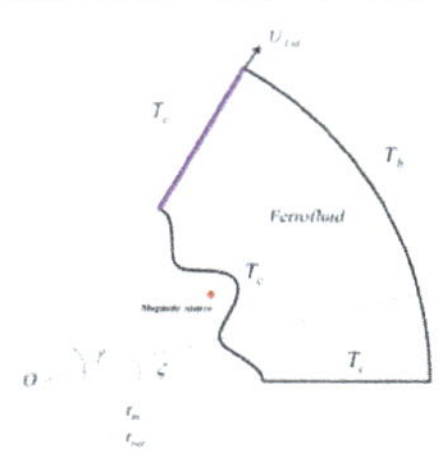	Control Volume based Finite Element Method	Fe_3O_4- water	$10 \le Re \le 600$ $0 \le Mn_F \le 10$ $0 \le Ha \le 10$ $0 \le \phi \le 0.04$ $Pr = 6.8$	Effects of Kelvin forces are more pronounced for high Reynolds number. Heat transfer enhancement has direct relationship with the Reynolds number and the magnetic number; while it has inverse relationship with the Hartmann number
Sheikholeslami et al. (2016)	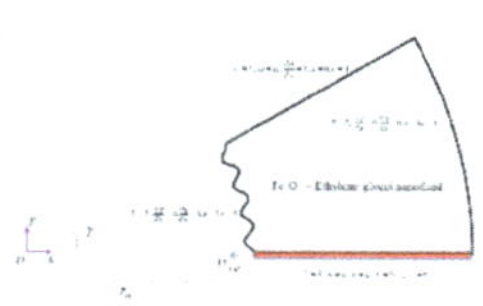	Control Volume based Finite Element Method	Fe_3O_4- Ethylene glycol	$3000 \le Re \le 6000$ $0 \le \Delta\varphi \le 10$ $0 \le \phi \le 0.04$ $Pr = 149.54$	Effect of electric filed on heat transfer is more pronounce at low Reynolds number.

continued on following page

Table 9. Continued

Sheikholeslami et al. (2016)	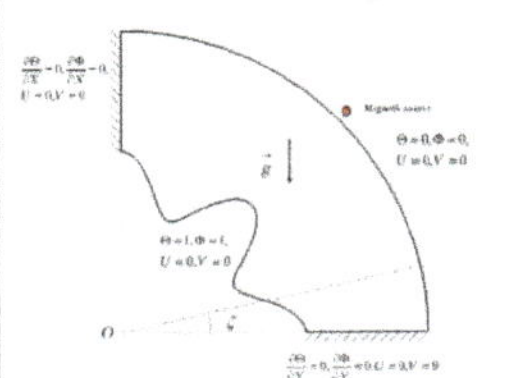	Control Volume based Finite Element Method	Fe_3O_4- water	$10^3 \le Ra \le 10^5$ $0.1 \le Nr \le 4$ $0 \le Ha \le 10$ $2 \le Le \le 4$ $Pr = 6.85$	Nusselt number has direct relationship with Rayleigh number, buoyancy ratio number and Lewis number while it has reverse relationship with Hartmann number
Sheikholeslami, and Rashidi (2015)	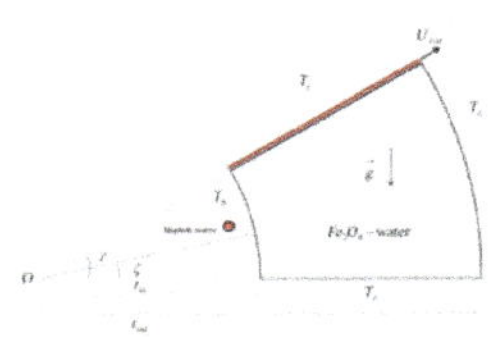	Control Volume based Finite Element Method	Fe_3O_4- water	$0.001 \le Ri \le 10$ $0 \le \delta^* \le 0.6$ $0 \le Mn_F \le 10$ $0 \le Ha \le 10$ $Re = 100$	Nusselt number has direct relationship with Richardson number, nanoparticle volume fraction while it has reverse relationship with Hartmann number and Magnetic number

et al. (2014). Sheikholeslami et al. (2014) studied free convection heat transfer in a nanofluid filled inclined L-shaped enclosure. Ferrohydrodynamic and Magnetohydrodynamic effects on ferrofluid flow and convective heat transfer has been investigated by Sheikholeslami and Ganji (2014). They found that Nusselt number increases with augment of Rayleigh number and nanoparticle volume fraction but it decreases with increase of Hartmann number. Magnetic number has different effect on Nusselt number corresponding to Rayleigh number.

Sheikholeslami et al. (2015) considered the effect of thermal radiation on ferrofluid flow and heat transfer in a semi annulus enclosure in the presence of magnetic source. Sheikholeslami Kandelousi (2014) studied the effect of spatially variable magnetic field on ferrofluid flow and heat transfer considering constant heat flux boundary condition. Free convection heat transfer in a nanofluid filled enclosure with elliptic inner cylinder has been presented by Sheikholeslami et al. (2014). Sheikholeslami and Rashidi (2015) studied the effect of space dependent magnetic field on free convection of Fe3O4-water nanofluid. Effect of non-uniform magnetic field on forced convection heat transfer of nanofluid has been studied by Sheikholeslami et al. (2015). Electrohydrodynamic nanofluid hydrothermal treatment in an enclosure with sinusoidal upper wall has been investigated by Sheikholeslami and Ellahi (2015). Sheikholeslami et al. (2015) applied two phase model for magnetic nanofluid forced convective heat transfer in existence of variable magnetic field. Sheikholeslami et al. (2016) investigated forced convection heat transfer in a semi annulus under the influence of a variable magnetic field. Sheikholeslami et al. (2016)studied Effect of electric field on hydrothermal behavior of nanofluid in a complex geometry. They found that effect of electric filed on heat transfer is more pronounce at low Reynolds number. Sheikholeslami et al. (2016) investigated non-uniform magnetic field effect on nanofluid hydrothermal treatment considering Brownian motion and thermophoresis effects. Sheikholeslami, and Rashidi (2015) studied Ferrofluid heat transfer treatment in the presence of variable magnetic field. They found that Nusselt number has direct relationship with Richardson number, nanoparticle volume fraction while it has reverse relationship with Hartmann number and Magnetic number. Table 9 shows the summary of the Control Volume based Finite Element Method studies on nanofluid.

Table 10. Summary of the Lattice Boltzmann method studies on nanofluid

Authors	Gometry of study	Method	Type of nanofluid	Properties	Remarks
Sheikholeslami et al. (2014)	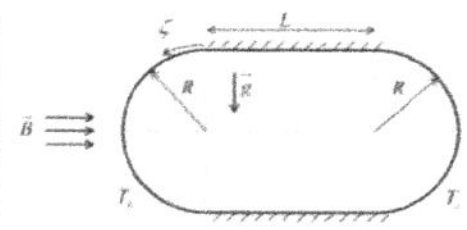	Lattice Boltzmann Method	Al_2O_3-water	$10^4 \le Ra \le 10^6$ $0 \le \phi \le 0.04$ $0 \le Ha \le 60$ $R / L = 0.5$ $\Pr = 6.8$	Enhancement in heat transfer increases with increase of Hartmann number except for $Ra = 10^4$ in which $Ha = 40$ roles as a critical Hartmann number
Sheikholeslami et al. (2013)	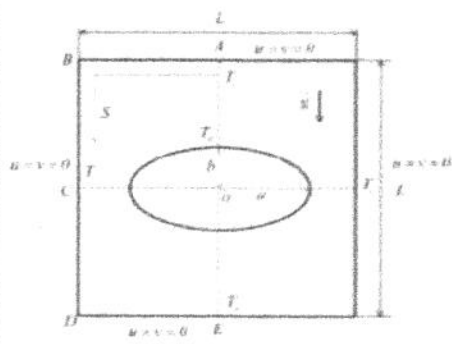	Lattice Boltzmann Method	Cu-water	$10^3 \le Ra \le 10^6$ $0 \le \phi \le 0.06$ $0.65 \le \varepsilon \le 0.95$ $a = 0.4$ $\Pr = 6.8$	The minimum value of enhancement of heat transfer occurs at $\varepsilon = 0.95$ for $Ra = 10^5$ but for other values of Rayleigh number it obtained at $\varepsilon = 0.65$.
Sheikholeslami et al. (2012)	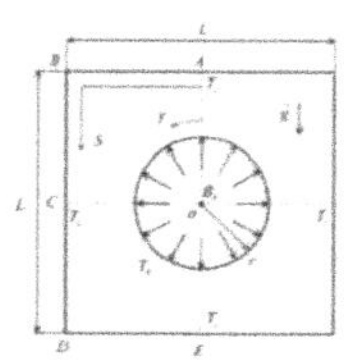	Lattice Boltzmann Method	Cu-water	$10^4 \le Ra \le 10^6$ $0 \le \phi \le 0.06$ $0 \le Ha \le 60$ $\Pr = 6.8$	As the nanoparticle volume fraction and Rayleigh number increase average Nusselt number increases but opposite trends are observed when the Hartmann number increases.
Sheikholeslami et al. (2013)	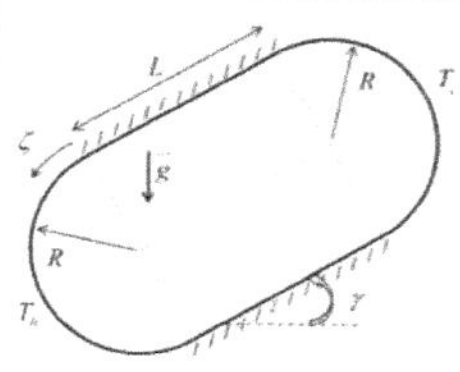	Lattice Boltzmann Method	Cu-water	$10^4 \le Ra \le 10^6$ $0 \le \phi \le 0.04$ $-60^\circ \le \gamma \le 60^\circ$ $R / L = 0.5$ $\Pr = 6.8$	The change of inclination angle has a significant impact on the thermal and hydrodynamic flow field
Sheikholeslami et al. (2013)	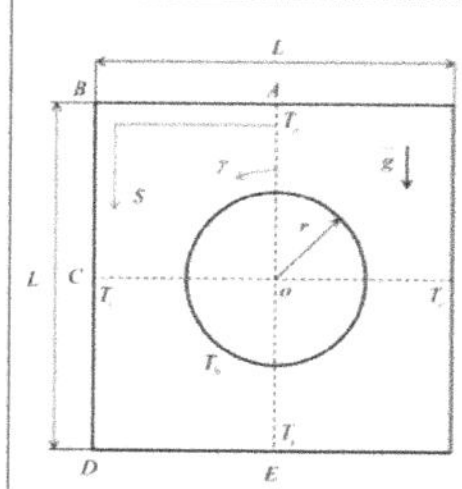	Lattice Boltzmann Method	Cu-water Ag-water Al_2O_3-water TiO_2--water	$10^3 \le Ra \le 10^6$ $0 \le \phi \le 0.04$ $1.5 \le \lambda \le 4.5$ $\Pr = 6.8$	Choosing copper as the nanoparticle leads to obtain the highest enhancement for this problem.

continued on following page

Table 10. Continued

Ashorynejad et al. (2013)	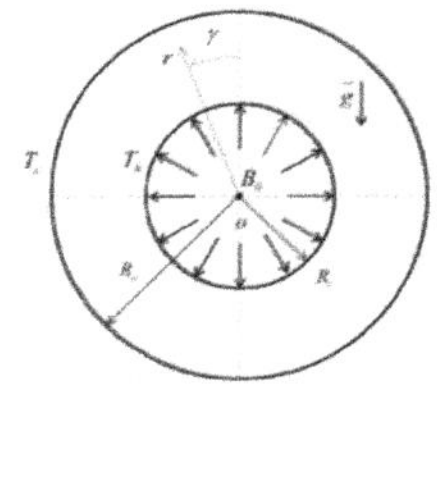	Lattice Boltzmann Method	Ag-water	$10^3 \le Ra \le 10^6$ $0 \le \phi \le 0.04$ $0 \le Ha \le 60$ $\Pr = 6.8$	Flow oscillations can be suppressed effectively by imposing an external radial magnetic field.
Sheikholeslami et al. (2013)	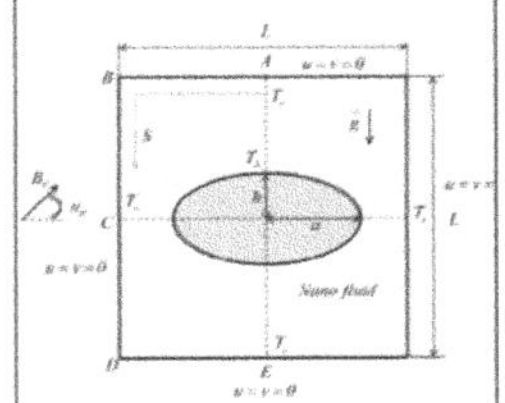	Lattice Boltzmann Method	Al_2O_3-water	$10^4 \le Ra \le 10^6$ $0 \le \phi \le 0.04$ $0 \le Ha \le 100$ $\Pr = 6.8$	The enhancement in heat transfer increases as Hartmann number increases but it decreases with increase of Rayleigh number
		Lattice Boltzmann Method			
Sheikholeslami et al. [120]	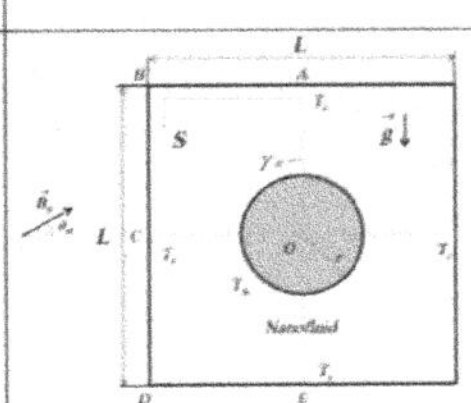	Lattice Boltzmann Method	Cu-water	$10^4 \le Ra \le 10^6$ $0 \le \phi \le 0.06$ $0 \le Ha \le 40$ $1.5 \le \lambda \le 4.5$ $\Pr = 6.8$	Enhancement ratio increases with decrease of Rayleigh number and it increases with augment of Hartmann number
Sheikholeslami and Gorji (2014)	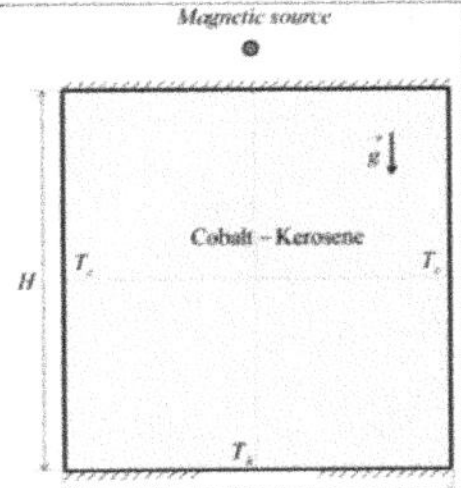	Lattice Boltzmann Method	Cobalt-kerosene	$10^4 \le Ra \le 10^6$ $0 \le \phi \le 0.04$ $0.675 \le A \le 1.325$ $0.2 \le \varepsilon \le 0.8$ $\Pr = 6.8$	Particles with a smaller size have better ability to dissipate heat, and a larger volume fraction would provide a stronger driving force which leads to increase in temperature profile
Sheikholeslami et al. (2014)	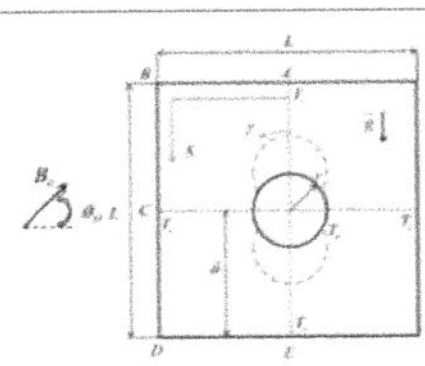	Lattice Boltzmann Method	Cu-water	$10^4 \le Ra \le 10^6$ $0 \le \phi \le 0.06$ $0 \le Ha \le 40$ $0.2 \le \frac{\delta}{L} \le 0.8$ $\lambda = 3.5$ $\Pr = 6.8$	Nusselt number has direct relationship with nanoparticle volume fraction and Rayleigh number but it has inverse relationship with Hartmann number and position of inner cylinder at high Rayleigh number

continued on following page

Table 10. Continued

Sheikholeslami et al. (2014)	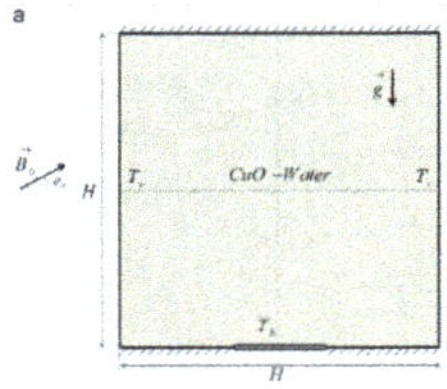	Lattice Boltzmann Method	CuO-water	$10^3 \le Ra \le 10^5$ $0 \le \phi \le 0.04$ $0 \le Ha \le 100$ $0.4 \le \varepsilon \le 0.8$ $Pr = 6.8$	Enhancement in heat transfer increases as Hartmann number and heat source length increase but it decreases with increase of Rayleigh number. Also it can be found that effect of Hartmann number and heat source length is more pronounced at high Rayleigh number.
Sheikholeslami and Ganji (2015)	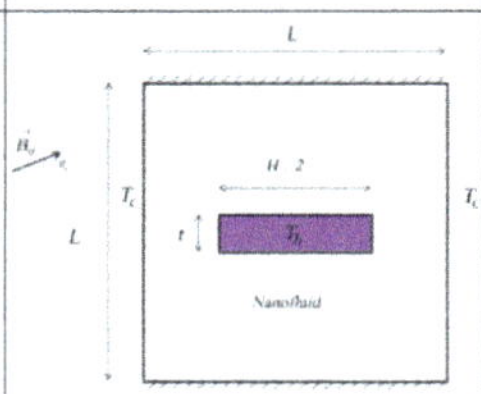	Lattice Boltzmann Method	CuO-water	$10^3 \le Ra \le 10^5$ $0 \le \phi \le 0.04$ $0 \le Ha \le 100$ $Pr = 6.8$	Heat transfer rate and Dimensionless entropy generation number increase with increase of the Rayleigh number and nanoparticle volume fraction but it decreases with increase of the Hartmann number
Sheikholeslami et al. (2015)	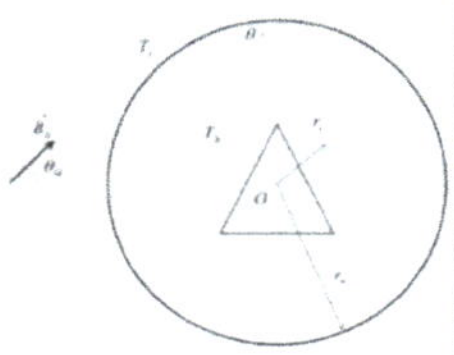	Lattice Boltzmann Method	Al_2O_3-water	$10^3 \le Ra \le 10^5$ $0 \le \phi \le 0.04$ $0 \le Ha \le 100$ $Pr = 6.8$	Lattice Boltzmann method with double-population is a powerful approach for the simulation of natural convection heat transfer in nanofluids in regions with curved boundaries
Sheikholeslami and Ellahi (2015)		Lattice Boltzmann Method	Fe_3O_4-Plasma	$50 \le Re \le 400$ $0 \le Mn_F \le 100$ $\phi = 0.04$ $Ha = 20$	Back flow occurs near the region where the magnetic source is located. Also it can be found that skin friction coefficient is a decreasing function of Reynolds number and Magnetic number
Sheikholeslami et al. (2015)	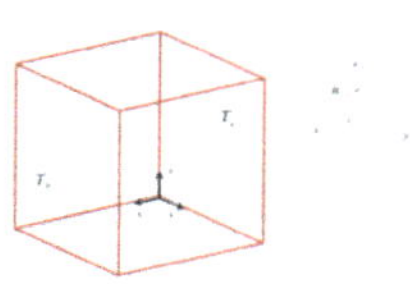	Lattice Boltzmann Method	Al_2O_3-water	$10^3 \le Ra \le 10^5$ $0 \le \phi \le 0.04$ $0 \le Ha \le 100$ $Pr = 6.8$	Nusselt number increases with increase of nanoparticle volume fraction and Rayleigh number while it decreases with increase of Hartmann number.

continued on following page

Table 10. Continued

Sheikholeslami and Ellahi (2015)	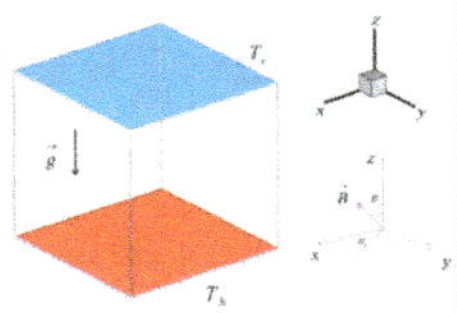	Lattice Boltzmann Method	Al_2O_3-water	$10^3 \le Ra \le 10^5$ $0 \le \phi \le 0.04$ $0 \le Ha \le 100$ $Pr = 6.8$	Applying magnetic field results in a force opposite to the flow direction that leads to drag the flow and then reduces the convection currents by reducing the velocities
Sheikholeslami et al. (2016)	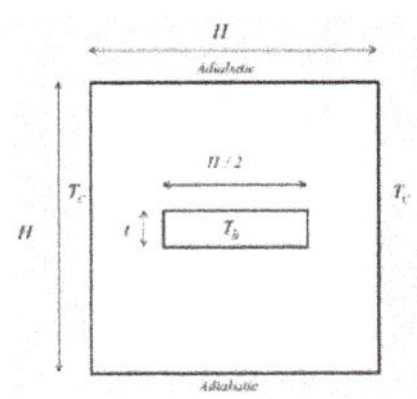	Lattice Boltzmann Method	Cu-water Ag-water Al_2O_3-water TiO_2--water	$10^3 \le Ra \le 10^5$ $0 \le \phi \le 0.04$ $2 \le H/t \le 9$ $Pr = 6.8$	The effect of nanoparticle volume fraction is found to be more pronounced for low Rayleigh number as compared to high Rayleigh number
Kefayati (2013)	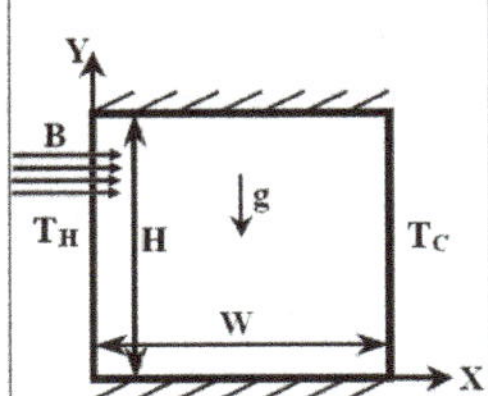	lattice Boltzmann method	Cu-water	$10^3 \le Ra \le 10^5$ $0 \le \phi \le 0.06$ $0 \le Ha \le 90$ $0.5 \le A \le 2$	The heat transfer decreases by increment of Hartmann number. Heat transfer decreases with growth of the aspect ratio but this growth causes the effect of nanoparticles to increase. Magnetic field augments the effect of nanoparticles at High Rayleigh numbers (Ra=10^5). The effect of nanoparticles rises for high Hartmann numbers when the aspect ratio increases.
Mejri et al. (2013)	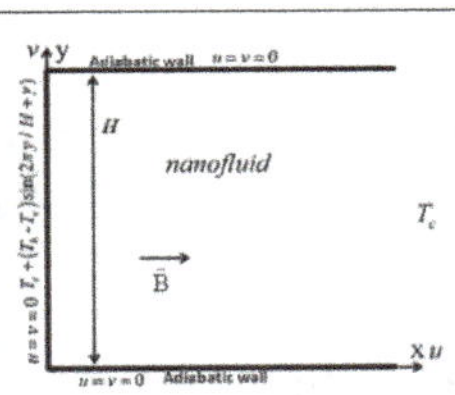	lattice Boltzmann method	Al_2O_3 -water	$10^3 \le Ra \le 10^6$ $0 \le \phi \le 0.06$ $0 \le Ha \le 60$ $\gamma = 0, \pi/4, \pi/2, 3\pi/4, \pi$	The heat transfer rate decreases with the increase of Hartmann number and increases with the rise of Rayleigh number.

2.7. Lattice Boltzmann Method

Investigation of nanofluid flow and heat transfer in presence of magnetic field using KKL model has been studied by Sheikholeslami et al. (2014). Nanofluid hydrothermal behaviors in an enclosure with curve boundaries have been studied by Sheikholeslami et al. (2013; 2012). Ashorynejad et al. (2013) studied magnetic field effects on natural convection flow of a nanofluid in a horizontal cylindrical annulus. They found that flow oscillations can be suppressed effectively by imposing an external radial magnetic field. MHD effects on nanofluid flow and heat transfer in a semi-annulus enclosure has been studied by Sheikholeslami et al. (2013; 2014). They showed that the enhancement in heat transfer increases as Hartmann number increases but it decreases with increase of Rayleigh number. Free convection of ferrofluid in a cavity heated from below in the presence of an external magnetic field has been investigated by Sheikholeslami and Gorji (2014). They found that particles with a smaller size have better ability to dissipate heat, and a larger volume fraction would provide a stronger driving force which leads to increase in temperature profile.

Sheikholeslami et al. (2014) studied MHD free convection in an eccentric semi-annulus filled with nanofluid. Sheikholeslami et al. (2014) simulated MHD CuO–water nanofluid flow and convective heat transfer considering Lorentz forces. Entropy generation of nanofluid in presence of magnetic field was studied by Sheikholeslami and Ganji (2015). Sheikholeslami et al. (2015) simulated magnetohydrodynamic natural convection heat transfer of Al_2O_3-water nanofluid in a horizontal cylindrical enclosure with an inner triangular cylinder. Sheikholeslami and Ellahi (2015) studied ferrofluid flow for magnetic drug targeting. They showed that back flow occurs near the region where the magnetic source is located. Sheikholeslami et al. (2015) simulated the magnetic field effect on hydrothermal behavior of nanofluid in a cubic cavity. Three dimensional mesoscopic simulation of magnetic field effect on natural convection of nanofluid has been studied by Sheikholeslami and Ellahi (2015). Nanofluid heat transfer enhancement and entropy generation has been studied by Sheikholeslami et al. (2016). Effect of a magnetic source on free convection in a cavity subjugated to nanofluid has been investigated (Kefavati, 2013). Table 10 shows the summary of the Lattice Boltzmann Method studies on nanofluid. Recently several authors apply new methods for simulation of hydrothermal behavior.

REFERENCES

Abu-Nada, E., & Chamkha, A. J. (2010). Effect of nanofluid variable properties on natural convection in enclosures filled with a CuO–EG–Water nanofluid. *International Journal of Thermal Sciences*, *49*(12), 2339–2352. doi:10.1016/j.ijthermalsci.2010.07.006

Abu-Nada, E., & Oztop, H. F. (2009). Effects of inclination angle on natural convection in enclosures filled with Cu–water nanofluid. *International Journal of Heat and Fluid Flow*, *30*(4), 669–678. doi:10.1016/j.ijheatfluidflow.2009.02.001

Akbarinia, A., & Laur, R. (2009). Investigation the diameter of solid particles effects on a laminar nanofluid flow in a curved tube using a two phase approach. *International Journal of Heat and Fluid Flow*, *30*(4), 706–714. doi:10.1016/j.ijheatfluidflow.2009.03.002

Angue Mintsa, H., Roy, G., Nguyen, C. T., & Doucet, D. (2009). New temperature dependent thermal conductivity data for water-based nanofluids. *International Journal of Thermal Sciences*, *48*(2), 363–371. doi:10.1016/j.ijthermalsci.2008.03.009

Ashorynejad, H. R., Mohamad, A. A., & Sheikholeslami, M. (2013). Magnetic field effects on natural convection flow of a nanofluid in a horizontal cylindrical annulus using Lattice Boltzmann method. *International Journal of Thermal Sciences*, *64*, 240–250. doi:10.1016/j.ijthermalsci.2012.08.006

Ashorynejad, H. R., Sheikholeslami, M., Pop, I., & Ganji, D. D. (2013). Nanofluid flow and heat transfer due to a stretching cylinder in the presence of magnetic field. *Heat and Mass Transfer*, *49*(3), 427–436. doi:10.100700231-012-1087-6

Behzadmehr, A., Saffar-Avval, M., & Galanis, N. (2007). Prediction of turbulent forced convection of a nanofluid in a tube with uniform heat flux using a two phase approach. *International Journal of Heat and Fluid Flow*, *28*(2), 211–219. doi:10.1016/j.ijheatfluidflow.2006.04.006

Brinkman, H. (1952). The viscosity of concentrated suspensions and solutions. *The Journal of Chemical Physics*, *20*(4), 571. doi:10.1063/1.1700493

Buongiorno, J. (2006). Convective transport in nanofluids. ASME. *Journal of Heat Transfer*, *128*(3), 240–250. doi:10.1115/1.2150834

Chamkha, A. J., & Aly, A. M. (2011). MHD Free Convection Flow of a Nanofluid past a Vertical Plate in the Presence of Heat Generation or Absorption Effects. *Chemical Engineering Communications*, *198*(3), 425–441. doi:10.1080/00986445.2010.520232

Chamkha, A. J., & Rashad, A. M. (2014). MHD Forced Convection Flow of a Nanofluid Adjacent to a Non-Isothermal Wedge, Computational. *Thermal Science*, *6*(1), 27–39. doi:10.1615/ComputThermalScien.2014005800

Charuyakorn, P., Sengupta, S., & Roy, S. K. (1991). Forced convection heat transfer in micro encapsulated phase change material slurries. *International Journal of Heat and Mass Transfer*, *34*, 819–833. doi:10.1016/0017-9310(91)90128-2

Chon, C. H., Kihm, K. D., Lee, S. P., & Choi, S. U. S. (2005). Empirical correlation finding the role of temperature and particle size for nanofluid (Al2O3) thermal conductivity enhancement. *Applied Physics Letters*, *87*(15), 153107. doi:10.1063/1.2093936

Das, M. K., & Ohal, P. S. (2009). Natural convection heat transfer augmentation in a partially heated and partially cooled square cavity utilizing nanofluids. *International Journal of Numerical Methods for Heat & Fluid Flow*, *19*(3/4), 411–431. doi:10.1108/09615530910938353

Ding, Y., & Wen, D. (2005). Particle migration in a flow of nanoparticle suspensions. *Powder Technology*, *149*(2–3), 84–92. doi:10.1016/j.powtec.2004.11.012

Domairry, Sheikholeslami, Ashorynejad, Gorla, & Khani. (2012). Natural convection flow of a non-Newtonian nanofluid between two vertical flat plates. *Proc IMechE Part N:J Nanoengineering and Nanosystems, 225*(3), 115–122. doi:10.1177/1740349911433468

Eastman, J. A., Choi, S. U. S., Li, S., Soyez, G., Thompson, L. J., & Di Melfi, R. J. (1999). Novel thermal properties of nanostructured materials. *Materials Science Forum*, *312–314*, 629–634. doi:10.4028/www.scientific.net/MSF.312-314.629

Eastman, J. A., Phillpot, S. R., Choi, U. S., & Keblinski, P. (2000). Thermal transport in nanofluids. *Annual Review of Materials Research*, *34*(1), 219–246. doi:10.1146/annurev.matsci.34.052803.090621

Einstein, A. (1956). *Investigation on the Theory of Brownian Motion*. New York: Dover.

Garoosi, F., Garoosi, S., & Hooman, K. (2014). Numerical simulation of natural convection and mixed convection of the nanofluid in a square cavity using Buongiorno model. *Powder Technology*, *268*(December), 279–292. doi:10.1016/j.powtec.2014.08.006

Garoosi, F., & Hoseininejad, F. (2016). Numerical study of natural and mixed convection heat transfer between differentially heated cylinders in an adiabatic enclosure filled with nanofluid. *Journal of Molecular Liquids*, *215*, 1–17. doi:10.1016/j.molliq.2015.12.016

Garoosi, F., Rohani, B., & Rashidi, M. M. (2015). Two-phase mixture modeling of mixed convection of nanofluids in a square cavity with internal and external heating. *Powder Technology*, *275*(May), 304–321. doi:10.1016/j.powtec.2015.02.015

Ghalambaz, M., Sheremet, M. A., & Pop, I. (2015). Free convection in a parallelogrammic porous cavity filled with a nanofluid using Tiwari and Das' nanofluid model. *PLoS One*, *10*(5), e0126486. doi:10.1371/journal.pone.0126486 PMID:25993540

Gherasim, I., Roy, G., Nguyen, C. T., & Vo-Ngoc, D. (2009). Experimental investigation of nanofluids in confined laminar radial flows. *International Journal of Thermal Sciences*, *48*(8), 1486–1493. doi:10.1016/j.ijthermalsci.2009.01.008

Hayat, T., Imtiaz, M., Alsaedi, A., & Alzahrani, F. (2016). Effects of homogeneous–heterogeneous reactions in flow of magnetite-Fe3O4 nanoparticles by a rotating disk. *Journal of Molecular Liquids*, *216*(April), 845–855. doi:10.1016/j.molliq.2016.01.089

Hsiao, K.-L. (2014, November). Nanofluid flow with multimedia physical features for conjugate mixed convection and radiation. *Computers & Fluids*, *104*(20), 1–8. doi:10.1016/j.compfluid.2014.08.001

Jang, S. P., & Choi, S. U. S. (2004). The role of Brownian motion in the enhanced thermal conductivity of nanofluids. *Applied Physics Letters*, *84*(21), 4316–4318. doi:10.1063/1.1756684

Jang, S. P., Lee, J. H., Hwang, K. S., & Choi, S. U. S. (2007). Particle concentration and tube size dependence of viscosities of Al2O3-water nanofluids flowing through microand minitubes. *Applied Physics Letters*, *91*(24), 243112. doi:10.1063/1.2824393

Junaid Ahmad Khan, M. (2015). Three-Dimensional Flow of Nanofluid Induced by an Exponentially Stretching Sheet: An Application to Solar Energy. *PLoS One*, *10*(3), e0116603. doi:10.1371/journal.pone.0116603 PMID:25785857

Kefayati, G. H. R. (2013). Lattice Boltzmann simulation of natural convection in nanofluid-filled 2D long enclosures at presence of magnetic source. *Theoretical and Computational Fluid Dynamics*, *27*(6), 865–883. doi:10.100700162-012-0290-x

Kefayati, G. H. R. (2013). Lattice Boltzmann simulation of natural convection in a nanouid-filled inclined square cavity at presence of magnetic field. *Scientia Iranica*, *20*(5), 1517–1527.

Khan, M., & Malik, R. (2016). Forced convective heat transfer to Sisko nanofluid past a stretching cylinder in the presence of variable thermal conductivity. *Journal of Molecular Liquids*, *218*(June), 1–7. doi:10.1016/j.molliq.2016.02.024

Khanafer, K., Vafai, K., & Lightstone, M. (2003). Buoyancy-driven heat transfer enhancement in a two-dimensional enclosure utilizing nanofluids. *International Journal of Heat and Mass Transfer*, *46*(19), 3639–3653. doi:10.1016/S0017-9310(03)00156-X

Koo, J., & Kleinstreuer, C. (2004). A new thermal conductivity model for nanofluids. *Journal of Nanoparticle Research*, *6*(6), 577–588. doi:10.100711051-004-3170-5

Maı¨ga, S. E. B., Palm, S. M., Nguyen, C. T., Roy, G., & Galanis, N. (2005). Heat transfer enhancement by using nanofluids in forced convection flows. *International Journal of Heat and Fluid Flow*, *26*(4), 530–546. doi:10.1016/j.ijheatfluidflow.2005.02.004

Malvandi, A. (2015). The unsteady flow of a nanofluid in the stagnation point region of a time-dependent rotating sphere. *Thermal Science*, *19*(5), 1603–1612. doi:10.2298/TSCI121020079M

Malvandi, A., Hedayati, F., & Ganji, D. D. (2014). Slip effects on unsteady stagnation point flow of a nanofluid over a stretching sheet. *Powder Technology*, *253*, 377–384. doi:10.1016/j.powtec.2013.11.049

Masoumi, N., Sohrabi, N., & Behzadmehr, A. (2009). A new model for calculating the effective viscosity of nanofluids. *Journal of Physics. D, Applied Physics*, *42*(5), 055501. doi:10.1088/0022-3727/42/5/055501

Maxwell, J. C. (1904). *A Treatise on Electricity and Magnetism* (2nd ed.). Oxford, UK: Oxford University Press.

Mirmasoumi, S., & Behzadmehr, A. (2008). Numerical study of laminar mixed convection of a nanofluid in a horizontal tube using two-phase mixture model. *Applied Thermal Engineering*, *28*(7), 717–727. doi:10.1016/j.applthermaleng.2007.06.019

Namburu, P. K., Kulkarni, D. P., Misra, D., & Das, D. K. (2007). Viscosity of copper oxide nanoparticles dispersed in ethylene glycol and water mixture. *Experimental Thermal and Fluid Science*, *32*(2), 397–402. doi:10.1016/j.expthermflusci.2007.05.001

Nguyen, C. T., Desgranges, F., Galanis, N., Roy, G., Mare, T., Boucher, S., & Mintsa, H. A. (2008). Viscosity data for Al2O3-water nanofluid-hysteresis: Is heat transfer enhancement using nanofluids reliable? *International Journal of Thermal Sciences*, *47*(2), 103–111. doi:10.1016/j.ijthermalsci.2007.01.033

Orozco, D. (2005). Hydrodynamic behavior of suspension of polar particles. *Encyclopedia Surface Colloid Science*, *4*, 2375–2396.

Oztop, H. F., Abu-Nada, E., Varol, Y., & Al-Salem, K. (2011). Computational analysis of non- isothermal temperature distribution on natural convection in nanofluid filled enclosures. *Superlattices and Microstructures*, *49*(4), 453–467. doi:10.1016/j.spmi.2011.01.002

Pak, B. C., & Cho, Y. (1998). Hydrodynamic and heat transfer study of dispersed fluids with submicron metallic oxide particle. *Experimental Heat Transfer*, *11*(2), 151–170. doi:10.1080/08916159808946559

Palm, S. J., Roy, G., & Nguyen, C. T. (2006). Heat transfer enhancement with the use of nanofluids in radial flow cooling systems considering temperature dependent properties. *Applied Thermal Engineering*, *26*(17-18), 2209–2218. doi:10.1016/j.applthermaleng.2006.03.014

Patel, H. E., Sundarrajan, T., Pradeep, T., Dasgupta, A., Dasgupta, N., & Das, S. K. (2005). A micro-convection model for thermal conductivity of nanofluid. *Pramana Journal of Physics*, *65*(5), 863–869. doi:10.1007/BF02704086

Santra, A. K., Sen, S., & Chakraborty, N. (2008). Study of heat transfer augmentation in a differentially heated square cavity using copper-water nanofluid. Interna- tional. *Journal of Thermal Science*, *47*(9), 1113–1122. doi:10.1016/j.ijthermalsci.2007.10.005

Sasmal, C., & Nirmalkar, N. (2016). Momentum and heat transfer characteristics from heated spheroids in water based nanofluids. *International Journal of Heat and Mass Transfer*, *96*(May), 582–601. doi:10.1016/j.ijheatmasstransfer.2016.01.054

Selimefendigil, F., & Öztop, H. F. (2014). MHD mixed convection of nanofluid filled partially heated triangular enclosure with a rotating adiabatic. *Journal of the Taiwan Institute of Chemical Engineers*, *45*(5), 2150–2162. doi:10.1016/j.jtice.2014.06.018

Selimefendigil, F., & Öztop, H. F. (2014). Pulsating nanofluids jet impingement cooling of a heated horizontal surface. *International Journal of Heat and Mass Transfer*, *69*, 54–65. doi:10.1016/j.ijheatmasstransfer.2013.10.010

Selimefendigil, F., & Öztop, H. F. (2014). Numerical study of MHD mixed convection in a nanofluid filled lid driven square enclosure with a rotating cylinder. *International Journal of Heat and Mass Transfer*, *78*, 741–754. doi:10.1016/j.ijheatmasstransfer.2014.07.031

Selimefendigil, F., & Öztop, H. F. (2015). Natural convection and entropy generation of nanofluid filled cavity having different shaped obstacles under the influence of magnetic field and internal heat generation. *Journal of the Taiwan Institute of Chemical Engineers*, *56*, 42–56. doi:10.1016/j.jtice.2015.04.018

Selimefendigil, F., & Öztop, H. F. (2015). Numerical investigation and reduced order model of mixed convection at a backward facing step with a rotating cylinder subjected to nanofluid. *Computers & Fluids*, *109*, 27–37. doi:10.1016/j.compfluid.2014.12.007

Selimefendigil, F., & Öztop, H. F. (2016). Conjugate natural convection in a cavity with a conductive partition and filled with different nanofluids on different sides of the partition. *Journal of Molecular Liquids*, *216*, 67–77. doi:10.1016/j.molliq.2015.12.102

Selimefendigil, F., Öztop, H. F., & Abu-Hamdeh, N. (2016). Mixed convection due to rotating cylinder in an internally heated and flexible walled cavity filled with SiO2–water nanofluids: Effect of nanoparticle shape. *International Communications in Heat and Mass Transfer*, *71*, 9–19. doi:10.1016/j.icheatmasstransfer.2015.12.007

Sheikholeslami. (2014). Effect of spatially variable magnetic field on ferrofluid flow and heat transfer considering constant heat flux boundary condition. *The European Physical Journal Plus*, 129–248.

Sheikholeslami, Ashorynejad, Domairry, & Hashim. (2012b). *Flow and Heat Transfer of Cu-Water Nanofluid between a Stretching Sheet and a Porous Surface in a Rotating System*. Hindawi Publishing Corporation. doi:. doi:10.1155/2012/421320

Sheikholeslami, M. (2015). Effect of uniform suction on nanofluid flow and heat transfer over a cylinder. *Journal of the Brazilian Society of Mechanical Sciences and Engineering*, *37*(6), 1623–1633. doi:10.100740430-014-0242-z

Sheikholeslami, M., & Abelman, S. (2015). Two phase simulation of nanofluid flow and heat transfer in an annulus in the presence of an axial magnetic field. *IEEE Transactions on Nanotechnology*, *14*(3), 561–569. doi:10.1109/TNANO.2015.2416318

Sheikholeslami, M., Abelman, S., & Ganji, D. D. (2014). Numerical simulation of MHD nanofluid flow and heat transfer considering viscous dissipation. *International Journal of Heat and Mass Transfer*, *79*, 212–222. doi:10.1016/j.ijheatmasstransfer.2014.08.004

Sheikholeslami, M., Ashorynejad, H. R., & Rana, P. (2016). Lattice Boltzmann simulation of nanofluid heat transfer enhancement and entropy generation. *Journal of Molecular Liquids*, *214*, 86–95. doi:10.1016/j.molliq.2015.11.052

Sheikholeslami, M., Bandpay, M. G., & Ganji, D. D. (2012). Magnetic field effects on natural convection around a horizontal circular cylinder inside a square enclosure filled with nanofluid. *International Communications in Heat and Mass Transfer*, *39*(7), 978–986. doi:10.1016/j.icheatmasstransfer.2012.05.020

Sheikholeslami, M., Bandpy, M. G., & Ashorynejad, H. R. (2015). Lattice Boltzmann Method for simulation of magnetic field effect on hydrothermal behavior of nanofluid in a cubic cavity. *Physica A*, *432*, 58–70. doi:10.1016/j.physa.2015.03.009

Sheikholeslami, M., Bandpy, M. G., Ellahi, R., & Zeeshan, A. (2014). R. Ellahi, A. Zeeshan, Simulation of MHD CuO–water nanofluid flow and convective heat transfer considering Lorentz forces. *Journal of Magnetism and Magnetic Materials*, *369*, 69–80. doi:10.1016/j.jmmm.2014.06.017

Sheikholeslami, M., & Ellahi, R. (2015). Electrohydrodynamic nanofluid hydrothermal treatment in an enclosure with sinusoidal upper wall. *Appl. Sci.*, *5*(3), 294–306. doi:10.3390/app5030294

Sheikholeslami, M., & Ellahi, R. (2015). Simulation of ferrofluid flow for magnetic drug targeting using Lattice Boltzmann method, Journal of Zeitschrift Fur Naturforschung A. *Verlag der Zeitschrift für Naturforschung*, *70*(2), 115–124.

Sheikholeslami, M., & Ellahi, R. (2015). Three dimensional mesoscopic simulation of magnetic field effect on natural convection of nanofluid. *International Journal of Heat and Mass Transfer*, *89*, 799–808. doi:10.1016/j.ijheatmasstransfer.2015.05.110

Sheikholeslami, M., Ellahi, R., Ashorynejad, H. R., Domairry, G., & Hayat, T. (2014). Effects of Heat Transfer in Flow of Nanofluids Over a Permeable Stretching Wall in a Porous Medium. *Journal of Computational and Theoretical Nanoscience*, *11*(2), 1–11. doi:10.1166/jctn.2014.3384

Sheikholeslami, M., Ellahi, R., Hassan, M., & Soleimani, S. (2014). Mohsan Hassan, Soheil Soleimani, A study of natural convection heat transfer in a nanofluid filled enclosure with elliptic inner cylinder. *International Journal of Numerical Methods for Heat & Fluid Flow*, *24*(8), 8. doi:10.1108/HFF-07-2013-0225

Sheikholeslami, M., & Ganji, D. D. (2013c). Heat transfer of Cu-water nanofluid flow between parallel plates. *Powder Technology*, *235*, 873–879. doi:10.1016/j.powtec.2012.11.030

Sheikholeslami, M., & Ganji, D. D. (2014). Magnetohydrodynamic flow in a permeable channel filled with nanofluid. *Scientia IranicaB*, *21*(1), 203–212.

Sheikholeslami, M., & Ganji, D. D. (2014). Ferrohydrodynamic and Magnetohydrodynamic effects on ferrofluid flow and convective heat transfer. *Energy*, *75*, 400–410. doi:10.1016/j.energy.2014.07.089

Sheikholeslami, M., & Ganji, D. D. (2014a). Heated permeable stretching surface in a porous medium using Nanofluids. *Journal of Applied Fluid Mechanics*, *7*(3), 535–542.

Sheikholeslami, M., & Ganji, D. D. (2014b). Three dimensional heat and mass transfer in a rotating system using nanofluid. *Powder Technology*, *253*, 789–796. doi:10.1016/j.powtec.2013.12.042

Sheikholeslami, M., & Ganji, D. D. (2014c). Numerical investigation for two phase modeling of nanofluid in a rotating system with permeable sheet. *Journal of Molecular Liquids*, *194*, 13–19. doi:10.1016/j.molliq.2014.01.003

Sheikholeslami, M., & Ganji, D. D. (2015). Nanofluid flow and heat transfer between parallel plates considering Brownian motion using DTM. *Computer Methods in Applied Mechanics and Engineering*, *283*, 651–663. doi:10.1016/j.cma.2014.09.038

Sheikholeslami, M., & Ganji, D. D. (2015). Unsteady nanofluid flow and heat transfer in presence of magnetic field considering thermal radiation. *Journal of the Brazilian Society of Mechanical Sciences and Engineering*, *37*(3), 895–902. doi:10.100740430-014-0228-x

Sheikholeslami, M., & Ganji, D. D. (2015). M. Younus Javed, R. Ellahi, Effect of thermal radiation on magnetohydrodynamics nanofluid flow and heat transfer by means of two phase model. *Journal of Magnetism and Magnetic Materials*, *374*, 36–43. doi:10.1016/j.jmmm.2014.08.021

Sheikholeslami, M., & Ganji, D. D. (2015). Entropy generation of nanofluid in presence of magnetic field using Lattice Boltzmann Method. *Physica A*, *417*, 273–286. doi:10.1016/j.physa.2014.09.053

Sheikholeslami, M., Ganji, D. D., & Ashorynejad, H. R. (2013b). Investigation of squeezing unsteady nanofluid flow using ADM. *Powder Technology*, *239*, 259–265. doi:10.1016/j.powtec.2013.02.006

Sheikholeslami, M., Ganji, D. D., Ashorynejad, H. R., & Houman, B. (2012a). Rokni, Analytical investigation of Jeffery-Hamel flow with high magnetic field and nano particle by Adomian decomposition method, Appl. Math. Mech.-. *Engl. Ed.*, *33*(1), 1553–1564.

Sheikholeslami, M., Ganji, D. D., Gorji-Bandpy, M., & Soleimani, S. (2014). Soheil Soleimani, Magnetic field effect on nanofluid flow and heat transfer using KKL model. *Journal of the Taiwan Institute of Chemical Engineers*, *45*(3), 795–807. doi:10.1016/j.jtice.2013.09.018

Sheikholeslami, M., Ganji, D. D., & Rashidi, M. M. (2015). Ferrofluid flow and heat transfer in a semi annulus enclosure in the presence of magnetic source considering thermal radiation. *Journal of the Taiwan Institute of Chemical Engineers*, *47*, 6–17. doi:10.1016/j.jtice.2014.09.026

Sheikholeslami, M., Gorji-Bandpay, M., & Ganji, D. D. (2014). Investigation of nanofluid flow and heat transfer in presence of magnetic field using KKL model. *Arabian Journal for Science and Engineering*, *39*(6), 5007–5016. doi:10.100713369-014-1060-4

Sheikholeslami, M., & Gorji-Bandpy, M. (2014). Free convection of ferrofluid in a cavity heated from below in the presence of an external magnetic field. *Powder Technology*, *256*, 490–498. doi:10.1016/j.powtec.2014.01.079

Sheikholeslami, M., Gorji-Bandpy, M., & Domairry, G. (2013). Free convection of nanofluid filled enclosure using lattice Boltzmann method (LBM), Appl. Math. Mech. -. *Engl. Ed.*, *34*(7), 1–15.

Sheikholeslami, M., Gorji Bandpy, M., Ellahi, R., Hassan, M., & Soleimani, S. (2014). Mohsan Hassan, Soheil Soleimani, Effects of MHD on Cu-water nanofluid flow and heat transfer by means of CVFEM. *Journal of Magnetism and Magnetic Materials*, *349*, 188–200. doi:10.1016/j.jmmm.2013.08.040

Sheikholeslami, M., Gorji-Bandpy, M., & Ganji, D. D. (2013). Natural convection in a nanofluid filled concentric annulus between an outer square cylinder and an inner elliptic cylinder, Scientia Iranica, Transaction B. *Mechanical Engineering (New York, N.Y.)*, *20*(4), 1–13.

Sheikholeslami, M., Gorji-Bandpy, M., & Ganji, D. D. (2013). Numerical investigation of MHD effects on Al2O3-water nanofluid flow and heat transfer in a semi-annulus enclosure using LBM. *Energy*, *60*, 501–510. doi:10.1016/j.energy.2013.07.070

Sheikholeslami, M., Gorji-Bandpy, M., & Ganji, D. D. (2014). Soheil Soleimani, Natural convection heat transfer in a nanofluid filled inclined L-shaped enclosure, IJST. *Transactions of Mechanical Engineering*, *38*, 217–226.

Sheikholeslami, M., Gorji-Bandpy, M., & Ganji, D. D. (2014). Lattice Boltzmann method for MHD natural convection heat transfer using nanofluid. *Powder Technology*, *254*, 82–93. doi:10.1016/j.powtec.2013.12.054

Sheikholeslami, M., Gorji-Bandpy, M., & Ganji, D. D. (2014). MHD free convection in an eccentric semi-annulus filled with nanofluid. *Journal of the Taiwan Institute of Chemical Engineers*, *45*(4), 1204–1216. doi:10.1016/j.jtice.2014.03.010

Sheikholeslami, M., Gorji-Bandpy, M., Ganji, D. D., Rana, P., & Soleimani, S. (2014). Soheil Soleimani, Magnetohydrodynamic free convection of Al2O3-water nanofluid considering Thermophoresis and Brownian motion effects. *Computers & Fluids*, *94*, 147–160. doi:10.1016/j.compfluid.2014.01.036

Sheikholeslami, M., Gorji-Bandpy, M., Ganji, D. D., & Soheil Soleimani, S. M. (2012). Seyyedi, Natural convection of nanofluids in an enclosure between a circular and a sinusoidal cylinder in the presence of magnetic field. *International Communications in Heat and Mass Transfer*, *39*(9), 1435–1443. doi:10.1016/j.icheatmasstransfer.2012.07.026

Sheikholeslami, M., Gorji-Bandpy, M., Ganji, D. D., & Soleimani, S. (2013). Soheil Soleimani, Effect of a magnetic field on natural convection in an inclined half-annulus enclosure filled with Cu–water nanofluid using CVFEM. *Advanced Powder Technology*, *24*(6), 980–991. doi:10.1016/j.apt.2013.01.012

Sheikholeslami, M., Gorji-Bandpy, M., Ganji, D. D., & Soleimani, S. (2014). Soheil Soleimani, Natural convection heat transfer in a cavity with sinusoidal wall filled with CuO-water nanofluid in presence of magnetic field. *Journal of the Taiwan Institute of Chemical Engineers*, *45*(1), 40–49. doi:10.1016/j.jtice.2013.04.019

Sheikholeslami, M., Gorji-Bandpy, M., Ganji, D. D., & Soleimani, S. (2014). Soheil Soleimani, MHD natural convection in a nanofluid filled inclined enclosure with sinusoidal wall using CVFEM. *Neural Computing & Applications*, *24*(3-4), 873–882. doi:10.100700521-012-1316-4

Sheikholeslami, M., Gorji-Bandpy, M., Ganji, D. D., & Soleimani, S. (2014). Soheil Soleimani, Heat flux boundary condition for nanofluid filled enclosure in presence of magnetic field. *Journal of Molecular Liquids*, *193*, 174–184. doi:10.1016/j.molliq.2013.12.023

Sheikholeslami, M., Gorji-Bandpy, M., Ganji, D. D., & Soleimani, S. (2014). Soheil Soleimani, Thermal management for free convection of nanofluid using two phase model. *Journal of Molecular Liquids*, *194*, 179–187. doi:10.1016/j.molliq.2014.01.022

Sheikholeslami, M., Gorji-Bandpy, M., Seyyedi, S. M., Ganji, D. D., & Houman, B. (2013). Rokni, Soheil Soleimani, Application of LBM in simulation of natural convection in a nanofluid filled square cavity with curve boundaries. *Powder Technology*, *247*, 87–94. doi:10.1016/j.powtec.2013.06.008

Sheikholeslami, M., Gorji-Bandpy, M., & Soleimani, S. (2013). Soheil Soleimani, Two phase simulation of nanofluid flow and heat transfer using heatline analysis. *International Communications in Heat and Mass Transfer*, *47*, 73–81. doi:10.1016/j.icheatmasstransfer.2013.07.006

Sheikholeslami, M., Gorji-Bandpy, M., & Vajravelu, K. (2015). Lattice Boltzmann Simulation of Magnetohydrodynamic Natural Convection Heat Transfer of Al2O3-water Nanofluid in a Horizontal Cylindrical Enclosure with an Inner Triangular Cylinder. *International Journal of Heat and Mass Transfer*, *80*, 16–25. doi:10.1016/j.ijheatmasstransfer.2014.08.090

Sheikholeslami, M., Hatami, M., & Domairry, G. (2015). Numerical simulation of two phase unsteady nanofluid flow and heat transfer between parallel plates in presence of time dependent magnetic field. *Journal of the Taiwan Institute of Chemical Engineers*, *46*, 43–50. doi:10.1016/j.jtice.2014.09.025

Sheikholeslami, M., Hatami, M., & Ganji, D. D. (2013a). Analytical investigation of MHD nanofluid flow in a Semi-Porous Channel. *Powder Technology*, *246*, 327–336. doi:10.1016/j.powtec.2013.05.030

Sheikholeslami, M., Hatami, M., & Ganji, D. D. (2014). Nanofluid flow and heat transfer in a rotating system in the presence of a magnetic field. *Journal of Molecular Liquids*, *190*, 112–120. doi:10.1016/j.molliq.2013.11.002

Sheikholeslami, M., Hatami, M., & Ganji, D. D. (2015). Numerical investigation of nanofluid spraying on an inclined rotating disk for cooling process. *Journal of Molecular Liquids*, *211*, 577–583. doi:10.1016/j.molliq.2015.07.006

Sheikholeslami, M., & Rashidi, M. M. (2015). Effect of space dependent magnetic field on free convection of Fe3O4-water nanofluid. *Journal of the Taiwan Institute of Chemical Engineers*, *56*, 6–15. doi:10.1016/j.jtice.2015.03.035

Sheikholeslami, M., & Rashidi, M. M. (2015). Ferrofluid heat transfer treatment in the presence of variable magnetic field. *The European Physical Journal Plus*, *130*(6), 115. doi:10.1140/epjp/i2015-15115-4

Sheikholeslami, M., & Rashidi, M. M. (2016). Non-uniform magnetic field effect on nanofluid hydrothermal treatment considering Brownian motion and thermophoresis effects. *Journal of the Brazilian Society of Mechanical Sciences and Engineering*, *38*(4), 1171–1184. doi:10.100740430-015-0459-5

Sheikholeslami, M., Rashidi, M. M., & Ganji, D. D. (2015). Effect of non-uniform magnetic field on forced convection heat transfer of Fe3O4- water nanofluid. *Computer Methods in Applied Mechanics and Engineering*, *294*, 299–312. doi:10.1016/j.cma.2015.06.010

Sheikholeslami, M., Rashidi, M. M., & Ganji, D. D. (2015). Numerical investigation of magnetic nanofluid forced convective heat transfer in existence of variable magnetic field using two phase model. *Journal of Molecular Liquids*, *212*, 117–126. doi:10.1016/j.molliq.2015.07.077

Sheikholeslami, M., & Soheil Soleimani, D. D. (2016). Ganji, Effect of electric field on hydrothermal behavior of nanofluid in a complex geometry. *Journal of Molecular Liquids*, *213*, 153–161. doi:10.1016/j.molliq.2015.11.015

Sheikholeslami, M., Vajravelu, K., & Rashidi, M. M. (2016). Forced convection heat transfer in a semi annulus under the influence of a variable magnetic field. *International Journal of Heat and Mass Transfer*, *92*, 339–348. doi:10.1016/j.ijheatmasstransfer.2015.08.066

Sheikholeslami Kandelousi, M. (2014). KKL correlation for simulation of nanofluid flow and heat transfer in a permeable channel. *Physics Letters. [Part A]*, *378*(45), 3331–3339. doi:10.1016/j.physleta.2014.09.046

Sheikholeslami, Mustafa, & Ganji. (2015). Nanofluid flow and heat transfer over a stretching porous cylinder considering thermal radiation. *Iranian Journal of Science & Technology, 39*(3), 433-440.

Sheikholeslami, Rashidi, Al Saad, Firouzi, Rokni, & Domairry. (2016). Steady nanofluid flow between parallel plates considering Thermophoresis and Brownian effects. *Journal of King Saud University - Science, 28*(4), 380-389.

Sheremet, M. A., Dinarvand, S., & Pop, I. (2015). Effect of thermal stratification on free convection in a square porous cavity filled with a nanofluid using Tiwari and Das' nanofluid model. *Physica E, Low-Dimensional Systems and Nanostructures*, *69*, 332–341. doi:10.1016/j.physe.2015.02.005

Sheremet, M. A., & Pop, I. (2014). Conjugate natural convection in a square porous cavity filled by a nanofluid using Buongiorno's mathematical model. *International Journal of Heat and Mass Transfer*, *79*, 137–145. doi:10.1016/j.ijheatmasstransfer.2014.07.092

Sheremet, M. A., & Pop, I. (2015a). Free convection in a triangular cavity filled with a porous medium saturated by a nanofluid: Buongiorno's mathematical model. *International Journal of Numerical Methods for Heat & Fluid Flow*, *25*(5), 1138–1161. doi:10.1108/HFF-06-2014-0181

Sheremet, M. A., & Pop, I. (2015b). Natural convection in a horizontal cylindrical annulus filled with a porous medium saturated by a nanofluid using Tiwari and Das' nanofluid model. *The European Physical Journal Plus*, *130*(6), 107. doi:10.1140/epjp/i2015-15107-4

Sheremet, M. A., Pop, I., & Ishak, A. (2015). Double-diffusive mixed convection in a porous open cavity filled with a nanofluid using Buongiorno's model. *Transport in Porous Media*, *109*(1), 131–145. doi:10.100711242-015-0505-x

Sheremet, M. A., Pop, I., & Rahman, M. M. (2015). Three-dimensional natural convection in a porous enclosure filled with a nanofluid using Buongiorno's mathematical model. *International Journal of Heat and Mass Transfer*, *82*, 396–405. doi:10.1016/j.ijheatmasstransfer.2014.11.066

Sheremet, M. A., Pop, I., & Rosca, N. C. (2016). Magnetic field effect on the unsteady natural convection in a wavy-walled cavity filled with a nanofluid: Buongiorno's mathematical model. *J. Taiwan Institute of Chemical Engineers*, *61*, 211–222. doi:10.1016/j.jtice.2015.12.015

Soheil Soleimani, M. (2012). Sheikholeslami, D.D. Ganji and M. Gorji-Bandpay, Natural convection heat transfer in a nanofluid filled semi-annulus enclosure. *International Communications in Heat and Mass Transfer*, *39*(4), 565–574. doi:10.1016/j.icheatmasstransfer.2012.01.016

Teamah, M. A., & El-Maghlany, W. M. (2012). Augmentation of natural convective heat transfer in square cavity by utilizing nanofluids in the presence of magnetic field and uniform heat generation/absorption. *International Journal of Thermal Sciences*, *58*, 130–142. doi:10.1016/j.ijthermalsci.2012.02.029

Wang, B. X., Zhou, L. R., & Peng, X. F. (2003). A fractal model for predicting the effective thermal conductivity of liquid with suspension of nanoparticles. *International Journal of Heat and Mass Transfer*, *46*(14), 2665–2672. doi:10.1016/S0017-9310(03)00016-4

Xuan, Y., & Li, Q. (2000). Heat transfer enhancement of nanofluids. *International Journal of Heat and Fluid Flow*, *21*(1), 58–64. doi:10.1016/S0142-727X(99)00067-3

Xuan, Y., & Roetzel, W. (2000). Conceptions for heat transfer correlations of nanofluids. *International Journal of Heat and Mass Transfer*, *43*(19), 3701–3707. doi:10.1016/S0017-9310(99)00369-5

Yu, W., & Choi, S. U. S. (2003). The role of interfacial layers in the enhanced thermal conductivity of nanofluids: A renovated Maxwell model. *Journal of Nanoparticle Research*, *5*(1/2), 167–171. doi:10.1023/A:1024438603801

ADDITIONAL READING

Jafaryar, M., Sheikholeslami, M., Li, M., & Moradi, R. (2018). Nanofluid turbulent flow in a pipe under the effect of twisted tape with alternate axis. *Journal of Thermal Analysis and Calorimetry*. doi:10.100710973-018-7093-2

Li, Z., Shehzad, S. A., & Sheikholeslami, M. (2018). An application of CVFEM for nanofluid heat transfer intensification in a porous sinusoidal cavity considering thermal non-equilibrium model. *Computer Methods in Applied Mechanics and Engineering*, *339*, 663–680. doi:10.1016/j.cma.2018.05.015

Li, Z., Sheikholeslami, M., Chamkha, A. J., Raizah, Z. A., & Saleem, S. (2018). Control Volume Finite Element Method for nanofluid MHD natural convective flow inside a sinusoidal annulus under the impact of thermal radiation. *Computer Methods in Applied Mechanics and Engineering*, *338*, 618–633. doi:10.1016/j.cma.2018.04.023

Li, Z., Sheikholeslami, M., Jafaryar, M., Shafee, A., & Chamkha, A. J. (2018). Investigation of nanofluid entropy generation in a heat exchanger with helical twisted tapes. *Journal of Molecular Liquids*, *266*, 797–805. doi:10.1016/j.molliq.2018.07.009

Li, Z., Sheikholeslami, M., Samandari, M., & Shafee, A. (2018). Nanofluid unsteady heat transfer in a porous energy storage enclosure in existence of Lorentz forces. *International Journal of Heat and Mass Transfer*, *127*, 914–926. doi:10.1016/j.ijheatmasstransfer.2018.06.101

Sheikholeslami, M. (2017a). Lattice Boltzmann Method simulation of MHD non-Darcy nanofluid free convection. *Physica B, Condensed Matter*, *516*, 55–71. doi:10.1016/j.physb.2017.04.029

Sheikholeslami, M. (2017b). CuO-water nanofluid free convection in a porous cavity considering Darcy law. *The European Physical Journal Plus*, *132*(1), 55. doi:10.1140/epjp/i2017-11330-3

Sheikholeslami, M. (2017c). Influence of Lorentz forces on nanofluid flow in a porous cavity by means of Non- Darcy model. *Engineering Computations*, *34*(8), 2651–2667. doi:10.1108/EC-01-2017-0008

Sheikholeslami, M. (2017d). Magnetic field influence on nanofluid thermal radiation in a cavity with tilted elliptic inner cylinder. *Journal of Molecular Liquids*, *229*, 137–147. doi:10.1016/j.molliq.2016.12.024

Sheikholeslami, M. (2017e). Numerical simulation of magnetic nanofluid natural convection in porous media. *Physics Letters. [Part A]*, *381*(5), 494–503. doi:10.1016/j.physleta.2016.11.042

Sheikholeslami, M. (2017f). Influence of Lorentz forces on nanofluid flow in a porous cylinder considering Darcy model. *Journal of Molecular Liquids*, *225*, 903–912. doi:10.1016/j.molliq.2016.11.022

Sheikholeslami, M. (2017g). Influence of Coulomb forces on Fe3O4-H2O nanofluid thermal improvement. *International Journal of Hydrogen Energy*, *42*(2), 821–829. doi:10.1016/j.ijhydene.2016.09.185

Sheikholeslami, M. (2017h). Numerical investigation of MHD nanofluid free convective heat transfer in a porous tilted enclosure. *Engineering Computations*, *34*(6), 1939–1955. doi:10.1108/EC-08-2016-0293

Sheikholeslami, M. (2017i). Magnetic field influence on CuO -H2O nanofluid convective flow in a permeable cavity considering various shapes for nanoparticles. *International Journal of Hydrogen Energy*, *42*(31), 19611–19621. doi:10.1016/j.ijhydene.2017.06.121

Sheikholeslami, M. (2018a). Magnetic source impact on nanofluid heat transfer using CVFEM. *Neural Computing & Applications*, *30*(4), 1055–1064. doi:10.100700521-016-2740-7

Sheikholeslami, M. (2018b). Numerical simulation for external magnetic field influence on Fe3O4-water nanofluid forced convection. *Engineering Computations*, *35*(4), 1639–1654. doi:10.1108/EC-06-2017-0200

Sheikholeslami, M. (2018c). Finite element method for PCM solidification in existence of CuO nanoparticles. *Journal of Molecular Liquids*, *265*, 347–355. doi:10.1016/j.molliq.2018.05.132

Sheikholeslami, M. (2018d). Solidification of NEPCM under the effect of magnetic field in a porous thermal energy storage enclosure using CuO nanoparticles. *Journal of Molecular Liquids*, *263*, 303–315. doi:10.1016/j.molliq.2018.04.144

Sheikholeslami, M. (2018e). Influence of magnetic field on Al2O3-H2O nanofluid forced convection heat transfer in a porous lid driven cavity with hot sphere obstacle by means of LBM. *Journal of Molecular Liquids*, *263*, 472–488. doi:10.1016/j.molliq.2018.04.111

Sheikholeslami, M. (2018f). Numerical simulation for solidification in a LHTESS by means of Nano-enhanced PCM. *Journal of the Taiwan Institute of Chemical Engineers*, *86*, 25–41. doi:10.1016/j.jtice.2018.03.013

Sheikholeslami, M. (2018g). Numerical modeling of Nano enhanced PCM solidification in an enclosure with metallic fin. *Journal of Molecular Liquids*, *259*, 424–438. doi:10.1016/j.molliq.2018.03.006

Sheikholeslami, M. (2018h). Numerical investigation of nanofluid free convection under the influence of electric field in a porous enclosure. *Journal of Molecular Liquids*, *249*, 1212–1221. doi:10.1016/j.molliq.2017.11.141

Sheikholeslami, M. (2018i). CuO-water nanofluid flow due to magnetic field inside a porous media considering Brownian motion. *Journal of Molecular Liquids*, *249*, 921–929. doi:10.1016/j.molliq.2017.11.118

Sheikholeslami, M. (2018j). Application of Darcy law for nanofluid flow in a porous cavity under the impact of Lorentz forces. *Journal of Molecular Liquids*, *266*, 495–503. doi:10.1016/j.molliq.2018.06.083

Sheikholeslami, M., Barzegar Gerdroodbary, M., Valiallah Mousavi, S., Ganji, D. D., & Moradi, R. (2018). Heat transfer enhancement of ferrofluid inside an 90o elbow channel by non-uniform magnetic field. *Journal of Magnetism and Magnetic Materials*, *460*, 302–311. doi:10.1016/j.jmmm.2018.03.070

Sheikholeslami, M., & Bhatti, M. M. (2017). Active method for nanofluid heat transfer enhancement by means of EHD. *International Journal of Heat and Mass Transfer*, *109*, 115–122. doi:10.1016/j.ijheatmasstransfer.2017.01.115

Sheikholeslami, M., Darzi, M., & Li, Z. (2018). Experimental investigation for entropy generation and exergy loss of nano-refrigerant condensation process. *International Journal of Heat and Mass Transfer*, *125*, 1087–1095. doi:10.1016/j.ijheatmasstransfer.2018.04.155

Sheikholeslami, M., Darzi, M., & Sadoughi, M. K. (2018). Heat transfer improvement and Pressure Drop during condensation of refrigerant-based Nanofluid; An Experimental Procedure. *International Journal of Heat and Mass Transfer*, *122*, 643–650. doi:10.1016/j.ijheatmasstransfer.2018.02.015

Sheikholeslami, M., & Ghasemi, A. (2018). Solidification heat transfer of nanofluid in existence of thermal radiation by means of FEM. *International Journal of Heat and Mass Transfer*, *123*, 418–431. doi:10.1016/j.ijheatmasstransfer.2018.02.095

Sheikholeslami, M., Ghasemi, A., Li, Z., Shafee, A., & Saleem, S. (2018). Influence of CuO nanoparticles on heat transfer behavior of PCM in solidification process considering radiative source term. *International Journal of Heat and Mass Transfer*, *126*, 1252–1264. doi:10.1016/j.ijheatmasstransfer.2018.05.116

Sheikholeslami, M., Hayat, T., & Alsaedi, A. (2018). Numerical simulation for forced convection flow of MHD CuO-H2O nanofluid inside a cavity by means of LBM. *Journal of Molecular Liquids*, *249*, 941–948. doi:10.1016/j.molliq.2017.10.099

Sheikholeslami, M., Hayat, T., Muhammad, T., & Alsaedi, A. (2018). MHD forced convection flow of nanofluid in a porous cavity with hot elliptic obstacle by means of Lattice Boltzmann method. *International Journal of Mechanical Sciences*, *135*, 532–540. doi:10.1016/j.ijmecsci.2017.12.005

Sheikholeslami, M., Jafaryar, M., Ganji, D. D., & Li, Z. (2018). Exergy loss analysis for nanofluid forced convection heat transfer in a pipe with modified turbulators. *Journal of Molecular Liquids*, *262*, 104–110. doi:10.1016/j.molliq.2018.04.077

Sheikholeslami, M., Jafaryar, M., & Li, Z. (2018a). Second law analysis for nanofluid turbulent flow inside a circular duct in presence of twisted tape turbulators. *Journal of Molecular Liquids*, *263*, 489–500. doi:10.1016/j.molliq.2018.04.147

Sheikholeslami, M., Jafaryar, M., & Li, Z. (2018b). Nanofluid turbulent convective flow in a circular duct with helical turbulators considering CuO nanoparticles. *International Journal of Heat and Mass Transfer*, *124*, 980–989. doi:10.1016/j.ijheatmasstransfer.2018.04.022

Sheikholeslami, M., Jafaryar, M., Saleem, S., Li, Z., Shafee, A., & Jiang, Y. (2018). Nanofluid heat transfer augmentation and exergy loss inside a pipe equipped with innovative turbulators. *International Journal of Heat and Mass Transfer*, *126*, 156–163. doi:10.1016/j.ijheatmasstransfer.2018.05.128

Sheikholeslami, M., Jafaryar, M., Shafee, A., & Li, Z. (2018). Investigation of second law and hydrothermal behavior of nanofluid through a tube using passive methods. *Journal of Molecular Liquids*, *269*, 407–416. doi:10.1016/j.molliq.2018.08.019

Sheikholeslami, M., Li, Z., & Shafee, A. (2018a). Lorentz forces effect on NEPCM heat transfer during solidification in a porous energy storage system. *International Journal of Heat and Mass Transfer*, *127*, 665–674. doi:10.1016/j.ijheatmasstransfer.2018.06.087

Sheikholeslami, M., Li, Z., & Shamlooei, M. (2018). Nanofluid MHD natural convection through a porous complex shaped cavity considering thermal radiation. *Physics Letters. [Part A]*, *382*(24), 1615–1632. doi:10.1016/j.physleta.2018.04.006

Sheikholeslami, M., & Rokni, H. B. (2017). Simulation of nanofluid heat transfer in presence of magnetic field: A review. *International Journal of Heat and Mass Transfer*, *115*, 1203–1233. doi:10.1016/j.ijheatmasstransfer.2017.08.108

Sheikholeslami, M., & Rokni, H. B. (2018a). CVFEM for effect of Lorentz forces on nanofluid flow in a porous complex shaped enclosure by means of Non-equilibrium model. *Journal of Molecular Liquids*, *254*, 446–462. doi:10.1016/j.molliq.2018.01.130

Sheikholeslami, M., Rokni, H.B. (2018b). Magnetic nanofluid flow and convective heat transfer in a porous cavity considering Brownian motion effects. *Physics of Fluids*, *30*(1), doi:10.1063/1.5012517

Sheikholeslami, M., & Sadoughi, M. K. (2017). Mesoscopic method for MHD nanofluid flow inside a porous cavity considering various shapes of nanoparticles. *International Journal of Heat and Mass Transfer*, *113*, 106–114. doi:10.1016/j.ijheatmasstransfer.2017.05.054

Sheikholeslami, M., & Sadoughi, M. K. (2018). Simulation of CuO- water nanofluid heat transfer enhancement in presence of melting surface. *International Journal of Heat and Mass Transfer*, *116*, 909–919. doi:10.1016/j.ijheatmasstransfer.2017.09.086

Sheikholeslami, M., & Seyednezhad, M. (2018). Simulation of nanofluid flow and natural convection in a porous media under the influence of electric field using CVFEM. *International Journal of Heat and Mass Transfer*, *120*, 772–781. doi:10.1016/j.ijheatmasstransfer.2017.12.087

Sheikholeslami, M., Shafee, A., Ramzan, M., & Li, Z. (2018). Investigation of Lorentz forces and radiation impacts on nanofluid treatment in a porous semi annulus via Darcy law. *Journal of Molecular Liquids*, *272*, 8–14. doi:10.1016/j.molliq.2018.09.016

Sheikholeslami, M., Shamlooei, M., & Moradi, R. (2018). Numerical simulation for heat transfer intensification of nanofluid in a porous curved enclosure considering shape effect of Fe3O4 nanoparticles. *Chemical Engineering & Processing: Process Intensification*, *124*, 71–82. doi:10.1016/j.cep.2017.12.005

Sheikholeslami, M., & Shehzad, S. A. (2017a). Thermal radiation of ferrofluid in existence of Lorentz forces considering variable viscosity. *International Journal of Heat and Mass Transfer*, *109*, 82–92. doi:10.1016/j.ijheatmasstransfer.2017.01.096

Sheikholeslami, M., & Shehzad, S. A. (2017b). Magnetohydrodynamic nanofluid convective flow in a porous enclosure by means of LBM. *International Journal of Heat and Mass Transfer*, *113*, 796–805. doi:10.1016/j.ijheatmasstransfer.2017.05.130

Sheikholeslami, M., & Shehzad, S. A. (2018a). Numerical analysis of Fe3O4 –H2O nanofluid flow in permeable media under the effect of external magnetic source. *International Journal of Heat and Mass Transfer*, *118*, 182–192. doi:10.1016/j.ijheatmasstransfer.2017.10.113

Sheikholeslami, M., & Shehzad, S. A. (2018b). CVFEM simulation for nanofluid migration in a porous medium using Darcy model. *International Journal of Heat and Mass Transfer*, *122*, 1264–1271. doi:10.1016/j.ijheatmasstransfer.2018.02.080

Sheikholeslami, M., & Shehzad, S. A. (2018c). Simulation of water based nanofluid convective flow inside a porous enclosure via Non-equilibrium model. *International Journal of Heat and Mass Transfer*, *120*, 1200–1212. doi:10.1016/j.ijheatmasstransfer.2017.12.132

Sheikholeslami, M., & Shehzad, S. A. (2018d). Non-Darcy free convection of Fe3O4-water nanoliquid in a complex shaped enclosure under impact of uniform Lorentz force. *The Chinese Journal of Physiology*, *56*(1), 270–281. doi:10.1016/j.cjph.2017.12.017

Sheikholeslami, M., Shehzad, S. A., Abbasi, F. M., & Li, Z. (2018). Nanofluid flow and forced convection heat transfer due to Lorentz forces in a porous lid driven cubic enclosure with hot obstacle. *Computer Methods in Applied Mechanics and Engineering*, *338*, 491–505. doi:10.1016/j.cma.2018.04.020

Sheikholeslami, M., Shehzad, S. A., & Li, Z. (2018a). Nanofluid heat transfer intensification in a permeable channel due to magnetic field using Lattice Boltzmann method. *Physica B, Condensed Matter*, *542*, 51–58. doi:10.1016/j.physb.2018.03.036

Sheikholeslami, M., Shehzad, S. A., & Li, Z. (2018b). Water based nanofluid free convection heat transfer in a three dimensional porous cavity with hot sphere obstacle in existence of Lorenz forces. *International Journal of Heat and Mass Transfer*, *125*, 375–386. doi:10.1016/j.ijheatmasstransfer.2018.04.076

Sheikholeslami, M., Shehzad, S. A., Li, Z., & Shafee, A. (2018). Numerical modeling for Alumina nanofluid magnetohydrodynamic convective heat transfer in a permeable medium using Darcy law. *International Journal of Heat and Mass Transfer*, *127*, 614–622. doi:10.1016/j.ijheatmasstransfer.2018.07.013

Sheikholeslami, M., & Vajravelu, K. (2017). Forced convection heat transfer in Fe3O4- ethylene glycol nanofluid under the influence of Coulomb force. *Journal of Molecular Liquids*, *233*, 203–210. doi:10.1016/j.molliq.2017.03.026

Sheikholeslami, M., & Zeeshan, A. (2017). Analysis of flow and heat transfer in water based nanofluid due to magnetic field in a porous enclosure with constant heat flux using CVFEM. *Computer Methods in Applied Mechanics and Engineering*, *320*, 68–81. doi:10.1016/j.cma.2017.03.024

Sheikholeslami, M., Zeeshan, A., & Majeed, A. (2018). Control volume based finite element simulation of magnetic nanofluid flow and heat transport in non-Darcy medium. *Journal of Molecular Liquids*, *268*, 354–364. doi:10.1016/j.molliq.2018.07.031

Chapter 2
Uniform Lorenz Forces Impact on Nanoparticles Transportation

ABSTRACT

Natural convection under the influence of a magnetic field has great importance in many industrial applications such as crystal growth, metal casting, and liquid metal cooling blankets for fusion reactors. The existence of a magnetic field has a noticeable effect on heat transfer reduction under natural convection while in many engineering applications increasing heat transfer from solid surfaces is a goal. At this circumstance, the use of nanofluids with higher thermal conductivity can be considered as a promising solution. In this chapter, the influence of Lorentz forces on hydrothermal behavior is studied.

1. ENTROPY GENERATION OF NANOFLUID IN PRESENCE OF MAGNETIC FIELD USING LATTICE BOLTZMANN METHOD

1.1. Problem Definition

The considered physical geometry with related parameters and coordinates are shown in Figure 1(a). A rectangular body with height *t* and width $H / 2$ is placed in the center of the enclosure, is supposed to be isothermal at higher temperature than two vertical isothermal walls while the top and bottom walls are insulated. Also, it is also assumed that the uniform magnetic field ($\vec{B} = B_x \vec{e_x} + B_y \vec{e_y}$) of constant magnitude $B = \sqrt{B_x^2 + B_y^2}$ is applied, where $\vec{e_x}$ and $\vec{e_y}$ are unit vectors in the Cartesian coordinate system. The orientation of the magnetic field form an angle θ_M with horizontal axis such that $\theta_M = B_x / B_y$. The electric current *J* and the electromagnetic force *F* are defined by $J = \sigma\left(\vec{V}\times \vec{B}\right)$ and $F = \sigma\left(\vec{V}\times \vec{B}\right)\times \vec{B}$, respectively. In this section, θ_M is equal to zero (Sheikholeslami & Ganji, 2015).

DOI: 10.4018/978-1-5225-7595-5.ch002

Figure 1. (a) Geometry of the problem; (b) Discrete velocity set of two-dimensional nine-velocity (D_2Q_9) model

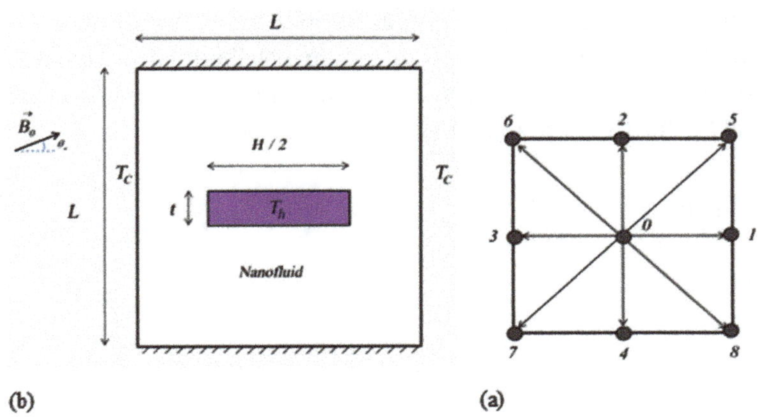

One of the novel computational fluid dynamics (CFD) methods which is solved Boltzmann equation to simulate the flow instead of solving the Navier–Stokes equations is called Lattice Boltzmann methods (LBM) (or Thermal Lattice Boltzmann methods (TLBM)). LBM has several advantages such as simple calculation procedure and efficient implementation for parallel computation, over other conventional CFD methods, because of its particulate nature and local dynamics. The thermal LB model utilizes two distribution functions, *f* and *g*, for the flow and temperature fields, respectively. It uses modeling of movement of fluid particles to capture macroscopic fluid quantities such as velocity, pressure and temperature. In this approach, the fluid domain discretized to uniform Cartesian cells. Each cell holds a fixed number of distribution functions, which represent the number of fluid particles moving in these discrete directions. The D_2Q_9 model was used and values of $w_0 = 4/9$ for $|c_0| = 0$ (for the static particle), $w_{1-4} = 1/9$ for $|c_{1-4}| = 1$ and $w_{5-9} = 1/36$ for $|c_{5-9}| = \sqrt{2}$ are assigned in this model (Figure 1(b)). The density and distribution functions i.e. the *f* and *g*, are calculated by solving the lattice Boltzmann equation, which is a special discretization of the kinetic Boltzmann equation. After introducing BGK approximation, the general form of lattice Boltzmann equation with external force is as follow:

For the flow field:

$$f_i(x + c_i\Delta t, t + \Delta t) = f_i(x,t) + \frac{\Delta t}{\tau_v}[f_i^{eq}(x,t) - f_i(x,t)] + \Delta t c_i F_k \tag{1}$$

For the temperature field:

$$g_i(x + c_i\Delta t, t + \Delta t) = g_i(x,t) + \frac{\Delta t}{\tau_C}[g_i^{eq}(x,t) - g_i(x,t)] \tag{2}$$

where Δt denotes lattice time step, c_i is the discrete lattice velocity in direction i, F_k is the external force in direction of lattice velocity, τ_v and τ_C denotes the lattice relaxation time for the flow and temperature fields. The kinetic viscosity υ and the thermal diffusivity α, are defined in terms of their respective relaxation times, i.e. $\upsilon = c_s^2(\tau_\upsilon - 1/2)$ and $\alpha = c_s^2(\tau_C - 1/2)$, respectively. Note that the limitation $0.5 < \tau$ should be satisfied for both relaxation times to ensure that viscosity and thermal diffusivity are positive. Furthermore, the local equilibrium distribution function determines the type of problem that needs to be solved. It also models the equilibrium distribution functions, which are calculated with Eqs. (3) and (4) for flow and temperature fields, respectively.

$$f_i^{eq} = w_i\rho\left[1 + \frac{c_i.u}{c_s^2} + \frac{1}{2}\frac{(c_i.u)^2}{c_s^4} - \frac{1}{2}\frac{u^2}{c_s^2}\right] \tag{3}$$

$$g_i^{eq} = w_i T\left[1 + \frac{c_i.u}{c_s^2}\right] \tag{4}$$

where w_i is a weighting factor and ρ is the lattice fluid density.

In order to incorporate buoyancy forces and magnetic forces in the model, the force term in the Equation (2) need to calculate as below (Sheikholeslami & Ganji, 2015):

$$\begin{aligned} F &= F_x + F_y \\ F_x &= 3w_i\rho\left[A\left(\upsilon \sin\left(\theta_M\right)\, cos\left(\theta_M\right)\right) - \left(u \sin^2\left(\theta_M\right)\right)\right], \\ F_y &= 3w_i\rho\left[g_y\beta\left(T - T_m\right) + A\left(u \sin\left(\theta_M\right)\, cos\left(\theta_M\right)\right) - \left(\upsilon \cos^2\left(\theta_M\right)\right)\right] \end{aligned} \tag{5}$$

where A is $A = \frac{Ha^2\mu}{L^2}$, $Ha = LB_0\sqrt{\frac{\sigma}{\mu}}$ is Hartmann number and θ_M is the direction of the magnetic field.

For natural convection, the Boussinesq approximation is applied and radiation heat transfer is negligible. To ensure that the code works in near incompressible regime, the characteristic velocity of the flow for natural ($V_{natural} \equiv \sqrt{\beta g_y \Delta TH}$) regime must be small compared with the fluid speed of sound. In the present study, the characteristic velocity selected as 0.1 of sound speed.

Finally, macroscopic variables calculate with the following formula:

$$\begin{aligned} &\text{Flow density} : \rho = \sum_i f_i, \\ &\text{Momentum} : \rho u = \sum_i \mathrm{c}_i f_i, \\ &\text{Temperature} : T = \sum_i g_i. \end{aligned} \tag{6}$$

According to Bejan (1982), one can find the volumetric entropy generation rate as

$$S^{\bullet}_{gen} = HTI + FFI \tag{7}$$

where *HTI* is the irreversibility due to heat transfer in the direction of finite temperature gradients and *FFI* is the contribution of fluid friction irreversibility to the total generated entropy.

In terms of the primitive variables, *HTI* and *FFI* become

$$\begin{aligned} HTI &= \frac{k\left(\nabla T.\nabla T\right)}{T^2} \\ FFI &= \frac{\mu\phi}{T} \end{aligned} \tag{8}$$

One can also define the Bejan number, Be, as

$$Be = \frac{HTI}{HTI + FFI} \tag{9}$$

Note that a Be value more/less than 0.5 shows that the contribution of *HTI* to the total entropy generation is higher/lower than that of *FFI*. The limiting value of $Be = 1$ shows that the only active entropy generation mechanism is *HTI* while $Be = 0$ represents no HTI contribution.

The dimensionless form of entropy generation rate, Ns, is defined as

$$Ns = \frac{\left(\frac{H}{\Omega}\right)^2 S^{\bullet}_{gen}}{k} \tag{10}$$

one finds that

$$Ns = \frac{\left(\frac{\partial\theta}{\partial x}\right)^2 + \left(\frac{\partial\theta}{\partial y}\right)^2}{\Omega^2\left(\Omega^{-1} + \theta\right)^2} + \frac{Ge\,\varphi}{Ra\Omega^2\left(\Omega^{-1} + \theta\right)} \tag{11}$$

where the dimensionless temperature difference is defined as

$$\Omega = \frac{T_h - T_c}{T_c} \tag{12}$$

The dimensionless viscous dissipation function, addressed in Equation (11), takes the following form

$$\varphi = 2\left(\frac{\partial u}{\partial x}\right)^2 + 2\left(\frac{\partial v}{\partial x}\right)^2 + \left(\frac{\partial u}{\partial x} + \frac{\partial v}{\partial x}\right)^2 \tag{13}$$

Here, Ge is the Gebhart number which is defined as (Sheikholeslami, Ashorynejad, & Rana, 2016).

$$Ge = \frac{g\beta H}{C_p} \tag{14}$$

Average Ns is denoted by $< Ns >$, where the angle brackets show an average taken over the area, as:

$$< Ns >= \int_A \frac{Ns}{A} dA \tag{15}$$

Selecting the fluid, trapped between the heated plate and the cavity, as the thermodynamic system, one observes that the amount of heat entered through the heated plate is equal to the one transferred to the surroundings via the isothermal walls. Moreover, one notes that the total volumetric entropy generation rate is obtainable as

$$< S^{\bullet}_{gen} >= \frac{q''}{H}\left(\frac{1}{T_c} - \frac{1}{T_H}\right) \tag{16}$$

where, in terms of Nu, it reads

$$< S^{\bullet}_{gen} >= 4Nu \frac{k}{H^2} \frac{\Omega^2}{1+\Omega} \tag{17}$$

Applying perturbation techniques for small values of Ω, say $\Omega << 1$, one has

$$< S^{\bullet}_{gen} >\approx \frac{4\, Nu\, \Omega^2 \left(1 - \Omega\right) k}{H^2} \tag{18}$$

The dimensionless entropy generation number can be obtained as

$$< Ns >= \frac{4\, Nu}{1+\Omega} \tag{19}$$

In order to simulate the nanofluid by the lattice Boltzmann method, because of the interparticle potentials and other forces on the nanoparticles, the nanofluid behaves differently from the pure liquid from the mesoscopic point of view and is of higher efficiency in energy transport as well as better stabilization than the common solid-liquid mixture. For modeling the nanofluid because of changing in the fluid thermal conductivity, density, heat capacitance and thermal expansion, some of the governed equations should change. The effective density $\left(\rho_{nf}\right)$, the effective heat capacity $\left(\rho C_p\right)_{nf}$, thermal expansion $\left(\rho\beta\right)_{nf}$ and electrical conductivity $\left(\sigma\right)_{nf}$ of the nanofluid are defined as:

$$\rho_{nf} = \rho_f(1-\phi) + \rho_s\phi, \tag{20}$$

$$\left(\rho C_p\right)_{nf} = \left(\rho C_p\right)_f(1-\phi) + \left(\rho C_p\right)_s\phi \tag{21}$$

$$\left(\rho\beta\right)_{nf} = \left(\rho\beta\right)_f(1-\phi) + \left(\rho\beta\right)_s\phi \tag{22}$$

$$\frac{\sigma_{nf}}{\sigma_f} = 1 + \frac{3\left(\frac{\sigma_s}{\sigma_f} - 1\right)\phi}{\left(\frac{\sigma_s}{\sigma_f} + 2\right) - \left(\frac{\sigma_s}{\sigma_f} - 1\right)\phi} \tag{23}$$

where ϕ is the solid volume fraction of the nanoparticles and subscripts f, nf and s stand for base fluid, nanofluid and solid, respectively.

$\left(k_{nf}\right)$ and $\left(\mu_{nf}\right)$ are obtained according to Koo–Kleinstreuer–Li (KKL) model (Sheikholeslami & Kandelousi, 2014):

$$\begin{aligned}
k_{nf} &= 1 + \frac{3\left(\frac{k_p}{k_f} - 1\right)\phi}{\left(\frac{k_p}{k_f} + 2\right) - \left(\frac{k_p}{k_f} - 1\right)\phi} + 5\times 10^4 g'(\phi, T, d_p)\phi\rho_f c_{p,f}\sqrt{\frac{\kappa_b T}{\rho_p d_p}} \\
g'\left(\phi, T, d_p\right) &= \left(a_6 + a_7 Ln\left(d_p\right) + a_8 Ln\left(\phi\right) + a_9 Ln\left(\phi\right)\ln\left(d_p\right) + a_{10} Ln\left(d_p\right)^2\right) \\
&+ Ln\left(T\right)\left(a_1 + a_2 Ln\left(d_p\right) + a_3 Ln\left(\phi\right) + a_4 Ln\left(\phi\right)\ln\left(d_p\right) + a_5 Ln\left(d_p\right)^2\right) \\
R_f &+ d_p / k_p = d_p / k_{p,eff}, R_f = 4\times 10^{-8} km^2 / W
\end{aligned} \tag{24}$$

$$\mu_{nf} = \frac{\mu_f}{\left(1-\phi\right)^{2.5}} + \frac{k_{Brownian}}{k_f} \times \frac{\mu_f}{\Pr} \tag{25}$$

In order to compare total heat transfer rate, Nusselt number is used. The average Nusselt numbers are defined as follows:

$$Nu_{ave} = \frac{k_{nf}}{k_f} \int_0^1 \frac{1}{4} \left(\left. \frac{\partial T}{\partial x} \right|_{X=0} + \left. \frac{\partial T}{\partial x} \right|_{X=1} \right) dy. \tag{26}$$

1.2. Effects of Active Parameters

In this study Lattice Boltzmann method (LBM) is used to investigate the natural convection in a square enclosure filled with CuO-water nanofluid in presence of magnetic field. A body is placed at the center of enclosure. Calculations are made for various values of Hartmann number (*Ha=0 to 100*), volume fraction of nanoparticle ($\phi = 0$ to 0.04) and Rayleigh number ($Ra = 10^3, 10^4$ and 10^5) when height of rectangular body ($L / t = 10$) and Prandtl number ($\Pr = 6.8$).

The effects of nanoparticles on streamlines and isotherms are shown in Figure 2. By adding nanoparticle the absolute values of stream functions indicate that the strength of flow decreases. Although

Figure 2. Streamlines (left) and isotherms (right) contours between CuO-water nanofluid ($\phi = 0.04$) ($- \cdot\cdot -$) and pure fluid ($\phi = 0$)(—) when $Ra = 10^4$, $L / t = 10$, $Ha = 0$

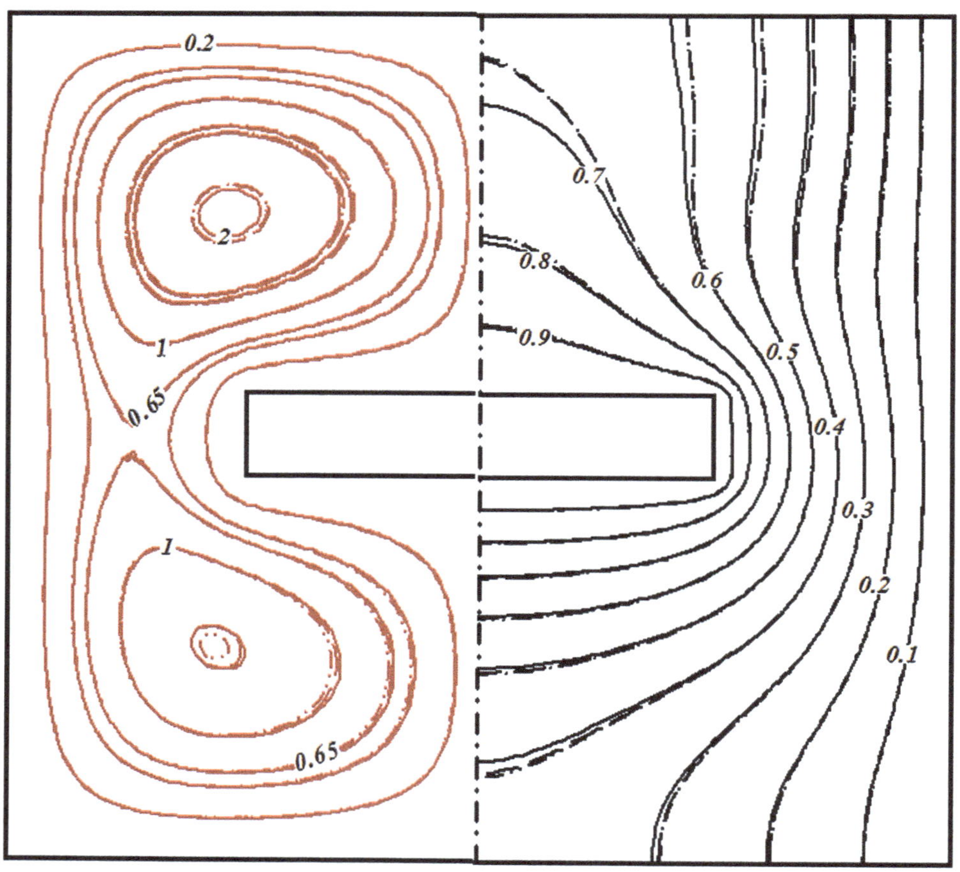

thermal boundary layer thickness decreases with adding nanoparticle in the base fluid, Nusselt number increases with increases of nanofluid volume fraction because of increment in thermal conductivity. Figure 3 shows the isotherms and streamlines for different Rayleigh numbers. In general, the heated lighter fluid is lifted and moves upward along the hot surface of the body and the vertical symmetry line until it encounters the cold top wall. Then the fluid becomes gradually colder and denser while it moves horizontally outward in contact with the cold top wall. Consequently, the cooled denser fluid descends along the cold side walls. For $Ra = 10^3$, the heat transfer in the enclosure is mainly dominated by the conduction mode. The circulation of the flow shows two overall rotating symmetric eddies with two inner vortices respectively as shown in Figure 3 for the streamlines. At $Ra = 10^4$, the patterns of the isotherms

Figure 3. Effects of Rayleigh numbers on streamlines (red) and isotherms (black) contours when $L\,/\,t = 10$, $Ha = 0$ *and* $\phi = 0.04$

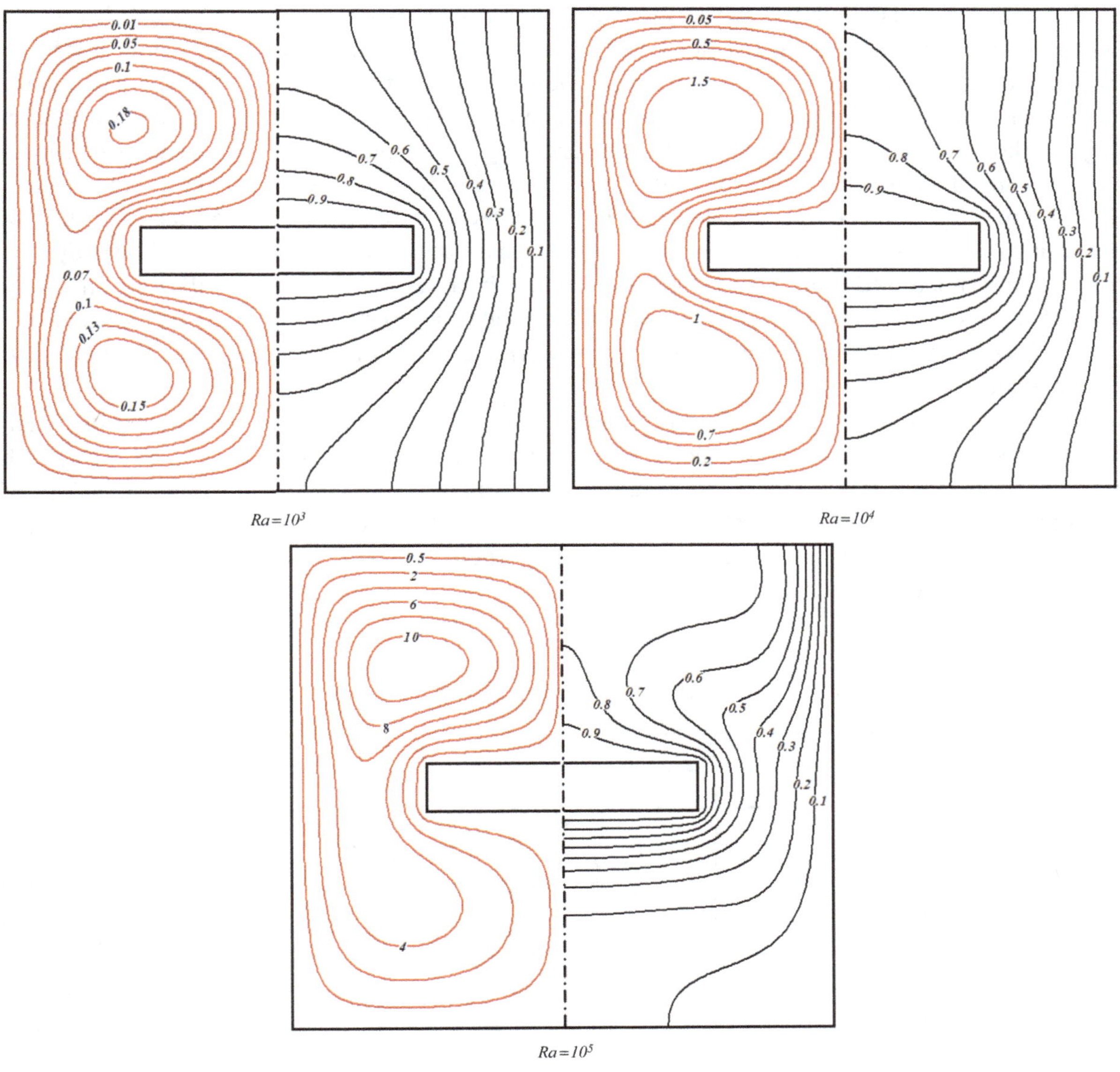

and streamlines are about the same as those for $Ra = 10^3$. However, a careful observation indicates that the thermal boundary layer on the bottom part of the body is thinner than that on the opposite side and the inner lower vortex slightly becomes smaller in size and weaker in strength compared with the upper one, because the effect of convection on heat transfer and flow increases with increasing the Rayleigh number. As the Rayleigh number increases up to 10^5, the role of convection in heat transfer becomes more significant and consequently the thermal boundary layer on the surface of the body becomes thinner. Also, a plume starts to appear on the top of the body and as a result the isotherms move upward, giving rise to a stronger thermal gradient in the upper part of the enclosure and a much lower thermal gradient in the lower part. In consequence, the dominant flow is in the upper half of the enclosure, and correspondingly the core of the recirculating eddies is located only in the upper half. At this Rayleigh number, the flow filed undergoes a bifurcation where two inner vorticies merge. The flow at the bottom of the enclosure is very weak compared with that at the middle and top regions, which suggests stratification effects in the lower regions of the enclosure.

Effects of Hartmann number on the streamlines and isotherms are shown in Figure 4. Hartmann number is the ratio of electromagnetic force to the viscous force. Variation of Hartmann number leads to the variation of the Lorentz force due to magnetic field and the Lorentz force produces more resistance to transport phenomena. Increase of the Hartmann number causes the flow strength decreases considerably. As the Hartmann number increases the primary eddy divide into two secondary eddies which are rotate in same direction. Pattern of the isotherms is affected strongly by changing intensity of magnetic field. There is high temperature gradient at the bottom of the body in the absent of magnetic field. With increasing Hartmann number thermal boundary layer thickness increases at the bottom of the body. The convection is suppressed at the higher Hartmann number. It causes that the plume on the top of the body wall disappears and isothermal lines become concentric and parallel between the body and the enclosure. It is shown that convection heat transfer becomes weaker and causes the heat transfer mostly dominated by conduction between the cylinders.

As shown in Table 1, at $Ra=10^3$, for all values of Hartmann number, the absolute values of stream function decreases with increase of nanoparticle volume fraction, whereas for higher value of Rayleigh number, i.e. ($Ra=10^4$ and $Ra=10^5$), the effect of nanoparticle volume fraction on absolute values of stream function is depended on the value of Hartmann number. In these Rayleigh numbers, when $Ha \leq 50$ absolute values of stream function decreases as nanoparticle volume fraction decreases but opposite trend is observed when $Ha > 50$.

Table 1. Effects of the nanoparticle volume fraction and Hartmann number on $|\psi_{\max}|$ *when* $L/t = 10$

Ha	φ	Ra		
		10^3	10^4	10^5
0	0	0.188952	2.05299	11.48
0	0.04	0.18672	2.0787	11.4451
50	0	0.018267	0.142036	1.57514
50	0.04	0.016833	0.145037	1.55233
100	0	0.016523	0.046635	0.444481
100	0.04	0.01025	0.044569	0.446956

Figure 4. Effect of Hartmann number on the streamlines (red) and isotherms (black) when $L / t = 10$*,* $Ra = 10^5$ *and* $\phi = 0.04$

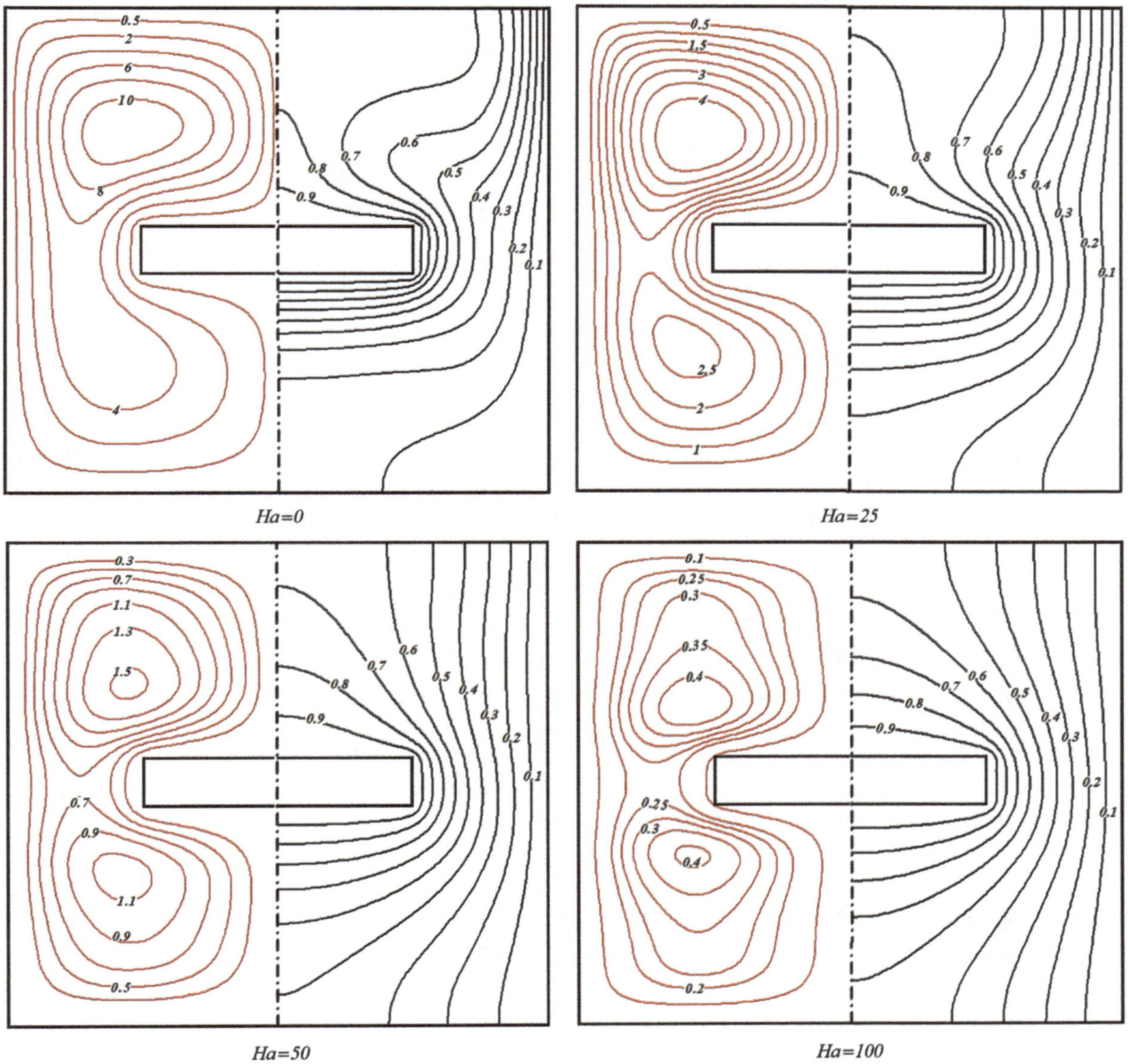

Figure 5 shows the effects of the nanoparticle volume fraction, Rayleigh number and Hartmann number for Cu-water nanofluids on average Nusselt number ratio ($Nu^* = \left(Nu_{ave}\big|_{\phi=0.04}\right) / \left(Nu_{ave}\big|_{\phi=0}\right)$). It can be found that the effect of nanoparticles is more pronounced at low Rayleigh numbers than at high Rayleigh numbers because of greater values of average Nusselt number ratios. This observation can be explained by noting that at low Rayleigh numbers the heat transfer is dominant by conduction. Therefore, the addition of high thermal conductivity nanoparticles will increase the conduction and therefore make

Figure 5. Effects of the active parameters on average Nusselt number ratio when $L / t = 10$

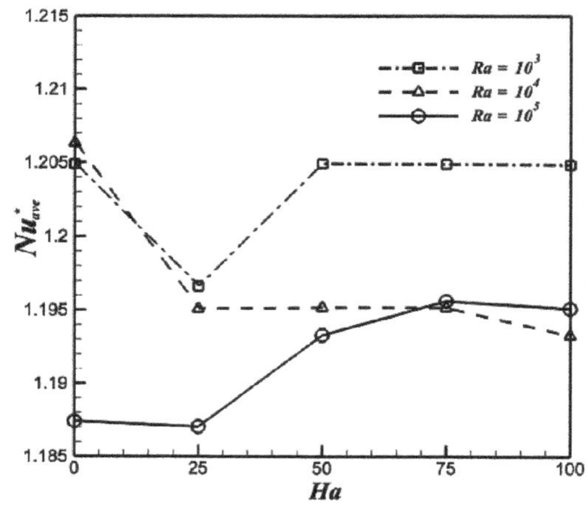

the enhancement more effective. When $Ha < 75$ minimum values of average Nusselt number ratio are obtained for $Ra=10^5$ but when $Ha > 75$ minimum values of this ratio occurs for $Ra=10^4$.

Effects of Hartmann number, the nanoparticle volume fraction, Rayleigh number and dimensionless temperature difference for CuO-water on Dimensionless entropy generation number are shown in Figure 6. The Dimensionless entropy generation number increases with increasing nanoparticle volume fraction and Rayleigh number. As seen in Figure 6, $< Ns >$ decreases with increase of Ω. This fact is in line with the predictions of our Equation (11). This also makes physical sense since, as $\Omega = T_h / T_c - 1$, higher values of Ω imply a greater temperature difference (leading to higher heat transfer rates) and consequently boosted HTI values. So decrease in $< Ns >$, according to $S^{\bullet}_{gen} = Ns.k\left(\Omega / L\right)^2$ leads to increase in the total entropy generation. Also these figures indicate that increasing Hartmann number leads to Dimensionless entropy generation number decreases.

2. MHD NATURAL CONVECTION IN A NANOFLUID FILLED INCLINED ENCLOSURE WITH SINUSOIDAL WALL USING CVFEM

2.1. Problem Definition

Schematic of the problem and the related boundary conditions as well as the mesh of enclosure which is used in the present CVFEM program are shown in Figure 7 (Sheikholeslami, 2018). The enclosure has a width /height aspect ratio of two. The two sidewalls with length H are thermally insulated whereas the lower flat and upper sinusoidal walls are maintained at constant temperatures T_h and T_c, respectively. Under all circumstances $T_h > T_c$ condition is maintained. The shape of the upper sinusoidal wall profile is assumed to mimic the following pattern

$$Y = H - \left\{a\left(H + \sin\left(\pi x - \pi / 2\right)\right)\right\} \tag{27}$$

Figure 6. Effects of the nanoparticle volume fraction, Hartmann number, Rayleigh number and dimensionless temperature difference for CuO-water on Dimensionless entropy generation number (Ns) (a) when $Ra = 10^5$, $\Omega = 0.06$; (b) $\phi = 0.04$, $\Omega = 0.06$; (c) $\phi = 0.04$, $Ra = 10^5$ and $L / t = 10$

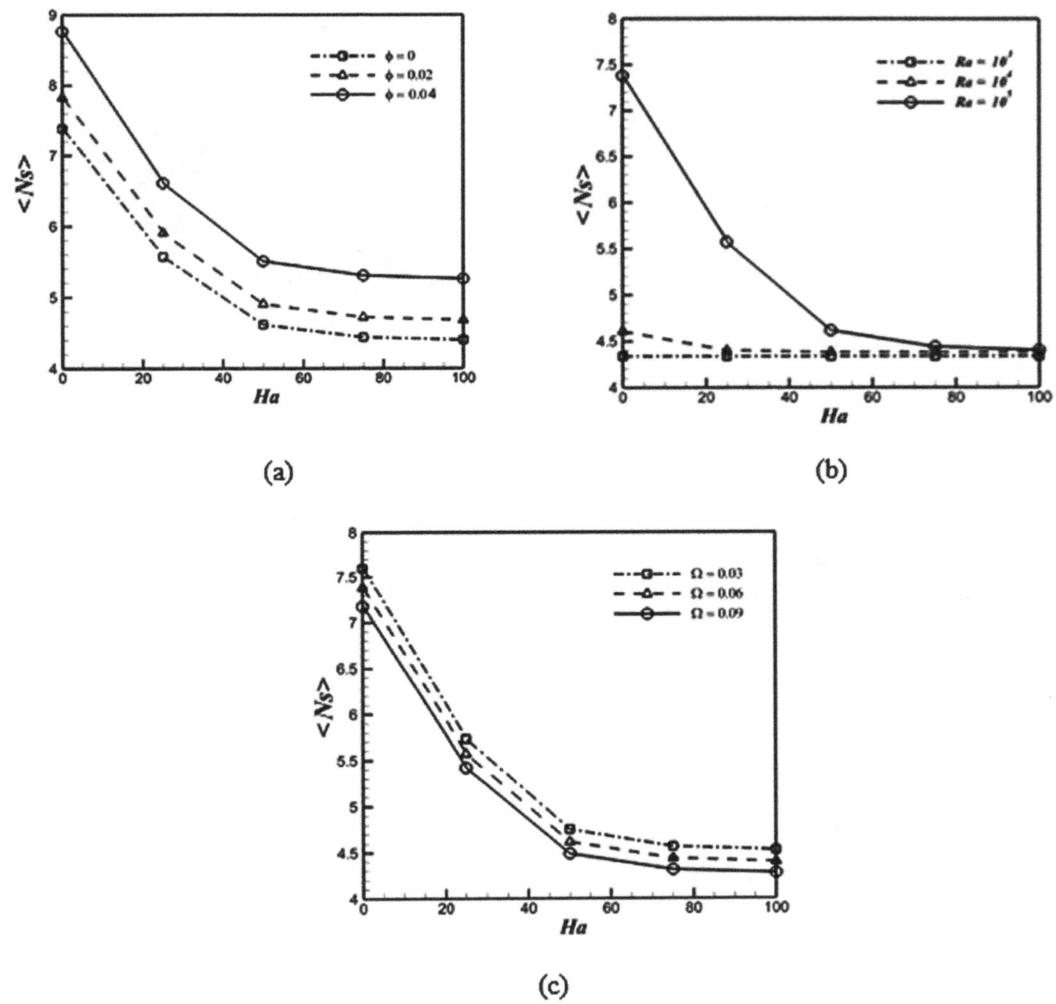

where a is the dimensionless amplitude of the sinusoidal wall. It is also assumed that the uniform magnetic field ($\vec{B} = B_x \vec{e_x} + B_y \vec{e_y}$) of constant magnitude $B = \sqrt{B_x^2 + B_y^2}$ is applied, where $\vec{e_x}$ and $\vec{e_y}$ are unit vectors in the Cartesian coordinate system. The orientation of the magnetic field form an angle γ with horizontal axis such that $\gamma = B_x / B_y$. The electric current J and the electromagnetic force F are defined by $J = \sigma \left(\vec{V} \times \vec{B} \right)$ and $F = \sigma \left(\vec{V} \times \vec{B} \right) \times \vec{B}$ respectively.

The flow is two-dimensional, laminar and incompressible. The radiation, viscous dissipation, induced electric current and Joule heating are neglected. The magnetic Reynolds number is assumed to be small so that the induced magnetic field can be neglected compared to the applied magnetic field. The flow

Figure 7. (a) Geometry and the boundary conditions, (b) the mesh of enclosure considered in this work

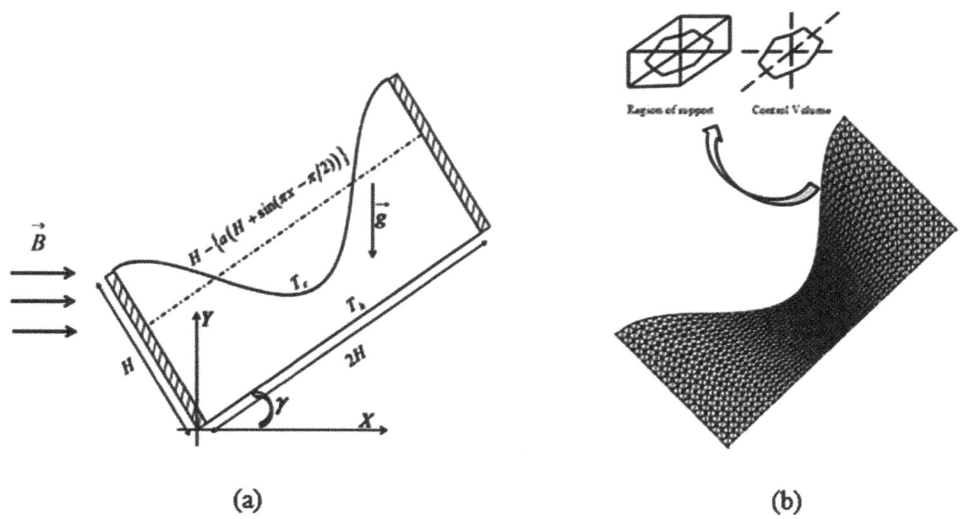

is considered to be steady, two dimensional and laminar. Neglecting displacement currents, induced magnetic field, and using the Boussinesq approximation, the governing equations of heat transfer and fluid flow for nanofluid can be obtained as follows (Sheikholeslami et al., 2018):

$$\frac{\partial u}{\partial x}+\frac{\partial v}{\partial y}=0 \tag{28}$$

$$u\frac{\partial u}{\partial x}+v\frac{\partial u}{\partial y}=-\frac{1}{\rho_{nf}}\frac{\partial P}{\partial x}+\upsilon_{nf}\left(\frac{\partial^2 u}{\partial x^2}+\frac{\partial^2 u}{\partial y^2}\right)+\frac{\sigma_{nf}B^2}{\rho_{nf}}\left(v\sin\lambda\cos\lambda-u\sin^2\lambda\right) \tag{29}$$

$$u\frac{\partial v}{\partial x}+v\frac{\partial v}{\partial y}=-\frac{1}{\rho_{nf}}\frac{\partial P}{\partial y}+\upsilon_{nf}\left(\frac{\partial^2 v}{\partial x^2}+\frac{\partial^2 v}{\partial y^2}\right)+\beta_{nf}g\left(T-T_c\right)+\frac{\sigma_{nf}B^2}{\rho_{nf}}\left(u\sin\lambda\cos\lambda-v\cos^2\lambda\right) \tag{30}$$

$$u\frac{\partial T}{\partial x}+v\frac{\partial T}{\partial y}=\alpha_{nf}\left(\frac{\partial^2 T}{\partial x^2}+\frac{\partial^2 T}{\partial y^2}\right) \tag{31}$$

where the effective density (ρ_{nf}) and heat capacitance $\left(\rho C_p\right)_{nf}$ of the nanofluid are defined as:

$$\rho_{nf}=\rho_f(1-\phi)+\rho_s\phi \tag{32}$$

$$\left(\rho C_p\right)_{nf} = \left(\rho C_p\right)_f (1-\phi) + \left(\rho C_p\right)_s \phi \tag{33}$$

where ϕ is the solid volume fraction of nanoparticles. Thermal diffusivity of the nanofluids is

$$\alpha_{nf} = \frac{k_{nf}}{\left(\rho C_p\right)_{nf}} \tag{34}$$

and the thermal expansion coefficient of the nanofluids can be determined as

$$\beta_{nf} = \beta_f (1-\phi) + \beta_s \phi \tag{35}$$

The dynamic viscosity of the nanofluids given is

$$\mu_{nf} = \frac{\mu_f}{(1-\phi)^{2.5}} \tag{36}$$

The effective thermal conductivity of the nanofluid can be approximated as:

$$\frac{k_{nf}}{k_f} = \frac{k_s + 2k_f - 2\phi(k_f - k_s)}{k_s + 2k_f + \phi(k_f - k_s)} \tag{37}$$

and the effective electrical conductivity of nanofluid was presented as below:

$$\frac{\sigma_{nf}}{\sigma_f} = 1 + \frac{3\left(\sigma_s / \sigma_f - 1\right)\phi}{\left(\sigma_s / \sigma_f + 2\right) - \left(\sigma_s / \sigma_f - 1\right)\phi} \tag{38}$$

The stream function and vorticity are defined as:

$$u = \frac{\partial \psi}{\partial y}, \quad v = -\frac{\partial \psi}{\partial x}, \quad \omega = \frac{\partial v}{\partial x} - \frac{\partial u}{\partial y} \tag{39}$$

The stream function satisfies the continuity Equation (28). The vorticity equation is obtained by eliminating the pressure between the two momentum equations, i.e. by taking y-derivative of Equation (29) and subtracting from it the x-derivative of Equation (30). This gives:

$$\frac{\partial\psi}{\partial y}\frac{\partial\omega}{\partial x}-\frac{\partial\psi}{\partial x}\frac{\partial\omega}{\partial y}=\upsilon_{nf}\left(\frac{\partial^2\omega}{\partial x^2}+\frac{\partial^2\omega}{\partial y^2}\right)+\beta_{nf}g\left(\frac{\partial T}{\partial x}\right)$$
$$+\frac{\sigma_{nf}B^2}{\rho_{nf}}\left(-\frac{\delta v}{\delta y}\sin\lambda\cos\lambda+\frac{\delta u}{\delta y}\sin^2\lambda+\frac{\delta u}{\delta x}\sin\lambda\cos\lambda-\frac{\delta v}{\delta x}\cos^2\lambda\right) \tag{40}$$

$$\frac{\partial\psi}{\partial y}\frac{\partial T}{\partial x}-\frac{\partial\psi}{\partial x}\frac{\partial T}{\partial y}=\alpha_{nf}\left(\frac{\partial^2 T}{\partial x^2}+\frac{\partial^2 T}{\partial y^2}\right) \tag{41}$$

$$\frac{\partial^2\psi}{\partial x^2}+\frac{\partial^2\psi}{\partial y^2}=-\omega \tag{42}$$

By introducing the following non-dimensional variables:

$$X=\frac{x}{L},Y=\frac{y}{L},\Omega=\frac{\omega L^2}{\alpha_f},\Psi=\frac{\psi}{\alpha_f},\Theta=\frac{T-T_c}{T_h-T_c},U=\frac{uL}{\alpha_f},V=\frac{vL}{\alpha_f} \tag{43}$$

where in Equation (43) $L=r_{out}-r_{in}=r_{in}$. Using the dimensionless parameters, the equations now become:

$$\frac{\partial\Psi}{\partial Y}\frac{\partial\Omega}{\partial X}-\frac{\partial\Psi}{\partial X}\frac{\partial\Omega}{\partial Y}=\left[\frac{\Pr_f}{\left(1-\phi\right)^{2.5}\left(\left(1-\phi\right)+\phi\frac{\rho_s}{\rho_f}\right)}\right]\left(\frac{\partial^2\Omega}{\partial X^2}+\frac{\partial^2\Omega}{\partial Y^2}\right)$$
$$+Ra_f\Pr_f\left[\left(1-\phi\right)+\phi\frac{\beta_s}{\beta_f}\right]\left(\frac{\partial\Theta}{\partial X}\right)$$
$$+Ha^2\Pr_f\left(1+\frac{3\left(\frac{\sigma_s}{\sigma_f}-1\right)\phi}{\left(\frac{\sigma_s}{\sigma_f}+2\right)-\left(\frac{\sigma_s}{\sigma_f}-1\right)\phi}\right)\left(\frac{1}{\left(1-\phi\right)+\phi\frac{\rho_s}{\rho_f}}\right)\left(-\frac{\delta V}{\delta Y}\tan\lambda+\frac{\delta U}{\delta Y}\tan^2\lambda+\frac{\delta U}{\delta X}\tan\lambda-\frac{\delta V}{\delta X}\right) \tag{44}$$

$$\frac{\partial \Psi}{\partial Y}\frac{\partial \Theta}{\partial X}-\frac{\partial \Psi}{\partial X}\frac{\partial \Theta}{\partial Y}=\left[\frac{\frac{k_{nf}}{k_f}}{\left((1-\phi)+\phi\frac{(\rho Cp)_s}{(\rho Cp)_f}\right)}\right]\left(\frac{\partial^2 \Theta}{\partial X^2}+\frac{\partial^2 \Theta}{\partial Y^2}\right) \tag{45}$$

$$\frac{\partial^2 \Psi}{\partial X^2}+\frac{\partial^2 \Psi}{\partial Y^2}=-\Omega \tag{46}$$

where $Ra_f = g\beta_f L^3 (T_h - T_c)/(\alpha_f \upsilon_f)$ is the Rayleigh number for the base fluid, $Ha = LB_x\sqrt{\sigma_f / \mu_f}$ is the Hartmann number and $\Pr_f = \upsilon_f / \alpha_f$ is the Prandtl number for the base fluid. The boundary conditions as shown in Figure 7 are:

$\Theta = 1.0$ on the hot wall

$\Theta = 0.0$ on the cold wall

$\partial\Theta/\partial n = 0.0$ on the two other insulation boundaries

$\Psi = 0.0$ on all solid boundaries (47)

The values of vorticity on the boundary of the enclosure can be obtained using the stream function formulation and the known velocity conditions during the iterative solution procedure.

The local Nusselt number of the nanofluid along the hot wall can be expressed as:

$$Nu_{loc} = \frac{k_{nf}}{k_f}\frac{\partial \Theta}{\partial n}\bigg|_{hot\ wall} \tag{48}$$

where *n* is normal to surface. The average Nusselt number on the hot wall is evaluated as:

$$Nu_{ave} = \frac{1}{2}\int_0^2 Nu_{loc}\, dS \tag{49}$$

2.2. Effects of Active Parameters

Numerical simulations of natural convection nanofluid flow in an enclosure with one sinusoidal wall in presence of magnetic field were performed using CVFEM. Calculations are made for various values of Hartmann number ($Ha = 0, 20, 60$ and 100), Rayleigh number ($Ra = 10^3, 10^4$ and 10^5), volume fraction of nanoparticles ($\phi = 0\%, 2\%, 4\%$ and 6%) and inclination angle ($\gamma = 0^{\circ}, 30^{\circ}, 60^{\circ}$ and 90°) at constant dimensionless amplitude of the sinusoidal wall ($a = 0.3$) and Prandtl number ($\Pr = 6.2$).

Figures 8 and 9 show isotherms (up) and streamlines (down) contours for different values of Rayleigh number, Hartmann number and inclination angle. The figures show that the absolute value of stream function increases as Rayleigh number enhances and it decreases as Hartmann number increases. Also it can be seen that maximum values of $|\Psi_{max}|$ are observed at $\gamma = 90^{\circ}$ for $Ra = 10^3, 10^4$ while it is obtained at $\gamma = 60^{\circ}$ for $Ra = 10^5$. At $Ra = 10^3$, for all inclination angles, the isotherms are nearly

Figure 8a. Isotherms (up) and streamlines (down) contours for different values of Rayleigh number, Hartmann number and inclination angle at $a = 0.3, \phi = 0.06$ *and* $\Pr = 6.2$

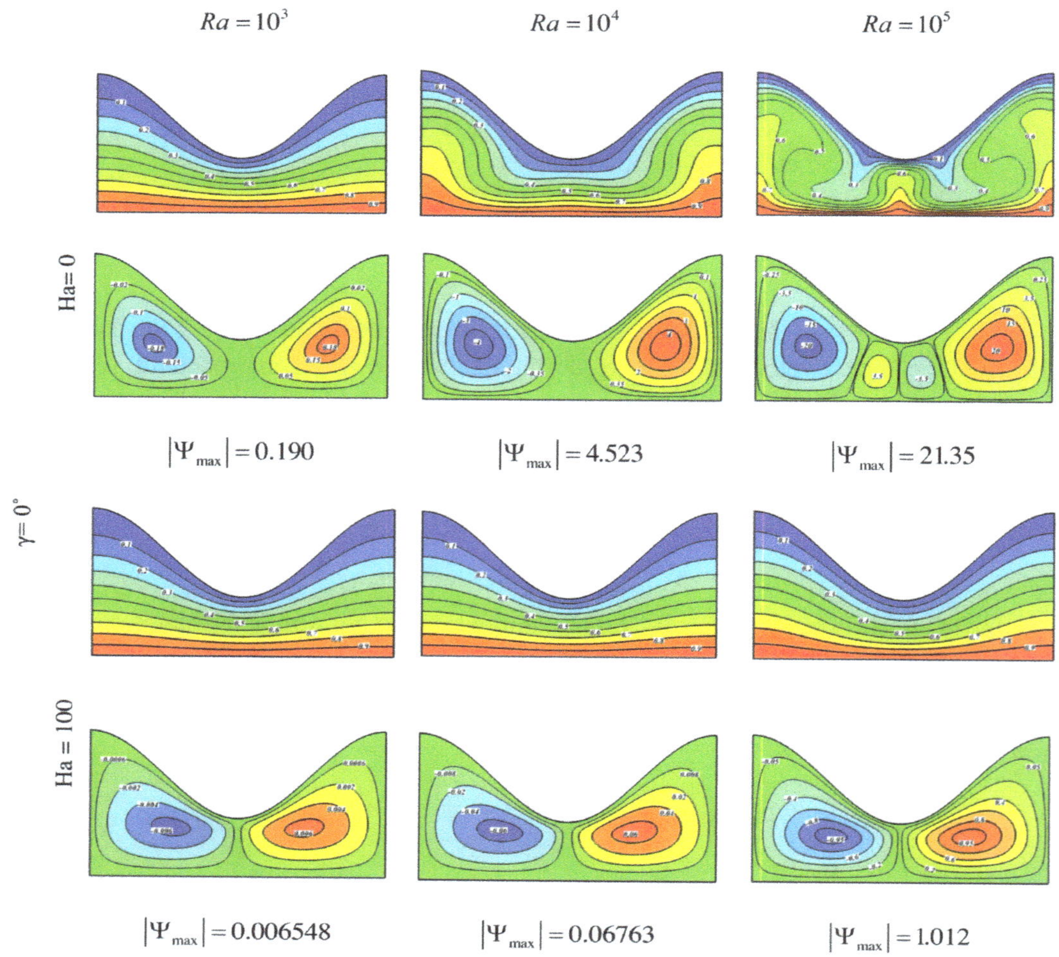

Figure 8b. Isotherms (up) and streamlines (down) contours for different values of Rayleigh number, Hartmann number and inclination angle at $\alpha = 0.3, \phi = 0.06$ *and* $\Pr = 6.2$

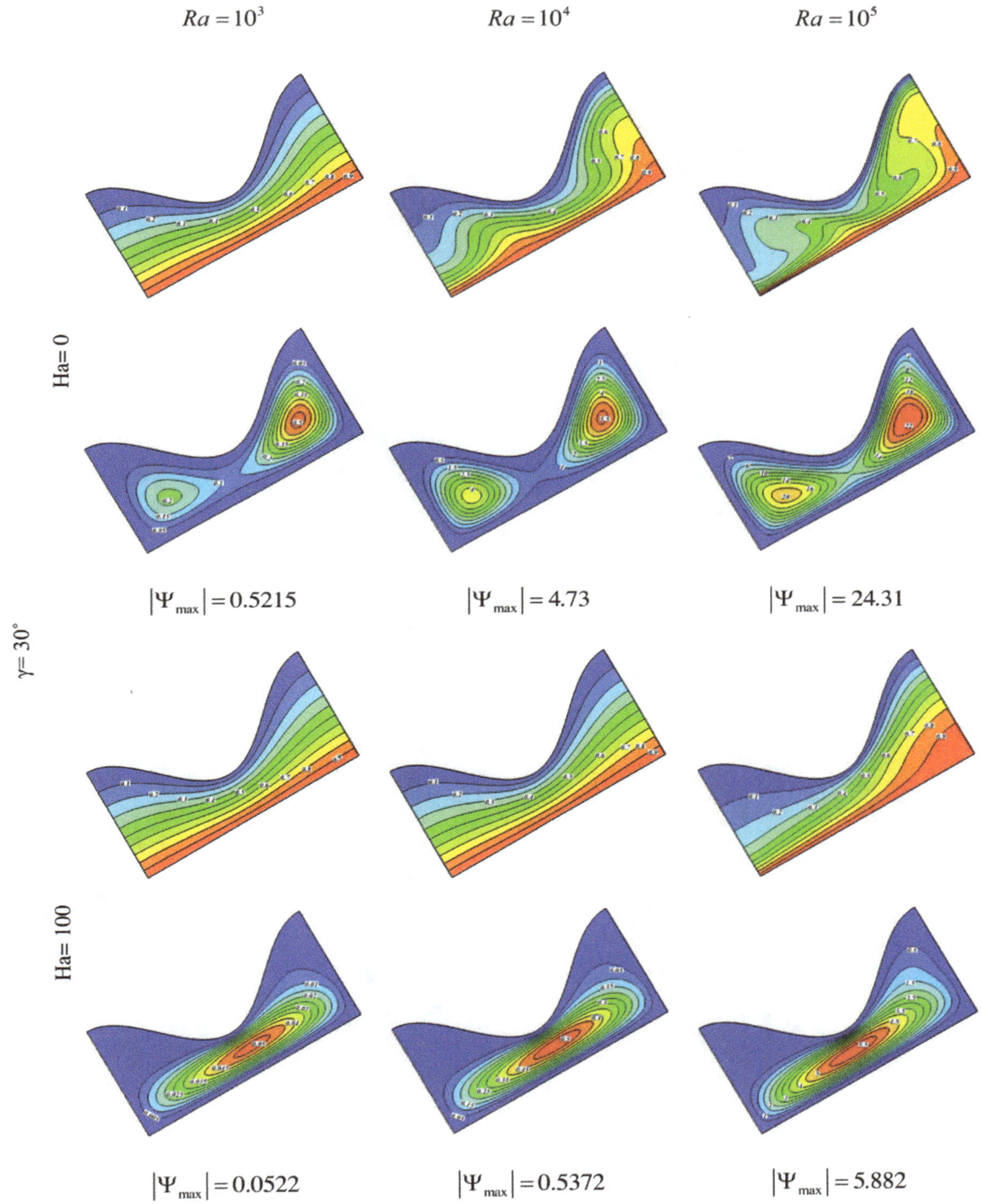

Figure 9a. Isotherms (up) and streamlines (down) contours for different values of Rayleigh number, inclination angle and inclination angle at $a = 0.3, \phi = 0.06$ *and* $\Pr = 6.2$

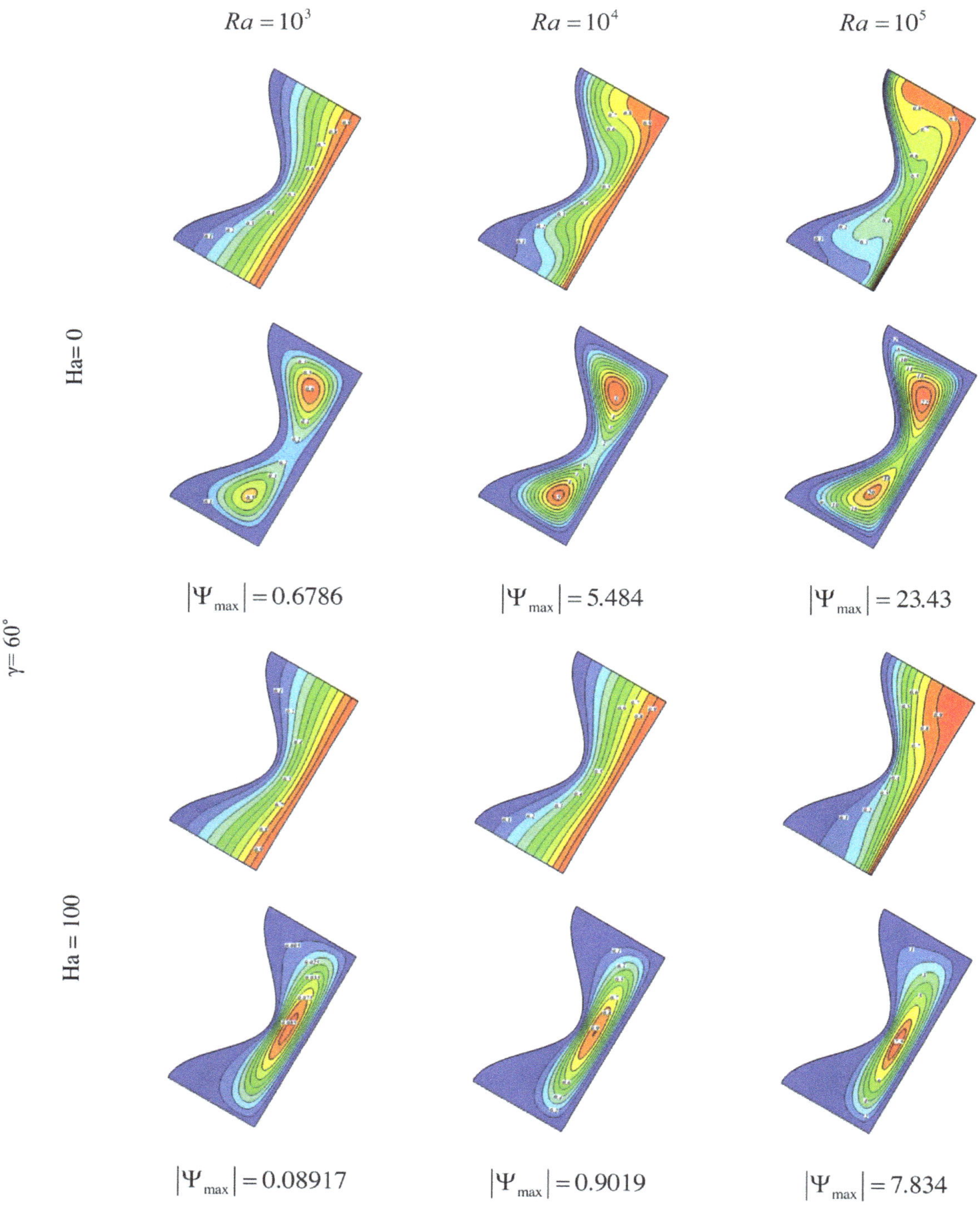

Figure 9b. Isotherms (up) and streamlines (down) contours for different values of Rayleigh number, inclination angle and inclination angle at $a = 0.3, \phi = 0.06$ *and* $\Pr = 6.2$

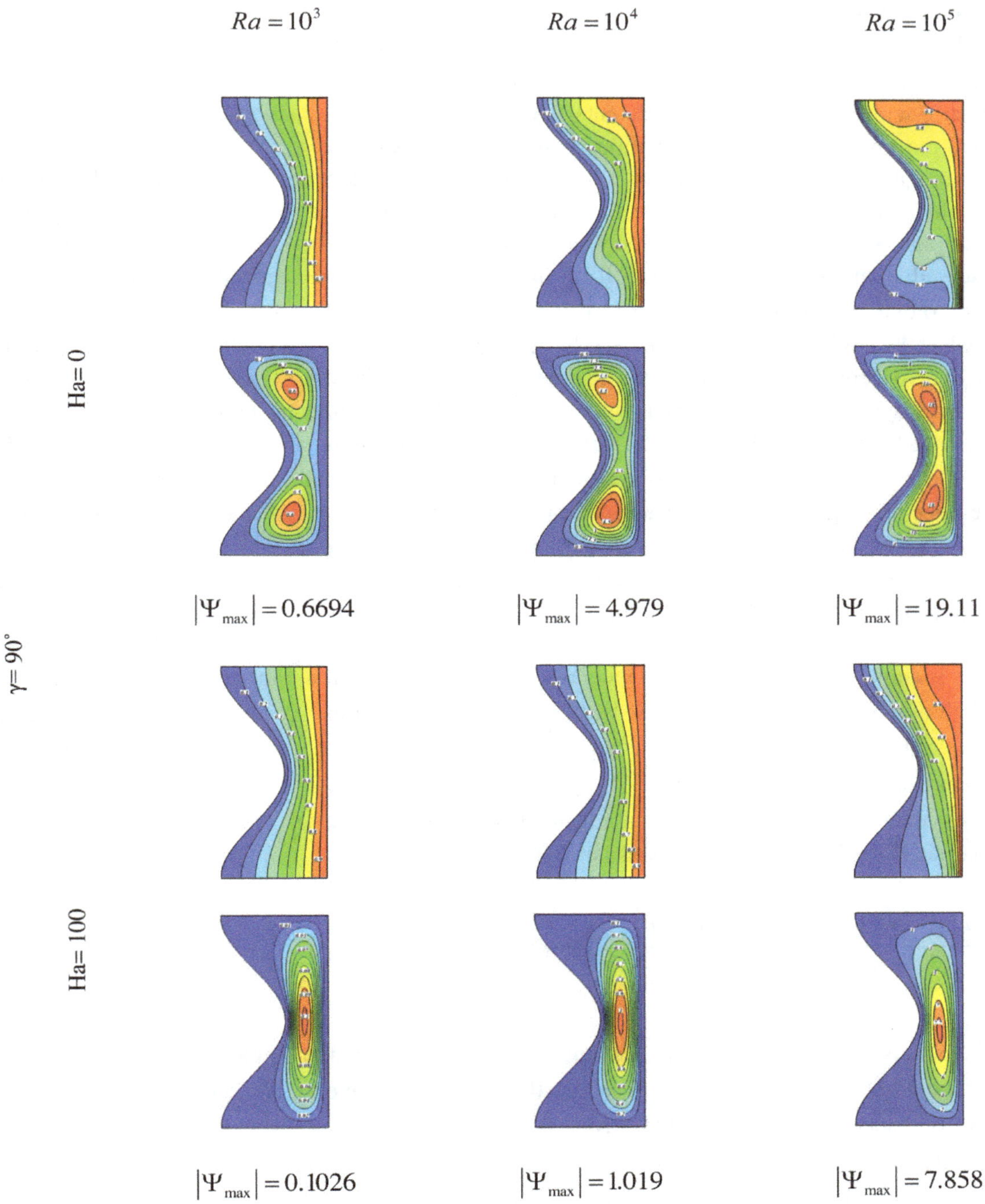

smooth curves and nearly parallel to each other which follow the geometry of the sinusoidal surfaces; this pattern is the characteristic of conduction dominant mechanism of heat transfer at low Rayleigh numbers. At $\gamma = 0^{\circ}$ two counter rotating vortices cores are observed. This bi-cellular flow pattern divides the cavity into two symmetric parts respect to vertical centerline of the enclosure. By increasing inclination angle, at first the upper vortex becomes stronger and then at $\gamma = 90^{\circ}$, streamlines become symmetric with respect to horizontal centerline of the enclosure. In general, as Rayleigh number increases, the buoyancy-driven circulations inside the enclosure become stronger as it is clear from greater magnitudes of stream function and more distortion appears in the isotherms. When the magnetic field is imposed on the enclosure, the velocity field suppressed owing to the retarding effect of the Lorenz force. So intensity of convection weakens significantly. The braking effect of the magnetic field is observed from the maximum stream function value. Increase of Hartmann number merge two vortexes in to one except for $\gamma = 0^{\circ}$. Also magnetic field disappear the thermal plume over the hot wall and makes the isotherms parallel to each other due to domination of conduction mode of heat transfer.

Figure 10 depicts the effects of the nanoparticle volume fraction, Hartmann number, inclination angle and Rayleigh number on Local Nusselt number. Generally, increasing the nanoparticles volume fraction and Rayleigh number leads to an increase in local Nusselt number. In absence of magnetic field, at $\gamma = 0^{\circ}$ the local Nusselt profile is symmetric respect to the vertical center line of the enclosure. But in presence of magnetic field, because of domination of conduction mechanism, maximum value of local Nusselt number occurs at vertical center line. At $\gamma = 90^{\circ}$ the local Nusselt decreases with increase of *S* and increasing Hartmann number leads to decrease in Nusselt number. When $Ha = 0$, the number of extermum in in the local Nusselt number profile is corresponding to exist of thermal plume.

Effects of the Hartmann number, Rayleigh number and inclination angle on the average Nusselt number is shown in Figure 11(a, b). Generally, the average Nusselt number increases with increase of Rayleigh number while it decreases as Hartmann number increases. At $Ra = 10^5$ in absence of magnetic field maximum value of average Nusselt number is obtained at $\gamma = 0^{\circ}$ but for higher values of Hartmann number maximum value of Nu_{ave} occurs at $\gamma = 90^{\circ}$.

To estimate the enhancement of heat transfer between the case of $\phi = 0.06$ and the pure fluid (base fluid) case, the enhancement is defined as:

$$E = \frac{Nu(\phi = 0.06) - Nu(basefluid)}{Nu(basefluid)} \times 100 \tag{50}$$

The effects of Hartmann number, Rayleigh number and inclination angle on heat transfer enhancement ratio is shown in Figure 11(c). At $Ra = 10^3$, maximum value of enhancement for low Hartmann number is observed at $\gamma = 0^{\circ}$, but for $Ha > 20$ maximum values of it occur for $\gamma = 90^{\circ}$. Also it can be seen that for $Ra = 10^4$ and 10^5 maximum value of *E* are obtained for $\gamma = 60^{\circ}$ and $\gamma = 0^{\circ}$, respectively. It is an interesting observation that at $Ra = 10^5$ the enhancement in heat transfer for case of $\gamma = 0^{\circ}$ increases with increase of Hartmann number when $Ha < 60$ while opposite trend is observed for $Ha > 60$. For other value of inclination angles, enhancement in heat transfer is an increasing function of Hartmann number.

Figure 10. Effects of the nanoparticle volume fraction, Hartmann number, inclination angle and Rayleigh number on Local Nusselt number when (a), (b) $Ra = 10^5, \gamma = 90^{\circ}$ *; (c),(d)* $Ra = 10^5, \phi = 0.06$ *; (e),(f)* $Ra = 10^5, \phi = 0.06$ *; (g),(h)* $\phi = 0.06, \gamma = 90^{\circ}$

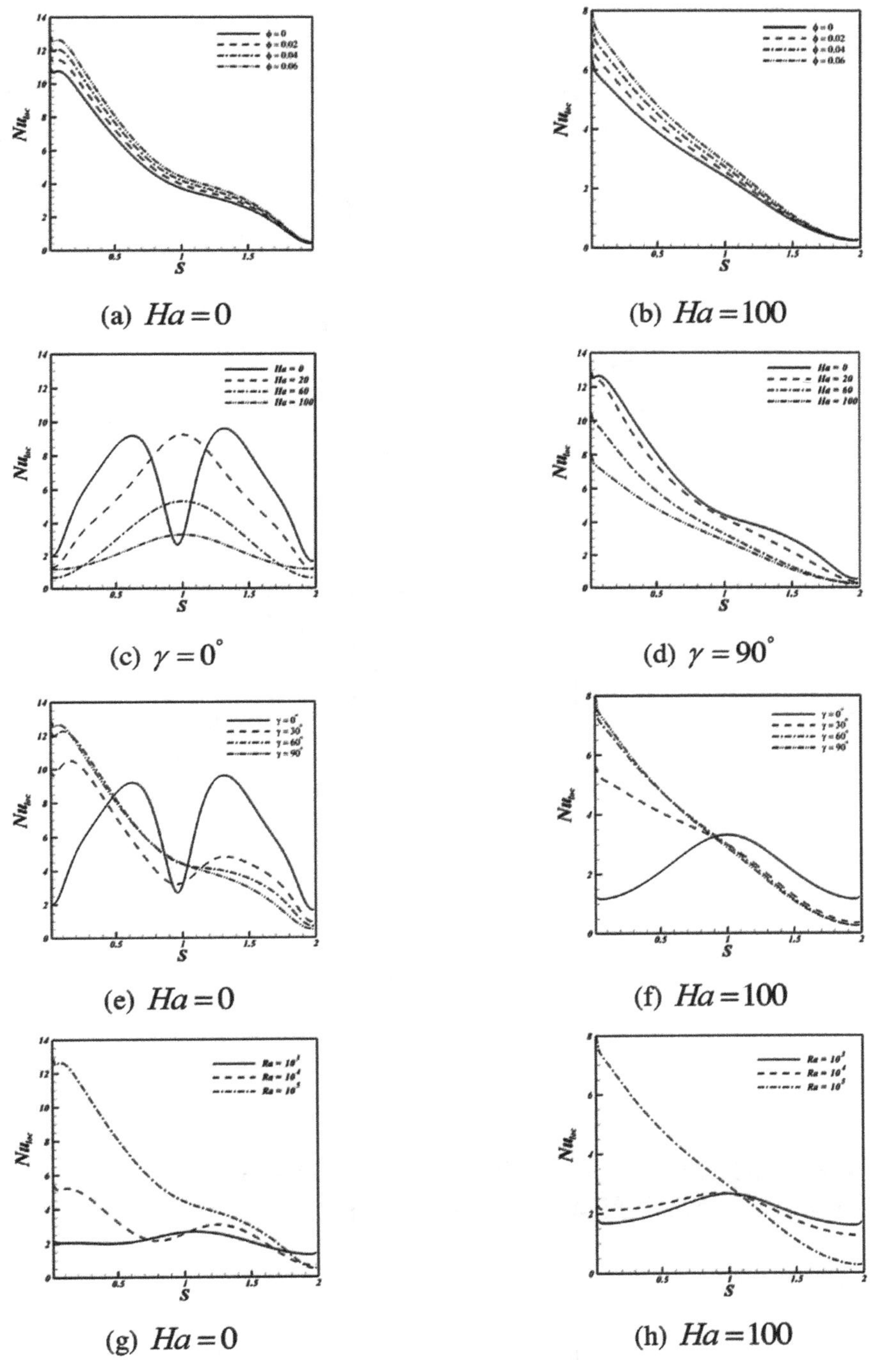

(a) $Ha = 0$ (b) $Ha = 100$

(c) $\gamma = 0^{\circ}$ (d) $\gamma = 90^{\circ}$

(e) $Ha = 0$ (f) $Ha = 100$

(g) $Ha = 0$ (h) $Ha = 100$

Figure 11. Effects of the Hartmann number, Rayleigh number and inclination angle on the average Nusselt number when (a) $Ra = 10^5$; (b) $\gamma = 90^\circ$ at $a = 0.3$ and $\phi = 0.06$; (c)Effects of Hartmann number, Rayleigh number and inclination angle on the ratio of heat transfer enhancement due to addition of nanoparticles $a = 0.3$

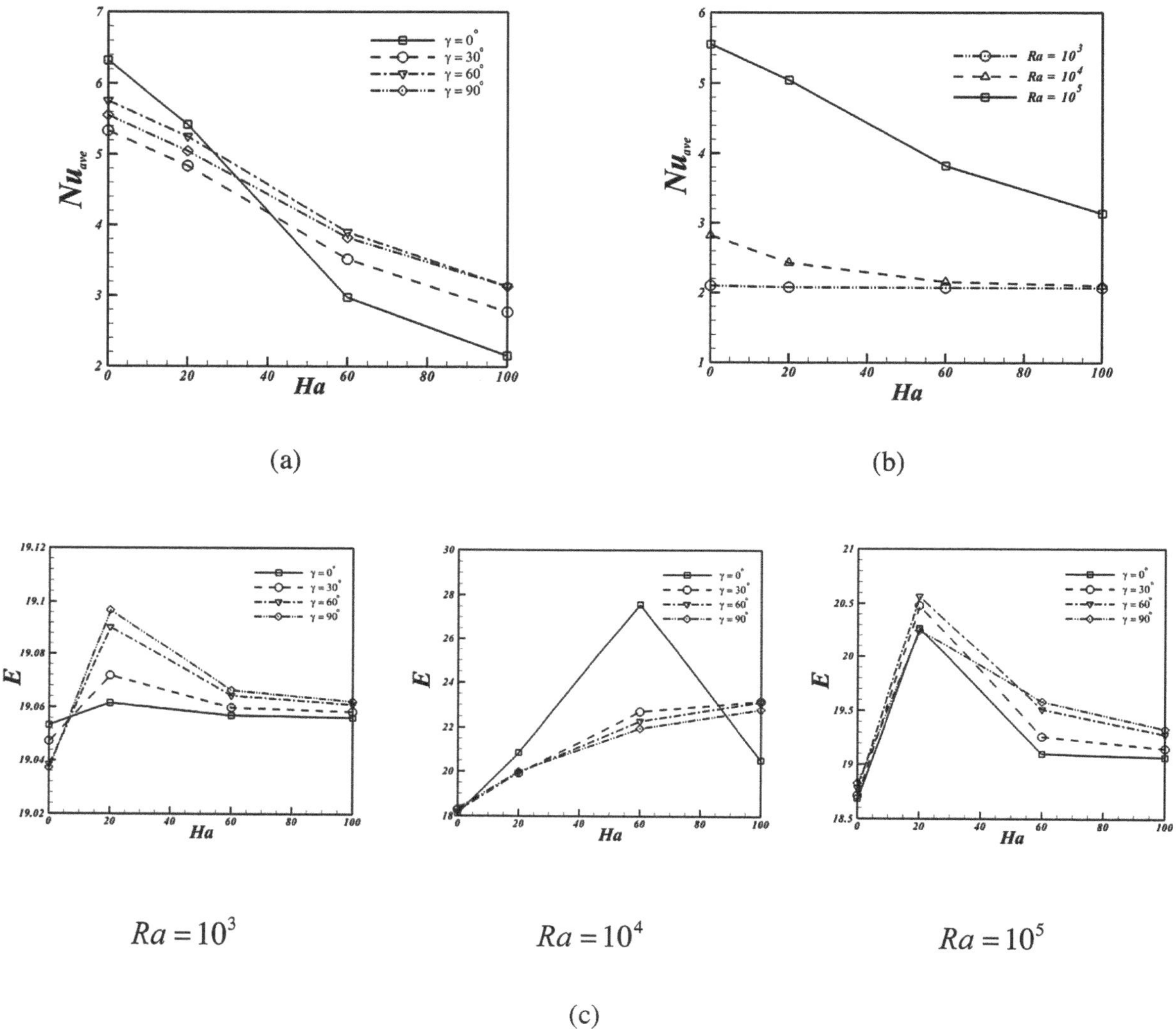

3. EFFECTS OF MHD ON CU-WATER NANOFLUID FLOW AND HEAT TRANSFER BY MEANS OF CVFEM

3.1. Problem Definition

The schematic diagram and the mesh of the semi-annulus enclosure used in the present CVFEM program are shown in Figure 12(a) (Sheikholeslami et al., 2018). The system consists of a circular enclosure with radius of r_{out}, within which an inclined elliptic cylinder is located and rotates from $\gamma = 0^\circ$ to 90°. T_h

Figure 12. (a)Geometry and the boundary conditions with (b) the mesh of enclosure considered in this work; (c) A sample triangular element and its corresponding control volume

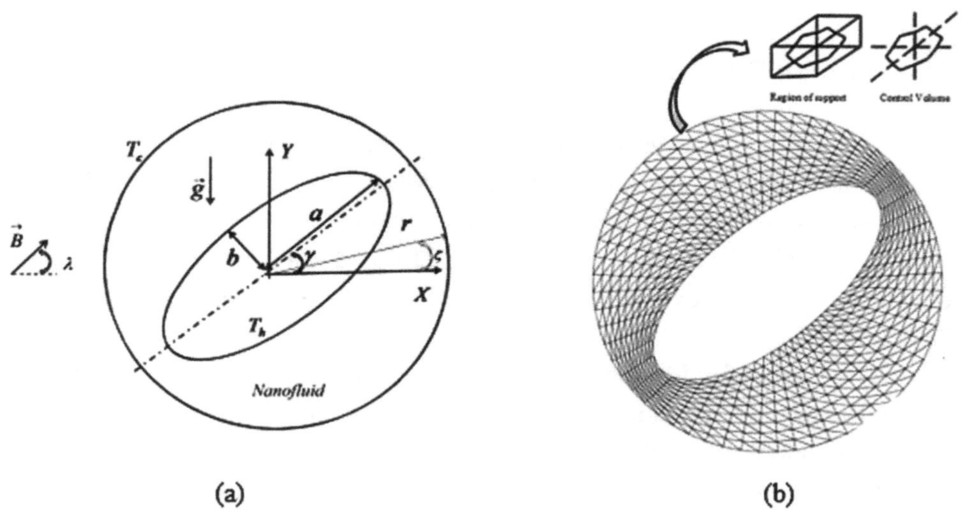

and T_c are the constant temperatures of the inner and outer cylinders, respectively ($T_h > T_c$). Setting a as the major axis and b as the minor axis of elliptic cylinder, the eccentricity (ε) for the inner cylinder is defined as (Sheikholeslami et al., 2014):

$$\varepsilon = \sqrt{a^2 - b^2}\Big/a \text{ or } b = \sqrt{1 - \varepsilon^2}.a \tag{51}$$

In this study, for the inner ellipse, the eccentricity and the major axis are 0.9 and $0.8L$, respectively.

Also, it is also assumed that the uniform magnetic field ($\vec{B} = B_x \vec{e}_x + B_y \vec{e}_y$) of constant magnitude $B = \sqrt{B_x^2 + B_y^2}$ is applied, where $\vec{e}_x$ and $\vec{e}_y$ are unit vectors in the Cartesian coordinate system. The orientation of the magnetic field form an angle λ with horizontal axis such that $\lambda = B_x / B_y$. The electric current J and the electromagnetic force F are defined by $J = \sigma\left(\vec{V} \times \vec{B}\right)$ and $F = \sigma\left(\vec{V} \times \vec{B}\right) \times \vec{B}$, respectively. In this study λ is equal to zero.

The governing equations are similar to those of exist in section (2.1). The local Nusselt number of the nanofluid along the cold wall can be expressed as:

$$Nu_{loc} = \left(\frac{k_{nf}}{k_f}\right)\frac{\partial \Theta}{\partial r} \tag{52}$$

where r is the radial direction. The average Nusselt number on the cold circular wall is evaluated as:

$$Nu_{ave} = \frac{1}{2\pi}\int_0^{2\pi} Nu_{loc}\left(\zeta\right)d\zeta \tag{53}$$

3.2. Effects of Active Parameters

MHD natural convection heat transfer between a circular enclosure and an elliptic cylinder filled with nanofluid is investigated numerically using the Control Volume based Finite Element Method. Calculations are carried out for constant eccentricity ($\varepsilon = 0.9$), major axis ($a = 0.8L$) and Prandtl (*Pr*=6.2) at different values of Rayleigh number ($Ra = 10^3, 10^4, 10^5$), Hartmann number ($Ha = 0, 20, 60$ and 100) and inclined angle of inner cylinder ($\gamma = 0^\circ, 30^\circ, 60^\circ, 90^\circ$) and volume fraction of nanoparticles ($\phi = 0\%, 2\%, 4\%$ and 6%).

Isotherms and streamlines for different values of Ra, γ and Ha are shown in Figures 13 and 14. At $Ra = 10^3$ the isotherms are parallel to each other and take the form of inner and outer wall and the stream function magnitude is relatively small which indicates the domination of conduction heat transfer mechanism. Increasing inclination angle leads to increase in the absolute value of maximum stream function ($\left|\Psi_{max}\right|$) at this Rayleigh number. At $\gamma = 0^\circ$ the stream lines and isotherms are symmetric respect to the vertical center line of the enclosure. Each pair cells have two cells. The top vortex is stronger because at this area the hot surface is located beneath of the cold one which helps the flow circulation whereas the arrangement of the existence of cold wall under the hot cylinder at the bottom of the enclosure resists the flow circulation. As $\gamma = 0^\circ$ increases these two pair cells merge together and form two single cell different locations inside the enclosure. At $\gamma = 90^\circ$ again the stream lines pattern become symmetric with respect to the vertical centerline of the enclosure. At $Ra = 10^4$ the thermal plumbs start to appear over the hot elliptic cylinder. Besides, the stream function values start to growth which show that the convection heat transfer mechanism become comparable with conduction. At $\gamma = 0^\circ$ increase pattern off stream line is similar to that of $Ra = 10^3$ but the size of the vortices at the bottom of the enclosure. In addition, the temperature contours becomes stratified beneath the hot cylinder. With increase of γ to 30° a secondary vortex appears at the top of the enclosure and the thermal plumb slant to left because of more available space at this area. As the inclination angle of the inner cylinder increases furthermore this secondary vortex disappears and the stream lines show two main vortices in the enclosure. Also it can be found that effect of increasing inclination angle on $\left|\Psi_{max}\right|$ become less pronounced at $\gamma > 30^\circ$. When Rayleigh number increases up to $Ra = 10^5$ isotherms are totally distorted at the top of the enclosure while it is stratified at the bottom of the enclosure which shows the heat transfer mechanism is dominated by convection. In the area above the thermal plumb completely formed which impinging the hot fluid to the cold wall of the enclosure, these results in the thermal boundary layer over the cold wall of the enclosure. As seen the secondary vortex exist at $\gamma = 30^\circ$ and 60° which can the thermal plumb slant to left. Also, as seen maximum value of $\left|\Psi_{max}\right|$ occurs at $\gamma = 60^\circ$. It is worthwhile mentioning that the effect of magnetic field is to decrease the value of the velocity magnitude throughout the enclosure because the presence of magnetic field introduces a force called the Lorentz force, which acts against the flow if the magnetic field is applied in the normal direction. This type of resisting force slows down the fluid velocity. Increase of Hartmann number make the core of vortices move toward the horizontal centerline. Also magnetic field causes the thermal plume to disap-

Figure 13a. Isotherms (up) and streamlines (down) contours for different values of $Ra, \gamma = 0°, 30°, 45°$ and Ha at $\phi = 0.06$

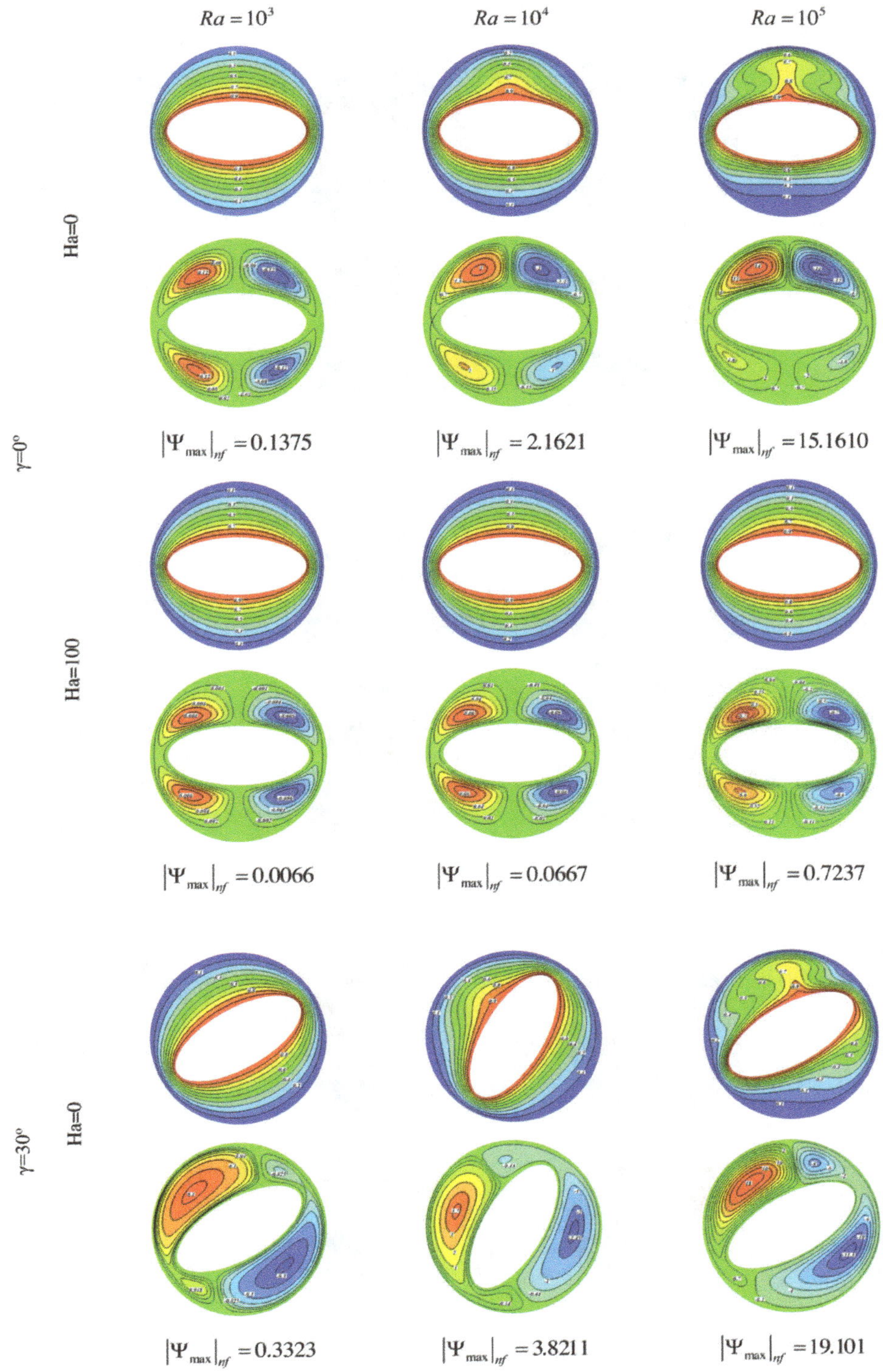

Figure 13b. Isotherms (up) and streamlines (down) contours for different values of $Ra, \gamma = 0^\circ, 30^\circ, 45^\circ$ and Ha at $\phi = 0.06$

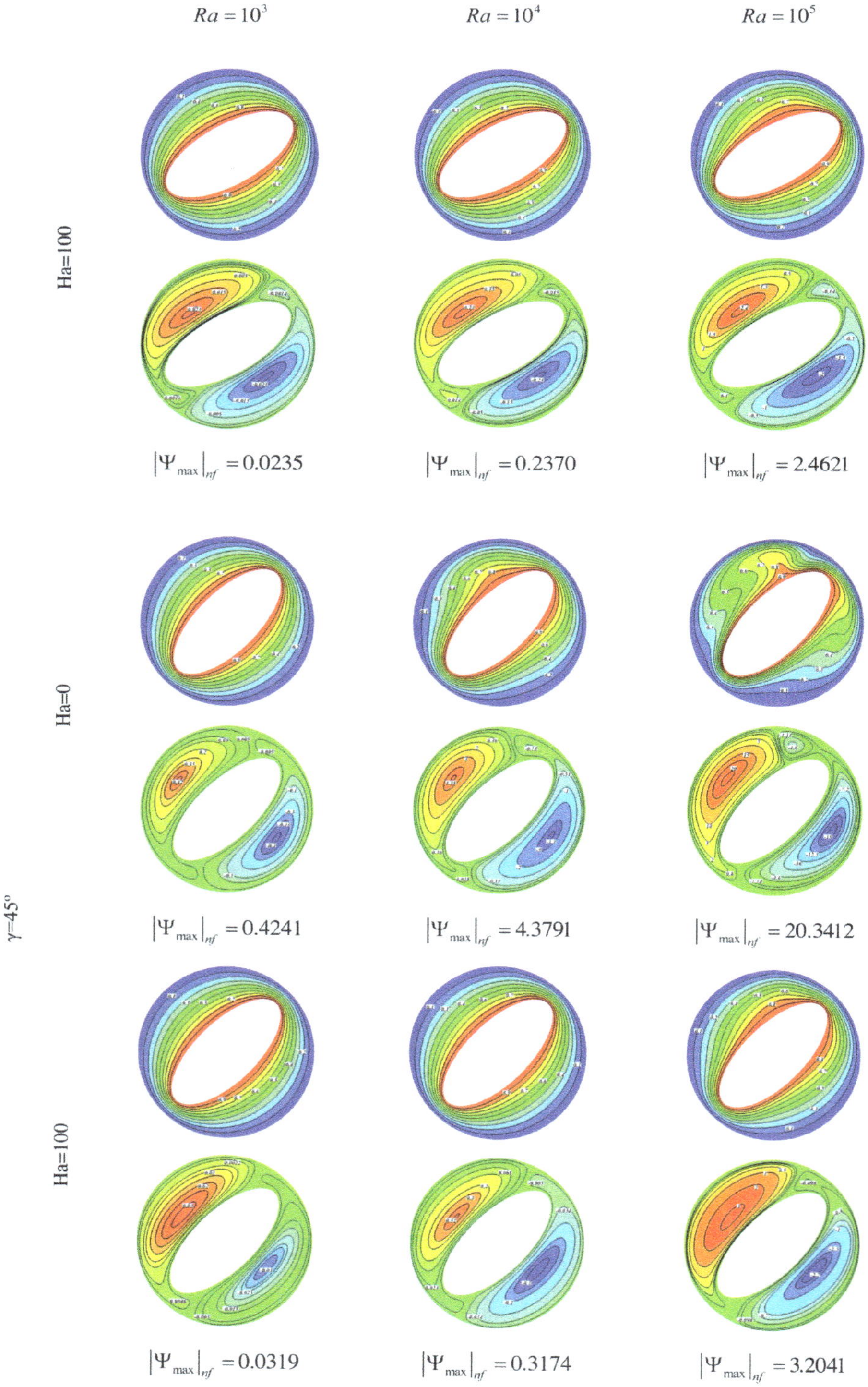

Figure 14a. Isotherms (up) and streamlines (down) contours for different values of $Ra, \gamma = 60°, 90°$ and Ha at $\phi = 0.06$

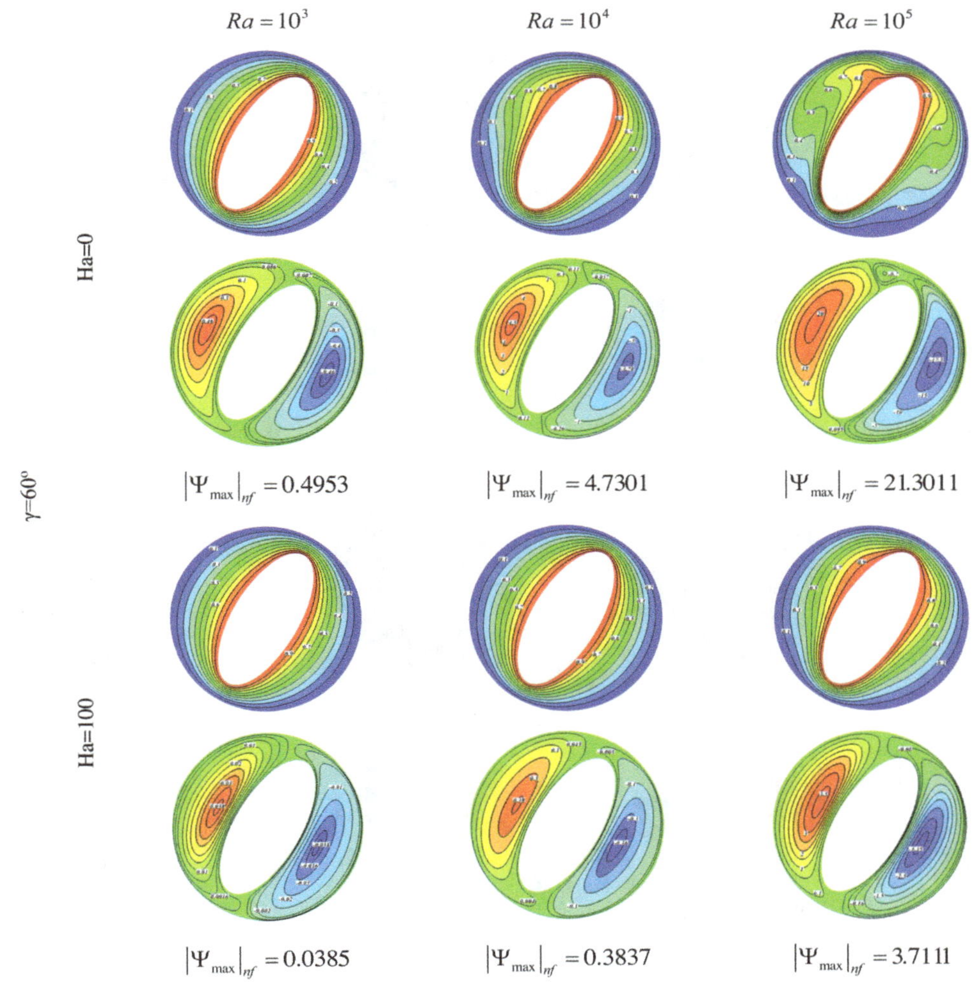

pear and makes the isotherms parallel to each other due to domination of conduction mode of heat transfer.

Figure 15 shows the distribution of local Nusselt numbers along the surface of the outer circular wall for different inclination angle, Rayleigh number and Hartmann number. As Rayleigh number increases the local Nusselt number increases due to increment of convection effect. At $Ra = 10^3$ the Nu_{loc} profile is nearly symmetry respect to the horizontal center line. As Rayleigh number enhances (e.g. $Ra = 10^4$ and $Ra = 10^5$), the Nu_{loc} profile is no longer symmetry and local Nusselt number is considerably small over the bottom wall of the enclosure. Increasing Hartmann number causes local Nusselt number to decreases. These local Nusselt number profiles are more complex due to the presence of thermal plume at the vicinity of the top wall of the enclosure.

The corresponding polynomial representation of such model for Nusselt number is as the following:

Figure 14b. Isotherms (up) and streamlines (down) contours for different values of $Ra, \gamma = 60^{\circ}, 90^{\circ}$ *and* Ha *at* $\phi = 0.06$

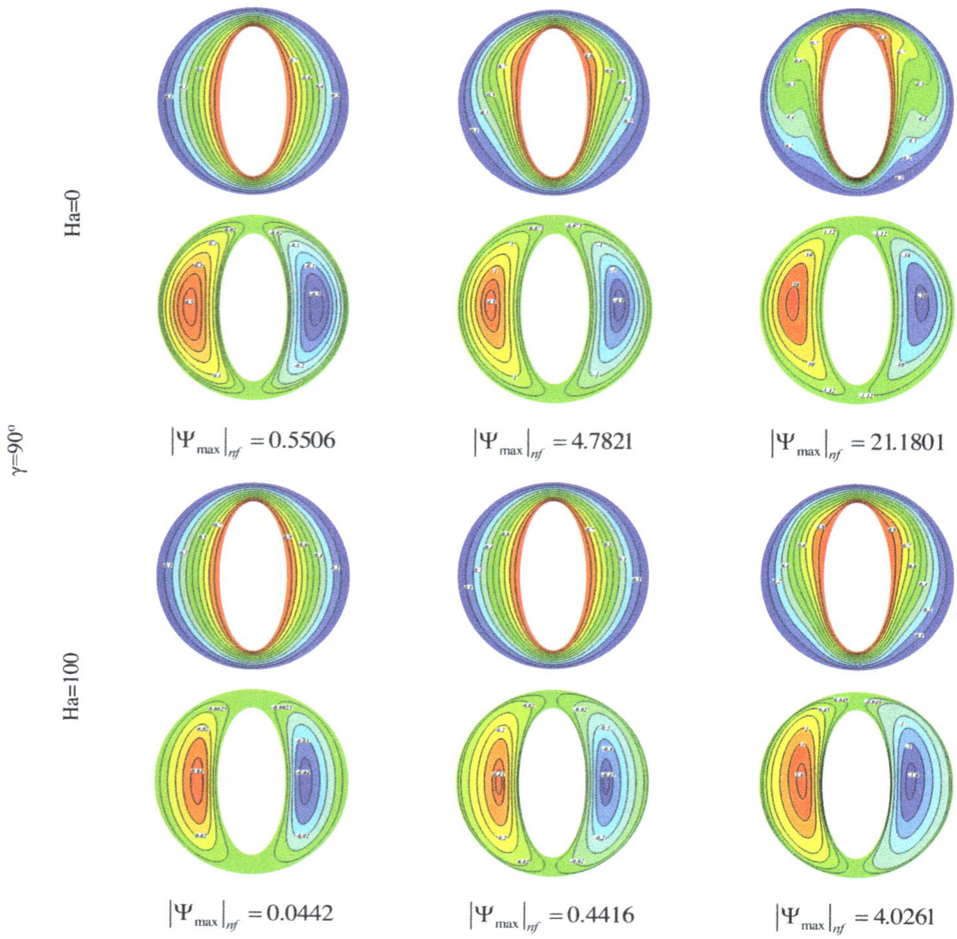

$$\begin{aligned} Nu &= a_{13} + a_{23}Y_1 + a_{33}Y_2 + a_{43}Y_1^2 + a_{53}Y_2^2 + a_{63}Y_1Y_2 \\ Y_1 &= a_{11} + a_{21}Ra^* + a_{31}Ha^* + a_{41}Ra^{*2} + a_{51}Ha^{*2} + a_{61}Ha^*Ra^* \\ Y_2 &= a_{12} + a_{22}\gamma + a_{32}\phi + a_{42}\gamma^2 + a_{52}\phi^2 + a_{62}\gamma\phi \end{aligned} \tag{54}$$

Also a_{ij} can be found in Table 2 for example a_{21} equals to (-1.52352).

Table 2. Constant coefficient for using Equation (54)

a_{ij}	i=1	i=2	i=3	i=4	i=5	i=6
j=1	4.302581	-1.52352	1.51045	0.267076	0.380695	-0.59522
j=2	2.170725	-0.05626	6.043703	0.071014	-2.55729	-0.09138
j=3	3.084162	-2.27497	-0.06891	0.556024	0.155194	0.131637

Figure 15. Effects of the inclination angle, Hartmann number and Rayleigh number for Cu-water ($\phi = 0.06$) nanofluids on Local Nusselt number

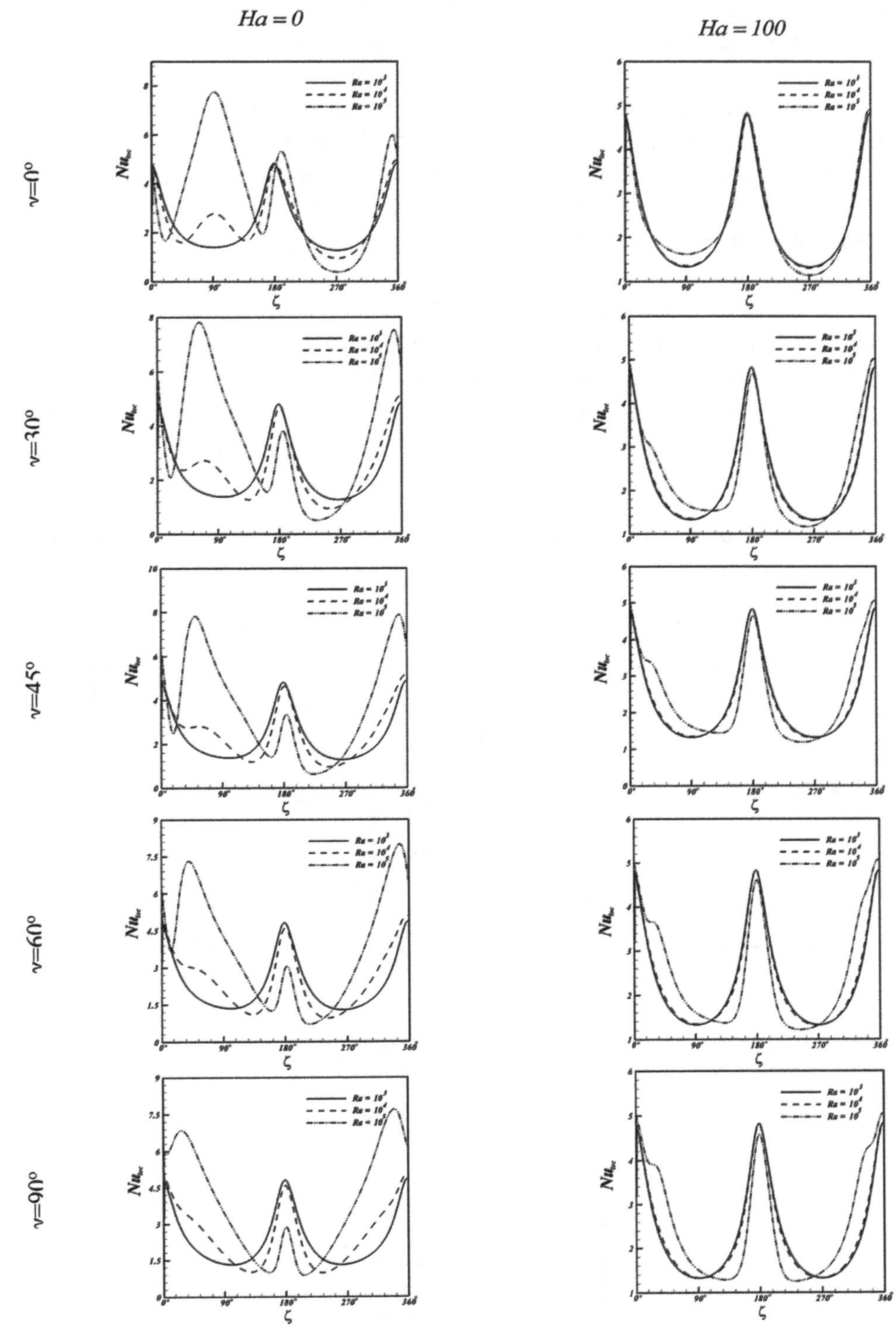

Effects of the volume fraction of nanoparticles, inclination angle, Hartmann number and Rayleigh number on average Nusselt number are shown in Figures 16 and 17. Adding nanoparticles leads to increase thermal boundary layer thickness, Nusselt number increases because Nusselt number is a multiplication of temperature gradient and the thermal conductivity ratio and reduction in temperature gradient due to the presence of nanoparticles is much smaller than the thermal conductivity ratio. Increasing Rayleigh number is associated with an increase in the heat transfer and the Nusselt number. This is due to stronger convective heat transfer for higher Rayleigh number. Increasing Hartmann number causes Lorenz force to increase and leads to a substantial suppression of the convection. So Nusselt number has reveres relationship with Hartmann number. Also these figures show that inclination angle has direct relationship with average Nusselt number. As seen in Figure 16 inclination angle has no significant effect on average Nusselt number at high Hartmann number.

Heat transfer enhancement ratio due to addition of nanoparticles for different values of Ha, γ and Ra is shown in Figure 18. Generally, increasing Rayleigh number causes heat transfer enhancement to decrease because of domination of conduction mechanism in low Rayleigh number. Also Hartmann number is an increasing function of En. As inclination angle increases En decreases except for $Ra = 10^4$. At $Ra = 10^4$ maximum values of En occur at $\gamma = 30°$ and $45°$ for $Ha = 0$ and 100, respectively.

Figure 16. Effects of the inclination angle, Hartmann number and Rayleigh number for Cu-water ($\phi = 0.06$) nanofluids on Average Nusselt number

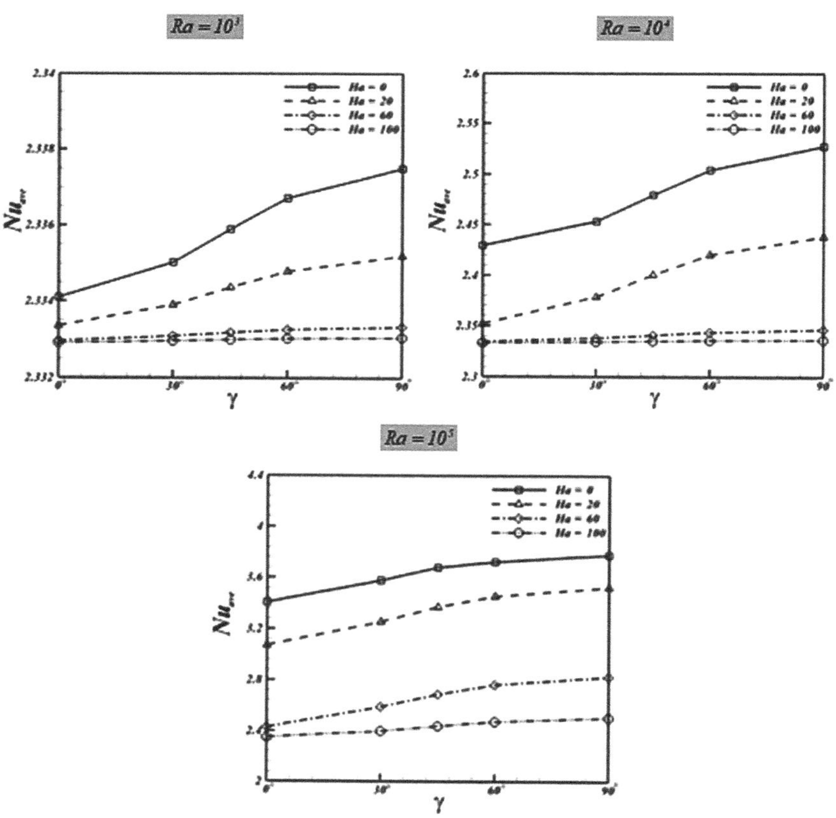

Figure 17. Variation of Nu_{ave} for various input parameters

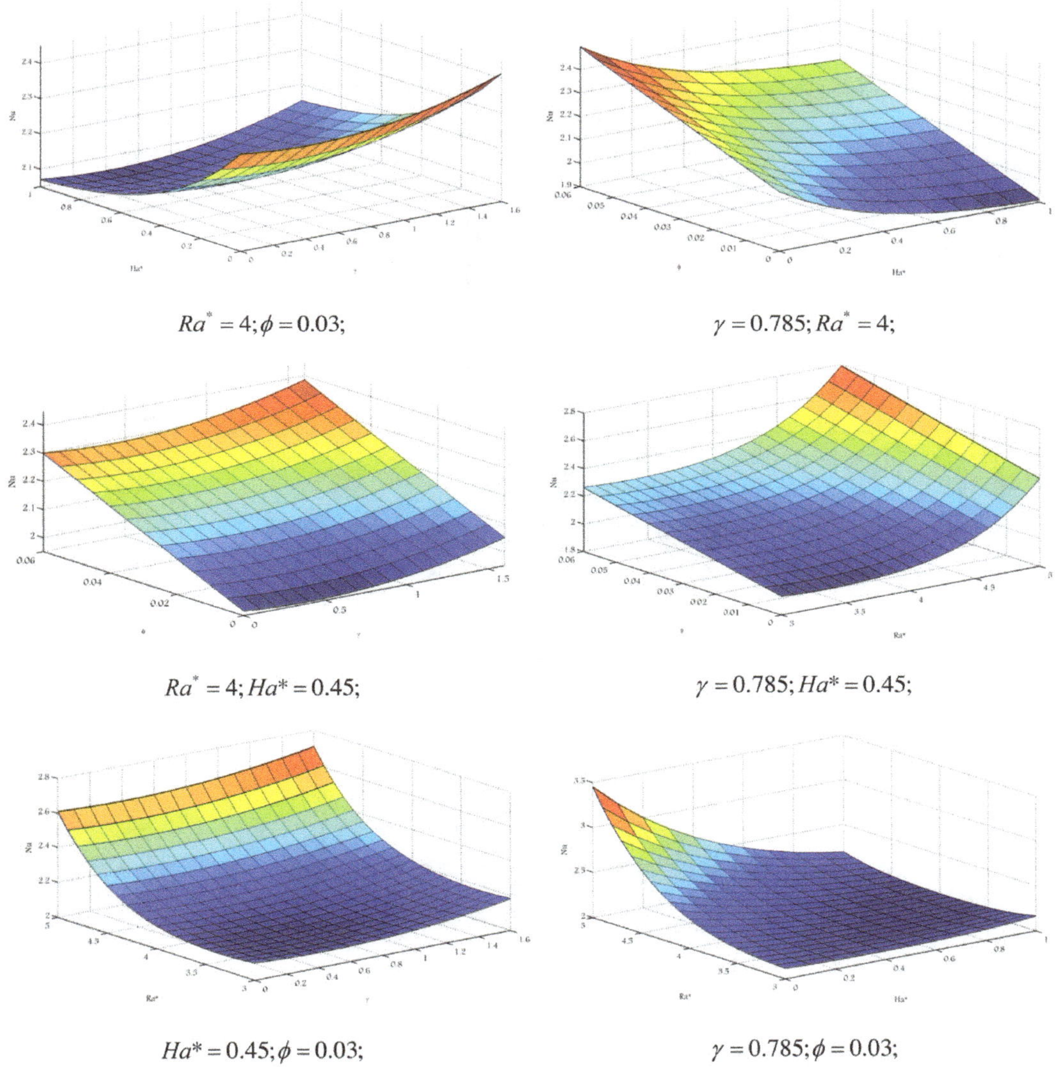

4. HEAT FLUX BOUNDARY CONDITION FOR NANOFLUID FILLED ENCLOSURE IN PRESENCE OF MAGNETIC FIELD

4.1. Problem Definition

The physical model along with the important geometrical parameters is as shown in Figure 19(a). The width and height of the enclosure is L. The outer cylinder is maintained at constant cold temperature T_c, whereas the inner circular wall is under constant heat flux (Sheikholeslami, Jafaryar, Shafee, & Li, 2018). To assess the shape of inner circular and outer rectangular boundary which consists of the right and top walls, a supper elliptic function can be used as follows

Figure 18. Effects of the inclination angle, Hartmann number and Rayleigh number on the ratio of heat transfer enhancement due to addition of nanoparticles when $\mathrm{Pr} = 6.2$ *(Cu-Water case)*

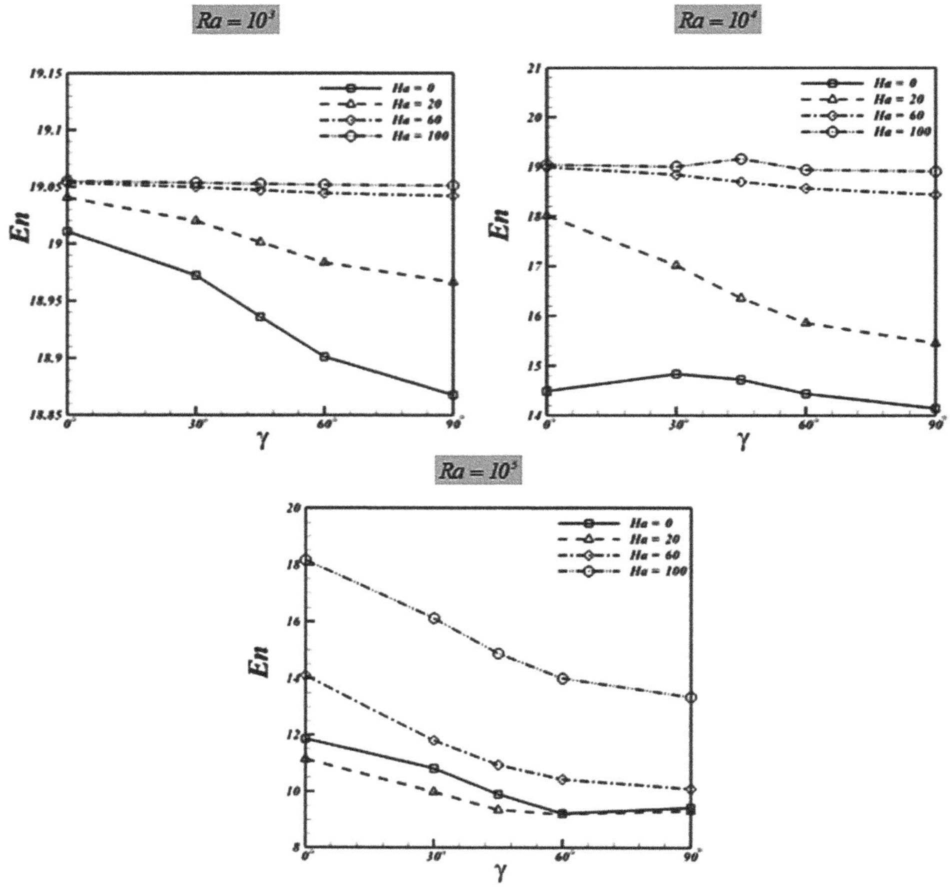

$$\left(\frac{X}{a}\right)^{2\hat{n}} + \left(\frac{Y}{b}\right)^{2\hat{n}} = 1 \tag{55}$$

When $a = b$ and $\hat{n} = 1$ the geometry becomes a circle. As $\hat{n}$ increases from 1 the geometry would approach a rectangle for $a \neq b$ and square for $a = b$. It is also assumed that the uniform magnetic field ($\vec{B} = B_x \vec{e}_x + B_y \vec{e}_y$) of constant magnitude $B = \sqrt{B_x^2 + B_y^2}$ is applied, where $\vec{e}_x$ and $\vec{e}_y$ are unit vectors in the Cartesian coordinate system. The orientation of the magnetic field form an angle λ with horizontal axis such that $\lambda = B_x / B_y$. In this study, λ equals to zero. The electric current *J* and the electromagnetic force F are defined by $J = \sigma\left(\vec{V} \times \vec{B}\right)$ and $F = \sigma\left(\vec{V} \times \vec{B}\right) \times \vec{B}$, respectively.

The flow is steady, two-dimensional, laminar and incompressible. The radiation, viscous dissipation, induced electric current and Joule heating are neglected. The magnetic Reynolds number is assumed to

Figure 19. (a) Geometry and the boundary conditions with (b) the mesh of Geometry considered in this work

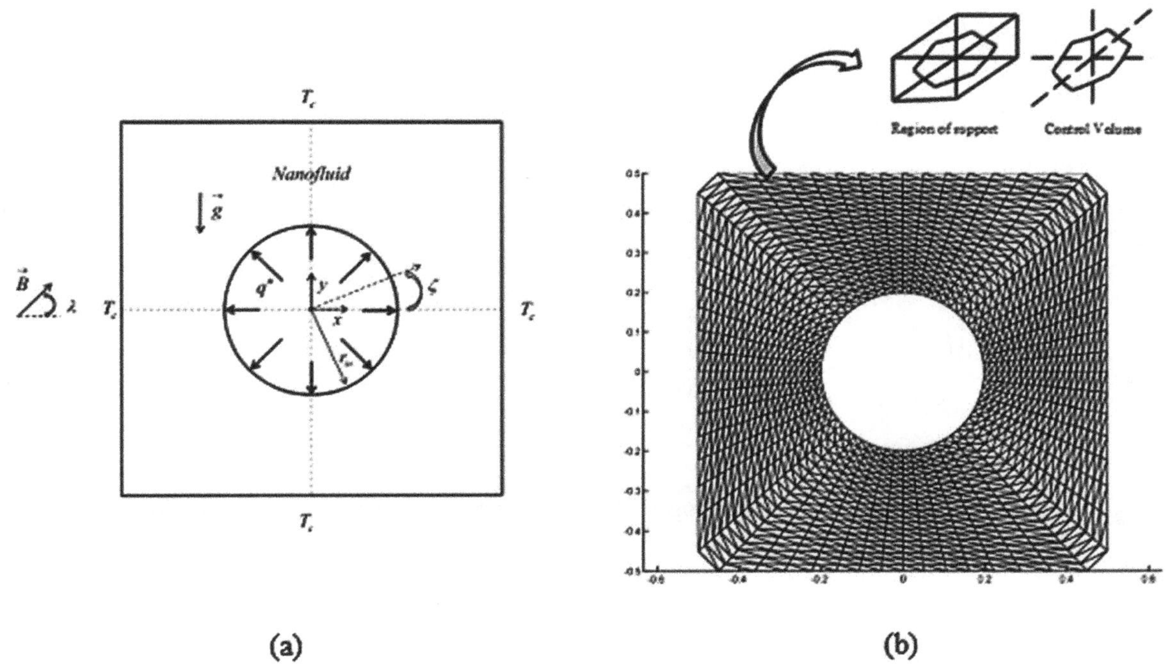

be small so that the induced magnetic field can be neglected compared to the applied magnetic field. Neglecting displacement currents, induced magnetic field, and using the Boussinesq approximation, the governing equations of heat transfer and fluid flow for nanofluid can be obtained as follows:

$$\frac{\partial u}{\partial x}+\frac{\partial v}{\partial y}=0 \tag{56}$$

$$u\frac{\partial u}{\partial x}+v\frac{\partial u}{\partial y}=-\frac{1}{\rho_{nf}}\frac{\partial P}{\partial x}+\upsilon_{nf}\left(\frac{\partial^2 u}{\partial x^2}+\frac{\partial^2 u}{\partial y^2}\right)+\frac{\sigma_{nf}B^2}{\rho_{nf}}\left(v\sin\lambda\cos\lambda-u\sin^2\lambda\right) \tag{57}$$

$$u\frac{\partial v}{\partial x}+v\frac{\partial v}{\partial y}=-\frac{1}{\rho_{nf}}\frac{\partial P}{\partial y}+\upsilon_{nf}\left(\frac{\partial^2 v}{\partial x^2}+\frac{\partial^2 v}{\partial y^2}\right)+\beta_{nf}g\left(T-T_c\right)+\frac{\sigma_{nf}B^2}{\rho_{nf}}\left(u\sin\lambda\cos\lambda-v\cos^2\lambda\right) \tag{58}$$

$$u\frac{\partial T}{\partial x}+v\frac{\partial T}{\partial y}=\alpha_{nf}\left(\frac{\partial^2 T}{\partial x^2}+\frac{\partial^2 T}{\partial y^2}\right) \tag{59}$$

where (ρ_{nf}), (β_{nf}), $\left(\rho C_p\right)_{nf}$ and $\left(\sigma_{nf}\right)$ are defined as:

$$\rho_{nf} = \rho_f(1-\phi) + \rho_s\phi$$

$$\beta_{nf} = \beta_f(1-\phi) + \beta_s\phi$$

$$\left(\rho C_p\right)_{nf} = \left(\rho C_p\right)_f (1-\phi) + \left(\rho C_p\right)_s \phi \tag{60}$$

$$\frac{\sigma_{nf}}{\sigma_f} = 1 + \frac{3\left(\sigma_s / \sigma_f - 1\right)\phi}{\left(\sigma_s / \sigma_f + 2\right) - \left(\sigma_s / \sigma_f - 1\right)\phi} \tag{61}$$

$\left(k_{nf}\right)$ and $\left(\mu_{nf}\right)$ are obtained according to Koo–Kleinstreuer–Li (KKL) model (Sheikholeslami, 2014):

$$\begin{aligned} k_{nf} &= 1 + \frac{3\left(\frac{k_p}{k_f} - 1\right)\phi}{\left(\frac{k_p}{k_f} + 2\right) - \left(\frac{k_p}{k_f} - 1\right)\phi} + 5\times 10^4 g'(\phi, T, d_p)\phi\rho_f c_{p,f}\sqrt{\frac{\kappa_b T}{\rho_p d_p}} \\ g'\left(\phi, T, d_p\right) &= \left(a_6 + a_7 Ln\left(d_p\right) + a_8 Ln\left(\phi\right) + a_9 Ln\left(\phi\right)\ln\left(d_p\right) + a_{10} Ln\left(d_p\right)^2\right) \\ &+ Ln\left(T\right)\left(a_1 + a_2 Ln\left(d_p\right) + a_3 Ln\left(\phi\right) + a_4 Ln\left(\phi\right)\ln\left(d_p\right) + a_5 Ln\left(d_p\right)^2\right) \\ R_f + d_p / k_p &= d_p / k_{p,eff}, R_f = 4\times 10^{-8} km^2 / W \end{aligned} \tag{62}$$

$$\mu_{nf} = \frac{\mu_f}{\left(1-\phi\right)^{2.5}} + \frac{k_{Brownian}}{k_f} \times \frac{\mu_f}{\Pr} \tag{63}$$

The stream function and vorticity are defined as:

$$u = \frac{\partial \psi}{\partial y},\ v = -\frac{\partial \psi}{\partial x},\ \omega = \frac{\partial v}{\partial x} - \frac{\partial u}{\partial y} \tag{64}$$

The stream function satisfies the continuity Equation (56). The vorticity equation is obtained by eliminating the pressure between the two momentum equations, i.e. by taking y-derivative of Equation (57) and subtracting from it the x-derivative of Equation (58). This gives:

$$\frac{\partial\psi}{\partial y}\frac{\partial\omega}{\partial x}-\frac{\partial\psi}{\partial x}\frac{\partial\omega}{\partial y}=\upsilon_{nf}\left(\frac{\partial^2\omega}{\partial x^2}+\frac{\partial^2\omega}{\partial y^2}\right)+\beta_{nf}g\left(\frac{\partial T}{\partial x}\right)$$
$$+\frac{\sigma_{nf}B^2}{\rho_{nf}}\left(-\frac{\delta v}{\delta y}\sin\lambda\cos\lambda+\frac{\delta u}{\delta y}\sin^2\lambda+\frac{\delta u}{\delta x}\sin\lambda\cos\lambda-\frac{\delta v}{\delta x}\cos^2\lambda\right) \tag{65}$$

$$\frac{\partial\psi}{\partial y}\frac{\partial T}{\partial x}-\frac{\partial\psi}{\partial x}\frac{\partial T}{\partial y}=\alpha_{nf}\left(\frac{\partial^2 T}{\partial x^2}+\frac{\partial^2 T}{\partial y^2}\right) \tag{66}$$

$$\frac{\partial^2\psi}{\partial x^2}+\frac{\partial^2\psi}{\partial y^2}=-\omega \tag{67}$$

By introducing the following non-dimensional variables:

$$X=\frac{x}{L},\ Y=\frac{y}{L}\ ,\Omega=\frac{\omega L^2}{\alpha_f},\Psi=\frac{\psi}{\alpha_f},\Theta=\frac{T-T_c}{\left(q''L/k_f\right)},U=\frac{uL}{\alpha_f},V=\frac{vL}{\alpha_f} \tag{68}$$

Using the dimensionless parameters, the equations now become:

$$\frac{\partial\Psi}{\partial Y}\frac{\partial\Omega}{\partial X}-\frac{\partial\Psi}{\partial X}\frac{\partial\Omega}{\partial Y}=\Pr_f/\left(\left(1-\phi\right)^{2.5}\left(\left(1-\phi\right)+\phi\frac{\rho_s}{\rho_f}\right)\right)\left(\frac{\partial^2\Omega}{\partial X^2}+\frac{\partial^2\Omega}{\partial Y^2}\right)$$
$$+Ra_f\Pr_f\left[\left(1-\phi\right)+\phi\frac{\beta_s}{\beta_f}\right]\left(\frac{\partial\Theta}{\partial X}\right)$$
$$+Ha^2\Pr_f\left(1+3\left(\sigma_s/\sigma_f-1\right)\phi/\left(\left(\sigma_s/\sigma_f+2\right)-\left(\sigma_s/\sigma_f-1\right)\phi\right)\right)$$
$$\times 1/\left(\left(1-\phi\right)+\phi\frac{\rho_s}{\rho_f}\right)\left(-\frac{\delta V}{\delta Y}\tan\lambda+\frac{\delta U}{\delta Y}\tan^2\lambda+\frac{\delta U}{\delta X}\tan\lambda-\frac{\delta V}{\delta X}\right) \tag{69}$$

$$\frac{\partial\Psi}{\partial Y}\frac{\partial\Theta}{\partial X}-\frac{\partial\Psi}{\partial X}\frac{\partial\Theta}{\partial Y}=\frac{k_{nf}}{k_f}/\left(\left(1-\phi\right)+\phi\frac{\left(\rho Cp\right)_s}{\left(\rho Cp\right)_f}\right)\left(\frac{\partial^2\Theta}{\partial X^2}+\frac{\partial^2\Theta}{\partial Y^2}\right) \tag{70}$$

$$\frac{\partial^2\Psi}{\partial X^2}+\frac{\partial^2\Psi}{\partial Y^2}=-\Omega \tag{71}$$

where $Ra_f = g\beta_f L^4 q'' / \left(k_f \alpha_f \nu_f\right)$ is the Rayleigh number for the base fluid, $Ha = LB_x \sqrt{\sigma_f / \mu_f}$ is the Hartmann number and $\Pr_f = \upsilon_f / \alpha_f$ is the Prandtl number for the base fluid. The boundary conditions as shown in Figure 19 are:

$$\frac{\partial \Theta}{\partial n} = -1 \text{ on the inner circular boundary}$$

$$\Theta = 0.0 \text{ on the outer walls boundary}$$

$$\Psi = 0.0 \text{ on all solid boundaries} \tag{72}$$

The values of vorticity on the boundary of the enclosure can be obtained using the stream function formulation and the known velocity conditions during the iterative solution procedure. The local Nusselt number of the nanofluid along the hot wall can be expressed as:

$$Nu_{local} = \left(\frac{k_{nf}}{k_f}\right) \frac{1}{\theta}\bigg|_{inner\ wall} \tag{73}$$

where *r* is the radial direction. The average Nusselt number on hot circular wall is evaluated as:

$$Nu_{ave} = \frac{1}{2\pi} \int_0^{2\pi} Nu_{loc}\, d\zeta \tag{74}$$

To estimate the enhancement of heat transfer between the case of $\phi = 0.04$ and the pure fluid (base fluid) case, the enhancement is defined as:

$$E = \frac{Nu\left(\phi = 0.04\right) - Nu\left(basefluid\right)}{Nu\left(basefluid\right)} \times 100 \tag{75}$$

The heatlines are adequate tools for visualization and analysis of 2D convection heat transfer, through an extension of the heat flux line concept to include the advection terms. Heat function (*H*) are defined in terms of the energy equation as

$$\frac{\partial H}{\partial Y} = U\Theta - \frac{\partial \Theta}{\partial X},\ -\frac{\partial H}{\partial X} = V\Theta - \frac{\partial \Theta}{\partial Y} \tag{76}$$

4.2. Effects of Active Parameters

CVFEM is applied to solve the problem of natural convection in an enclosure filled with Al_2O_3-water nanofluid in presence of magnetic field. The effective thermal conductivity and viscosity of nanofluid are calculated by KKL (Koo-Kleinstreuer-Li) correlation. Calculations are made for various values of volume fraction of nanoparticles ($\phi = 0$ to 4%), Rayleigh number ($Ra = 10^3, 10^4$ and 10^5), aspect ratio ($r_{in} / L = 0.2, 0.3$ and 0.4) and Hartmann number ($Ha = 0$ to 100) at constant Prandtl number ($\mathrm{Pr} = 6.2$).

Comparisons of the isotherms, streamlines and heatlines for different values of Hartmann number, aspect ratio and Rayleigh number are shown in Figures 20 and 21. By increasing Rayleigh number the prominent heat transfer mechanism is turn from conduction to convection. Also it can be seen that increasing aspect ratio leads to decrease thermal boundary layer thickness and intensity of convection because of domination of conduction heat transfer. When the magnetic field is imposed on the enclosure, the velocity field suppressed owing to the retarding effect of the Lorenz force. So intensity of convection weakens significantly. The braking effect of the magnetic field is observed from the maximum stream function value. The core vortex is shift upward vertically as the Hartmann number increases. Also imposing magnetic field leads to omit the thermal plume over the inner wall. At high Hartmann number the conduction heat transfer mechanism is more pronounced. For this reason the isotherms are parallel to each other.

The heat flow within the enclosure is displayed using the heat function obtained from conductive heat fluxes ($\partial\Theta / \partial X, \partial\Theta / \partial Y$) as well as convective heat fluxes ($V\Theta, U\Theta$). Heatlines emanate from hot regimes and end on cold regimes illustrating the path of heat flow. The domination of conduction heat transfer in low Rayleigh number and high Hartmann number can be observed from the heatline patterns since no passive area exists. The increase of Ra causes the clustering of heatlines from hot to the cold wall and generates passive heat transfer area in which heat is rotated without having significant effect on heat transfer between walls.

Distribution of local Nusselt numbers along the surface of the inner circular wall for different aspect ratio, Rayleigh number and Hartmann number are shown in Figure 22. Increasing Rayleigh number and aspect ratio lead to an increase in local Nusselt number but increasing Hartmann number causes local Nusselt number to decrease.

Figure 23 shows the effects of aspect ratio, Rayleigh number and Hartmann number on average Nusselt number. Increasing Hartmann number causes Lorenz force to increase and leads to a substantial suppression of the convection. So Nusselt number has reveres relationship with Hartmann number. As aspect ratio increases, space to accelerate the flow inside the cavity decreases. So thermal boundary layer thickness decreases and in turn $1 / \theta$ increases. As volume fraction of nanoparticle increases, thermal diffusivity increases. So the high values of thermal diffusivity cause the boundary thickness to increase and accordingly decrease $1 / \theta$. Nusselt number is function of $1 / \theta$ and k_{nf} / k_f. Because of reduction in $1 / \theta$ due to the presence of nanoparticles is much smaller than thermal conductivity ratio therefore an augment in Nusselt number is taken place by increasing the volume fraction of nanoparticles. The distance between cold and hot walls decreases with increase of aspect ratio. So Nusselt number decreases with increase of r_{in} / L.

Figure 20. Comparison of the isotherms, streamlines, heatlines for different values of Hartmann number and Raleigh number at $r_{in} / L = 0.2, \phi = 0.04$ *and* $\Pr = 6.2$

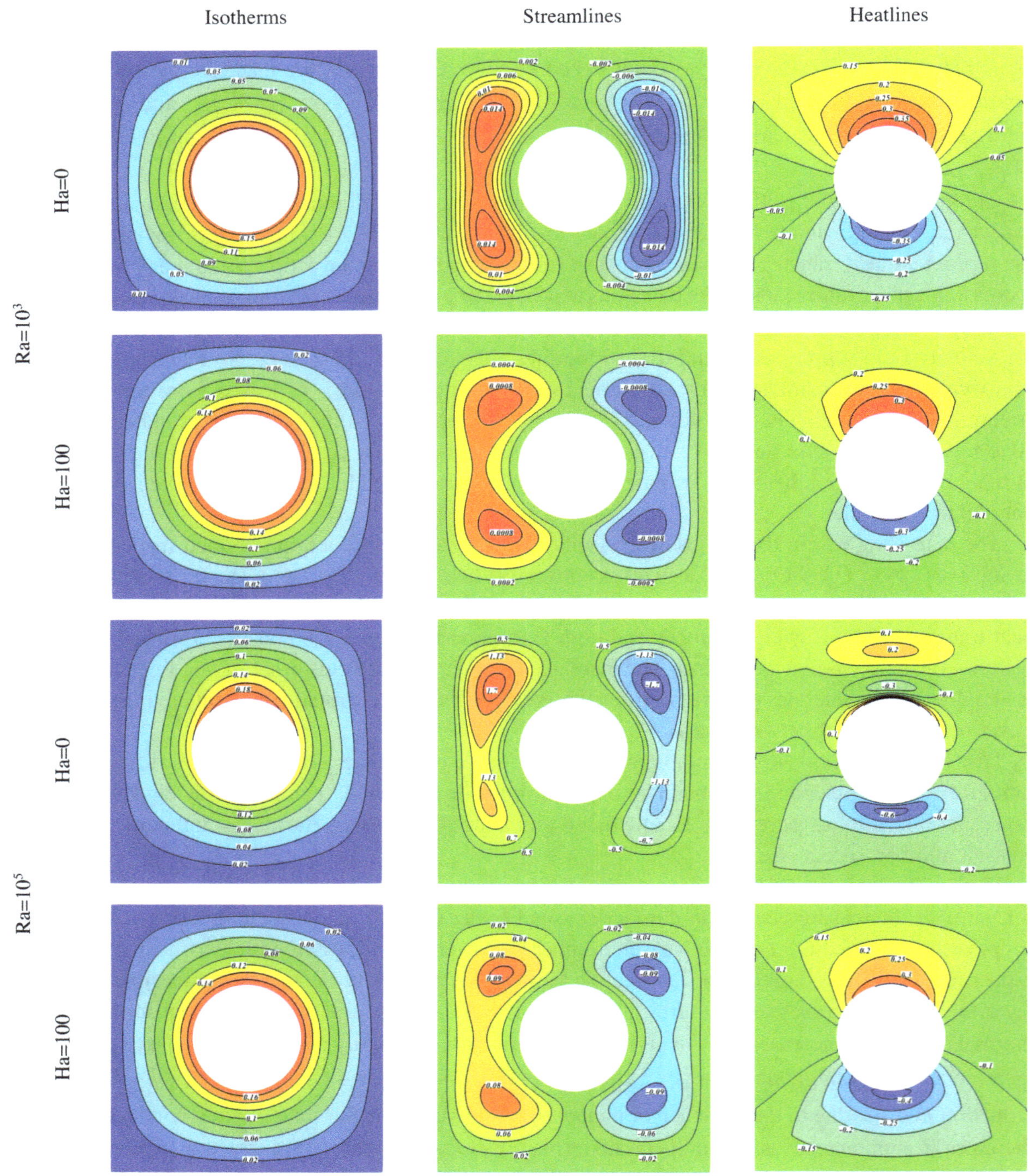

Figure 21. Comparison of the isotherms, streamlines, heatlines for different values of Hartmann number and Raleigh number at $r_{in} / L = 0.4, \phi = 0.04$ *and* $\mathrm{Pr} = 6.2$

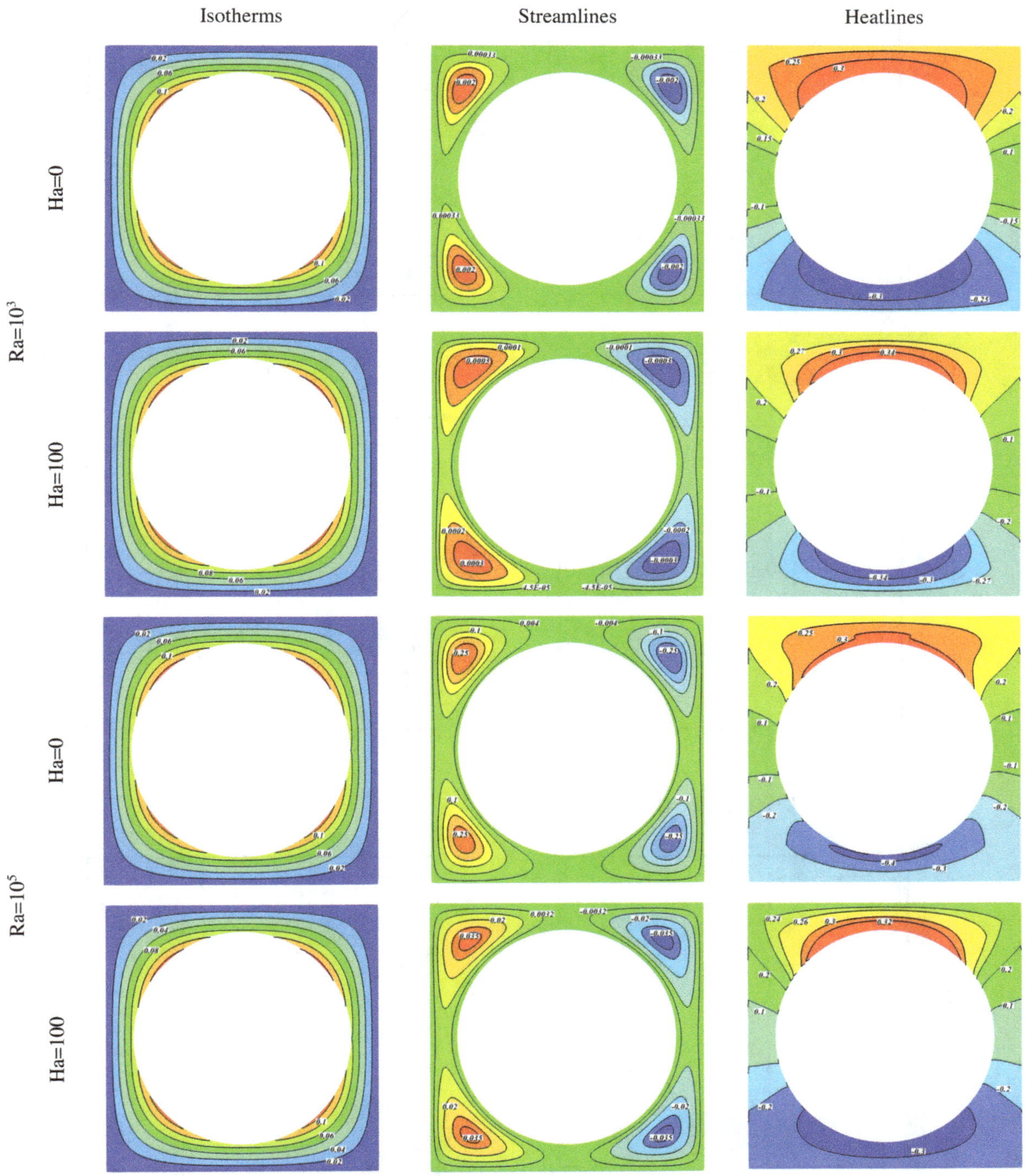

Figure 22. Effects of aspect ratio, Rayleigh number and Hartmann number on local Nusselt number

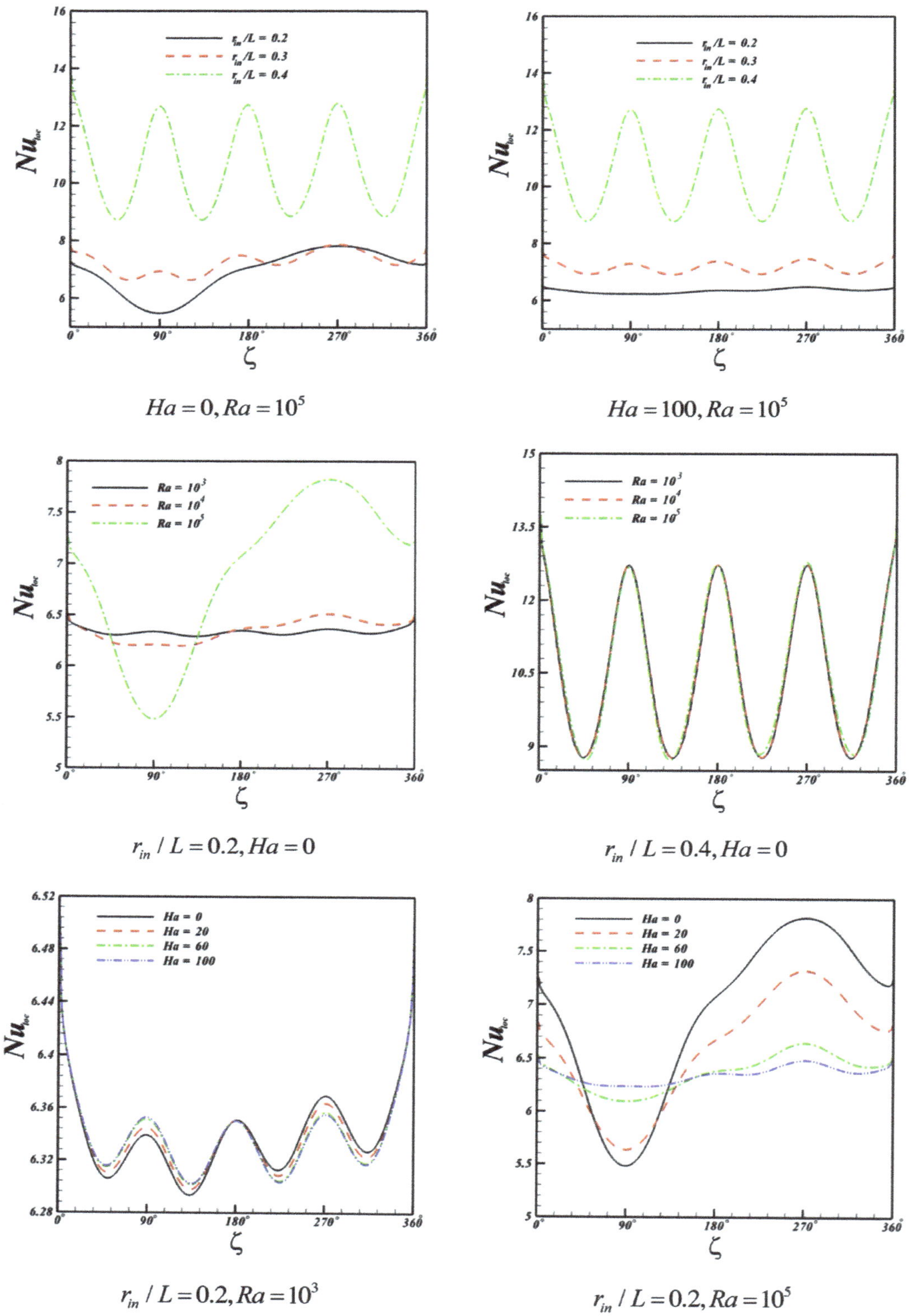

Figure 23. Effects of aspect ratio, Rayleigh number and Hartmann number on average Nusselt number

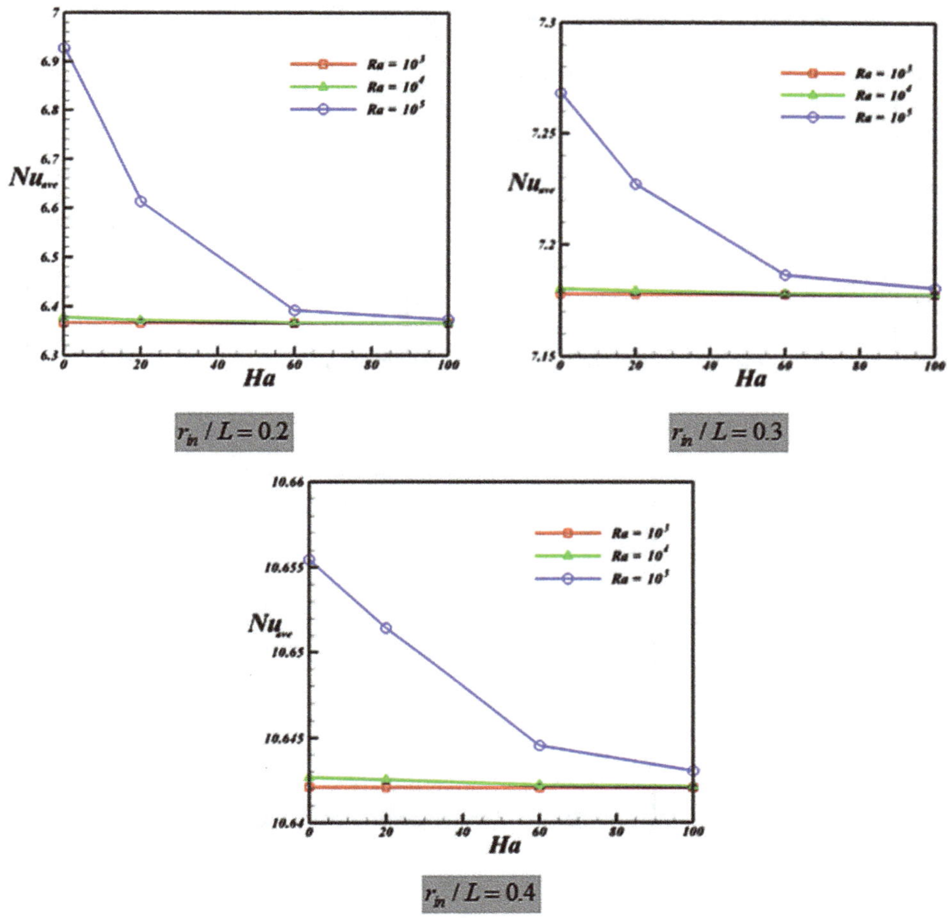

The heat transfer enhancement ratio due to addition of nanoparticles for different values of r_{in} / L, Ha and Ra is shown in Figure 24. Heat transfer enhancement ratio has direct relationship with Hartmann number and aspect ratio but has reverse relationship with Rayleigh number. This observation is due to domination of conduction heat transfer in low Rayleigh number and high Hartmann number or aspect ratio. Therefore, the addition of high thermal conductivity nanoparticles will increase the conduction and make the enhancement more effective.

5. MAGNETIC FIELD EFFECT ON NANOFLUID FLOW AND HEAT TRANSFER USING KKL MODEL

5.1. Problem Definition

The physical model along with the important geometrical parameters and the mesh of the enclosure used in the present CVFEM program are shown in Figure 25 (Li et al., 2018). The width and height of the

Figure 24. Effects of aspect ratio, Rayleigh number and Hartmann number on heat transfer enhancement

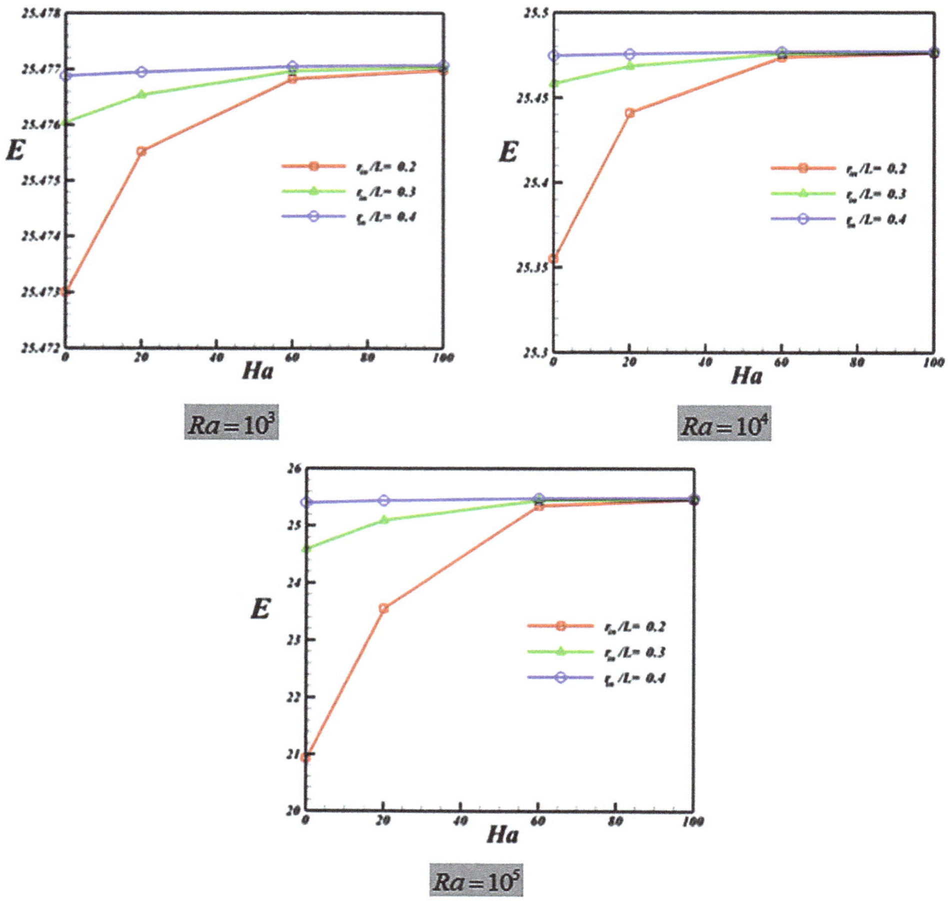

enclosure is H. The right and top wall of the enclosure are maintained at constant cold temperatures T_c whereas the inner circular hot wall is maintained at constant hot temperature T_h and the two bottom and left walls with the length of $H / 2$ are thermally isolated. Under all cases $T_h > T_c$ condition is maintained. In this section $r_{in} / r_{out} = 0.75$. It is also assumed that the uniform magnetic field ($\vec{B} = B_x \vec{e}_x + B_y \vec{e}_y$) of constant magnitude $B = \sqrt{B_x^2 + B_y^2}$ is applied, where $\vec{e}_x$ and $\vec{e}_y$ are unit vectors in the Cartesian coordinate system (Sheikholeslami, 2018). The governing equations are similar to those of exist in section (4.1). The local Nusselt number of the nanofluid along the hot wall can be expressed as:

$$Nu_{local} = \left(\frac{k_{nf}}{k_f}\right)\frac{\partial \Theta}{\partial r} \tag{77}$$

where *r* is the radial direction. The average Nusselt number on hot circular wall is evaluated as:

Figure 25. (a) Geometry and the boundary conditions with (b) the mesh of enclosure considered in this work

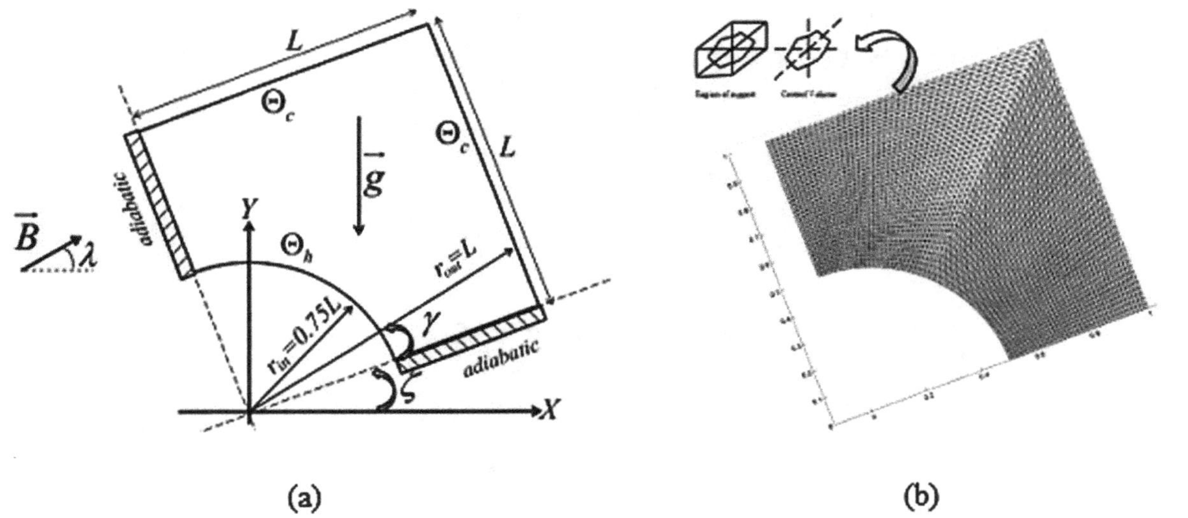

$$Nu_{ave} = \frac{1}{\gamma}\int_{0}^{\gamma} Nu_{loc}\left(\zeta\right)d\zeta \tag{78}$$

5.2. Effects of Active Parameters

In this study, MHD effect on natural convection heat transfer in an L-shape inclined enclosure filled with nanofluid is investigated numerically using the Control Volume based Finite Element Method. The fluid in the enclosure is Al_2O_3-water nanofluid. Calculations are made for various values of Hartmann number ($Ha = 0, 20, 60$ and 100), volume fraction of nanoparticle ($\phi = 0\%$ and 4%), Rayleigh number ($Ra = 10^3, 10^4$ and 10^5), inclination angle ($\gamma = -90°, -60°, -30°$ and $0°$) and constant Prandtl number ($\Pr = 6.2$).

Comparisons of the isotherms, streamlines and heatlines for different values of Hartmann number, inclined angles and Rayleigh number are shown in Figures 26 and 27. At $Ra = 10^3$ the isotherms are parallel to each other and take the shape of enclosure which is the main characteristic of conduction heat transfer mechanism. As Rayleigh number increases the isotherms become more distorted and the stream function values enhance which is due to the domination of convective heat transfer mechanism at higher Rayleigh numbers. At $Ra = 10^5$, thermal plume appears on the hot circular wall and three vortices exist in streamline. By increasing $|\zeta|$, these vortices merge to one eddy. As $|\zeta|$ increases, the hot wall locates on the cold wall. So convective heat transfer becomes weak and in turn Nusselt number decreases with increase of $|\zeta|$.

When the magnetic field is imposed on the enclosure, the velocity field suppressed owing to the retarding effect of the Lorenz force. So intensity of convection weakens significantly. The braking effect of the magnetic field is observed from the maximum stream function value. The core vortex is shift

Figure 26a. Comparison of the isotherms, streamlines and heatlines for different values of Hartmann number and inclination angle at $Ra = 10^3, \phi = 0.04$ *and* $\Pr = 6.2$

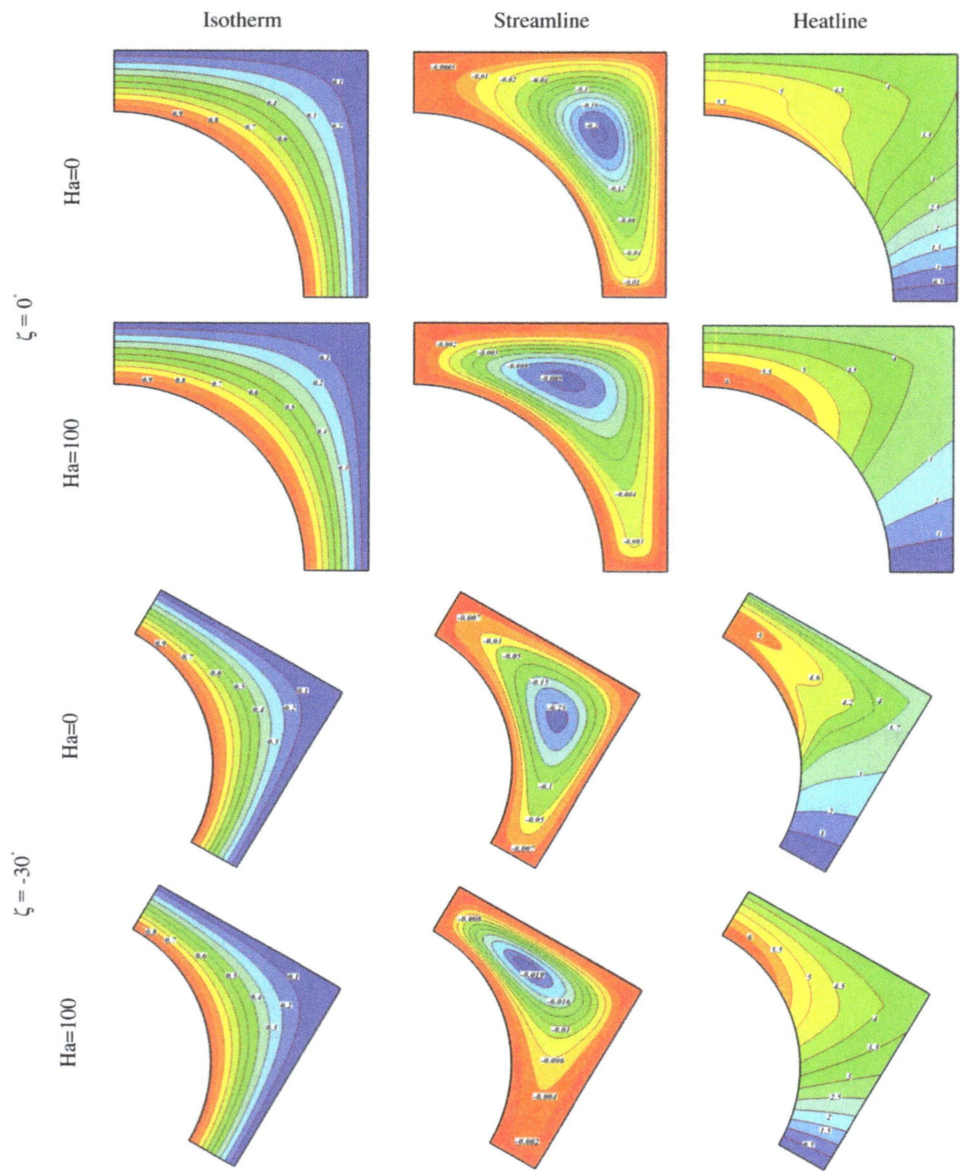

upward vertically as the Hartmann number increases. Also imposing magnetic field leads to omit the thermal plume over the inner wall. At high Hartmann number the conduction heat transfer mechanism is more pronounced. For this reason the isotherms are parallel to each other.

The heat flow within the enclosure is displayed using the heat function obtained from conductive heat fluxes ($\partial\Theta / \partial X, \partial\Theta / \partial Y$) as well as convective heat fluxes ($V\Theta, U\Theta$). Heatlines emanate from hot regimes and end on cold regimes illustrating the path of heat flow. The domination of conduction

Figure 26b. Comparison of the isotherms, streamlines and heatlines for different values of Hartmann number and inclination angle at $Ra = 10^3, \phi = 0.04$ *and* $\Pr = 6.2$

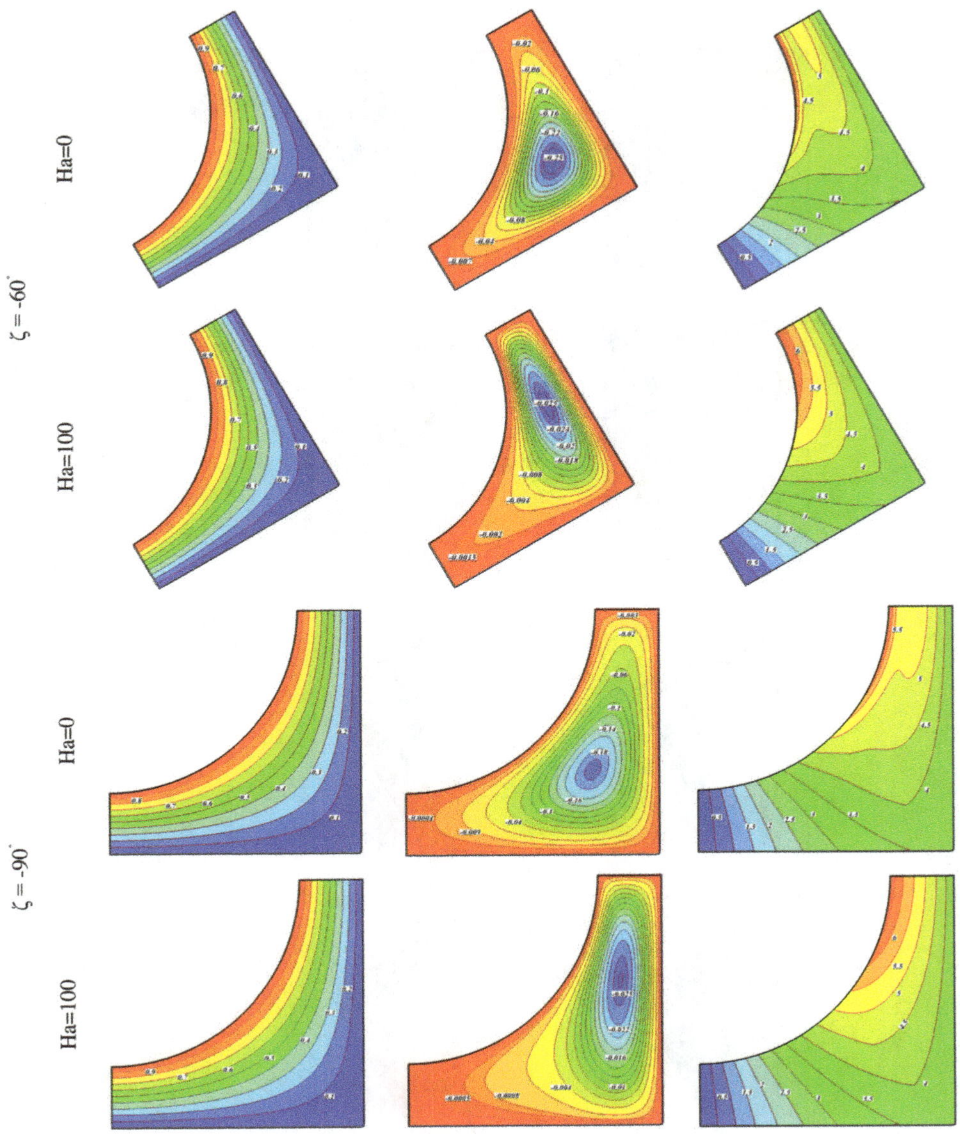

heat transfer in low Rayleigh number and high Hartmann number can be observed from the heatline patterns since no passive area exists. The increase of Ra causes the clustering of heatlines from hot to the cold wall and generates passive heat transfer area in which heat is rotated without having significant effect on heat transfer between walls.

Figure 28 shows the distribution of local Nusselt numbers along the surface of the inner circular wall for different inclination angle, Rayleigh number and Hartmann number. Increasing Rayleigh number leads to an increase in local Nusselt number but increasing Hartmann number and inclination angle cause

Figure 27a. Comparison of the isotherms, streamlines and heatlines for different values of Hartmann number and inclination angle at $Ra = 10^5, \phi = 0.04$ *and* $\Pr = 6.2$

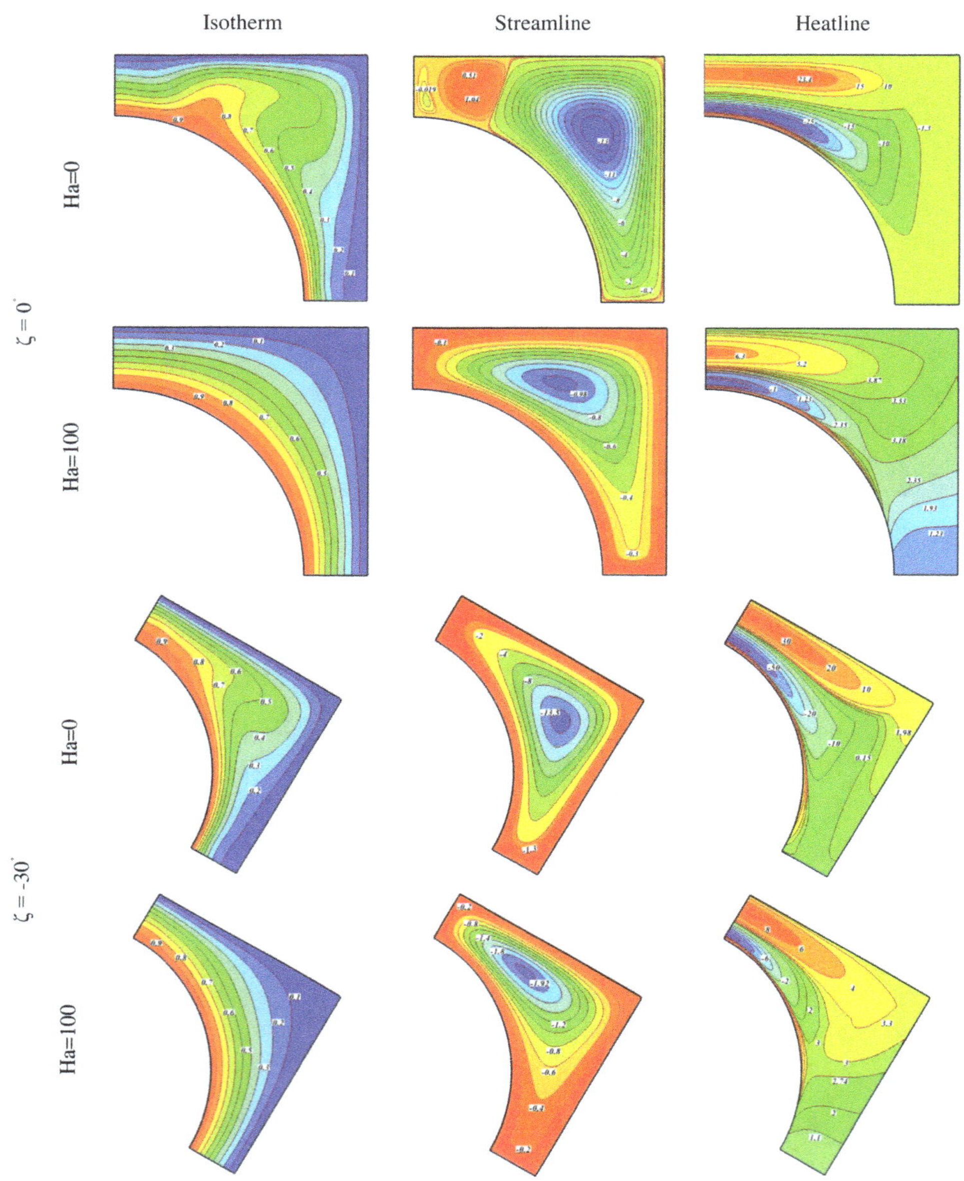

local Nusselt number to decrease. At $Ra = 10^3$, because of domination conduction heat transfer mechanism, the distribution of the local Nusselt numbers along the surface of inner circular shows the symmetric shape. It is interesting to notice that at high Rayleigh number the local Nusselt number profiles are more complex due to the presence of thermal plume. In all cases expect for $Ra = 10^5, \gamma = -90^{\circ}$, one minimum point exist in local Nusselt number profile which is occurred in lower values of $|\gamma|$ with

Figure 27b. Comparison of the isotherms, streamlines and heatlines for different values of Hartmann number and inclination angle at $Ra = 10^5, \phi = 0.04$ *and* $\Pr = 6.2$

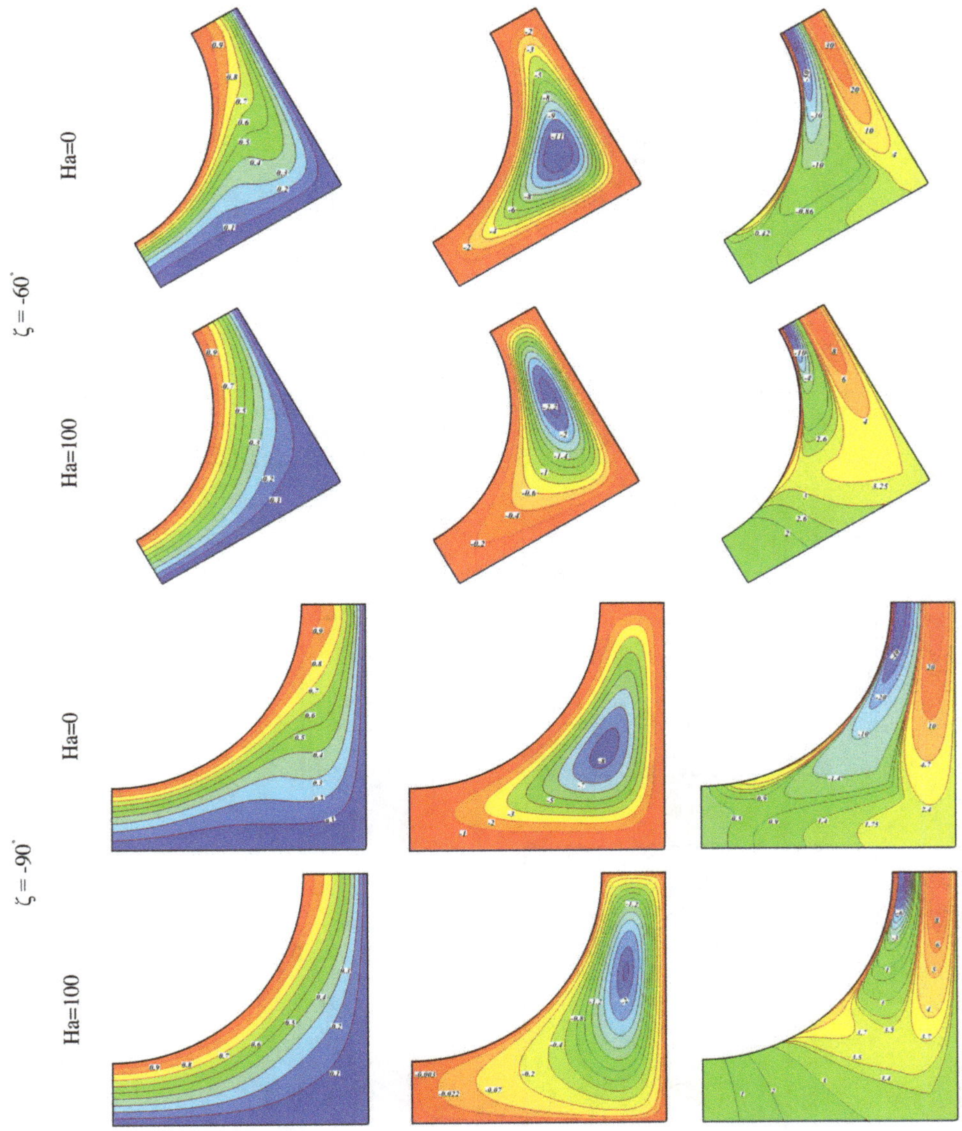

increase of Hartmann number. Effects of Rayleigh number, Hartmann number and inclination angle on average Nusselt number in shown in Figure 29. Nusselt number is an increasing function of Rayleigh number but it is a decreasing function of Hartmann number and inclined angle. Also it can be found that effect of Hartmann number on Nusselt number is more pronounced at $\zeta = 0^\circ$.

The heat transfer enhancement ratio due to addition of nanoparticles for different values of ζ, Ha and Ra is shown in Figure 30. In general, it can be found that the effect of nanoparticles is more pro-

Figure 28. Effects of the Hartmann number, Rayleigh number and inclination angle for Cu-water nanofluids on Local Nusselt number

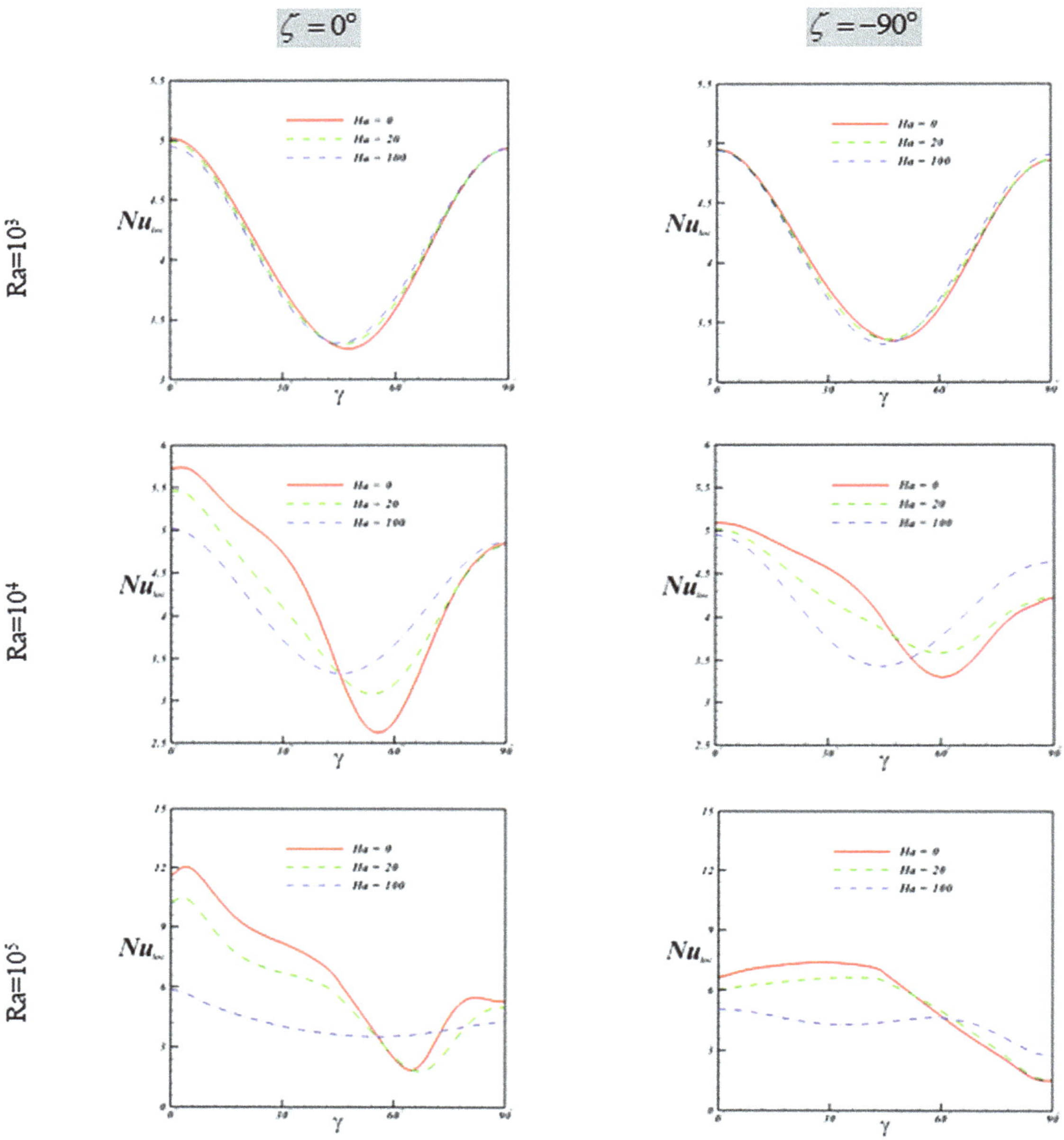

nounced at low Rayleigh number and high values of Hartmann number because of the greater enhancement rate. This observation can be explained by noting that at low Rayleigh number the heat transfer is dominant by conduction. Therefore, the addition of high thermal conductivity nanoparticles will increase the conduction and make the enhancement more effective. Inclination angle has no significant effect on rate on enhancement when $Ra = 10^3, 10^4$ while E increases with increase of $|\zeta|$ when $Ra = 10^5$. Fi-

Figure 29. Effects of Rayleigh number, Hartmann number and inclination angle on average Nusselt number at $\phi = 0.04$

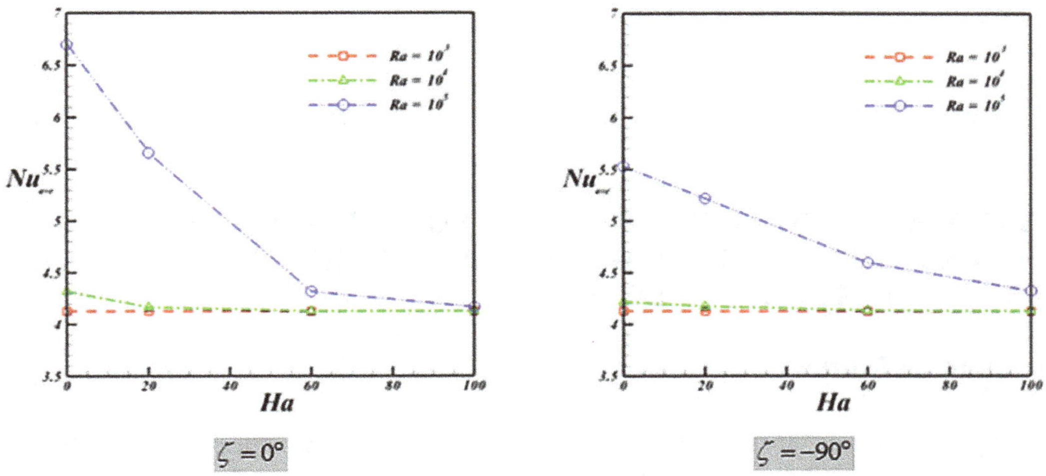

Figure 30. Effects of Rayleigh number, Hartmann number and inclination angle on ratio of heat transfer enhancement due to addition of nanoparticles

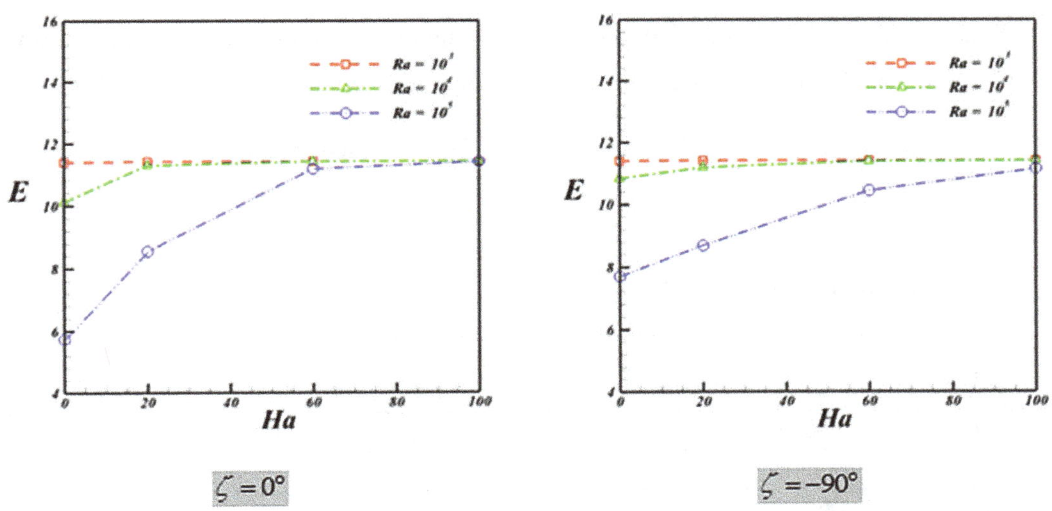

nally, the corresponding polynomial representation of such model for each of Nusselt number and rate of enhancement are presented as follows:

$$\begin{aligned} Nu &= a_{13} + a_{23}Y_1 + a_{33}Y_2 + a_{43}Y_1^2 + a_{53}Y_2^2 + a_{63}Y_1Y_2 \\ Y_1 &= a_{11} + a_{21}\zeta + a_{31}\phi + a_{41}\zeta^2 + a_{51}\phi^2 + a_{61}\zeta\,\phi \\ Y_2 &= a_{12} + a_{22}Ra^* + a_{32}Ha^* + a_{42}Ra^{*2} + a_{52}Ha^{*2} + a_{62}Ra^*\,Ha^* \end{aligned} \tag{79}$$

$$\begin{aligned} E &= b_{13} + b_{23}Y_1 + b_{33}Y_2 + b_{43}Y_1^2 + b_{53}Y_2^2 + b_{63}Y_1Y_2 \\ Y_1 &= b_{11} + b_{21}\zeta + b_{31}Ha^* + b_{41}\zeta^2 + b_{51}Ha^{*2} + b_{61}\zeta Ha^* \\ Y_2 &= b_{12} + b_{22}Ra^* + b_{32}\zeta + b_{42}Ra^{*2} + b_{52}\zeta^2 + b_{62}Ra^*\zeta \end{aligned} \quad (80)$$

where $Ra^* = \log\left(Ra\right)$ and $Ha^* = Ha / 100$. Also a_{ij} and b_{ij} can be found in Tables 3 and 4.

6. MAGNETOHYDRODYNAMIC FREE CONVECTION OF AL2O3-WATER NANOFLUID CONSIDERING THERMOPHORESIS AND BROWNIAN MOTION EFFECTS

6.1. Problem Definition

The schematic diagram and the mesh of the semi-annulus enclosure used in the present CVFEM program are shown in Figure 31 (Sheikholeslami & Gorji-Bandpy, 2014). The inner and outer walls are maintained at constant temperatures T_h and T_c, respectively while the two other walls are thermally insulated. Also the boundary conditions of concentration are similar to temperature. It is also assumed that the uniform magnetic field ($\vec{B} = B_x \vec{e}_x + B_y \vec{e}_y$) of constant magnitude $B = \sqrt{B_x^2 + B_y^2}$ is applied, where $\vec{e}_x$ and $\vec{e}_y$ are unit vectors in the Cartesian coordinate system. The orientation of the magnetic field form an angle λ with horizontal axis such that $\lambda = B_x / B_y$. The electric current J and the electromagnetic force F are defined by $J = \sigma\left(\vec{V} \times \vec{B}\right)$ and $F = \sigma\left(\vec{V} \times \vec{B}\right) \times \vec{B}$, respectively.

Table 3. Constant coefficient for using Equation (70)

a_{ij}	*i=1*	*i=2*	*i=3*	*i=4*	*i=5*	*i=6*
j=1	4.080931	-0.10849	10.44936	-0.10891	0.417974	-0.03188
j=2	7.809771	-2.67057	2.414376	0.452514	0.790997	-0.97843
j=3	8.465516	0.484181	-4.13987	-0.01217	0.4827	0.145098

Table 4. Constant coefficient for using Equation (80)

b_{ij}	*i=1*	*i=2*	*i=3*	*i=4*	*i=5*	*i=6*
j=1	9.333563	-0.00506	4.537482	0.195157	-2.33844	0.508573
j=2	3.204931	5.085197	0.530967	-0.77932	0.195157	-0.07679
j=3	-134.848	14.51322	11.03078	-0.10825	0.060356	-1.0671

Figure 31. (a) Geometry and the boundary conditions; (b) the mesh of enclosure considered in this work

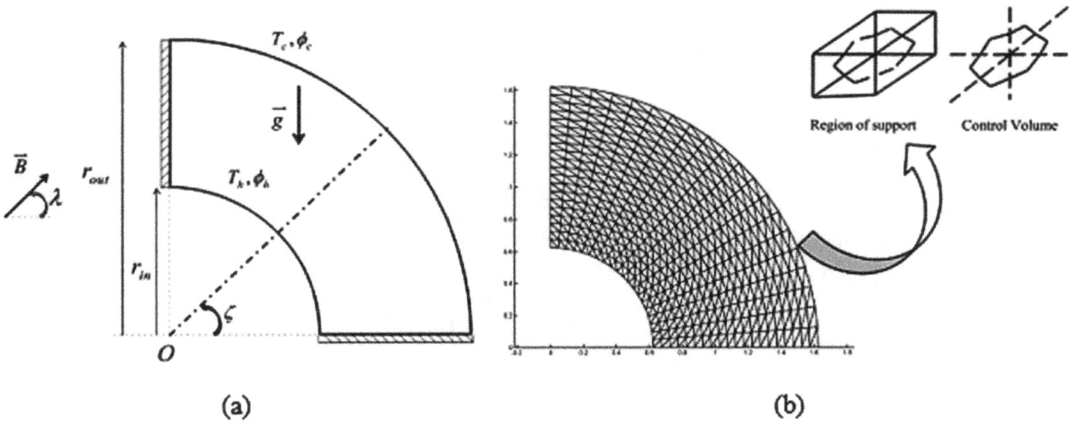

The nanofluid's density, ρ is

$$\begin{aligned}\rho &= \phi\rho_p + (1-\phi)\rho_f \\ &\cong \phi\rho_p + (1-\phi)\left\{\rho_{f_0}\left(1-\beta(T-T_c)\right)\right\}\end{aligned} \tag{81}$$

where ρ_f, is the base fluid's density, T_c, is a reference temperature, ρ_{f_0} is the base fluid's density at the reference temperature, β is the volumetric coefficient of expansion. Taking the density of base fluid as that of the nanofluid, as adopted by (Sheikholeslami & Ganji, 2014), the density ρ in Equation (81), thus becomes

$$\rho \cong \phi\rho_p + (1-\phi)\left\{\rho_0\left(1-\beta(T-T_c)\right)\right\} \tag{82}$$

ρ_0 is the nanofluid's density at the reference temperature.

The continuity, momentum under Boussinesq approximation and energy equations for the laminar and steady state natural convection in a two-dimensional enclosure can be written in dimensional form as follows:

$$\frac{\partial u}{\partial x} + \frac{\partial v}{\partial y} = 0 \tag{83}$$

$$\rho_f\left\{u\frac{\partial u}{\partial x} + v\frac{\partial u}{\partial y}\right\} = -\frac{\partial P}{\partial x} + \mu\left(\frac{\partial^2 u}{\partial x^2} + \frac{\partial^2 u}{\partial y^2}\right) + \sigma B^2\left(v\sin\lambda\cos\lambda - u\sin^2\lambda\right) \tag{84}$$

$$\rho_f\left\{u\frac{\partial v}{\partial x}+v\frac{\partial v}{\partial y}\right\}=-\frac{\partial P}{\partial y}+\mu\left(\frac{\partial^2 v}{\partial x^2}+\frac{\partial^2 v}{\partial y^2}\right)-\left(\phi-\phi_c\right)\left(\rho_p-\rho_{f_0}\right)g$$
$$+(1-\phi_c)\rho_{f_0}\left(T-T_c\right)g+\sigma B^2\left(u\sin\lambda\cos\lambda-v\cos^2\lambda\right) \tag{85}$$

$$u\frac{\partial T}{\partial x}+v\frac{\partial T}{\partial y}=\alpha\left(\frac{\partial^2 T}{\partial x^2}+\frac{\partial^2 T}{\partial y^2}\right)+\frac{\left(\rho c\right)_p}{\left(\rho c\right)_f}\left[\begin{array}{l}D_B\left\{\frac{\partial \phi}{\partial x}.\frac{\partial T}{\partial x}+\frac{\partial \phi}{\partial y}.\frac{\partial T}{\partial y}\right\}\\ +(D_T/T_c)\left\{\left(\frac{\partial T}{\partial x}\right)^2+\left(\frac{\partial T}{\partial y}\right)^2\right\}\end{array}\right] \tag{86}$$

$$u\frac{\partial \phi}{\partial x}+v\frac{\partial \phi}{\partial y}=D_B\left\{\frac{\partial^2 \phi}{\partial x^2}+\frac{\partial^2 \phi}{\partial y^2}\right\}+\left(\frac{D_T}{T_c}\right)\left\{\frac{\partial^2 T}{\partial x^2}+\frac{\partial^2 T}{\partial y^2}\right\} \tag{87}$$

Boundary conditions are

$$\begin{array}{ll} T=T_h,\quad \phi=\phi_h & \text{on the inner circular boundary}\\ T=T_c,\quad \phi=\phi_c & \text{on the outer circular boundary}\\ \partial T/\partial n=\partial\phi/\partial n=0.0 & \text{on two other insulation boundaries}\\ \psi=0, & \text{all the solid boundaries}\end{array} \tag{88}$$

The stream function and vorticity are defined as follows:

$$u=\frac{\partial \psi}{\partial y},\ v=-\frac{\partial \psi}{\partial x},\ \omega=\frac{\partial v}{\partial x}-\frac{\partial u}{\partial y} \tag{89}$$

The stream function satisfies the continuity Equation (83). The vorticity equation is obtained by eliminating the pressure between the two momentum equations, i.e. by taking y-derivative of Equation (84) and subtracting from it the *x*-derivative of Equation (85). Also the following non-dimensional variables should be introduced:

$$X=\frac{x}{L},Y=\frac{y}{L},\Omega=\frac{\omega L^2}{\alpha},\Psi=\frac{\psi}{\alpha},\Theta=\frac{T-T_c}{T_h-T_c},\Phi=\frac{\phi-\phi_c}{\phi_h-\phi_c},U=\frac{uL}{\alpha},V=\frac{vL}{\alpha} \tag{90}$$

where in Equation (90) $L=r_{out}-r_{in}=r_{in}$.

By using these dimensionless parameters the equations become:

$$\left[\frac{\partial\Psi}{\partial Y}\frac{\partial\Omega}{\partial X}-\frac{\partial\Psi}{\partial X}\frac{\partial\Omega}{\partial Y}\right]=\Pr\left(\frac{\partial^2\Omega}{\partial X^2}+\frac{\partial^2\Omega}{\partial Y^2}\right)+\Pr Ra\left(\frac{\partial\Theta}{\partial X}-Nr\frac{\partial\Phi}{\partial X}\right)$$
$$+Ha^2\Pr\left(-\frac{\delta V}{\delta Y}\tan\lambda+\frac{\delta U}{\delta Y}\tan^2\lambda+\frac{\delta U}{\delta X}\tan\lambda-\frac{\delta V}{\delta X}\right) \tag{91}$$

$$\frac{\partial\Psi}{\partial Y}\frac{\partial\Theta}{\partial X}-\frac{\partial\Psi}{\partial X}\frac{\partial\Theta}{\partial Y}=\left(\frac{\partial^2\Theta}{\partial X^2}+\frac{\partial^2\Theta}{\partial Y^2}\right)+Nb\left(\frac{\partial\Phi}{\partial X}\frac{\partial\Theta}{\partial X}+\frac{\partial\Phi}{\partial Y}\frac{\partial\Theta}{\partial Y}\right)+Nt\left(\left(\frac{\partial\Theta}{\partial X}\right)^2+\left(\frac{\partial\Theta}{\partial Y}\right)^2\right) \tag{92}$$

$$\frac{\partial\Psi}{\partial Y}\frac{\partial\Phi}{\partial X}-\frac{\partial\Psi}{\partial X}\frac{\partial\Phi}{\partial Y}=\frac{1}{Le}\left(\frac{\partial^2\Phi}{\partial X^2}+\frac{\partial^2\Phi}{\partial Y^2}\right)+\frac{Nt}{Nb\ Le}\left(\frac{\partial^2\Theta}{\partial X^2}+\frac{\partial^2\Theta}{\partial Y^2}\right) \tag{93}$$

$$\frac{\partial^2\Psi}{\partial X^2}+\frac{\partial^2\Psi}{\partial Y^2}=-\Omega \tag{94}$$

where thermal Rayleigh number, the buoyancy ratio number, Prandtl number, the Brownian motion parameter, the thermophoretic parameter, Lewis number and Hartmann number of nanofluid are defined as:

$$Ra=\left(1-\phi_c\right)\rho_{f_0}g\beta L^3\left(T_h-T_c\right)/\left(\mu\alpha\right),$$

$$Nr=\left(\rho_p-\rho_0\right)\left(\phi_h-\phi_c\right)/\left[\left(1-\phi_c\right)\rho_{f_0}\beta L\left(T_h-T_c\right)\right],$$

$$\Pr=\mu/\rho_f\alpha,$$

$$Nb=(\rho c)_p D_B(\phi_h-\phi_c)/\left((\rho c)_f\alpha\right),$$

$$Nt=(\rho c)_p D_T(T_h-T_c)/[(\rho c)_f\alpha T_c],$$

$$Le=\alpha/D_B$$

and

$$Ha = LB_x\sqrt{\sigma / \mu}\,,$$

respectively.

The equation (94) has been obtained using small temperature gradient in a dilute suspension of nanoparticles. The boundary conditions as shown in Figure 31 are:

$$\Theta = 1.0,\ \Phi = 1.0 \text{ on the inner circular boundary} \tag{95}$$

$$\Theta = 0.0, \Phi = 0.0 \text{ on the outer circular boundary}$$

$$\partial\Theta/\partial n = \partial\Phi/\partial n = 0.0 \text{ on two other insulation boundaries}$$

$$\Psi = 0.0 \text{ on all solid boundaries}$$

The values of vorticity on the boundary of the enclosure can be obtained using the stream function formulation and the known velocity conditions during the iterative solution procedure.

The local Nusselt number on the hot circular wall can be expressed as:

$$Nu_{loc} = -\frac{\partial\Theta}{\partial n} \tag{96}$$

where *n* is the direction normal to the inner cylinder surface. The average number on the hot circular wall is evaluated as:

$$Nu_{ave} = \frac{1}{0.5\pi}\int_0^{0.5\pi} Nu_{loc}\left(\zeta\right)d\zeta \tag{97}$$

The heatlines are adequate tools for visualization and analysis of 2D convection heat transfer, through an extension of the heat flux line concept to include the advection terms. Heat function (*H*) are defined in terms of the energy equation as

$$\frac{\partial H}{\partial Y} = U\Theta - \frac{\partial\Theta}{\partial X}, \tag{98}$$

$$-\frac{\partial H}{\partial X} = V\Theta - \frac{\partial\Theta}{\partial Y} \tag{99}$$

6.2. Effects of Active Parameters

MHD effect on natural convection heat transfer in an enclosure filled with nanofluid is investigated numerically using CVFEM. Effects of Hartmann number ($Ha = 0, 30, 60$ and 100), buoyancy ratio number ($Nr = 0.1$ to 4) and Lewis number ($Le = 2, 4, 6$ and 8) on flow and heat transfer characteristics are examined. Brownian motion parameter of nanofluids ($Nb = 0.5$), thermophoretic parameter of nanofluids ($Nt = 0.5$), thermal Rayleigh number ($Ra = 10^5$) and Prandtl number ($\Pr = 10$) are fixed.

Effects of Hartmann number, Lewis number and buoyancy ratio number on isotherms, streamlines, isoconcentration and heatline contours are shown in Figures 32-35. When the magnetic field is imposed on the enclosure, the velocity field suppressed owing to the retarding effect of the Lorenz force. So intensity of convection weakens significantly. The braking effect of the magnetic field is observed from the maximum stream function value. The core of vortex is shift downward vertically as the Hartmann number increases. Also imposing magnetic field leads to omit the thermal plume over the inner wall. At high Hartmann number, the conduction heat transfer mechanism is more pronounced. For this reason the isotherms are parallel to each other. Increasing Hartmann number causes the concentration boundary layer thickness near inner wall to increase.

The heat flow within the enclosure is displayed using the heat function obtained from conductive heat fluxes ($\partial\Theta / \partial X, \partial\Theta / \partial Y$) as well as convective heat fluxes ($V\Theta, U\Theta$). Heatlines emanate from hot regimes and end on cold regimes illustrating the path of heat flow. As seen heatline has two regions rotate in different direction. The lower one is greater than another which means that more heat transfer occurs in this region. As Hartmann number increases heatlines become weaker because of reduction of heat transfer rate by applying magnetic field. The domination of conduction heat transfer in high Hartmann number can be observed from the heatline patterns since no passive area exists.

It should be mentioned that negative *Nr* values (opposing buoyancy forces) showed more complex and interesting flow patterns, such as multi-cells, which is worthy for presentation and discussion. So in this study results for positive *Nr* (aiding buoyancy forces), where the temperature and species induced buoyancy forces aides each other, are not considered. For $Nr = 0$, the species induced buoyancy force has no effect on flow; the flow is solely driven by the thermal buoyancy force. However, the effect of species induced buoyancy increases as *Nr* value increases, and reaches a certain value, where the effect of thermally induced buoyancy becomes negligible in comparison to the solutal one. For small *Nr* value, the flow is mainly driven by the thermal buoyancy force. When *Nr* increases a reverse thermal plume appears at $\zeta = 90^{\circ}$. This phenomena is due to existing one counter clockwise eddy at this region. By increasing Hartmann number, the two main eddies merged into one counterclockwise eddy. The isconcentrations are more distorted with increase of solutal forces. As *Nr* increases the upper region of healine counters divided into two smaller one and this new region disappear with increase of Hartmann number.

The mass flow is given by $\psi_{\max} \approx \delta_s v$, where the solutal boundary layer thickness is given by $\delta_s \approx \left(RaLeNr\right)^{-1/4}$ and $v \approx \left(RaLeNr\right)^{1/2}$ so $\psi_{\max} \approx \left(RaLeNr\right)^{1/4}$. The solutal boundary layer, δ_s, becomes thinner by increasing of Le. Heatlines are found to be more distorted as Le augments.

Effects of Hartmann number, buoyancy ratio number and Lewis number on Local Nusselt number are shown in Figure 36. As seen Local Nusselt number increases as buoyancy ratio number increases but it decreases with increase of Hartmann number and Lewis number. Generally as ζ increases local

Figure 32. Comparison of the isotherms, streamlines, isoconcentration and heatline contours for different values of thermal Hartmann number when $Nr = 0.1, Le = 8, Nt = Nb = 0.5, Ra = 10^5$ *and* $\Pr = 10$

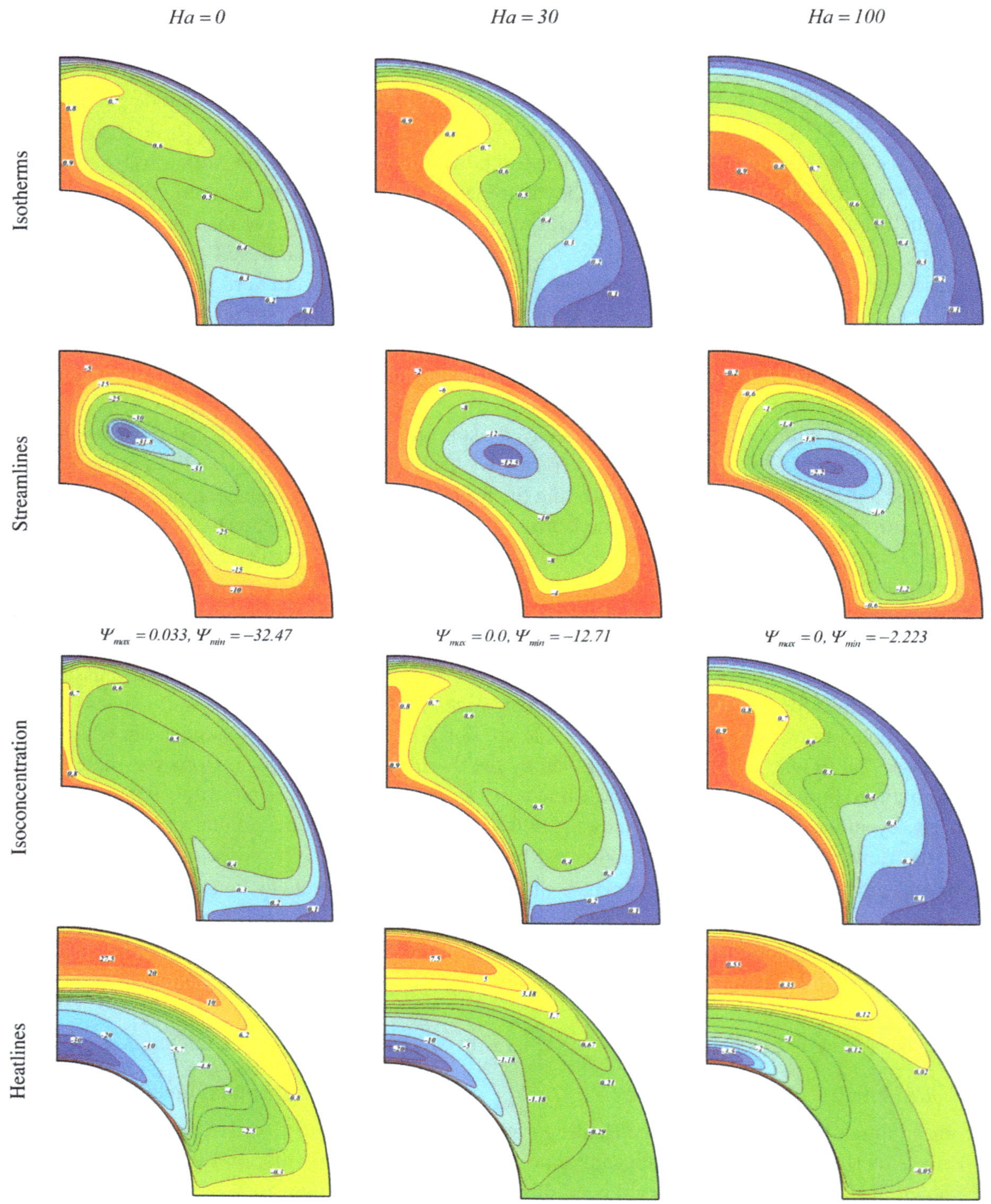

Figure 33. Comparison of the isotherms, streamlines, isoconcentration and heatline contours for different values of thermal Hartmann number when $Nr = 4, Le = 8, Nt = Nb = 0.5, Ra = 10^5$ *and* $\Pr = 10$

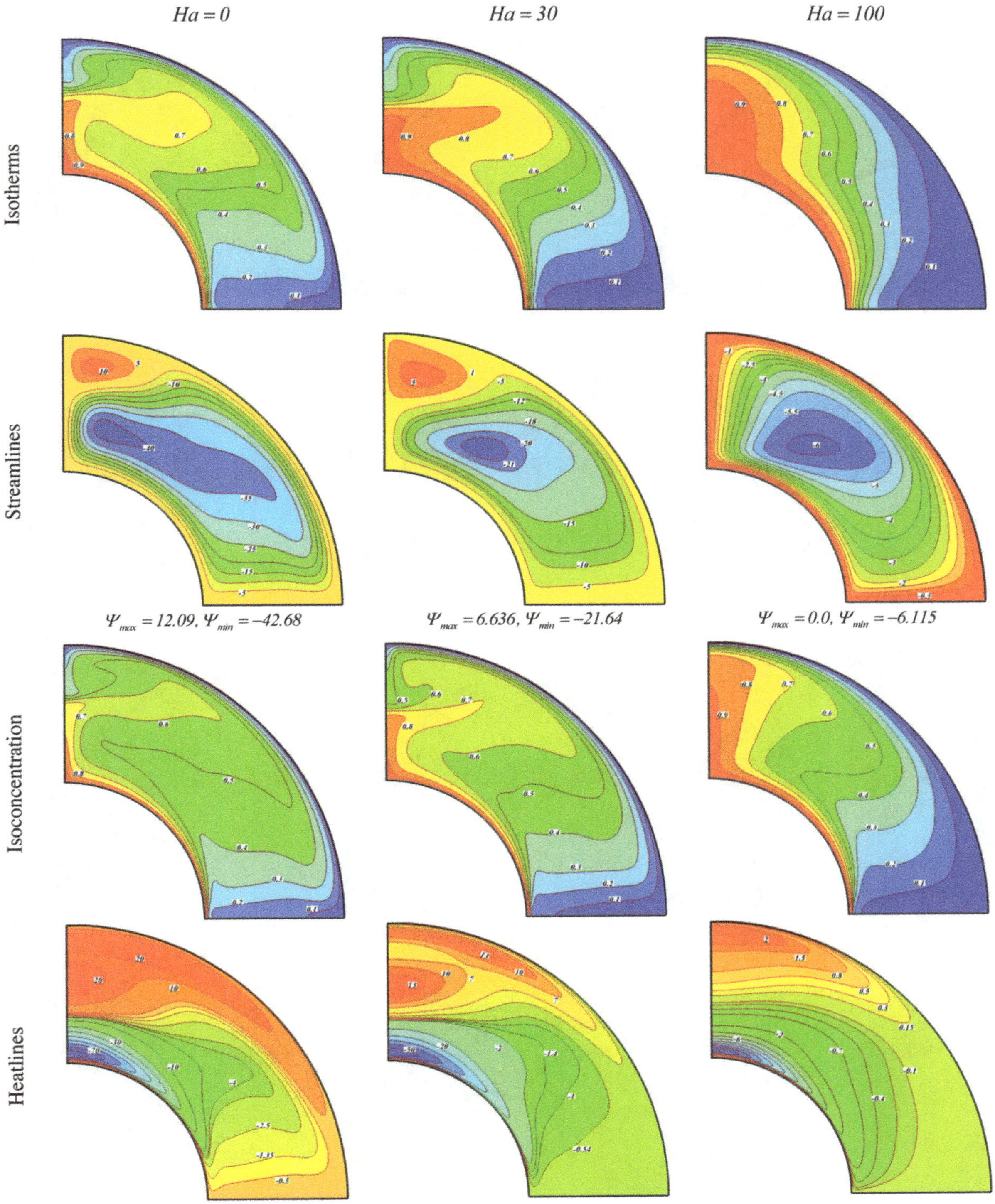

Figure 34. Comparison of the isotherms, streamlines, isoconcentration and heatline contours for different values of thermal Hartmann number when $Nr = 4, Le = 2, Nt = Nb = 0.5, Ra = 10^5$ *and* $\Pr = 10$

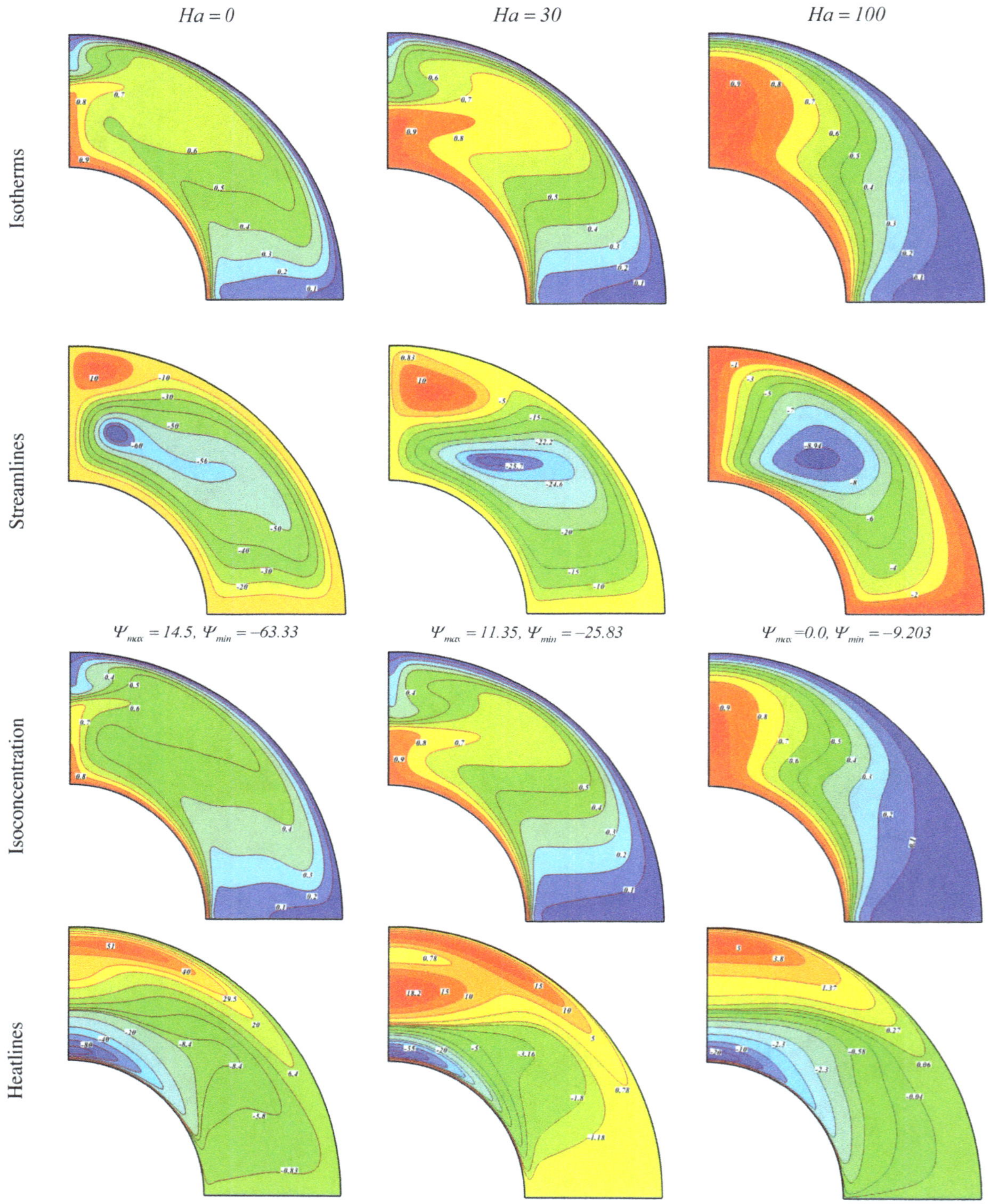

Figure 35. Comparison of the isotherms, streamlines, isoconcentration and heatline contours for different values of thermal Hartmann number when $Nr = 0.1, Le = 2, Nt = Nb = 0.5, Ra = 10^5$ and $\Pr = 10$

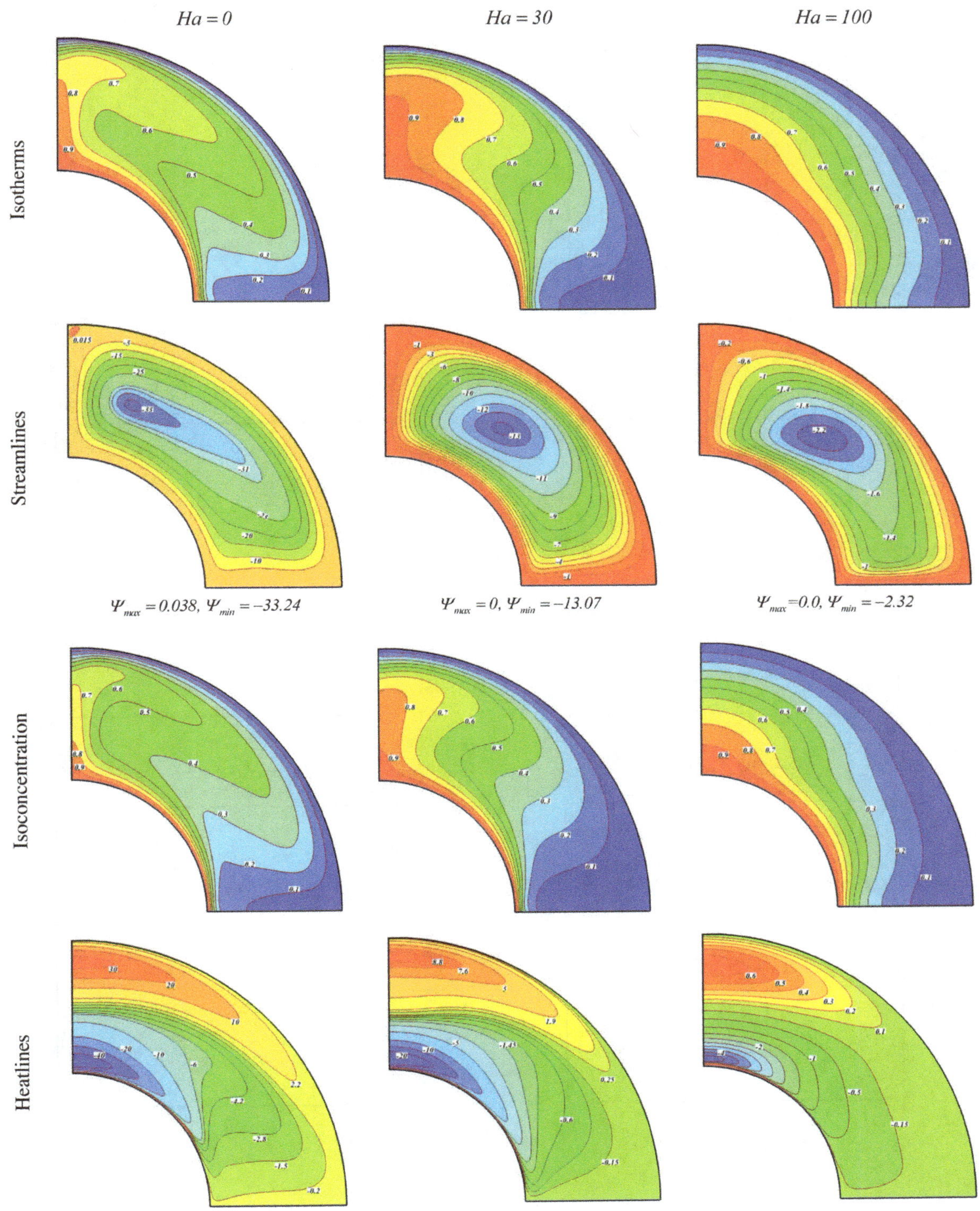

Figure 36. Effects of Hartmann number, buoyancy ratio number and Lewis number on Local Nusselt number at $Nt = Nb = 0.5, Ra = 10^5$ *and* $\Pr = 10$

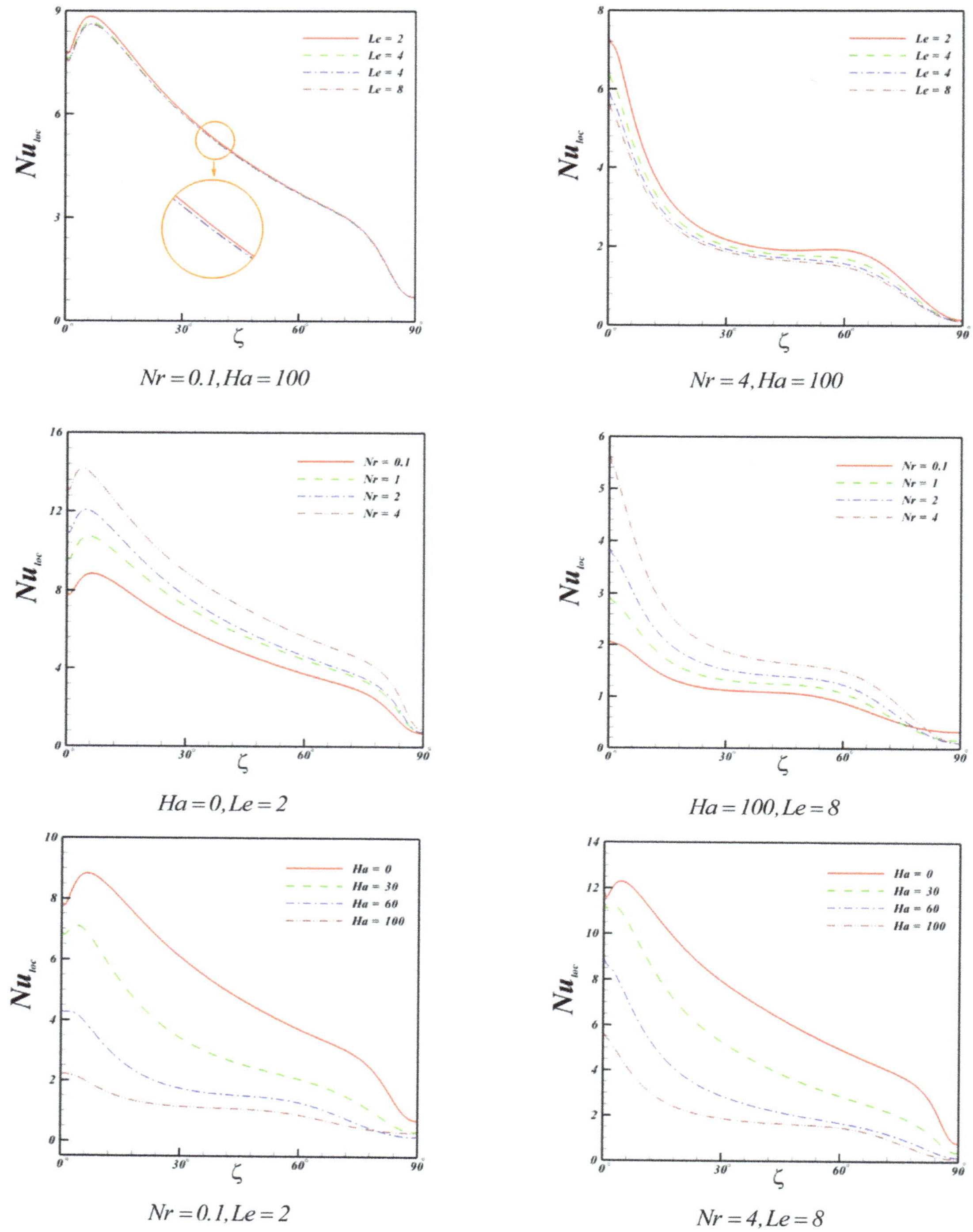

Nusselt number decreases due to increment of thermal boundary layer thickness. In absence of magnetic field Nu_{loc} has one local maximum near the bottom wall. The occurrence of maxima for Nu_{loc} is due to dense heatlines based on conductive heat transport occurring at this portion. This point disappears at high Hartmann number. Local Nusselt number profile has minimum point at $\zeta = 90^{\circ}$ because of existing of thermal plume at this region. Also this figure shows that effect of Lewis number on Nu_{loc} is negligible at low buoyancy ratio number.

The corresponding polynomial representation of such model for Nusselt number is as follows:

$$\begin{aligned} Nu &= a_{13} + a_{23}Y_1 + a_{33}Y_2 + a_{43}Y_1^2 + a_{53}Y_2^2 + a_{63}Y_1Y_2 \\ Y_1 &= a_{11} + a_{21}Le + a_{31}Nr + a_{41}Le^2 + a_{51}Nr^2 + a_{61}Le\,Nr \\ Y_2 &= a_{12} + a_{22}Ha^* + a_{32}Nr + a_{42}Ha^{*2} + a_{52}Nr^2 + a_{62}Ha^*\,Nr \end{aligned} \tag{100}$$

where a_{ij} can be found in Table 5 for example a_{21} equals to (-0.16499).

Effects of Hartmann number, buoyancy ratio number and Lewis number on average Nusselt number are shown in Figures 37 and 38. Presence of magnetic field leads to disappear the thermal plume over inner wall and makes the isotherms parallel to each other due to domination of conduction mode of heat transfer. Therefore average Nusselt number decreases with increase of Hartmann number. As Lewis number increases, thermal boundary layer thickness increases and in turn Nusselt number decreases. Effect of buoyancy ratio number on Nu_{ave} is in contrast with Ha and Le. Also it can be found that effect of Hartmann number and Lewis number are more pronounced at higher values of buoyancy ratio number.

7. SIMULATION OF MHD CUO–WATER NANOFLUID FLOW AND CONVECTIVE HEAT TRANSFER CONSIDERING LORENTZ FORCES

7.1. Problem Definition

The numerical model consists in a two-dimensional square cavity with side equal to *H* which represents the characteristic dimension of the problem (see Figure 39(a)) (Sheikholeslami, 2018). The heat source is centrally located on the bottom surface and its length 1 varied from 2/5 to 4/5 of *H*; the ratio *1/H* is called ε. The cooling is achieved by the two vertical walls. The heat source has a temperature T_h while

Table 5. Constant coefficient for using Equation (100)

a_{ij}	i=1	i=2	i=3	i=4	i=5	i=6
j=1	3.069651	-0.16499	0.578896	0.012949	-0.01115	-0.02536
j=2	4.906119	-7.59508	0.585453	3.681993	-0.01115	-0.28078
j=3	0.819671	-0.56997	0.947823	0.094258	-0.00422	0.021492

Figure 37. Effects of Hartmann number, buoyancy ratio number and Lewis number on average Nusselt number at $Nt = Nb = 0.5, Ra = 10^5$ *and* $\Pr = 10$

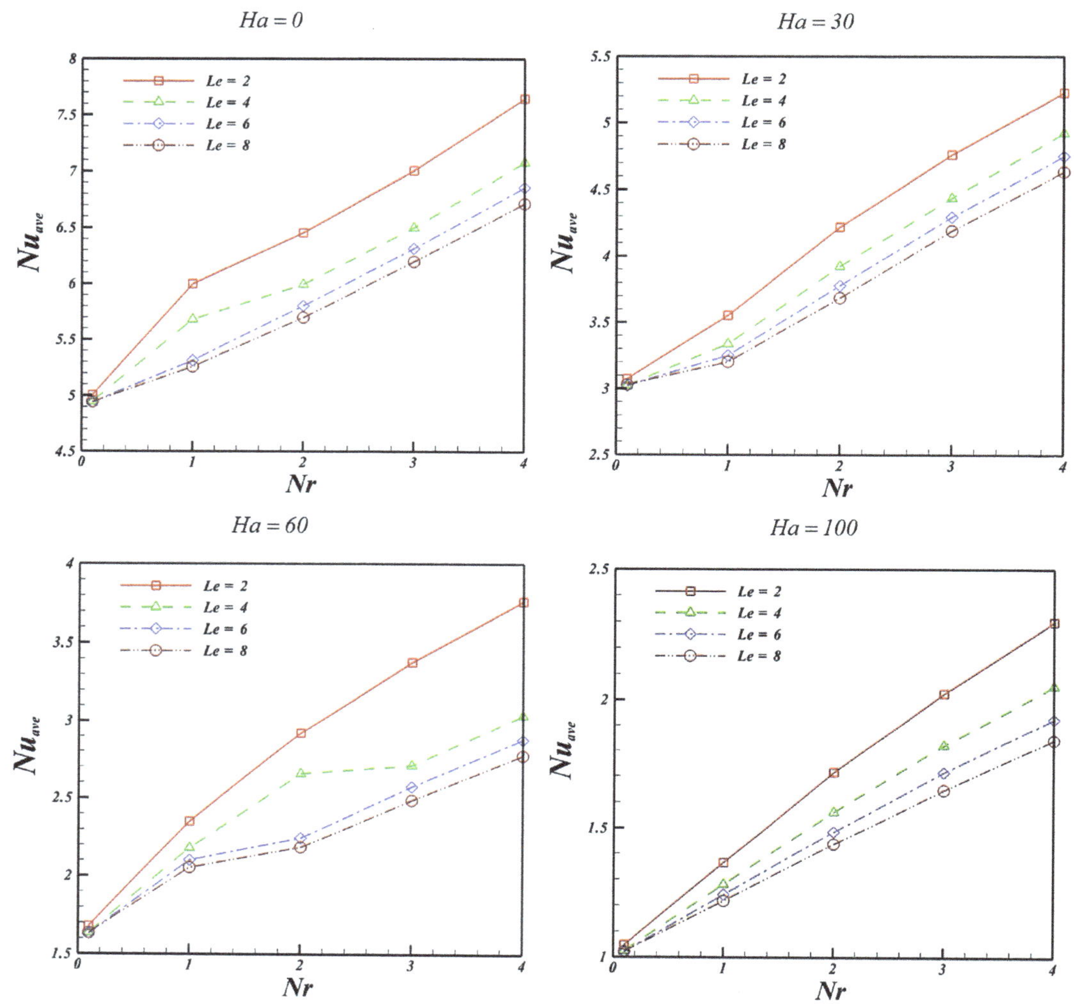

the cooling walls have a temperature T_c; all the other surfaces are adiabatic $\left(T_h > T_c\right)$. Also, it is also assumed that the uniform magnetic field ($\vec{B} = B_x \vec{e}_x + B_y \vec{e}_y$) of constant magnitude $B = \sqrt{B_x^2 + B_y^2}$ is applied, where $\vec{e}_x$ and $\vec{e}_y$ are unit vectors in the Cartesian coordinate system. The orientation of the magnetic field form an angle θ_M with horizontal axis such that $\theta_M = \cot^{-1}\left(B_x / B_y\right)$. The electric current J and the electromagnetic force F are defined by $J = \sigma\left(\vec{V} \times \vec{B}\right)$ and $F = \sigma\left(\vec{V} \times \vec{B}\right) \times \vec{B}$, respectively. The governing equations are similar to those of exist in section (1.1). In order to compare total heat transfer rate, Nusselt number is used. The local and average Nusselt numbers on cold enclosure are defined as follows:

Figure 38. Variation of Nu_{ave} for various input parameters

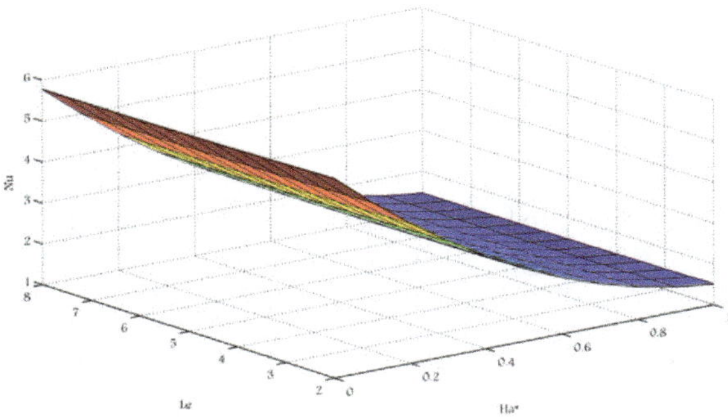

$Nr = 1.69$

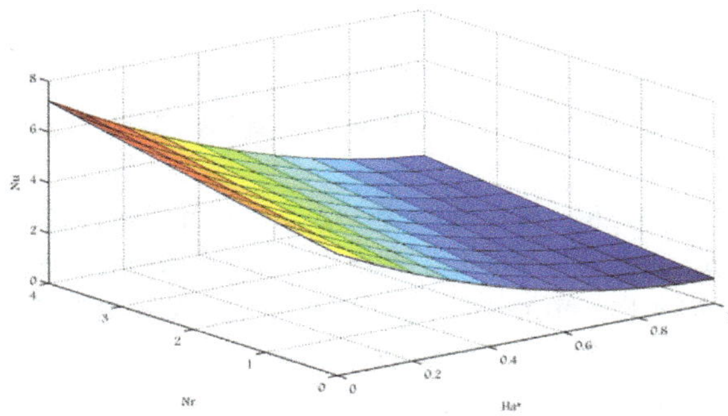

$Le = 5$

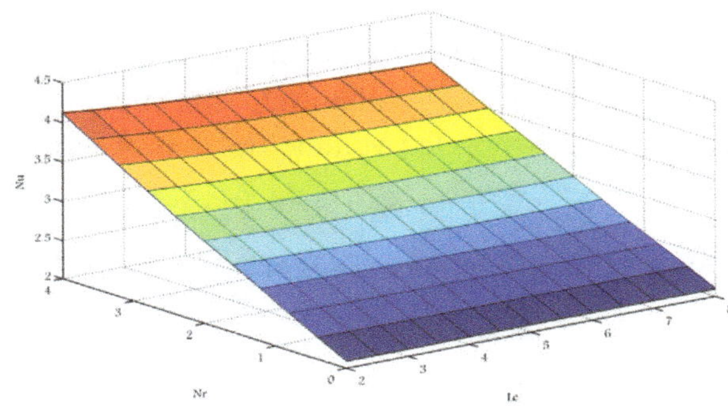

$Ha^{*} = 0.45$

Figure 39. (a) Geometry of the problem; (b) Discrete velocity set of two-dimensional nine-velocity (D_2Q_9) model

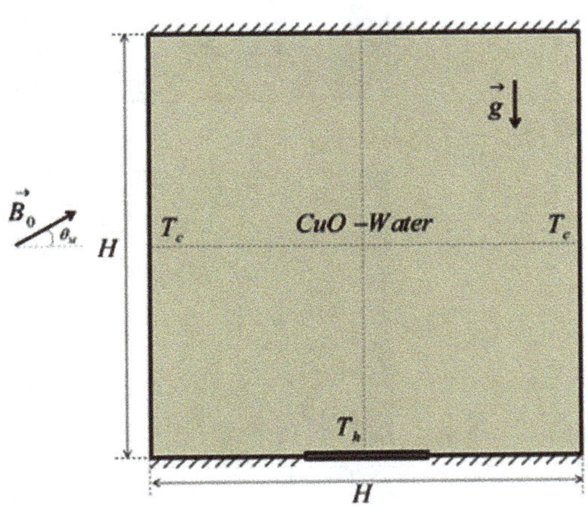

(a)

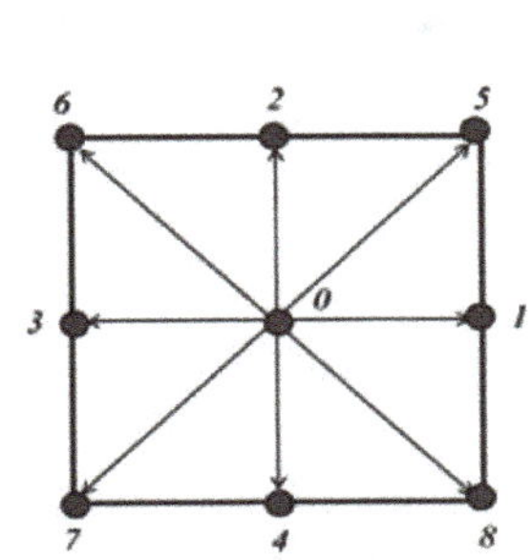

(b)

$$Nu = \frac{k_{nf}}{k_f} \frac{\partial T}{\partial Y}\bigg|_{Y=0} \text{ and } Nu_{ave} = \int_{\frac{1-\varepsilon}{2}}^{\frac{1+\varepsilon}{2}} Nu dX \tag{101}$$

7.2. Effects of Active Parameters

Effect of horizontal magnetic field on development of the heat transfer phenomenon in a square enclosure partially heated from below is investigated. Lattice Boltzmann method scheme was utilized to obtain the numerical simulation in a cavity filled with CuO-water. The present numerical solution is validated by comparing the present code results against the results of Calcagni et al. (2005) for viscous flow ($\phi = 0$) (see Figure 40). Effect of active parameters such as: Rayleigh number ($Ra = 10^3, 10^4$ and 10^5), Hartmann number ($Ha = 0, 20, 60$ and 100), heat source length ($\varepsilon = 1 / H = 0.4, 0.6$ and 0.8) and volume fraction of nanoparticle ($\phi = 0$ and 0.04) on flow and heat transfer are examined.

Figure 41 depicts the effect of volume fraction of nanoparticle on streamlines and isotherms. As seen the velocity components increase with increase of nanoparticles volume fraction which enhances the energy transport within the fluid. Thus, the absolute values of stream functions which indicate that the strength of flow increase with increasing the volume fraction of nanoparticle. The sensitivity of

Figure 40. Comparison of isotherms between the present work and experimental and numerical study of Calcagni et al. (2005)

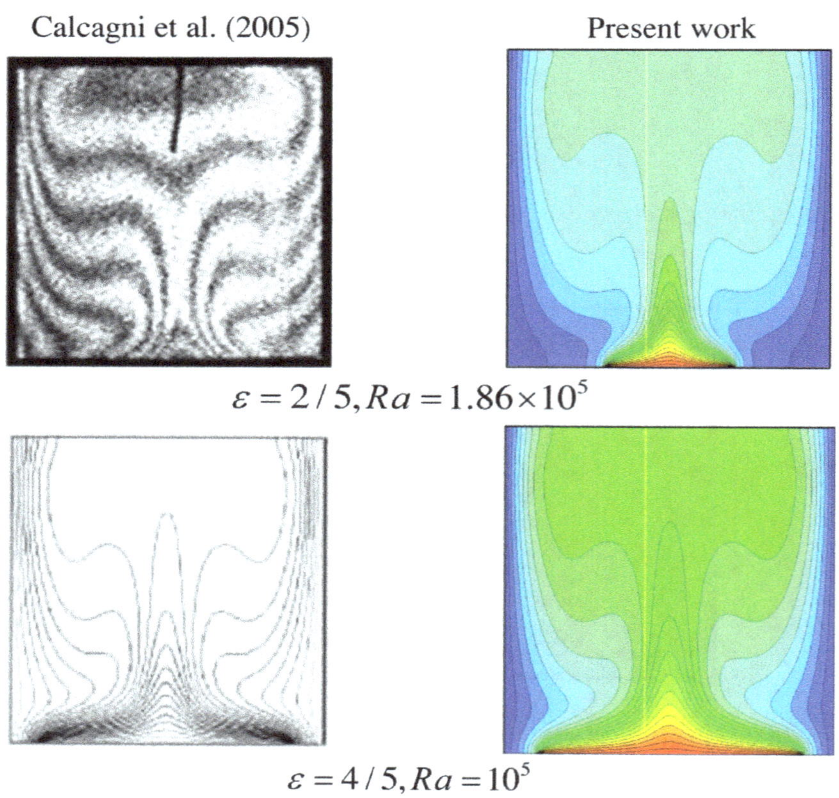

Figure 41. Comparison of the isotherms (left) and streamlines (right) contours between nanofluid ($\phi = 0.04$) (- - -) and pure fluid ($\phi = 0$) (—) when $\varepsilon = 0.8, Ra = 10^5$

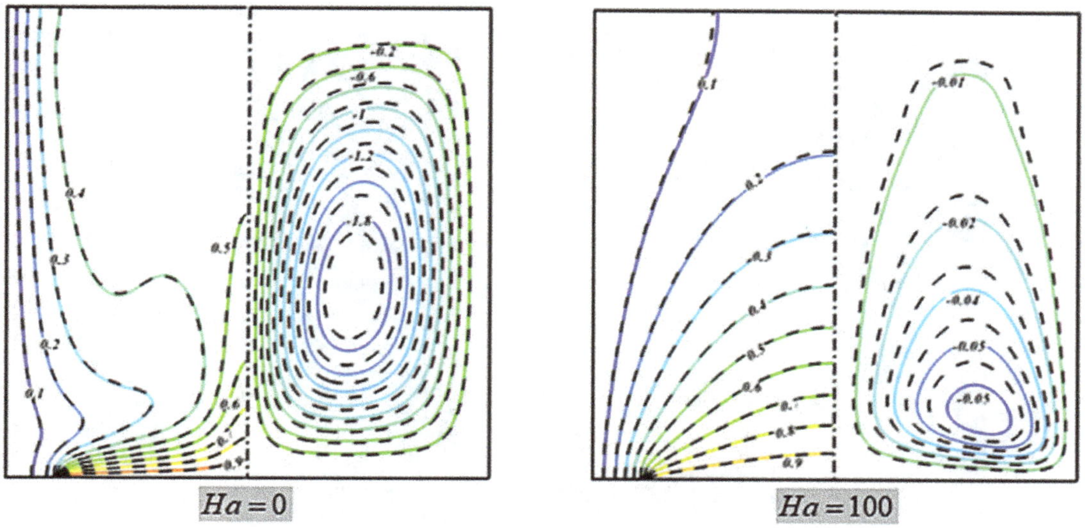

thermal boundary layer thickness to volume fraction of nanoparticles is related to the increased thermal conductivity of the nanofluid. In fact, higher values of thermal conductivity are accompanied by higher values of thermal diffusivity. The high value of thermal diffusivity causes a drop in the temperature gradients and accordingly increases the boundary layer thickness. Although adding nanoparticles leads to increase thermal boundary layer thickness, Nusselt number increases because it is a multiplication of temperature gradient and the thermal conductivity ratio. Reduction in temperature gradient is much smaller than the thermal conductivity ratio.

Effects of Rayleigh number, heat source length and Hartmann number on isotherms (left) and streamlines (right) contours are shown in Figures 42, 43 and 44. By increasing Rayleigh number the buoyancy forces increases and overcome the viscous forces and the heat transfer is dominated by convection at high Rayleigh number. Moreover, the isotherms are more distorted at higher Rayleigh numbers due to the stronger convection effects. Increasing Hartmann number causes Lorenz force to increase and leads to a substantial suppression of the convection. The core of main cell move downward with increase of Hartmann number. As Lorentz forces increase, the conduction heat transfer mechanism is more marked and isotherms are parallel to each other. Also it can be seen that the absolute values of stream function increases with increase of heat source length. Thermal boundary layer thickness near the bottom wall decreases with augment of ε. Effects of heat source length and Hartmann number on flow and heat transfer are more sensible for high Rayleigh number.

Figure 45 shows the effects of Rayleigh number, Hartmann number and heat source length on average Nusselt number. Nusselt number increases with increase of Rayleigh number and heat source length while it decreases with increase of Hartmann number. Effects of Rayleigh number, heat source length and Hartmann number on heat transfer enhancement are depicted in Figure 46. It can be found that the effect of nanoparticles is more obvious at low Rayleigh number than at high Rayleigh number. This observa-

Figure 42. Effects of heat source length and Hartmann number on isotherms (left) and streamlines (right) contours when $\phi = 0.04, \mathrm{Ra} = 10^3$

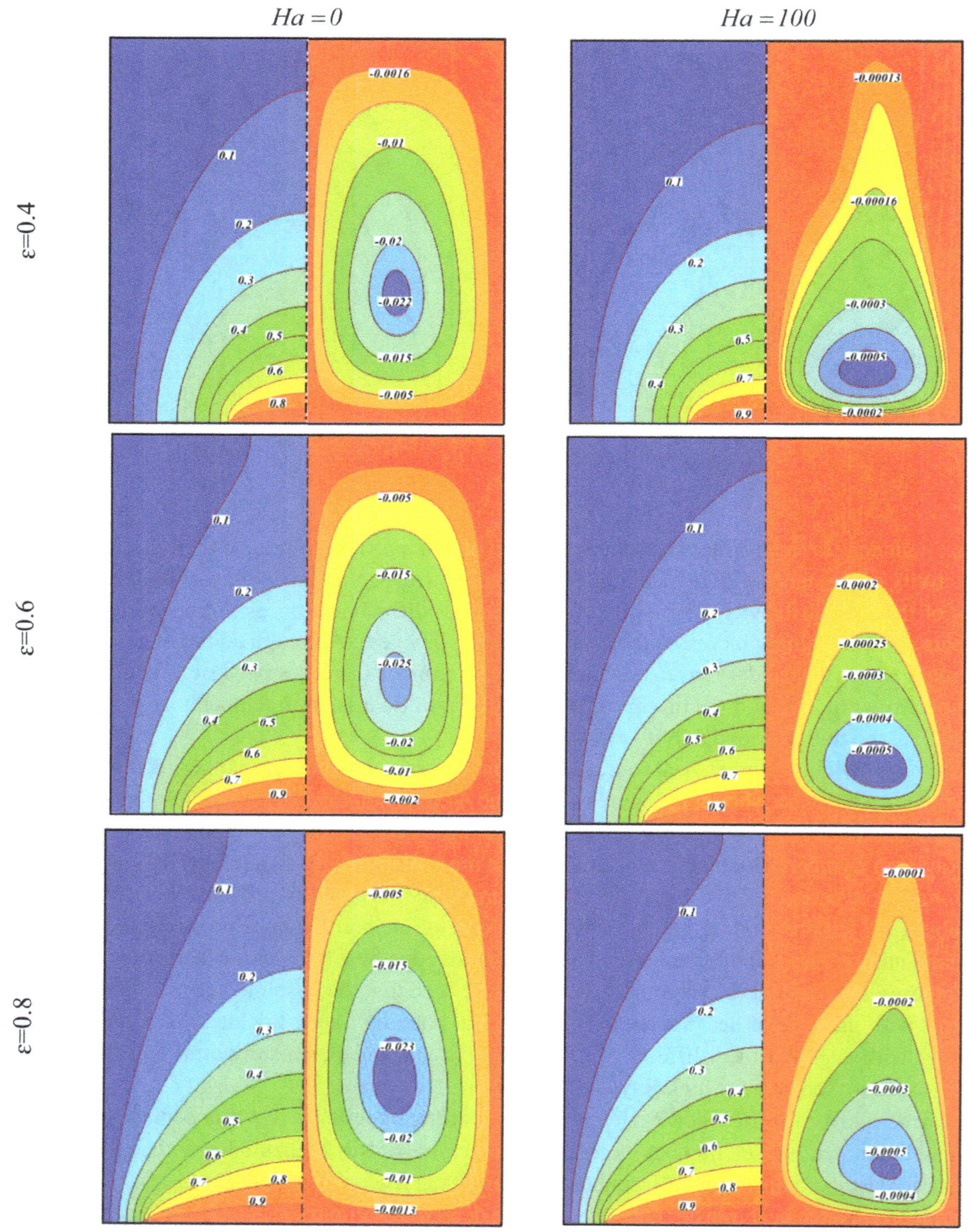

Figure 43. Effects of heat source length and Hartmann number on isotherms (left) and streamlines (right) contours when $\phi = 0.04, \mathrm{Ra} = 10^4$

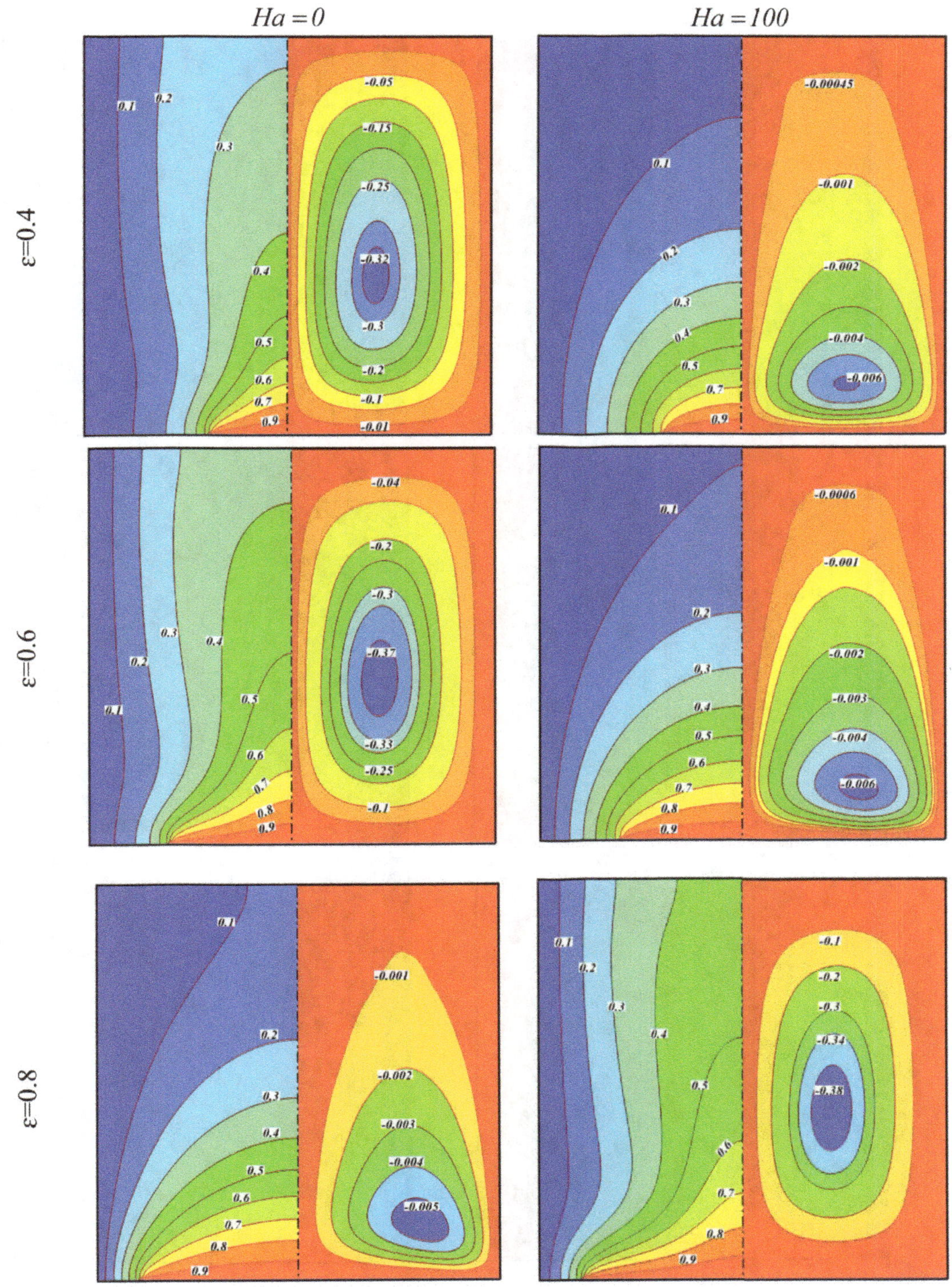

Figure 44. Effects of heat source length and Hartmann number on isotherms (left) and streamlines (right) contours when $\phi = 0.04, \mathrm{Ra} = 10^5$

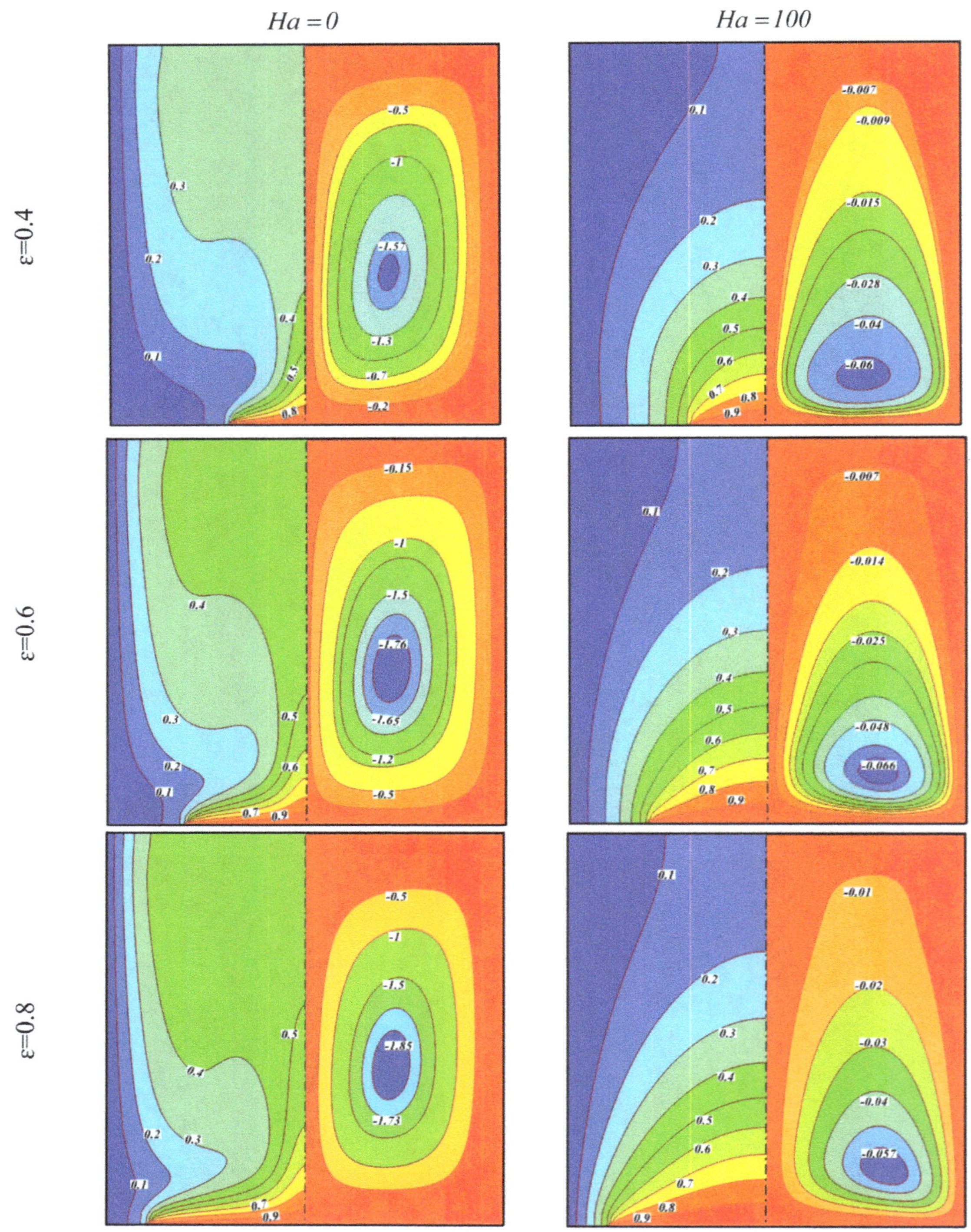

Figure 45. Effects of Rayleigh number, heat source length and Hartmann number on average Nusselt number when $\phi = 0.04$

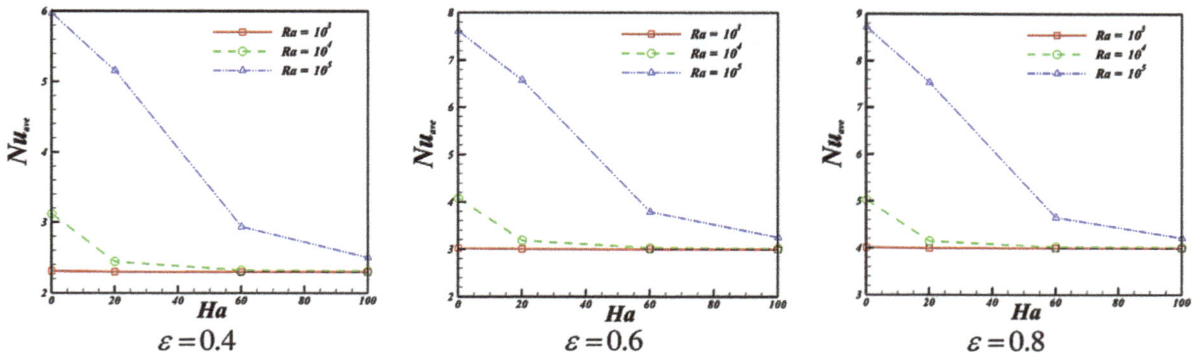

Figure 46. Effects of Rayleigh number, heat source length and Hartmann number on heat transfer enhancement

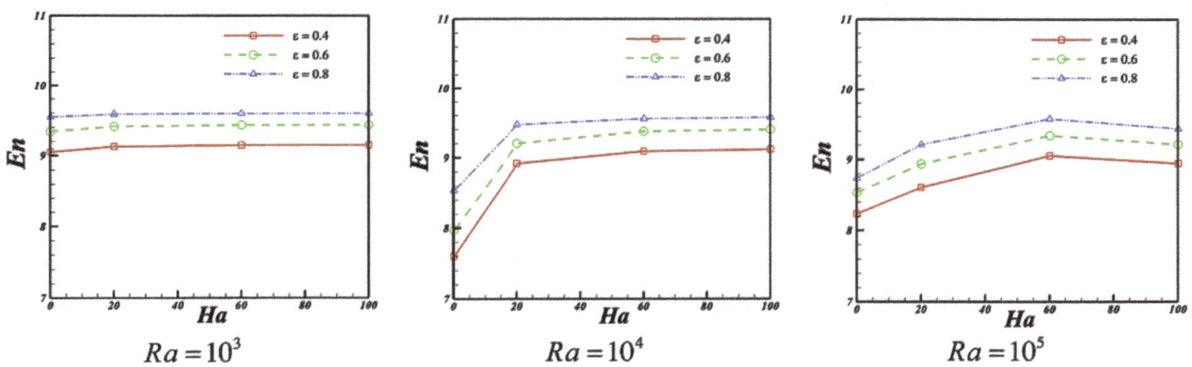

tion can be clarified by noting that at low Rayleigh number the heat transfer is dominant by conduction. Therefore, the addition of high thermal conductivity nanoparticles will increase the conduction and so make the enhancement more effective. Also this figure depicts that increasing Hartmann number leads to increase in heat transfer enhancement. Furthermore, rate of enhancement increases with increases of heat source length.

8. THREE DIMENSIONAL MESOSCOPIC SIMULATION OF MAGNETIC FIELD EFFECT ON NATURAL CONVECTION OF NANOFLUID

8.1. Problem Definition

The numerical model consists in a three-dimensional square cavity with side equal to *L* which represents the characteristic dimension of the problem (see Figure 47) (Sheikholeslami & Ellahi, 2015). The hot and cold wall were located at $Z = z / L = 0$ and $Z = z / L = 1$, respectively. All the other surfaces

Figure 47. Geometry of the problem

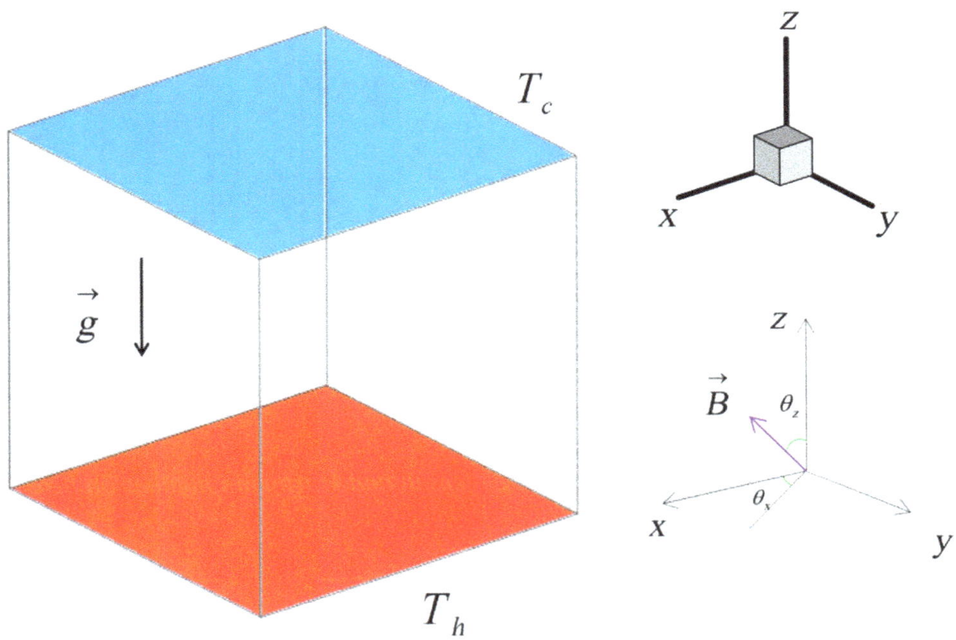

are adiabatic. Also, it is also assumed that the uniform magnetic field ($\vec{B} = B_x \vec{e_x} + B_y \vec{e_y} + B_z \vec{e_z}$) of constant magnitude $B = \sqrt{B_x^2 + B_y^2 + B_z^2}$ is applied, where $\vec{e_x}, \vec{e_y}$ and $\vec{e_z}$ are unit vectors in the Cartesian coordinate system. The orientation of the magnetic field forms an angle θ_z with z axis and θ_x with x axis. The electric current J and the electromagnetic force F are defined by $J = \sigma\left(\vec{V} \times \vec{B}\right)$ and $F = \sigma\left(\vec{V} \times \vec{B}\right) \times \vec{B}$, respectively. In this study, $\theta_x = \theta_z = 90^\circ$.

The thermal LB model utilizes two distribution functions, *f* and *g*, for the flow and temperature fields, respectively. It uses modeling of movement of fluid particles to capture macroscopic fluid quantities such as velocity, pressure and temperature. In this approach, the fluid domain discretized to uniform Cartesian cells. Each cell holds a fixed number of distribution functions, which represent the number of fluid particles moving in these discrete directions. The density and distribution functions i.e. the *f* and *g*, are calculated by solving the lattice Boltzmann equation, which is a special discretization of the kinetic Boltzmann equation. After introducing BGK approximation, the general form of lattice Boltzmann equation with external force is as follow:

For the flow field:

$$f_i(x + c_i\Delta t, t + \Delta t) = f_i(x,t) + \frac{\Delta t}{\tau_v}[f_i^{eq}(x,t) - f_i(x,t)] + \Delta t c_i F_k \tag{102}$$

For the temperature field:

$$g_i(x + c_i\Delta t, t + \Delta t) = g_i(x,t) + \frac{\Delta t}{\tau_C}[g_i^{eq}(x,t) - g_i(x,t)] \tag{103}$$

where Δt denotes lattice time step, c_i is the discrete lattice velocity in direction i, F_k is the external force in direction of lattice velocity, τ_v and τ_C denotes the lattice relaxation time for the flow and temperature fields. The kinetic viscosity υ and the thermal diffusivity α, are defined in terms of their respective relaxation times, i.e. $\upsilon = c_s^2(\tau_v - 1/2)$ and $\alpha = c_s^2(\tau_C - 1/2)$, respectively. Note that the limitation $0.5 < \tau$ should be satisfied for both relaxation times to ensure that viscosity and thermal diffusivity are positive. Furthermore, the local equilibrium distribution function determines the type of problem that needs to be solved. The D_3Q_{19} model was used (Figure 48). These 19 velocities are given as follows (Sheikholeslami et al., 2015).

$$c_i = \begin{pmatrix} 0 & 0 & 0 & 0 & -1 & -1 & -1 & -1 & 1 & 1 & 1 & 0 & 0 & 0 & 0 & -1 & 1 & 0 \\ -1 & -1 & 1 & 1 & 0 & -1 & -1 & 1 & 0 & -1 & 1 & 0 & 0 & -1 & 1 & 0 & 0 & 0 \\ -1 & 1 & -1 & 1 & -1 & 0 & 0 & 0 & 1 & 0 & 0 & -1 & 1 & 0 & 0 & 0 & 0 & 0 \end{pmatrix} \tag{104}$$

It also models the equilibrium distribution functions, which are calculated with Eqs. (105) and (106) for flow and temperature fields, respectively.

$$f_i^{eq} = w_i\rho\left[1 + \frac{c_i.u}{c_s^2} + \frac{1}{2}\frac{(c_i.u)^2}{c_s^4} - \frac{1}{2}\frac{u^2}{c_s^2}\right] \tag{105}$$

Figure 48. Discrete velocity for the D_3Q_{19} model

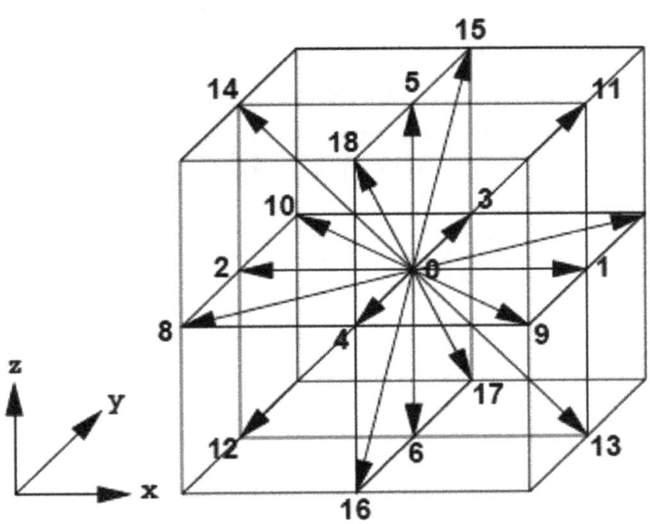

$$g_i^{eq} = w_i T\left[1 + \frac{c_i.u}{c_s^2}\right] \tag{106}$$

where ρ, T are the lattice fluid density and temperature.

Weighting factor $\left(w_i\right)$ is defined as follows:

$$w_i = \begin{cases} 1/3 & i = 0 \\ 1/18 & i = 1:6 \\ 1/36 & i = 7:18 \end{cases} \tag{107}$$

In order to incorporate buoyancy forces and magnetic forces in the model, the force term in the Equation (102) need to calculate as below (Sheikholeslami & Ellahi, 2015):

$$\begin{aligned}
&F = F_x + F_y + F_z \\
&F_x = 3w_i\rho A\Big[-\cos\left(\theta_z\right)\left(u\cos\left(\theta_z\right) + w\sin\left(\theta_z\right)\cos\left(\theta_x\right)\right) \\
&-\sin\left(\theta_z\right)\sin\left(\theta_x\right)\left(u\sin\left(\theta_z\right)\sin\left(\theta_x\right) - v\sin\left(\theta_z\right)\cos\left(\theta_x\right)\right)\Big], \\
&F_y = 3w_i\rho A\Big[\sin\left(\theta_z\right)\cos\left(\theta_x\right)\left(u\sin\left(\theta_z\right)\sin\left(\theta_x\right) - v\sin\left(\theta_z\right)\cos\left(\theta_x\right)\right) \\
&-\cos\left(\theta_z\right)\left(v\cos\left(\theta_z\right) - w\sin\left(\theta_z\right)\sin\left(\theta_x\right)\right)\Big] \\
&F_z = 3w_i\rho\Big[g_z\beta\left(T - T_m\right) + A\sin\left(\theta_z\right)\sin\left(\theta_x\right)\left(v\cos\left(\theta_z\right) - w\sin\left(\theta_z\right)\sin\left(\theta_x\right)\right) \\
&+A\sin\left(\theta_z\right)\cos\left(\theta_x\right)\left(u\cos\left(\theta_z\right) - w\sin\left(\theta_z\right)\cos\left(\theta_x\right)\right)\Big]
\end{aligned} \tag{108}$$

where A is $A = \frac{Ha^2\mu}{L^2}$, $Ha = LB_0\sqrt{\frac{\sigma}{\mu}}$ is Hartmann number.

For natural convection, the Boussinesq approximation is applied and radiation heat transfer is negligible. To ensure that the code works in near incompressible regime, the characteristic velocity of the flow for natural $(V_{natural} \equiv \sqrt{\beta g_z \Delta TL})$ regime must be small compared with the fluid speed of sound. In the present study, the characteristic velocity selected as 0.1 of sound speed.

Finally, macroscopic variables calculate with the following formula:

$$\begin{aligned}
&\text{Flow density}: \ \rho = \sum_i f_i, \\
&\text{Momentum}: \ \rho u = \sum_i c_i f_i, \\
&\text{Temperature}: \ T = \sum_i g_i.
\end{aligned} \tag{109}$$

In order to simulate the nanofluid by the lattice Boltzmann method, because of the interparticle potentials and other forces on the nanoparticles, the nanofluid behaves differently from the pure liquid from the mesoscopic point of view and is of higher efficiency in energy transport as well as better stabilization than the common solid-liquid mixture. For modeling the nanofluid because of changing in the fluid thermal conductivity, density, heat capacitance and thermal expansion, some of the governed equations should change. The effective density $\left(\rho_{nf}\right)$, the effective heat capacity $\left(\rho C_p\right)_{nf}$, thermal expansion $\left(\rho\beta\right)_{nf}$ and electrical conductivity $\left(\sigma\right)_{nf}$ of the nanofluid are defined as:

$$\rho_{nf} = \rho_f(1-\phi) + \rho_s\phi \tag{110}$$

$$\left(\rho C_p\right)_{nf} = \left(\rho C_p\right)_f (1-\phi) + \left(\rho C_p\right)_s \phi \tag{111}$$

$$\left(\rho\beta\right)_{nf} = \left(\rho\beta\right)_f (1-\phi) + \left(\rho\beta\right)_s \phi \tag{112}$$

$$\frac{\sigma_{nf}}{\sigma_f} = 1 + \frac{3\left(\frac{\sigma_s}{\sigma_f} - 1\right)\phi}{\left(\frac{\sigma_s}{\sigma_f} + 2\right) - \left(\frac{\sigma_s}{\sigma_f} - 1\right)\phi} \tag{113}$$

where ϕ is the solid volume fraction of the nanoparticles and subscripts f, nf and s stand for base fluid, nanofluid and solid, respectively.

$\left(k_{nf}\right)$ and $\left(\mu_{nf}\right)$ are obtained according to Koo–Kleinstreuer–Li (KKL) model (Sheikholeslami, 2014):

$$\begin{gathered}
k_{nf} = 1 + \frac{3\left(\frac{k_p}{k_f} - 1\right)\phi}{\left(\frac{k_p}{k_f} + 2\right) - \left(\frac{k_p}{k_f} - 1\right)\phi} + 5\times 10^4 g'(\phi, T, d_p)\phi\rho_f c_{p,f}\sqrt{\frac{\kappa_b T}{\rho_p d_p}} \\
g'\left(\phi, T, d_p\right) = \left(a_6 + a_7 Ln\left(d_p\right) + a_8 Ln\left(\phi\right) + a_9 Ln\left(\phi\right)\ln\left(d_p\right) + a_{10} Ln\left(d_p\right)^2\right) \\
+ Ln\left(T\right)\left(a_1 + a_2 Ln\left(d_p\right) + a_3 Ln\left(\phi\right) + a_4 Ln\left(\phi\right)\ln\left(d_p\right) + a_5 Ln\left(d_p\right)^2\right) \\
R_f + d_p / k_p = d_p / k_{p,eff}, R_f = 4\times 10^{-8} km^2 / W
\end{gathered} \tag{114}$$

$$\mu_{nf} = \frac{\mu_f}{\left(1-\phi\right)^{2.5}} + \frac{k_{Brownian}}{k_f} \times \frac{\mu_f}{\Pr} \tag{115}$$

In order to compare total heat transfer rate, Nusselt number is used. The local and average Nusselt numbers on hot wall are defined as follows:

$$Nu = \frac{k_{nf}}{k_f}\frac{\partial T}{\partial Z}\bigg|_{Z=0} \quad and \quad Nu_{ave} = \int_0^1 \int_0^1 Nu\, dY dX \tag{116}$$

To estimate the enhancement of heat transfer between the case of $\phi = 0.04$ and the pure fluid (base fluid) case, the heat transfer enhancement is defined as:

$$En = \frac{Nu_{ave}\left(\phi = 0.04\right) - Nu_{ave}\left(basefluid\right)}{Nu_{ave}\left(basefluid\right)} \times 100 \tag{117}$$

The kinetic energy of the fluid is defined as follows.

$$E_c = 0.5\left[\left(U\right)^2 + \left(V\right)^2 + \left(W\right)^2\right] \tag{118}$$

8.2. Effects of Active Parameters

A numerical investigation is presented for free convection heat transfer of Al_2O_3-water nanofluid in presence of magnetic field parallel to gravity in a cubic cavity heated from below. The governing equations are solved via Lattice Boltzmann method. Computations are carried out for different values of Rayleigh number ($Ra = 10^3, 10^4$ and 10^5), Hartmann number (*Ha*=0,20,40 and 60) and volume fraction of nanoparticle ($\phi = 0$ and 0.04).

Effects of Hartmann number and Rayleigh number on isotherm, streamlines and isokinetic energy are shown in Figures 49, 50, and 51. All the contours are symmetric with respect to the vertical midplane of the enclosure due to symmetrical boundary conditions. In general, natural convection of nanofluid in the enclosure is affected by buoyancy and Lorentz forces. The buoyancy force has an aiding effect on natural convection, but the Lorentz force has an opposing effect. Either of the two forces is important, when $Ha^2 / Ra \approx 1$. The buoyancy force is dominant when $Ha^2 / Ra << 1$ and the Lorentz force is dominant when $Ha^2 / Ra >> 1$.

At low Rayleigh number, conduction is the main heat transfer mechanism number. So the shape of the isotherms tends to follow the geometry of the enclosure. The maximum value of stream function can be viewed as a measure of the intensity of natural convection in the cavity. It is evident from the figures that by increasing the Ra, the maximum value of the stream function increases; this means that the flow move faster as natural convection is stronger and the isotherm will be distorted. For high Rayleigh number, where buoyancy forces are more dominant than the viscous forces, the convection currents inside

Figure 49. Effect of Hartmann number on (a) isotherm, (b) streamlines and (c) isokinetic energy at $Y = y / L = 0.5$ *when* $Ra = 10^3, \phi = 0.04$

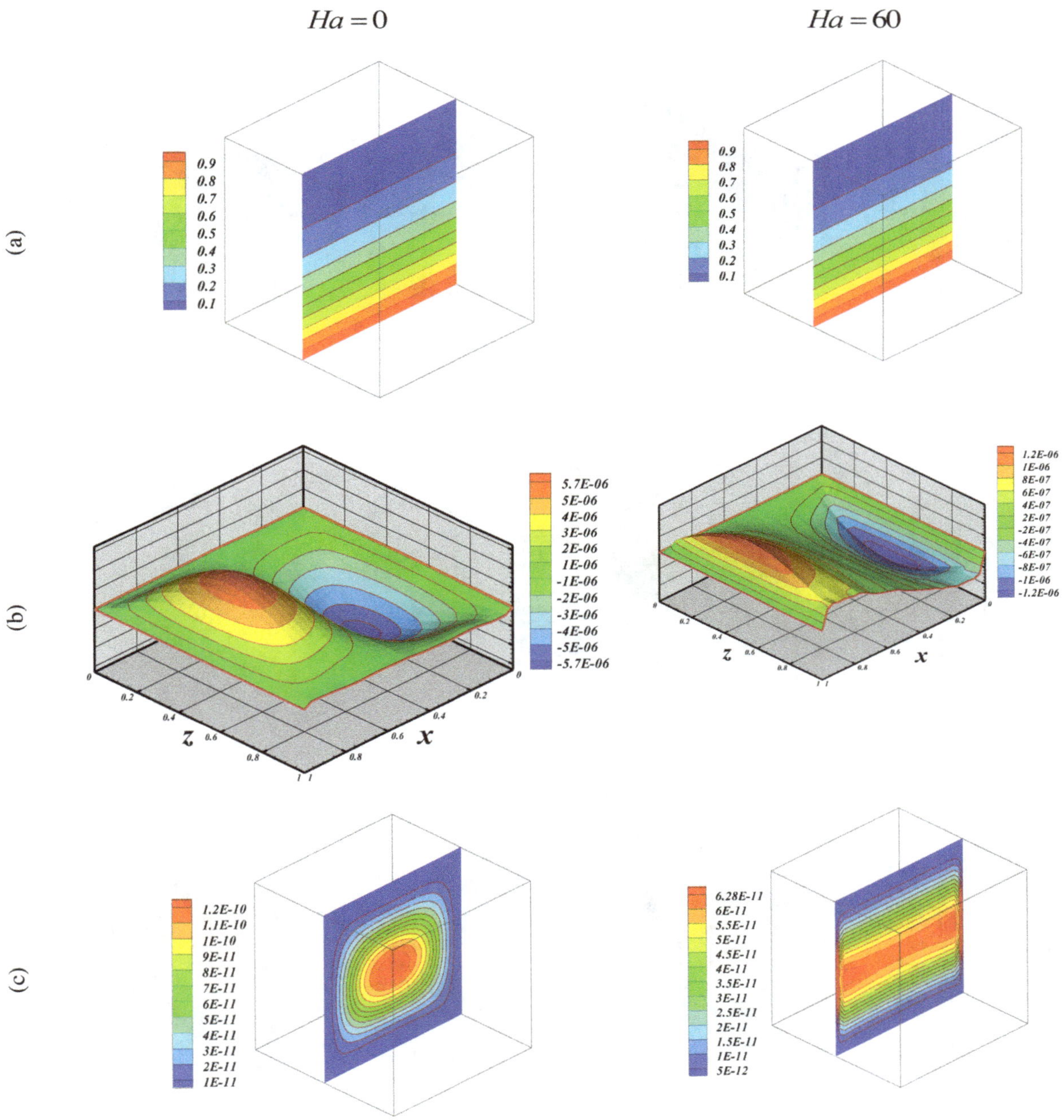

the enclosure become very strong. Since the cold fluid has a downward motion with an increase of circular flow, the convection becomes the basic mode of heat transfer. For all Rayleigh numbers by increasing the Hartmann number have a pure conduction regime. Because Lorentz force interacts with the buoyancy force and suppresses the convection flow by reducing the velocities. Furthermore, it is shown that as the Rayleigh number is increased, the convective heat transfer is increased, so that for suppression

Figure 50. Effect of Hartmann number on (a) isotherm, (b) streamlines and (c) isokinetic energy at $Y = y / L = 0.5$ *when* $Ra = 10^4, \phi = 0.04$

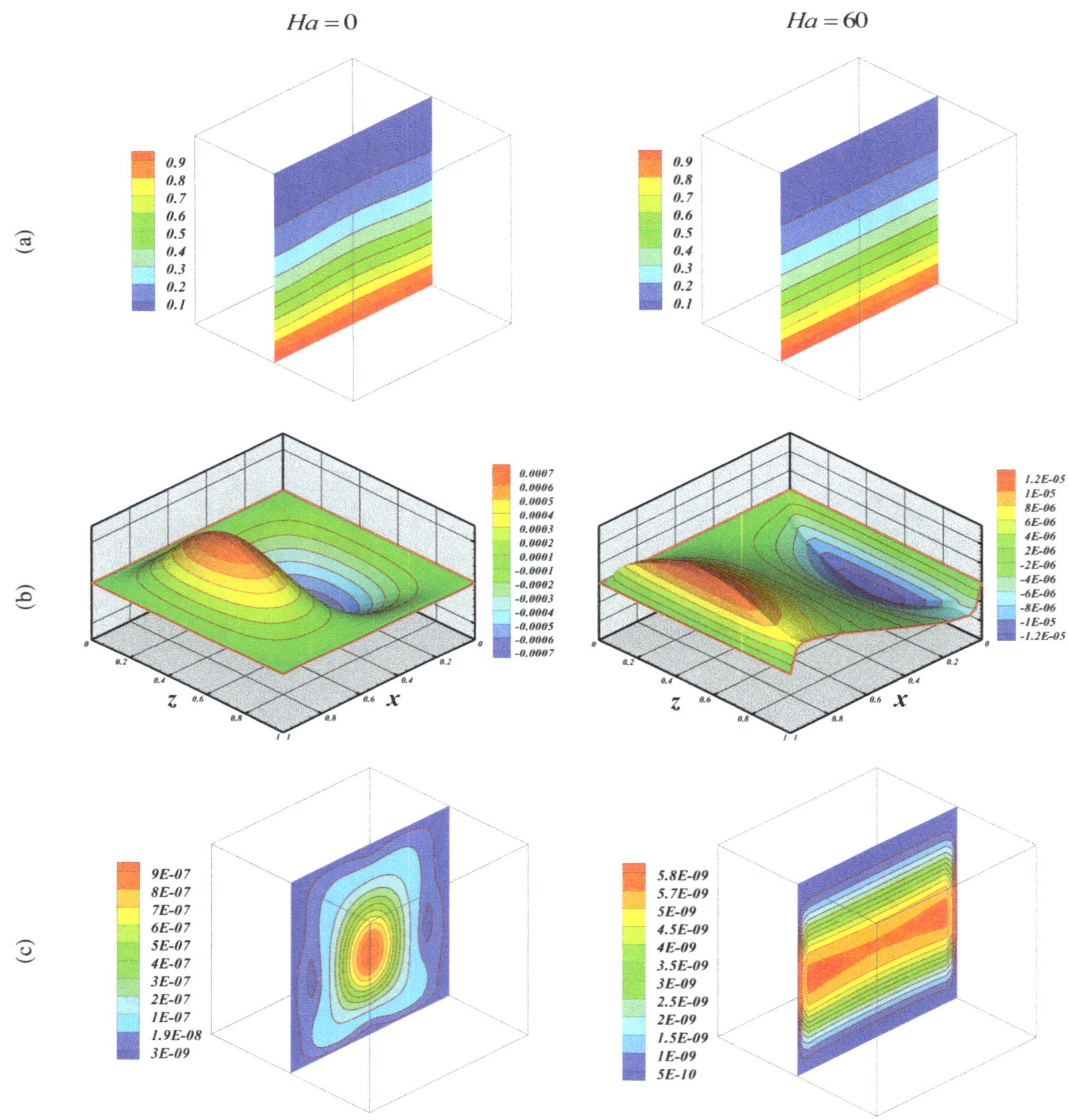

of convection is needed higher magnetic field. By increasing magnetic field, the alteration in the temperature distribution seems to be more serious. The thermal boundary layers at the two isothermal walls die out, indicating the weakened role of the convection in the heat transfer mechanism. Also it can be found that the maximum value of stream function decreases with increase of Hartmann number. At $Ra = 10^3$, contour of kinetic energy shows a single cell. Lorentz forces elongated this cell along the

Figure 51. Effect of Hartmann number on (a) isotherm, (b) streamlines and (c) isokinetic energy at $Y = y / L = 0.5$ *when* $Ra = 10^5, \phi = 0.04$

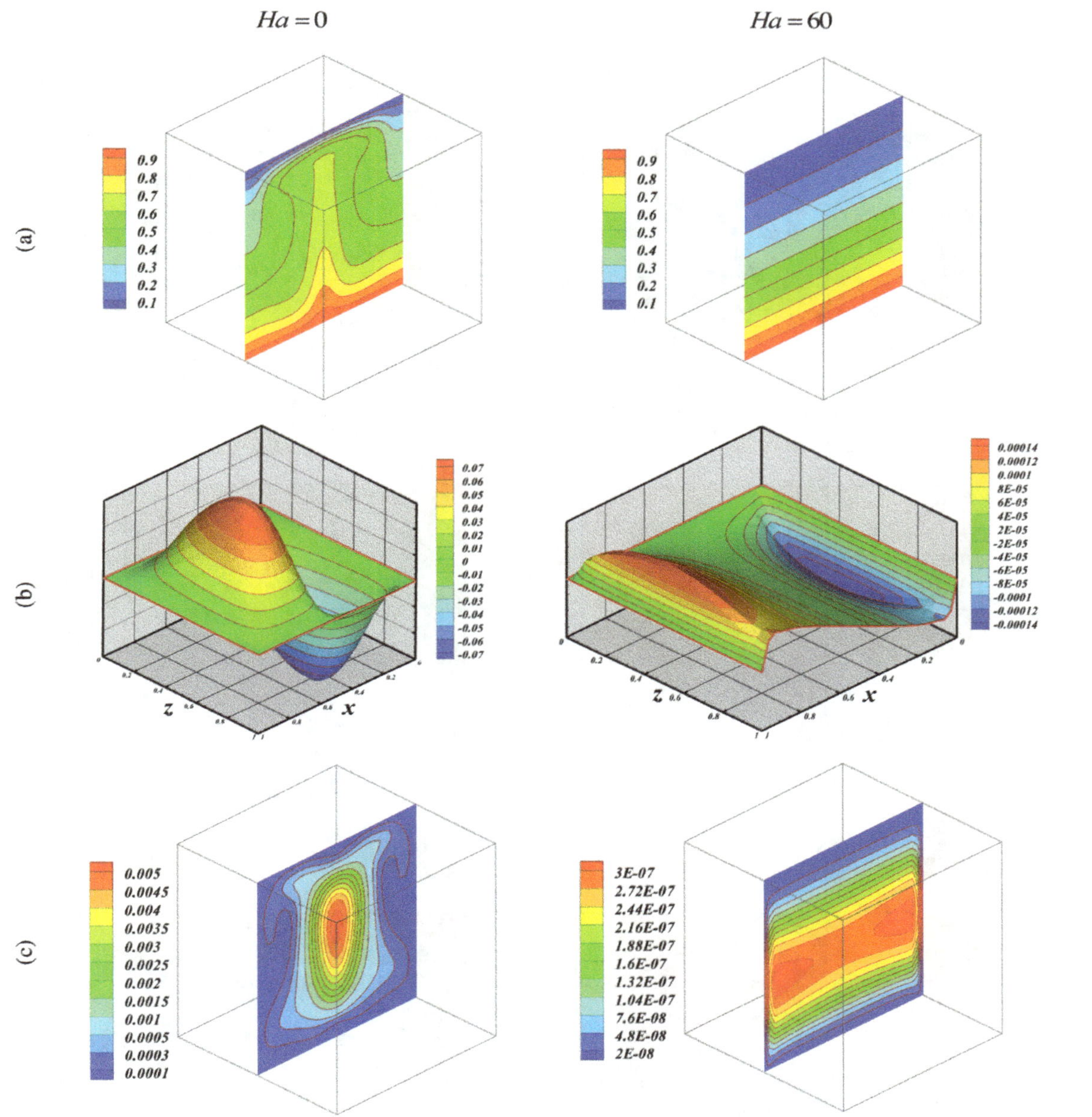

horizontal axis. As Rayleigh number increases up to 10^4, two smaller cells appear near the vertical wall. At $Ra = 10^5$, the main cell of isokinetic energy is elongated along the vertical axis. Also it can be found that kinetic energy increases with increase of Rayleigh number but it decreases with augment of Hartmann number.

Effects of Rayleigh number and Hartmann number on local and average Nusselt number on hot wall are depicted in Figures 52 and 53(a). Thermal boundary layer thickness increases with augment

Figure 52. Effects of Rayleigh number and Hartmann number on local Nusselt number on hot wall when $\phi = 0.04$

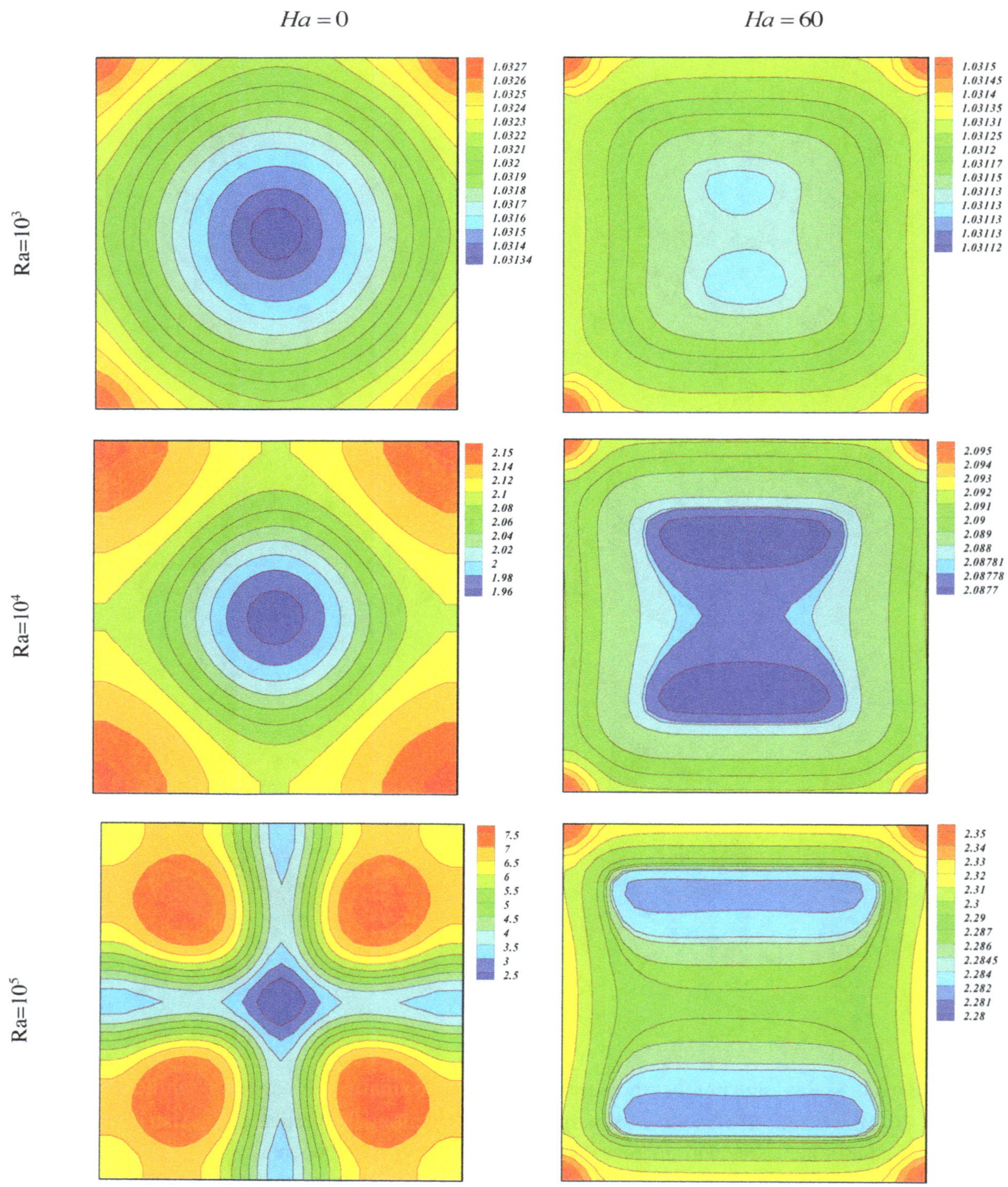

Figure 53. Effects of Rayleigh number and Hartmann number on average Nusselt and heat transfer enhancement

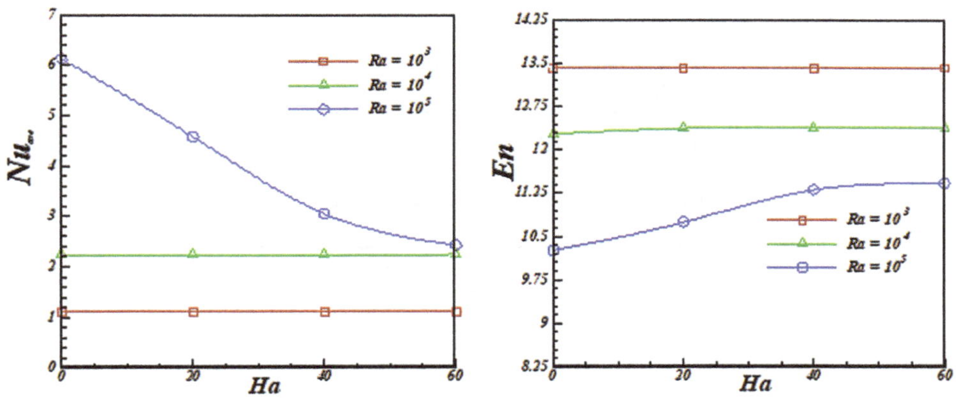

(a) $\phi = 0.04$ (b)

of Hartmann number while it decreases with increase of Rayleigh number. So, Nusselt number has direct relationship with Rayleigh number but it has inverse relationship with Hartmann number. Figure 53 (b) shows the effects of Rayleigh number and Hartmann number on heat transfer enhancement. It is noteworthy that the heat transfer enhancement becomes to its maximum value at low Raleigh number and high Hartmann number. This observation is due to domination of conduction in this case. So the addition of high thermal conductivity nanoparticles will increase the conduction and therefore make the enhancement more effective.

9. TWO PHASE SIMULATION OF NANOFLUID FLOW AND HEAT TRANSFER IN AN ANNULUS IN THE PRESENCE OF AN AXIAL MAGNETIC FIELD

9.1. Problem Definition

Flow is assumed to be steady, laminar and unidirectional; therefore the radial and axial components of the velocity and the derivatives of the velocity with respect to θ and z are zero (Figure 54) (Sheikholeslami & Abelman, 2015). Under these assumptions and in cylindrical coordinates, the governing equations for nanofluid flow, heat and mass transfer following the azimuthal direction can be written as follows:

$$\mu\left(\frac{\partial^2 v}{\partial r^2}+\frac{1}{r}\frac{\partial v}{\partial r}-\frac{v}{r^2}\right)-\sigma v B_0^{\,2}=\rho_f v\frac{\partial v}{\partial r} \tag{119}$$

$$\frac{k}{r}\frac{\partial}{\partial r}\left(r\frac{\partial T}{\partial r}\right)+\mu\left(\frac{\partial v}{\partial r}-\frac{v}{r}\right)^2+\left(\rho C_p\right)_p\left[D_B\frac{\partial C}{\partial r}\frac{\partial T}{\partial r}+\frac{D_T}{T_1}\left(\frac{\partial T}{\partial r}\right)^2\right]=\left(\rho C_p\right)_f v\frac{\partial T}{\partial r} \tag{120}$$

Figure 54. Geometry of the problem

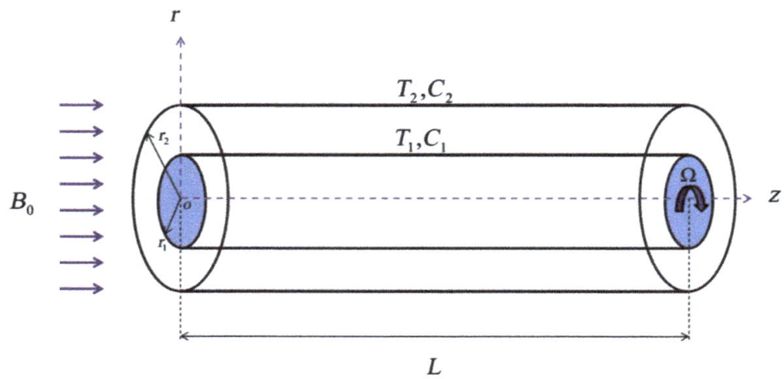

$$\frac{D_B}{r}\frac{\partial}{\partial r}\left(r\frac{\partial C}{\partial r}\right)+\frac{D_T}{T_1}\frac{\partial}{\partial r}\left(r\frac{\partial T}{\partial r}\right)=v\frac{\partial C}{\partial r} \tag{121}$$

$$\begin{aligned} r=r_1 &: v(r)=\Omega_1 r_1, T=T_1, C=C_1 \\ r=r_2 &: v(r)=0, T=T_2, C=C_2 \end{aligned} \tag{122}$$

The governing equation and boundary conditions, Equation (119) to (122), which are in non-dimensional form, become:

$$\frac{\partial^2 v^*}{\partial r^{*2}}+\frac{1}{r^*}\frac{\partial v^*}{\partial r^*}-\left(\frac{Ha^2}{(1-\eta)^2}+\frac{1}{r^{*2}}\right)v^*-\mathrm{Re}\,v^*\frac{\partial v^*}{\partial r^*}=0 \tag{123}$$

$$\frac{1}{r^*}\frac{\partial}{\partial r^*}\left(r^*\frac{\partial \theta}{\partial r^*}\right)+Ec\,\mathrm{Pr}\left(\frac{\partial v^*}{\partial r^*}-\frac{v^*}{r^*}\right)^2-\mathrm{Pr}\,\mathrm{Re}\,v^*\frac{\partial \theta}{\partial r^*}+Nb\frac{\partial \theta}{\partial r^*}\frac{\partial \phi}{\partial r^*}+Nt\left(\frac{\partial \theta}{\partial r^*}\right)^2=0 \tag{124}$$

$$\frac{1}{r^*}\frac{\partial}{\partial r^*}\left(r^*\frac{\partial \phi}{\partial r^*}\right)-\mathrm{Re}\,Sc\,v^*\frac{\partial \phi}{\partial r^*}+\frac{Nt}{Nb}\left(\frac{\partial^2 \theta}{\partial r^{*2}}+\frac{1}{r^*}\frac{\partial \theta}{\partial r^*}\right)=0 \tag{125}$$

$$\begin{aligned} r^*=\eta &: v^*(r^*)=1, \theta=1, \phi=1 \\ r^*=1 &: v^*(r^*)=0, \theta=0, \phi=0 \end{aligned} \tag{126}$$

where

$$r^* = \frac{r}{r_2}, v^* = \frac{v}{\Omega_1 r_1}, \eta = \frac{r_1}{r_2}, Ha = B_0\left(r_2 - r_1\right)\sqrt{\frac{\sigma}{\mu}}, \theta = \frac{T - T_2}{T_1 - T_2}, \phi = \frac{C - C_2}{C_1 - C_2}, \mathrm{Re} = \frac{\rho_f \Omega r_1 r_2}{\mu}$$
$$\mathrm{Pr} = \frac{\mu}{\rho_f \alpha}, \alpha = \frac{k}{\left(\rho C_p\right)_f}, \ Ec = \frac{\rho_f\left(\Omega_1 r_1\right)^2}{\left(\rho C_p\right)_f \Delta T}, Nb = \frac{\left(\rho C_p\right)_p D_B \Delta C}{\left(\rho C_p\right)_f \alpha}, \tag{127}$$
$$Nt = \frac{\left(\rho C_p\right)_p D_T \Delta T}{\left(\rho C_p\right)_f \alpha T_1}, Sc = \frac{\mu}{\rho_f D_B}$$

Nusselt number *Nu* along the inner wall is defined as

$$Nu = -A_4 \left.\frac{\partial \theta}{\partial r^*}\right|_{r^*=\eta} \tag{128}$$

9.2. Effects of Active Parameters

Flow, heat and mass transfer of Al_2O_3-water nanofluid between two horizontal coaxial cylinders in presence of an axial magnetic field is investigated. Two phase model is used for simulation of nanofluid flow and heat transfer. ODEs along with initial conditions are solved using the fourth order Runge-Kutta integration technique (Ashorynejad, Sheikholeslami, Pop, & Ganji, 2013). The effects of Hartmann number, Reynolds number, Schmidt number, Brownian parameter, thermophoresis parameter, Eckert number and aspect ratio on flow, heat and mass transfer characteristics are examined.

Effects of Hartmann number and Reynolds number on velocity, temperature and concentration profiles are shown in Figures 55 and 56. Hartmann number is the ratio of electromagnetic force to the viscous force. Lorentz forces increases with increase of Hartmann number and in turn the velocity decrease with increase of Hartmann number. As Hartmann number increases temperature profiles decrease but the concentration profile increases. Reynolds number is the ratio of inertial forces to viscous forces. Increasing Reynolds number leads to increase velocity and temperature profiles but opposite trend is observed for concentration profile.

Table 6. Effect of Schmidt number on concentration when $\mathrm{Pr} = 10, \eta = 0.5, Ec = 0.01, \mathrm{Re} = 1, Nb = 0.1, Nt = 0.1$.

Ha	Sc		
	0.5	1	2
0	0.343909	0.344245	0.344933
2	0.461028	0.461393	0.46213
4	0.628474	0.628794	0.629434
6	0.705387	0.705624	0.706094

Figure 55. Effect of Hartmann number on velocity, temperature and concentration profiles when $\Pr = 10, \eta = 0.5, Ec = 0.01, \mathrm{Re} = 1, Sc = 1, Nb = 0.1, Nt = 0.1$.

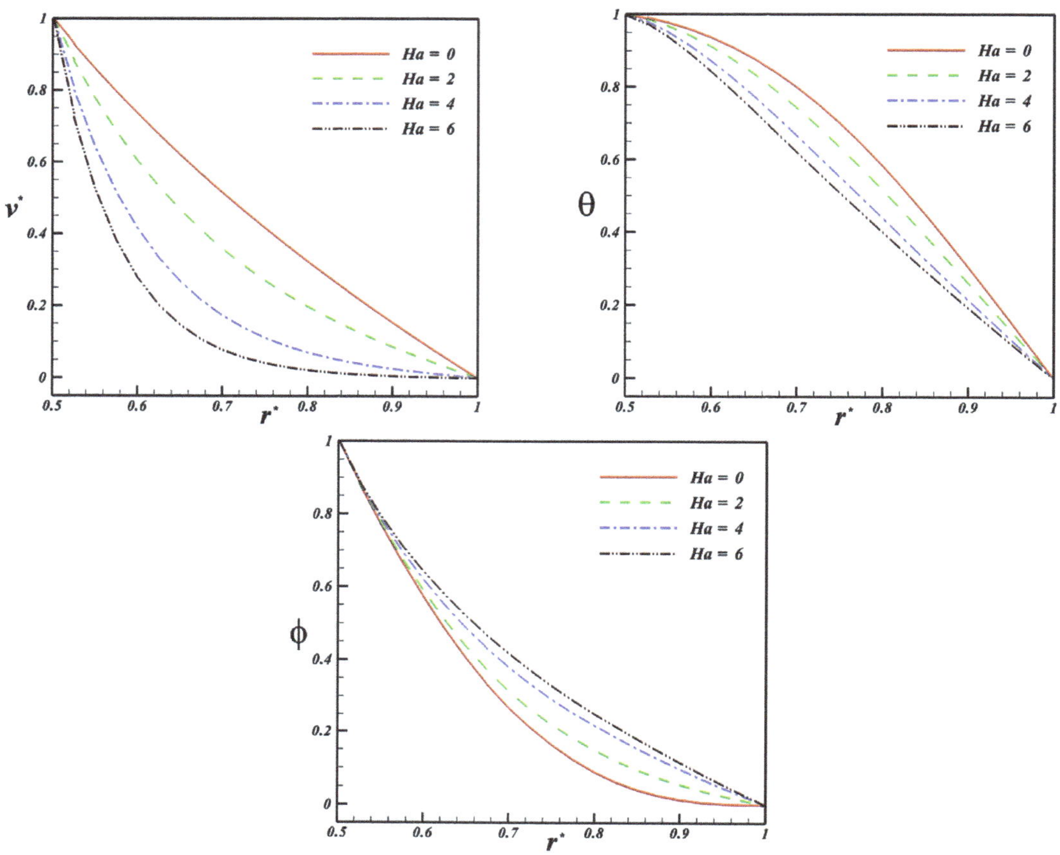

Figure 57 shows the effects of Hartmann number and Reynolds number on Nusselt number. Increasing the Hartmann number leads to increase in temperature boundary layer thickness while the opposite trend is observed for the Reynolds number. So, Nusselt number increases with increasing Hartmann number while it decreases with increasing Reynolds number.

Figure 58 depicts the effect of the Brownian parameter on temperature, concentration profiles and Nusselt number. By increasing Brownian motion parameter, thermal diffusivity decreases and in turn thermal boundary layer thickness increase. So, rate of heat transfer decreases with increase of Brownian motion parameter. Also it can be found that concentration boundary layer increases with increase of Brownian motion parameter. Figure 59 depicts the effect of the thermophoresis parameter on the concentration profiles and Nusselt number. Effect of the thermophoresis parameter on the Nusselt number is similar to that of the Brownian parameter. Increasing the thermophoresis parameter leads to increase in the concentration boundary layer. The Nusselt number decreases as the thermophoresis parameter increases.

Figure 60 shows the effect of the Eckert number on the temperature profile and Nusselt number. As the Eckert number increases, viscous dissipation increases and in turn the temperature increases and Nusselt

Figure 56. Effect of Reynolds number on velocity, temperature and concentration profiles when $\Pr = 10, \eta = 0.5, Ha = 1, Ec = 0.01,\ Sc = 1, Nb = 0.1, Nt = 0.1$.

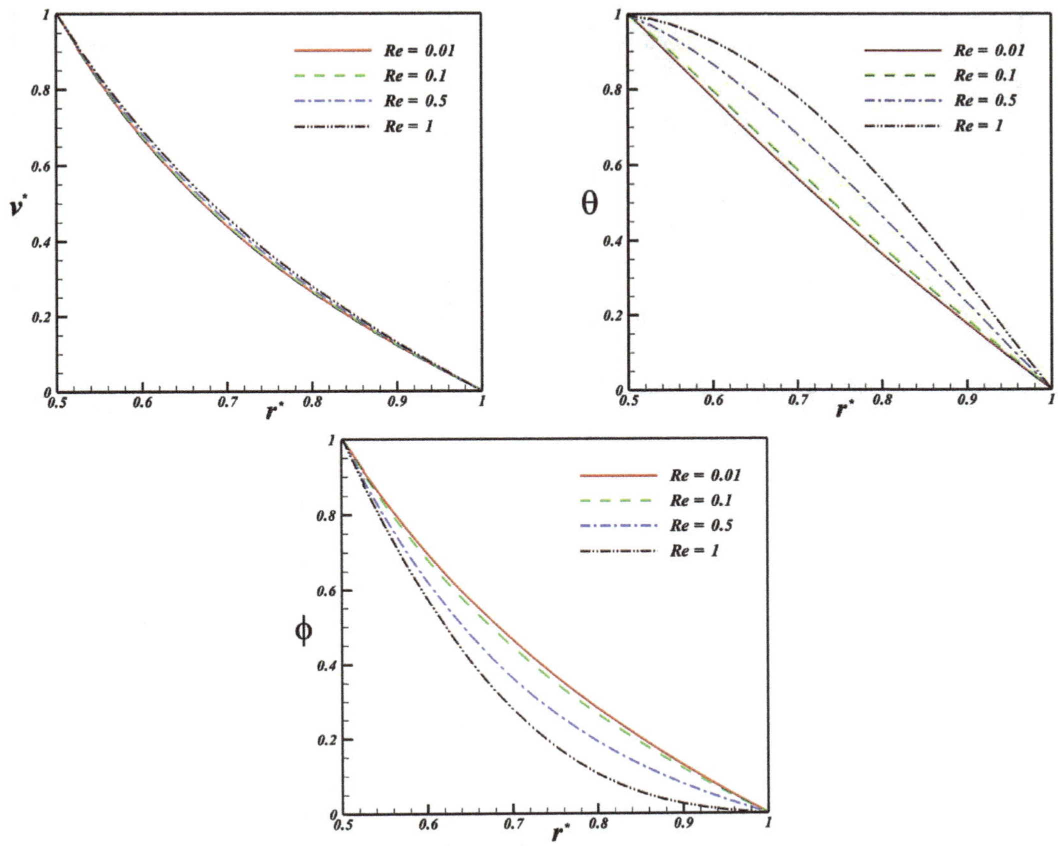

Figure 57. Effect of Reynolds number and Hartmann number on Nusselt number when $\Pr = 10, \eta = 0.5, Ec = 0.01, Sc = 1, Nb = 0.1, Nt = 0.1$.

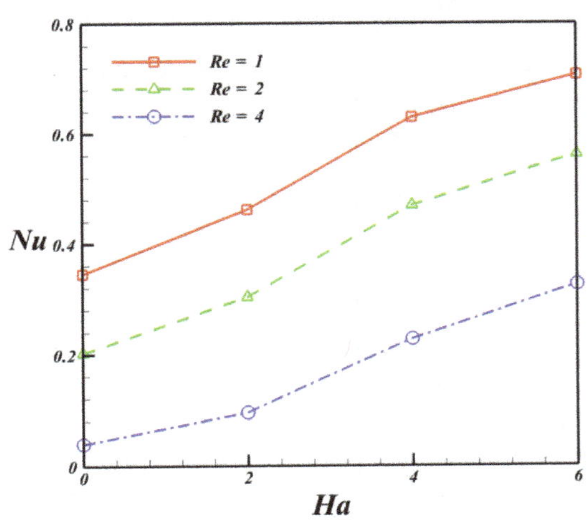

Figure 58. Effect of Brownian parameter on temperature, concentration profiles and Nusselt number when $\Pr = 10, \eta = 0.5, Ec = 0.01, \mathrm{Re} = 1, Sc = 1, Nt = 0.1$.

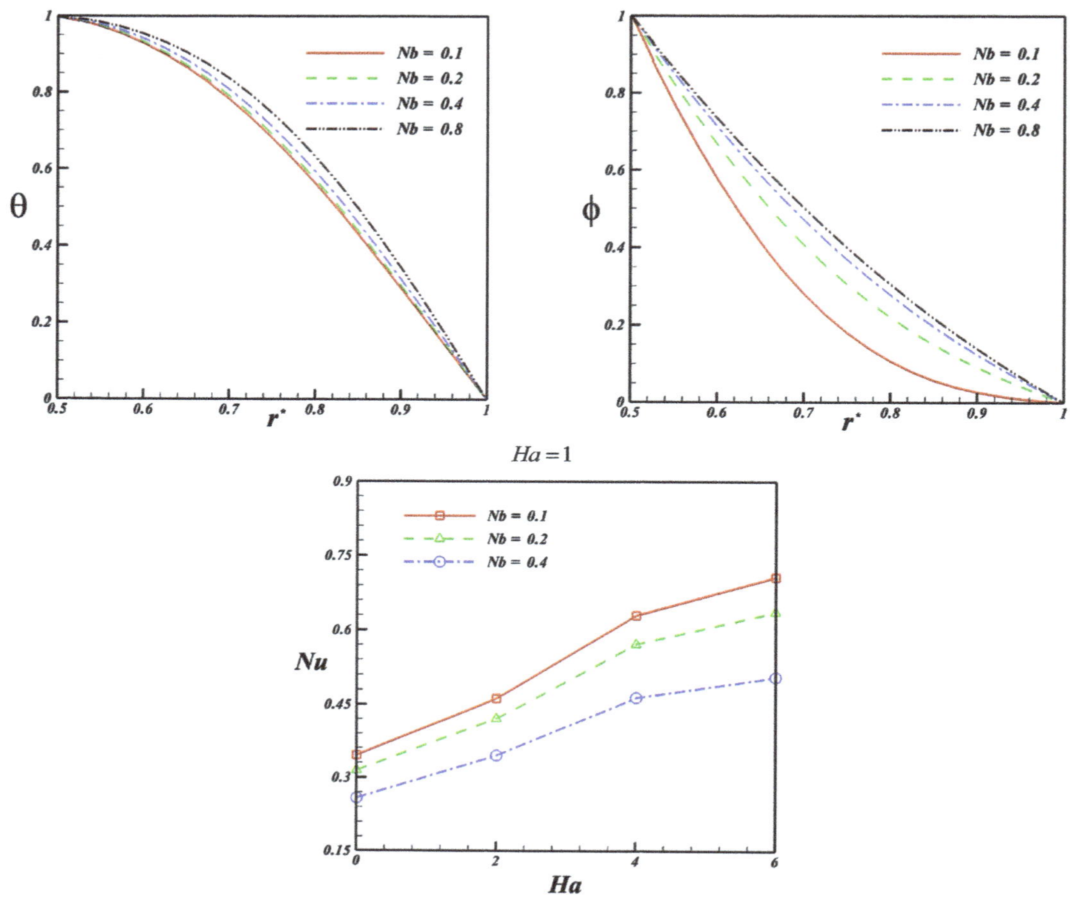

Figure 59. Effect of thermophoresis parameter on concentration profile and Nusselt number when $\Pr = 10, \eta = 0.5, Ec = 0.01, \mathrm{Re} = 1, Sc = 1, Nb = 0.1$.

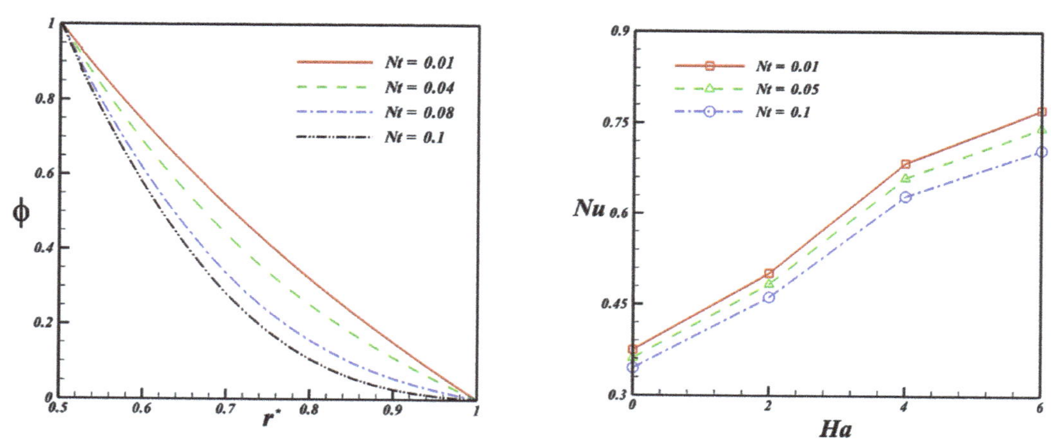

$Ha = 1$

Figure 60. Effect of Eckert number on temperature profile and Nusselt number when $\Pr = 10, \eta = 0.5, \mathrm{Re} = 1, Sc = 1, Nb = 0.1, Nt = 0.1$.

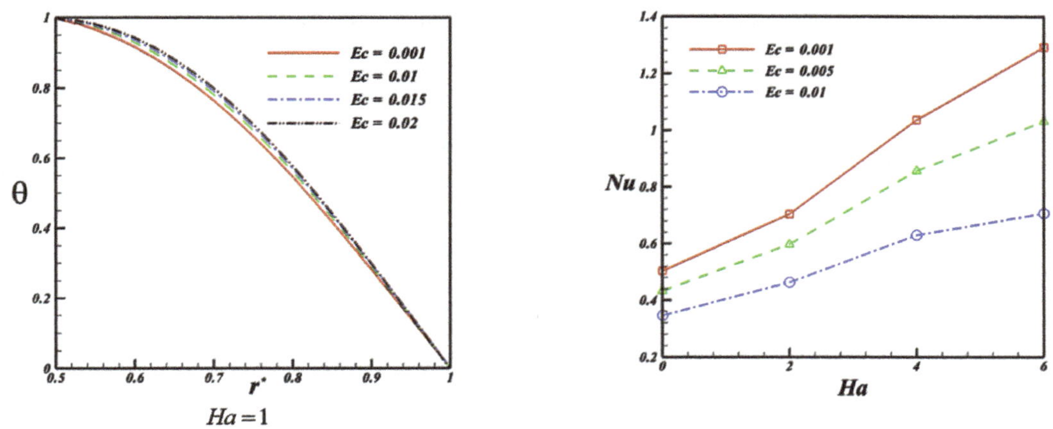

number decreases. Effect of the Schmidt number on the concentration profile is depicted in Figure 61 and Table Concentration increases with increase of Schmidt number, while the thermal boundary layer thickness decreases. So Nusselt number increases with increase of Schmidt number. Figure 62 shows the effect of the aspect ratio on the Nusselt number. As the aspect ratio increases, the distance between the hot and cold walls of the horizontal coaxial cylinders decreases and the Nusselt number increases.

The corresponding polynomial representation of such model for Nusselt number is as the following:

Figure 61. Effect of Schmidt number on concentration profile when $\Pr = 10, \eta = 0.5, Ha = 1, Ec = 0.01, \mathrm{Re} = 1, Nb = 0.1, Nt = 0.1$.

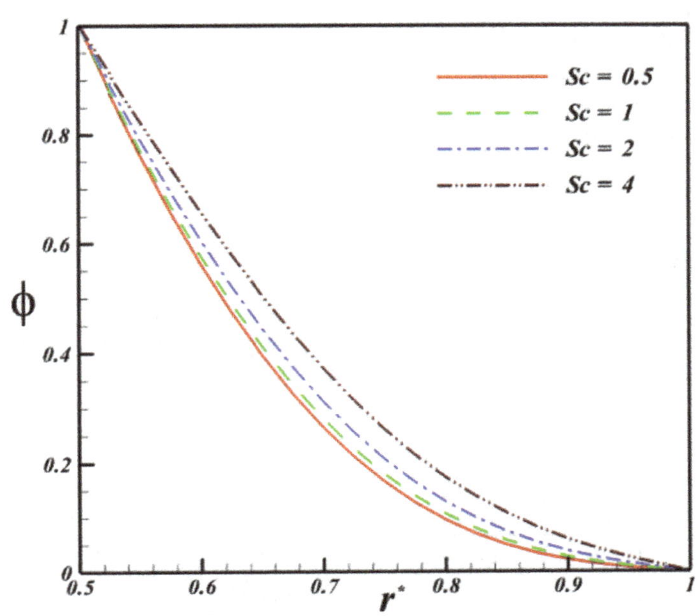

Figure 62. Effect of aspect ratio on Nusselt number when $\Pr = 10, Ec = 0.01, \mathrm{Re} = 1, Sc = 1, Nb = 0.1, Nt = 0.1$.

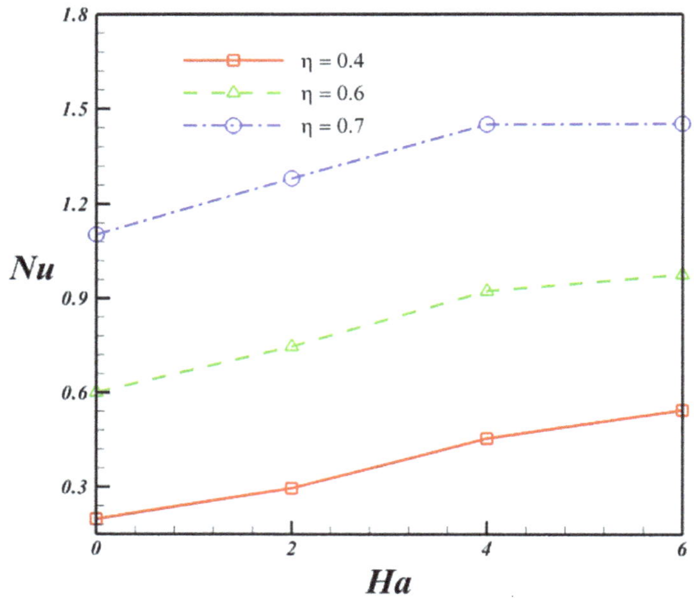

$$\begin{aligned}
Nu_{ave} &= a_{16} + a_{26}Y_5 + a_{36}Y_6 + a_{46}Y_5^2 + a_{56}Y_6^2 + a_{66}Y_5Y_6 \\
Y_1 &= a_{11} + a_{21}Sc + a_{31}Nb + a_{41}Sc^2 + a_{51}Nb^2 + a_{61}Sc\,Nb \\
Y_2 &= a_{12} + a_{22}\,\mathrm{Re} + a_{32}Nt + a_{42}\,\mathrm{Re}^2 + a_{52}Nt^2 + a_{62}\,\mathrm{Re}\,Nt \\
Y_3 &= a_{13} + a_{23}\eta + a_{33}Ha + a_{43}\eta^2 + a_{53}Ha^2 + a_{63}\eta\,Ha \\
Y_4 &= a_{14} + a_{24}Y_1 + a_{34}Y_2 + a_{44}Y_1^2 + a_{54}Y_2^2 + a_{64}Y_1Y_2 \\
Y_5 &= a_{15} + a_{25}Ec + a_{35}Y_4 + a_{45}Ec^2 + a_{55}Y_4^2 + a_{65}Ec\,Y_4
\end{aligned} \qquad (129)$$

Also a_{ij} can be found in Table 7.

Table 7. Constant coefficient for using Equation (129)

a_{ij}	*i=1*	*i=2*	*i=3*	*i=4*	*i=5*	*i=6*
j=1	0.306009	1.185703	-2.24994	-0.38755	6.581819	-2.16197
j=2	4.757492	-6.27872	-1.12292	2.132729	46.88285	-2.22654
j=2	2.756484	-11.9752	0.124053	14.13088	-0.00312	-0.07635
j=3	-0.37891	-0.55728	0.378953	1.631652	0.788778	0.06663
j=4	-0.16013	34.84572	2.02581	-1219.24	0.107653	-128.364
j=5	-0.69095	2.318391	1.015165	-1.70229	0.019379	-0.10102
j=6	0.306009	1.185703	-2.24994	-0.38755	6.581819	-2.16197

10. MAGNETIC FIELD EFFECT ON UNSTEADY NANOFLUID FLOW AND HEAT TRANSFER USING BUONGIORNO MODEL

10.1. Problem Definition and Semi-Analytical Method

10.1.1. Problem Statement

Heat and mass transfer analysis in the unsteady two-dimensional squeezing flow of nanofluid between the infinite parallel plates is considered (Figure 63) (Sheikholeslami, Ganji, & Rashidi, 2016). The two plates are placed at $\ell(1-\beta t)^{1/2}=h(t)$. When $\beta>0$ the two plates are squeezed until they touch $t=1/\beta$ and for $\beta<0$ the two plates are separated. The viscous dissipation effect, the generation of heat due to friction caused by shear in the flow, is retained. Also, it is also assumed that the uniform magnetic field ($\vec{B}=B\vec{e}_y$) is applied, where $\vec{e}_y$ is unit vectors in the Cartesian coordinate system. The electric current J and the electromagnetic force F are defined by $J=\sigma\left(\vec{V}\times\vec{B}\right)$ and $F=\sigma\left(\vec{V}\times\vec{B}\right)\times\vec{B}$, respectively.

The governing equations for mass, momentum, energy and mass transfer in unsteady two dimensional flow of nanofluid are:

Figure 63. Geometry of problem

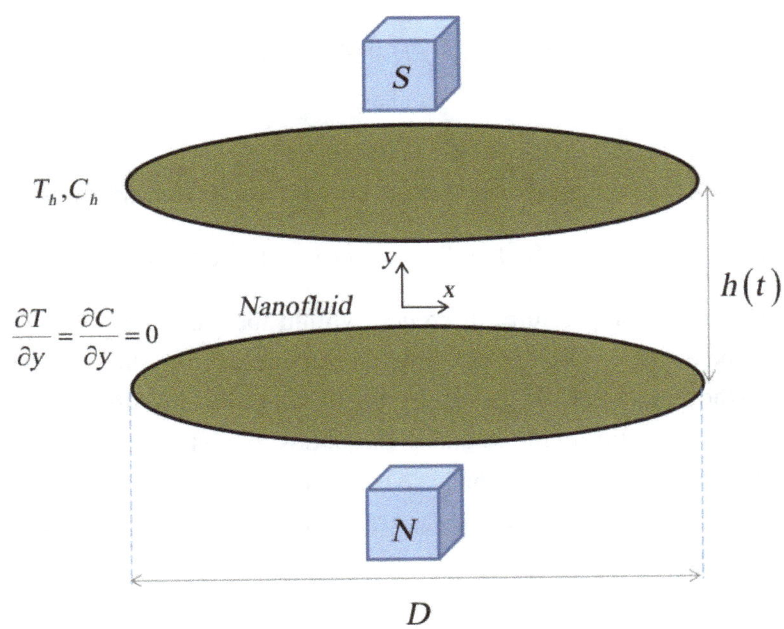

$$\frac{\partial u}{\partial x}+\frac{\partial v}{\partial y}=0, \tag{130}$$

$$\rho_f\left(\frac{\partial u}{\partial t}+u\frac{\partial u}{\partial v}+v\frac{\partial u}{\partial y}\right)=-\frac{\partial p}{\partial x}+\mu(\frac{\partial^2 u}{\partial x^2}+\frac{\partial^2 u}{\partial y^2})-\sigma B^2 u, \tag{131}$$

$$\rho_f\left(\frac{\partial v}{\partial t}+u\frac{\partial v}{\partial x}+v\frac{\partial v}{\partial y}\right)=-\frac{\partial p}{\partial y}+\mu(\frac{\partial^2 v}{\partial x^2}+\frac{\partial^2 v}{\partial y^2}), \tag{132}$$

$$\begin{aligned}\frac{\partial T}{\partial t}+u\frac{\partial T}{\partial x}+v\frac{\partial T}{\partial y}&=\alpha(\frac{\partial^2 T}{\partial x^2}+\frac{\partial^2 T}{\partial y^2})+\frac{\mu}{\left(\rho c_p\right)_f}(4(\frac{\partial u}{\partial x})^2)\\&+\frac{\left(\rho c_P\right)_p}{\left(\rho c_P\right)_f}\left[\begin{array}{l}D_B\left\{\frac{\partial C}{\partial x}.\frac{\partial T}{\partial x}+\frac{\partial C}{\partial y}.\frac{\partial T}{\partial y}\right\}\\+(D_T/T_c)\left\{\left(\frac{\partial T}{\partial x}\right)^2+\left(\frac{\partial T}{\partial y}\right)^2\right\}\end{array}\right]-\frac{1}{\left(\rho c_p\right)_f}\frac{\partial q_r}{\partial y},\end{aligned} \tag{133}$$

$$\frac{\partial C}{\partial t}+u\frac{\partial C}{\partial x}+v\frac{\partial C}{\partial y}=D_B(\frac{\partial^2 C}{\partial x^2}+\frac{\partial^2 C}{\partial y^2})+\left(\frac{D_T}{T_c}\right)\left\{\frac{\partial^2 T}{\partial x^2}+\frac{\partial^2 T}{\partial y^2}\right\} \tag{134}$$

where the radiation heat flux q_r is considered according to Rosseland approximation such that $q_r=-\frac{4\sigma_e}{3\beta_R}\frac{\partial T^4}{\partial y}$ where σ_e, β_R are the Stefan–Boltzmann constant and the mean absorption coefficient, respectively. The fluid-phase temperature differences within the flow are assumed to be sufficiently small so that T^4 may be expressed as a linear function of temperature. This is done by expanding T^4 in a Taylor series about the temperature T_c and neglecting higher order terms to yield, $T^4 \cong 4T_c^3 T-3T_c^4$.

Here u and v are the velocities in the x and y directions respectively, T is the temperature, C is the concentration, P is the pressure, ρ_f is the base fluid's density, μ is the dynamic viscosity, k is the thermal conductivity, c_p is the specific heat of nanofluid, D_B is the diffusion coefficient of the diffusing species. The relevant boundary conditions are:

$$\begin{aligned} &C = 0, v = v_w = dh / dt, T = T_H, C = C_H \; at \; y = h(t), \\ &v = \partial u / \partial y = \partial T / \partial y = \partial C / \partial y = 0 \; at \; y = 0. \end{aligned} \tag{135}$$

We introduce these parameters:

$$\begin{aligned} &\eta = \frac{y}{[\ell(1-\gamma t)^{1/2}]}, \\ &u = \frac{\gamma x}{[2(1-\gamma t)]} f'(\eta), \\ &v = -\frac{\gamma \ell}{[2(1-\gamma t)^{1/2}]} f(\eta), \\ &\theta = \frac{T}{T_H}, \\ &\phi = \frac{C}{C_h}. \end{aligned} \tag{136}$$

Substituting the above variables into (131) and (132) and then eliminating the pressure gradient from the resulting equations give:

$$f^{iv} - S\left(\eta f''' + 3f'' + f'f'' - ff'''\right) - Ha^2 f'' = 0, \tag{137}$$

Using (136), Equations (133) and (134) reduce to the following differential equations:

$$\left(1 + \frac{4}{3} Rd\right)\theta'' + \Pr S\left(f\theta' - \eta\theta'\right) + \Pr Ec\left(f''^2\right) + Nb\,\phi'\theta' + Nt\,\theta'^2 = 0, \tag{138}$$

$$\phi'' + S.Sc\left(f\phi' - \eta\phi'\right) + \frac{Nt}{Nb}\theta'' = 0, \tag{139}$$

With these boundary conditions:

$$\begin{aligned} &f(0) = 0, f''(0) = 0, \theta'(0) = 0, \\ &\phi'(0) = 0, f(1) = 1, f'(1) = 0, \\ &\theta(1) = \phi(1) = 1, \end{aligned} \tag{140}$$

where S is the squeeze number, $\Pr$ is the Prandtl number, Ec is the Eckert number, Sc is the Schmidt number, Ha is Hartman number of nanofluid, Nb is the Brownian motion parameter, Nt is the thermophoretic parameter and Rd is the Radiation parameter, which are defined as:

$$\begin{aligned}
&S = \frac{\beta l^2}{2\mu}\rho_f, \Pr = \frac{\mu}{\rho_f \alpha}, Ec = \frac{1}{c_p}\left(\frac{\beta x}{2\left(1-\beta t\right)}\right)^2 \\
&Sc = \frac{\mu}{\rho_f D}, Ha = \ell B\sqrt{\frac{\sigma}{\mu}\left(1-\beta t\right)}, \\
&Nb = (\rho c)_p D_B (C_h) / \left((\rho c)_f \alpha\right), \\
&Nt = (\rho c)_p D_T (T_H) / [(\rho c)_f \alpha T_c], \\
&Rd = 4\sigma_e T_c^3 / \left(\beta_R k\right).
\end{aligned} \tag{141}$$

Skin friction coefficient and Nusselt number are defined as:

$$C_f^* = -f''(1) \tag{142}$$

$$Nu^* = -\theta'\left(1\right) \tag{143}$$

10.1.2. DTM Solution

Applying the differential transforms of Eqs. (137), (138) and (139) gives:

$$\begin{aligned}
&(k+1)(k+2)(k+3)(k+4)F[k+4] + S\sum_{m=0}^{k}\left(\Delta[k-m-1](m+1)(m+2)(m+3)F[m+3]\right) \\
&-3S(k+1)(k+2)F[k+2] - S\sum_{m=0}^{k}\left(\left(k-m+1\right)F[k-m+1](m+1)(m+2)F[m+2]\right) \\
&+S\sum_{m=0}^{k}\left(F[k-m](m+1)(m+2)(m+3)F[m+3]\right) - Ha^2(k+1)(k+2)F[k+2] = 0, \\
&\Delta[m] = \begin{cases} 1 & m=1 \\ 0 & m \neq 1 \end{cases}
\end{aligned} \tag{144}$$

$$F[0] = 0, F[1] = a_1, F[2] = 0, F[3] = a_2 \tag{145}$$

$$\left(1+\frac{4}{3}Rd\right)(k+1)(k+2)\Theta[k+2]+\Pr.S.\sum_{m=0}^{k}\left(F[k-m](m+1)\Theta[m+1]\right)$$
$$-\Pr.S.\sum_{m=0}^{k}\left(\Delta[k-m](m+1)\Theta[m+1]\right)$$
$$+\Pr Ec\sum_{m=0}^{k}\left(\left(k-m+1\right)\left(k-m+2\right)F[k-m+2](m+1)(m+2)F[m+2]\right) \tag{146}$$
$$+Nb.\sum_{m=0}^{k}\left(\Phi[k-m](m+1)\Theta[m+1]\right)+Nt.\sum_{m=0}^{k}\left(\Theta[k-m](m+1)\Theta[m+1]\right)=0$$
$$\Delta[m]=\begin{cases}1 & m=1\\ 0 & m\neq 1\end{cases}$$

$$\Theta[0]=a_3,\Theta[1]=0 \tag{147}$$

$$(k+1)(k+2)\Phi[k+2]+Sc.S.\sum_{m=0}^{k}\left(F[k-m](m+1)\Phi[m+1]\right)$$
$$+Sc.S.\sum_{m=0}^{k}\left(\Delta[k-m](m+1)\Phi[m+1]\right)+\frac{Nt}{Nb}(k+1)(k+2)\Theta[k]=0 \tag{148}$$
$$\Delta[m]=\begin{cases}1 & m=1\\ 0 & m\neq 1\end{cases}$$

$$\Phi[0]=a_4,\Phi[1]=0 \tag{149}$$

By solving the above equations:

$$F[0]=0,\ F[1]=a_1,\ F[2]=0,\ F[3]=a_2,F[4]=0$$
$$F[5]=\frac{3}{20}Sa_2+\frac{1}{20}Sa_1a_2+\frac{1}{20}a_1a_2+\frac{1}{20}Ha^2\,a_2,\ldots \tag{150}$$

$$\Theta[0]=a_3,\ \Theta[1]=0,\Theta[2]=0,$$
$$\Theta[3]=0,$$
$$\Theta[4]=-\left(9\Pr Ec\,a_2^2\right)/\left(3+4Rd\right), \tag{151}$$
$$\Theta[5]=0,\ldots$$

$$\begin{aligned} &\Phi[0] = a_4,\ \Phi[1] = 0, \Phi[2] = 0, \Phi[3] = 0, \\ &\Phi[4] = \frac{9\,Nt \Pr Ec\, a_2^3}{Nb\left(3 + 4Rd\right)}, \\ &\Phi[5] = 0, \ldots \end{aligned} \tag{152}$$

Finally we have:

$$F\left(\eta\right) = a_1\eta + a_2\eta^3 + \left(\frac{3}{20}Sa_2 + \frac{1}{20}Sa_1a_2 + \frac{1}{20}a_1a_2 + \frac{1}{20}Ha^2\, a_2\right)\eta^5 + \ldots \tag{153}$$

$$\Theta\left(\eta\right) = a_3 - \left(9\Pr Ec\, a_2^2\right) / \left(3 + 4Rd\right)\eta^4 + \ldots \tag{154}$$

$$\Phi\left(\eta\right) = a_4 + \frac{9\,Nt \Pr Ec\, a_2^3}{Nb\left(3 + 4Rd\right)}\eta^4 + \ldots \tag{155}$$

According to Eqs. (140), (153), (154) and (155) it can be obtained the values of a_1, a_2, a_3, a_4 . By substituting obtained a_1, a_2, a_3, a_4 into Equations (153), (154) and (155), it can be obtained the expression of $F\left(\eta\right), \Theta\left(\eta\right)$ and $\Phi\left(\eta\right)$.

10.2. Effects of Active Parameters

Unsteady nanofluid flow and heat transfer between parallel sheets is studied considering Buongiorno model using DTM. The influences of the squeeze number, radiation parameter, Hartmann number, Brownian motion parameter, thermophoretic parameter and Eckert number on heat and mass characteristics are examined.

Effects of the Hartmann and squeeze numbers on the velocity profiles and skin friction coefficient are demonstrated in Figure 64. It is significant to reminder that the squeeze number $\left(S\right)$ defines the movement of the sheets ($S > 0$ corresponds to the plates moving apart, while $S < 0$ corresponds to the plates moving together). In this paper, positive values of squeeze number are examined. Squeeze number has different effect on vertical velocity profile near each sheet. f' increases with increases S of when $\eta > 0.5$ but opposite style is detected when $\eta < 0.5$. It is valuable saying that the impact of magnetic field is to reduce the value of the velocity magnitude near the lower plat because the existence of magnetic field presents a force called the Lorentz force, which acts in contradiction of the flow. This type of resisting force slows down the fluid velocity. Also it can be concluded that skin friction coefficient augments with enhance of Hartmann number and squeeze number.

Figure 64. Effects of the squeeze number and Hartmann number on the velocity profile and skin friction coefficient

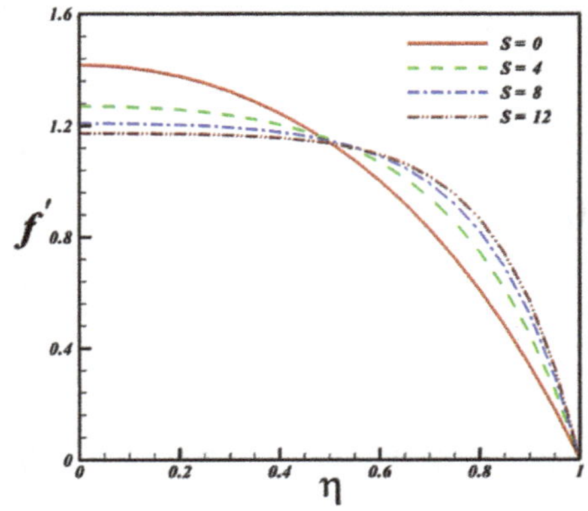

(a) $Ha = 2$

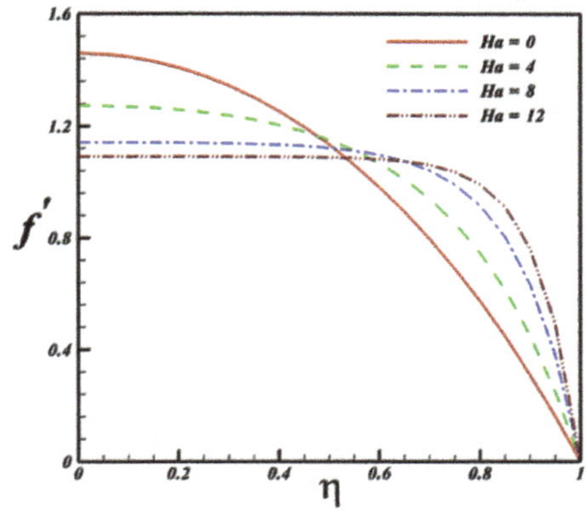

(b) $S = 0.5$

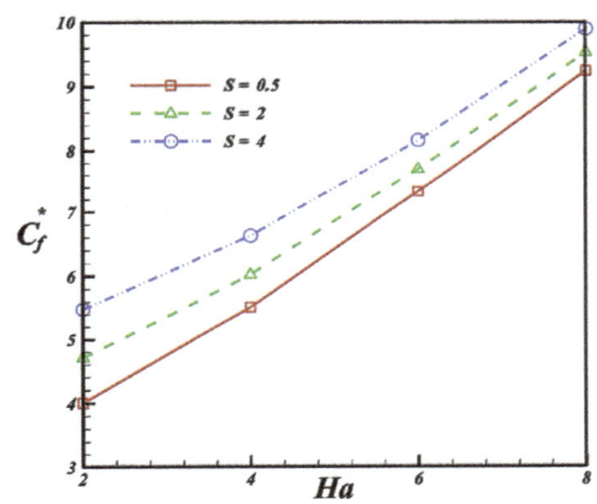

(c)

Figure 65 shows the effect of the radiation parameter, squeeze number, Hartmann number and Nt on the temperature profile. Effects of active parameters on Nusselt number are depicted in Figures 66 and 67. The corresponding polynomial representation for Nusselt number is as the following:

$$\begin{aligned} Nu^* &= -0.14575 + 0.16685S - 0.015376Ha - 0.084727Rd + 4.41505Ec \\ &+4.57118Nt - 2.2554Nb - 4.305\times10^{-4} S\,Ha - 1.12733\times10^{-3} S\,Rd \\ &+0.14115\,S\,Ec - 1.04122\,S\,Nt - 0.026763\,S\,Nb - 8.9126\times10^{-3} Ha\,Rd \\ &+0.84028Ha\,Ec + 0.042273Ha\,Nt + 0.096047\,Ha\,Nb - 2.60638\,Rd\,Ec \\ &+0.11069Rd\,Nt - 0.62086\,Rd\,Nb - 1.19695\,Ec\,Nt + 68.06014\,Ec\,Nb \\ &+2.62781\,Nt\,Nb + 4.5899\times10^{-5} S^2 + 1.60079\,Ha^2 + 0.025525\,Rd^2 \\ &-15.10193\,Ec^2 + 1.65357\,Nt^2 + 1.65357\,Nb^2 \end{aligned} \tag{156}$$

Figure 65. Effects of the radiation parameter, squeeze number, Hartmann number and Nt *on the temperature profile when* $Sc = 0.5, Nb = 0.1, Ec = 0.1$ *and* $\Pr = 10$

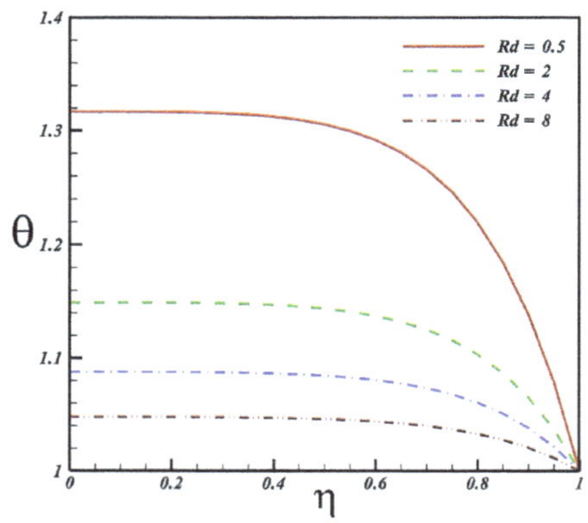

(a) $S = 0.5, Ha = 2, Nt = 0.1$

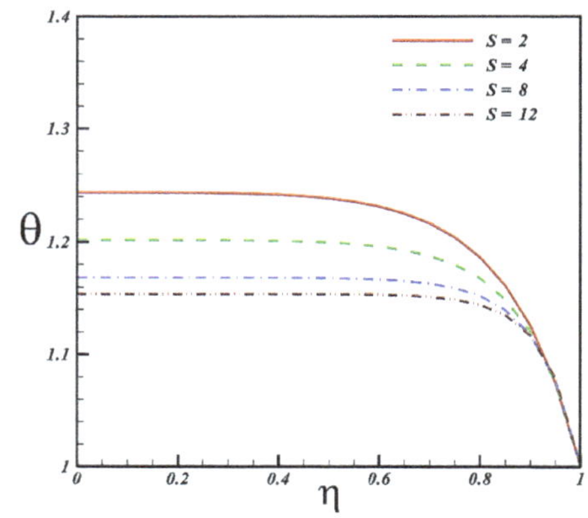

(b) $Ha = 2, Rd = 0.5, Nt = 0.1$

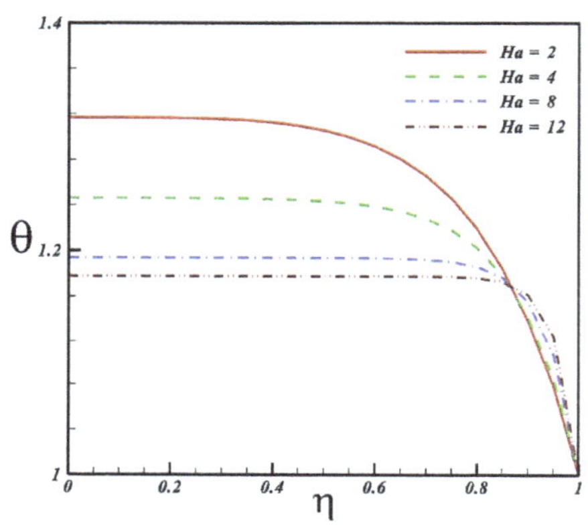

(c) $S = 0.5, Rd = 0.5, Nt = 0.1$

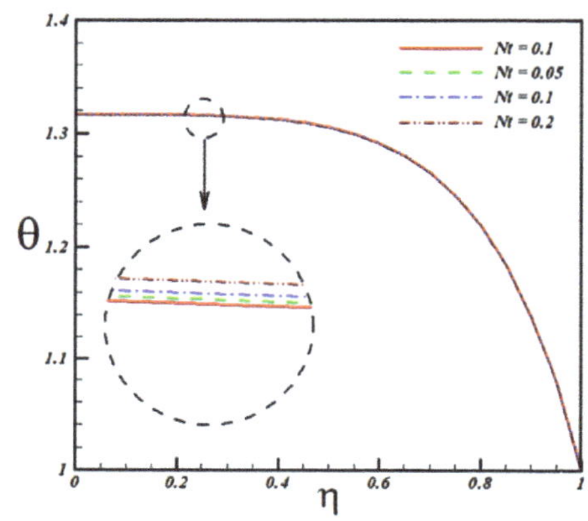

(d) $S = 0.5, Ha = 2, Rd = 0.5$

Figure 66. Effects of the radiation parameter, squeeze number and Hartmann number on Nusselt number when $Sc = 0.5, Nb = 0.1, Ec = 0.1, Nt = 0.1$ *and* $\Pr = 10$.

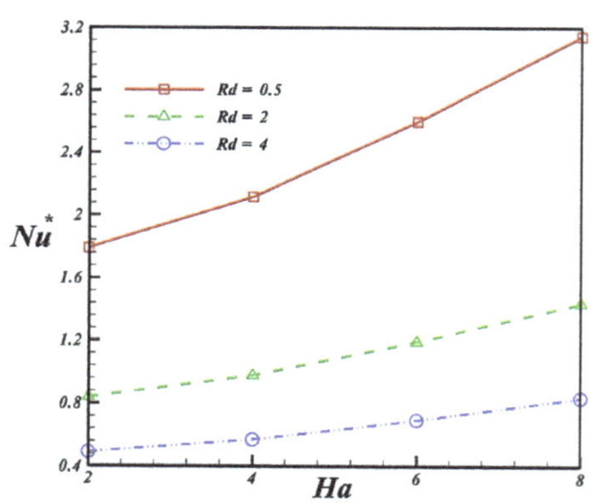

(a) $S = 0.5$

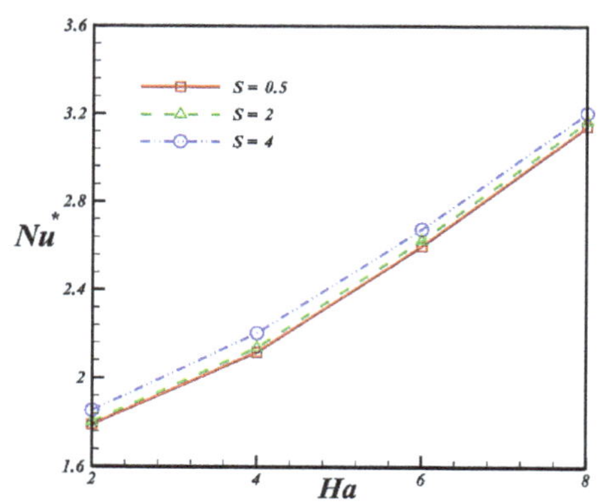

(b) $Rd = 0.5$

Figure 67a. Effect of active parameters on Nusselt number

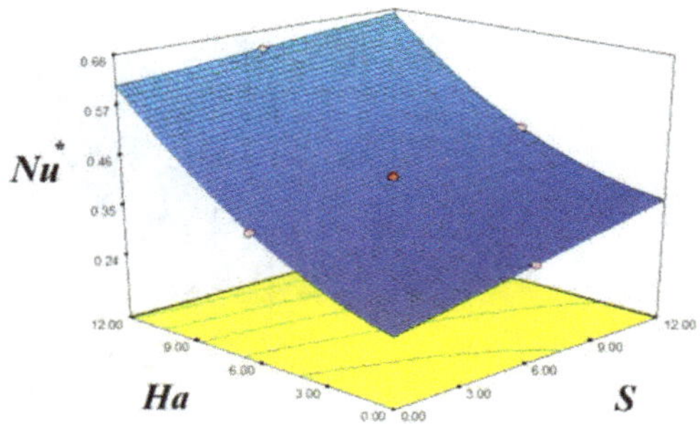

$Rd = 4.5, Ec = 0.06, Nb = 0.15, Nt = 0.15, Sc = 0.5$

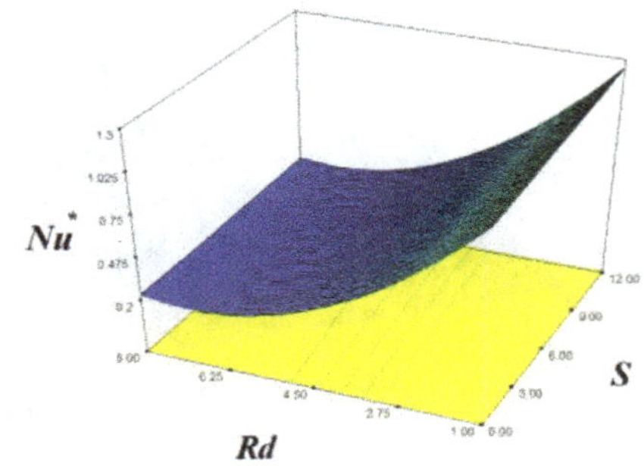

$Ha = 6, Ec = 0.06, Nb = 0.15, Nt = 0.15, Sc = 0.5$

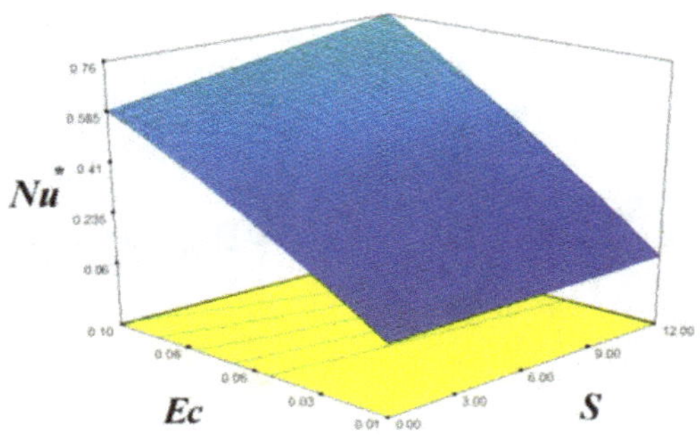

$Ha = 6, Rd = 4.5, Nb = 0.15, Nt = 0.15, Sc = 0.5$

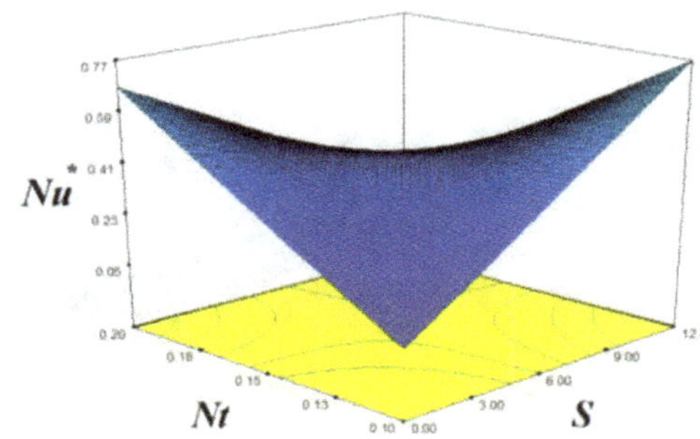

$Ha = 6, Rd = 4.5, Ec = 0.06, Nb = 0.15, Sc = 0.5$

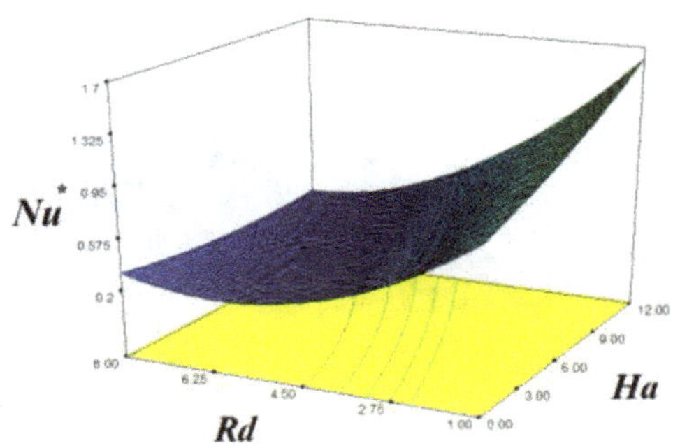

$S = 6, Ec = 0.06, Nb = 0.15, Nt = 0.15, Sc = 0.5$

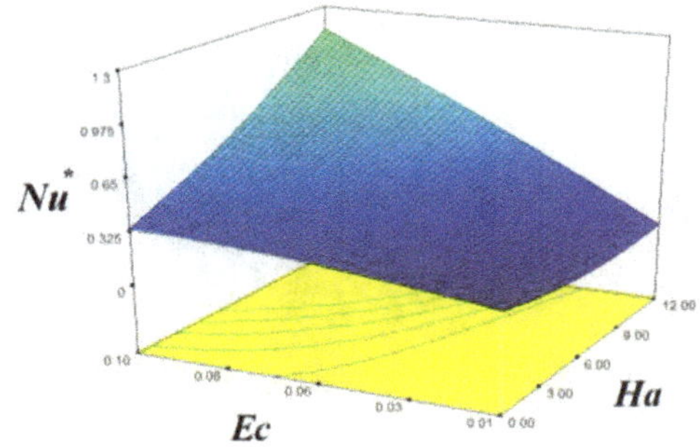

$S = 6, Rd = 4.5, Nb = 0.15, Nt = 0.15, Sc = 0.5$

As radiation parameter augments thermal boundary layer thickness near the upper wall augments. So, Nusselt number reduces with augment of this parameter. Enhance in the squeeze number which can be associated with reducing in the kinematic viscosity, an augment in the distance between the plates and an augment in the speed at which the plates move. Thermal boundary layer thickness near the upper wall reduces as the squeeze number augments. Temperature profiles have meeting point near $\eta = 0.82$ for different values of Hartmann number. Increasing Hartmann number leads to augment in temperature

Figure 67b. Effect of active parameters on Nusselt number

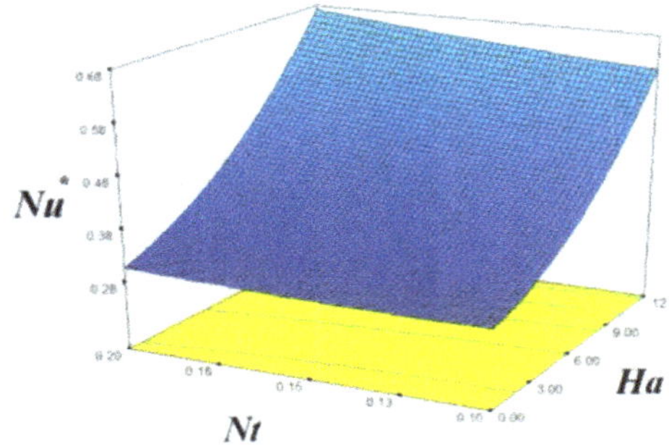

$S = 6, Rd = 4.5, Ec = 0.06, Nb = 0.15, Sc = 0.5$

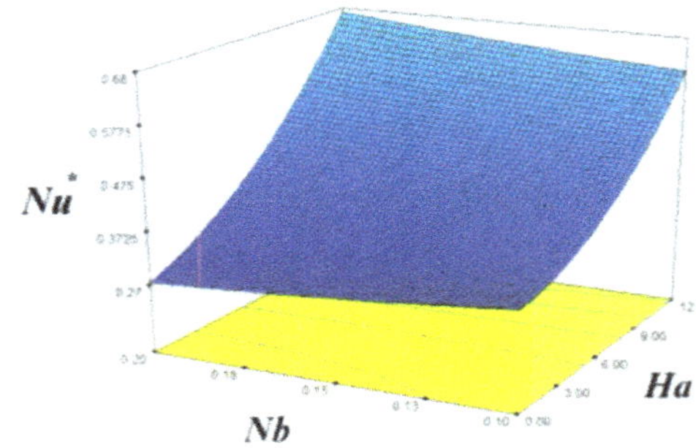

$S = 6, Rd = 4.5, Ec = 0.06, Nt = 0.15, Sc = 0.5$

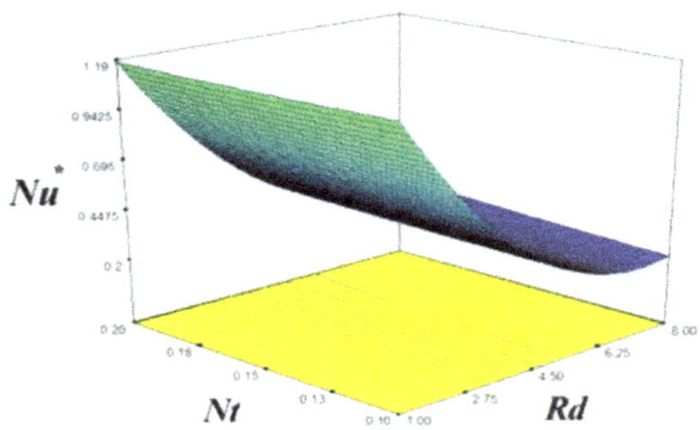

$S = 6, Ha = 6, Ec = 0.06, Nb = 0.15, Sc = 0.5$

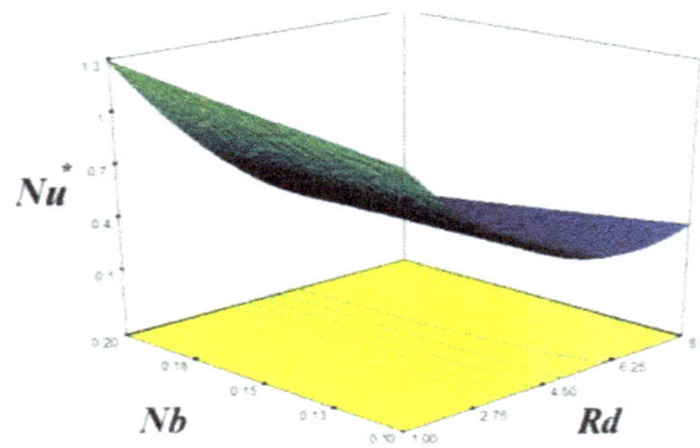

$S = 6, Ha = 6, Ec = 0.06, Nt = 0.15, Sc = 0.5$

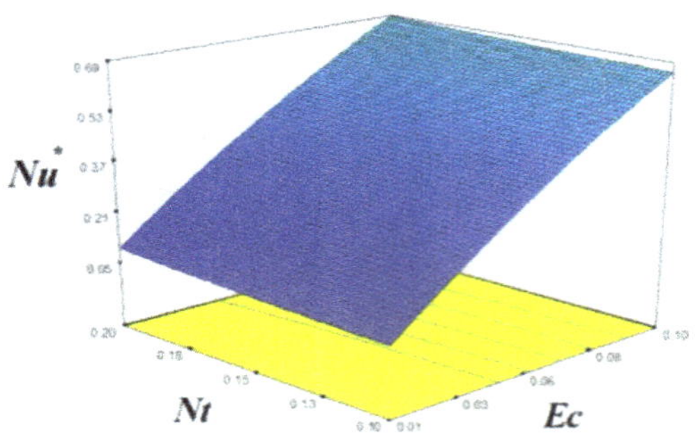

$S = 6, Ha = 6, Rd = 4.5, Nb = 0.15, Sc = 0.5$

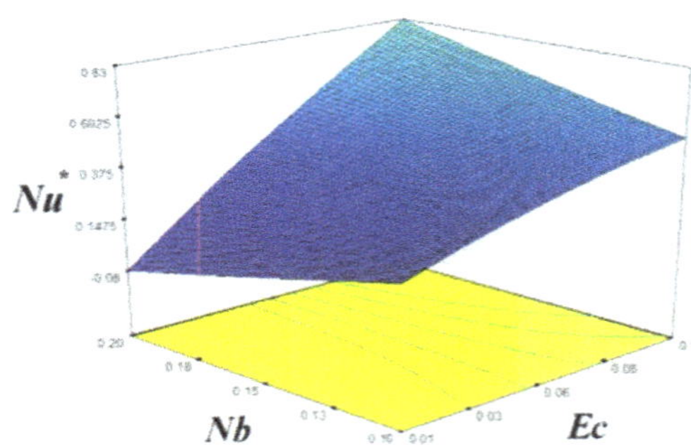

$S = 6, Ha = 6, Rd = 4.5, Nt = 0.15, Sc = 0.5$

profile gradient near the hot plate. Nusselt number enhances with rise of Hartmann number and squeeze number. Nusselt number augments slightly as Nt increase due to decrease in thermal boundary thickness near the upper plate.

Effects of the radiation parameter, squeeze number, Hartmann number, Nt and Nb on the concentration profile are displayed in Figure 68. Concentration profile enhances with augment of squeeze

Figure 68. Effects of the radiation parameter, squeeze number, Hartmann number Nt *and* Nb *on the concentration profile when* $Sc = 0.5, Ec = 0.1$ *and* $\Pr = 10$.

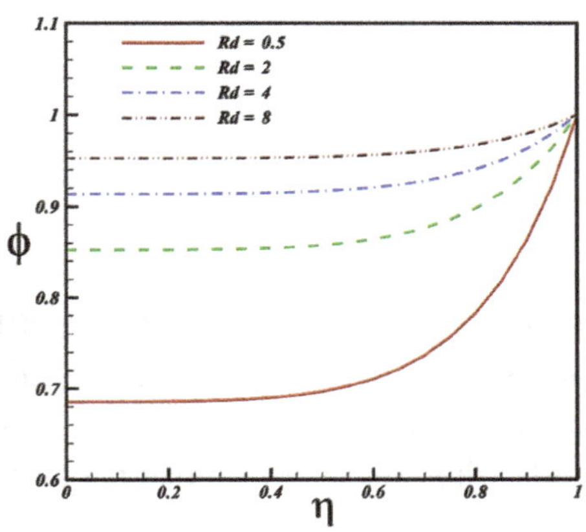

(a) $S = 0.5, Ha = 2, Nt = 0.1, Nb = 0.1$

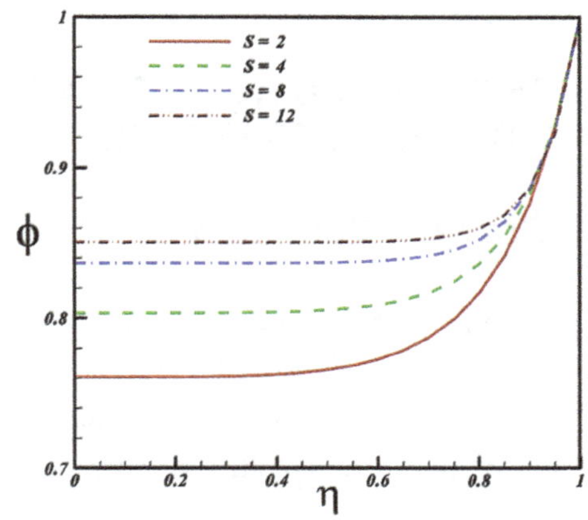

(b) $Ha = 2, Rd = 0.5, Nt = 0.1, Nb = 0.1$

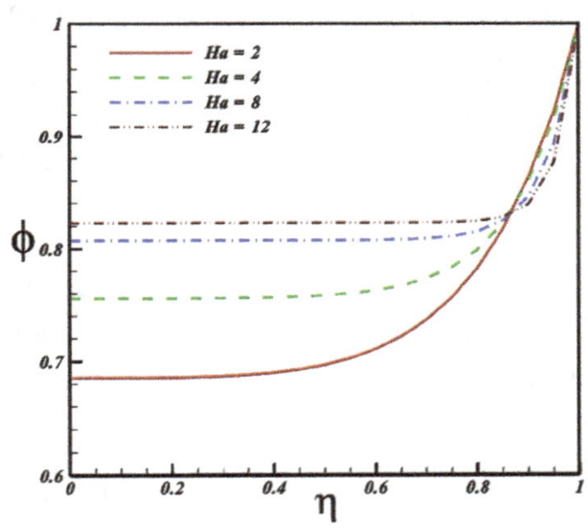

(c) $S = 0.5, Rd = 0.5, Nt = 0.1, Nb = 0.1$

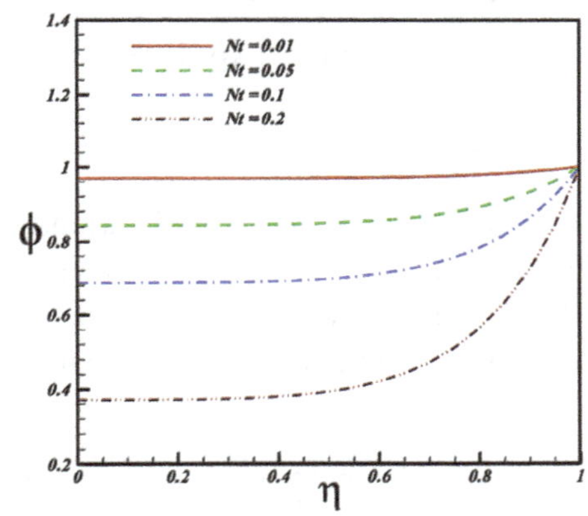

(d) $S = 0.5, Ha = 2, Rd = 0.5, Nb = 0.1$

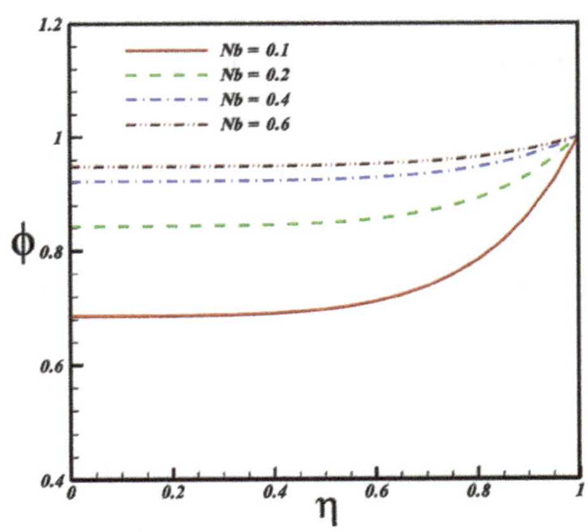

(e) $S = 0.5, Ha = 2, Rd = 0.5, Nt = 0.1$

number, radiation parameter and Nb but it reduces with increase of Nt. As Hartmann number increases, concentration enhances when $\eta < 0.82$ but opposite manners observed when $\eta > 0.82$.

11. FREE CONVECTION OF MAGNETIC NANOFLUID CONSIDERING MFD VISCOSITY EFFECT

11.1. Problem Definition

The geometry of this section is shown in Figure 69(a) (Sheikholeslami, Rashidi, Hayat, & Ganji, 2016). The heat source is centrally located on the bottom surface and its length L/3. The cooling is achieved by the two vertical walls. The heat source has constant heat flux q'' while the cooling walls have a constant temperature T_c; all the other surfaces are adiabatic. Also, it is also assumed that the uniform magnetic field ($\vec{B} = B_x \vec{e}_x + B_y \vec{e}_y$) of constant magnitude $B = \sqrt{B_x^2 + B_y^2}$ is applied, where $\vec{e}_x$ and $\vec{e}_y$ are unit vectors in the Cartesian coordinate system. The orientation of the magnetic field form an angle θ_M with horizontal axis such that $\theta_M = \cot^{-1}\left(B_x / B_y\right)$. The electric current J and the electromagnetic force F are defined by $J = \sigma\left(\vec{V} \times \vec{B}\right)$ and $F = \sigma\left(\vec{V} \times \vec{B}\right) \times \vec{B}$, respectively. Several authors investigated about nanofluid applications.

The flow is steady, two-dimensional, laminar and incompressible. The induced electric current and Joule heating are neglected. The magnetic Reynolds number is assumed to be small so that the induced magnetic field can be neglected compared to the applied magnetic field. Neglecting displacement cur-

Figure 69. (a) Geometry and the boundary conditions; (b) a sample triangular element and its corresponding control volume

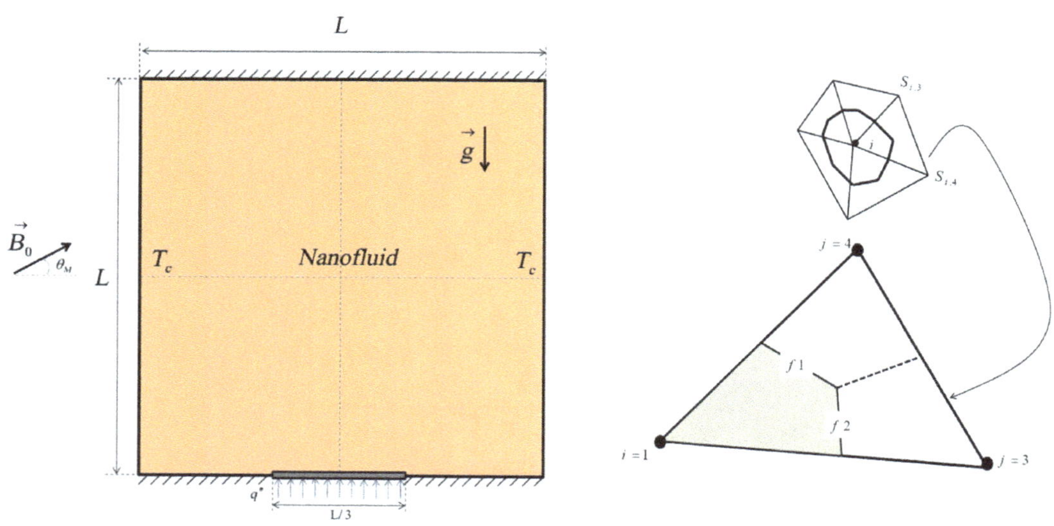

(a) (b)

rents, induced magnetic field, and using the Boussinesq approximation, the governing equations of heat transfer and fluid flow for nanofluid can be obtained as follows:

$$\frac{\partial u}{\partial x}+\frac{\partial v}{\partial y}=0 \tag{157}$$

$$\rho_{nf}\left(u\frac{\partial u}{\partial x}+v\frac{\partial u}{\partial y}\right)=-\frac{\partial P}{\partial x}+\eta\left(\frac{\partial^2 u}{\partial x^2}+\frac{\partial^2 u}{\partial y^2}\right) +\sigma_{nf}B_0^2\left(v\sin\theta_M\cos\theta_M-u\sin^2\theta_M\right) \tag{158}$$

$$\rho_{nf}\left(u\frac{\partial v}{\partial x}+v\frac{\partial v}{\partial y}\right)=-\frac{\partial P}{\partial y}+\eta\left(\frac{\partial^2 v}{\partial x^2}+\frac{\partial^2 v}{\partial y^2}\right)+\rho_{nf}\beta_{nf}g\left(T-T_c\right) +\sigma_{nf}B_0^2\left(u\sin\theta_M\cos\theta_M-v\cos^2\theta_M\right) \tag{159}$$

$$u\frac{\partial T}{\partial x}+v\frac{\partial T}{\partial y}=\alpha_{nf}\left(\frac{\partial^2 T}{\partial x^2}+\frac{\partial^2 T}{\partial y^2}\right) \tag{160}$$

where $\eta=(1+\vec{\delta}.\vec{B})\mu_{nf}$, the variation of MFD viscosity (δ) has been taken to be isotropic, $\delta_1=\delta_2=\delta_3=\delta$. The effective density (ρ_{nf}), the thermal expansion coefficient (β_{nf}), heat capacitance $\left(\rho C_p\right)_{nf}$ and electrical conductivity of nanofluid $\left(\sigma_{nf}\right)$ of the nanofluid are defined as:

$$\rho_{nf}=\rho_f(1-\phi)+\rho_s\phi$$

$$\beta_{nf}=\beta_f(1-\phi)+\beta_s\phi$$

$$\left(\rho C_p\right)_{nf}=\left(\rho C_p\right)_f(1-\phi)+\left(\rho C_p\right)_s\phi \tag{161}$$

$$\frac{\sigma_{nf}}{\sigma_f}=1+\frac{3\left(\sigma_s/\sigma_f-1\right)\phi}{\left(\sigma_s/\sigma_f+2\right)-\left(\sigma_s/\sigma_f-1\right)\phi} \tag{162}$$

Koo-Kleinstreuer-Li (KKL) is used to simulate thermal conductivity of nanofluid (Sheikholeslami, 2014):

$$k_{eff} = k_{static} + k_{Brownian} \tag{163}$$

$$\frac{k_{static}}{k_f} = 1 + \frac{3\left(k_p / k_f - 1\right)\phi}{\left(k_p / k_f + 2\right) - \left(k_p / k_f - 1\right)\phi} \tag{164}$$

$$k_{Brownian} = 5 \times 10^4 \phi \rho_f c_{p,f} \sqrt{\frac{\kappa_b T}{\rho_p d_p}} g'(T, \phi, d_p) \tag{165}$$

$$R_f + \frac{d_p}{k_p} = \frac{d_p}{k_{p,eff}}, R_f = 4 \times 10^{-8} km^2 / W \tag{166}$$

$$\begin{aligned} g'\left(T, \phi, d_p\right) = &\left(a_1 + a_2 \ln\left(d_p\right) + a_3 \ln\left(\phi\right) + a_4 \ln\left(\phi\right)\ln\left(d_p\right) + a_5 \ln\left(d_p\right)^2\right)\ln\left(T\right) \\ &+ \left(a_6 + a_7 \ln\left(d_p\right) + a_8 \ln\left(\phi\right) + a_9 \ln\left(\phi\right)\ln\left(d_p\right) + a_{10} \ln\left(d_p\right)^2\right) \end{aligned} \tag{167}$$

The effective viscosity due to micro mixing in suspensions, can be obtained as follows (Sheikholeslami, 2014):

$$\mu_{eff} = \mu_{static} + \mu_{Brownian} = \mu_{static} + \frac{k_{Brownian}}{k_f} \times \frac{\mu_f}{\Pr_f} \tag{168}$$

where $\mu_{static} = \dfrac{\mu_f}{\left(1 - \phi\right)^{2.5}}$ is viscosity of the nanofluid, as given originally by Brinkman.

The stream function and vorticity are defined as:

$$u = \frac{\partial \psi}{\partial y},\ v = -\frac{\partial \psi}{\partial x},\ \omega = \frac{\partial v}{\partial x} - \frac{\partial u}{\partial y} \tag{169}$$

The stream function satisfies the continuity Equation (157). The vorticity equation is obtained by eliminating the pressure between the two momentum equations, i.e. by taking y-derivative of Equation (158) and subtracting from it the x-derivative of Equation (159). This gives:

$$\begin{aligned}\frac{\partial \psi}{\partial y}\frac{\partial \omega}{\partial x}-\frac{\partial \psi}{\partial x}\frac{\partial \omega}{\partial y} &= \upsilon_{nf}\left(1+\delta B_0\left(\cos\theta_M+\sin\theta_M\right)\right)\left(\frac{\partial^2\omega}{\partial x^2}+\frac{\partial^2\omega}{\partial y^2}\right)\\ &+\beta_{nf}g\left(\frac{\partial T}{\partial x}\right)+\frac{\sigma_{nf}B_0^2}{\rho_{nf}}\begin{pmatrix}-\frac{\delta v}{\delta y}\sin\theta_M\cos\theta_M+\frac{\delta u}{\delta y}\sin^2\theta_M\\ +\frac{\delta u}{\delta x}\sin\theta_M\cos\theta_M-\frac{\delta v}{\delta x}\cos^2\theta_M\end{pmatrix}\end{aligned} \tag{170}$$

$$\frac{\partial \psi}{\partial y}\frac{\partial T}{\partial x}-\frac{\partial \psi}{\partial x}\frac{\partial T}{\partial y}=\alpha_{nf}\left(\frac{\partial^2 T}{\partial x^2}+\frac{\partial^2 T}{\partial y^2}\right) \tag{171}$$

$$\frac{\partial^2\psi}{\partial x^2}+\frac{\partial^2\psi}{\partial y^2}=-\omega \tag{172}$$

By introducing the following non-dimensional variables:

$$X=\frac{x}{L},Y=\frac{y}{L},\Omega=\frac{\omega L^2}{\alpha_f},\Psi=\frac{\psi}{\alpha_f},\Theta=\frac{T-T_c}{\left(q''L/k_f\right)},U=\frac{uL}{\alpha_f},V=\frac{vL}{\alpha_f} \tag{173}$$

Using the dimensionless parameters, the equations now become:

$$\begin{aligned}\frac{\partial \Psi}{\partial Y}\frac{\partial \Omega}{\partial X}-\frac{\partial \Psi}{\partial X}\frac{\partial \Omega}{\partial Y} &= \Pr_f\left(\frac{\mu_{nf}}{\mu_f}\frac{\rho_f}{\rho_{nf}}\frac{k_f}{k_{nf}}\frac{\left(\rho C_p\right)_{nf}}{\left(\rho C_p\right)_f}\right)\left(1+\delta^*\left(\cos\theta_M+\sin\theta_M\right)\right)\left(\frac{\partial^2\Omega}{\partial X^2}+\frac{\partial^2\Omega}{\partial Y^2}\right)\\ &+Ra_f\Pr_f\frac{\beta_{nf}}{\beta_f}\left(\frac{\partial\Theta}{\partial X}\right)+Ha^2\Pr_f\frac{\sigma_{nf}}{\sigma_f}\frac{\rho_f}{\rho_{nf}}\left(-\frac{\delta V}{\delta Y}\tan\theta_M+\frac{\delta U}{\delta Y}\tan^2\theta_M+\frac{\delta U}{\delta X}\tan\theta_M-\frac{\delta V}{\delta X}\right)\end{aligned} \tag{174}$$

$$\frac{\partial \Psi}{\partial Y}\frac{\partial \Theta}{\partial X}-\frac{\partial \Psi}{\partial X}\frac{\partial \Theta}{\partial Y}=\frac{k_{nf}}{k_f}\frac{\left(\rho C_p\right)_f}{\left(\rho C_p\right)_{nf}}\left(\frac{\partial^2\Theta}{\partial X^2}\right) \tag{175}$$

$$\frac{\partial^2 \Psi}{\partial X^2} + \frac{\partial^2 \Psi}{\partial Y^2} = -\Omega \tag{176}$$

where $Ra_f = g\beta_f L^4 q'' / \left(k_f \, \alpha_f \upsilon_f\right)$ is the Rayleigh number for the base fluid, $Ha = LB_x \sqrt{\sigma_f / \mu_f}$ is the Hartmann number and $\Pr_f = \upsilon_f / \alpha_f$ is the Prandtl number for the base fluid. Also $\delta^* = \delta B_0$ is viscosity parameter. The boundary conditions as shown in Figure 69 are:

$\partial\Theta/\partial n = -1.0$ on the heat source

$\Theta = 0.0$ on the left and right

$\partial\Theta/\partial n = 0.0$ on all the other adiabatic surfaces

$\Psi = 0.0$ on all solid boundaries (177)

The values of vorticity on the boundary of the enclosure can be obtained using the stream function formulation and the known velocity conditions during the iterative solution procedure. The local Nusselt number of the nanofluid along the heat source can be expressed as:

$$Nu_{local} = \left(\frac{k_{nf}}{k_f}\right) \frac{1}{\theta}\bigg|_{L/3<X<2L/3, Y=0} \tag{178}$$

The average Nusselt number is evaluated as:

$$Nu_{ave} = \frac{1}{L/3} \int_{L/3}^{2L/3} Nu_{loc}\left(X\right) dX \tag{179}$$

Nusselt ratio is defined as follows:

$$Nu_{ratio} = \frac{Nu_{ave}\big|_{\delta^*=1}}{Nu_{ave}\big|_{\delta^*=0}} \tag{180}$$

11.2. Effects of Active Parameters

In this section, magnetohydrodynamic natural convective heat transfer in a cooling system of electronic components is investigated. Magnetic field dependent (MFD) viscosity effect is taken in to account. Brownian motion effect is considered for simulating effective thermal conductivity and viscosity. Influence of active parameters such as: Rayleigh number ($Ra = 10^3, 10^4$ and 10^5), Hartmann number

(*Ha*=0,15,30 and 60) and viscosity parameter ($\delta^* = 0$, 1) on flow and heat transfer are examined when $\phi = 0.04, \Pr = 6.2$.

Effects viscosity parameter, Hartmann number and Rayleigh number on isotherms and streamlines are shown in Figures 70, 71 and 72. At low Rayleigh number, the streamlines take the enclosure geometry. By increasing Rayleigh number the prominent heat transfer mechanism is turn from conduction to convection. When the magnetic field is imposed on the enclosure, the velocity field suppressed owing to the

Figure 70. Effects of viscosity parameter ($\delta^* = 1$ (− − −) and $\delta^* = 0$ (—)) and Hartmann number on isotherms (Right) and streamlines (left) when $\phi = 0.04, Ra = 10^3$.

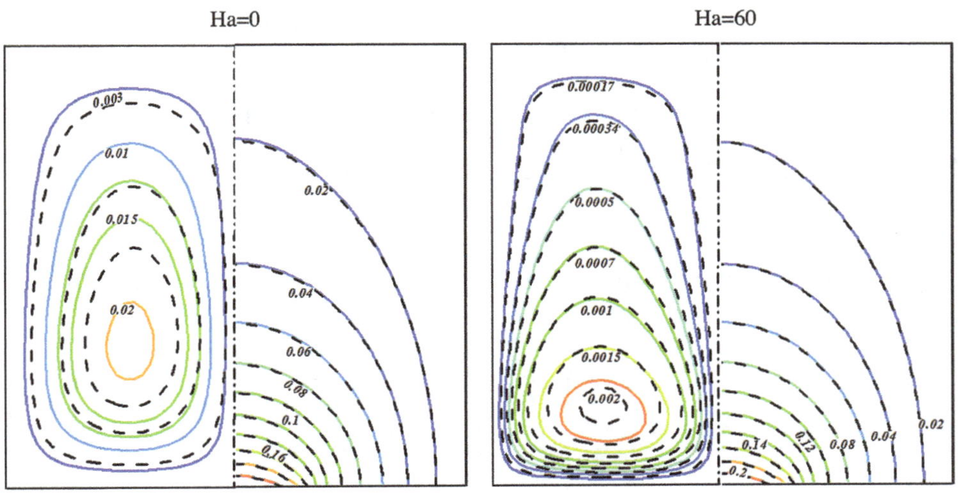

Figure 71. Effects of viscosity parameter ($\delta^* = 1$ (− − −) and $\delta^* = 0$ (—)) and Hartmann number on isotherms (Right) and streamlines (left) when $\phi = 0.04, Ra = 10^4$.

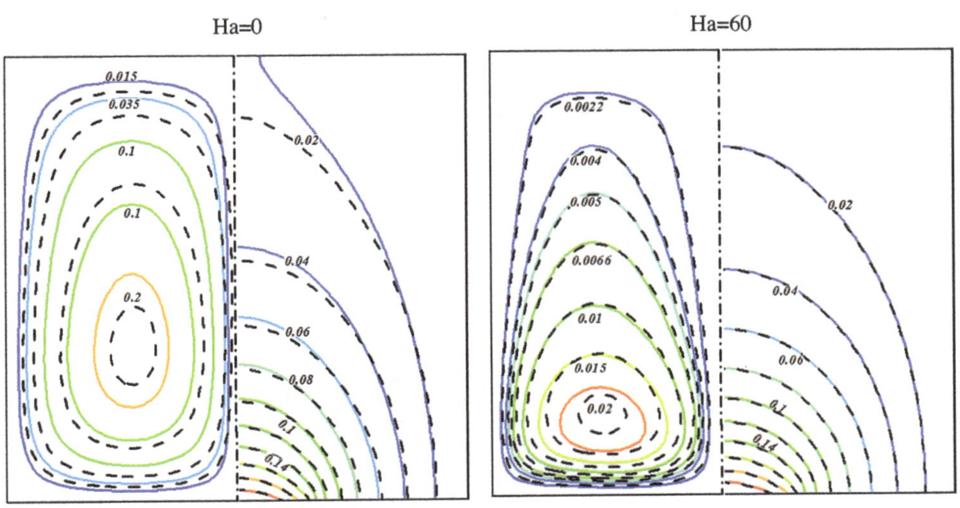

Figure 72. Effects of viscosity parameter ($\delta^* = 1$ (– – –) and $\delta^* = 0$ (—)) and Hartmann number on isotherms (Right) and streamlines (left) when $\phi = 0.04, Ra = 10^5$.

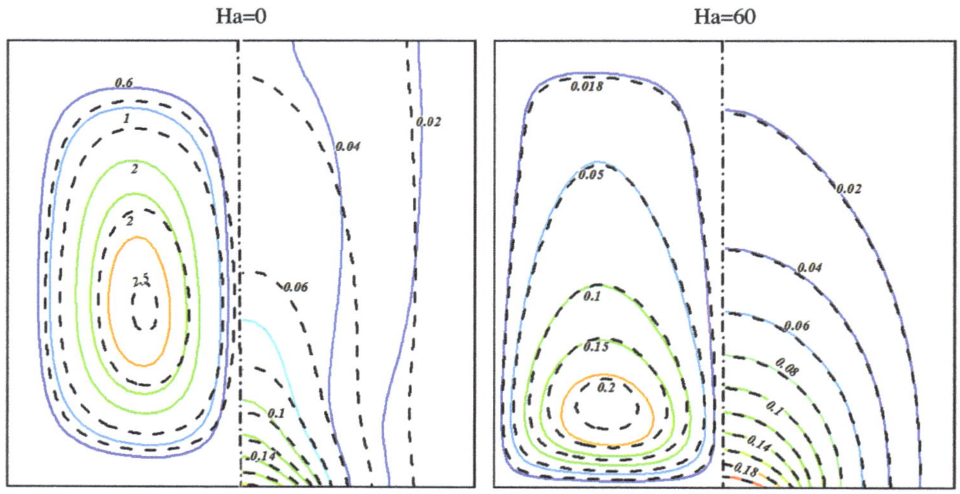

Figure 73. Effects of Hartmann number and Rayleigh number on Local Nusselt number when $\phi = 0.04, \delta^* = 1, \Pr = 6.2$.

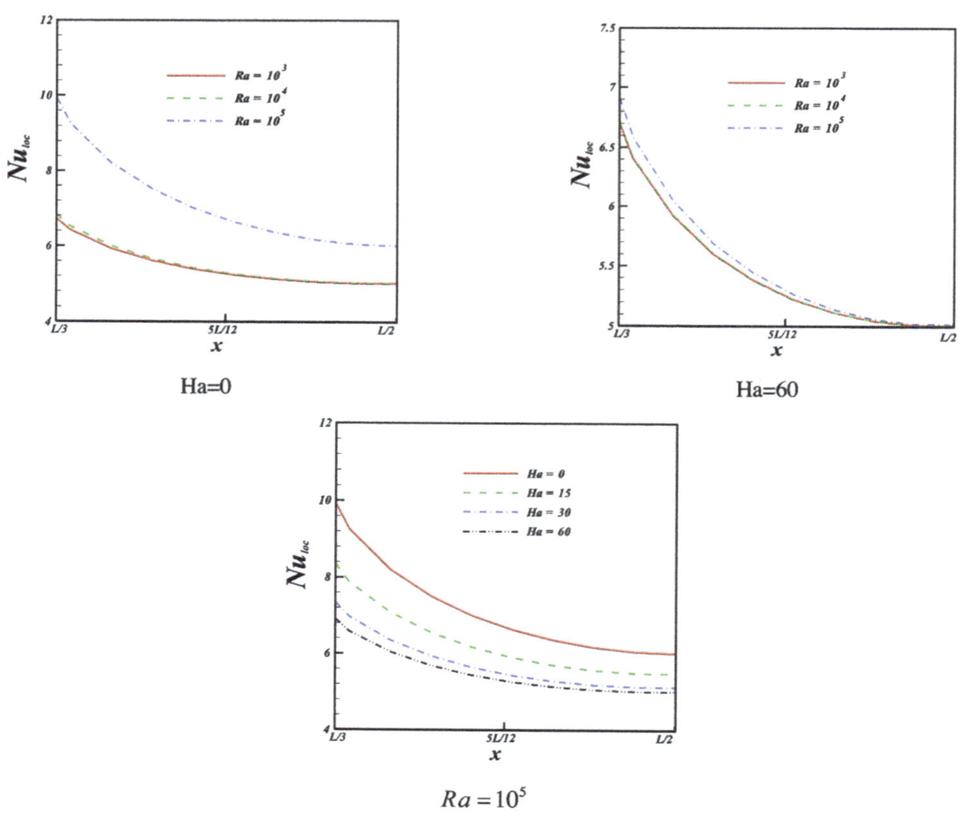

retarding effect of the Lorenz force. So intensity of convection weakens significantly. The decelerating effect of the magnetic field is observed from the maximum stream function value. The core vortex is shift downward vertically as the Hartmann number increases. Also imposing magnetic field leads to omit the thermal plume over the bottom wall. At high Hartmann number the conduction heat transfer mechanism is more pronounced. For this reason the isotherms are parallel to each other. Considering MFD viscosity

Figure 74. Effects of Hartmann number and Rayleigh number on average Nusselt number when $\phi = 0.04, \delta^* = 1, \Pr = 6.2$.

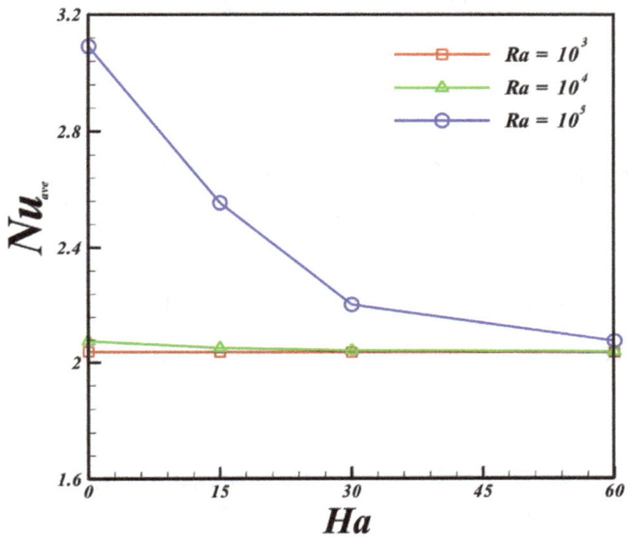

Figure 75. Effects of Hartmann number and Rayleigh number on Nusselt number ratio when $\phi = 0.04, \Pr = 6.2$.

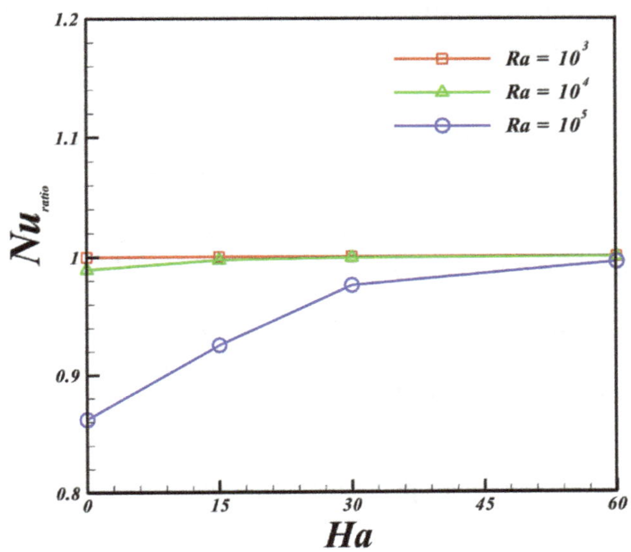

effect increases the thermal boundary layer thickness and in turn Nusselt number decreases with increase of viscosity parameter. Also it can be seen that flow circulation increases with increase of this parameter.

Effects of Hartmann number and Rayleigh number on local and average Nusselt number are shown in Figures 73 and 74. Increasing Rayleigh number leads the Nusselt number to enhance. Increasing Lorentz forces make thermal boundary layer thickness to increase and in turn Nusselt number decreases with enhance of Hartmann number. Figure 75 depicts the effect of Hartmann number and Rayleigh number on Nusselt number ratio. As viscosity parameter increases the rate of heat transfer decreases. This reduction is more pronounced for higher values of Rayleigh number and lower values of Hartmann number. It is interesting observation that increasing Lorentz forces increase the effect of MFD viscosity of Nusselt number. At low Rayleigh number, considering MFD viscosity has no significant effect on Nusselt number.

REFERENCES

Ashorynejad, H. R., Sheikholeslami, M., Pop, I., & Ganji, D. D. (2013). Nanofluid flow and heat transfer due to a stretching cylinder in the presence of magnetic field. *Heat and Mass Transfer*, *49*(3), 427–436. doi:10.100700231-012-1087-6

Bejan, A. (1982). *Entropy generation through heat and fluid flow*. New York: Wiley.

Calcagni, B., Marsili, F., & Paroncini, M. (2005). Natural convective heat transfer in square enclosures heated from below. *Applied Thermal Engineering*, *25*(16), 2522–2531. doi:10.1016/j.applthermaleng.2004.11.032

Li, Z., Sheikholeslami, M., Jafaryar, M., Shafee, A., & Chamkha, A. J. (2018). Investigation of nanofluid entropy generation in a heat exchanger with helical twisted tapes. *Journal of Molecular Liquids*, *266*, 797–805. doi:10.1016/j.molliq.2018.07.009

Sheikholeslami, M. (2014). KKL correlation for simulation of nanofluid flow and heat transfer in a permeable channel. *Physics Letters. [Part A]*, *378*(45), 3331–3339. doi:10.1016/j.physleta.2014.09.046

Sheikholeslami, M. (2018). Numerical simulation for external magnetic field influence on Fe3O4-water nanofluid forced convection. *Engineering Computations*, *35*(4), 1639–1654. doi:10.1108/EC-06-2017-0200

Sheikholeslami, M. (2018). Application of Darcy law for nanofluid flow in a porous cavity under the impact of Lorentz forces. *Journal of Molecular Liquids*, *266*, 495–503. doi:10.1016/j.molliq.2018.06.083

Sheikholeslami, M. (2018). Magnetic source impact on nanofluid heat transfer using CVFEM. *Neural Computing & Applications*, *30*(4), 1055–1064. doi:10.100700521-016-2740-7

Sheikholeslami, M., & Abelman, S. (2015). Two phase simulation of nanofluid flow and heat transfer in an annulus in the presence of an axial magnetic field. *IEEE Transactions on Nanotechnology*, *14*(3), 561–569. doi:10.1109/TNANO.2015.2416318

Sheikholeslami, M., Ashorynejad, H. R., & Rana, P. (2016). Lattice Boltzmann simulation of nanofluid heat transfer enhancement and entropy generation. *Journal of Molecular Liquids*, *214*, 86–95. doi:10.1016/j.molliq.2015.11.052

Sheikholeslami, M., Bandpy, M. G., & Ashorynejad, H. R. (2015). Lattice Boltzmann Method for simulation of magnetic field effect on hydrothermal behavior of nanofluid in a cubic cavity. *Physica A*, *432*, 58–70. doi:10.1016/j.physa.2015.03.009

Sheikholeslami, M., & Ellahi, R. (2015). Three dimensional mesoscopic simulation of magnetic field effect on natural convection of nanofluid. *International Journal of Heat and Mass Transfer*, *89*, 799–808. doi:10.1016/j.ijheatmasstransfer.2015.05.110

Sheikholeslami, M., Ellahi, R., Hassan, M., & Soleimani, S. (2014). A study of natural convection heat transfer in a nanofluid filled enclosure with elliptic inner cylinder. *International Journal of Numerical Methods for Heat & Fluid Flow*, *24*(8), 8. doi:10.1108/HFF-07-2013-0225

Sheikholeslami, M., & Ganji, D. D. (2014). Three dimensional heat and mass transfer in a rotating system using nanofluid. *Powder Technology*, *253*, 789–796. doi:10.1016/j.powtec.2013.12.042

Sheikholeslami, M., & Ganji, D. D. (2015). Entropy generation of nanofluid in presence of magnetic field using Lattice Boltzmann Method. *Physica A*, *417*, 273–286. doi:10.1016/j.physa.2014.09.053

Sheikholeslami, M., Ganji, D. D., & Rashidi, M. M. (2016). Magnetic field effect on unsteady nanofluid flow and heat transfer using Buongiorno model. *Journal of Magnetism and Magnetic Materials*, *416*, 164–173. doi:10.1016/j.jmmm.2016.05.026

Sheikholeslami, M., Gorji-Bandpy, M., Ganji, D. D., Rana, P., & Soleimani, S. (2014). Soheil Soleimani, Magnetohydrodynamic free convection of Al2O3-water nanofluid considering Thermophoresis and Brownian motion effects. *Computers & Fluids*, *94*, 147–160. doi:10.1016/j.compfluid.2014.01.036

Sheikholeslami, M., Jafaryar, M., Shafee, A., & Li, Z. (2018). Investigation of second law and hydrothermal behavior of nanofluid through a tube using passive methods. *Journal of Molecular Liquids*, *269*, 407–416. doi:10.1016/j.molliq.2018.08.019

Sheikholeslami, M., Rashidi, M. M., Hayat, T., & Ganji, D. D. (2016). Free convection of magnetic nanofluid considering MFD viscosity effect. *Journal of Molecular Liquids*, *218*, 393–399. doi:10.1016/j.molliq.2016.02.093

Sheikholeslami, M., Shehzad, S. A., Li, Z., & Shafee, A. (2018). Numerical modeling for Alumina nanofluid magnetohydrodynamic convective heat transfer in a permeable medium using Darcy law. *International Journal of Heat and Mass Transfer*, *127*, 614–622. doi:10.1016/j.ijheatmasstransfer.2018.07.013

Sheikholeslami, M., Zeeshan, A., & Majeed, A. (2018). Control volume based finite element simulation of magnetic nanofluid flow and heat transport in non-Darcy medium. *Journal of Molecular Liquids*, *268*, 354–364. doi:10.1016/j.molliq.2018.07.031

ADDITIONAL READING

Jafaryar, M., Sheikholeslami, M., Li, M., & Moradi, R. (2018). Nanofluid turbulent flow in a pipe under the effect of twisted tape with alternate axis. *Journal of Thermal Analysis and Calorimetry*. doi:10.100710973-018-7093-2

Li, Z., Shehzad, S. A., & Sheikholeslami, M. (2018). An application of CVFEM for nanofluid heat transfer intensification in a porous sinusoidal cavity considering thermal non-equilibrium model. *Computer Methods in Applied Mechanics and Engineering*, *339*, 663–680. doi:10.1016/j.cma.2018.05.015

Li, Z., Sheikholeslami, M., Chamkha, A. J., Raizah, Z. A., & Saleem, S. (2018). Control Volume Finite Element Method for nanofluid MHD natural convective flow inside a sinusoidal annulus under the impact of thermal radiation. *Computer Methods in Applied Mechanics and Engineering*, *338*, 618–633. doi:10.1016/j.cma.2018.04.023

Li, Z., Sheikholeslami, M., Jafaryar, M., Shafee, A., & Chamkha, A. J. (2018). Investigation of nanofluid entropy generation in a heat exchanger with helical twisted tapes. *Journal of Molecular Liquids*, *266*, 797–805. doi:10.1016/j.molliq.2018.07.009

Li, Z., Sheikholeslami, M., Samandari, M., & Shafee, A. (2018). Nanofluid unsteady heat transfer in a porous energy storage enclosure in existence of Lorentz forces. *International Journal of Heat and Mass Transfer*, *127*, 914–926. doi:10.1016/j.ijheatmasstransfer.2018.06.101

Sheikholeslami, M. (2017a). Lattice Boltzmann Method simulation of MHD non-Darcy nanofluid free convection. *Physica B, Condensed Matter*, *516*, 55–71. doi:10.1016/j.physb.2017.04.029

Sheikholeslami, M. (2017b). CuO-water nanofluid free convection in a porous cavity considering Darcy law. *The European Physical Journal Plus*, *132*(1), 55. doi:10.1140/epjp/i2017-11330-3

Sheikholeslami, M. (2017c). Influence of Lorentz forces on nanofluid flow in a porous cavity by means of Non- Darcy model. *Engineering Computations*, *34*(8), 2651–2667. doi:10.1108/EC-01-2017-0008

Sheikholeslami, M. (2017d). Magnetic field influence on nanofluid thermal radiation in a cavity with tilted elliptic inner cylinder. *Journal of Molecular Liquids*, *229*, 137–147. doi:10.1016/j.molliq.2016.12.024

Sheikholeslami, M. (2017e). Numerical simulation of magnetic nanofluid natural convection in porous media. *Physics Letters. [Part A]*, *381*(5), 494–503. doi:10.1016/j.physleta.2016.11.042

Sheikholeslami, M. (2017f). Influence of Lorentz forces on nanofluid flow in a porous cylinder considering Darcy model. *Journal of Molecular Liquids*, *225*, 903–912. doi:10.1016/j.molliq.2016.11.022

Sheikholeslami, M. (2017g). Influence of Coulomb forces on Fe3O4-H2O nanofluid thermal improvement. *International Journal of Hydrogen Energy*, *42*(2), 821–829. doi:10.1016/j.ijhydene.2016.09.185

Sheikholeslami, M. (2017h). Numerical investigation of MHD nanofluid free convective heat transfer in a porous tilted enclosure. *Engineering Computations*, *34*(6), 1939–1955. doi:10.1108/EC-08-2016-0293

Sheikholeslami, M. (2017i). Magnetic field influence on CuO -H2O nanofluid convective flow in a permeable cavity considering various shapes for nanoparticles. *International Journal of Hydrogen Energy*, *42*(31), 19611–19621. doi:10.1016/j.ijhydene.2017.06.121

Sheikholeslami, M. (2018a). Magnetic source impact on nanofluid heat transfer using CVFEM. *Neural Computing & Applications*, *30*(4), 1055–1064. doi:10.100700521-016-2740-7

Sheikholeslami, M. (2018b). Numerical simulation for external magnetic field influence on Fe3O4-water nanofluid forced convection. *Engineering Computations*, *35*(4), 1639–1654. doi:10.1108/EC-06-2017-0200

Sheikholeslami, M. (2018c). Finite element method for PCM solidification in existence of CuO nanoparticles. *Journal of Molecular Liquids*, *265*, 347–355. doi:10.1016/j.molliq.2018.05.132

Sheikholeslami, M. (2018d). Solidification of NEPCM under the effect of magnetic field in a porous thermal energy storage enclosure using CuO nanoparticles. *Journal of Molecular Liquids*, *263*, 303–315. doi:10.1016/j.molliq.2018.04.144

Sheikholeslami, M. (2018e). Influence of magnetic field on Al2O3-H2O nanofluid forced convection heat transfer in a porous lid driven cavity with hot sphere obstacle by means of LBM. *Journal of Molecular Liquids*, *263*, 472–488. doi:10.1016/j.molliq.2018.04.111

Sheikholeslami, M. (2018f). Numerical simulation for solidification in a LHTESS by means of Nano-enhanced PCM. *Journal of the Taiwan Institute of Chemical Engineers*, *86*, 25–41. doi:10.1016/j.jtice.2018.03.013

Sheikholeslami, M. (2018g). Numerical modeling of Nano enhanced PCM solidification in an enclosure with metallic fin. *Journal of Molecular Liquids*, *259*, 424–438. doi:10.1016/j.molliq.2018.03.006

Sheikholeslami, M. (2018h). Numerical investigation of nanofluid free convection under the influence of electric field in a porous enclosure. *Journal of Molecular Liquids*, *249*, 1212–1221. doi:10.1016/j.molliq.2017.11.141

Sheikholeslami, M. (2018i). CuO-water nanofluid flow due to magnetic field inside a porous media considering Brownian motion. *Journal of Molecular Liquids*, *249*, 921–929. doi:10.1016/j.molliq.2017.11.118

Sheikholeslami, M. (2018j). Application of Darcy law for nanofluid flow in a porous cavity under the impact of Lorentz forces. *Journal of Molecular Liquids*, *266*, 495–503. doi:10.1016/j.molliq.2018.06.083

Sheikholeslami, M., Barzegar Gerdroodbary, M., Valiallah Mousavi, S., Ganji, D. D., & Moradi, R. (2018). Heat transfer enhancement of ferrofluid inside an 90o elbow channel by non-uniform magnetic field. *Journal of Magnetism and Magnetic Materials*, *460*, 302–311. doi:10.1016/j.jmmm.2018.03.070

Sheikholeslami, M., & Bhatti, M. M. (2017). Active method for nanofluid heat transfer enhancement by means of EHD. *International Journal of Heat and Mass Transfer*, *109*, 115–122. doi:10.1016/j.ijheatmasstransfer.2017.01.115

Sheikholeslami, M., Darzi, M., & Li, Z. (2018). Experimental investigation for entropy generation and exergy loss of nano-refrigerant condensation process. *International Journal of Heat and Mass Transfer*, *125*, 1087–1095. doi:10.1016/j.ijheatmasstransfer.2018.04.155

Sheikholeslami, M., Darzi, M., & Sadoughi, M. K. (2018). Heat transfer improvement and Pressure Drop during condensation of refrigerant-based Nanofluid; An Experimental Procedure. *International Journal of Heat and Mass Transfer*, *122*, 643–650. doi:10.1016/j.ijheatmasstransfer.2018.02.015

Sheikholeslami, M., & Ghasemi, A. (2018). Solidification heat transfer of nanofluid in existence of thermal radiation by means of FEM. *International Journal of Heat and Mass Transfer*, *123*, 418–431. doi:10.1016/j.ijheatmasstransfer.2018.02.095

Sheikholeslami, M., Ghasemi, A., Li, Z., Shafee, A., & Saleem, S. (2018). Influence of CuO nanoparticles on heat transfer behavior of PCM in solidification process considering radiative source term. *International Journal of Heat and Mass Transfer*, *126*, 1252–1264. doi:10.1016/j.ijheatmasstransfer.2018.05.116

Sheikholeslami, M., Hayat, T., & Alsaedi, A. (2018). Numerical simulation for forced convection flow of MHD CuO-H2O nanofluid inside a cavity by means of LBM. *Journal of Molecular Liquids*, *249*, 941–948. doi:10.1016/j.molliq.2017.10.099

Sheikholeslami, M., Hayat, T., Muhammad, T., & Alsaedi, A. (2018). MHD forced convection flow of nanofluid in a porous cavity with hot elliptic obstacle by means of Lattice Boltzmann method. *International Journal of Mechanical Sciences*, *135*, 532–540. doi:10.1016/j.ijmecsci.2017.12.005

Sheikholeslami, M., Jafaryar, M., Ganji, D. D., & Li, Z. (2018). Exergy loss analysis for nanofluid forced convection heat transfer in a pipe with modified turbulators. *Journal of Molecular Liquids*, *262*, 104–110. doi:10.1016/j.molliq.2018.04.077

Sheikholeslami, M., Jafaryar, M., & Li, Z. (2018a). Second law analysis for nanofluid turbulent flow inside a circular duct in presence of twisted tape turbulators. *Journal of Molecular Liquids*, *263*, 489–500. doi:10.1016/j.molliq.2018.04.147

Sheikholeslami, M., Jafaryar, M., & Li, Z. (2018b). Nanofluid turbulent convective flow in a circular duct with helical turbulators considering CuO nanoparticles. *International Journal of Heat and Mass Transfer*, *124*, 980–989. doi:10.1016/j.ijheatmasstransfer.2018.04.022

Sheikholeslami, M., Jafaryar, M., Saleem, S., Li, Z., Shafee, A., & Jiang, Y. (2018). Nanofluid heat transfer augmentation and exergy loss inside a pipe equipped with innovative turbulators. *International Journal of Heat and Mass Transfer*, *126*, 156–163. doi:10.1016/j.ijheatmasstransfer.2018.05.128

Sheikholeslami, M., Jafaryar, M., Shafee, A., & Li, Z. (2018). Investigation of second law and hydrothermal behavior of nanofluid through a tube using passive methods. *Journal of Molecular Liquids*, *269*, 407–416. doi:10.1016/j.molliq.2018.08.019

Sheikholeslami, M., Li, Z., & Shafee, A. (2018a). Lorentz forces effect on NEPCM heat transfer during solidification in a porous energy storage system. *International Journal of Heat and Mass Transfer*, *127*, 665–674. doi:10.1016/j.ijheatmasstransfer.2018.06.087

Sheikholeslami, M., Li, Z., & Shamlooei, M. (2018). Nanofluid MHD natural convection through a porous complex shaped cavity considering thermal radiation. *Physics Letters. [Part A]*, *382*(24), 1615–1632. doi:10.1016/j.physleta.2018.04.006

Sheikholeslami, M., & Rokni, H. B. (2017). Simulation of nanofluid heat transfer in presence of magnetic field: A review. *International Journal of Heat and Mass Transfer*, *115*, 1203–1233. doi:10.1016/j.ijheatmasstransfer.2017.08.108

Sheikholeslami, M., & Rokni, H. B. (2018a). CVFEM for effect of Lorentz forces on nanofluid flow in a porous complex shaped enclosure by means of Non-equilibrium model. *Journal of Molecular Liquids*, *254*, 446–462. doi:10.1016/j.molliq.2018.01.130

Sheikholeslami, M., Rokni, H.B. (2018b). Magnetic nanofluid flow and convective heat transfer in a porous cavity considering Brownian motion effects. *Physics of Fluids*, *30*(1), doi:10.1063/1.5012517

Sheikholeslami, M., & Sadoughi, M. K. (2017). Mesoscopic method for MHD nanofluid flow inside a porous cavity considering various shapes of nanoparticles. *International Journal of Heat and Mass Transfer*, *113*, 106–114. doi:10.1016/j.ijheatmasstransfer.2017.05.054

Sheikholeslami, M., & Sadoughi, M. K. (2018). Simulation of CuO- water nanofluid heat transfer enhancement in presence of melting surface. *International Journal of Heat and Mass Transfer*, *116*, 909–919. doi:10.1016/j.ijheatmasstransfer.2017.09.086

Sheikholeslami, M., & Seyednezhad, M. (2018). Simulation of nanofluid flow and natural convection in a porous media under the influence of electric field using CVFEM. *International Journal of Heat and Mass Transfer*, *120*, 772–781. doi:10.1016/j.ijheatmasstransfer.2017.12.087

Sheikholeslami, M., Shafee, A., Ramzan, M., & Li, Z. (2018). Investigation of Lorentz forces and radiation impacts on nanofluid treatment in a porous semi annulus via Darcy law. *Journal of Molecular Liquids*, *272*, 8–14. doi:10.1016/j.molliq.2018.09.016

Sheikholeslami, M., Shamlooei, M., & Moradi, R. (2018). Numerical simulation for heat transfer intensification of nanofluid in a porous curved enclosure considering shape effect of Fe3O4 nanoparticles. *Chemical Engineering & Processing: Process Intensification*, *124*, 71–82. doi:10.1016/j.cep.2017.12.005

Sheikholeslami, M., & Shehzad, S. A. (2017a). Thermal radiation of ferrofluid in existence of Lorentz forces considering variable viscosity. *International Journal of Heat and Mass Transfer*, *109*, 82–92. doi:10.1016/j.ijheatmasstransfer.2017.01.096

Sheikholeslami, M., & Shehzad, S. A. (2017b). Magnetohydrodynamic nanofluid convective flow in a porous enclosure by means of LBM. *International Journal of Heat and Mass Transfer*, *113*, 796–805. doi:10.1016/j.ijheatmasstransfer.2017.05.130

Sheikholeslami, M., & Shehzad, S. A. (2018a). Numerical analysis of Fe3O4 –H2O nanofluid flow in permeable media under the effect of external magnetic source. *International Journal of Heat and Mass Transfer*, *118*, 182–192. doi:10.1016/j.ijheatmasstransfer.2017.10.113

Sheikholeslami, M., & Shehzad, S. A. (2018b). CVFEM simulation for nanofluid migration in a porous medium using Darcy model. *International Journal of Heat and Mass Transfer*, *122*, 1264–1271. doi:10.1016/j.ijheatmasstransfer.2018.02.080

Sheikholeslami, M., & Shehzad, S. A. (2018c). Simulation of water based nanofluid convective flow inside a porous enclosure via Non-equilibrium model. *International Journal of Heat and Mass Transfer*, *120*, 1200–1212. doi:10.1016/j.ijheatmasstransfer.2017.12.132

Sheikholeslami, M., & Shehzad, S. A. (2018d). Non-Darcy free convection of Fe3O4-water nanoliquid in a complex shaped enclosure under impact of uniform Lorentz force. *The Chinese Journal of Physiology*, *56*(1), 270–281. doi:10.1016/j.cjph.2017.12.017

Sheikholeslami, M., Shehzad, S. A., Abbasi, F. M., & Li, Z. (2018). Nanofluid flow and forced convection heat transfer due to Lorentz forces in a porous lid driven cubic enclosure with hot obstacle. *Computer Methods in Applied Mechanics and Engineering*, *338*, 491–505. doi:10.1016/j.cma.2018.04.020

Sheikholeslami, M., Shehzad, S. A., & Li, Z. (2018a). Nanofluid heat transfer intensification in a permeable channel due to magnetic field using Lattice Boltzmann method. *Physica B, Condensed Matter*, *542*, 51–58. doi:10.1016/j.physb.2018.03.036

Sheikholeslami, M., Shehzad, S. A., & Li, Z. (2018b). Water based nanofluid free convection heat transfer in a three dimensional porous cavity with hot sphere obstacle in existence of Lorenz forces. *International Journal of Heat and Mass Transfer*, *125*, 375–386. doi:10.1016/j.ijheatmasstransfer.2018.04.076

Sheikholeslami, M., Shehzad, S. A., Li, Z., & Shafee, A. (2018). Numerical modeling for Alumina nanofluid magnetohydrodynamic convective heat transfer in a permeable medium using Darcy law. *International Journal of Heat and Mass Transfer*, *127*, 614–622. doi:10.1016/j.ijheatmasstransfer.2018.07.013

Sheikholeslami, M., & Vajravelu, K. (2017). Forced convection heat transfer in Fe3O4- ethylene glycol nanofluid under the influence of Coulomb force. *Journal of Molecular Liquids*, *233*, 203–210. doi:10.1016/j.molliq.2017.03.026

Sheikholeslami, M., & Zeeshan, A. (2017). Analysis of flow and heat transfer in water based nanofluid due to magnetic field in a porous enclosure with constant heat flux using CVFEM. *Computer Methods in Applied Mechanics and Engineering*, *320*, 68–81. doi:10.1016/j.cma.2017.03.024

Sheikholeslami, M., Zeeshan, A., & Majeed, A. (2018). Control volume based finite element simulation of magnetic nanofluid flow and heat transport in non-Darcy medium. *Journal of Molecular Liquids*, *268*, 354–364. doi:10.1016/j.molliq.2018.07.031

Chapter 3
Space-Dependent Lorenz Forces Influence on Nanofluid Behavior

ABSTRACT

Magnetic nanofluid (Ferrofluid) is a magnetic colloidal suspension consisting of base liquid and magnetic nanoparticles with a size range of 5–15 nm in diameter coated with a surfactant layer. The effect of magnetic field on fluids is worth investigating due to its numerous applications in a wide range of fields. The study of interaction of the magnetic field or the electromagnetic field with fluids have been documented (e.g., among nuclear fusion, chemical engineering, medicine, and transformer cooling). The goal of nanofluid is to achieve the highest possible thermal properties at the smallest possible concentrations by uniform dispersion and stable suspension of nanoparticles in host fluids. In this chapter, the influence of external magnetic field on ferrofluid flow and heat transfer is investigated. Both effects of Ferrohydrodynamic (FHD) and Magnetohydrodynamic (MHD) have been taken in to account. So, the effects of Lorentz and Kelvin forces on hydrothermal behavior are examined.

1. EFFECT OF SPACE DEPENDENT MAGNETIC FIELD ON FREE CONVECTION OF FE_3O_4-WATER NANOFLUID

1.1. Problem Definition

The physical model along with the important geometrical parameters and the mesh of the enclosure used in the present CVFEM program are shown in Figure 1 (Sheikholeslami, 2014). The inner circular wall is maintained at constant hot temperature T_h, the two horizontal walls are thermally isolated and other walls are maintained at constant cold temperature T_c ($T_h > T_c$). To reach the shape of inner circular and outer rectangular boundary which consists of the right and top walls, a supper elliptic function can be used as follows;

DOI: 10.4018/978-1-5225-7595-5.ch003

Figure 1. (a) Geometry and the boundary conditions; (b) the mesh of enclosure considered in this work.

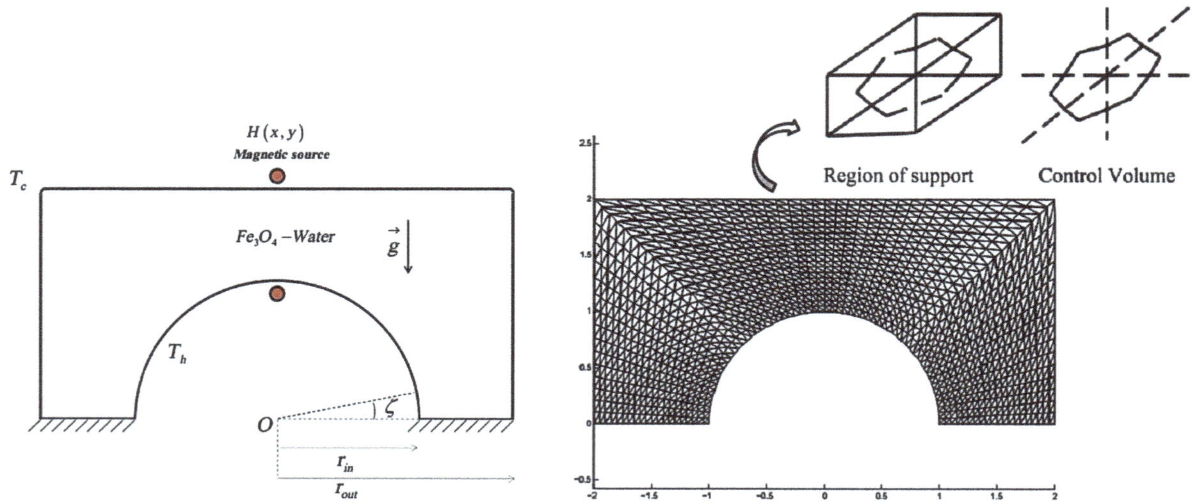

(a) (b)

$$\left(\frac{X}{a}\right)^{2n} + \left(\frac{Y}{b}\right)^{2n} = 1 \tag{1}$$

When $a = b$ and $n = 1$ the geometry becomes a circle. As n increases from 1 the geometry would approach a rectangle for $a \neq b$ and square for $a = b$. In this study we assumed that $r_{in} / r_{out} = 0.6$.

For the expression of the magnetic field strength it can be considered that two magnetic sources at points $\left(\overline{a}_1, \overline{b}_1\right)$ and $\left(\overline{a}_2, \overline{b}_2\right)$. The components of the magnetic field intensity ($\overline{H}_x, \overline{H}_y$) and the magnetic field strength ($\overline{H}$) can be considered as (Sheikholeslami and Rashidi, 2015):

$$\overline{H}_x = \frac{\gamma}{2\pi} \frac{1}{\left(x - \overline{a}_1\right)^2 + \left(y - \overline{b}_1\right)^2} \left(y - \overline{b}_1\right) - \frac{\gamma}{2\pi} \frac{1}{\left(x - \overline{a}_2\right)^2 + \left(y - \overline{b}_2\right)^2} \left(y - \overline{b}_2\right) \tag{2}$$

$$\overline{H}_y = -\frac{\gamma}{2\pi} \frac{1}{\left(x - \overline{a}_1\right)^2 + \left(y - \overline{b}_1\right)^2} \left(x - \overline{a}_1\right) - \frac{\gamma}{2\pi} \frac{1}{\left(x - \overline{a}_2\right)^2 + \left(y - \overline{b}_2\right)^2} \left(x - \overline{a}_2\right) \tag{3}$$

$$\overline{H} = \sqrt{\overline{H}_x^2 + \overline{H}_y^2} \tag{4}$$

where γ the magnetic field strength at the source (of the wire). The contours of the magnetic field strength are shown in Figure 2. In this study magnetic source is located at

$$\left(\overline{a_1} = 1.05cols, \overline{b_1} = 0.5rows\right)$$

and

$$\left(\overline{a_2} = -0.05cols, \overline{b_2} = 0.5rows\right).$$

The flow is two-dimensional, laminar and incompressible. The magnetic Reynolds number is assumed to be small so that the induced magnetic field can be neglected compared to the applied magnetic field. Using the Boussinesq approximation, the governing equations of heat transfer and fluid flow for nanofluid can be obtained as follows:

$$\frac{\partial u}{\partial x} + \frac{\partial v}{\partial y} = 0 \tag{5}$$

Figure 2. Contours of the (a) magnetic field strength H; (b) magnetic field intensity component in x direction Hx; (c) magnetic field intensity component in y direction Hy.

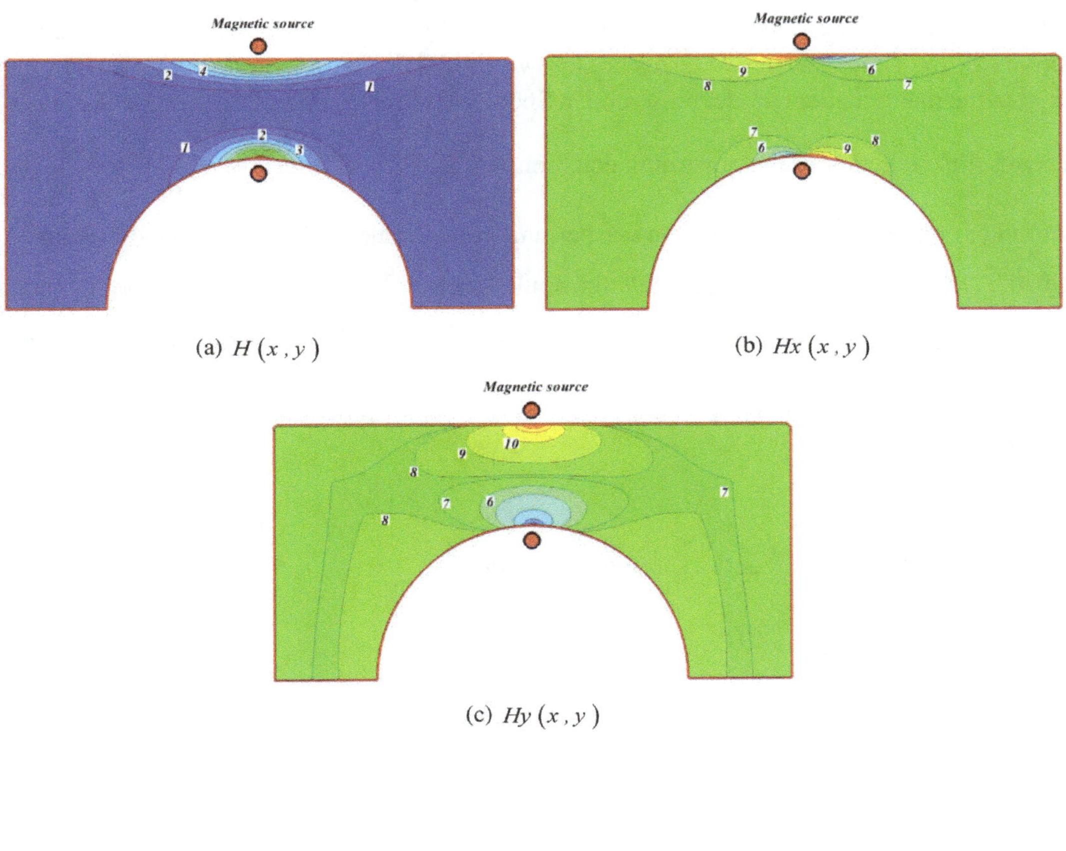

(a) $H(x, y)$

(b) $Hx(x, y)$

(c) $Hy(x, y)$

$$\rho_{nf}\left(u\frac{\partial u}{\partial x}+v\frac{\partial u}{\partial y}\right)=-\frac{\partial P}{\partial x}+\mu_{nf}\left(\frac{\partial^2 u}{\partial x^2}+\frac{\partial^2 u}{\partial y^2}\right)+\mu_0 M\frac{\partial \overline{H}}{\partial x}-\sigma_{nf}B_y^2 u+\sigma_{nf}B_x B_y v \tag{6}$$

$$\rho_{nf}\left(u\frac{\partial v}{\partial x}+v\frac{\partial v}{\partial y}\right)=-\frac{\partial P}{\partial y}+\mu_{nf}\left(\frac{\partial^2 v}{\partial x^2}+\frac{\partial^2 v}{\partial y^2}\right)+\mu_0 M\frac{\partial \overline{H}}{\partial y}-\sigma_{nf}B_x^2 v+\sigma_{nf}B_x B_y u$$
$$+\rho_{nf}\beta_{nf}g\left(T-T_c\right) \tag{7}$$

$$\left(\rho C_p\right)_{nf}\left(u\frac{\partial T}{\partial x}+v\frac{\partial T}{\partial y}\right)=k_{nf}\left(\frac{\partial^2 T}{\partial x^2}+\frac{\partial^2 T}{\partial y^2}\right)+\sigma_{nf}\left(uB_y-vB_x\right)^2$$
$$-\mu_0 T\frac{\partial M}{\partial T}\left(u\frac{\partial \overline{H}}{\partial x}+v\frac{\partial \overline{H}}{\partial y}\right)+\mu_{nf}\left\{2\left(\frac{\partial u}{\partial x}\right)^2+2\left(\frac{\partial v}{\partial y}\right)^2+\left(\frac{\partial u}{\partial y}+\frac{\partial v}{\partial x}\right)^2\right\} \tag{8}$$

The terms $\mu_0 M\frac{\partial \overline{H}}{\partial x}$ and $\mu_0 M\frac{\partial \overline{H}}{\partial y}$ in (6) and (7), respectively, represent the components of magnetic force, per unit volume, and depend on the existence of the magnetic gradient on the corresponding *x* and *y* directions. These two terms are well known from FHD which is the so-called the Kelvin force. The term s $-\sigma_{nf}B_y^2 u+\sigma_{nf}B_x B_y v$ and $-\sigma_{nf}B_x^2 v+\sigma_{nf}B_x B_y u$ appearing in (6) and (7), respectively, represent the Lorentz force per unit volume towards the *x* and *y* directions and arise due to the electrical conductivity of the fluid. These two terms are known in MHD. The principles of MHD and FHD are combined in the mathematical model presented in Sheikholeslami and Ganji (2014) and the above mentioned terms arise together in the governing equations (6) and (7). The term $\mu_0 T\frac{\partial M}{\partial T}\left(u\frac{\partial \overline{H}}{\partial x}+v\frac{\partial \overline{H}}{\partial y}\right)$ in *Equation* (8) represents the thermal power per unit volume due to the magneto caloric effect. Also the term $\sigma_{nf}\left(uB_y-vB_x\right)^2$ in (8) represents the Joule heating. For the variation of the magnetization M, with the magnetic field intensity $\overline{H}$ and temperature T, the following relation is used Loukopoulos and Tzirtzilakis (2004):

$$M=K'\overline{H}\left(T_c'-T\right) \tag{9}$$

where K' is a constant and T_c' is the Curie temperature.

In the above equations, μ_0 is the magnetic permeability of vacuum $\left(4\pi \times 10^{-7} Tm / A\right)$, $\overline{H}$ is the magnetic field strength, $\overline{B}$ is the magnetic induction $\left(\overline{B} = \mu_0 \overline{H}\right)$ and the bar above the quantities denotes that they are dimensional. The effective density (ρ_{nf}) and heat capacitance $\left(\rho C_p\right)_{nf}$ of the nanofluid are defined as:

$$\rho_{nf} = \rho_f(1-\phi) + \rho_s\phi \tag{10}$$

$$\left(\rho C_p\right)_{nf} = \left(\rho C_p\right)_f (1-\phi) + \left(\rho C_p\right)_s \phi \tag{11}$$

where ϕ is the solid volume fraction of nanoparticles. Thermal diffusivity of the nanofluid is

$$\alpha_{nf} = \frac{k_{nf}}{\left(\rho C_p\right)_{nf}} \tag{12}$$

and the thermal expansion coefficient of the nanofluid can be determined as

$$\beta_{nf} = \beta_f(1-\phi) + \beta_s\phi \tag{13}$$

The dynamic viscosity of the nanofluids is given is

$$\mu_{nf} = \frac{\mu_f}{(1-\phi)^{2.5}} \tag{14}$$

The effective thermal conductivity of the nanofluid can be approximated by the Maxwell–Garnetts (MG) model as (Sheikholeslami, 2014):

$$\frac{k_{nf}}{k_f} = \frac{k_s + 2k_f - 2\phi(k_f - k_s)}{k_s + 2k_f + \phi(k_f - k_s)} \tag{15}$$

and the effective electrical conductivity of nanofluid was presented as below:

$$\frac{\sigma_{nf}}{\sigma_f} = 1 + \frac{3\left(\frac{\sigma_s}{\sigma_f} - 1\right)\phi}{\left(\frac{\sigma_s}{\sigma_f} + 2\right) - \left(\frac{\sigma_s}{\sigma_f} - 1\right)\phi} \tag{16}$$

By introducing the following non-dimensional variables:

$$\begin{aligned} & a = \frac{\bar{a}}{L}, b = \frac{\bar{b}}{L}, X = \frac{x}{L}, Y = \frac{y}{L}, \Theta = \frac{T - T_c}{T_h - T_c}, U = \frac{uL}{\alpha_f}, V = \frac{vL}{\alpha_f}, \\ & H = \frac{\overline{H}}{\overline{H_0}}, H_x = \frac{\overline{H_x}}{\overline{H_0}}, H_y = \frac{\overline{H_y}}{\overline{H_0}}, P = \frac{p}{\rho_f \left(\alpha_f / \mathrm{L}\right)^2} \end{aligned} \tag{17}$$

where in *Equation* (17) $\overline{H_0} = \overline{H}\left(\bar{a}, 0\right) = \frac{\gamma}{2\pi\left|b\right|}$ and $L = r_{out} - r_{in}$. Using the dimensionless parameters, the equations now become:

$$\frac{\partial U}{\partial X} + \frac{\partial V}{\partial Y} = 0 \tag{18}$$

$$\begin{aligned} & U\frac{\partial U}{\partial X} + V\frac{\partial U}{\partial Y} = -\frac{\partial P}{\partial X} + \Pr\left[\frac{\mu_{nf} / \mu_f}{\rho_{nf} / \rho_f}\right]\left(\frac{\partial^2 U}{\partial X^2} + \frac{\partial^2 U}{\partial Y^2}\right) \\ & + Mn_F \Pr\left(\frac{\rho_f}{\rho_{nf}}\right)\left(\varepsilon_2 - \varepsilon_1 - \Theta\right) H \frac{\partial H}{\partial X} - Ha^2 \Pr\left[\frac{\sigma_{nf} / \sigma_f}{\rho_{nf} / \rho_f}\right]\left(H_y^2 U - H_x H_y V\right) \end{aligned} \tag{19}$$

$$\begin{aligned} & U\frac{\partial V}{\partial X} + V\frac{\partial V}{\partial Y} = -\frac{\partial P}{\partial X} + \Pr\left[\frac{\mu_{nf} / \mu_f}{\rho_{nf} / \rho_f}\right]\left(\frac{\partial^2 V}{\partial X^2} + \frac{\partial^2 V}{\partial Y^2}\right) \\ & + Mn_F \Pr\left(\frac{\rho_f}{\rho_{nf}}\right)\left(\varepsilon_2 - \varepsilon_1 - \Theta\right) H \frac{\partial H}{\partial Y} - Ha^2 \Pr\left[\frac{\sigma_{nf} / \sigma_f}{\rho_{nf} / \rho_f}\right]\left(H_x^2 V - H_x H_y U\right) + Ra \Pr\left[\frac{\beta_{nf}}{\beta_f}\right]\Theta \end{aligned} \tag{20}$$

$$U\frac{\partial \Theta}{\partial X}+V\frac{\partial \Theta}{\partial Y}=\left[\frac{\frac{k_{nf}}{k_f}}{\frac{(\rho C_P)_{nf}}{(\rho C_P)_f}}\right]\left(\frac{\partial^2\Theta}{\partial X^2}+\frac{\partial^2\Theta}{\partial Y^2}\right)+Ha^2\,Ec\left[\frac{\frac{\sigma_{nf}}{\sigma_f}}{\frac{(\rho C_P)_{nf}}{(\rho C_P)_f}}\right]\left\{U H_y - V H_x\right\}^2$$
$$+Mn_F Ec\frac{(\rho C_P)_f}{(\rho C_P)_{nf}}\left\{U\frac{\partial H}{\partial X}+V\frac{\partial H}{\partial Y}\right\}H\left(\varepsilon_1+\Theta\right) \tag{21}$$
$$+Ec\left[\frac{\frac{\mu_{nf}}{\mu_f}}{\frac{(\rho C_P)_{nf}}{(\rho C_P)_f}}\right]\left\{2\left(\frac{\partial U}{\partial X}\right)^2+2\left(\frac{\partial V}{\partial Y}\right)^2+\left(\frac{\partial U}{\partial Y}+\frac{\partial V}{\partial X}\right)^2\right\}$$

where

$$Ra_f = g\beta_f L^3\left(T_h - T_c\right)/\left(\alpha_f \upsilon_f\right), \mathrm{Pr}_f = \upsilon_f / \alpha_f, Ha = L\mu_0 H_0\sqrt{\sigma_f / \mu_f}, \varepsilon_1 = T_1 / \Delta T$$

$$\varepsilon_2 = T_c' / \Delta T, Ec = \left(\mu_f \alpha_f\right)/\left[\left(\rho C_P\right)_f \Delta T L^2\right]$$

and

$$Mn_F = \mu_0 H_0^2 K'\left(T_h - T_c\right)L^2 / \left(\mu_f \alpha_f\right)$$

are the Rayleigh number, Prandtl number, Hartmann number arising from MHD, temperature number, curie temperature number, Eckert number and Magnetic number arising from FHD the for the base fluid, respectively. The thermo physical properties of the nanofluid are given in Table 1 (Sheikholeslami and Ganji, 2014).

Table 1. Thermo physical properties of water and nanoparticles

	$\rho(kg / m^3)$	$C_p(j / kgk)$	$k(W / m.k)$	$\beta\times 10^5(K^{-1})$	$\sigma\left(\Omega\cdot \mathrm{m}\right)^{-1}$
Pure water	997.1	4179	0.613	21	0.05
Fe_3O_4	5200	670	6	1.3	25000

(Sheikholeslami and Ganji, 2014)

The stream function and vorticity are defined as:

$$u = \frac{\partial \psi}{\partial y}, v = -\frac{\partial \psi}{\partial x}, \omega = \frac{\partial v}{\partial x} - \frac{\partial u}{\partial y}$$
$$\Omega = \frac{\omega L^2}{\alpha_f}, \Psi = \frac{\psi}{\alpha_f} \tag{22}$$

The stream function satisfies the continuity *Equation* (18). The vorticity equation is obtained by eliminating the pressure between the two momentum equations, i.e. by taking y-derivative of *Equation* (20) and subtracting from it the x-derivative of *Equation* (19).

The boundary conditions as shown in Figure 1 are:

$\Theta = 1.0$ on the inner circular boundary

$\Theta = 0.0$ on the outer wall boundary

$\Psi = 0.0$ on all solid boundaries (23)

The values of vorticity on the boundary of the enclosure can be obtained using the stream function formulation and the known velocity conditions during the iterative solution procedure.

The local Nusselt number of the nanofluid along the hot wall can be expressed as:

$$Nu_{loc} = \left(\frac{k_{nf}}{k_f}\right)\frac{\partial \Theta}{\partial r} \tag{24}$$

where *r* is the radial direction. The average Nusselt number on the hot circular wall is evaluated as:

$$Nu_{ave} = \frac{1}{\gamma}\int_0^{\gamma} Nu_{loc}(\zeta)\, d\zeta \tag{25}$$

To estimate the enhancement of heat transfer between the case of $\phi = 0.04$ and the pure fluid (base fluid) case, the heat transfer enhancement is defined as:

$$En = \frac{Nu(\phi = 0.04) - Nu(basefluid)}{Nu(basefluid)} \times 100 \tag{26}$$

1.2. Effects of Active Parameters

In this work, effects of Ferrohydrodynamics and Magnetohydrodynamics on Fe_3O_4- water flow and heat transfer is investigated numerically using the Control Volume based Finite Element Method. Calculations are made for various values of volume fraction of nanoparticles ($\phi = 0.0$ and 0.04), Rayleigh numbers ($Ra = 10^3, 10^4$ and 10^5), Magnetic number arising from FHD ($Mn_F = 0, 20, 60$ and 100) and Hartmann number arising from MHD ($Ha = 0, 5$ and 10). In all calculations, the Prandtl number ($\Pr$), temperature number (ε_1) and Eckert number (Ec) are set to 6.8, 0.0 and 1.0e-6, respectively.

Figure 3 shows the comparison of the isotherms and streamlines between nanofluid and pure fluid. The velocity components of nanofluid are enhanced because of an augment in the energy transport in the fluid with the increasing of volume fraction. Thus, the absolute values of stream functions increase with increasing the volume fraction of nanofluid. Also it can be seen that thermal boundary layer thickness around the hot wall increases with increase of nanoparticle volume fraction.

Figure 4 depicts the effects of Rayleigh number, Hartmann number and Magnetic number on isotherms and streamlines. $|\Psi_{max}|$ increases with increase of Rayleigh number while it decreases with augment of Hartmann number and Magnetic number. When $Ra = 10^3$, the heat transfer in the enclosure is mainly dominated by the conduction mode. At absence of magnetic field the streamlines show one rotating eddy. When Magnetic parameter increases, another vortex appears near the centerline which is rotates in reverse direction in comparison with the primary vortex. Lorentz force has no significant effect on isotherm and streamline in low Rayleigh number. As the Rayleigh number increases up to 10^4, the role of convection in heat transfer becomes more important and consequently the thermal boundary layer on the surface of the inner wall becomes thinner. In addition, a plume starts to appear on the top of the inner circular wall. It is interesting remark that at $Mn_F = 100, Ha = 10$ the smaller primary vortex turns to two smaller eddies near the magnetic sources. At $Ra = 10^5$ two reverse plumes are observed in absence of FHD and MHD effects. As Hartmann number increases one of the plumes disappears due to Lorentz forces.

Figure 5 shows the effects of Magnetic number, Hartmann number and Rayleigh number on local Nusselt number along the hot wall. As Hartmann number increases Nusselt number decreases but opposite

Figure 3. Comparison of the isotherms and streamlines between nanofluid ($\phi = 0.04$) (– – –) and pure fluid ($\phi = 0$) (—) when $\Pr = 6.8$.

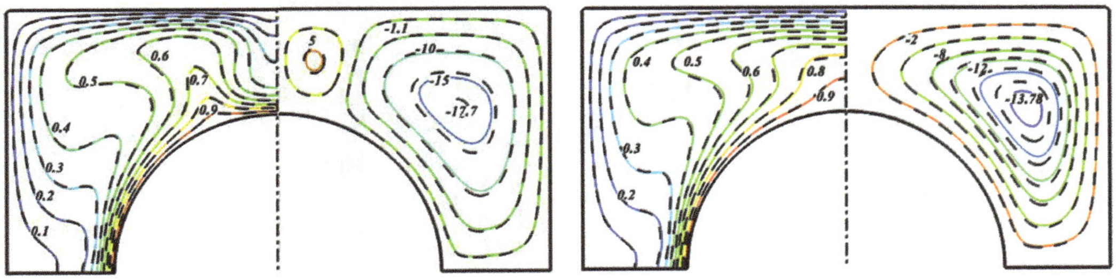

(a) $Ra = 10^5, Mn_F = 0, Ha = 0$ (b) $Ra = 10^5, Mn_F = 100, Ha = 10$

Figure 4. Isotherms (left) and streamlines (right) contours for different values of Rayleigh number, Hartmann number and Magnetic number when $\Pr = 6.8, \phi = 0.04$.

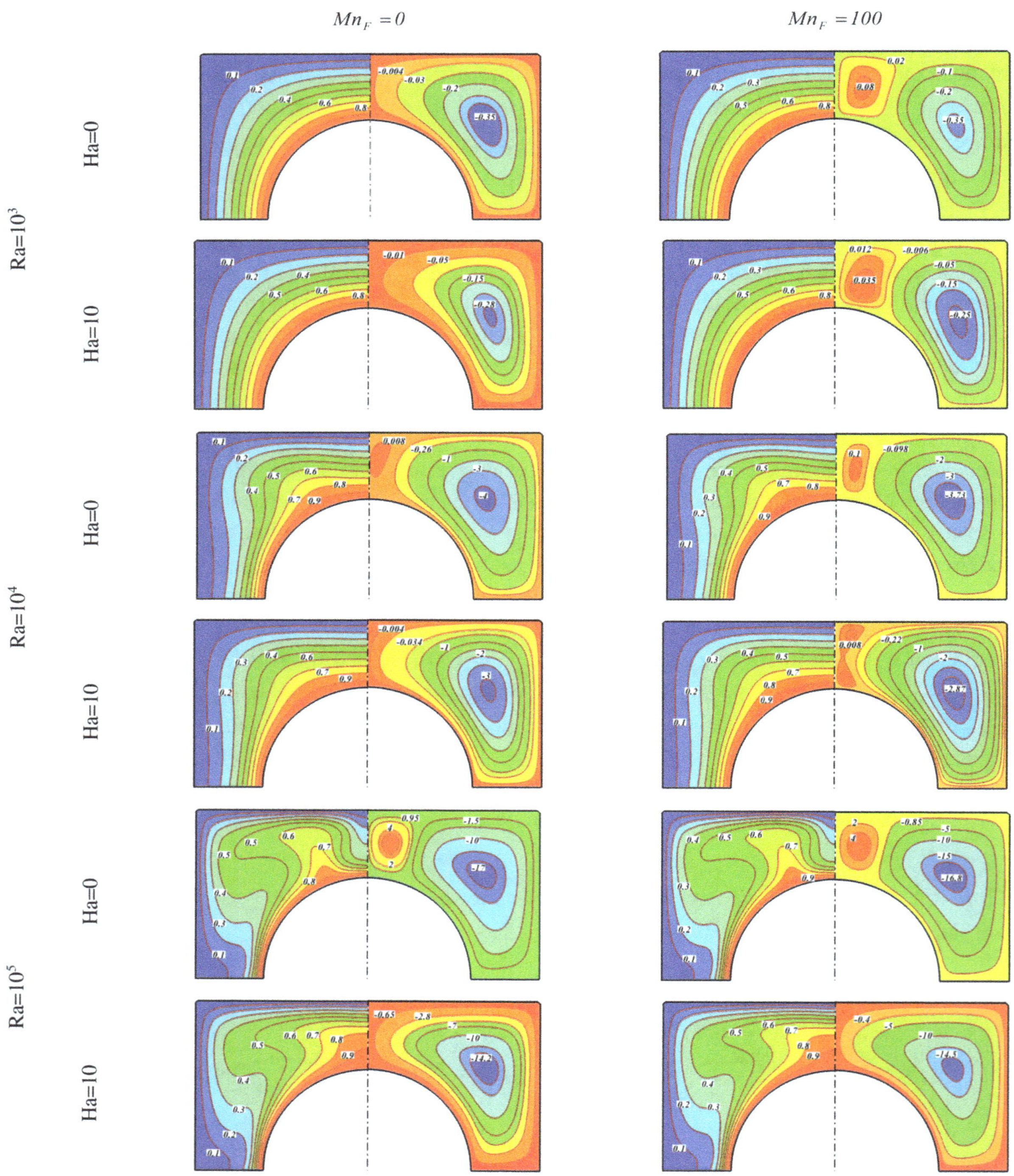

Figure 5. Effects of Magnetic number, Hartmann number and Rayleigh number on local Nusselt number Nu_{loc} *along hot wall when* $\Pr = 6.8, \phi = 0.04$

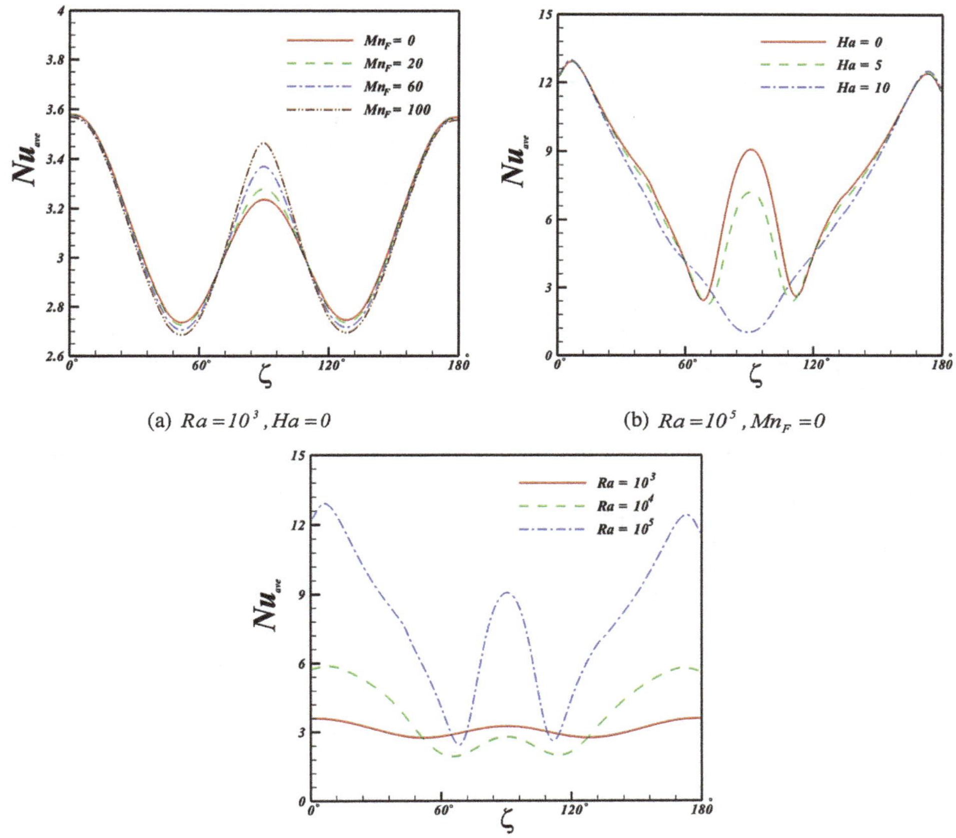

(a) $Ra = 10^3, Ha = 0$ (b) $Ra = 10^5, Mn_F = 0$

(c) $Mn_F = 0, Ha = 0$

trend is observed for Magnetic number and Rayleigh number. Existence of maximum and minimum points in local Nusselt number profile is corresponding to presence of thermal plumes. Figure 6 (a-c) shows the effects of Hartmann number and Rayleigh number on average Nusselt number along hot wall. Effects of Magnetic number, Rayleigh number and Hartmann number on heat transfer enhancement are shown in Figure 6 (d-f). The corresponding polynomial representations for Nusselt number and Enhancement in heat transfer are as the following:

$$
\begin{aligned}
Nu_{ave} &= a_{15} + a_{25}Y_3 + a_{35}Y_4 + a_{45}Y_3^2 + a_{55}Y_4^2 + a_{65}Y_4Y_3 \\
Y_1 &= a_{11} + a_{21}Ra^* + a_{31}Mn_F^* + a_{41}Ra^{*2} + a_{51}Mn_F^{*2} + a_{61}Ra^*\,Mn_F^* \\
Y_2 &= a_{12} + a_{22}\phi + a_{32}Ra^* + a_{42}\phi^2 + a_{52}Ra^{*2} + a_{62}\phi Ra^* \\
Y_3 &= a_{13} + a_{23}Y_1 + a_{33}Ha^* + a_{43}Y_1^2 + a_{53}Ha^{*2} + a_{63}Y_1Ha^* \\
Y_4 &= a_{14} + a_{24}Ha^* + a_{34}Y_2 + a_{44}Ha^{*2} + a_{54}Y_2^2 + a_{64}Ha^*\,Y_2
\end{aligned}
\tag{27}
$$

Figure 6. (a-c) Effects of Magnetic parameter, Hartmann number and Rayleigh number on average Nusselt number Nu_{ave} *along hot wall when* $\Pr = 6.8, \phi = 0.04$*; (d-f) Effects of Magnetic number, Rayleigh number and Hartmann number on heat transfer enhancement when* $\Pr = 6.8$.

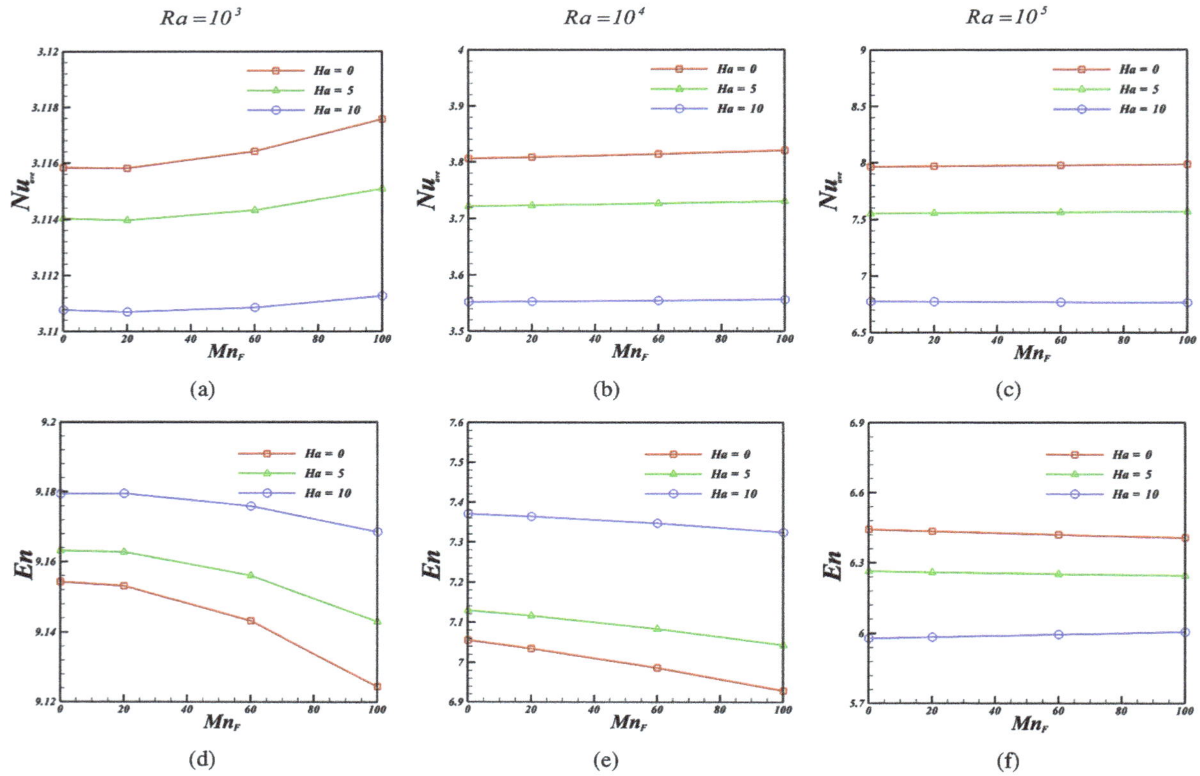

$$\begin{aligned} En &= b_{13} + b_{23}Y_1 + b_{33}Y_2 + b_{43}Y_1^2 + b_{53}Y_2^{\,2} + b_{63}Y_1Y_2 \\ Y_1 &= b_{11} + b_{21}Ra^* + b_{31}Ha^* + b_{41}Ra^{*2} + b_{51}Ha^{*2} + b_{61}Ra^*\,Ha^* \\ Y_2 &= b_{12} + b_{22}Ra^* + b_{32}Mn_F^{\ *} + b_{42}Ra^{*2} + b_{52}Mn_F^{\ *2} + b_{62}\operatorname{Re} Mn_F^{\ *} \end{aligned} \tag{28}$$

where $Ra^* = \log(Ra), Ha^* = Ha/10, Mn_F^{\ *} = Mn_F/100$. Also a_{ij}, b_{ij} can be found in Table 2 and 3.

Table 2. Constant coefficient for using Equation (27)

a_{ij}	i=1	i=2	i=3	i=4	i=5	i=6
j=1	19.51647	-10.0868	-0.00804	1.525112	0.001086	0.003349
j=2	19.53085	-0.87952	-10.1291	-0.03518	1.525112	2.186806
j=3	-0.29704	1.107821	0.654435	-0.00017	0.041509	-0.24967
j=4	-0.15202	0.647229	1.043281	0.044365	0.005999	-0.24883
j=5	-0.07156	0.161803	0.861233	1.818237	1.846604	-3.66821

Table 3. Constant coefficient for using Equation (28)

b_{ij}	i=1	i=2	i=3	i=4	i=5	i=6
j=1	21.33634	-5.71464	0.798586	0.542492	0.064253	-0.21888
j=2	21.72597	-5.81042	-0.0488	0.542492	-0.01392	0.006107
j=3	0.061649	4.86387	-3.87977	-0.38679	0.13164	0.25619

As Buoyancy force (due to temperature gradient) and Kelvin force (due to external non-uniform magnetic field) increases, the rate of heat transfer increases while opposite trend is observed for Lorentz forces. It means that increasing Lorentz forces lead to decrease of Nusselt number. Also it can be concluded that the effect of Magnetic parameter on Nu_{ave} is more pounced at low Rayleigh number. At low Rayleigh number, heat transfer enhancement increases with increase of Hartmann number while opposite trend is observed for Magnetic parameter. At high Rayleigh number Magnetic parameter has no significant effect on enhancement but increasing Hartmann number causes the heat transfer enhancement to decrease. The effect of nanoparticles is more pronounced at low Rayleigh number than at high Rayleigh number due to domination of conduction heat transfer mechanism.

2. SIMULATION OF FERROFLUID FLOW FOR MAGNETIC DRUG TARGETING USING LATTICE BOLTZMANN METHOD

2.1. Problem Definition

The steady, two-dimensional, incompressible, laminar flow is considered taking place in a channel. The width of channel is D and the length of the channel is L, such that $L / D = 25$. The flow at the entrance is assumed to be fully developed ($\frac{u}{u_{in}} = 4\frac{y}{D}(1-\frac{y}{D})$) (see Figure 7(a)) (Sheikholeslami and Ellahi, 2015). The flow is subjected to a magnetic source, which is placed very close to the lower plate and below it ($a = 0.25L, b = -0.1D$) (see Figure 8). The fluid in vessel is considered a homogenous mixture of Fe_3O_4-Blood plasma. Thermophysical properties of the nanoscale ferromagnetic particle and blood plasma are assumed to be constant (Table 4).

The components of the magnetic field intensity (H_x, H_y) and the magnetic field strength (H) can be considered as (Sheikholeslami and Ganji, 2014):

$$H_x = \frac{|b|}{(x-a)^2 + (y-b)^2}(y-b) \tag{29}$$

$$H_y = -\frac{|b|}{(x-a)^2 + (y-b)^2}(x-a) \tag{30}$$

Figure 7. (a) Geometry of the vessel; (b) Discrete velocity set of two-dimensional nine-velocity

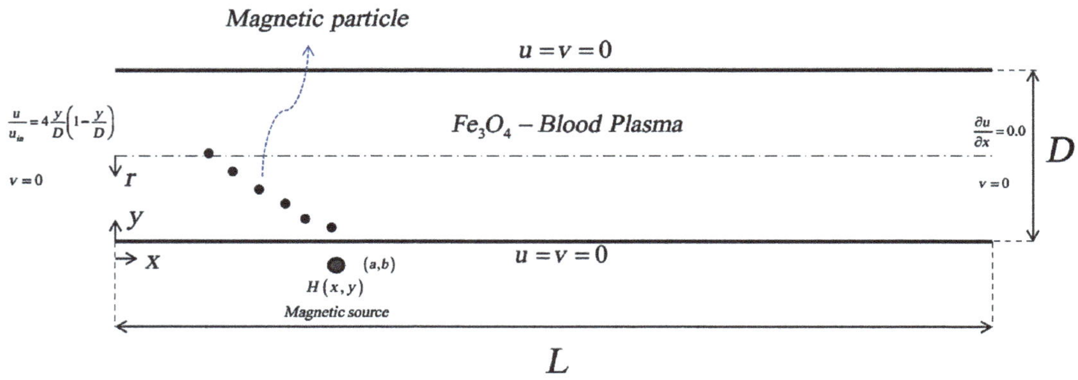

(a)

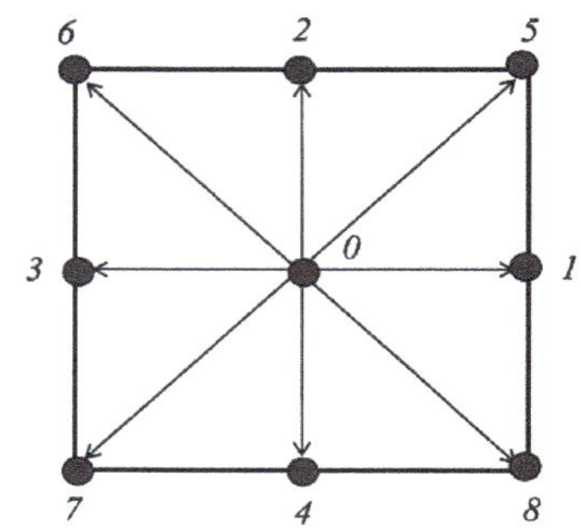

(b)

Figure 8. Contours of the magnetic field strength H.

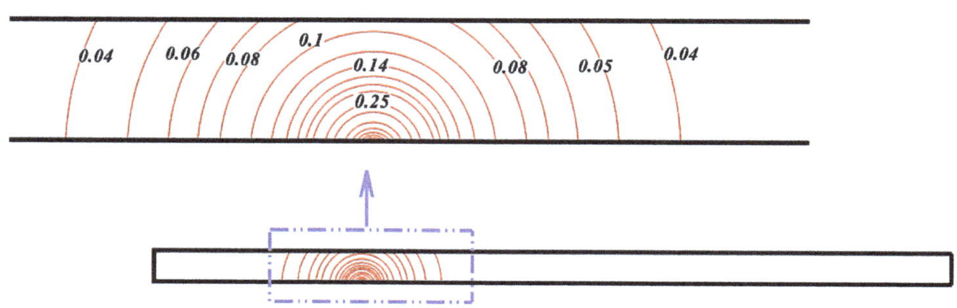

Table 4. Thermo physical properties of Blood plasma and Magnetite

	$\rho(kg/m^3)$	$\sigma(\Omega/m)$	$\mu(Pa.s)$
Blood plasma	1025	0.7	0.0015045
Magnetite (Fe_3O_4)	5000	25000	-

$$H = \sqrt{H_x^2 + H_y^2} = \frac{|b|}{\sqrt{(x-a)^2 + (y-b)^2}} \tag{31}$$

One of the novel computational fluid dynamics (CFD) methods which is solved Boltzmann equation to simulate the flow instead of solving the Navier–Stokes equations is called Lattice Boltzmann methods (LBM). LBM has several advantages such as simple calculation procedure and efficient implementation for parallel computation, over other conventional CFD methods, because of its particulate nature and local dynamics. The LB model uses one distribution function *f* for the flow. It uses modeling of movement of fluid particles to capture macroscopic fluid quantities such as velocity and pressure. In this approach, the fluid domain discretized to uniform Cartesian cells. Each cell holds a fixed number of distribution functions, which represent the number of fluid particles moving in these discrete directions. The D_2Q_9 model [6] was used and values of $w_0 = 4/9$ for $|c_0| = 0$ (for the static particle), $w_{1-4} = 1/9$ for $|c_{1-4}| = 1$ and $w_{5-9} = 1/36$ for $|c_{5-9}| = \sqrt{2}$ are assigned in this model (Figure 7(b)). The density and distribution functions are calculated by solving the lattice Boltzmann equation, which is a special discretization of the kinetic Boltzmann equation. After introducing BGK approximation, the general form of lattice Boltzmann equation with external force is as follow:

For the flow field:

$$f_i(x + c_i\Delta t, t + \Delta t) = f_i(x,t) + \frac{\Delta t}{\tau_v}[f_i^{eq}(x,t) - f_i(x,t)] + \Delta t c_i F_k \tag{32}$$

where Δt denotes lattice time step, c_i is the discrete lattice velocity in direction i, F_k is the external force in direction of lattice velocity, τ_v denotes the lattice relaxation time for the flow. The kinetic viscosity υ is defined in terms of its respective relaxation times, i.e. $\upsilon = c_s^2(\tau_v - 1/2)$. Note that the limitation $0.5 < \tau$ should be satisfied for both relaxation times to ensure that viscosity and thermal diffusivity are positive. Furthermore, the local equilibrium distribution function determines the type of problem that needs to be solved. It also models the equilibrium distribution functions, which are calculated with Equation (32) for flow field, respectively.

$$f_i^{eq} = w_i \rho \left[1 + \frac{c_i.u}{c_s^2} + \frac{1}{2}\frac{(c_i.u)^2}{c_s^4} - \frac{1}{2}\frac{u^2}{c_s^2}\right] \tag{33}$$

where w_i is a weighting factor and ρ is the lattice fluid density.

In order to incorporate buoyancy forces and magnetic forces in the model, the force term in the Equation (32) need to calculate as below:

$$F = F_x + F_y$$
$$F_x = 3w_i\rho\left[A\left(-H_y^2u + H_xH_yv\right) + B\left(H\frac{\partial H}{\partial x}\right)\right], \tag{34}$$
$$F_y = 3w_i\rho\left[A\left(-H_x^2v + H_xH_yu\right) + B\left(H\frac{\partial H}{\partial y}\right)\right]$$

where A and B are $A = \frac{Ha^2 v}{L^2}$ and $B = Mn_F u_{in}^2$, respectively. $\mathrm{Re} = \frac{Du_{in}}{v}$, $Ha = \mu_0 H_0 D\sqrt{\frac{\sigma}{\mu}}$ and $Mn_F = \frac{\mu_0 \chi H_0^2}{\rho u_{in}^2}$ are Reynolds number, Hartmann number and Magnetic number arising from FHD (Sheikholeslami and Ganji, 2014).

Finally, macroscopic variables calculate with the following formula:

$$\text{Flow density}: \ \rho = \sum_i f_i, \qquad \text{Momentum}: \ \rho u = \sum_i c_i f_i, \tag{35}$$

In order to simulate the nanofluid by the lattice Boltzmann method, because of the interparticle potentials and other forces on the nanoparticles, the nanofluid behaves differently from the pure liquid from the mesoscopic point of view and is of higher efficiency in energy transport as well as better stabilization than the common solid-liquid mixture. For modeling the nanofluid because of changing in the fluid density, viscosity and electrical conductivity some of the governed equations should change. The effective density (ρ_{nf}), the effective viscosity (μ_{nf}) and electrical conductivity $(\sigma)_{nf}$ of the nanofluid are defined as:

$$\rho_{nf} = \rho_f(1-\phi) + \rho_s\phi \tag{36}$$

$$\mu_{nf} = \frac{\mu_f}{(1-\phi)^{2.5}} \tag{37}$$

$$\frac{\sigma_{nf}}{\sigma_f} = 1 + \frac{3\left(\frac{\sigma_s}{\sigma_f} - 1\right)\phi}{\left(\frac{\sigma_s}{\sigma_f} + 2\right) - \left(\frac{\sigma_s}{\sigma_f} - 1\right)\phi} \tag{38}$$

with ϕ being the volume fraction of the nanoscale ferromagnetic particle (Fe_3O_4) and subscripts *nf, s* and *f* stand for the mixture, nanoscale ferromagnetic particle (Fe_3O_4) and the carrier fluid (Blood plasma), respectively.

The Local and average skin friction coefficient is defined as follows:

$$C_f = \frac{2}{\mathrm{Re}} \frac{\rho_f}{\rho_{nf}} \frac{1}{\left(1-\phi\right)^{2.5}} \frac{\partial U}{\partial Y}\bigg|_{X=0} \text{ and } \overline{C_f} = \int_0^L C_f dX \tag{39}$$

2.2. Effects of Active Parameters

Targeted drug delivery system under the influence of an applied magnetic field is studied. The working fluid is homogenies mixture of Magnetite in Blood plasma. Lattice Boltzmann method scheme was utilized to obtain the numerical simulation. Effect of Reynolds number ($\mathrm{Re} = 50, 100, 200$ and 400), Magnetic number ($Mn_F = 0, 20, 60$ and 100) on streamline, velocity profile and skin friction coefficient are examined for constant values of volume fraction of nanoparticle ($\phi = 0.04$) and Hartmann number ($Ha = 20$).

Effects of Magnetic number and Reynolds number on streamlines are shown in Figures 9, 10, 11 and 12. In the absence of the magnetic field $\left(Mn_F = 0\right)$ the streamlines are straight lines. As Reynolds number increases, velocity boundary layer thickness decreases. In the presence of the magnetic field a clock wise vortex appears at the area where the magnetic source is placed. This vortex causes perturbation in the flow pattern and nanofluid flow recirculation may occur in this region. This phenomenon is more pronounced for higher values of Magnetic number.

Figure 13 shows the effects of Magnetic number and Reynolds number on velocity profile at $x = 0.25L$. At $x = 0.25L$ flow is separated from the bottom wall and then reattaches with the lower plate. Finally, the flow at the outlet is again reverted to fully developed. This phenomenon is more pronounced for higher values of Reynolds number. Effects of Magnetic number and Reynolds number on local skin friction coefficient along the lower and upper walls are shown in Figure 14. The value of local skin friction coefficient the lower wall increases rapidly near $x = 0.2L$, where it reaches its maximum value. Very close to the region where the source is located $x = 0.25L$, a corresponding decrement takes place and at $x = 0.25L$ this parameter takes its minimum negative value. $C_f\big|_{Y=1}$ increases near the area of the magnetic source in a smoother way and it's sign does not change as happens with the lower plate where reverse of the flow occurs. The wall shear is more influenced on the lower plate, below of which the magnetic source is located. It is an interesting observation that there are two points with zero values of skin friction for lower plate. In these points there is no skin friction and this result may be exciting in the case of creation of fibrinoid. Figure 15 shows the effects of Magnetic number and Reynolds number on average skin friction coefficient. The drag acting on the upper plate is greater than that of the lower. Average skin friction coefficient decreases with increase of Reynolds number and Magnetic number.

Figure 9. Effect of Magnetic number on streamlines when $\mathrm{Re} = 50, Ha = 20, \phi = 0.04$.

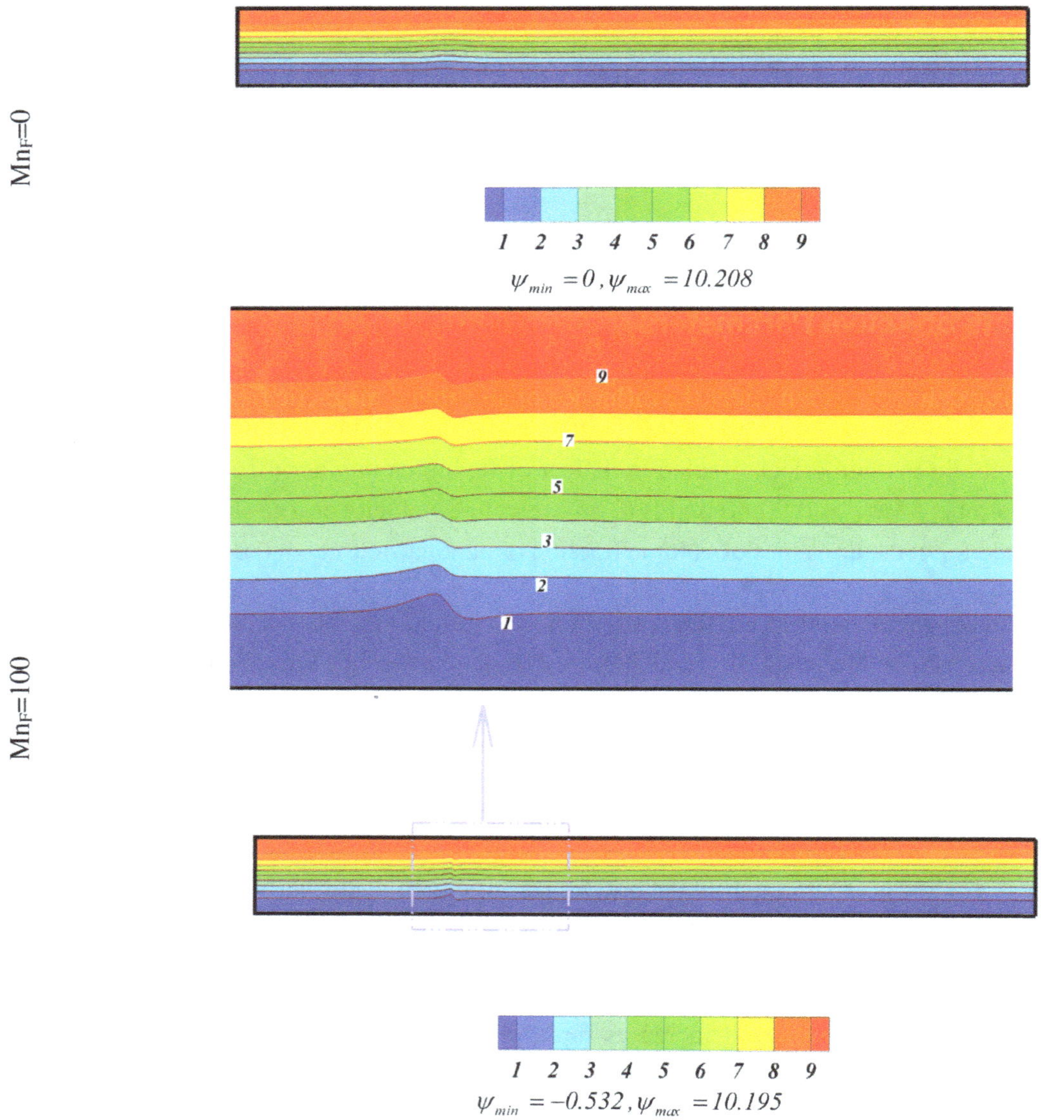

3. MAGNETIC NANOFLUID FORCED CONVECTIVE HEAT TRANSFER IN EXISTENCE OF VARIABLE MAGNETIC FIELD USING TWO PHASE MODEL

3.1. Problem Definition

The schematic diagram and the mesh of the semi-annulus enclosure used in the present CVFEM program are shown in Figure 16 (Sheikholeslami, Rashidi and Ganji, 2015). The inner wall is maintained at constant temperatures T_h and the other walls are maintained at constant temperature $T_c \left(T_h > T_c\right)$. The boundary conditions for concentration are similar to those of temperature. For the expression of the

Figure 10. Effect of Magnetic number on streamlines when $\mathrm{Re} = 100, Ha = 20, \phi = 0.04$.

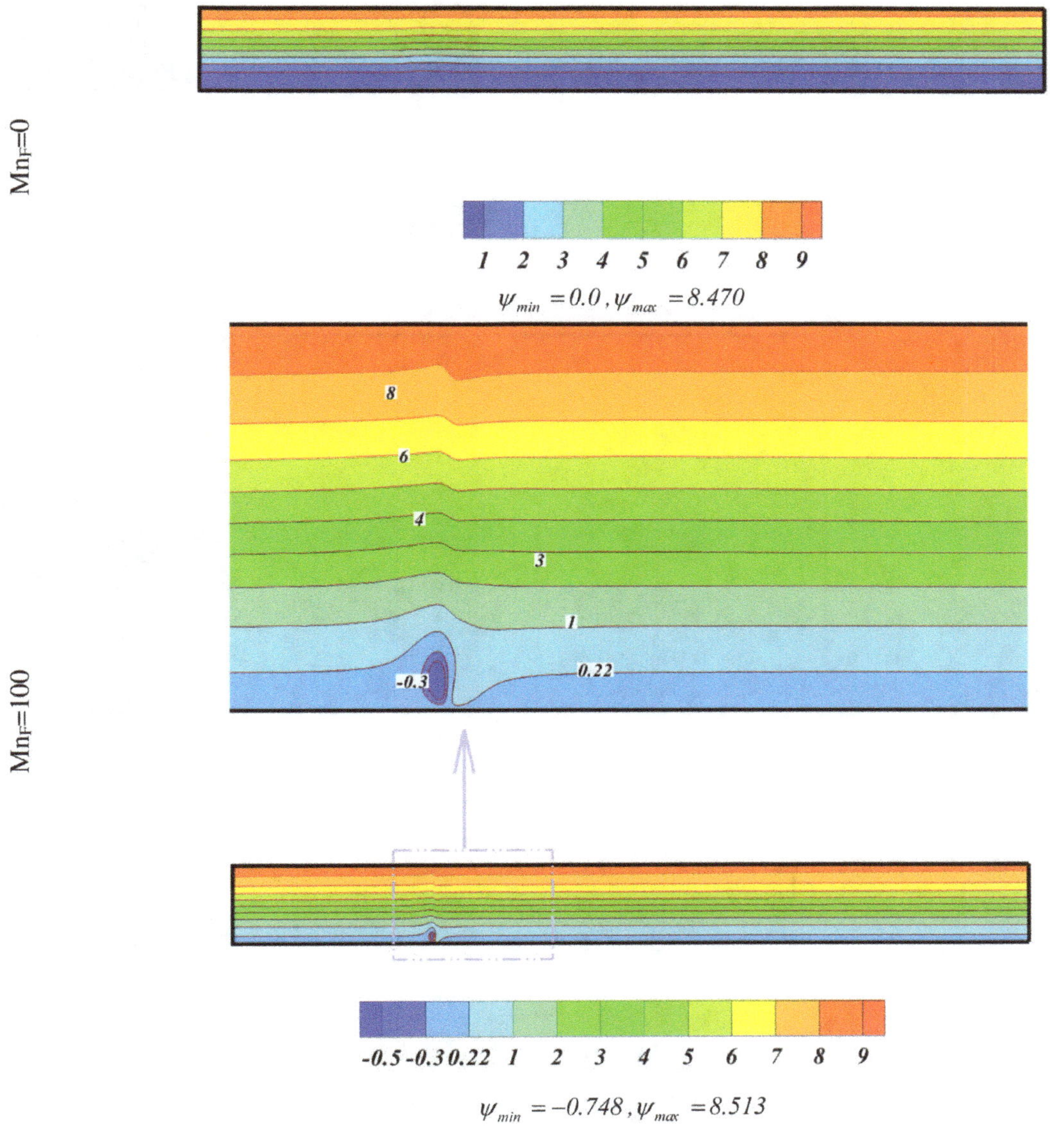

magnetic field strength it can be considered that the magnetic source represents a magnetic wire placed vertically to thex-y plane at the point $\left(\bar{a}, \bar{b}\right)$. The components of the magnetic field intensity ($\overline{H_x}, \overline{H_y}$) and the magnetic field strength ($\overline{H}$) can be considered as (Sheikholeslami and Ganji, 2014):

$$\overline{H_x} = \frac{\gamma'}{2\pi} \frac{1}{\left(x - \bar{a}\right)^2 + \left(y - \bar{b}\right)^2} \left(y - \bar{b}\right) \tag{40}$$

Figure 11. Effect of Magnetic number on streamlines when $\mathrm{Re} = 200, Ha = 20, \phi = 0.04$.

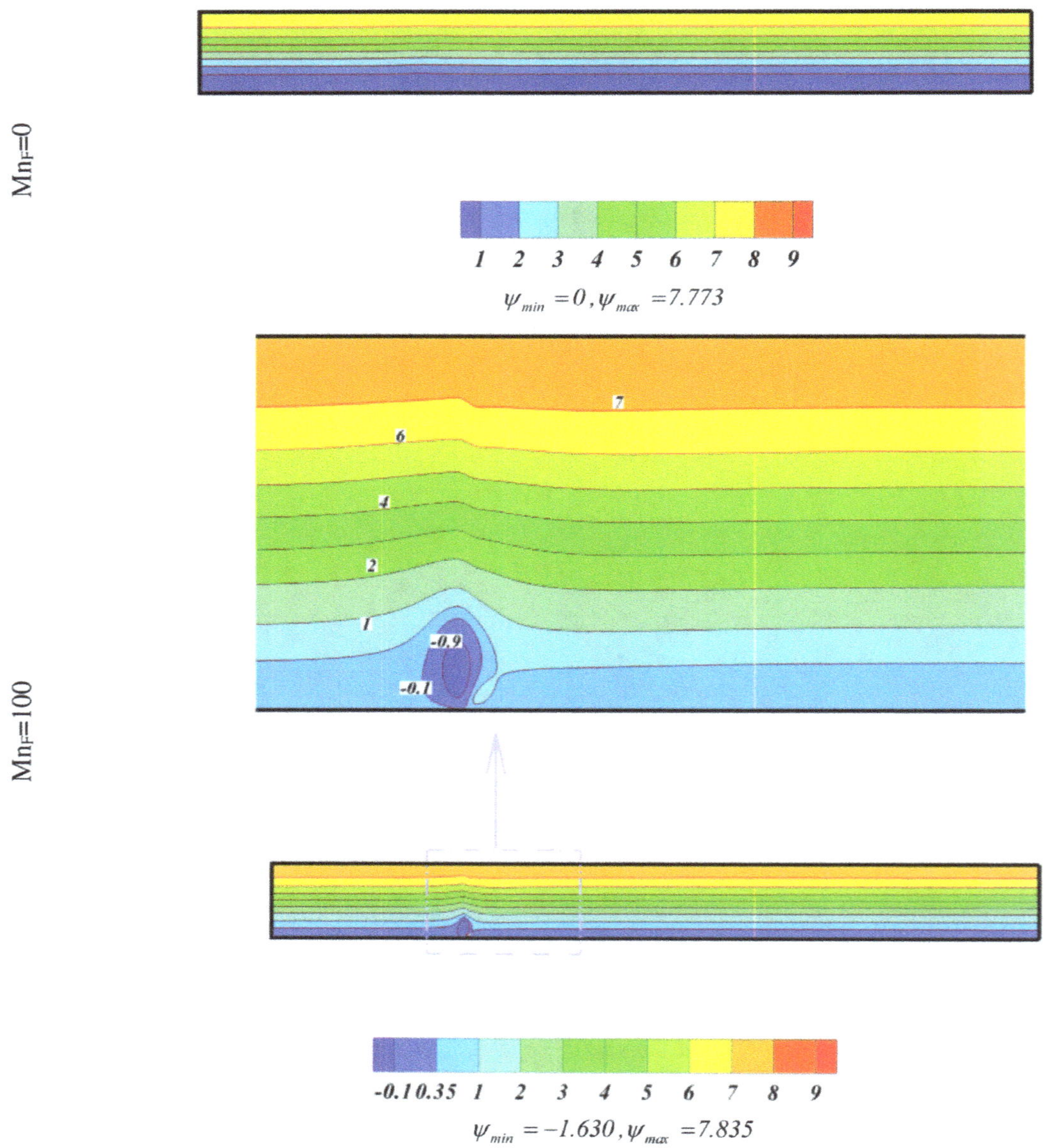

$$\overline{H}_y = -\frac{\gamma'}{2\pi}\frac{1}{\left(x-\bar{a}\right)^2+\left(y-\bar{b}\right)^2}\left(x-\bar{a}\right) \tag{41}$$

$$\overline{H} = \sqrt{\overline{H}_x^2+\overline{H}_y^2} = \frac{\gamma'}{2\pi}\frac{1}{\sqrt{\left(x-\bar{a}\right)^2+\left(y-\bar{b}\right)^2}} \tag{42}$$

Figure 12. Effect of Magnetic number on streamlines when $\mathrm{Re} = 400, Ha = 20, \phi = 0.04$.

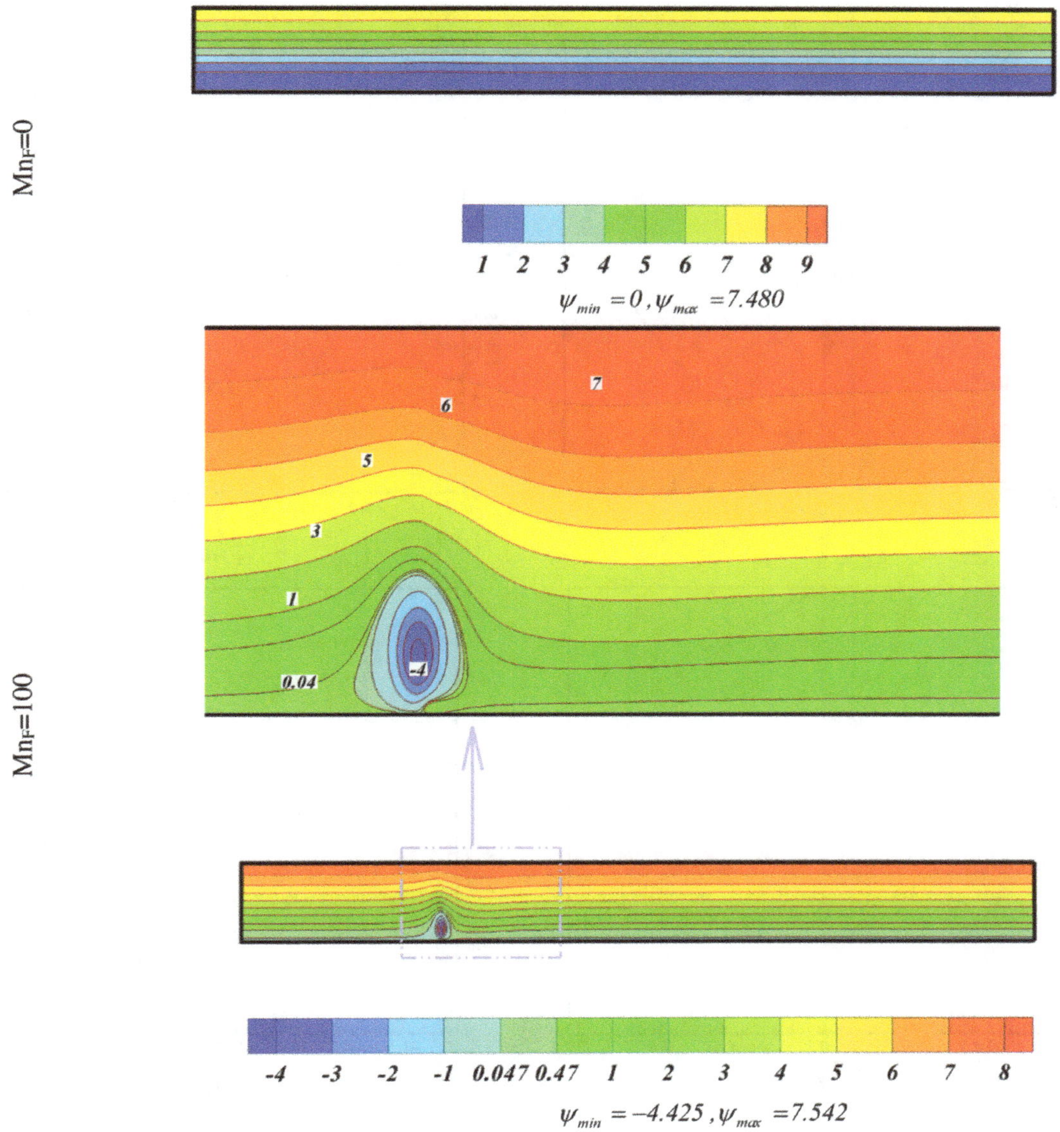

where γ' the magnetic field strength at the source (of the wire) and $\left(\bar{a},\bar{b}\right)$ is the position where the source is located. The contours of the magnetic field strength are shown in Figure 17. In this study magnetic source is located at $\left(-0.01\,\mathrm{cols}, 0.5\,\mathrm{rows}\right)$. The upper wall is Lid driven with velocity of U_{Lid}. In this section γ is equal to 45°.

The continuity, momentum under Boussinesq approximation and energy equations for the laminar and steady state natural convection in a two-dimensional enclosure can be written in dimensional form as follows:

Figure 13. Effects of Magnetic number and Reynolds number on velocity profile at $x = 0.25L$ *when* $Ha = 20, \phi = 0.04$.

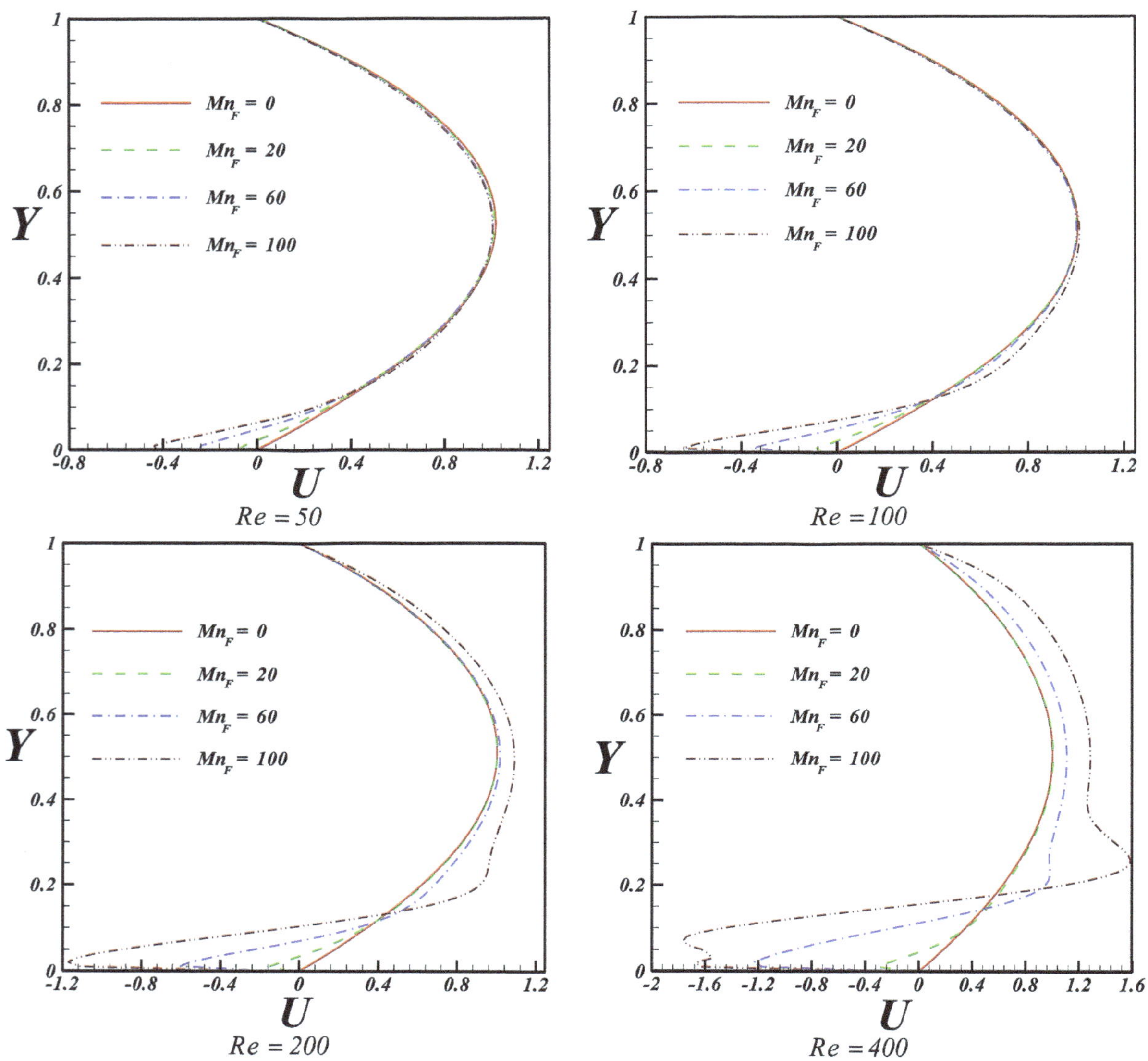

$$\frac{\partial u}{\partial x} + \frac{\partial v}{\partial y} = 0 \tag{43}$$

$$\rho_f \left\{ u \frac{\partial u}{\partial x} + v \frac{\partial u}{\partial y} \right\} = -\frac{\partial P}{\partial x} + \mu \left(\frac{\partial^2 u}{\partial x^2} + \frac{\partial^2 u}{\partial y^2} \right) - \sigma B_y^2 u + \sigma B_x B_y v \tag{44}$$

Figure 14. Effects of Magnetic number and Reynolds number on local skin friction coefficient along the (a) lower and (b) upper plates when $Ha = 20, \phi = 0.04$.

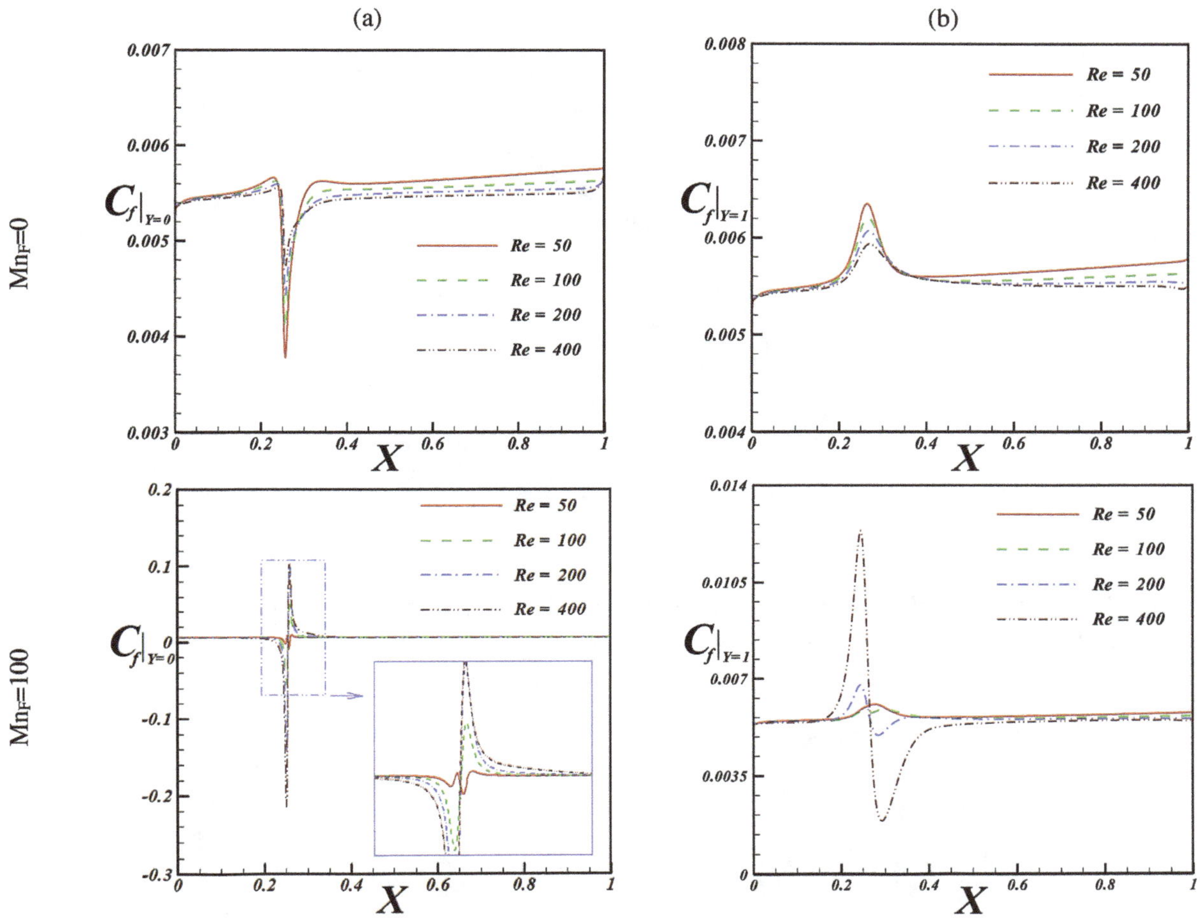

Figure 15. Effects of Magnetic number and Reynolds number on average skin friction coefficient along the (a) lower and (b) upper plates when $Ha = 20, \phi = 0.04$.

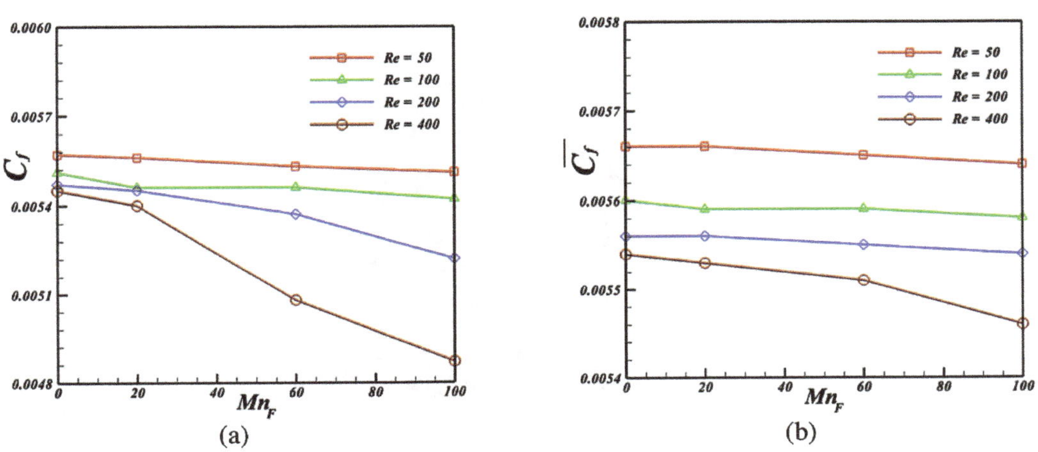

Figure 16. Geometry and the boundary conditions

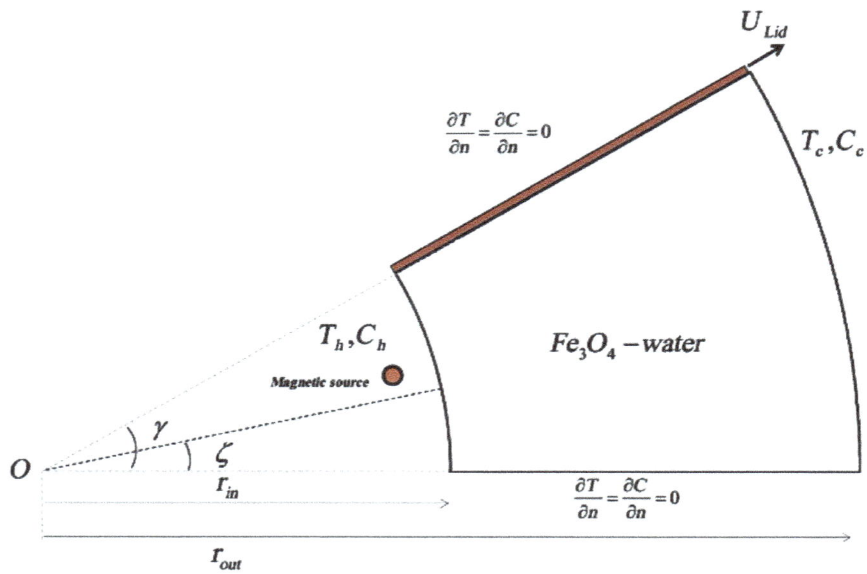

Figure 17. Contours of the (a) magnetic field strength H; (b) magnetic field intensity component in x direction Hx; (c) magnetic field intensity component in y direction Hy.

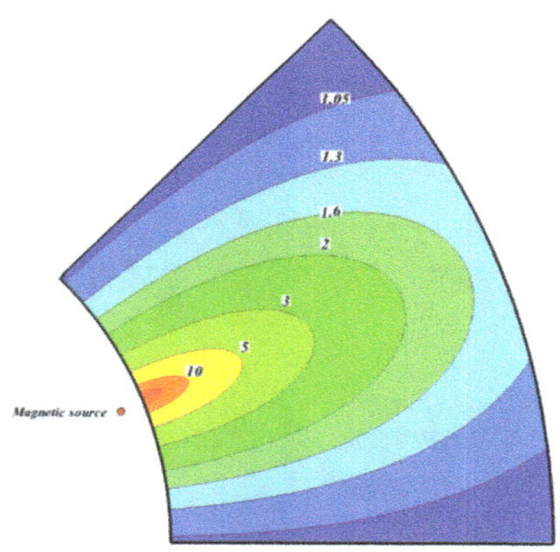

(a) $H(x, y)$

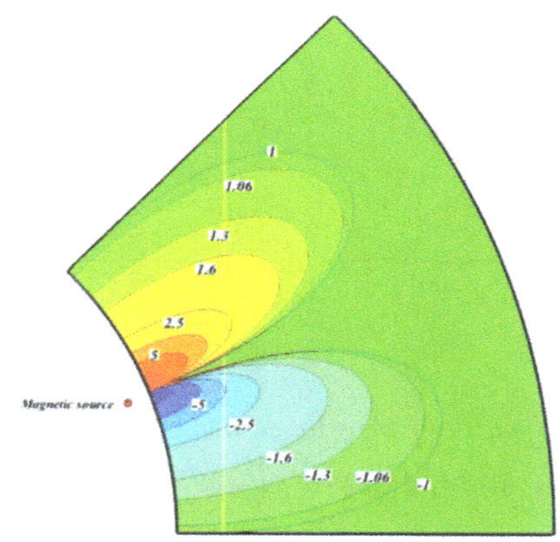

(b) $Hx(x, y)$

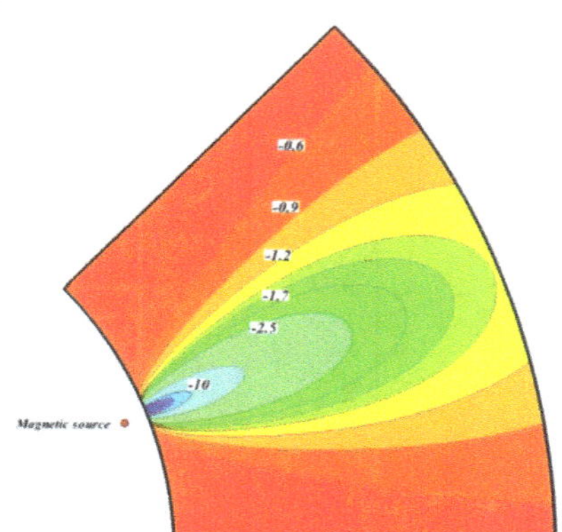

(c) $Hy(x, y)$

$$\rho_f\left\{u\frac{\partial v}{\partial x}+v\frac{\partial v}{\partial y}\right\}=-\frac{\partial P}{\partial y}+\mu\left(\frac{\partial^2 v}{\partial x^2}+\frac{\partial^2 v}{\partial y^2}\right)-\sigma B_x^2 v+\sigma B_x B_y u \tag{45}$$

$$u\frac{\partial T}{\partial x}+v\frac{\partial T}{\partial y}=\alpha\left(\frac{\partial^2 T}{\partial x^2}+\frac{\partial^2 T}{\partial y^2}\right)+\frac{(\rho c)_p}{(\rho c)_f}\left[\begin{array}{l}D_B\left\{\frac{\partial \phi}{\partial x}.\frac{\partial T}{\partial x}+\frac{\partial \phi}{\partial y}.\frac{\partial T}{\partial y}\right\}\\+(D_T/T_c)\left\{\left(\frac{\partial T}{\partial x}\right)^2+\left(\frac{\partial T}{\partial y}\right)^2\right\}\end{array}\right]$$
$$+\sigma\left(uB_y-vB_x\right)^2+\mu\left\{2\left(\frac{\partial u}{\partial x}\right)^2+2\left(\frac{\partial v}{\partial y}\right)^2+\left(\frac{\partial u}{\partial y}+\frac{\partial v}{\partial x}\right)^2\right\} \tag{46}$$

$$u\frac{\partial \phi}{\partial x}+v\frac{\partial \phi}{\partial y}=D_B\left\{\frac{\partial^2 \phi}{\partial x^2}+\frac{\partial^2 \phi}{\partial y^2}\right\}+\left(\frac{D_T}{T_c}\right)\left\{\frac{\partial^2 T}{\partial x^2}+\frac{\partial^2 T}{\partial y^2}\right\} \tag{47}$$

The stream function and vorticity are defined as follows:

$$u=\frac{\partial \psi}{\partial y},\ v=-\frac{\partial \psi}{\partial x},\ \omega=\frac{\partial v}{\partial x}-\frac{\partial u}{\partial y} \tag{48}$$

The following non-dimensional variables should be introduced:

$$X=\frac{x}{L},\ Y=\frac{y}{L},\ \Omega=\frac{\omega L}{U_{Lid}},\Psi=\frac{\psi}{U_{Lid}L},\Theta=\frac{T-T_c}{T_h-T_c},\ \Phi=\frac{\phi-\phi_c}{\phi_h-\phi_c},$$
$$U=\frac{u}{U_{Lid}},V=\frac{v}{U_{Lid}},H=\frac{H}{H_0},H_x=\frac{H_x}{H_0},H_y=\frac{H_y}{H_0} \tag{49}$$

By using these dimensionless parameters the equations become:

$$\left[\frac{\partial \Psi}{\partial Y}\frac{\partial \Omega}{\partial X}-\frac{\partial \Psi}{\partial X}\frac{\partial \Omega}{\partial Y}\right]=\frac{1}{\mathrm{Re}}\left(\frac{\partial^2 \Omega}{\partial X^2}+\frac{\partial^2 \Omega}{\partial Y^2}\right)$$
$$-\frac{Ha^2}{\mathrm{Re}}\times\left\{\frac{\partial V}{\partial X}H_x^2+V\left(2H_x\frac{\partial H_x}{\partial X}\right)-\frac{\partial U}{\partial X}H_xH_y-U\frac{\partial H_x}{\partial X}H_y-U\frac{\partial H_y}{\partial X}H_x\right.$$
$$\left.-\frac{\partial U}{\partial Y}H_y^2-U\left(2H_y\frac{\partial H_y}{\partial Y}\right)+\frac{\partial V}{\partial Y}H_xH_y+V\frac{\partial H_x}{\partial Y}H_y+V\frac{\partial H_y}{\partial Y}H_x\right\} \tag{50}$$

$$\frac{\partial \Psi}{\partial Y}\frac{\partial \Theta}{\partial X}-\frac{\partial \Psi}{\partial X}\frac{\partial \Theta}{\partial Y}=\frac{1}{\mathrm{Re}\,\mathrm{Pr}}\left(\frac{\partial^2 \Theta}{\partial X^2}+\frac{\partial^2 \Theta}{\partial Y^2}\right)+\frac{Nb}{\mathrm{Re}}\left(\frac{\partial \Phi}{\partial X}\frac{\partial \Theta}{\partial X}+\frac{\partial \Phi}{\partial Y}\frac{\partial \Theta}{\partial Y}\right)+\frac{Nt}{\mathrm{Re}}\left(\left(\frac{\partial \Theta}{\partial X}\right)^2+\left(\frac{\partial \Theta}{\partial Y}\right)^2\right)$$
$$+Ha^2\,Ec\left\{U\,H_y-V\,H_x\right\}^2+Ec\left[2\left(\frac{\partial U}{\partial X}\right)^2+2\left(\frac{\partial V}{\partial Y}\right)^2+\left(\frac{\partial U}{\partial Y}+\frac{\partial V}{\partial X}\right)^2\right] \tag{51}$$

$$\frac{\partial \Psi}{\partial Y}\frac{\partial \Phi}{\partial X}-\frac{\partial \Psi}{\partial X}\frac{\partial \Phi}{\partial Y}=\frac{1}{Le\,\mathrm{Re}}\left(\frac{\partial^2 \Phi}{\partial X^2}+\frac{\partial^2 \Phi}{\partial Y^2}\right)+\frac{Nt}{Nb\;Le\,\mathrm{Re}}\left(\frac{\partial^2 \Theta}{\partial X^2}+\frac{\partial^2 \Theta}{\partial Y^2}\right) \tag{52}$$

$$\frac{\partial^2 \Psi}{\partial X^2}+\frac{\partial^2 \Psi}{\partial Y^2}=-\Omega \tag{53}$$

where Prandtl number, the Brownian motion parameter, the thermophoretic parameter, Lewis number, Hartmann number, Eckert number and Reynolds number are defined as:

$$\mathrm{Pr}=\mu / \rho_f \alpha\,,$$

$$Nb=(\rho c)_p D_B(\phi_h-\phi_c)/\left((\rho c)_f \alpha\right), Nt=(\rho c)_p D_T(T_h-T_c)/[(\rho c)_f \alpha T_c], Le=\alpha / D_B,$$

$$Ha=LH_0\mu_0\sqrt{\sigma / \mu},$$

$$\mathrm{Ec}=\left(U_{Lid}^2\right)/\left[\left(C_P\right)\Delta T\right]$$

and

$$\mathrm{Re}=\frac{\rho_f L U_{Lid}}{\mu},$$

respectively.

The local Nusselt number of the nanofluid along the hot wall can be expressed as:

$$Nu_{loc} = \frac{\partial \Theta}{\partial r} \tag{54}$$

where *r* is the radial direction. The average Nusselt number on the hot circular wall is evaluated as:

$$Nu_{ave} = \frac{1}{\gamma}\int_{0}^{\gamma} Nu_{loc}(\zeta)\,d\zeta \tag{55}$$

The heatlines are adequate tools for visualization and analysis of 2D convection heat transfer, through an extension of the heat flux line concept to include the advection terms. Heat function (*H*) are defined in terms of the energy equation as

$$\frac{\partial H}{\partial Y} = U\Theta - \frac{\partial \Theta}{\partial X},\ -\frac{\partial H}{\partial X} = V\Theta - \frac{\partial \Theta}{\partial Y} \tag{56}$$

3.2. Effects of Active Parameters

In this section, forced convection heat transfer of ferrofluid in presence of variable magnetic field is investigated using CVFEM. Two phase model is used to simulate nanofluid. Calculations are made for various values of Reynolds numbers ($\mathrm{Re} = 10, 100$ and 500), Lewis number ($Le = 2, 4$ and 8) and Hartmann number ($Ha = 0,\ 5, 10$ and 20). In all calculations, the Prandtl number (Pr), temperature number (ε_1) and Eckert number (Ec) are set to 6.8, 0.0 and 10^{-5}, respectively.

Effects of Reynolds number and Hartmann number on isotherms, streamlines, isoconcentration and heatline contours are shown in Figures 18, 19 and 20. At low Reynolds number, one main eddy exists in streamline. By increasing Reynolds number, another small eddy generates near the bottom wall. As Reynolds number increases up to 500, the second eddy become stronger. So the enclosure divides into two region respect to $\zeta = 22.5^\circ$.

Due to existing two eddies which are rotates in reverse direction, thermal plume appear generate near the hot wall. As Hartmann number increases, Lorentz forces suppress the flow and diminish the thermal plume. Isoconcentration becomes more distributed in high Reynolds number and this effect is reduced in presence of magnetic field. Heat lines contours has two regions in low Reynolds number while it has three regions in high Reynolds number. Applying magnetic field leads to generate a small passive region at bottom right corner.

Figures 21 and 22 show the effects of Hartmann number, Reynolds number and Lewis number on local and average Nusselt number. As Reynolds number increase thermal boundary layer thickness near the hot wall decreases and in turn rate of heat transfer increases with increase of Reynolds number. As Hartmann number increases Lorentz forces becomes stronger and suppress the flow. So thermal boundary layer thickness increases with increase of Hartmann number and in turn Nusselt number decreases with increase of Hartmann number. Increasing Lewis number leads to decrease in rate of heat transfer. Due to existing thermal plume, maximum or minimum point appears in local Nusselt number profile.

Figure 18. Comparison of the isotherms, streamlines, isoconcentration and heatline contours for different values of Hartmann number when $Le = 4, Nt = Nb = 0.5, \mathrm{Re} = 10$ *and* $\mathrm{Pr} = 6.85$.

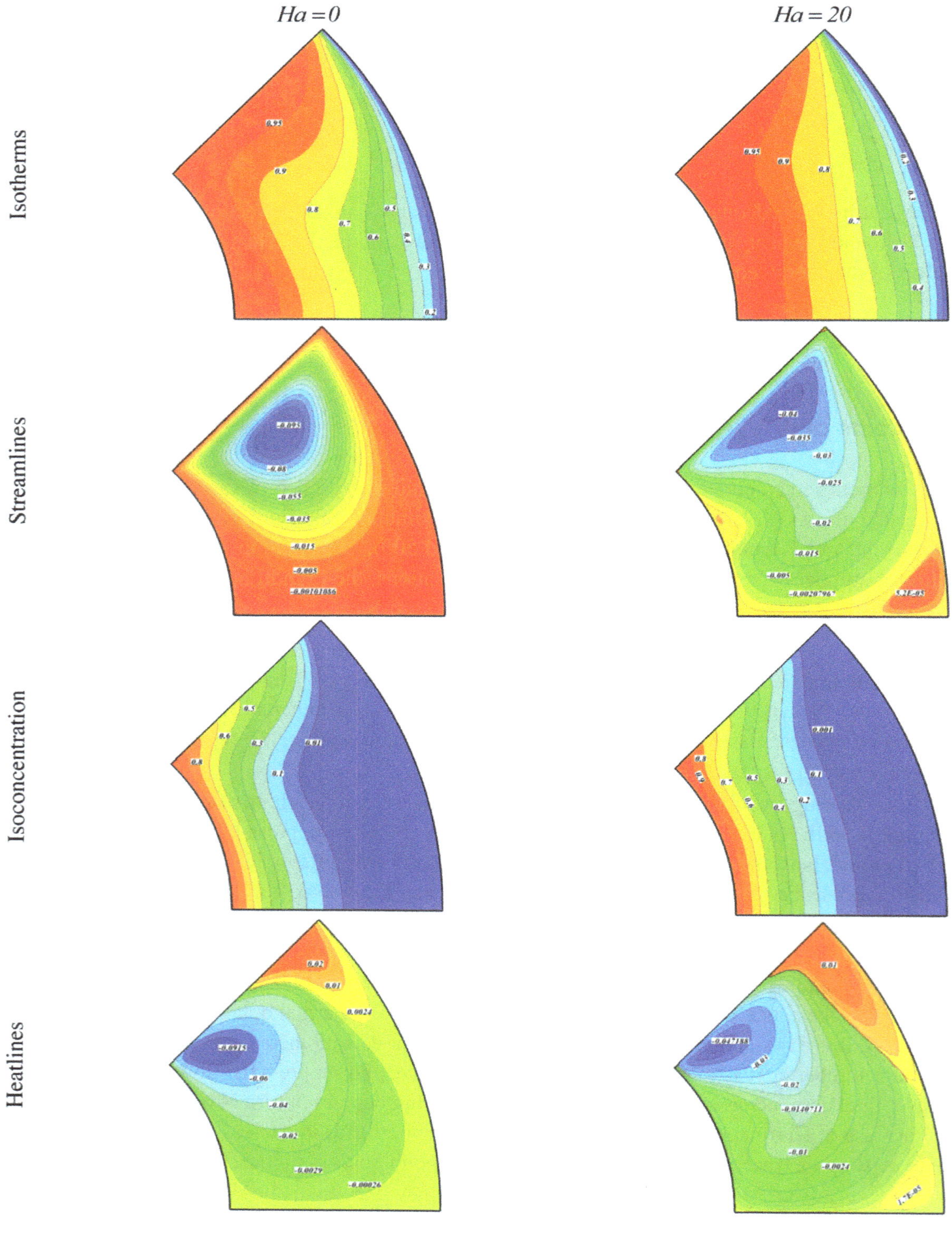

Figure 19. Comparison of the isotherms, streamlines, isoconcentration and heatline contours for different values of Hartmann number when $\mathrm{Le}=4, Nt=Nb=0.5, \mathrm{Re}=100$ *and* $\mathrm{Pr}=6.85$.

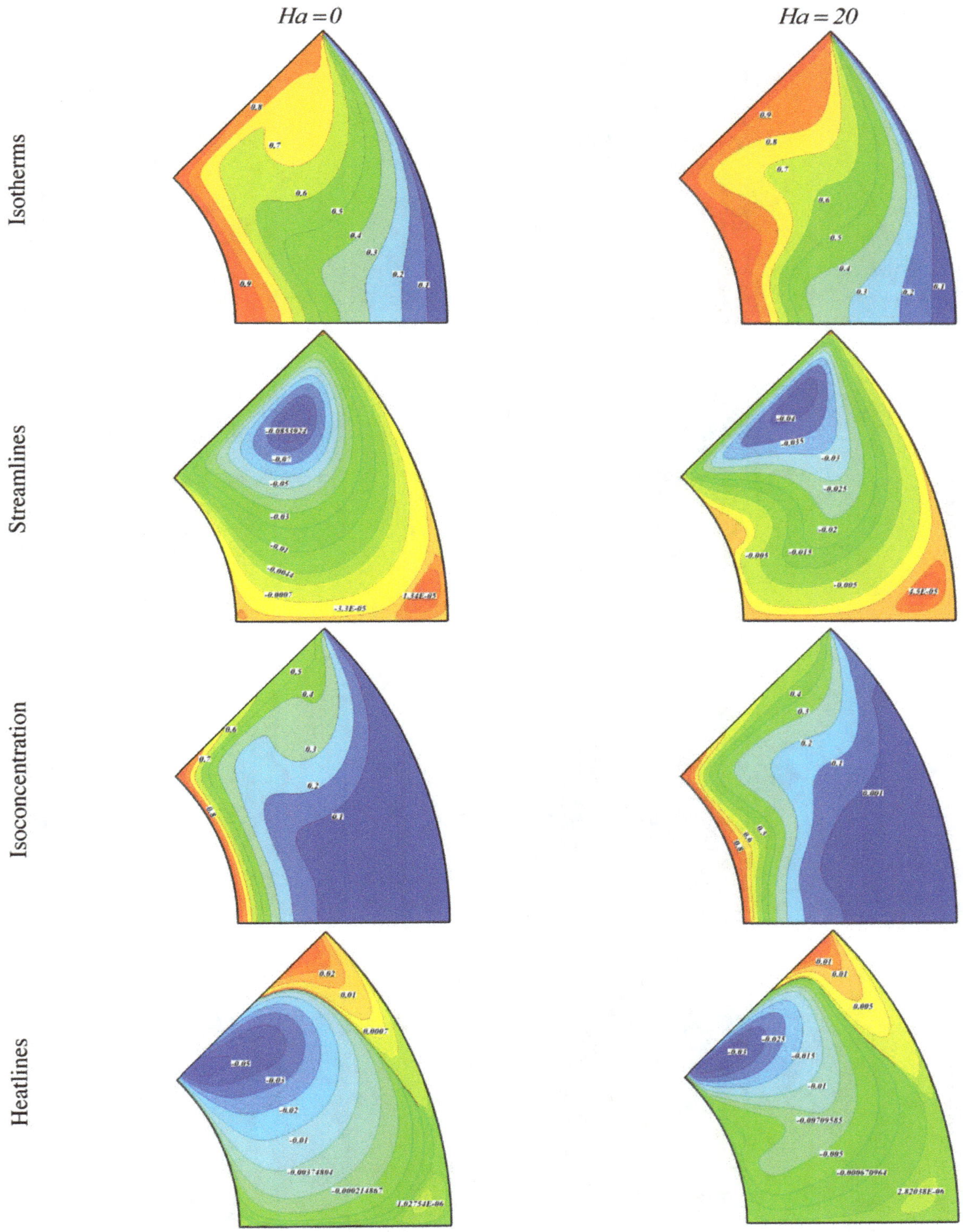

Figure 20. Comparison of the isotherms, streamlines, isoconcentration and heatline contours for different values of Hartmann number when $Le = 4, Nt = Nb = 0.5, \mathrm{Re} = 500$ *and* $\mathrm{Pr} = 6.85$.

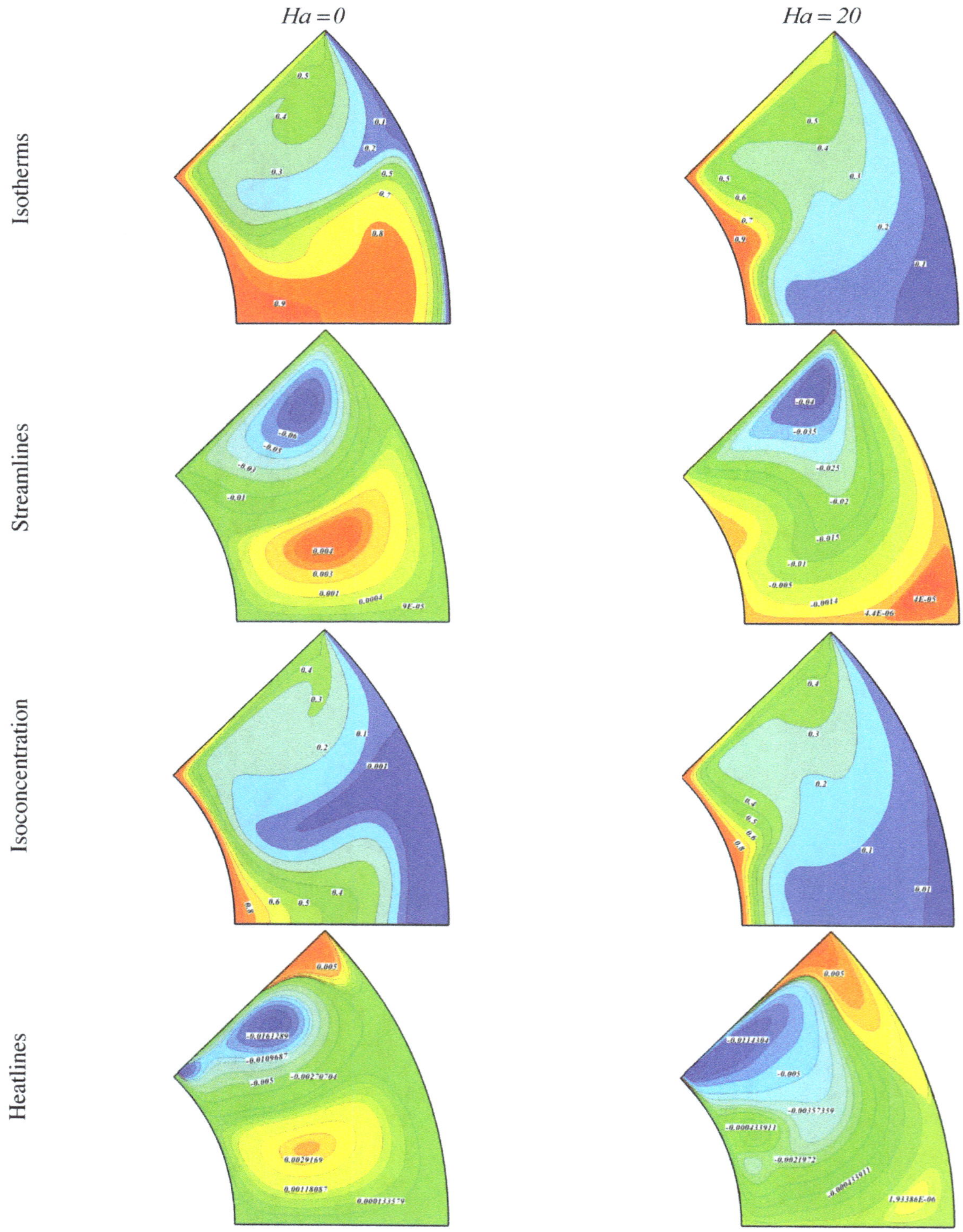

Figure 21. Effects of Hartmann number, Reynolds number and Lewis number on local Nusselt number Nu_{loc} along hot wall

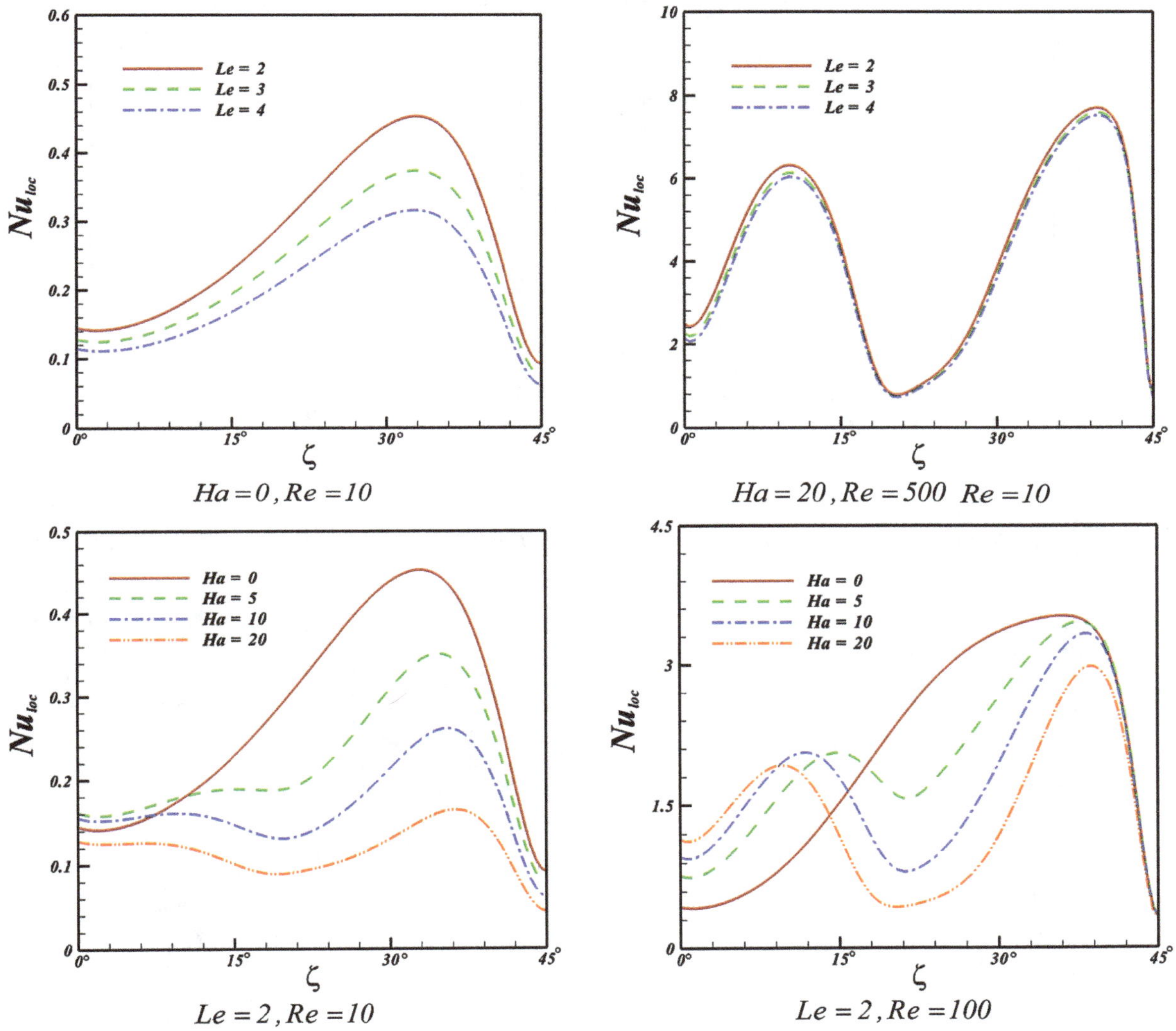

4. NON-UNIFORM MAGNETIC FIELD EFFECT ON NANOFLUID HYDROTHERMAL TREATMENT CONSIDERING BROWNIAN MOTION AND THERMOPHORESIS EFFECTS

4.1. Problem Definition

The physical model along with the important geometrical parameters and the mesh of the enclosure used in the present CVFEM program is shown in Figure 23 (Sheikholeslami and Rashidi, 2016). The inner wall is maintained at constant heat flux. The outer wall is maintained at constant temperature T_h. The

Figure 22. Effects of Hartmann number, Reynolds number and Lewis number on average Nusselt number Nu_{ave} along hot wall.

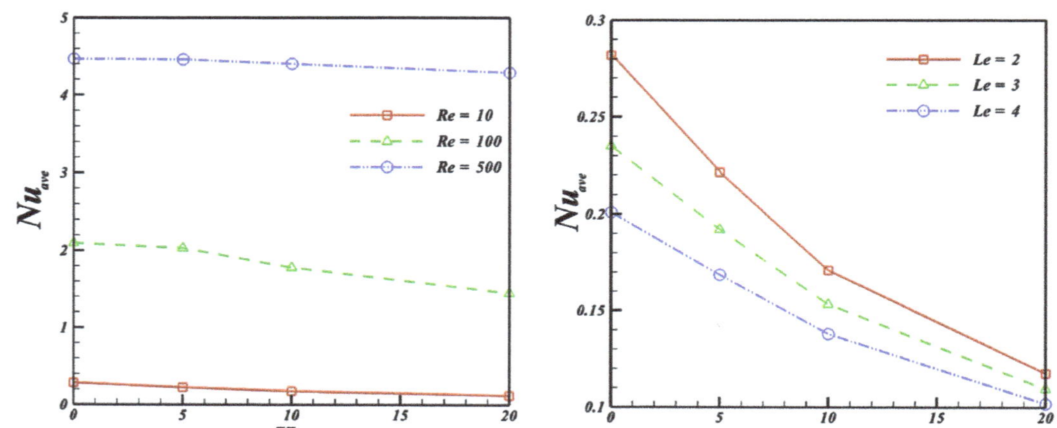

Le = 2 *Re = 10*

Figure 23. (a)Geometry and the boundary conditions; (b) the mesh of enclosure considered in this work.

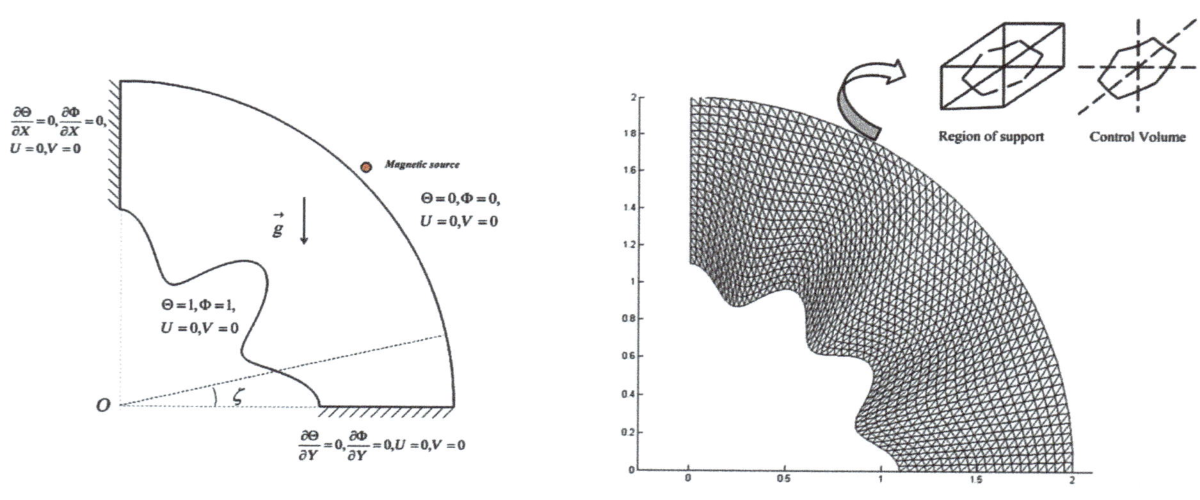

(a) (b)

inner and outer walls are maintained at constant concentration C_h and C_c respectively. The shape of inner cylinder profile is assumed to mimic the following pattern

$$r = r_{in} + A\cos\left(N\left(\zeta\right)\right) \tag{57}$$

in which r_{in} is the base circle radius, r_{out} is the radius of outer cylinder, A and N are amplitude and number of undulations, respectively, ζ is the rotation angle. In this study A and N equal to 0.2 and 4, respectively. The contours of the magnetic field strength are shown in Figure 24. In this study magnetic source is located at $\left(-0.05cols, 0.5rows\right)$.

The nanofluid's density, ρ is

$$\begin{aligned} \rho &= \phi\rho_p + (1-\phi)\rho_f \\ &\cong \phi\rho_p + (1-\phi)\left\{\rho_{f_0}\left(1-\beta(T-T_c)\right)\right\} \end{aligned} \tag{58}$$

where ρ_f is the base fluid's density, T_c is a reference temperature, ρ_{f_0} is the base fluid's density at the reference temperature, β is the volumetric coefficient of expansion. Taking the density of base fluid as that of the nanofluid, the density becomes

$$\rho \cong \phi\rho_p + (1-\phi)\left\{\rho_0\left(1-\beta(T-T_c)\right)\right\} \tag{59}$$

ρ_0 is the nanofluid's density at the reference temperature.

Figure 24. Contours of the (a) magnetic field strength H; (b) magnetic field intensity component in x direction Hx; (c) magnetic field intensity component in y direction Hy.

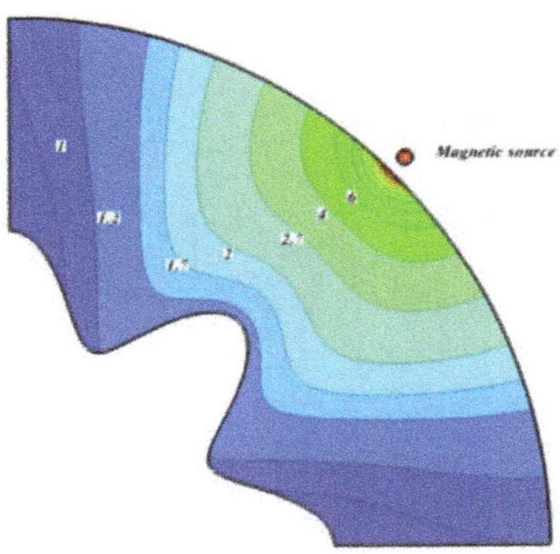

(a) $H(x,y)$

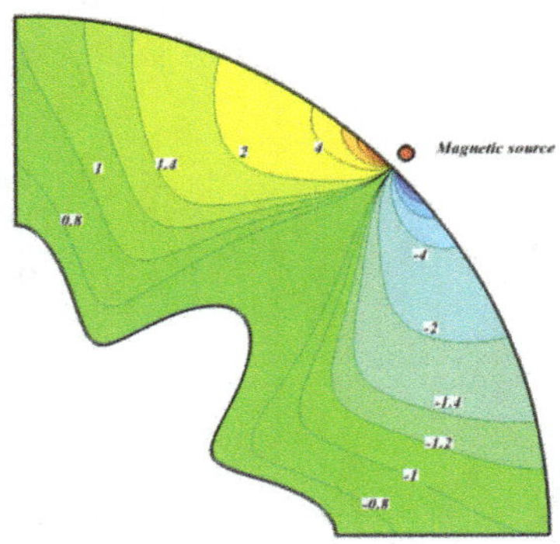

(b) $Hx(x,y)$

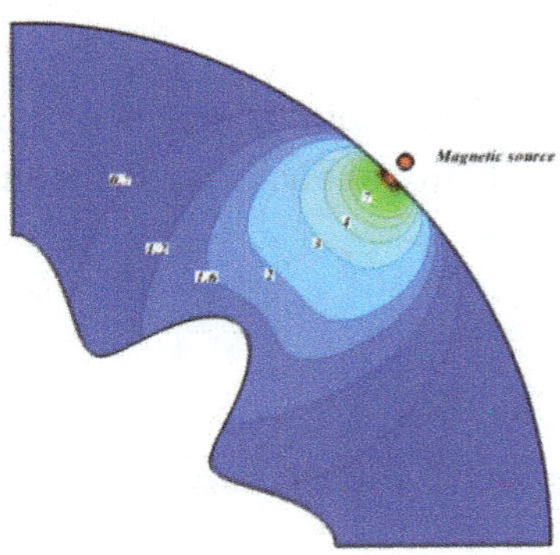

(c) $Hy(x,y)$

The continuity, momentum under Boussinesq approximation and energy equations for the laminar and steady state natural convection in a two-dimensional enclosure can be written in dimensional form as follows (Sheikholeslami and Rashidi, 2016):

$$\frac{\partial u}{\partial x}+\frac{\partial v}{\partial y}=0 \tag{60}$$

$$\rho_f\left\{u\frac{\partial u}{\partial x}+v\frac{\partial u}{\partial y}\right\}=-\frac{\partial P}{\partial x}+\mu\left(\frac{\partial^2 u}{\partial x^2}+\frac{\partial^2 u}{\partial y^2}\right)+\mu_0 M\frac{\partial \overline{H}}{\partial x}-\sigma B_y^2 u+\sigma B_x B_y v \tag{61}$$

$$\begin{aligned}\rho_f\left\{u\frac{\partial v}{\partial x}+v\frac{\partial v}{\partial y}\right\}&=-\frac{\partial P}{\partial y}+\mu\left(\frac{\partial^2 v}{\partial x^2}+\frac{\partial^2 v}{\partial y^2}\right)-\left(\phi-\phi_c\right)\left(\rho_p-\rho_{f_0}\right)g\\&+(1-\phi_c)\rho_{f_0}\left(T-T_c\right)g+\mu_0 M\frac{\partial \overline{H}}{\partial y}-\sigma B_x^2 v+\sigma B_x B_y u\end{aligned} \tag{62}$$

$$\begin{aligned}u\frac{\partial T}{\partial x}+v\frac{\partial T}{\partial y}&=\alpha\left(\frac{\partial^2 T}{\partial x^2}+\frac{\partial^2 T}{\partial y^2}\right)+\frac{(\rho c)_p}{(\rho c)_f}\left[\begin{array}{l}D_B\left\{\frac{\partial \phi}{\partial x}.\frac{\partial T}{\partial x}+\frac{\partial \phi}{\partial y}.\frac{\partial T}{\partial y}\right\}\\+(D_T/T_c)\left\{\left(\frac{\partial T}{\partial x}\right)^2+\left(\frac{\partial T}{\partial y}\right)^2\right\}\end{array}\right]\\&+\left(uB_y-vB_x\right)^2-\mu_0 T\frac{\partial M}{\partial T}\left(u\frac{\partial \overline{H}}{\partial x}+v\frac{\partial \overline{H}}{\partial y}\right)+\mu\left\{2\left(\frac{\partial u}{\partial x}\right)^2+2\left(\frac{\partial v}{\partial y}\right)^2+\left(\frac{\partial u}{\partial y}+\frac{\partial v}{\partial x}\right)^2\right\}\end{aligned} \tag{63}$$

$$u\frac{\partial \phi}{\partial x}+v\frac{\partial \phi}{\partial y}=D_B\left\{\frac{\partial^2 \phi}{\partial x^2}+\frac{\partial^2 \phi}{\partial y^2}\right\}+\left(\frac{D_T}{T_c}\right)\left\{\frac{\partial^2 T}{\partial x^2}+\frac{\partial^2 T}{\partial y^2}\right\} \tag{64}$$

For the variation of the magnetization M, with the magnetic field intensity $\overline{H}$ and temperature T, the following relation derived experimentally in Sheikholeslami and Ganji (2014) is considered:

$$M=K'\overline{H}\left(T_c'-T\right) \tag{65}$$

where K' is a constant and T_c' is the Curie temperature.

In the above equations, μ_0 is the magnetic permeability of vacuum $\left(4\pi \times 10^{-7} Tm / A\right)$, $\overline{H}$ is the magnetic field strength, $\overline{B}$ is the magnetic induction $\left(\overline{B} = \mu_0 \overline{H}\right)$ and the bar above the quantities denotes that they are dimensional.

The stream function and vorticity are defined as follows:

$$u = \frac{\partial \psi}{\partial y},\ v = -\frac{\partial \psi}{\partial x},\ \omega = \frac{\partial v}{\partial x} - \frac{\partial u}{\partial y} \tag{66}$$

Also the following non-dimensional variables should be introduced:

$$\begin{aligned} & X = \frac{x}{L}, Y = \frac{y}{L}, \Omega = \frac{\omega L^2}{\alpha}, \Psi = \frac{\psi}{\alpha}, \Theta = \frac{T - T_c}{T_h - T_c}, \Phi = \frac{\phi - \phi_c}{\phi_h - \phi_c}, \\ & U = \frac{uL}{\alpha}, V = \frac{vL}{\alpha}, H = \frac{\overline{H}}{\overline{\overline{H_0}}}, H_x = \frac{\overline{H_x}}{\overline{\overline{H_0}}}, H_y = \frac{\overline{H_y}}{\overline{\overline{H_0}}} \end{aligned} \tag{67}$$

By using these dimensionless parameters the equations become:

$$\begin{aligned} & \left[\frac{\partial \Psi}{\partial Y}\frac{\partial \Omega}{\partial X} - \frac{\partial \Psi}{\partial X}\frac{\partial \Omega}{\partial Y}\right] = \Pr\left(\frac{\partial^2 \Omega}{\partial X^2} + \frac{\partial^2 \Omega}{\partial Y^2}\right) + \Pr Ra\left(\frac{\partial \Theta}{\partial X} - Nr\frac{\partial \Phi}{\partial X}\right) \\ & + Mn_F \Pr\left\{\frac{\partial H}{\partial X}\frac{\partial \Theta}{\partial Y} - \frac{\partial H}{\partial Y}\frac{\partial \Theta}{\partial X}\right\} H \\ & - Ha^2 \Pr\times\left\{\frac{\partial V}{\partial X} H_x^2 + V\left(2H_x \frac{\partial H_x}{\partial X}\right) - \frac{\partial U}{\partial X} H_x H_y - U\frac{\partial H_x}{\partial X} H_y - U\frac{\partial H_y}{\partial X} H_x \right. \\ & \left. -\frac{\partial U}{\partial Y} H_y^2 - U\left(2H_y \frac{\partial H_y}{\partial Y}\right) + \frac{\partial V}{\partial Y} H_x H_y + V\frac{\partial H_x}{\partial Y} H_y + V\frac{\partial H_y}{\partial Y} H_x \right\} \end{aligned} \tag{68}$$

$$\begin{aligned} & \frac{\partial \Psi}{\partial Y}\frac{\partial \Theta}{\partial X} - \frac{\partial \Psi}{\partial X}\frac{\partial \Theta}{\partial Y} = \left(\frac{\partial^2 \Theta}{\partial X^2} + \frac{\partial^2 \Theta}{\partial Y^2}\right) + Nb\left(\frac{\partial \Phi}{\partial X}\frac{\partial \Theta}{\partial X} + \frac{\partial \Phi}{\partial Y}\frac{\partial \Theta}{\partial Y}\right) + Nt\left(\left(\frac{\partial \Theta}{\partial X}\right)^2 + \left(\frac{\partial \Theta}{\partial Y}\right)^2\right) \\ & + Ha^2\, Ec\left\{U H_y - V H_x\right\}^2 + Mn_F Ec\left\{U\frac{\partial H}{\partial X} + V\frac{\partial H}{\partial Y}\right\} H\left(\varepsilon_1 + \Theta\right) \\ & + Ec\left\{2\left(\frac{\partial U}{\partial X}\right)^2 + 2\left(\frac{\partial V}{\partial Y}\right)^2 + \left(\frac{\partial U}{\partial Y} + \frac{\partial V}{\partial X}\right)^2\right\} \end{aligned} \tag{69}$$

$$\frac{\partial \Psi}{\partial Y}\frac{\partial \Phi}{\partial X}-\frac{\partial \Psi}{\partial X}\frac{\partial \Phi}{\partial Y}=\frac{1}{Le}\left(\frac{\partial^2 \Phi}{\partial X^2}+\frac{\partial^2 \Phi}{\partial Y^2}\right)+\frac{Nt}{Nb\ Le}\left(\frac{\partial^2 \Theta}{\partial X^2}+\frac{\partial^2 \Theta}{\partial Y^2}\right) \tag{70}$$

$$\frac{\partial^2 \Psi}{\partial X^2}+\frac{\partial^2 \Psi}{\partial Y^2}=-\Omega \tag{71}$$

where thermal Rayleigh number, the buoyancy ratio number, Prandtl number, the Brownian motion parameter, the thermophoretic parameter, Lewis number, Hartmann number, Eckert number and Magnetic number arising from FHD of nanofluid are defined as:

$$Ra=\left(1-\phi_c\right)\rho_{f_0} g\beta L^3\left(T_h-T_c\right)/\left(\mu\alpha\right),$$

$$Nr=\left(\rho_p-\rho_0\right)\left(\phi_h-\phi_c\right)/\left[\left(1-\phi_c\right)\rho_{f_0}\beta L\left(T_h-T_c\right)\right],$$

$$\Pr=\mu/\rho_f\alpha,$$

$$Nb=(\rho c)_p D_B(\phi_h-\phi_c)/\left((\rho c)_f\alpha\right),$$

$$Nt=(\rho c)_p D_T(T_h-T_c)/[(\rho c)_f\alpha T_c],$$

$$Le=\alpha/D_B, Ha=LH_0\mu_0\sqrt{\sigma/\mu}, \mathrm{Ec}=\left(\alpha\mu\right)/\left[\left(\rho C_P\right)\Delta TL^2\right]$$

and

$$Mn_F=\mu_0 H_0^2 K'\left(T_h-T_c\right)L^2/\left(\mu\alpha\right),$$

respectively.

The local Nusselt number of the nanofluid along the inner wall can be expressed as:

$$Nu_{local}=\left(\frac{k_{nf}}{k_f}\right)\frac{1}{\theta}\bigg|_{inner\ wall} \tag{72}$$

The average Nusselt number on the hot wall is evaluated as:

$$Nu_{ave} = \frac{1}{0.5\pi} \int_{0}^{0.5\pi} Nu_{loc} \, d\gamma \tag{73}$$

4.2. Effects of Active Parameters

Natural convection heat transfer in an enclosure filled with nanofluid external magnetic field is investigated numerically. Effects of Rayleigh number ($Ra = 10^3, 10^4, 10^5$), buoyancy ratio number ($Nr = 0.1$ and 4), Hartmann number ($Ha = 0, 2, 6$ and 10) and Lewis number ($Le = 2$ and 4) on flow and heat transfer characteristics are examined. In all calculations, the Prandtl number ($\Pr$), temperature number (ε_1), Eckert number (Ec), Brownian motion parameter (Nb) and thermophoretic parameter of nanofluid (Nt) are set to 6.85, $0.0, 10^{-6}, 0.5$ and 0.5, respectively.

Effects of Hartmann number, Rayleigh number, buoyancy ratio number and Lewis number on isotherms, streamlines, isoconcentration and heatline contours are shown in Figure 25, 26, 27 and 28. At $Ra = 10^3$ the conduction heat transfer mechanism is more pronounced. For this reason the isotherms are parallel to each other. As Ra increases, the distribution of isotherm contours increases. At $Ra=10^3$, two equal eddies exist which are symmetric respect to $\zeta = 45°$. The strength of upper eddy increases with increase of Hartmann number.

With increase of Ra, the role of convection in heat transfer becomes more significant. Also it can be seen that as Ra increases the distribution of isoconcentration contours increases. The heat flow within the enclosure is displayed using the heat function obtained from conductive heat fluxes ($\partial\Theta / \partial X, \partial\Theta / \partial Y$) as well as convective heat fluxes ($V\Theta, U\Theta$). Heatlines emanate from hot regimes and end on cold regimes illustrating the path of heat flow. The domination of conduction heat transfer in low Rayleigh number can be observed from the heatline patterns since no passive area exists. The increase of Ra causes the clustering of heatlines from hot to the cold wall and generates passive heat transfer area in which heat is rotated without having significant effect on heat transfer between walls. By increasing buoyancy ratio, a small eddy which is rotates clock wise appears near the vertical wall. This eddy disappears by increasing Hartmann number. As Hartmann number increases, the Lorentz forces increase and in turn the nanofluid flow suppressed. So, thermal boundary layer thickness increases with increase of Lorentz forces. As Rayleigh number increases, the buoyancy forces increases and in turn thermal boundary layer thickness near the hot wall decreases. Similar trend is observed for buoyancy ratio number and Lewis number.

Effects of Hartmann number, Rayleigh number, buoyancy ratio number and Lewis number on local Nusselt number are shown in Figure 29. The profiles of the Nu_{loc} profiles have local extremes, which are related to the thermal plumes and crests over the inner cylinder. Also it can be found that local Nusselt number decreases with increase of Hartmann number. Figure 30 shows the effect of Hartmann number, Rayleigh number, buoyancy ratio number and Lewis number on average Nusselt number. Average Nusselt number increases with increase of Rayleigh number, buoyancy ratio number and Lewis number while it decreases with increase of Hartmann number.

Figure 25. Comparison of the isotherms, streamlines, isoconcentration and heatline contours for different values of Hartmann number when $Nr = 0.1, Le = 2, Nt = Nb = 0.5, Ra = 10^3, Mn_F = 5$ *and* $\Pr = 6.85$.

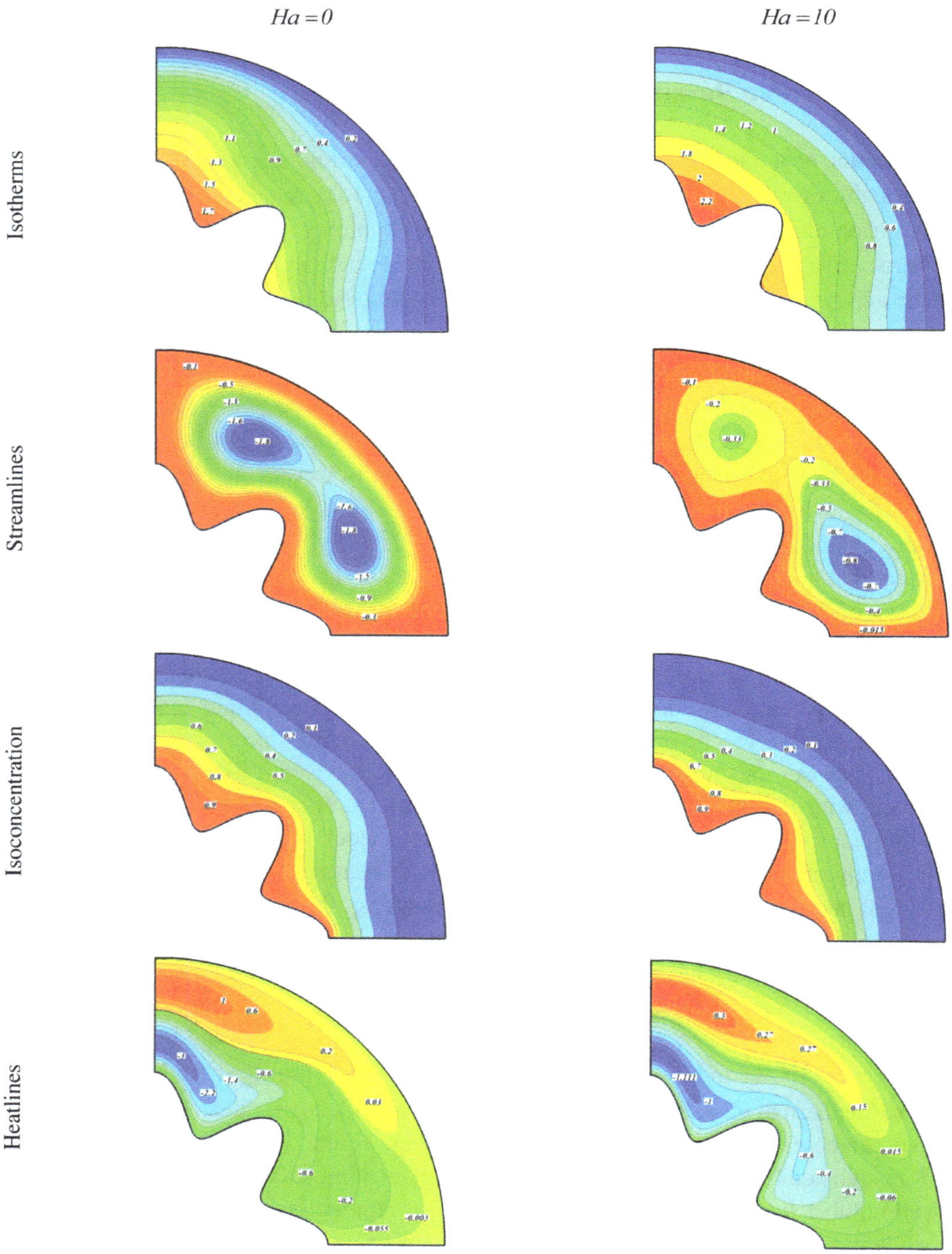

Figure 26. Comparison of the isotherms, streamlines, isoconcentration and heatline contours for different values of Hartmann number when $Nr = 0.1, Le = 2, Nt = Nb = 0.5, Ra = 10^5, Mn_F = 5$ *and* $\Pr = 6.85$.

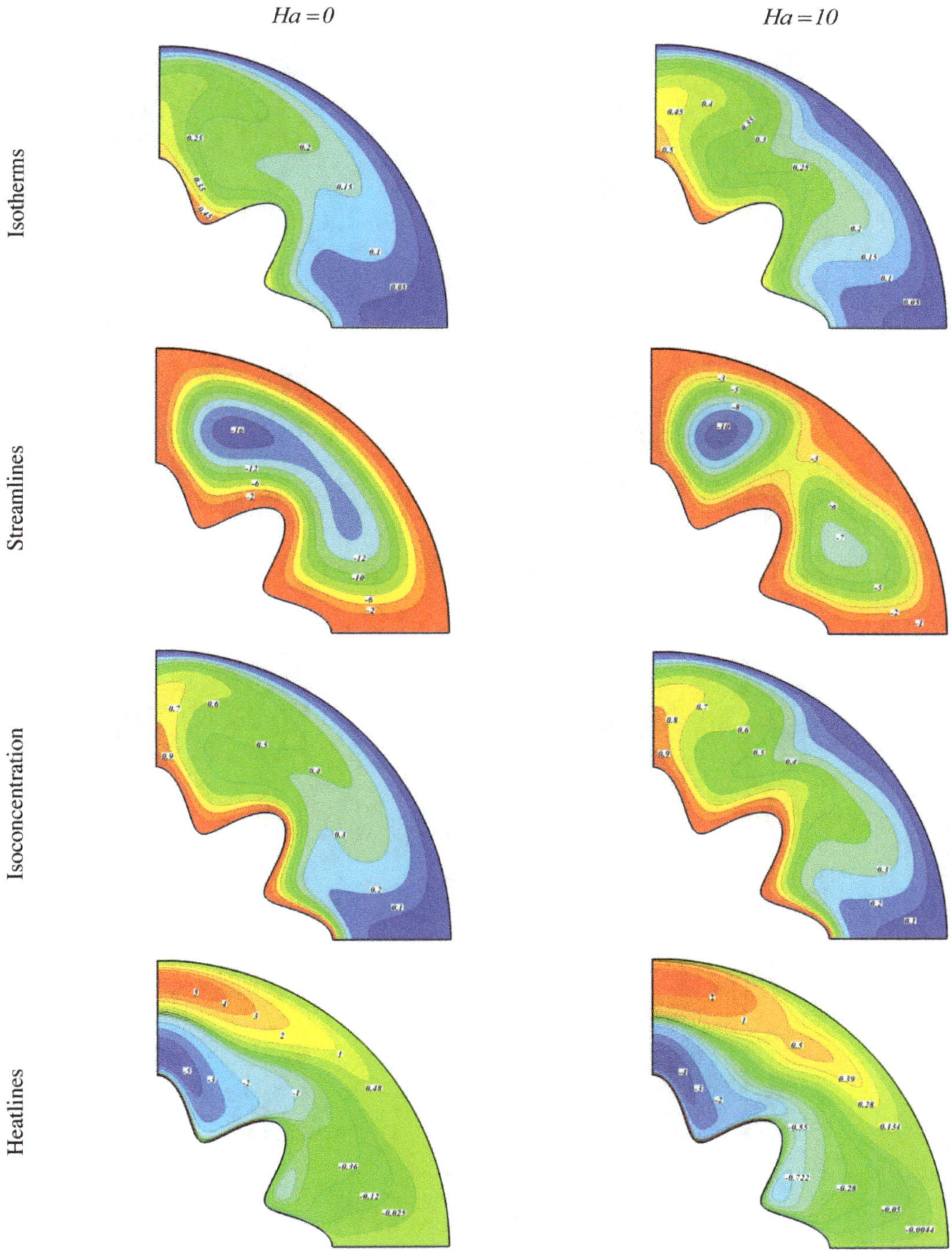

Figure 27. Comparison of the isotherms, streamlines, isoconcentration and heatline contours for different values of Hartmann number when $Nr = 4, Le = 2, Nt = Nb = 0.5, Ra = 10^5, Mn_F = 5$ *and* $\Pr = 6.85$.

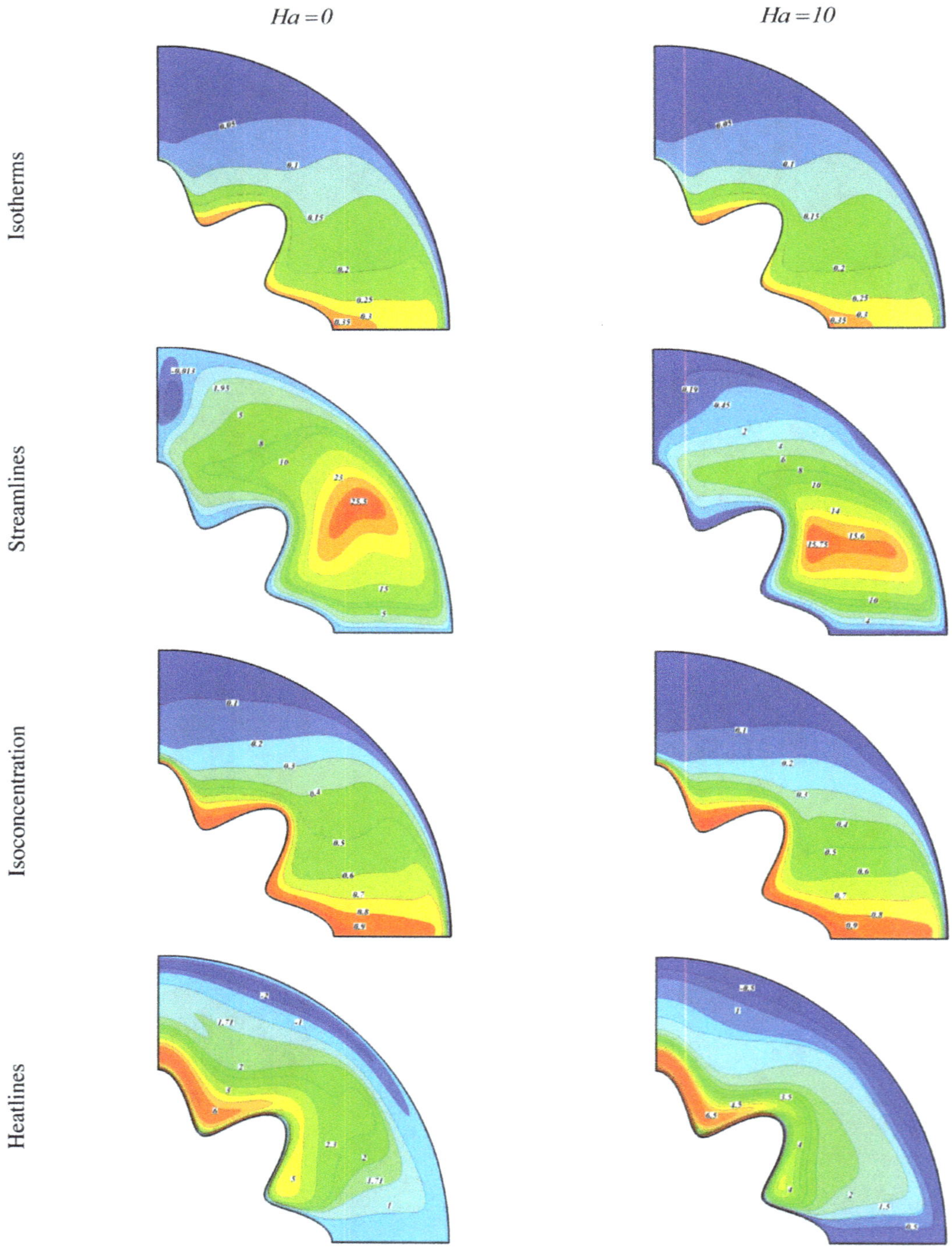

Figure 28. Comparison of the isotherms, streamlines, isoconcentration and heatline contours for different values of Hartmann number when $Nr = 0.1, Le = 4, Nt = Nb = 0.5, Ra = 10^5, Mn_F = 5$ *and* $\Pr = 6.85$.

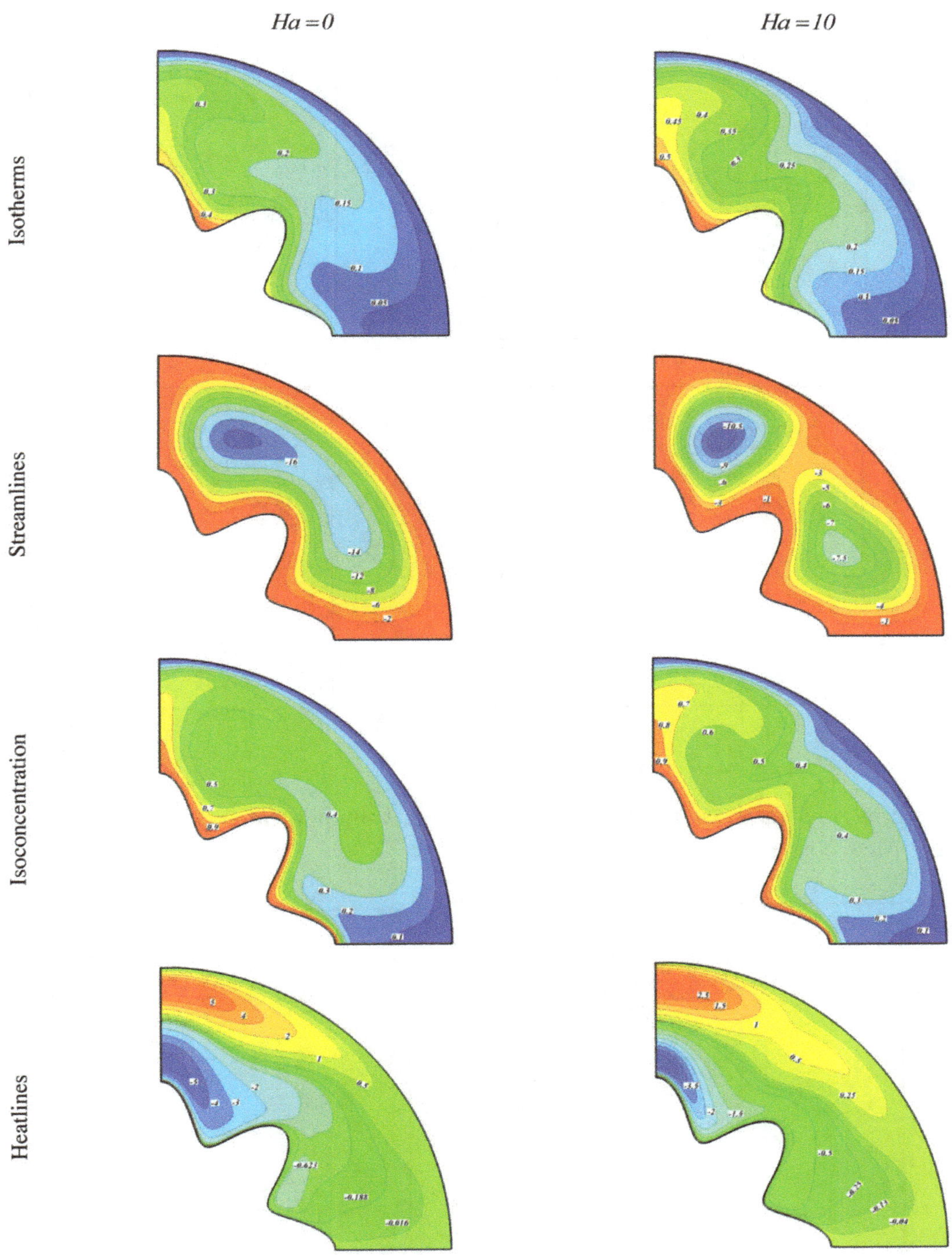

Figure 29. Effects of Hartmann number, Rayleigh number, buoyancy ratio number and Lewis number on local Nusselt number Nu_{loc} along cold wall

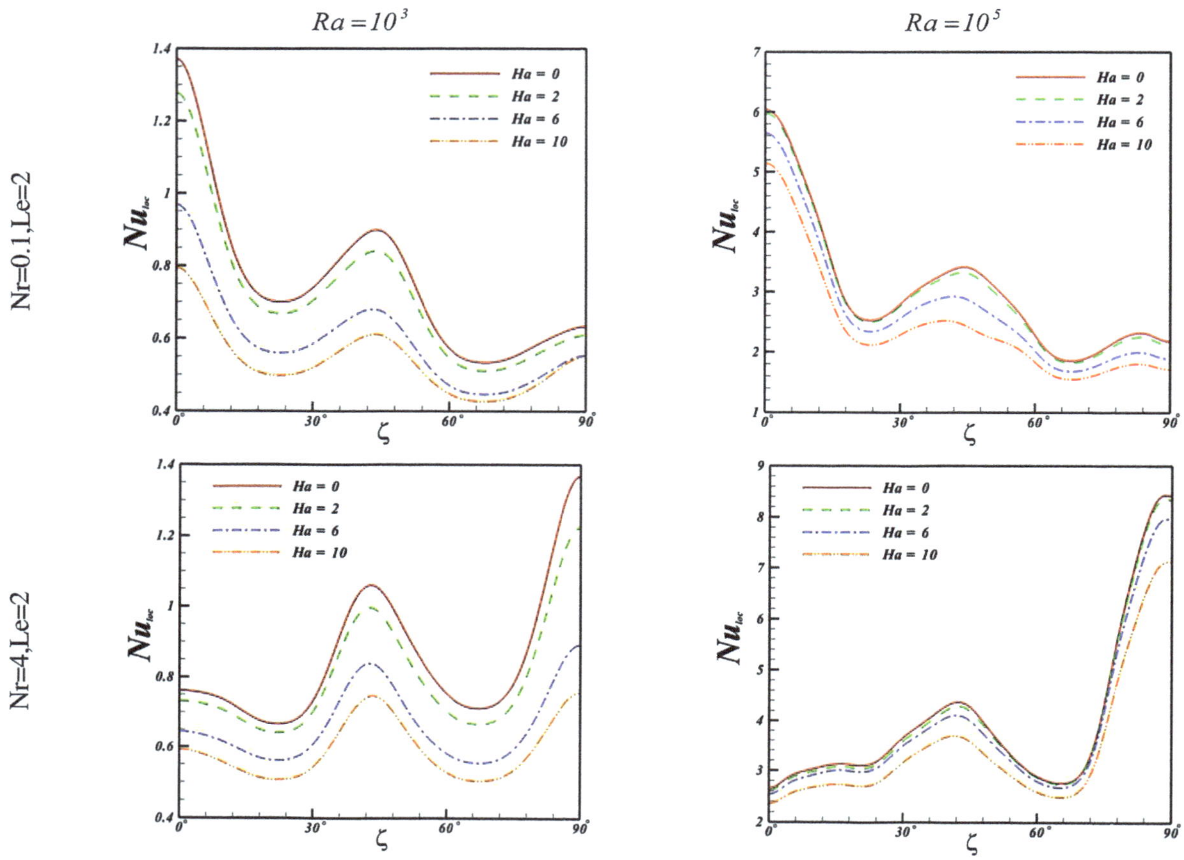

Figure 30. Effects of Hartmann number, Rayleigh number, buoyancy ratio number and Lewis number on average Nusselt number

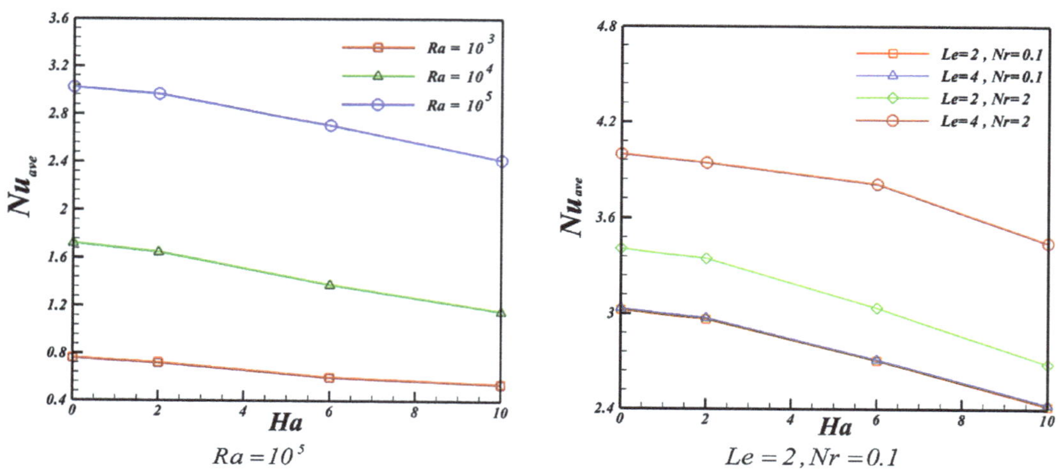

5. FERROFLUID MIXED CONVECTION HEAT TRANSFER IN EXISTENCE OF VARIABLE MAGNETIC FIELD

5.1. Problem Definition

The schematic diagram and the mesh of the semi-annulus enclosure used in the present CVFEM program are shown in Figure 31 (Sheikholeslami and Rashidi, 2015). The inner wall is maintained at constant temperatures T_h and the other walls are maintained at constant temperature $T_c\left(T_h > T_c\right)$. For the expression of the magnetic field strength it can be considered that the magnetic source represents a magnetic wire placed vertically to the x-y plane at the point $\left(\bar{a}, \bar{b}\right)$. The contours of the magnetic field strength are shown in Figure 32. In this study magnetic source is located at $\left(-0.01\,\mathrm{cols}, 0.5\,\mathrm{rows}\right)$. The upper wall is Lid driven with velocity of U_{Lid}.

The flow is two-dimensional, laminar and incompressible. The magnetic Reynolds number is assumed to be small so that the induced magnetic field can be neglected compared to the applied magnetic field. The flow is considered to be steady, two dimensional and laminar. Using the Boussinesq approximation, the governing equations of heat transfer and fluid flow for nanofluid can be obtained as follows:

$$\frac{\partial u}{\partial x} + \frac{\partial v}{\partial y} = 0 \tag{74}$$

Figure 31. (a)Geometry and the boundary conditions; (b) the mesh of enclosure considered in this work.

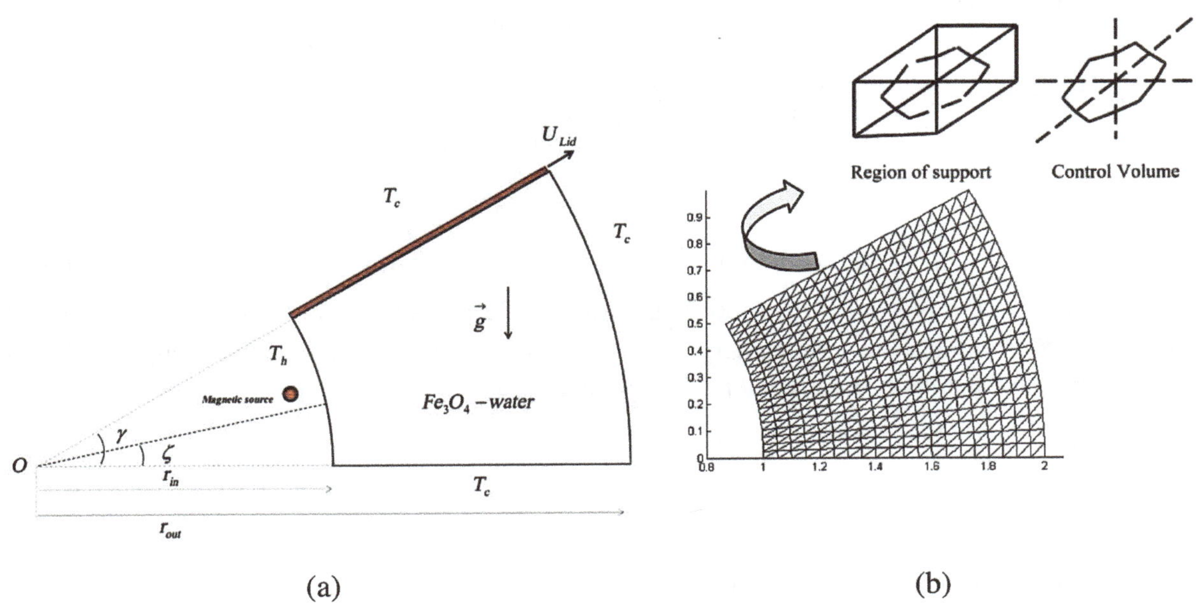

Figure 32. Contours of the (a) magnetic field strength H; (b) magnetic field intensity component in x direction Hx; (c) magnetic field intensity component in y direction Hy.

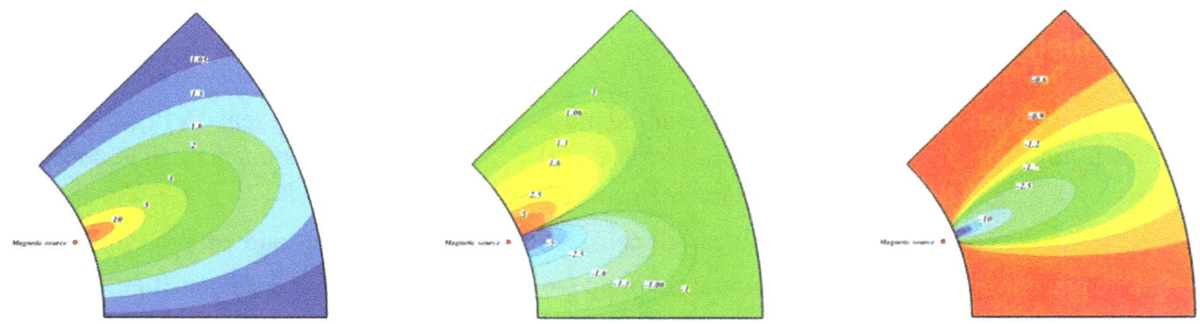

(a) $H(x,y)$ (b) $Hx(x,y)$ (c) $Hy(x,y)$

$$\rho_{nf}\left(u\frac{\partial u}{\partial x}+v\frac{\partial u}{\partial y}\right)=-\frac{\partial P}{\partial x}+\eta\left(\frac{\partial^2 u}{\partial x^2}+\frac{\partial^2 u}{\partial y^2}\right)+\mu_0 M\frac{\partial \overline{H}}{\partial x}-\sigma_{nf}B_y^2 u+\sigma_{nf}B_x B_y v \tag{75}$$

$$\rho_{nf}\left(u\frac{\partial v}{\partial x}+v\frac{\partial v}{\partial y}\right)=-\frac{\partial P}{\partial y}+\eta\left(\frac{\partial^2 v}{\partial x^2}+\frac{\partial^2 v}{\partial y^2}\right)+\mu_0 M\frac{\partial \overline{H}}{\partial y}-\sigma_{nf}B_x^2 v+\sigma_{nf}B_x B_y u+\rho_{nf}\beta_{nf}g\left(T-T_c\right) \tag{76}$$

$$\begin{gathered}\left(\rho C_p\right)_{nf}\left(u\frac{\partial T}{\partial x}+v\frac{\partial T}{\partial y}\right)=k_{nf}\left(\frac{\partial^2 T}{\partial x^2}+\frac{\partial^2 T}{\partial y^2}\right)+\sigma_{nf}\left(uB_y-vB_x\right)^2\\-\mu_0 T\frac{\partial M}{\partial T}\left(u\frac{\partial \overline{H}}{\partial x}+v\frac{\partial \overline{H}}{\partial y}\right)+\eta\left\{2\left(\frac{\partial u}{\partial x}\right)^2+2\left(\frac{\partial v}{\partial y}\right)^2+\left(\frac{\partial u}{\partial y}+\frac{\partial v}{\partial x}\right)^2\right\}\end{gathered} \tag{77}$$

where $\eta=(1+\vec{\delta}.\vec{B})\mu_{nf}$, the variation of MFD viscosity (δ) has been taken to be isotropic, $\delta_1=\delta_2=\delta_3=\delta$. For the variation of the magnetization M, with the magnetic field intensity $\overline{H}$ and temperature T, the following relation derived experimentally in Sheikholeslami and Ganji (2014) and also used in Loukopoulos and Tzirtzilakis (2004) is considered:

$$M=K'\overline{H}\left(T_c'-T\right) \tag{78}$$

where K' is a constant and T_c' is the Curie temperature.

In the above equations, μ_0 is the magnetic permeability of vacuum $\left(4\pi \times 10^{-7} Tm / A\right)$, $\overline{H}$ is the magnetic field strength, $\overline{B}$ is the magnetic induction $\left(\overline{B} = \mu_0 \overline{H}\right)$ and the bar above the quantities denotes that they are dimensional.

The effective density, heat capacitance, thermal diffusivity, thermal expansion coefficient, dynamic viscosity and effective electrical conductivity of the nanofluid are defined as:

$$\rho_{nf} = \rho_f(1-\phi) + \rho_s\phi \tag{79}$$

$$\left(\rho C_p\right)_{nf} = \left(\rho C_p\right)_f (1-\phi) + \left(\rho C_p\right)_s \phi \tag{80}$$

$$\alpha_{nf} = k_{nf} / \left(\rho C_p\right)_{nf} \tag{81}$$

$$\beta_{nf} = \beta_f(1-\phi) + \beta_s\phi \tag{82}$$

$$\mu_{nf} = \mu_f(1-\phi)^{-2.5} \tag{83}$$

$$\frac{k_{nf}}{k_f} = \frac{k_s + 2k_f - 2\phi(k_f - k_s)}{k_s + 2k_f + \phi(k_f - k_s)} \tag{84}$$

$$\frac{\sigma_{nf}}{\sigma_f} = 1 + 3\left(\sigma_s / \sigma_f - 1\right)\phi / \left[\left(\sigma_s / \sigma_f + 2\right) - \left(\sigma_s / \sigma_f - 1\right)\phi\right] \tag{85}$$

By introducing the following non-dimensional variables:

$$X = \frac{x}{L}, Y = \frac{y}{L}, \Theta = \frac{T - T_c}{T_h - T_c}, U = \frac{u}{U_{Lid}}, V = \frac{v}{U_{Lid}}, H = \frac{\overline{H}}{\overline{H_0}}, H_x = \frac{\overline{H_x}}{\overline{H_0}}, H_y = \frac{\overline{H_y}}{\overline{H_0}} \tag{86}$$

where in Equation (86) $\overline{H_0} = \overline{H}\left(\overline{a}, 0\right) = \frac{\gamma}{2\pi\left|\overline{b}\right|}$ and $L = r_{out} - r_{in} = r_{in}$.

Using the dimensionless parameters, the equations now become:

$$\frac{\partial U}{\partial X}+\frac{\partial V}{\partial Y}=0 \tag{87}$$

$$\begin{aligned}U\frac{\partial U}{\partial X}+V\frac{\partial U}{\partial Y}&=-\frac{\partial P}{\partial X}+\frac{1}{\mathrm{Re}}\left[\frac{\mu_{nf}/\mu_f}{\rho_{nf}/\rho_f}\right]\left(1+\delta^*\left(H_x+H_y\right)\right)\left(\frac{\partial^2 U}{\partial X^2}+\frac{\partial^2 U}{\partial Y^2}\right)\\&+Mn_F\left(\frac{\rho_f}{\rho_{nf}}\right)\left(\varepsilon_2-\varepsilon_1-\Theta\right)H\frac{\partial H}{\partial X}-\frac{Ha^2}{\mathrm{Re}}\left[\frac{\sigma_{nf}/\sigma_f}{\rho_{nf}/\rho_f}\right]\left(H_y^2U-H_xH_yV\right)\end{aligned} \tag{88}$$

$$\begin{aligned}U\frac{\partial V}{\partial X}+V\frac{\partial V}{\partial Y}&=-\frac{\partial P}{\partial X}+\frac{1}{\mathrm{Re}}\left[\frac{\mu_{nf}/\mu_f}{\rho_{nf}/\rho_f}\right]\left(1+\delta^*\left(H_x+H_y\right)\right)\left(\frac{\partial^2 V}{\partial X^2}+\frac{\partial^2 V}{\partial Y^2}\right)\\&+Mn_F\left(\frac{\rho_f}{\rho_{nf}}\right)\left(\varepsilon_2-\varepsilon_1-\Theta\right)H\frac{\partial H}{\partial Y}-\frac{Ha^2}{\mathrm{Re}}\left[\frac{\sigma_{nf}/\sigma_f}{\rho_{nf}/\rho_f}\right]\left(H_x^2V-H_xH_yU\right)+\frac{Gr}{\mathrm{Re}^2}\left[\frac{\beta_{nf}}{\beta_f}\right]\Theta\end{aligned} \tag{89}$$

$$\begin{aligned}U\frac{\partial \Theta}{\partial X}+V\frac{\partial \Theta}{\partial Y}&=\frac{1}{\mathrm{Pr\,Re}}\left[\frac{\dfrac{k_{nf}}{k_f}}{\dfrac{(\rho C_P)_{nf}}{(\rho C_P)_f}}\right]\left(\frac{\partial^2 \Theta}{\partial X^2}+\frac{\partial^2 \Theta}{\partial Y^2}\right)\\&+Ha^2\frac{Ec}{\mathrm{Re}}\left[\frac{\dfrac{\sigma_{nf}}{\sigma_f}}{\dfrac{(\rho C_P)_{nf}}{(\rho C_P)_f}}\right]\left\{U\,H_y-V\,H_x\right\}^2+Mn_F Ec\frac{(\rho C_P)_f}{(\rho C_P)_{nf}}\left\{U\frac{\partial H}{\partial X}+V\frac{\partial H}{\partial Y}\right\}H\left(\varepsilon_1+\Theta\right)\\&+\frac{Ec}{\mathrm{Re}}\left[\frac{\dfrac{\mu_{nf}}{\mu_f}}{\dfrac{(\rho C_P)_{nf}}{(\rho C_P)_f}}\right]\left(1+\delta^*\left(H_x+H_y\right)\right)\left\{2\left(\frac{\partial U}{\partial X}\right)^2+2\left(\frac{\partial V}{\partial Y}\right)^2+\left(\frac{\partial U}{\partial Y}+\frac{\partial V}{\partial X}\right)^2\right\}\end{aligned} \tag{90}$$

where

$$Re = \frac{\rho_f L U_{Lid}}{\mu_f}, Gr = g\beta\Delta T L^3 / \upsilon^2, Ha = LH_0\mu_0\sqrt{\sigma_f / \mu_f}, \varepsilon_1 = T_1 / \Delta T$$

$$Ec = \left(\rho_f U_{Lid}^{\ 2}\right) / \left[\left(\rho C_P\right)_f \Delta T\right] Ri = {Gr}/{\mathrm{Re}^2}, \delta^* = \delta\mu_0 H_0$$

and

$$Mn_F = \mu_0 H_0^2 K'\left(T_h - T_c\right) / \left(\rho_f U_{Lid}^{\ 2}\right)$$

are the Reynolds number, Grashof number, Hartmann number arising from MHD, temperature number, Eckert number, Richardson number, viscosity parameter and Magnetic number arising from FHD the for the base fluid, respectively. The stream function and vorticity are defined as:

$$u = \frac{\partial\psi}{\partial y}, v = -\frac{\partial\psi}{\partial x}, \omega = \frac{\partial v}{\partial x} - \frac{\partial u}{\partial y}$$
$$\Omega = \frac{\omega L}{U_{Lid}}, \Psi = \frac{\psi}{L U_{Lid}} \quad (91)$$

The stream function satisfies the continuity Equation (87). The vorticity equation is obtained by eliminating the pressure between the two momentum equations, i.e. by taking y-derivative of Equation (88) and subtracting from it the x-derivative of Equation (89). The boundary conditions as shown in Figure 31 are:

$\Theta = 1.0$ on the inner circular boundary

$\Theta = 0.0$ on the other walls (92)

The values of vorticity on the boundary of the enclosure can be obtained using the stream function formulation and the known velocity conditions during the iterative solution procedure. The local Nusselt number of the nanofluid along the hot wall can be expressed as:

$$Nu_{loc} = \left(\frac{k_{nf}}{k_f}\right)\frac{\partial\Theta}{\partial r} \quad (93)$$

where *r* is the radial direction. The average Nusselt number on the hot circular wall is evaluated as:

$$Nu_{ave} = \frac{1}{\gamma}\int_{0}^{\gamma} Nu_{loc}\left(\zeta\right)d\zeta \tag{94}$$

5.2. Effects of Active Parameters

In this section, mixed convection heat transfer of ferrofluid in presence of variable magnetic field is investigated using CVFEM. Also the effect of magnetic field dependent (MFD) viscosity on hydrothermal behavior is considered. Calculations are made for various values of volume fraction of nanoparticles ($\phi = 0\%$ and 4%), Richardson numbers ($Ri = 0.001, 1$ and 10), Magnetic number ($Mn_F = 0, 2, 4, 6$ and 10), Hartmann number ($Ha = 0,\ 5$ and 10) and viscosity parameter ($\delta^* = 0, 0.2, 0.4$ and 0.6). In all calculations, the Prandtl number ($\Pr$), temperature number (ε_1), Eckert number (Ec) and Reynolds number (Re) are set to 6.8, 0.0, 10^{-5} and 100, respectively.

Comparison of the streamlines between nanofluid and pure fluid is shown in Figure 33. The velocity components of nanofluid are increased because of an increase in the energy transport in the fluid with the increasing of volume fraction. Thermal boundary layer thickness decreases with increase of nanofluid volume fraction. Isotherms and streamlines contours for different values of viscosity parameter, Richardson, Hartmann and Magnetic numbers are shown in Figures 34, 35 and 36. When $Ri = 0.01$, the heat transfer in the enclosure is mainly dominated by the conduction mode. At $Mn_F = 0, Ha = 0$ the streamlines show one rotating eddy.

When Hartmann number increases a small vortex generates near the magnetic source, so thermal plume appears in this region. As magnetic number increases, the main vortex turns in to two smaller vortexes and in turn reverse thermal plume appears near the location of magnetic source. As Richardson number increases, the role of convection in heat transfer becomes more significant. At $Ri = 10$, the center of main vortex move downward and thermal boundary layer thickness near the inner wall becomes thinner. When Hartmann and Magnetic numbers increase up to 10, the main vortex turns in to three

Figure 33. Comparison of the streamlines between nanofluid ($\phi = 0.04$) ($-\cdot-\cdot-$) and pure fluid ($\phi = 0$) (—) when $Ri = 10, \mathrm{Re} = 100,\ Mn_F = 10,\ Ha = 10, Ec = 10^{-5}, \varepsilon_1 = 0, \delta^* = 0, \Pr = 6.8$.

Figure 34. Isotherms (left) and streamlines (right) contours for different values of Richardson number and Magnetic number when $Ha = 0, \delta^* = 0$.

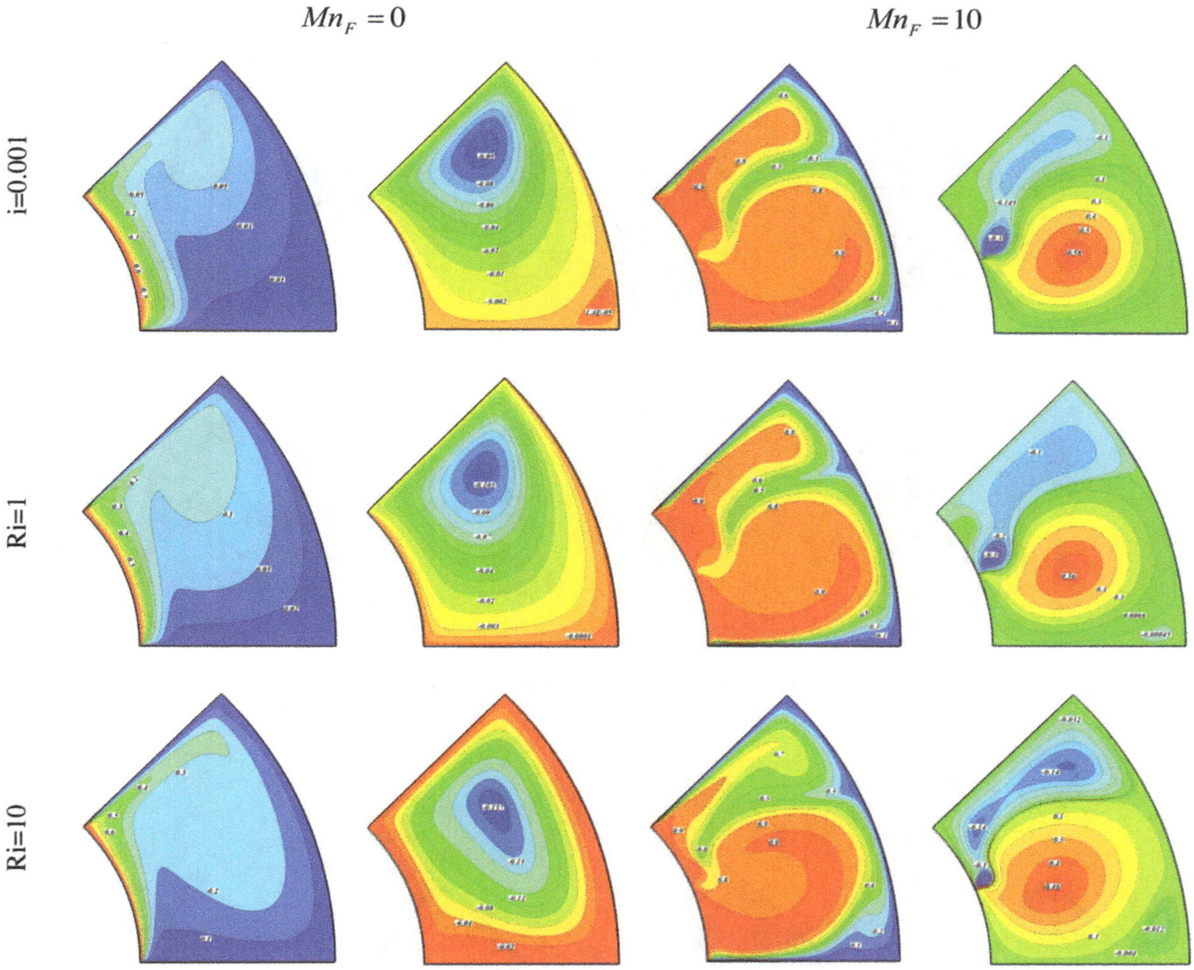

eddies. Three thermal plumes generate over the inner wall due to existence of eddies which are rotated in different direction. Also it can be seen that as viscosity parameter increases thermal boundary layer thickness decreases and one counter clock wise eddy generates at center of the enclosure.

Figure 37 depicts that the effects of Magnetic number, Hartmann number, viscosity parameter and Richardson number on average Nusselt number. Thermal boundary layer thickness increases with increase of Magnetic number, Hartmann number while it decreases with augment of Richardson number. So, average Nusselt number increases with increase of Richardson number while it decreases with increase of Hartmann number and Magnetic number. Also it can be concluded that by considering the effect of magnetic field on viscosity of the fluid, rate of heat transfer increases.

Figure 35. Isotherms (up) and streamlines (down) contours for different values of Richardson number when $Mn_F = 0, \mathrm{Ha} = 10, \delta^* = 0$.

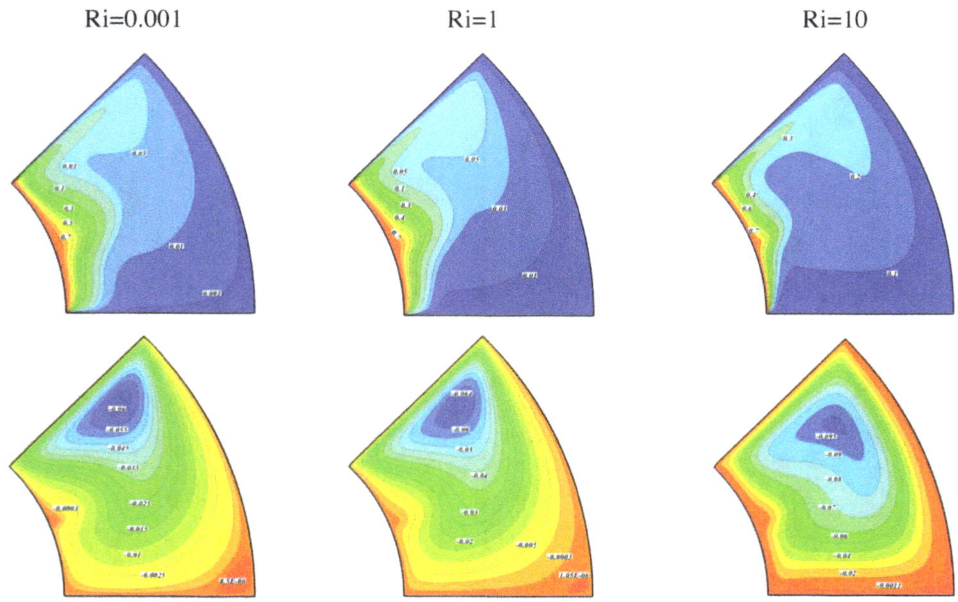

Figure 36. Isotherms (left) and streamlines (right) contours for different values of δ^*, Ha *and* Mn_F *when* $Ri = 10$.

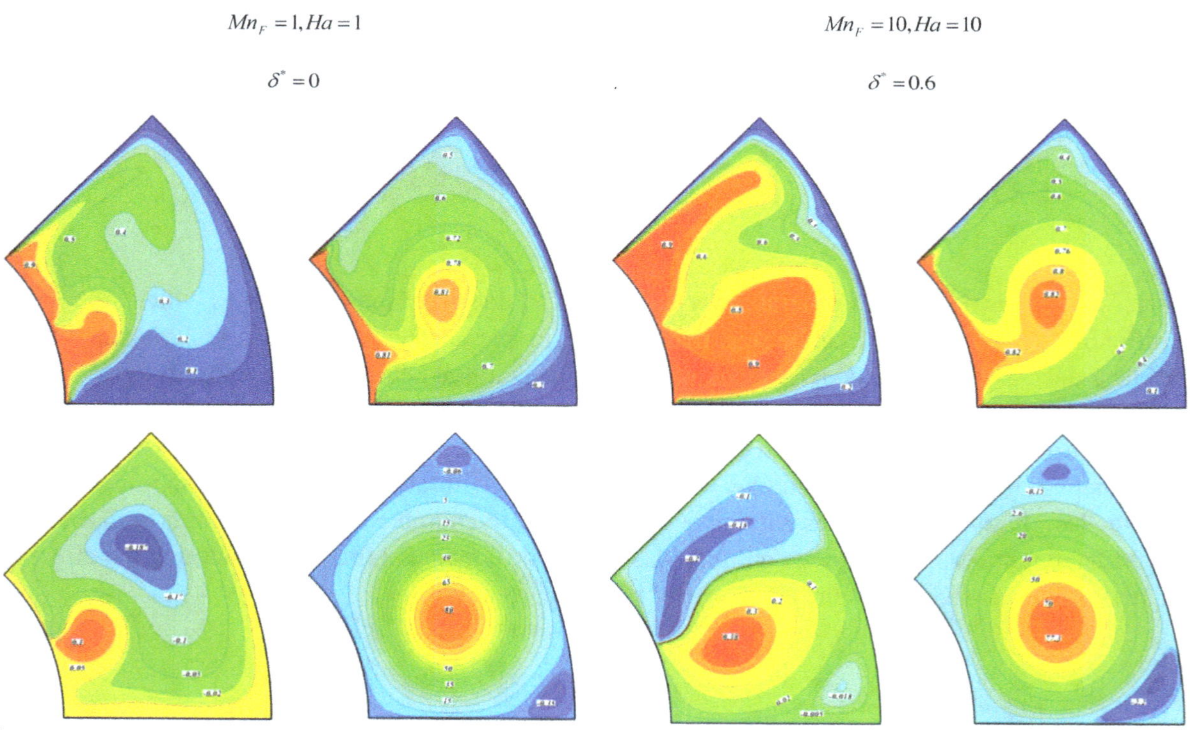

Figure 37. Effects of Ha, Mn_F, Ri and δ^* on average Nusselt number Nu_{ave} along hot wall

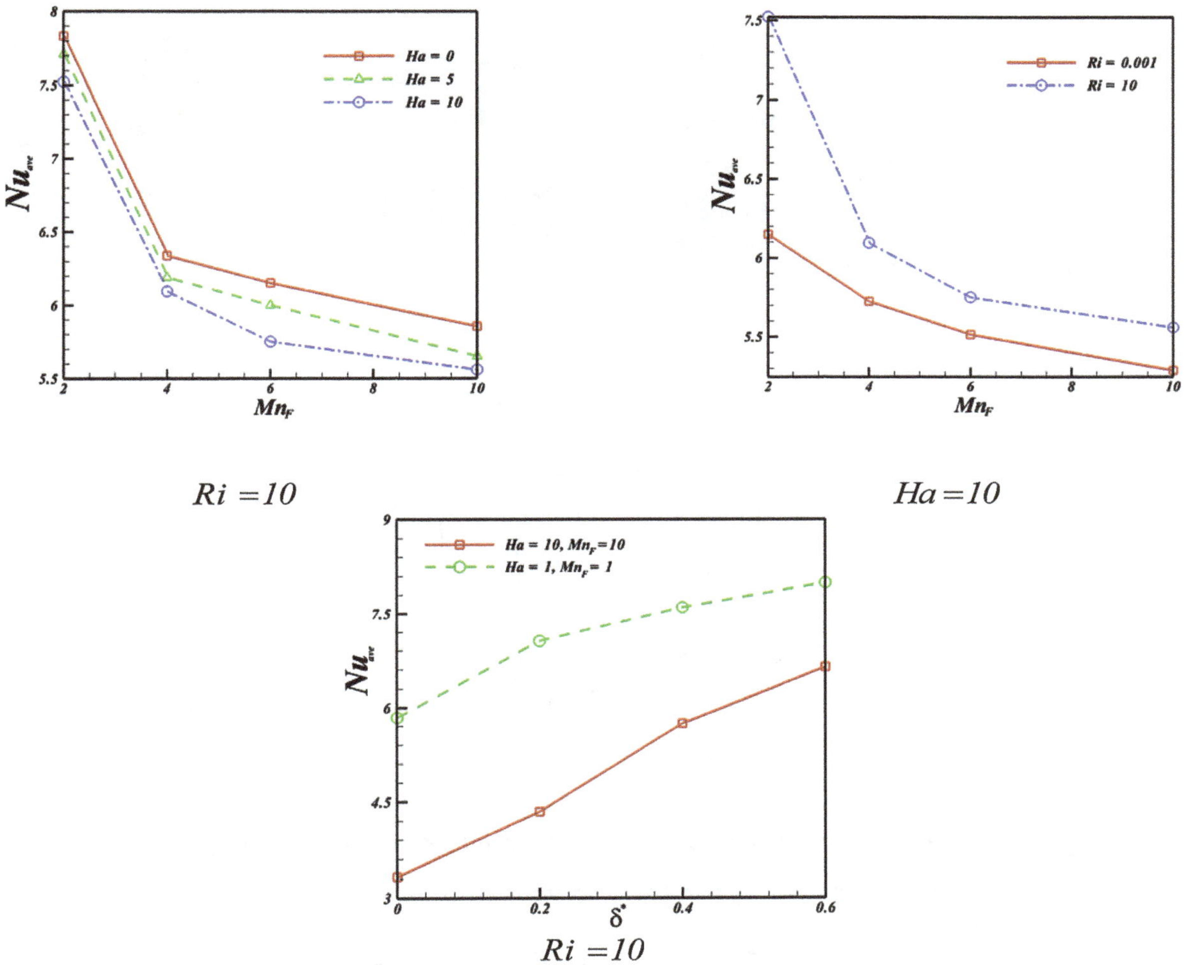

6. INFLUENCE OF MAGNETIC FIELD ON HEAT TRANSFER OF MAGNETIC NANOFLUID IN A SINUSOIDAL DOUBLE PIPE HEAT EXCHANGER

6.1. Problem Definition

Figure 38(a) illustrates the three dimensional schematic of the sinusoidal double pipe heat exchanger in the presence of the wire parallel to axis of heat exchanger (Sheikholeslami, Gerdroodbary, Mousavi, Ganji and Moradi, 2018). The magnetic field is generated by an electric current going through a thin and straight wire oriented parallel to the longitudinal axis (z) at the position (a, b) and the current in the wire flows in the direction of positive z-axis. The structure of the magnetic field in vicinity of the wire is depicted in the Figure 38(b). Figure 39 illustrates the geometry of a sinusoidal two-tube heat exchanger with length L and the height inner tube d_i, amplitude of (δ) and wavelength of (L_w). The total length of wavy wall is six wavelengths, i.e. there are six waves along wavy wall. For the current study,

Figure 38. (a) Three-dimensional sinusoidal double-tube heat exchanger with magnetic field carrying wire, (b)The effect of magnetic field intensity on the ferrofluid of inside inner pipe of heat exchanger

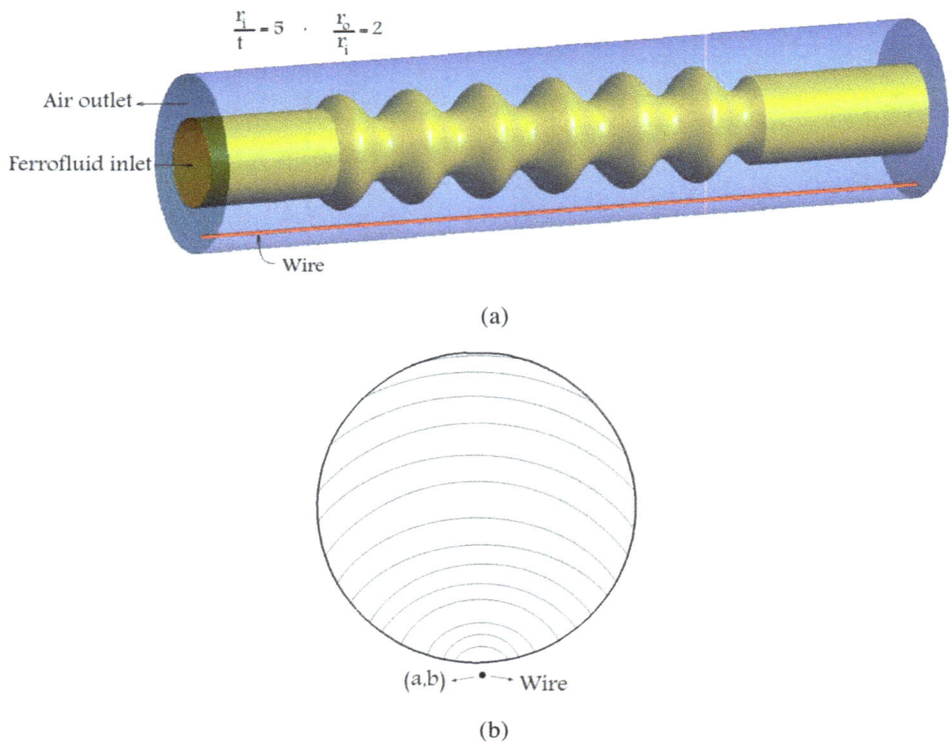

Figure 39. Two-dimensional sinusoidal double-tube heat exchanger without magnetic field carrying wire

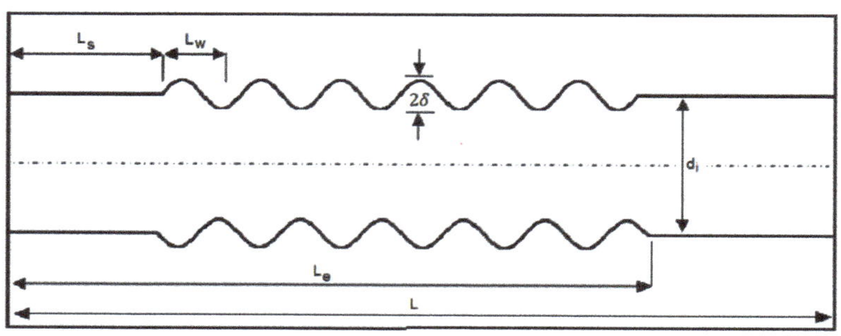

the following dimensionless geometric parameters are applied; wavy amplitude (A) is assumed 0, 0.1, 0.2 and 0.3 in this study. In addition, the profile of the lower wavy-wall can be represented by:

$$S(z) = -\frac{d_i}{2} - \delta \sin\left[\frac{2\pi(z - L_s)}{L_w}\right], L_s \leq z \leq L_e \tag{95}$$

In this section, Navier–Stokes equations and energy equations are coupled to obtain heat transfer inside the sinusoidal double-tube heat exchanger. To investigate the influence of the magnetic field, the components of the magnetic field should be account in the momentum equations. Moreover, it is assumed that physical properties of the fluid are constant. The effects of magnetic fields on the viscosity and the thermal conductivity of the ferrofluid have been assumed to be negligible. It should be mentioned that the non-uniform transverse magnetic field has negligible effect in MHD, and the Lorentz force is also considered negligible compared to the magnetic force due to the electrical conductivity.

Considering these assumptions the dimensional conservation equations for steady state condition are as follows:

Continuity equation:

$$\frac{\partial u}{\partial x}+\frac{\partial v}{\partial y}+\frac{\partial w}{\partial z}=0 \tag{96}$$

Momentum equation:

$$\rho_m\left(u\frac{\partial u}{\partial x}+v\frac{\partial u}{\partial y}+w\frac{\partial u}{\partial z}\right)=-\frac{\partial p}{\partial x}+\mu_m\left(\frac{\partial^2 u}{\partial x^2}+\frac{\partial^2 u}{\partial y^2}+\frac{\partial^2 u}{\partial z^2}\right)+F_K\left(x\right) \tag{97}$$

$$\rho_m\left(u\frac{\partial v}{\partial x}+v\frac{\partial v}{\partial y}+w\frac{\partial v}{\partial z}\right)=-\frac{\partial p}{\partial y}+\mu_m\left(\frac{\partial^2 v}{\partial x^2}+\frac{\partial^2 v}{\partial y^2}+\frac{\partial^2 v}{\partial z^2}\right)+F_K\left(y\right) \tag{98}$$

$$\rho_m\left(u\frac{\partial w}{\partial x}+v\frac{\partial w}{\partial y}+w\frac{\partial w}{\partial z}\right)=-\frac{\partial p}{\partial y}+\mu_m\left(\frac{\partial^2 w}{\partial x^2}+\frac{\partial^2 w}{\partial y^2}+\frac{\partial^2 w}{\partial z^2}\right) \tag{99}$$

Energy equation:

$$\left(\rho_m C_p\right)_m\left(u\frac{\partial T}{\partial x}+v\frac{\partial T}{\partial y}+w\frac{\partial T}{\partial z}\right)=k_m\left(\frac{\partial^2 T}{\partial x^2}+\frac{\partial^2 T}{\partial y^2}+\frac{\partial^2 T}{\partial z^2}\right) \tag{100}$$

The terms $F_K\left(x\right)$ and $F_K\left(y\right)$ is related to FHD due to the existence of the magnetic gradient and is called the Kelvin force. $\mu_0 M\frac{\partial H}{\partial x}$ and $\mu_0 M\frac{\partial H}{\partial y}$ are the components of Kelvin force in the x and y directions, respectively. They are resulted from the electric current flowing through the wire. Therefore, it is needed to define the magnetic field of electric current. The components of the magnetic field H_x, H_y in the x and y directions are calculated as follows (Sheikholeslami and Ganji, 2014):

$$H_x\left(x,y\right)=\frac{I}{2\pi}\frac{\left(x-a\right)}{\left(x-a\right)^2+\left(y-b\right)^2} \tag{101}$$

$$H_y\left(x,y\right)=\frac{I}{2\pi}\frac{\left(y-b\right)}{\left(x-a\right)^2+\left(y-b\right)^2} \tag{102}$$

The magnetic field strength is given by

$$H\left(x,y,z\right)=\frac{I}{2\pi}\frac{1}{\sqrt{\left(x-a\right)^2+\left(y-b\right)^2}} \tag{103}$$

It is needed to be mentioned that the term $\mu_0 M\frac{\partial H}{\partial z}$ is should be added to the momentum equation in the z direction when axial non-uniform magnetic gradient is existed in the domain. M is the magnetization and is defined as (Yamaguchi, 2008):

$$M=\frac{6m_p}{\pi d_p^3}\left[coth\left(\xi\right)-\frac{1}{\xi}\right] \tag{104}$$

The unit cell of the crystal structure of magnetite has a volume of about 730 A^3 and contains 8 molecules of Fe_3O_4, each of them having a magnetic moment of 4 μ_B (Kittel, 1967). Therefore the particle magnetic moment for the magnetite particles is obtained as

$$m_p=\frac{4\mu_B\pi d_p^3}{6\times 91.25\times 10^{-30}} \tag{105}$$

Also ξ is the Langevin parameter and is defined as (Yamaguchi, 2008)

$$\xi=\frac{\mu_0 m_p H}{k_B T} \tag{106}$$

It is also noted that dimensionless magnet number (Mn) is used to measure and the effect of the magnetic field intensity. Magnetic number (Mn) is dependent to the magnetic field intensity. This means that *Mn* increases with an increase in the magnetic field intensity.

$$Mn = \frac{\mu_0 \chi H_r^2 h^2}{\rho_m \alpha_m^2} \tag{107}$$

where χ is the magnetic susceptibility of ferrofluid. As mentioned earlier, the magnetic susceptibility of a ferrofluid containing 4 Vol% with a mean diameter of 10 nm is in the order of $\chi = 0.348586$ (Kittel, 1967). H_r is the characteristic of magnetic field strength and calculated by $H_r = H\left(a, 0\right) = \frac{I}{2\pi b}$. The mixture physical properties in the above equations are calculated as below.

The Reynolds number and Nusselt number as two main non dimensional numbers is calculated by the following equation

$$Re_m = \frac{\rho_m v_m d_i}{\mu_m} \tag{108}$$

$$Nu_{hot} = \frac{q'' D_{h_i}}{k_m \left(T_w - T_b\right)} \tag{109}$$

where T_b is bulk temperature and D_{h_i} is expressed by following equation:

$$D_{h_i} = d_i + 2\delta \tag{110}$$

In the present study, the second order upwind numerical scheme decoupling with the SIMPLEC algorithm is used, and all the governing equations are solved through a finite volume CFD in-house code.

The inflow conditions of ferrofluid are equivalent to $Re_m = 100$, 50 and the Reynolds number of air flow is $Re_{air} = 2300$. Also, for studying heat transfer, results of Nusselt number have been presented for water based ferrofluid containing of 4 Vol% Fe3O4 spherical shape particles with 10 nm mean diameter. Boundary conditions were applied to the ferrofluid inflow (inlet velocity) with constant temperature ($T_{hot,in} = T_0$) and air flow as cold gas (inlet velocity) with constant temperature ($T_{cold,in} = T_0$). Influences of external forces on nanofluid behavior were demonstrated by several authors.

6.2. Effects of Active Parameters

Effects of different parameters like Reynolds number, magnetic number and geometric shape coefficient on the heat transfer of ferrofluid are comprehensively studied. The effects of different geometric configurations (4 types) of inner pipe with ferrofluid flow are investigated in the non-uniform magnetic field. Figure 40 illustrates streamlines for two cavities in different geometric shapes. Production of ed-

Figure 40. Streamlines for two cavities of sinusoidal pipe in Re_m *=50. (a) A=0.1, (b) A=0.2, (c) A=0.3*

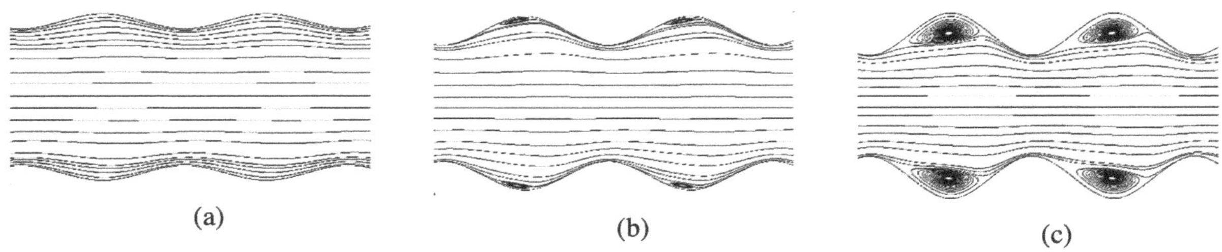

dies as a result of separation in cavities results in an increase in the heat transfer rate. Since an adverse pressure gradient is formed as a result of geometric nonuniformity (i.e. Sinusoidal shape), the separation is clearly discerned in the domain.

Figure 41 represents the Nusselt number along the inner tube for various geometrical shapes under different magnetic number. It can be seen that non-uniform magnetic field enhances velocity gradient near the wall and hence results in an increase in Nusselt number. This observation is due to retarting flow because of existence of Lorentz forces.

Figure 42 the Nusselt number for geometric coefficient and different Reynolds numbers. The figure shows that the effects of Reynolds number are more in maximum Nusselt number. Moreover, an increase in Reynolds number results in an increase in Nusselt number. Nusselt has a periodically decreasing behavior from the beginning of the sinusoidal part of the pipe. Also, increasing Reynolds number the moves separation point toward the crest of the sinus wave in diverging part of the pipe. Electrical wire produces a non-uniform magnetic field in x and y directions. This field is perpendicular to ferrofliud flow direction. As a magnetic field intensity is increased, the force to flow in cross planes increases. This increase results in a secondary flow, which appears as two eddies. Figure 43 illustrates streamlines in the presence of non-uniform magnetic field at plane on $A = 0$, $z^* = 15.5$. These eddies diffuse ferrofluid towards the wall in x-y plane. Two eddies are symmetric with respect to the y axis. It can be seen that streamlines recede from electrical wire as a result of kelvin force. Figure 44 illustrates the temperature distribution on plane at $z^* = 15.5$ for simple double pipe heat exchanger ($A = 0$) with $\mathrm{Re}_m = 50$. It can be observed that magnetic field causes cold boundary layer to extend towards the center of inner pipe and the magnetic field intensification increases this extension.

Figure 45 shows the temperature distribution for sinusoidal pipe ($A = 0.1$) at the plane $z^* = 15.5$. The comparison of Figures 44 and 45 indicates that the effect of magnetic field in $A = 0.1$ is less than $A = 0$. According to Equation (107), the intensity of a magnetic field has an inverse relation with the distance from electrical wire. In A=0.1 due to a sinusoidal wall of the pipe, the distance between the wire and the centerline of the pipe increases and the magnetic field intensity decreases. Figure 46 compares the effect of geometric shape factor on heat flux (Nusselt number) for $\mathrm{Re}_m = 50$. In converging section of the inner tube, Nusselt number increases due to the increase in temperature gradient. On the contrary, Nusselt number decreases in the diverging section due to a reduction in temperature gradient.

Figure 41. Effect Nusselt number for Re_m *=50 and values of geometric factor. (a) A=0, (b) A=0.1, (c) A=0.2, (d) A=0.3*

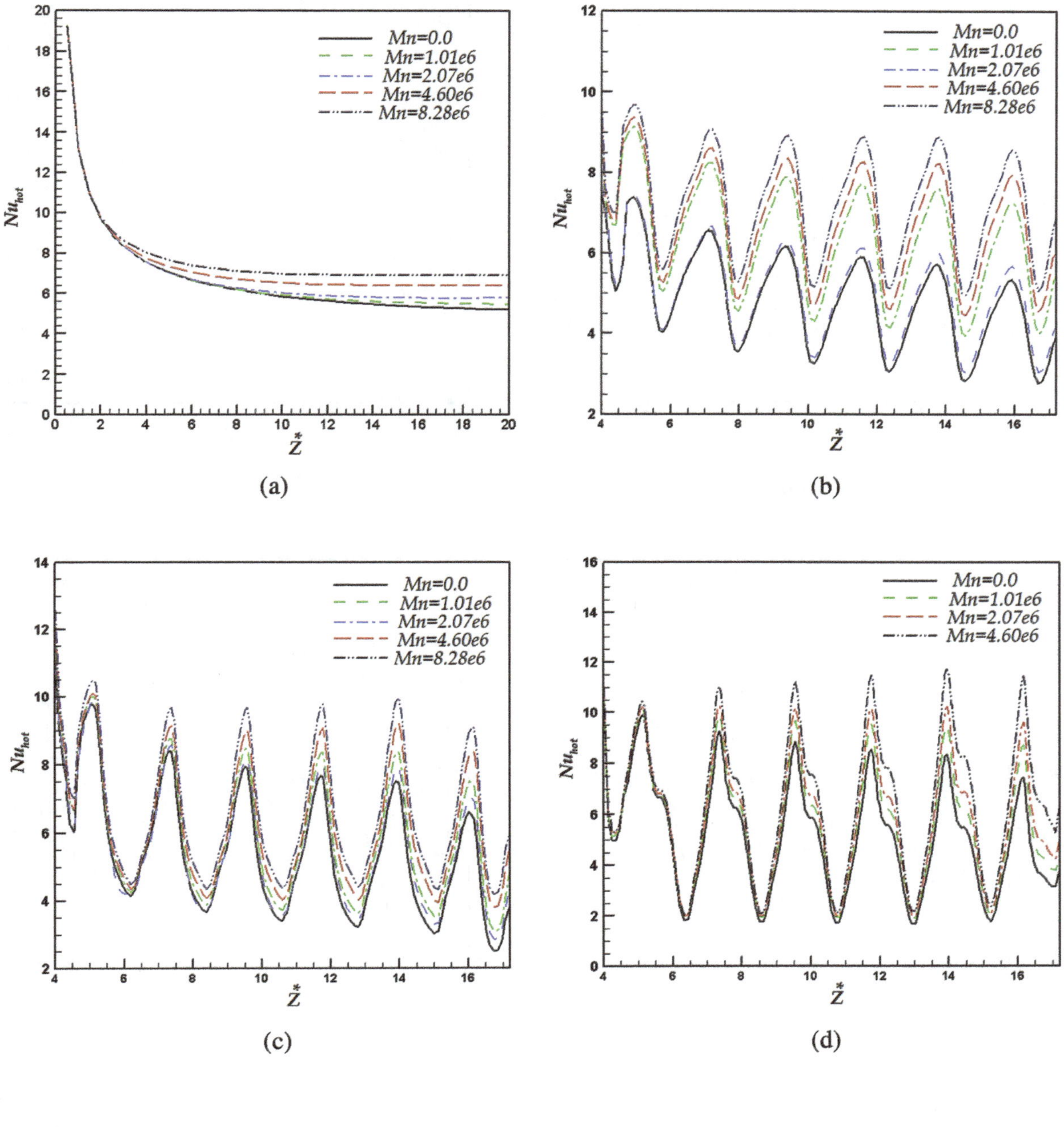

Figure 42. Effect of Nusselt number for different Reynolds in geometric factor. (a) A=0, (b) A=0.1, (c) A=0.2, (d) A=0.3

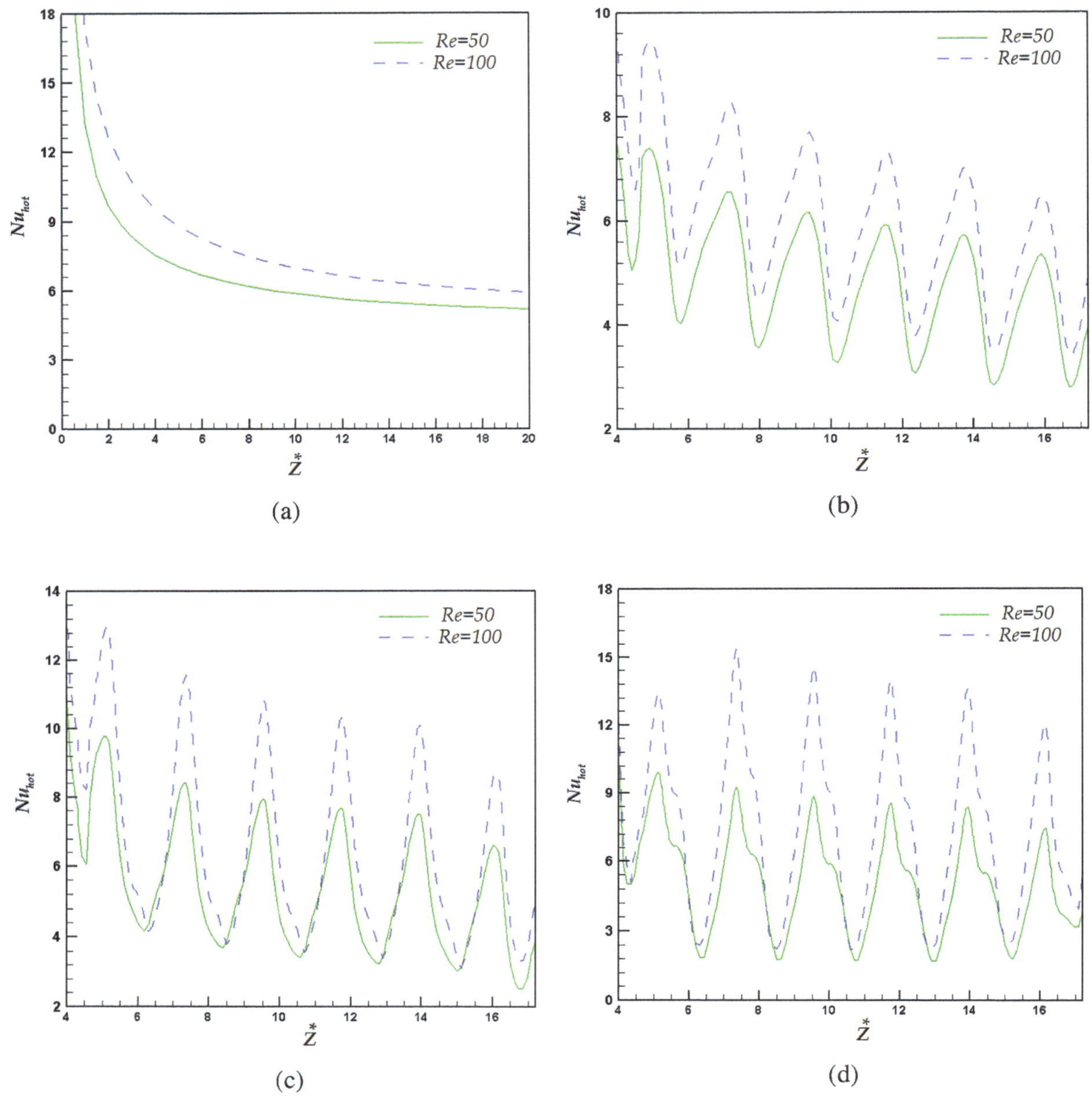

Figure 47 illustrates the variation of friction coefficient along the inner tube at $\mathrm{Re}_m = 100$. The friction coefficient increases due to increase of magnetic field intensity along the tube. In higher shape coefficient, this increase is not as intense as Nusselt number increase due to the magnetic field intensity. Figure 48 presents axial velocity distribution at $A = 0$, $z^* = 15.5$ and for different magnetic field in-

Figure 43. Streamlines for Re_m *=50 and geometric factor A=0 in* $z^* = 15.5$

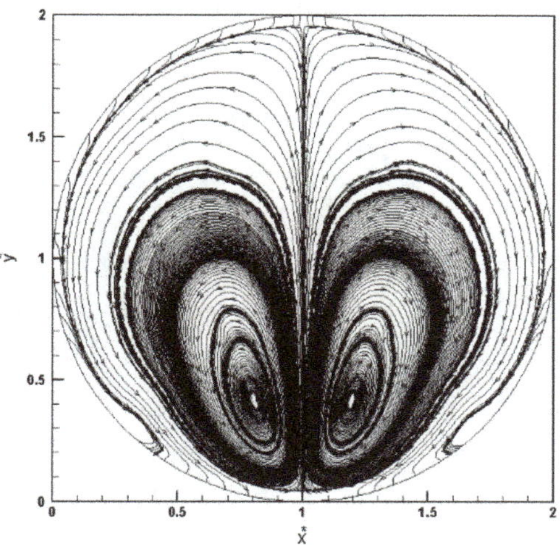

Figure 44. Non- dimensional temperature profile in Re_m *=50,* $z^* = 15.5$ *and geometric factor* $A = 0$*. (a)* $Mn = 0$*, (b)* $Mn = 1.01 \times 10^6$*, (c)* $Mn = 2.07 \times 10^6$*, (d)* $Mn = 4.60 \times 10^6$*, (e)* $Mn = 8.28 \times 10^6$*, (f)* $Mn = 18.64 \times 10^6$

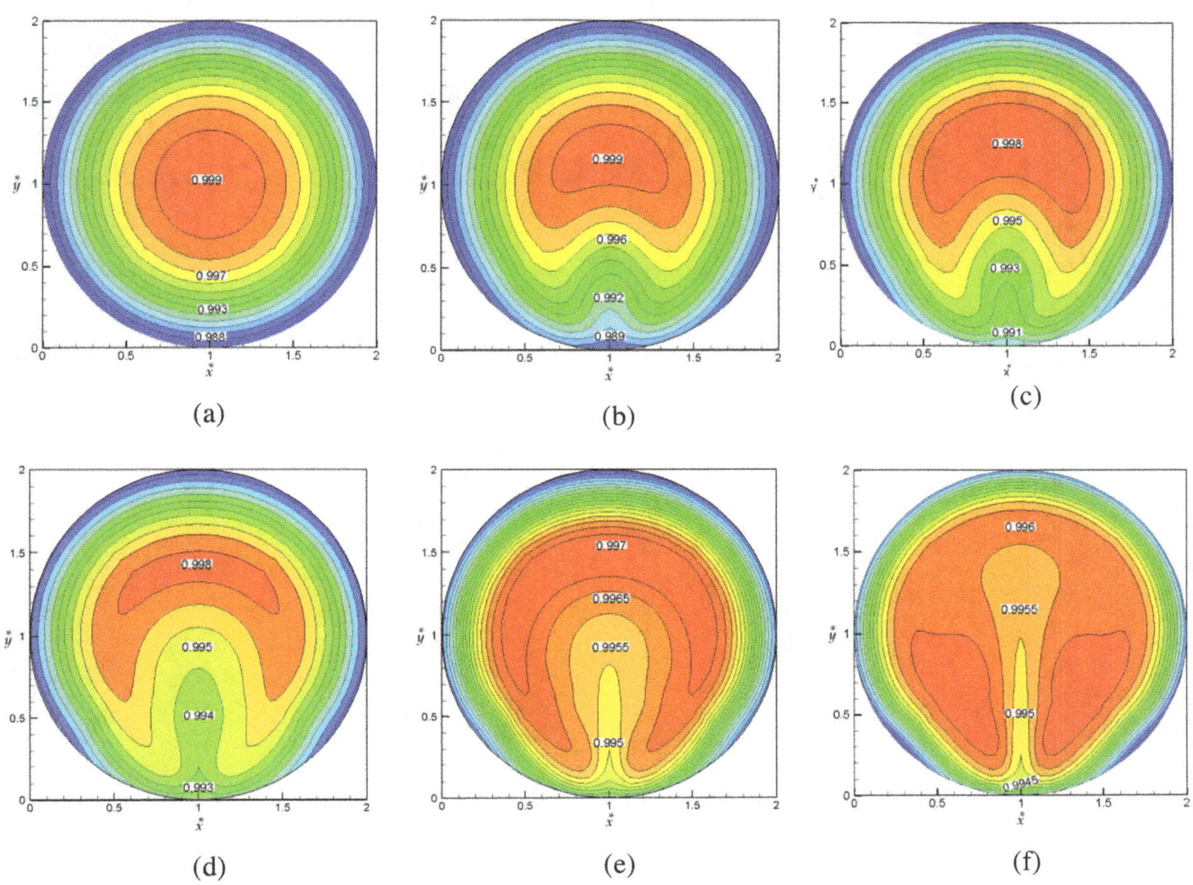

Figure 45. Non- dimensional temperature profile in Re_m *=50,* $z^* = 15.5$ *and geometric factor* $A = 0.1$*. (a)* $Mn = 0$*, (b)* $Mn = 1.01\times10^6$*, (c)* $Mn = 2.07\times10^6$*, (d)* $Mn = 4.60\times10^6$*, (e)* $Mn = 8.28\times10^6$*, (f)* $Mn = 18.64\times10^6$

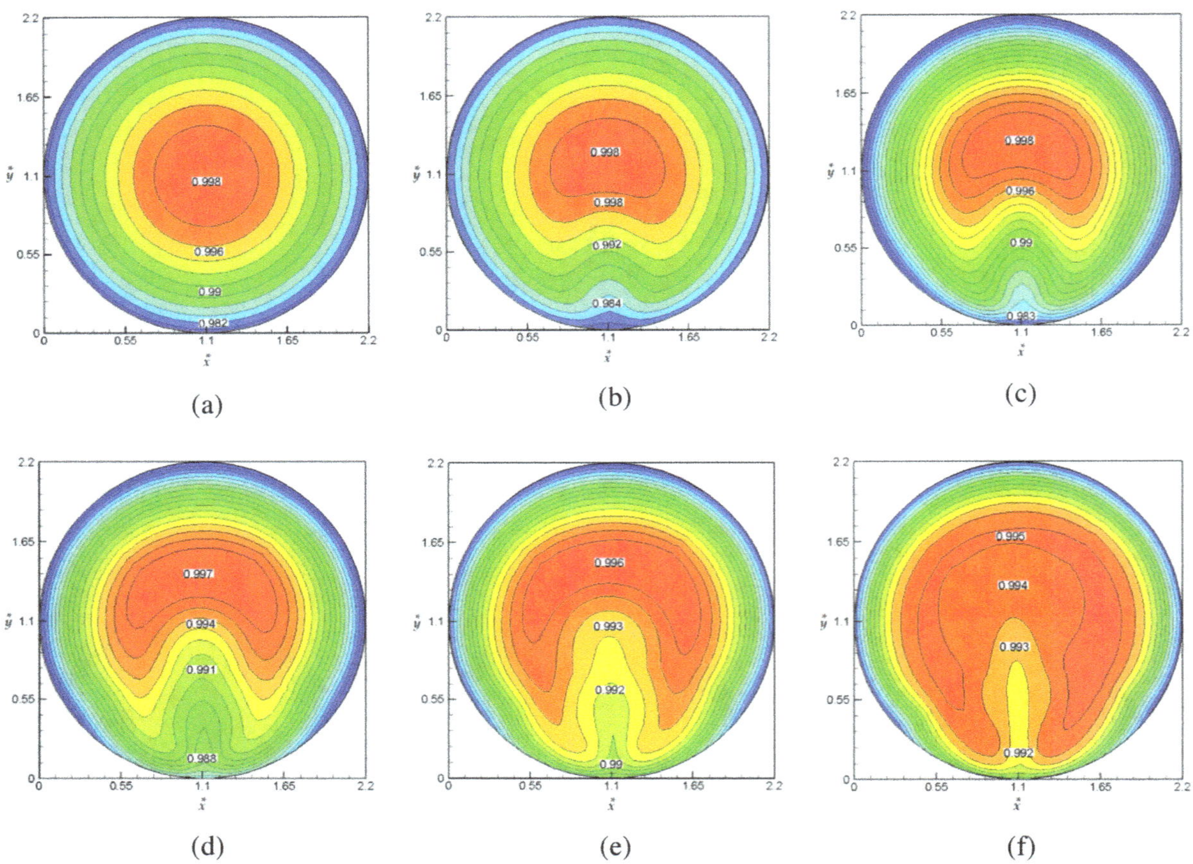

Figure 46. Effect Nusselt number in different values geometric factor in case $Mn = 2.07\times10^6$ *and* $\mathrm{Re_m} = 50$

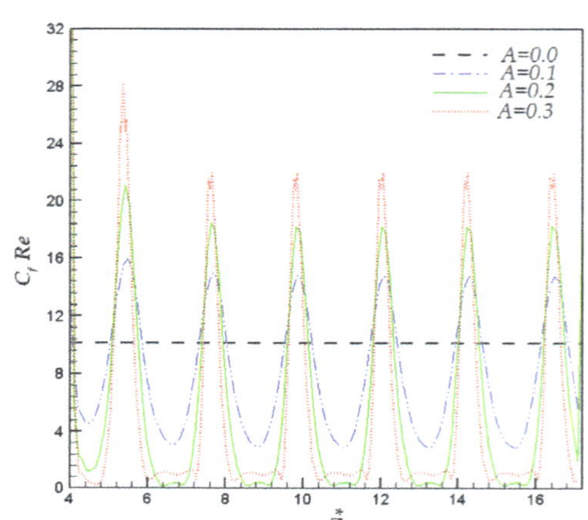

Figure 47. Effect friction factor in Reynolds number $\mathrm{Re}_m = 100$ *and different values of geometric factor. (a) A=0, (b) A=0.1, (c) A=0.2, (d) A=0.3*

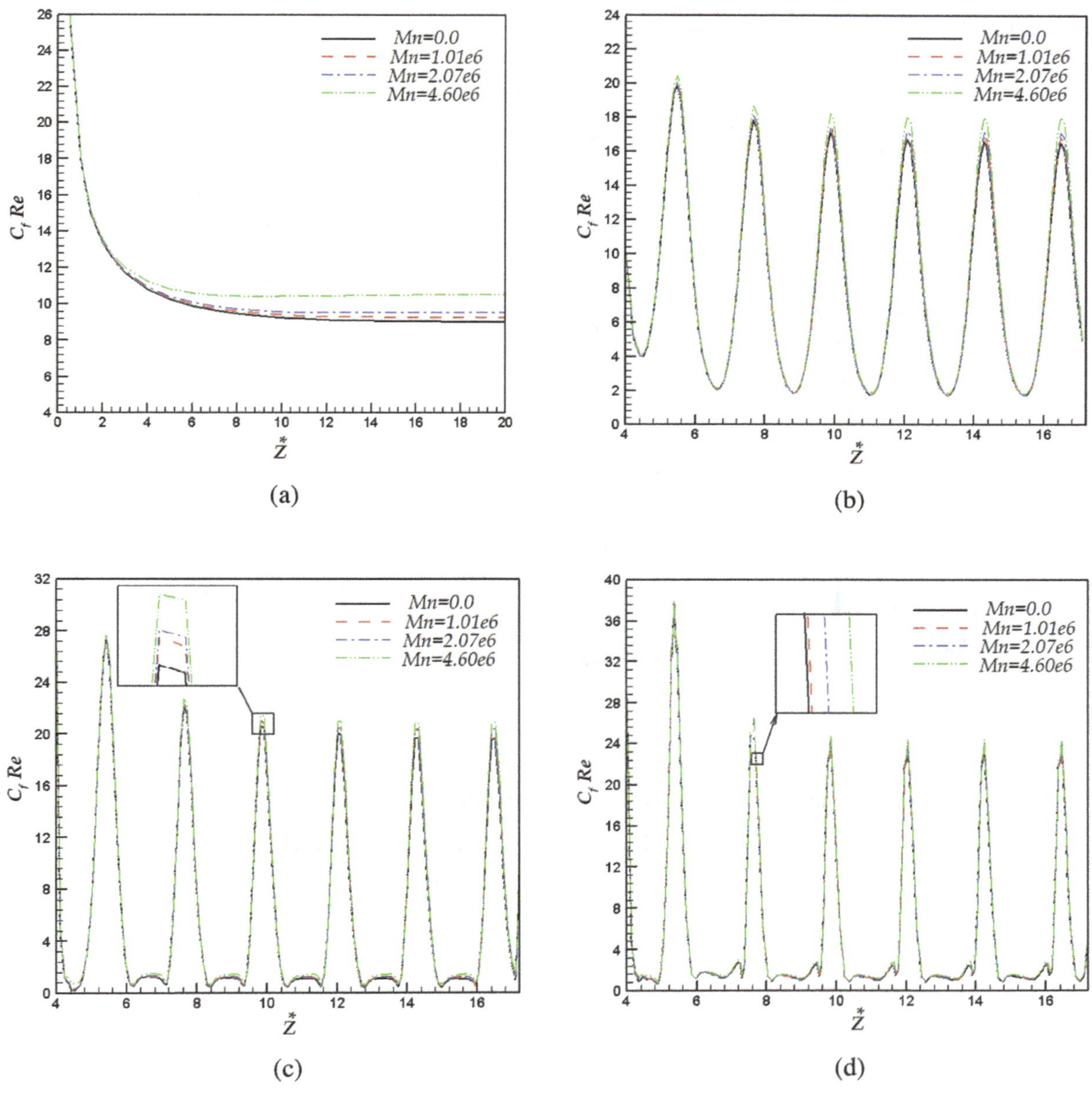

tensities in $\mathrm{Re}_m = 50$. It can be seen that increase in magnetic field causes Ferrofluid to go toward the inner tube wall.

The effect of geometric shape on average Nusselt is presented in Figure 49 for Mn=0 and $\mathrm{Re}_m = 100$. As expected, with increase in shape coefficient, averaged Nusselt increases due to the sinusoidal shape of inner tube. Figure 50 illustrates the variation temperature distribution along the x direction for differ-

Figure 48. The effect of non-uniform crossover magnetic field on non- dimensional axial velocity distribution of ferrofluid(three dimension) for Re_m *=50,* $z^* = 15.5$ *and geometric factor* $A = 0$ *. (a)* $Mn = 0$*, (b)* $Mn = 1.01 \times 10^6$*, (c)* $Mn = 2.07 \times 10^6$*, (d)* $Mn = 4.60 \times 10^6$*, (e)* $Mn = 8.28 \times 10^6$ *, (f)* $Mn = 18.64 \times 10^6$

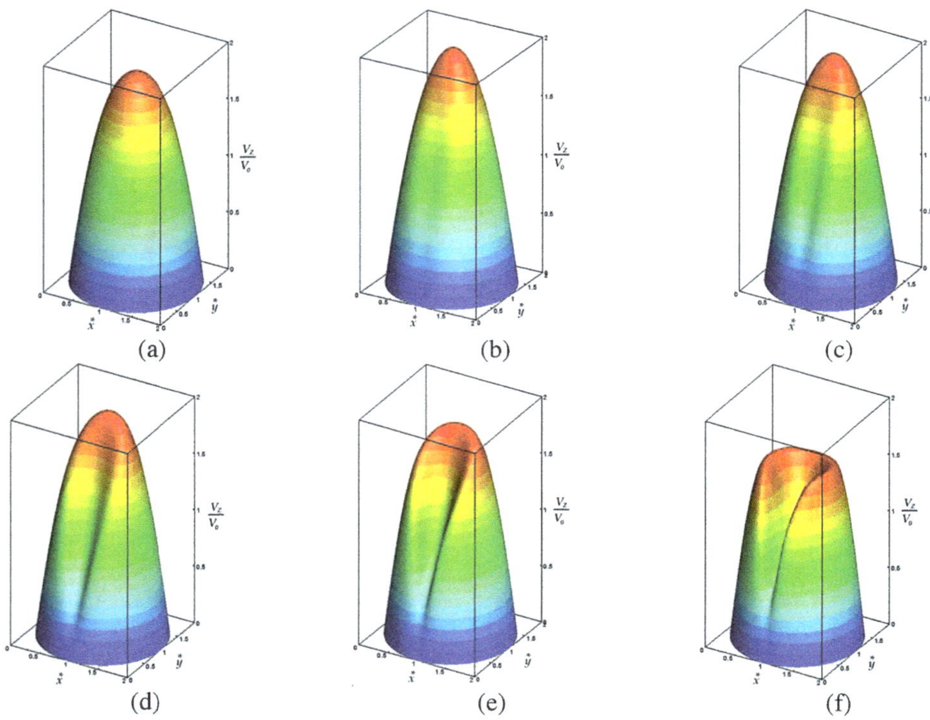

Figure 49. The effect of geometric form factor on average Nusselt in Re_m *=100*

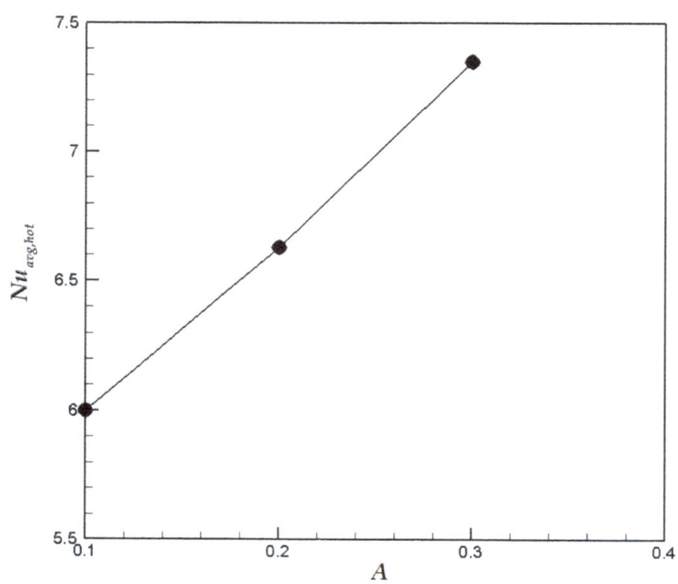

Figure 50. The effect of geometric of magnetic field in Dimensionless temperature for Re_m *=50,* $z^* = 15.5$*,* $y^* = 1$ *and geometric facto(a) A=0, (b) A=0.1, (c) A=0.2, (d) A=0.*

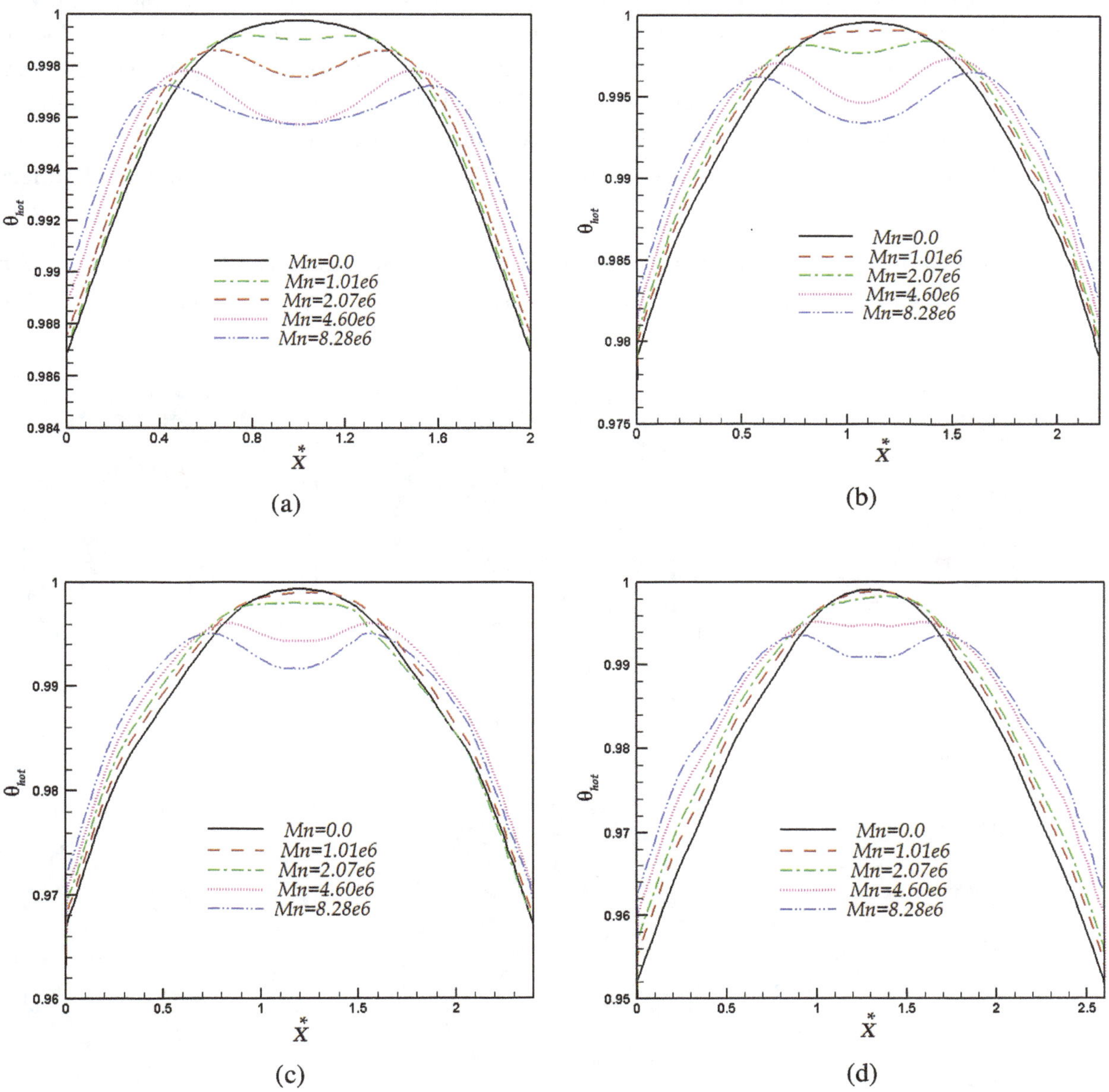

ent shape coefficients at $\mathrm{Re}_m = 100$. It is found that magnetic field increases the velocity gradient in the vicinity of the tube wall and this increase results in the extension of the cold boundary layer towards ferrofluid. So, ferrofluids temperature decreases at the centerline of inner tube and heat transfer improves. Figure 51 depicts non-dimensional temperature contour in various sections along the sinusoidal double pipe (A=0.2) in the presence of magnetic field ($Mn = 2.07 \times 10^6$). In the axial direction, cold boundary layer extends toward centerline of the pipe and intensifies heat transfer. Figure 52 indicates non-dimensional temperature along the heat exchanger for different shape coefficient at $\mathrm{Re}_m = 100$. The results show that the magnetic field decreases the outlet temperature of ferrofluid. Figure 53 compares the effect

Figure 51. Non- dimensional temperature contour in 6 sinusiodal wave sections inside of inner tube, for A=0.2 and $Mn = 2.07 \times 10^6$ *(a)* $z^* = 4.55$, *(b)* $z^* = 6.75$, *(c)* $z^* = 8.95$, *(d)* $z^* = 11.15$, *(e)* $z^* = 13.35$, *(f)* $z^* = 15.5$

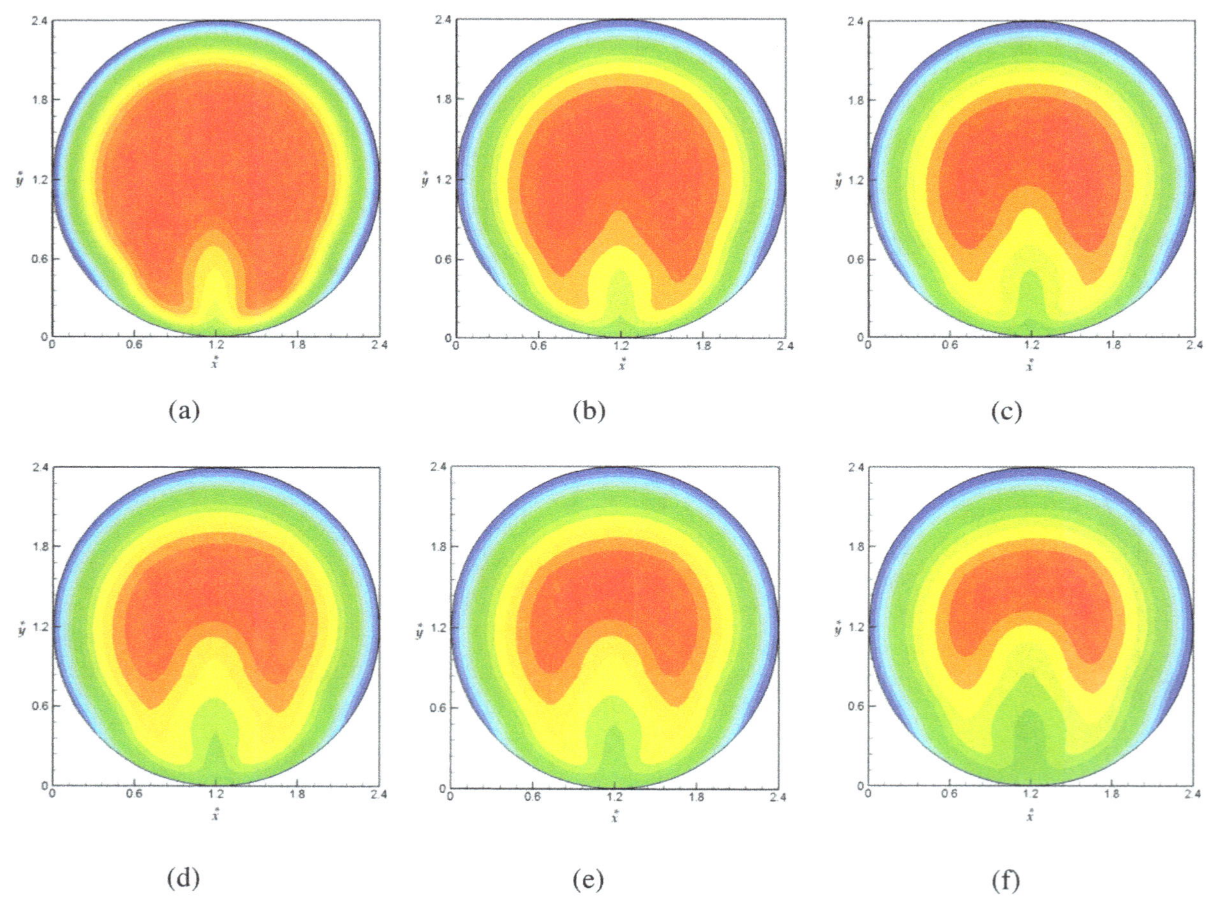

Figure 52. The effect of magnetic field in non- dimensional temperature in along the axis of heat exchanger for Re_m *=50 with different values of geometric factor (a) A=0, (b) A=0.1, (c) A=0.2, (d) A=0.3*

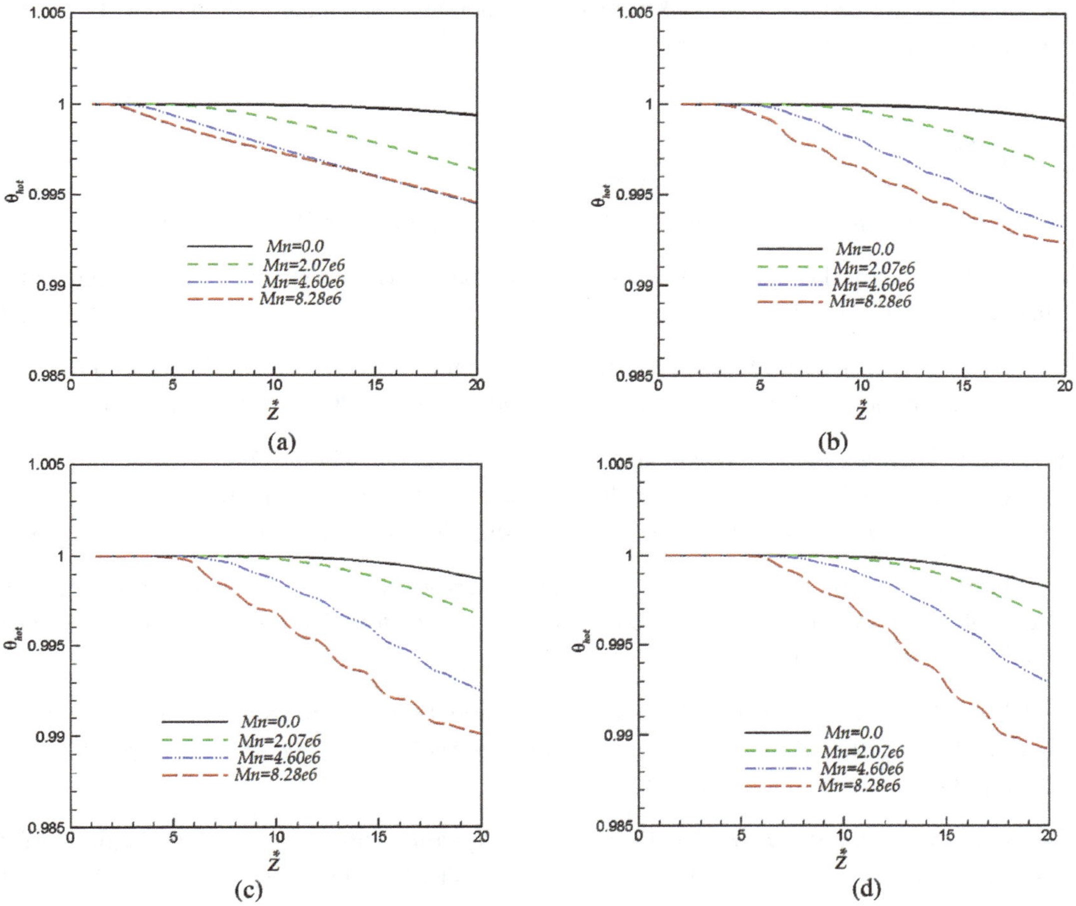

Figure 53. The ratio of average Nusselt number of ferrofluid for Re_m *=50 in various intensities of magnetic field with different geometric factors*

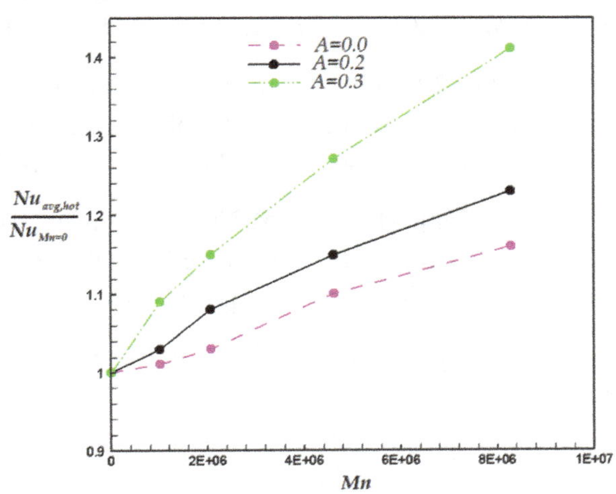

of magnetic field on the ratio of the Nusselt number in different shape coefficients. This ratio is defined as the ratio of mean Nusselt number in the presence of a magnetic field to the Nusselt number without magnetic field. It is noticed that this ratio significantly increases as magnetic field intensity increases in ferrofluid in constant Reynolds number.

REFERENCES

Kittel, C. (1967). *Introduction to Solid State Physics*. New York: John Wiley & Sons.

Loukopoulos, V., & Tzirtzilakis, E. (2004). Biomagnetic channel flow in spatially varying magnetic field. *International Journal of Engineering Science*, *42*(5-6), 571–590. doi:10.1016/j.ijengsci.2003.07.007

Sheikholeslami. (2014). Effect of spatially variable magnetic field on ferrofluid flow and heat transfer considering constant heat flux boundary condition. *The European Physical Journal Plus*, 129–248.

Sheikholeslami, M., Barzegar Gerdroodbary, M., Mousavi, S. V., Ganji, D. D., & Moradi, R. (2018, August). Heat transfer enhancement of ferrofluid inside an 90° elbow channel by non-uniform magnetic field. *Journal of Magnetism and Magnetic Materials*, *460*, 302–311. doi:10.1016/j.jmmm.2018.03.070

Sheikholeslami, M., & Ellahi, R. (2015). Simulation of ferrofluid flow for magnetic drug targeting using Lattice Boltzmann method. *Journal of Zeitschrift Fur Naturforschung A. Verlag der Zeitschrift für Naturforschung*, *70*(2), 115–124.

Sheikholeslami, M., & Ellahi, R. (2015). Three dimensional mesoscopic simulation of magnetic field effect on natural convection of nanofluid. *International Journal of Heat and Mass Transfer*, *89*, 799–808. doi:10.1016/j.ijheatmasstransfer.2015.05.110

Sheikholeslami, M., & Ganji, D. D. (2014). Ferrohydrodynamic and Magnetohydrodynamic effects on ferrofluid flow and convective heat transfer. *Energy*, *75*, 400–410. doi:10.1016/j.energy.2014.07.089

Sheikholeslami, M., & Rashidi, M. M. (2015). Effect of space dependent magnetic field on free convection of Fe3O4-water nanofluid. *Journal of the Taiwan Institute of Chemical Engineers*, *56*, 6–15. doi:10.1016/j.jtice.2015.03.035

Sheikholeslami, M., & Rashidi, M. M. (2015). Ferrofluid heat transfer treatment in the presence of variable magnetic field. *The European Physical Journal Plus*, *130*(6), 115. doi:10.1140/epjp/i2015-15115-4

Sheikholeslami, M., & Rashidi, M. M. (2016). Non-uniform magnetic field effect on nanofluid hydrothermal treatment considering Brownian motion and thermophoresis effects. *Journal of the Brazilian Society of Mechanical Sciences and Engineering*, *38*(4), 1171–1184. doi:10.100740430-015-0459-5

Sheikholeslami, M., Rashidi, M. M., & Ganji, D. D. (2015). Numerical investigation of magnetic nanofluid forced convective heat transfer in existence of variable magnetic field using two phase model. *Journal of Molecular Liquids*, *212*, 117–126. doi:10.1016/j.molliq.2015.07.077

Yamaguchi, H. (2008). *Engineering Fluid Mechanics*. Springer Science.This is a document created to process the author's additional reading section.

ADDITIONAL READING

Jafaryar, M., Sheikholeslami, M., Li, M., & Moradi, R. (2018). Nanofluid turbulent flow in a pipe under the effect of twisted tape with alternate axis. *Journal of Thermal Analysis and Calorimetry*. doi:10.100710973-018-7093-2

Li, Z., Shehzad, S. A., & Sheikholeslami, M. (2018). An application of CVFEM for nanofluid heat transfer intensification in a porous sinusoidal cavity considering thermal non-equilibrium model. *Computer Methods in Applied Mechanics and Engineering*, *339*, 663–680. doi:10.1016/j.cma.2018.05.015

Li, Z., Sheikholeslami, M., Chamkha, A. J., Raizah, Z. A., & Saleem, S. (2018). Control Volume Finite Element Method for nanofluid MHD natural convective flow inside a sinusoidal annulus under the impact of thermal radiation. *Computer Methods in Applied Mechanics and Engineering*, *338*, 618–633. doi:10.1016/j.cma.2018.04.023

Li, Z., Sheikholeslami, M., Jafaryar, M., Shafee, A., & Chamkha, A. J. (2018). Investigation of nanofluid entropy generation in a heat exchanger with helical twisted tapes. *Journal of Molecular Liquids*, *266*, 797–805. doi:10.1016/j.molliq.2018.07.009

Li, Z., Sheikholeslami, M., Samandari, M., & Shafee, A. (2018). Nanofluid unsteady heat transfer in a porous energy storage enclosure in existence of Lorentz forces. *International Journal of Heat and Mass Transfer*, *127*, 914–926. doi:10.1016/j.ijheatmasstransfer.2018.06.101

Sheikholeslami, M. (2017a). Lattice Boltzmann Method simulation of MHD non-Darcy nanofluid free convection. *Physica B, Condensed Matter*, *516*, 55–71. doi:10.1016/j.physb.2017.04.029

Sheikholeslami, M. (2017b). CuO-water nanofluid free convection in a porous cavity considering Darcy law. *The European Physical Journal Plus*, *132*(1), 55. doi:10.1140/epjp/i2017-11330-3

Sheikholeslami, M. (2017c). Influence of Lorentz forces on nanofluid flow in a porous cavity by means of Non- Darcy model. *Engineering Computations*, *34*(8), 2651–2667. doi:10.1108/EC-01-2017-0008

Sheikholeslami, M. (2017d). Magnetic field influence on nanofluid thermal radiation in a cavity with tilted elliptic inner cylinder. *Journal of Molecular Liquids*, *229*, 137–147. doi:10.1016/j.molliq.2016.12.024

Sheikholeslami, M. (2017e). Numerical simulation of magnetic nanofluid natural convection in porous media. *Physics Letters. [Part A]*, *381*(5), 494–503. doi:10.1016/j.physleta.2016.11.042

Sheikholeslami, M. (2017f). Influence of Lorentz forces on nanofluid flow in a porous cylinder considering Darcy model. *Journal of Molecular Liquids*, *225*, 903–912. doi:10.1016/j.molliq.2016.11.022

Sheikholeslami, M. (2017g). Influence of Coulomb forces on Fe3O4-H2O nanofluid thermal improvement. *International Journal of Hydrogen Energy*, *42*(2), 821–829. doi:10.1016/j.ijhydene.2016.09.185

Sheikholeslami, M. (2017h). Numerical investigation of MHD nanofluid free convective heat transfer in a porous tilted enclosure. *Engineering Computations*, *34*(6), 1939–1955. doi:10.1108/EC-08-2016-0293

Sheikholeslami, M. (2017i). Magnetic field influence on CuO -H2O nanofluid convective flow in a permeable cavity considering various shapes for nanoparticles. *International Journal of Hydrogen Energy*, *42*(31), 19611–19621. doi:10.1016/j.ijhydene.2017.06.121

Sheikholeslami, M. (2018a). Magnetic source impact on nanofluid heat transfer using CVFEM. *Neural Computing & Applications*, *30*(4), 1055–1064. doi:10.100700521-016-2740-7

Sheikholeslami, M. (2018b). Numerical simulation for external magnetic field influence on Fe3O4-water nanofluid forced convection. *Engineering Computations*, *35*(4), 1639–1654. doi:10.1108/EC-06-2017-0200

Sheikholeslami, M. (2018c). Finite element method for PCM solidification in existence of CuO nanoparticles. *Journal of Molecular Liquids*, *265*, 347–355. doi:10.1016/j.molliq.2018.05.132

Sheikholeslami, M. (2018d). Solidification of NEPCM under the effect of magnetic field in a porous thermal energy storage enclosure using CuO nanoparticles. *Journal of Molecular Liquids*, *263*, 303–315. doi:10.1016/j.molliq.2018.04.144

Sheikholeslami, M. (2018e). Influence of magnetic field on Al2O3-H2O nanofluid forced convection heat transfer in a porous lid driven cavity with hot sphere obstacle by means of LBM. *Journal of Molecular Liquids*, *263*, 472–488. doi:10.1016/j.molliq.2018.04.111

Sheikholeslami, M. (2018f). Numerical simulation for solidification in a LHTESS by means of Nano-enhanced PCM. *Journal of the Taiwan Institute of Chemical Engineers*, *86*, 25–41. doi:10.1016/j.jtice.2018.03.013

Sheikholeslami, M. (2018g). Numerical modeling of Nano enhanced PCM solidification in an enclosure with metallic fin. *Journal of Molecular Liquids*, *259*, 424–438. doi:10.1016/j.molliq.2018.03.006

Sheikholeslami, M. (2018h). Numerical investigation of nanofluid free convection under the influence of electric field in a porous enclosure. *Journal of Molecular Liquids*, *249*, 1212–1221. doi:10.1016/j.molliq.2017.11.141

Sheikholeslami, M. (2018i). CuO-water nanofluid flow due to magnetic field inside a porous media considering Brownian motion. *Journal of Molecular Liquids*, *249*, 921–929. doi:10.1016/j.molliq.2017.11.118

Sheikholeslami, M. (2018j). Application of Darcy law for nanofluid flow in a porous cavity under the impact of Lorentz forces. *Journal of Molecular Liquids*, *266*, 495–503. doi:10.1016/j.molliq.2018.06.083

Sheikholeslami, M., Barzegar Gerdroodbary, M., Valiallah Mousavi, S., Ganji, D. D., & Moradi, R. (2018). Heat transfer enhancement of ferrofluid inside an 90o elbow channel by non-uniform magnetic field. *Journal of Magnetism and Magnetic Materials*, *460*, 302–311. doi:10.1016/j.jmmm.2018.03.070

Sheikholeslami, M., & Bhatti, M. M. (2017). Active method for nanofluid heat transfer enhancement by means of EHD. *International Journal of Heat and Mass Transfer*, *109*, 115–122. doi:10.1016/j.ijheatmasstransfer.2017.01.115

Sheikholeslami, M., Darzi, M., & Li, Z. (2018). Experimental investigation for entropy generation and exergy loss of nano-refrigerant condensation process. *International Journal of Heat and Mass Transfer*, *125*, 1087–1095. doi:10.1016/j.ijheatmasstransfer.2018.04.155

Sheikholeslami, M., Darzi, M., & Sadoughi, M. K. (2018). Heat transfer improvement and Pressure Drop during condensation of refrigerant-based Nanofluid; An Experimental Procedure. *International Journal of Heat and Mass Transfer*, *122*, 643–650. doi:10.1016/j.ijheatmasstransfer.2018.02.015

Sheikholeslami, M., & Ghasemi, A. (2018). Solidification heat transfer of nanofluid in existence of thermal radiation by means of FEM. *International Journal of Heat and Mass Transfer*, *123*, 418–431. doi:10.1016/j.ijheatmasstransfer.2018.02.095

Sheikholeslami, M., Ghasemi, A., Li, Z., Shafee, A., & Saleem, S. (2018). Influence of CuO nanoparticles on heat transfer behavior of PCM in solidification process considering radiative source term. *International Journal of Heat and Mass Transfer*, *126*, 1252–1264. doi:10.1016/j.ijheatmasstransfer.2018.05.116

Sheikholeslami, M., Hayat, T., & Alsaedi, A. (2018). Numerical simulation for forced convection flow of MHD CuO-H2O nanofluid inside a cavity by means of LBM. *Journal of Molecular Liquids*, *249*, 941–948. doi:10.1016/j.molliq.2017.10.099

Sheikholeslami, M., Hayat, T., Muhammad, T., & Alsaedi, A. (2018). MHD forced convection flow of nanofluid in a porous cavity with hot elliptic obstacle by means of Lattice Boltzmann method. *International Journal of Mechanical Sciences*, *135*, 532–540. doi:10.1016/j.ijmecsci.2017.12.005

Sheikholeslami, M., Jafaryar, M., Ganji, D. D., & Li, Z. (2018). Exergy loss analysis for nanofluid forced convection heat transfer in a pipe with modified turbulators. *Journal of Molecular Liquids*, *262*, 104–110. doi:10.1016/j.molliq.2018.04.077

Sheikholeslami, M., Jafaryar, M., & Li, Z. (2018a). Second law analysis for nanofluid turbulent flow inside a circular duct in presence of twisted tape turbulators. *Journal of Molecular Liquids*, *263*, 489–500. doi:10.1016/j.molliq.2018.04.147

Sheikholeslami, M., Jafaryar, M., & Li, Z. (2018b). Nanofluid turbulent convective flow in a circular duct with helical turbulators considering CuO nanoparticles. *International Journal of Heat and Mass Transfer*, *124*, 980–989. doi:10.1016/j.ijheatmasstransfer.2018.04.022

Sheikholeslami, M., Jafaryar, M., Saleem, S., Li, Z., Shafee, A., & Jiang, Y. (2018). Nanofluid heat transfer augmentation and exergy loss inside a pipe equipped with innovative turbulators. *International Journal of Heat and Mass Transfer*, *126*, 156–163. doi:10.1016/j.ijheatmasstransfer.2018.05.128

Sheikholeslami, M., Jafaryar, M., Shafee, A., & Li, Z. (2018). Investigation of second law and hydrothermal behavior of nanofluid through a tube using passive methods. *Journal of Molecular Liquids*, *269*, 407–416. doi:10.1016/j.molliq.2018.08.019

Sheikholeslami, M., Li, Z., & Shafee, A. (2018a). Lorentz forces effect on NEPCM heat transfer during solidification in a porous energy storage system. *International Journal of Heat and Mass Transfer*, *127*, 665–674. doi:10.1016/j.ijheatmasstransfer.2018.06.087

Sheikholeslami, M., Li, Z., & Shamlooei, M. (2018). Nanofluid MHD natural convection through a porous complex shaped cavity considering thermal radiation. *Physics Letters. [Part A]*, *382*(24), 1615–1632. doi:10.1016/j.physleta.2018.04.006

Sheikholeslami, M., & Rokni, H. B. (2017). Simulation of nanofluid heat transfer in presence of magnetic field: A review. *International Journal of Heat and Mass Transfer*, *115*, 1203–1233. doi:10.1016/j.ijheatmasstransfer.2017.08.108

Sheikholeslami, M., & Rokni, H. B. (2018a). CVFEM for effect of Lorentz forces on nanofluid flow in a porous complex shaped enclosure by means of Non-equilibrium model. *Journal of Molecular Liquids*, *254*, 446–462. doi:10.1016/j.molliq.2018.01.130

Sheikholeslami, M., Rokni, H.B. (2018b). Magnetic nanofluid flow and convective heat transfer in a porous cavity considering Brownian motion effects. *Physics of Fluids*, *30*(1), doi:10.1063/1.5012517

Sheikholeslami, M., & Sadoughi, M. K. (2017). Mesoscopic method for MHD nanofluid flow inside a porous cavity considering various shapes of nanoparticles. *International Journal of Heat and Mass Transfer*, *113*, 106–114. doi:10.1016/j.ijheatmasstransfer.2017.05.054

Sheikholeslami, M., & Sadoughi, M. K. (2018). Simulation of CuO- water nanofluid heat transfer enhancement in presence of melting surface. *International Journal of Heat and Mass Transfer*, *116*, 909–919. doi:10.1016/j.ijheatmasstransfer.2017.09.086

Sheikholeslami, M., & Seyednezhad, M. (2018). Simulation of nanofluid flow and natural convection in a porous media under the influence of electric field using CVFEM. *International Journal of Heat and Mass Transfer*, *120*, 772–781. doi:10.1016/j.ijheatmasstransfer.2017.12.087

Sheikholeslami, M., Shafee, A., Ramzan, M., & Li, Z. (2018). Investigation of Lorentz forces and radiation impacts on nanofluid treatment in a porous semi annulus via Darcy law. *Journal of Molecular Liquids*, *272*, 8–14. doi:10.1016/j.molliq.2018.09.016

Sheikholeslami, M., Shamlooei, M., & Moradi, R. (2018). Numerical simulation for heat transfer intensification of nanofluid in a porous curved enclosure considering shape effect of Fe3O4 nanoparticles. *Chemical Engineering & Processing: Process Intensification*, *124*, 71–82. doi:10.1016/j.cep.2017.12.005

Sheikholeslami, M., & Shehzad, S. A. (2017a). Thermal radiation of ferrofluid in existence of Lorentz forces considering variable viscosity. *International Journal of Heat and Mass Transfer*, *109*, 82–92. doi:10.1016/j.ijheatmasstransfer.2017.01.096

Sheikholeslami, M., & Shehzad, S. A. (2017b). Magnetohydrodynamic nanofluid convective flow in a porous enclosure by means of LBM. *International Journal of Heat and Mass Transfer*, *113*, 796–805. doi:10.1016/j.ijheatmasstransfer.2017.05.130

Sheikholeslami, M., & Shehzad, S. A. (2018a). Numerical analysis of Fe3O4 –H2O nanofluid flow in permeable media under the effect of external magnetic source. *International Journal of Heat and Mass Transfer*, *118*, 182–192. doi:10.1016/j.ijheatmasstransfer.2017.10.113

Sheikholeslami, M., & Shehzad, S. A. (2018b). CVFEM simulation for nanofluid migration in a porous medium using Darcy model. *International Journal of Heat and Mass Transfer*, *122*, 1264–1271. doi:10.1016/j.ijheatmasstransfer.2018.02.080

Sheikholeslami, M., & Shehzad, S. A. (2018c). Simulation of water based nanofluid convective flow inside a porous enclosure via Non-equilibrium model. *International Journal of Heat and Mass Transfer*, *120*, 1200–1212. doi:10.1016/j.ijheatmasstransfer.2017.12.132

Sheikholeslami, M., & Shehzad, S. A. (2018d). Non-Darcy free convection of Fe3O4-water nanoliquid in a complex shaped enclosure under impact of uniform Lorentz force. *The Chinese Journal of Physiology*, *56*(1), 270–281. doi:10.1016/j.cjph.2017.12.017

Sheikholeslami, M., Shehzad, S. A., Abbasi, F. M., & Li, Z. (2018). Nanofluid flow and forced convection heat transfer due to Lorentz forces in a porous lid driven cubic enclosure with hot obstacle. *Computer Methods in Applied Mechanics and Engineering*, *338*, 491–505. doi:10.1016/j.cma.2018.04.020

Sheikholeslami, M., Shehzad, S. A., & Li, Z. (2018a). Nanofluid heat transfer intensification in a permeable channel due to magnetic field using Lattice Boltzmann method. *Physica B, Condensed Matter*, *542*, 51–58. doi:10.1016/j.physb.2018.03.036

Sheikholeslami, M., Shehzad, S. A., & Li, Z. (2018b). Water based nanofluid free convection heat transfer in a three dimensional porous cavity with hot sphere obstacle in existence of Lorenz forces. *International Journal of Heat and Mass Transfer*, *125*, 375–386. doi:10.1016/j.ijheatmasstransfer.2018.04.076

Sheikholeslami, M., Shehzad, S. A., Li, Z., & Shafee, A. (2018). Numerical modeling for Alumina nanofluid magnetohydrodynamic convective heat transfer in a permeable medium using Darcy law. *International Journal of Heat and Mass Transfer*, *127*, 614–622. doi:10.1016/j.ijheatmasstransfer.2018.07.013

Sheikholeslami, M., & Vajravelu, K. (2017). Forced convection heat transfer in Fe3O4- ethylene glycol nanofluid under the influence of Coulomb force. *Journal of Molecular Liquids*, *233*, 203–210. doi:10.1016/j.molliq.2017.03.026

Sheikholeslami, M., & Zeeshan, A. (2017). Analysis of flow and heat transfer in water based nanofluid due to magnetic field in a porous enclosure with constant heat flux using CVFEM. *Computer Methods in Applied Mechanics and Engineering*, *320*, 68–81. doi:10.1016/j.cma.2017.03.024

Sheikholeslami, M., Zeeshan, A., & Majeed, A. (2018). Control volume based finite element simulation of magnetic nanofluid flow and heat transport in non-Darcy medium. *Journal of Molecular Liquids*, *268*, 354–364. doi:10.1016/j.molliq.2018.07.031

Chapter 4
Discharging of Nano-Enhanced PCM via Finite Element Method

ABSTRACT

Latent heat thermal energy storage systems (LHTESS), which work based on energy storage and retrieval during solid-liquid phase change, is used to establish balance between energy supply and demand. LHTESS stores and retrieves thermal energy during solid-liquid phase change, while in SHTESS phase change doesn't occur during the energy storage and retrieval process. LHTESS has a lot of advantages in comparison to SHTESS. The most important one is storing a large amount of energy during phase change process, which makes the energy storage density in LHTESS much higher than SHTESS. Because of this property, LHTESS have a wide application in different cases, such as solar air dryer, HVAC systems, electronic chip cooling, and engine heat recovery. The main restriction for these systems is thermal conductivity weakness of common PCMs. In this chapter, the method of adding nanoparticles to pure PCM and making nano-enhanced phase change material (NEPCM) and using fin with suitable array are presented to accelerate solidification process. The numerical approach which is used in this chapter is standard Galerkin finite element method.

1. DISCHARGING PROCESS EXPEDITION OF NEPCM IN Y-SHAPED FIN-ASSISTED LATENT HEAT THERMAL ENERGY STORAGE SYSTEM

1.1. Problem Definition and Mathematical Model

1.1.1. Problem Definition

The main geometry of present study is a Fin-assisted LHTESS, which HTF exists in the inner tube, and in the space between inner and outer tube, is filled by PCM. Y-shaped fin is connected to the outer side of HTF tube to improve heat transfer from HTF to PCM (Figure 1) (Lohrasbi, Sheikholeslami, & Ganji, 2016). The two-dimensional solution domain is presented in Figure 2. The angle between fin branches and axes is assumed to be $\beta = \pi / 16, 2\pi / 16, 3\pi / 16, 4\pi / 16$, Fin length and thickness assumed

DOI: 10.4018/978-1-5225-7595-5.ch004

Figure 1. Fin assisted LHTESS

Figure 2. Two dimensional solution domain used in present study

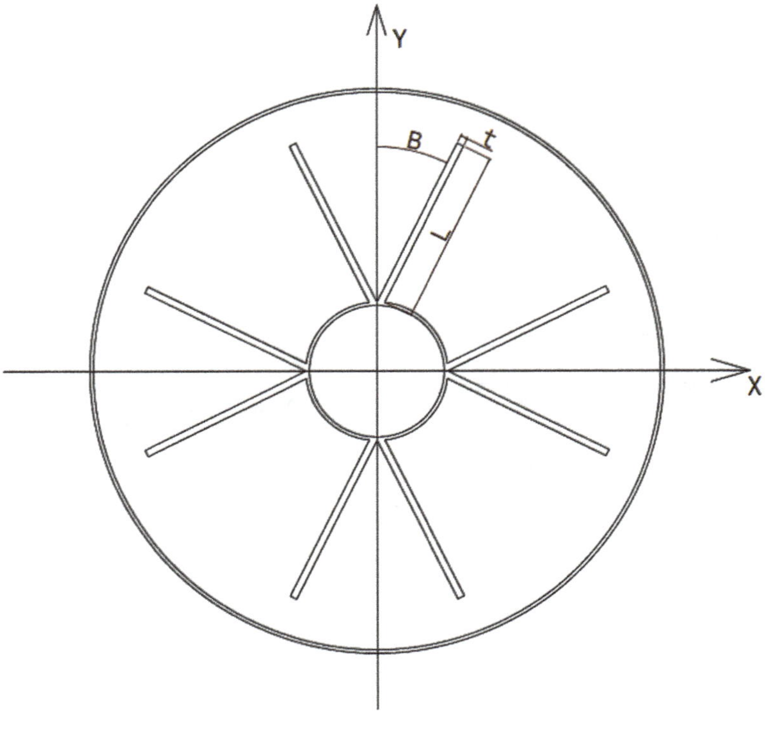

to be equal to $L = 2{,}2.4{,}2.8\,cm$ and $t = 0.5{,}0.75{,}1\,mm$ respectively. The constant temperature boundary condition equal to 240K is applied to inner tube and the initial temperature of liquid PCM is assumed equal to 278K. The PCM, Nanoparticles and fin properties are listed in Table 1.

The transient Governing equations of conduction-dominated solidification process are presented as below:

$$\rho C_p \frac{dT}{dt} = \nabla\left(k\nabla T\right) + L_f \frac{dS}{dt} \tag{1}$$

$$\begin{aligned} S &= 1 && T < T_m - T_0 \\ S &= 0 && T > T_m + T_0 \\ S &= \frac{\left(T_m + T_0 / 2 - T\right)}{T_0} && T_m - T_0 < T < T_m + T_0 \end{aligned} \tag{2}$$

When pure PCM is applied in LHTESS, pure PCM properties are used in governing equations but when copper nanoparticles are dispersed in PCM, the NEPCM properties should be obtained by using the following equations and be applied in governing equations (Sheikholeslami, 2014):

$$\rho_{nf} = \left(1-\phi\right)\rho_{bf} + \phi\rho_p \tag{3}$$

$$\left(\rho C_p\right)_{nf} = \left(1-\phi\right)\left(\rho C_p\right)_{bf} + \phi\left(\rho C_p\right)_p \tag{4}$$

$$\left(\rho L_f\right)_{bf} = \left(1-\phi\right)\left(\rho L\right)_{bf} \tag{5}$$

Table 1. The physical properties of water as PCM, Copper as nanoparticles and aluminum fin

Property	PCM	Nanoparticles	Fin
$\rho\left[kg / m^3\right]$	997	8954	2700
$C_p\left[j / kg\,K\right]$	4179	385	902
$k\left[w / mK\right]$	0.6	400	200
$L_f\left[j / kg\right]$	335000	-	-

In present study, thermal conductivity of NEPCM is evaluated by the following equation based on Maxwell (Sheikholeslami, 2014) work.

$$\frac{k_{nf}}{k_{bf}} = \frac{k_p + 2k_{bf} - 2\phi\left(k_{bf} - k_p\right)}{k_p + 2k_{bf} + 2\phi\left(k_{bf} - k_p\right)} \tag{6}$$

where ϕ, k_p and k_{bf} are nanoparticle volume fraction, nanoparticle and base fluid thermal conductivity respectively.

Total energy density released during discharging process, is calculated from the following equation:

$$E_{density} = \frac{\rho}{V}\int\left(c_p T + \left(1 - s\right)L_f\right)dV \tag{7}$$

1.1.2. Numerical Method

Standard Galerkin Finite Element Method with cubic interpolation over triangles is implemented to solve the present phase change problem. Nodal values are placed on corners and sides of the grid cells. The Galerkin equations are formed by symbolic analysis, which substitutes definitions, segregates dependencies on variables, applies integration by parts, integrates over cells, and ultimately differentiates the resulting system with respect to system variables to form the coupling matrix. Equations are solved simultaneously by an iterative method. For nonlinear systems Newton-Raphson iteration process with back-tracking is used. For time dependent problems, such as solidification problem, an implicit Backward Difference Method for integration in time is used. Variables are approximated by quadratic polynomials in time, and the time step is controlled to keep the cubic term smaller than the required value of error. The residual Galerkin integral over a patch of cells surrounding each mesh node is minimized by the Finite Element Equations. Then the residuals in each cell independently are analyzed as a measure of compliance, and subdivide each cell in which the required error tolerance is exceeded. Any cell thus split can be re-merged whenever the cell error drops to of the splitting tolerance. Adaptive grid refinement is used to simulate solidification process in present work. When the initial mesh generation is performed, code estimates the error and refines mesh in order to reach to the desired accuracy. In unsteady problems, this procedure also has to be applied to the initial values of the variables in order to refine the mesh where rapid change in variables occur (Figure 3). Comparison between the present code based on Galerkin Finite Element Method and experimental results obtained by Ismaeil et al. (2001) indicates good agreement, which validates the present code (see Figure 4). Moreover it proves that ignoring natural convection in numerical simulation of solidification phenomenon leads to results close to reality.

1.2. Effects of Active Parameters

1.2.1. Applying Y-Shaped Fin to LHTESS

In this section the effect of adding Y-shaped fins to LHTESS containing pure PCM during solidification process and expedition of energy retrieval will be investigated. For LHTESS without fin, solidification

Figure 3. Adaptive grid refinement procedure for $\beta = 3\pi / 16, L = 2.4cm, t = 1mm$

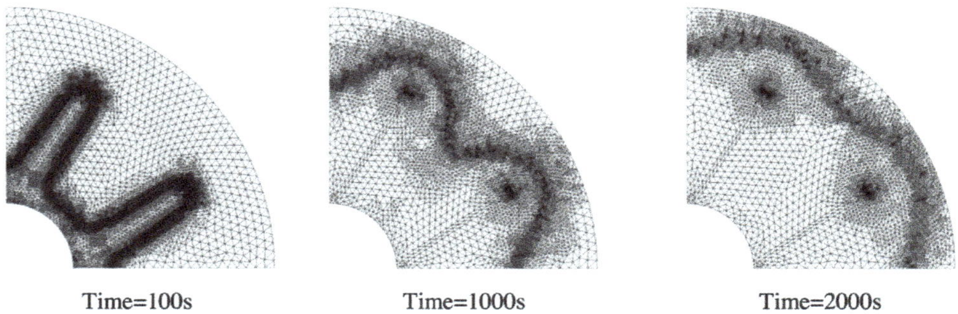

Figure 4. Comparison between solidification front in fin-assisted LHTESS in present study and experimental work by Ismaeil et al. (2001)

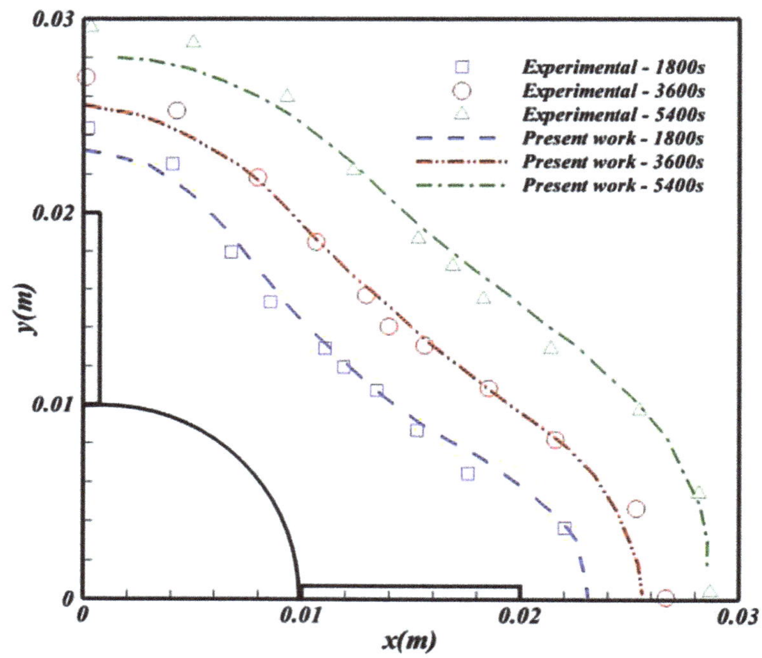

process begins only in the region adjacent to the inner tube containing HTF, but by attaching fins with high thermal conductivity to the inner tube, solidification also begins in the regions adjacent to the fins surfaces, and as a result, solidification rate will be enhanced. In order to achieve this purpose, Y-shaped fin array as can be seen from Figures 1 and 2 for different values of branch angle, thickness and length will be applied to the system. In Table 2 the mentioned geometry parameters are listed.

According to Figures 5 and 6 it is obvious that for all values of fin length and thickness, the best performance of LHTESS during solidification occurs when the branch angle is $\beta = 3\pi / 16$ among the investigated values. Because when the value of β is small, branches of Y-shaped fin are close to the axes, therefore in the region between two branches in each quadrant of solution domain, the amount of

Table 2. Geometry parameters of Y-shaped fins

Parameter	Different Values Used in Simulation
β	$\frac{\pi}{16},\frac{2\pi}{16},\frac{3\pi}{16},\frac{4\pi}{16}$
L	$2,2.4,2.8\,cm$
t	$0.5,0.75,1\,mm$

Figure 5. Full solidification time for different values of fin geometry parameters, (a) L=2cm, (b) L=2.4, (c) L=2.8

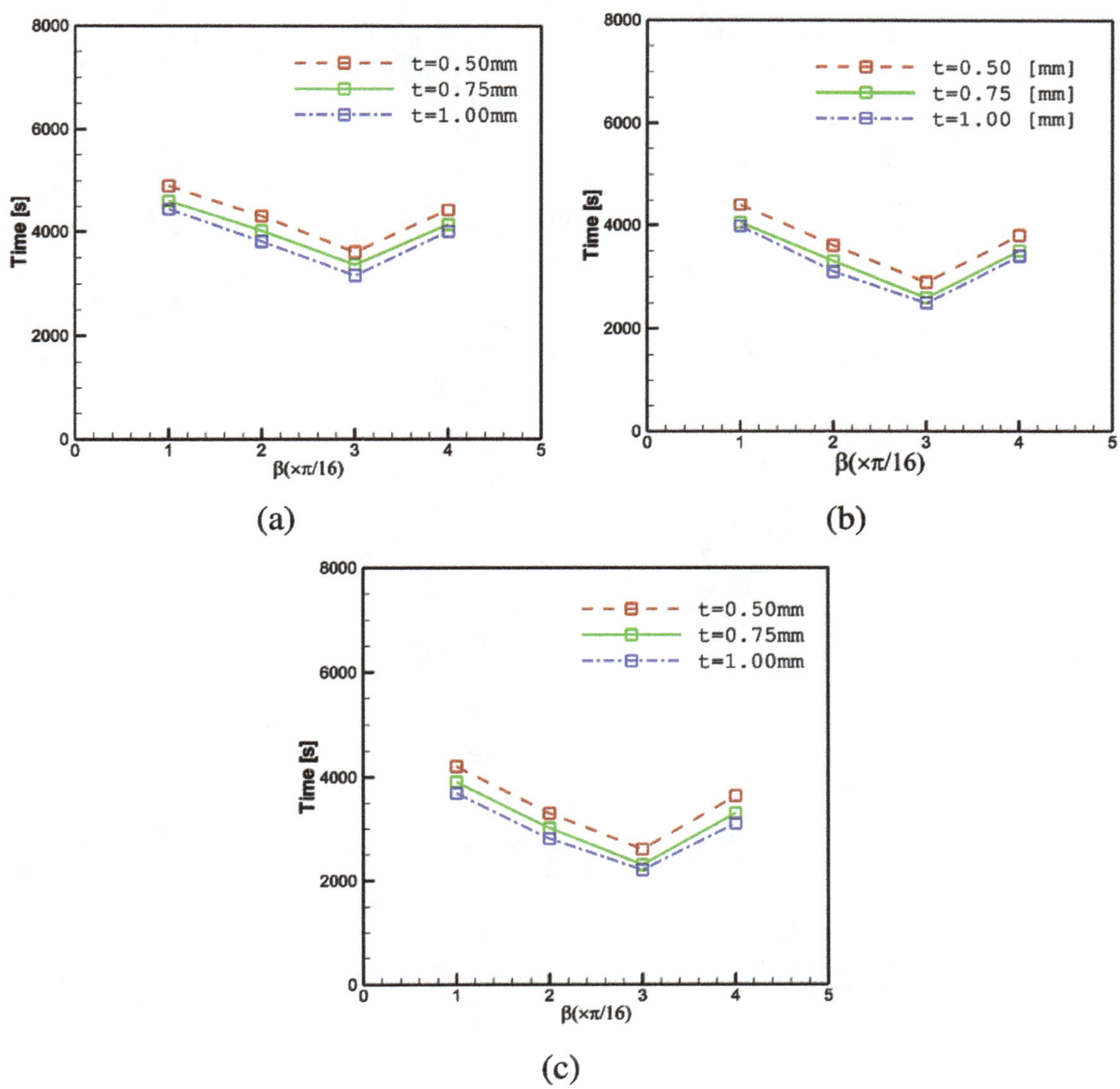

Figure 6. Effect of fin branch angle on temperature (left side) and solid fraction (right side) contours. L=2.8 [cm], t=0.75 [mm]

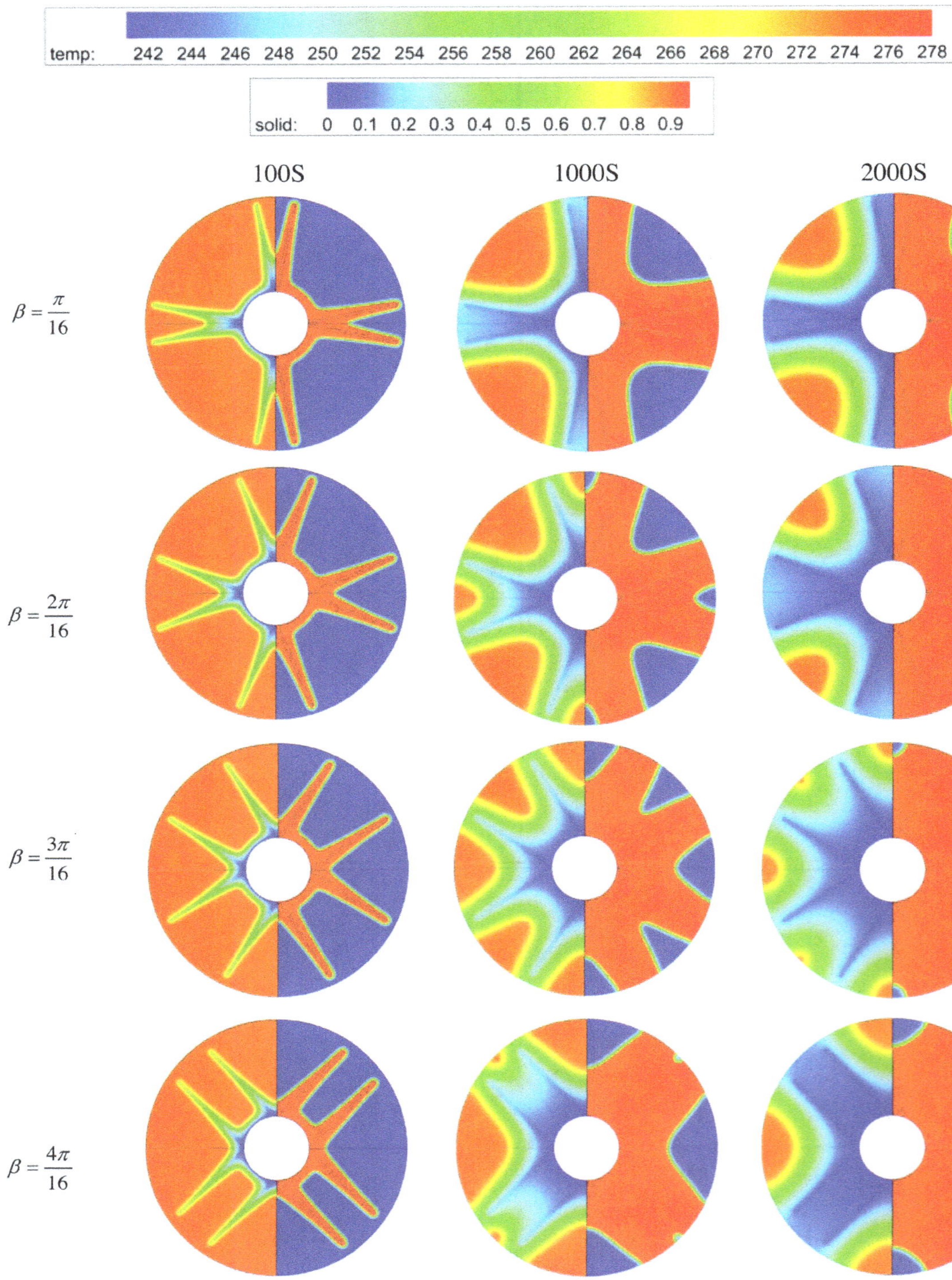

PCM without fin is too high and because in this region, thermal penetration depth is not enhanced by fins, Solidification rate is too low, which increases full solidification time. Similarly when the value of fin branch angle is too high, in region between two branches in each quadrant of solution domain, penetration depth is enhanced but in the regions between branches and axes, solidification rate is too low. So it is obvious that the best choice for β is a value between $\pi / 16, 4\pi / 16$, which in the values discussed in this paper, $\beta = 3\pi / 16$ demonstrates the lowest full solidification time and the most uniform solidification process, which is obvious from related contour and diagrams. As can be observed in Figures 5 and 7 full solidification time for three values of fin thickness are presented. These diagrams in-

Figure 7. Effect of fin branch thickness on temperature (left side) and solid fraction (right side) contours $\beta=3\pi/16$ [Rad], L=2.8 [cm]

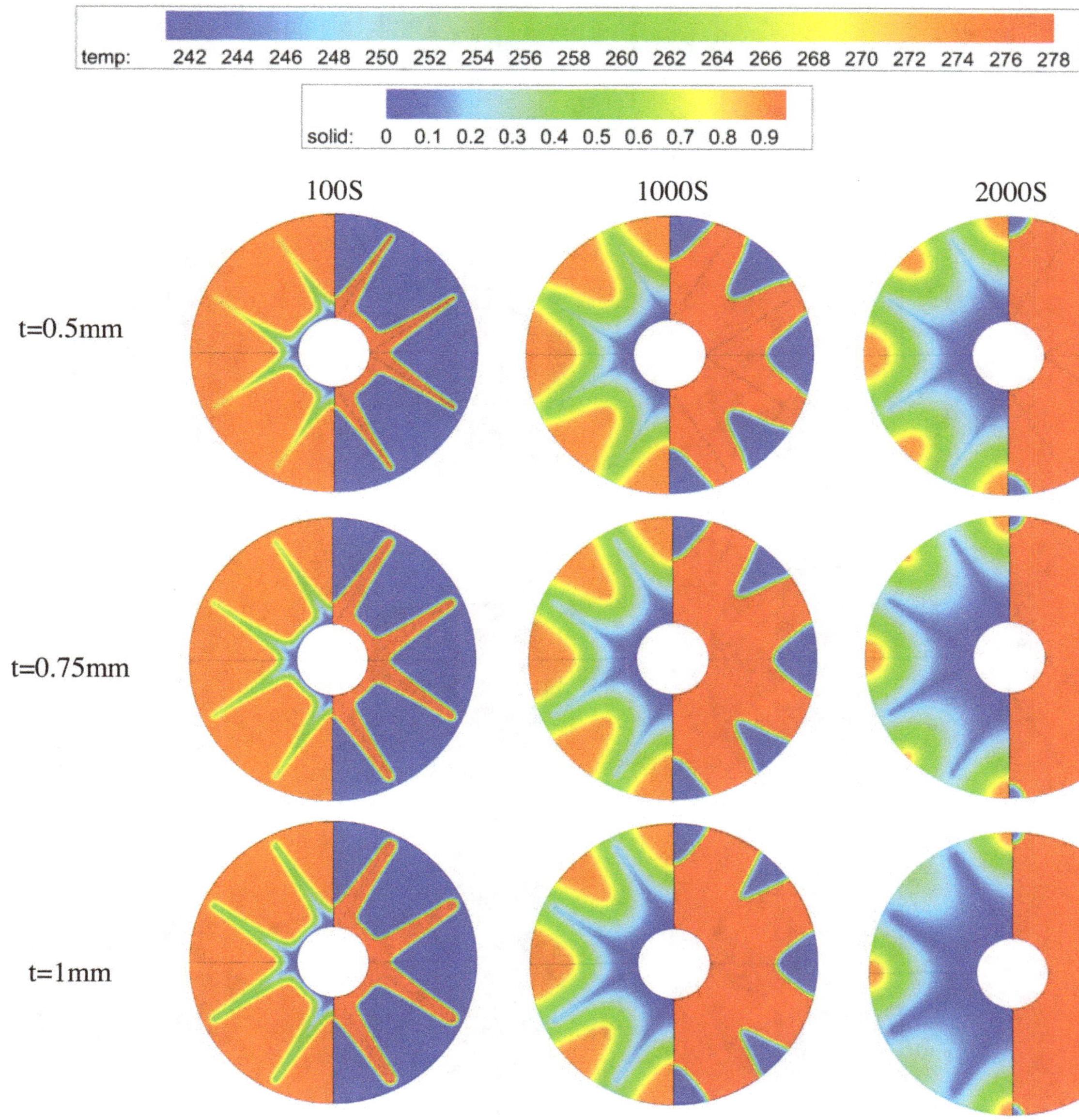

dicate that by increasing fin thickness from the value of $0.5 to 0.75mm$, solidification rate enhancement is considerable but by changing the value to $1mm$, solidification rate enhancement is not significant. Moreover, since increasing the fin thickness value leads to increasing fin volume in LHTESS and decreasing the volume of employed PCM and as a result, maximum energy storage capacity will decrease, so the best value of fin thickness according to reducing full solidification time and avoiding the reduction of maximum energy storage capacity is $t = 0.75mm$.

According to Figures 5 and 8, by increasing in fin length for all values of branch angles and thickness, full solidification time is decreased considerably and because of this significant enhancement,

Figure 8. Effect of fin length temperature (left side) and solid fraction (right side) contours.
β=3π/16 [Rad], t=0.75 [mm]

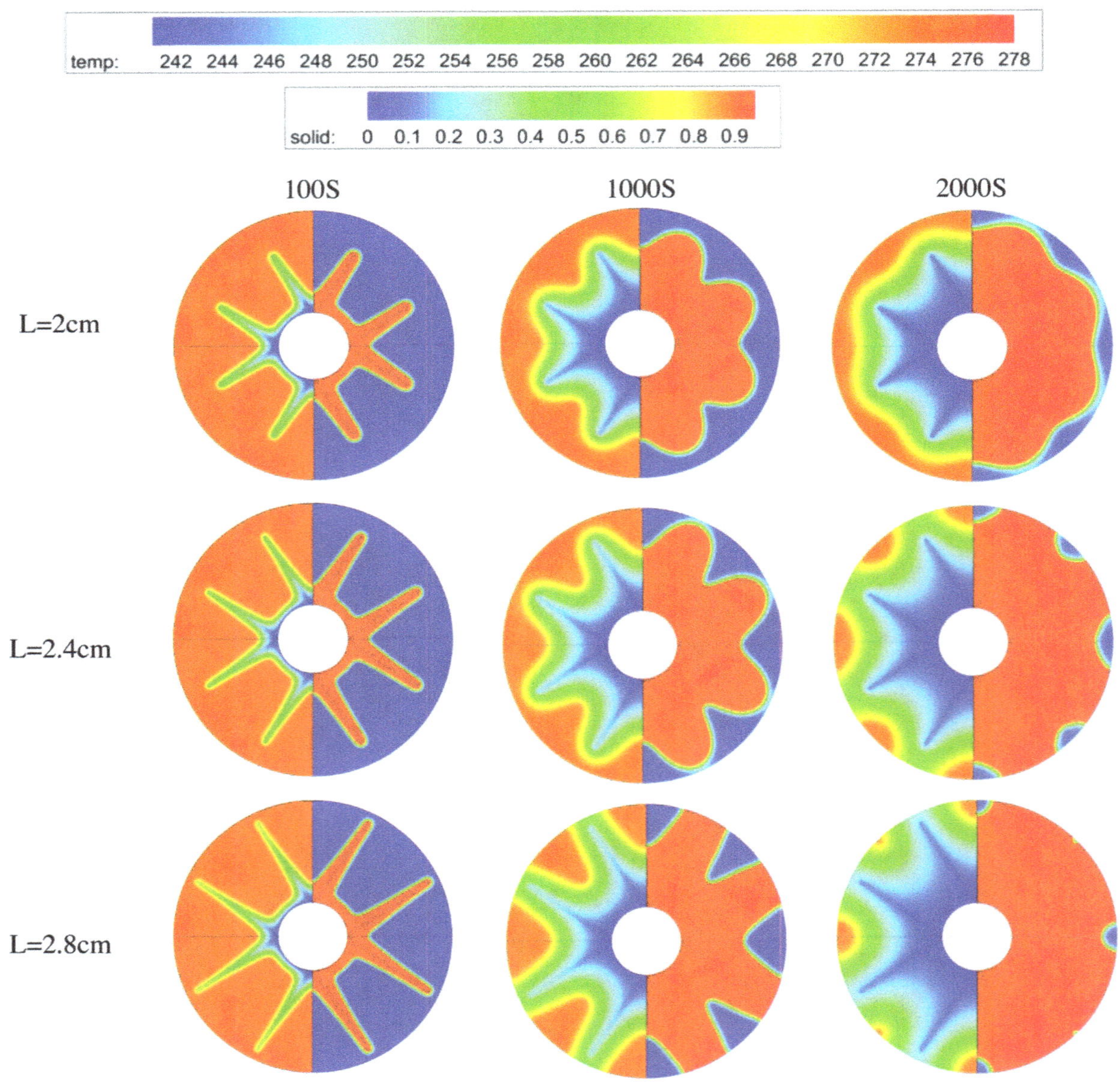

decrease in value of maximum energy storage capacity can be ignored. The main reason is the augmentation of penetration depth due to increase in fin length. So the best choice for fin length among discussed values is $L = 2.8cm$. Therefore between all discussed cases, the best choice for geometry parameters is $\beta = \pi / 16, t = 0.75mm L = 2.8cm$. From solid fraction contour, this case shows the fastest and most uniform solidification process.

According to Figure 6, the temperature distribution and solid fraction contours has been illustrated to study the effect of fin branch angle on solidification process in LHTESS. As can be seen, in all time steps, average temperature for the case of $\beta = 3\pi / 16$ is lower in comparison to the other cases, and the value of solid fraction is higher, which means that for the mentioned branch angle, solidification rate is higher. According to Figure 7, the temperature distribution and solid fraction contours has been illustrated to study the effect of fin thickness on solidification process in LHTESS. As can be seen, in all time steps, average temperature for the case of t=1mm is lower and solid fraction value is higher, but the enhancement difference for the case of t=1 [mm] in comparison to t=0.75 [mm] is insignificant. According to Figure 8, the temperature distribution and solid fraction contours has been illustrated to study the effect of fin length on solidification process in LHTESS. As can be observed, in all time steps, by increasing in fin length, average temperature is lower and the value of solid fraction in higher. It can be observed that when the solidification front moves away a little from fin tips, phase change process is not enhanced by fins anymore. By employing longer fins, the region affected by them will be wider and heat transfer enhancement will be more considerable.

1.2.2. Adding Nanoparticles to LHTESS

The effect of adding Copper particles as high thermal conductivity nanoparticles, to water as PCM and making NEPCM, on LHTESS without Y-shaped fins performance during solidification process is illustrated in Table 3 and Figure 9 it can be observed that adding nanoparticles to LHTESS has moderately considerable effect on solidification process, this is because in conduction dominated phase change procedures, adding nanoparticles has much more effect in comparison to natural convection dominated procedures (Dhaidan, Khodadadi, Al-Hattab, & Al-Mashat, 2013), therefore since in solidification process, conduction is predominant heat transfer mechanism, increasing of nanoparticles volume fraction shows rather significant enhancement in process rate.

According to Figure 9, it can be observed that at the beginning of the process the enhancement of solidification rate, by adding nanoparticles is insignificant, but as time progresses, the enhancement

Table 3. Effect of nanoparticles volume fraction on full solidification time and improvement

Nanoparticle Volume Fraction (ϕ)	Full Solidification Time (s)	Increment in Rate of Solidification (%)
0.00	13000	-
0.025	11847	8.8
0.050	10800	16.9

Figure 9. Effect of nanoparticles volume fraction on solidification front position

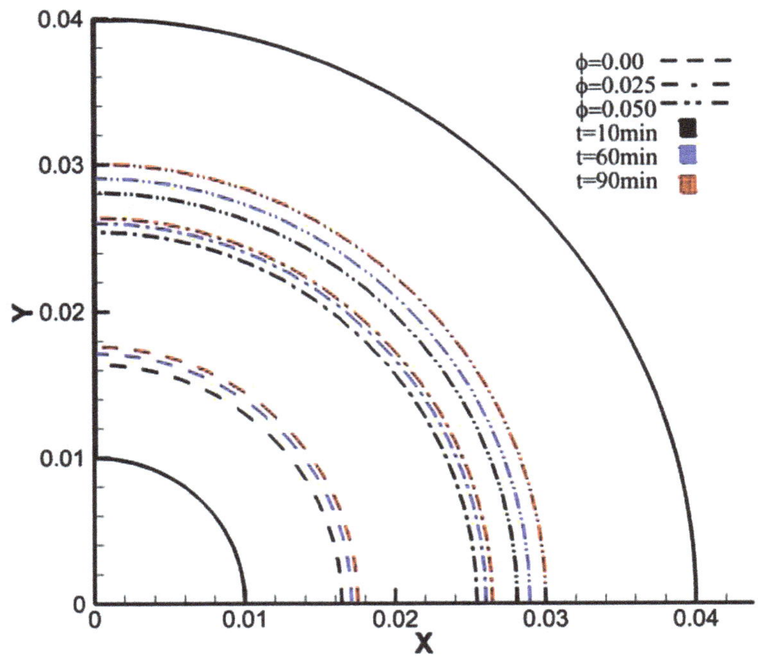

of the process rate because of nanoparticles dispersion augments. Because of the enhancement in the discharging rate by adding 5% copper nanoparticles, this volume fraction value can be reported as the best choice for nanoparticles volume fraction in this system.

1.2.3. Comparison Between Adding Nanoparticles and Applying Y-Shaped Fins

Adding nanoparticles decreases maximum energy storage capacity because of decreasing the value of heat capacity and latent heat, while applying fin in LHTESS decreases maximum energy storage capacity because of decreasing in PCM mass. But both of these enhancement methods enhance solidification rate. In order to compare these techniques from the viewpoint of both parameters of solidification rate and maximum energy storage capacity, by using Equation (7), Figures 10 and 11 have been produced.

According to Figure 10 it is obvious that in Y-shaped fin-assisted LHTESS, discharging rate is significantly higher than the case of nanoparticles dispersion and as a result, full solidification time is significantly lower. Moreover, according to Figure 11, comparison between the Maximum energy storage capacity in the case of adding nanoparticles to the system and adding Y-shaped fin, indicate that by employing fin in the system, Maximum energy storage capacity, which is equal to total energy at the beginning of discharging process, is almost the same as the LHTESS containing pure PCM without fin, and the LHTESS containing NEPCM, But the discharging rate enhancement is significantly higher, therefore, adding fin to LHTESS is a better enhancement technique in comparison to Nanoparticle dispersion from the viewpoint of either discharging rate or maximum energy storage capacity.

Figure 10. Solid fraction during discharging process

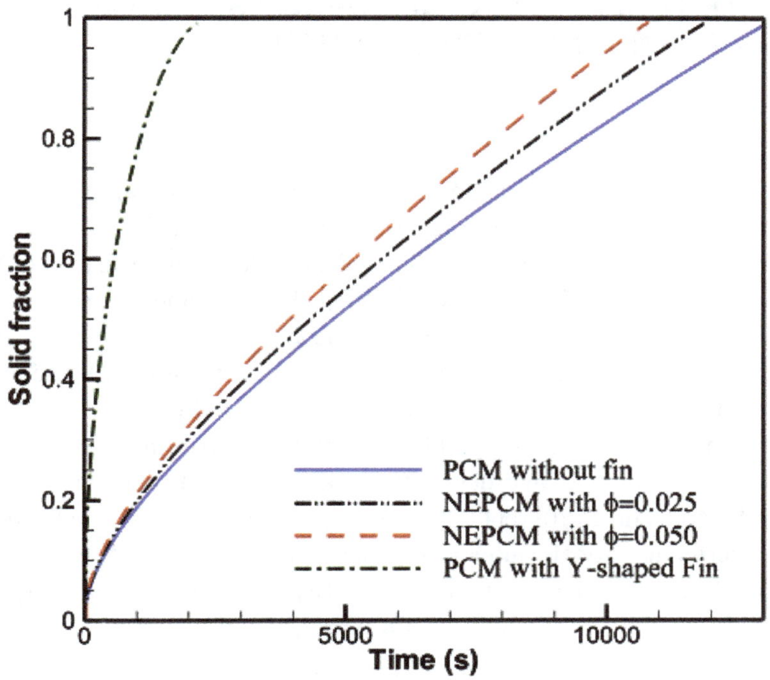

Figure 11. Total energy released during discharging process

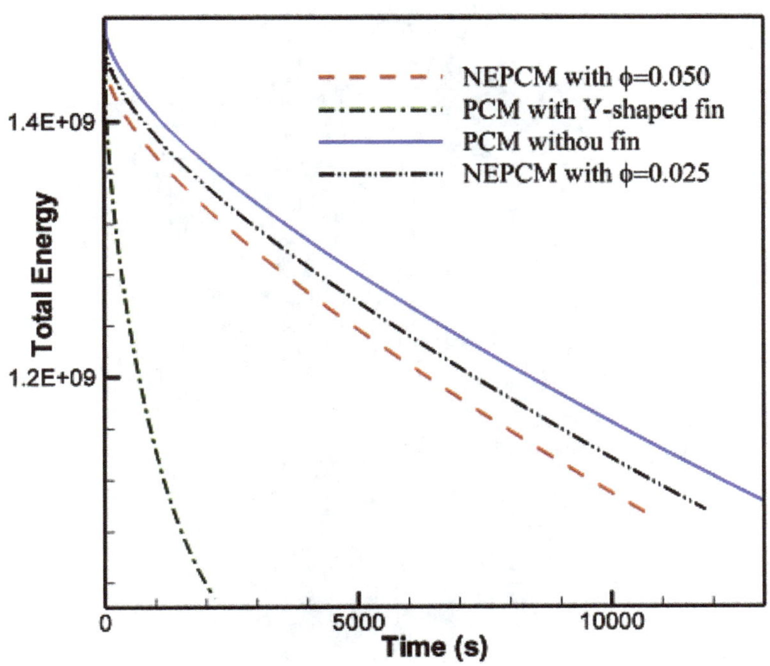

2. SNOWFLAKE SHAPED FIN FOR EXPEDITING DISCHARGING PROCESS IN LATENT HEAT THERMAL ENERGY STORAGE SYSTEM CONTAINING NANO-ENHANCED PHASE CHANGE MATERIAL

2.1. Problem Statement

The main geometry of the present study is a Snowflake shaped fin-assisted LHTESS, which has been illustrated in Figure 12 (Sheikholeslami, Lohrasbi, & Ganji, 2016). HTF flows in the inner tube, and the space between inner tube and storage container shell, which is square shaped (Khillarkar, Gong, & Mujumdar, 2000), is filled with PCM. The snowflake shaped fin is connected to the outer side of the HTF tube in order to enhance thermal penetration depth into the space filled with PCM. The two-dimensional solution domain is illustrated in Figure 13. The fin geometry parameters are listed in Table 4. Constant temperature equal to 240K is applied to the inner tube and the liquid PCM initial temperature is assumed equal to 285K. In Figure 14, solidification front for different values of fin branch angles and length have been illustrated at the same time step, in order to study the behavior of solidification process at different values of fin branch length. Three different values for each branch length have been investigated, from this figure it can be inferred that for all values of branches direction, thermal penetration depth increases and solidification rate has strictly increasing behavior. Therefore fin branches length has been chosen

Figure 12. Three dimensional view of Snowflake shaped fin-assisted LHTESS

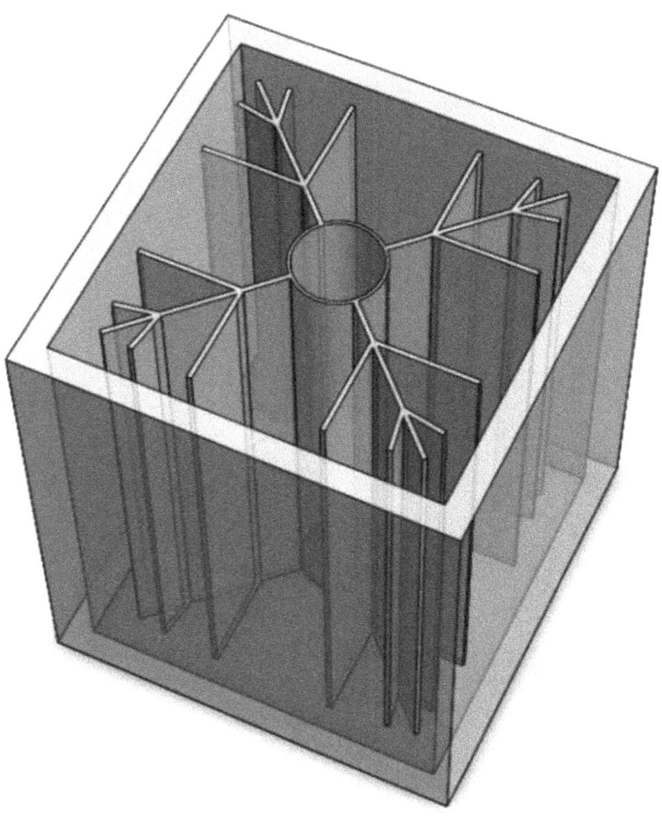

Figure 13. Solution domain and geometry parameters

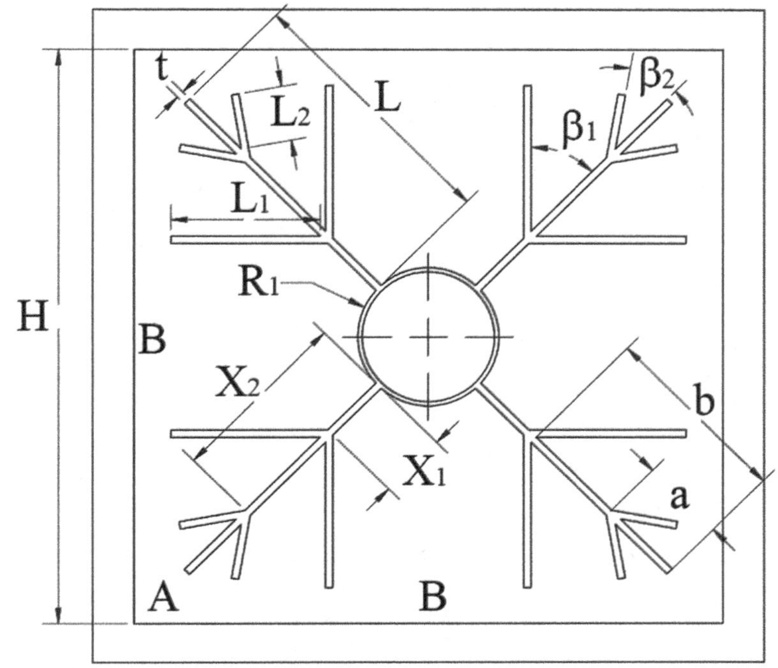

Table 4. Geometry parameters of Snowflake shaped fin structure

Geometry Parameters	Values
β_1	$[2\pi / 12 - 5\pi/12][Rad]$
β_2	$[2\pi / 12 - 5\pi/12][Rad]$
X_1	$0.1L, 0.2L$
X_2	$0.6L, 0.7L$

long enough and constant, which for longer branches, they will interfere or exceed of the calculation domain, for different values of branches directions. This choice reduces geometry parameters and simplifies the analysis of the other parameters, which have unknown effect on the performance of LHTESS. Fin thickness has been changed in order to keep constant the cross section area of the fin. This choice has been made because among the three geometry parameters in the present study including β, L and w, the effect of fin thickness on solid-liquid phase change is not significant (Ismail, Alves, & Modesto, 2001), therefore its effect on solidification process has not been investigated. On the other hand, if the volume of the employed fin varies for different cases, the value of energy storage capacity will be different, so the value of fin volume-in two dimensional simulation cross section of the fin system, has to

Figure 14. Phase change front at t=1000s, for Snowflake shaped fin assisted LHTESS with $\beta_1 = \beta_2 = 2\pi / 12, 4\pi / 12 \, and \, 4\pi / 12$.

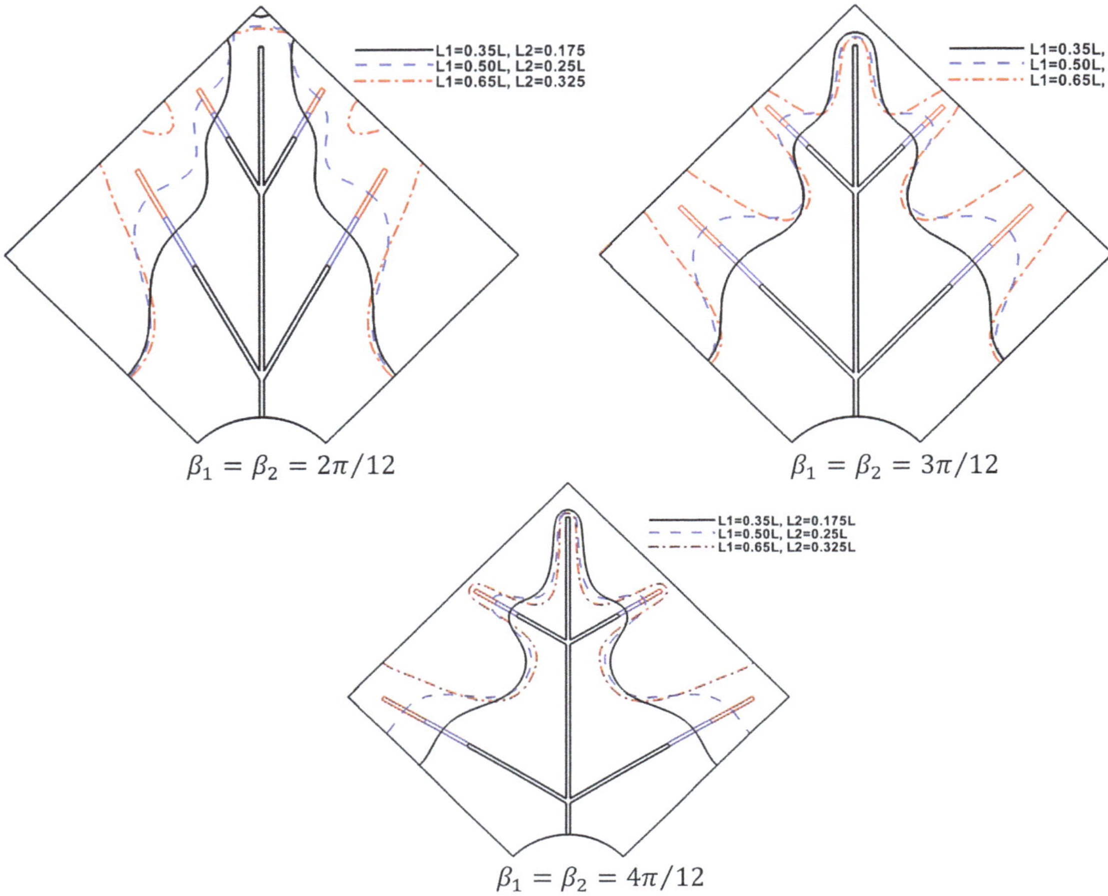

be the same in order to study the effect of fin geometry parameters on solidification rate, at the same value of energy storage capacity. To do so, fin length and direction is changed to study their effect on the system performance, and fin thickness which has negligible effect on the process has been changed in order to keep constant the cross section area of the fin and energy storage capacity.

2.2. Numerical Results

In this section, the reason of employing Snowflake shaped fin structure in square shaped LHTESS will be investigated. Solid fraction-temperature contour plots for LHTESS containing pure PCM without fin are illustrated in Figure 15. The contour plots indicate that in square shaped LHTESS, solidification process at the corners happens too slowly and it increases the full process time. As can be observed in Figure 15 at 15000s, if the LHTESS shape had been cylindrical, the solidification process would have been completed by this time step, but for square shape LHTESS the remaining PCM at the corners slow-down the process significantly, which full solidification is achieved after 22500 seconds from the begin-

Figure 15. Solid fraction (left side) and temperature (right side) contour plots at three different time steps during solidification process of LHTESS without fins

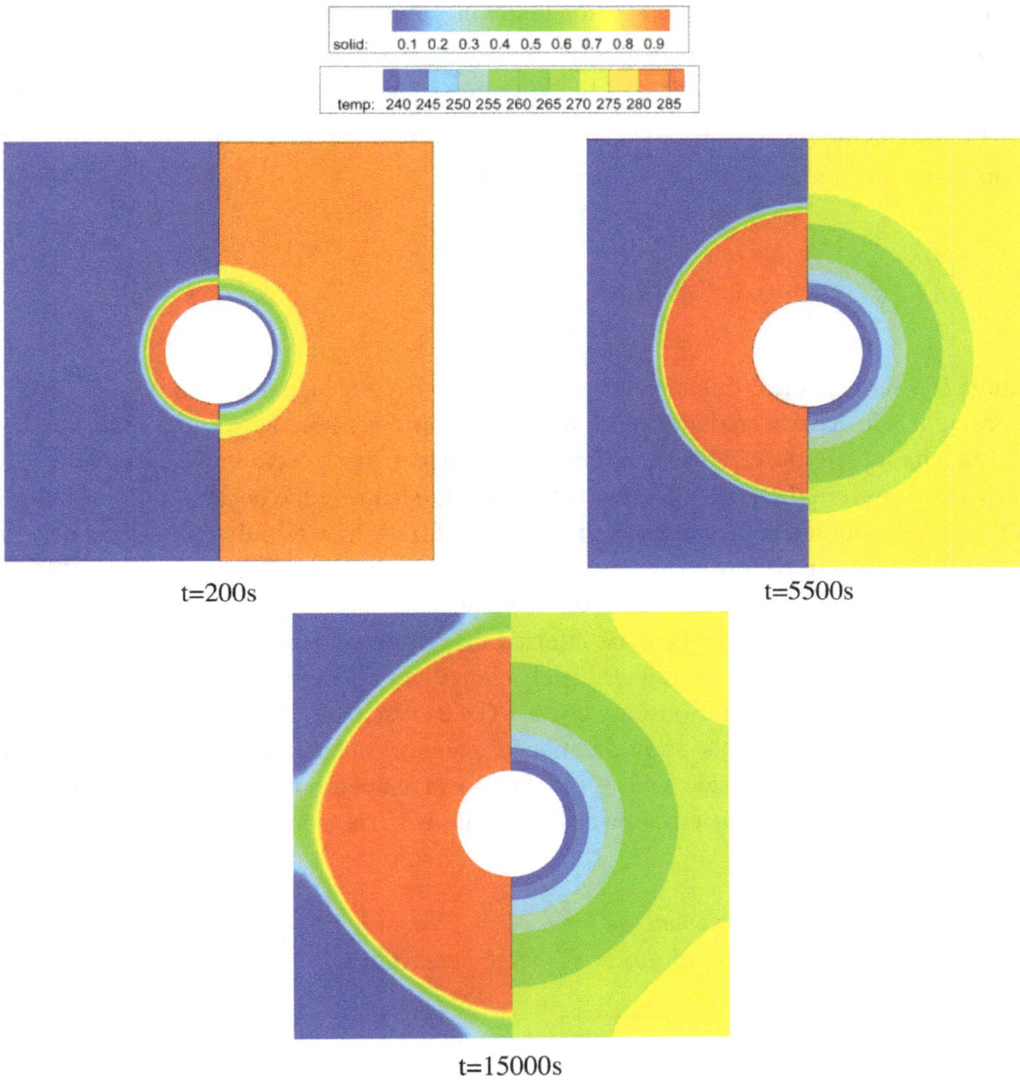

ning of the process. This issue shows the requirement of increasing thermal penetration depth at the corners of the square shaped LHTESS, which in this paper is carried out by employing fin with an innovative array in order to increase the solidification rate of PCM. In order to increase the thermal penetration depth at the corners of the square shaped LHTESS, we will use an innovative fin configuration which we have named Snowflake shaped fin structure as illustrated in Figure 13. In this structure, each fin has four branches, the main question is that the bigger branch should be near the cold wall or at the right side after the smaller branch. To answer this, a secondary problem is considered in order to justify the reason of special snowflake crystal structure from the viewpoint of heat transfer analysis. In this problem, conduction heat transfer through the Snowflake shaped fin with thermal conductivity equal to $200W / mK$, with constant temperature equal to 273K in one side, zero heat flux at the end side, surrounding temperature equal to 298K and convection heat transfer coefficient equal to $15W / m^2K$ have been simulated. It should be noted that the boundary condition at the surfaces are applied by convection, and this is not relevant to the natural convection effect on solidification process of the main problem of this paper, To investigate the performance of fin branches, efficiency is measured as indicated in Equation (8) (Bergman, Incropera, & Lavine, 2011). Where q_f is the actual heat transfer from the fin surface area, h is the convection heat transfer coefficient, A_f is the fin surface area and θ_b is the difference between fin base temperature and ambient temperature.

$$\eta \equiv \frac{q_f}{q_{max}} = \frac{q_f}{hA_f\theta_b} \tag{8}$$

In Figures 16a and 16b the efficiency of two branches, when the bigger branch is in left side-near the cold wall- or right side, have been illustrated. Comparison between these two figures indicates that, when the bigger branch is at left, the efficiency values of the two branches are close to each other for all values of branches distance from the end side. This will be used to achieve the desired uniform solidification process in LHTESS. But when the bigger branch locates at right side, for all values of distance from the end side, its efficiency will be lower and since bigger branch gets more volume in the LHTESS instead of PCM in comparison with smaller branch, the lower efficiency of bigger branch is not reasonable in this application. The other result that can be inferred from Figures 6a and 6b is that the closer the two branches locates to the cold wall, the higher the efficiency of both branches.

It can be claimed that the mysteriousness of snowflake crystal structure (Figure 17) can be justified by heat transfer reasons, which in the abovementioned sentences, it has been indicated that this structure, with smaller branch near the end side, causes more uniform temperature distribution on the branches in comparison with other arrays, which in snowflakes in nature, the reason of this structure can be prevent-

Figure 16. Fin branches efficiency when the bigger branch is in left position-near the cold wall (a) - or in the right (b) versus the distance of branches from the end side

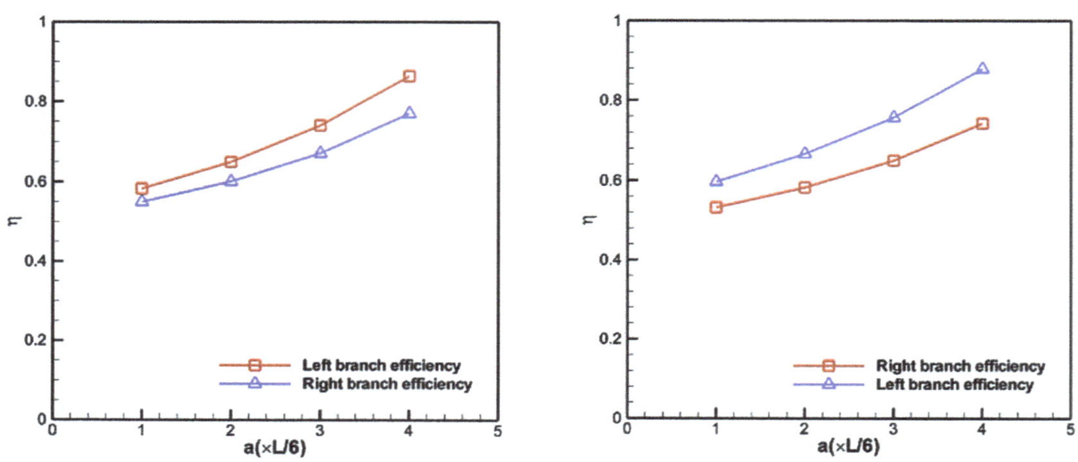

(a) (b)

Figure 17. Snowflake crystal structure

ing the snowflakes from melting and destruction. This concept can be applied to fin structure to achieve more uniform temperature distribution on the fin surface to improve heat transfer rate between the fin surface area and surroundings. Snowflake shaped fin structure is firstly proposed in this work and has not been investigated before. It is obvious that by adding branches to the main fin, the ability of fin to widespread heat into the PCM will be improved. But the configuration of branches and the effect of geometry parameters have to be investigated. As mentioned above, snowflake structure is able to achieve uniform temperature distribution on its surface, and this feature can be used to achieve uniform solidification by immersing this fin structure into the LHTESS. Therefore, solidification process simulation in Snowflake shaped fin-assisted LHTESS has been carried out for different values of geometry parameters to obtain the best fin structure. In Figure 18, full solidification time for different geometry parameters are illustrated. From Figure 18, it can be observed that, changing the direction of smaller branch (β_2) doesn't have considerable effect on the solidification rate of PCM, but changing the bigger branch direction (β_1) has a significant effect on the process rate. According to Figure 18, the best case among the investigated cases of geometry parameters is $X_1 = 0.1L$,$X_2 = 0.6L$,$\beta_1 = 45°$,$\beta_2 = 75°$.

2.2.1. Effect of Branches Direction on Solidification Rate

In Figure 19, the effect of changing bigger branch direction on solid fraction of PCM during solidification process, for the best values for the position of branches among the investigated cases, which are $X_1 = 0.1L$ and $X_2 = 0.6L$ have been illustrated. According to this figure, changing the bigger branch direction has a rather considerable effect on the solidification rate of PCM. For all values of β_2, when the direction of bigger branch is $\beta_1 = 5\pi / 12$, solidification rate, which is equal to the slope of solid fraction diagram versus time, is the lowest, and among the other values of second branch direction, $\beta_1 = 3\pi / 12$ has the highest rate. The reason that can be explained here is that, in the bigger branch direction, two parameters should be considered, the first one is the increasing thermal penetration depth in the direction of the corners of the quarter of the solution domain which are named in Figure 13 as A and B. The second one is keeping the optimized distance from the small branch in order to keep the amount of PCM in the space between branches in optimized value. Therefore the best value of bigger

Figure 18. Full solidification time for different values of geometry parameters of Snowflake shaped fin-assisted LHTESS

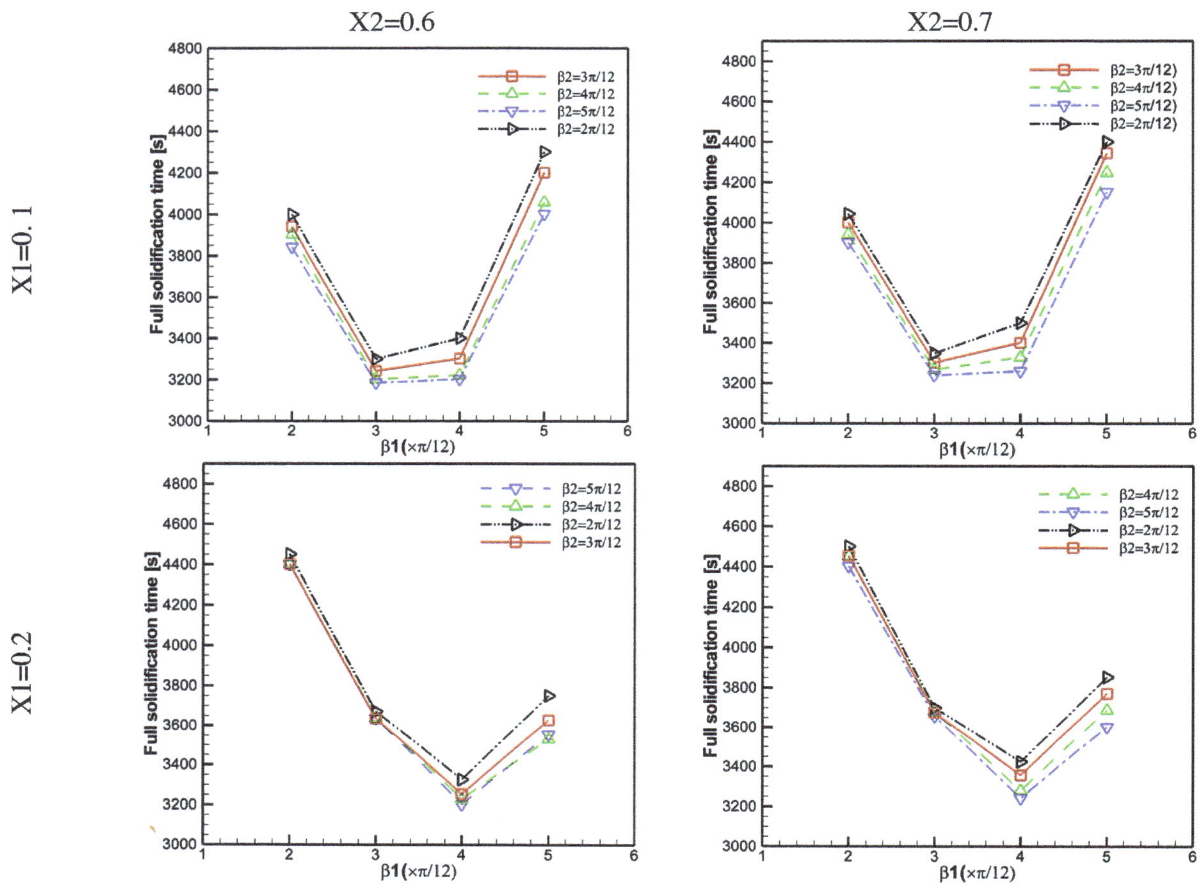

branch direction according to both of abovementioned parameters is $\beta_1 = 3\pi / 12$. Also, it can be observed that, the behavior of the solid fraction rate doesn't change significantly by changing in smaller branch direction.

2.2.2. The Effect of Changing the Distance Between the Branches on Solidification Rate

In Figure 20 the effect of changing the distance between the branches on solid fraction of PCM during discharging process are illustrated. It can be observed that the best case for the branches distance from cold wall is $X_1 = 0.1L, X_2 = 0.6L$, which is the closest case to the cold wall for both branches. As mentioned before, when the branches are closer to the cold wall, both branches efficiency values are higher. Therefore, we have chosen X_1 so close to the cold wall and inevitably chosen X_2 Equal to 0.6L in order to increase the thermal penetration depth at the corners of LHTESS. In Figures 21 and 22 solid fraction and temperature contour plots are illustrated for LHTESS with snowflake shaped fin with best

Figure 19. The effect of changing bigger branch direction on solid fraction of PCM during solidification process for $X_1 = 0.1L$ and $X_2 = 0.6L$.

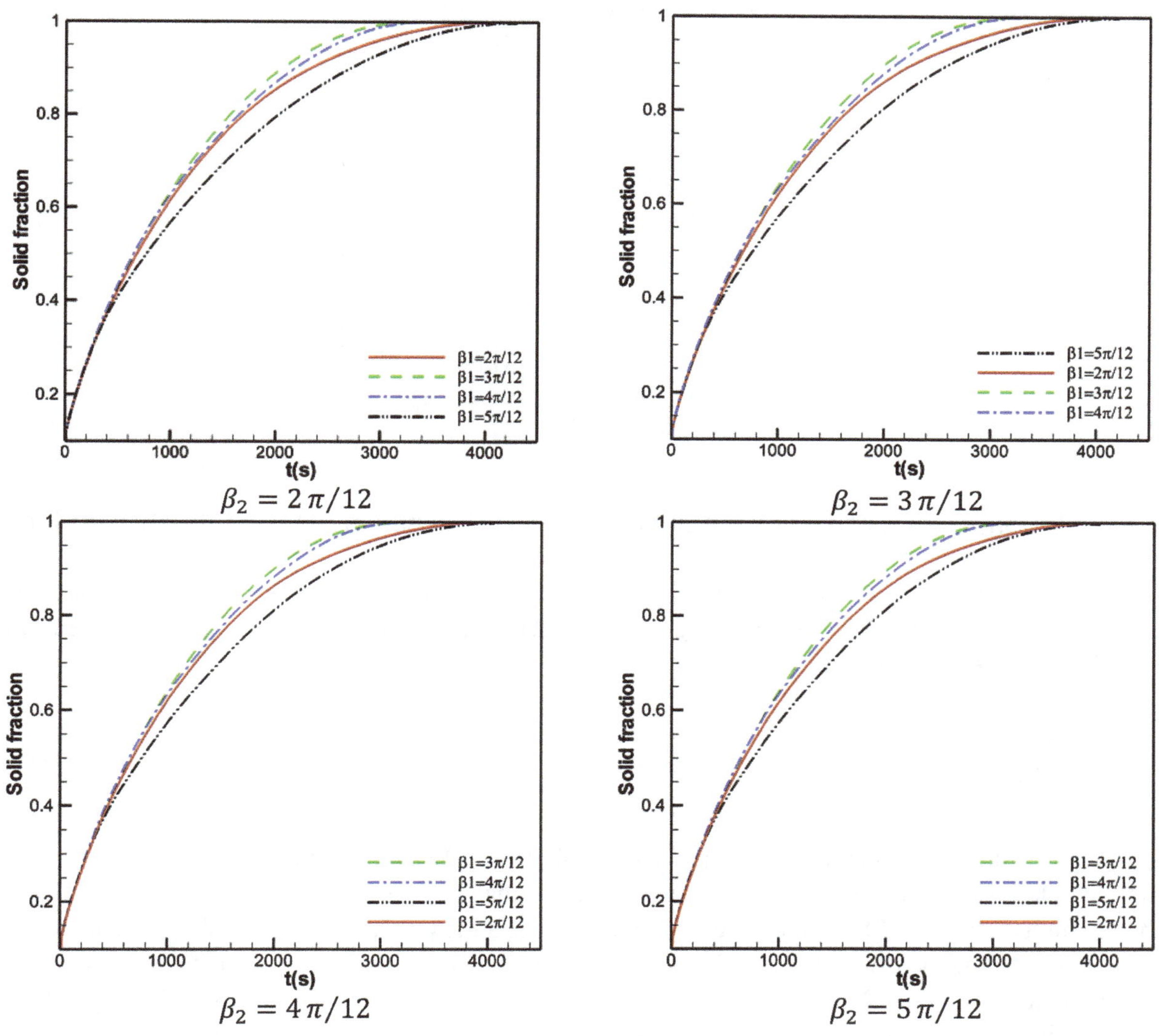

branch structure among the investigated cases and simple longitudinal fin with the same value of cross section area, in three time steps, which are 200, 1500 and 3200 seconds after the beginning of solidification process. It should be mentioned that 3200s is the time that full solidification is achieved for the case of Snowflake shaped fin-assisted LHTESS with the best branches array, and these time steps are used in longitudinal fin assisted LHTESS in order to simplify the comparison. Comparison between these cases and LHTESS without fin indicates that solidification rate in Snowflake shaped fin-assisted LHTESS is so higher than other cases. Although, longitudinal fin has the best direction, which increases thermal penetration depth at the corners of square shaped LHTESS, solidification rate in Snowflake shaped fin-assisted LHTESS is higher approximately 36% than the case of longitudinal fin-assisted LHTESS.

Figure 20. The effect of changing the distance between the branches on the solid fraction of PCM during discharging process in Snowflake shaped fin-assisted LHTESS

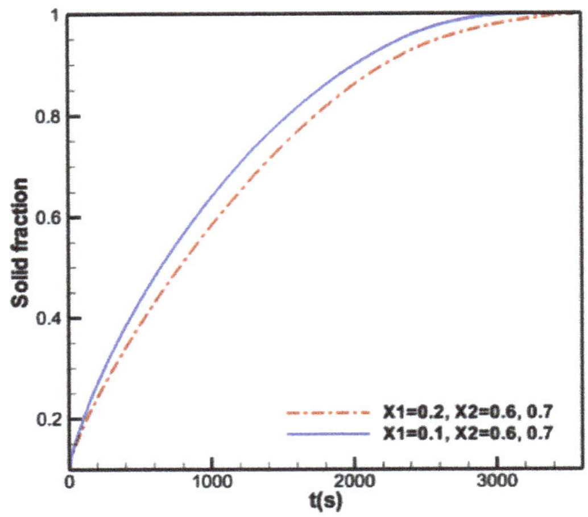

2.2.3. Performance Enhancement of Discharging Process in LHTESS by Adding Fins

In Figure 23, average temperature over the whole domain during discharging process has been illustrated in order to analyze the efficiency of adding fin to the LHTESS. The average temperature has been calculated using the following equation:

$$T_{ave} = \frac{\int TdA}{\int dA} \quad (9)$$

According to temperature contours illustrated in Figure 15, 21 and 22, and average temperature over the computational domain in Figure 23, it can be observed that the average temperature for the Snowflake shaped fin-assisted LHTESS is the lowest, which means that the whole computational domain temperature is closer to the cold wall and the fin system is more efficient in enhancing penetration depth into the PCM. Also in Figure 23, full solidification time for LHTESS without fin, with simple longitudinal fin and with snowflake shaped fin are illustrated, comparison between them indicates that adding fins to LHTESS is a highly efficient technique for expediting the discharging process in LHTESS, and the snowflake fin structure has the best performance among the investigated cases, although the simple longitudinal fin has the best direction which improves thermal penetration depth at the corners of the square shaped LHTESS. Moreover, the comparison between Fin-assisted and without fin LHTESS indicates that employing fin of any configuration is an efficient method for increasing the solidification rate of PCM in LHTESS without changing the thermos-physical properties of PCM.

2.2.4. Optimization of the Snowflake-Shaped Fin Configuration

The optimization method which has been employed in present study is RSM. This approach is consisted of mathematical methods which are appropriate for optimization studies. It is so efficient in problems

Figure 21. Solid fraction (left side) and temperature (right side) contour plots at three different time steps during solidification process of Snowflake shaped fin-assisted LHTESS

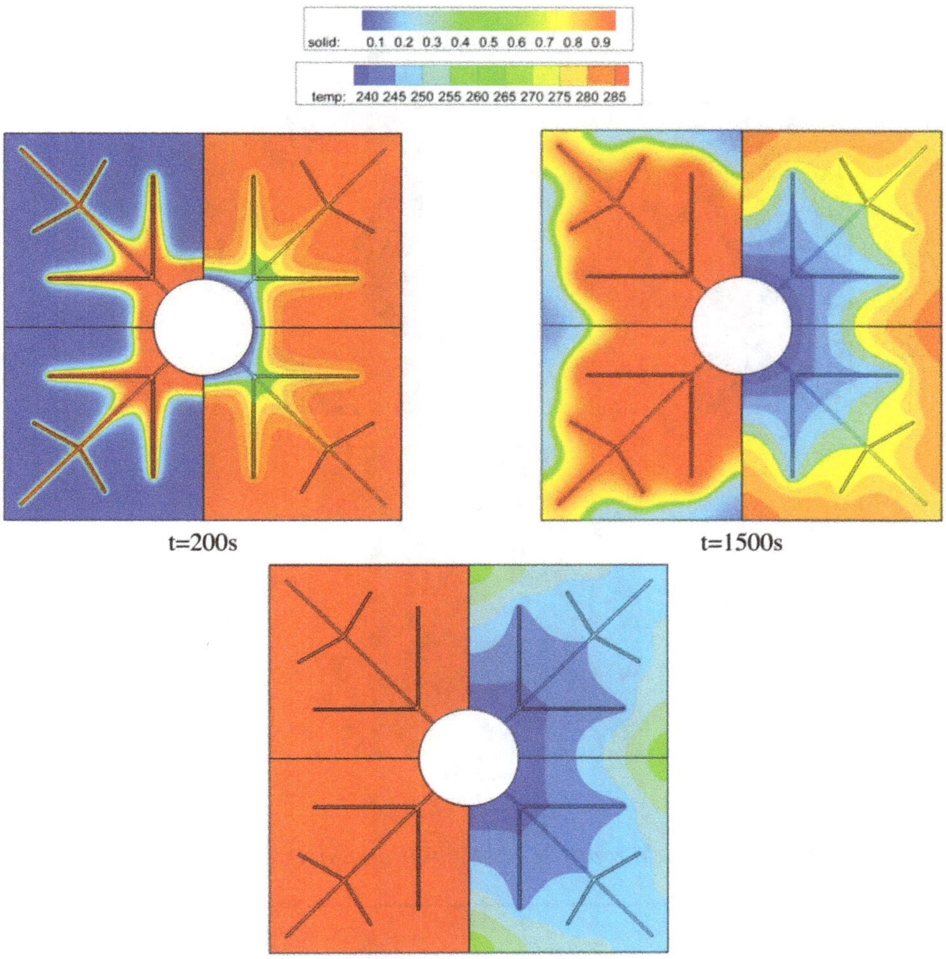

t=200s t=1500s t=3200s

where several design parameters affect the response of the system. In the optimization procedure by RSM, the responses which have been obtained by a limited number of cases properly chosen in the design space are characterized (Khuri & Cornell, 1996). The main purpose of optimization procedure in this paper is to find the best configuration of Snowflake shaped fin based on solidification expedition. The reason of employing this array in square container is to cover the space filled with PCM, in order to reduce thermal resistance as possible. The value of thermal resistance has an indirect relationship with thermal conductivity of PCM and direct relationship with the space between fins (Bergman, Incropera, & Lavine, 2011). Therefore thermal resistance of liquid PCM, due to its low thermal conductivity is so high that it slows down the energy retrieval process of LHTESS. It is obvious that by immersing fin into the system, thermal properties of PCM doesn't change, and thermal resistance is controlled just by the distance between fin branches. If there is no fin in the system, thermal resistance will be so high because

Figure 22. Solid fraction (left side) and temperature (right side) contour plots at three different time steps during solidification process of longitudinal fin-assisted LHTESS

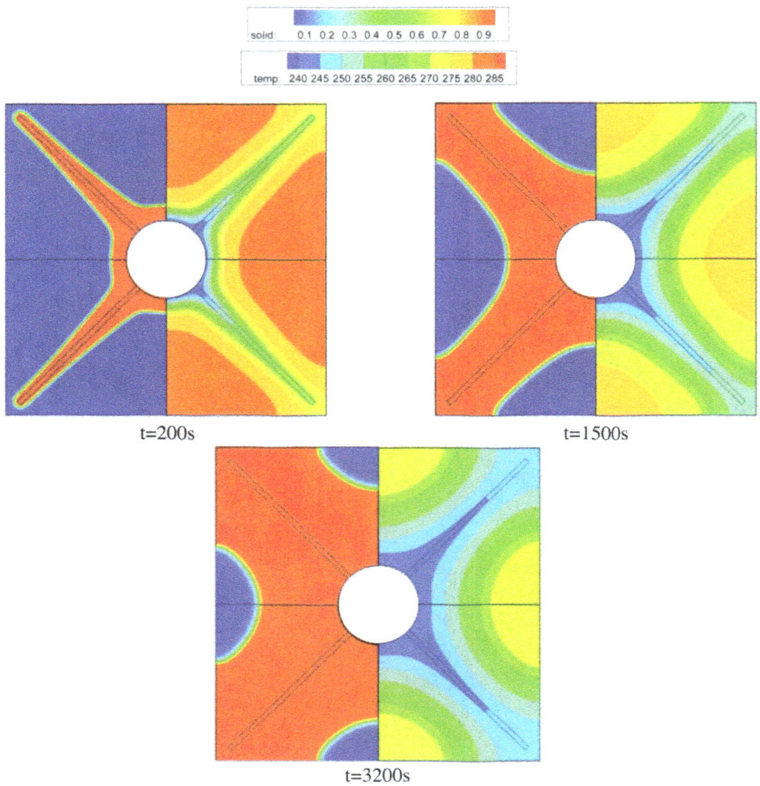

Figure 23. Average temperature variations over computational domain during solidification process

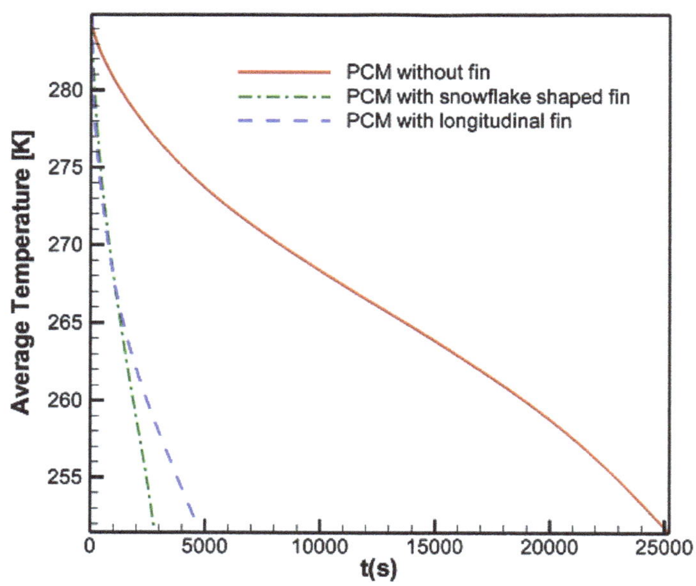

the distance between the cold wall and the liquid PCM far from cold wall, is so high that leads to lower discharging rate. By applying fin into the system, because of its high thermal conductivity, solidification procedure begins not only in the region adjacent to the cold wall, but also in the regions adjacent to the fin system, therefore it resembles the case that the distance between the cold wall and PCM decreases. This distance can be optimized to obtain minimum thermal resistance and more discharging rate.

2.2.4.1. Effect of Bigger Branch Direction on Solidification Rate

From Figure 18, it can be observed that the effect of changing bigger branch direction on solidification rate of PCM is significant. With the increase in the bigger branch direction, first the full solidification rate decreases, then increases. The reason of this behavior is that in the bigger branch direction investigation, two parameters should be considered, the first one is the increase of thermal penetration depth toward the corners of the solution domain which are marked by (B) in Figure 13. The second one is to keep the optimized distance from the smaller branch in order to achieve optimized mass distribution in the space between branches during solidification process. Therefore the best value of bigger branch direction according to both of the abovementioned parameters is $\beta_1 = 3.69\pi / 12$.

2.2.4.2. Effect of Smaller Branch Direction on Solidification Rate

According to Figure 18, with the increase in smaller branch direction, full solidification time decreases, because at the corners of the domain, which are marked by (A) in Figure 13 thermal penetration depth is increased by the main fin branch as illustrated in Figure 13, therefore the need for the increase of thermal penetration depth by the second branch at the corners is insignificant. But with the increase in second branch direction, mass distribution in the space between the bigger and smaller branches are controlled and optimized. According to the mentioned points, the optimized smaller branch direction is $\beta_2 = 5\pi / 12$. It should be noted that the existence of smaller branch is necessary, because the distance between cold wall and the corners of container is so high that causes high thermal resistance. Therefore this branch covers the space filled with PCM at the corners and lowers thermal resistance and expedites energy retrieval of LHTESS.

2.2.4.3. Effect of Bigger Branch Distance From Cold Wall on Solidification Rate

With the increase in bigger branch distance from cold wall, full solidification time doesn't have absolute ascending or descending trend, first it decreases and then increases (Figure 24). Because if X_1 is lower, the bigger branch is closer to the cold wall and therefore the average temperature on its surface is closer to the cold wall temperature, thus the bigger branch efficiency is higher according to Equation 10, which leads to heat transfer enhancement and higher solidification rate. On the other hand, if the value of X_1 is higher, the distance between bigger and smaller branch is lower and mass distribution in the space between branches is more uniform. Based on the abovementioned reasons, the optimized value of bigger branch distance from cold wall is a value between the biggest and smallest value of X_1, which is $X_1 = 0.16L$.

Figure 24. Time surfaces for different geometry parameters of Snowflake shaped fin

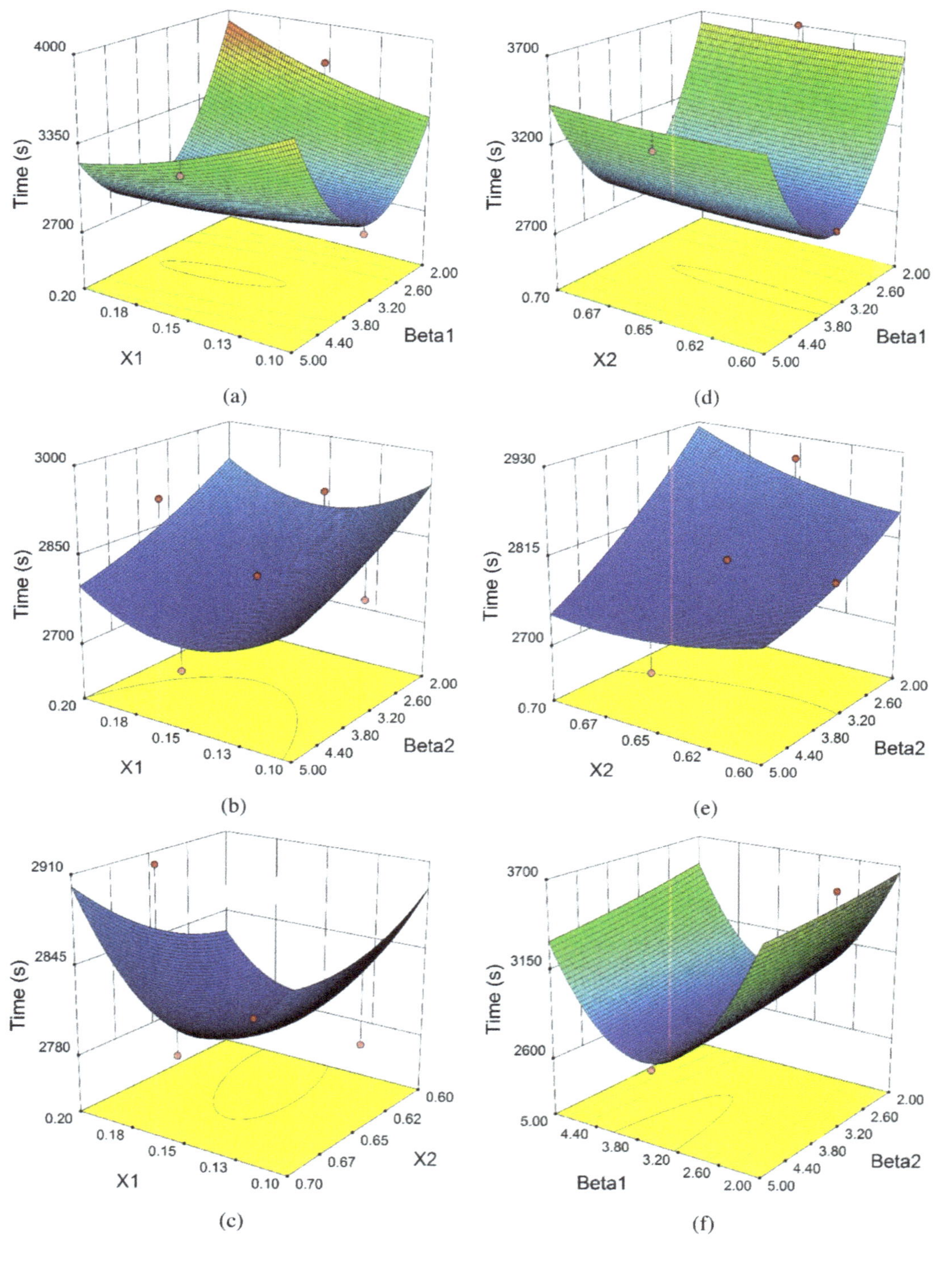

2.2.4.4. Effect of Smaller Branch Distance From Cold Wall on Solidification Rate

If smaller branch is close to the corners of the solution domain, far from cold wall, thermal penetration depth at the corners will be enhanced, but on the other hand, it should be closer to the cold wall in order to control the mass distribution in the space between smaller and bigger branches (Sheikholeslami, Lohrasbi, & Ganji, 2016). The interaction between these two factors, makes the optimized value of smaller branch distance from cold wall be $X_2 = 0.66L$.

In Figures 25 and 26 solid-temperature contours are illustrated for LHTESS with simple longitudinal fin and with snowflake shaped fin with optimized structure in three time steps including 200, 1500 and 2800 seconds after the beginning of solidification process. It should be mentioned that 2800s is the time that full solidification is achieved for the case of optimized Snowflake shaped fin-assisted LHTESS, and these time steps are used in longitudinal fin assisted LHTESS in order to simplify the comparison. Comparison between these cases and LHTESS without fin indicates that solidification rate in Snowflake

Figure 25. Solid-Temperature contours at three different times during solidification process of Snowflake shaped fin-assisted LHTESS

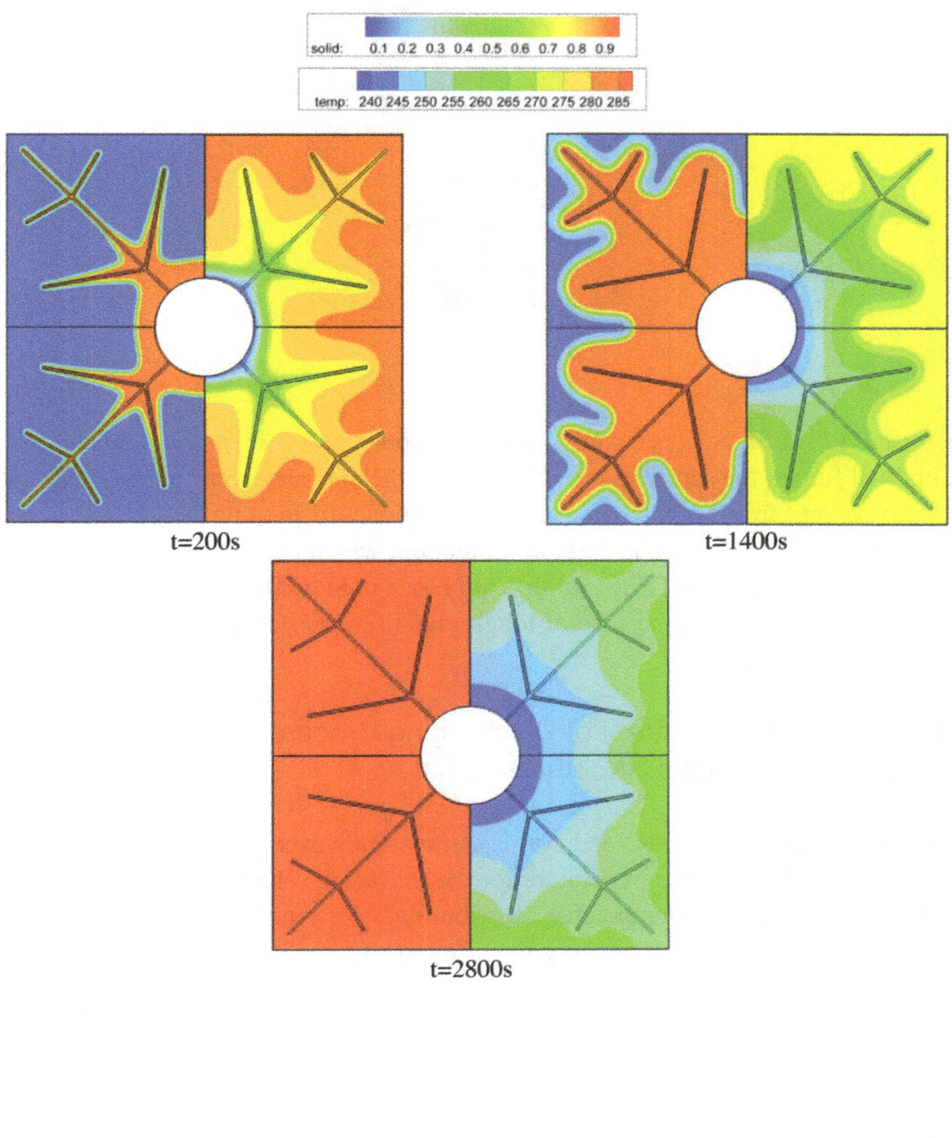

Figure 26. Solid-Temperature contours at three different time steps during solidification process of longitudinal fin-assisted LHTESS

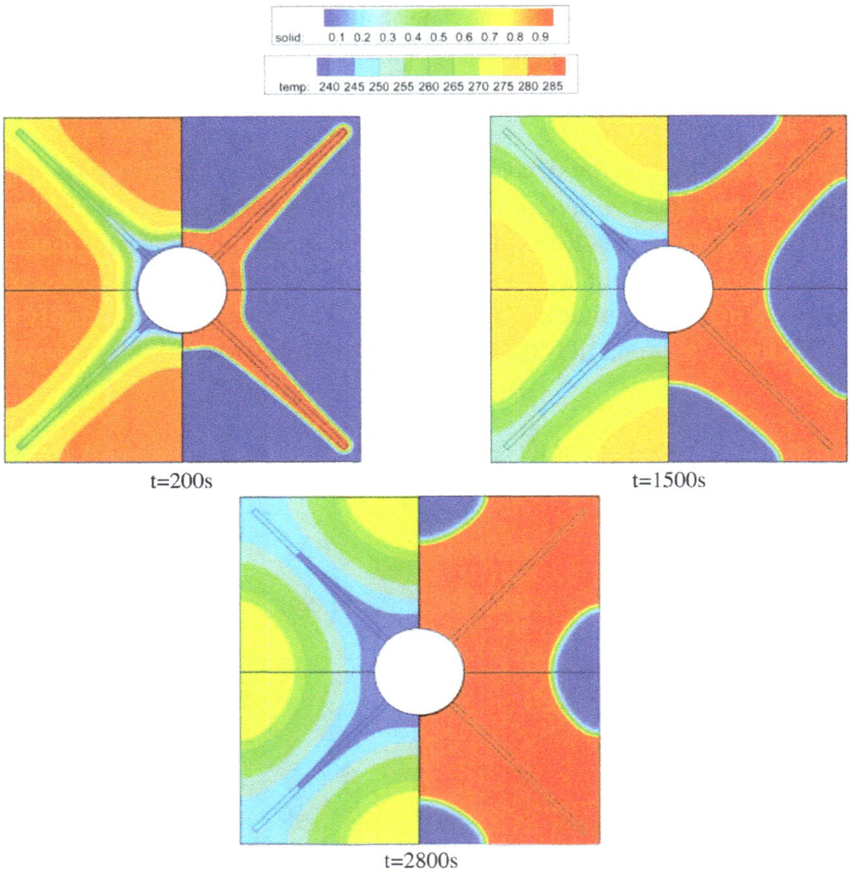

shaped fin-assisted LHTESS is significantly higher than other cases. It should be noted that the investigated longitudinal fin has the same cross section area as the optimized Snowflake shaped fin, in order to compare these systems from the viewpoint of solidification rate, in a constant value of energy storage capacity. Although, longitudinal fin has the best direction, which increases thermal penetration depth at the corners of square shaped LHTESS, solidification rate in Snowflake shaped fin-assisted LHTESS is higher approximately 49% than the case of longitudinal fin-assisted LHTESS. Nanotechnology has been used in various applications.

2.2.5. Performance Enhancement of Discharging Process in LHTESS by Adding Various Structures of Fin

According to temperature contour plots and average temperature over the solution domain in Figure 27, it can be observed that the average temperature for the Snowflake shaped fin-assisted LHTESS in the lowest and it has the most uniform temperature distribution, which means that the solution domain temperature is closer to the cold wall and the fin system is more efficient in enhancing penetration depth into

Figure 27. Average temperature variations over the solution domain during solidification process

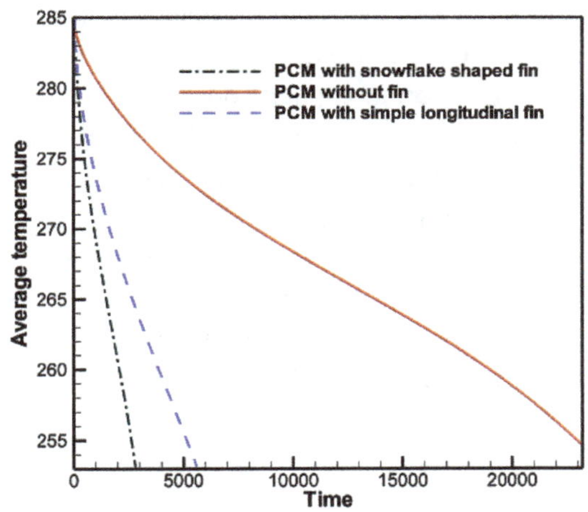

the PCM. Also in Figure 27, full solidification time for LHTESS without fin, with simple longitudinal fin and with snowflake shaped fin are illustrated, comparison between these cases indicates that adding fin of any structure to LHTESS, is a highly efficient technique for expediting the discharging process in LHTESS without changing the physical properties of PCM, and the snowflake shaped fin configuration has the best performance among the investigated cases, although the simple longitudinal fin has the best direction which improves thermal penetration depth at the corners of the square shaped LHTESS.

2.2.6. Comparison Between the Enhancement Techniques

In Figure 28, total energy released during discharging process of LHTESS has been illustrated in order to investigate two parameters. The first parameter is maximum energy storage capacity, which is equal

Figure 28. Total energy released during the discharging process in LHTESS containing PCM without a fin, PCM with Snowflake shaped fin and NEPCM with $\phi = 2.5, 5.0\%$.

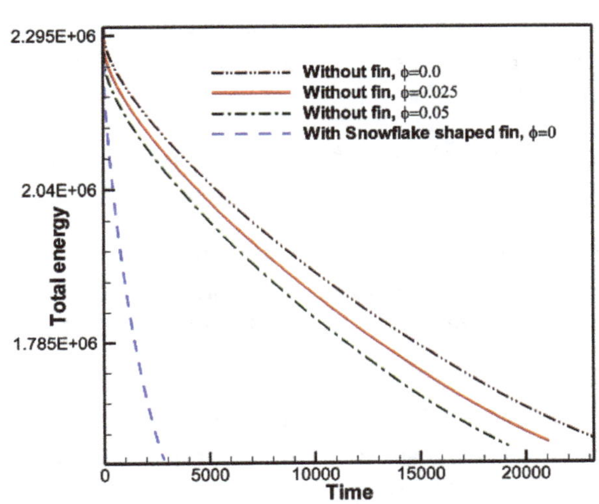

to the sum of sensible and latent heat, and the second one is full solidification time. The reason that can be explained as the importance of these factors in present paper is that in the case of adding fins to the systems, energy storage capacity reduces because less amount of PCM is used in the system. But for the case of nanoparticles dispersion, energy storage capacity decreases because of a decrease in latent heat of fusion and heat capacity, which can be observed according to Equation 4 and 5. The effect of nanoparticles dispersion and making NEPCM, and adding fin to LHTESS has been investigated considering both of the abovementioned parameters. According to Figure 28, full solidification time for Snowflake shaped fin-assisted LHTESS is the lowest, moreover maximum energy storage density is almost the same for all of these cases. Therefore, adding snowflake shaped fin, enhances the solidification rate considerably without decreasing maximum energy storage capacity significantly, which verifies the efficiency of adding snowflake shaped fin configuration to LHTESS.

REFERENCES

Bergman, T. L., Incropera, F. P., & Lavine, A. S. (2011). *Fundamentals of heat and mass transfer*. John Wiley & Sons.

Dhaidan, N. S., Khodadadi, J. M., Al-Hattab, T. A., & Al-Mashat, S. M. (2013). Experimental and numerical investigation of melting of phase change material/nanoparticle suspensions in a square container subjected to a constant heat flux. *International Journal of Heat and Mass Transfer*, *66*, 672–683. doi:10.1016/j.ijheatmasstransfer.2013.06.057

Ismail, K. A., Alves, C. L., & Modesto, M. S. (2001). Numerical and experimental study on the solidification of PCM around a vertical axially finned isothermal cylinder. *Applied Thermal Engineering*, *21*(1), 53–77. doi:10.1016/S1359-4311(00)00002-8

Khillarkar, D. B., Gong, Z. X., & Mujumdar, A. S. (2000). Melting of a phase change material in concentric horizontal annuli of arbitrary cross-section. *Applied Thermal Engineering*, *20*(10), 893–912. doi:10.1016/S1359-4311(99)00058-7

Khuri, A.I., & Cornell, J.A. (1996). *Response surfaces: designs and analyses*. CRC Press.

Lohrasbi, S., Sheikholeslami, M., & Ganji, D. D. (2016). Discharging process expedition of NEPCM in fin-assisted Latent Heat Thermal Energy Storage System. *Journal of Molecular Liquids*, *221*, 833–841. doi:10.1016/j.molliq.2016.06.044

Sheikholeslami. (2014). Effect of spatially variable magnetic field on ferrofluid flow and heat transfer considering constant heat flux boundary condition. *The European Physical Journal Plus*, 129–248.

Sheikholeslami, M., Lohrasbi, S., & Ganji, D. D. (2016). Numerical analysis of discharging process acceleration in LHTESS by immersing innovative fin configuration using finite element method. *Applied Thermal Engineering*, *107*, 154–166. doi:10.1016/j.applthermaleng.2016.06.158

Sheikholeslami, M., Lohrasbi, S., & Ganji, D. D. (2016). Response surface method optimization of innovative fin structure for expediting discharging process in latent heat thermal energy storage system containing nano-enhanced phase change material. *Journal of the Taiwan Institute of Chemical Engineers*, *67*, 115–125. doi:10.1016/j.jtice.2016.08.019

ADDITIONAL READING

Jafaryar, M., Sheikholeslami, M., Li, M., & Moradi, R. (2018). Nanofluid turbulent flow in a pipe under the effect of twisted tape with alternate axis. *Journal of Thermal Analysis and Calorimetry*. doi:10.100710973-018-7093-2

Li, Z., Shehzad, S. A., & Sheikholeslami, M. (2018). An application of CVFEM for nanofluid heat transfer intensification in a porous sinusoidal cavity considering thermal non-equilibrium model. *Computer Methods in Applied Mechanics and Engineering*, *339*, 663–680. doi:10.1016/j.cma.2018.05.015

Li, Z., Sheikholeslami, M., Chamkha, A. J., Raizah, Z. A., & Saleem, S. (2018). Control Volume Finite Element Method for nanofluid MHD natural convective flow inside a sinusoidal annulus under the impact of thermal radiation. *Computer Methods in Applied Mechanics and Engineering*, *338*, 618–633. doi:10.1016/j.cma.2018.04.023

Li, Z., Sheikholeslami, M., Jafaryar, M., Shafee, A., & Chamkha, A. J. (2018). Investigation of nanofluid entropy generation in a heat exchanger with helical twisted tapes. *Journal of Molecular Liquids*, *266*, 797–805. doi:10.1016/j.molliq.2018.07.009

Li, Z., Sheikholeslami, M., Samandari, M., & Shafee, A. (2018). Nanofluid unsteady heat transfer in a porous energy storage enclosure in existence of Lorentz forces. *International Journal of Heat and Mass Transfer*, *127*, 914–926. doi:10.1016/j.ijheatmasstransfer.2018.06.101

Sheikholeslami, M. (2017a). Lattice Boltzmann Method simulation of MHD non-Darcy nanofluid free convection. *Physica B, Condensed Matter*, *516*, 55–71. doi:10.1016/j.physb.2017.04.029

Sheikholeslami, M. (2017b). CuO-water nanofluid free convection in a porous cavity considering Darcy law. *The European Physical Journal Plus*, *132*(1), 55. doi:10.1140/epjp/i2017-11330-3

Sheikholeslami, M. (2017c). Influence of Lorentz forces on nanofluid flow in a porous cavity by means of Non- Darcy model. *Engineering Computations*, *34*(8), 2651–2667. doi:10.1108/EC-01-2017-0008

Sheikholeslami, M. (2017d). Magnetic field influence on nanofluid thermal radiation in a cavity with tilted elliptic inner cylinder. *Journal of Molecular Liquids*, *229*, 137–147. doi:10.1016/j.molliq.2016.12.024

Sheikholeslami, M. (2017e). Numerical simulation of magnetic nanofluid natural convection in porous media. *Physics Letters. [Part A]*, *381*(5), 494–503. doi:10.1016/j.physleta.2016.11.042

Sheikholeslami, M. (2017f). Influence of Lorentz forces on nanofluid flow in a porous cylinder considering Darcy model. *Journal of Molecular Liquids*, *225*, 903–912. doi:10.1016/j.molliq.2016.11.022

Sheikholeslami, M. (2017g). Influence of Coulomb forces on Fe3O4-H2O nanofluid thermal improvement. *International Journal of Hydrogen Energy*, *42*(2), 821–829. doi:10.1016/j.ijhydene.2016.09.185

Sheikholeslami, M. (2017h). Numerical investigation of MHD nanofluid free convective heat transfer in a porous tilted enclosure. *Engineering Computations*, *34*(6), 1939–1955. doi:10.1108/EC-08-2016-0293

Sheikholeslami, M. (2017i). Magnetic field influence on CuO -H2O nanofluid convective flow in a permeable cavity considering various shapes for nanoparticles. *International Journal of Hydrogen Energy*, *42*(31), 19611–19621. doi:10.1016/j.ijhydene.2017.06.121

Sheikholeslami, M. (2018a). Magnetic source impact on nanofluid heat transfer using CVFEM. *Neural Computing & Applications*, *30*(4), 1055–1064. doi:10.100700521-016-2740-7

Sheikholeslami, M. (2018b). Numerical simulation for external magnetic field influence on Fe3O4-water nanofluid forced convection. *Engineering Computations*, *35*(4), 1639–1654. doi:10.1108/EC-06-2017-0200

Sheikholeslami, M. (2018c). Finite element method for PCM solidification in existence of CuO nanoparticles. *Journal of Molecular Liquids*, *265*, 347–355. doi:10.1016/j.molliq.2018.05.132

Sheikholeslami, M. (2018d). Solidification of NEPCM under the effect of magnetic field in a porous thermal energy storage enclosure using CuO nanoparticles. *Journal of Molecular Liquids*, *263*, 303–315. doi:10.1016/j.molliq.2018.04.144

Sheikholeslami, M. (2018e). Influence of magnetic field on Al2O3-H2O nanofluid forced convection heat transfer in a porous lid driven cavity with hot sphere obstacle by means of LBM. *Journal of Molecular Liquids*, *263*, 472–488. doi:10.1016/j.molliq.2018.04.111

Sheikholeslami, M. (2018f). Numerical simulation for solidification in a LHTESS by means of Nano-enhanced PCM. *Journal of the Taiwan Institute of Chemical Engineers*, *86*, 25–41. doi:10.1016/j.jtice.2018.03.013

Sheikholeslami, M. (2018g). Numerical modeling of Nano enhanced PCM solidification in an enclosure with metallic fin. *Journal of Molecular Liquids*, *259*, 424–438. doi:10.1016/j.molliq.2018.03.006

Sheikholeslami, M. (2018h). Numerical investigation of nanofluid free convection under the influence of electric field in a porous enclosure. *Journal of Molecular Liquids*, *249*, 1212–1221. doi:10.1016/j.molliq.2017.11.141

Sheikholeslami, M. (2018i). CuO-water nanofluid flow due to magnetic field inside a porous media considering Brownian motion. *Journal of Molecular Liquids*, *249*, 921–929. doi:10.1016/j.molliq.2017.11.118

Sheikholeslami, M. (2018j). Application of Darcy law for nanofluid flow in a porous cavity under the impact of Lorentz forces. *Journal of Molecular Liquids*, *266*, 495–503. doi:10.1016/j.molliq.2018.06.083

Sheikholeslami, M., Barzegar Gerdroodbary, M., Valiallah Mousavi, S., Ganji, D. D., & Moradi, R. (2018). Heat transfer enhancement of ferrofluid inside an 90o elbow channel by non-uniform magnetic field. *Journal of Magnetism and Magnetic Materials*, *460*, 302–311. doi:10.1016/j.jmmm.2018.03.070

Sheikholeslami, M., & Bhatti, M. M. (2017). Active method for nanofluid heat transfer enhancement by means of EHD. *International Journal of Heat and Mass Transfer*, *109*, 115–122. doi:10.1016/j.ijheatmasstransfer.2017.01.115

Sheikholeslami, M., Darzi, M., & Li, Z. (2018). Experimental investigation for entropy generation and exergy loss of nano-refrigerant condensation process. *International Journal of Heat and Mass Transfer*, *125*, 1087–1095. doi:10.1016/j.ijheatmasstransfer.2018.04.155

Sheikholeslami, M., Darzi, M., & Sadoughi, M. K. (2018). Heat transfer improvement and Pressure Drop during condensation of refrigerant-based Nanofluid; An Experimental Procedure. *International Journal of Heat and Mass Transfer*, *122*, 643–650. doi:10.1016/j.ijheatmasstransfer.2018.02.015

Sheikholeslami, M., & Ghasemi, A. (2018). Solidification heat transfer of nanofluid in existence of thermal radiation by means of FEM. *International Journal of Heat and Mass Transfer*, *123*, 418–431. doi:10.1016/j.ijheatmasstransfer.2018.02.095

Sheikholeslami, M., Ghasemi, A., Li, Z., Shafee, A., & Saleem, S. (2018). Influence of CuO nanoparticles on heat transfer behavior of PCM in solidification process considering radiative source term. *International Journal of Heat and Mass Transfer*, *126*, 1252–1264. doi:10.1016/j.ijheatmasstransfer.2018.05.116

Sheikholeslami, M., Hayat, T., & Alsaedi, A. (2018). Numerical simulation for forced convection flow of MHD CuO-H2O nanofluid inside a cavity by means of LBM. *Journal of Molecular Liquids*, *249*, 941–948. doi:10.1016/j.molliq.2017.10.099

Sheikholeslami, M., Hayat, T., Muhammad, T., & Alsaedi, A. (2018). MHD forced convection flow of nanofluid in a porous cavity with hot elliptic obstacle by means of Lattice Boltzmann method. *International Journal of Mechanical Sciences*, *135*, 532–540. doi:10.1016/j.ijmecsci.2017.12.005

Sheikholeslami, M., Jafaryar, M., Ganji, D. D., & Li, Z. (2018). Exergy loss analysis for nanofluid forced convection heat transfer in a pipe with modified turbulators. *Journal of Molecular Liquids*, *262*, 104–110. doi:10.1016/j.molliq.2018.04.077

Sheikholeslami, M., Jafaryar, M., & Li, Z. (2018a). Second law analysis for nanofluid turbulent flow inside a circular duct in presence of twisted tape turbulators. *Journal of Molecular Liquids*, *263*, 489–500. doi:10.1016/j.molliq.2018.04.147

Sheikholeslami, M., Jafaryar, M., & Li, Z. (2018b). Nanofluid turbulent convective flow in a circular duct with helical turbulators considering CuO nanoparticles. *International Journal of Heat and Mass Transfer*, *124*, 980–989. doi:10.1016/j.ijheatmasstransfer.2018.04.022

Sheikholeslami, M., Jafaryar, M., Saleem, S., Li, Z., Shafee, A., & Jiang, Y. (2018). Nanofluid heat transfer augmentation and exergy loss inside a pipe equipped with innovative turbulators. *International Journal of Heat and Mass Transfer*, *126*, 156–163. doi:10.1016/j.ijheatmasstransfer.2018.05.128

Sheikholeslami, M., Jafaryar, M., Shafee, A., & Li, Z. (2018). Investigation of second law and hydrothermal behavior of nanofluid through a tube using passive methods. *Journal of Molecular Liquids*, *269*, 407–416. doi:10.1016/j.molliq.2018.08.019

Sheikholeslami, M., Li, Z., & Shafee, A. (2018a). Lorentz forces effect on NEPCM heat transfer during solidification in a porous energy storage system. *International Journal of Heat and Mass Transfer*, *127*, 665–674. doi:10.1016/j.ijheatmasstransfer.2018.06.087

Sheikholeslami, M., Li, Z., & Shamlooei, M. (2018). Nanofluid MHD natural convection through a porous complex shaped cavity considering thermal radiation. *Physics Letters. [Part A]*, *382*(24), 1615–1632. doi:10.1016/j.physleta.2018.04.006

Sheikholeslami, M., & Rokni, H. B. (2017). Simulation of nanofluid heat transfer in presence of magnetic field: A review. *International Journal of Heat and Mass Transfer*, *115*, 1203–1233. doi:10.1016/j.ijheatmasstransfer.2017.08.108

Sheikholeslami, M., & Rokni, H. B. (2018a). CVFEM for effect of Lorentz forces on nanofluid flow in a porous complex shaped enclosure by means of Non-equilibrium model. *Journal of Molecular Liquids*, *254*, 446–462. doi:10.1016/j.molliq.2018.01.130

Sheikholeslami, M., Rokni, H.B. (2018b). Magnetic nanofluid flow and convective heat transfer in a porous cavity considering Brownian motion effects. *Physics of Fluids*, *30*(1), doi:10.1063/1.5012517

Sheikholeslami, M., & Sadoughi, M. K. (2017). Mesoscopic method for MHD nanofluid flow inside a porous cavity considering various shapes of nanoparticles. *International Journal of Heat and Mass Transfer*, *113*, 106–114. doi:10.1016/j.ijheatmasstransfer.2017.05.054

Sheikholeslami, M., & Sadoughi, M. K. (2018). Simulation of CuO- water nanofluid heat transfer enhancement in presence of melting surface. *International Journal of Heat and Mass Transfer*, *116*, 909–919. doi:10.1016/j.ijheatmasstransfer.2017.09.086

Sheikholeslami, M., & Seyednezhad, M. (2018). Simulation of nanofluid flow and natural convection in a porous media under the influence of electric field using CVFEM. *International Journal of Heat and Mass Transfer*, *120*, 772–781. doi:10.1016/j.ijheatmasstransfer.2017.12.087

Sheikholeslami, M., Shafee, A., Ramzan, M., & Li, Z. (2018). Investigation of Lorentz forces and radiation impacts on nanofluid treatment in a porous semi annulus via Darcy law. *Journal of Molecular Liquids*, *272*, 8–14. doi:10.1016/j.molliq.2018.09.016

Sheikholeslami, M., Shamlooei, M., & Moradi, R. (2018). Numerical simulation for heat transfer intensification of nanofluid in a porous curved enclosure considering shape effect of Fe3O4 nanoparticles. *Chemical Engineering & Processing: Process Intensification*, *124*, 71–82. doi:10.1016/j.cep.2017.12.005

Sheikholeslami, M., & Shehzad, S. A. (2017a). Thermal radiation of ferrofluid in existence of Lorentz forces considering variable viscosity. *International Journal of Heat and Mass Transfer*, *109*, 82–92. doi:10.1016/j.ijheatmasstransfer.2017.01.096

Sheikholeslami, M., & Shehzad, S. A. (2017b). Magnetohydrodynamic nanofluid convective flow in a porous enclosure by means of LBM. *International Journal of Heat and Mass Transfer*, *113*, 796–805. doi:10.1016/j.ijheatmasstransfer.2017.05.130

Sheikholeslami, M., & Shehzad, S. A. (2018a). Numerical analysis of Fe3O4 –H2O nanofluid flow in permeable media under the effect of external magnetic source. *International Journal of Heat and Mass Transfer*, *118*, 182–192. doi:10.1016/j.ijheatmasstransfer.2017.10.113

Sheikholeslami, M., & Shehzad, S. A. (2018b). CVFEM simulation for nanofluid migration in a porous medium using Darcy model. *International Journal of Heat and Mass Transfer*, *122*, 1264–1271. doi:10.1016/j.ijheatmasstransfer.2018.02.080

Sheikholeslami, M., & Shehzad, S. A. (2018c). Simulation of water based nanofluid convective flow inside a porous enclosure via Non-equilibrium model. *International Journal of Heat and Mass Transfer*, *120*, 1200–1212. doi:10.1016/j.ijheatmasstransfer.2017.12.132

Sheikholeslami, M., & Shehzad, S. A. (2018d). Non-Darcy free convection of Fe3O4-water nanoliquid in a complex shaped enclosure under impact of uniform Lorentz force. *The Chinese Journal of Physiology*, *56*(1), 270–281. doi:10.1016/j.cjph.2017.12.017

Sheikholeslami, M., Shehzad, S. A., Abbasi, F. M., & Li, Z. (2018). Nanofluid flow and forced convection heat transfer due to Lorentz forces in a porous lid driven cubic enclosure with hot obstacle. *Computer Methods in Applied Mechanics and Engineering*, *338*, 491–505. doi:10.1016/j.cma.2018.04.020

Sheikholeslami, M., Shehzad, S. A., & Li, Z. (2018a). Nanofluid heat transfer intensification in a permeable channel due to magnetic field using Lattice Boltzmann method. *Physica B, Condensed Matter*, *542*, 51–58. doi:10.1016/j.physb.2018.03.036

Sheikholeslami, M., Shehzad, S. A., & Li, Z. (2018b). Water based nanofluid free convection heat transfer in a three dimensional porous cavity with hot sphere obstacle in existence of Lorenz forces. *International Journal of Heat and Mass Transfer*, *125*, 375–386. doi:10.1016/j.ijheatmasstransfer.2018.04.076

Sheikholeslami, M., Shehzad, S. A., Li, Z., & Shafee, A. (2018). Numerical modeling for Alumina nanofluid magnetohydrodynamic convective heat transfer in a permeable medium using Darcy law. *International Journal of Heat and Mass Transfer*, *127*, 614–622. doi:10.1016/j.ijheatmasstransfer.2018.07.013

Sheikholeslami, M., & Vajravelu, K. (2017). Forced convection heat transfer in Fe3O4- ethylene glycol nanofluid under the influence of Coulomb force. *Journal of Molecular Liquids*, *233*, 203–210. doi:10.1016/j.molliq.2017.03.026

Sheikholeslami, M., & Zeeshan, A. (2017). Analysis of flow and heat transfer in water based nanofluid due to magnetic field in a porous enclosure with constant heat flux using CVFEM. *Computer Methods in Applied Mechanics and Engineering*, *320*, 68–81. doi:10.1016/j.cma.2017.03.024

Sheikholeslami, M., Zeeshan, A., & Majeed, A. (2018). Control volume based finite element simulation of magnetic nanofluid flow and heat transport in non-Darcy medium. *Journal of Molecular Liquids*, *268*, 354–364. doi:10.1016/j.molliq.2018.07.031

Chapter 5
Nanoparticle Transportation in a Porous Medium

ABSTRACT

The study of convective heat transfer in fluid-saturated porous media has many important applications in technology geothermal energy recovery such as oil recovery, food processing, fiber and granular insulation, porous burner and heater, combustion of low-calorific fuels to diesel engines, and design of packed bed reactors. Also, the flow in porous tubes or channels has been under considerable attention in recent years because of its various applications in biomedical engineering, transpiration cooling boundary layer control, and gaseous diffusion. Nanofluids are produced by dispersing the nanometer-scale solid particles into base liquids with low thermal conductivity such as water, ethylene glycol (EG), and oils. In this chapter, nanofluid hydrothermal behavior in porous media has been investigated.

1. NANOFLUID HEAT TRANSFER OVER A PERMEABLE STRETCHING WALL IN A POROUS MEDIUM

1.1. Problem Definition

A steady, constant property, two-dimensional flow of an incompressible nanofluid through a homogenous porous medium with permeability of *K*, over a stretching surface with linear velocity distribution, i.e., $u_w = \frac{u_0 x}{L}$ is assumed (Figure 1) (Sheikholeslami, Ellahi, Ashorynejad, Domairry and Hayat, 2014).

The fluid is a water based nanofluid containing different types of nanoparticles: Cu, Al_2O_3, Ag and TiO_2. It is assumed that the base fluid and the nanoparticles are in thermal equilibrium and no slip occurs between them. The thermo physical properties of the nanofluid are given in Table 1.

The transport properties of the medium can be considered independent from the temperature when the temperature difference between wall and ambient is not significant (Starov and Zhdanov, 2001). The origin is kept fixed while the wall is stretching and the *y*-axis is perpendicular to the surface. Using the above-mentioned assumptions, the continuity equation is:

DOI: 10.4018/978-1-5225-7595-5.ch005

Figure 1. Schematic theme of the problem geometry

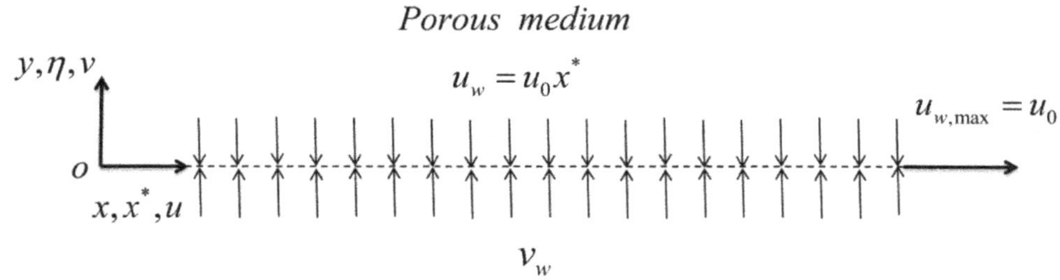

Table 1. Thermo physical properties of water and nanoparticles

	$\rho(kg / m^3)$	$C_p(j / kgk)$	$k(W / m.k)$	$\beta \times 10^5(K^{-1})$
Pure water	997.1	4179	0.613	21
Copper(Cu)	8933	385	401	1.67
Silver(Ag)	10 500	235	429	1.89
Alumina(Al_2O_3)	3970	765	40	0.85
Titanium Oxide(TiO_2)	4250	686.2	8.9538	0.9

$$\frac{\partial u}{\partial x} + \frac{\partial v}{\partial y} = 0 \tag{1}$$

where u and v are velocity components in the x and y directions, respectively. The Brinkman model x-momentum equation reads:

$$\rho_{nf}(u\frac{\partial u}{\partial x} + v\frac{\partial v}{\partial y}) = \mu_{eff_{nf}} \frac{\partial^2 u}{\partial y^2} - \frac{\mu_{nf}}{K} u \tag{2}$$

$$u\frac{\partial T}{\partial x} + v\frac{\partial T}{\partial y} = \frac{k_{nf}}{(\rho C_p)_{nf}} \frac{\partial^2 T}{\partial y^2} \tag{3}$$

where $\mu_{eff_{nf}}$ is the effective viscosity which for simplicity in the present study is considered to be identical to the dynamic viscosity, μ_{nf}. This assumption is reasonable for packed beds of particles (Hooman, Gurgenci and Merrikh, 2007)

The effective density ρ_{nf}, the effective dynamic viscosity μ_{nf}, the heat capacitance $(\rho C_p)_{nf}$ and the thermal conductivity k_{nf} of the nanofluid are given as:

$$\rho_{nf} = \rho_f(1-\phi) + \rho_s\phi \tag{4}$$

$$\mu_{nf} = \frac{\mu_f}{(1-\phi)^{2.5}} \tag{5}$$

$$\left(\rho C_p\right)_{nf} = \left(\rho C_p\right)_f (1-\phi) + \left(\rho C_p\right)_s \phi \tag{6}$$

$$\frac{k_{nf}}{k_f} = \frac{k_s + 2k_f - 2\phi(k_f - k_s)}{k_s + 2k_f + 2\phi(k_f - k_s)} \tag{7}$$

Here, ϕ is the solid volume fraction.

The hydrodynamic boundary conditions are:

$$u(x^*,0) = u_0 x^*, v(x^*,0) = v_w, u(x^*,\infty) = 0 \tag{8}$$

where $x^* = \frac{x}{L}$ is the non-dimensional *x*-coordinate and *L* is the length of the porous plate.

The following thermal boundary conditions are considered:

$$T(x^*,0) = T_\infty + T_0(x^*)^n, T(x^*,\infty) = T_\infty \tag{9}$$

$$-k_{nf}\left.\frac{\partial T}{\partial y}\right|_{(x^*,0)} = q_0(x^*)^n, T(x^*,\infty) = T_\infty \tag{10}$$

The power-law temperature and heat flux distribution, described in Equations (9) and (10), resent a wider range of thermal boundary conditions including isoflux and isothermal cases. For example, by setting *n* equal to zero, Equations (9) and (10) yield isothermal and isoflux, respectively.

Second law of thermodynamics analysis of porous media is found to be more complicated compared to the clear fluid counterpart due to increased number of variables in governing equations (Nield and Kuznetsov, 2005). In the non-Darcian regime, there are three alternative models for the fluid friction term which are the clear-fluid compatible model, the Darcy model, and the Nield model or the power of drag model. Following the entropy generation function introduced by Nield and Kuznetsov (2005) the volumetric entropy generation rate, $S^{\bullet}_{gen}$, reads:

$$S_{gen} = \frac{k_{nf}}{T^2}[(\frac{\partial T}{\partial x})^2 + (\frac{\partial T}{\partial y})^2] + \frac{\mu_{nf}}{T}\{2[(\frac{\partial u}{\partial x})^2 + (\frac{\partial u}{\partial y})^2] + (\frac{\partial u}{\partial x} + \frac{\partial u}{\partial y})^2\} + \frac{\mu_{nf}}{TK}(u^2 + v^2) \tag{11}$$

Using boundary layer approximations, Equation (11) reduces to:

$$S^{\bullet}_{gen} = \frac{k_{nf}}{T^2}(\frac{\partial T}{\partial y})^2 + \frac{\mu_{nf}}{T}(\frac{\partial u}{\partial y})^2 + \frac{\mu_{nf}}{TK}u^2 \tag{12}$$

Using the stream function, $\psi(x,y)$, the continuity equation is satisfied:

$$u = \frac{\partial \psi}{\partial y}, v = \frac{\partial \psi}{\partial x} \tag{13}$$

The hydrodynamic boundary layer thickness scales with $\sqrt{K}$. This can be found through a scale analysis between the first and the second terms on the right hand side of Equation (2), i.e., the viscous and the Darcy terms. Therefore, instead of the other similarity parameters reported in the literature, the following dimensionless similarity parameter is defined

$$\eta = \frac{y}{\sqrt{K}} \tag{14}$$

The u-velocity is assumed to be correlated to $f(\eta)$, a dimensionless similarity function as:

$$u = u_0 x^* f'(\eta) \tag{15}$$

where $f'(\eta)$ is $\frac{df}{d\eta}$. Using stream function definition, Equation (15), the stream function and the v-velocity take the following forms:

$$v = -\frac{u_0}{L}\sqrt{K} f(\eta), v = u_0 x^* \sqrt{K} f(\eta) \tag{16}$$

Substituting from u and v into Equations (2) and (4), one will find the following differential equation for the u-momentum equation:

$$f''' + \mathrm{Re}\, A_1(ff'' - (f')^2) - f' = 0, \mathrm{Re} = \frac{\rho_f u_0 K}{L\mu_f} \tag{17}$$

where A_1 is a parameter having the following form:

$$A_1 = (1-\phi) + \frac{\rho_s}{\rho_f}\phi \tag{18}$$

where *Re* is the Reynolds number. Equation (18) should be solved subjected to the following boundary conditions:

$$f(0) = \frac{-v_w L}{u_0\sqrt{K}} = f_w, f'(0) = 1, f'(\infty) = 0 \tag{19}$$

f_w is the injection parameter. Positive/negative values of f_w show suction/injection into/from the porous surface, respectively.

The wall shear stress is the driving force that drags fluid flow along the stretching wall. The wall shear stress term can then be found, in terms of the similarity function, as:

$$\tau_w = -\mu_{nf}\left.\frac{\partial u}{\partial y}\right|_{y=0} = \frac{-\mu_{nf}u_0 x^* f''(0)}{2\sqrt{K}} \tag{20}$$

Introducing a similarity function, θ, as:

$$T - T_\infty = T_{ref}(x^*)^n\theta(\eta) \tag{21}$$

where T_{ref} is T_0 and $q_0\frac{\sqrt{K}}{k}$ for the power-law temperature and heat flux boundary conditions, respectively. The thermal energy equation reads:

$$\theta'' + \mathrm{Re}.\mathrm{Pr}.\frac{A_1}{A_3}.A_2.(1-\phi)^{2.5}(f\theta' - nf'\theta)) = 0, \mathrm{Pr} = \frac{\mu_f(\rho C_p)_f}{\rho_f k_f}, \mathrm{Re} = \frac{\rho_f u_0 K}{L\mu_f} \tag{22}$$

where A_2, A_3 are parameters having the following form:

$$A_2 = (1-\phi) + \frac{(\rho C_p)_s}{(\rho C_p)_f}\phi \tag{23}$$

$$A_3 = \frac{k_{nf}}{k_f} = \frac{k_s + 2k_f - 2\phi(k_f - k_s)}{k_s + 2k_f + 2\phi(k_f - k_s)}$$

Which are subjected to the following boundary conditions:

$$\theta(0) = 1, \theta(\infty) = 0 \text{ Power-law temperature} \tag{24}$$

$$\theta'(0) = -1, \theta(\infty) = 0 \text{ Power-law heat flux}$$

For power-law temperature and heat flux boundary conditions, respectively. Employing the definition of convective heat transfer coefficient, the local Nusselt numbers, become:

$$Nu_x = \frac{hx}{k} = \begin{cases} \dfrac{-\theta'(0)x}{\sqrt{K}} & Power-law\ temperature \\ \dfrac{q_w x}{K(T_w - T_\infty)} = \dfrac{x}{\theta(0)\sqrt{K}} & Power-law\ heat\ flux \end{cases} \tag{25}$$

Finally, the local volumetric entropy generation rate for the above cases, respectively, reads:

$$S^{\bullet}_{gen} = HTI + FFI \tag{26}$$

where HTI is the heat transfer irreversibility due to heat transfer in the direction of finite temperature gradients. HTI is common in all types of thermal engineering applications.

The last term (FFI) is the contribution of fluid friction irreversibility to the total entropy generation. Not only the wall and fluid layer shear stress but also the momentum exchange at the solid boundaries (pore level) contributes to FFI . In terms of the primitive variables, HTI and FFI become

$$HTI = \begin{cases} \dfrac{A_3 . k_f}{(\theta T_0 (x^*)^n + T_\infty)^2} (\dfrac{\theta' T_0 (x^*)^n}{\sqrt{K}})^2 & Power-law\ temperature \\ \dfrac{A_3 . k_f}{(\theta \sqrt{K} q_0 (x^*)^n / k + T_\infty)^2} (\dfrac{\theta' q_0 (x^*)^n}{A_3 . k_f})^2 & Power-law\ heat\ flux \end{cases} \tag{27}$$

$$FFI = \begin{cases} \dfrac{\mu_f . (1-\phi)^{2.5}}{(\theta T_0 (x^*)^n + T_\infty)} [(\dfrac{f'' u_0 x^*}{\sqrt{K}})^2 + (\dfrac{f' u_0 x^*}{\sqrt{K}})^2] & Power-law\ temperature \\ \dfrac{\mu_f . (1-\phi)^{2.5}}{(\theta \sqrt{K} q_0 (x^*)^n / k + T_\infty)} [(\dfrac{f'' u_0 x^*}{\sqrt{K}})^2 + (\dfrac{f' u_0 x^*}{\sqrt{K}})^2] & Power-law\ heat\ flux \end{cases} \tag{28}$$

where T_∞ and T_0 are measured in degrees of Kelvin.

One can also define the Bejan number, Be, as

$$Be = \frac{HTI}{HTI + FFI} \tag{29}$$

The Bejan number shows the ratio of entropy generation due to heat transfer irreversibility to the total entropy generation so that a *Be* value more/less than 0.5 shows that the contribution of HTI to the total entropy generation is higher/less than that of FFI. The limiting value of $Be = 1$ shows that the only active entropy generation mechanism is *HTI* while $Be = 0$ represents no HTI contribution to the total entropy production.

1.2. Semi Analytical Method

In the heart of all different engineering sciences, everything shows itself in the mathematical relation that most of these problems and phenomena are modeled by ordinary or partial differential equations. Since there are some limitations with the common perturbation method, and also because of this fact that the basis of the common perturbation method is upon the existence of a small parameter, developing this method for different applications is very difficult. Therefore, some different methods have recently introduced some ways to eliminate the small parameter, such as the Homotopy Perturbation Method (Sheikholeslami, Ganji, and Rokni, 2013; Sheikholeslami, Ashorynejad, Ganji, Yıldırım, 2012; Sheikholeslami and Ganji, 2013; Sheikholeslami, Ashorynejad, Ganji and Kolahdooz, 2011), Differential Transformation Method (Sheikholeslami, Ganji and Rashidi, 2016; Sheikholeslami and Ganji, 2015; Sheikholeslami, Ashorynejad, Barari, and Soleimani, 2013; Domairry, Sheikholeslami, Reza, Ashorynejad, Subba, Gorla, and Khani, 2012; Sheikholeslami, Azimi and Ganji, 2015; Sheikholeslami, Rashidi, Al Saad, Firouzi, Houman, Rokni and Domairry, 2015), Homotopy Analysis Method (Sheikholeslami, Ashorynejad, Domairry and Hashim, 2012), Adomian Decomposition Method (Sheikholeslami and Ganji, 2016; Sheikholeslami, Ganji, Ashorynejad and Rokni, 2012; Sheikholeslami, Ganji and Ashorynejad, 2013) and Optimal Homotopy Asymptotic Method (OHAM) (Sheikholeslami, Ashorynejad, Domairry and Hashim, 2012). Sample codes for new semi analytical methods are presented in *appendix*.

1.2.1. BASIC Idea of HAM

Let us assume the following nonlinear differential equation in form of:

$$N[u(\tau)] = 0, \tag{30}$$

where N is a nonlinear operator, τ is an independent variable and $u(\tau)$ is the solution of equation. We define the function, $\phi(\tau, p)$ as follows:

$$\lim_{p \to 0} \phi(\tau, p) = u_0(\tau) \tag{31}$$

where, $p \in [0,1]$ and $u_0(\tau)$ is the initial guess which satisfies the initial or boundary condition,

$$\lim_{p\to 1}\phi(\tau,p)=u_0(\tau) \tag{32}$$

And by using the generalized homotopy method, Liao's so-called zeroth-order deformation equation will be:

$$(1-p)L[\phi(\tau,p)-u_0(\tau)]=p\hbar H(\tau)N[\phi(\tau,p)] \tag{33}$$

where $\hbar$ is the auxiliary parameter which helps us increase the results convergence, H(τ) is the auxiliary function and L is the linear operator. It should be noted that there is a great freedom to choose the auxiliary parameter $\hbar$, the auxiliary function H(τ), the initial guess $u_0(\tau)$ and the auxiliary linear operator L. This freedom plays an important role in establishing the keystone of validity and flexibility of HAM as shown in this paper. Thus, when p increases from 0 to 1 the solution $\phi(\tau,p)$ changes between the initial guess $u_0(\tau)$ and the solution $u(\tau)$. The Taylor series expansion of $\phi(\tau,p)$ with respect to p is:

$$\phi(\tau,p)=u_0(\tau)+\sum_{m=1}^{\infty}u_m(\tau)p^m \tag{34}$$

and

$$u_0^{[m]}(\tau)=\left.\frac{\partial^m\phi(\tau,p)}{\partial p^m}\right|_{p=0} \tag{35}$$

where $u_0^{[m]}(\tau)$ for brevity is called the m th order of deformation derivation which reads:

$$u_m(\tau)=\frac{u_0^{[m]}}{m!}=\left.\frac{1}{m!}\frac{\partial^m\phi(\tau,p)}{\partial p^m}\right|_{p=0} \tag{36}$$

It's clear that if the auxiliary parameter is $\hbar=-1$ and the auxiliary function is determined to be H(τ)=1, Equation (33)will be:

$$(1-p)L[\phi(\tau,p)-u_0(\tau)]+p(\tau)N[\phi(\tau,p)]=0 \tag{37}$$

This statement is commonly used in the HAM procedure. Indeed, in HAM we solve the nonlinear differential equation by separating any Taylor expansion term. Now we define the vector of:

$$\overrightarrow{u_m}=\{\overrightarrow{u_1},\overrightarrow{u_2},\overrightarrow{u_3},\ldots,\overrightarrow{u_n}\} \tag{38}$$

According to the definition in Equation (36), the governing equation and the corresponding initial condition of $u_m(\tau)$ can be deduced from zeroth-order deformation Equation(9.30). Differentiating Equation (30) for m times with respect to the embedding parameter p and setting p=0and finally dividing by m!, we will have the so-called m th order deformation equation in the form,

$$L[u_m(\tau) - \chi_m u_{m-1}(\tau)] = \hbar H(\tau) R(\vec{u}_{m-1}) \tag{39}$$

where,

$$R(\vec{u}_{m-1}) = \frac{1}{(m-1)!} \frac{\partial^m \phi(\tau,p)}{\partial p^m}\bigg|_{p=0} \tag{40}$$

And

$$\chi_m = \begin{cases} 0 & m \le 1 \\ 1 & m > 1 \end{cases} \tag{41}$$

1.2.2. Application of HAM

For HAM solutions of the governing equations, we choose the initial approximations of $f(\eta)$ and $\theta(\eta)$ as follow:

$$f_0(\eta) = -\exp(-\eta) + 1 + f_w \tag{42}$$

$$\theta_0(\eta) = \exp(-\eta) \tag{43}$$

And the auxiliary linear operators are:

$$L_1(f) = f''' + f'' \tag{44}$$

$$L_2(\theta) = \theta'' + \theta' \tag{45}$$

These auxiliary linear operators satisfy:

$$L_1(C_1 + C_2 + C_3 \exp(-\eta)) \tag{46}$$

$$L_2(C_4 + C_5 \exp(-\eta)) \quad (47)$$

where C_i *(i=1,2,3,4,5,6)* are constants. Introducing a non-zero auxiliary parameters $\hbar_1$ and $\hbar_2$, we develop the zeroth-order deformation problems as follow:

$$(1-p)L[f(\eta;p) - f_0(\eta)] = p\hbar_1 N[f(\eta;p)] \quad (48)$$

$$f(0;p) = f_w, f^{'}(0;p) = 1, f^{'}(\infty;p) = 0 \quad (49)$$

$$(1-p)L[\theta(\eta;p) - \theta_0(\eta)] = p\hbar_2 N[\theta(\eta;p)] \quad (50)$$

$$\theta(0;p) = 1, \theta(\infty;p) = 0 \text{ Power-law temperature} \quad (51)$$

$$\theta^{'}(0;p) = -1, \theta(\infty;p) = 0 \text{ Power-law heat flux} \quad (52)$$

where nonlinear operators, N_1 and N_2 are defined as:

$$N_1[f(\eta;p),\theta(\eta;p)] = \frac{\partial^3 f(\eta;p)}{\partial\eta^3} + \operatorname{Re} A_1(f(\eta;p)\frac{\partial^2 f(\eta;p)}{\partial\eta^2} - (\frac{\partial f(\eta;p)}{\partial\eta})^2) - \frac{\partial f(\eta;p)}{\partial\eta} \quad (53)$$

$$N_2[f(\eta;p),\theta(\eta;p)] = \frac{\partial^2\theta(\eta;p)}{\partial\eta^2} + \operatorname{Re}\Pr\frac{A_1}{A_3}A_2(f(\eta;p)\frac{\partial\theta(\eta;p)}{\partial\eta}) - n\operatorname{Re}\Pr\frac{A_1}{A_3}A_2(\theta(\eta;p)\frac{\partial f(\eta;p)}{\partial\eta}) \quad (54)$$

For *p=0* and *p=1* we, respectively, have:

$$f(\eta;0) = f_0(\eta) f(\eta;1) = f(\eta) \quad (55)$$

$$\theta(\eta;0) = \theta_0(\eta)\theta(\eta;1) = \theta(\eta) \quad (56)$$

As *p* increases from 0 to 1, $f(\eta;p)$ and $\theta(\eta;p)$ vary, respectively, from $f_0(\eta)$ and $\theta_0(\eta)$ to $f(\eta)$ and $\theta(\eta)$. By Taylor's theorem, $f(\eta)$ and $\theta(\eta)$ can be expanded in a power series of *p* as follows:

$$f(\eta;p)=f_0(\eta)+\sum_{m=1}^{\infty}f_m(\eta)p^m \tag{57a}$$

$$f_m(\tau)=\frac{1}{m!}\frac{\partial^m f(\eta,p)}{\partial p^m} \tag{57b}$$

And

$$\theta(\eta;p)=\theta_0(\eta)+\sum_{m=1}^{\infty}\theta_m(\eta)p^m \tag{58a}$$

$$\theta_m(\tau)=\frac{1}{m!}\frac{\partial^m \theta(\eta,p)}{\partial p^m} \tag{58b}$$

In which $\hbar_1$ and $\hbar_2$ are chosen in such a way that these series are convergent at p=1. Convergence of the series (57a) and (58a) depends on the auxiliary parameters $\hbar_1$ and $\hbar_2$.

Assume that $\hbar_1$ and $\hbar_2$ are selected such that the series (57a) and (58a) are convergent at p=1, then due to Equations (48) and (50) we have:

$$f(\eta)=f_0(\eta)+\sum_{m=1}^{\infty}f_m(\eta) \tag{59}$$

$$\theta(\eta)=\theta_0(\eta)+\sum_{m=1}^{\infty}\theta_m(\eta) \tag{60}$$

Differentiating the zeroth-order deformation Equations (36)and(38) m times with respect to p and then dividing them by m! and finally setting p=0, we have the following mth-order deformation problem:

$$L_1[f_m(\eta)-\chi_m f_{m-1}(\eta)]=\hbar_1 R_m^f(\eta) \tag{61}$$

$$f_m(0)=0, f_m^{'}(0;p)=0, f_m^{'}(\infty;p)=0 \tag{62}$$

$$L_2[\theta_m(\eta)-\chi_m\theta_{m-1}(\eta)]=\hbar_2 R_m^{\theta}(\eta) \tag{63}$$

$$\theta_m(0) = 0, \theta_m(\infty; p) = 0 \text{ Power-law temperature} \tag{64}$$

$$\theta'_m(0; p) = 0, \theta_m(\infty; p) = 0 \text{ Power-law heat flux} \tag{65}$$

$$R_m^f(\eta) = f'''_{m-1} + \mathrm{Re}.A_1.(\sum_{n=0}^{m-1} f_{m-1-n} f''_n - \sum_{n=0}^{m-1} f'_{m-1-n} f'_n) - f'_{m-1} \tag{66}$$

$$R_m^\theta(\eta) = \theta''_{m-1} + \mathrm{Re}.\mathrm{Pr}.\frac{A_1}{A_3}.A_2.(\sum_{n=0}^{m-1} f_{m-1-n}\theta'_n) - n.\mathrm{Re}.\mathrm{Pr}.\frac{A_1}{A_3}.A_2.(\sum_{n=0}^{m-1} \theta_{m-1-n} f'_n) \tag{67}$$

We use MAPLE software to obtain the solution of these equations. Two first deformations of the coupled solutions are presented as follow:

$$f_1(\eta) = -\hbar.\mathrm{Re}.A_1.\exp(-\eta)[1 + f_w + \eta + f_w\eta - (1 + f_w)\exp(\eta) \tag{68}$$

$$\begin{aligned} \theta_1(\eta) = & -0.5\hbar \Pr \mathrm{Re} \frac{A_1}{A_3} A_2 \exp(-2\eta) + \hbar \Pr \mathrm{Re} \frac{A_1}{A_3} A_2 \exp(-\eta) \\ & -0.5\hbar \Pr \mathrm{Re} \frac{A_1}{A_3} A_2 \exp(-\eta) + \hbar \Pr \mathrm{Re} f_w \frac{A_1}{A_3} A_2 \exp(-\eta) - \hbar\eta \exp(-\eta) \text{ Power-law temperature} \\ & -0.5n\hbar \Pr \mathrm{Re} \frac{A_1}{A_3} A_2 \exp(-2\eta) + 0.5n\hbar \Pr \mathrm{Re} \frac{A_1}{A_3} A_2 \exp(-\eta) \end{aligned} \tag{69}$$

$$\begin{aligned} \theta_1(\eta) = & -0.5\hbar \frac{A_1}{A_3} A_2[-\Pr \mathrm{Re} \exp(-\eta) - 2\Pr \mathrm{Re}\,\eta \\ & -2\Pr \mathrm{Re}\,\eta f_w + 2\eta + 2 + n \Pr \mathrm{Re} \exp(-\eta) - 2n \Pr \mathrm{Re}] \end{aligned} \quad \text{Power-law heat flux} \tag{70}$$

The solutions $f_2(\eta)$ *and* $\theta_2(\eta)$ were too long to be mentioned here, therefore, they are shown graphically.

The convergence and the rate of approximation for the HAM solution strongly depend on the values of auxiliary parameter $\hbar$. $\hbar$ curves for (a) f (b) $\theta_{isothermal}$ (c) $\theta_{isoflux}$ in $\mathrm{Re} = f_w = \Pr = 1, \phi = 0.01$ and $n = 0$ are shown in Figure 2. Using the $\hbar$ -curve, we can easily choose the value of auxiliary parameter $\hbar$ to guarantee the convergence. For this problem $\hbar = -0.21$ in step 20 has good accuracy.

Figure 2. h curves for (a) f *(b)* $\theta_{isothermal}$ *(c)* $\theta_{isoflux}$ *in* $\mathrm{Re} = f_w = \mathrm{Pr} = 1, \phi = 0.01$ *and* $n = 0$

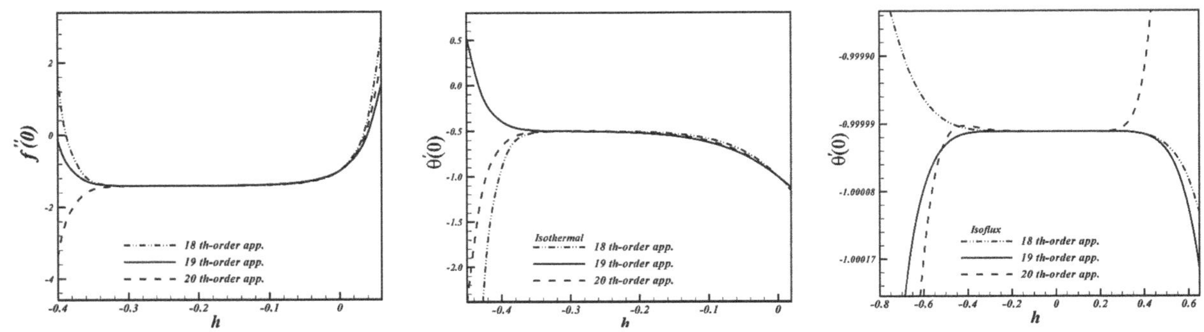

(a) (b) (c)

1.3. Effects of Active Parameters

Results are given for the velocity, temperature distribution, wall shear stress and Nusselt number and entropy generation for different non-dimensional numbers.

Figures 3 and 4 are presented to show the effect of the volume fraction of nanoparticles (Cu) on velocity profiles and temperature distribution (a) Power-law temperature, (b) Power-law heat flux, respectively when $\mathrm{Pr} = 6.2, f_w = 1, \mathrm{Re} = 1, n = 2$. When the volume of fraction for the nanoparticles increases from 0 to 0.2, all boundary layers thicknesses decrease. This agrees with the physical behavior,

Figure 3. Effect of nanoparticle volume fraction on velocity profiles when $f_w = 1, \mathrm{Re} = 1$.

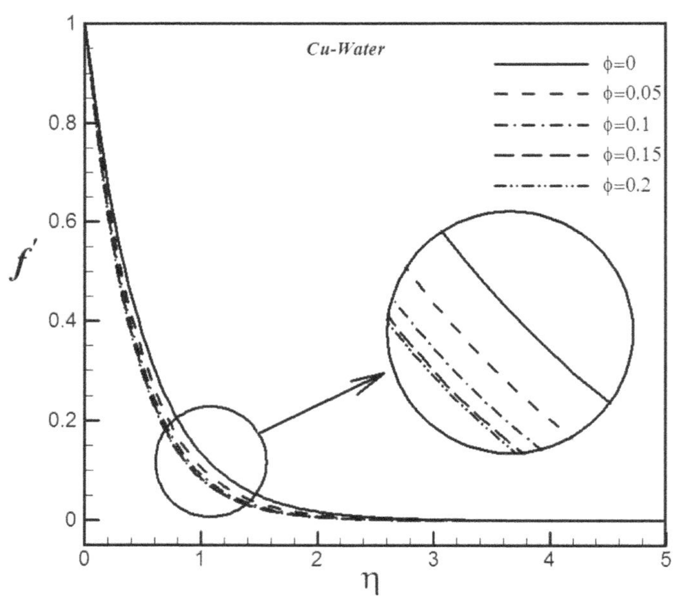

Figure 4. Effect of nanoparticle volume fraction on temperature distribution (a) Power-law temperature (b) Power-law heat flux when $\mathrm{Pr} = 6.2, f_w = 1, \mathrm{Re} = 1, n = 2$.

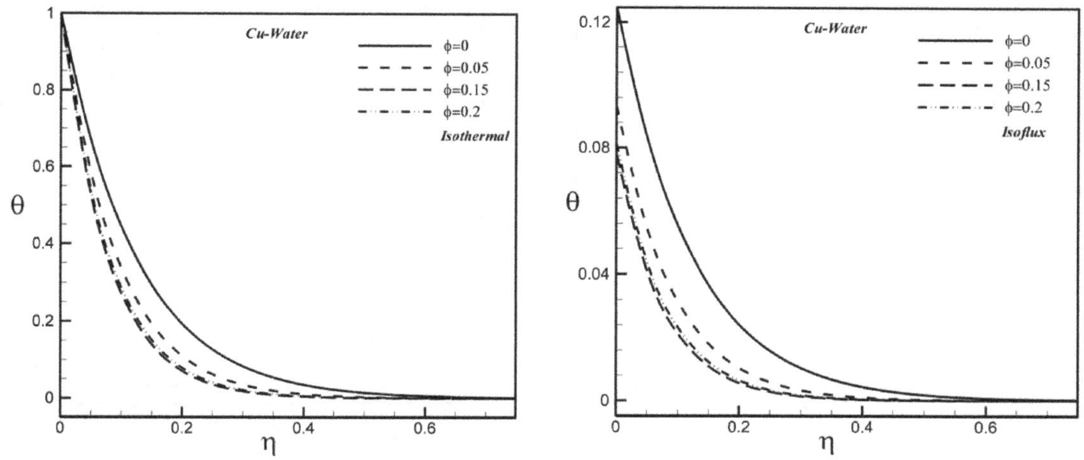

(a) (b)

when the volume of copper nanoparticles increases the thermal conductivity increases and then the thermal boundary layer thickness decreases.

Figures 5 and 6 display the behavior of the velocity and the temperature profiles using different nanofluids When $\mathrm{Pr} = 6.2, \phi = 0.1, \mathrm{Re} = 1, f_w = 1, n = 2$. The tables show that by using different types of nanofluids the values of the velocity and temperature change. When Silver is chosen as the nanopar-

Figure 5. Velocity for different types of nanofluids when $\phi = 0.1, \mathrm{Re} = 1, f_w = 1$.

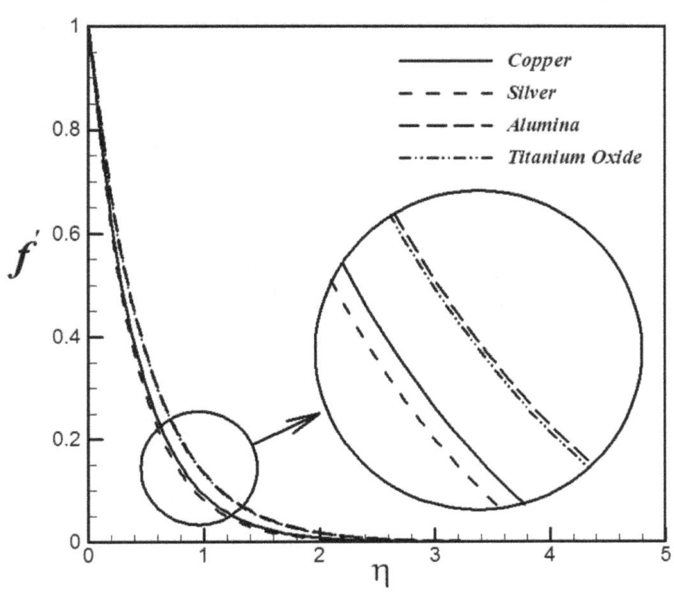

Figure 6. Temperature profiles (a) Power-law temperature (b) Power-law heat flux for different types of nanofluids when $\Pr = 6.2, \phi = 0.1, \mathrm{Re} = 1, f_w = 1, n = 2$.

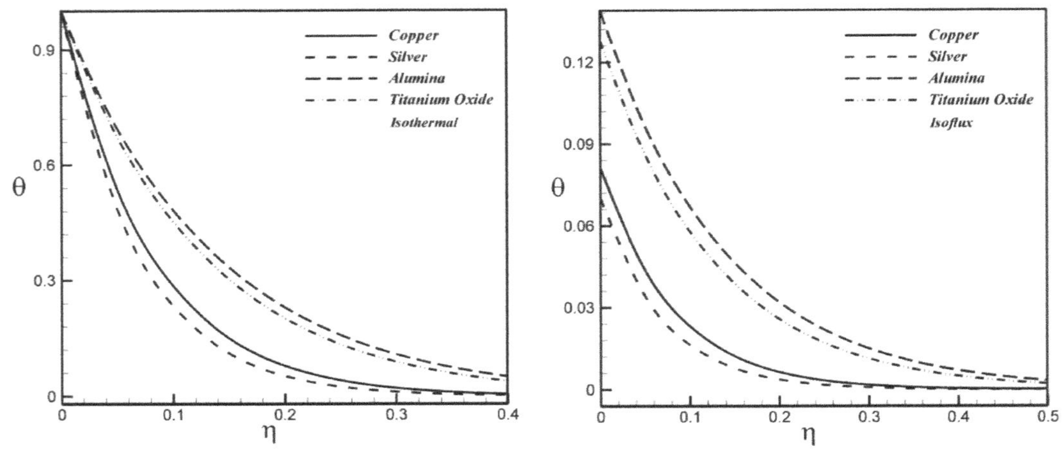

(a) (b)

ticle, the maximum amount of all boundary layer thicknesses observed, while minimum amount of those amounts observed by choosing Alumina. (Figures 5 and 6).

Figure 7 shows variation of skin friction coefficients ($-f''(0)$) versus (a) Re and (b) f_w for selected values of the nanoparticles volume parameter in the case of Cu-water. For both suction and injection, it is observed that skin friction increases as ϕ increases. Also this change occurs when Re or f_w increases. It should be noticed that the changes are more noticeable for higher values of ϕ when values of Re and f_w are greater.

Figure 7. Effects of the nanoparticle volume fraction ϕ, Reynolds number and wall injection/suction Parameter on skin friction coefficient when (a) $f_w = 1$ *(b)* $\mathrm{Re} = 1$.

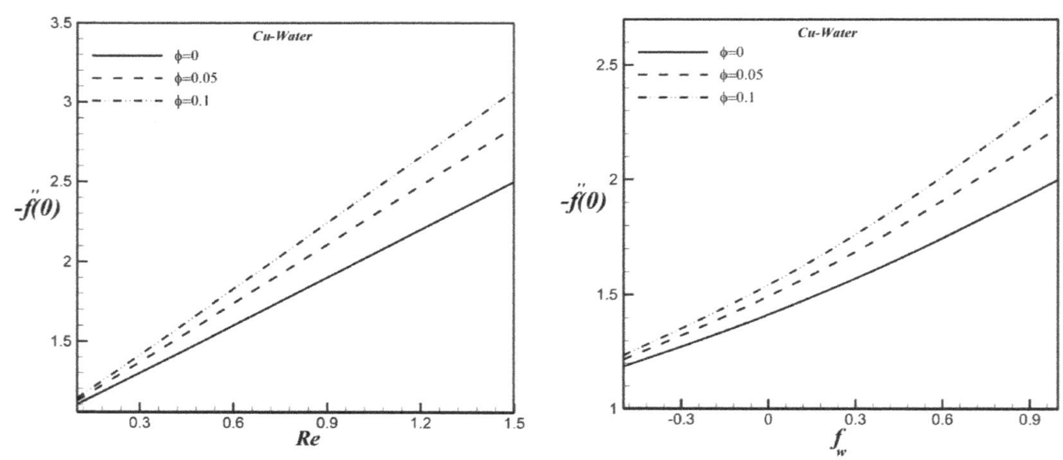

(a) (b)

Figure 8 shows variation of Nusselt number for Power-law temperature case ($-\theta'(0)$) versus (a) Re, (b) f_w and (c) n for selected values of the nanoparticles volume parameter in the case of Cu-water.

It is obvious from Figure 8 that the heat transfer rates increase with the increase of the nanoparticles volume fraction (ϕ), Re, f_w and n .The change in the Nusselt number is found to be upper for higher values of the parameter ϕ, and this change are more noticeable with the increase of Re, f_w and n.

Table 2 shows the effects of the nanoparticle volume fraction ϕ for different types of nanofluids on skin friction coefficient when $\mathrm{Re}=1, f_w=1$. Table 3 and 4 show the effects of the nanoparticle volume fraction ϕ for different types of nanofluids on Nusselt number for Power-law temperature case, Power-law heat flux case respectively when $\mathrm{Pr}=6.2, n=2, \mathrm{Re}=1, f_w=1$. These tables show that the values of $-f''(0), -\theta'(0)$ and $1/\theta(0)$ change with nanofluid changes, i.e. we can say that the shear stress and rate of hate transfer change by using different types of nanofluid. This means that the nanofluids will be important in the cooling and heating processes.

Choosing Silver as the nanoparticle leads to the maximum amount of skin friction coefficient and rate of hate transfer, while selecting Alumina leads to the minimum amount of those values (Tables 2, 3 and 4). Figure 9 shows the effect of nanoparticle volume fraction on (a) HTI, (b) FFI, (c) S_{gen} and (d) Be when $\mathrm{Pr}=6.2, \mathrm{Re}=1, n=2, x^*=0.5, f_w=1, u_0=1m/s, T_\infty=T_0=10K$ and $K=0.001$. Figure 10 shows the effect of different types of nanofluids on (a) HTI, (b) FFI, (c) S_{gen} and (d) Be

Figure 8. Effects of the nanoparticle volume fraction ϕ, Reynolds number, wall injection/suction Parameter and power of temperature/heat flux distribution on Nusselt number (Power-law temperature) when (a) $\mathrm{Pr}=6.2, f_w=1, n=2$ (b) $\mathrm{Pr}=6.2, \mathrm{Re}=1, n=2$ (c) $\mathrm{Pr}=6.2, f_w=1, \mathrm{Re}=1$.

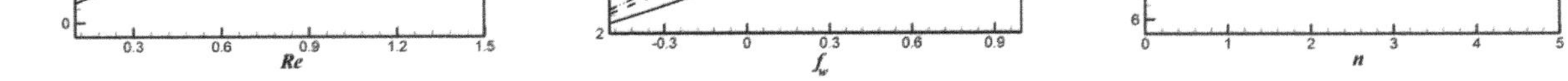

(a) (b) (c)

Table 2. Effects of the nanoparticle volume fraction for different types of nanofluids on skin friction coefficient when $\mathrm{Re}=1, f_w=1$.

ϕ	Nanoparticles			
	Cu	Ag	Al_2O_3	TiO_2
0.05	2.229703	2.298824	2.010783788	2.023135
0.1	2.380028	2.500792	1.9975454	2.019124
0.2	2.483631	2.663553	1.913780762	1.94593

Table 3. Effects of the nanoparticle volume fraction for different types of nanofluids on Nusselt number (Power-law temperature) when $\Pr = 6.2, n = 2, \mathrm{Re} = 1, f_w = 1$.

ϕ	**Nanoparticles**			
	Cu	**Ag**	**Al_2O_3**	**TiO_2**
0.05	10.71299	11.77269	7.741767	8.085111
0.1	12.31904	14.26799	7.201689	7.822867
0.2	12.30637	15.04481	5.630545	6.602276

Figure 9. Effect of nanoparticle volume fraction on (a) HTI*, (b)* FFI*, (c)* $S^{\bullet}_{gen}$ *and (d)* Be *When* $\Pr = 6.2, \mathrm{Re} = 1, n = 2, x^* = 0.5, f_w = 1, u_0 = 1m/s, T_\infty = T_0 = 10K$ *and* $K = 0.001$

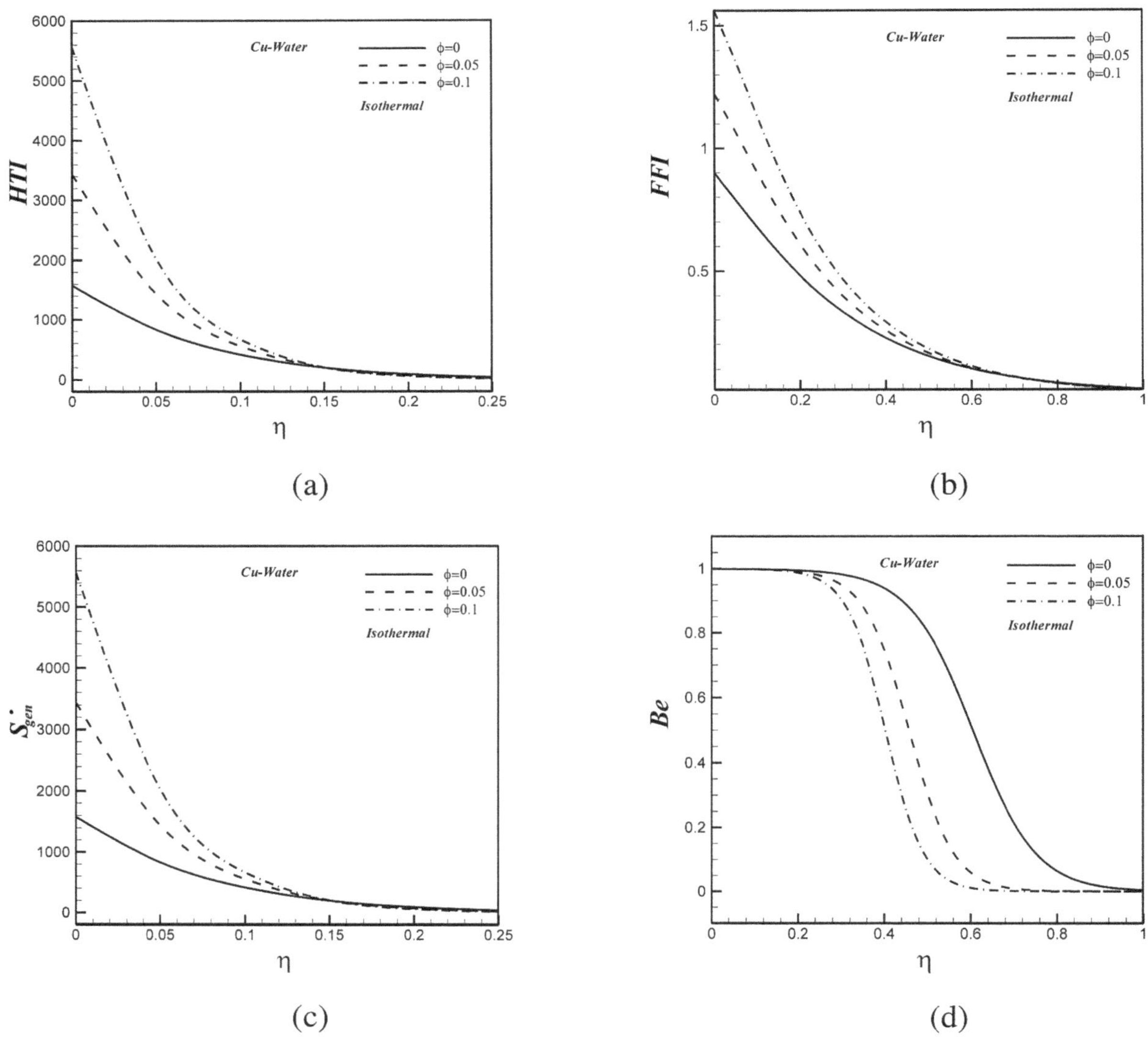

Table 4. Effects of the nanoparticle volume fraction for different types of nanofluids on Nusselt number (Power-law heat flux) when $\mathrm{Pr} = 6.2, n = 2, \mathrm{Re} = 1, f_w = 1$.

ϕ	**Nanoparticles**			
	Cu	**Ag**	**Al_2O_3**	**TiO_2**
0.05	10.71299	11.77268758	7.741767427	8.085111261
0.1	12.31904	14.26799218	7.201688959	7.822867028
0.2	12.30637	15.04480717	5.630544587	6.60227588

Figure 10. Effect of different types of nanofluids on (a) HTI*, (b)* FFI*, (c)* $S^{\bullet}_{gen}$ *and (d)* Be *When* $\mathrm{Pr} = 6.2, \mathrm{Re} = 1, n = 2, \phi = 0.1, x^* = 0.5, f_w = 1, u_0 = 1m/s, T_\infty = T_0 = 10K$ *and* $K = 0.001$

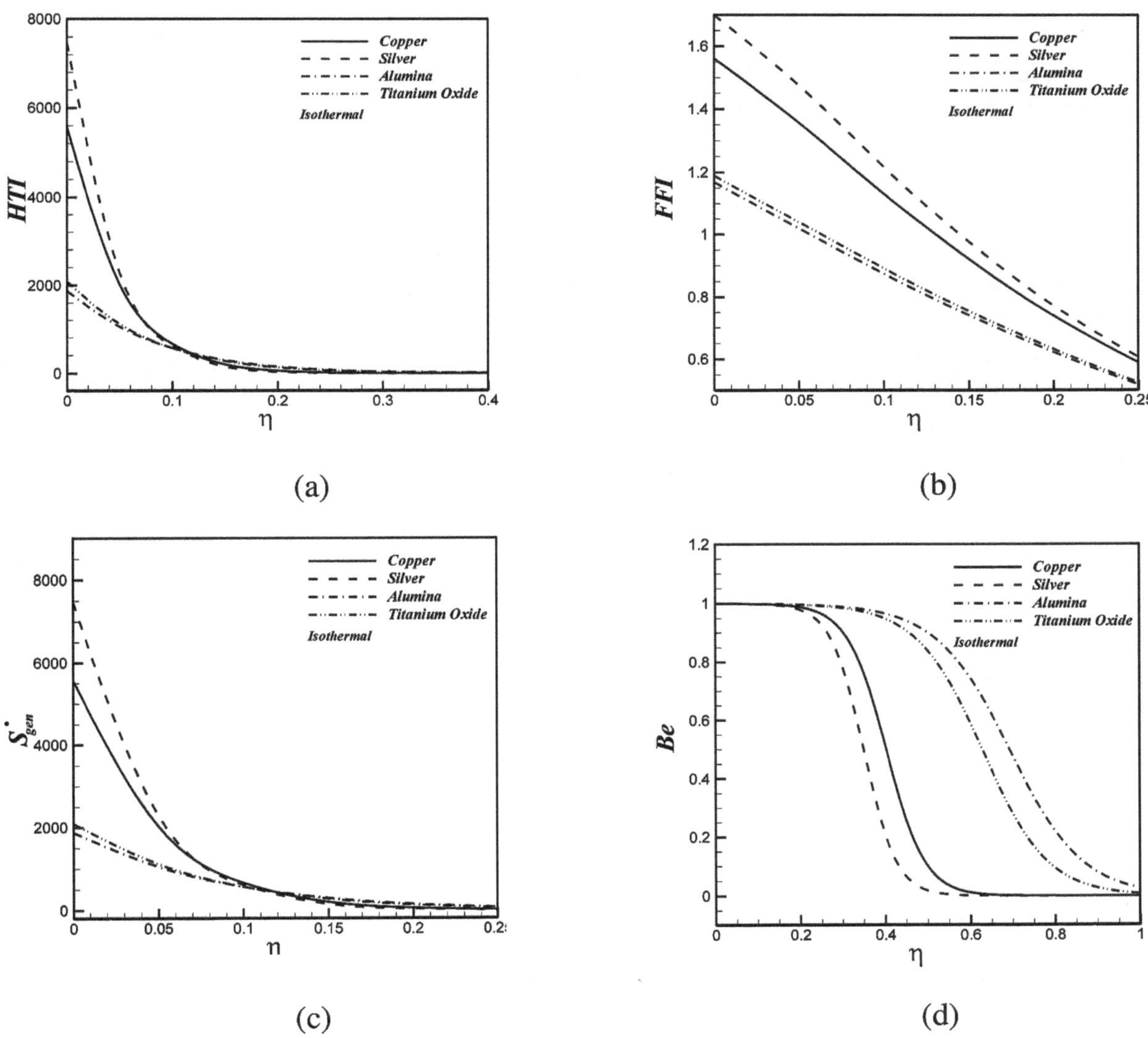

when $\Pr = 6.2, \mathrm{Re} = 1, n = 2, \phi = 0.1, x^* = 0.5, f_w = 1, u_0 = 1m / s, T_\infty = T_0 = 10K$ and $K = 0.001$. It is observed that, regardless of the boundary condition, increasing percentage of nanoparticles (ϕ) leads to the increase in the heat transfer irreversibility due to heat transfer in the direction of finite temperature gradients (HTI) and the contribution of fluid friction irreversibility to the total entropy generation (FFI) whereas the increase in HTI is more than increase in FFI. The entropy generation rate (S_{gen}) reduces while we get farter from the surface of the porous plate and HTI reduction in higher nanoparticles percentages occurs in farter distances. With studying the Bejan number we can see that HTI beats FFI near the surface of porous plate. Adding Alumina nanoparticles leads to the minimum amount of heat loss while the maximum amount of heat loss occurs when we use Silver as the nanoparticles and there will be the same for HTI and FFI. By choosing Silver and Copper as the nanoparticle HTI beats FFI close to the porous plate while selecting Alumina and Titanium Oxide as the nanoparticle cause this fact occurs farter from the porous plate.

2. MAGNETOHYDRODYNAMIC FLOW IN A PERMEABLE CHANNEL FILLED WITH NANOFLUID

2.1. Problem Definition

The laminar two-dimensional stationary nanofluid flow in a semi-porous channel made by a long rectangular plate with length of L_x in uniform translation in x^* direction and an infinite porous plate is considered. The distance between the two plates is h. We observe a normal velocity q on the porous wall. A uniform magnetic field B is assumed to be applied towards direction y^* (Figure 11) (Sheikholeslami and Ganji, 2014)

In the case of a short circuit to neglect the electrical field and perturbations to the basic normal field and without any gravity forces, the governing equations are:

$$\frac{\partial u^*}{\partial x^*} + \frac{\partial v^*}{\partial y^*} = 0, \tag{71}$$

Figure 11. Schematic diagram of the system

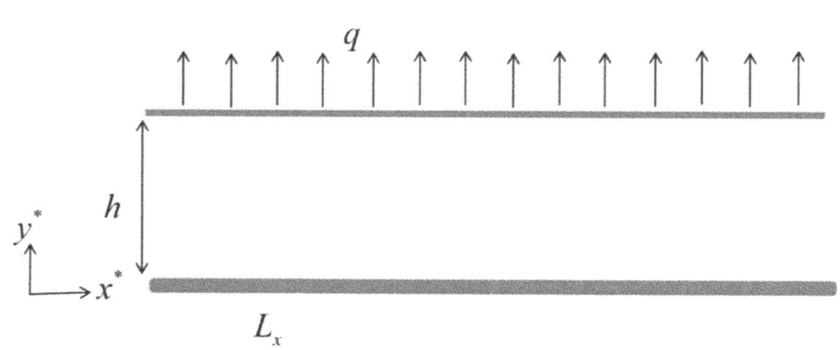

$$u*\frac{\partial u*}{\partial x*}+v*\frac{\partial u*}{\partial y*}=-\frac{1}{\rho_{nf}}\frac{\partial P*}{\partial x*}+\frac{\mu_{nf}}{\rho_{nf}}\left(\frac{\partial^2 u*}{\partial x*^2}+\frac{\partial^2 u*}{\partial y*^2}\right)-u*\frac{\sigma_{nf}B^2}{\rho_{nf}}, \tag{72}$$

$$u*\frac{\partial v*}{\partial x*}+v*\frac{\partial v*}{\partial y*}=-\frac{1}{\rho_{nf}}\frac{\partial P*}{\partial y*}+\frac{\mu_{nf}}{\rho_{nf}}\left(\frac{\partial^2 v*}{\partial x*^2}+\frac{\partial^2 v*}{\partial y*^2}\right), \tag{73}$$

The appropriate boundary conditions for the velocity are:

$$y*=0: u*=u_0*, v*=0, \tag{74}$$

$$y*=h: u*=0, v*=-q, \tag{75}$$

Calculating a mean velocity U by the relation:

$$y*=0: u*=u_0*, v*=0, \tag{76}$$

We consider the following transformations:

$$x=\frac{x*}{L_x}; y=\frac{y*}{h}, \tag{77}$$

$$u=\frac{u*}{U}; v=\frac{v*}{q}, P_y=\frac{P*}{\rho_f . q^2} \tag{78}$$

Then, we can consider two dimensionless numbers: the Hartman number Ha for the description of magnetic forces and the Reynolds number Re for dynamic forces:

$$Ha=Bh\sqrt{\frac{\sigma_f}{\rho_f . \upsilon_f}}, \tag{79}$$

$$\mathrm{Re}=\frac{hq}{\mu_{nf}}\rho_{nf}. \tag{80}$$

where the effective density (ρ_{nf}) is defined as:

$$\rho_{nf} = \rho_f(1-\phi) + \rho_s\phi \tag{81}$$

where ϕ is the solid volume fraction of nanoparticles. The dynamic viscosity, thermal conductivity and effective electrical conductivity of the nanofluid are defined as

$$\mu_{nf} = \frac{\mu_f}{(1-\phi)^{2.5}} \tag{82}$$

$$\frac{k_{nf}}{k_f} = \frac{k_s + 2k_f - 2\phi(k_f - k_s)}{k_s + 2k_f + \phi(k_f - k_s)} \tag{83}$$

$$\frac{\sigma_{nf}}{\sigma_f} = 1 + \frac{3\left(\frac{\sigma_s}{\sigma_f} - 1\right)\phi}{\left(\frac{\sigma_s}{\sigma_f} + 2\right) - \left(\frac{\sigma_s}{\sigma_f} - 1\right)\phi} \tag{84}$$

Substituting Equations (76) and (80) into Equations (71) and (73) leads to the dimensionless equations:

$$\frac{\partial u}{\partial x} + \frac{\partial v}{\partial y} = 0, \tag{85}$$

$$u\frac{\partial u}{\partial x} + v\frac{\partial u}{\partial y} = -\varepsilon^2\frac{\partial P_y}{\partial x} + \frac{\mu_{nf}}{\rho_{nf}}\frac{1}{hq}\left(\varepsilon^2\frac{\partial^2 u}{\partial x^2} + \frac{\partial^2 u}{\partial y^2}\right) - u\frac{Ha^2}{\mathrm{Re}}\frac{B^*}{A^*}, \tag{86}$$

$$u\frac{\partial v}{\partial x} + v\frac{\partial v}{\partial y} = -\frac{\partial P_y}{\partial x} + \frac{\mu_{nf}}{\rho_{nf}}\frac{1}{hq}\left(\varepsilon^2\frac{\partial^2 v}{\partial x^2} + \frac{\partial^2 v}{\partial y^2}\right). \tag{87}$$

where A^* and B^* are constant parameters:

$$A^* = (1-\phi) + \frac{\rho_s}{\rho_f}\phi, B^* = 1 + \frac{3\left(\frac{\sigma_s}{\sigma_f}-1\right)\phi}{\left(\frac{\sigma_s}{\sigma_f}+2\right)-\left(\frac{\sigma_s}{\sigma_f}-1\right)\phi} \tag{88}$$

Quantity of ε is defined as the aspect ratio between distance h and a characteristic length L_x of the slider. This ratio is normally small. Berman's similarity transformation is used to be free from the aspect ratio of ε :

$$v = -V\left(y\right); u = \frac{u*}{U} = u_0 U\left(y\right) + x\frac{dV}{dy}. \tag{89}$$

Introducing Equation (89) in the second momentum equation (87) shows that quantity $\partial P_y / \partial y$ does not depend on the longitudinal variable x. With the first momentum equation, we also observe that $\partial^2 P_y / \partial x^2$ is independent of x. We omit asterisks for simplicity. Then a separation of variables leads to:

$$V'^2 - VV'' - \frac{1}{\mathrm{Re}}\frac{1}{A^*\left(1-\phi\right)^{2.5}}V''' + \frac{Ha^2}{\mathrm{Re}}\frac{B^*}{A^*}V' = \varepsilon^2\frac{\partial^2 P_y}{\partial x^2} = \varepsilon^2\frac{1}{x}\frac{\partial P_y}{\partial x}, \tag{90}$$

$$UV' - VU' = \frac{1}{\mathrm{Re}}\frac{1}{A^*\left(1-\phi\right)^{2.5}}\left[U'' - Ha^2 B^*\left(1-\phi\right)^{2.5}U\right]. \tag{91}$$

The right-hand side of Equation (90) is constant. So, we derive this equation with respect to x. This gives:

$$V^{IV} = Ha^2 B^*\left(1-\phi\right)^{2.5}V'' + \mathrm{Re}\,A^*\left(1-\phi\right)^{2.5}\left[V'V'' - VV'''\right], \tag{92}$$

Where primes denote differentiation with respect to y and asterisks have been omitted for simplicity. The dynamic boundary conditions are:

$$y = 0:\ U = 1; V = 0; V' = 0, \tag{93}$$

$$y = 1:\ U = 0; V = 1; V' = 0. \tag{94}$$

2.2. Semi Analytical Method

2.2.1. Basic Idea of OHAM

Following differential equation is considered:

$$L(u(t)) + N(u(t)) + g(t) = 0, B(u)=0, \tag{95}$$

where *L* is a linear operator, τ is an independent variable, $u(t)$ is an unknown function, $g(t)$ is a known function, $N(u(t))$ is a nonlinear operator and *B* is a boundary operator. By means of OHAM one first constructs a set of equations:

$$\begin{aligned}&(1-p)[L(\phi(\tau,p)) + g(\tau)] - H(p)[L(\phi(\tau,p)) + g(\tau) + N(\phi(\tau,p))] = 0\\&B(\phi(\tau,p)) = 0,\end{aligned} \tag{96}$$

where $p \in [0,1]$ is an embedding parameter, $H(p)$ denotes a nonzero auxiliary function for $p \neq 0$ and $H(0) = 0$, $\phi(\tau,p)$ is an unknown function. Obviously, when $p = 0$ and $p = 1$, it holds that:

$$\phi(\tau,0) = u_0(\tau),\ \phi(\tau,1) = u(\tau). \tag{97}$$

Thus, as *p* increases from *0* to *1*, the solution $\phi(\tau,p)$ varies from $u_0(\tau)$ to the solution $u(\tau)$, where $u_0(\tau)$ is obtained from Equation (96) for $p = 0$:

$$L(u_0(\tau)) + g(\tau) = 0,\ B(u_0) = 0. \tag{98}$$

We choose the auxiliary function $H(p)$ in the form:

$$H(p) = pC_1 + p_2C_2 + \ldots \tag{99}$$

where $C_1, C_2, \ldots$ are constants which can be determined later.

Expanding $\phi(\tau,p)$ in a series with respect to *p*, one has:

$$\phi(\tau,p,C_i) = u_0(\tau) + \sum_{k\geq 1} u_k(\tau,C_i)p_k,\ i = 1,2,\ldots \tag{100}$$

Substituting Equation (100) into Equation (86), collecting the same powers of *p*, and equating each coefficient of *p* to zero, we obtain set of differential equation with boundary conditions. Solving differential equations by boundary conditions $u_0(\tau), u_1(\tau,C_1), u_2(\tau,C_2), \ldots$ are obtained. Generally speaking, the solution of Equation (95) can be determined approximately in the form:

$$\tilde{u}^{(m)} = u_0(\tau) + \sum_{k=1}^{m} u_k(\tau, C_i). \tag{101}$$

Note that the last coefficient C_m can be function of τ. Substituting Equation (98) into Equation (95), there results the following residual:

$$R(\tau, C_i) = L(\tilde{u}^{(m)}(\tau, C_i)) + g(\tau) + N(\tilde{u}^{(m)}(\tau, C_i)). \tag{102}$$

If $R(\tau, C_i) = 0$ then $\tilde{u}^{(m)}(\tau, C_i)$ happens to be the exact solution. Generally such a case will not arise for nonlinear problems, but we can minimize the functional by Galerkin method (GM):

$$w_i = \frac{\partial R(\tau, C_1, C_2, \ldots, C_m)}{\partial C_i}, i = 1, 2, \ldots, m, \tag{103}$$

The unknown constants $C_i (i = 1, 2, \ldots, m)$ can be identified from the conditions:

$$J(C_1, C_2) = \int_a^b w_i . R(\tau, C_1, C_2, \ldots, C_m) d\tau = 0, \tag{104}$$

where *a* and *b* are two values, depending on the given problem. With these constants, the approximate solution (of order *m*) (Equation (101)) is well determined. It can be observed that the method proposed in this work generalizes these two methods using the special (more general) auxiliary function $H(p)$.

2.2.2. Application of OHAM

In this section, OHAM is applied to nonlinear ordinary differential Equations (91) and (92). According to the OHAM, we have:

$$\begin{aligned} V^{IV} &= Ha^2 B^* (1-\phi)^{2.5} V'' + \operatorname{Re} A^* (1-\phi)^{2.5} \left[V'V'' - VV''' \right], \\ UV' - VU' &= \frac{1}{\operatorname{Re}} \frac{1}{A^* (1-\phi)^{2.5}} \left[U'' - Ha^2 B^* (1-\phi)^{2.5} U \right]. \end{aligned} \tag{105}$$

We consider $V, U, H_1(p)$ and $H_2(p)$ as following:

$$V = V_0 + pV_1 + p^2 V_2,$$

$$U = U_0 + pU_1 + p^2 U_2,$$

$$H_1(p) = pC_{11} + p^2C_{12},$$

$$H_2(p) = pC_{21} + p^2C_{22}. \tag{106}$$

Substituting $V, U, H_1(p)$ and $H_2(p)$ from Equation (106) into Equation (105) and some simplification and rearranging based on powers of p-terms, we have:

$$p^0 : V^{IV} = 0,$$

$$U'' = 0,$$

$$V_0(0) = 0, V_0'(0) = 0, V_0(1) = 1, V_0'(1) = 0$$

$$U_0(0) = 1, U_0(1) = 0 \tag{107}$$

$$p^1 :$$

$$V_1^{IV} + C_{11} V_0^{IV} - C_{11} \operatorname{Re} A^* \left(1-\phi\right)^{2.5} V_0'' V_0' + C_{11} \operatorname{Re} A^* \left(1-\phi\right)^{2.5} V_0''' V_0$$
$$-C_{11} Ha^2 B^* \left(1-\phi\right)^{2.5} V_0'' - V_0^{IV} = 0,$$

$$U_1'' - C_{21} \operatorname{Re} A^* \left(1-\phi\right)^{2.5} V_0' U_0 + C_{22} U_0'' - U_0'' + C_{21} \operatorname{Re} A^* \left(1-\phi\right)^{2.5} U_0' V_0$$
$$-C_{21} Ha^2 B^* \left(1-\phi\right)^{2.5} U_0 = 0,$$

$$V_1(0) = 0, \ V_1'(0) = 0, \ V_1(1) = 0, \ V_1'(1) = 0$$

$$U_1(0) = 0, \ U_1(1) = 0 \tag{108}$$
$$\vdots$$

Solving Equations (107) and(108) with boundary conditions:

$$V_0(y) = -2y^3 + 3y^2,$$

$$U_0(y) = -y + 1, \tag{109}$$

$$V_1(y) = C_{11}\,(0.05714285714\,\mathrm{Re}\,A^*\left(1-\phi\right)^{2.5} y^7 - 0.2y^6 - 0.1Ha^2\,B^*\left(1-\phi\right)^{2.5} y^5$$
$$+0.25Ha^2B^*\left(1-\phi\right)^{2.5} y^4 - 0.3857142857\,\mathrm{Re}\,A^*\left(1-\phi\right)^{2.5} y^3$$
$$-0.2\,Ha^2B^*\left(1-\phi\right)^{2.5} y^3 + (0.22855714286\,\mathrm{Re}\,A^*\left(1-\phi\right)^{2.5} + 0.05)\,y^2),$$

$$U(y) = C_{21}\,(0.2\,\mathrm{Re}\,y^5 - 0.75\,\mathrm{Re}\,A^*\left(1-\phi\right)^{2.5} y^4 + \mathrm{Re}\,A^*\left(1-\phi\right)^{2.5} y^3$$
$$-0.1667Ha^2\,B^*\left(1-\phi\right)^{2.5} y^3 + 0.5Ha^2B^*\left(1-\phi\right)^{2.5} y^2 \qquad (110)$$
$$-0.45\,\mathrm{Re}\,A^*\left(1-\phi\right)^{2.5} y - 0.3333Ha^2B^*\left(1-\phi\right)^{2.5} y)$$

The terms of $V_2(y)$ and $U_2(y)$ are too large that mentioned graphically. Therefore final expression for $V(y)$ and $U(y)$ is:

$$V(y) = V_0(y) + V_1(y) + V_2(y)$$
$$U(y) = U_0(y) + U_1(y) + U_2(y). \qquad (111)$$

By Substituting $V(y)$ and $U(y)$ into Equation (105), $R_1(\eta, C_{11}, C_{12})$ and $R_2(\eta, C_{21}, C_{22})$ are obtained then J_1 and J_2 are obtained in the flowing manner:

$$J(C_{11}, C_{12}) = \int_a^b w_i\,.\,R_1\,d\tau = 0, \qquad (112)$$

$$J(C_{21}, C_{22}) = \int_a^b w_i\,.\,R_2 d\tau = 0. \qquad (113)$$

The constants C_{11}, C_{12}, C_{21} and C_{22} obtain from Equations (112) and (113). By substituting these constants into Equation (111), an expression for $V(y)$ and $U(y)$ is obtained.

2.3. Effects of Active Parameters

In the present laminar nanofluid flow in a permeable channel in the presence of uniform magnetic field is studied (Figure 11). Optimal Homotopy Asymptotic Method (using Galerkin method to minimize the residual) is used in order to solve this problem.

Figure 12 shows the effects of various values of Hartmann number on $V(y)$ and $U(y)$. Generally, when the magnetic field is imposed on the enclosure, the velocity field suppressed owing to the retarding effect of the Lorenz force. For low Reynolds number, as Hartmann number increases $V(y)$ decreases for $y > y_m$ but opposite trend is observed for $y < y_m$, y_m is a meeting point that all curves joint together at this point. When Reynolds number increases this meeting point shifts to the solid wall and

Figure 12. Effect of various values of Hartmann numbers (Ha) on $V(y)$ and $U(y)$, when $\phi = 0.06$.

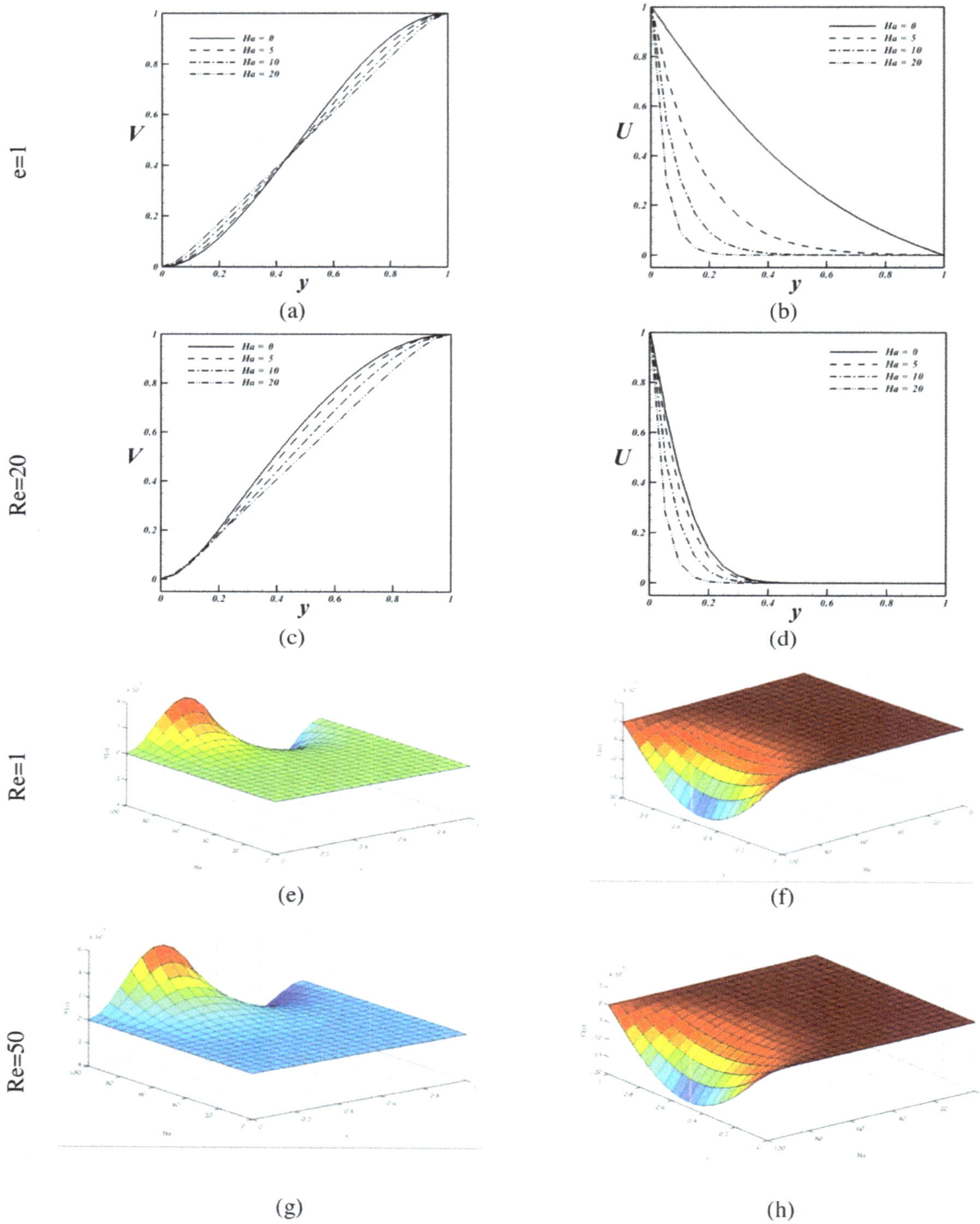

it can be seen that $V(y)$ decreases with increase of Hartmann number. As Hartmann number increases $U(y)$ decreases for all values of Reynolds number. Besides, this figure shows that this change is more pronounced for low Reynolds numbers.

Figure 13 shows the effects of various values of Reynolds numbers (Re) on $V(y)$ and $U(y)$. It is worth to mention that the Reynolds number indicates the relative significance of the inertia effect compared to the viscous effect. Thus, velocity profile decreases as Re increases and in turn increasing Re leads to increase in the magnitude of the skin friction coefficient. With increasing Reynolds number, $V(y)$ and $U(y)$ increase. These effects become less at higher Hartmann numbers because of retarding flow owing to Lorenz forces. Also it shows that increasing Hartmann number leads to increasing the curve of velocity profile.

3. HEATED PERMEABLE STRETCHING SURFACE IN A POROUS MEDIUM USING NANOFLUID

3.1. Problem Definition

Consider the steady, two-dimensional flow of a nanofluid near the stagnation point on a stretching sheet saturated at a porous surface (highly permeable) as shown in Figure 14 (Sheikholeslami and Ganji, 2014). The stretching velocity $U_w(x)$ and the free stream velocity $U_\infty(x)$ are assumed to vary proportional to the distance x from the stagnation point, i.e. $U_w(x) = ax$ and $U_\infty(x) = bx$, where a and b are constants with $a > 0$ and $b \geq 0$. It is assumed that the temperature at the stretching surface takes the constant values T_w, while the temperature of the ambient nanofluid, attained as y tends to infinity, takes the constant values T_∞. The fluid is a water based nanofluid containing different types of nanoparticles: Cu, Al_2O_3, Ag and TiO_2. It is assumed that the base fluid and the nanoparticles are in thermal equilibrium and no slip occurs between them.

Under these assumptions:

$$\frac{\partial u}{\partial x} + \frac{\partial v}{\partial y} = 0, \tag{114}$$

$$\rho_{nf}\left(u\frac{\partial u}{\partial x} + v\frac{\partial u}{\partial y} - U_\infty \frac{dU_\infty}{dx}\right) = \mu_{nf}\frac{\partial^2 u}{\partial y^2} + \frac{\mu_{nf}}{K}\left(U_\infty - u\right), \tag{115}$$

$$\left(\rho C_p\right)_{nf}\left(u\frac{\partial T}{\partial x} + v\frac{\partial T}{\partial y}\right) = \frac{\partial}{\partial y}\left(k_{nf}^*\frac{\partial T}{\partial y}\right), k_{nf}^* = k_{nf}(1 + \varepsilon\theta) \tag{116}$$

Subject to the boundary conditions

Figure 13. Effects of various values of Reynolds numbers (Re *) on* $V(y)$ *and* $U(y)$*, when* $\phi = 0.06$ *.*

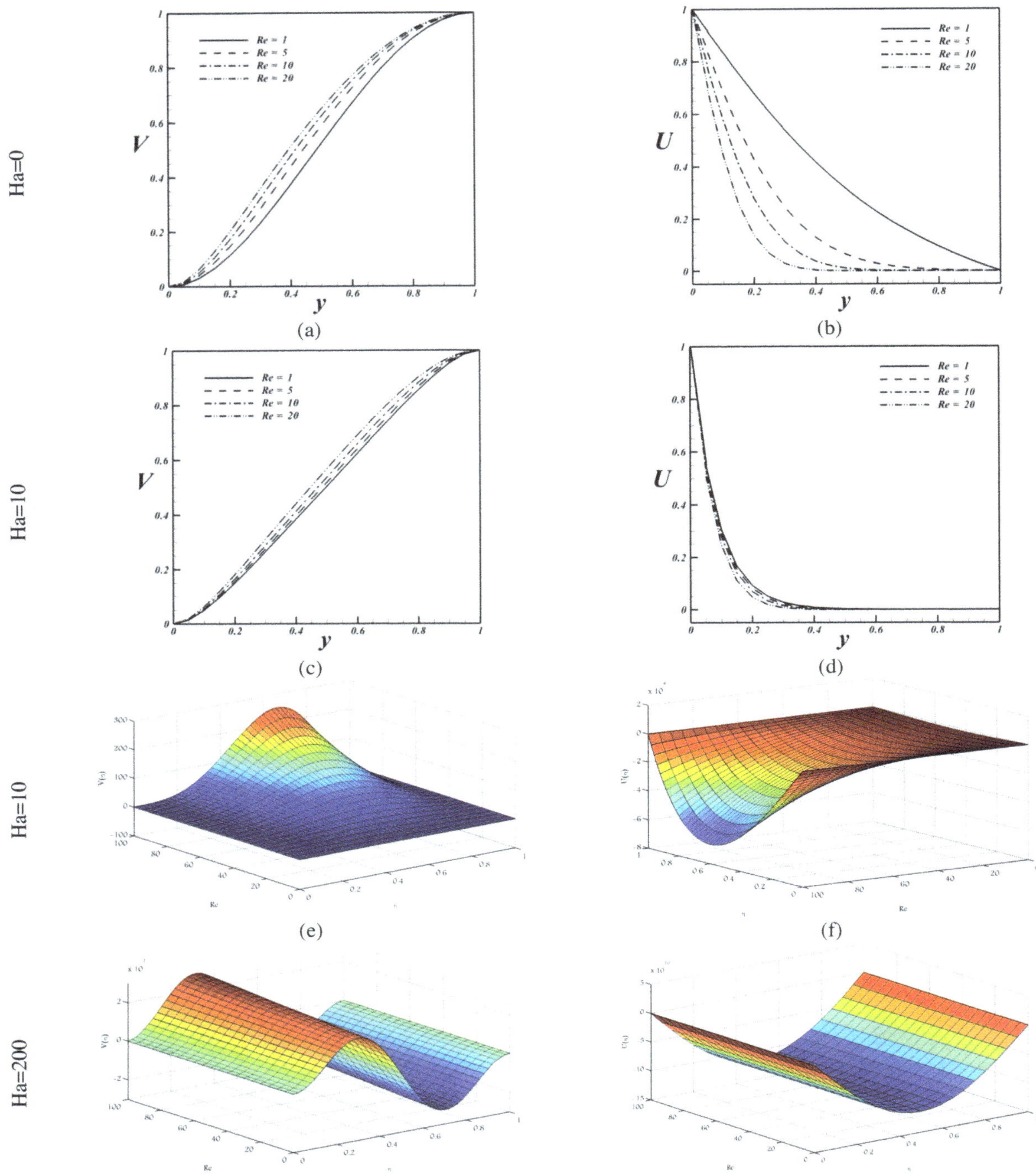

Figure 14. Figure of geometry

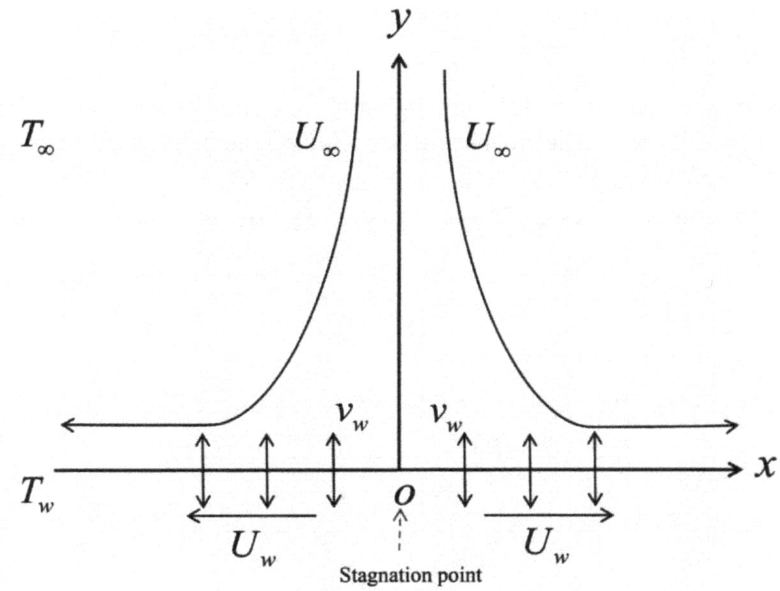

$$\begin{aligned} &y = 0 : u = U_w\left(x\right), v = v_w, T = T_w \\ &y \to \infty : u \to U_\infty\left(x\right), T \to T_\infty \end{aligned} \tag{117}$$

where u and v are the velocity components along the *x* and *y* axes, respectively, T is fluid temperature, ε is Thermal conductivity parameter and K is the permeability of the porous medium. Also, v_w is the wall mass flux with $v_w < 0$ for suctions and $v_w > 0$ for injection, respectively.

The effective density ρ_{nf}, the effective dynamic viscosity μ_{nf}, the heat capacitance $(\rho C_p)_{nf}$ and the are given as:

$$\rho_{nf} = \rho_f(1-\phi) + \rho_s\phi \tag{118}$$

$$\mu_{nf} = \frac{\mu_f}{(1-\phi)^{2.5}} \tag{119}$$

$$\left(\rho C_p\right)_{nf} = \left(\rho C_p\right)_f(1-\phi) + \left(\rho C_p\right)_s\phi \tag{120}$$

Effective thermal conductivity (k_{nf}) can be incorporated from the following expression (Sheikholeslami and Ganji, 2014):

$$\frac{k_{nf}}{k_f} = \frac{k_s + (n-1)k_f - (n-1)\phi(k_f - k_s)}{k_s + (n-1)k_f + \phi(k_f - k_s)} \tag{121}$$

where n is the empirical shape factor for the nanoparticle. In particular, $n = 3$ for spherical shaped nanoparticles and $n = 3/2$ for cylindrical ones. Models of nanofluid based on different formulas for thermal conductivity and dynamic viscosity is shown in Table 5.

The continuity equation (114) is satisfied by introducing a stream function ψ such that

$$u = \frac{\partial \psi}{\partial y}$$

and

$$v = -\frac{\partial \psi}{\partial x}. \tag{122}$$

The momentum and energy equations can be transformed into the corresponding ordinary differential equations by the following transformation:

$$\eta = \left(\frac{a}{v}\right)^{1/2} y, f(\eta) = \frac{\psi}{\left(av\right)^{1/2} x}, \theta(\eta) = \frac{T - T_\infty}{T_w - T_\infty}. \tag{123}$$

Using model *I*, the transformed ordinary differential equations are:

$$f''' + ff'' - f'^2 + \lambda^2 + \frac{K_1}{A_1 \cdot (1-\phi)^{2.5}}\left(\lambda - f'\right) = 0, \tag{124}$$

Table 5. Models of nanofluid based on different formulas for thermal conductivity and dynamic viscosity.

Model	Shape of Nanoparticles	Thermal Conductivity	Dynamic Viscosity
I	Spherical	$\frac{k_{nf}}{k_f} = \frac{k_s + 2k_f - 2\phi(k_f - k_s)}{k_s + 2k_f + \phi(k_f - k_s)}$	$\mu_{nf} = \frac{\mu_f}{(1-\phi)^{2.5}}$
II	Cylindrical (nanotubes)	$\frac{k_{nf}}{k_f} = \frac{k_s + (1/2)k_f - (1/2)\phi(k_f - k_s)}{k_s + (1/2)k_f + \phi(k_f - k_s)}$	$\mu_{nf} = \frac{\mu_f}{(1-\phi)^{2.5}}$

$$\frac{1}{\Pr}\cdot\frac{A_1 . A_3 .\left(1-\phi\right)^{2.5}}{A_2}\left(\left(1+\varepsilon\theta\right)\theta''+\varepsilon\left(\theta'\right)^2\right)+f\theta'=0. \tag{125}$$

subject to the boundary conditions (4) which become

$$\begin{gathered} f(0)=\frac{-v_w}{\left(av\right)^{0.5}}=\gamma, f'(0)=1, \theta(0)=1, \\ f'\left(\infty\right)\to\lambda, \theta\left(\infty\right)\to 0. \end{gathered} \tag{126}$$

Here prime denote differentiation with respect to η, $\lambda=b/a$ is the Velocity ratio parameter, $\Pr=\left(\mu_f\left(\rho C_p\right)_f\right)/\left(\rho_f k_f\right)$ is the Prandtl number and $K_1=\mu_f/\left(\rho_f K a\right)$ is the Permeability parameter and A_1, A_2, A_3 are parameters having the following form:

$$A_1=(1-\phi)+\frac{\rho_s}{\rho_f}\phi \tag{127}$$

$$A_2=(1-\phi)+\frac{(\rho C_p)_s}{(\rho C_p)_f}\phi \tag{128}$$

$$A_3=\frac{k_{nf}}{k_f}=\frac{k_s+2k_f-2\varphi(k_f-k_s)}{k_s+2k_f+\varphi(k_f-k_s)} \tag{129}$$

The quantities of practical interest in this study are the skin friction coefficient $\left(C_f\right)$ and the local Nusselt number $\left(Nu_x\right)$, which are defined as

$$C_f=\frac{\mu_{nf}}{\rho_f U_w^2}\left(\frac{\partial u}{\partial y}\right)_{y=0}, Nu=\frac{x\,k_{nf}}{k_f\left(T_w-T_\infty\right)}\left(-\frac{\partial T}{\partial y}\right)_{y=0}, \tag{130}$$

with k_{nf} being the thermal conductivity of the nanofluid.

Therefore, the skin friction coefficient and the local Nusselt number can be expressed as

$$\tilde{C}_f=\frac{1}{2}C_f \mathrm{Re}_x^{0.5}=\frac{1}{\left(1-\phi\right)^{2.5}}f''(0), \tilde{N}u_x=Nu_x \mathrm{Re}_x^{-0.5}=-\frac{k_{nf}}{k_f}\theta'(0), \tag{131}$$

where $\mathrm{Re}_x = \rho_f U_w x / \mu_f$ is the local Reynolds number based on the stretching velocity. The quantities $\tilde{C}_f$ and $\tilde{N}u_x$ are referred as the reduced skin friction coefficient and reduced local Nusselt number, respectively.

3.2. Effects of Active Parameters

The ordinary differential equations with the boundary conditions have been solved numerically for some values of the governing parameters Thermal conductivity parameter, volume fraction of the nanoparticles, Permeability parameter, suction/injection parameter, Velocity ratio parameter and different kinds of nanoparticles using the fourth order Runge–Kutta.

Effects of Thermal conductivity parameter (ε) and nanoparticle volume fraction (ϕ) on Temperature distribution and Nusselt number when $K_1 = 0.1$ and $\mathrm{Pr} = 6.2$ is shown in Figure 15. Because of increasing in the thermal conductivity due to increase in ε , the thermal boundary layer thickness increases, thus decreases heat transfer rate at the surface. When the volume fraction of the nanoparticles increases from 0 to 0.2, the thermal boundary layer is increased. This agrees with the physical behavior in that when the volume fraction of copper increases the thermal conductivity increases, and then the thermal boundary layer thickness increases, hence increases heat transfer rate at the surface.

Figure 16 graphical representation of the effect of the suction/injection parameter (γ) for Cu-water on Velocity profile and Temperature distribution for two different values of Velocity ratio parameter ($\lambda = 0.1$ and $\lambda = 2$, when $K_1 = 0.1, \phi = 0.1, \varepsilon = 0.1$ and $\mathrm{Pr} = 6.2$ .We know that the effect of suction is to bring the fluid closer to the surface and, therefore, to reduce the thermal boundary layer thickness and in turn increases the Nusselt number, but opposite trend is observed for injection. When stronger injection is provided, the heated Cu-water is pushed less from the wall than for a regular fluid ($\phi = 0$), i.e. the existence of the nanoparticle leads to a small increase of the velocity profiles, but for suction, it is noted an opposite behavior. It is clear that the thermal boundary layer thickness for the injection case is greater than for suction. When $\lambda > 1$, the flow has a boundary layer structure and when $\lambda < 1$ the flow has an inverted boundary layer structure, which results from the fact that when $\left(b / a\right) < 1$, the stretching velocity ax of the surface exceeds the velocity bx of the external stream. It is to be noted that no momentum boundary layer is formed when $\lambda = 1$. It is seen from Figure 16, the thermal boundary layer thickness for $\lambda < 1$ is greater than for $\lambda > 1$. For $\lambda = 0.1$, the effect of suction is to decrease the velocity profile, whereas the effect of injection is to increase this profile. Also it can be found that for both cases, all boundary layer thicknesses increase with increasing values of suction/injection parameter.

Table 6 shows that the effects of Permeability parameter (K_1), Suction/injection parameter (γ), Velocity ratio parameter (λ) and nanoparticle volume fraction (ϕ) on skin friction coefficient when $\varepsilon = 0.1$ and $\mathrm{Pr} = 6.2$. Physically, negative sign of $\tilde{C}_f$ implies that the stretching tube exerts a dragging force on the fluid and positive sign implies the opposite. For both suction ($\gamma > 0$) or injection ($\gamma < 0$) and both cases of λ ($\lambda = 0.1$ or $\lambda = 2$) it can be seen that the absolute values of skin friction increases due to increase in Permeability parameter and Suction/injection parameter, while it decreases due to decrease in nanoparticle volume fraction.

Figure 15. Effect of Thermal conductivity parameter (ε) and nanoparticle volume fraction (ϕ) on Temperature distribution and Nusselt number when $K_1 = 0.1$ and $\Pr = 6.2$.

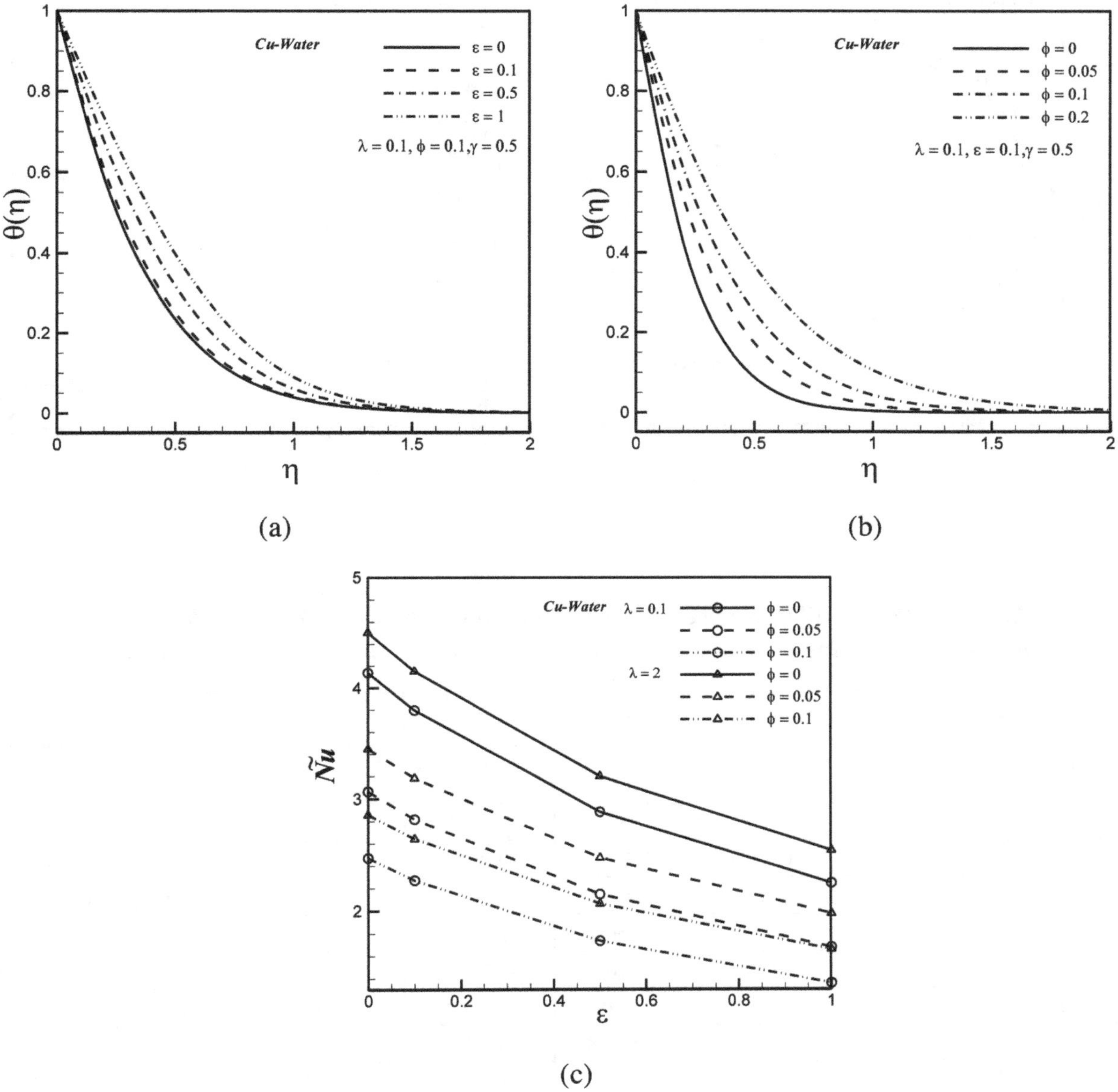

Table 7 shows that the effects of Permeability parameter (K_1), Suction/injection parameter (γ), Velocity ratio parameter (λ) and nanoparticle volume fraction (ϕ) on Nusselt number when $\varepsilon = 0.1$ and $\Pr = 6.2$ for Cu-Water. When $\lambda = 0.1$ and $\lambda = 2$, for suction Nusselt number decreases with increasing in nanoparticle volume fraction, whereas the opposite trend is observed for injection. For both suction or injection, Nusselt number increases due to increase the Permeability parameter when $\lambda = 2$, but opposite behavior is noted when $\lambda = 0.1$. However, increasing in γ leads to increase in Nusselt number for both the cases of λ.

Figure 16. Effect of Suction or injection parameter (γ) and nanoparticle volume fraction (ϕ) on Velocity profile and Temperature distribution when $K_1 = 0.1, \phi = 0.1, \varepsilon = 0.1$ and $\Pr = 6.2$.

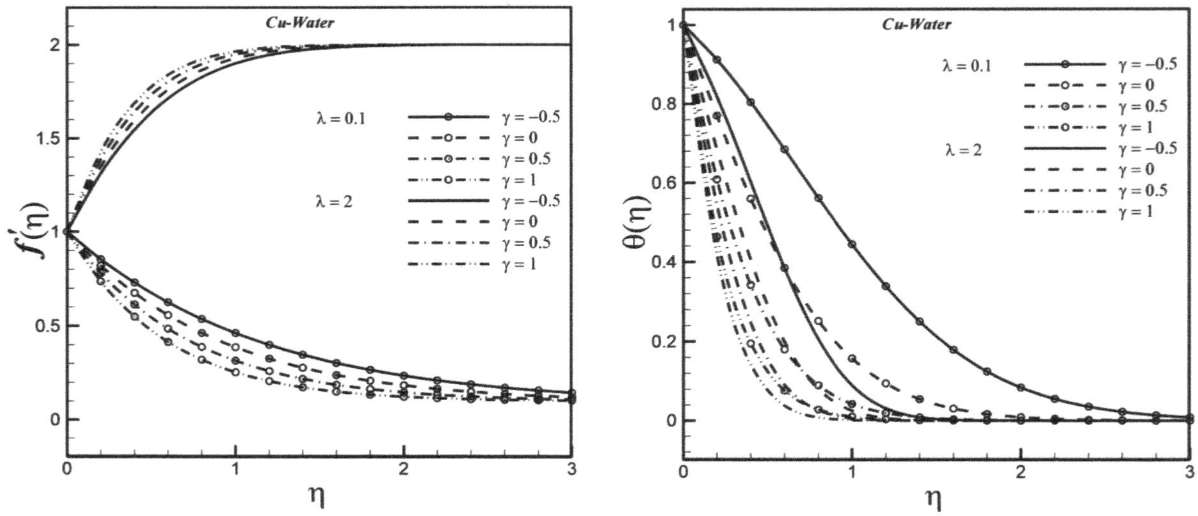

(a) (b)

Table 6. Effects of Permeability parameter, Suction/injection parameter, Velocity ratio parameter and nanoparticle volume fraction on skin friction coefficient when $\varepsilon = 0.1$ and $\Pr = 6.2$ for Cu-Water.

K_1	γ	$\lambda = 0.1$			$\lambda = 2$		
		ϕ					
		0	0.05	0.1	0	0.05	0.1
0.1	-0.5	-0.80524	-0.79792	-0.79443	1.778452	1.773849	1.771663
0.1	0	-1.01007	-1.00264	-0.9991	2.041795	2.037279	2.035133
0.1	0.5	-1.26387	-1.25665	-1.25322	2.333717	2.329335	2.327254
0.5	0.5	-1.40827	-1.37611	-1.36056	2.425567	2.404448	2.394366
1	0.5	-1.5677	-1.5104	-1.48234	2.535495	2.495021	2.475585
0.5	-0.5	-0.95151	-0.91898	-0.90323	1.874645	1.852573	1.842026
1	-0.5	-1.11245	-1.05467	-1.02635	1.989145	1.947061	1.926822

Table 8 shows that the effects of Velocity ratio parameter (λ) and nanoparticle volume fraction (ϕ) on skin friction coefficient and Nusselt number when $\varepsilon = 0.1, K_1 = 0.1$ and $\Pr = 6.2$ for Cu-Water. For both suction ($\gamma > 0$) and injection ($\gamma < 0$), when $\lambda < 1$ absolute values of skin friction decreases with increasing in λ, while it increases when $\lambda > 1$. Also, for both suction and injection and any values of λ it can be found that Nusselt number increases as Velocity ratio parameter increases.

Table 7. Effects of Permeability parameter, Suction/injection parameter and nanoparticle volume fraction on Nusselt number when $\varepsilon = 0.1$ *and* $\Pr = 6.2$ *for Cu-Water.*

K_1	γ	$\lambda = 0.1$			$\lambda = 2$		
		ϕ					
		0	0.05	0.1	0	0.05	0.1
0.1	-0.5	0.336449	0.374097	0.383887	0.740698	0.783077	0.788656
0.1	0	1.657345	1.351415	1.16013	2.121312	1.796571	1.591408
0.1	0.5	3.797954	2.817711	2.272061	4.151551	3.183882	2.643174
0.5	0.5	3.782502	2.804163	2.259577	4.156554	3.18765	2.646277
1	0.5	3.765819	2.789282	2.245761	4.16241	3.1921	2.649959
0.5	-0.5	0.305268	0.346851	0.359213	0.747181	0.78797	0.792681
1	-0.5	0.271571	0.316794	0.331823	0.754629	0.793661	0.79739

Table 8. Effects of Velocity ratio parameter and nanoparticle volume fraction on skin friction coefficient and Nusselt number when $\varepsilon = 0.1, K_1 = 0.1$ *and* $\Pr = 6.2$ *for Cu-Water.*

γ	λ	$\tilde{C}_f$			$\tilde{Nu}_x$		
		ϕ					
		0	0.05	0.1	0	0.05	0.1
0.5	0.1	-1.26387	-1.25665	-1.25322	3.797954	2.817711	2.272061
	0.2	-1.17913	-1.17312	-1.17025	3.810096	2.831477	2.287149
	1.7	1.552826	1.54958	1.548038	4.089999	3.123066	2.583777
	2	2.333717	2.329335	2.327254	4.151551	3.183882	2.643174
-0.5	0.1	-0.80524	-0.79792	-0.79443	0.336449	0.374097	0.383887
	0.2	-0.76547	-0.75929	-0.75635	0.350128	0.38932	0.400184
	1.7	1.165817	1.162405	1.160785	0.67334	0.717759	0.725932
	2	1.778452	1.773849	1.771663	0.740698	0.783077	0.788656

Table 9 shows variation Variation in Nusselt number with different base fluids (Water $\Pr = 6.2$ and Ethylene glycol $\Pr = 203.6$) when $K_1 = 0.1, \gamma = 0.5, \lambda = 0.1$. For fixed values of nanoparticle volume fraction, selecting Ethylene glycol instead of Water as base fluid lead to decrease thermal boundary layer thickness, so, Cu-Ethylene glycol has higher Nusselt number than Cu- Water due to lower thermal conductivity. Tables 10 and 11 display the behavior of the skin friction coefficient and Nusselt number using different nanofluids. Figure 17 (a) shows that temperature distribution for different types of nanofluids when $M = 1, S = 0.1, n = 1, \lambda = 0.1, \phi = 0.1$ and $\Pr = 6.2$. These tables show that by using different types of nanofluid the value of $\tilde{C}_f$, $\tilde{Nu}_x$ and temperature profile change. This means that the

Table 9. Variation in Nusselt number with different base fluids (Water $\Pr = 6.2$ *and Ethylene glycol* $\Pr = 203.6$ *) when* $K_1 = 0.1, \gamma = 0.5, \lambda = 0.1$.

ε	Cu - Water			Cu – Ethylene glycol		
	ϕ					
	0	**0.05**	**0.1**	**0**	**0.05**	**0.1**
0	4.138303	3.066625	2.470897	103.7016	76.98598	62.00115
0.1	3.797954	2.817711	2.272061	94.3584	70.06946	56.44514
0.5	2.883762	2.147712	1.735969	69.44076	51.62293	41.62623
1	2.245959	1.678457	1.359372	52.307	38.93695	31.43358

Table 10. Effects of the nanoparticle volume fraction for different types of nanofluids on skin friction coefficient when $M = 1$ *and* $\lambda = 0.1$.

λ	ϕ	Nanoparticles			
		Cu	**Ag**	**Al_2O_3**	**TiO_2**
0.1	0.05	-1.26387	-1.25498	-1.26346	-1.263
	0.1	-1.25322	-1.25095	-1.26396	-1.26315
2	0.05	2.329335	2.328318	2.333467	2.333187
	0.1	2.327254	2.325883	2.333774	2.333277

Table 11. Effects of the nanoparticle volume fraction for different types of nanofluids on Nusselt number when $S = 0.1, n = 1, M = 1, \lambda = 0.1$ *and* $\Pr = 6.2$.

λ	ϕ	Nanoparticles			
		Cu	**Ag**	**Al_2O_3**	**TiO_2**
0.1	0.05	3.797954	2.668673	3.312571	3.329105
	0.1	2.272061	2.082538	2.961854	2.995371
2	0.05	3.183882	3.036321	3.67321	3.689447
	0.1	2.643174	2.45449	3.32729	3.36026

nanofluids will be important in the cooling and heating processes. Choosing Titanium oxide as the nanoparticle leads to the maximum amount rate of hate transfer, while selecting Silver leads to the minimum amount of it. Also, choosing Alumina as the nanoparticle leads to the maximum amount skin friction coefficient, while selecting Silver leads to the minimum amount of it. In Figure 17(b), the Nusselt numbers versus the volume fraction of nanoparticle is shown when different models of nanofluid based on different formulas for thermal conductivity and dynamic viscosity is used. For all amount of λ, model *II* for nanotubes has higher Nusselt number than model *I* for spherical shaped nanoparticles due to lower thermal conductivity.

Figure 17. (a) Temperature distribution for different types of nanofluids when $K_1 = 0.1, \phi = 0.1, \varepsilon = 0.1, \gamma = 0.5, \lambda = 0.1$ *and* $\Pr = 6.2$ *; (b) Variation in Nusselt number with nanoparticle volume fraction for different models* $K_1 = 0.1, \varepsilon = 0.1, \gamma = 0.5$ *and* $\Pr = 6.2$.

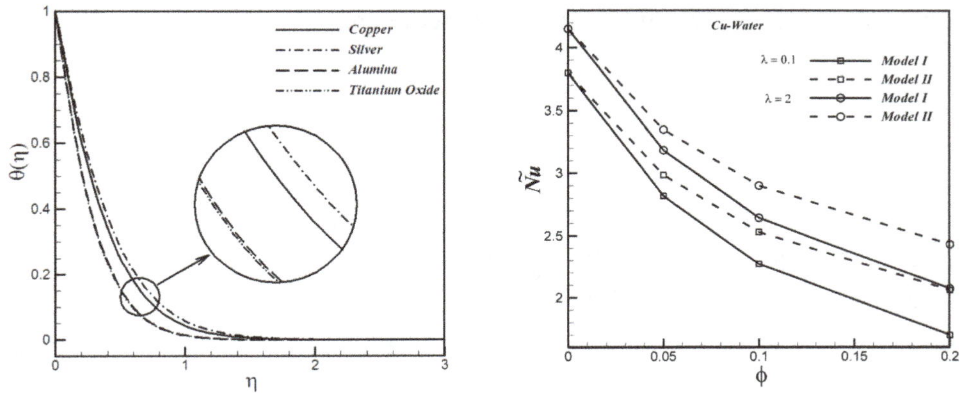

4. TWO PHASE MODELING OF NANOFLUID IN A ROTATING SYSTEM WITH PERMEABLE SHEET

4.1. Problem Definition

Consider the steady nanofluid flow between two horizontal parallel plates when the fluid and the plates rotate together around the *y*-axis which is normal to the plates with an angular velocity. A cartesian coordinate system is considered as follows: the *x*-axis is along the plate, the y-axis is perpendicular to it and the *z*-axis is normal to the *x y* plane (see Figure 18) (Sheikholeslami and Ganji, 2014). The upper plate is subjected to a constant wall injection velocity $v_0 (> 0)$, respectively. The plates are located at

Figure 18. Geometry of problem

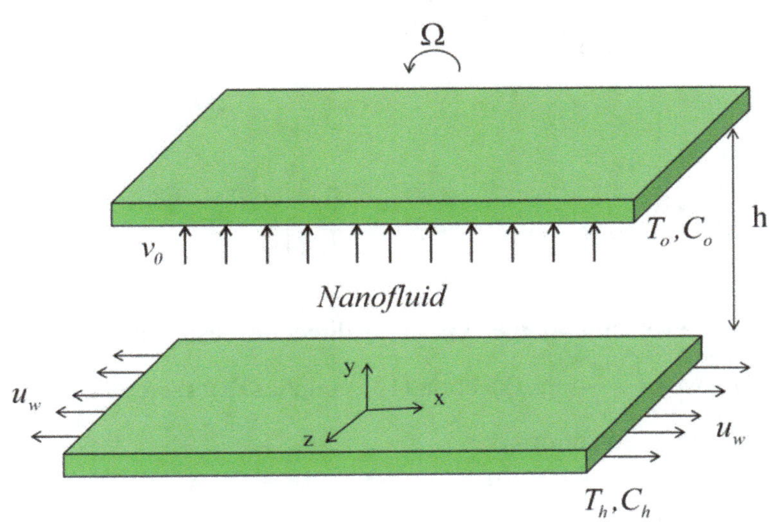

$y=0$ and $y=h$.The lower plate is being stretched by two equal and opposite forces so that the position of the point $\left(0,0,0\right)$ remains unchanged. The upper plate is subjected to a constant flow injection with a velocity v_0.

The governing equations in a rotating frame of reference are:

$$\frac{\partial u}{\partial x}+\frac{\partial v}{\partial y}+\frac{\partial w}{\partial z}=0 \tag{132}$$

$$\rho_f\left(u\frac{\partial u}{\partial x}+\nu\frac{\partial u}{\partial y}+2\Omega w\right)=-\frac{\partial p^*}{\partial x}+\mu\left(\frac{\partial^2 u}{\partial x^2}+\frac{\partial^2 u}{\partial y^2}\right), \tag{133}$$

$$\rho_f\left(u\frac{\partial v}{\partial y}\right)=-\frac{\partial p^*}{\partial y}+\mu\left(\frac{\partial^2 v}{\partial x^2}+\frac{\partial^2 v}{\partial y^2}\right), \tag{134}$$

$$\rho_f\left(u\frac{\partial w}{\partial x}+\nu\frac{\partial w}{\partial y}-2\Omega w\right)=\mu\left(\frac{\partial^2 w}{\partial x^2}+\frac{\partial^2 w}{\partial y^2}\right), \tag{135}$$

$$\begin{aligned} u\frac{\partial T}{\partial x}+v\frac{\partial T}{\partial y}+w\frac{\partial T}{\partial z}&=\alpha\left(\frac{\partial^2 T}{\partial x^2}+\frac{\partial^2 T}{\partial y^2}+\frac{\partial^2 T}{\partial z^2}\right)+\\ &+\frac{\left(\rho c_P\right)_p}{\left(\rho c_P\right)_f}\left[\begin{array}{l} D_B\left\{\frac{\partial C}{\partial x}\cdot\frac{\partial T}{\partial x}+\frac{\partial C}{\partial y}\cdot\frac{\partial T}{\partial y}+\frac{\partial C}{\partial z}\cdot\frac{\partial T}{\partial z}\right\} \\ +(D_T/T_c)\left\{\left(\frac{\partial T}{\partial x}\right)^2+\left(\frac{\partial T}{\partial y}\right)^2+\left(\frac{\partial T}{\partial z}\right)^2\right\}\end{array}\right], \end{aligned} \tag{136}$$

$$u\frac{\partial C}{\partial x}+v\frac{\partial C}{\partial y}+w\frac{\partial C}{\partial z}=D_B\left(\frac{\partial^2 C}{\partial x^2}+\frac{\partial^2 C}{\partial y^2}+\frac{\partial^2 C}{\partial z^2}\right)+\left(\frac{D_T}{T_0}\right)\left\{\frac{\partial^2 T}{\partial x^2}+\frac{\partial^2 T}{\partial y^2}+\frac{\partial^2 T}{\partial z^2}\right\} \tag{137}$$

Here u,v and z are the velocities in the x,y and z directions respectively. Also p^* is the modified fluid pressure. The absence of $\frac{\partial p^*}{\partial z}$ in Equation (135) implies that there is a net cross-flow along the z-axis. The relevant boundary conditions are:

$$u = ax, v = 0, w = 0, T = T_h, C = C_h \ at \ y = 0 \tag{138}$$

$$u = 0, v = v_0, w = 0, T = T_0, C = C_0 \ at \ y = +h$$

The following non-dimensional variables are introduced:

$$\eta = \frac{y}{h}, u = axf'(\eta), \nu = -ah\, f(\eta), w = axg(\eta)$$
$$\theta(\eta) = \frac{T - T_h}{T_0 - T_h}, \phi(\eta) = \frac{C - C_h}{C_0 - C_h} \tag{139}$$

where a prime denotes differentiation with respect to η.

By substituting (139) in equations (132)–(135), we have:

$$-\frac{1}{\rho h}\frac{\partial p^*}{\partial \eta} = a^2 x [f' - ff'' - \frac{f'''}{R} + \frac{2K_r}{R} g], \tag{140}$$

$$-\frac{1}{\rho h}\frac{\partial p^*}{\partial \eta} = a^2 h [ff' + \frac{1}{R} f''], \tag{141}$$

$$g'' - R(f'g - fg') + 2K_r f' = 0 \tag{142}$$

and the non-dimensional quantities are defined.

$$R = \frac{ah^2}{v}, K_r = \frac{\Omega h^2}{v} \tag{143}$$

Equation (139) with the help of (140) can be written as:

$$f''' - R[f'^2 - ff''] - 2K_r^2 g = A \tag{144}$$

Differentiation of Equation (142) with respect to η gives:

$$f^{iv} - R(f' f'' - ff'') - 2K_r g' = 0 \tag{145}$$

Therefore, the governing equations and boundary conditions for this case in non-dimensional form are given by:

$$f^{iv} - R(f'f'' - ff'') - 2K_r g' = 0 \tag{146}$$

$$g'' - R(f'g - fg') + 2K_r f' = 0 \tag{147}$$

Also equations (136) and (137) turn to:

$$\theta'' + \Pr Rf\theta' + Nb\phi'\theta' + Nt\theta'^2 = 0, \tag{148}$$

$$\phi'' + R.Scf\phi' + \frac{Nt}{Nb}\theta'' = 0, \tag{149}$$

With these boundary conditions:

$$\begin{aligned} &f = 0, f' = 1, g = 0, \theta = 1, \phi = 1 \, at \, \eta = 0 \\ &f = \lambda, f' = 0, g = 0, \theta = 0, \phi = 0 \, at \, \eta = 1 \end{aligned} \tag{150}$$

Other non-dimensional quantities are defined as:

$$\begin{aligned} &\lambda = v_0 / (a\,h), \Pr = \frac{\mu}{\rho_f \alpha}, Sc = \frac{\mu}{\rho_f D}, \\ &Nb = (\rho c)_p D_B (C_h) / \left((\rho c)_f \alpha\right), \\ &Nt = (\rho c)_p D_T (T_H) / [(\rho c)_f \alpha T_c]. \end{aligned} \tag{151}$$

Skin friction coefficient $\left(C_f\right)$ along the stretching wall and Nusselt number $\left(Nu\right)$ along the stretching wall are defined as

$$\tilde{C}_f = \left(Rx / h\right) C_f = f''\left(0\right), Nu = -\theta'(0) \tag{152}$$

4.2. Effects of Active Parameters

In this study, three dimensional two phase simulation of nanofluid flow and heat transfer is investigated. Fourth order Runge–Kutta integration scheme featuring a shooting technique with the MAPLE package is used to solve this problem.

Effect of injection parameter on velocity, temperature, concentration profiles is shown in Figure 19. As injection parameter increases, velocity profiles increase. Temperature profile decreases with increase of injection parameter while opposite trend is observed for concentration profile.

Figure 20 shows the effects of Reynolds number on velocity, temperature and concentration profiles. Effects of injection parameter and Reynolds number on Nusselt number is shown in Figure 21. It is worth to mention that the Reynolds number indicates the relative significance of the inertia effect compared to the viscous effect. Thus, both velocity and temperature boundary layer thicknesses decrease with increase of Reynolds number and in turn increasing Reynolds number leads to increase in the magnitude of the skin friction coefficient and Nusselt number. Also it can be seen that concentration profile increases with augment of Reynolds number. Also it can be seen that Nusselt number increases with increase of injection parameter.

Effects of Rotation parameter on velocity profile and Nusselt number are depicted in Figure 22 and Table 12, respectively. With increasing Rotation parameter, the transverse velocity increases. As Rota-

Figure 19. Effect of injection parameter on velocity profiles $\left(f, g\right)$, temperature profile $\left(\theta\right)$ and concentration profile $\left(\phi\right)$ when $R = 0.5, K_r = 0.5, Sc = 0.5, Nb = Nt = 0.1$ and $\Pr = 10$

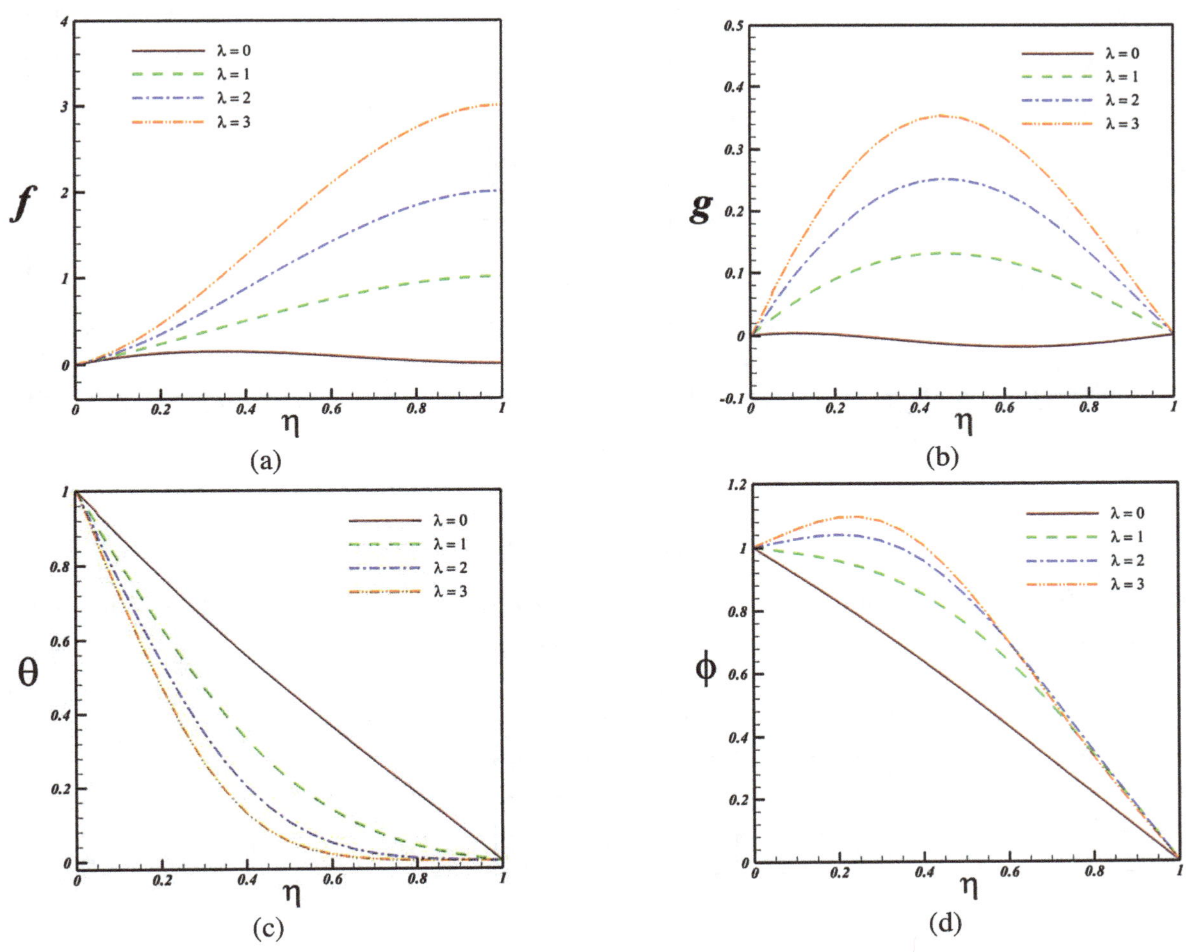

Figure 20. Effect of Reynolds number on velocity profiles (f,g), *temperature profile* (θ) *and concentration profile* (ϕ) *when* $\lambda = 0.5, K_r = 0.5, Sc = 0.5, Nb = Nt = 0.1$ *and* $\Pr = 10$

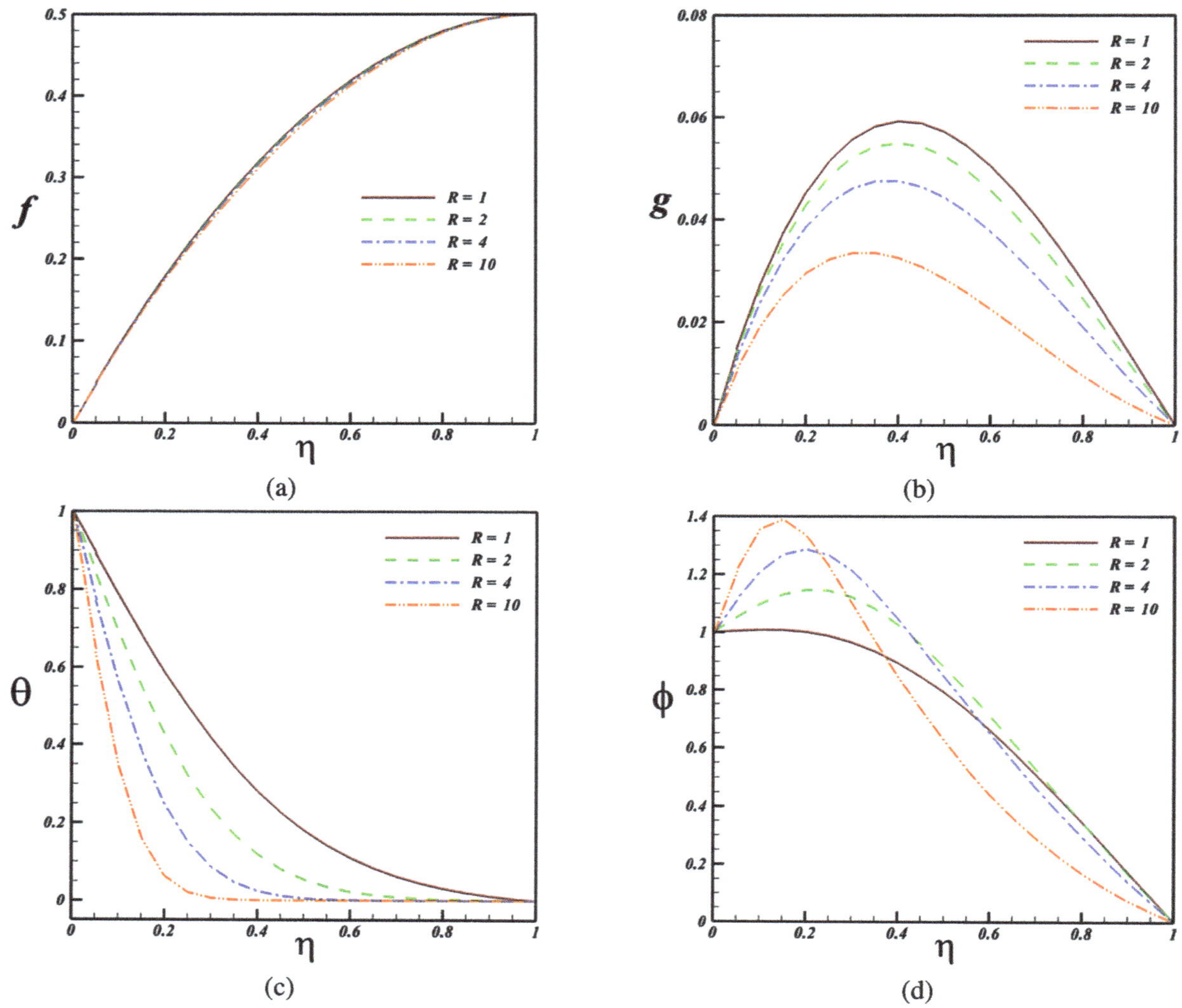

Table 12. Effect of Rotation parameter on Nusselt number when $Nt = 0.1, Nb = 0.1, Sc = 0.1, R = 1$

λ	K_r			
	0.5	**2**	**4**	**6**
1	2.633504	2.633507	2.633518	2.634296
2	3.271111	3.271753	3.274186	3.279422
3	3.745806	3.746079	3.747427	3.751176

Figure 21. Effects of injection parameter and Reynolds number on Nusselt number when $K_r = 0.5, Nt = 0.1, Nb = 0.1, Sc = 0.1$ *and* $\Pr = 10$

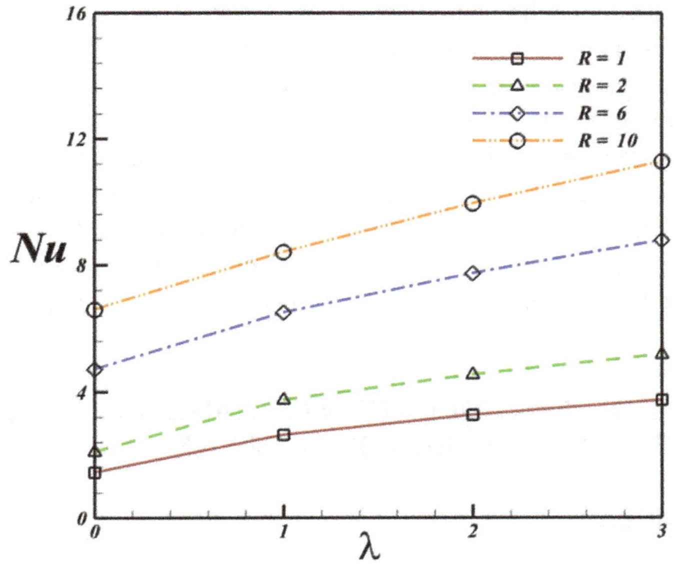

Figure 22. Effect of Rotation parameter on velocity profile when $R = 0.5, \lambda = 0.5, Sc = 0.5, Nb = Nt = 0.1.$

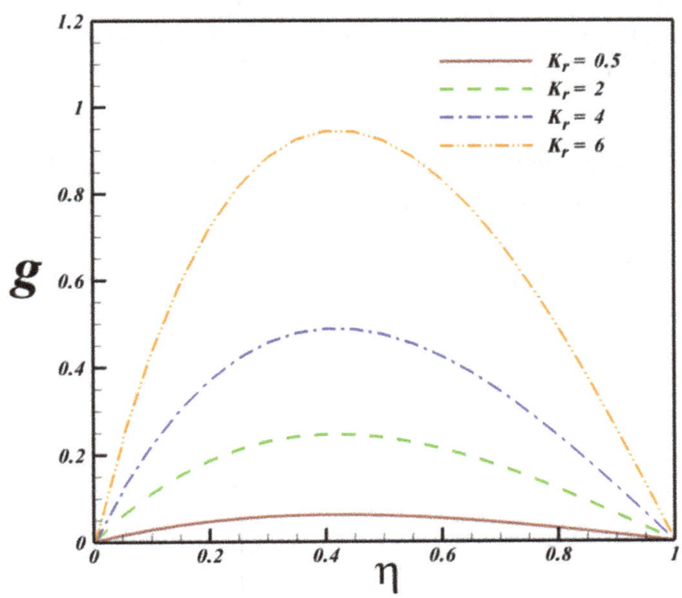

tion parameter increases thermal boundary layer thickness decreases and in turn Nusselt number increases with increase of K_r.

Figure 23 shows the effect of Schmidt number on concentration profile and Nusselt number. Schmidt number is a dimensionless number defined as the ratio of momentum diffusivity (viscosity) and mass diffusivity. So concentration profile decreases as Schmidt number increases. Also it can be concluded that increasing Schmidt number causes a slight decrease in rate of heat transfer.

Effects of Brownian parameter and Thermophoretic parameter on temperature, concentration profiles and Nusselt number are shown in Figure 24.Theses active parameters have similar effects on heat and mass transfer characteristics. It means that temperature boundary layer thickness increases with increase of them while opposite trend is observed for concentration boundary layer thickness. Nusselt number is a decreasing function of Thermophoretic parameter and Brownian parameter.

5. KKL CORRELATION FOR SIMULATION OF NANOFLUID FLOW AND HEAT TRANSFER IN A PERMEABLE CHANNEL

5.1. Problem Definition

The unsteady flow between two parallel flat plates is considered as shown in Figure 25 (Sheikholeslami, 2014). The wall, which coincides with the x axis, is stationary and heated externally. In order to cool the heated wall, cooled fluid is injected with velocity v_w uniformly from the other plate, which expands or contracts at a time-dependent rate $a\left(t\right)$. Take y to be perpendicular to the plates and assume u and v to be the velocity components in the x and y directions respectively. In this perspective the flow field may be assumed to be stagnation flow. The nanofluid is a two component mixture with the follow-

Figure 23. Effect of Schmidt number on concentration profile and Nusselt number when (a) $R = 0.5, \lambda = 0.5, K_r = 0.5, Nb = Nt = 0.1$; (b) $K_r = 0.5, Nt = 0.1, Nb = 0.1, R = 1$ and $\Pr = 10$

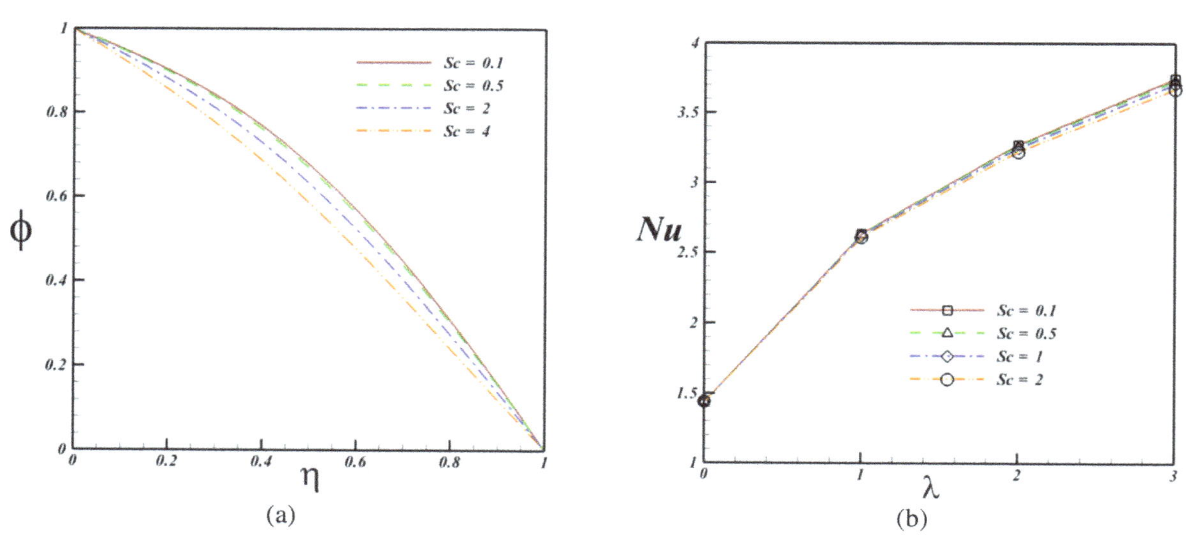

(a) (b)

Figure 24. Effects of Brownian parameter and thermophoretic parameter on temperature, concentration profileandNusseltnumberwhen(a,b) $R = 0.5, \lambda = 0.5, K_r = 0.5, Sc = 0.5$ *;(c)* $K_r = 0.5, Sc = 0.1, R = 1$ *and* $\Pr = 10$

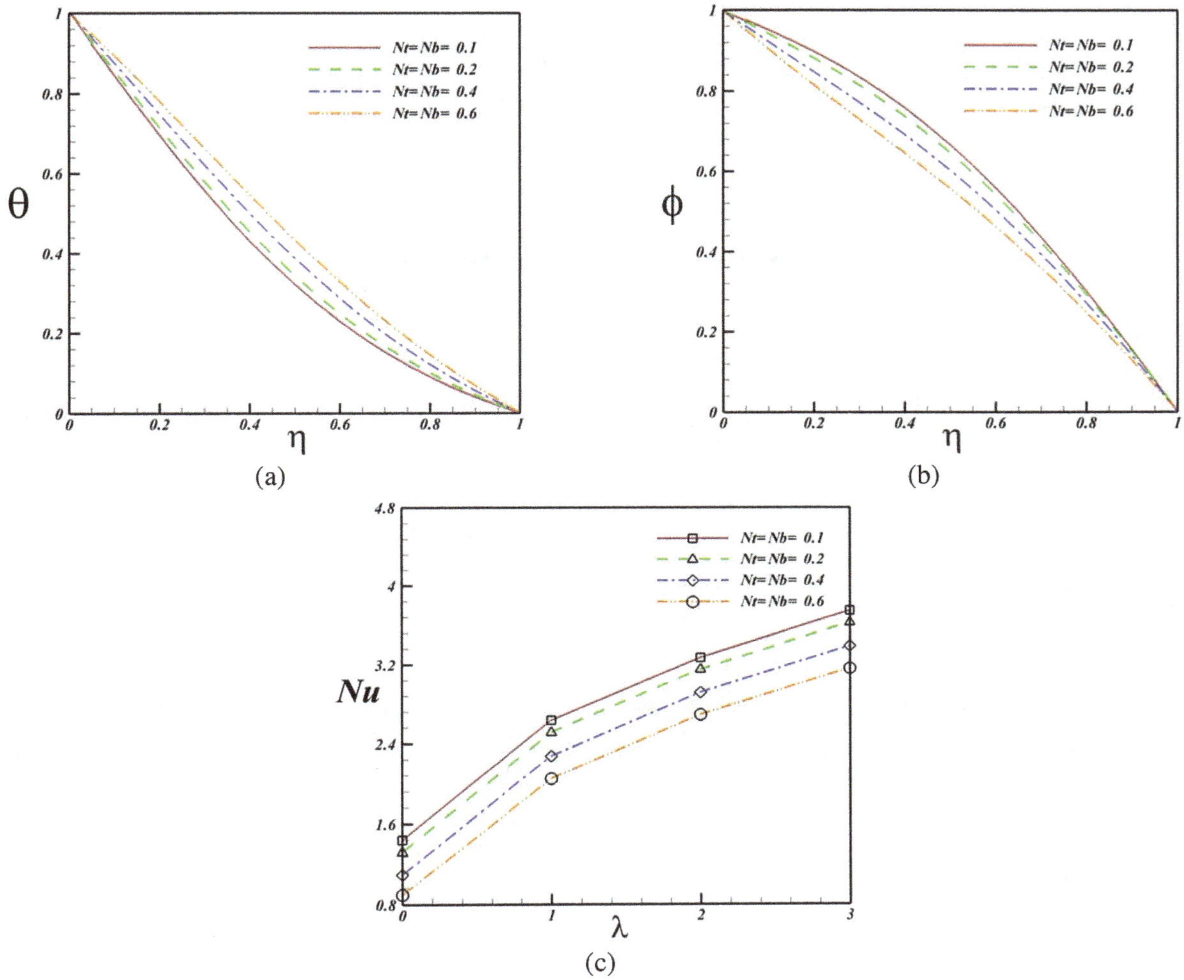

ing assumptions: Incompressible; No-chemical reaction; Negligible radiative heat transfer; Nano-solid-particles and the base fluid are in thermal equilibrium and no slip occurs between them. Under these assumptions, the Navier–Stokes equations are:

$$\frac{\partial u}{\partial x} + \frac{\partial v}{\partial y} = 0, \tag{153}$$

$$\rho_{nf}\left(\frac{\partial u}{\partial t} + u\frac{\partial u}{\partial v} + v\frac{\partial u}{\partial y}\right) = -\frac{\partial p}{\partial x} + \mu_{nf}\left(\frac{\partial^2 u}{\partial x^2} + \frac{\partial^2 u}{\partial y^2}\right), \tag{154}$$

Figure 25. Geometry of problem

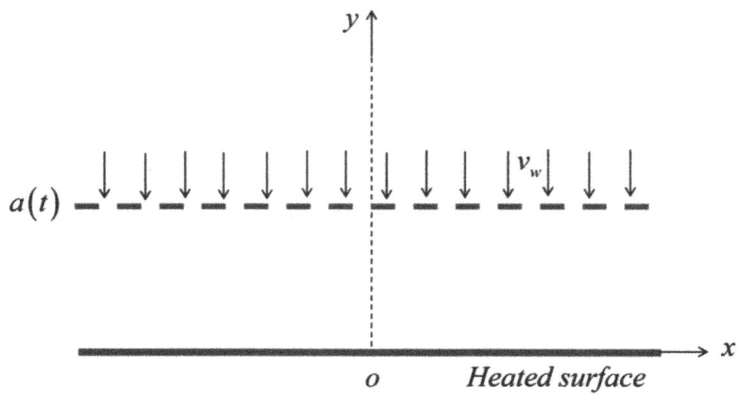

$$\rho_{nf}\left(\frac{\partial v}{\partial t}+u\frac{\partial v}{\partial v}+v\frac{\partial v}{\partial y}\right)=-\frac{\partial p}{\partial y}+\mu_{nf}(\frac{\partial^2 v}{\partial x^2}+\frac{\partial^2 v}{\partial y^2}), \tag{155}$$

$$\frac{\partial T}{\partial t}+u\frac{\partial T}{\partial x}+v\frac{\partial T}{\partial y}=\frac{k_{nf}}{\left(\rho C_p\right)_{nf}}(\frac{\partial^2 T}{\partial x^2}+\frac{\partial^2 T}{\partial y^2}). \tag{156}$$

Here u and v are the velocities in the x and y directions respectively, T is the temperature, P is the pressure, effective density ρ_{nf} and the effective heat capacity $\left(\rho C_p\right)_{nf}$ of the nanofluid are defined as:

$$\begin{aligned}&\rho_{nf}=(1-\phi)\rho_f+\phi\rho_p\\&(\rho C_p)_{nf}=(1-\phi)(\rho C_p)_f+\phi(\rho C_p)_p\end{aligned} \tag{157}$$

$\left(k_{nf}\right)$ and $\left(\mu_{nf}\right)$ are obtained according to Koo–Kleinstreuer–Li (KKL) model (Li, 2008):

$$\begin{aligned}&k_{nf}=1+\frac{3\left(\frac{k_p}{k_f}-1\right)\phi}{\left(\frac{k_p}{k_f}+2\right)-\left(\frac{k_p}{k_f}-1\right)\phi}+5\times10^4 g'(\phi,T,d_p)\phi\rho_f c_{p,f}\sqrt{\frac{\kappa_b T}{\rho_p d_p}}\\&g'\left(\phi,T,d_p\right)=\left(a_6+a_7Ln\left(d_p\right)+a_8Ln\left(\phi\right)+a_9Ln\left(\phi\right)\ln\left(d_p\right)+a_{10}Ln\left(d_p\right)^2\right)\\&+Ln\left(T\right)\left(a_1+a_2Ln\left(d_p\right)+a_3Ln\left(\phi\right)+a_4Ln\left(\phi\right)\ln\left(d_p\right)+a_5Ln\left(d_p\right)^2\right)\\&R_f+d_p/k_p=d_p/k_{p,eff},R_f=4\times10^{-8}km^2/W\end{aligned} \tag{158}$$

$$\mu_{nf} = \frac{\mu_f}{\left(1-\phi\right)^{2.5}} + \frac{k_{Brownian}}{k_f} \times \frac{\mu_f}{\Pr} \quad (159)$$

All needed coefficients and properties are illustrated in Tables 13 and 14 (Li, 2008).

The relevant boundary conditions are:

Table 13. Thermo physical properties of water and nanoparticles

	$\rho(kg/m^3)$	$C_p(J/kgK)$	$k(W/m.K)$	$d_p(nm)$
Pure water	997.1	4179	0.613	-
Al_2O_3	3970	765	25	47

(Li, 2008)

Table 14. The coefficient values of Al_2O_3 – Water nanofluids

Coefficient Values	Al_2O_3 – Water
a_1	52.813488759
a_2	6.115637295
a_3	0.6955745084
a_4	4.17455552786E-02
a_5	0.176919300241
a_6	-298.19819084
a_7	-34.532716906
a_8	-3.9225289283
a_9	-0.2354329626
a_{10}	-0.999063481

(Li, 2008)

$$u = 0, v = -v_w = -Aa^{\bullet}, T(a) = T_0 \; at \; y = a(t),$$
$$u = 0, v = 0; at \; y = 0. \tag{160}$$

where $A = v_w / a^{\bullet}$ is the measure of wall permeability, and T_0 is the temperature of the porous plate ($y = a$) which has the same temperature as that of the incoming coolant.

Introducing the stream function

$$\psi = \frac{vx}{a} F(\eta, t), \tag{161}$$

where $\eta = y / a$.

Substituting ψ into Equations (153)–(155) and eliminating the pressure term from the momentum equation, the following expression can be obtained:

$$\psi = \frac{v_{nf} x}{a} F(\eta, t), \tag{162}$$

We introduce these parameters:

$$F_{\eta\eta\eta\eta} + \frac{A_1}{A_2} \alpha \left(\eta F_{\eta\eta\eta} + 3F_{\eta\eta} \right) + FF_{\eta\eta\eta} - F_\eta F_{\eta\eta} - \frac{A_1}{A_2} a^2 v_f^{-1} F_{\eta\eta t} = 0, \tag{163}$$

where α is the wall expansion ratio defined by

Note that the expansion ratio will be positive for expansion and negative for contraction.

$$\alpha = \frac{aa^{\bullet}}{v_f} \tag{164}$$

And A_1 and A_2 are constant parameters that are defined as:

$$A_1 = \frac{\rho_{nf}}{\rho_f}, A_2 = \frac{\mu_{nf}}{\mu_f}, \tag{165}$$

The boundary conditions are

$$F_\eta(0) = 0, F(0) = 0, F_\eta(1) = 0, F(1) = \frac{A_1}{A_2} R, \tag{166}$$

where R is the Reynolds number defined by $R = av_w / v_f$. Note that R is positive for injection and negative for suction. For this model, we only consider the case that R is positive.

Let

$$f = \frac{F}{R}, l = \frac{x}{a} \tag{167}$$

A similar solution with respect to both space and time can be developed following the transformation (Dauenhauer and Majdalani, 2003; Majdalani, Zhou and Dawson, 2002). This can be accomplished by considering in the case: α is a constant and $f = f(\eta)$. It leads to $f_{\eta\eta t} = 0$. To realize this condition, the value of expansion ratio α must be specified by its initial value

$$\alpha = \frac{aa^{\bullet}}{v_f} = \frac{a_0 a_0^{\bullet}}{v_f} = cte, \tag{168}$$

where a_0 and $a_0^{\bullet}$ denote the initial channel height and expansion ratio, respectively. Integrating Equation (168) with respect to time, the similar solution can be achieved. The result is

$$\frac{a}{a_0} = \sqrt{1 + 2v_f \alpha t a_0^{-2}}. \tag{169}$$

For a physical setting in which the injection coefficient A is constant (Majdalani, Zhou and Dawson, 2002). Since $v_w = Aa^{\bullet}$, an expression for the injection velocity variation can be determined. From Equations (168) and (169), it is clear that

$$\frac{a}{a_0} = \frac{v_w(0)}{v_w(t)} = \sqrt{1 + 2v_f \alpha t a_0^{-2}}. \tag{170}$$

Under these assumptions, Equation (163) becomes

$$f^{iv} + \frac{A_1}{A_2}\left(\alpha\left(\eta f''' + 3f''\right) + R\left(ff''' - f' f''\right)\right) = 0, \tag{171}$$

with boundary conditions

$$f'(0) = 0, f(0) = 0, f'(1) = 0, f(1) = 1, \tag{172}$$

At a distance η form the wall, the temperature of the fluid can be expressed as:

$$T = T_0 + \sum C_m \left(x / a\right)^m q_m \left(0\right) \tag{173}$$

and the temperature of the heated wall can be expressed as:

$$\frac{A_3}{A_4}\left(-\alpha \Pr\left(mq_m + \eta q_m{'}\right) + \Pr R\left(mf'q_m - fq_m{'}\right)\right) = q_m{''} \tag{174}$$

where A_3 and A_4 are constant parameters that are defined as:

$$A_3 = \frac{\left(\rho C_p\right)_{nf}}{\left(\rho C_p\right)_f}, A_4 = \frac{k_{nf}}{k_f}. \tag{175}$$

with boundary conditions

$$q_m\left(0\right) = 1, q_m\left(1\right) = 0. \tag{176}$$

However, it is not possible to get a single value for the heat transfer coefficient along the heated wall if the wall temperature follows a polynomial variation, unless the temperature along the heated surface is expressed by a single term in Equation (173), i.e.

$$T_w = T_0 + C_m \left(x / a\right)^m q_m \left(0\right) \tag{177}$$

In this case the non-dimensional Nusselt number is obtained as:

$$Nu = -\frac{k_{nf}}{k_f}\frac{\partial T}{\partial \eta} / \left(T_w - T_0\right) = -q_m{'}\left(0\right). \tag{178}$$

5.2. Numerical Method

Before employing the Runge-Kutta integration scheme, first we reduce the governing differential equations into a set of first order ODEs (Sheikholeslami, 2014; 2015).

Let $x_1 = \eta, x_2 = f, x_3 = f', x_4 = f'', x_5 = f''', x_6 = q_m, x_7 = q_m{'}$. We obtain the following system:

$$\begin{pmatrix} x_1' \\ x_2' \\ x_3' \\ x_4' \\ x_5' \\ x_6' \\ x_7' \end{pmatrix} = \begin{pmatrix} 1 \\ x_3 \\ x_4 \\ x_5 \\ -\frac{A_1}{A_2}\left(\alpha\left(x_1 x_5 + 3x_4\right) + R\left(x_2 x_5 - x_3 x_4\right)\right) \\ x_7 \\ \frac{A_3}{A_4}\left(-\alpha \Pr\left(m\, x_6 + x_1\, x_7\right) + \Pr R\left(m\, x_3 x_6 - x_2 x_7\right)\right) \end{pmatrix} \tag{179}$$

and the corresponding initial conditions are

$$\begin{pmatrix} x_1 \\ x_2 \\ x_3 \\ x_4 \\ x_5 \\ x_6 \\ x_7 \end{pmatrix} = \begin{pmatrix} 0 \\ 0 \\ 0 \\ u_1 \\ u_2 \\ 1 \\ u_3 \end{pmatrix} \tag{180}$$

The above nonlinear coupled ODEs along with initial conditions are solved using fourth Order Runge-Kutta integration technique. Suitable values of the unknown initial conditions u_1, u_2, u_3 and u_4 are approximated through Newton's method until the boundary conditions at $f'\left(1\right) = 0, f\left(1\right) = 1, q_m\left(1\right) = 0$ are satisfied. The computations have been performed by using MAPLE. The maximum value of $x = 1$, to each group of parameters is determined when the values of unknown boundary conditions at $x = 0$ do not change to a successful loop with error less than 10^{-6}.

5.3. Effects of Active Parameters

Flow and heat transfer of nanofluid fluid between two parallel plates is studied numerically. One of plates is externally heated, and the other plate, through which coolant fluid is injected, expands or contracts with time. Effect of Reynolds number and expansion ratio on the velocity profiles is shown in Figure 26. As expansion ratio increases f. This figure shows that there is a maximum point for f' between the two plates. Increasing expansion ratio leads to shift maximum velocity point of f' to the solid wall. Also it can be seen that maximum values of f' increases with increase of expansion ratio. Effect of Reynolds number on velocity profiles is similar to expansion ratio. Nanoparticles transportation in various mediums was investigated by several authors.

Figure 26. Effect of Reynolds number and expansion ratio on the velocity profiles at (a) $\phi = 0.04, R = 1$; *(b)* $\phi = 0.04, \alpha = 1$.

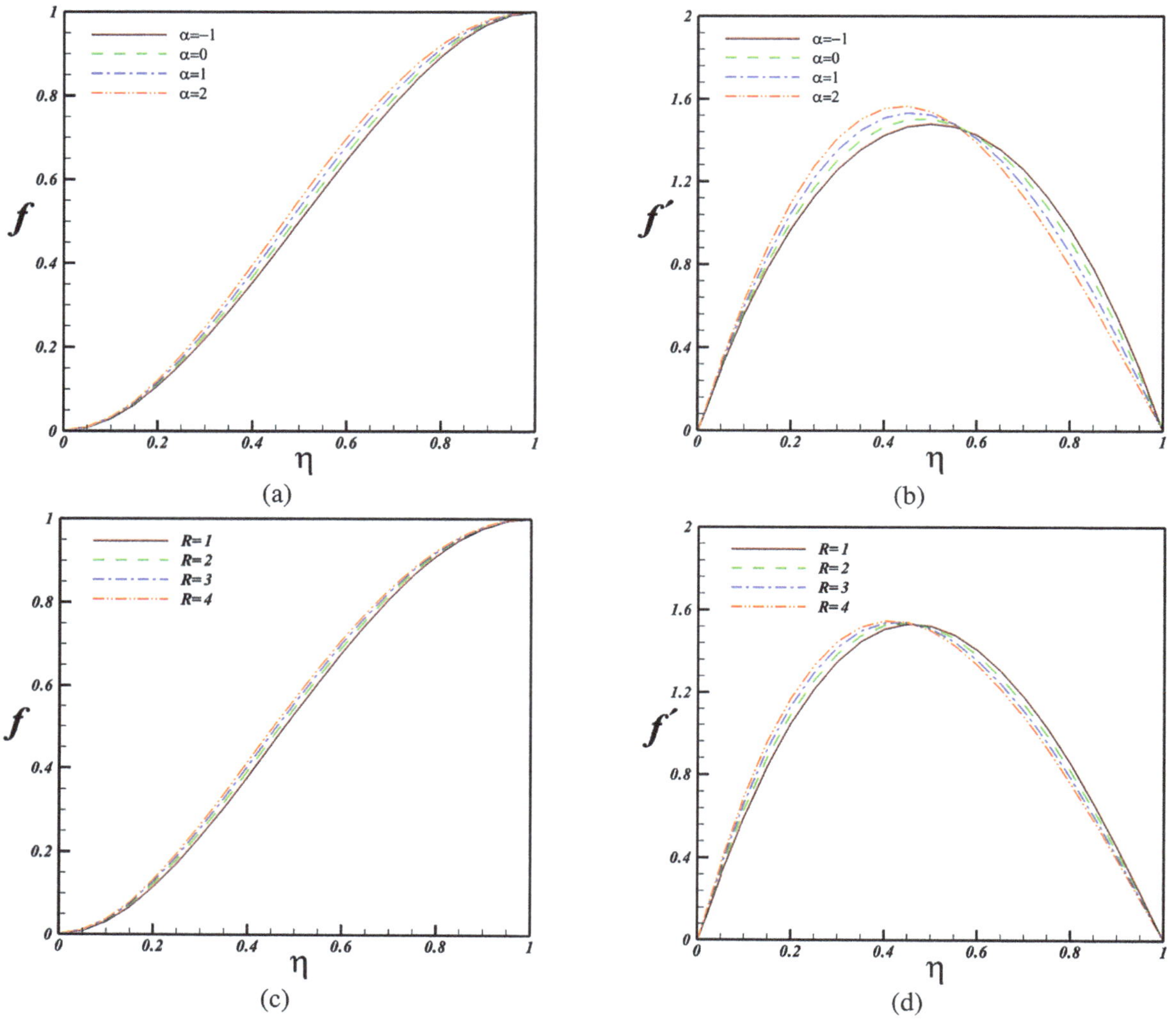

Figure 27 shows the effect of volume fraction of nanofluid on the temperature profile. The sensitivity of thermal boundary layer thickness to volume fraction of nanoparticles is related to the increased thermal conductivity of the nanofluid. In fact, higher values of thermal conductivity are accompanied by higher values of thermal diffusivity. The high values of thermal diffusivity cause a fall in the temperature gradients and accordingly increase the boundary thickness. This increase in thermal boundary layer thickness reduces the Nusselt number; however, the Nusselt number is a multiplication of temperature gradient and the thermal conductivity ratio (conductivity of the nanofluid to the conductivity of the base fluid). Since the reduction in temperature gradient due to the presence of nanoparticles is much smaller than thermal conductivity ratio therefore an enhancement in Nusselt is taken place by increasing the volume fraction of nanoparticles.

Figure 27. Effect of volume fraction of nanofluid on the temperature profile when $\alpha = 1, R = 1, m = 1$ *and* $\mathrm{Pr} = 6.2$.

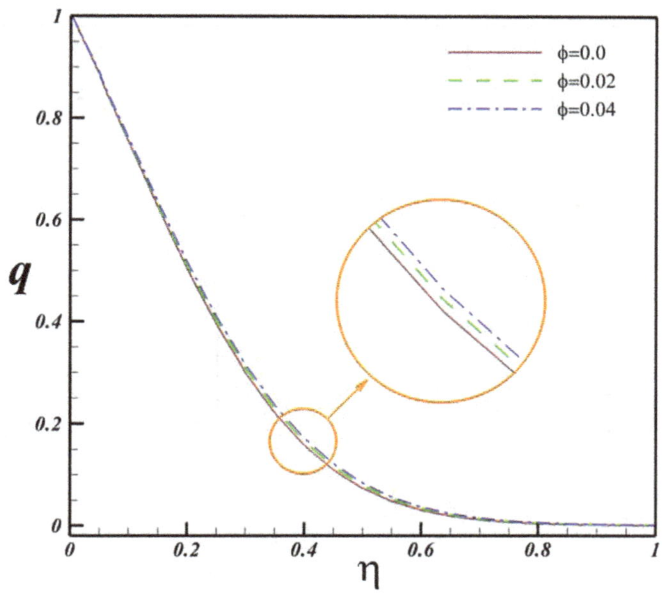

Figure 28 demonstrates the effect of expansion ratio on temperature profile and Nusselt number. Increasing expansion ratio causes thermal boundary layer thickness to increase; therefore Nusselt number decreases with increase of expansion ratio. Effect of Reynolds number on temperature profile and Nusselt number is shown in Figure 29. Reynolds number indicates the relative significance of the iner-

Figure 28. Effect of expansion ratio on (a) the temperature profile when $\mathrm{R} = 1, m = 1, \phi = 0.04$; *(b) Nusselt number when* $R = 1, m = 1$ *and* $\mathrm{Pr} = 6.2$.

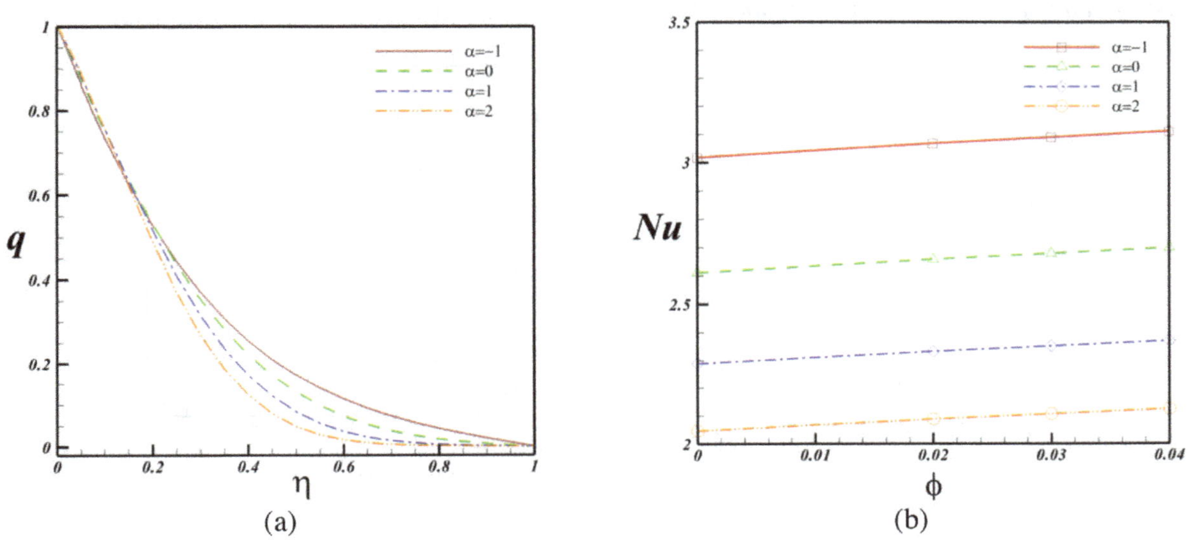

Figure 29. Effect of Reynolds number on (a) the temperature profile when $\alpha = 1, m = 1, \phi = 0.04$ *; (b) Nusselt number when* $\alpha = 1, m = 1$ *and* $\mathrm{Pr} = 6.2$.

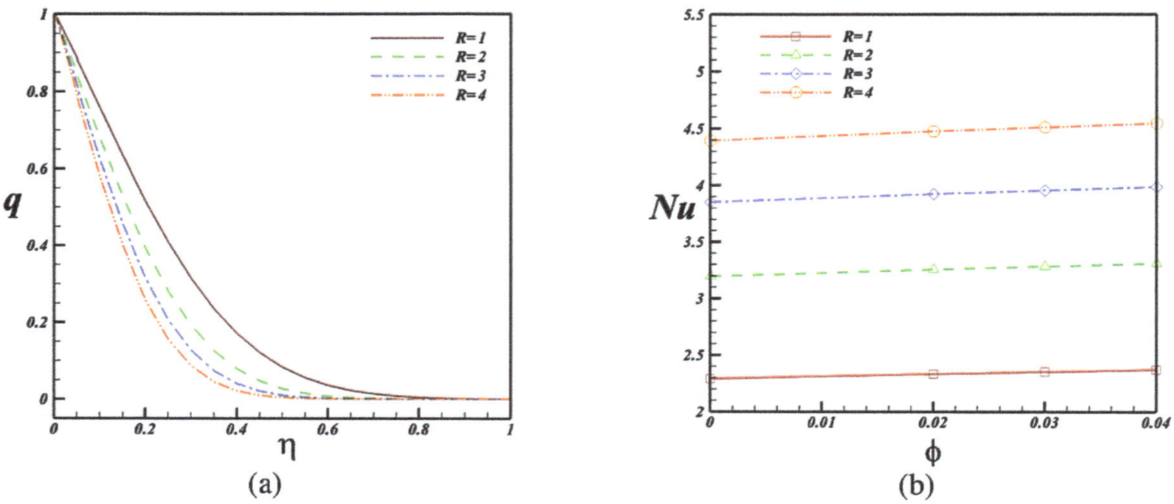

tia effect compared to the viscous effect. Thus, thermal boundary layer thickness decreases as R increases and in turn increasing Reynolds number leads to increase in Nusselt number. Figure 30 shows the effect of power law index on temperature profile and Nusselt number. Temperature gradients near the solid wall increases with augment of power law index. So Nusselt number is an increasing function of power law index.

The enhancement of heat transfer between the case of $\phi = 0.04$ and the pure fluid (base fluid) case is defined as:

Figure 30. Effect of power law index on (a) the temperature profile when $\alpha = 1, \mathrm{R} = 1, \phi = 0.04$ *; (b) Nusselt number when* $\alpha = 1, \mathrm{R} = 1$ *and* $\mathrm{Pr} = 6.2$.

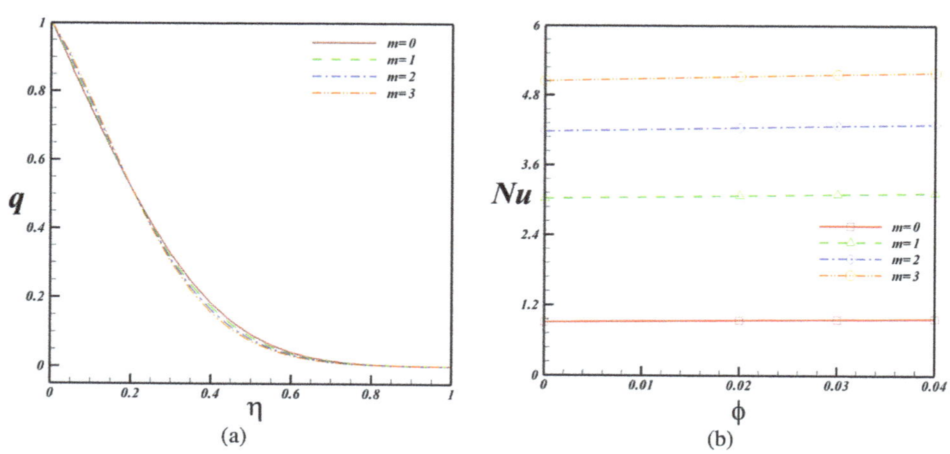

Figure 31. Effects of the Reynolds number, expansion ratio and power law index on enhancement heat transfer when $\Pr = 6.2$.

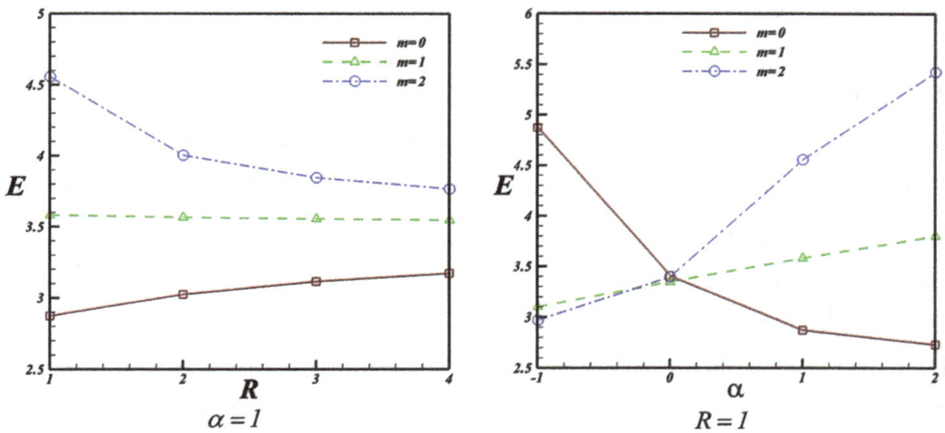

$$E = \frac{Nu(\phi = 0.04) - Nu(basefluid)}{Nu(basefluid)} \times 100 \quad (181)$$

Heat transfer enhancement due to addition of nanoparticles for different values of Reynolds number, expansion ratio and power law index is shown in Figure 31. When $m = 0$, enhancement of heat transfer increases as Reynolds number increases but it decrease with increase of R when $m > 0$. Also values of enhancement for $m > 0$ are greater than $m = 0$. The enhancement of heat transfer increases with augment of expansion ratio when $m > 0$ but opposite trend is observed for $m = 0$. It is interesting observation that for all values of power law index, values of E has no significant changes when $\alpha = 0$.

REFERENCES

Dauenhauer, C. E., & Majdalani, J. (2003). Exact self-similarity solution of the Navier–Stokes equations for a porous channel with orthogonally moving walls. *Physics of Fluids, 151*, 485–1495.

Domairry, Sheikholeslami, Ashorynejad, Gorla, & Khani. (2012). Natural convection flow of a non-Newtonian nanofluid between two vertical flat plates. *J Nanoengineering and Nanosystems, 225*(3), 115–122. Doi:10.1177/1740349911433468

Hooman, K., Gurgenci, H., & Merrikh, A. A. (2007). Heat transfer and entropy generation optimization of forced convection in porous-saturated ducts of rectangular cross-section, Int. *J. Heat MassTransf., 50*(11-12), 2051–2059. doi:10.1016/j.ijheatmasstransfer.2006.11.015

Li, J. (2008). *Computational analysis of nanofluid flow in micro channels with applications to micro-heat sinks and bio-MEMS* (PhD Thesis). NC State University, Raleigh, NC.

Majdalani, J., Zhou, C., & Dawson, C. A. (2002). Two-dimensional viscous flows between slowly expanding or contracting walls with weak permeability. *Journal of Biomechanics*, *35*(10), 1399–1403. doi:10.1016/S0021-9290(02)00186-0 PMID:12231285

Nield, D. A., & Kuznetsov, A. V. (2005). Forced convection in porous media: transverse heterogeneity effects and thermal development. Taylor and Francis.

Sheikholeslami. (2014). Effect of spatially variable magnetic field on ferrofluid flow and heat transfer considering constant heat flux boundary condition. *The European Physical Journal Plus*, 129–248.

Sheikholeslami, Ashorynejad, Domairry, & Hashim. (2012). *Flow and Heat Transfer of Cu-Water Nanofluid between a Stretching Sheet and a Porous Surface in a Rotating System*. Hindawi Publishing Corporation. doi:10.1155/2012/421320

Sheikholeslami, Ashorynejad, Ganji, & Kolahdooz. (2011). *Investigation of Rotating MHD Viscous Flow and Heat Transfer between Stretching and Porous Surfaces Using Analytical Method*. Hindawi Publishing Corporation. doi:10.1155/2011/258734

Sheikholeslami, Rashidi, Al Saad, Firouzi, Rokni, & Domairry. (2015). Steady nanofluid flow between parallel plates considering Thermophoresis and Brownian effects. *Journal of King Saud University - Science*. doi:10.1016/j.jksus.2015.06.003

Sheikholeslami, M. (2015). Effect of uniform suction on nanofluid flow and heat transfer over a cylinder. *Journal of the Brazilian Society of Mechanical Sciences and Engineering*, *37*(6), 1623–1633. doi:10.100740430-014-0242-z

Sheikholeslami, M., Ashorynejad, H. R., Barari, A., & Soleimani, S. (2013). Soheil Soleimani, Investigation of heat and mass transfer of rotating MHD viscous flow between a stretching sheet and a porous surface. *Engineering Computations*, *30*(3), 357–378. doi:10.1108/02644401311314330

Sheikholeslami, M., Ashorynejad, H. R., Domairry, D., & Hashim, I. (2012). Investigation of the Laminar Viscous Flow in a Semi-Porous Channel in the Presence of Uniform Magnetic Field using Optimal Homotopy Asymptotic Method. *Sains Malaysiana*, *41*(10), 1177–1229.

Sheikholeslami, M., Ashorynejad, H. R., Ganji, D. D., & Yıldırım, A. (2012). Homotopy perturbation method for three-dimensional problem of condensation film on inclined rotating disk. *Scientia Iranica B*, *19*(3), 437–442. doi:10.1016/j.scient.2012.03.006

Sheikholeslami, M., Azimi, M., & Ganji, D. D. (2015). Application of Differential Transformation Method for Nanofluid Flow in a Semi-Permeable Channel Considering Magnetic Field Effect. *International Journal for Computational Methods in Engineering Science and Mechanics*, *16*(4), 246–255. do i:10.1080/15502287.2015.1048384

Sheikholeslami, M., Ellahi, R., Ashorynejad, H. R., Domairry, G., & Hayat, T. (2014). Effects of Heat Transfer in Flow of Nanofluids Over a Permeable Stretching Wall in a Porous Medium. *Journal of Computational and Theoretical Nanoscience*, *11*(2), 1–11. doi:10.1166/jctn.2014.3384

Sheikholeslami, M., & Ganji, D. D. (2013). Heat transfer of Cu-water nanofluid flow between parallel plates. *Powder Technology*, *235*, 873–879. doi:10.1016/j.powtec.2012.11.030

Sheikholeslami, M., & Ganji, D. D. (2014). Magnetohydrodynamic flow in a permeable channel filled with nanofluid. *Scientia IranicaB*, *21*(1), 203–212.

Sheikholeslami, M., & Ganji, D. D. (2014). Heated permeable stretching surface in a porous medium using Nanofluids. *Journal of Applied Fluid Mechanics*, *7*(3), 535–542.

Sheikholeslami, M., & Ganji, D. D. (2014). Numerical investigation for two phase modeling of nanofluid in a rotating system with permeable sheet. *Journal of Molecular Liquids*, *194*, 13–19. doi:10.1016/j.molliq.2014.01.003

Sheikholeslami, M., & Ganji, D. D. (2015). Nanofluid flow and heat transfer between parallel plates considering Brownian motion using DTM. *Computer Methods in Applied Mechanics and Engineering*, *283*, 651–663. doi:10.1016/j.cma.2014.09.038

Sheikholeslami, M., & Ganji, D. D. (2016). *New semianalytical methods: application for MHD nanofluid hydrothermal behavior*. External Magnetic Field Effects on Hydrothermal Treatment of Nanofluid.

Sheikholeslami, M., Ganji, D. D., & Ashorynejad, H. R. (2013). Investigation of squeezing unsteady nanofluid flow using ADM. *Powder Technology*, *239*, 259–265. doi:10.1016/j.powtec.2013.02.006

Sheikholeslami, M., Ganji, D. D., Ashorynejad, H. R., & Houman, B. (2012). Rokni, Analytical investigation of Jeffery-Hamel flow with high magnetic field and nano particle by Adomian decomposition method, Appl. Math. Mech.-. *Engl. Ed.*, *33*(1), 1553–1564.

Sheikholeslami, M., Ganji, D. D., & Houman, B. (2013). Rokni, Nanofluid Flow in a Semi-Porous Channel in the Presence of Uniform Magnetic Field. *IJE Transactions C. Aspects*, *26*(6), 653–662.

Sheikholeslami, M., Ganji, D. D., & Rashidi, M. M. (2016). Magnetic field effect on unsteady nanofluid flow and heat transfer using Buongiorno model. *Journal of Magnetism and Magnetic Materials*, *416*, 164–173. doi:10.1016/j.jmmm.2016.05.026

Sheikholeslami Kandelousi, M. (2014). KKL correlation for simulation of nanofluid flow and heat transfer in a permeable channel. *Physics Letters. [Part A]*, *378*(45), 3331–3339. doi:10.1016/j.physleta.2014.09.046

Starov, V. M., & Zhdanov, V. G. (2001). Effective viscosity and permeability of porous media. *Colloids and Surfaces. A, Physicochemical and Engineering Aspects*, *192*(1-3), 363–375. doi:10.1016/S0927-7757(01)00737-3

ADDITIONAL READING

Jafaryar, M., Sheikholeslami, M., Li, M., & Moradi, R. (2018). Nanofluid turbulent flow in a pipe under the effect of twisted tape with alternate axis. *Journal of Thermal Analysis and Calorimetry*. doi:10.100710973-018-7093-2

Li, Z., Shehzad, S. A., & Sheikholeslami, M. (2018). An application of CVFEM for nanofluid heat transfer intensification in a porous sinusoidal cavity considering thermal non-equilibrium model. *Computer Methods in Applied Mechanics and Engineering*, *339*, 663–680. doi:10.1016/j.cma.2018.05.015

Li, Z., Sheikholeslami, M., Chamkha, A. J., Raizah, Z. A., & Saleem, S. (2018). Control Volume Finite Element Method for nanofluid MHD natural convective flow inside a sinusoidal annulus under the impact of thermal radiation. *Computer Methods in Applied Mechanics and Engineering*, *338*, 618–633. doi:10.1016/j.cma.2018.04.023

Li, Z., Sheikholeslami, M., Jafaryar, M., Shafee, A., & Chamkha, A. J. (2018). Investigation of nanofluid entropy generation in a heat exchanger with helical twisted tapes. *Journal of Molecular Liquids*, *266*, 797–805. doi:10.1016/j.molliq.2018.07.009

Li, Z., Sheikholeslami, M., Samandari, M., & Shafee, A. (2018). Nanofluid unsteady heat transfer in a porous energy storage enclosure in existence of Lorentz forces. *International Journal of Heat and Mass Transfer*, *127*, 914–926. doi:10.1016/j.ijheatmasstransfer.2018.06.101

Sheikholeslami, M. (2017a). Lattice Boltzmann Method simulation of MHD non-Darcy nanofluid free convection. *Physica B, Condensed Matter*, *516*, 55–71. doi:10.1016/j.physb.2017.04.029

Sheikholeslami, M. (2017b). CuO-water nanofluid free convection in a porous cavity considering Darcy law. *The European Physical Journal Plus*, *132*(1), 55. doi:10.1140/epjp/i2017-11330-3

Sheikholeslami, M. (2017c). Influence of Lorentz forces on nanofluid flow in a porous cavity by means of Non- Darcy model. *Engineering Computations*, *34*(8), 2651–2667. doi:10.1108/EC-01-2017-0008

Sheikholeslami, M. (2017d). Magnetic field influence on nanofluid thermal radiation in a cavity with tilted elliptic inner cylinder. *Journal of Molecular Liquids*, *229*, 137–147. doi:10.1016/j.molliq.2016.12.024

Sheikholeslami, M. (2017e). Numerical simulation of magnetic nanofluid natural convection in porous media. *Physics Letters. [Part A]*, *381*(5), 494–503. doi:10.1016/j.physleta.2016.11.042

Sheikholeslami, M. (2017f). Influence of Lorentz forces on nanofluid flow in a porous cylinder considering Darcy model. *Journal of Molecular Liquids*, *225*, 903–912. doi:10.1016/j.molliq.2016.11.022

Sheikholeslami, M. (2017g). Influence of Coulomb forces on Fe3O4-H2O nanofluid thermal improvement. *International Journal of Hydrogen Energy*, *42*(2), 821–829. doi:10.1016/j.ijhydene.2016.09.185

Sheikholeslami, M. (2017h). Numerical investigation of MHD nanofluid free convective heat transfer in a porous tilted enclosure. *Engineering Computations*, *34*(6), 1939–1955. doi:10.1108/EC-08-2016-0293

Sheikholeslami, M. (2017i). Magnetic field influence on CuO -H2O nanofluid convective flow in a permeable cavity considering various shapes for nanoparticles. *International Journal of Hydrogen Energy*, *42*(31), 19611–19621. doi:10.1016/j.ijhydene.2017.06.121

Sheikholeslami, M. (2018a). Magnetic source impact on nanofluid heat transfer using CVFEM. *Neural Computing & Applications*, *30*(4), 1055–1064. doi:10.100700521-016-2740-7

Sheikholeslami, M. (2018b). Numerical simulation for external magnetic field influence on Fe3O4-water nanofluid forced convection. *Engineering Computations*, *35*(4), 1639–1654. doi:10.1108/EC-06-2017-0200

Sheikholeslami, M. (2018c). Finite element method for PCM solidification in existence of CuO nanoparticles. *Journal of Molecular Liquids*, *265*, 347–355. doi:10.1016/j.molliq.2018.05.132

Sheikholeslami, M. (2018d). Solidification of NEPCM under the effect of magnetic field in a porous thermal energy storage enclosure using CuO nanoparticles. *Journal of Molecular Liquids*, *263*, 303–315. doi:10.1016/j.molliq.2018.04.144

Sheikholeslami, M. (2018e). Influence of magnetic field on Al2O3-H2O nanofluid forced convection heat transfer in a porous lid driven cavity with hot sphere obstacle by means of LBM. *Journal of Molecular Liquids*, *263*, 472–488. doi:10.1016/j.molliq.2018.04.111

Sheikholeslami, M. (2018f). Numerical simulation for solidification in a LHTESS by means of Nano-enhanced PCM. *Journal of the Taiwan Institute of Chemical Engineers*, *86*, 25–41. doi:10.1016/j.jtice.2018.03.013

Sheikholeslami, M. (2018g). Numerical modeling of Nano enhanced PCM solidification in an enclosure with metallic fin. *Journal of Molecular Liquids*, *259*, 424–438. doi:10.1016/j.molliq.2018.03.006

Sheikholeslami, M. (2018h). Numerical investigation of nanofluid free convection under the influence of electric field in a porous enclosure. *Journal of Molecular Liquids*, *249*, 1212–1221. doi:10.1016/j.molliq.2017.11.141

Sheikholeslami, M. (2018i). CuO-water nanofluid flow due to magnetic field inside a porous media considering Brownian motion. *Journal of Molecular Liquids*, *249*, 921–929. doi:10.1016/j.molliq.2017.11.118

Sheikholeslami, M. (2018j). Application of Darcy law for nanofluid flow in a porous cavity under the impact of Lorentz forces. *Journal of Molecular Liquids*, *266*, 495–503. doi:10.1016/j.molliq.2018.06.083

Sheikholeslami, M., Barzegar Gerdroodbary, M., Valiallah Mousavi, S., Ganji, D. D., & Moradi, R. (2018). Heat transfer enhancement of ferrofluid inside an 90o elbow channel by non-uniform magnetic field. *Journal of Magnetism and Magnetic Materials*, *460*, 302–311. doi:10.1016/j.jmmm.2018.03.070

Sheikholeslami, M., & Bhatti, M. M. (2017). Active method for nanofluid heat transfer enhancement by means of EHD. *International Journal of Heat and Mass Transfer*, *109*, 115–122. doi:10.1016/j.ijheatmasstransfer.2017.01.115

Sheikholeslami, M., Darzi, M., & Li, Z. (2018). Experimental investigation for entropy generation and exergy loss of nano-refrigerant condensation process. *International Journal of Heat and Mass Transfer*, *125*, 1087–1095. doi:10.1016/j.ijheatmasstransfer.2018.04.155

Sheikholeslami, M., Darzi, M., & Sadoughi, M. K. (2018). Heat transfer improvement and Pressure Drop during condensation of refrigerant-based Nanofluid; An Experimental Procedure. *International Journal of Heat and Mass Transfer*, *122*, 643–650. doi:10.1016/j.ijheatmasstransfer.2018.02.015

Sheikholeslami, M., & Ghasemi, A. (2018). Solidification heat transfer of nanofluid in existence of thermal radiation by means of FEM. *International Journal of Heat and Mass Transfer*, *123*, 418–431. doi:10.1016/j.ijheatmasstransfer.2018.02.095

Sheikholeslami, M., Ghasemi, A., Li, Z., Shafee, A., & Saleem, S. (2018). Influence of CuO nanoparticles on heat transfer behavior of PCM in solidification process considering radiative source term. *International Journal of Heat and Mass Transfer*, *126*, 1252–1264. doi:10.1016/j.ijheatmasstransfer.2018.05.116

Sheikholeslami, M., Hayat, T., & Alsaedi, A. (2018). Numerical simulation for forced convection flow of MHD CuO-H2O nanofluid inside a cavity by means of LBM. *Journal of Molecular Liquids*, *249*, 941–948. doi:10.1016/j.molliq.2017.10.099

Sheikholeslami, M., Hayat, T., Muhammad, T., & Alsaedi, A. (2018). MHD forced convection flow of nanofluid in a porous cavity with hot elliptic obstacle by means of Lattice Boltzmann method. *International Journal of Mechanical Sciences*, *135*, 532–540. doi:10.1016/j.ijmecsci.2017.12.005

Sheikholeslami, M., Jafaryar, M., Ganji, D. D., & Li, Z. (2018). Exergy loss analysis for nanofluid forced convection heat transfer in a pipe with modified turbulators. *Journal of Molecular Liquids*, *262*, 104–110. doi:10.1016/j.molliq.2018.04.077

Sheikholeslami, M., Jafaryar, M., & Li, Z. (2018a). Second law analysis for nanofluid turbulent flow inside a circular duct in presence of twisted tape turbulators. *Journal of Molecular Liquids*, *263*, 489–500. doi:10.1016/j.molliq.2018.04.147

Sheikholeslami, M., Jafaryar, M., & Li, Z. (2018b). Nanofluid turbulent convective flow in a circular duct with helical turbulators considering CuO nanoparticles. *International Journal of Heat and Mass Transfer*, *124*, 980–989. doi:10.1016/j.ijheatmasstransfer.2018.04.022

Sheikholeslami, M., Jafaryar, M., Saleem, S., Li, Z., Shafee, A., & Jiang, Y. (2018). Nanofluid heat transfer augmentation and exergy loss inside a pipe equipped with innovative turbulators. *International Journal of Heat and Mass Transfer*, *126*, 156–163. doi:10.1016/j.ijheatmasstransfer.2018.05.128

Sheikholeslami, M., Jafaryar, M., Shafee, A., & Li, Z. (2018). Investigation of second law and hydrothermal behavior of nanofluid through a tube using passive methods. *Journal of Molecular Liquids*, *269*, 407–416. doi:10.1016/j.molliq.2018.08.019

Sheikholeslami, M., Li, Z., & Shafee, A. (2018a). Lorentz forces effect on NEPCM heat transfer during solidification in a porous energy storage system. *International Journal of Heat and Mass Transfer*, *127*, 665–674. doi:10.1016/j.ijheatmasstransfer.2018.06.087

Sheikholeslami, M., Li, Z., & Shamlooei, M. (2018). Nanofluid MHD natural convection through a porous complex shaped cavity considering thermal radiation. *Physics Letters. [Part A]*, *382*(24), 1615–1632. doi:10.1016/j.physleta.2018.04.006

Sheikholeslami, M., & Rokni, H. B. (2017). Simulation of nanofluid heat transfer in presence of magnetic field: A review. *International Journal of Heat and Mass Transfer*, *115*, 1203–1233. doi:10.1016/j.ijheatmasstransfer.2017.08.108

Sheikholeslami, M., & Rokni, H. B. (2018a). CVFEM for effect of Lorentz forces on nanofluid flow in a porous complex shaped enclosure by means of Non-equilibrium model. *Journal of Molecular Liquids*, *254*, 446–462. doi:10.1016/j.molliq.2018.01.130

Sheikholeslami, M., Rokni, H.B. (2018b). Magnetic nanofluid flow and convective heat transfer in a porous cavity considering Brownian motion effects. *Physics of Fluids*, *30*(1), doi:10.1063/1.5012517

Sheikholeslami, M., & Sadoughi, M. K. (2017). Mesoscopic method for MHD nanofluid flow inside a porous cavity considering various shapes of nanoparticles. *International Journal of Heat and Mass Transfer*, *113*, 106–114. doi:10.1016/j.ijheatmasstransfer.2017.05.054

Sheikholeslami, M., & Sadoughi, M. K. (2018). Simulation of CuO- water nanofluid heat transfer enhancement in presence of melting surface. *International Journal of Heat and Mass Transfer, 116*, 909–919. doi:10.1016/j.ijheatmasstransfer.2017.09.086

Sheikholeslami, M., & Seyednezhad, M. (2018). Simulation of nanofluid flow and natural convection in a porous media under the influence of electric field using CVFEM. *International Journal of Heat and Mass Transfer, 120*, 772–781. doi:10.1016/j.ijheatmasstransfer.2017.12.087

Sheikholeslami, M., Shafee, A., Ramzan, M., & Li, Z. (2018). Investigation of Lorentz forces and radiation impacts on nanofluid treatment in a porous semi annulus via Darcy law. *Journal of Molecular Liquids, 272*, 8–14. doi:10.1016/j.molliq.2018.09.016

Sheikholeslami, M., Shamlooei, M., & Moradi, R. (2018). Numerical simulation for heat transfer intensification of nanofluid in a porous curved enclosure considering shape effect of Fe3O4 nanoparticles. *Chemical Engineering & Processing: Process Intensification, 124*, 71–82. doi:10.1016/j.cep.2017.12.005

Sheikholeslami, M., & Shehzad, S. A. (2017a). Thermal radiation of ferrofluid in existence of Lorentz forces considering variable viscosity. *International Journal of Heat and Mass Transfer, 109*, 82–92. doi:10.1016/j.ijheatmasstransfer.2017.01.096

Sheikholeslami, M., & Shehzad, S. A. (2017b). Magnetohydrodynamic nanofluid convective flow in a porous enclosure by means of LBM. *International Journal of Heat and Mass Transfer, 113*, 796–805. doi:10.1016/j.ijheatmasstransfer.2017.05.130

Sheikholeslami, M., & Shehzad, S. A. (2018a). Numerical analysis of Fe3O4 –H2O nanofluid flow in permeable media under the effect of external magnetic source. *International Journal of Heat and Mass Transfer, 118*, 182–192. doi:10.1016/j.ijheatmasstransfer.2017.10.113

Sheikholeslami, M., & Shehzad, S. A. (2018b). CVFEM simulation for nanofluid migration in a porous medium using Darcy model. *International Journal of Heat and Mass Transfer, 122*, 1264–1271. doi:10.1016/j.ijheatmasstransfer.2018.02.080

Sheikholeslami, M., & Shehzad, S. A. (2018c). Simulation of water based nanofluid convective flow inside a porous enclosure via Non-equilibrium model. *International Journal of Heat and Mass Transfer, 120*, 1200–1212. doi:10.1016/j.ijheatmasstransfer.2017.12.132

Sheikholeslami, M., & Shehzad, S. A. (2018d). Non-Darcy free convection of Fe3O4-water nanoliquid in a complex shaped enclosure under impact of uniform Lorentz force. *The Chinese Journal of Physiology, 56*(1), 270–281. doi:10.1016/j.cjph.2017.12.017

Sheikholeslami, M., Shehzad, S. A., Abbasi, F. M., & Li, Z. (2018). Nanofluid flow and forced convection heat transfer due to Lorentz forces in a porous lid driven cubic enclosure with hot obstacle. *Computer Methods in Applied Mechanics and Engineering, 338*, 491–505. doi:10.1016/j.cma.2018.04.020

Sheikholeslami, M., Shehzad, S. A., & Li, Z. (2018a). Nanofluid heat transfer intensification in a permeable channel due to magnetic field using Lattice Boltzmann method. *Physica B, Condensed Matter, 542*, 51–58. doi:10.1016/j.physb.2018.03.036

Sheikholeslami, M., Shehzad, S. A., & Li, Z. (2018b). Water based nanofluid free convection heat transfer in a three dimensional porous cavity with hot sphere obstacle in existence of Lorenz forces. *International Journal of Heat and Mass Transfer*, *125*, 375–386. doi:10.1016/j.ijheatmasstransfer.2018.04.076

Sheikholeslami, M., Shehzad, S. A., Li, Z., & Shafee, A. (2018). Numerical modeling for Alumina nanofluid magnetohydrodynamic convective heat transfer in a permeable medium using Darcy law. *International Journal of Heat and Mass Transfer*, *127*, 614–622. doi:10.1016/j.ijheatmasstransfer.2018.07.013

Sheikholeslami, M., & Vajravelu, K. (2017). Forced convection heat transfer in Fe3O4- ethylene glycol nanofluid under the influence of Coulomb force. *Journal of Molecular Liquids*, *233*, 203–210. doi:10.1016/j.molliq.2017.03.026

Sheikholeslami, M., & Zeeshan, A. (2017). Analysis of flow and heat transfer in water based nanofluid due to magnetic field in a porous enclosure with constant heat flux using CVFEM. *Computer Methods in Applied Mechanics and Engineering*, *320*, 68–81. doi:10.1016/j.cma.2017.03.024

Sheikholeslami, M., Zeeshan, A., & Majeed, A. (2018). Control volume based finite element simulation of magnetic nanofluid flow and heat transport in non-Darcy medium. *Journal of Molecular Liquids*, *268*, 354–364. doi:10.1016/j.molliq.2018.07.031

Chapter 6
Influence of Shape of Nanoparticles on Nanofluid Hydrothermal Behavior

ABSTRACT

The shape of nanoparticles can change the thermal conductivity of nanofluid. So, the effect of shape factor on nanofluid flow and heat transfer has been reported in this chapter. Governing equations are presented in vorticity stream function formulation. Control volume-based finite element method (CVFEM) is utilized to obtain the results. Results indicate that platelet shape has the highest rate of heat transfer.

1. INTRODUCTION

Most conventional heat transfer fluids, such as water, Ethylene Glycol and engine oil, have limited capabilities in term of thermal properties, which in turn, may impose severe restrictions in many thermal applications. And in spite of considerable research and efforts deployed, a clear and urgent need does exist to date to develop new strategies in order to improve the effective thermal behaviours of these fluids. On the other hand, most solids, in particular metals, have thermal conductivities much higher, say by 1–3 orders of magnitude, compared to that of liquids. Hence, one can then expect that fluid containing solid particles may significantly increase its conductivity. Sheikholeslami (2017) investigated the effect of nanoparticle shape on convective flow in a permeable cavity. Sheikholeslami and Bhatti (2017) studied the forced convection of nanofluid in presence of constant magnetic field considering shape effects of nanoparticles. Sheikholeslami and Shamlooei (2017) reported the magnetic source influence on nanofluid flow in porous medium considering shape factor effect. Sheikholeslami and Ganji (2017) simulated the shape factor effect on magnetohydrodynamic CuO -water transportation inside a porous cavity. In recent decade, various methods have been utilized to show nanofluid behavior.

DOI: 10.4018/978-1-5225-7595-5.ch006

2. FORCED CONVECTION OF NANOFLUID IN PRESENCE OF CONSTANT MAGNETIC FIELD CONSIDERING SHAPE EFFECTS OF NANOPARTICLES

2.1. Problem Definition

Figure 1 depicts the geometry, boundary condition and sample element. The lower wall has the velocity of U_{Lid} and others are stationary. The lower wall has constant temperature T_h and the temperature

Figure 1. (a) Geometry and the boundary conditions with (b) the mesh of geometry considered in this work; (c) a sample triangular element and its corresponding control volume.

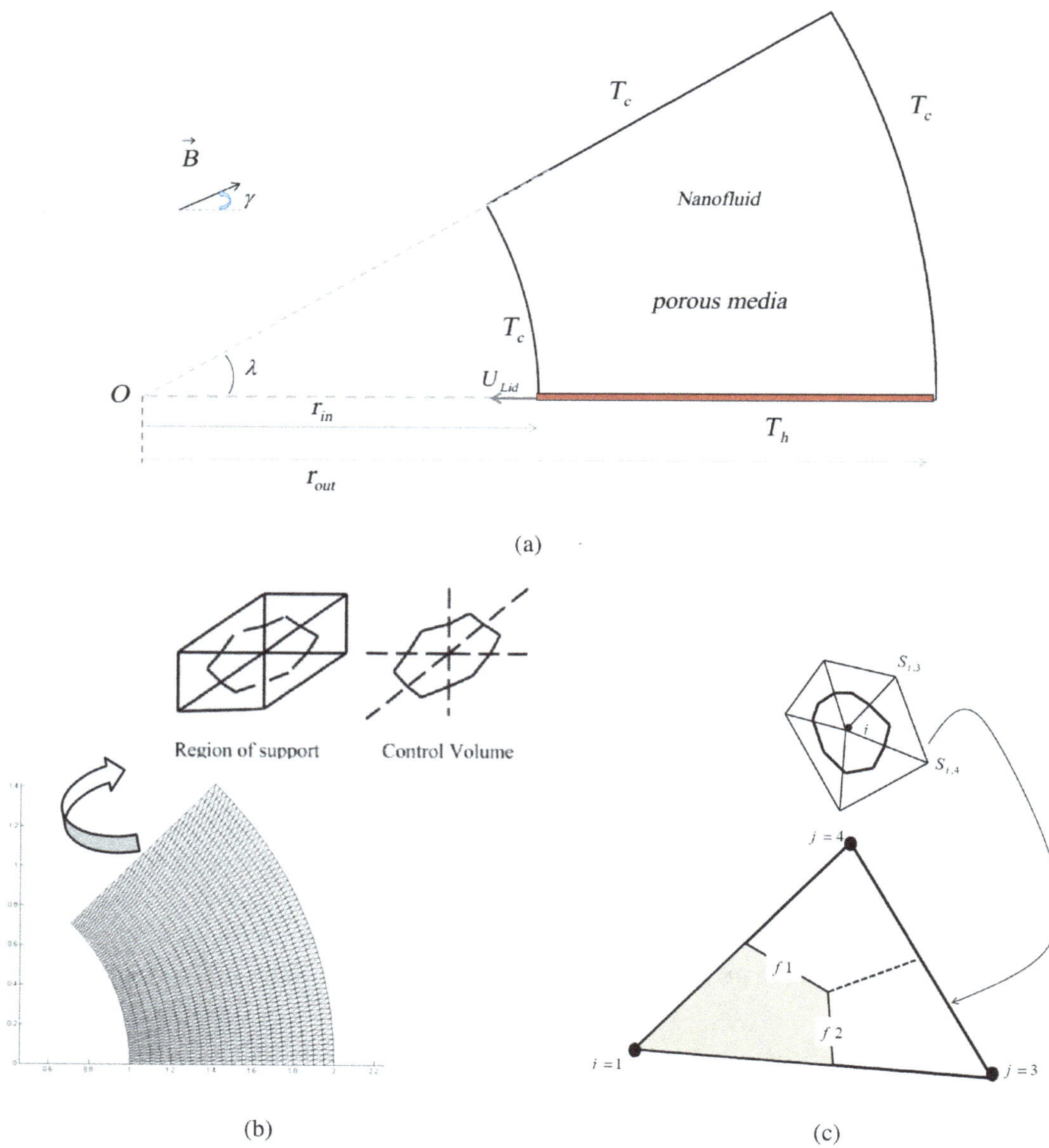

of other walls is T_c. Horizontal magnetic field has been applied. Nanofluid forced convection heat transfer in a porous semi annulus is investigated.

2.2. Governing Equation

2D steady convective flow of nanofluid in a porous media is considered in existence of constant magnetic field. The PDEs equations are:

$$\frac{\partial u}{\partial x}+\frac{\partial v}{\partial y}=0 \tag{1}$$

$$\rho_{nf}\left(v\frac{\partial u}{\partial y}+\frac{\partial u}{\partial x}u\right)=\left[-\sigma_{nf}B_y^2u+\sigma_{nf}B_xB_yv+\left(\frac{\partial^2 u}{\partial y^2}+\frac{\partial^2 u}{\partial x^2}\right)\mu_{nf}-\frac{\partial P}{\partial x}\right]-\frac{\mu_{nf}}{K}u \tag{2}$$

$$\rho_{nf}\left(\frac{\partial v}{\partial x}u+\frac{\partial v}{\partial y}v\right)=+\mu_{nf}\left(\frac{\partial^2 v}{\partial x^2}+\frac{\partial^2 v}{\partial y^2}\right)-\frac{\partial P}{\partial y}+B_y\sigma_{nf}B_xu-B_x\sigma_{nf}B_xv-\frac{\mu_{nf}}{K}v,$$
$$B_x=B_o\cos\lambda,\ B_y=B_o\sin\lambda \tag{3}$$

$$\left(\rho C_p\right)_{nf}\left(v\frac{\partial T}{\partial y}+u\frac{\partial T}{\partial x}\right)=k_{nf}\left(\frac{\partial^2 T}{\partial x^2}+\frac{\partial^2 T}{\partial y^2}\right), \tag{4}$$

$\left(\rho C_p\right)_{nf}$, ρ_{nf}, and σ_{nf} are defined as:

$$\left(\rho C_p\right)_{nf}=\left(\rho C_p\right)_f(1-\phi)+\left(\rho C_p\right)_s\phi \tag{5}$$

$$\rho_{nf}=\rho_f(1-\phi)+\rho_s\phi \tag{6}$$

$$\frac{\sigma_{nf}}{\sigma_f}=1+\frac{3\left(\frac{\sigma_s}{\sigma_f}-1\right)\phi}{\left(\frac{\sigma_s}{\sigma_f}+2\right)-\left(\frac{\sigma_s}{\sigma_f}-1\right)\phi} \tag{7}$$

The KKL (Koo-Kleinstreuer-Li) correlation has been utilized for viscosity of nanofluid

$$\begin{aligned}
&\mu_{eff} = \mu_{static} + \mu_{Brownian} = \mu_{static} + \frac{k_{Brownian}}{k_f} \times \frac{\mu_f}{\Pr_f} \\
&k_{Brownian} = 5 \times 10^4 \phi \rho_f c_{p,f} \sqrt{\frac{\kappa_b T}{\rho_p d_p}} g'(T,\phi,d_p) \\
&g'(T,\phi,d_p) = \left(a_1 + a_2 \ln(d_p) + a_3 \ln(\phi) + a_4 \ln(\phi)\ln(d_p) + a_5 \ln(d_p)^2\right)\ln(T) \\
&+\left(a_6 + a_7 \ln(d_p) + a_8 \ln(\phi) + a_9 \ln(\phi)\ln(d_p) + a_{10} \ln(d_p)^2\right)
\end{aligned} \tag{8}$$

The related coefficient and properties of Cuo-water nanofluid is presented in Table 1 and 2. Maxwell model and Hamilton–Crosser model for irregular particle geometries by introducing a shape factor can be expressed as

$$\frac{k_{nf}}{k_f} = \frac{k_p + (m+1)k_f - (m+1)\phi(k_f - k_p)}{k_p + (m+1)k_f + \phi(k_f - k_p)} \tag{9}$$

Table 1. The coefficient values of CuO – Water nanofluid

Coefficient Values	CuO – Water
a_1	-26.5933108
a_2	-0.403818333
a_3	-33.3516805
a_4	-1.915825591
a_5	6.421858E-02
a_6	48.40336955
a_7	-9.787756683
a_8	190.245610009
a_9	10.9285386565
a_{10}	-0.72009983664

Table 2. Thermo physical properties of water and nanoparticles

	$\rho(kg/m^3)$	$C_p(j/kgk)$	$k(W/m.k)$	$\beta\times10^5(K^{-1})$	$d_p(nm)$	$\sigma(\Omega\cdot\mathrm{m})^{-1}$
Water	997.1	4179	0.613	21	-	0.05
CuO	6500	540	18	29	45	10^{-10}

in which k_p and k_f are the conductivities of the particle material and the base fluid. In this equation "m" is shaper factor. Table 3 shows the different values of shape factors for various shapes of nanoparticles. Vorticity and stream function should be used to eliminate pressure source terms:

$$\omega+\frac{\partial u}{\partial y}-\frac{\partial v}{\partial x}=0,\ \frac{\partial\psi}{\partial x}=-v,\ \frac{\partial\psi}{\partial y}=u \tag{10}$$

Introducing dimensionless quantities:

$$U=\frac{u}{U_{Lid}},V=\frac{v}{U_{Lid}},\theta=\frac{T-T_c}{\Delta T},\Delta T=T_h-T_c,\left(X,Y\right)=\frac{\left(x,y\right)}{L},\Psi=\frac{\psi L}{U_{Lid}},\Omega=\frac{\omega}{LU_{Lid}} \tag{11}$$

The final formulae are:

$$\frac{\partial^2\Psi}{\partial Y^2}+\frac{\partial^2\Psi}{\partial X^2}=-\Omega, \tag{12}$$

Table 3. The values of shape factor of different shapes of nanoparticles

m	Spherical		3
	Platelet		5.7
	Cylinder		4.8
	Brick	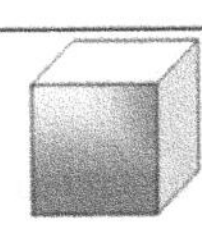	3.7

$$U\frac{\partial\Omega}{\partial X}+\frac{\partial\Omega}{\partial Y}V=\frac{1}{\mathrm{Re}}\frac{A_5}{A_1}\left(\frac{\partial^2\Omega}{\partial Y^2}+\frac{\partial^2\Omega}{\partial X^2}\right)+\frac{Ha^2}{\mathrm{Re}}\frac{A_6}{A_1}\left(\frac{\partial U}{\partial X}B_xB_y-\frac{\partial V}{\partial X}B_x^{\ 2}+\frac{\partial U}{\partial Y}B_y^2-\frac{\partial V}{\partial Y}B_xB_y\right)-\frac{1}{\mathrm{Re}\,Da}\frac{A_5}{A_1}\Omega, \tag{13}$$

$$\frac{\partial\theta}{\partial X}U+\frac{\partial\theta}{\partial Y}V=\frac{1}{\mathrm{Re}\,\mathrm{Pr}}\frac{A_4}{A_2}\left(\frac{\partial^2\theta}{\partial X^2}+\frac{\partial^2\theta}{\partial Y^2}\right) \tag{14}$$

where dimensionless and constants parameters are illustrated as:

$$\mathrm{Re}=\frac{\rho_f U_{Lid} L}{\mu_f}, Ha=LB_0\sqrt{\sigma_f/\mu_f}, A_1=\frac{\rho_{nf}}{\rho_f}, A_2=\frac{(\rho C_p)_{nf}}{(\rho C_p)_f}, A_4=\frac{k_{nf}}{k_f}, A_5=\frac{\mu_{nf}}{\mu_f}, A_6=\frac{\sigma_{nf}}{\sigma_f}, Da=\frac{K}{L^2} \tag{15}$$

and boundary conditions are:

$\theta=1.0$ on bottom wall

$\theta=0.0$ on other walls

$\Psi=0.0$ on all walls (16)

Local and average Nusselt over the hot wall can calculate as:

$$Nu_{loc}=\frac{\partial\theta}{\partial y}\left(\frac{k_{nf}}{k_f}\right) \tag{17}$$

$$Nu_{ave}=\frac{1}{L}\int_{r_{in}}^{r_{out}} Nu_{loc}\,dx \tag{18}$$

2.3. Effects of Active Parameters

In this paper, effect of uniform magnetic field on nanofluid forced convection in a porous semi annulus is investigated using Control volume based finite element method. Numerical outputs are illustrated for various values of Hartmann number ($Ha=0$ to 40), Darcy number ($Da=0.01$ to 100), Reynolds number ($\mathrm{Re}=100$ to 600) and volume fraction of CuO ($\phi=0\%$ to 4%). Table 4 shows the shape of nanoparticle effect on Nusselt number. Maximum values of Nusselt number is obtained if Platelet shape is selected for nanoparticles. So, Platelet shape selected for further investigation.

Figure 2 demonstrates the effect of nanofluid volume fraction on isotherms and streamlines. Augmenting nanofluid volume fraction leads to augment thermal boundary layer thickness. The nanofluid velocity increases due to increase in the solid movements. Figures 3, 4 and 5 demonstrate the influence of Darcy, Hartmann and Reynolds numbers on hydrothermal treatment. In absence of magnetic field, when Darcy

Figure 2. Influence of nanofluid volume fraction on streamlines (left) and isotherms (right) contours (nanofluid ($\phi=0.04$)(—) and pure fluid ($\phi=0$) (- - -)) when $\mathrm{Re}=600, Da=100$

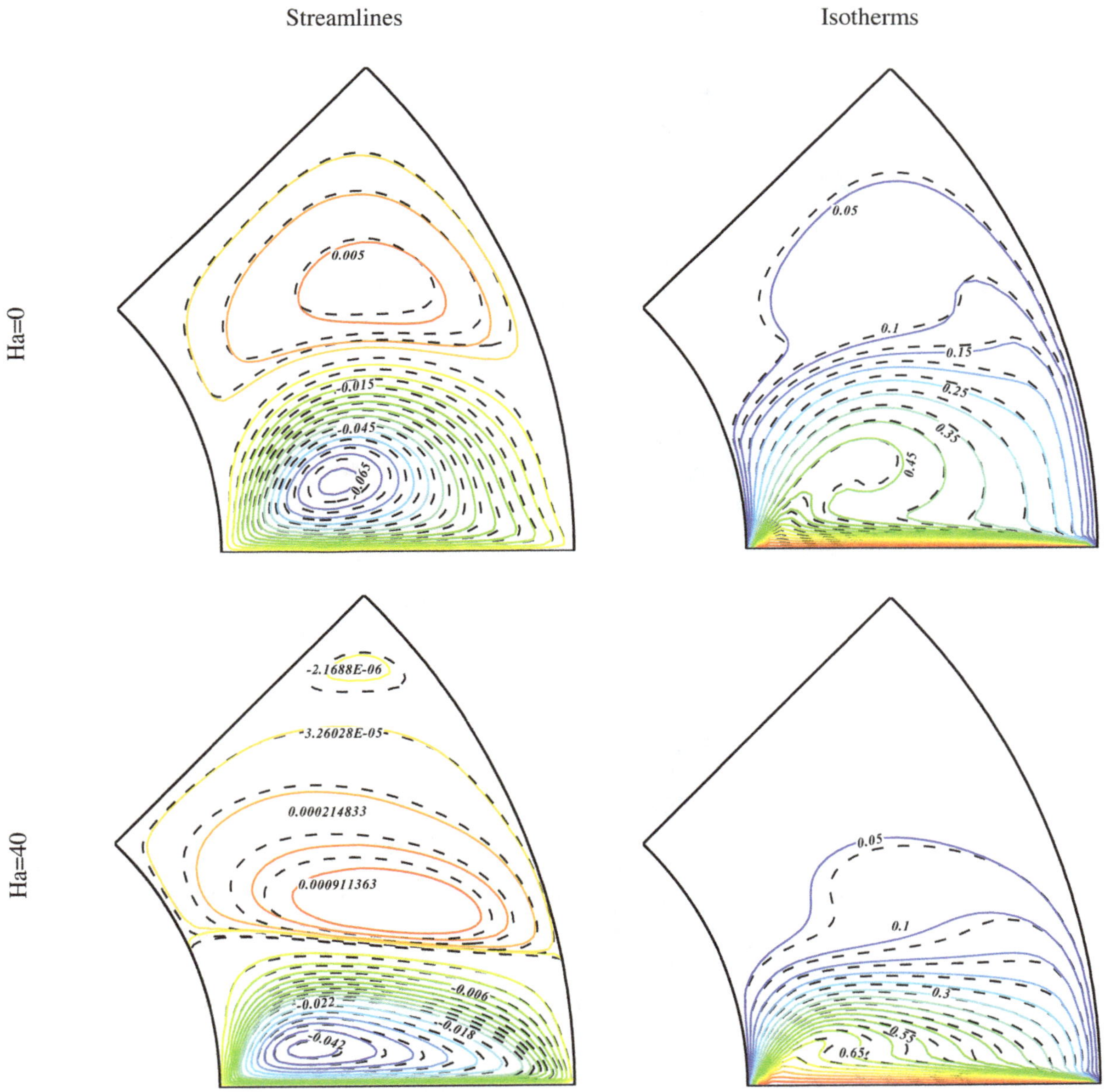

Figure 3. Effects of Darcy and Hartmann numbers on Isotherms (left) and streamlines (right) contours when $\phi = 0.04, \mathrm{Re} = 100$

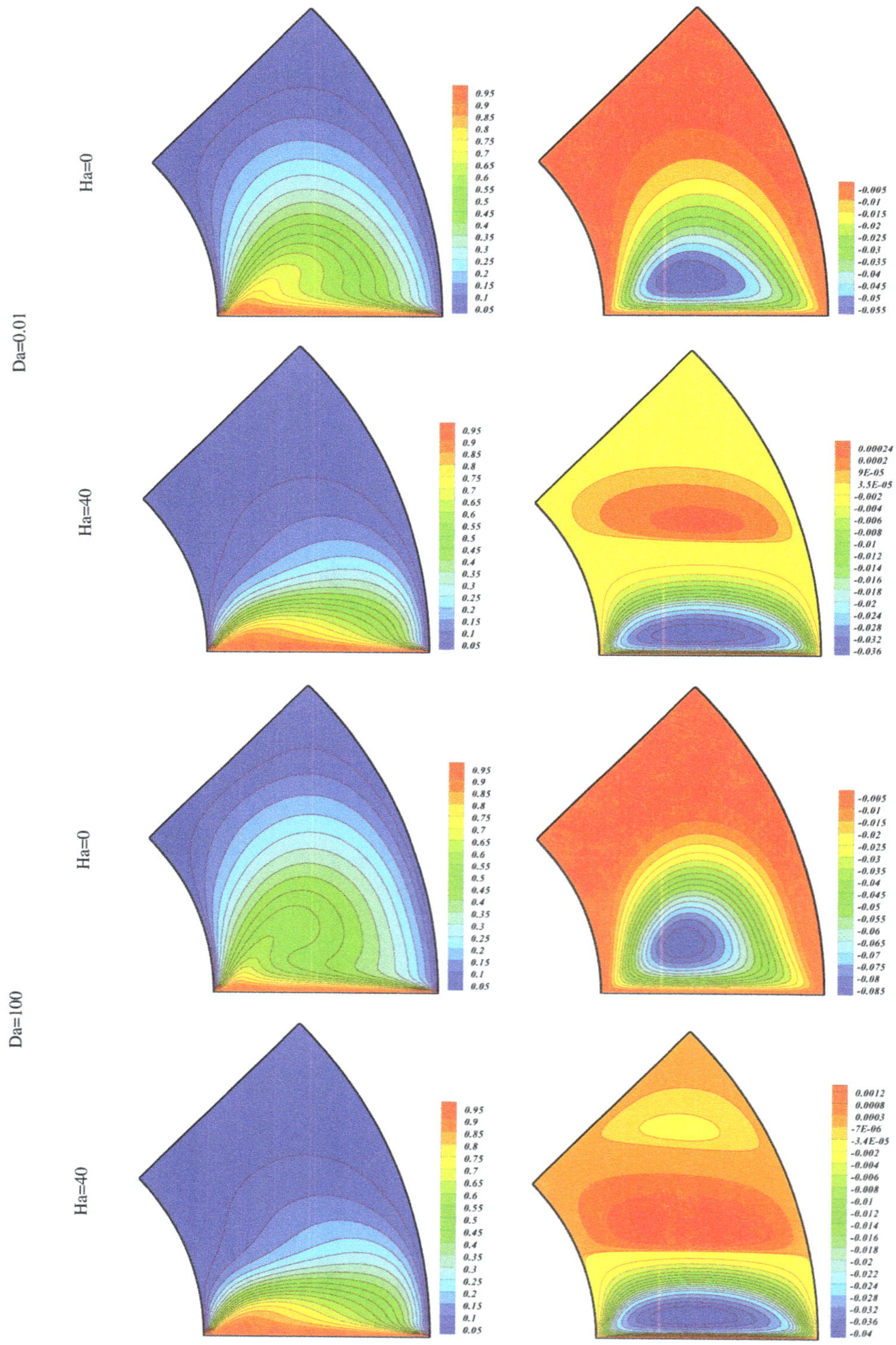

Figure 4. Effects of Darcy and Hartmann numbers on Isotherms (left) and streamlines (right) contours when $\phi = 0.04, \mathrm{Re} = 300$

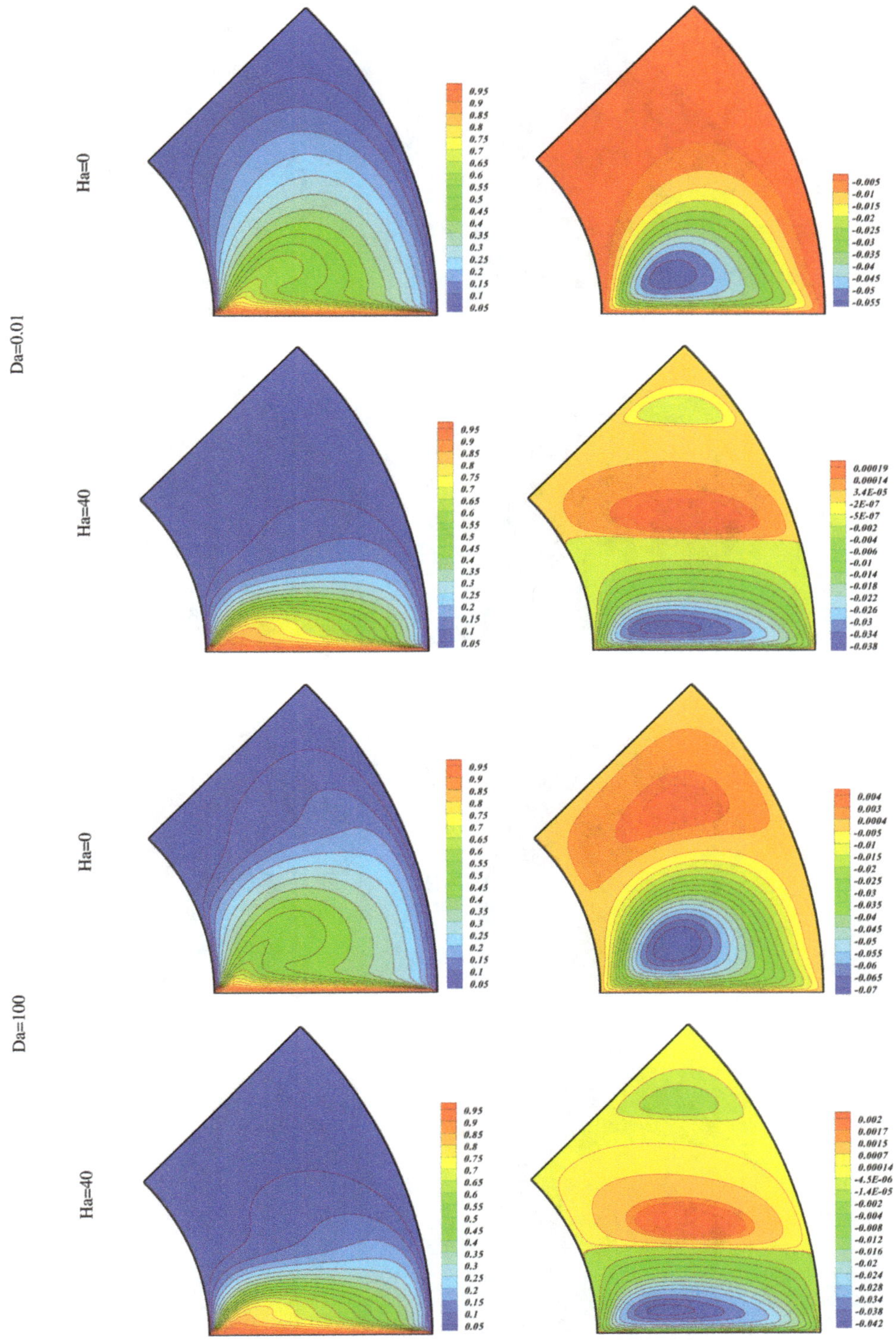

Figure 5. Effects of Darcy and Hartmann numbers on Isotherms (left) and streamlines (right) contours when $\phi = 0.04, \mathrm{Re} = 600$

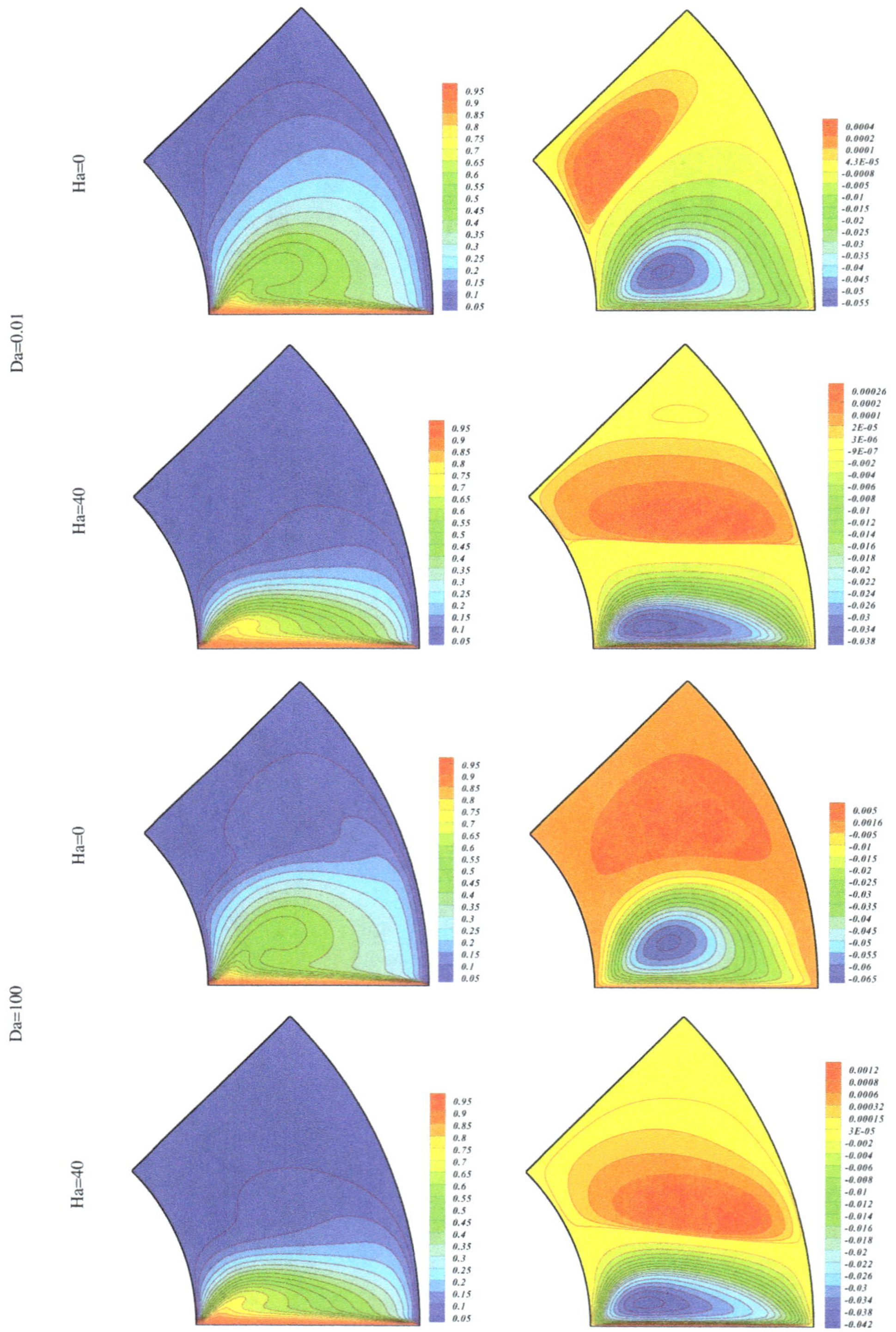

Table 4. Nusselt number for various shapes of nanoparticles when $\mathrm{Re} = 600, \phi = 0.04$

Da	Ha	Spherical	Brick	Cylinder	Platelet
0.01	0	14.45798	14.77039	15.25743	15.65244
0.01	40	12.44094	12.70772	13.12349	13.46058
100	0	15.1156	15.44261	15.95241	16.3659
100	40	12.71177	12.98315	13.40598	13.74869

and Reynolds number are low, only one eddy exists in streamline and isotherms are parallel together. As Hartmann number increases, the main eddy converts to two smaller ones. As Darcy number increases, the convective heat transfer enhances and three layers generates for streamlines in existence of magnetic field. As Reynolds number enhances, isotherms become denser to lid wall due to increment of convective mode. Also the absolute values of stream function augment with rise of Reynolds number. At Re=600, two eddies generates in absence of magnetic field. The upper one stretches to right side with increase of Darcy number. These eddies convert to three ones in presence of magnetic field. Increasing Lorentz forces make the isotherms becomes less dense. Also velocity decreases with rise of Hartmann number.

Influence of important parameters on average Nusselt number is shown in Figure 6. The correlation for average Nusselt number is as follows:

$$\begin{aligned} Nu_{ave} &= 10.74 + 0.7Da^{*} + 1.9\mathrm{Re}^{*} - 0.34Ha^{*} - 0.045Da^{*}\,\mathrm{Re}^{*} - 0.028Da^{*}\,Ha^{*} \\ &+ 0.05\,\mathrm{Re}^{*}\,Ha^{*} + 0.28Da^{*2} - 0.18\,\mathrm{Re}^{*2} - 0.09Ha^{*2} \end{aligned} \tag{19}$$

where $\mathrm{Re}^{*} = 0.01\,\mathrm{Re}, Da^{*} = 0.01Da, Ha^{*} = 0.1Ha$. Rate of heat transfer enhances with rise of Reynolds and Darcy numbers because of increment in convective heat transfer. Adding nanoparticles causes Nusselt number to increases due to increase in thermal conductivity of nanofluid. As Hartmann number increases the temperature gradient decreases and in turn Nusselt number reduces with rise of Lorentz forces.

3. EFFECT OF SHAPE FACTOR ON FE_3O_4-WATER NANOFLUID FORCED CONVECTION IN EXISTENCE OF EXTERNAL MAGNETIC FIELD

3.1. Problem Definition

Sample element, boundary condition and Geometry are demonstrated in Figure 7. The bottom wall is hot and other walls are cold. Also the bottom wall can move horizontally. Porous cavity filled with nanofluid and affected by horizontal magnetic field.

3.2. Governing Equation

2D steady convective non-Darcy flow of nanofluid is considered in existence of constant magnetic field. The PDEs are:

Figure 6. Influences of the Darcy, Reynolds number and Hartmann numbers on average Nusselt number

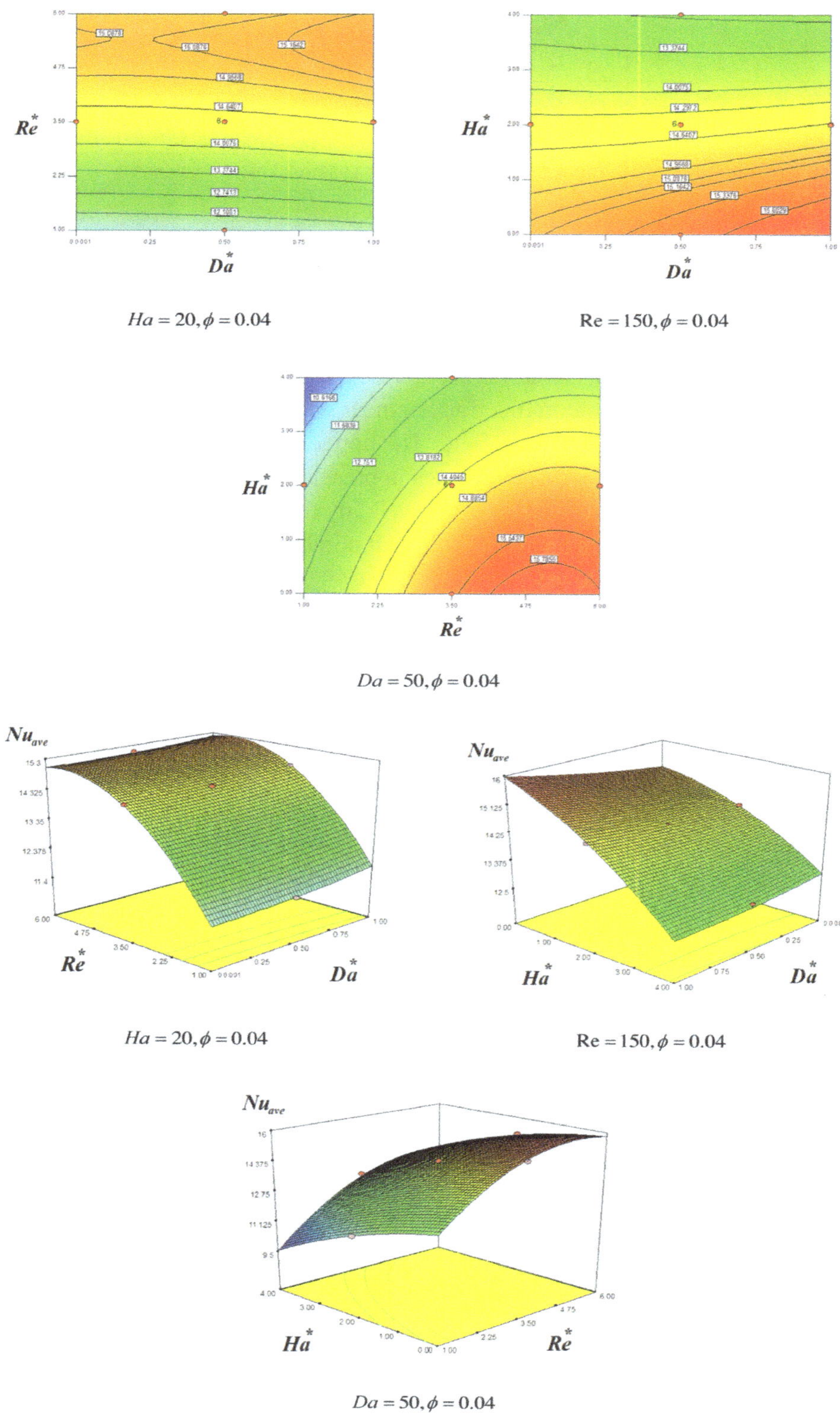

$Ha = 20, \phi = 0.04$

$\mathrm{Re} = 150, \phi = 0.04$

$Da = 50, \phi = 0.04$

$Ha = 20, \phi = 0.04$

$\mathrm{Re} = 150, \phi = 0.04$

$Da = 50, \phi = 0.04$

Figure 7. (a) Geometry and the boundary conditions with (b) the mesh of geometry considered in this work; (c) a sample triangular element and its corresponding control volume

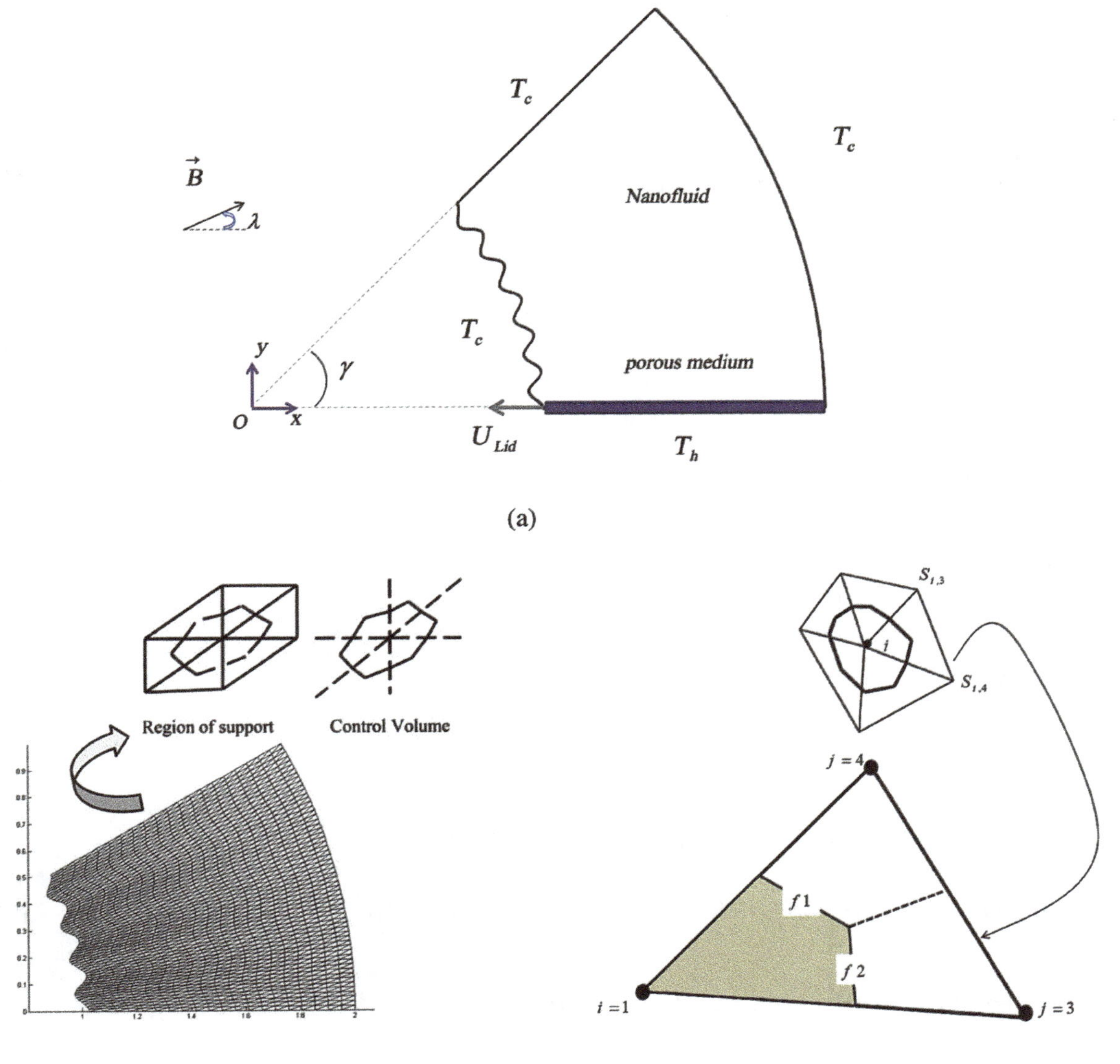

$$\frac{\partial v}{\partial y} + \frac{\partial u}{\partial x} = 0 \tag{20}$$

$$\left[-\sigma_{nf} B_y^2 u + \sigma_{nf} B_x B_y v + \left(\frac{\partial^2 u}{\partial y^2} + \frac{\partial^2 u}{\partial x^2}\right)\mu_{nf} - \frac{\partial P}{\partial x}\right] - \frac{\mu_{nf}}{K} u = \rho_{nf}\left(v\frac{\partial u}{\partial y} + \frac{\partial u}{\partial x} u\right) \tag{21}$$

$$\mu_{nf}\left(\frac{\partial^2 v}{\partial y^2}+\frac{\partial^2 v}{\partial x^2}\right)-\frac{\partial P}{\partial y}-B_x\sigma_{nf}B_x v+B_y\sigma_{nf}B_x u-\frac{\mu_{nf}}{K}v=\rho_{nf}\left(\frac{\partial v}{\partial x}u+\frac{\partial v}{\partial y}v\right), \quad (22)$$
$$B_x=B_o\cos\lambda,\ B_y=B_o\sin\lambda$$

$$\left(\rho C_p\right)_{nf}\left(v\frac{\partial T}{\partial y}+u\frac{\partial T}{\partial x}\right)=k_{nf}\left(\frac{\partial^2 T}{\partial x^2}+\frac{\partial^2 T}{\partial y^2}\right), \quad (23)$$

$\left(\rho C_p\right)_{nf}$, ρ_{nf}, k_{nf} and σ_{nf} are defined as:

$$\left(\rho C_p\right)_{nf}=\left(\rho C_p\right)_f(1-\phi)+\left(\rho C_p\right)_s\phi \quad (24)$$

$$\rho_{nf}=\rho_f(1-\phi)+\rho_s\phi \quad (25)$$

$$\frac{\sigma_{nf}}{\sigma_f}=1+\frac{3\left(\frac{\sigma_s}{\sigma_f}-1\right)\phi}{\left(\frac{\sigma_s}{\sigma_f}+2\right)-\left(\frac{\sigma_s}{\sigma_f}-1\right)\phi} \quad (26)$$

According to experimental outputs, μ_{nf} is obtained as follows:

$$\mu_{nf}=\left(0.035B^2+3.1B-27886.4807\phi^2+4263.02\phi+316.0629\right)e^{-0.01T} \quad (27)$$

Properties of nanoparticles and base fluid are provided table 5.

Table 5. Thermo physical properties of water and nanoparticles

	$\rho(kg/m^3)$	$C_p(j/kgk)$	$k(W/m.k)$	$d_p(nm)$	$\sigma\left(\Omega\cdot\mathrm{m}\right)^{-1}$
Pure water	997.1	4179	0.613	-	0.05
Fe_3O_4	5200	670	6	7.5	25000

k_{nf} can be expressed as:

$$\frac{k_{nf}}{k_f} = \frac{-m\left(k_f - k_p\right)\phi + \left(k_p - k_f\right)\phi + mk_f + k_p + k_f}{mk_f + \left(k_f - k_p\right)\phi + k_f + k_p} \tag{28}$$

Different values of shape factors for various shapes of nanoparticles are illustrated in table 3. Vorticity and stream function should be employed to eliminate pressure source terms:

$$\omega + \frac{\partial u}{\partial y} - \frac{\partial v}{\partial x} = 0, \frac{\partial \psi}{\partial x} = -v, \frac{\partial \psi}{\partial y} = u \tag{29}$$

Introducing dimensionless quantities:

$$\begin{gathered} U = \frac{u}{U_{Lid}}, V = \frac{v}{U_{Lid}}, \theta = \frac{T - T_c}{\Delta T}, \Delta T = T_h - T_c, \\ \left(X, Y\right) = \frac{\left(x, y\right)}{L}, \Psi = \frac{\psi L}{U_{Lid}}, \Omega = \frac{\omega}{LU_{Lid}} \end{gathered} \tag{30}$$

The final formulae are:

$$\frac{\partial^2 \Psi}{\partial X^2} + \Omega + \frac{\partial^2 \Psi}{\partial Y^2} = 0, \tag{31}$$

$$\begin{gathered} \frac{\partial \Omega}{\partial Y} V + \frac{\partial \Omega}{\partial X} U = \frac{1}{\mathrm{Re}} \frac{A_5}{A_1} \left(\frac{\partial^2 \Omega}{\partial Y^2} + \frac{\partial^2 \Omega}{\partial X^2} \right) + \frac{Ha^2}{\mathrm{Re}} \frac{A_6}{A_1} \left(\frac{\partial U}{\partial X} B_x B_y - \frac{\partial V}{\partial X} B_x^{\ 2} + \frac{\partial U}{\partial Y} B_y^2 - \frac{\partial V}{\partial Y} B_x B_y \right) \\ - \frac{1}{\mathrm{Re}\, Da} \frac{A_5}{A_1} \Omega, \end{gathered} \tag{32}$$

$$\frac{A_4}{A_2} \left(\frac{\partial^2 \theta}{\partial Y^2} + \frac{\partial^2 \theta}{\partial X^2} \right) = \Pr \mathrm{Re} \left(\frac{\partial \theta}{\partial X} U + \frac{\partial \theta}{\partial Y} V \right) \tag{33}$$

where dimensionless and constants parameters are illustrated as:

$$\mathrm{Re} = \frac{\rho_f U_{Lid} L}{\mu_f}, Ha = LB_0\sqrt{\sigma_f / \mu_f},$$
$$A_1 = \frac{\rho_{nf}}{\rho_f}, A_5 = \frac{\mu_{nf}}{\mu_f}, \tag{34}$$
$$A_2 = \frac{\left(\rho C_p\right)_{nf}}{\left(\rho C_p\right)_f}, A_4 = \frac{k_{nf}}{k_f}, A_6 = \frac{\sigma_{nf}}{\sigma_f}$$

and boundary conditions are:

$\theta = 1.0$ on bottom wall

$\theta = 0.0$ on other walls

$\Psi = 0.0$ on all walls (35)

Nu_{loc} and Nu_{ave} over the hot wall can calculate as:

$$Nu_{loc} = \frac{\partial \theta}{\partial y}\left(\frac{k_{nf}}{k_f}\right) \tag{36}$$

$$Nu_{ave} = \frac{1}{L}\int_{r_{in}}^{r_{out}} Nu_{loc}\, dx \tag{37}$$

3.3. Effects of Active Parameters

In this article, the impact of Lorentz forces on forced convection of nanofluid in a lid driven porous cavity is presented. The working fluid is Fe_3O_4-water and its viscosity is a function of nanofluid volume fraction and magnetic field. CVFEM is utilized to find the effects of Darcy number ($Da = 0.01$ to 100), Reynolds number ($\mathrm{Re} = 100$ to 500), volume fraction of Fe_3O_4 ($\phi = 0\%$ to 4%), shape of nanoparticle and Hartmann number ($Ha = 0$ to 40).

Impacts of shape of the nanoparticles on Nu_{ave} are discussed in Table 6. The maximum Nu is caused by Platelet, followed by Cylinder, Brick and Spherical. Therefore, Platelet nanoparticle has been selected for further investigation.

Impact of adding nanoparticles in the water on velocity and temperature contours is depicted in Figure 8. Temperature gradient reduces with augment of ϕ. $|\Psi_{max}|$ augments with adding nanoparticles because of increment in the solid movements.

Figures 9, 10, and 11 demonstrate the impact of Da, Re, Ha on isotherms and stream lines. At $Da = 0.01, \mathrm{Re} = 100, Ha = 0$, only one eddy exists in streamline. By applying magnetic field, two new

Table 6. Effect of shape of nanoparticles on Nusselt number when $\mathrm{Re} = 600, \phi = 0.04$

Da	Ha	Spherical	Brick	Cylinder	Platelet
0.01	0	12.90812	13.19201	13.635	13.99464
0.01	40	11.22557	11.47157	11.85536	12.16689
100	0	13.42214	13.71721	14.1776	14.55133
100	40	11.43061	11.67944	12.06752	12.38241

Figure 8. Influence of nanofluid volume fraction on streamlines (left) and isotherms (right) contours (nanofluid ($\phi = 0.04$ *)(—) and pure fluid(* $\phi = 0$ *) (- - -)) when* $\mathrm{Re} = 500, Da = 100$

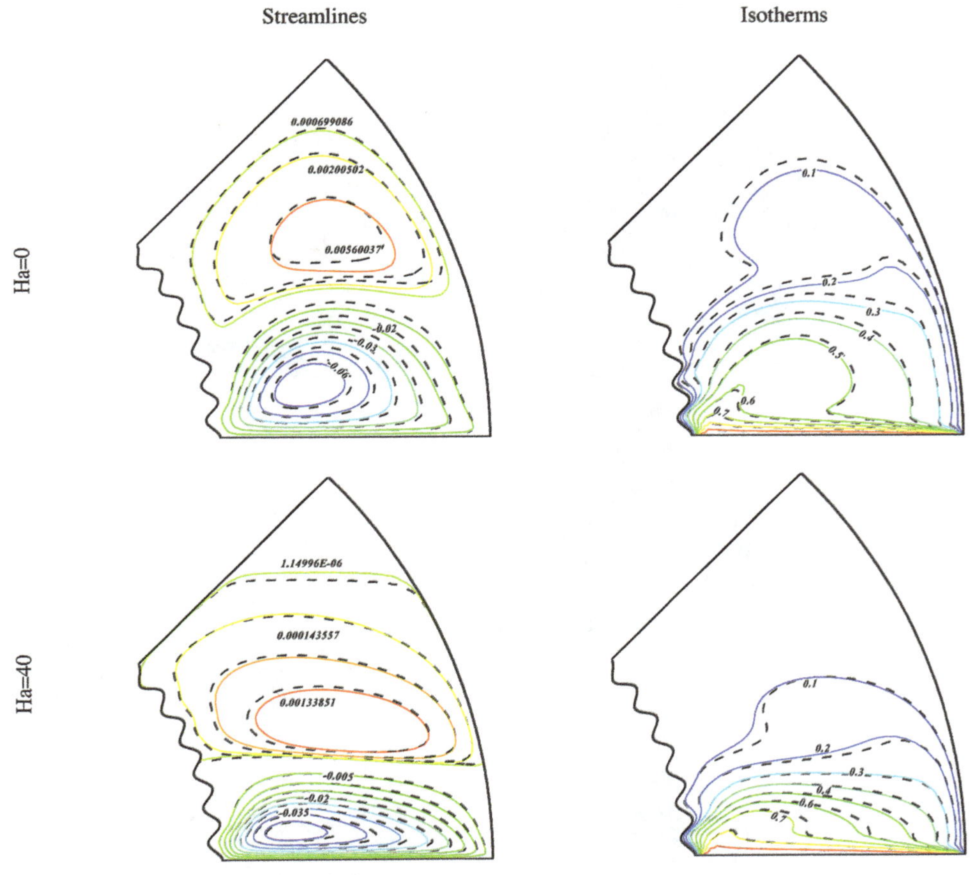

eddies with opposite direction generate above the main eddy. The distortion of isotherms becomes less than before. As Darcy number increases, convective mode becomes stronger and the primary eddy converts to two eddies. Increasing Hartmann number leads to generate the third eddy. As Re augments, the strength of new eddy enhances. In absence of magnetic field at Re=500, a counter clock wise eddy appears above the main eddy. Impacts of Darcy and Hartmann number on isotherms and streamlines at Re=500 is similar to those of Re=100 except the stream function values. It means that the absolute stream function values augment with augment of Reynolds number.

Figure 9. Effect of Darcy and Hartmann numbers on isotherms (left) and streamlines (right) contours when $\phi = 0.04, \mathrm{Re} = 100$

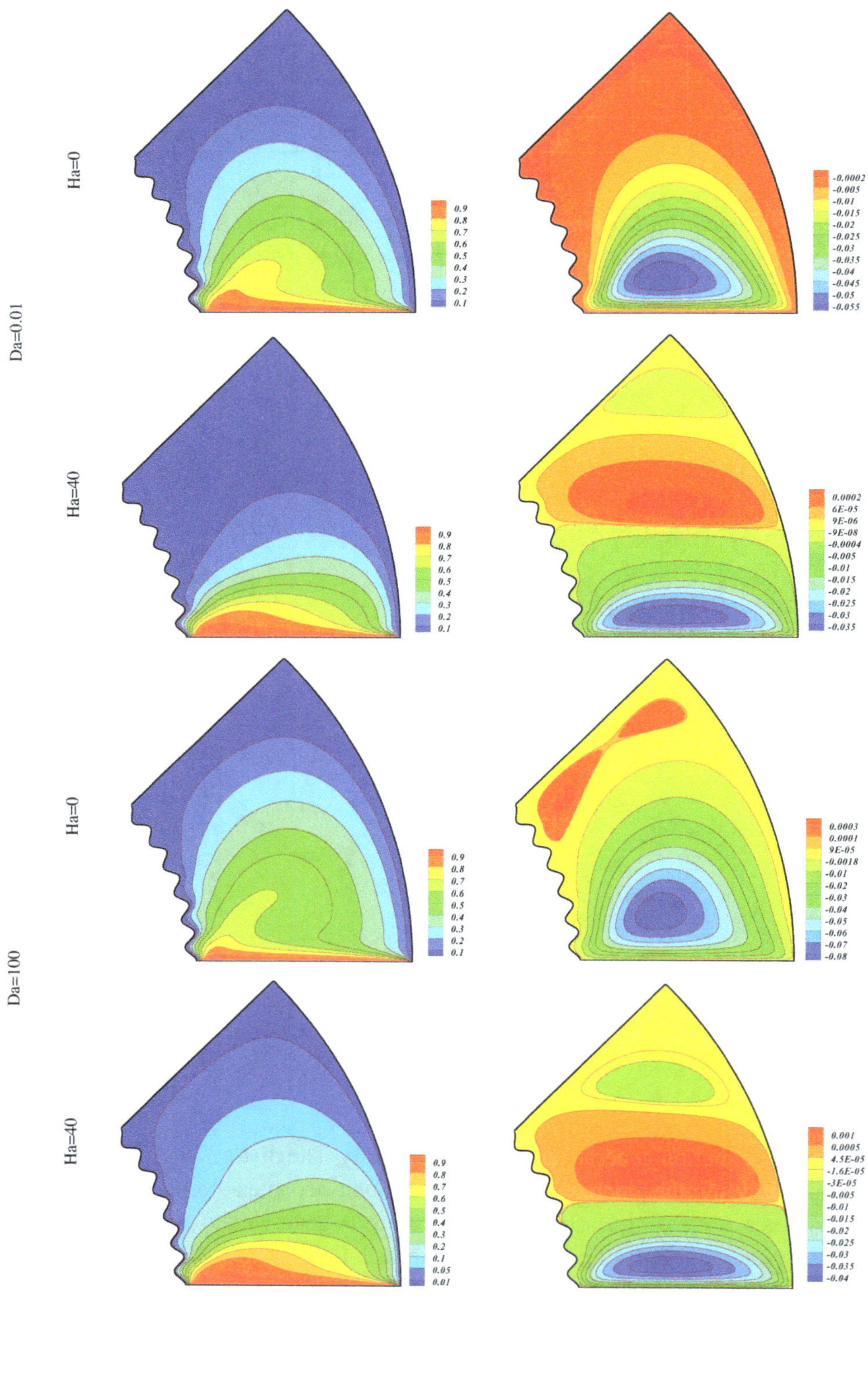

Figure 10. Effect of Darcy and Hartmann numbers on isotherms (left) and streamlines (right) contours when $\phi = 0.04, \mathrm{Re} = 300$

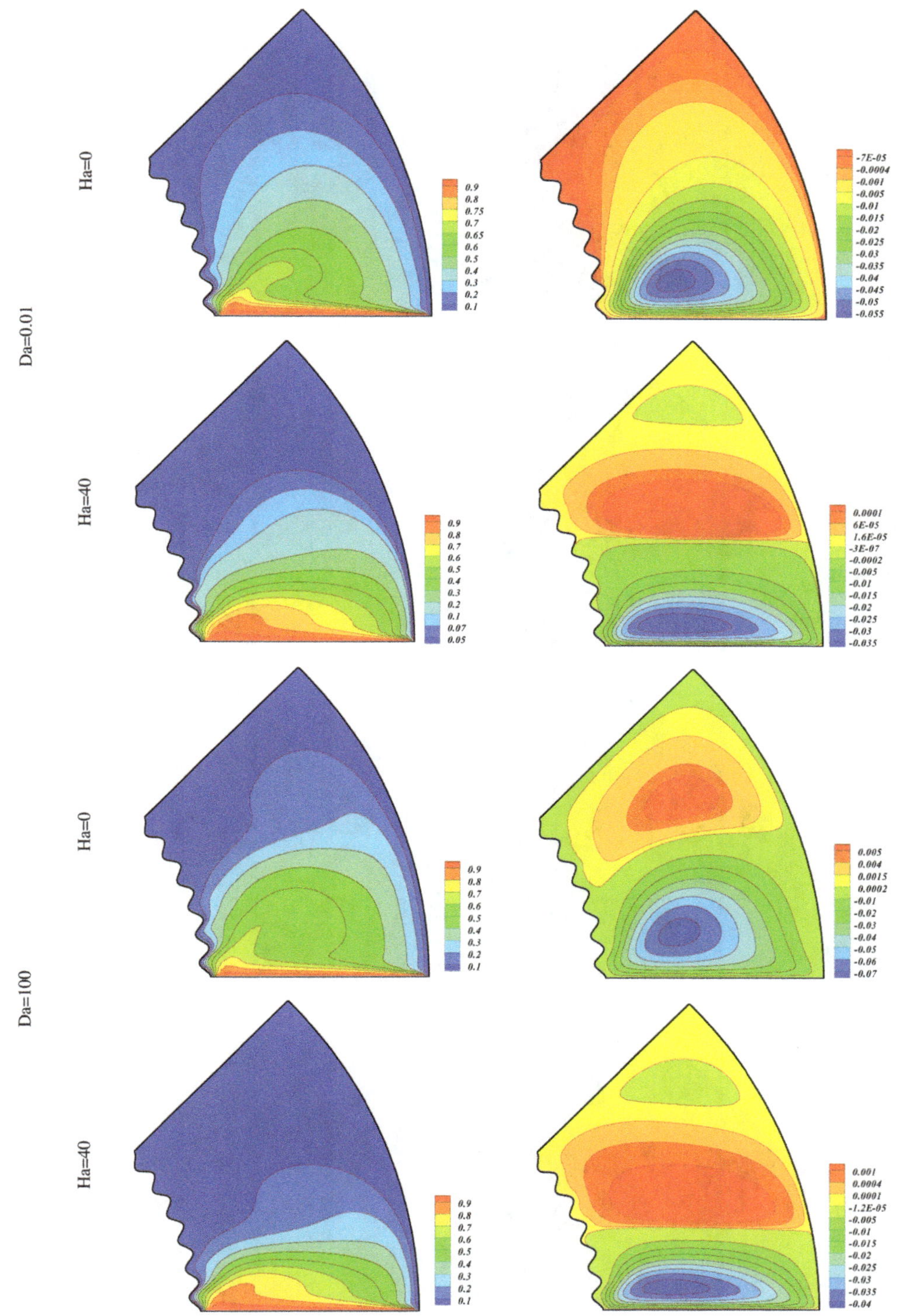

Figure 11. Effect of Darcy and Hartmann numbers on isotherms (left) and streamlines (right) contours when $\phi = 0.04, \mathrm{Re} = 500$

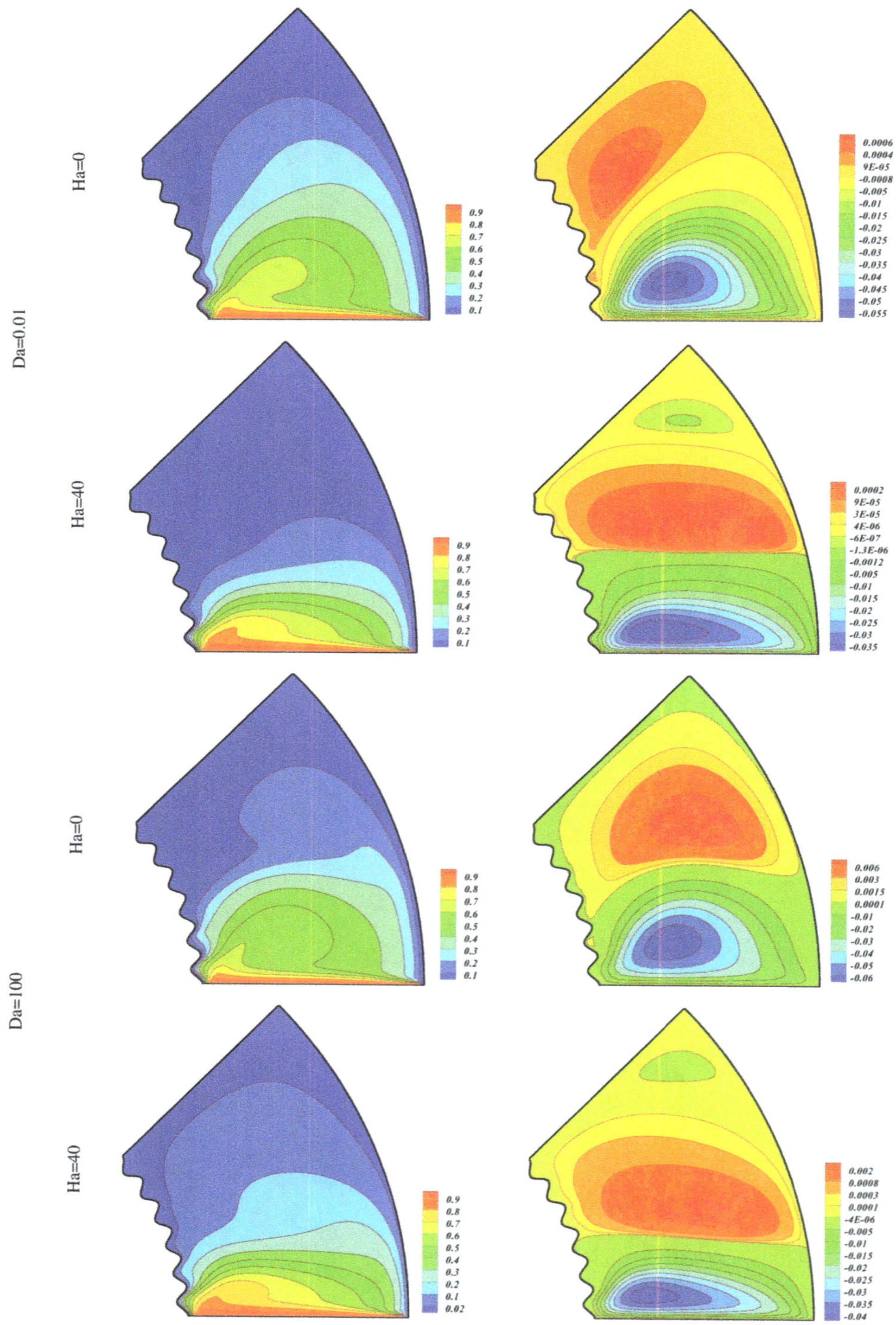

Influence of significant parameters on Nu_{ave} is illustrated in Figure 12. The correlation for Nu_{ave} is as follows:

$$\begin{aligned}Nu_{ave} &= 9.94 + 1.69\,\mathrm{Re}^* + 0.66Da^* - 0.18Ha^* - 0.057Da^*\,\mathrm{Re}^* + 0.051\,\mathrm{Re}^*\,Ha^* \\ &-0.13Da^*\,Ha^* - 0.18\,\mathrm{Re}^{*2} + 0.49Da^{*2} - 0.13Ha^{*2}\end{aligned} \quad (38)$$

where $\mathrm{Re}^* = 0.01\,\mathrm{Re}, Ha^* = 0.1Ha, Da^* = 0.01Da$. Increasing Reynolds number can enhance the heat transfer rate because of reduction in thermal boundary layer thickness. Similar trend is seen for Darcy number. Temperature gradient decreases with augment of Lorentz forces. So, Nu_{ave} decreases with rise of Hartmann number.

4. MAGNETIC SOURCE EFFECT ON NANOFLUID FLOW IN POROUS MEDIUM CONSIDERING VARIOUS SHAPE OF NANOPARTICLES

4.1. Problem Definition

Geometry, boundary condition and sample element are illustrated (see Figure 13). External magnetic source is applied (see Figure 14). $\overline{H}, \overline{H}_x, \overline{H}_y$ are:

$$\overline{H}_y = \left(\bar{a} - x\right)\frac{\gamma}{2\pi}\left[\left(\bar{b} - y\right)^2 + \left(\bar{a} - x\right)^2\right]^{-1}, \quad (39)$$

$$\overline{H}_x = \left(y - \bar{b}\right)\frac{\gamma}{2\pi}\left[\left(\bar{b} - y\right)^2 + \left(\bar{a} - x\right)^2\right]^{-1}, \quad (40)$$

$$\overline{H} = \left(\overline{H}_y^2 + \overline{H}_x^2\right)^{0.5}. \quad (41)$$

4.2. Governing Equation

2D steady convective non-Darcy flow of nanofluid is considered in existence of variable magnetic field. The PDEs are:

$$\frac{\partial u}{\partial x} + \frac{\partial v}{\partial y} = 0, \quad (42)$$

Figure 12. Influences of the Darcy, Reynolds number, and Hartmann numbers on average Nusselt number

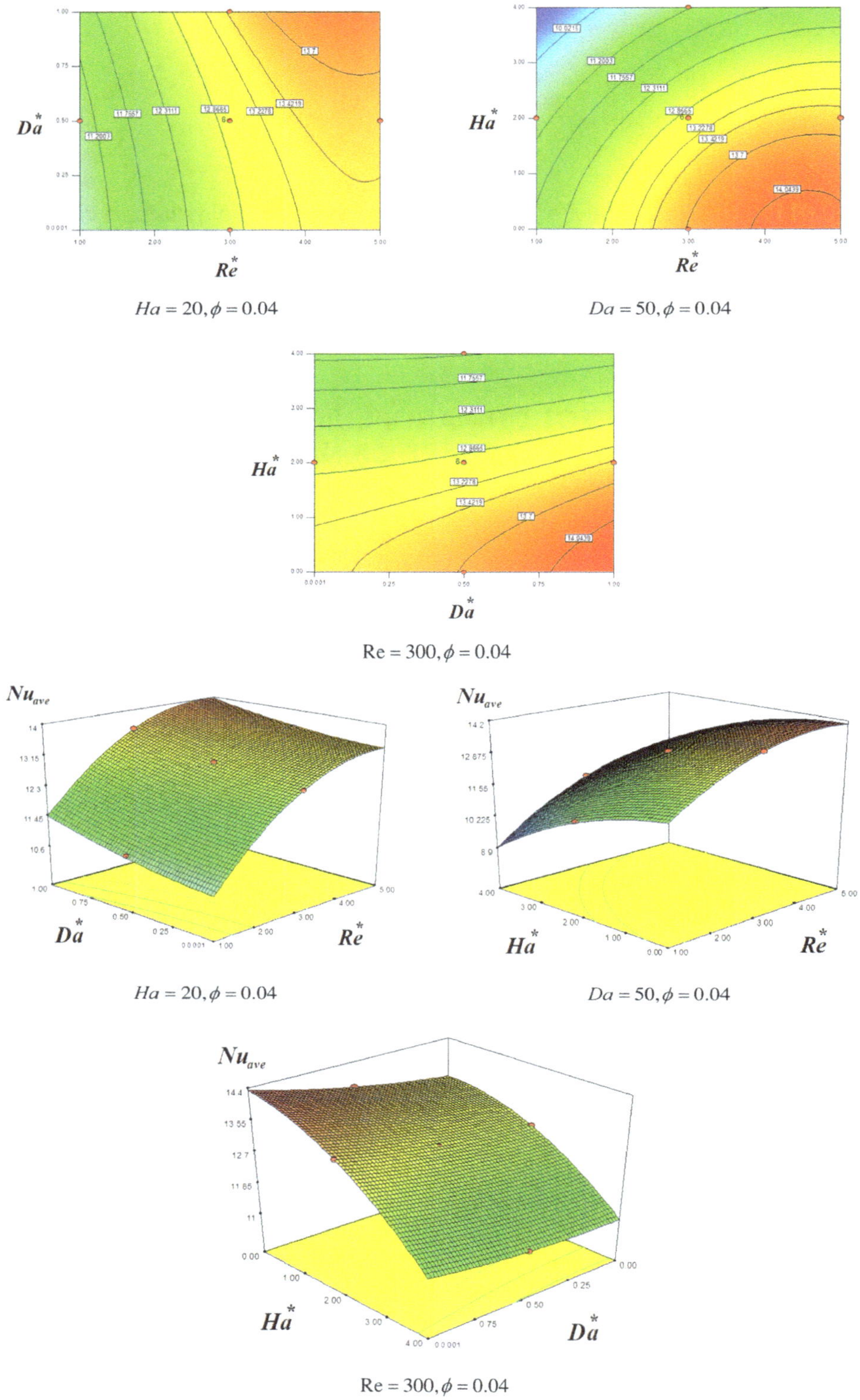

Figure 13. (a) Geometry and the boundary conditions with; (b) sample for mesh; (c) a sample triangular element and its corresponding control volume

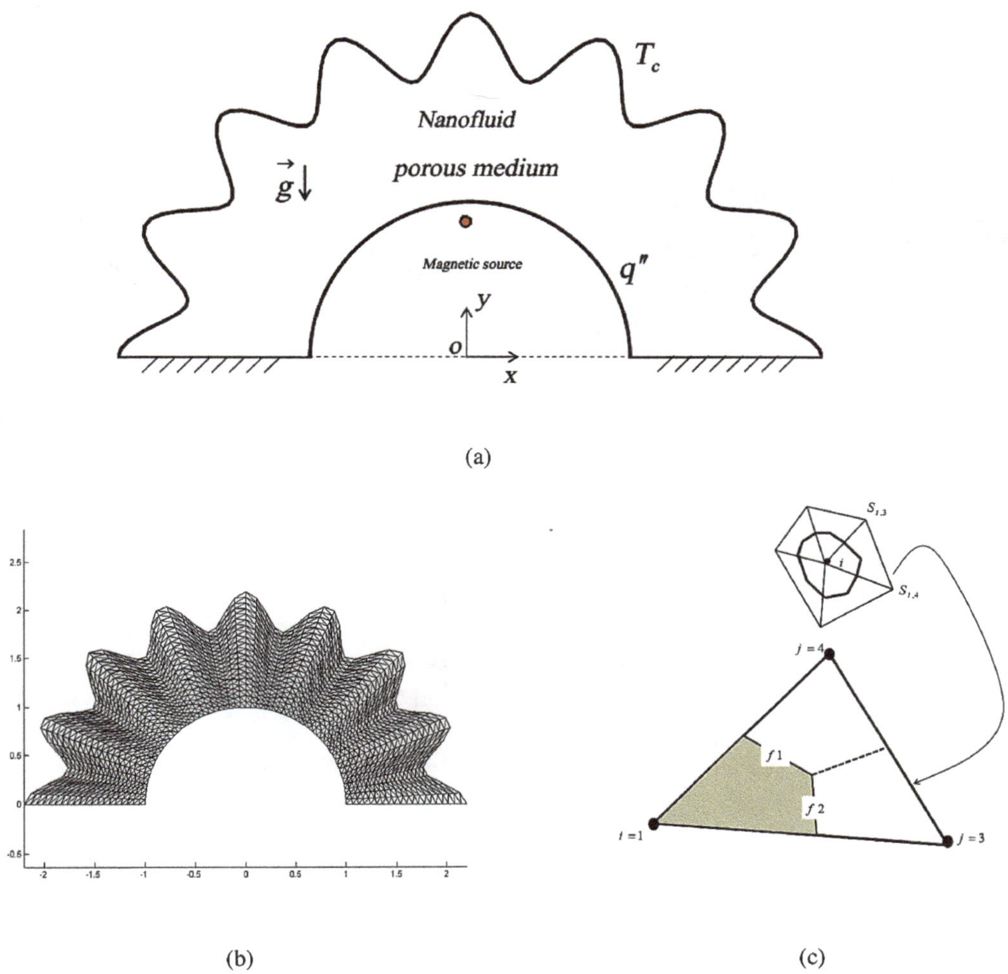

$$\left(\frac{\partial^2 u}{\partial y^2}+\frac{\partial^2 u}{\partial x^2}\right)\mu_{nf}-\frac{\partial P}{\partial x}-\mu_0^2\sigma_{nf}\bar{H}_y^2 u+\sigma_{nf}\mu_0^2\bar{H}_x\bar{H}_y v$$
$$-\frac{\mu_{nf}}{K}u=\left(\rho_{nf}\right)\left(\frac{\partial u}{\partial x}u+\frac{\partial u}{\partial y}v\right), \tag{43}$$

$$\mu_{nf}\left(\frac{\partial^2 v}{\partial x^2}+\frac{\partial^2 v}{\partial y^2}\right)-\frac{\partial P}{\partial y}+\mu_0^2\bar{H}_y\sigma_{nf}\bar{H}_x u-\mu_0^2\bar{H}_x\sigma_{nf}\bar{H}_x v-\frac{\mu_{nf}}{K}v$$
$$+\left(T-T_c\right)\beta_{nf}g\rho_{nf}=\rho_{nf}\left(\frac{\partial v}{\partial x}u+\frac{\partial v}{\partial y}v\right), \tag{44}$$

Figure 14. Contours of the (a) magnetic field strength H; (b) magnetic field intensity component in x direction Hx; (c) magnetic field intensity component in y direction Hy

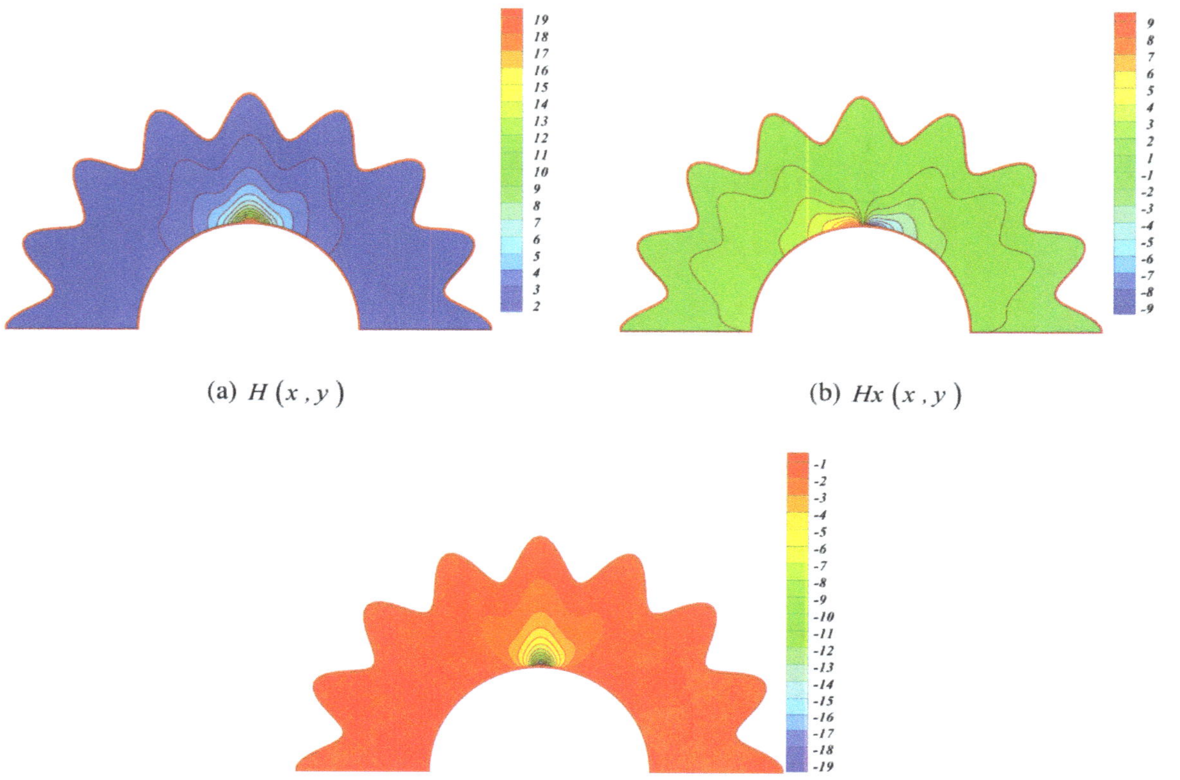

(a) $H(x,y)$ (b) $Hx(x,y)$

(c) $Hy(x,y)$

$$k_{nf}\left(\frac{\partial^2 T}{\partial y^2}+\frac{\partial^2 T}{\partial x^2}\right)-\frac{\partial q_r}{\partial y}=\left(v\frac{\partial T}{\partial y}+u\frac{\partial T}{\partial x}\right)\left(\rho C_p\right)_{nf},$$
$$\left[q_r=-\frac{4\sigma_e}{3\beta_R}\frac{\partial T^4}{\partial y},\quad T^4\cong 4T_c^3T-3T_c^4\right]. \tag{45}$$

$\left(\rho C_p\right)_{nf}$, $\left(\rho\beta\right)_{nf}$, ρ_{nf}, k_{nf} and σ_{nf} are calculated as:

$$\left(\rho C_p\right)_{nf}=\left(\rho C_p\right)_f(1-\phi)+\left(\rho C_p\right)_s\phi \tag{46}$$

$$\rho_{nf}=\rho_f(1-\phi)+\rho_s\phi \tag{47}$$

$$\left(\rho\beta\right)_{nf} = \left(\rho\beta\right)_{f}(1-\phi) + \left(\rho\beta\right)_{s}\phi, \tag{48}$$

$$\sigma_{nf} = \sigma_{f}\left[\frac{3\phi\left(\sigma 1 - 1\right)}{\left(1-\sigma 1\right)\phi + \left(2+\sigma 1\right)} + 1\right], \sigma 1 = \sigma_{s} / \sigma_{f}. \tag{49}$$

μ_{nf} is obtained as follows:

$$\mu_{nf} = \left(0.035\mu_0^2 H^2 + 3.1\mu_0 H - 27886.4807\phi^2 + 4263.02\phi + 316.0629\right)e^{-0.01T} \tag{50}$$

k_{nf} can be expressed as:

$$\frac{k_{nf}}{k_f} = \frac{-m\left(k_f - k_p\right)\phi + \left(k_p - k_f\right)\phi + mk_f + k_p + k_f}{mk_f + \left(k_f - k_p\right)\phi + k_f + k_p} \tag{51}$$

Vorticity and stream function should be used to eliminate pressure source terms:

$$\omega + \frac{\partial u}{\partial y} - \frac{\partial v}{\partial x} = 0, \frac{\partial \psi}{\partial x} = -v, \frac{\partial \psi}{\partial y} = u. \tag{52}$$

Dimensionless parameters are defined as:

$$\begin{gathered}\left(H_y, H_x, H\right) = \frac{\left(\overline{H_y}, \overline{H_x}, \overline{H}\right)}{\overline{H_0}}, \left(b, a\right) = \frac{\left(\bar{b}, \bar{a}\right)}{L}, \\ U = \frac{uL}{\alpha_{nf}}, V = \frac{vL}{\alpha_{nf}}, \left(X, Y\right) = \frac{\left(x, y\right)}{L}, \theta = \frac{T - T_c}{\Delta T}, \Delta T = q''L / k_f, \\ \Psi = \frac{\psi}{\alpha_{nf}}, \Omega = \frac{\omega L^2}{\alpha_{nf}}.\end{gathered} \tag{53}$$

So equations change to:

$$\frac{\partial^2 \Psi}{\partial Y^2} + \Omega + \frac{\partial^2 \Psi}{\partial X^2} = 0, \tag{54}$$

$$\frac{\partial \Omega}{\partial X}U + V\frac{\partial \Omega}{\partial Y} = \Pr\frac{A_5}{A_1}\frac{A_2}{A_4}\left(\frac{\partial^2 \Omega}{\partial Y^2} + \frac{\partial^2 \Omega}{\partial X^2}\right)$$
$$+\Pr Ha^2\frac{A_6}{A_1}\frac{A_2}{A_4}\left(\frac{\partial U}{\partial X}H_y H_x - \frac{\partial V}{\partial X}H_x^{\ 2} + \frac{\partial U}{\partial Y}H_y^2 - \frac{\partial V}{\partial Y}H_y H_x\right) \tag{55}$$
$$+\Pr Ra\frac{A_3 A_2^2}{A_1 A_4^2}\frac{\partial \theta}{\partial X} - \frac{\Pr}{Da}\frac{A_5}{A_1}\frac{A_2}{A_4},$$

$$\frac{\partial \theta}{\partial X}U + V\frac{\partial \theta}{\partial Y} = \left(\frac{\partial^2 \theta}{\partial X^2} + \frac{\partial^2 \theta}{\partial Y^2}\right) + \frac{4}{3}\frac{1}{A_4}Rd\frac{\partial^2 \theta}{\partial Y^2}. \tag{56}$$

and dimensionless parameters are:

$$Ra_f = g\beta_f L^3 \Delta T / \left(\alpha_f \upsilon_f\right), \Pr_f = \upsilon_f / \alpha_f, Ha = L\mu_0 H_0\sqrt{\sigma_f / \mu_f},$$
$$Da = K / L^2, \mathrm{Ec} = \left(\mu_f \alpha_f\right) / \left[\left(\rho C_P\right)_f \Delta T\, L^2\right],$$
$$A_1 = \frac{\rho_{nf}}{\rho_f}, A_2 = \frac{\left(\rho C_p\right)_{nf}}{\left(\rho C_p\right)_f}, A_3 = \frac{\left(\rho\beta\right)_{nf}}{\left(\rho\beta\right)_f}, A_4 = \frac{k_{nf}}{k_f}, A_5 = \frac{\mu_{nf}}{\mu_f}, A_6 = \frac{\sigma_{nf}}{\sigma_f}, \tag{57}$$
$$Rd = 4\sigma_e T_c^3 / \left(\beta_R k_f\right)$$

and boundary conditions are:

on inner wall $\frac{\partial \theta}{\partial n} = 1.0$

on outer wall $\theta = 0.0$

on other walls $\frac{\partial \theta}{\partial n} = 0$

on all walls $\Psi = 0.0$ (58)

Nu_{loc}, Nu_{ave} over the hot wall can be calculated as:

$$Nu_{loc} = \left(\frac{k_{nf}}{k_f}\right)\left(1 + \left(\frac{k_{nf}}{k_f}\right)^{-1}\frac{4Rd}{3}\right)\frac{1}{\theta}, \tag{59}$$

$$Nu_{ave} = \frac{1}{\pi}\int_0^{\pi} Nu_{loc}\, d\xi. \tag{60}$$

4.3. Effects of Active Parameters

Fe_3O_4-water nanofluid flow in a permeable enclosure in existence of variable Lorentz force is investigated numerically. Shape factor and magnetic field influences on nanofluid properties have been considered. CVFEM is employed to find the influences of radiation parameter $\left(Rd = 0\, to\, 0.8\right)$, Darcy number ($Da = 0.01$ to 100), Rayleigh number ($Ra = 10^3, 10^4, 10^5$), volume fraction of Fe_3O_4 ($\phi = 0\%$ to 4%), shape of nanoparticle and Hartmann number ($Ha = 0$ to 10).

Table 7 demonstrates the impact of shape factor on rate of heat transfer. The maximum and minimum values of Nusselt number are obtained for Platelet and Spherical shapes, respectively. So, Platelet shape is selected for further investigation. Influence of adding nanoparticles in to water on flow and heat transfer is depicted in Figure 15. $\left|\Psi_{max}\right|$ increases with adding nanoparticles because of increment in the solid movements. Isotherms become less dense with increase of ϕ. Figure 16 demonstrates the influence of Rd on streamlines and isotherms contours. Thermal boundary layer thickness increases with augment of radiation parameter. By increasing Lorentz forces, impact of radiation parameter on streamlines becomes no significant. Figures 17, 18 and 19 demonstrate the impact of Da, Ra, Ha on isotherms and stream lines. Only one eddy appears in streamline. By applying magnetic field, the main eddy moves downward. The distortion of isotherms becomes less than before. As *Ra* enhances, thermal plume generates near the vertical symmetric line. Increasing Lorentz forces makes $\left|\Psi_{max}\right|$ to decrease and shifts the thermal plume to left side. As Darcy number increases, convective mode becomes stronger and the primary eddy converts to two eddies.

Effects of important parameters on Nu_{ave} are demonstrated in Figure 20. The correlation for Nu_{ave} is as follows:

$$\begin{aligned} Nu_{ave} &= 27.97 - 3.84Rd - 13.2\log\left(Ra\right) - 0.6Da^* + 1.5Ha^* \\ &+ 2.32Rd\log\left(Ra\right) + 0.5Rd\,Da^* - 0.79Rd\,Ha^* \\ &+ 0.25Da^*\log\left(Ra\right) - 0.4Ha^*\log\left(Ra\right) - 0.64Da^*Ha^* \\ &- 3.11Rd^2 + 1.72\left(\log\left(Ra\right)\right)^2 + 0.33\left(Da^*\right)^2 - 0.22\left(Ha^*\right)^2 \end{aligned} \tag{61}$$

Table 7. Effect of shape of nanoparticles on Nusselt number when $Da = 100, Ra = 10^5, Rd = 0.8, \phi = 0.04$

	Ha	
	0	**10**
Spherical	10.80306	8.304584
Brick	10.87294	8.344276
Cylinder	10.98493	8.407705
Platelet	11.07831	8.460315

Figure 15. Impact of nanofluid volume fraction on streamlines (up) and isotherms (bottom) contours (nanofluid ($\phi = 0.04$)(—) and pure fluid ($\phi = 0$) ($- \cdot -$)) when $Ra = 10^5, Da = 100, Rd = 0.8$

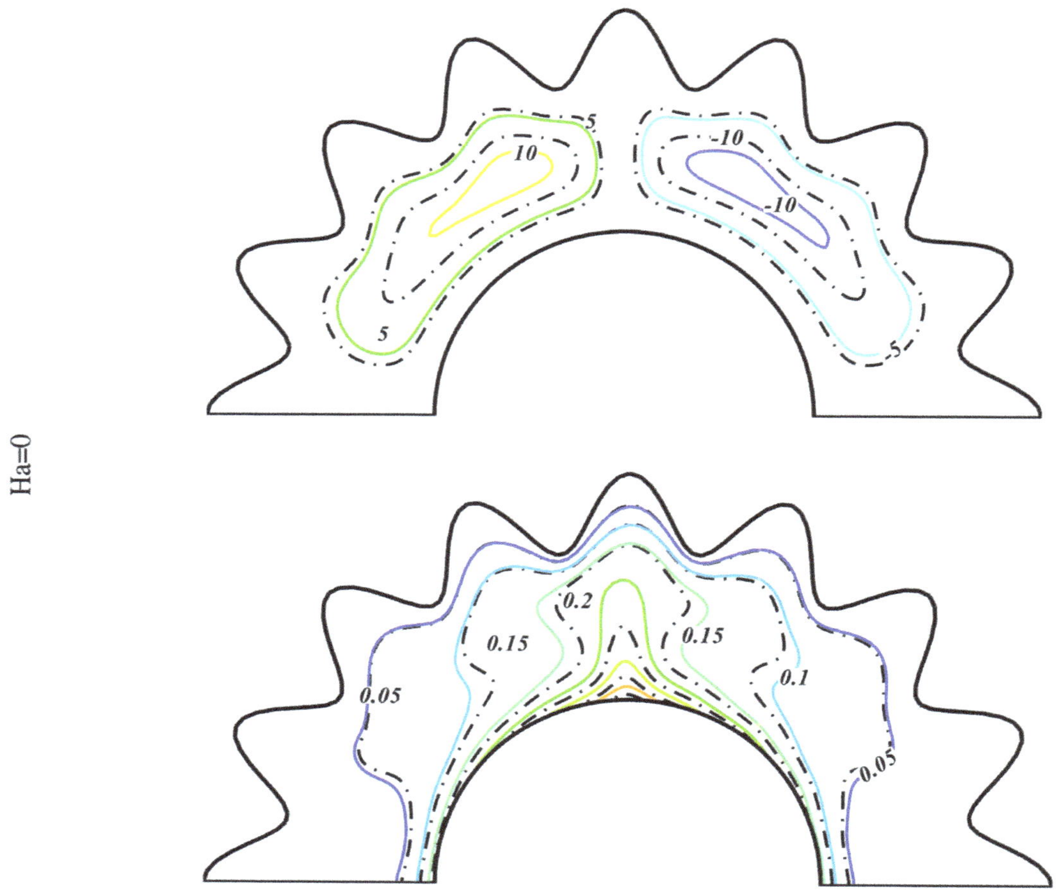

where $Da^* = 0.01Da, Ha^* = 0.1Ha$. As convective heat transfer augments, Nusselt number increases. So, Nusselt number has direct relationship with permeability of porous media and buoyancy forces. Lorentz forces make the conduction mechanism to improve. So Nusselt number has reverse relationship with Hartmann number.

5. MAGNETOHYDRODYNAMIC CUO-WATER TRANSPORTATION INSIDE A POROUS CAVITY CONSIDERING SHAPE FACTOR EFFECT

5.1. Problem Definition

Geometry, boundary condition and sample element are depicted (see Figure 21). The bottom wall is hot and other walls are cold. Also the bottom wall can move horizontally. Porous cavity filled with nanofluid and affected by horizontal magnetic field.

Figure 16. Impact of radiation parameter on streamlines (up) and isotherms (bottom) contours ($Rd = 0.8$ (—), $Rd = 0$ (- - -)) when $Ra = 10^5, Da = 100, \phi = 0.04$

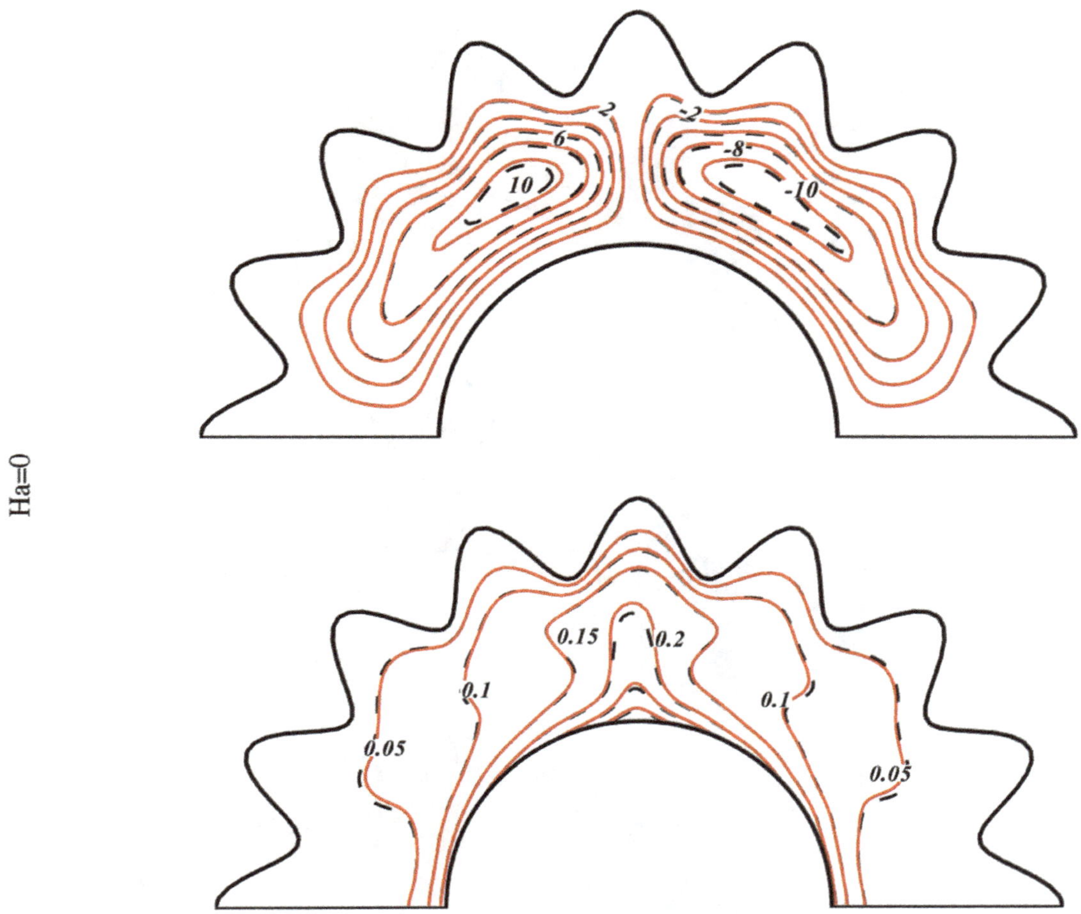

5.2. Governing Equation

2D steady convective non-Darcy flow of nanofluid is considered in existence of constant magnetic field. The PDEs are:

$$\frac{\partial v}{\partial y} + \frac{\partial u}{\partial x} = 0 \tag{62}$$

$$\left[-\sigma_{nf} B_y^2 u + \sigma_{nf} B_x B_y v + \left(\frac{\partial^2 u}{\partial y^2} + \frac{\partial^2 u}{\partial x^2}\right)\mu_{nf} - \frac{\partial P}{\partial x}\right] - \frac{\mu_{nf}}{K} u = \rho_{nf}\left(v\frac{\partial u}{\partial y} + \frac{\partial u}{\partial x} u\right) \tag{63}$$

Figure 17. Influence of Da, Ha *on streamlines (left) and isotherms (right) contours when* $\phi = 0.04, Ra = 10^3$

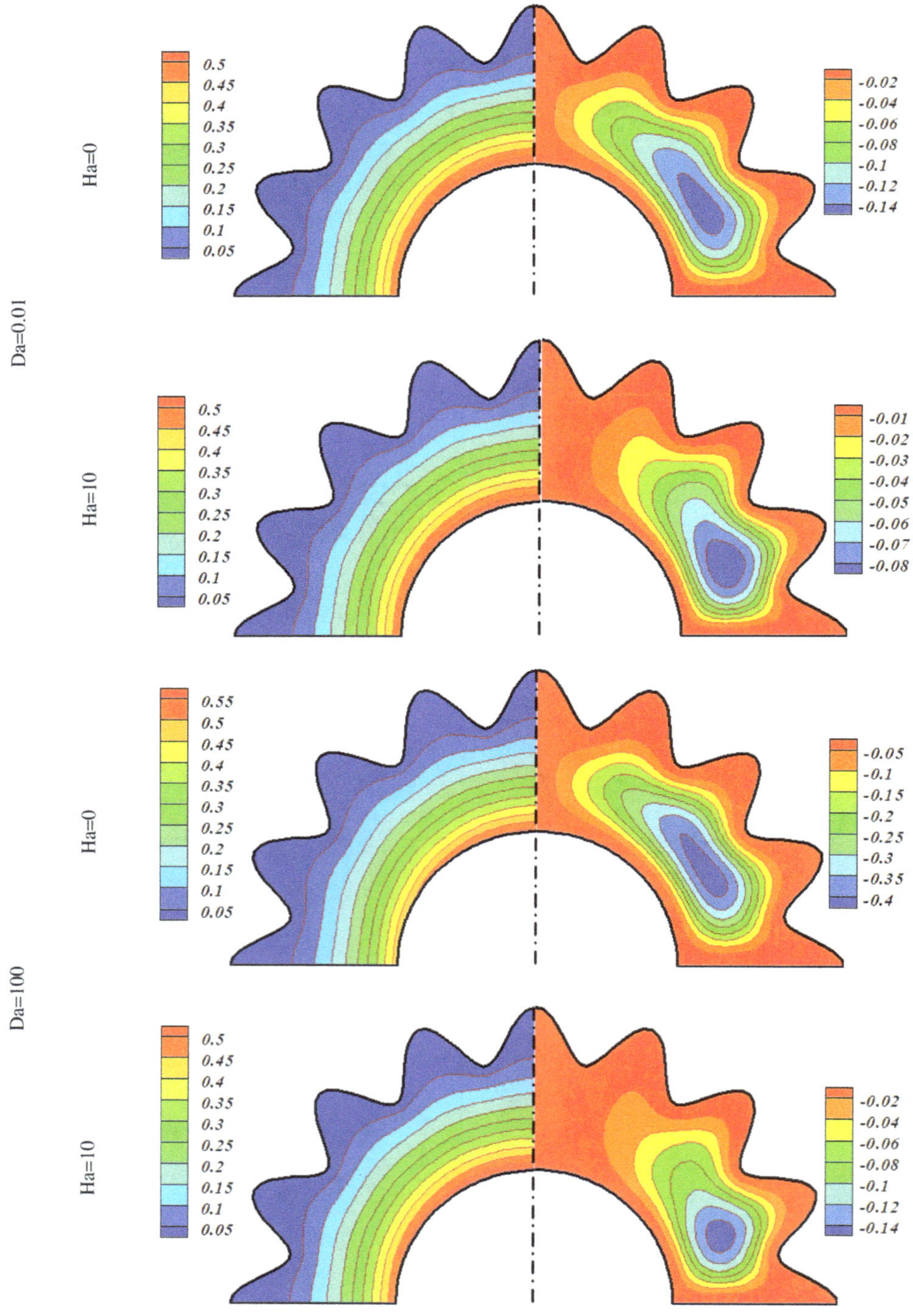

Figure 18. Influence of Da, Ha *on streamlines (left) and isotherms (right) contours when* $\phi = 0.04, Ra = 10^4$

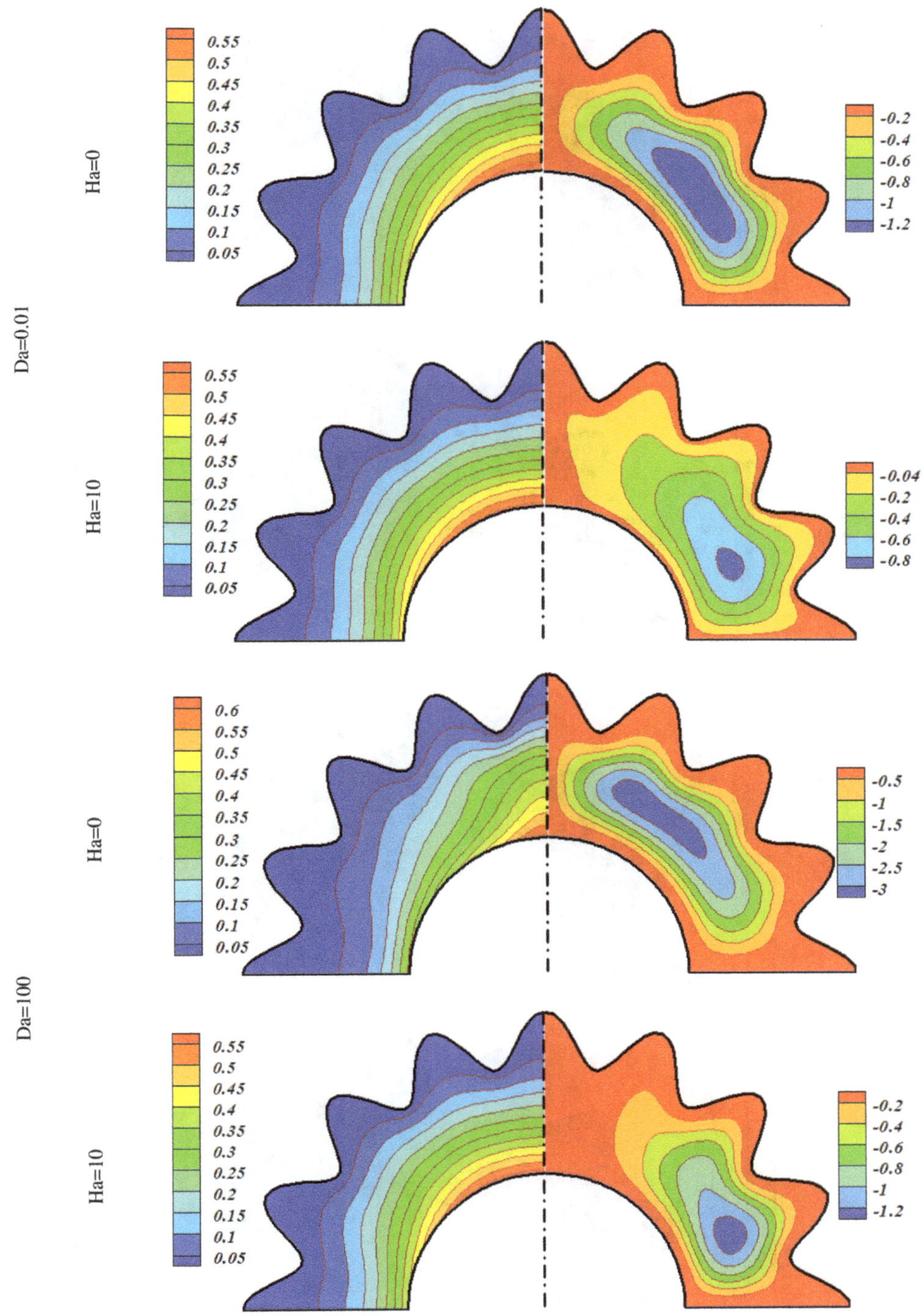

Figure 19. Influence of Da, Ha *on streamlines (left) and isotherms (right) contours when* $\phi = 0.04, Ra = 10^5$

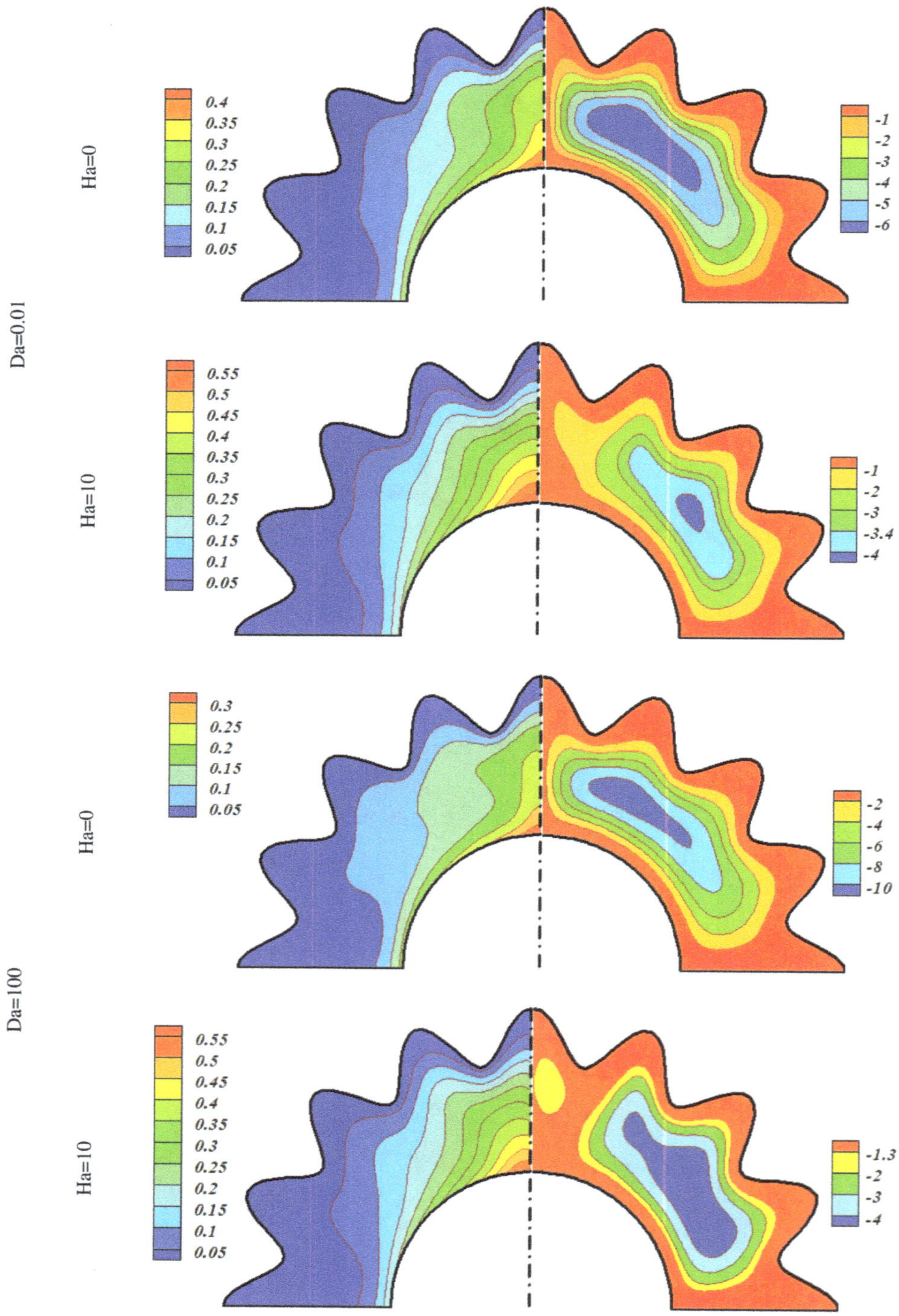

Figure 20a. Effects of Da, Ha *and* Ra *on average Nusselt number*

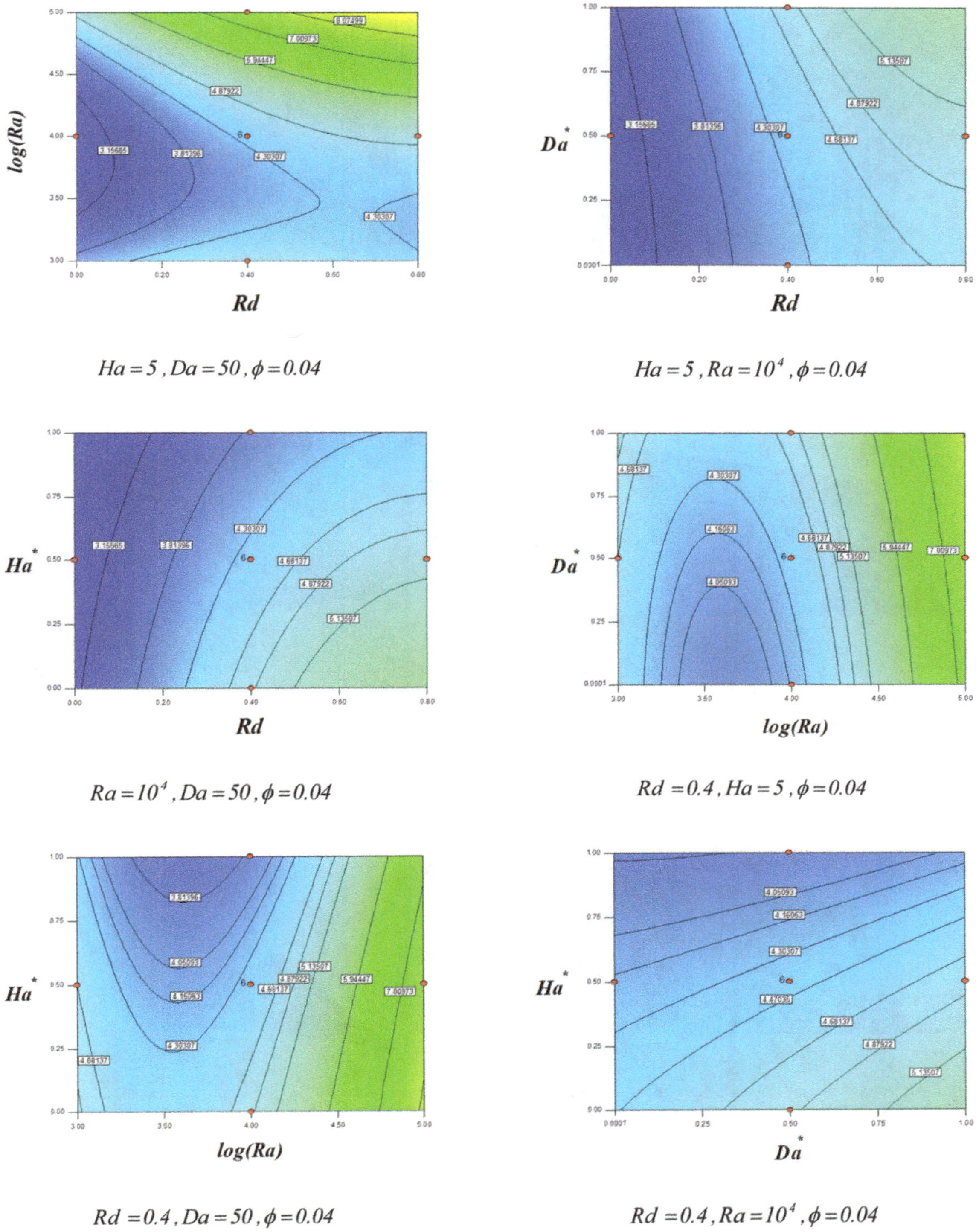

Figure 20b. Effects of Da, Ha, and Ra on average Nusselt number

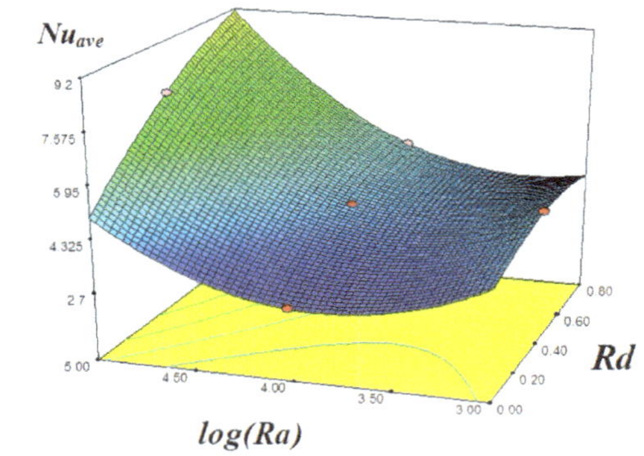

$Ha = 5, Da = 50, \phi = 0.04$

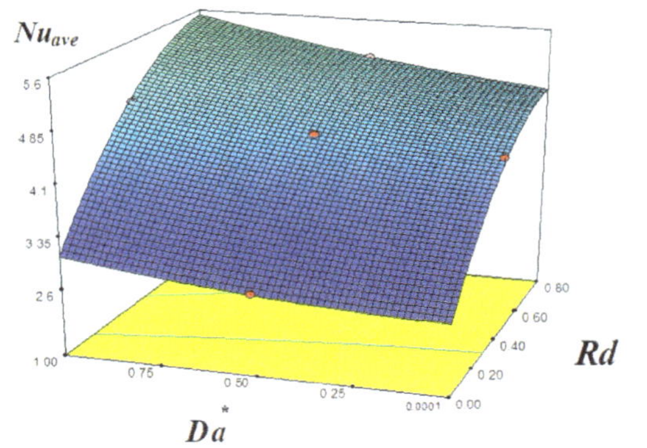

$Ha = 5, Ra = 10^4, \phi = 0.04$

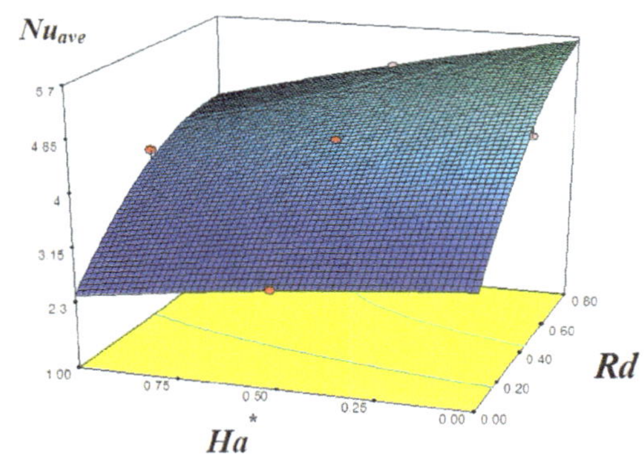

$Ra = 10^4, Da = 50, \phi = 0.04$

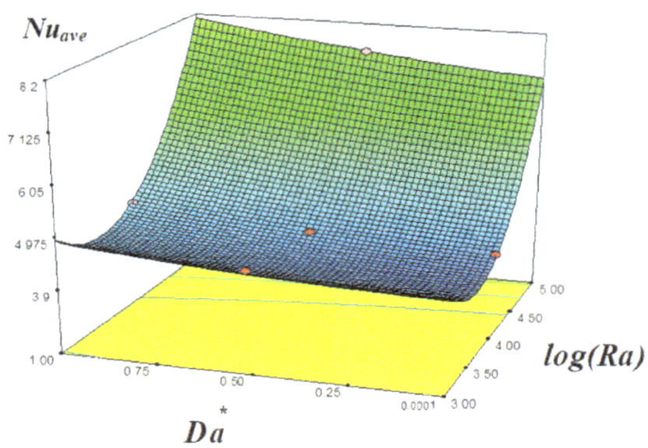

$Rd = 0.4, Ha = 5, \phi = 0.04$

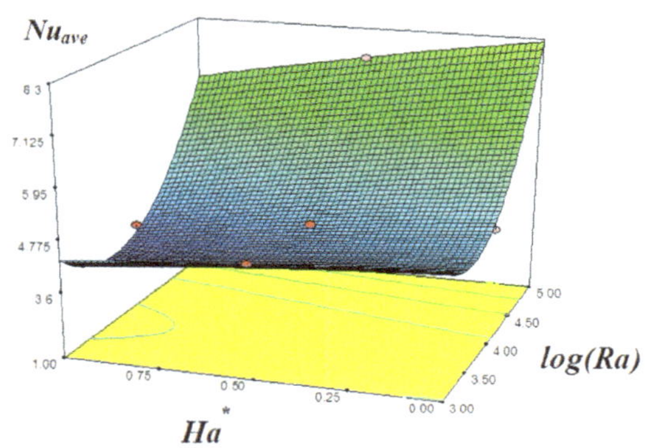

$Rd = 0.4, Da = 50, \phi = 0.04$

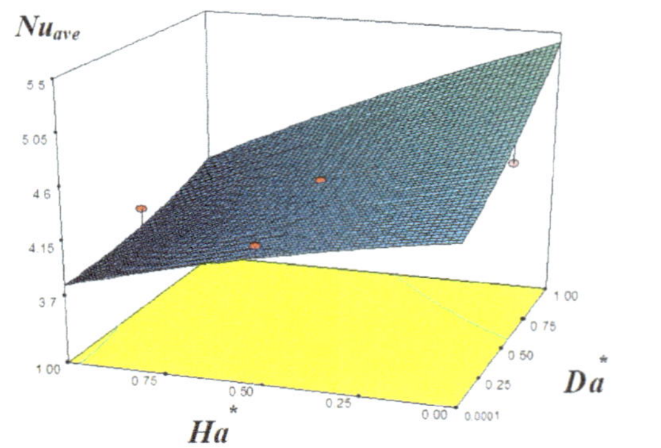

$Rd = 0.4, Ra = 10^4, \phi = 0.04$

Figure 21. (a) Geometry and the boundary conditions with (b) a sample triangular element and its corresponding control volume

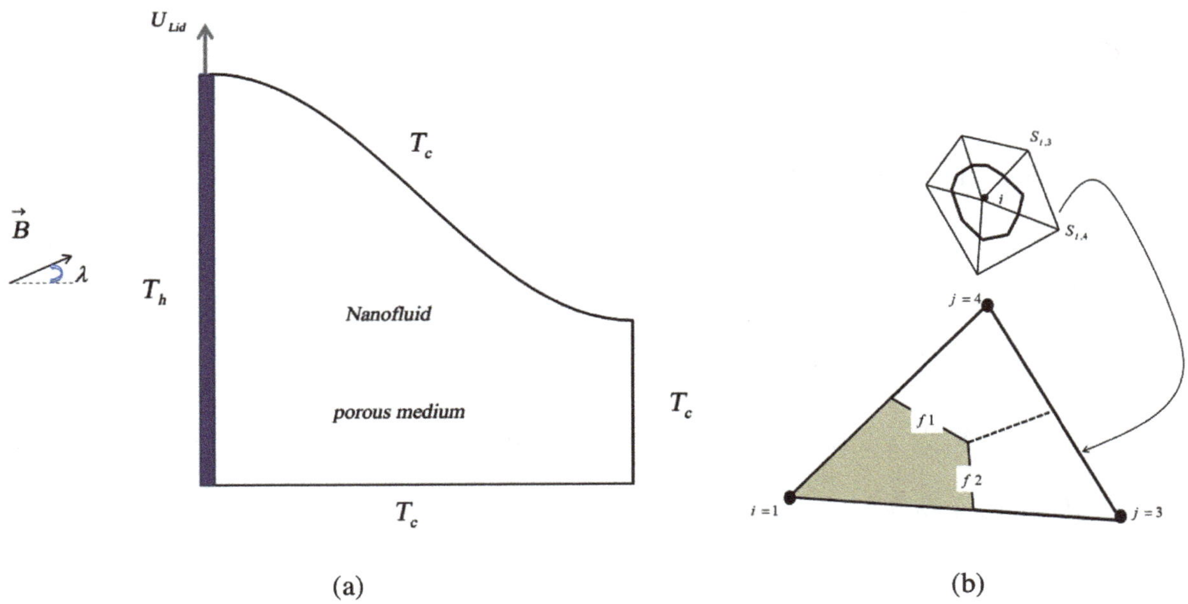

$$\mu_{nf}\left(\frac{\partial^2 v}{\partial y^2}+\frac{\partial^2 v}{\partial x^2}\right)-\frac{\partial P}{\partial y}-B_x\sigma_{nf}B_x v+B_y\sigma_{nf}B_x u-\frac{\mu_{nf}}{K}v=\rho_{nf}\left(\frac{\partial v}{\partial x}u+\frac{\partial v}{\partial y}v\right), \tag{64}$$
$$B_x=B_o\cos\lambda,\ B_y=B_o\sin\lambda$$

$$\left(\rho C_p\right)_{nf}\left(v\frac{\partial T}{\partial y}+u\frac{\partial T}{\partial x}\right)=k_{nf}\left(\frac{\partial^2 T}{\partial x^2}+\frac{\partial^2 T}{\partial y^2}\right), \tag{65}$$

$\left(\rho C_p\right)_{nf}$, ρ_{nf} and σ_{nf} are defined as:

$$\left(\rho C_p\right)_{nf}=\left(\rho C_p\right)_f(1-\phi)+\left(\rho C_p\right)_s\phi \tag{66}$$

$$\rho_{nf}=\rho_f(1-\phi)+\rho_s\phi \tag{67}$$

$$\frac{\sigma_{nf}}{\sigma_f} = 1 + \frac{3\left(\frac{\sigma_s}{\sigma_f} - 1\right)\phi}{\left(\frac{\sigma_s}{\sigma_f} + 2\right) - \left(\frac{\sigma_s}{\sigma_f} - 1\right)\phi} \tag{68}$$

μ_{nf} is calculated via KKL model:

$$\begin{aligned} &\mu_{eff} = \mu_{static} + \mu_{Brownian} = \mu_{static} + \frac{k_{Brownian}}{k_f} \times \frac{\mu_f}{\Pr_f} \\ &k_{Brownian} = 5\times 10^4 g'(\phi, T, d_p)\phi\rho_f c_{p,f}\sqrt{\frac{\kappa_b T}{d_p \rho_p}} \\ &g'(\phi, d_p, T) = Ln(T)\left(a_1 + a_3 Ln(\phi) + a_2 Ln(d_p) + a_5 Ln(d_p)^2 + a_4 \ln(d_p) Ln(\phi)\right) \\ &+\left(a_6 + a_8 Ln(\phi) + a_7 Ln(d_p) + a_{10} Ln(d_p)^2 + a_9 Ln(\phi)\ln(d_p)\right) \end{aligned} \tag{69}$$

k_{nf} can be expressed as:

$$\frac{k_{nf}}{k_f} = \frac{-m(k_f - k_p)\phi + (k_p - k_f)\phi + mk_f + k_p + k_f}{mk_f + (k_f - k_p)\phi + k_f + k_p} \tag{70}$$

Vorticity and stream function should be used to eliminate pressure source terms:

$$\omega + \frac{\partial u}{\partial y} - \frac{\partial v}{\partial x} = 0, \frac{\partial \psi}{\partial x} = -v, \frac{\partial \psi}{\partial y} = u \tag{71}$$

Introducing dimensionless quantities:

$$\begin{aligned} &U = \frac{u}{U_{Lid}}, V = \frac{v}{U_{Lid}}, \theta = \frac{T - T_c}{\Delta T}, \Delta T = T_h - T_c, (X, Y) = \frac{(x, y)}{L} \\ &, \Psi = \frac{\psi L}{U_{Lid}}, \Omega = \frac{\omega}{L U_{Lid}} \end{aligned} \tag{72}$$

The final formulae are:

$$\frac{\partial^2 \Psi}{\partial X^2} + \Omega + \frac{\partial^2 \Psi}{\partial Y^2} = 0, \tag{73}$$

$$\begin{aligned}&\frac{\partial \Omega}{\partial Y}V+\frac{\partial \Omega}{\partial X}U=\frac{1}{\mathrm{Re}}\frac{A_5}{A_1}\left(\frac{\partial^2 \Omega}{\partial Y^2}+\frac{\partial^2 \Omega}{\partial X^2}\right)\\&+\frac{Ha^2}{\mathrm{Re}}\frac{A_6}{A_1}\left(\frac{\partial U}{\partial X}B_xB_y-\frac{\partial V}{\partial X}B_x^{\ 2}+\frac{\partial U}{\partial Y}B_y^2-\frac{\partial V}{\partial Y}B_xB_y\right)\\&-\frac{1}{\mathrm{Re}\,Da}\frac{A_5}{A_1}\Omega,\end{aligned} \tag{74}$$

$$\frac{A_4}{A_2}\left(\frac{\partial^2 \theta}{\partial Y^2}+\frac{\partial^2 \theta}{\partial X^2}\right)=\Pr\mathrm{Re}\left(\frac{\partial \theta}{\partial X}U+\frac{\partial \theta}{\partial Y}V\right) \tag{75}$$

where dimensionless and constants parameters are illustrated as:

$$\begin{aligned}&\mathrm{Re}=\frac{\rho_f U_{Lid} L}{\mu_f},Ha=LB_0\sqrt{\sigma_f/\mu_f},\\&A_1=\frac{\rho_{nf}}{\rho_f},A_5=\frac{\mu_{nf}}{\mu_f},\\&A_2=\frac{\left(\rho C_p\right)_{nf}}{\left(\rho C_p\right)_f},A_4=\frac{k_{nf}}{k_f},A_6=\frac{\sigma_{nf}}{\sigma_f}\end{aligned} \tag{76}$$

and boundary conditions are:

$\theta = 1.0$ on bottom wall

$\theta = 0.0$ on other walls

$\Psi = 0.0$ on all walls (77)

Nu_{loc} and Nu_{ave} over the hot wall can calculate as:

$$Nu_{loc}=\frac{\partial \theta}{\partial y}\left(\frac{k_{nf}}{k_f}\right) \tag{78}$$

$$Nu_{ave}=\frac{1}{L}\int_{r_{in}}^{r_{out}} Nu_{loc}\,dx \tag{79}$$

5.3. Effects of Active Parameters

In this research, the impact of Lorentz forces on forced convection of nanofluid in a lid driven porous cavity is illustrated. μ_{nf} of CuO-water nanofluid are estimated by means of KKL model. CVFEM is utilized to find the effects of Darcy number ($Da = 0.01$ to 100), Reynolds number ($\mathrm{Re} = 100$ to 500), volume fraction of CuO ($\phi = 0\%$ to 4%), shape of nanoparticle and Hartmann number ($Ha = 0$ to 40).

Influences of shape of the nanoparticles on Nu_{ave} are discussed in Table 8. The maximum Nu is caused by Platelet, followed by Cylinder, Brick and Spherical. So, Platelet nanoparticle has been selected for further investigation.

Impact of adding nanoparticles in the base fluid on velocity and temperature contours is depicted in Figure 22. Temperature gradient decreases with augment of ϕ. $\left|\Psi_{\max}\right|$ augments with adding nanoparticles because of increment in the solid movements.

Figures 23, 24 and 25 demonstrate the impact of Da, Re, Ha on isotherms and stream lines. At $Da = 0.01, \mathrm{Re} = 100, Ha = 0$, only one eddy exists in streamline. By applying magnetic field, two new eddies with opposite direction generate above the main eddy. The distortion of isotherms becomes less than before. As Darcy number increases, convective mode becomes stronger and the primary eddy converts to two eddies. Increasing Hartmann number leads to generate the third eddy. As Re augments, the strength of secondary eddy enhances. In absence of magnetic field at Re=500, a counter clock wise eddy appears above the main eddy. Impacts of Darcy and Hartmann number on isotherms and streamlines at Re=500 is similar to those of Re=100 except the stream function values. It means that the absolute stream function values augment with increase of Reynolds number.

Influence of significant parameters on Nu_{ave} is depicted in Figure 26. The correlation for Nu_{ave} is as follows:

$$\begin{aligned} Nu_{ave} &= -0.25 + 0.29\,\mathrm{Re}^* - 0.1Da^* + 0.2Ha^* + 0.12Da^*\,\mathrm{Re}^* \\ &\quad -0.07\,\mathrm{Re}^*\,Ha^* - 0.12Da^*\,Ha^* + 0.006\,\mathrm{Re}^{*2} + 0.26Da^{*2} - 0.023Ha^{*2} \end{aligned} \tag{80}$$

where $\mathrm{Re}^* = 0.01\,\mathrm{Re}, Ha^* = 0.1Ha, Da^* = 0.01Da$. Increasing Reynolds number can enhance the heat transfer rate because of reduction in thermal boundary layer thickness. Similar trend is seen for Darcy

Table 8. Effect of shape of nanoparticles on Nusselt number when $Da = 100, \mathrm{Re} = 600, \phi = 0.04$

	Ha	
	0	**40**
Spherical	2.587961	0.572717
Brick	2.61641	0.579895
Cylinder	2.659818	0.59086
Platelet	2.694209	0.599562

Figure 22. Influence of nanofluid volume fraction on streamlines (left) and isotherms (right) contours (nanofluid ($\phi = 0.04$)(—) and pure fluid($\phi = 0$) (- - -)) when $\mathrm{Re} = 500, Da = 100$

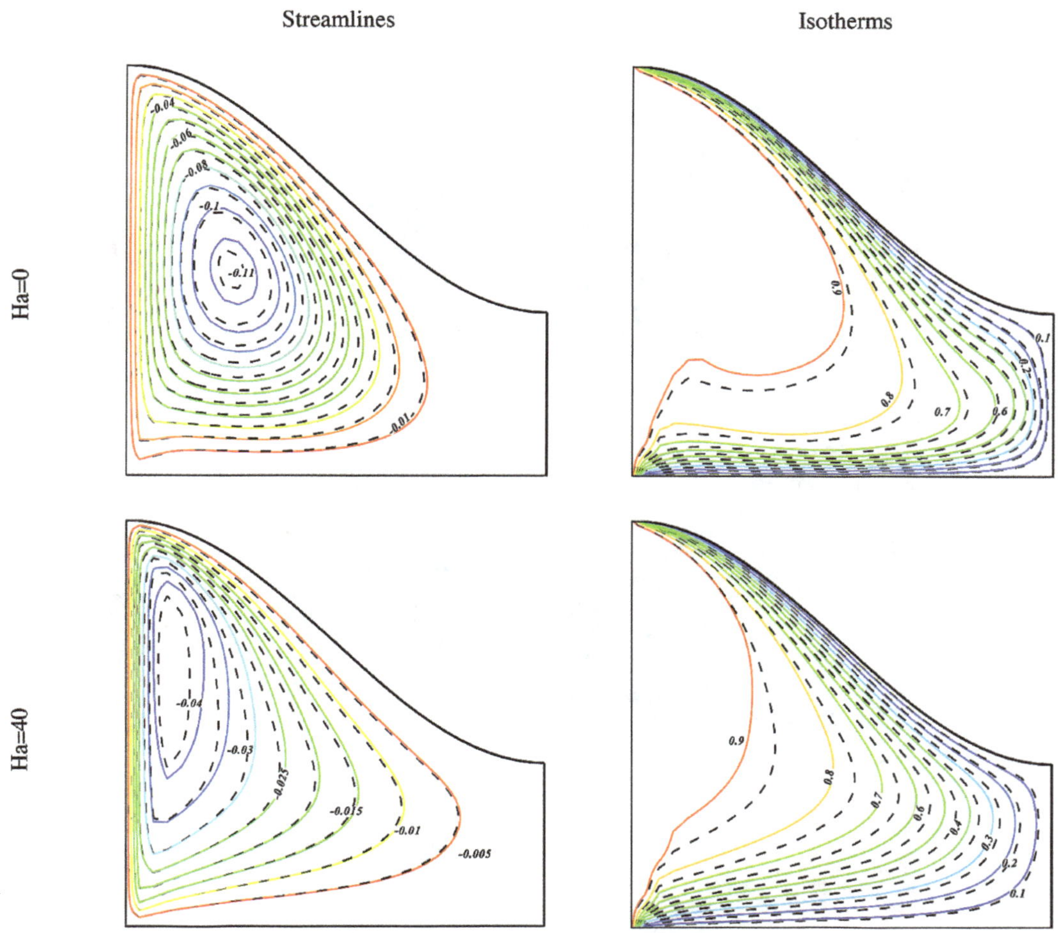

number. Temperature gradient decreases with augment of Lorentz forces. So, Nu_{ave} decreases with increase of Hartmann number.

6. MAGNETIC FIELD INFLUENCE ON CUO -H_2O NANOFLUID CONVECTIVE FLOW IN A PERMEABLE CAVITY CONSIDERING VARIOUS SHAPES FOR NANOPARTICLES

6.1. Problem Definition

Geometry, boundary conditions and sample element are depicted (see Figure 27). The bottom wall is hot and other walls are cold. Also the bottom wall can move horizontally. Porous cavity filled with nanofluid and affected by horizontal magnetic field.

Figure 23. Effect of Darcy and Hartmann numbers on Isotherms (left) and streamlines (right) contours when $\phi = 0.04, \mathrm{Re} = 100$

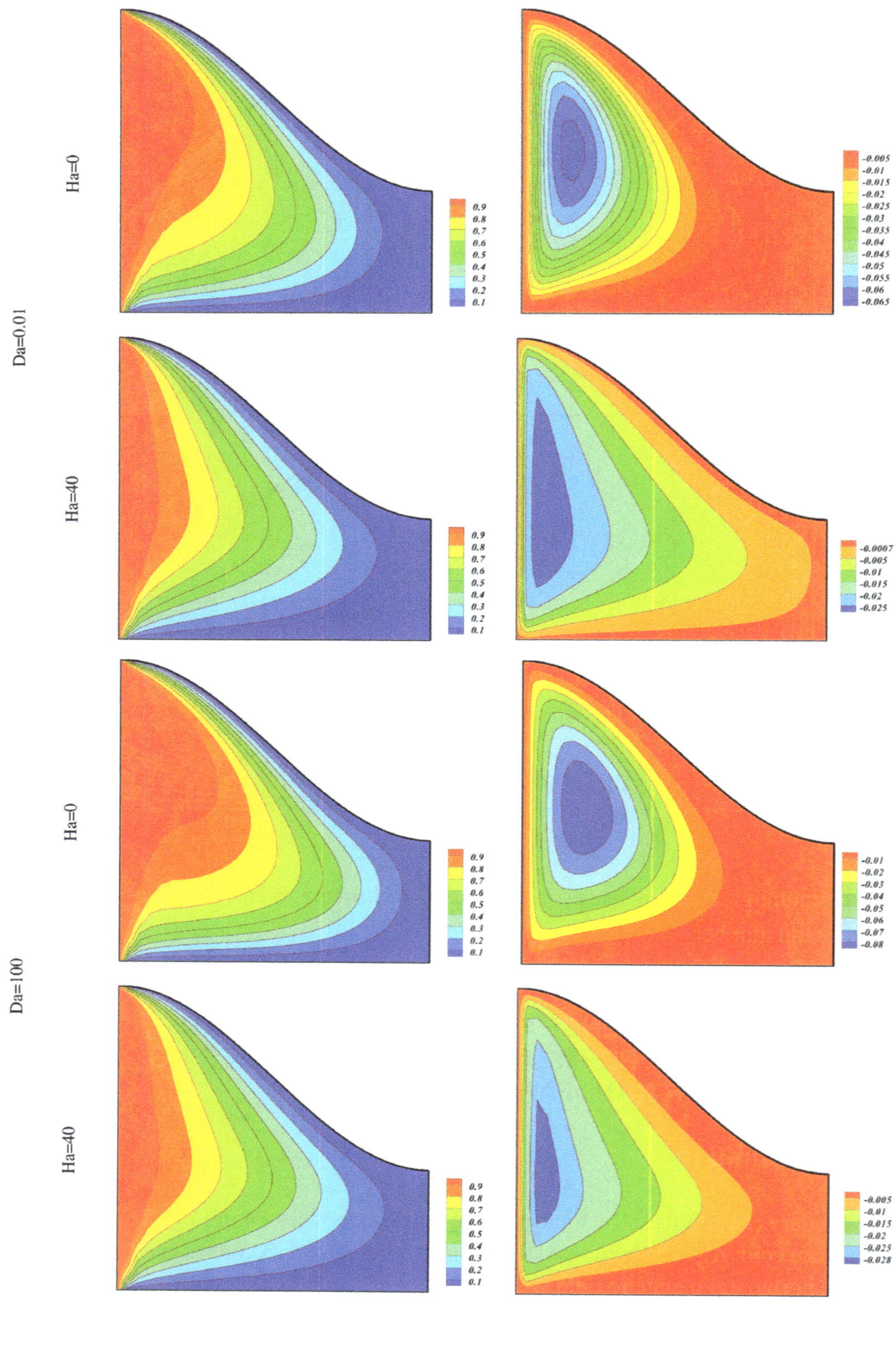

Figure 24. Effect of Darcy and Hartmann numbers on Isotherms (left) and streamlines (right) contours when $\phi = 0.04, \mathrm{Re} = 300$

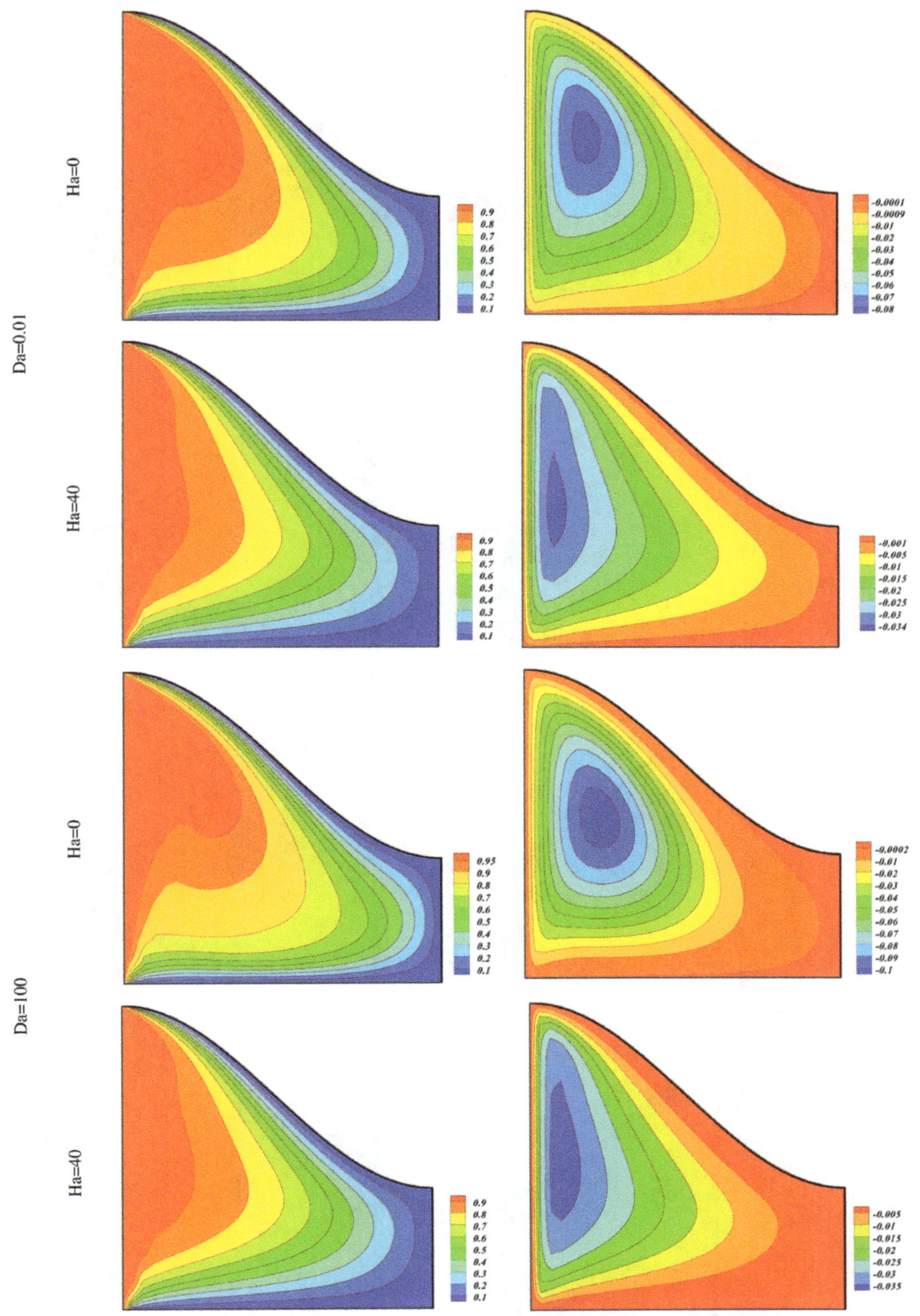

Figure 25. Effect of Darcy and Hartmann numbers on Isotherms (left) and streamlines (right) contours when $\phi = 0.04, \mathrm{Re} = 500$

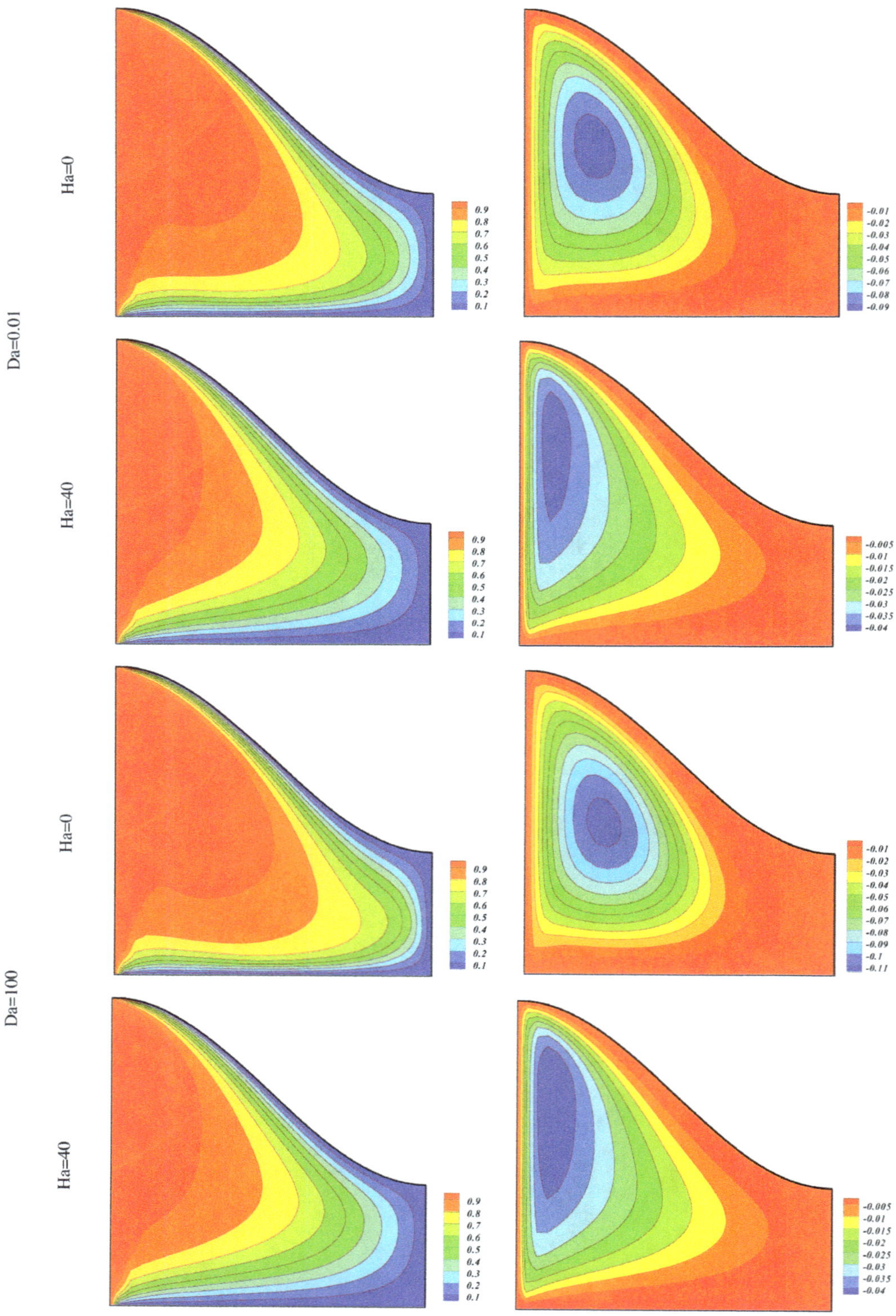

Figure 26. Influences of the Darcy, Reynolds number and Hartmann numbers on average Nusselt number

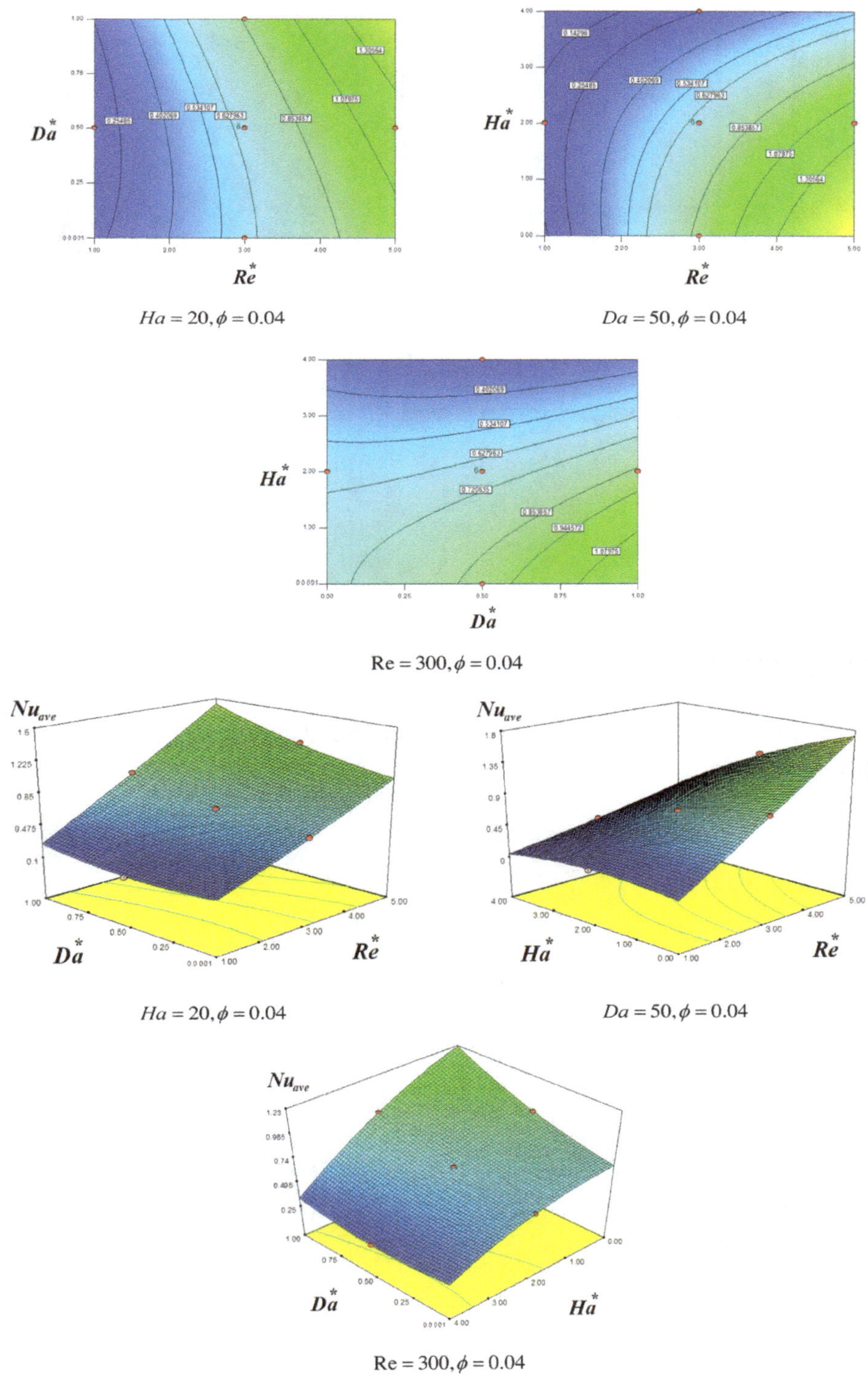

Figure 27. (a) Geometry and the boundary conditions with (b) a sample triangular element and its corresponding control volume

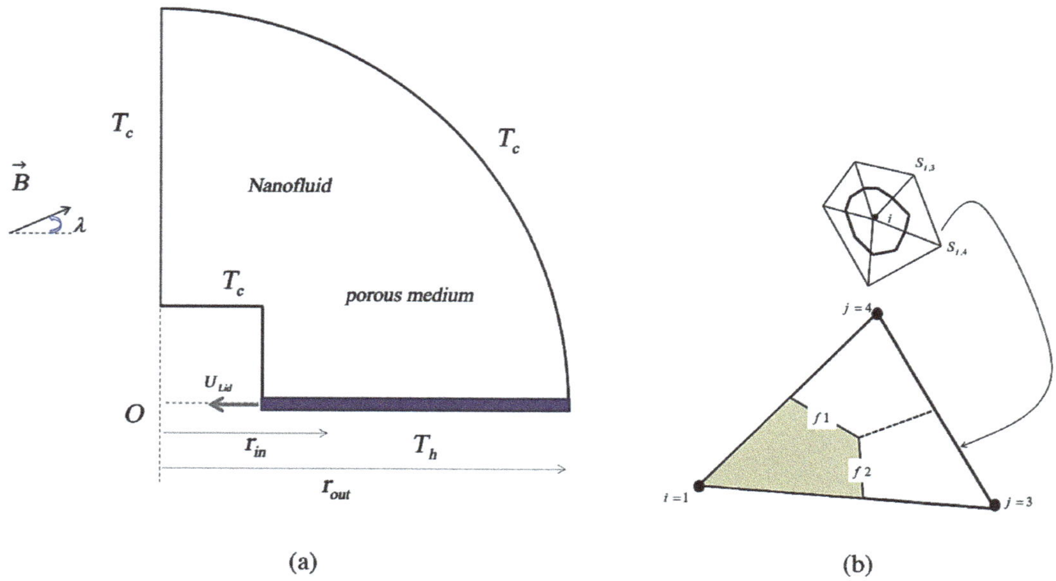

6.2. Governing Equation

2D steady convective non-Darcy flow of nanofluid is considered in existence of constant magnetic field. The PDEs equations are:

$$\frac{\partial v}{\partial y}+\frac{\partial u}{\partial x}=0, \tag{81}$$

$$\left[-\sigma_{nf}B_y^2 u+\sigma_{nf}B_x B_y v+\left(\frac{\partial^2 u}{\partial y^2}+\frac{\partial^2 u}{\partial x^2}\right)\mu_{nf}-\frac{\partial P}{\partial x}\right]-\frac{\mu_{nf}}{K}u=\rho_{nf}\left(v\frac{\partial u}{\partial y}+\frac{\partial u}{\partial x}u\right), \tag{82}$$

$$\mu_{nf}\left(\frac{\partial^2 v}{\partial y^2}+\frac{\partial^2 v}{\partial x^2}\right)-\frac{\partial P}{\partial y}-B_x\sigma_{nf}B_x v+B_y\sigma_{nf}B_x u-\frac{\mu_{nf}}{K}v=\rho_{nf}\left(\frac{\partial v}{\partial x}u+\frac{\partial v}{\partial y}v\right),$$
$$B_x=B_o\cos\lambda,\ B_y=B_o\sin\lambda, \tag{83}$$

$$\left(\rho C_p\right)_{nf}\left(v\frac{\partial T}{\partial y}+u\frac{\partial T}{\partial x}\right)=k_{nf}\left(\frac{\partial^2 T}{\partial x^2}+\frac{\partial^2 T}{\partial y^2}\right). \tag{84}$$

$\left(\rho C_p\right)_{nf}$, ρ_{nf} and σ_{nf} are defined as:

$$\left(\rho C_p\right)_{nf}=\left(\rho C_p\right)_f(1-\phi)+\left(\rho C_p\right)_s\phi, \tag{85}$$

$$\rho_{nf}=\rho_f(1-\phi)+\rho_s\phi, \tag{86}$$

$$\frac{\sigma_{nf}}{\sigma_f}=1+\frac{3\left(\frac{\sigma_s}{\sigma_f}-1\right)\phi}{\left(\frac{\sigma_s}{\sigma_f}+2\right)-\left(\frac{\sigma_s}{\sigma_f}-1\right)\phi}, \tag{87}$$

μ_{nf} is calculated via KKL model:

$$\begin{aligned}
&\mu_{eff}=\mu_{static}+\mu_{Brownian}=\mu_{static}+\frac{k_{Brownian}}{k_f}\times\frac{\mu_f}{\Pr_f},\\
&k_{Brownian}=5\times10^4 g'(\phi,T,d_p)\phi\rho_f c_{p,f}\sqrt{\frac{\kappa_b T}{d_p\rho_p}},\\
&g'\left(\phi,d_p,T\right)=Ln\left(T\right)\left(a_1+a_3Ln\left(\phi\right)+a_2Ln\left(d_p\right)+a_5Ln\left(d_p\right)^2+a_4\ln\left(d_p\right)Ln\left(\phi\right)\right)\\
&+\left(a_6+a_8Ln\left(\phi\right)+a_7Ln\left(d_p\right)+a_{10}Ln\left(d_p\right)^2+a_9Ln\left(\phi\right)\ln\left(d_p\right)\right).
\end{aligned} \tag{88}$$

k_{nf} can be expressed as:

$$\frac{k_{nf}}{k_f}=\frac{-m\left(k_f-k_p\right)\phi+\left(k_p-k_f\right)\phi+mk_f+k_p+k_f}{mk_f+\left(k_f-k_p\right)\phi+k_f+k_p} \tag{89}$$

Vorticity and stream function should be used to eliminate pressure source terms:

$$\omega + \frac{\partial u}{\partial y} - \frac{\partial v}{\partial x} = 0, \frac{\partial \psi}{\partial x} = -v, \frac{\partial \psi}{\partial y} = u \tag{90}$$

Introducing dimensionless quantities:

$$U = \frac{u}{U_{Lid}}, V = \frac{v}{U_{Lid}}, \theta = \frac{T - T_c}{\Delta T}, \Delta T = T_h - T_c, \left(X, Y\right) = \frac{\left(x, y\right)}{L}, \Psi = \frac{\psi L}{U_{Lid}}, \Omega = \frac{\omega}{L U_{Lid}}. \tag{91}$$

The final formulae are:

$$\frac{\partial^2 \Psi}{\partial X^2} + \Omega + \frac{\partial^2 \Psi}{\partial Y^2} = 0, \tag{92}$$

$$\begin{aligned} &\frac{\partial \Omega}{\partial Y} V + \frac{\partial \Omega}{\partial X} U = \frac{1}{\mathrm{Re}} \frac{A_5}{A_1} \left(\frac{\partial^2 \Omega}{\partial Y^2} + \frac{\partial^2 \Omega}{\partial X^2} \right) \\ &+ \frac{Ha^2}{\mathrm{Re}} \frac{A_6}{A_1} \left(\frac{\partial U}{\partial X} B_x B_y - \frac{\partial V}{\partial X} B_x^{\ 2} + \frac{\partial U}{\partial Y} B_y^2 - \frac{\partial V}{\partial Y} B_x B_y \right) \\ &- \frac{1}{\mathrm{Re}\, Da} \frac{A_5}{A_1} \Omega, \end{aligned} \tag{93}$$

$$\frac{A_4}{A_2} \left(\frac{\partial^2 \theta}{\partial Y^2} + \frac{\partial^2 \theta}{\partial X^2} \right) = \Pr \mathrm{Re} \left(\frac{\partial \theta}{\partial X} U + \frac{\partial \theta}{\partial Y} V \right). \tag{94}$$

where dimensionless and constants parameters are illustrated as:

$$\begin{aligned} &\mathrm{Re} = \frac{\rho_f U_{Lid} L}{\mu_f}, Ha = L B_0 \sqrt{\sigma_f / \mu_f}, \\ &A_1 = \frac{\rho_{nf}}{\rho_f}, A_5 = \frac{\mu_{nf}}{\mu_f}, \\ &A_2 = \frac{\left(\rho C_p\right)_{nf}}{\left(\rho C_p\right)_f}, A_4 = \frac{k_{nf}}{k_f}, A_6 = \frac{\sigma_{nf}}{\sigma_f}. \end{aligned} \tag{95}$$

and boundary conditions are:

$\theta = 1.0$ on bottom wall

$\theta = 0.0$ on other walls

$\Psi = 0.0$ on all walls (96)

Nu_{loc} and Nu_{ave} over the hot wall can calculate as:

$$Nu_{loc} = \frac{\partial \theta}{\partial y}\left(\frac{k_{nf}}{k_f}\right) \quad (97)$$

$$Nu_{ave} = \frac{1}{L}\int_{r_{in}}^{r_{out}} Nu_{loc}\, dx \quad (98)$$

6.3. Effects of Active Parameters

In this paper, the effect of magnetic field on forced convection of CuO-H_2O nanofluid in a lid driven porous cavity is demonstrated. μ_{nf} of CuO- H_2O nanofluid are estimated by means of KKL model. CVFEM is utilized to find the effects of Darcy number ($Da = 0.01$ to 100), Reynolds number ($\mathrm{Re} = 100$ to 500), volume fraction of CuO ($\phi = 0\%$ to 4%), shape of nanoparticle and Hartmann number ($Ha = 0$ to 40).

Influences of shape of the nanoparticles on Nu_{ave} are discussed in Table 9. The maximum Nu is caused by Platelet, followed by Cylinder, Brick and Spherical. Therefore, Platelet nanoparticle has been selected for further investigation.

Table 9. Effect of shape of nanoparticles on Nusselt number when $Da = 100, \mathrm{Re} = 600, \phi = 0.04$

	Ha	
	0	**40**
Spherical	8.654815	7.765007
Brick	8.85174	7.947995
Cylinder	9.159554	8.234346
Platelet	9.409924	8.467536

Impact of adding nanoparticles in the base fluid on velocity and temperature contours is depicted in Figure 28. Temperature gradient decreases with augment of ϕ. $|\Psi_{max}|$ augments with adding nanoparticles because of increment in the solid movements.

Figures 29, 30, and 31 demonstrate the impact of Da, Re, Ha on isotherms and stream lines. At $Da = 0.01, \mathrm{Re} = 100, Ha = 0$, only one eddy exists in streamline. By applying magnetic field, two new eddies with opposite direction generate above the main eddy. The distortion of isotherms becomes less than before. As Darcy number increases, convective mode becomes stronger and the primary eddy converts to two eddies. Increasing Hartmann number leads to generate the third eddy. As Re augments, the strength of secondary eddy which is generated due to increase of Darcy number, enhances. In absence of magnetic field at Re=500, a counter clock wise eddy appears above the main eddy. Impacts of Darcy and Hartmann numbers on isotherms and streamlines at Re=500 is similar to those of Re=100 except the stream function values. It means that the absolute stream function values augment with rise of Reynolds number.

Influence of significant parameters on Nu_{ave} is illustrated in Figure 32. The correlation for Nu_{ave} is as follows:

Figure 28. Influence of nanofluid volume fraction on streamlines (left) and isotherms (right) contours (nanofluid ($\phi = 0.04$)(—) and pure fluid($\phi = 0$) (- - -)) when $\mathrm{Re} = 500, Da = 100$

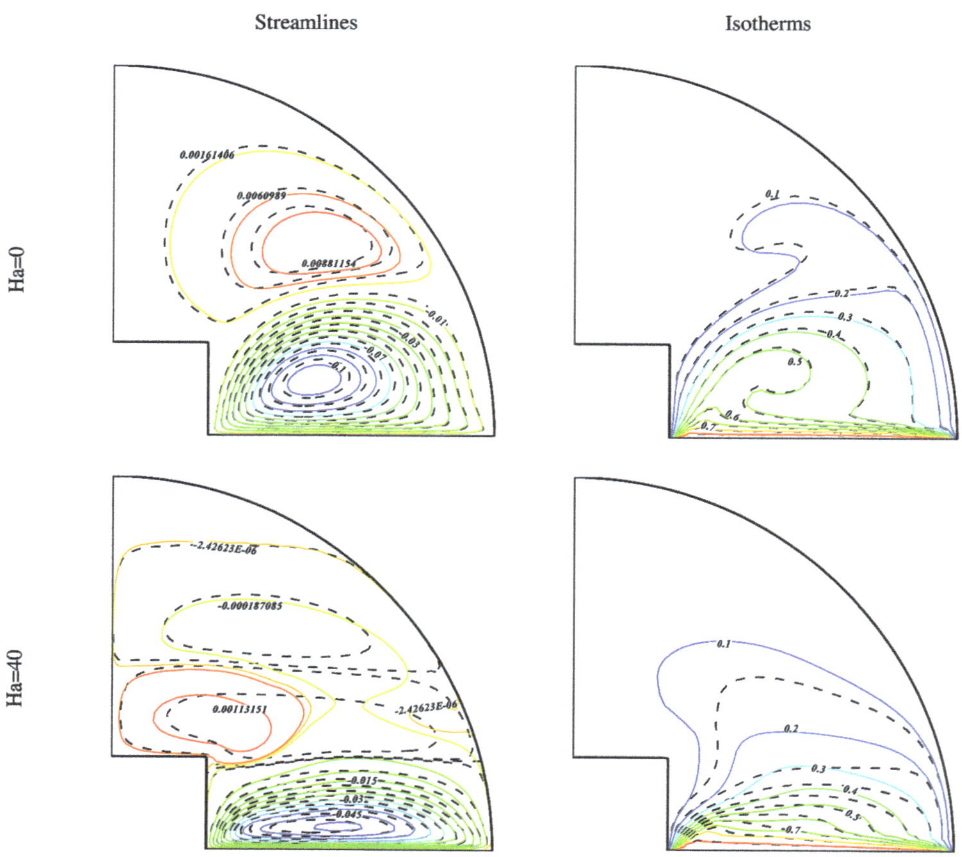

Figure 29. Effect of Darcy and Hartmann numbers on Isotherms (left) and streamlines (right) contours when $\phi = 0.04, \mathrm{Re} = 100$

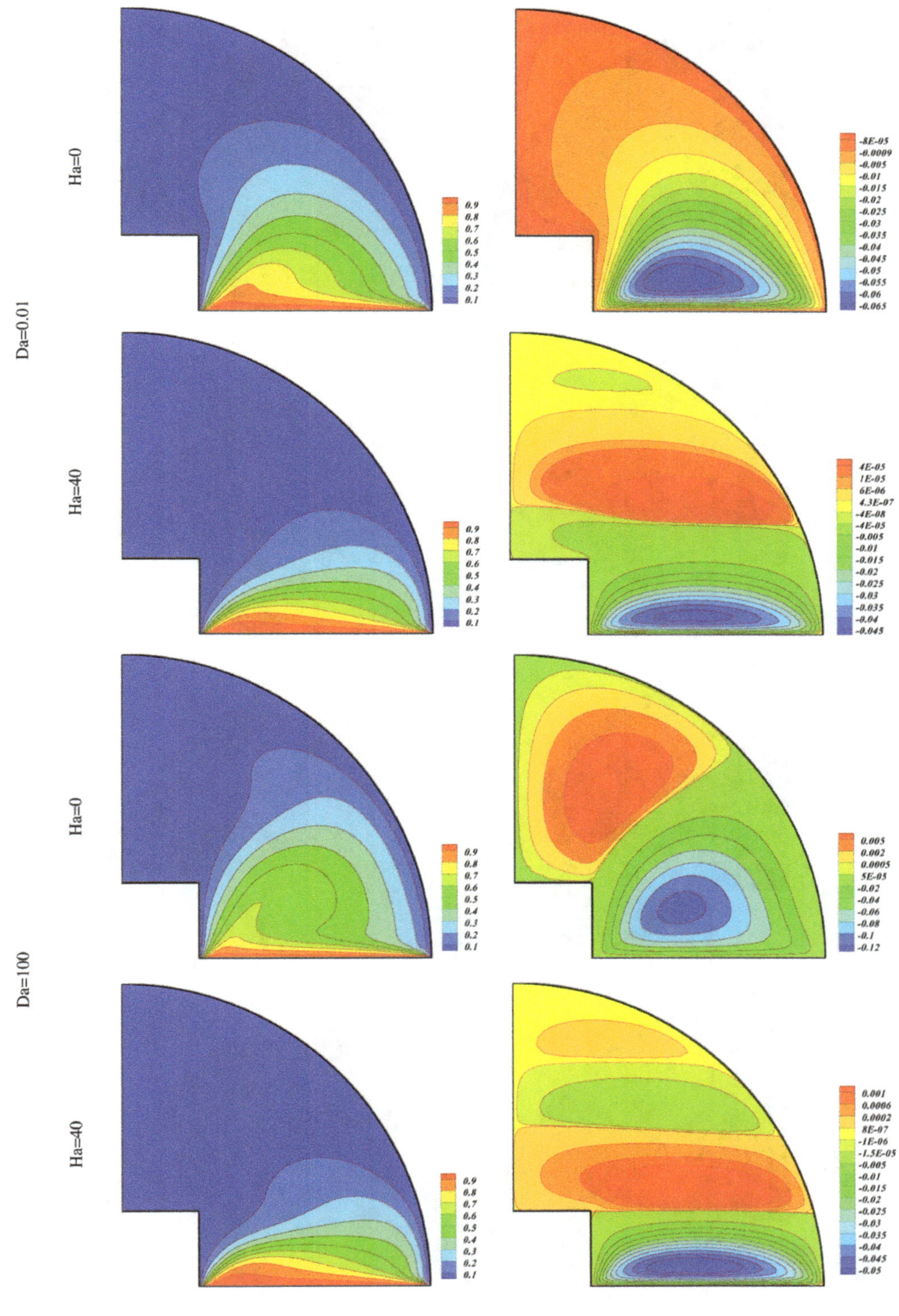

Figure 30. Effect of Darcy and Hartmann numbers on Isotherms (left) and streamlines (right) contours when $\phi = 0.04, \mathrm{Re} = 300$

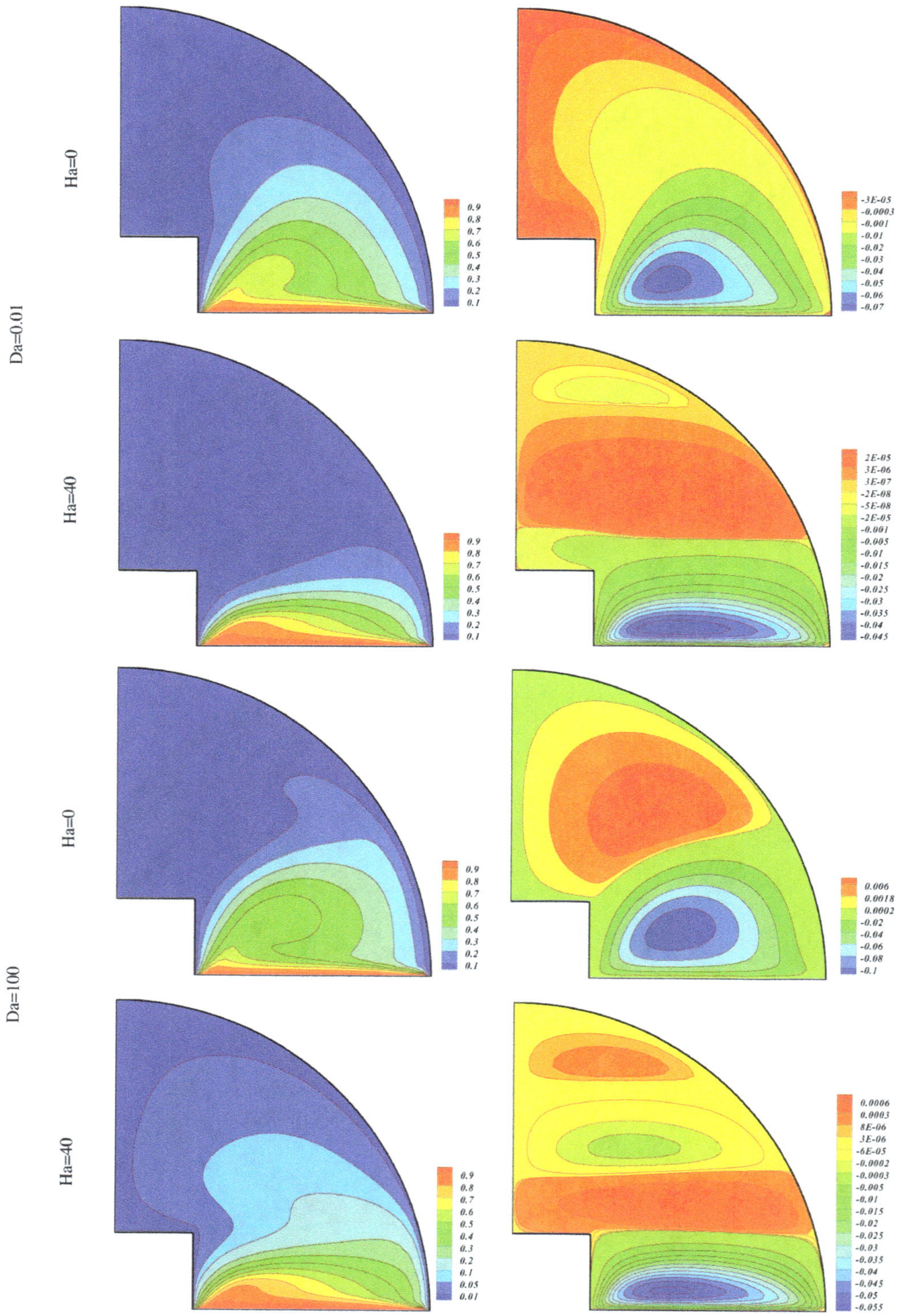

Figure 31. Effect of Darcy and Hartmann numbers on Isotherms (left) and streamlines (right) contours when $\phi = 0.04, \mathrm{Re} = 500$

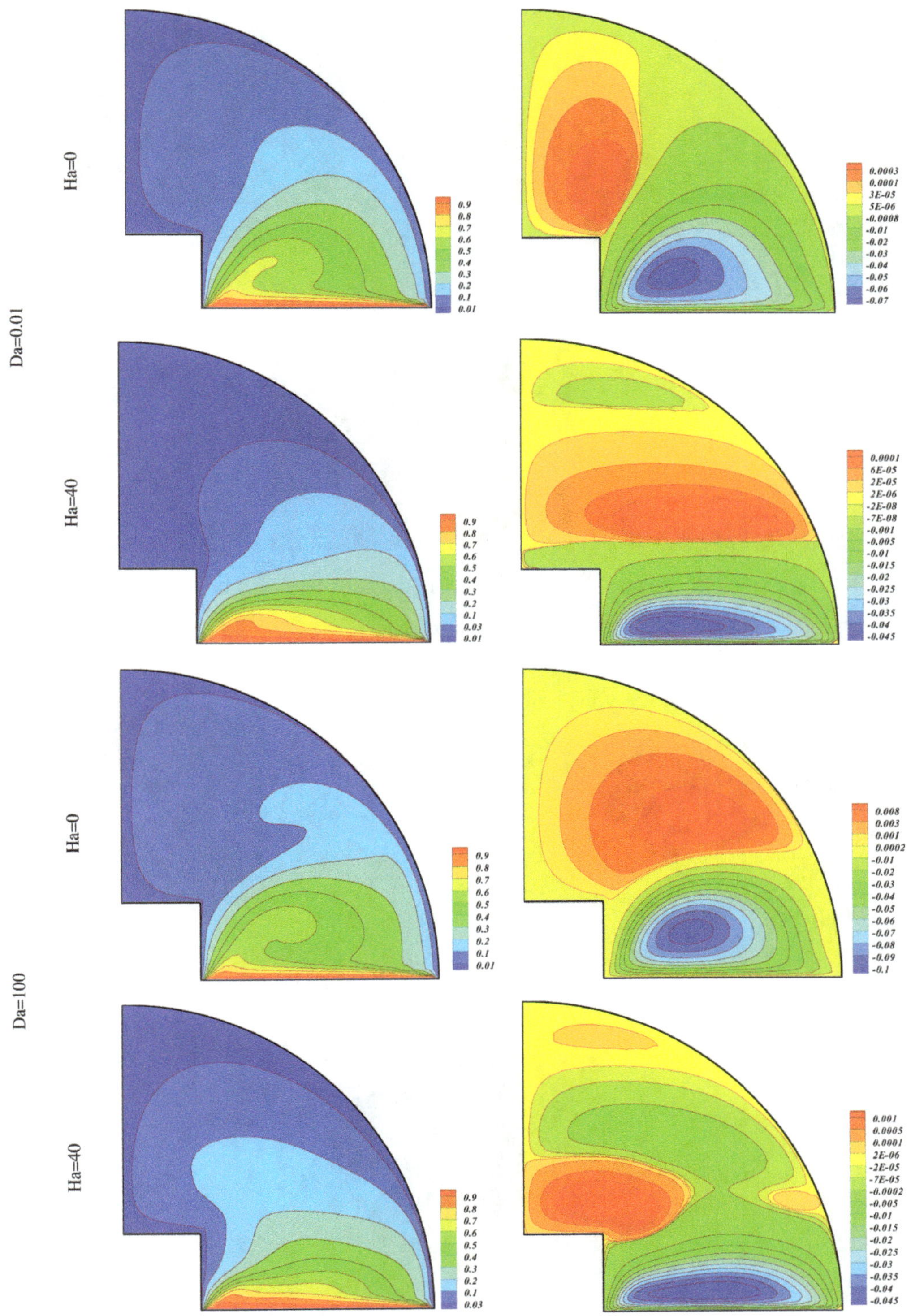

Figure 32. Influences of the Darcy, Reynolds number and Hartmann numbers on average Nusselt number

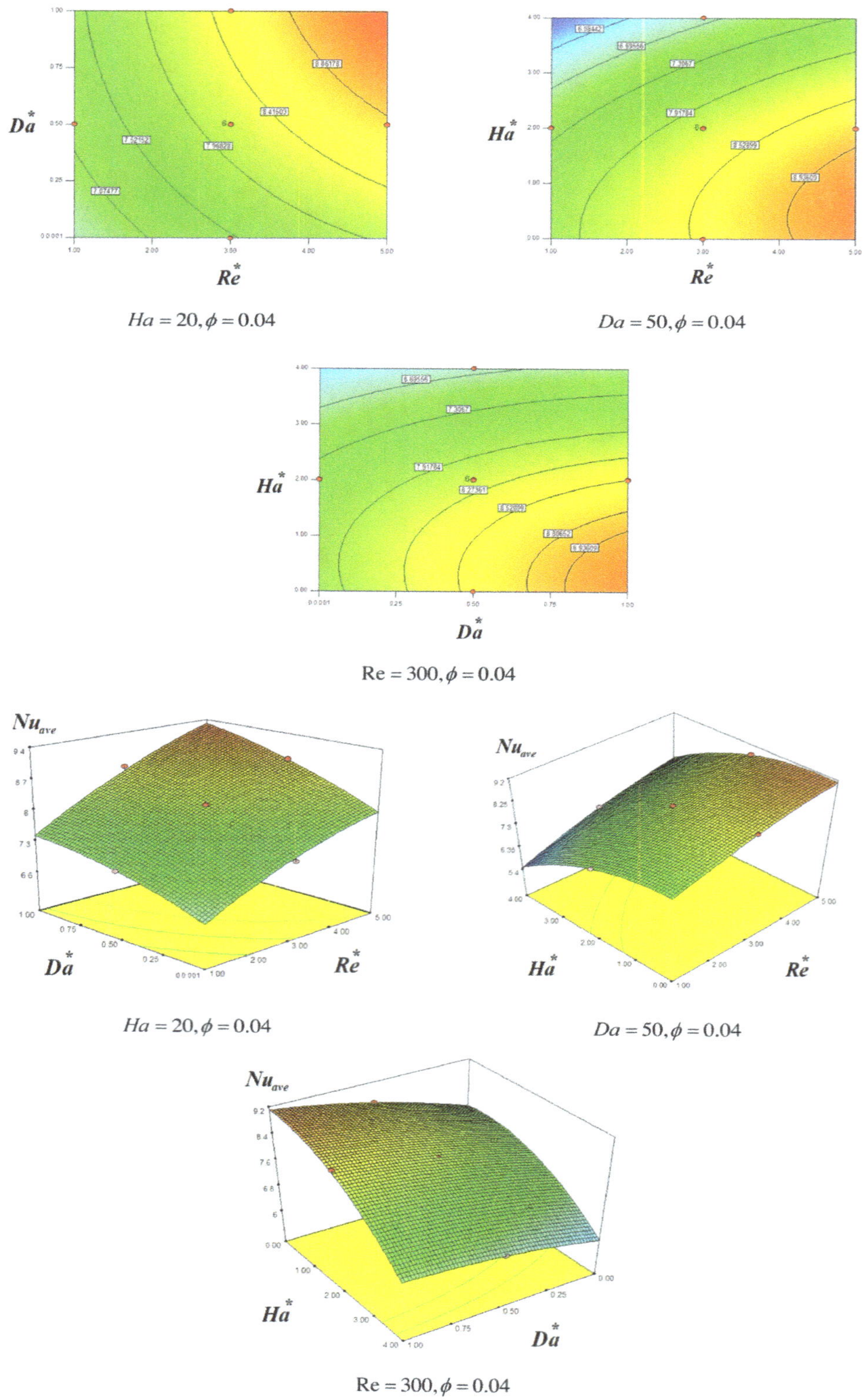

$Ha = 20, \phi = 0.04$ $Da = 50, \phi = 0.04$

$\mathrm{Re} = 300, \phi = 0.04$

$Ha = 20, \phi = 0.04$ $Da = 50, \phi = 0.04$

$\mathrm{Re} = 300, \phi = 0.04$

$$\begin{aligned} Nu_{ave} &= 6.54 + 0.52\,\mathrm{Re}^* + 1.56Da^* + 0.05Ha^* + 0.12Da^*\,\mathrm{Re}^* \\ &+0.03\,\mathrm{Re}^*\,Ha^* - 0.15Da^*\,Ha^* - 0.04\,\mathrm{Re}^{*2} + 0.58Da^{*2} - 0.14Ha^{*2} \end{aligned} \tag{99}$$

where $\mathrm{Re}^* = 0.01\,\mathrm{Re}, Ha^* = 0.1Ha, Da^* = 0.01Da$. Increasing Reynolds number can enhance the heat transfer rate because of reduction in thermal boundary layer thickness. Similar trend is seen for Darcy number. Temperature gradient decreases with augment of Lorentz forces. So, Nu_{ave} decreases with rise of Hartmann number.

REFERENCES

Sheikholeslami, M. (2017). Magnetic field influence on CuO -H2O nanofluid convective flow in a permeable cavity considering various shapes for nanoparticles, International Journal of Hydrogen Energy. *International Journal of Hydrogen Energy*, *42*(31), 19611–19621. doi:10.1016/j.ijhydene.2017.06.121

Sheikholeslami, M., & Bhatti, M. M. (2017). Forced convection of nanofluid in presence of constant magnetic field considering shape effects of nanoparticles. *International Journal of Heat and Mass Transfer*, *111*, 1039–1049. doi:10.1016/j.ijheatmasstransfer.2017.04.070

Sheikholeslami, M., & Ganji, D. D. (2017). Numerical modeling of magnetohydrodynamic CuO -water transportation inside a porous cavity considering shape factor effect, Colloids and Surfaces A: Physicochemical and Engineering Aspects. *Colloids and Surfaces. A, Physicochemical and Engineering Aspects*, *529*, 705–714. doi:10.1016/j.colsurfa.2017.06.046

Sheikholeslami, M., & Shamlooei, M. (2017). Magnetic source influence on nanofluid flow in porous medium considering shape factor effect. *Physics Letters. [Part A]*, *381*(36), 3071–3078. doi:10.1016/j.physleta.2017.07.028

ADDITIONAL READING

Jafaryar, M., Sheikholeslami, M., Li, M., & Moradi, R. (2018). Nanofluid turbulent flow in a pipe under the effect of twisted tape with alternate axis. *Journal of Thermal Analysis and Calorimetry*. doi:10.100710973-018-7093-2

Li, Z., Shehzad, S. A., & Sheikholeslami, M. (2018). An application of CVFEM for nanofluid heat transfer intensification in a porous sinusoidal cavity considering thermal non-equilibrium model. *Computer Methods in Applied Mechanics and Engineering*, *339*, 663–680. doi:10.1016/j.cma.2018.05.015

Li, Z., Sheikholeslami, M., Chamkha, A. J., Raizah, Z. A., & Saleem, S. (2018). Control Volume Finite Element Method for nanofluid MHD natural convective flow inside a sinusoidal annulus under the impact of thermal radiation. *Computer Methods in Applied Mechanics and Engineering*, *338*, 618–633. doi:10.1016/j.cma.2018.04.023

Li, Z., Sheikholeslami, M., Jafaryar, M., Shafee, A., & Chamkha, A. J. (2018). Investigation of nanofluid entropy generation in a heat exchanger with helical twisted tapes. *Journal of Molecular Liquids*, *266*, 797–805. doi:10.1016/j.molliq.2018.07.009

Li, Z., Sheikholeslami, M., Samandari, M., & Shafee, A. (2018). Nanofluid unsteady heat transfer in a porous energy storage enclosure in existence of Lorentz forces. *International Journal of Heat and Mass Transfer*, *127*, 914–926. doi:10.1016/j.ijheatmasstransfer.2018.06.101

Sheikholeslami, M. (2017a). Lattice Boltzmann Method simulation of MHD non-Darcy nanofluid free convection. *Physica B, Condensed Matter*, *516*, 55–71. doi:10.1016/j.physb.2017.04.029

Sheikholeslami, M. (2017b). CuO-water nanofluid free convection in a porous cavity considering Darcy law. *The European Physical Journal Plus*, *132*(1), 55. doi:10.1140/epjp/i2017-11330-3

Sheikholeslami, M. (2017c). Influence of Lorentz forces on nanofluid flow in a porous cavity by means of Non- Darcy model. *Engineering Computations*, *34*(8), 2651–2667. doi:10.1108/EC-01-2017-0008

Sheikholeslami, M. (2017d). Magnetic field influence on nanofluid thermal radiation in a cavity with tilted elliptic inner cylinder. *Journal of Molecular Liquids*, *229*, 137–147. doi:10.1016/j.molliq.2016.12.024

Sheikholeslami, M. (2017e). Numerical simulation of magnetic nanofluid natural convection in porous media. *Physics Letters. [Part A]*, *381*(5), 494–503. doi:10.1016/j.physleta.2016.11.042

Sheikholeslami, M. (2017f). Influence of Lorentz forces on nanofluid flow in a porous cylinder considering Darcy model. *Journal of Molecular Liquids*, *225*, 903–912. doi:10.1016/j.molliq.2016.11.022

Sheikholeslami, M. (2017g). Influence of Coulomb forces on Fe3O4-H2O nanofluid thermal improvement. *International Journal of Hydrogen Energy*, *42*(2), 821–829. doi:10.1016/j.ijhydene.2016.09.185

Sheikholeslami, M. (2017h). Numerical investigation of MHD nanofluid free convective heat transfer in a porous tilted enclosure. *Engineering Computations*, *34*(6), 1939–1955. doi:10.1108/EC-08-2016-0293

Sheikholeslami, M. (2017i). Magnetic field influence on CuO -H2O nanofluid convective flow in a permeable cavity considering various shapes for nanoparticles. *International Journal of Hydrogen Energy*, *42*(31), 19611–19621. doi:10.1016/j.ijhydene.2017.06.121

Sheikholeslami, M. (2018a). Magnetic source impact on nanofluid heat transfer using CVFEM. *Neural Computing & Applications*, *30*(4), 1055–1064. doi:10.100700521-016-2740-7

Sheikholeslami, M. (2018b). Numerical simulation for external magnetic field influence on Fe3O4-water nanofluid forced convection. *Engineering Computations*, *35*(4), 1639–1654. doi:10.1108/EC-06-2017-0200

Sheikholeslami, M. (2018c). Finite element method for PCM solidification in existence of CuO nanoparticles. *Journal of Molecular Liquids*, *265*, 347–355. doi:10.1016/j.molliq.2018.05.132

Sheikholeslami, M. (2018d). Solidification of NEPCM under the effect of magnetic field in a porous thermal energy storage enclosure using CuO nanoparticles. *Journal of Molecular Liquids*, *263*, 303–315. doi:10.1016/j.molliq.2018.04.144

Sheikholeslami, M. (2018e). Influence of magnetic field on Al2O3-H2O nanofluid forced convection heat transfer in a porous lid driven cavity with hot sphere obstacle by means of LBM. *Journal of Molecular Liquids*, *263*, 472–488. doi:10.1016/j.molliq.2018.04.111

Sheikholeslami, M. (2018f). Numerical simulation for solidification in a LHTESS by means of Nano-enhanced PCM. *Journal of the Taiwan Institute of Chemical Engineers*, *86*, 25–41. doi:10.1016/j.jtice.2018.03.013

Sheikholeslami, M. (2018g). Numerical modeling of Nano enhanced PCM solidification in an enclosure with metallic fin. *Journal of Molecular Liquids*, *259*, 424–438. doi:10.1016/j.molliq.2018.03.006

Sheikholeslami, M. (2018h). Numerical investigation of nanofluid free convection under the influence of electric field in a porous enclosure. *Journal of Molecular Liquids*, *249*, 1212–1221. doi:10.1016/j.molliq.2017.11.141

Sheikholeslami, M. (2018i). CuO-water nanofluid flow due to magnetic field inside a porous media considering Brownian motion. *Journal of Molecular Liquids*, *249*, 921–929. doi:10.1016/j.molliq.2017.11.118

Sheikholeslami, M. (2018j). Application of Darcy law for nanofluid flow in a porous cavity under the impact of Lorentz forces. *Journal of Molecular Liquids*, *266*, 495–503. doi:10.1016/j.molliq.2018.06.083

Sheikholeslami, M., Barzegar Gerdroodbary, M., Valiallah Mousavi, S., Ganji, D. D., & Moradi, R. (2018). Heat transfer enhancement of ferrofluid inside an 90o elbow channel by non-uniform magnetic field. *Journal of Magnetism and Magnetic Materials*, *460*, 302–311. doi:10.1016/j.jmmm.2018.03.070

Sheikholeslami, M., & Bhatti, M. M. (2017). Active method for nanofluid heat transfer enhancement by means of EHD. *International Journal of Heat and Mass Transfer*, *109*, 115–122. doi:10.1016/j.ijheatmasstransfer.2017.01.115

Sheikholeslami, M., Darzi, M., & Li, Z. (2018). Experimental investigation for entropy generation and exergy loss of nano-refrigerant condensation process. *International Journal of Heat and Mass Transfer*, *125*, 1087–1095. doi:10.1016/j.ijheatmasstransfer.2018.04.155

Sheikholeslami, M., Darzi, M., & Sadoughi, M. K. (2018). Heat transfer improvement and Pressure Drop during condensation of refrigerant-based Nanofluid; An Experimental Procedure. *International Journal of Heat and Mass Transfer*, *122*, 643–650. doi:10.1016/j.ijheatmasstransfer.2018.02.015

Sheikholeslami, M., & Ghasemi, A. (2018). Solidification heat transfer of nanofluid in existence of thermal radiation by means of FEM. *International Journal of Heat and Mass Transfer*, *123*, 418–431. doi:10.1016/j.ijheatmasstransfer.2018.02.095

Sheikholeslami, M., Ghasemi, A., Li, Z., Shafee, A., & Saleem, S. (2018). Influence of CuO nanoparticles on heat transfer behavior of PCM in solidification process considering radiative source term. *International Journal of Heat and Mass Transfer*, *126*, 1252–1264. doi:10.1016/j.ijheatmasstransfer.2018.05.116

Sheikholeslami, M., Hayat, T., & Alsaedi, A. (2018). Numerical simulation for forced convection flow of MHD CuO-H2O nanofluid inside a cavity by means of LBM. *Journal of Molecular Liquids*, *249*, 941–948. doi:10.1016/j.molliq.2017.10.099

Sheikholeslami, M., Hayat, T., Muhammad, T., & Alsaedi, A. (2018). MHD forced convection flow of nanofluid in a porous cavity with hot elliptic obstacle by means of Lattice Boltzmann method. *International Journal of Mechanical Sciences*, *135*, 532–540. doi:10.1016/j.ijmecsci.2017.12.005

Sheikholeslami, M., Jafaryar, M., Ganji, D. D., & Li, Z. (2018). Exergy loss analysis for nanofluid forced convection heat transfer in a pipe with modified turbulators. *Journal of Molecular Liquids*, *262*, 104–110. doi:10.1016/j.molliq.2018.04.077

Sheikholeslami, M., Jafaryar, M., & Li, Z. (2018a). Second law analysis for nanofluid turbulent flow inside a circular duct in presence of twisted tape turbulators. *Journal of Molecular Liquids*, *263*, 489–500. doi:10.1016/j.molliq.2018.04.147

Sheikholeslami, M., Jafaryar, M., & Li, Z. (2018b). Nanofluid turbulent convective flow in a circular duct with helical turbulators considering CuO nanoparticles. *International Journal of Heat and Mass Transfer*, *124*, 980–989. doi:10.1016/j.ijheatmasstransfer.2018.04.022

Sheikholeslami, M., Jafaryar, M., Saleem, S., Li, Z., Shafee, A., & Jiang, Y. (2018). Nanofluid heat transfer augmentation and exergy loss inside a pipe equipped with innovative turbulators. *International Journal of Heat and Mass Transfer*, *126*, 156–163. doi:10.1016/j.ijheatmasstransfer.2018.05.128

Sheikholeslami, M., Jafaryar, M., Shafee, A., & Li, Z. (2018). Investigation of second law and hydrothermal behavior of nanofluid through a tube using passive methods. *Journal of Molecular Liquids*, *269*, 407–416. doi:10.1016/j.molliq.2018.08.019

Sheikholeslami, M., Li, Z., & Shafee, A. (2018a). Lorentz forces effect on NEPCM heat transfer during solidification in a porous energy storage system. *International Journal of Heat and Mass Transfer*, *127*, 665–674. doi:10.1016/j.ijheatmasstransfer.2018.06.087

Sheikholeslami, M., Li, Z., & Shamlooei, M. (2018). Nanofluid MHD natural convection through a porous complex shaped cavity considering thermal radiation. *Physics Letters. [Part A]*, *382*(24), 1615–1632. doi:10.1016/j.physleta.2018.04.006

Sheikholeslami, M., & Rokni, H. B. (2017). Simulation of nanofluid heat transfer in presence of magnetic field: A review. *International Journal of Heat and Mass Transfer*, *115*, 1203–1233. doi:10.1016/j.ijheatmasstransfer.2017.08.108

Sheikholeslami, M., & Rokni, H. B. (2018a). CVFEM for effect of Lorentz forces on nanofluid flow in a porous complex shaped enclosure by means of Non-equilibrium model. *Journal of Molecular Liquids*, *254*, 446–462. doi:10.1016/j.molliq.2018.01.130

Sheikholeslami, M., Rokni, H.B. (2018b). Magnetic nanofluid flow and convective heat transfer in a porous cavity considering Brownian motion effects. *Physics of Fluids*, *30*(1), doi:10.1063/1.5012517

Sheikholeslami, M., & Sadoughi, M. K. (2017). Mesoscopic method for MHD nanofluid flow inside a porous cavity considering various shapes of nanoparticles. *International Journal of Heat and Mass Transfer*, *113*, 106–114. doi:10.1016/j.ijheatmasstransfer.2017.05.054

Sheikholeslami, M., & Sadoughi, M. K. (2018). Simulation of CuO- water nanofluid heat transfer enhancement in presence of melting surface. *International Journal of Heat and Mass Transfer, 116*, 909–919. doi:10.1016/j.ijheatmasstransfer.2017.09.086

Sheikholeslami, M., & Seyednezhad, M. (2018). Simulation of nanofluid flow and natural convection in a porous media under the influence of electric field using CVFEM. *International Journal of Heat and Mass Transfer, 120*, 772–781. doi:10.1016/j.ijheatmasstransfer.2017.12.087

Sheikholeslami, M., Shafee, A., Ramzan, M., & Li, Z. (2018). Investigation of Lorentz forces and radiation impacts on nanofluid treatment in a porous semi annulus via Darcy law. *Journal of Molecular Liquids, 272*, 8–14. doi:10.1016/j.molliq.2018.09.016

Sheikholeslami, M., Shamlooei, M., & Moradi, R. (2018). Numerical simulation for heat transfer intensification of nanofluid in a porous curved enclosure considering shape effect of Fe3O4 nanoparticles. *Chemical Engineering & Processing: Process Intensification, 124*, 71–82. doi:10.1016/j.cep.2017.12.005

Sheikholeslami, M., & Shehzad, S. A. (2017a). Thermal radiation of ferrofluid in existence of Lorentz forces considering variable viscosity. *International Journal of Heat and Mass Transfer, 109*, 82–92. doi:10.1016/j.ijheatmasstransfer.2017.01.096

Sheikholeslami, M., & Shehzad, S. A. (2017b). Magnetohydrodynamic nanofluid convective flow in a porous enclosure by means of LBM. *International Journal of Heat and Mass Transfer, 113*, 796–805. doi:10.1016/j.ijheatmasstransfer.2017.05.130

Sheikholeslami, M., & Shehzad, S. A. (2018a). Numerical analysis of Fe3O4 –H2O nanofluid flow in permeable media under the effect of external magnetic source. *International Journal of Heat and Mass Transfer, 118*, 182–192. doi:10.1016/j.ijheatmasstransfer.2017.10.113

Sheikholeslami, M., & Shehzad, S. A. (2018b). CVFEM simulation for nanofluid migration in a porous medium using Darcy model. *International Journal of Heat and Mass Transfer, 122*, 1264–1271. doi:10.1016/j.ijheatmasstransfer.2018.02.080

Sheikholeslami, M., & Shehzad, S. A. (2018c). Simulation of water based nanofluid convective flow inside a porous enclosure via Non-equilibrium model. *International Journal of Heat and Mass Transfer, 120*, 1200–1212. doi:10.1016/j.ijheatmasstransfer.2017.12.132

Sheikholeslami, M., & Shehzad, S. A. (2018d). Non-Darcy free convection of Fe3O4-water nanoliquid in a complex shaped enclosure under impact of uniform Lorentz force. *The Chinese Journal of Physiology, 56*(1), 270–281. doi:10.1016/j.cjph.2017.12.017

Sheikholeslami, M., Shehzad, S. A., Abbasi, F. M., & Li, Z. (2018). Nanofluid flow and forced convection heat transfer due to Lorentz forces in a porous lid driven cubic enclosure with hot obstacle. *Computer Methods in Applied Mechanics and Engineering, 338*, 491–505. doi:10.1016/j.cma.2018.04.020

Sheikholeslami, M., Shehzad, S. A., & Li, Z. (2018a). Nanofluid heat transfer intensification in a permeable channel due to magnetic field using Lattice Boltzmann method. *Physica B, Condensed Matter, 542*, 51–58. doi:10.1016/j.physb.2018.03.036

Sheikholeslami, M., Shehzad, S. A., & Li, Z. (2018b). Water based nanofluid free convection heat transfer in a three dimensional porous cavity with hot sphere obstacle in existence of Lorenz forces. *International Journal of Heat and Mass Transfer*, *125*, 375–386. doi:10.1016/j.ijheatmasstransfer.2018.04.076

Sheikholeslami, M., Shehzad, S. A., Li, Z., & Shafee, A. (2018). Numerical modeling for Alumina nanofluid magnetohydrodynamic convective heat transfer in a permeable medium using Darcy law. *International Journal of Heat and Mass Transfer*, *127*, 614–622. doi:10.1016/j.ijheatmasstransfer.2018.07.013

Sheikholeslami, M., & Vajravelu, K. (2017). Forced convection heat transfer in Fe3O4- ethylene glycol nanofluid under the influence of Coulomb force. *Journal of Molecular Liquids*, *233*, 203–210. doi:10.1016/j.molliq.2017.03.026

Sheikholeslami, M., & Zeeshan, A. (2017). Analysis of flow and heat transfer in water based nanofluid due to magnetic field in a porous enclosure with constant heat flux using CVFEM. *Computer Methods in Applied Mechanics and Engineering*, *320*, 68–81. doi:10.1016/j.cma.2017.03.024

Sheikholeslami, M., Zeeshan, A., & Majeed, A. (2018). Control volume based finite element simulation of magnetic nanofluid flow and heat transport in non-Darcy medium. *Journal of Molecular Liquids*, *268*, 354–364. doi:10.1016/j.molliq.2018.07.031

Chapter 7
Influence of Electric Field on Nanofluid Forced Convection Heat Transfer

ABSTRACT

In this chapter, the effect of electric field on forced convection heat transfer of nanofluid is presented. The governing equations are derived and presented in vorticity stream function formulation. Control volume-based finite element method (CVFEM) is employed to solve the final equations. Results indicate that the flow style depends on supplied voltage, and this effect is more sensible for low Reynolds number.

1. INTRODUCTION

Thermal conductivity of fluid can be improved by adding Nano size metal particle in to the bas fluid. This method can enhance heat transfer. Electric field is one of the effective active methods of enhancing heat transfer. Shu and Lai (1995) applied electric field on induced flow. Kasayapanand et al. (2002) investigated the effect of electrode arrangements on Nusselt number. Sheikholeslami et al. (2017) presented HD nanofluid force convective heat transfer. They considered electric field dependent viscosity. Sheikholeslami and Ellahi (2015) investigated electrohydrodynamic nanofluid hydrothermal treatment in an enclosure with sinusoidal upper wall. Sheikholeslami et al. (2016) studied the effect of electric field on nanofluid flow in a complex geometry. Sheikholeslami (2017) investigated Fe3O4-H2O nanofluid thermal improvement inexistence of Coulomb forces. Sheikholeslami and Ganji (2017) considered variable viscosity for simulation of nanofluid forced convection in existence of electric field. Sheikholeslami, and Bhatti (2017) presented active method for nanofluid heat transfer enhancement by means of EHD. Sheikholeslami and Rokni (2017) presented the impact of EFD viscosity on nanofluid forced convection in a cavity with sinusoidal wall. Sheikholeslami and Vajravelu (2017)] reported forced convection heat transfer in Fe3O4- ethylene glycol nanofluid under the influence of Coulomb force. In recent decade, applications of nanofluid for various engineering problems were presented.

DOI: 10.4018/978-1-5225-7595-5.ch007

2. EHD NANOFLUID FORCE CONVECTIVE HEAT TRANSFER CONSIDERING ELECTRIC FIELD DEPENDENT VISCOSITY

2.1. Problem Definition

Geometry and boundary conditions are shown in Figure 1. All walls are stationary except the lower one. The lid wall is hot wall and the others are cold. Distribution of electric density is depicted in Figure 2.

Figure 1. (a) Geometry of the problem and boundary conditions; (b) the mesh of enclosure considered in this work; (c) A sample triangular element and its corresponding control volume.

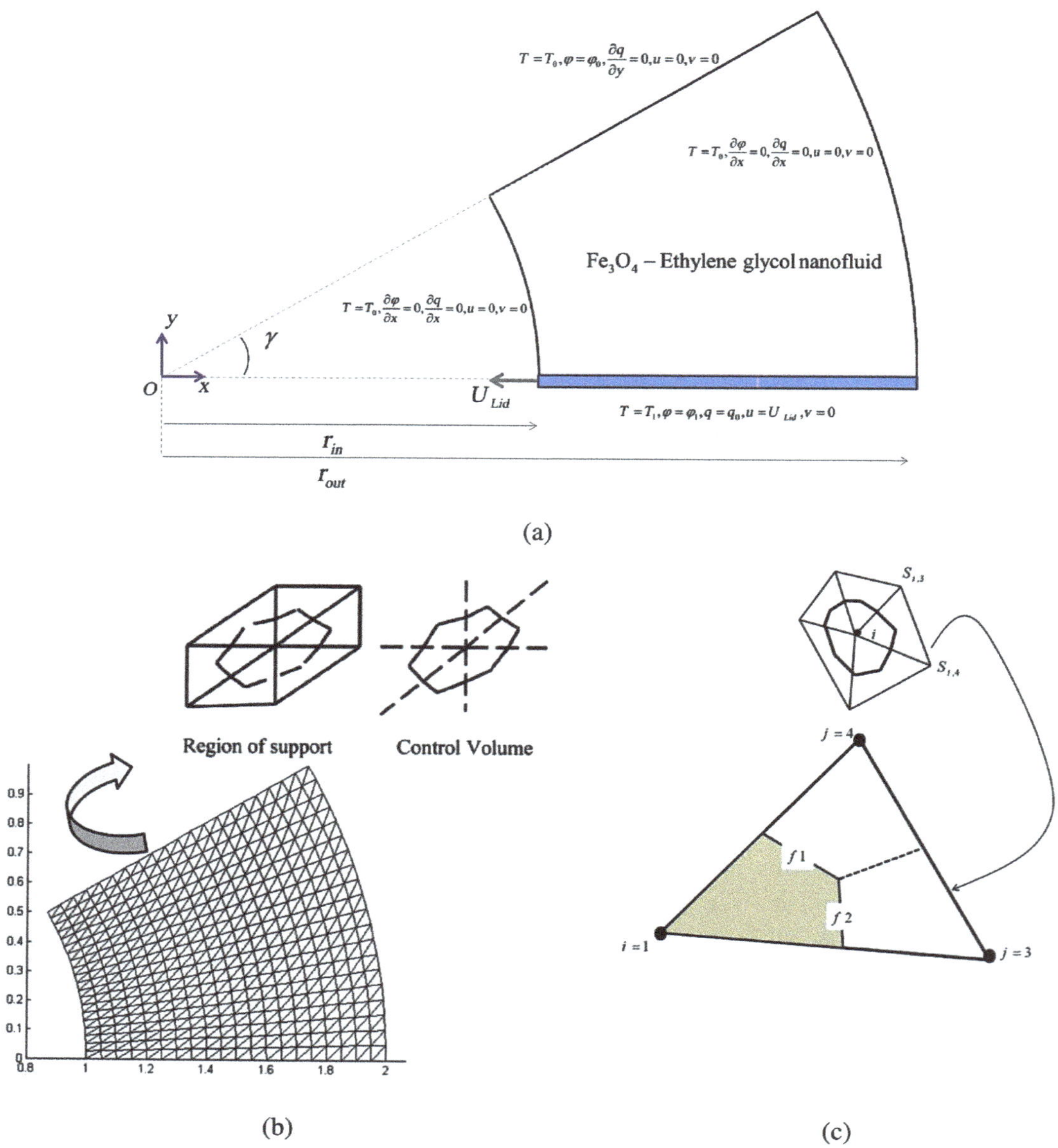

Figure 2. Electric density distribution injected by the bottom electrode

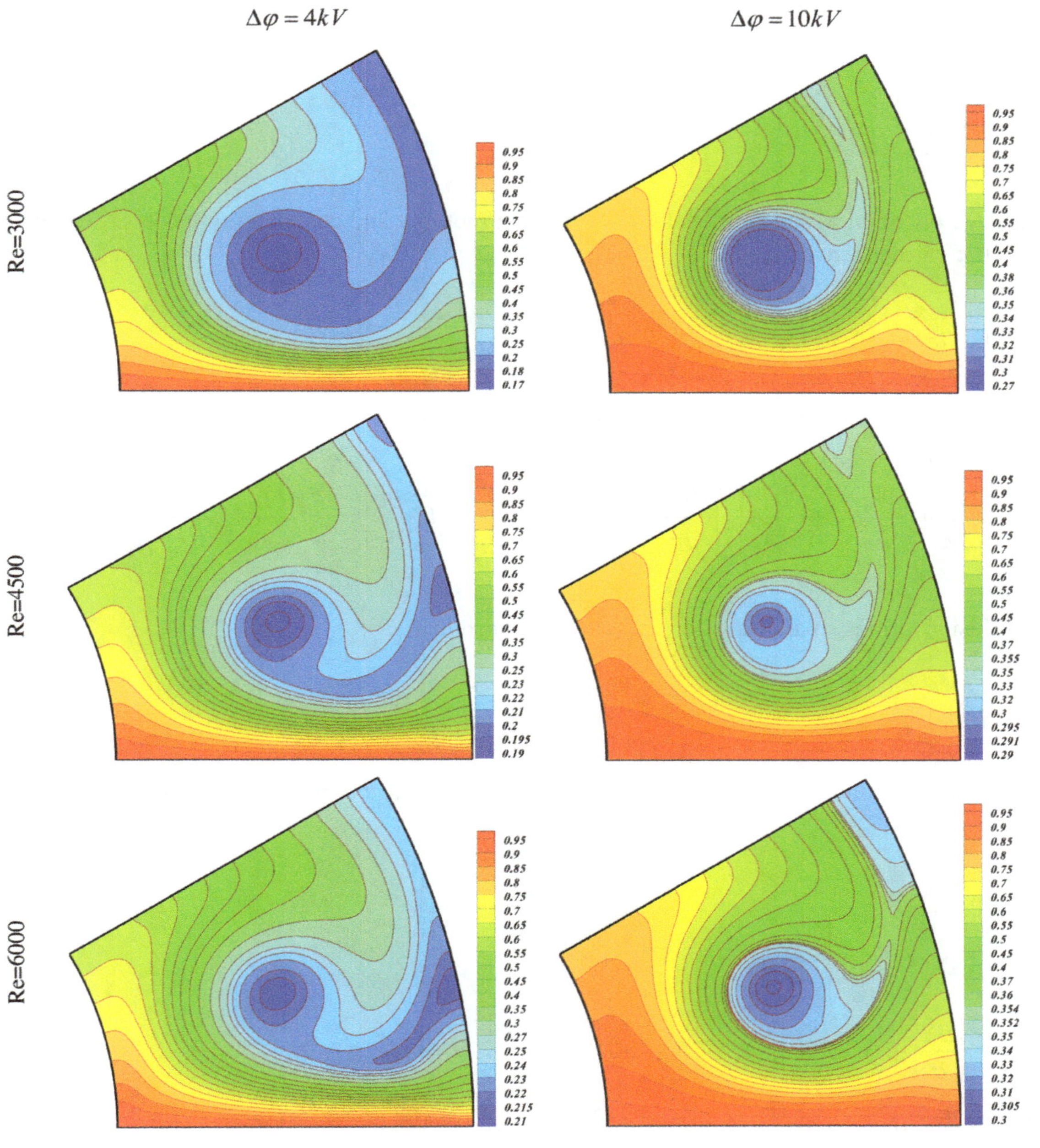

2.2. Governing Equation

Electric field equations should be combined with hydrothermal equations. Electric field equations are:

$$q = \nabla.\left(\vec{E}\,\varepsilon\right) \tag{1}$$

$$\vec{E} = \left(-\nabla \varphi\right) \tag{2}$$

$$\frac{\partial q}{\partial t} = -\nabla . \vec{J} \tag{3}$$

Charge distribution can be modeled in two ways: mobility and conductivity models. The equation of electric current density is:

$$\vec{J} = q\vec{V} + \sigma \vec{E} - D\nabla q \tag{4}$$

Using the above equations leads to the following equation:

$$\frac{\partial \bar{q}}{\partial \bar{t}} + \bar{u}\frac{\partial \bar{q}}{\partial \bar{x}} + \bar{v}\frac{\partial \bar{q}}{\partial \bar{y}} + \frac{1}{\mathrm{Re}\,\mathrm{Pr}_E}\left[\bar{q}\left(\frac{\partial \overline{E_y}}{\partial \bar{y}} + \frac{\partial \overline{E_x}}{\partial \bar{x}}\right) + \overline{E_y}\frac{\partial \bar{q}}{\partial \bar{y}} + \overline{E_x}\frac{\partial \bar{q}}{\partial \bar{x}}\right] = \frac{\rho_{nf}/\rho_f}{\mu_{nf}/\mu_f}\frac{1}{\mathrm{Re}\,D_e}\left(\frac{\partial^2 \bar{q}}{\partial \bar{y}^2} + \frac{\partial^2 \bar{q}}{\partial \bar{x}^2}\right) \tag{5}$$

Diffusion term is small, Equation (4) can be changed to:

$$\vec{J} = qb\vec{E} + q\vec{V} \tag{6}$$

In existence of electric field, Coulomb forces should be added to momentum equations:

$$\begin{cases} \nabla . \vec{V} = 0 \\ \rho_{nf}\left(\frac{\partial \vec{V}}{\partial t} + \left(\vec{V} . \nabla\right)\vec{V}\right) = -\nabla p + \mu_{nf}\nabla^2 \vec{V} + q\vec{E} \\ \left(\rho C_p\right)_{nf}\left(\frac{\partial T}{\partial t} + \left(\vec{V} . \nabla\right)T\right) = k_{nf}\nabla^2 T + \vec{J} . \vec{E} \\ \nabla . \vec{J} + \frac{\partial q}{\partial t} = 0 \\ \nabla . \varepsilon \vec{E} = q \\ \vec{E} = -\nabla \varphi \end{cases} \tag{7}$$

$\rho_{nf}, \left(\rho C_p\right)_{nf}, \alpha_{nf}, \beta_{nf}, \mu_{nf}$ and k_{nf} are defined as:

$$\rho_{nf} = \rho_s\phi + \rho_f(1-\phi) \tag{8}$$

$$\left(\rho C_p\right)_{nf} = \left(\rho C_p\right)_s \phi + \left(\rho C_p\right)_f (1-\phi) \tag{9}$$

$$\alpha_{nf} = k_{nf} / \left(\rho C_p\right)_{nf} \tag{10}$$

$$\beta_{nf} = \beta_s\phi + \beta_f(1-\phi) \tag{11}$$

$$k_{nf} = k_f \frac{-2\phi(k_f - k_s) + 2k_f + k_s}{+\phi(k_f - k_s)_s + 2k_f + k} \tag{12}$$

Table 1 illustrate the properties of the base fluid and nanoparticles. Effect of electric field on viscosity of nanofluid has been taken into account:

$$\mu = A_1 + A_2\left(\Delta\varphi\right) + A_3\left(\Delta\varphi\right)^2 + A_4\left(\Delta\varphi\right)^3 \tag{13}$$

Table 2 shows the coefficient values of this equation. Non-dimensional parameters are presented as follow:

$$\begin{gathered} \bar{t} = \frac{tU_{Lid}}{L}, \bar{p} = \frac{P}{\rho U_{Lid}^{\;2}}, \bar{y} = \frac{y}{L}, \bar{x} = \frac{x}{L}, \bar{v} = \frac{v}{U_{Lid}}, \bar{u} = \frac{u}{U_{Lid}}, \\ \theta = \frac{T - T_0}{\nabla \mathrm{T}}, \nabla \mathrm{T} = T_1 - T_0, \bar{\varphi} = \frac{\varphi - \varphi_0}{\nabla\varphi}, \nabla\varphi = \varphi_1 - \varphi_0, \bar{q} = \frac{q}{q_0}, \bar{E} = \frac{E}{E_0} \end{gathered} \tag{14}$$

where ∇T and $\nabla\varphi$ are $\left(T_1 - T_0\right)$ and $\left(\varphi_1 - \varphi_0\right)$, respectively. By eliminating the over bar, the equations are:

Table 1. Thermo physical properties of water and nanoparticles

	$\rho(kg / m^3)$	$C_p(j / kgk)$	$k(W / m.k)$	$\mu(Pa.s)$
Ethylene glycol	1110	2400	0.26	0.0162
Fe_2O_4	5200	670	6	-

Table 2. The coefficient values of Equation (13)

Coefficient Values	ϕ=0	ϕ=0.05
A_1	1.0603E+001	9.5331
A_2	-2.698E-003	-3.4119E-003
A_3	2.9082E-006	5.5228E-006
A_4	-1.1876E-008	-4.1344E-008

$$\begin{cases} \nabla.\left(\vec{V}\right) = 0 \\ \left(\dfrac{\partial \vec{V}}{\partial t} + \left(\vec{V}.\nabla\right)\vec{V}\right) = -\nabla p + \dfrac{\rho_{nf} / \rho_f}{\mu_{nf} / \mu_f} \dfrac{1}{\mathrm{Re}} \nabla^2 \vec{V} + \dfrac{S_E}{\rho_{nf} / \rho_f} q \vec{E} \\ \left(\dfrac{\partial \theta}{\partial t} + \left(\vec{V}.\nabla\right)\theta\right) = \dfrac{k_{nf} / k_f}{\left(\rho C_p\right)_{nf} / \left(\rho C_p\right)_f} \dfrac{1}{\mathrm{RePr}} \nabla^2 \theta + \dfrac{S_E Ec}{\left(\rho C_p\right)_{nf} / \left(\rho C_p\right)_f} \left(\vec{\mathrm{J}}.\vec{E}\right) \\ \nabla.\vec{\mathrm{J}} + \dfrac{\partial q}{\partial t} = 0 \\ \nabla.\varepsilon \vec{E} = q,\ \vec{E} = -\nabla \varphi \end{cases} \tag{15}$$

Stream function and vorticity can be defined as:

$$v = -\frac{\partial \psi}{\partial x}, u = \frac{\partial \psi}{\partial y}, \omega = \frac{\partial v}{\partial x} - \frac{\partial u}{\partial y},\ \Omega = \frac{\omega L}{U_{Lid}}, \Psi = \frac{\psi}{L U_{Lid}} \tag{16}$$

Stream function can satisfy the continuity equation. Vorticity equation can be derived by eliminating pressure sources.

Nu_{loc} and Nu_{ave} along the lid wall can be obtained as:

$$Nu_{loc} = \left(\frac{k_{nf}}{k_f}\right)\frac{\partial \Theta}{\partial Y} \tag{17}$$

$$Nu_{ave} = \frac{1}{L}\int_{r_{in}}^{r_{out}} Nu_{loc}\, dX \tag{18}$$

2.3. Effects of Active Parameters

Nanofluid flow in a semi lid driven enclosure in existence of electric field has been investigated. The working fluid is Ethylene glycol and Fe_3O_4. Viscosity of nanofluid relies on strength of electric field. Effects of volume fraction of nanoparticles ($\phi = 0\%$ and 5%), Reynolds number ($\mathrm{Re} = 3000, 4500$ and 6000) and supplied voltage ($\Delta\varphi = 0, 4, 6$ and $10kV$) are examined. In all cases Pr and Ec equal to 149.54 and10^{-6}, respectively.

Figures 3, 4 and 5 depict the impacts of supplied voltage and Reynolds number on isotherm and streamlines.

At low Reynolds number, streamline has one main eddy and two very small vortexes at upper corners of the cavity. By increasing electric field the main eddy convert to two smaller eddies. So thermal boundary layer thickness decreases near the moving wall and in turn temperature gradient reduces. As

Figure 3. Effect of supplied voltage on streamlines and isotherm when $\mathrm{Re} = 3000, \phi = 0.05$

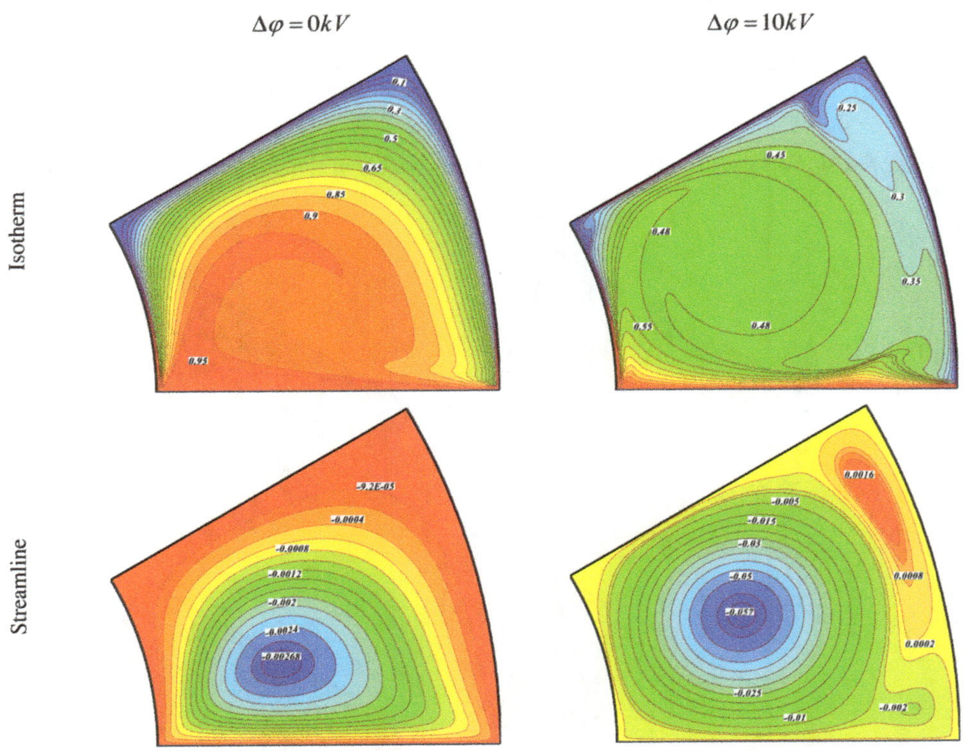

Figure 4. Effect of supplied voltage on streamlines and isotherm when $\mathrm{Re}=4500, \phi=0.05$.

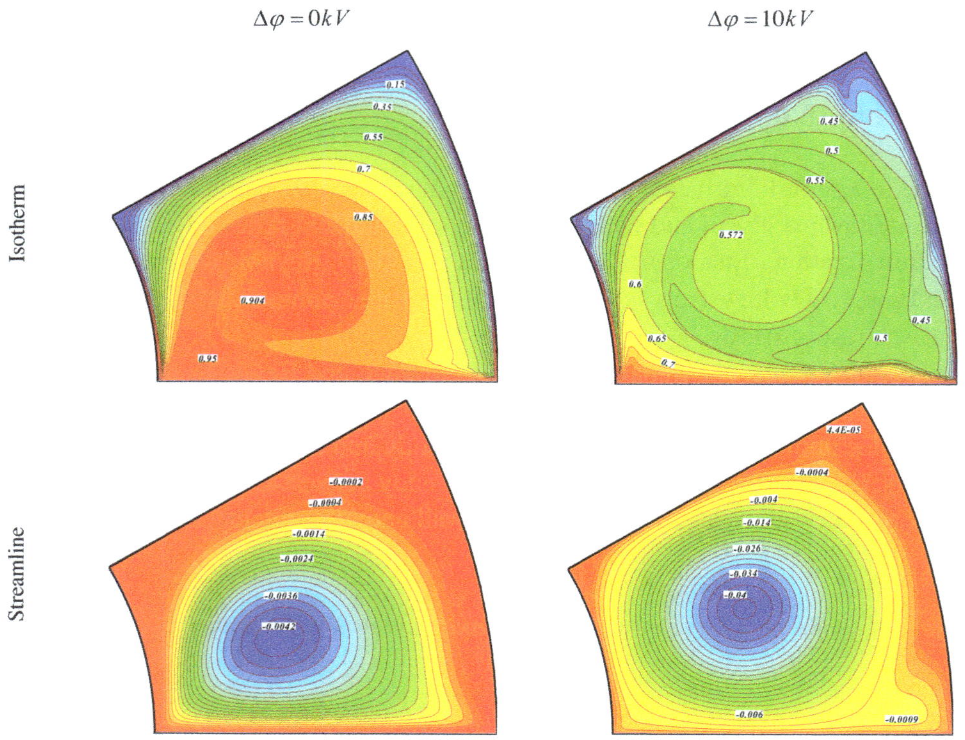

Figure 5. Effect of supplied voltage on streamlines and isotherm when $\mathrm{Re}=6000, \phi=0.05$.

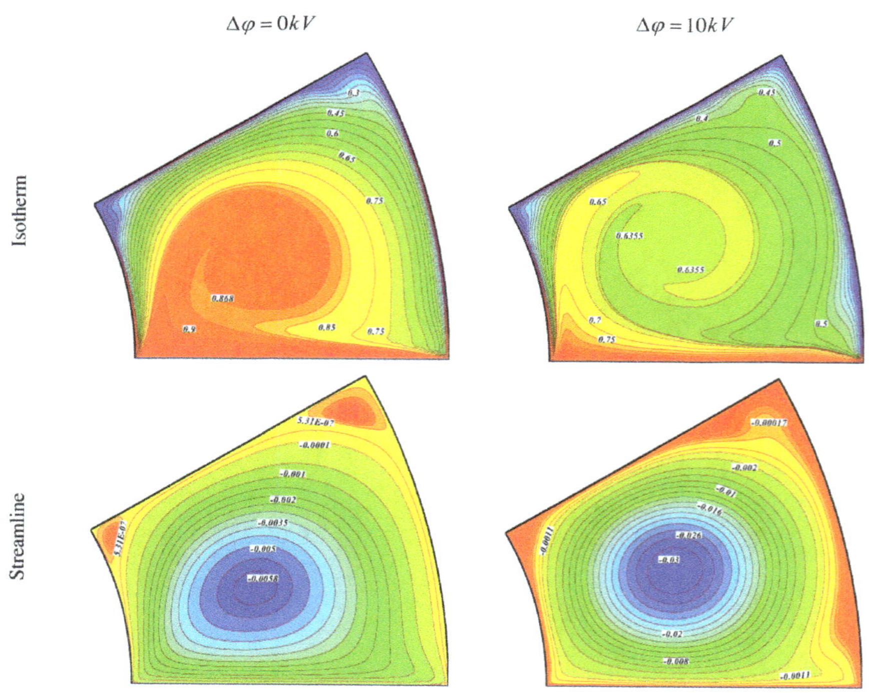

Reynolds number increases, thermal boundary layer thickness decreases. So rate of heat transfer enhances with rise of Reynolds number.

Effect of $\Delta\varphi$ and Re on local and average Nusselt number are shown in Figures 6 and 7. Enhancing supplied voltage and Reynolds number leads to more distortion of isotherms. So Nusselt number augments with rise of these parameters. By applying electric field stronger mixing leading to destruction of thermal boundary layer and create vortex flow, thinning the boundary layer, and therefore cause the rise in heat transfer rate. Existence of extremums in local Nusselt number is due to generation of thermal

Figure 6. Effects of Reynolds number and supplied voltage on local Nusselt number when $\phi = 0.05$.

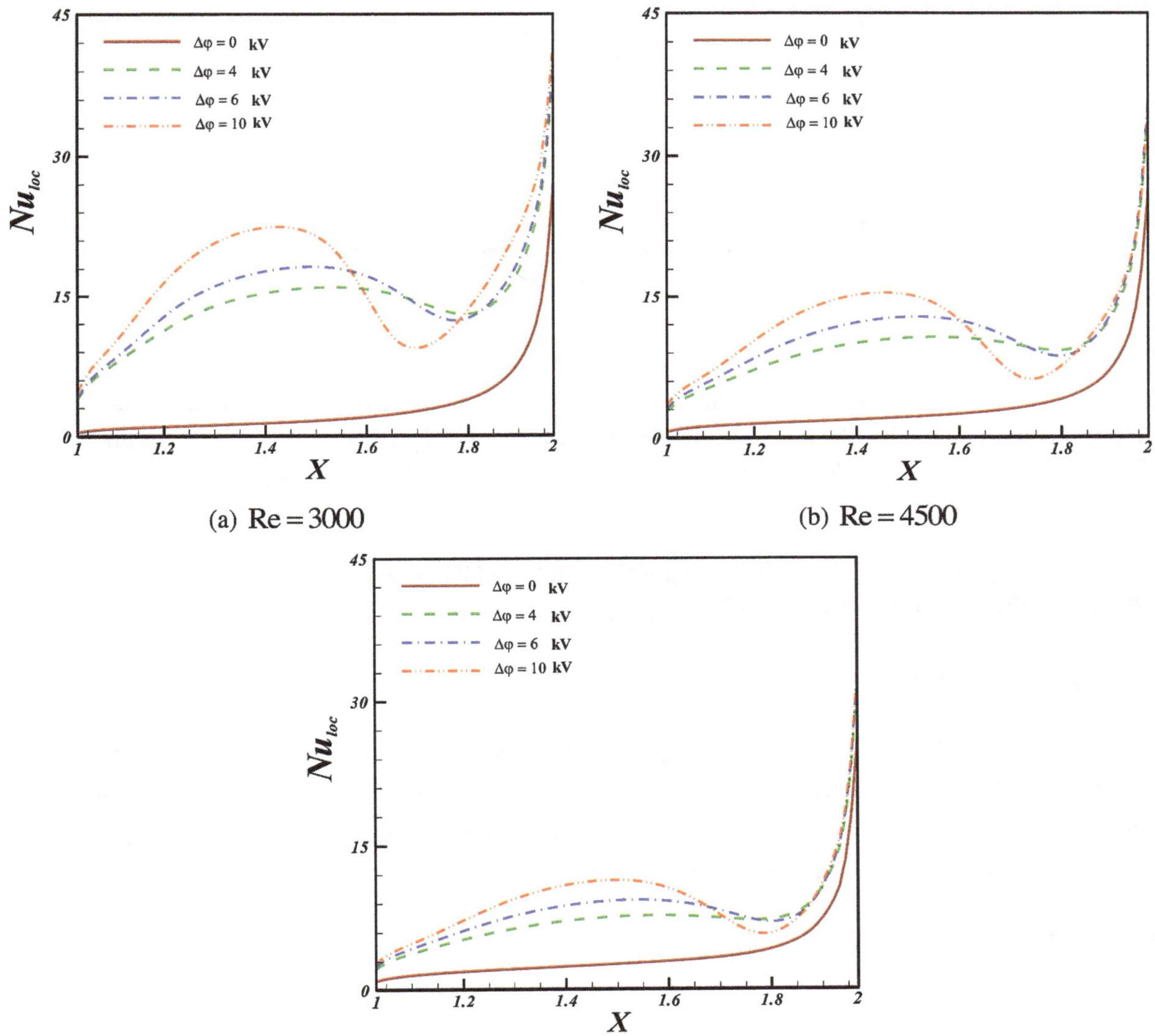

(a) $\mathrm{Re} = 3000$

(b) $\mathrm{Re} = 4500$

(c) $\mathrm{Re} = 6000$

Figure 7. Effects of Reynolds number and supplied voltage on average Nusselt number.

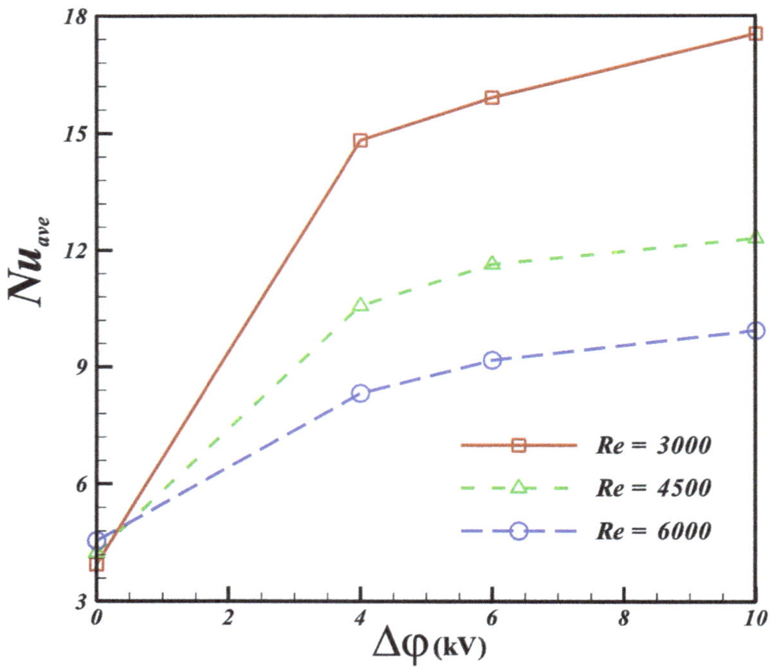

plumes. Effect of coulomb forces on rate of heat transfer is more sensible for lower Reynolds number. This is due to this fact that the boundary layer thickness in low Reynolds number is thicker than that of high Reynolds number. So electric field can more effective in low Reynolds number.

3. ELECTROHYDRODYNAMIC NANOFLUID HYDROTHERMAL TREATMENT IN AN ENCLOSURE WITH SINUSOIDAL UPPER WALL

3.1. Problem Definition

Figure 8 illustrates the physical geometry along with the important parameters and mesh of the enclosure. The lower wall has the velocity of U_{Lid} and others are stationary. The lower wall has constant temperature T_1 and the temperature of other walls is T_0. Also the retain boundary conditions are depicted in Figure 8(a). The formula of the upper sinusoidal wall is:

$$Y = 1 - \left\{0.03\left(1 + \sin\left(2\pi x - \pi / 2\right)\right)\right\} \tag{19}$$

Figure 9 depicts the distribution of electric density for different Reynolds number and supplied voltage.

Figure 8. (a) Geometry of the problem and boundary conditions; (b) the mesh of enclosure considered in this work.

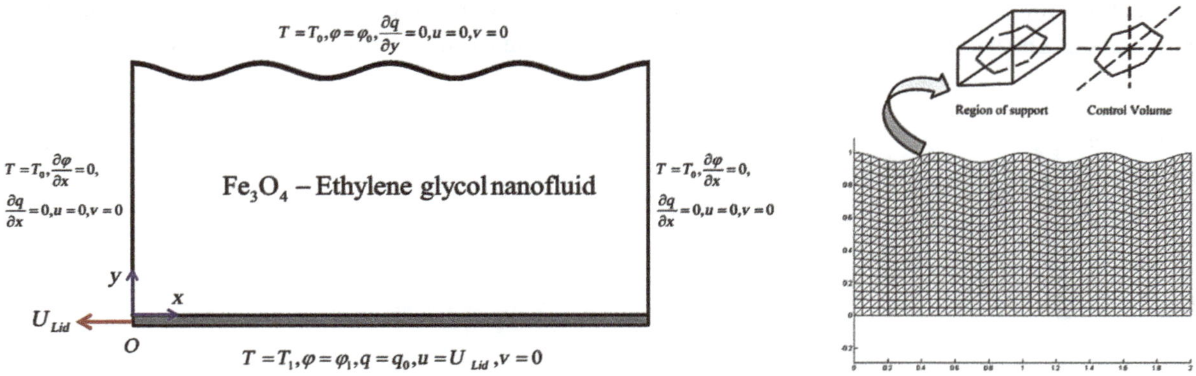

(a)

(b)

Figure 9. Electric density distribution injected by the bottom electrode

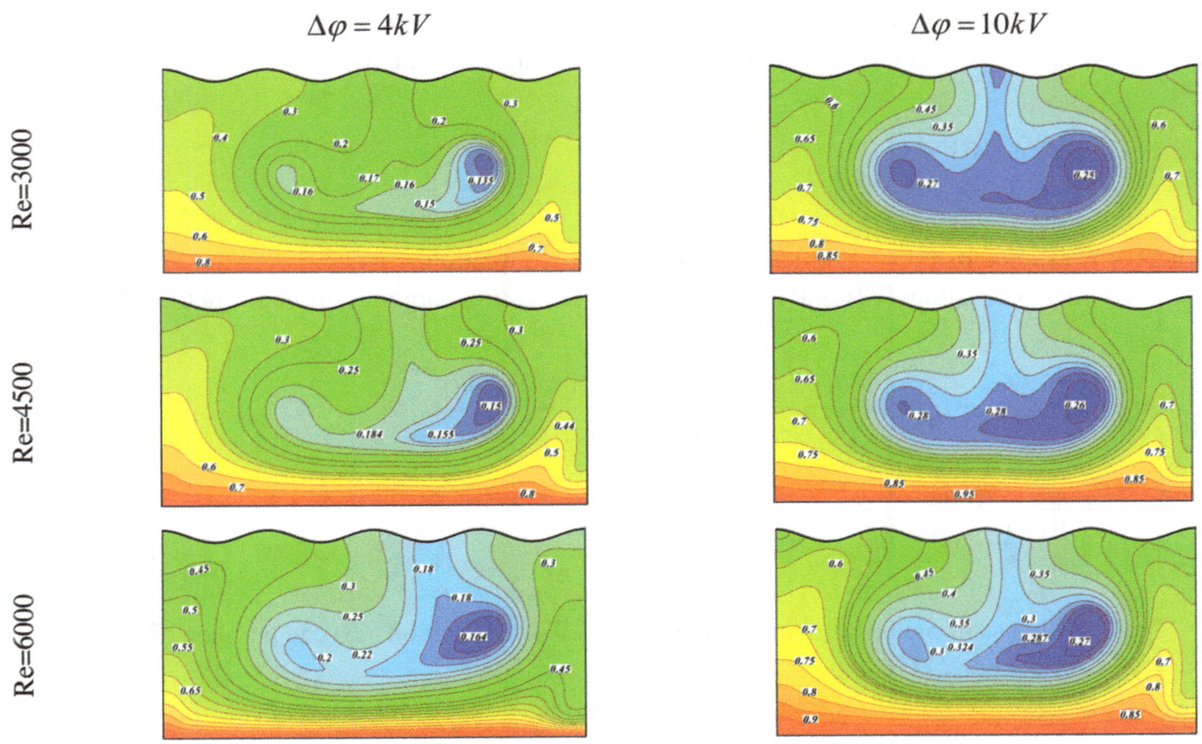

3.2. Governing Equation

In order to simulate nanofluid hydrothermal treatment in existence of electric field, we should combined equations of electric fields with those of hydrothermal. The formulas of electric field are:

$$\nabla.\left(\vec{E}\,\varepsilon\right) = q \tag{20}$$

$$\left(-\nabla\varphi\right) = \vec{E} \tag{21}$$

$$\nabla.\vec{J} + \frac{\partial q}{\partial t} = 0 \tag{22}$$

There exist two model for charge distribution: (1) conductivity model and (2) mobility model. In first model, electro-convection rely on temperature gradient. But in second model, electro-convection is independent of temperature gradient in the liquid. In the case of free charge origination, second model is more acceptable according to experimental results. Electric current density can be defined as:

$$\vec{J} = \sigma\,\vec{E} - D\nabla q + q\,\vec{V} \tag{23}$$

where $\sigma\,\vec{E}$ is ionic mobility, $D\nabla q$ is diffusion, $q\,\vec{V}$ is convection.

According to Equations (22) and (23), the equation for electric charge density can be obtained as follow:

$$\begin{aligned}&\frac{\partial \bar{q}}{\partial \bar{t}} + \bar{v}\frac{\partial \bar{q}}{\partial \bar{y}} + \bar{u}\frac{\partial \bar{q}}{\partial \bar{x}} + \frac{1}{\operatorname{Re}\Pr_E}\left[\bar{q}\left(\frac{\partial \overline{E_y}}{\partial \bar{y}} + \frac{\partial \overline{E_x}}{\partial \bar{x}}\right) + \overline{E_y}\frac{\partial \bar{q}}{\partial \bar{y}} + \overline{E_x}\frac{\partial \bar{q}}{\partial \bar{x}}\right] \\ &= \frac{\rho_{nf}/\rho_f}{\mu_{nf}/\mu_f}\frac{1}{\operatorname{Re}D_e}\left(\frac{\partial^2 \bar{q}}{\partial \bar{y}^2} + \frac{\partial^2 \bar{q}}{\partial \bar{x}^2}\right)\end{aligned} \tag{24}$$

The diffusion term can be taken negligible. Also $D\nabla q$ in Equation (23) can be taken negligible and $\sigma = bq$. So Equation (23) can be considered as:

$$\vec{J} = q\vec{V} + qb\,\vec{E} \tag{25}$$

In presence of electric field Coulomb forces should be added to momentum equation and Joule heating effect should be added in energy equation. So we have:

$$\begin{cases} \nabla.\vec{V} = 0 \\ \rho_{nf}\left(\frac{\partial \vec{V}}{\partial t} + \left(\vec{V}.\nabla\right)\vec{V}\right) = -\nabla p + \mu_{nf}\nabla^2\vec{V} + q\vec{E} \\ \left(\rho C_p\right)_{nf}\left(\frac{\partial T}{\partial t} + \left(\vec{V}.\nabla\right)T\right) = k_{nf}\nabla^2 T + \vec{J}.\vec{E} \\ \nabla.\vec{J} + \frac{\partial q}{\partial t} = 0 \\ \nabla.\varepsilon\vec{E} = q \\ \vec{E} = -\nabla\varphi \end{cases} \tag{26}$$

$\rho_{nf}, \left(\rho C_p\right)_{nf}, \alpha_{nf}, \mu_{nf}$ and k_{nf} are defined as:

$$\rho_{nf} = \rho_s\phi + \rho_f(1-\phi) \tag{27}$$

$$\left(\rho C_p\right)_{nf} = \left(\rho C_p\right)_s\phi + \left(\rho C_p\right)_f(1-\phi) \tag{28}$$

$$\alpha_{nf} = k_{nf} / \left(\rho C_p\right)_{nf} \tag{29}$$

$$\mu_{nf} = \frac{\mu_f}{(1-\phi)^{2.5}} \tag{30}$$

$$k_{nf} = k_f\frac{-2\phi(k_f - k_s) + 2k_f + k_s}{+\phi(k_f - k_s)_s + 2k_f + k} \tag{31}$$

The thermo physical properties of the working fluid are given in Table1. Non-dimensional parameters are introduced as follow:

$$\bar{t}=\frac{tU_{Lid}}{L},\bar{p}=\frac{P}{\rho U_{Lid}^{2}},\bar{y}=\frac{y}{L},\bar{x}=\frac{x}{L},$$
$$\bar{v}=\frac{v}{U_{Lid}},\bar{u}=\frac{u}{U_{Lid}},\theta=\frac{T-T_0}{\nabla \mathrm{T}},\nabla \mathrm{T}=T_1-T_0, \tag{32}$$
$$\bar{\varphi}=\frac{\varphi-\varphi_0}{\nabla\varphi},\nabla\varphi=\varphi_1-\varphi_0,\bar{q}=\frac{q}{q_0},\bar{E}=\frac{E}{E_0}$$

In order to reach clear formulation, over bar will be deleted in next equations. So, the governing equations can be considered as follows:

$$\begin{cases}\nabla.\left(\vec{V}\right)=0\\ \left(\frac{\partial \vec{V}}{\partial t}+\left(\vec{V}.\nabla\right)\vec{V}\right)=-\nabla p+\frac{\rho_{nf}/\rho_f}{\mu_{nf}/\mu_f}\frac{1}{\mathrm{Re}}\nabla^2\vec{V}+\frac{S_E}{\rho_{nf}/\rho_f}q\vec{E}\\ \left(\frac{\partial\theta}{\partial t}+\left(\vec{V}.\nabla\right)\theta\right)=\frac{k_{nf}/k_f}{\left(\rho C_p\right)_{nf}/\left(\rho C_p\right)_f}\frac{1}{\mathrm{RePr}}\nabla^2\theta+\frac{S_E Ec}{\left(\rho C_p\right)_{nf}/\left(\rho C_p\right)_f}\left(\vec{\mathrm{J}}.\vec{E}\right)\\ \nabla.\vec{\mathrm{J}}+\frac{\partial q}{\partial t}=0\\ \nabla.\varepsilon\vec{E}=q\\ \vec{E}=-\nabla\varphi\end{cases} \tag{33}$$

The formulas of vorticity and stream function are:

$$\omega=\frac{\partial v}{\partial x}-\frac{\partial u}{\partial y},$$
$$v=-\frac{\partial\psi}{\partial x},u=\frac{\partial\psi}{\partial y}, \tag{34}$$
$$\Omega=\frac{\omega L}{U_{Lid}},\Psi=\frac{\psi}{LU_{Lid}}$$

Continuity equation has been satisfied by the stream function. By eliminating pressure between x-momentum and y- momentum, the vorticity equation can be obtained.

Nu_{loc} and Nu_{ave} along the lid wall can be obtained as:

$$Nu_{loc}=\left(\frac{k_{nf}}{k_f}\right)\frac{\partial\Theta}{\partial Y} \tag{35}$$

$$Nu_{ave} = \frac{1}{L}\int_{0}^{L} Nu_{loc} dX \tag{36}$$

3.3. Effects of Active Parameters

Electric filed effect on hydrothermal behavior of nanofluid in an enclosure with moving lower and sinusoidal upper walls is presented. Table 1 illustrates the properties of Ethylene glycol and Fe_3O_4. Calculations are prepared for various values of supplied voltage ($\Delta\varphi = 0, 4, 6$ and $10kV$), volume fraction of nanoparticles ($\phi = 0\%$ and 4%) and Reynolds number numbers ($\mathrm{Re} = 3000, 4500$ and 6000). In all calculations, the Prandtl number ($\Pr$) and Eckert number (Ec) are set to 149.54, 0.0 and 1e-6.

Influence of Reynolds number and supplied voltage on streamlines and isotherm are shown in Figures 10, 11 and 12. At Re=3000, one main eddy and two very small eddied at upper corners of the enclosure exist in streamline. As electric filed applied the main eddy turn in to two smaller ones. Also isotherm becomes denser near the hot wall due to existence these eddies. As Reynolds number increases up to 6000, the upper right eddy becomes stronger and isotherms become denser near the bottom wall. Since convective heat transfer is helped more successfully at a greater flow rate, Nusselt number is improved at great Reynolds numbers. So isotherm becomes more distorted at greater values of Reynolds number. As electric filed increases, the main cell convert to two eddies which are rotates opposite direction.

Figures13 and 14 depict the influences of $\Delta\varphi$ and Re on Nu_{loc} and Nu_{ave} along the lid wall. As Reynolds number increases, Nusselt number increases due to decrease in thermal boundary layer thickness. Increasing supplied voltage make the isotherms more distorted. Local Nusselt number profiles have extremums at higher values of supplied voltage because of existence of thermal plumes. Nusselt

Figure 10. Effect of supplied voltage on streamlines and isotherm when $\mathrm{Re} = 3000, \phi = 0.04$.

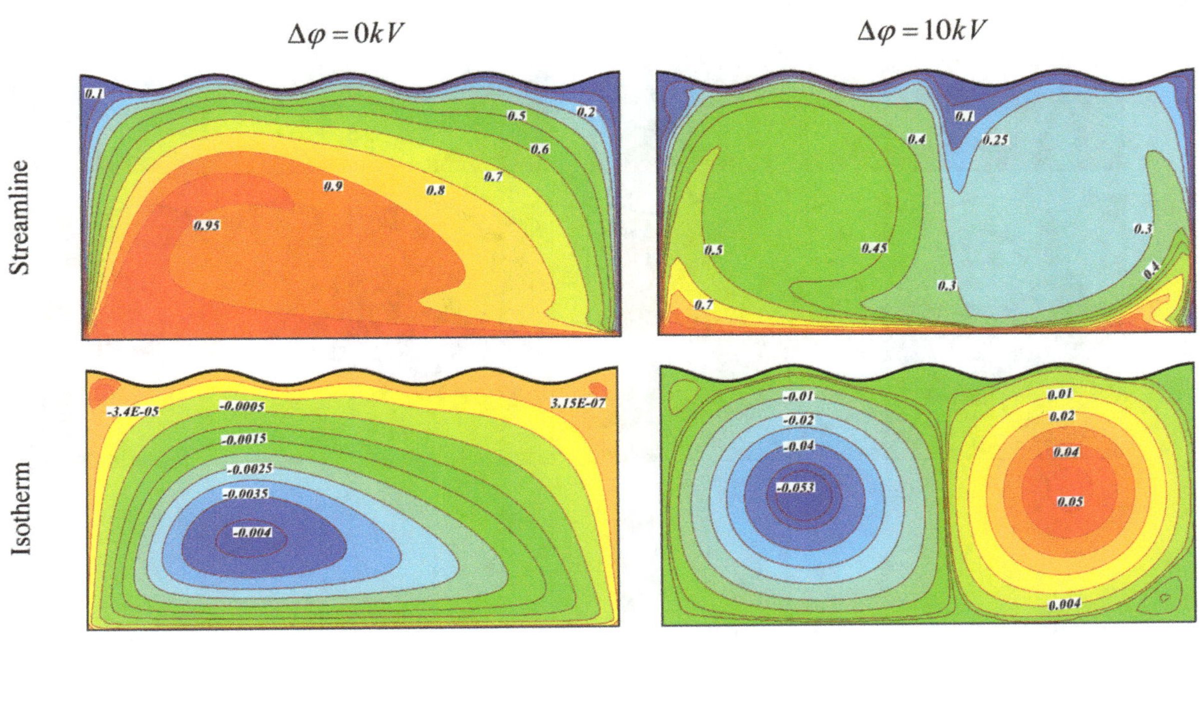

Figure 11. Effect of supplied voltage on streamlines and isotherm when $\mathrm{Re} = 4500, \phi = 0.04$.

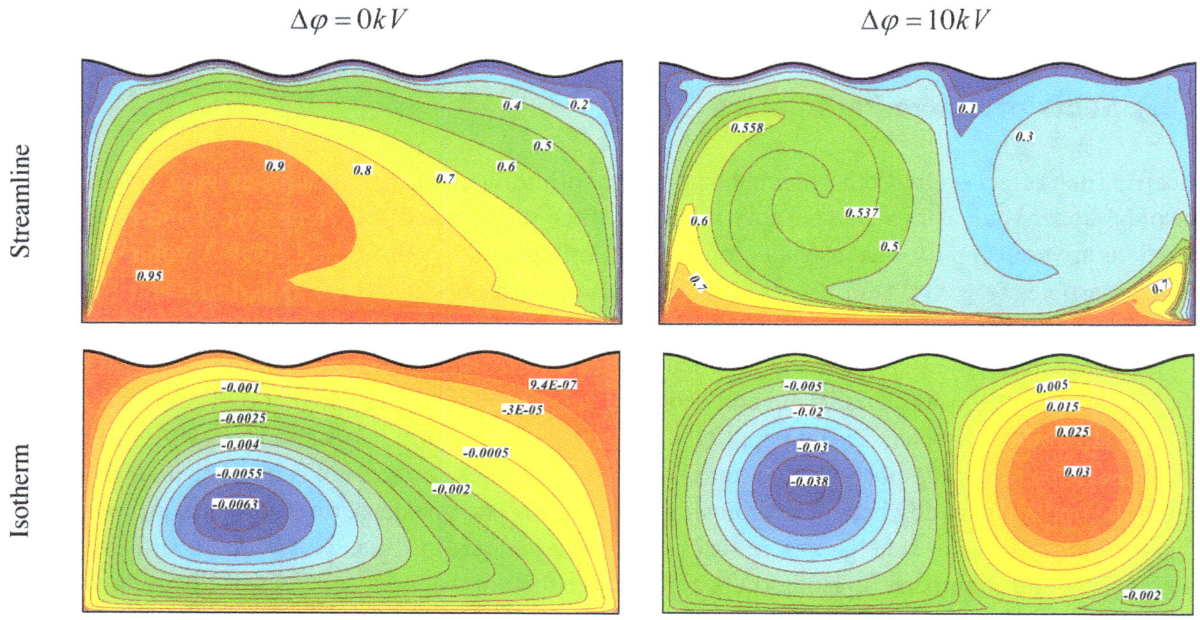

Figure 12. Effect of supplied voltage on streamlines and isotherm when $\mathrm{Re} = 6000, \phi = 0.04$.

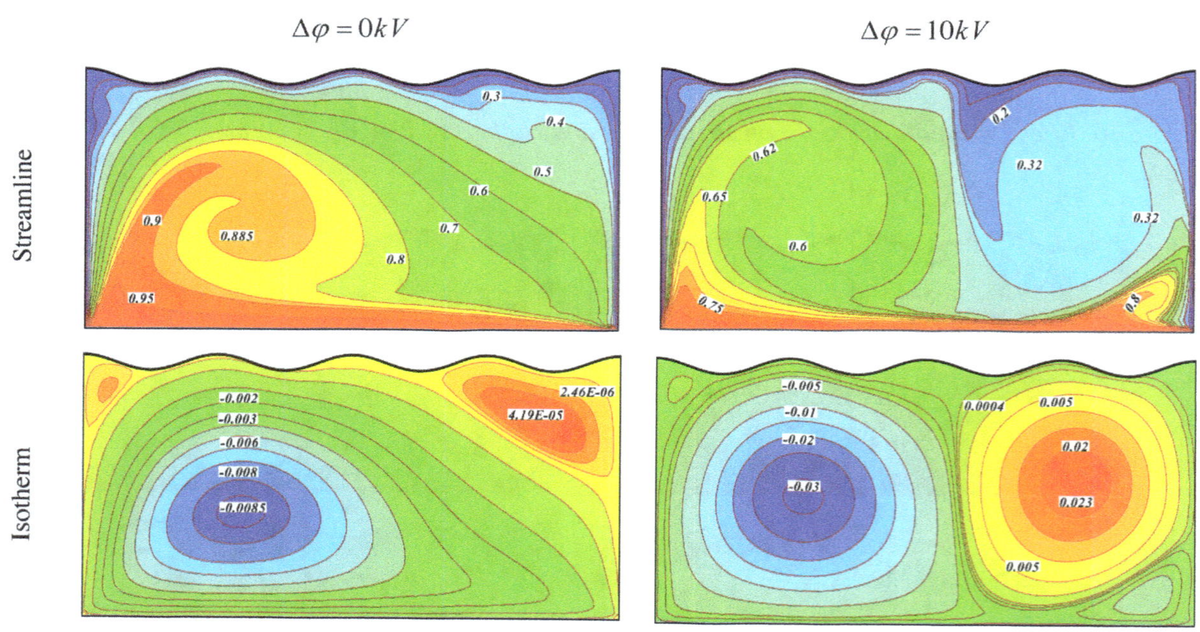

Figure 13. Effects of Reynolds number and supplied voltage on local Nusselt number when $\phi = 0.04$.

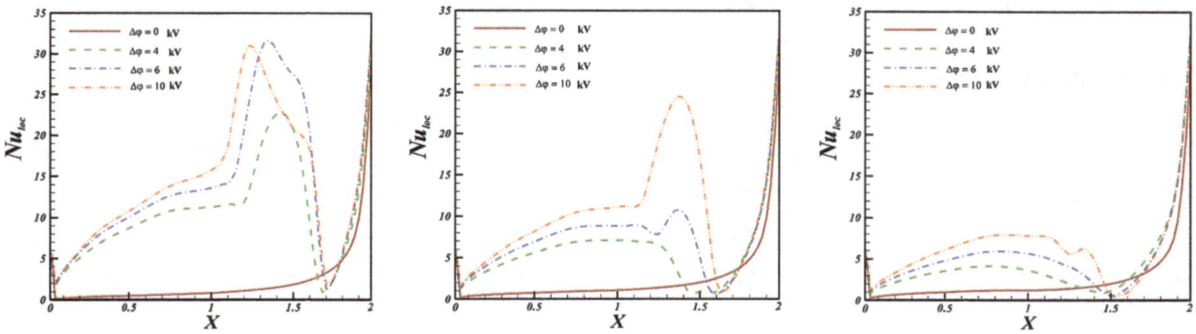

(a) Re = 3000 (b) Re = 4500 (c) Re = 6000

Figure 14. Effects of Reynolds number and supplied voltage on average Nusselt number.

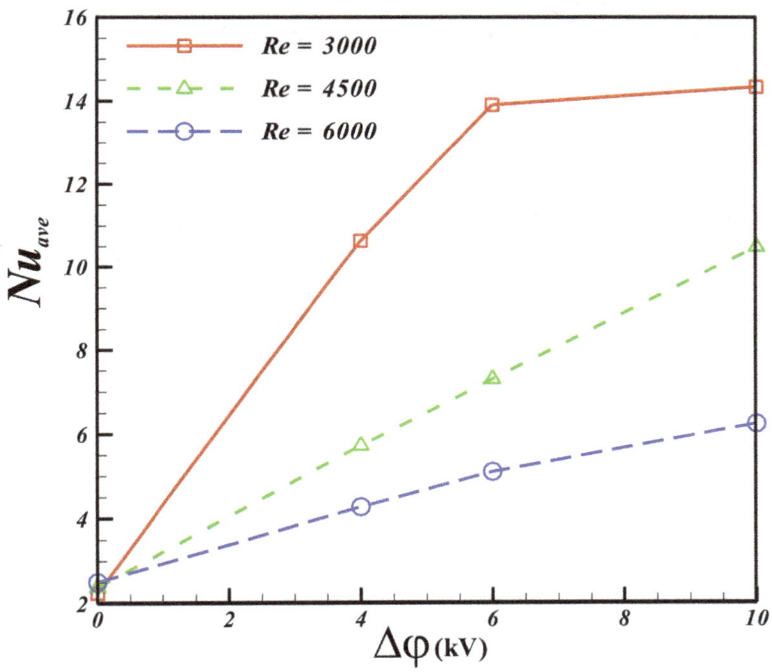

umber is an increasing function of supplied voltage. In absence of electric filed, Nusselt number for Re=6000 is 1.117 times higher than that of Re=3000 while in presence of electric filed $\left(\Delta\varphi = 10\right)$, Nusselt number for Re=6000 is 2.298385times lower than that of Re=3000. Also it can be concluded that Nusselt number at $\Delta\varphi = 10$ for Re=3000, 4500 and 6000 are times higher than those of obtained at $\Delta\varphi = 0$. This observation confirms that impact of electric filed is more marked for lower Reynolds number.

4. EFFECT OF ELECTRIC FIELD ON HYDROTHERMAL BEHAVIOR OF NANOFLUID IN A COMPLEX GEOMETRY

4.1. Problem Definition

Figure 15 illustrates the physical geometry along with the important parameters and mesh of the enclosure. The lower wall has the velocity of U_{Lid} and others are stationary. The lower wall has constant temperature T_1 and the temperature of other walls is T_0. Also the retain boundary conditions are depicted in Figure 15 (a). The shape of inner cylinder profile is assumed to mimic the following pattern

$$r = r_{in} + A\cos\left(N\left(\zeta\right)\right) \tag{37}$$

in which r_{in} is the base circle radius, r_{out} is the radius of outer cylinder, A and N are amplitude and number of undulations, respectively. ζ is the rotation angle. In this study A and N equal to 0.025 and 48, respectively. Figure 16 depicts the distribution of electric density for different Reynolds number and supplied voltage.

4.2. Governing Equation

In order to simulate nanofluid hydrothermal treatment in existence of electric field, we should combined equations of electric fields with those of hydrothermal. The formulas of electric field are:

$$\nabla.\left(\vec{E}\,\varepsilon\right) = q \tag{38}$$

$$\left(-\nabla\varphi\right) = \vec{E} \tag{39}$$

$$\nabla.\vec{\mathrm{J}} + \frac{\partial q}{\partial t} = 0 \tag{40}$$

$$\vec{\mathrm{J}} = \sigma\,\vec{E} - D\nabla q + q\vec{V} \tag{41}$$

Figure 15. (a) Geometry of the problem and boundary conditions; (b) the mesh of enclosure considered in this work; (c) A sample triangular element and its corresponding control volume.

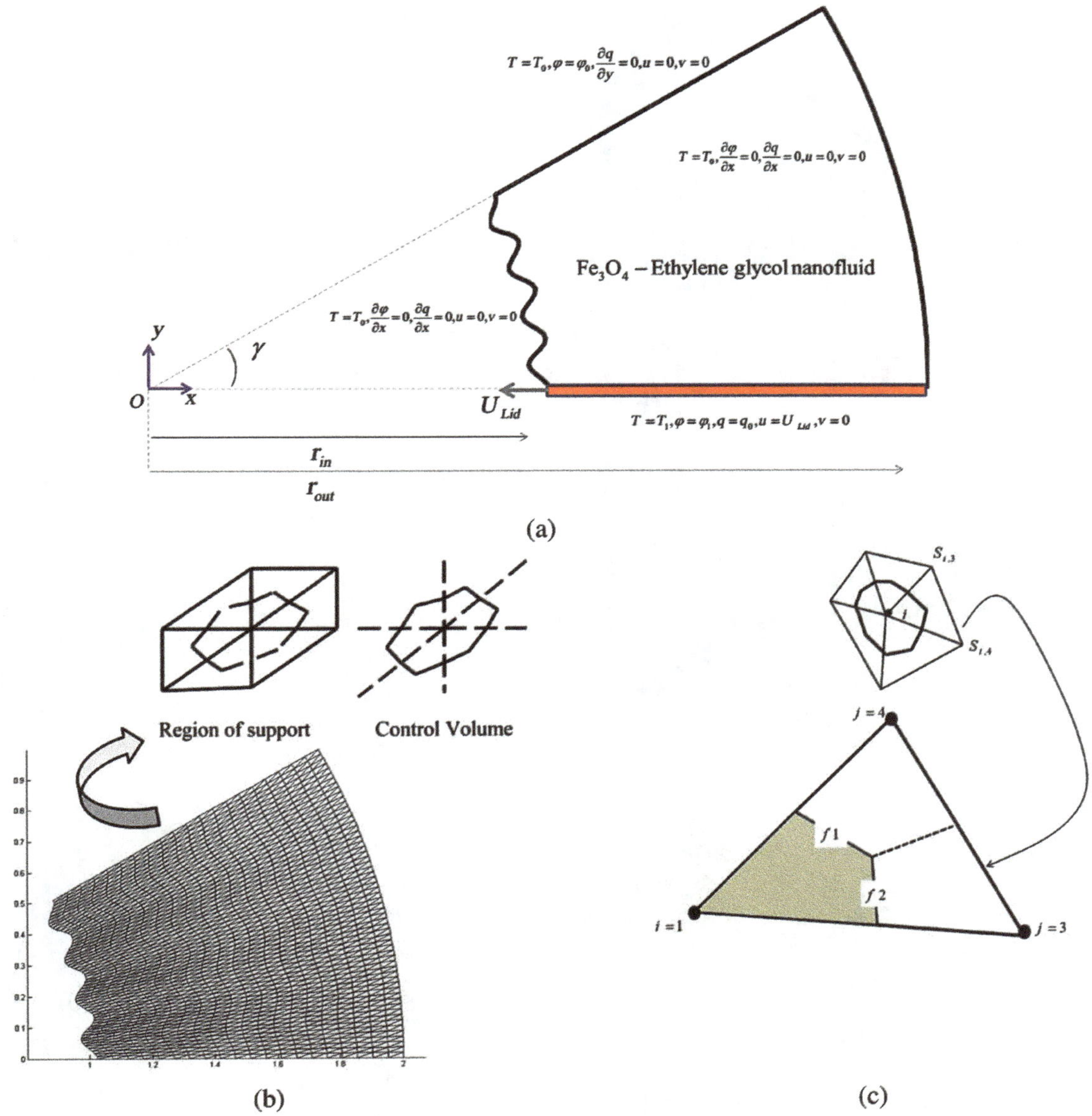

$$\frac{\partial \bar{q}}{\partial \bar{t}} + \bar{v}\frac{\partial \bar{q}}{\partial \bar{y}} + \bar{u}\frac{\partial \bar{q}}{\partial \bar{x}} + \frac{1}{\mathrm{Re}\,\mathrm{Pr}_E}\left[\bar{q}\left(\frac{\partial \bar{E}_y}{\partial \bar{y}} + \frac{\partial \bar{E}_x}{\partial \bar{x}}\right) + \bar{E}_y\frac{\partial \bar{q}}{\partial \bar{y}} + \bar{E}_x\frac{\partial \bar{q}}{\partial \bar{x}}\right] = \frac{\rho_{nf}/\rho_f}{\mu_{nf}/\mu_f}\frac{1}{\mathrm{Re}\,D_e}\left(\frac{\partial^2 \bar{q}}{\partial \bar{y}^2} + \frac{\partial^2 \bar{q}}{\partial \bar{x}^2}\right) \quad (42)$$

Diffusion term can be taken negligible. Also $D\nabla q$ in Equation (41) can be taken negligible and $\sigma = bq$ (Sheikholeslami & Ghasemi, 2018). So Equation (5) can be considered as:

Figure 16. Electric density distribution injected by the bottom electrode

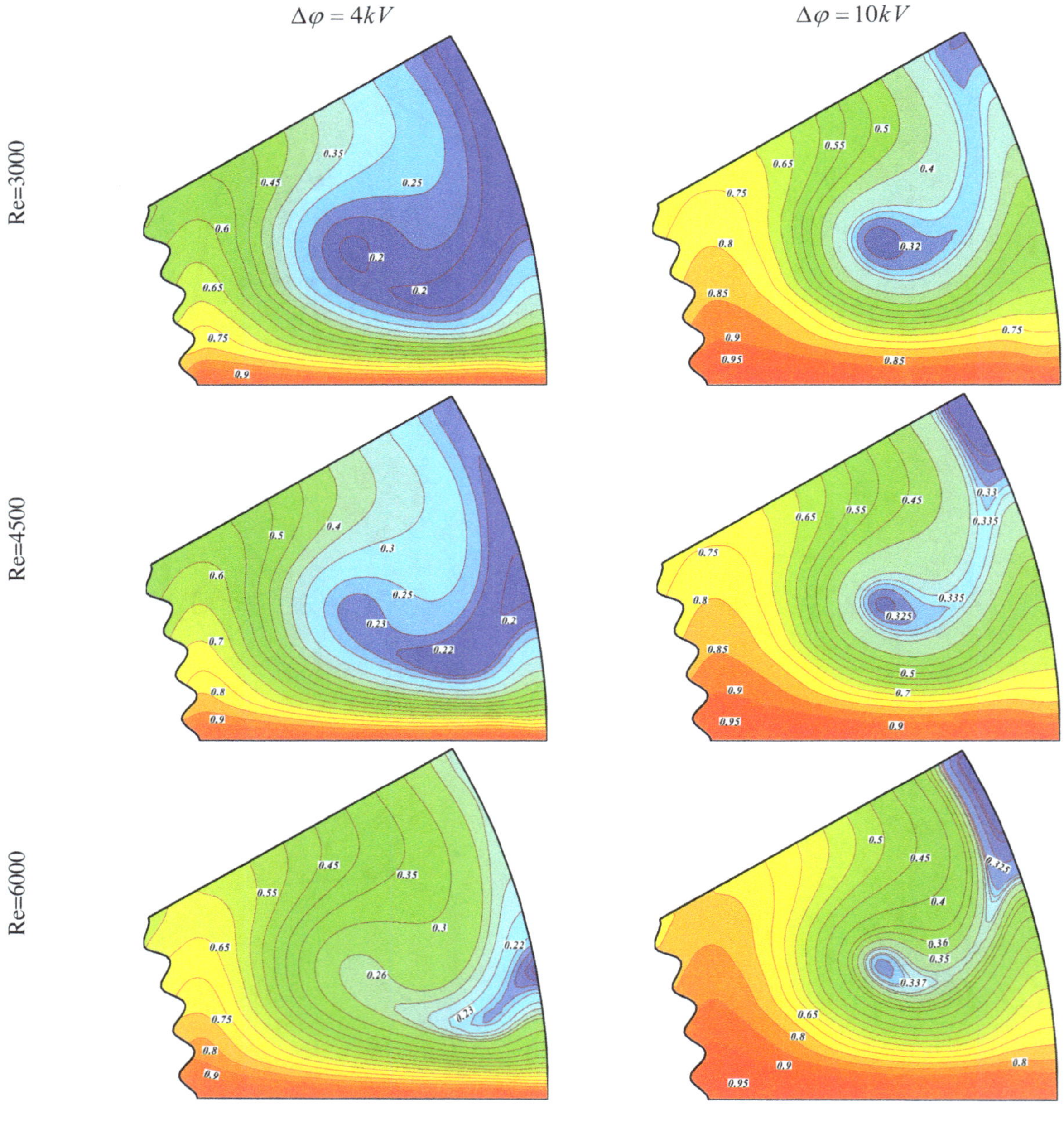

$$\vec{J} = q\vec{V} + qb\vec{E} \tag{43}$$

In presence of electric field Coulomb forces should be added to momentum equation and Joule heating effect should be added in energy equation. So we have:

$$\begin{cases} \nabla.\vec{V} = 0 \\ \rho_{nf}\left(\dfrac{\partial \vec{V}}{\partial t} + \left(\vec{V}.\nabla\right)\vec{V}\right) = -\nabla p + \mu_{nf}\nabla^2\vec{V} + q\vec{E} \\ \left(\rho C_p\right)_{nf}\left(\dfrac{\partial T}{\partial t} + \left(\vec{V}.\nabla\right)T\right) = k_{nf}\nabla^2 T + \vec{J}.\vec{E} \\ \nabla.\vec{J} + \dfrac{\partial q}{\partial t} = 0 \\ \nabla.\varepsilon\vec{E} = q \\ \vec{E} = -\nabla\varphi \end{cases} \tag{44}$$

$\rho_{nf}, \left(\rho C_p\right)_{nf}, \alpha_{nf}, k_{nf}$ and μ_{nf} are defined as:

$$\rho_{nf} = \rho_s\phi + \rho_f(1-\phi) \tag{45}$$

$$\left(\rho C_p\right)_{nf} = \left(\rho C_p\right)_s\phi + \left(\rho C_p\right)_f(1-\phi) \tag{46}$$

$$\alpha_{nf} = k_{nf} / \left(\rho C_p\right)_{nf} \tag{47}$$

$$k_{nf} = k_f\frac{-2\phi(k_f - k_s) + 2k_f + k_s}{+\phi(k_f - k_s)_s + 2k_f + k} \tag{48}$$

$$\mu_{nf} = \frac{\mu_f}{(1-\phi)^{2.5}} \tag{49}$$

Non-dimensional parameters are introduced as follow:

$$\begin{gathered} \bar{t} = \frac{tU_{Lid}}{L}, \bar{p} = \frac{P}{\rho U_{Lid}^{\;2}}, \bar{y} = \frac{y}{L}, \bar{x} = \frac{x}{L}, \bar{v} = \frac{v}{U_{Lid}}, \bar{u} = \frac{u}{U_{Lid}}, \\ \theta = \frac{T - T_0}{\nabla T}, \nabla T = T_1 - T_0, \bar{\varphi} = \frac{\varphi - \varphi_0}{\nabla\varphi}, \nabla\varphi = \varphi_1 - \varphi_0, \bar{q} = \frac{q}{q_0}, \bar{E} = \frac{E}{E_0} \end{gathered} \tag{50}$$

where ∇T and $\nabla\varphi$ are $\left(T_1 - T_0\right)$ and $\left(\varphi_1 - \varphi_0\right)$, respectively. In order to reach clear formulation, over bar will be deleted in next equations. So, the governing equations can be considered as follows:

$$\begin{cases} \nabla.\left(\vec{V}\right) = 0 \\ \left(\dfrac{\partial \vec{V}}{\partial t} + \left(\vec{V}.\nabla\right)\vec{V}\right) = -\nabla p + \dfrac{\rho_{nf} / \rho_f}{\mu_{nf} / \mu_f}\dfrac{1}{\text{Re}}\nabla^2\vec{V} + \dfrac{S_E}{\rho_{nf} / \rho_f} q\vec{E} \\ \left(\dfrac{\partial \theta}{\partial t} + \left(\vec{V}.\nabla\right)\theta\right) = \dfrac{k_{nf} / k_f}{\left(\rho C_p\right)_{nf} / \left(\rho C_p\right)_f}\dfrac{1}{\text{RePr}}\nabla^2\theta + \dfrac{S_E Ec}{\left(\rho C_p\right)_{nf} / \left(\rho C_p\right)_f}\left(\vec{J}.\vec{E}\right) \\ \nabla.\vec{\text{J}} + \dfrac{\partial q}{\partial t} = 0 \\ \nabla.\varepsilon\vec{E} = q \\ \vec{E} = -\nabla\varphi \end{cases} \tag{51}$$

The formulas of vorticity and stream function are:

$$\begin{aligned} &\omega = \frac{\partial v}{\partial x} - \frac{\partial u}{\partial y}, \\ &v = -\frac{\partial \psi}{\partial x}, u = \frac{\partial \psi}{\partial y}, \\ &\Omega = \frac{\omega L}{U_{Lid}}, \Psi = \frac{\psi}{L U_{Lid}} \end{aligned} \tag{52}$$

Continuity equation has been satisfied by the stream function. By eliminating pressure between x-momentum and y- momentum, the vorticity equation can be obtained.

Nu_{loc} and Nu_{ave} along the lid wall can be obtained as:

$$Nu_{loc} = \left(\frac{k_{nf}}{k_f}\right)\frac{\partial \Theta}{\partial Y} \tag{53}$$

$$Nu_{ave} = \frac{1}{L}\int_{r_{in}}^{r_{out}} Nu_{loc}\, dX \tag{54}$$

4.3. Effects of Active Parameters

Effect of non-uniform electric filed on nanofluid hydrothermal behavior in an enclosure with sinusoidal wall is presented. The working fluid is Ethylene glycol and Fe_3O_4. Calculations are prepared for various values of supplied voltage ($\Delta\varphi = 0, 4, 6$ and $10kV$), volume fraction of nanoparticles ($\phi = 0\%$ and 5%) and Reynolds number numbers ($\mathrm{Re} = 3000, 4500$ and 6000). In all calculations, the Prandtl number (Pr) and Eckert number (Ec) are set to 149.54 and 1e-6.

Influence of Reynolds number and supplied voltage on streamlines and isotherm are shown in Figures 17, 18 and 19. At Re=3000, one main eddy and two very small eddied at upper corners of the enclosure exist in streamline. As electric filed applied the main eddy turn in to two smaller ones. Also isotherm becomes denser near the hot wall due to existence these eddies. As Reynolds number increases up to 6000, the upper right eddy becomes stronger and isotherms become denser near the bottom wall. Since convective heat transfer is helped more successfully at a greater flow rate, Nusselt number is improved at great Reynolds numbers. So isotherm becomes more distorted at greater values of Reynolds number. As electric filed increases, the main cell convert to two eddies which are rotates opposite direction.

Figures 20 and 21 depict the influences of $\Delta\varphi$ and Re on Nu_{loc} and Nu_{ave} along the lid wall. As Reynolds number increases, Nusselt number increases due to decrease in thermal boundary layer thickness. Increasing supplied voltage make the isotherms more distorted. Local Nusselt number profiles

Figure 17. Effect of supplied voltage on streamlines and isotherm when $\mathrm{Re} = 3000, \phi = 0.05$.

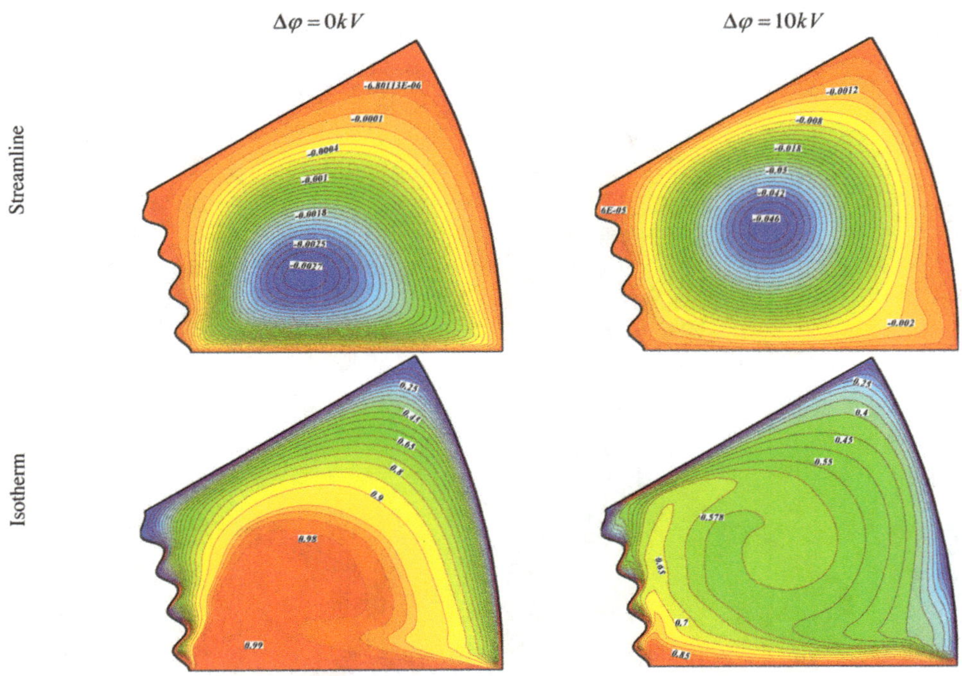

Figure 18. Effect of supplied voltage on streamlines and isotherm when $\mathrm{Re} = 4500, \phi = 0.05$.

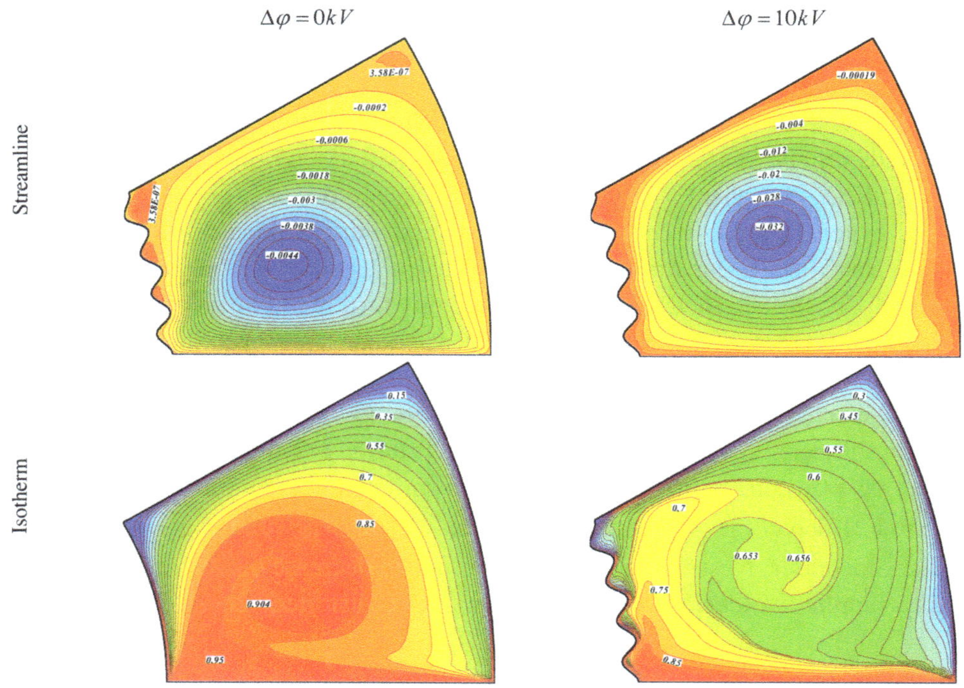

Figure 19. Effect of supplied voltage on streamlines and isotherm when $\mathrm{Re} = 6000, \phi = 0.05$.

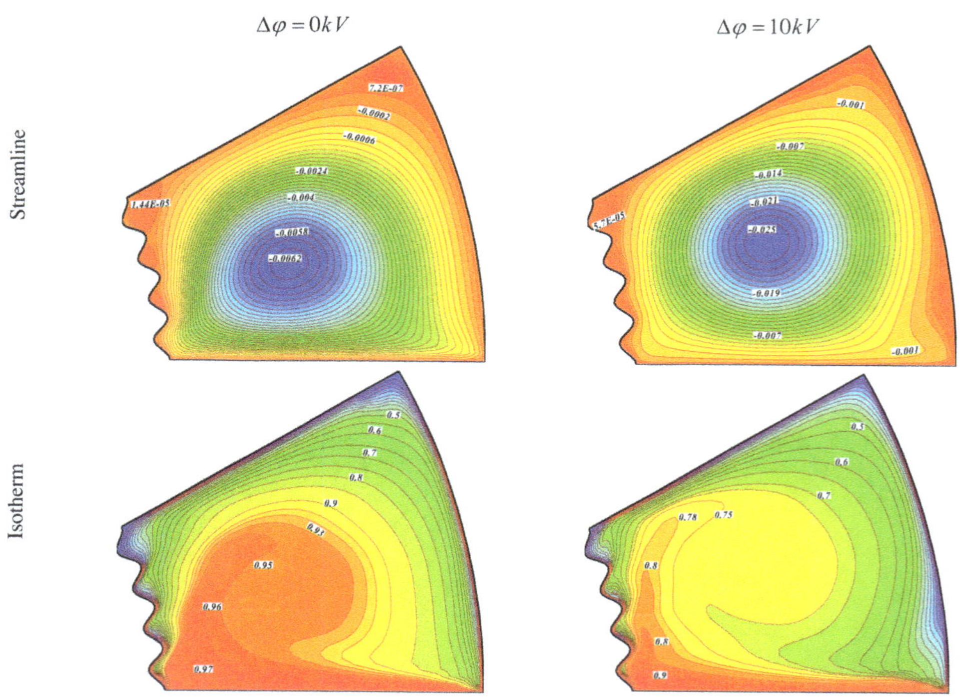

Figure 20. Effects of Reynolds number and supplied voltage on local Nusselt number when $\phi = 0.05$.

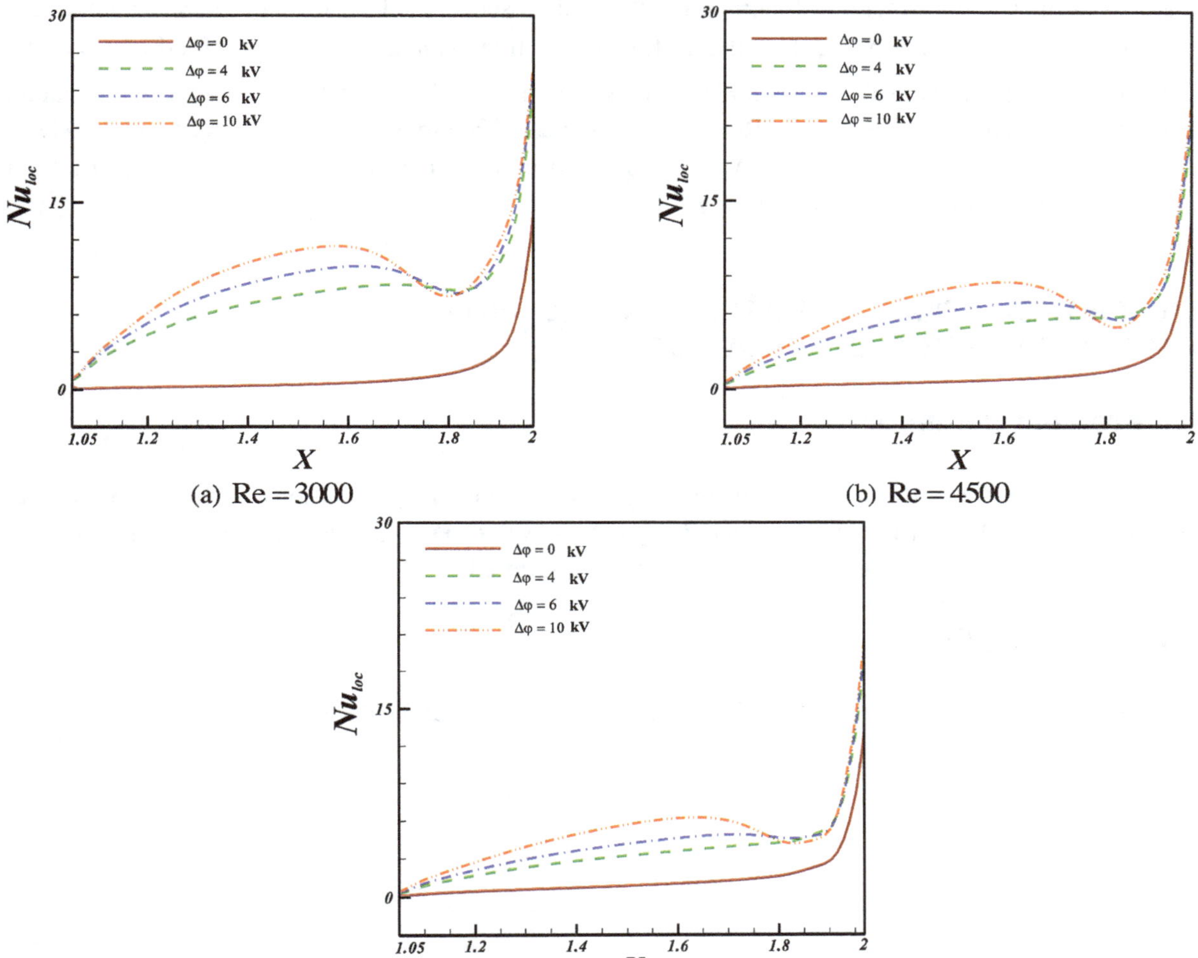

(a) Re = 3000

(b) Re = 4500

(c) Re = 6000

Figure 21. Effects of Reynolds number and supplied voltage on average Nusselt number.

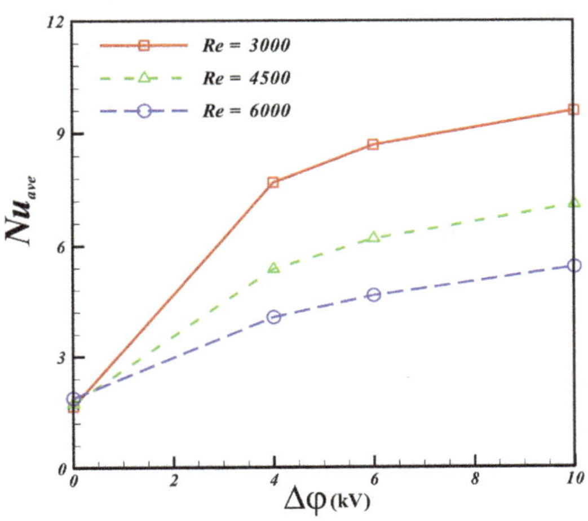

have extremums at higher values of supplied voltage because of existence of thermal plumes. Nusselt umber is an increasing function of supplied voltage. In absence of electric filed, Nusselt number for Re=6000 is 1.135921times higher than that of Re=3000 while in presence of electric filed $\left(\Delta\varphi = 10\right)$, Nusselt number for Re=6000 is 1.768835 times lower than that of Re=3000. Also it can be concluded that Nusselt number at $\Delta\varphi = 10$ for Re=3000, 4500 and 6000 are 5.816335, 4.112412 and 2.894771 times higher than those of obtained at $\Delta\varphi = 0$. This observation confirms that impact of electric filed is more marked for lower Reynolds number.

4. EFFECT OF COULOMB FORCES ON FE_3O_4-H2O NANOFLUID THERMAL IMPROVEMENT

4.1. Problem Definition

Figure 22 demonstrates the schematic of this paper and its boundary conditions. Influence of electric field on Fe_3O_4-H_2O nanofluid is considered. Only the top wall can move. The contour of electric density for various values of active parameters is depicted in Figure 23.

4.2. Governing Equation

According to Gauss's law and Maxwell's relation, electric field can be defined as:

$$q = \nabla.\varepsilon \vec{E} \tag{55}$$

$$\vec{E} = -\nabla\varphi \tag{56}$$

$$\nabla.\vec{J} + \frac{\partial q}{\partial t} = 0 \tag{57}$$

$$\vec{\mathrm{J}} = q\vec{V} - D\nabla q + \sigma\vec{E} \tag{58}$$

The governing equations are as follows:

Figure 22. (a) Geometry of the problem and boundary conditions; (b) the mesh of enclosure considered in this work; (c) A sample triangular element and its corresponding control volume.

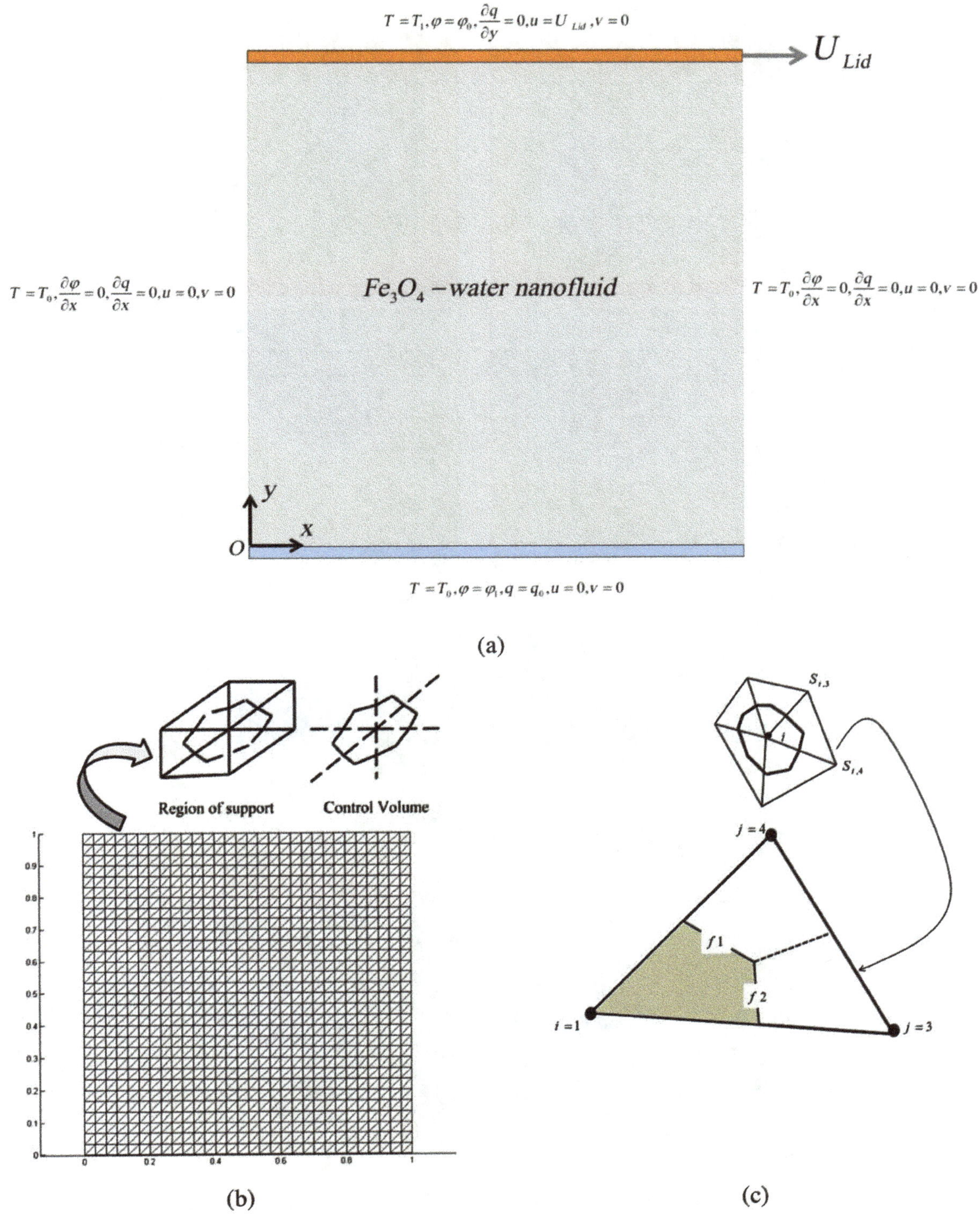

Figure 23. Electric density distribution injected by the bottom electrode

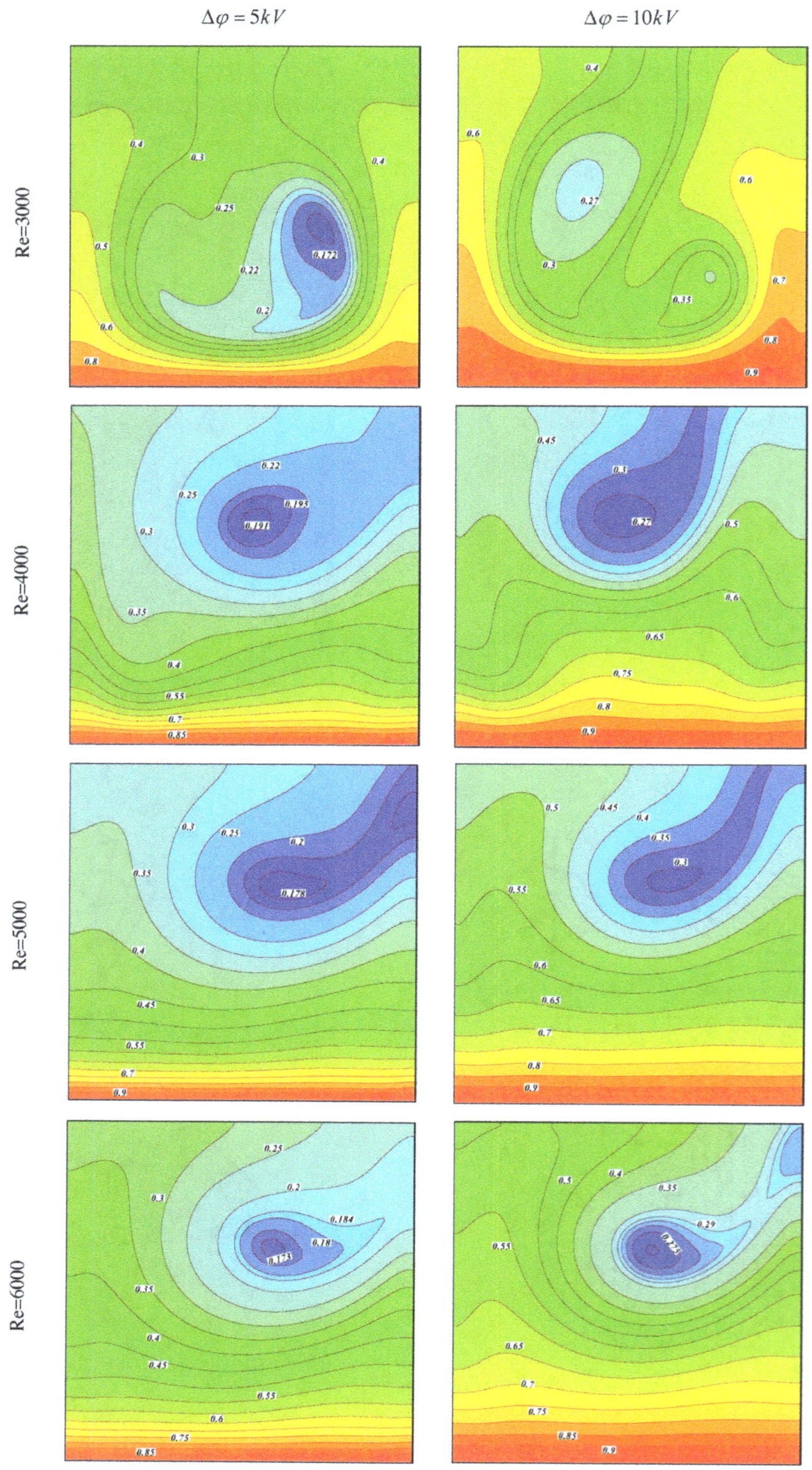

$$\begin{cases} \nabla.\vec{V} = 0 \\ \rho_{nf}\left(\left(\vec{V}.\nabla\right)\vec{V} + \frac{\partial \vec{V}}{\partial t}\right) = q\,\vec{E} + \mu_{nf}\nabla^2\vec{V} - \nabla p \\ \left(\left(\vec{V}.\nabla\right)T + \frac{\partial T}{\partial t}\right) = \frac{k_{nf}}{\left(\rho C_p\right)_{nf}}\nabla^2 T + \frac{\vec{J}.\vec{E}}{\left(\rho C_p\right)_{nf}} \\ \nabla.\vec{J} + \frac{\partial q}{\partial t} = 0 \\ \nabla.\varepsilon\,\vec{E} = q \\ \vec{E} = -\nabla\varphi \end{cases} \tag{59}$$

ρ_{nf}, $\left(\rho C_p\right)_{nf}$, μ_{nf} and k_{nf} can be obtained as:

$$\begin{aligned} &\rho_{nf} = \rho_f(1-\phi) + \rho_s\phi, \left(\rho C_p\right)_{nf} = \left(\rho C_p\right)_f(1-\phi) + \left(\rho C_p\right)_s\phi, \\ &\mu_{nf} = \frac{\mu_f}{(1-\phi)^{2.5}}, \frac{k_{nf}}{k_f} = \frac{k_s + 2k_f - 2\phi(k_f - k_s)}{k_s + 2k_f + \phi(k_f - k_s)} \end{aligned} \tag{60}$$

Properties of Fe_3O_4 and H_2O are illustrated in Table 3.

So, the final dimensionless equations in presence of electric field are:

$$\begin{cases} \nabla.\vec{V} = 0 \\ \left(\left(\vec{V}.\nabla\right)\vec{V} + \frac{\partial \vec{V}}{\partial t}\right) = \frac{S_E}{\rho_{nf}/\rho_f}q\,\vec{E} + \frac{\rho_{nf}/\rho_f}{\mu_{nf}/\mu_f}\frac{1}{\mathrm{Re}}\nabla^2\vec{V} - \nabla p \\ \left(\left(\vec{V}.\nabla\right)\theta + \frac{\partial \theta}{\partial t}\right) = \frac{k_{nf}/k_f}{\left(\rho C_p\right)_{nf}/\left(\rho C_p\right)_f}\frac{1}{\mathrm{RePr}}\nabla^2\theta + \frac{S_E Ec}{\left(\rho C_p\right)_{nf}/\left(\rho C_p\right)_f}\left(\vec{J}.\vec{E}\right) \\ \vec{E} = -\nabla\varphi \\ q = \nabla.\varepsilon\,\vec{E} \\ \nabla.\vec{J} = -\frac{\partial q}{\partial t} \end{cases} \tag{61}$$

where

Table 3. Thermo physical properties of water and nanoparticles

	$\rho(kg / m^3)$	$C_p(j / kgk)$	$k(W / m.k)$
H_2O	997.1	4179	0.613
Fe_3O_4	5200	670	6

$$\begin{aligned} &\bar{u} = \frac{u}{U_{Lid}}, \bar{v} = \frac{v}{U_{Lid}}, \bar{y} = \frac{y}{L}, \bar{x} = \frac{x}{L}, \\ &\bar{t} = \frac{tU_{Lid}}{L}, \bar{p} = \frac{P}{\rho U_{Lid}^{2}} \bar{q} = \frac{q}{q_0}, \bar{E} = \frac{E}{E_0}, \\ &\nabla \mathrm{T} = T_1 - T_0, \theta = \frac{T - T_0}{\nabla \mathrm{T}}, \nabla \varphi = \varphi_1 - \varphi_0, \bar{\varphi} = \frac{\varphi - \varphi_0}{\nabla \varphi} \end{aligned} \tag{62}$$

By eliminating pressure gradient, vorticity and stream function can be introduced as:

$$\omega = \frac{\partial v}{\partial x} - \frac{\partial u}{\partial y}, -\frac{\partial \psi}{\partial x} = v, \frac{\partial \psi}{\partial y} = u, \Psi = \frac{\psi}{\alpha_f}, \Omega = \frac{\omega L^2}{\alpha_f} \tag{63}$$

Nu_{loc} and Nu_{ave} along the left wall are defined as:

$$Nu_{loc} = \frac{\partial \Theta}{\partial X} \left(\frac{k_{nf}}{k_f} \right) \tag{64}$$

$$Nu_{ave} = \frac{1}{L} \int_0^L Nu_{loc} \, dY \tag{65}$$

4.3. Effects of Active Parameters

Coulomb forces impact on $Fe_3O_4 - H_2O$ hydrothermal behavior in a cavity with moving wall is studied. Various values of active parameters are examined such as Reynolds number ($\mathrm{Re} = 3000, 4000, 5000$ and 6000), volume fraction of solid particle ($\phi = 0\%$ and 4%) and supplied voltage ($\Delta\varphi = 0, 5, 8$ and $10 kV$). Impacts of $\Delta\varphi$ and Re on isotherm and streamlines are demonstrated in Figures 24, 25, 26 and 27. At low Re, one main vortex exists in streamline. As Coulomb forces augment main vortex turn in to three smaller vortexes and temperature gradient near the hot walls enhances.

As lid velocity increases, another eddy appears at right bottom of the enclosure. As voltage enhances, the previous cell alter to two vortexes which are revolves opposite direction. Further increase in Reynolds number leads to generate another small eddy at left bottom of the enclosure. Temperature gradient augments at high Reynolds numbers due to stronger convective heat transfer.

Figure 24. Effect of supplied voltage on streamlines and isotherm when $\mathrm{Re} = 3000, \phi = 0.04$.

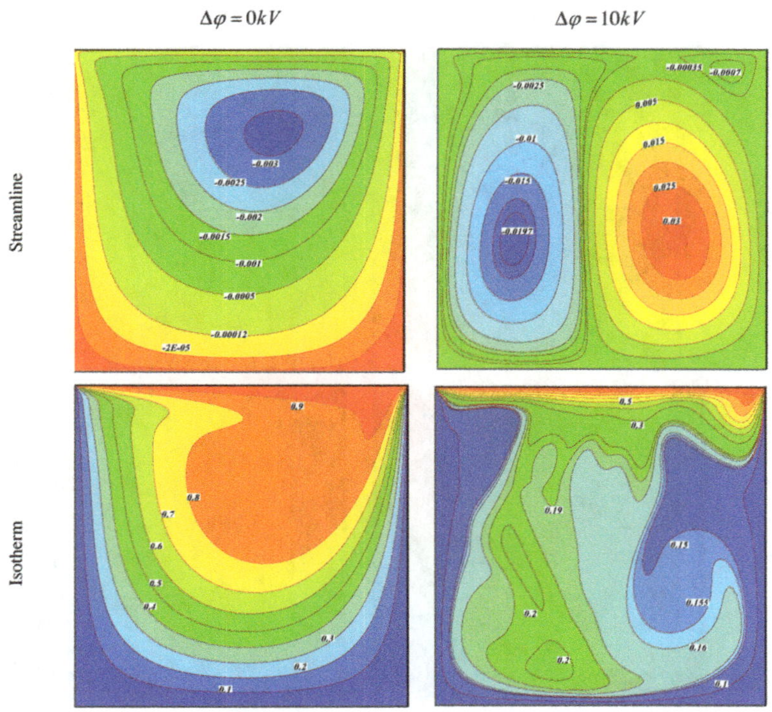

Figure 25. Effect of supplied voltage on streamlines and isotherm when $\mathrm{Re} = 4000, \phi = 0.04$.

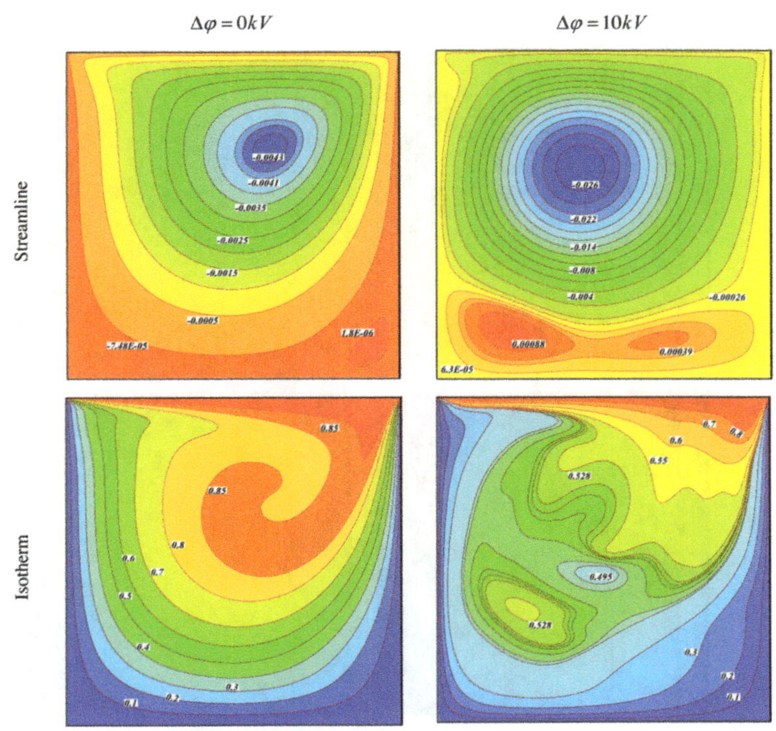

Figure 26. Effect of supplied voltage on streamlines and isotherm when $\mathrm{Re} = 5000, \phi = 0.04$.

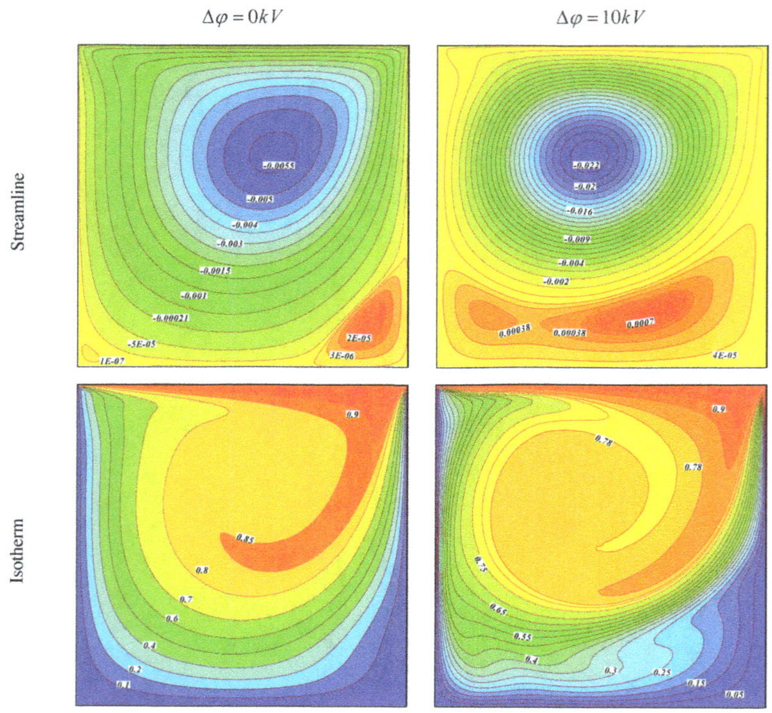

Figure 27. Effect of supplied voltage on streamlines and isotherm when $\mathrm{Re} = 6000, \phi = 0.04$.

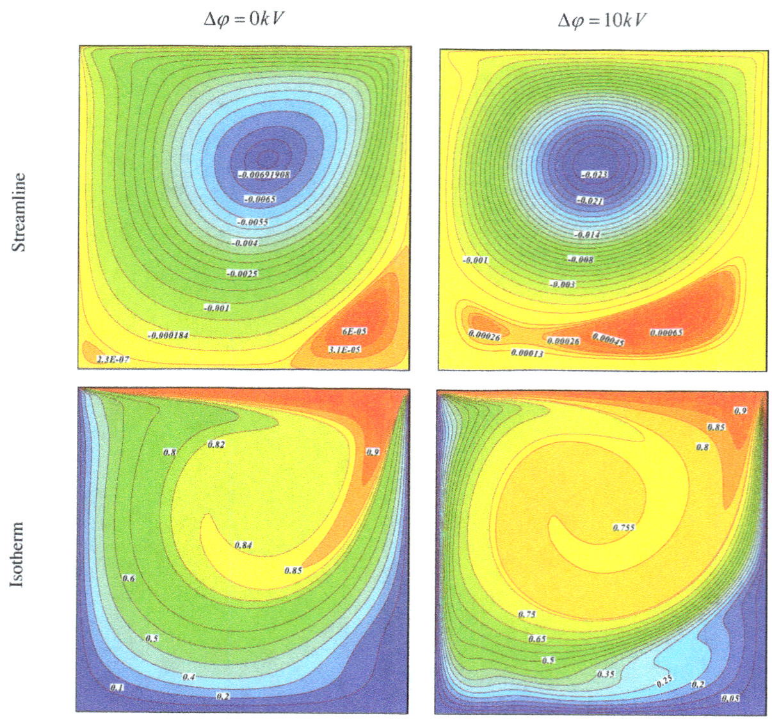

Impacts of $\Delta\varphi$ and Re on Nu_{loc} and Nu_{ave} are demonstrated in Figures 28 and 29. As lid wall velocity augments, Nusselt number enhances due to rise in temperature gradient. Augmenting Coulomb forces causes the isotherms near the left wall to be denser. Presence of thermal plumes for high values of supplied voltage leads to existence of extremums for Nu_{loc} profiles. The Nusselt numbers in the existence of electric field with supplied voltage $\Delta\varphi = 10kV$ of Re=3000, 4000, 5000 and 6000, are respectively 1.09, 2.15, 8.96and 1.76 times higher than those in absence of electric field.

Figure 28. Effects of Reynolds number and supplied voltage on local Nusselt number when $\phi = 0.04$.

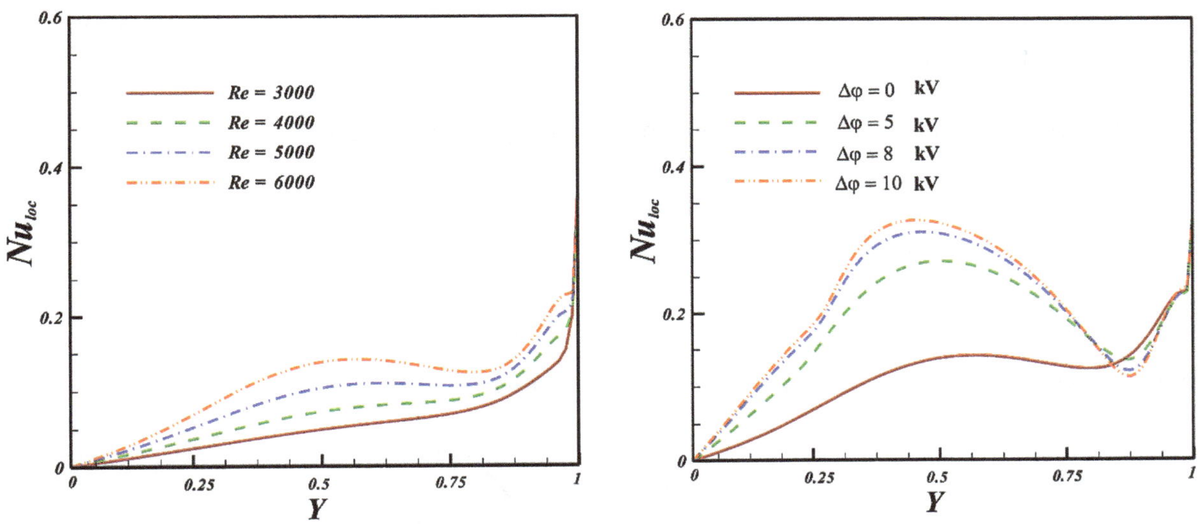

(a) without EHD (b) Re = 6000

Figure 29. Effects of Reynolds number and supplied voltage on average Nusselt number.

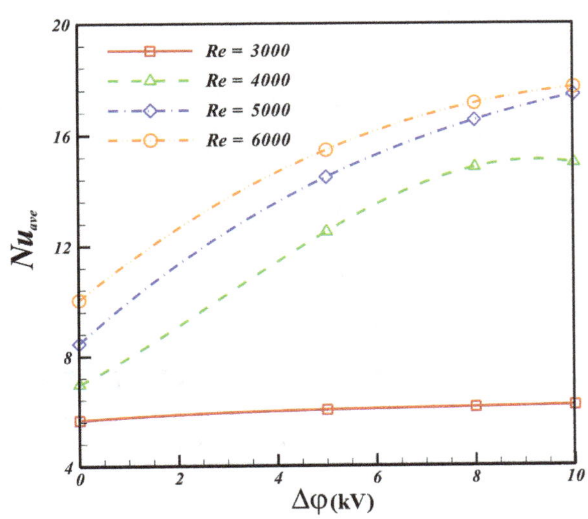

5. ACTIVE METHOD FOR NANOFLUID HEAT TRANSFER ENHANCEMENT BY MEANS OF EHD

5.1. Problem Definition

Figure 30 illustrates the schematic of this problem and its boundary conditions. Influence of electric field on Fe_3O_4- Ethylene glycol nanofluid is considered. Only the bottom wall can move. Figure 31 illustrates the contour of q f or various values of $\nabla\varphi$ and Re. As $\nabla\varphi$ increases the distortion of iso-electric density lines become more and one cell appear in right side. Influence of $\nabla\varphi$ on q is more sensible than Re.

5.2. Governing Equation

According to Gauss's law and Maxwell's relation, electric field can be defined as:

$$q = \nabla.\varepsilon \vec{E} \tag{66}$$

Figure 30. Geometry of the problem and boundary conditions

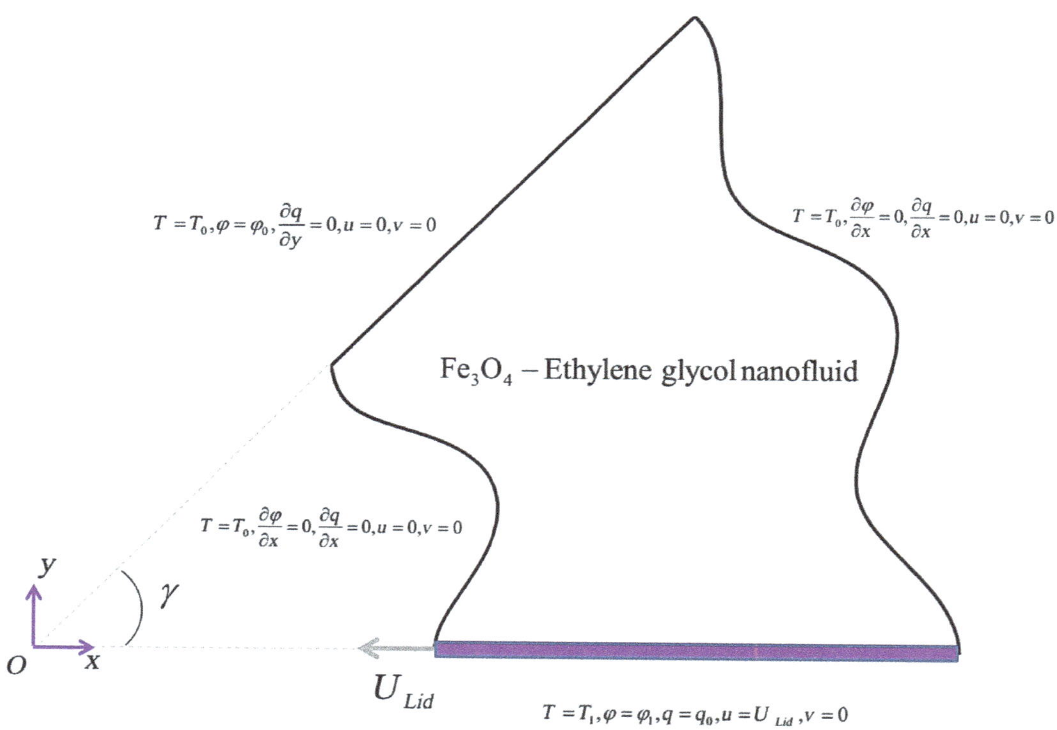

Figure 31. Electric density distribution injected by the bottom electrode

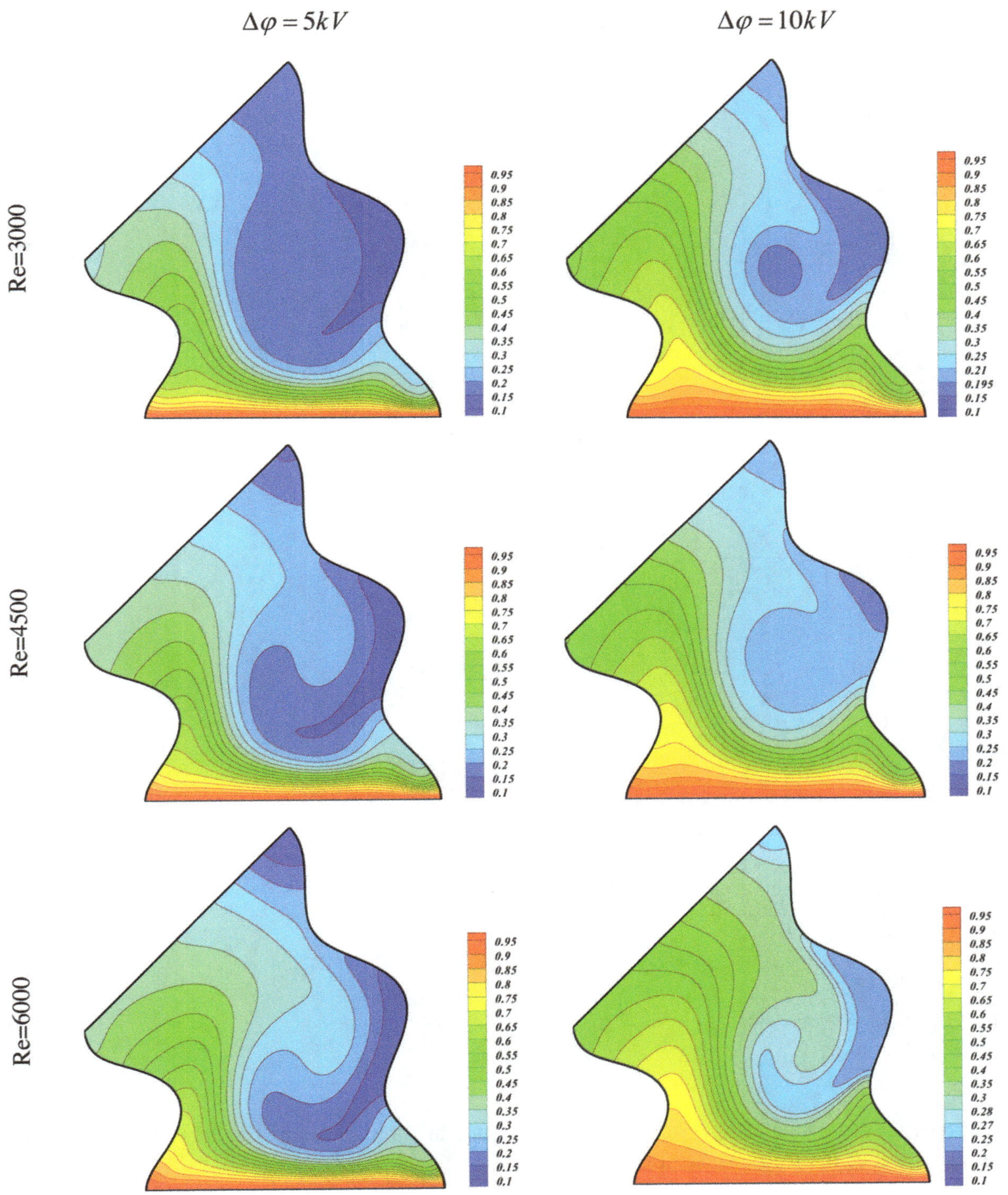

$$\vec{E} = -\nabla\varphi \tag{67}$$

$$\nabla.\vec{\mathrm{J}} + \frac{\partial q}{\partial t} = 0 \tag{68}$$

$$\vec{\mathrm{J}} = q\vec{V} - D\nabla q + \sigma\vec{E} \tag{69}$$

The governing equations are:

$$\begin{cases} \nabla.\vec{V} = 0 \\ \left(\left(\vec{V}.\nabla\right)\vec{V} + \dfrac{\partial\vec{V}}{\partial t}\right) = \dfrac{q\vec{E}}{\rho_{nf}} + \dfrac{\mu_{nf}}{\rho_{nf}}\nabla^2\vec{V} - \dfrac{\nabla p}{\rho_{nf}} \\ \left(\left(\vec{V}.\nabla\right)T + \dfrac{\partial T}{\partial t}\right) = \dfrac{k_{nf}}{\left(\rho C_p\right)_{nf}}\nabla^2 T + \dfrac{\vec{\mathrm{J}}.\vec{E}}{\left(\rho C_p\right)_{nf}} \\ \nabla\varphi = -\vec{E} \\ \dfrac{\partial q}{\partial t} = -\nabla.\vec{\mathrm{J}} \\ q = \nabla.\varepsilon\vec{E} \end{cases} \tag{70}$$

k_{nf}, $\left(\rho C_p\right)_{nf}$, μ_{nf} and ρ_{nf} can be obtained as:

$$\begin{aligned} &\frac{k_{nf}}{k_f} = \frac{-2\phi(k_f - k_s) + k_s + 2k_f}{\phi(k_f - k_s) + k_s + 2k_f}, \\ &\left(\rho C_p\right)_{nf} = \left(\rho C_p\right)_f(1-\phi) + \left(\rho C_p\right)_s\phi, \\ &\mu = A_1 + A_2\left(\Delta\varphi\right) + A_3\left(\Delta\varphi\right)^2 + A_4\left(\Delta\varphi\right)^3, \\ &\rho_{nf} = \rho_f(1-\phi) + \rho_s\phi \end{aligned} \tag{71}$$

Properties of Fe_3O_4 and ethylene glycol are illustrated in Table 1. So, the final dimensionless equations in presence of electric field are:

$$\begin{cases} \nabla.\vec{V} = 0 \\ \left(\left(\vec{V}.\nabla\right)\vec{V} + \dfrac{\partial \vec{V}}{\partial t}\right) = \dfrac{1}{\mathrm{Re}}\dfrac{\rho_{nf}/\rho_f}{\mu_{nf}/\mu_f}\nabla^2\vec{V} - \nabla p + \dfrac{S_E}{\rho_{nf}/\rho_f} q\vec{E} \\ \left(\left(\vec{V}.\nabla\right)\theta + \dfrac{\partial \theta}{\partial t}\right) = \dfrac{1}{\mathrm{Pr}\,\mathrm{Re}}\dfrac{k_{nf}/k_f}{\left(\rho C_p\right)_{nf}/\left(\rho C_p\right)_f}\nabla^2\theta + \dfrac{1}{\left(\rho C_p\right)_{nf}/\left(\rho C_p\right)_f} S_E Ec\left(\vec{\mathrm{J}}.\vec{E}\right) \\ \vec{E} = -\nabla\varphi \\ q = \nabla.\varepsilon\vec{E} \\ \nabla.\vec{\mathrm{J}} = -\dfrac{\partial q}{\partial t} \end{cases} \tag{72}$$

where

$$\begin{aligned} &\left(\bar{u},\bar{v}\right) = \frac{\left(u,v\right)}{U_{Lid}}, \bar{\varphi} = \frac{\varphi - \varphi_0}{\nabla\varphi}, \left(\bar{y},\bar{x}\right) = \frac{\left(y,x\right)}{L}, \theta = \frac{T - T_0}{\nabla \mathrm{T}}, \\ &\bar{t} = \frac{tU_{Lid}}{L}, \bar{p} = \frac{P}{\rho U_{Lid}^{2}}\bar{q} = \frac{q}{q_0}, \bar{E} = \frac{E}{E_0}, \\ &\nabla \mathrm{T} = T_1 - T_0, \nabla\varphi = \varphi_1 - \varphi_0. \end{aligned} \tag{73}$$

By eliminating pressure gradient, vorticity and stream function can be introduced as:

$$v = -\frac{\partial\psi}{\partial x}, \frac{\partial\psi}{\partial y} = u, \Psi = \frac{\psi}{\alpha_f}, \Omega = \frac{\omega L^2}{\alpha_f}, \omega = \frac{\partial v}{\partial x} - \frac{\partial u}{\partial y} \tag{74}$$

Nu_{loc} and Nu_{ave} along the hot wall are calculated as:

$$Nu_{loc} = \left(\frac{k_{nf}}{k_f}\right)\frac{\partial\Theta}{\partial X} \tag{75}$$

$$Nu_{ave} = \frac{1}{L}\int_0^L Nu_{loc}\, dY \tag{76}$$

5.3. Effects of Active Parameters

Fe_3O_4 – Ethylene glycol nanofluid forced convection in an enclosure is examined in existence of electric field. The bottom wall is moving lid. Roles of supplied voltage ($\Delta\varphi = 0$ to $10kV$), Reynolds number ($\mathrm{Re} = 3000$ to 6000) and volume fraction of Fe_3O_4 ($\phi = 0\%$ to 5%) are presented graphically.

Effects of $\Delta\varphi$ and Re on streamlines and isotherm are illustrated in Figures 32, 33 and 34. Two eddies appear in streamlines which are rotates in reverse direction. The counter clock wise eddy is stronger and located at left side upper the square obstacle. Existing such eddies leads to generate two thermal plumes at left and upper side of square obstacle. As electric field is applied, the clock wise eddy becomes stronger and the other one convert to two smaller counter clock wise eddies which are located at left side of enclosure. Isotherms become more disturb by augmenting $\Delta\varphi$. By increasing Re, the strength of rotating eddies enhances and the distortion of isotherms becomes more than before. Temperature gradient enhances at high Reynolds numbers due to stronger convective heat transfer. At high Reynolds number, Coulomb force is weaker than viscous force, so the impact of adding electric field becomes weaker with augment of Re.

Figure 32. Effect of supplied voltage on streamlines and isotherm when $\mathrm{Re} = 3000, \phi = 0.05$.

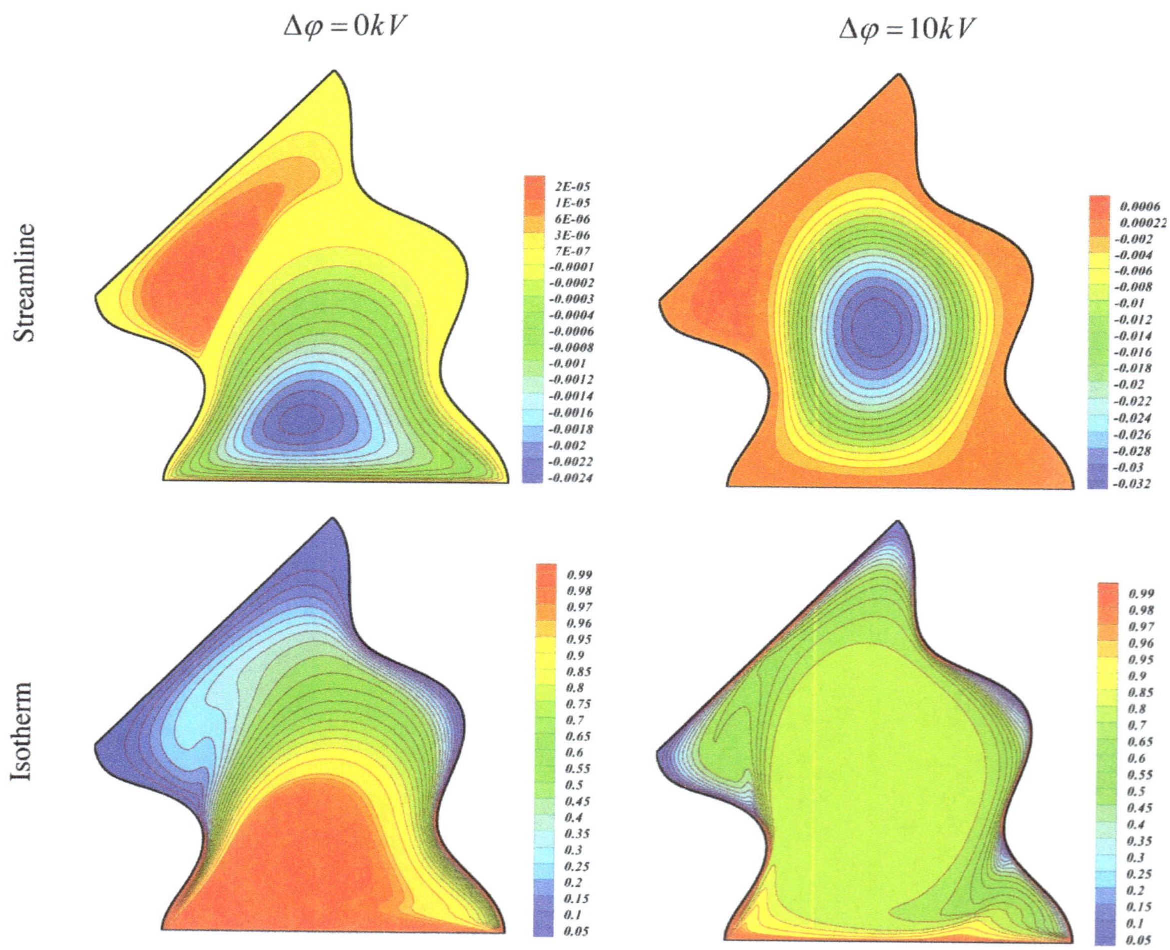

Figure 33. Effect of supplied voltage on streamlines and isotherm when $\mathrm{Re} = 4500, \phi = 0.05$.

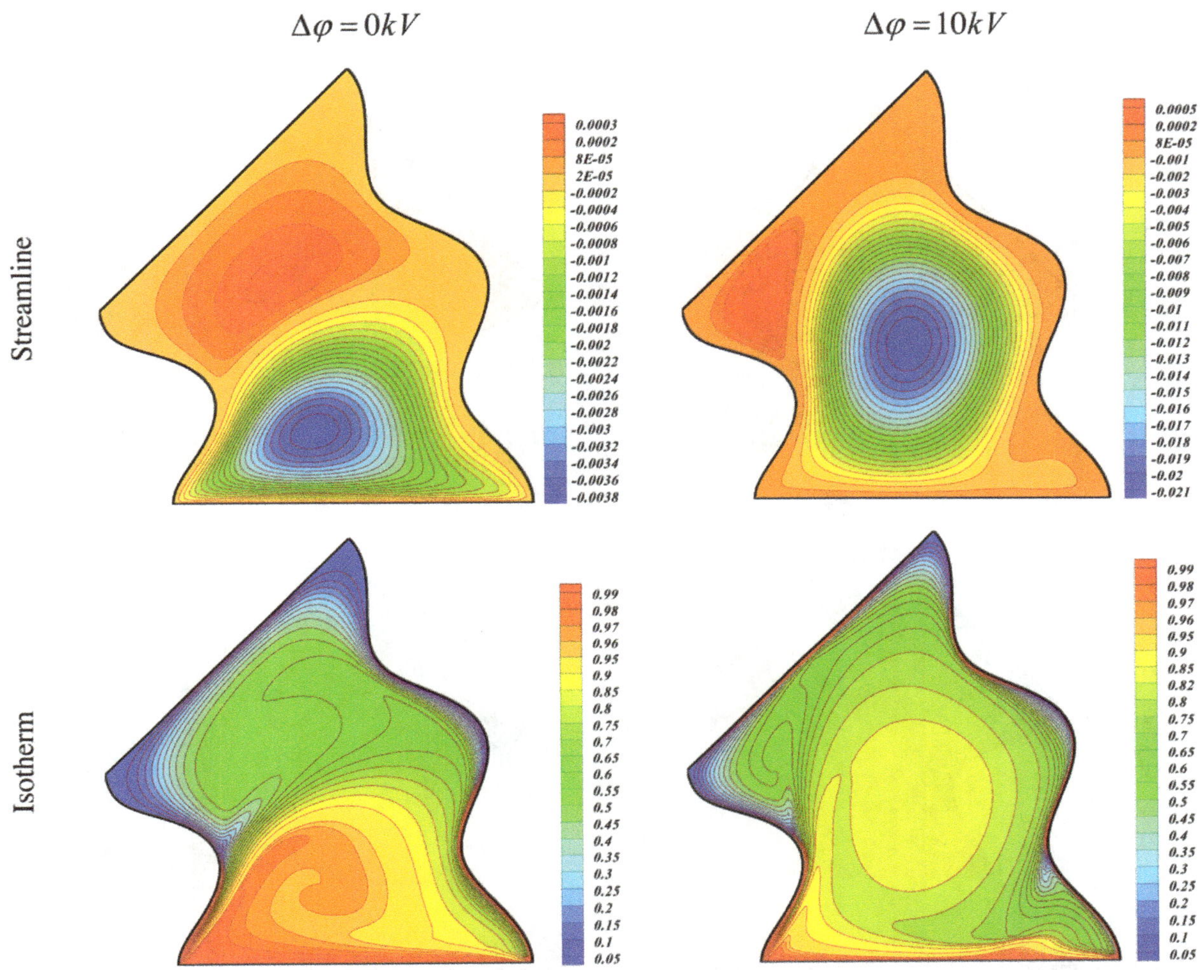

Influences of Re and $\Delta\varphi$ on Nu_{ave} are illustrated in Figure 35. A correlation for average Nusselt number is presented as:

$$Nu_{ave} = 4.86 - 1.52\,\mathrm{Re}^* + 1.01\Delta\varphi - 0.12\,\mathrm{Re}^*\,\Delta\varphi + 0.17\left(\mathrm{Re}^*\right)^2 - 0.02\left(\Delta\varphi\right)^2 \tag{77}$$

where $\mathrm{Re}^* = \mathrm{Re} \times 10^{-3}$ and $\Delta\varphi$ is voltage supply in Kilovolt. In absence of electric field, as velocity of the bottom wall augments, convective heat transfer becomes stronger and temperature gradient enhances with rise of Re. Coulomb force helps the convention mode to enhance. So Nu_{ave} augments with rise of Re and $\Delta\varphi$. In existence of high electric field enhancing *Re* shows reverse behavior and Nusselt number decreases with augment of Re. This outputs proved that using electric field is more useful in low Reynolds number.

Figure 34. Effect of supplied voltage on streamlines and isotherm when $\mathrm{Re} = 6000, \phi = 0.05$.

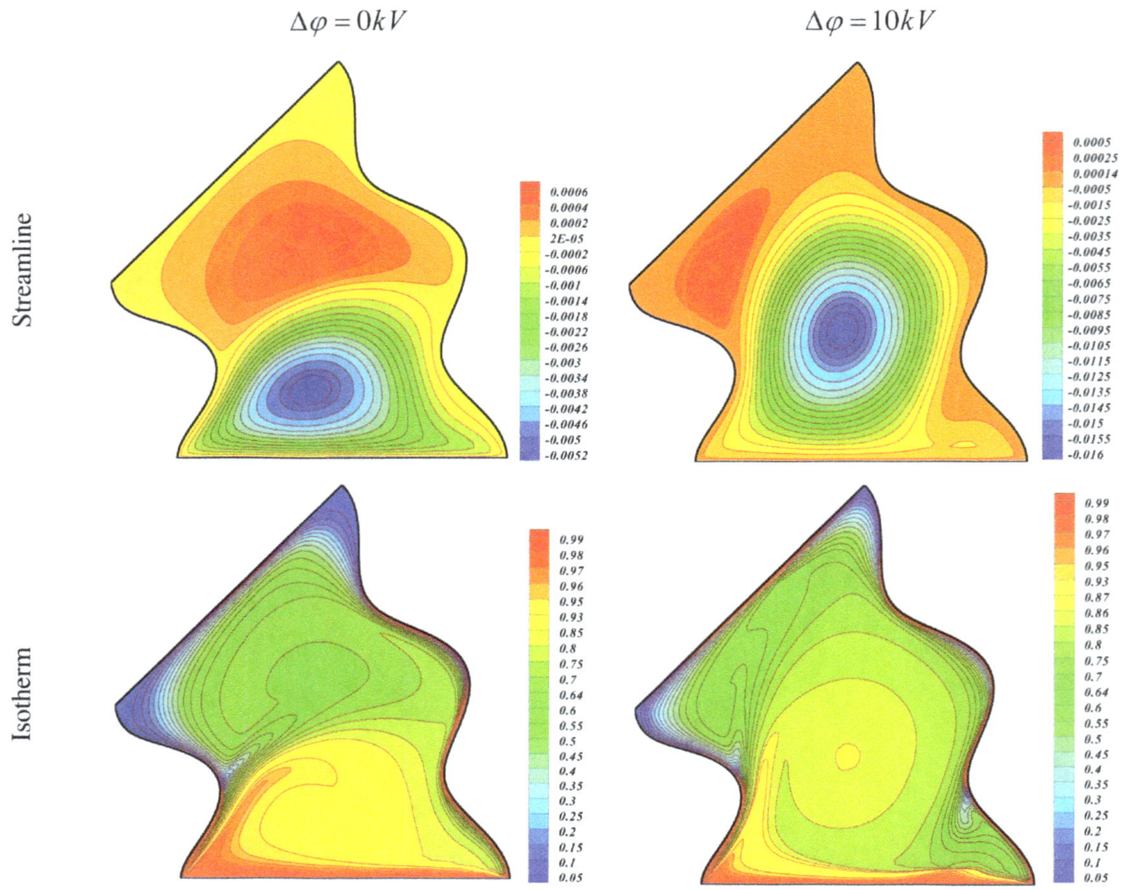

6. INFLUENCE OF ELECTRIC FIELD ON NANOFLUID FORCED CONVECTION IN A CAVITY WITH VARIABLE PROPERTIES

6.1. Problem Definition

Figure 36 demonstrates the schematic of this problem and its boundary conditions. Influence of electric field on Fe_3O_4- Ethylene glycol nanofluid is considered. Only the bottom wall can be moved. Figure 37 illustrates the contour of q for various values of $\nabla\varphi$ and Re. As $\nabla\varphi$ increases the distortion of iso-electric density lines become more and one cell appear in right side. Influence of $\nabla\varphi$ on q is more sensible than Re.

6.2. Governing Equation

According to Gauss's law and Maxwell's relation, electric field can be defined as:

Figure 35. Effects of Reynolds number and supplied voltage on average Nusselt number when $\phi = 0.05$.

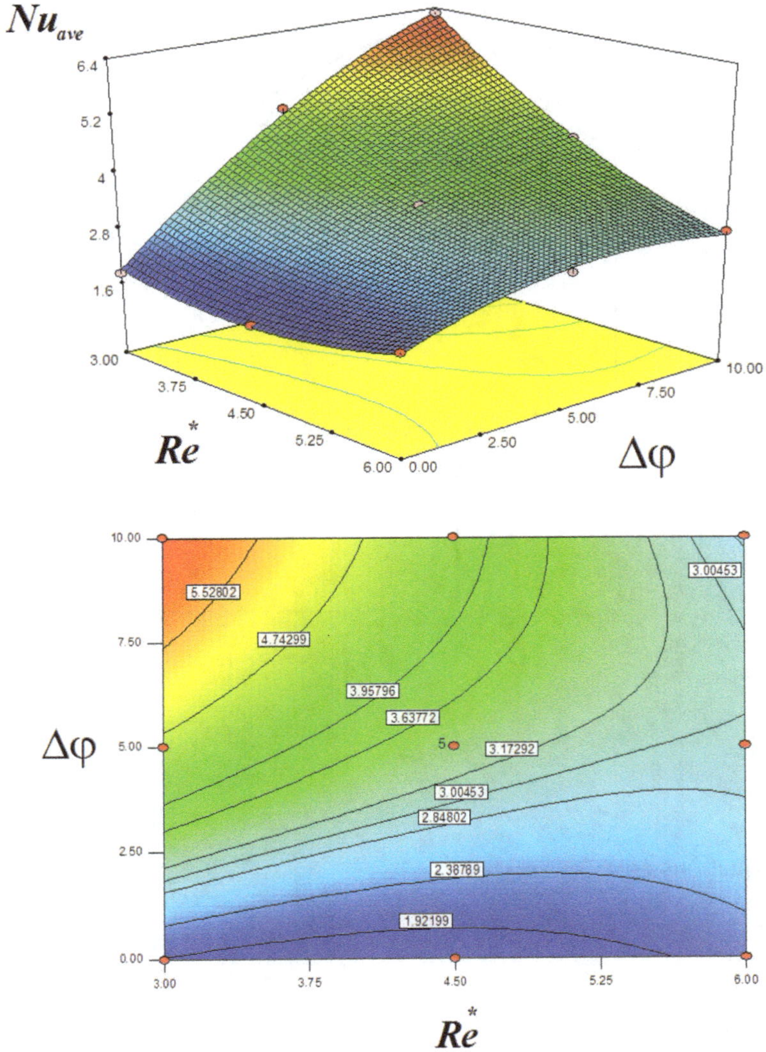

$$q = \nabla . \varepsilon \vec{E} \tag{78}$$

$$\vec{E} = -\nabla \varphi \tag{79}$$

$$\nabla . \vec{\mathrm{J}} + \frac{\partial q}{\partial t} = 0 \tag{80}$$

Figure 36. Geometry of the problem and boundary conditions

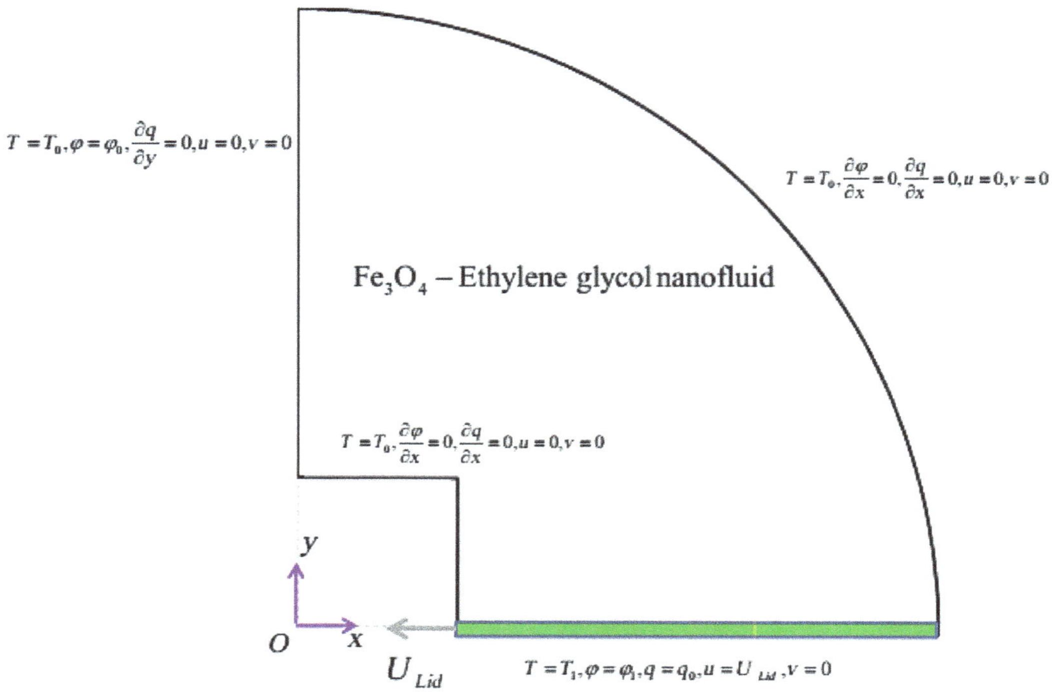

$$\vec{J} = q\vec{V} - D\nabla q + \sigma \vec{E} \tag{81}$$

The governing equations are:

$$\begin{cases} \nabla.\vec{V} = 0 \\ \left(\left(\vec{V}.\nabla\right)\vec{V} + \frac{\partial \vec{V}}{\partial t}\right) = \frac{q\vec{E}}{\rho_{nf}} + \frac{\mu_{nf}}{\rho_{nf}}\nabla^2\vec{V} - \frac{\nabla p}{\rho_{nf}} \\ \left(\left(\vec{V}.\nabla\right)T + \frac{\partial T}{\partial t}\right) = \frac{k_{nf}}{\left(\rho C_p\right)_{nf}}\nabla^2 T + \frac{\vec{J}.\vec{E}}{\left(\rho C_p\right)_{nf}} \\ \nabla\varphi = -\vec{E} \\ \frac{\partial q}{\partial t} = -\nabla.\vec{J} \\ q = \nabla.\varepsilon\vec{E} \end{cases} \tag{82}$$

The formulae of $\left(\rho C_p\right)_{nf}$ and ρ_{nf} are:

Figure 37. Electric density distribution injected by the bottom electrode

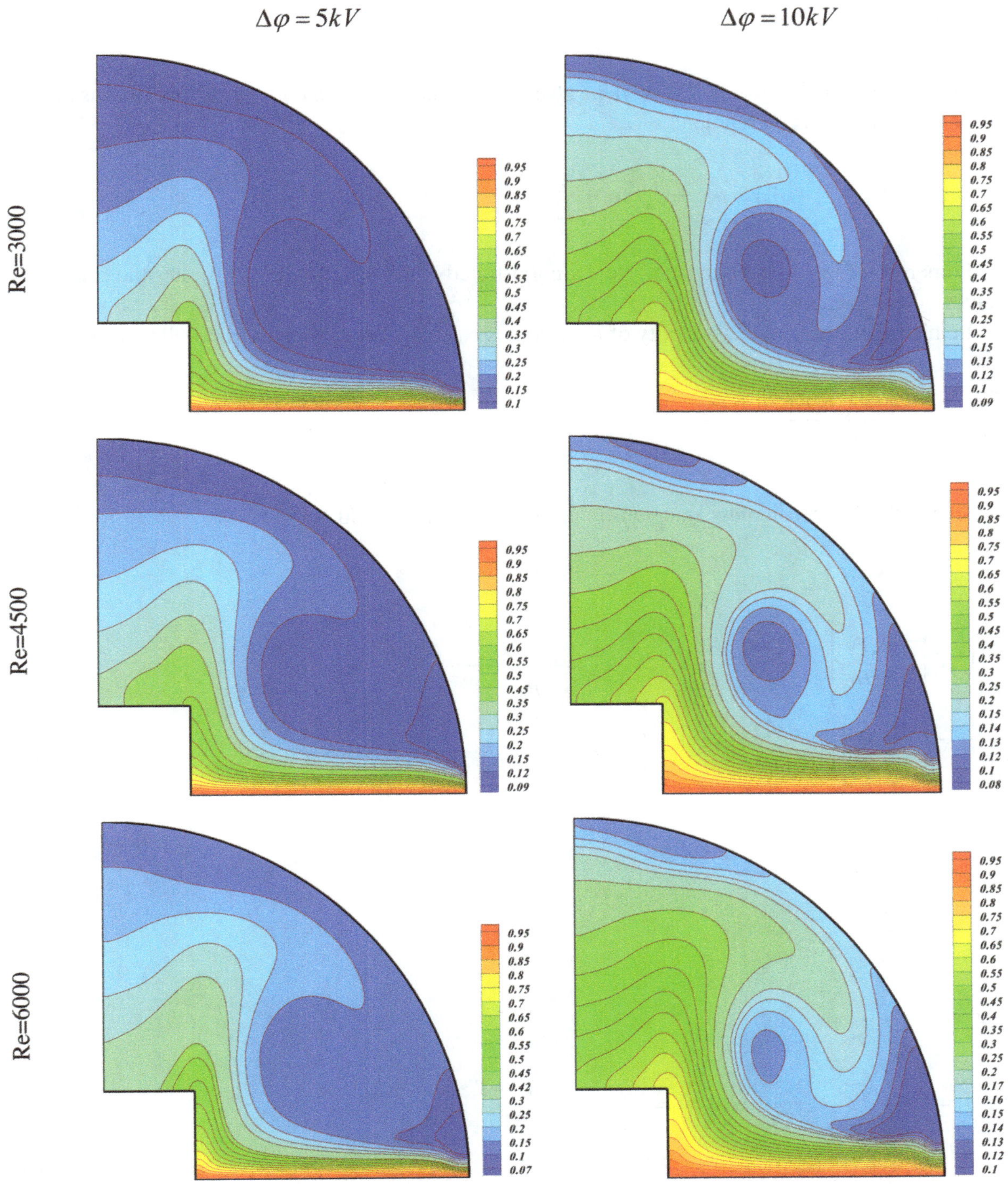

$$\begin{aligned}&\left(\rho C_p\right)_{nf}=\left(\rho C_p\right)_f(1-\phi)+\left(\rho C_p\right)_s\phi\\&\rho_{nf}=\rho_f(1-\phi)+\rho_s\phi\end{aligned}\tag{83}$$

In this paper, viscosity of Fe_3O_4-ethylene glycol is considered as a function of electric field as follows according to previous correlation:

$$\mu=A_1+A_2\left(\Delta\varphi\right)+A_3\left(\Delta\varphi\right)^2+A_4\left(\Delta\varphi\right)^3\tag{84}$$

Properties of Fe_3O_4 and ethylene glycol are illustrated in Table 1. Table 2 illustrates the coefficient values of this formula.

In this paper, thermal conductivity of nanofluid is calculated according to Maxwell method:

$$\frac{k_{nf}}{k_f}=\frac{-2\phi(k_f-k_s)+k_s+2k_f}{\phi(k_f-k_s)+k_s+2k_f}\tag{85}$$

So, the final dimensionless equations in presence of electric field are:

$$\begin{cases}\nabla.\vec{V}=0\\ \left(\left(\vec{V}.\nabla\right)\vec{V}+\dfrac{\partial\vec{V}}{\partial t}\right)=\dfrac{1}{\mathrm{Re}}\dfrac{\rho_{nf}/\rho_f}{\mu_{nf}/\mu_f}\nabla^2\vec{V}-\nabla p+\dfrac{S_E}{\rho_{nf}/\rho_f}q\vec{E}\\ \left(\left(\vec{V}.\nabla\right)\theta+\dfrac{\partial\theta}{\partial t}\right)=\dfrac{1}{\Pr\mathrm{Re}}\dfrac{k_{nf}/k_f}{\left(\rho C_p\right)_{nf}/\left(\rho C_p\right)_f}\nabla^2\theta+\dfrac{1}{\left(\rho C_p\right)_{nf}/\left(\rho C_p\right)_f}S_E Ec\left(\vec{\mathrm{J}}.\vec{E}\right)\\ \vec{E}=-\nabla\varphi\\ q=\nabla.\varepsilon\vec{E}\\ \nabla.\vec{\mathrm{J}}=-\dfrac{\partial q}{\partial t}\end{cases}\tag{86}$$

where

$$\begin{aligned}&\left(\bar{u},\bar{v}\right)=\frac{\left(u,v\right)}{U_{Lid}},\bar{\varphi}=\frac{\varphi-\varphi_0}{\nabla\varphi},\left(\bar{y},\bar{x}\right)=\frac{\left(y,x\right)}{L},\theta=\frac{T-T_0}{\nabla\mathrm{T}},\\&\bar{t}=\frac{tU_{Lid}}{L},\bar{p}=\frac{P}{\rho U_{Lid}^{\ 2}}\bar{q}=\frac{q}{q_0},\bar{E}=\frac{E}{E_0},\\&\nabla\mathrm{T}=T_1-T_0,\nabla\varphi=\varphi_1-\varphi_0,\end{aligned}\tag{87}$$

By eliminating pressure gradient, vorticity and stream function can be introduced as:

$$v = -\frac{\partial \psi}{\partial x}, \frac{\partial \psi}{\partial y} = u, \Psi = \frac{\psi}{\alpha_f}, \Omega = \frac{\omega L^2}{\alpha_f}, \omega = \frac{\partial v}{\partial x} - \frac{\partial u}{\partial y} \tag{88}$$

Nu_{loc} and Nu_{ave} along the bottom wall are calculated as:

$$Nu_{loc} = \left(\frac{k_{nf}}{k_f}\right)\frac{\partial \Theta}{\partial X} \tag{89}$$

$$Nu_{ave} = \frac{1}{L}\int_0^L Nu_{loc}\, dY \tag{90}$$

6.3. Effects of Active Parameters

Fe_3O_4 – Ethylene glycol nanofluid forced convection in an enclosure is studied in existence of electric filed. The bottom wall is moving. Roles of supplied voltage ($\Delta\varphi = 0$ to $10kV$), Reynolds number ($\mathrm{Re} = 3000$ to 6000) and volume fraction of Fe_3O_4 ($\phi = 0\%$ to 5%) are presented graphically.

Influences of $\Delta\varphi$ and Re on streamlines and isotherm are illustrated in Figures 38, 39 and 40. In absence of electric field, two eddies appear in streamlines which are rotates in reverse direction. The counter clock wise eddy is stronger and located at upper left side upper of the square obstacle. Existing such eddies leads to generate two thermal plumes at left and upper side of square obstacle. As electric field is applied, the clock wise eddy becomes stronger and the other one convert to two smaller counter clock wise eddies which are located at left side of enclosure. Isotherms become more disturb by augmenting voltage supply due to increment of Coulomb force. By increasing Re, the strength of rotating eddies enhances and the distortion of isotherms becomes more than before. Temperature gradient enhances at high Reynolds numbers due to stronger convective heat transfer. At high Re, Coulomb force is weaker than viscous force, so the impact of adding electric field becomes weaker with augment of Re.

Influences of Re and $\Delta\varphi$ on Nu_{ave} are illustrated in Figure 41. A correlation for average Nusselt number is presented as:

$$Nu_{ave} = 2.52 - 0.68\,\mathrm{Re}^* + 0.57\Delta\varphi - 0.058\,\mathrm{Re}^*\,\Delta\varphi + 0.08\left(\mathrm{Re}^*\right)^2 - 0.011\left(\Delta\varphi\right)^2 \tag{91}$$

where $\mathrm{Re}^* = \mathrm{Re} \times 10^{-3}$ and $\Delta\varphi$ is voltage supply in Kilovolt. In absence of electric field, as velocity of the bottom wall augments, convective heat transfer becomes stronger and temperature gradient enhances with rise of Re. As electric field augments, Coulomb force becomes stronger. Coulomb force helps the convective mode to enhance. So rate of heat transfer enhances with rise of voltage supply. In

Figure 38. Effect of supplied voltage on streamlines and isotherm when $\mathrm{Re} = 3000, \phi = 0.05$.

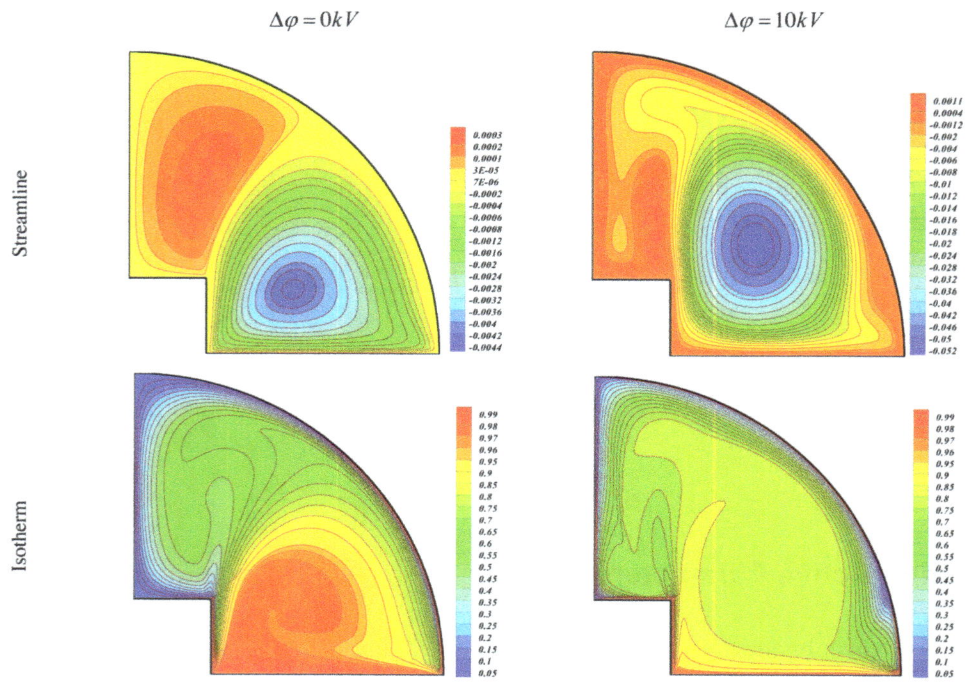

Figure 39. Effect of supplied voltage on streamlines and isotherm when $\mathrm{Re} = 4500, \phi = 0.05$.

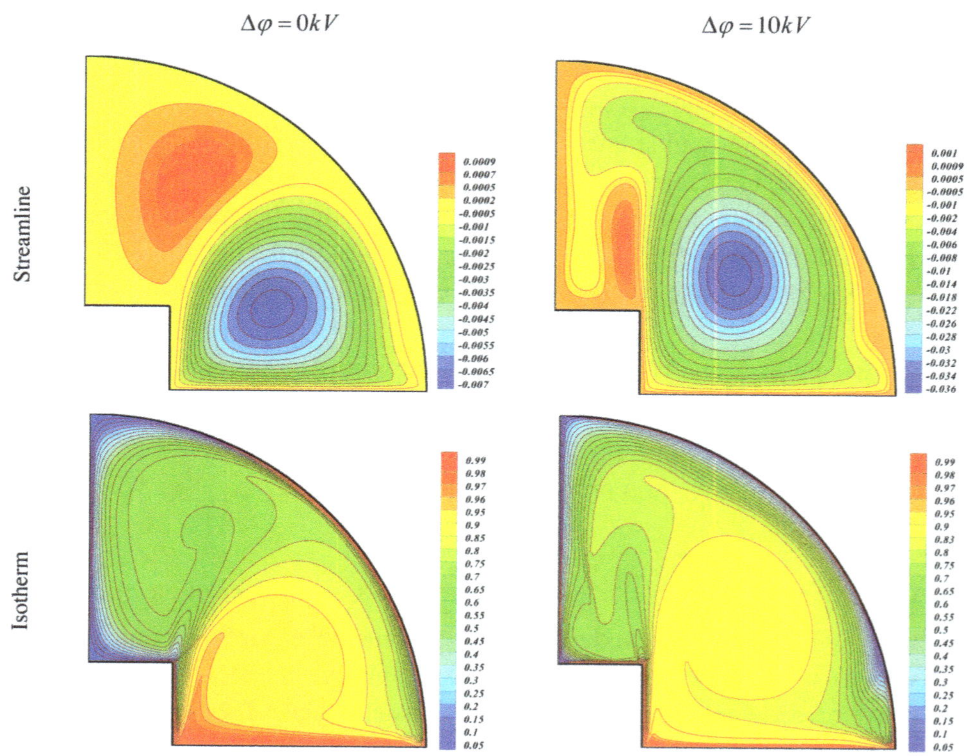

Figure 40. Effect of supplied voltage on streamlines and isotherm when $\mathrm{Re} = 6000, \phi = 0.05$.

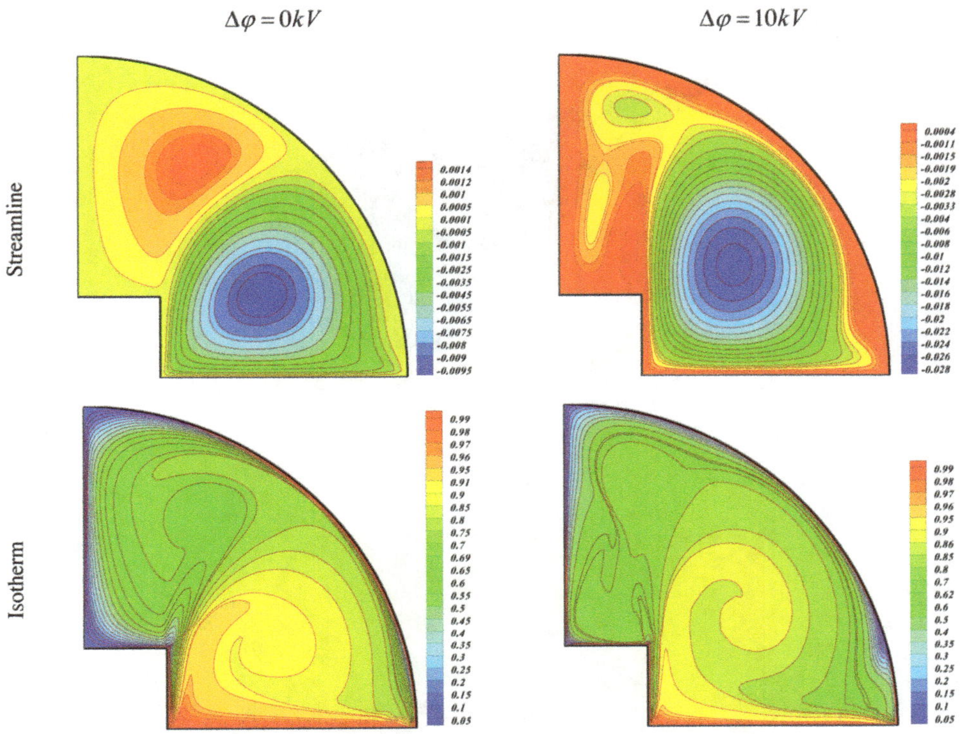

existence of high electric field, enhancing *Re* shows reverse behavior and Nusselt number reduces with rise of Re. This outputs proved that using electric field is more useful in low Reynolds number.

7. EFFECT OF EFD VISCOSITY ON NANOFLUID FORCED CONVECTION IN A CAVITY WITH SINUSOIDAL WALL

7.1. Problem Definition

Figure 42 illustrates the geometry of this problem and its boundary conditions. Impact of electric field on Fe_3O_4- Ethylene glycol nanofluid is considered. Figure 43 shows the contour of q for various values of $\nabla\varphi$ and Re. As $\nabla\varphi$ increases the distortion of isoelectric density lines become more and one cell appear in right side. Influence of $\nabla\varphi$ on q is more sensible than Re.

7.2. Governing Equation

According to Gauss's law and Maxwell's relation, electric field can be defined as:

$$q = \nabla.\varepsilon \vec{E} \tag{92}$$

Figure 41. Effects of Reynolds number and supplied voltage on average Nusselt number.

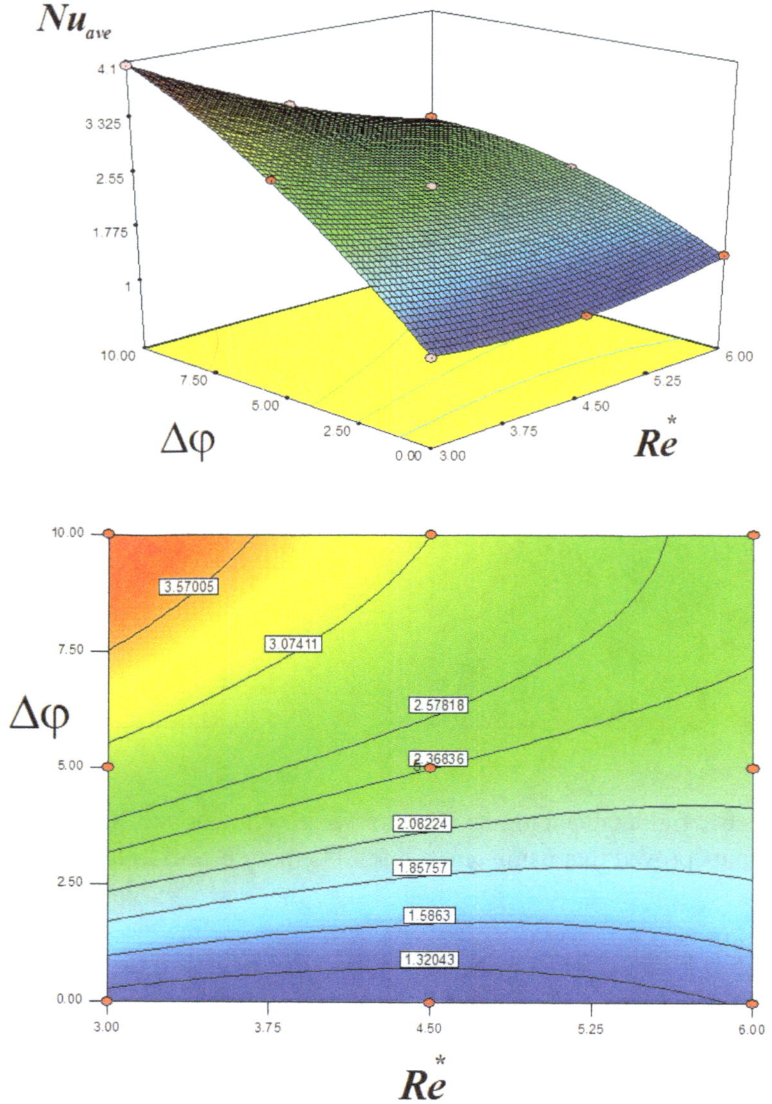

$$\vec{E} = -\nabla\varphi \tag{93}$$

$$\nabla.\vec{\mathrm{J}} + \frac{\partial q}{\partial t} = 0 \tag{94}$$

$$\vec{\mathrm{J}} = q\vec{V} - D\nabla q + \sigma\vec{E} \tag{95}$$

Figure 42. Geometry of the problem and boundary conditions

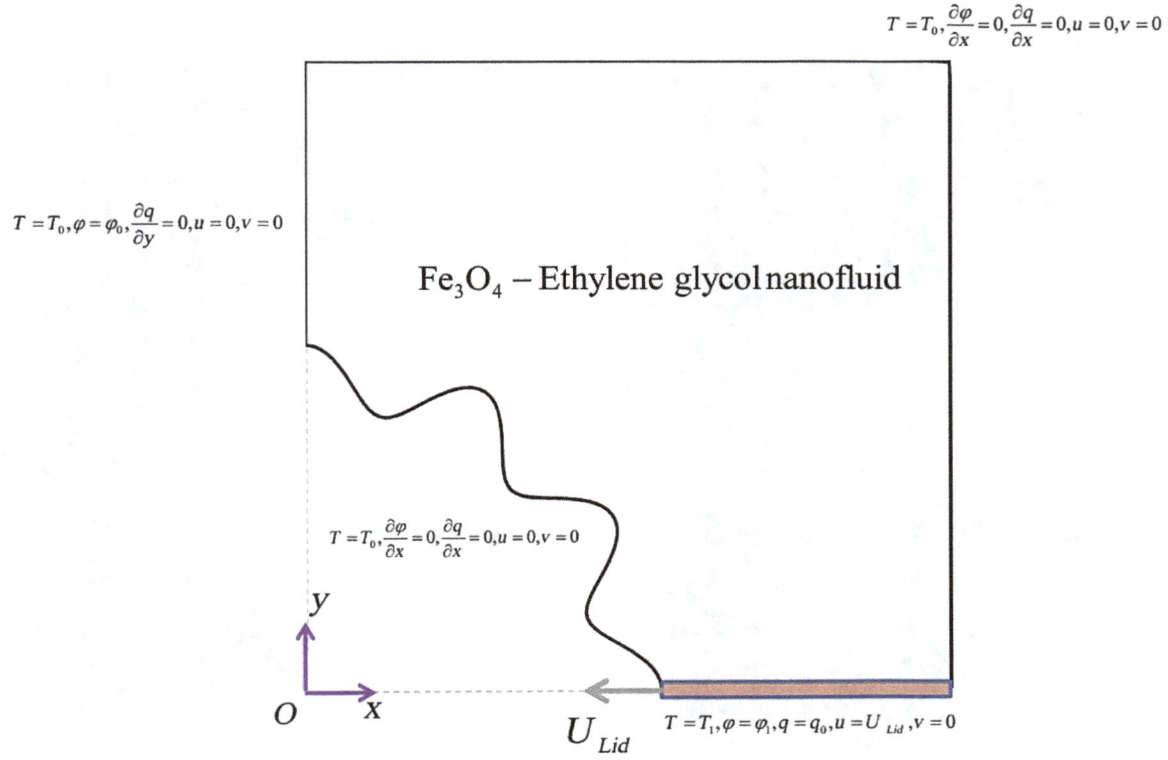

The governing equations are:

$$\begin{cases} \nabla.\vec{V} = 0 \\ \left(\left(\vec{V}.\nabla\right)\vec{V} + \dfrac{\partial \vec{V}}{\partial t}\right) = \dfrac{q\vec{E}}{\rho_{nf}} + \dfrac{\mu_{nf}}{\rho_{nf}}\nabla^2\vec{V} - \dfrac{\nabla p}{\rho_{nf}} \\ \left(\left(\vec{V}.\nabla\right)T + \dfrac{\partial T}{\partial t}\right) = \dfrac{k_{nf}}{\left(\rho C_p\right)_{nf}}\nabla^2 T + \dfrac{\vec{J}.\vec{E}}{\left(\rho C_p\right)_{nf}} \\ \nabla\varphi = -\vec{E} \\ \dfrac{\partial q}{\partial t} = -\nabla.\vec{J} \\ q = \nabla.\varepsilon\vec{E} \end{cases} \tag{96}$$

k_{nf}, $\left(\rho C_p\right)_{nf}$, μ_{nf} and ρ_{nf} can be obtained as:

Figure 43. Electric density distribution injected by the bottom electrode

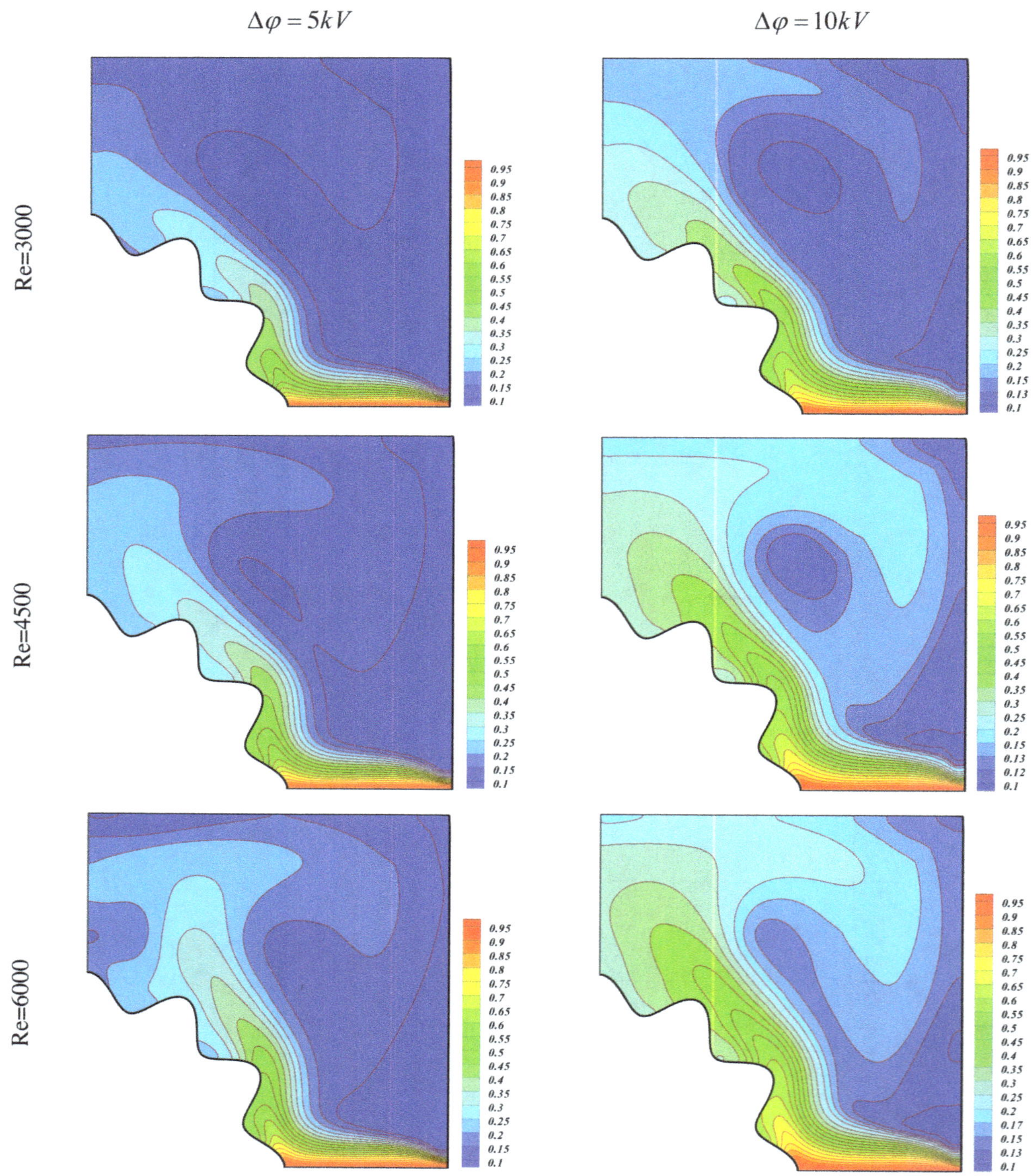

$$\frac{k_{nf}}{k_f} = \frac{-2\phi(k_f - k_s) + k_s + 2k_f}{\phi(k_f - k_s) + k_s + 2k_f}, \quad \left(\rho C_p\right)_{nf} = \left(\rho C_p\right)_f (1-\phi) + \left(\rho C_p\right)_s \phi,$$
$$\mu_{nf} = A_1 + A_2\left(\Delta\varphi\right) + A_3\left(\Delta\varphi\right)^2 + A_4\left(\Delta\varphi\right)^3 \tag{97}$$
$$\rho_{nf} = \rho_f(1-\phi) + \rho_s\phi$$

Properties of Fe_3O_4 and ethylene glycol are presented in Table 1. Table 2 illustrates the coefficient values of this formula.

So, the final dimensionless equations in presence of electric field are:

$$\begin{cases} \nabla.\vec{V} = 0 \\ \left(\left(\vec{V}.\nabla\right)\vec{V} + \dfrac{\partial \vec{V}}{\partial t}\right) = \dfrac{1}{\mathrm{Re}} \dfrac{\rho_{nf} / \rho_f}{\mu_{nf} / \mu_f} \nabla^2 \vec{V} - \nabla p + \dfrac{S_E}{\rho_{nf} / \rho_f} q\,\vec{E} \\ \left(\left(\vec{V}.\nabla\right)\theta + \dfrac{\partial \theta}{\partial t}\right) = \dfrac{1}{\Pr \mathrm{Re}} \dfrac{k_{nf} / k_f}{\left(\rho C_p\right)_{nf} / \left(\rho C_p\right)_f} \nabla^2\theta + \dfrac{1}{\left(\rho C_p\right)_{nf} / \left(\rho C_p\right)_f} S_E Ec\left(\vec{\mathrm{J}}.\vec{E}\right) \\ \vec{E} = -\nabla\varphi \\ q = \nabla.\varepsilon\,\vec{E} \\ \nabla.\vec{\mathrm{J}} = -\dfrac{\partial q}{\partial t} \end{cases} \tag{98}$$

where

$$\left(\bar{u}, \bar{v}\right) = \frac{\left(u, v\right)}{U_{Lid}}, \bar{\varphi} = \frac{\varphi - \varphi_0}{\nabla\varphi}, \left(\bar{y}, \bar{x}\right) = \frac{\left(y, x\right)}{L}, \theta = \frac{T - T_0}{\nabla \mathrm{T}},$$
$$\bar{t} = \frac{tU_{Lid}}{L}, \bar{p} = \frac{P}{\rho U_{Lid}^{\,2}} \bar{q} = \frac{q}{q_0}, \bar{E} = \frac{E}{E_0}, \tag{99}$$
$$\nabla \mathrm{T} = T_1 - T_0, \nabla\varphi = \varphi_1 - \varphi_0,$$

By eliminating pressure gradient, vorticity and stream function can be introduced as:

$$v = -\frac{\partial\psi}{\partial x}, \frac{\partial\psi}{\partial y} = u, \Psi = \frac{\psi}{\alpha_f}, \Omega = \frac{\omega L^2}{\alpha_f}, \omega = \frac{\partial v}{\partial x} - \frac{\partial u}{\partial y} \tag{100}$$

Nu_{loc} and Nu_{ave} along the left wall are calculated as:

$$Nu_{loc} = \left(\frac{k_{nf}}{k_f} \right) \frac{\partial \Theta}{\partial X} \tag{101}$$

$$Nu_{ave} = \frac{1}{L} \int_0^L Nu_{loc} \, dY \tag{102}$$

7.3. Effects of Active Parameters

Fe_3O_4 – Ethylene glycol nanofluid forced convection in an enclosure is simulated in existence of Coulomb forces. The bottom wall is moving lid. Roles of supplied voltage ($\Delta\varphi = 0$ to $10kV$), Reynolds number ($\mathrm{Re} = 3000$ to 6000) and volume fraction of Fe_3O_4 ($\phi = 0\%$ to 5%) are presented as graphs.

Effect of $\Delta\varphi$ and Re on streamlines and isotherms are shown in Figures 44, 45 and 46. In low Reynolds number, one clock wise eddy and three small counter clock wise eddies can be seen in streamline. As electric field increases eddies become stronger. By increasing Re, the strength of rotating eddies enhances and the distortion of isotherms becomes more than before. At high Reynolds number, the clock wise eddy converts to two smaller one. Also increasing Reynolds number makes isotherms more complicated. So, temperate gradient enhances with rise of Reynolds number. As electric field is

Figure 44. Effect of supplied voltage on streamlines and isotherm when $\mathrm{Re} = 3000, \phi = 0.05$.

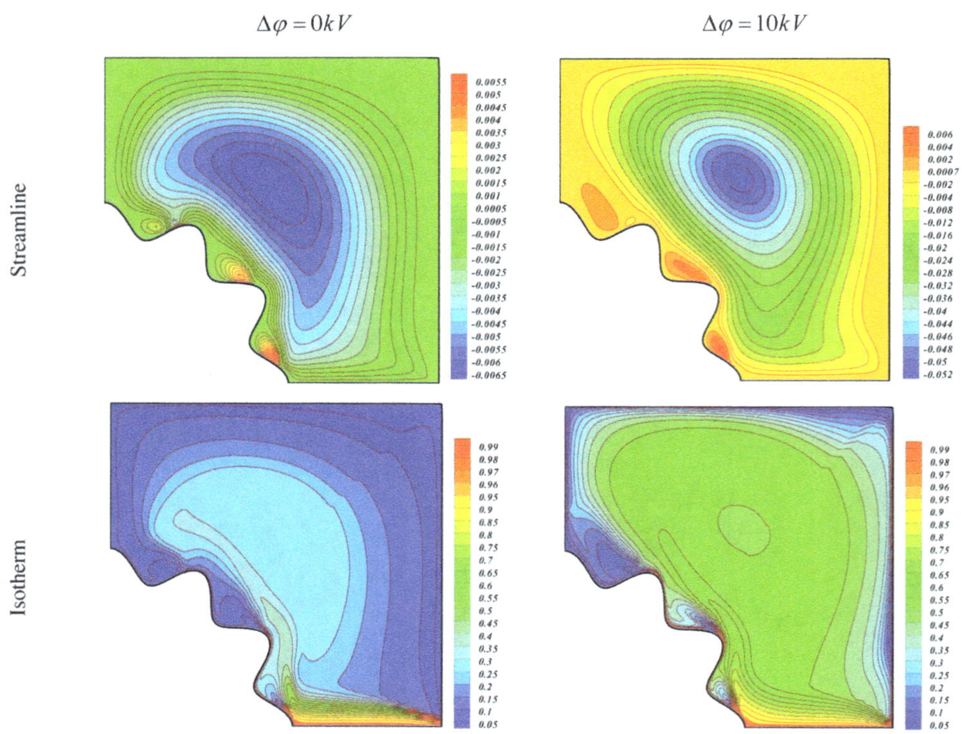

Figure 45. Effect of supplied voltage on streamlines and isotherm when $\mathrm{Re} = 4500, \phi = 0.05$.

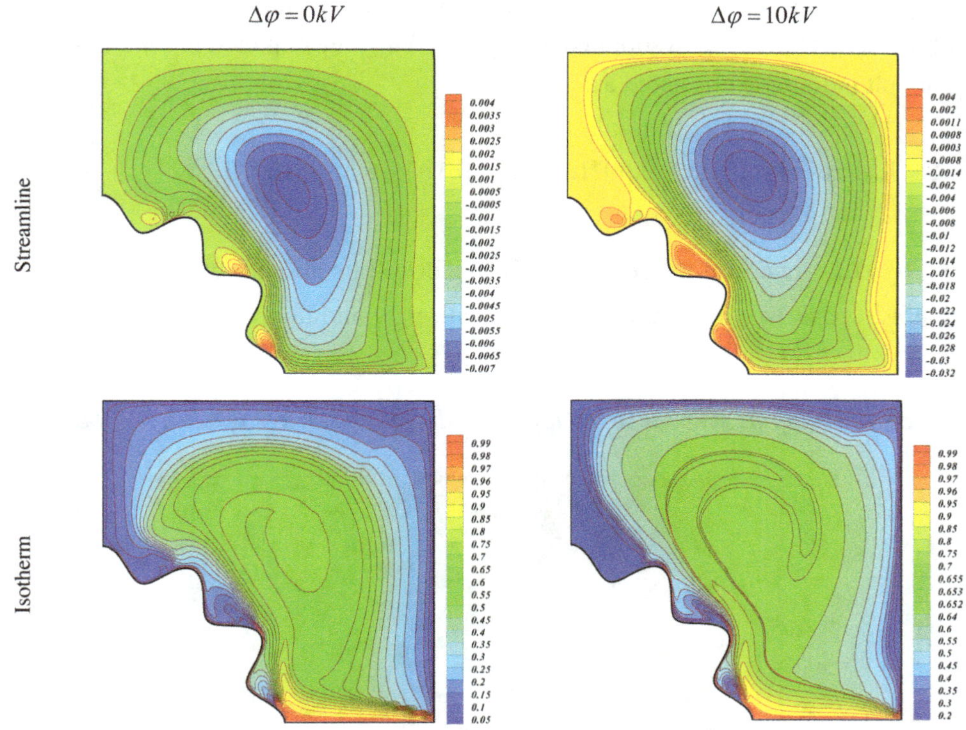

Figure 46. Effect of supplied voltage on streamlines and isotherm when $\mathrm{Re} = 6000, \phi = 0.05$.

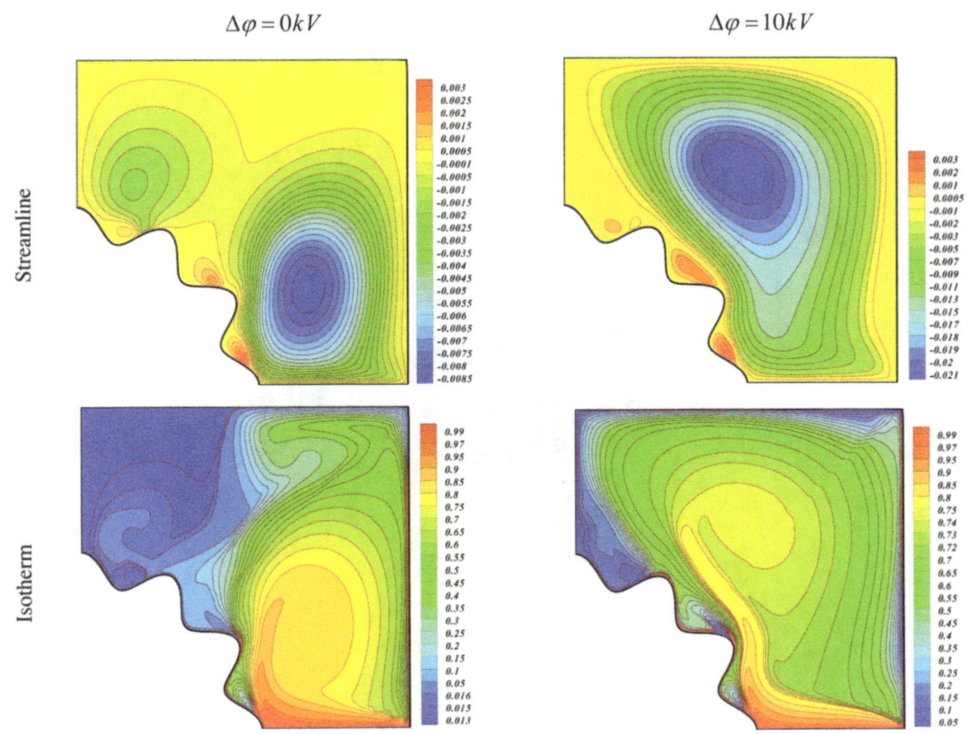

applied, the two clock wise eddies merge together. Isotherms become more disturb by augmenting $\Delta\varphi$. Effect of adding electric field is more pronounced in low Reynolds number.

Impacts of Reynolds number and supplied voltage on Nu_{ave} are depicted in Figure 47. A correlation for average Nusselt number is presented as:

$$Nu_{ave} = 5.24 - 1.66\,\mathrm{Re}^* + 0.9\Delta\varphi - 0.09\,\mathrm{Re}^*\,\Delta\varphi + 0.185\left(\mathrm{Re}^*\right)^2 - 0.029\left(\Delta\varphi\right)^2 \tag{103}$$

where $\mathrm{Re}^* = \mathrm{Re}\times 10^{-3}$ and $\Delta\varphi$ is voltage supply in Kilovolt. In absence of electric field, as Reynolds number augments, convective heat transfer becomes stronger and temperature gradient enhances with

Figure 47. Effects of Reynolds number and supplied voltage on average Nusselt number.

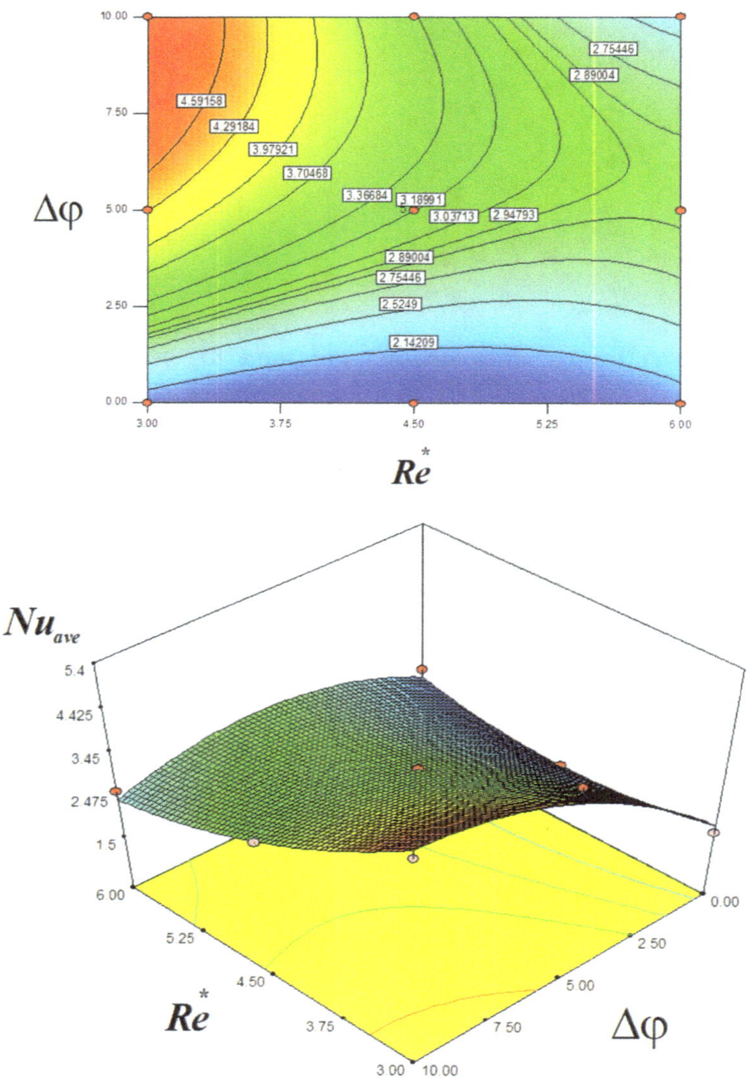

augment of Re. Electric field boosts the convention mode. So Nu_{ave} augments with augment $\Delta\varphi$. Effect of adding electric field becomes weaker in higher Reynolds number.

8. FORCED CONVECTION HEAT TRANSFER IN FE_3O_4-ETHYLENE GLYCOL NANOFLUID UNDER THE INFLUENCE OF COULOMB FORCE

8.1. Problem Definition

Figure 48 demonstrates the schematic of this paper and its boundary conditions. Influence of electric field on Fe_3O_4- Ethylene glycol nanofluid is considered. Only the bottom wall can be moved. Figure 49 illustrates the contour of q for various values of $\nabla\varphi$ and Re. As $\nabla\varphi$ increases the distortion of iso-electric density lines become more and one cell appear in right side. Influence of $\nabla\varphi$ on q is more sensible than Re.

8.2. Governing Equation

According to Gauss's law and Maxwell's relation, electric field can be defined as:

$$q = \nabla.\varepsilon \vec{E} \tag{104}$$

Figure 48. Geometry of the problem and boundary conditions

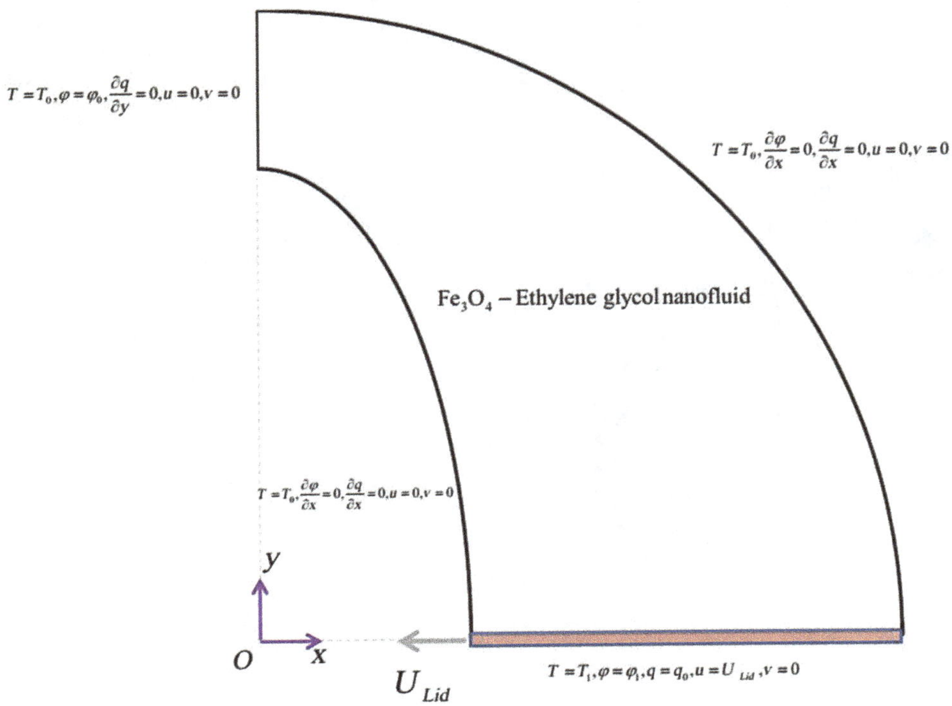

Figure 49. Electric density distribution injected by the bottom electrode

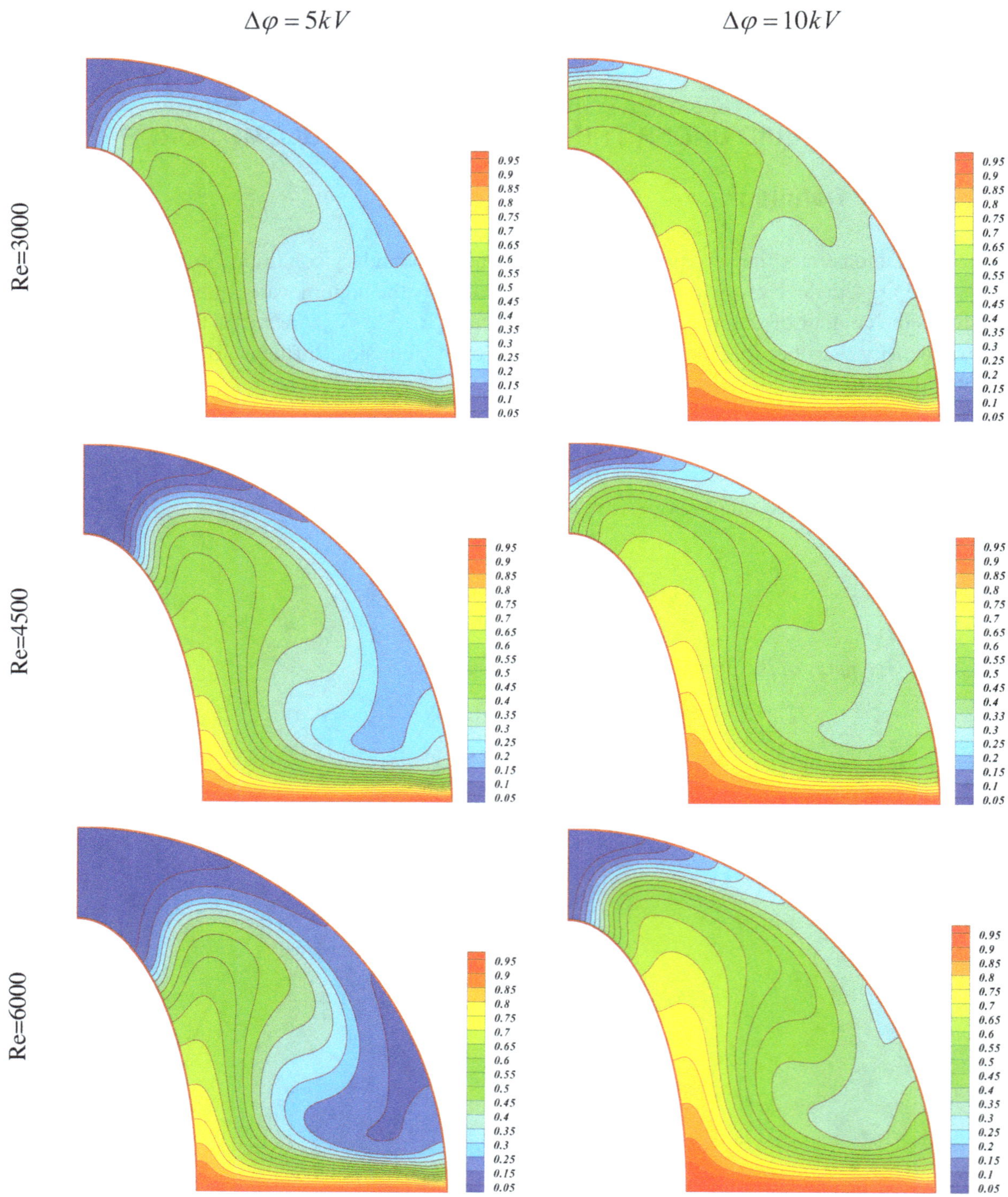

$$\vec{E} = -\nabla \varphi \tag{105}$$

$$\nabla . \vec{\mathrm{J}} + \frac{\partial q}{\partial t} = 0 \tag{106}$$

$$\vec{\mathrm{J}} = q\vec{V} - D\nabla q + \sigma \vec{E} \tag{107}$$

The governing equations are:

$$\begin{cases} \nabla . \vec{V} = 0 \\ \left(\left(\vec{V} . \nabla \right) \vec{V} + \dfrac{\partial \vec{V}}{\partial t} \right) = \dfrac{q\vec{E}}{\rho_{nf}} + \dfrac{\mu_{nf}}{\rho_{nf}} \nabla^2 \vec{V} - \dfrac{\nabla p}{\rho_{nf}} \\ \left(\left(\vec{V} . \nabla \right) T + \dfrac{\partial T}{\partial t} \right) = \dfrac{k_{nf}}{\left(\rho C_p \right)_{nf}} \nabla^2 T + \dfrac{\vec{\mathrm{J}} . \vec{E}}{\left(\rho C_p \right)_{nf}} \\ \nabla \varphi = -\vec{E} \\ \dfrac{\partial q}{\partial t} = -\nabla . \vec{\mathrm{J}} \\ q = \nabla . \varepsilon \vec{E} \end{cases} \tag{108}$$

k_{nf}, $\left(\rho C_p \right)_{nf}$, μ_{nf} and ρ_{nf} can be obtained as:

$$\begin{aligned} &\frac{k_{nf}}{k_f} = \frac{-2\phi(k_f - k_s) + k_s + 2k_f}{\phi(k_f - k_s) + k_s + 2k_f}, \quad \left(\rho C_p \right)_{nf} = \left(\rho C_p \right)_f (1 - \phi) + \left(\rho C_p \right)_s \phi, \\ &\mu = A_1 + A_2 \left(\Delta \varphi \right) + A_3 \left(\Delta \varphi \right)^2 + A_4 \left(\Delta \varphi \right)^3 \\ &\rho_{nf} = \rho_f (1 - \phi) + \rho_s \phi \end{aligned} \tag{109}$$

Properties of Fe_3O_4 and ethylene glycol are illustrated in Table 1 (Sheikoleslami & Ellahi, 2015). EFD viscosity is presented by Sheikoleslami and Ellahi (2015). Table 2 illustrates the coefficient values of this formula.

So, the final dimensionless equations in presence of electric field are:

$$\begin{cases} \nabla.\vec{V}=0 \\ \left(\left(\vec{V}.\nabla\right)\vec{V}+\dfrac{\partial\vec{V}}{\partial t}\right)=\dfrac{1}{\mathrm{Re}}\dfrac{\rho_{nf}/\rho_f}{\mu_{nf}/\mu_f}\nabla^2\vec{V}-\nabla p+\dfrac{S_E}{\rho_{nf}/\rho_f}q\vec{E} \\ \left(\left(\vec{V}.\nabla\right)\theta+\dfrac{\partial\theta}{\partial t}\right)=\dfrac{1}{\mathrm{Pr\,Re}}\dfrac{k_{nf}/k_f}{\left(\rho C_p\right)_{nf}/\left(\rho C_p\right)_f}\nabla^2\theta+\dfrac{1}{\left(\rho C_p\right)_{nf}/\left(\rho C_p\right)_f}S_E Ec\left(\vec{\mathrm{J}}.\vec{E}\right) \\ \vec{E}=-\nabla\varphi \\ q=\nabla.\varepsilon\vec{E} \\ \nabla.\vec{\mathrm{J}}=-\dfrac{\partial q}{\partial t} \end{cases} \tag{110}$$

where

$$\begin{gathered} \left(\bar{u},\bar{v}\right)=\frac{\left(u,v\right)}{U_{Lid}},\bar{\varphi}=\frac{\varphi-\varphi_0}{\nabla\varphi},\left(\bar{y},\bar{x}\right)=\frac{\left(y,x\right)}{L},\theta=\frac{T-T_0}{\nabla\mathrm{T}}, \\ \bar{t}=\frac{tU_{Lid}}{L},\bar{p}=\frac{P}{\rho U_{Lid}^{2}}\bar{q}=\frac{q}{q_0},\bar{E}=\frac{E}{E_0}, \\ \nabla\mathrm{T}=T_1-T_0,\nabla\varphi=\varphi_1-\varphi_0, \end{gathered} \tag{111}$$

By eliminating pressure gradient, vorticity and stream function can be introduced as:

$$v=-\frac{\partial\psi}{\partial x},\frac{\partial\psi}{\partial y}=u,\Psi=\frac{\psi}{\alpha_f},\Omega=\frac{\omega L^2}{\alpha_f},\omega=\frac{\partial v}{\partial x}-\frac{\partial u}{\partial y} \tag{112}$$

Nu_{loc} and Nu_{ave} along the left wall are calculated as:

$$Nu_{loc}=\left(\frac{k_{nf}}{k_f}\right)\frac{\partial\Theta}{\partial X} \tag{113}$$

$$Nu_{ave}=\frac{1}{L}\int_0^L Nu_{loc}\,dY \tag{114}$$

8.3. Effects of Active Parameters

Coulomb forces impact on nanofluid forced convective heat transfer is reported. Fe_3O_4 – Ethylene glycol is chosen as working fluid and viscosity of this nanofluid is taken into account as function of electric filed. The Lid wall is considered as positive electrode. Roles of supplied voltage ($\Delta\varphi = 0$ to $10kV$), Reynolds number ($\mathrm{Re} = 3000$ to 6000) and volume fraction of Fe_3O_4 ($\phi = 0\%$ to 5%) are presented graphically.

Impacts of $\Delta\varphi$ and Re on isotherm and streamlines are depicted in Figures 50, 51 and 52.

In absence of electric field, there are two vortexes in streamlines which are rotates in reverse direction. The upper one is counter clock wise vortex and it diminishes in existence of Coulomb force. Also this force makes to generate thermal plume over the lid wall. As Reynolds number augments the strength of vortexes enhance and thermal plume generates in absence of electric field, too. Isotherms become more disturb with increase of Coulomb force. So temperature gradient enhances near the positive electrode with enhance of Coulomb force and Reynolds number. Furthermore, it can be concluded from these figures that impact of Coulomb force of nanofluid heat transfer is more sensible in lower values of Reynolds number. This is due to this fact that at high Re, Coulomb force is weaker than viscous force; therefore the influence of applying electric field becomes weaker with rise of Re.

Figure 50. Effect of supplied voltage on streamlines and isotherm when $\mathrm{Re} = 3000, \phi = 0.05$.

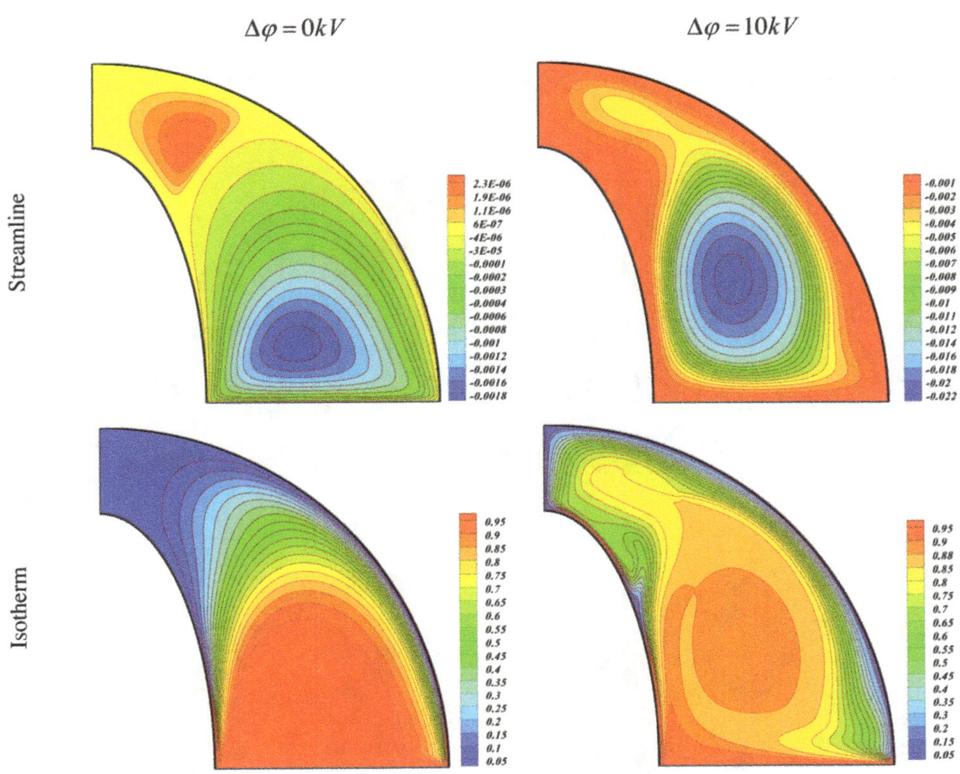

Figure 51. Effect of supplied voltage on streamlines and isotherm when $\mathrm{Re} = 4500, \phi = 0.05$.

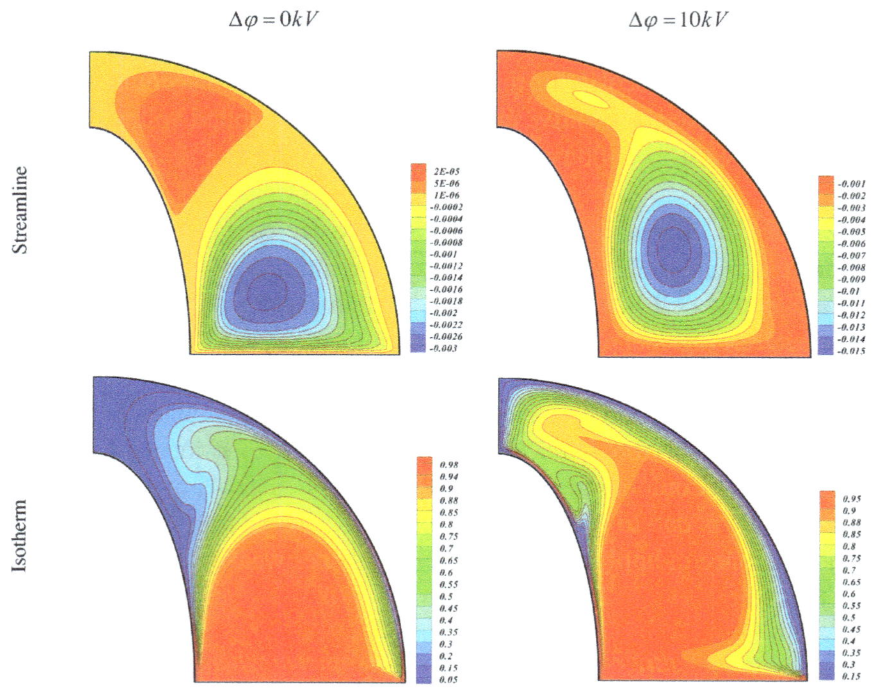

Figure 52. Effect of supplied voltage on streamlines and isotherm when $\mathrm{Re} = 6000, \phi = 0.05$.

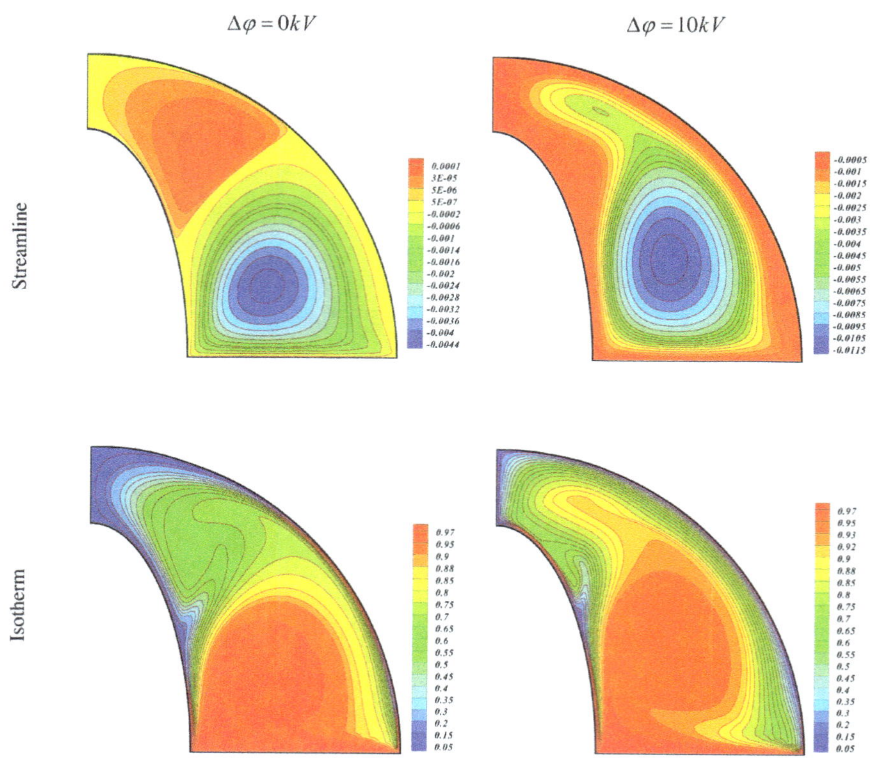

Figure 53 depicts the impacts of Re and $\Delta\varphi$ on Nu_{ave}. A formula for Nu_{ave} can be presented as:

$$Nu_{ave} = 4.19 - 1.08\,\mathrm{Re}^{*} + 1.05\Delta\varphi - 0.13\,\mathrm{Re}^{*}\,\Delta\varphi + 0.11\left(\mathrm{Re}^{*}\right)^{2} - 7.3\times10^{-3}\left(\Delta\varphi\right)^{2} \tag{115}$$

where $\mathrm{Re}^{*} = \mathrm{Re}\times10^{-3}$ and $\Delta\varphi$ is voltage supply in Kilovolt. In absence of Coulomb force, as Reynolds number augments, convective heat transfer becomes stronger and temperature gradient augments

Figure 53. Effects of Reynolds number and supplied voltage on average Nusselt number.

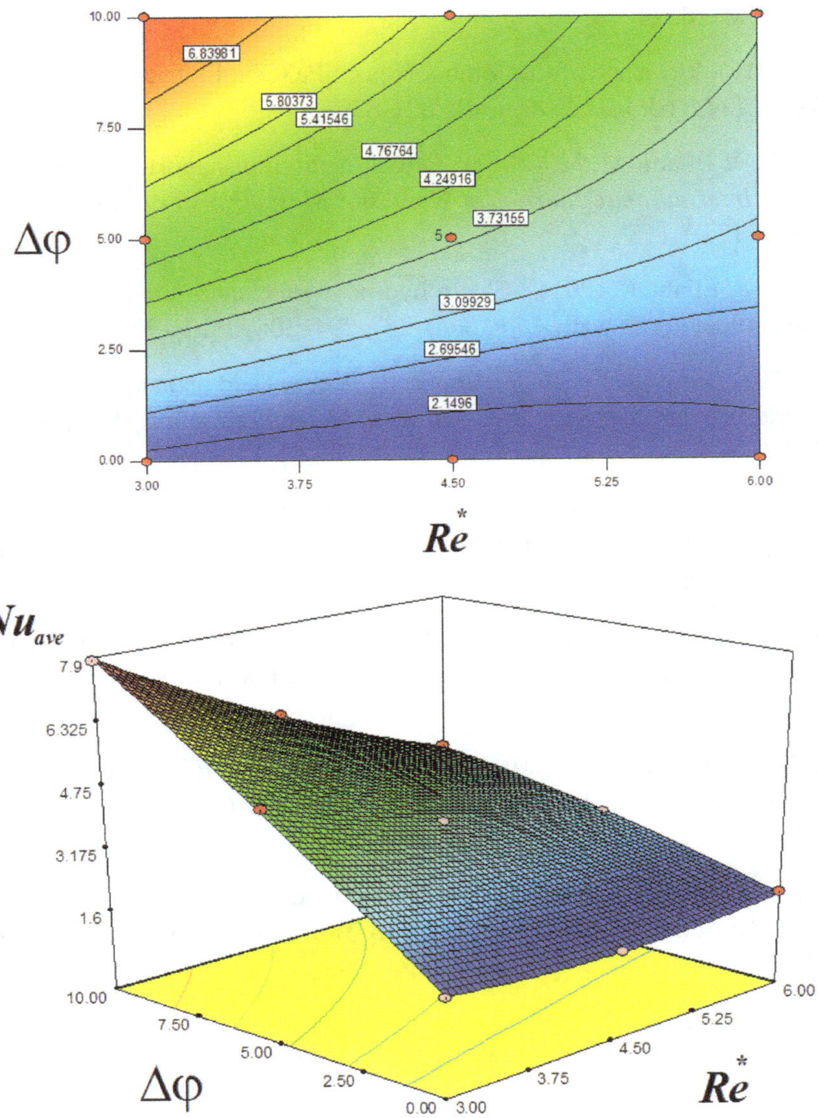

with rise of Re. Applying electric field improves the convention mechanism. Therefore Nu_{ave} augments with rise of Re and $\Delta\varphi$. In existence of high Coulomb force, increasing *Re* results reverse effect on rate of heat transfer. This observation indicates that using electric field is more useful in lower values of Re.

REFERENCES

Kasayapanand, N., Tiansuwan, J., Asvapoositkul, W., Vorayos, N., & Kiatsiriroat, T. (2002). Effect of the electrode arrangements in tube bank on the characteristic of electrohydrodynamic heat transfer enhancement: Low Reynolds number. *Journal of Enhanced Heat Transfer*, *9*(5-6), 229–242. doi:10.1080/10655130216024

Sheikholeslami. (2016). Influence of Coulomb forces on Fe3O4-H2O nanofluid thermal improvement. *International Journal of Hydrogen Energy, 42*, 821-829.

Sheikholeslami, M., & Bhatti, M. M. (2017). Active method for nanofluid heat transfer enhancement by means of EHD. *International Journal of Heat and Mass Transfer*, *109*, 115–122. doi:10.1016/j.ijheatmasstransfer.2017.01.115

Sheikholeslami, M., & Ellahi, R. (2015). Electrohydrodynamic nanofluid hydrothermal treatment in an enclosure with sinusoidal upper wall. *Appl. Sci.*, *5*(3), 294–306. doi:10.3390/app5030294

Sheikholeslami, M., & Ganji, D. D. (2017). Impact of electric field on nanofluid forced convection heat transfer with considering variable properties. *Journal of Molecular Liquids*, *229*, 566–573. doi:10.1016/j.molliq.2016.12.107

Sheikholeslami, M., & Ghasemi, A. (2018). Solidification heat transfer of nanofluid in existence of thermal radiation by means of FEM. *International Journal of Heat and Mass Transfer*, *123*, 418–431. doi:10.1016/j.ijheatmasstransfer.2018.02.095

Sheikholeslami, M., Hayat, T., Alsaedi, A., & Abelman, S. (2017). EHD nanofluid force convective heat transfer considering electric field dependent viscosity. *International Journal of Heat and Mass Transfer*, 108.

Sheikholeslami, M., & Rokni, H. B. (2017). Rokni, Influence of EFD viscosity on nanofluid forced convection in a cavity with sinusoidal wall. *Journal of Molecular Liquids*, *232*, 390–395. doi:10.1016/j.molliq.2017.02.042

Sheikholeslami, M., & Soheil Soleimani, D. D. (2016). Ganji, Effect of electric field on hydrothermal behavior of nanofluid in a complex geometry. *Journal of Molecular Liquids*, *213*, 153–161. doi:10.1016/j.molliq.2015.11.015

Sheikholeslami, M., & Vajravelu, K. (2017). Forced convection heat transfer in Fe3O4- ethylene glycol nanofluid under the influence of Coulomb force. *Journal of Molecular Liquids*, *233*, 203–210. doi:10.1016/j.molliq.2017.03.026

Shu, H. S., & Lai, F. C. (1995). Effect of Electrical Field on Buoyancy-Induced Flows in an Enclosure. *IEEE Industry Applications Society, 2*, 1465–1471.

ADDITIONAL READING

Jafaryar, M., Sheikholeslami, M., Li, M., & Moradi, R. (2018). Nanofluid turbulent flow in a pipe under the effect of twisted tape with alternate axis. *Journal of Thermal Analysis and Calorimetry*. doi:10.100710973-018-7093-2

Li, Z., Shehzad, S. A., & Sheikholeslami, M. (2018). An application of CVFEM for nanofluid heat transfer intensification in a porous sinusoidal cavity considering thermal non-equilibrium model. *Computer Methods in Applied Mechanics and Engineering*, *339*, 663–680. doi:10.1016/j.cma.2018.05.015

Li, Z., Sheikholeslami, M., Chamkha, A. J., Raizah, Z. A., & Saleem, S. (2018). Control Volume Finite Element Method for nanofluid MHD natural convective flow inside a sinusoidal annulus under the impact of thermal radiation. *Computer Methods in Applied Mechanics and Engineering*, *338*, 618–633. doi:10.1016/j.cma.2018.04.023

Li, Z., Sheikholeslami, M., Jafaryar, M., Shafee, A., & Chamkha, A. J. (2018). Investigation of nanofluid entropy generation in a heat exchanger with helical twisted tapes. *Journal of Molecular Liquids*, *266*, 797–805. doi:10.1016/j.molliq.2018.07.009

Li, Z., Sheikholeslami, M., Samandari, M., & Shafee, A. (2018). Nanofluid unsteady heat transfer in a porous energy storage enclosure in existence of Lorentz forces. *International Journal of Heat and Mass Transfer*, *127*, 914–926. doi:10.1016/j.ijheatmasstransfer.2018.06.101

Sheikholeslami, M. (2017a). Lattice Boltzmann Method simulation of MHD non-Darcy nanofluid free convection. *Physica B, Condensed Matter*, *516*, 55–71. doi:10.1016/j.physb.2017.04.029

Sheikholeslami, M. (2017b). CuO-water nanofluid free convection in a porous cavity considering Darcy law. *The European Physical Journal Plus*, *132*(1), 55. doi:10.1140/epjp/i2017-11330-3

Sheikholeslami, M. (2017c). Influence of Lorentz forces on nanofluid flow in a porous cavity by means of Non- Darcy model. *Engineering Computations*, *34*(8), 2651–2667. doi:10.1108/EC-01-2017-0008

Sheikholeslami, M. (2017d). Magnetic field influence on nanofluid thermal radiation in a cavity with tilted elliptic inner cylinder. *Journal of Molecular Liquids*, *229*, 137–147. doi:10.1016/j.molliq.2016.12.024

Sheikholeslami, M. (2017e). Numerical simulation of magnetic nanofluid natural convection in porous media. *Physics Letters. [Part A]*, *381*(5), 494–503. doi:10.1016/j.physleta.2016.11.042

Sheikholeslami, M. (2017f). Influence of Lorentz forces on nanofluid flow in a porous cylinder considering Darcy model. *Journal of Molecular Liquids*, *225*, 903–912. doi:10.1016/j.molliq.2016.11.022

Sheikholeslami, M. (2017g). Influence of Coulomb forces on Fe3O4-H2O nanofluid thermal improvement. *International Journal of Hydrogen Energy*, *42*(2), 821–829. doi:10.1016/j.ijhydene.2016.09.185

Sheikholeslami, M. (2017h). Numerical investigation of MHD nanofluid free convective heat transfer in a porous tilted enclosure. *Engineering Computations*, *34*(6), 1939–1955. doi:10.1108/EC-08-2016-0293

Sheikholeslami, M. (2017i). Magnetic field influence on CuO -H2O nanofluid convective flow in a permeable cavity considering various shapes for nanoparticles. *International Journal of Hydrogen Energy*, *42*(31), 19611–19621. doi:10.1016/j.ijhydene.2017.06.121

Sheikholeslami, M. (2018a). Magnetic source impact on nanofluid heat transfer using CVFEM. *Neural Computing & Applications*, *30*(4), 1055–1064. doi:10.100700521-016-2740-7

Sheikholeslami, M. (2018b). Numerical simulation for external magnetic field influence on Fe3O4-water nanofluid forced convection. *Engineering Computations*, *35*(4), 1639–1654. doi:10.1108/EC-06-2017-0200

Sheikholeslami, M. (2018c). Finite element method for PCM solidification in existence of CuO nanoparticles. *Journal of Molecular Liquids*, *265*, 347–355. doi:10.1016/j.molliq.2018.05.132

Sheikholeslami, M. (2018d). Solidification of NEPCM under the effect of magnetic field in a porous thermal energy storage enclosure using CuO nanoparticles. *Journal of Molecular Liquids*, *263*, 303–315. doi:10.1016/j.molliq.2018.04.144

Sheikholeslami, M. (2018e). Influence of magnetic field on Al2O3-H2O nanofluid forced convection heat transfer in a porous lid driven cavity with hot sphere obstacle by means of LBM. *Journal of Molecular Liquids*, *263*, 472–488. doi:10.1016/j.molliq.2018.04.111

Sheikholeslami, M. (2018f). Numerical simulation for solidification in a LHTESS by means of Nano-enhanced PCM. *Journal of the Taiwan Institute of Chemical Engineers*, *86*, 25–41. doi:10.1016/j.jtice.2018.03.013

Sheikholeslami, M. (2018g). Numerical modeling of Nano enhanced PCM solidification in an enclosure with metallic fin. *Journal of Molecular Liquids*, *259*, 424–438. doi:10.1016/j.molliq.2018.03.006

Sheikholeslami, M. (2018h). Numerical investigation of nanofluid free convection under the influence of electric field in a porous enclosure. *Journal of Molecular Liquids*, *249*, 1212–1221. doi:10.1016/j.molliq.2017.11.141

Sheikholeslami, M. (2018i). CuO-water nanofluid flow due to magnetic field inside a porous media considering Brownian motion. *Journal of Molecular Liquids*, *249*, 921–929. doi:10.1016/j.molliq.2017.11.118

Sheikholeslami, M. (2018j). Application of Darcy law for nanofluid flow in a porous cavity under the impact of Lorentz forces. *Journal of Molecular Liquids*, *266*, 495–503. doi:10.1016/j.molliq.2018.06.083

Sheikholeslami, M., Barzegar Gerdroodbary, M., Valiallah Mousavi, S., Ganji, D. D., & Moradi, R. (2018). Heat transfer enhancement of ferrofluid inside an 90o elbow channel by non-uniform magnetic field. *Journal of Magnetism and Magnetic Materials*, *460*, 302–311. doi:10.1016/j.jmmm.2018.03.070

Sheikholeslami, M., & Bhatti, M. M. (2017). Active method for nanofluid heat transfer enhancement by means of EHD. *International Journal of Heat and Mass Transfer*, *109*, 115–122. doi:10.1016/j.ijheatmasstransfer.2017.01.115

Sheikholeslami, M., Darzi, M., & Li, Z. (2018). Experimental investigation for entropy generation and exergy loss of nano-refrigerant condensation process. *International Journal of Heat and Mass Transfer*, *125*, 1087–1095. doi:10.1016/j.ijheatmasstransfer.2018.04.155

Sheikholeslami, M., Darzi, M., & Sadoughi, M. K. (2018). Heat transfer improvement and Pressure Drop during condensation of refrigerant-based Nanofluid; An Experimental Procedure. *International Journal of Heat and Mass Transfer*, *122*, 643–650. doi:10.1016/j.ijheatmasstransfer.2018.02.015

Sheikholeslami, M., & Ghasemi, A. (2018). Solidification heat transfer of nanofluid in existence of thermal radiation by means of FEM. *International Journal of Heat and Mass Transfer*, *123*, 418–431. doi:10.1016/j.ijheatmasstransfer.2018.02.095

Sheikholeslami, M., Ghasemi, A., Li, Z., Shafee, A., & Saleem, S. (2018). Influence of CuO nanoparticles on heat transfer behavior of PCM in solidification process considering radiative source term. *International Journal of Heat and Mass Transfer*, *126*, 1252–1264. doi:10.1016/j.ijheatmasstransfer.2018.05.116

Sheikholeslami, M., Hayat, T., & Alsaedi, A. (2018). Numerical simulation for forced convection flow of MHD CuO-H2O nanofluid inside a cavity by means of LBM. *Journal of Molecular Liquids*, *249*, 941–948. doi:10.1016/j.molliq.2017.10.099

Sheikholeslami, M., Hayat, T., Muhammad, T., & Alsaedi, A. (2018). MHD forced convection flow of nanofluid in a porous cavity with hot elliptic obstacle by means of Lattice Boltzmann method. *International Journal of Mechanical Sciences*, *135*, 532–540. doi:10.1016/j.ijmecsci.2017.12.005

Sheikholeslami, M., Jafaryar, M., Ganji, D. D., & Li, Z. (2018). Exergy loss analysis for nanofluid forced convection heat transfer in a pipe with modified turbulators. *Journal of Molecular Liquids*, *262*, 104–110. doi:10.1016/j.molliq.2018.04.077

Sheikholeslami, M., Jafaryar, M., & Li, Z. (2018a). Second law analysis for nanofluid turbulent flow inside a circular duct in presence of twisted tape turbulators. *Journal of Molecular Liquids*, *263*, 489–500. doi:10.1016/j.molliq.2018.04.147

Sheikholeslami, M., Jafaryar, M., & Li, Z. (2018b). Nanofluid turbulent convective flow in a circular duct with helical turbulators considering CuO nanoparticles. *International Journal of Heat and Mass Transfer*, *124*, 980–989. doi:10.1016/j.ijheatmasstransfer.2018.04.022

Sheikholeslami, M., Jafaryar, M., Saleem, S., Li, Z., Shafee, A., & Jiang, Y. (2018). Nanofluid heat transfer augmentation and exergy loss inside a pipe equipped with innovative turbulators. *International Journal of Heat and Mass Transfer*, *126*, 156–163. doi:10.1016/j.ijheatmasstransfer.2018.05.128

Sheikholeslami, M., Jafaryar, M., Shafee, A., & Li, Z. (2018). Investigation of second law and hydrothermal behavior of nanofluid through a tube using passive methods. *Journal of Molecular Liquids*, *269*, 407–416. doi:10.1016/j.molliq.2018.08.019

Sheikholeslami, M., Li, Z., & Shafee, A. (2018a). Lorentz forces effect on NEPCM heat transfer during solidification in a porous energy storage system. *International Journal of Heat and Mass Transfer*, *127*, 665–674. doi:10.1016/j.ijheatmasstransfer.2018.06.087

Sheikholeslami, M., Li, Z., & Shamlooei, M. (2018). Nanofluid MHD natural convection through a porous complex shaped cavity considering thermal radiation. *Physics Letters. [Part A]*, *382*(24), 1615–1632. doi:10.1016/j.physleta.2018.04.006

Sheikholeslami, M., & Rokni, H. B. (2017). Simulation of nanofluid heat transfer in presence of magnetic field: A review. *International Journal of Heat and Mass Transfer*, *115*, 1203–1233. doi:10.1016/j.ijheatmasstransfer.2017.08.108

Sheikholeslami, M., & Rokni, H. B. (2018a). CVFEM for effect of Lorentz forces on nanofluid flow in a porous complex shaped enclosure by means of Non-equilibrium model. *Journal of Molecular Liquids*, *254*, 446–462. doi:10.1016/j.molliq.2018.01.130

Sheikholeslami, M., Rokni, H.B. (2018b). Magnetic nanofluid flow and convective heat transfer in a porous cavity considering Brownian motion effects. *Physics of Fluids*, *30*(1), doi:10.1063/1.5012517

Sheikholeslami, M., & Sadoughi, M. K. (2017). Mesoscopic method for MHD nanofluid flow inside a porous cavity considering various shapes of nanoparticles. *International Journal of Heat and Mass Transfer*, *113*, 106–114. doi:10.1016/j.ijheatmasstransfer.2017.05.054

Sheikholeslami, M., & Sadoughi, M. K. (2018). Simulation of CuO- water nanofluid heat transfer enhancement in presence of melting surface. *International Journal of Heat and Mass Transfer*, *116*, 909–919. doi:10.1016/j.ijheatmasstransfer.2017.09.086

Sheikholeslami, M., & Seyednezhad, M. (2018). Simulation of nanofluid flow and natural convection in a porous media under the influence of electric field using CVFEM. *International Journal of Heat and Mass Transfer*, *120*, 772–781. doi:10.1016/j.ijheatmasstransfer.2017.12.087

Sheikholeslami, M., Shafee, A., Ramzan, M., & Li, Z. (2018). Investigation of Lorentz forces and radiation impacts on nanofluid treatment in a porous semi annulus via Darcy law. *Journal of Molecular Liquids*, *272*, 8–14. doi:10.1016/j.molliq.2018.09.016

Sheikholeslami, M., Shamlooei, M., & Moradi, R. (2018). Numerical simulation for heat transfer intensification of nanofluid in a porous curved enclosure considering shape effect of Fe3O4 nanoparticles. *Chemical Engineering & Processing: Process Intensification*, *124*, 71–82. doi:10.1016/j.cep.2017.12.005

Sheikholeslami, M., & Shehzad, S. A. (2017a). Thermal radiation of ferrofluid in existence of Lorentz forces considering variable viscosity. *International Journal of Heat and Mass Transfer*, *109*, 82–92. doi:10.1016/j.ijheatmasstransfer.2017.01.096

Sheikholeslami, M., & Shehzad, S. A. (2017b). Magnetohydrodynamic nanofluid convective flow in a porous enclosure by means of LBM. *International Journal of Heat and Mass Transfer*, *113*, 796–805. doi:10.1016/j.ijheatmasstransfer.2017.05.130

Sheikholeslami, M., & Shehzad, S. A. (2018a). Numerical analysis of Fe3O4 –H2O nanofluid flow in permeable media under the effect of external magnetic source. *International Journal of Heat and Mass Transfer, 118*, 182–192. doi:10.1016/j.ijheatmasstransfer.2017.10.113

Sheikholeslami, M., & Shehzad, S. A. (2018b). CVFEM simulation for nanofluid migration in a porous medium using Darcy model. *International Journal of Heat and Mass Transfer, 122*, 1264–1271. doi:10.1016/j.ijheatmasstransfer.2018.02.080

Sheikholeslami, M., & Shehzad, S. A. (2018c). Simulation of water based nanofluid convective flow inside a porous enclosure via Non-equilibrium model. *International Journal of Heat and Mass Transfer, 120*, 1200–1212. doi:10.1016/j.ijheatmasstransfer.2017.12.132

Sheikholeslami, M., & Shehzad, S. A. (2018d). Non-Darcy free convection of Fe3O4-water nanoliquid in a complex shaped enclosure under impact of uniform Lorentz force. *The Chinese Journal of Physiology, 56*(1), 270–281. doi:10.1016/j.cjph.2017.12.017

Sheikholeslami, M., Shehzad, S. A., Abbasi, F. M., & Li, Z. (2018). Nanofluid flow and forced convection heat transfer due to Lorentz forces in a porous lid driven cubic enclosure with hot obstacle. *Computer Methods in Applied Mechanics and Engineering, 338*, 491–505. doi:10.1016/j.cma.2018.04.020

Sheikholeslami, M., Shehzad, S. A., & Li, Z. (2018a). Nanofluid heat transfer intensification in a permeable channel due to magnetic field using Lattice Boltzmann method. *Physica B, Condensed Matter, 542*, 51–58. doi:10.1016/j.physb.2018.03.036

Sheikholeslami, M., Shehzad, S. A., & Li, Z. (2018b). Water based nanofluid free convection heat transfer in a three dimensional porous cavity with hot sphere obstacle in existence of Lorenz forces. *International Journal of Heat and Mass Transfer, 125*, 375–386. doi:10.1016/j.ijheatmasstransfer.2018.04.076

Sheikholeslami, M., Shehzad, S. A., Li, Z., & Shafee, A. (2018). Numerical modeling for Alumina nanofluid magnetohydrodynamic convective heat transfer in a permeable medium using Darcy law. *International Journal of Heat and Mass Transfer, 127*, 614–622. doi:10.1016/j.ijheatmasstransfer.2018.07.013

Sheikholeslami, M., & Vajravelu, K. (2017). Forced convection heat transfer in Fe3O4- ethylene glycol nanofluid under the influence of Coulomb force. *Journal of Molecular Liquids, 233*, 203–210. doi:10.1016/j.molliq.2017.03.026

Sheikholeslami, M., & Zeeshan, A. (2017). Analysis of flow and heat transfer in water based nanofluid due to magnetic field in a porous enclosure with constant heat flux using CVFEM. *Computer Methods in Applied Mechanics and Engineering, 320*, 68–81. doi:10.1016/j.cma.2017.03.024

Sheikholeslami, M., Zeeshan, A., & Majeed, A. (2018). Control volume based finite element simulation of magnetic nanofluid flow and heat transport in non-Darcy medium. *Journal of Molecular Liquids, 268*, 354–364. doi:10.1016/j.molliq.2018.07.031

Chapter 8
Nanofluid Treatment in Existence of Magnetic Field Using Non-Darcy Model for Porous Media

ABSTRACT

In this chapter, the non-Darcy model is employed for porous media filled with nanofluid. Both natural and forced convection heat transfer can be analyzed with this model. The governing equations in forms of vorticity stream function are derived and then they are solved via control volume-based finite element method (CVFEM). The effect of Darcy number on nanofluid flow and heat transfer is examined.

1. INTRODUCTION

The knowledge of free or forced convection heat transfer inside geometries of irregular shape (for example, wavy channel and pipe bend) for porous media has many significant engineering applications; for example, geothermal engineering, solar-collectors, performance of cold storage, and thermal insulation of buildings. A considerable number of published articles are available that deal with flow characteristics, heat transfer, flow and heat transfer instability, transition to turbulence, design aspects, etc. For non-Darcy porous medium, Kumar and Gupta (2003) reported the flow and thermal fields' characteristics in wavy cavities. Sheikholeslami (2017) investigated MHD nanofluid free convective heat transfer in a porous tilted enclosure by means of non-Darcy porous medium. Sheikholeslami and Ganji (2017a) studied the magnetic nanofluid flow in a porous cavity using CuO nanoparticles. Sheikholeslami and Ganji (2017b) investigated the nanofluid transportation in porous media under the influence of external magnetic source. Sheikholeslami and Rokni (2017) reported nanofluid convective heat transfer intensification in a porous circular cylinder. Sheikholeslami and Shamlooei (2017) utilized CVFEM for convective flow of nanofluid inside a lid driven porous cavity. Sheikholeslami and Seyednezhad (2017) simulated the nanofluid heat transfer in a permeable enclosure in presence of variable magnetic field. Sheikholeslami

DOI: 10.4018/978-1-5225-7595-5.ch008

(2017) demonstrated the influence of Lorentz forces on nanofluid flow in a porous cavity by means of Non- Darcy model. Sheikholeslami and Zeeshan (2017) presented the numerical simulation of Fe3O4 -water nanofluid flow in a non-Darcy porous media. Nanofluid flows in various mediums were studied in recent years.

2. MHD NANOFLUID FREE CONVECTIVE HEAT TRANSFER IN A POROUS TILTED ENCLOSURE

2.1. Problem Definition

Figure 1 illustrates the important geometric parameters of current geometry. Also sample mesh is presented. The inner and outer cylinders are considered as hot and cold walls, respectively. Horizontal magnetic field has been considered.

2.2. Governing Equation

2D steady convective flow of nanofluid in a porous media is considered in existence of constant magnetic field. The PDEs equations are:

$$\frac{\partial v}{\partial y}+\frac{\partial u}{\partial x}=0 \tag{1}$$

$$\begin{aligned}&\frac{\mu_{nf}}{\rho_{nf}}\left(\frac{\partial^2 u}{\partial x^2}+\frac{\partial^2 u}{\partial y^2}\right)-\frac{1}{\rho_{nf}}\frac{\partial P}{\partial x}-\frac{1}{\rho_{nf}}\frac{\mu_{nf}}{K}u-\left(T_c-T\right)\beta_{nf}g\sin\gamma\\&+\sigma_{nf}B_0^2\left[-u\left(\sin\lambda\right)^2+v\left(\sin\lambda\right)\left(\cos\lambda\right)\right]=v\frac{\partial u}{\partial y}+u\frac{\partial u}{\partial x}\end{aligned} \tag{2}$$

$$\begin{aligned}&\frac{\mu_{nf}}{\rho_{nf}}\left(\frac{\partial^2 v}{\partial x^2}+\frac{\partial^2 v}{\partial y^2}\right)-\left(T_c-T\right)\beta_{nf}g\cos\gamma-\frac{\partial P}{\partial y}\frac{1}{\rho_{nf}}-\frac{1}{\rho_{nf}}\frac{\mu_{nf}}{K}v\\&+\sigma_{nf}B_0^2\left[-v\left(\cos\lambda\right)^2+u\left(\sin\lambda\right)\left(\cos\lambda\right)\right]=v\frac{\partial v}{\partial y}+u\frac{\partial v}{\partial x}\end{aligned} \tag{3}$$

$$\left(\rho C_p\right)_{nf}\left(v\frac{\partial T}{\partial y}+u\frac{\partial T}{\partial x}\right)=k_{nf}\left(\frac{\partial^2 T}{\partial x^2}+\frac{\partial^2 T}{\partial y^2}\right) \tag{4}$$

Figure 1. (a)Geometry and the boundary conditions with (b) the mesh of half-annulus enclosure considered in this work; (c) a sample triangular element and its corresponding control volume.

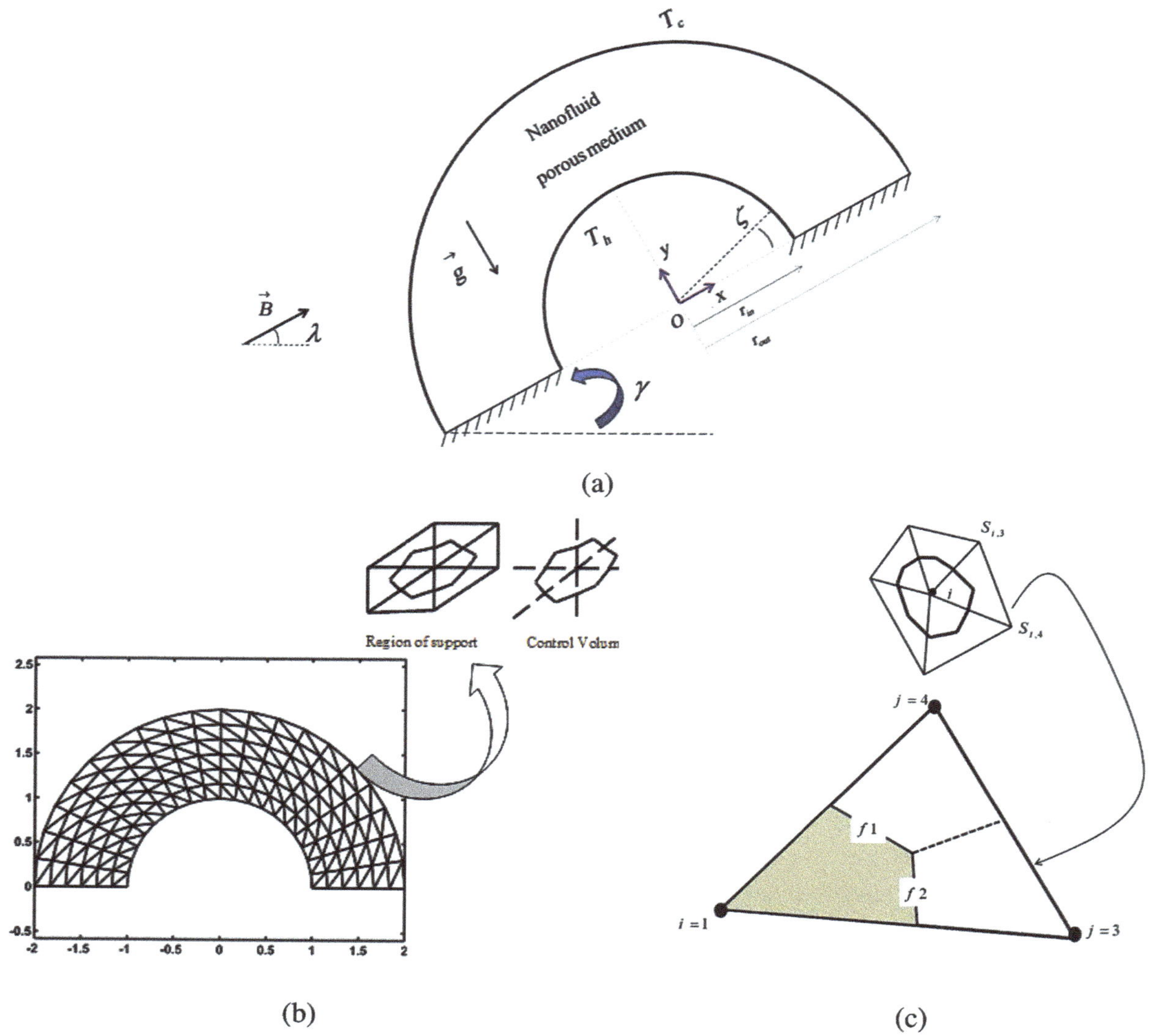

$\left(\rho C_p\right)_{nf}$, $\left(\rho\beta\right)_{nf}$, ρ_{nf} and σ_{nf} are defined as:

$$\left(\rho C_p\right)_{nf} = \left(\rho C_p\right)_f (1-\phi) + \left(\rho C_p\right)_s \phi \tag{5}$$

$$\left(\rho\beta\right)_{nf} = \left(\rho\beta\right)_f (1-\phi) + \left(\rho\beta\right)_s \phi, \tag{6}$$

$$\rho_{nf} = \rho_f(1-\phi) + \rho_s\phi \tag{7}$$

$$\frac{\sigma_{nf}}{\sigma_f} = 1 + \frac{3\left(\frac{\sigma_s}{\sigma_f} - 1\right)\phi}{\left(\frac{\sigma_s}{\sigma_f} + 2\right) - \left(\frac{\sigma_s}{\sigma_f} - 1\right)\phi} \tag{8}$$

k_{nf}, μ_{nf} are obtained according to Koo–Kleinstreuer–Li (KKL) model:

$$\begin{aligned}
&k_{nf} = k_{static} + k_{Brownian}\\
&\frac{k_{static}}{k_f} = 1 + \frac{3\left(\frac{k_p}{k_f} - 1\right)\phi}{\left(\frac{k_p}{k_f} + 2\right) - \left(\frac{k_p}{k_f} - 1\right)\phi}\\
&\frac{k_{Brownian}}{k_f} = 5\times10^4 g'(\phi, T, d_p)\phi\rho_f c_{p,f}\sqrt{\frac{\kappa_b T}{\rho_p d_p}}\\
&g'\left(\phi, T, d_p\right) = \left(a_1 + a_2 Ln\left(d_p\right) + a_3 Ln\left(\phi\right) + a_4 Ln\left(\phi\right)\ln\left(d_p\right) + a_5 Ln\left(d_p\right)^2\right) Ln\left(T\right)\\
&+\left(a_6 + a_7 Ln\left(d_p\right) + a_8 Ln\left(\phi\right) + a_9 \ln\left(d_p\right) Ln\left(\phi\right) + a_{10} Ln\left(d_p\right)^2\right)\\
&R_f = d_p / k_{p,eff} - d_p / k_p, R_f = 4\times10^{-8} km^2 / W
\end{aligned} \tag{9}$$

$$\mu_{nf} = \frac{\mu_f}{\left(1-\phi\right)^{2.5}} + \frac{k_{Brownian}}{k_f} \times \frac{\mu_f}{\Pr} \tag{10}$$

All required coefficients and properties are illustrated in Tables 1 and 2.

Vorticity and stream function should be used to eliminate pressure source terms:

$$\omega + \frac{\partial u}{\partial y} - \frac{\partial v}{\partial x} = 0, \frac{\partial \psi}{\partial x} = -v, \frac{\partial \psi}{\partial y} = u \tag{11}$$

Introducing dimensionless quantities:

Table 1. The coefficient values of CuO – Water nanofluid

Coefficient Values	CuO – Water
a_1	-26.5933108
a_2	-0.403818333
a_3	-33.3516805
a_4	-1.915825591
a_5	6.421858E-02
a_6	48.40336955
a_7	-9.787756683
a_8	190.245610009
a_9	10.9285386565
a_{10}	-0.72009983664

Table 2. Thermo physical properties of water and nanoparticles

	$\rho(kg / m^3)$	$C_p(j / kgk)$	$k(W / m.k)$	$\beta \times 10^5(K^{-1})$	$d_p(nm)$	$\sigma(\Omega \cdot m)^{-1}$
Water	997.1	4179	0.613	21	-	0.05
CuO	6500	540	18	29	45	10^{-10}

$$U = \frac{uL}{\alpha_{nf}}, V = \frac{vL}{\alpha_{nf}}, \theta = \frac{T - T_c}{\Delta T}, \Delta T = T_h - T_c, (X, Y) = \frac{(x, y)}{L}, \quad \Psi = \frac{\psi}{\alpha_{nf}}, \Omega = \frac{\omega L^2}{\alpha_{nf}} \tag{12}$$

The final formulae are:

$$\frac{\partial^2 \Psi}{\partial Y^2} + \frac{\partial^2 \Psi}{\partial X^2} = -\Omega, \tag{13}$$

$$U\frac{\partial \Omega}{\partial X}+\frac{\partial \Omega}{\partial Y}V=\Pr\frac{A_5}{A_1}\frac{A_2}{A_4}\left(\frac{\partial^2 \Omega}{\partial Y^2}+\frac{\partial^2 \Omega}{\partial X^2}\right)$$
$$+\Pr Ha^2\frac{A_6}{A_1}\frac{A_2}{A_4}\left(\frac{\partial U}{\partial X}\cos\lambda\sin\lambda-\frac{\partial V}{\partial X}\left(\cos\lambda\right)^2+\frac{\partial U}{\partial Y}\left(\sin\lambda\right)^2-\frac{\partial V}{\partial Y}\cos\lambda\sin\lambda\right) \quad (14)$$
$$+\Pr Ra\frac{A_3A_2^2}{A_1A_4^2}\left(\frac{\partial \theta}{\partial X}\cos\gamma-\frac{\partial \theta}{\partial Y}\sin\gamma\right)-\frac{\Pr}{Da}\frac{A_5}{A_1}\frac{A_2}{A_4}\Omega,$$

$$\frac{\partial \theta}{\partial X}U+\frac{\partial \theta}{\partial Y}V=\left(\frac{\partial^2 \theta}{\partial X^2}+\frac{\partial^2 \theta}{\partial Y^2}\right) \quad (15)$$

where dimensionless and constants parameters are defined as:

$$\Pr=\upsilon_f/\alpha_f, Ra=g\left(\rho\beta\right)_f\Delta TL^3/\left(\mu_f\alpha_f\right), Ha=LB_0\sqrt{\sigma_f/\mu_f}$$
$$A_1=\frac{\rho_{nf}}{\rho_f}, A_2=\frac{\left(\rho C_p\right)_{nf}}{\left(\rho C_p\right)_f}, A_3=\frac{\left(\rho\beta\right)_{nf}}{\left(\rho\beta\right)_f}, \quad (16)$$
$$A_4=\frac{k_{nf}}{k_f}, A_5=\frac{\mu_{nf}}{\mu_f}, A_6=\frac{\sigma_{nf}}{\sigma_f}$$

and boundary conditions are:

$\theta=1.0$ on inner wall

$\theta=0.0$ on outer wall

$\frac{\partial \theta}{\partial n}=0.0$ on other walls

$\Psi=0.0$ on all walls (17)

Local and average Nusselt over the hot wall can be calculated as:

$$Nu_{loc}=A_4\frac{\partial \theta}{\partial r} \quad (18)$$

$$Nu_{ave} = \frac{1}{0.5\pi} \int_{0}^{0.5\pi} Nu_{loc}(\zeta)\, d\zeta \tag{19}$$

2.3. Effects of Active Parameters

In this paper, magnetohydrodynamic nanofluid flow and convective heat transfer in a porous tilted annulus is investigated. CVFEM is utilized to obtain the outputs for various values of Hartmann number ($Ha = 0$ to 40), Rayleigh number ($Ra = 10^3, 10^4$ and 10^5), tilted angle ($\gamma = 0^\circ$ to 90°), Darcy number ($Da = 0.01$ to 100) and volume fraction of CuO ($\phi = 0\%$ and 4%).

Figures 2 and 3 demonstrate the influences of Hartmann, Rayleigh, Darcy numbers and tilted angle on hydrothermal behavior. At $\gamma = 0^\circ$, in conduction mode, there are two vortexes in streamline which are rotates in reverse direction. As buoyancy forces increases, vortexes become stronger and their centers move upward. Then thermal plume appears at $\xi = 90^\circ$. Appling magnetic field reduces the strength of vortexes and thermal plume. As tilted angle augments, the convective mode becomes less than before.

Figure 2a. Isotherms (down) and streamlines (up) contours for different values of Rayleigh number and Hartmann number when $Da = 0.01$

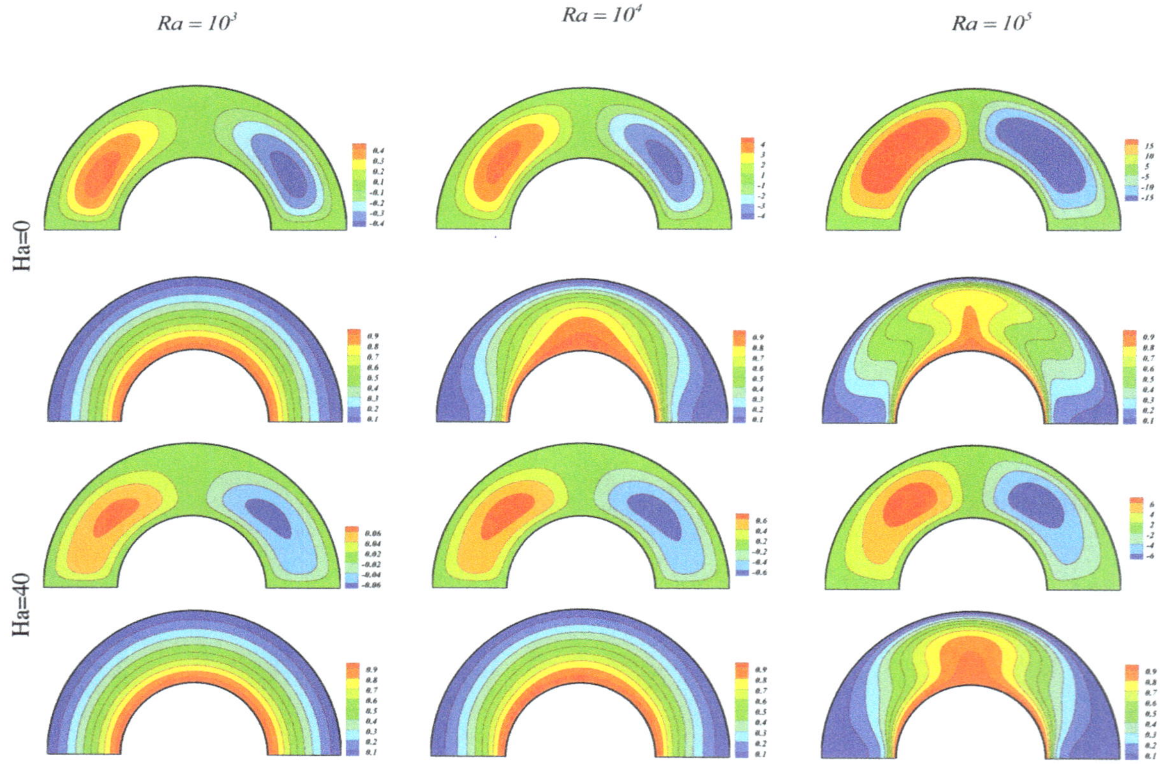

Figure 2b. Isotherms (down) and streamlines (up) contours for different values of Rayleigh number and Hartmann number when $Da = 0.01$

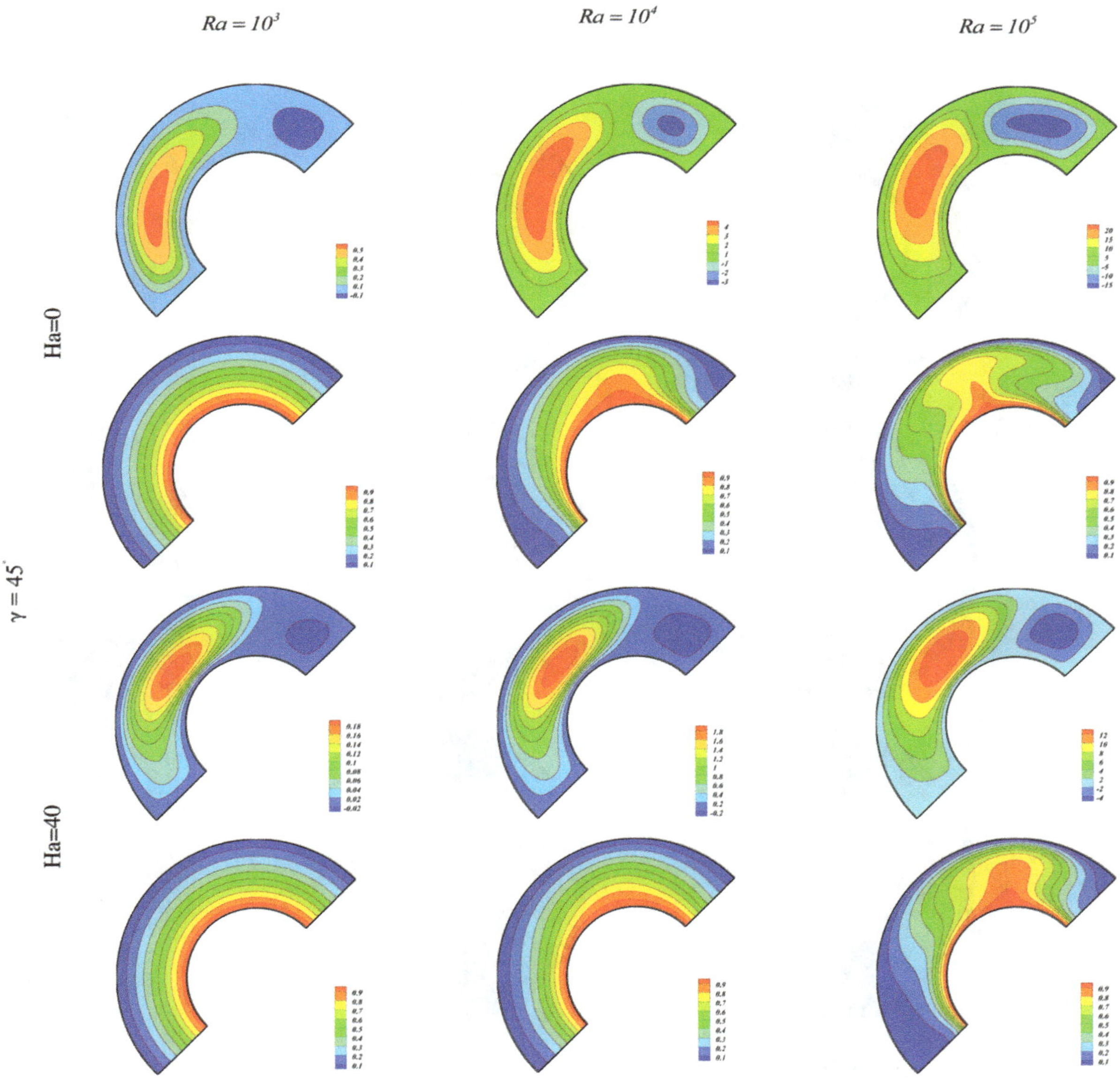

At $\gamma = 45^{\circ}$, the main vortexes convert to new two vortexes in which the upper one rotates clock wise. As Rayleigh number enhances the primary vortexes become stronger and thermal plume appears in the region between two vortexes. At $\gamma = 90^{\circ}$, only one vortex exists in absence of magnetic field. As buoyancy forces enhances, the main vortex convert to two vortexes and thermal plume generates. Increasing Darcy number makes the convective heat transfer to enhance. It is interesting observation that in pres-

Figure 2c. Isotherms (down) and streamlines (up) contours for different values of Rayleigh number and Hartmann number when $Da = 0.01$

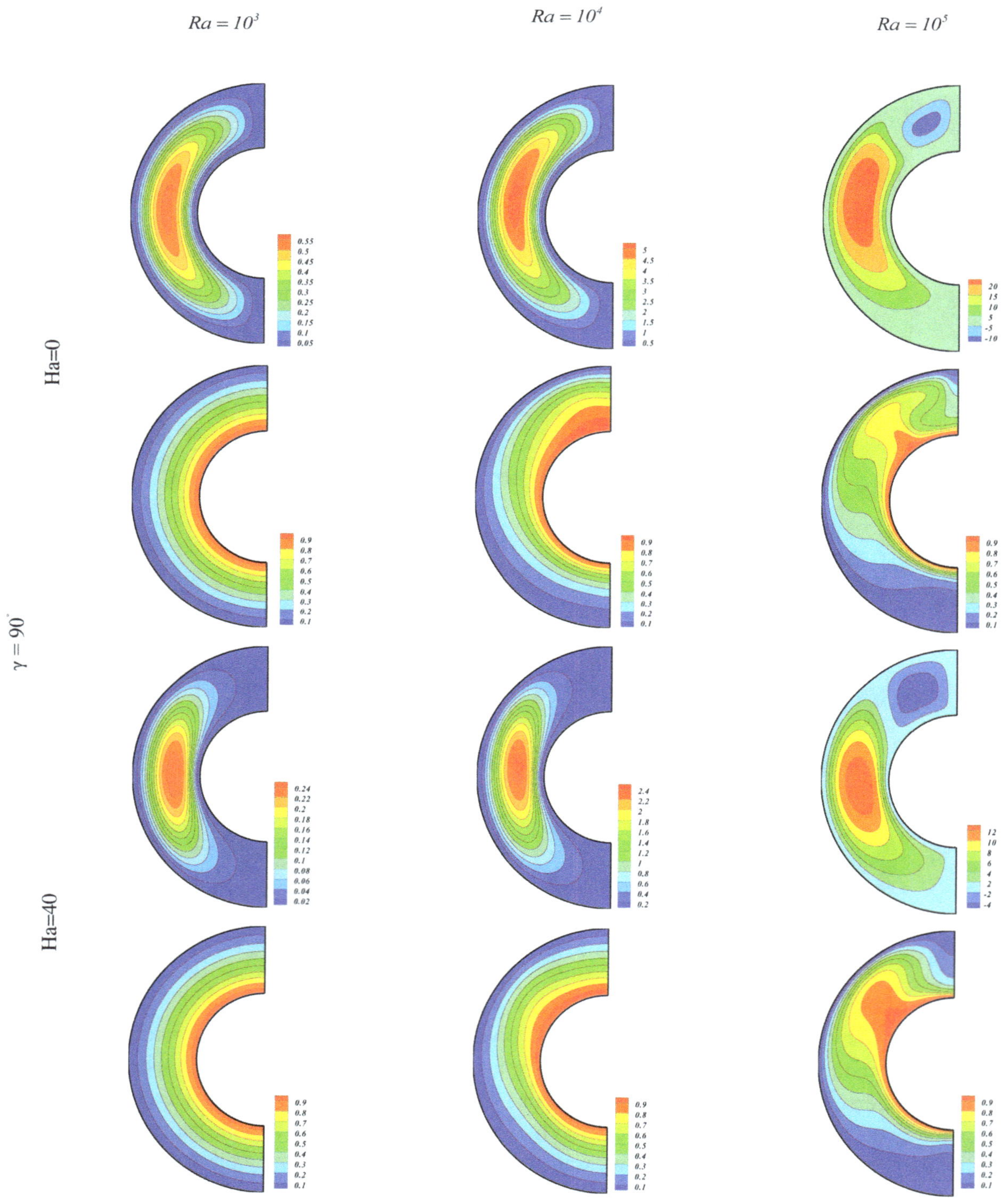

Figure 3a. Isotherms (down) and streamlines (up) contours for different values of Rayleigh number and Hartmann number when $Da = 100$

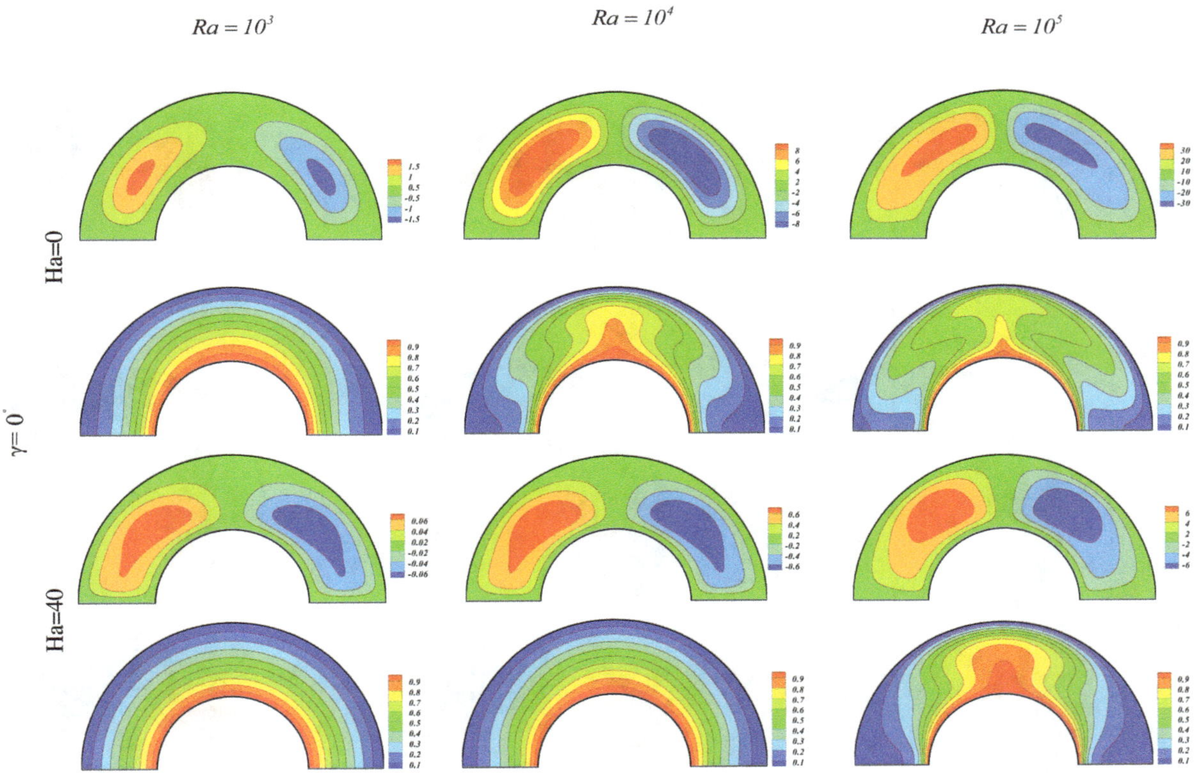

ence of magnetic field at high values of *Ra* and *Da*, two thermal plumes generates at upper region due to existence three rotating vortexes.

Impacts of significant parameters on Nu_{loc} and Nu_{ave} are illustrated in Figures 4 and 5. The correlation for Nu_{ave} is as follows:

$$\begin{aligned} Nu_{ave} &= 5.54 + 0.29\gamma - 0.47Da^{*} - 3.18\log\left(Ra\right) + 0.02Ha^{*} \\ &-0.02\gamma Da^{*} - 0.086\gamma Ha^{*} + 0.014\gamma\log\left(Ra\right) + 0.2Da^{*}Ha^{*} - 0.13\log\left(Ra\right)Ha^{*} \\ &-0.039\gamma^{2} - 0.015\left(Da^{*}\right)^{2} + 0.53\left(\log\left(Ra\right)\right)^{2} + 0.097\left(Ha^{*}\right)^{2} \end{aligned} \tag{20}$$

where $Ha^{*} = 0.1Ha, Da^{*} = 0.01Da$. Due to symmetric geometry and boundary conditions, Nu_{loc} profiles are symmetric respect to $\zeta = 90^{\circ}$ when $\gamma = 0^{\circ}$. Nusselt number enhances with augment of Darcy and Rayleigh numbers. Rate of heat transfer reduces with increase of tilted angle. Lorenz forces have reverse effect on Nusselt number due to increase in thermal boundary layer thickness with augment of Hartmann number.

Figure 3b. Isotherms (down) and streamlines (up) contours for different values of Rayleigh number and Hartmann number when $Da = 100$

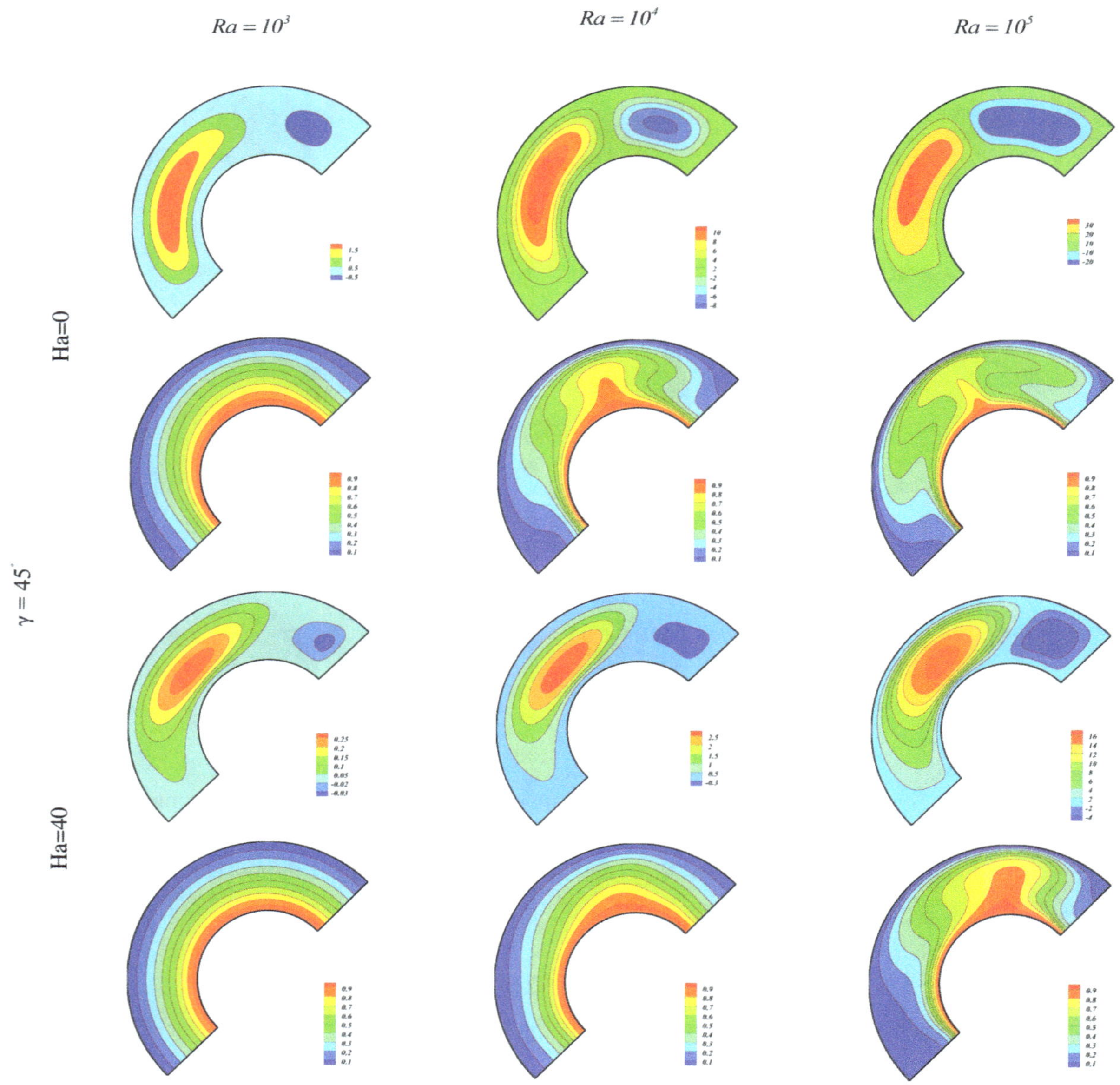

3. MAGNETIC NANOFLUID FLOW IN A POROUS CAVITY USING CUO NANOPARTICLES

3.1. Problem Definition

Figure 6 demonstrates the important geometric parameters of current geometry. Also sample mesh is presented. Constant heat flux is entered from inner wall. The outer wall is cold and

Figure 3c. Isotherms (down) and streamlines (up) contours for different values of Rayleigh number and Hartmann number when $Da = 100$

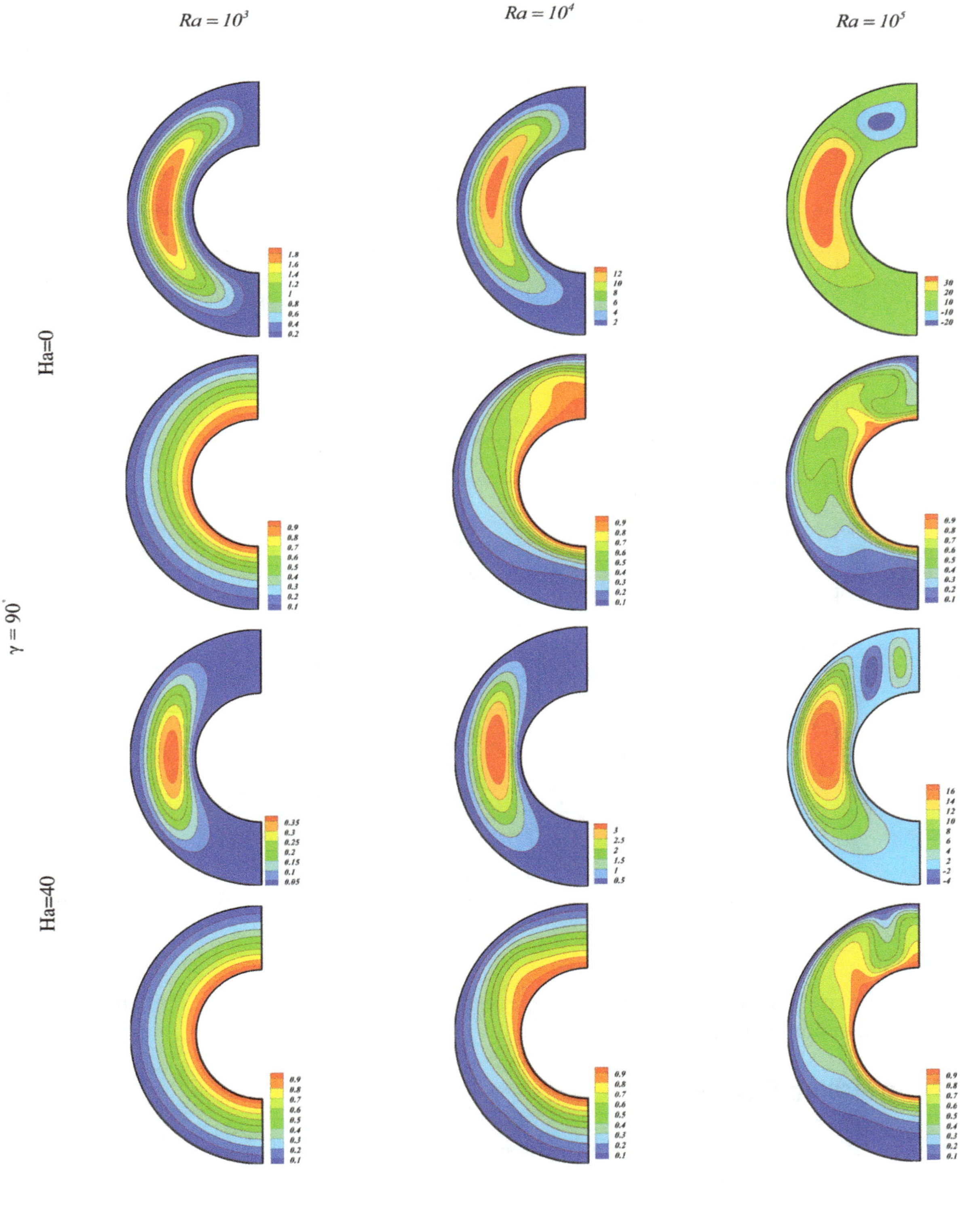

Figure 4. Effects of the Hartmann number, Rayleigh number and tilted angle on Local Nusselt number

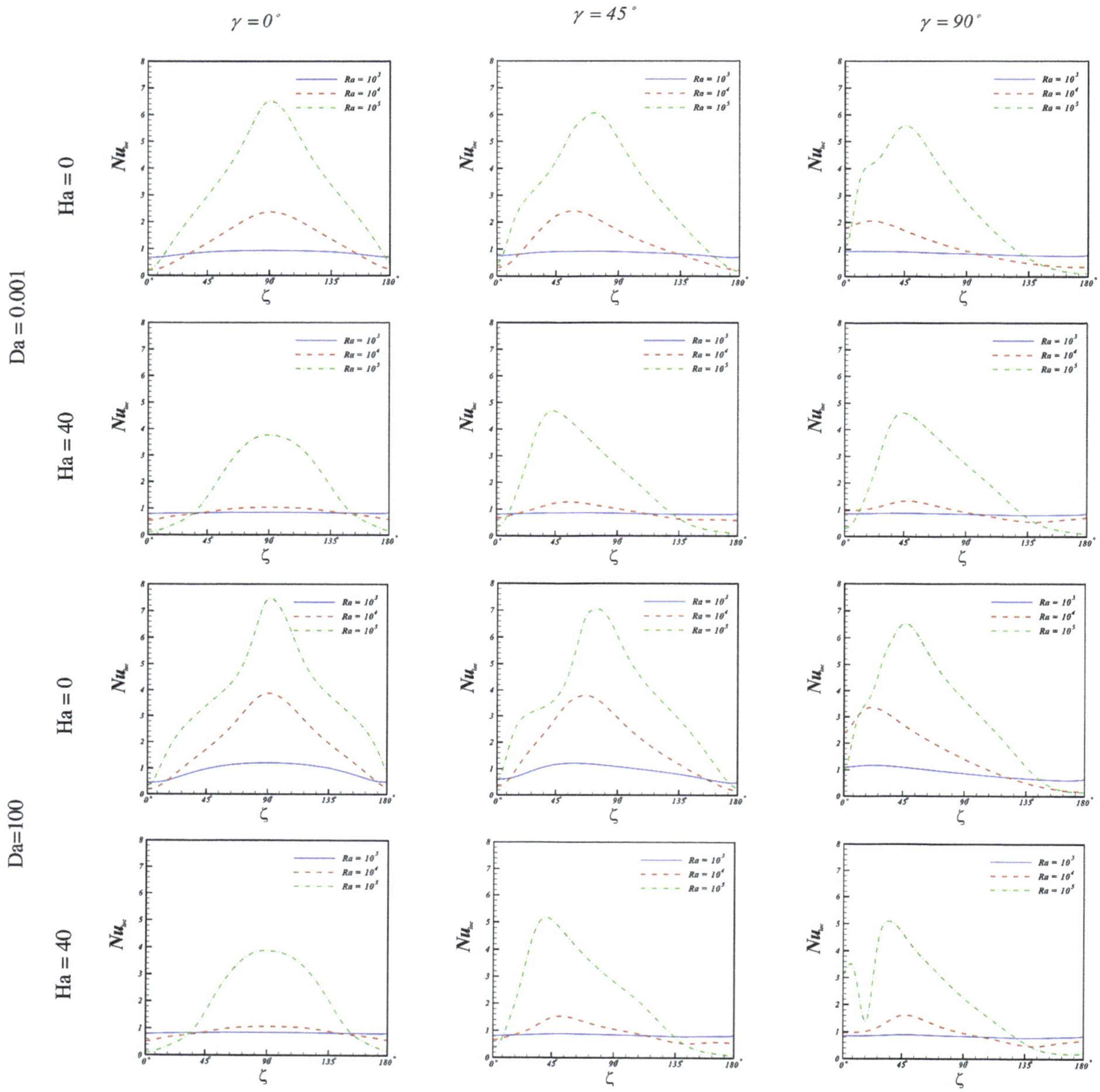

other walls are adiabatic. Horizontal magnetic field is taken into account. Radiation effect is considered in porous medium.

3.2. Governing Equation

Free convective MHD nanofluid flow in a porous media is considered. Non-Darcy model is used for porous media. The PDEs equations are:

Figure 5a. Effects of the Hartmann number, Rayleigh number and tilted angle on average Nusselt number

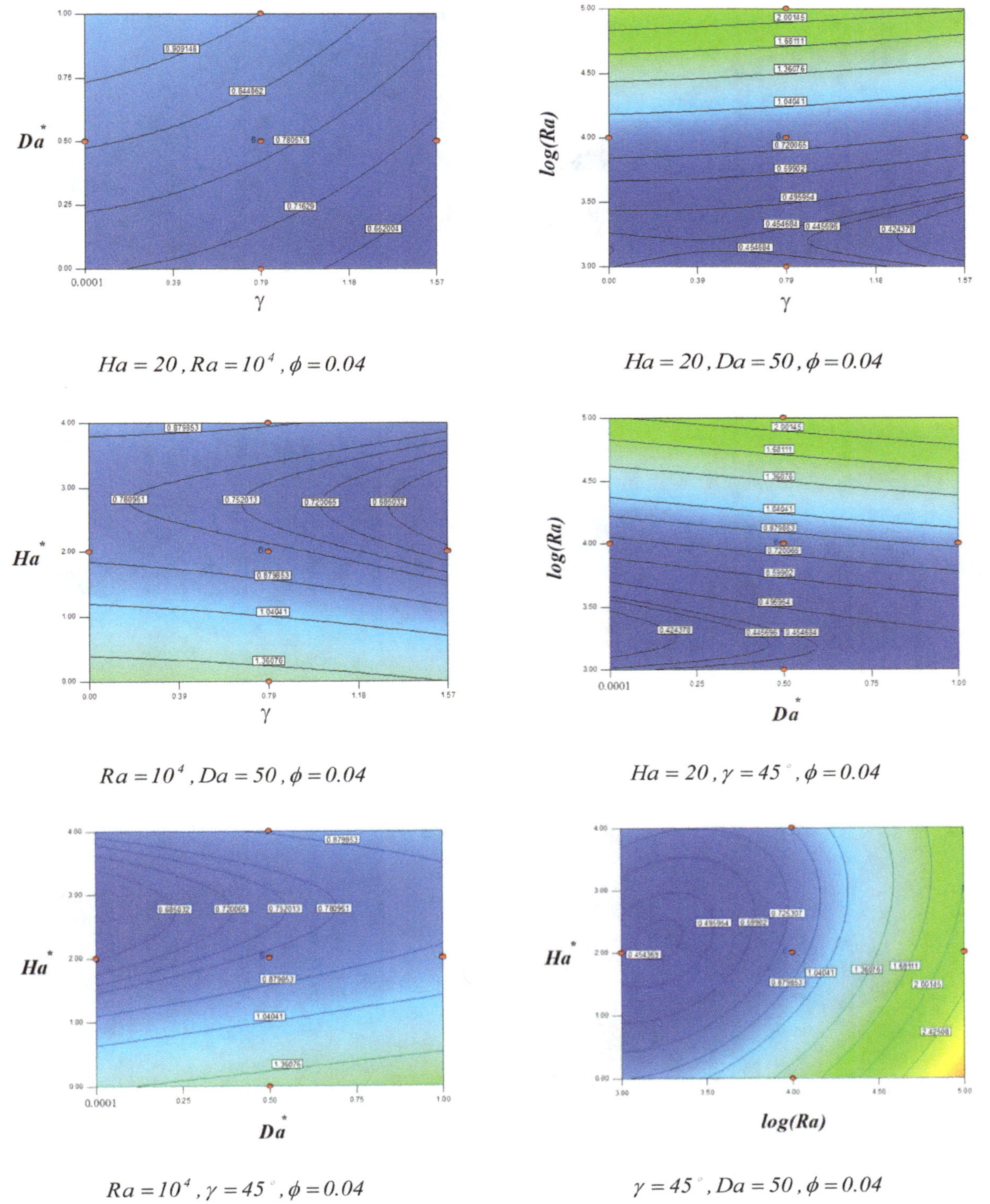

Figure 5b. Effects of the Hartmann number, Rayleigh number and tilted angle on average Nusselt number

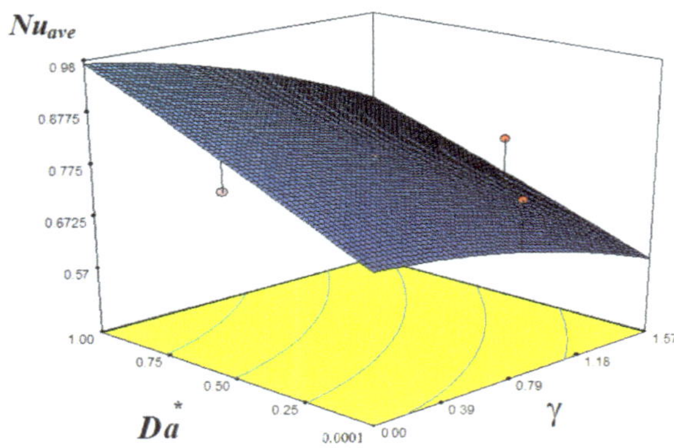

$Ha = 20, Ra = 10^4, \phi = 0.04$

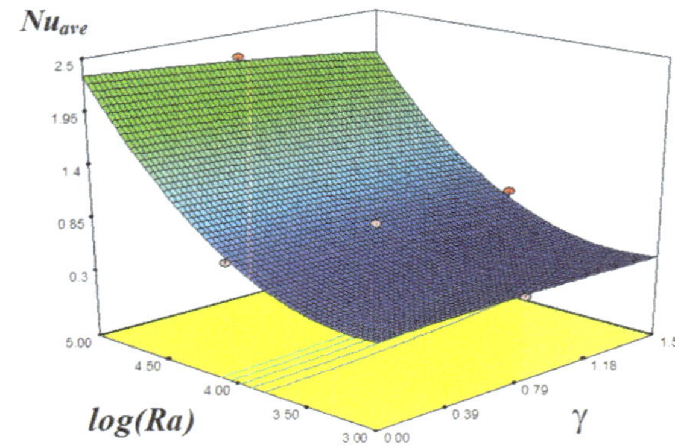

$Ha = 20, Da = 50, \phi = 0.04$

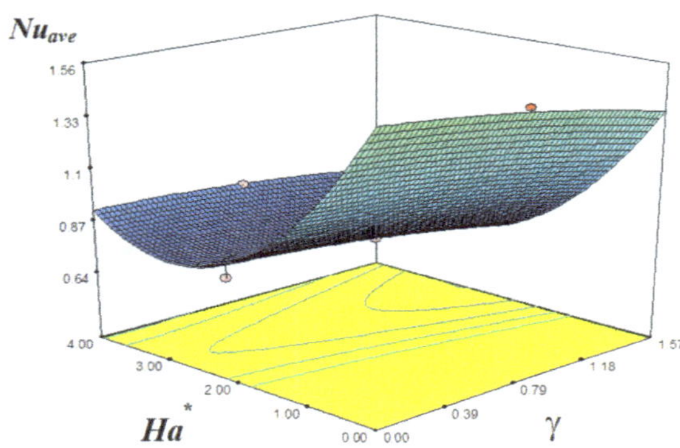

$Ra = 10^4, Da = 50, \phi = 0.04$

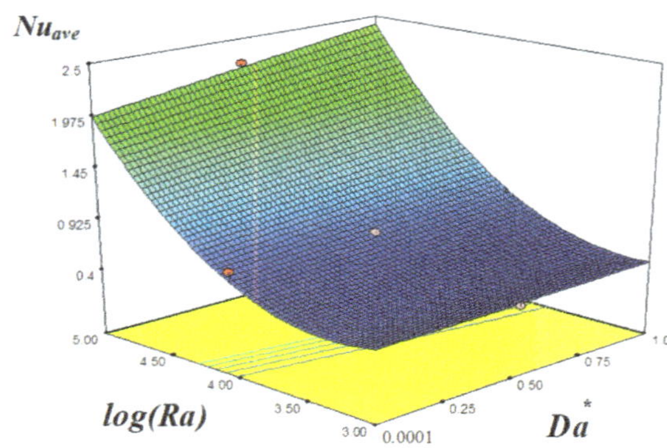

$Ha = 20, \gamma = 45^{\circ}, \phi = 0.04$

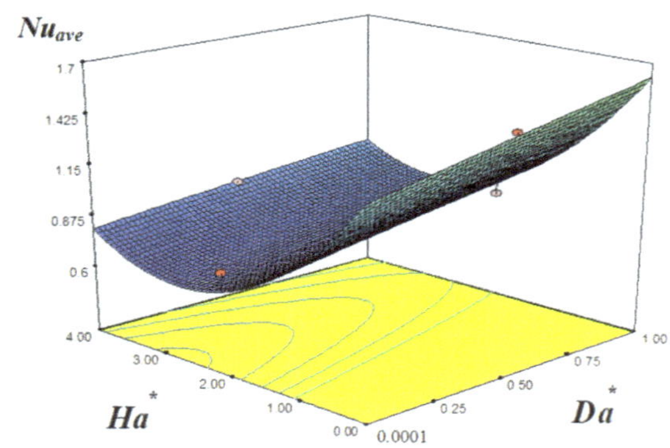

$Ra = 10^4, \gamma = 45^{\circ}, \phi = 0.04$

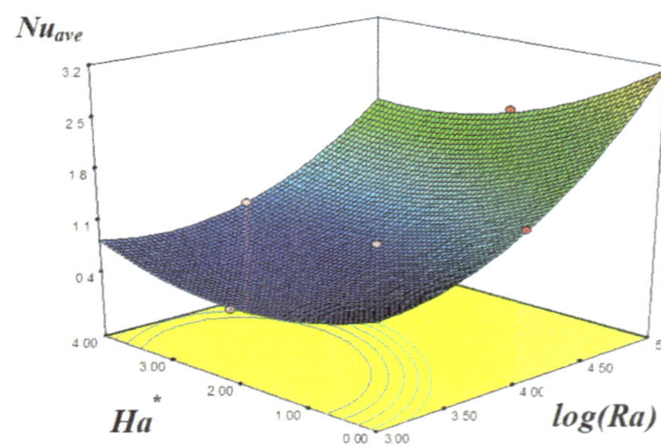

$\gamma = 45^{\circ}, Da = 50, \phi = 0.04$

Figure 6. (a) Geometry and the boundary conditions with (b) the mesh of geometry considered in this work; (c) a sample triangular element and its corresponding control volume

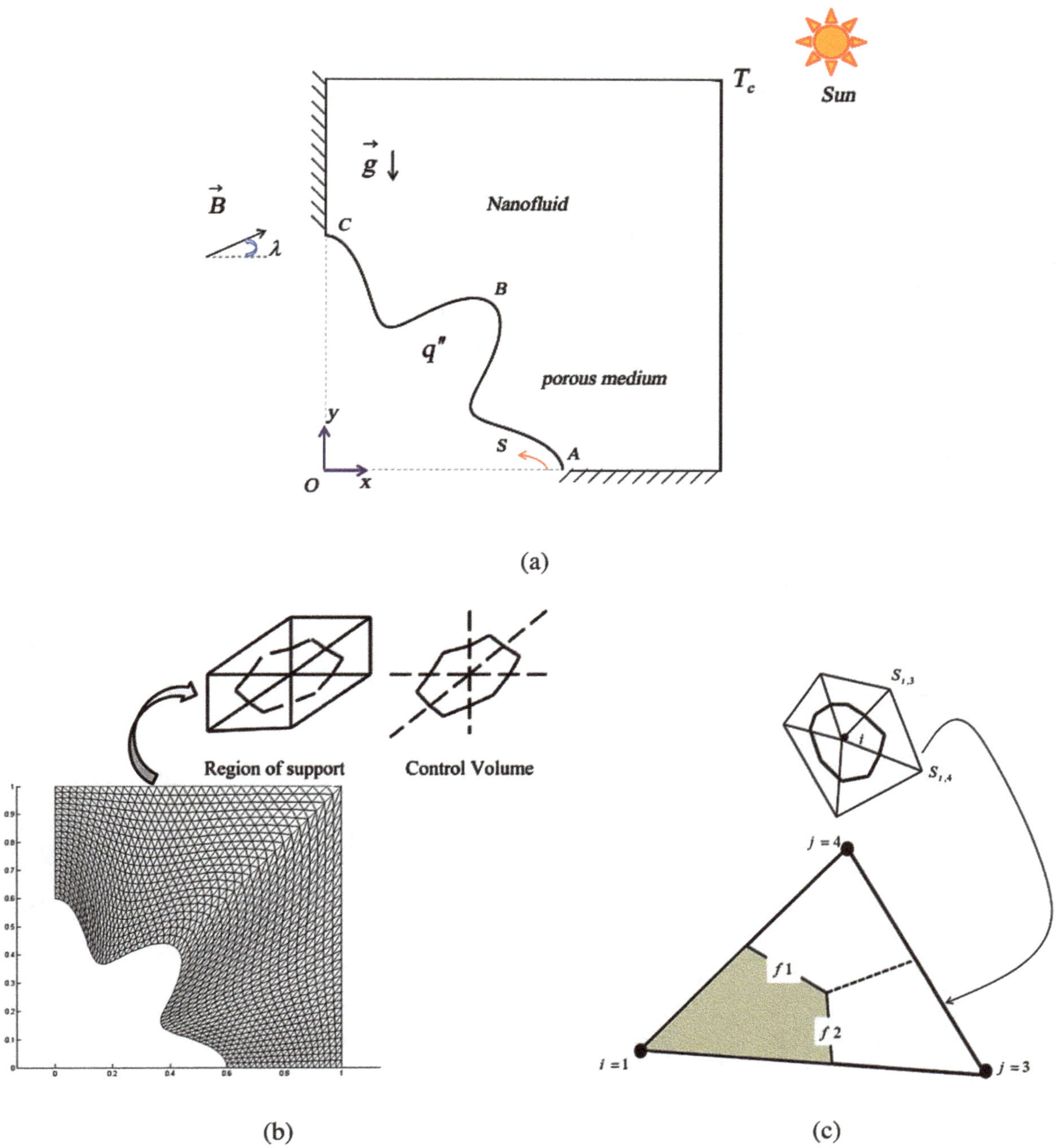

$$\frac{\partial v}{\partial y}+\frac{\partial u}{\partial x}=0 \tag{21}$$

$$\mu_{nf}\left(\frac{\partial^2 u}{\partial x^2}+\frac{\partial^2 u}{\partial y^2}\right)-\frac{\partial P}{\partial x}-\frac{\mu_{nf}}{K}u+\sigma_{nf}B_0^2\left[-u\left(\sin\lambda\right)^2+v\left(\sin\lambda\right)\left(\cos\lambda\right)\right]=\rho_{nf}\left(v\frac{\partial u}{\partial y}+u\frac{\partial u}{\partial x}\right) \tag{22}$$

$$\begin{gathered}\mu_{nf}\left(\frac{\partial^2 v}{\partial x^2}+\frac{\partial^2 v}{\partial y^2}\right)-\left(T_c-T\right)\left(\rho\beta\right)_{nf}g-\frac{\partial P}{\partial y}-\frac{\mu_{nf}}{K}v\\+\sigma_{nf}B_0^2\left[-v\left(\cos\lambda\right)^2+u\left(\sin\lambda\right)\left(\cos\lambda\right)\right]=\rho_{nf}\left(v\frac{\partial v}{\partial y}+u\frac{\partial v}{\partial x}\right)\end{gathered} \tag{23}$$

$$\left(\rho C_p\right)_{nf}\left(v\frac{\partial T}{\partial y}+u\frac{\partial T}{\partial x}\right)=k_{nf}\left(\frac{\partial^2 T}{\partial x^2}+\frac{\partial^2 T}{\partial y^2}\right)-\frac{\partial q_r}{\partial y},\left[q_r=-\frac{4\sigma_e}{3\beta_R}\frac{\partial T^4}{\partial y},T^4\cong 4T_c^3T-3T_c^4\right] \tag{24}$$

$\left(\rho C_p\right)_{nf},\left(\rho\beta\right)_{nf}$, ρ_{nf} and σ_{nf} are defined as:

$$\left(\rho C_p\right)_{nf}=\left(\rho C_p\right)_f(1-\phi)+\left(\rho C_p\right)_s\phi \tag{25}$$

$$\left(\rho\beta\right)_{nf}=\left(\rho\beta\right)_f(1-\phi)+\left(\rho\beta\right)_s\phi, \tag{26}$$

$$\rho_{nf}=\rho_f(1-\phi)+\rho_s\phi \tag{27}$$

$$\frac{\sigma_{nf}}{\sigma_f}=1+\frac{3\left(\frac{\sigma_s}{\sigma_f}-1\right)\phi}{\left(\frac{\sigma_s}{\sigma_f}+2\right)-\left(\frac{\sigma_s}{\sigma_f}-1\right)\phi} \tag{28}$$

k_{nf},μ_{nf} are obtained according to Koo–Kleinstreuer–Li (KKL) model:

$$k_{nf} = k_{static} + k_{Brownian}$$
$$\frac{k_{static}}{k_f} = 1 + \frac{3\left(\frac{k_p}{k_f} - 1\right)\phi}{\left(\frac{k_p}{k_f} + 2\right) - \left(\frac{k_p}{k_f} - 1\right)\phi}$$
$$\frac{k_{Brownian}}{k_f} = 5 \times 10^4 g'(\phi, T, d_p)\phi\rho_f c_{p,f}\sqrt{\frac{\kappa_b T}{\rho_p d_p}} \quad (29)$$
$$g'\left(\phi, T, d_p\right) = \left(a_1 + a_2 Ln\left(d_p\right) + a_3 Ln\left(\phi\right) + a_4 Ln\left(\phi\right)\ln\left(d_p\right) + a_5 Ln\left(d_p\right)^2\right) Ln\left(T\right)$$
$$+\left(a_6 + a_7 Ln\left(d_p\right) + a_8 Ln\left(\phi\right) + a_9 \ln\left(d_p\right) Ln\left(\phi\right) + a_{10} Ln\left(d_p\right)^2\right)$$
$$R_f = d_p / k_{p,eff} - d_p / k_p, R_f = 4 \times 10^{-8} km^2 / W$$

$$\mu_{nf} = \frac{\mu_f}{\left(1-\phi\right)^{2.5}} + \frac{k_{Brownian}}{k_f} \times \frac{\mu_f}{\Pr} \quad (30)$$

All required coefficients and properties are illustrated in Tables 1 and 2.

Vorticity and stream function should be used to eliminate pressure source terms:

$$\omega + \frac{\partial u}{\partial y} - \frac{\partial v}{\partial x} = 0, \frac{\partial \psi}{\partial x} = -v, \frac{\partial \psi}{\partial y} = u \quad (31)$$

Introducing dimensionless quantities:

$$U = \frac{uL}{\alpha_{nf}}, V = \frac{vL}{\alpha_{nf}}, \theta = \frac{T - T_c}{\Delta T}, \Delta T = q'' L / k_f, \left(X, Y\right) = \frac{\left(x, y\right)}{L}$$
$$, \Psi = \frac{\psi}{\alpha_{nf}}, \Omega = \frac{\omega L^2}{\alpha_{nf}} \quad (32)$$

The final formulae are:

$$\frac{\partial^2 \Psi}{\partial Y^2} + \frac{\partial^2 \Psi}{\partial X^2} = -\Omega, \quad (33)$$

$$
\begin{aligned}
& U\frac{\partial \Omega}{\partial X}+\frac{\partial \Omega}{\partial Y}V=\Pr\frac{A_5}{A_1}\frac{A_2}{A_4}\left(\frac{\partial^2 \Omega}{\partial Y^2}+\frac{\partial^2 \Omega}{\partial X^2}\right) \\
& +\Pr Ha^2\frac{A_6}{A_1}\frac{A_2}{A_4}\left(\frac{\partial U}{\partial X}\cos\lambda\sin\lambda-\frac{\partial V}{\partial X}\left(\cos\lambda\right)^2+\frac{\partial U}{\partial Y}\left(\sin\lambda\right)^2-\frac{\partial V}{\partial Y}\cos\lambda\sin\lambda\right) \\
& +\Pr Ra\frac{A_3 A_2^2}{A_1 A_4^2}\left(\frac{\partial \theta}{\partial X}\right)-\frac{\Pr}{Da}\frac{A_5}{A_1}\frac{A_2}{A_4}\Omega,
\end{aligned}
\tag{34}
$$

$$
\frac{\partial \theta}{\partial X}U+\frac{\partial \theta}{\partial Y}V=\left(\frac{\partial^2 \theta}{\partial X^2}+\frac{\partial^2 \theta}{\partial Y^2}\right)+\frac{4}{3}\frac{1}{A_4}Rd\frac{\partial^2 \theta}{\partial Y^2}
\tag{35}
$$

where dimensionless and constants parameters are defined as:

$$
\begin{aligned}
& \Pr=\upsilon_f/\alpha_f, Ra=g\left(\rho\beta\right)_f\Delta T L^3/\left(\mu_f\alpha_f\right), Ha=LB_0\sqrt{\sigma_f/\mu_f}, Rd=4\sigma_e T_c^3/\left(\beta_R k_f\right) \\
& A_1=\frac{\rho_{nf}}{\rho_f}, A_2=\frac{\left(\rho C_p\right)_{nf}}{\left(\rho C_p\right)_f}, A_3=\frac{\left(\rho\beta\right)_{nf}}{\left(\rho\beta\right)_f}, \\
& A_4=\frac{k_{nf}}{k_f}, A_5=\frac{\mu_{nf}}{\mu_f}, A_6=\frac{\sigma_{nf}}{\sigma_f}
\end{aligned}
\tag{36}
$$

and boundary conditions are:

$$\frac{\partial \theta}{\partial n}=1.0 \text{ on inner wall}$$

$$\theta=0.0 \text{ on outer wall}$$

$$\frac{\partial \theta}{\partial n}=0.0 \text{ on other walls}$$

$$\Psi=0.0 \text{ on all walls} \tag{37}$$

Local and average Nusselt number over the hot wall can be calculated as:

$$
Nu_{loc}=\frac{1}{\theta}\left(\frac{k_{nf}}{k_f}\right)\left(1+\frac{4}{3}Rd\left(\frac{k_{nf}}{k_f}\right)^{-1}\right)
\tag{38}
$$

$$Nu_{ave} = \frac{1}{S}\int_{0}^{s} Nu_{loc}\, ds \tag{39}$$

3.3. Effects of Active Parameters

Non-Darcy model is applied for natural convection of nanofluid in a porous enclosure. Influence of thermal radiation and magnetic field are taken into account. KKL model is utilized for estimating viscosity and thermal conductivity of CuO-water nanofluid. The numerical procedure is conducted by means of CVFEM. Effects of Darcy number ($Da = 0.01$ to 100), Radiation parameter ($Rd = 0$ to 0.8), Rayleigh number ($Ra = 10^3$ to 10^5), Hartmann number ($Ha = 0$ to 40) and volume fraction of nanofluid ($\phi = 0$ to 0.04) are examined.

Figure 7 demonstrates the effect of adding nanoparticle in to the base fluid on hydrothermal behavior. Thermal boundary layer thickness reduces with adding nanoparticles. So, rate of heat transfer en-

Figure 7. Influence of nanofluid volume fraction on streamlines (left) and isotherms (right) contours (nanofluid ($\phi = 0.04$)(—) and pure fluid($\phi = 0$) (- - -)) when $Ha = 40, Da = 100, Rd = 0.8$

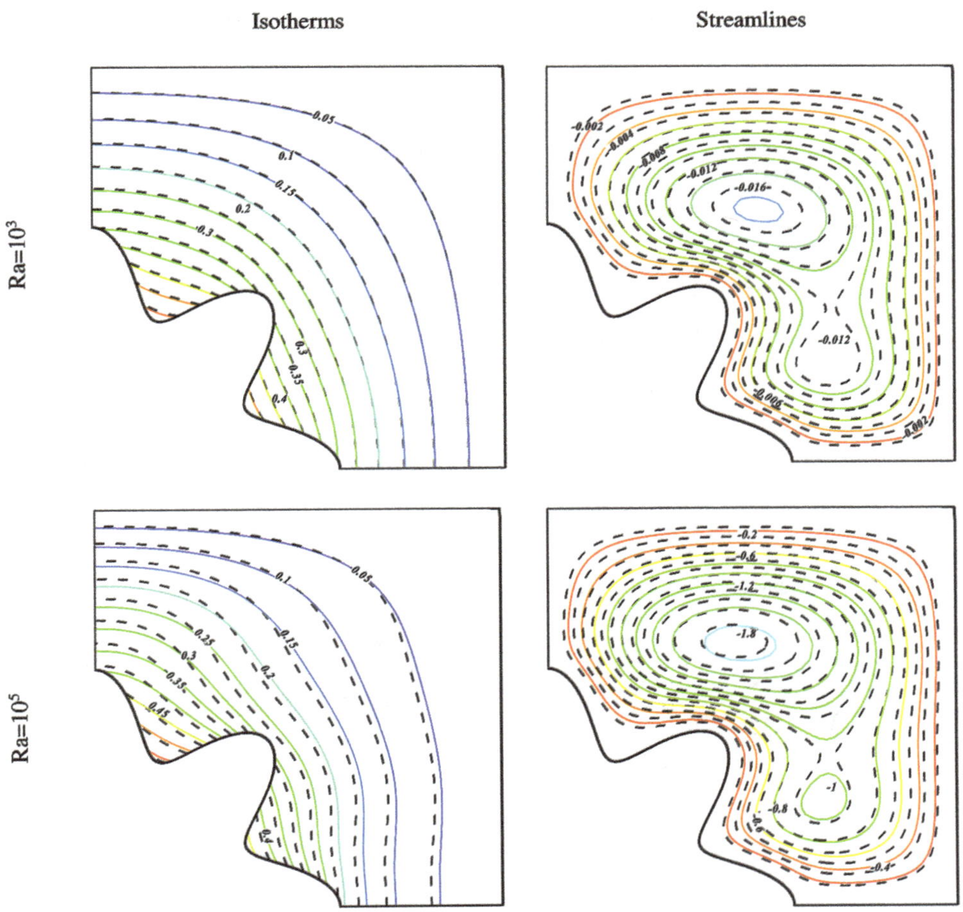

hances with augment of volume fraction of nanofluid. Also nanofluid velocity is greater than based fluid velocity due to increment in nanoparticles motion. Besides, influence of adding nanoparticles is more sensible in presence of magnetic field. Influence of radiation parameter on nanofluid flow and heat transfer is shown in Figure 8. As radiation parameter augments, thermal boundary layer thickness enhances. $|\Psi_{max}|$ augments with increase of radiation parameter.

Impacts of Darcy, Hartmann and Rayleigh numbers on hydrothermal behavior of nanofluid are reported in Figures 9, 10 and 11. In low Darcy and Rayleigh numbers, the conduction mechanism can be seen, so isotherms follow the shape of cylinders. As Rayleigh number increases, buoyancy forces enhance the convection heat transfer. So isotherms become more disturb with augment of Ra. Also $|\Psi_{max}|$ increases with rise of Ra. As Darcy number increase the permeability of medium augments and convective mechanism enhances. So rate of heat transfer and absolute values of stream function enhance with rise of Da. As magnetic field increases, Lorentz forces generate and these forces reduce the velocity of nanofluid. Also rate of heat transfer reduces with rise of Hartmann number.

Figure 8. Influence of radiation parameter on streamlines (left) and isotherms (right) contours (nanofluid ($Rd = 0.8$)(—) and pure fluid($Rd = 0$) ($- \cdot -$)) when $Ra = 10^5, Ha = 0$

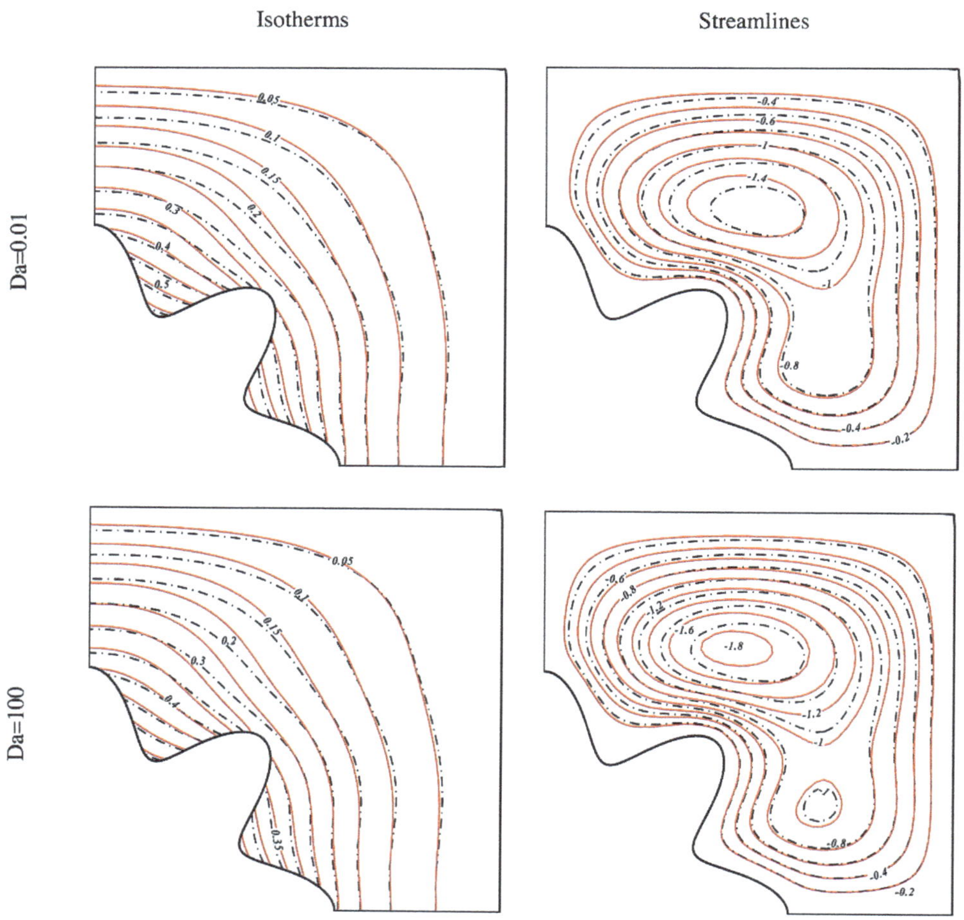

Figure 9. Isotherms (left) and streamlines (right) contours for different values of Darcy and Hartmann numbers when $Ra = 10^3, \phi = 0.04, Rd = 0.8$.

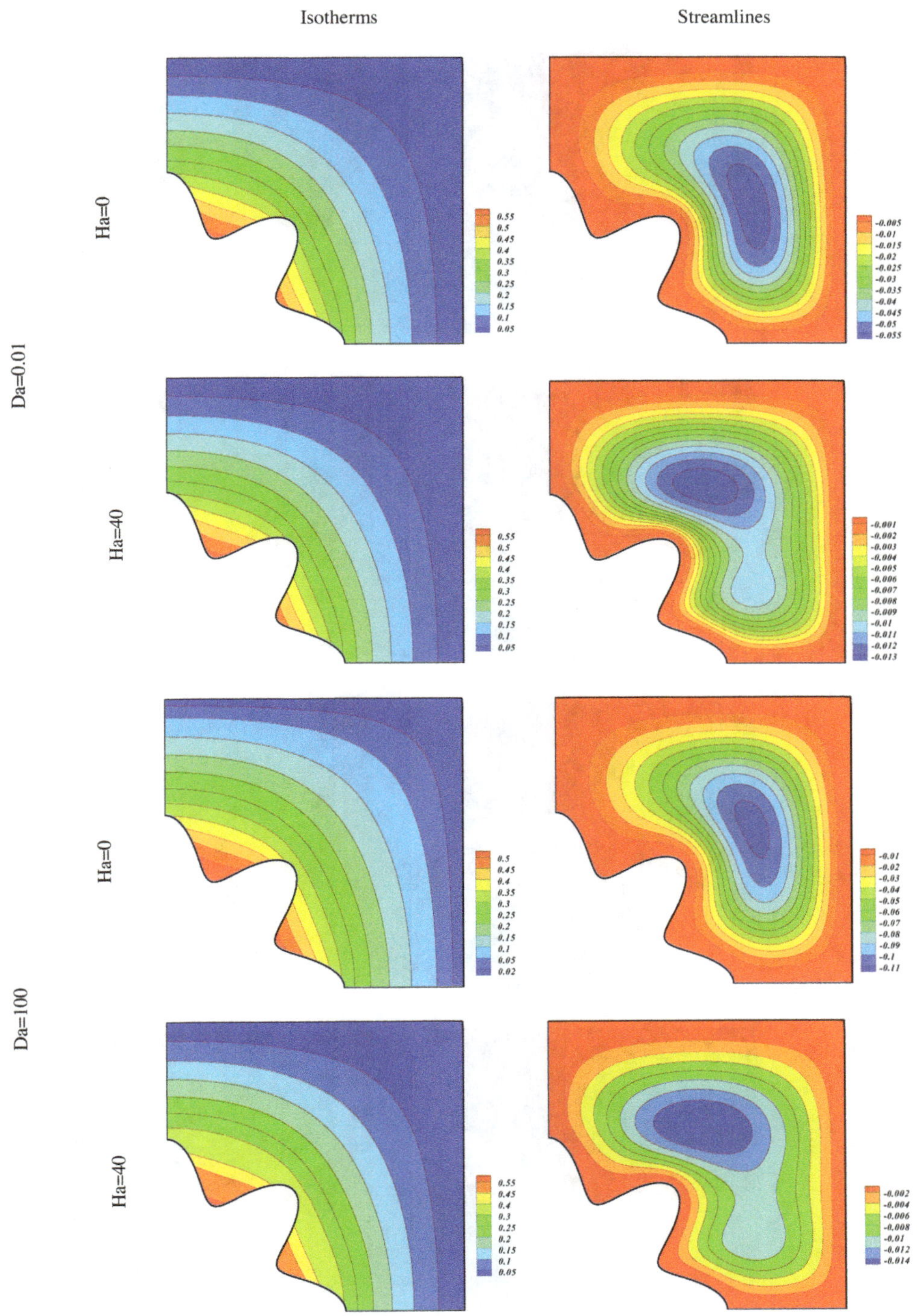

Figure 10. Isotherms (left) and streamlines (right) contours for different values of Darcy and Hartmann numbers when $Ra = 10^4, \phi = 0.04, Rd = 0.8$.

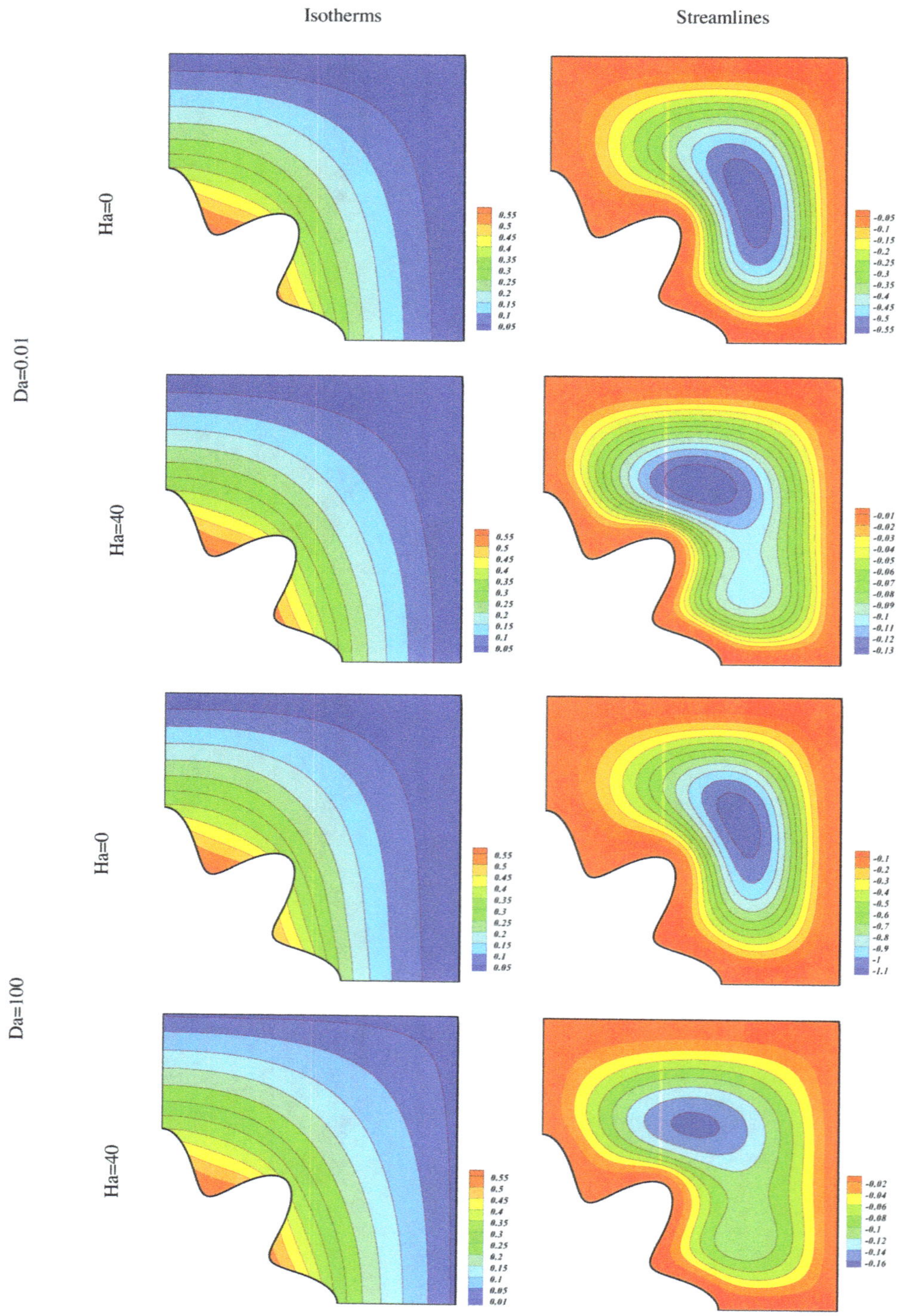

Figure 11. Isotherms (left) and streamlines (right) contours for different values of Darcy and Hartmann numbers when $Ra = 10^5, \phi = 0.04, Rd = 0.8$.

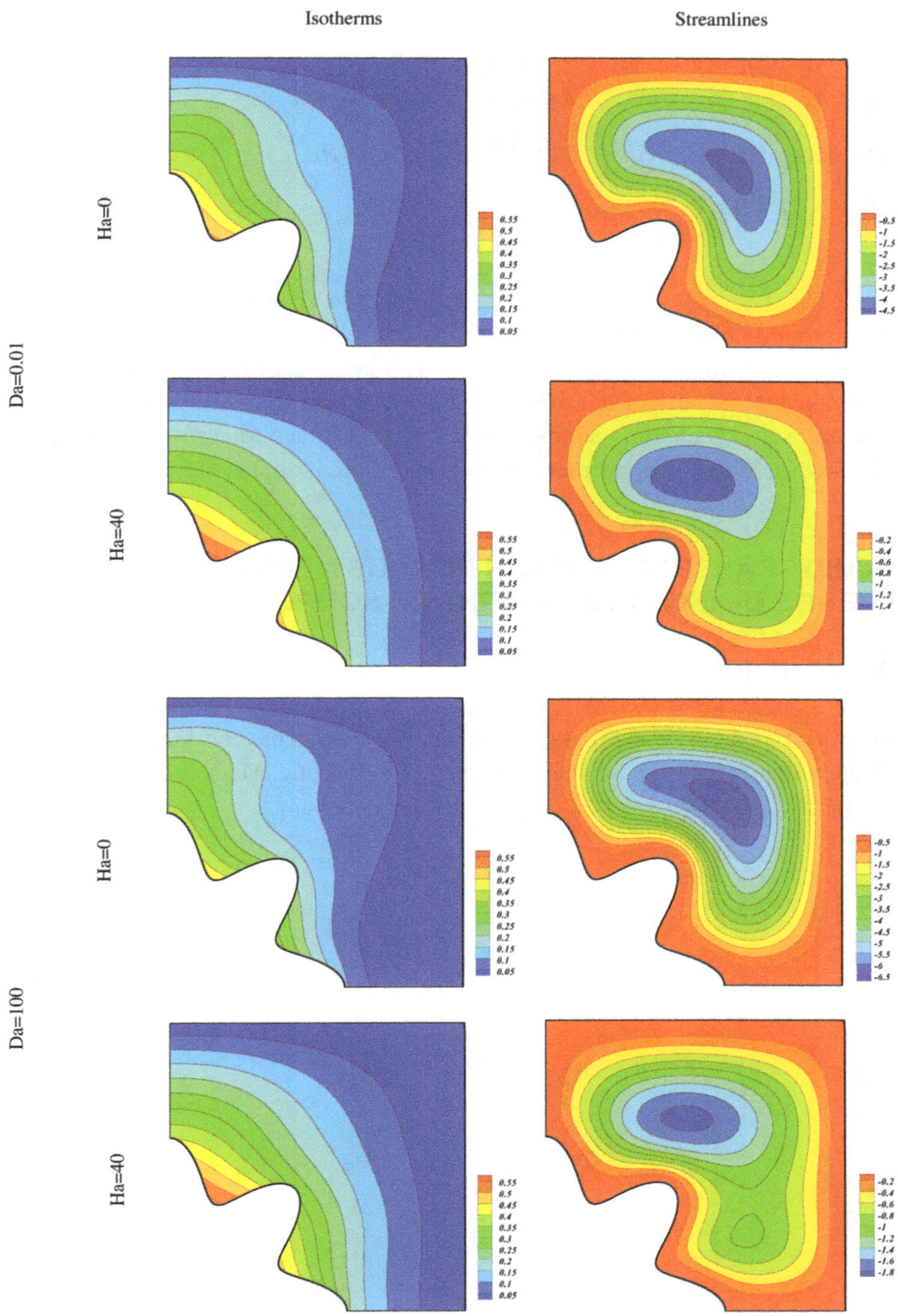

Figures 12 and 13 demonstrate the influence of Rd, Da, Ra and Ha on Nu_{loc}, Nu_{ave}. The formula for Nu_{ave} is:

$$\begin{aligned} Nu_{ave} &= 4.15 - 0.816Rd - 1.45\log\left(Ra\right) - 0.67Da^{*} + 0.79Ha^{*} \\ &+0.67Rd\log\left(Ra\right) + 0.14Rd\,Da^{*} - 0.16Rd\,Ha^{*} + 0.039\log\left(Ra\right)Da^{*} \\ &-0.28\log\left(Ra\right)Ha^{*} - 0.009Da^{*}Ha^{*} + 0.29Rd^{2} + 0.29\left(\log\left(Ra\right)\right)^{2} \\ &+0.66\left(Da^{*}\right)^{2} + 0.046\left(Ha^{*}\right)^{2} \end{aligned} \tag{40}$$

where $Ha^{*} = 0.1Ha, Da^{*} = 0.01Da$. The root mean squared error of this formula is equal to 0.98. Existence of extermum points on local Nusselt number is relevant to presence of undulation of inner wall and thermal plume. Nusselt number auagments with increase of Rayligh and Darcy numbers due to increment of convective heat transfer mechanism. Also Figure 13 indicates that rate of heat transfer increases with augment of radiation paramter. Futhermore, Lorentz forces reduce the convective heat transfer mode. So, Nusselt number decreases with enhance of Hartmann number.

4. NANOFLUID TRANSPORTATION IN POROUS MEDIA UNDER THE INFLUENCE OF EXTERNAL MAGNETIC SOURCE

4.1. Problem Definition

Boundary conditions are depicted in Figure 8. The inner elliptic wall has constant temperature considered as hot wall. Outer circular wall is cold wall, the others are adiabatic. Magnetic source has been considered as shown in Figure 15. $\overline{H_x}, \overline{H_y}, \overline{H}$ can be calculated as follow:

$$\overline{H_y} = \left[\left(\bar{b} - y\right)^2 + \left(\bar{a} - x\right)^2\right]^{-1} \frac{\gamma}{2\pi}\left(\bar{a} - x\right), \tag{41}$$

$$\overline{H_x} = \left[\left(\bar{b} - y\right)^2 + \left(\bar{a} - x\right)^2\right]^{-1} \frac{\gamma}{2\pi}\left(y - \bar{b}\right), \tag{42}$$

$$\overline{H} = \sqrt{\overline{H}_x^{2} + \overline{H}_y^{2}}. \tag{43}$$

Figure 12. Effects of radiation parameter, Darcy, Rayleigh number and Hartmann numbers on local Nusselt number at $\phi = 0.04$.

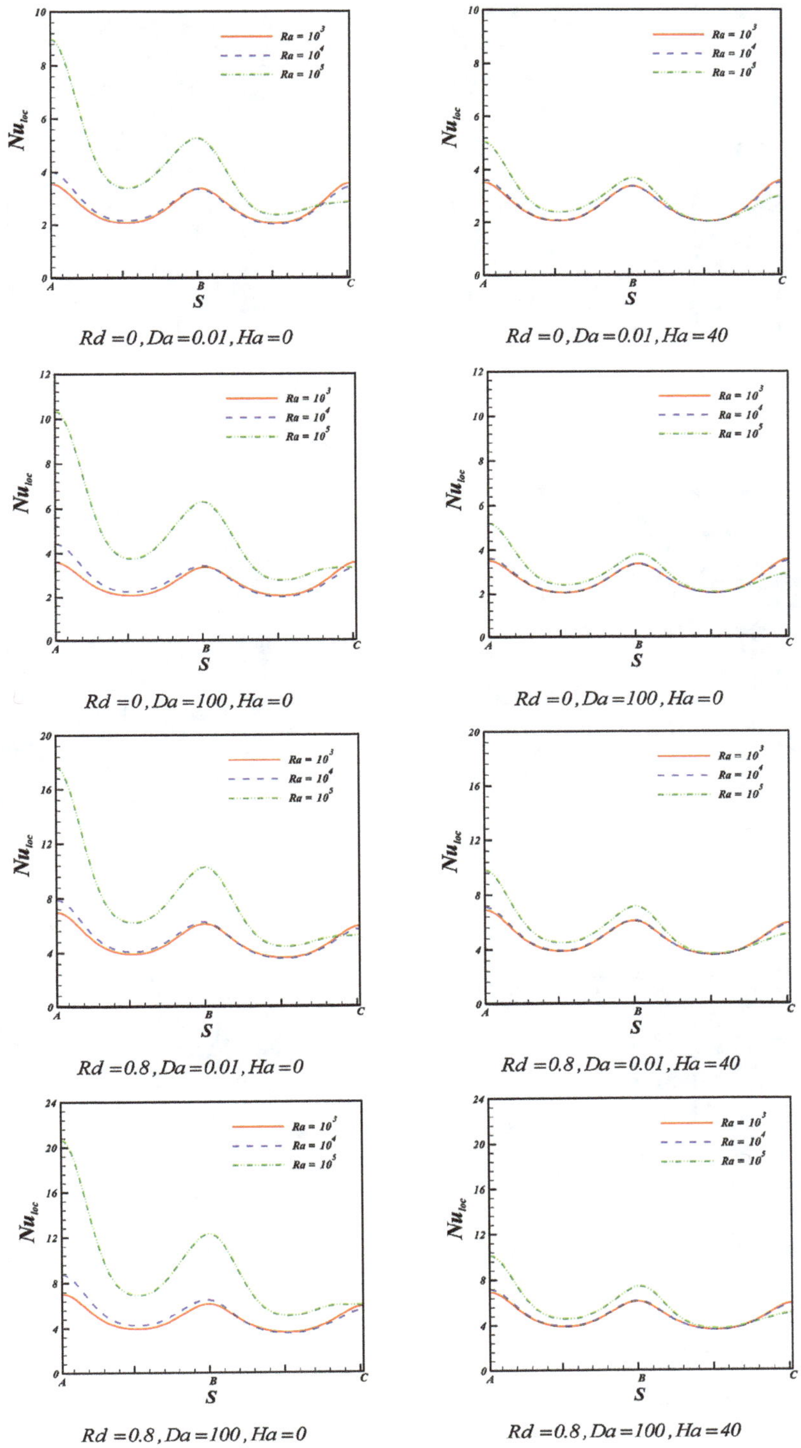

Figure 13a. Effects of radiation parameter, Darcy, Rayleigh number and Hartmann numbers on local Nusselt number at $\phi = 0.04$.

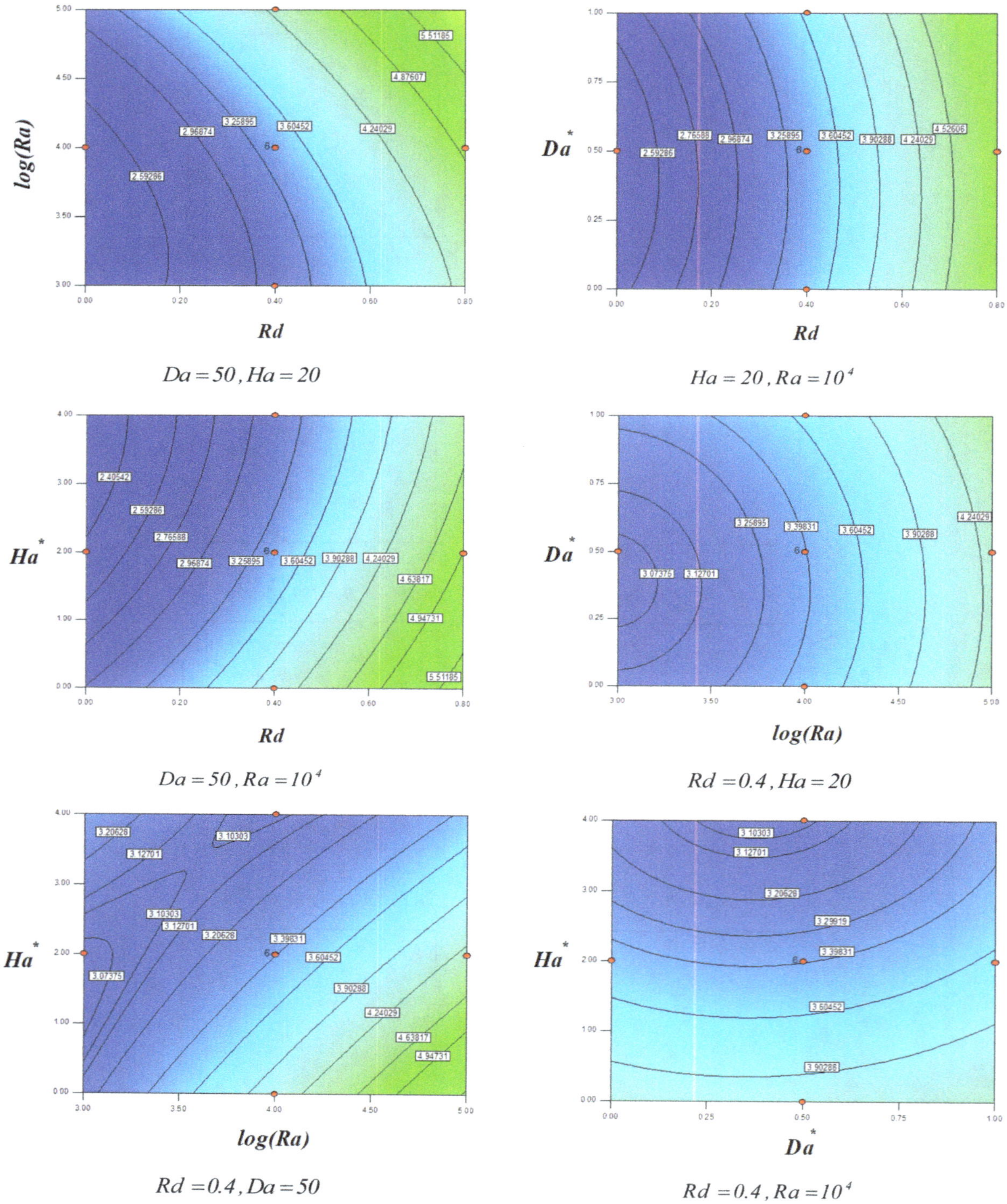

Figure 13b. Effects of radiation parameter, Darcy, Rayleigh number and Hartmann numbers on local Nusselt number at $\phi = 0.04$.

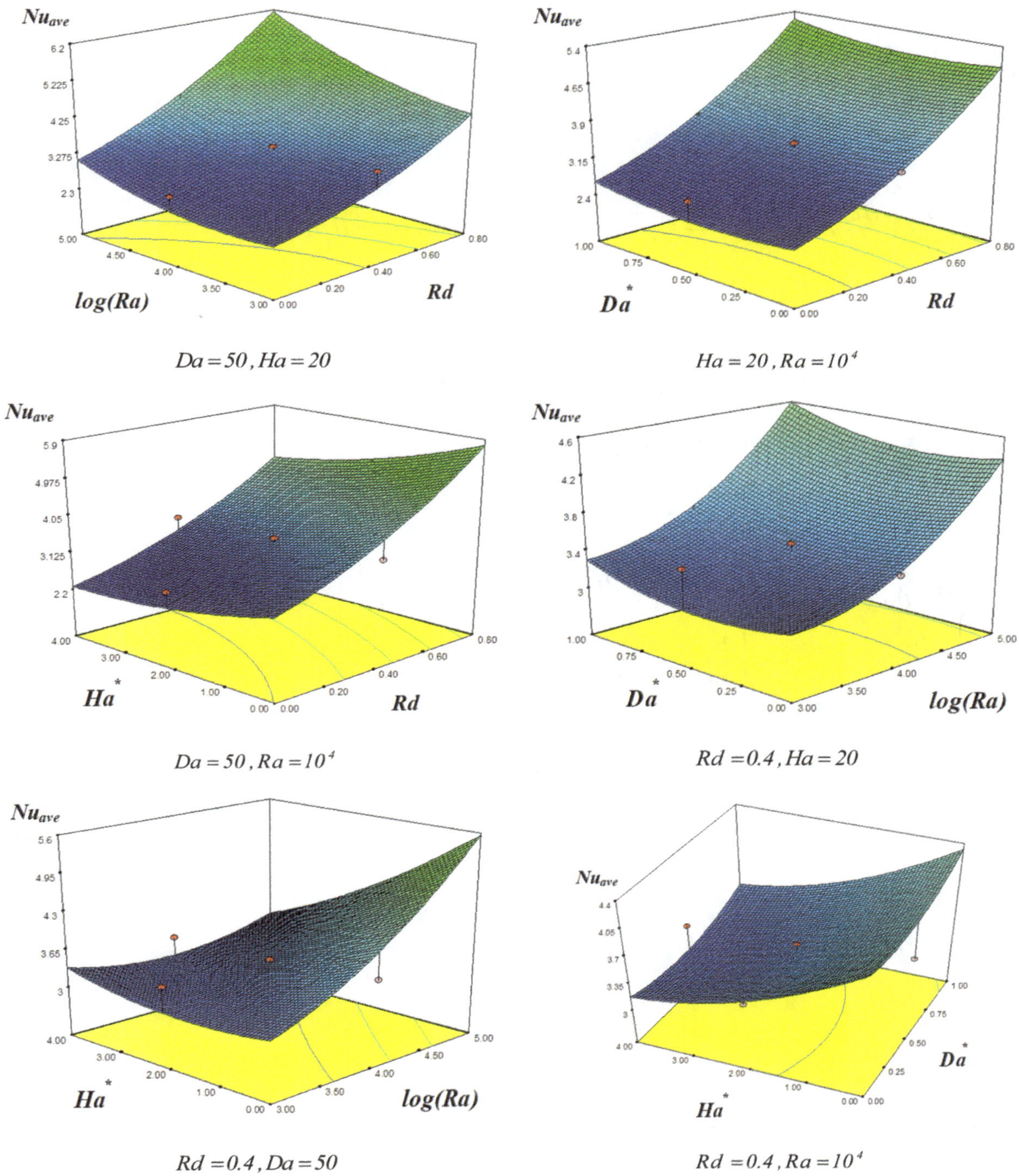

4.2. Governing Equation

2D laminar nanofluid flow and free convective heat transfer is taken into account. The governing PDEs are:

$$\frac{\partial v}{\partial y}+\frac{\partial u}{\partial x}=0, \tag{44}$$

$$\left(\rho_{nf}\right)\left(u\frac{\partial u}{\partial x}+\frac{\partial u}{\partial y}v\right)=\left(\frac{\partial^2 u}{\partial y^2}+\frac{\partial^2 u}{\partial x^2}\right)\mu_{nf}-\frac{\partial P}{\partial x}-\mu_0^2\sigma_{nf}H_y^2u+\sigma_{nf}\mu_0^2H_xH_yv-\frac{\mu_{nf}}{K}u, \tag{45}$$

$$\begin{aligned}\rho_{nf}\left(\frac{\partial v}{\partial x}u+\frac{\partial v}{\partial y}v\right)&=+\mu_{nf}\left(\frac{\partial^2 v}{\partial x^2}+\frac{\partial^2 v}{\partial y^2}\right)-\frac{\partial P}{\partial y}+\mu_0^2H_y\sigma_{nf}H_xu\\&-\mu_0^2H_x\sigma_{nf}H_xv-\frac{\mu_{nf}}{K}v+\left(T-T_c\right)\beta_{nf}g\rho_{nf},\end{aligned} \tag{46}$$

$$\begin{aligned}\left(\rho C_p\right)_{nf}\left(v\frac{\partial T}{\partial y}+u\frac{\partial T}{\partial x}\right)&=\sigma_{nf}\mu_0^2\left(H_xv-H_yu\right)^2+k_{nf}\left(\frac{\partial^2 T}{\partial x^2}+\frac{\partial^2 T}{\partial y^2}\right)\\&+\mu_{nf}\left\{2\left(\frac{\partial u}{\partial x}\right)^2+2\left(\frac{\partial v}{\partial y}\right)^2+\left(\frac{\partial u}{\partial y}+\frac{\partial v}{\partial x}\right)^2\right\},\end{aligned} \tag{47}$$

$\rho_{nf},\left(\rho C_p\right)_{nf},\beta_{nf},k_{nf}$ and σ_{nf} are calculated as

$$\rho_{nf}=\rho_f(1-\phi)+\rho_s\phi, \tag{48}$$

$$\left(\rho C_p\right)_{nf}=\left(\rho C_p\right)_f(1-\phi)+\left(\rho C_p\right)_s\phi, \tag{49}$$

$$\beta_{nf}=\beta_f(1-\phi)+\beta_s\phi, \tag{50}$$

$$k_{nf}=k_f\left(\frac{k_s-2\phi(k_f-k_s)+2k_f}{k_s+\phi(k_f-k_s)+2k_f}\right), \tag{51}$$

$$\sigma_{nf} = \sigma_f \left[\frac{3(\sigma 1 - 1)\phi}{(\sigma 1 + 2) - (\sigma 1 - 1)\phi} + 1 \right], \sigma 1 = \sigma_s / \sigma_f. \tag{52}$$

μ_{nf} is obtained as follows:

$$\mu_{nf} = \left(0.035\mu_0^2 H^2 + 3.1\mu_0 H - 27886.4807\phi^2 + 4263.02\phi + 316.0629\right) e^{-0.01T} \tag{53}$$

Dimensionless parameters are defined as:

$$\begin{gathered} (b, a) = \frac{(\bar{b}, \bar{a})}{L}, (H_y, H_x, H) = \frac{(\overline{H_y}, \overline{H_x}, \overline{H})}{\overline{\overline{H_0}}}, P = \frac{p}{\rho_f (\alpha_f / L)^2} \\ U = \frac{uL}{\alpha_f}, V = \frac{vL}{\alpha_f}, \Theta = \frac{T - T_c}{(T_h - T_c)}, (X, Y) = \frac{(x, y)}{L}. \end{gathered} \tag{54}$$

So equations change to:

$$\frac{\partial V}{\partial Y} + \frac{\partial U}{\partial X} = 0, \tag{55}$$

$$\begin{aligned} U\frac{\partial U}{\partial X} + \frac{\partial U}{\partial Y}V = {} & \Pr \left[\frac{\mu_{nf} / \mu_f}{\rho_{nf} / \rho_f} \right] \left(\frac{\partial^2 U}{\partial Y^2} + \frac{\partial^2 U}{\partial X^2} \right) \\ & -Ha^2 \Pr \left[\frac{\sigma_{nf} / \sigma_f}{\rho_{nf} / \rho_f} \right] \left(H_y^2 U - H_x H_y V \right) - \frac{\partial P}{\partial X} - \frac{\Pr}{Da} \left[\frac{\mu_{nf} / \mu_f}{\rho_{nf} / \rho_f} \right] U, \end{aligned} \tag{56}$$

$$\begin{aligned} V\frac{\partial V}{\partial Y} + U\frac{\partial V}{\partial X} = {} & \Pr \left(\frac{\partial^2 V}{\partial Y^2} + \frac{\partial^2 V}{\partial X^2} \right) \left[\frac{\mu_{nf} / \mu_f}{\rho_{nf} / \rho_f} \right] \\ & -Ha^2 \Pr \left[\frac{\sigma_{nf} / \sigma_f}{\rho_{nf} / \rho_f} \right] \left(H_x^2 V - H_x H_y U \right) \\ & -\frac{\partial P}{\partial Y} + Ra \Pr \left[\frac{\beta_{nf}}{\beta_f} \right] \Theta - \frac{\Pr}{Da} \left[\frac{\mu_{nf} / \mu_f}{\rho_{nf} / \rho_f} \right] V, \end{aligned} \tag{57}$$

$$
\begin{aligned}
&V\frac{\partial \Theta}{\partial Y}+U\frac{\partial \Theta}{\partial X}=\left[\frac{k_{nf}}{k_f}\frac{\left(\rho C_P\right)_f}{\left(\rho C_P\right)_{nf}}\right]\left(\frac{\partial^2\Theta}{\partial Y^2}+\frac{\partial^2\Theta}{\partial X^2}\right)\\
&+Ha^2\,Ec\left(\frac{\left(\rho C_P\right)_f}{\left(\rho C_P\right)_{nf}}\frac{\sigma_{nf}}{\sigma_f}\right)\left\{V\,H_x-U\,H_y\right\}^2\\
&+\left[\frac{\dfrac{\mu_{nf}}{\mu_f}}{\dfrac{\left(\rho C_P\right)_{nf}}{\left(\rho C_P\right)_f}}\right]Ec\left\{2\left(\frac{\partial U}{\partial X}\right)^2+2\left(\frac{\partial V}{\partial Y}\right)^2+\left(\frac{\partial U}{\partial Y}+\frac{\partial V}{\partial X}\right)^2\right\}
\end{aligned}
\tag{58}
$$

and dimensionless parameters are

$$
\begin{aligned}
&Ra_f=g\beta_f L^3\Delta T/\left(\alpha_f\upsilon_f\right),\Pr_f=\upsilon_f/\alpha_f,Ha=L\mu_0H_0\sqrt{\sigma_f/\mu_f},\\
&Ec=\left(\mu_f\alpha_f\right)/\left[\left(\rho C_P\right)_f\Delta T\,L^2\right],Da=K/L^2.
\end{aligned}
\tag{59}
$$

The thermo-physical properties of Fe_3O_4 and water are presented in Table 3. Pressure gradient source terms discard by vorticity stream function.

$$
\Omega=\frac{\omega L^2}{\alpha_f},\Psi=\frac{\psi}{\alpha_f},\omega=-\frac{\partial u}{\partial y}+\frac{\partial v}{\partial x},(u,v)=\left(\frac{\partial\psi}{\partial y},-\frac{\partial\psi}{\partial x}\right). \tag{60}
$$

According to Figure 14, boundary conditions are

on inner wall $\theta=1.0$

on outer wall $\theta=0.0$

on other walls $\dfrac{\partial\theta}{\partial y}=0$

Table 3. Thermo physical properties of water and nanoparticles

	$\rho(kg/m^3)$	$C_p(j/kgk)$	$k(W/m.k)$	$d_p(nm)$	$\sigma(\Omega\cdot m)^{-1}$
Pure water	997.1	4179	0.613	-	0.05
Fe_3O_4	5200	670	6	47	25000

Figure 14. (a)Geometry and the boundary conditions with (b) the mesh of Geometry considered in this work; (c) A sample triangular element and its corresponding control volume.

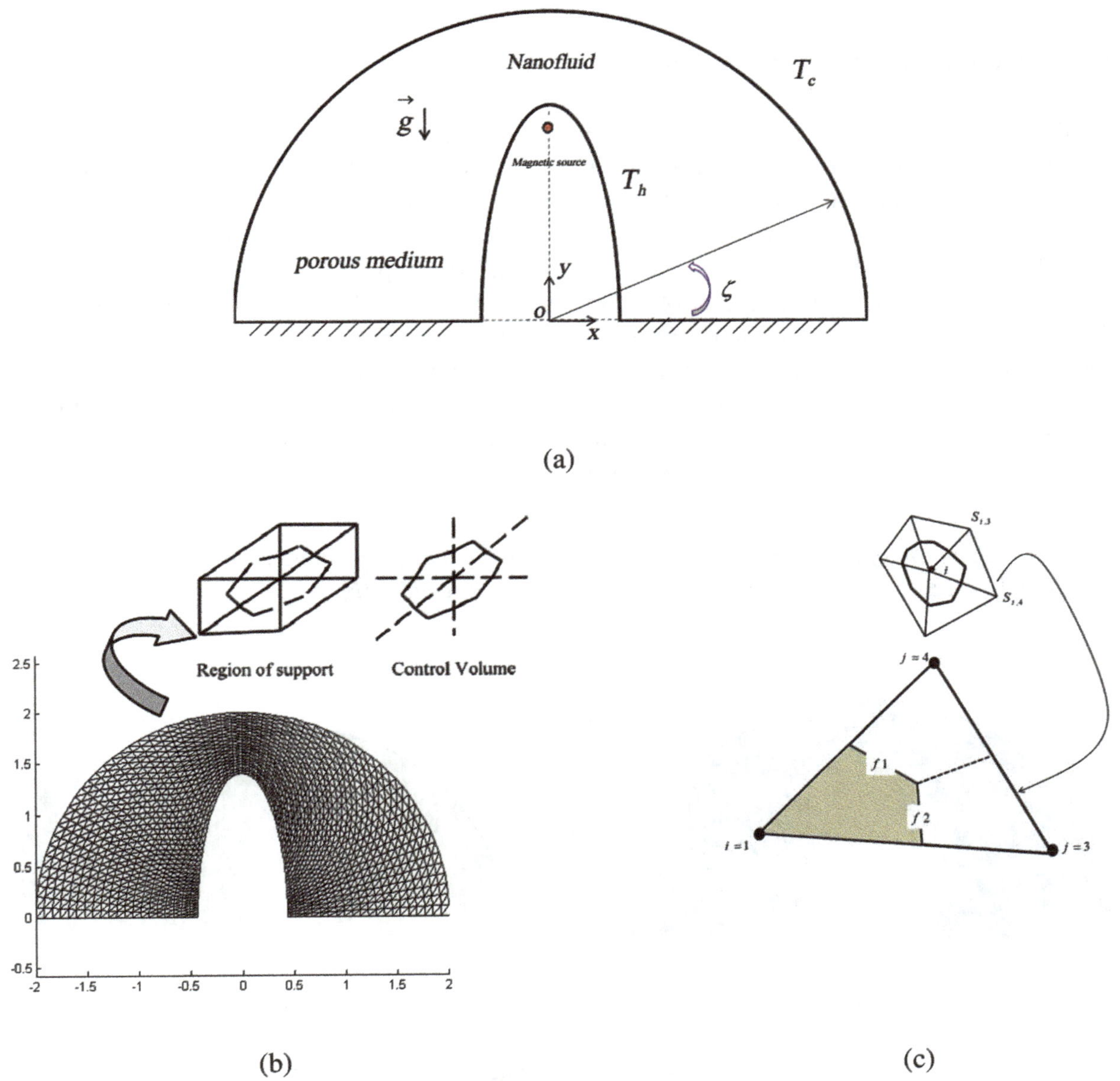

on all walls $\Psi = 0.0$ (61)

Nu_{loc}, Nu_{ave} along cold wall are:

$$Nu_{loc} = \left| \left(\frac{k_{nf}}{k_f} \right) \frac{\partial \Theta}{\partial r} \right|, \tag{62}$$

$$Nu_{ave} = \frac{1}{0.5\pi} \int_{0}^{0.5\pi} Nu_{loc}(\zeta)\, d\zeta, \tag{63}$$

4.3. Effects of Active Parameters

In this article, the influence of magnetic field on Fe_3O_4-water nanofluid in a porous enclosure is reported. The governing equations have been solved via Control volume based finite element method and the outputs are depicted in several plots for the influence of various parameters on the flow and heat transfer. These parameters are Darcy number (Da), Rayleigh number (Ra), Hartmann number (Ha) and volume fraction of Fe_3O_4 (ϕ). $\Pr$ and Ec are 6.8 and10^{-5}, respectively.

Figure 16 demonstrates the influence of adding Fe_3O_4 in to water on hydrothermal characteristic. This figure depicts that an increase in nanoparticle volume fraction results in increase in nanofluid velocity. It is also found that the thermal boundary layer thickness of water based nanofluid is higher than pure fluid.

Figures 17, 18 and 19 illustrate the impact of Darcy, Hartmann and Rayleigh numbers on isotherms and streamlines. In domination of conduction modes, one main clock wise cell appears in half of the enclosure. An augment in magnetic field results in generate secondary cell near the vertical center line.

Figure 15. Contours of the (a) magnetic field strength H ; (b) magnetic field intensity component in x direction Hx ; (c) magnetic field intensity component in y direction Hy .

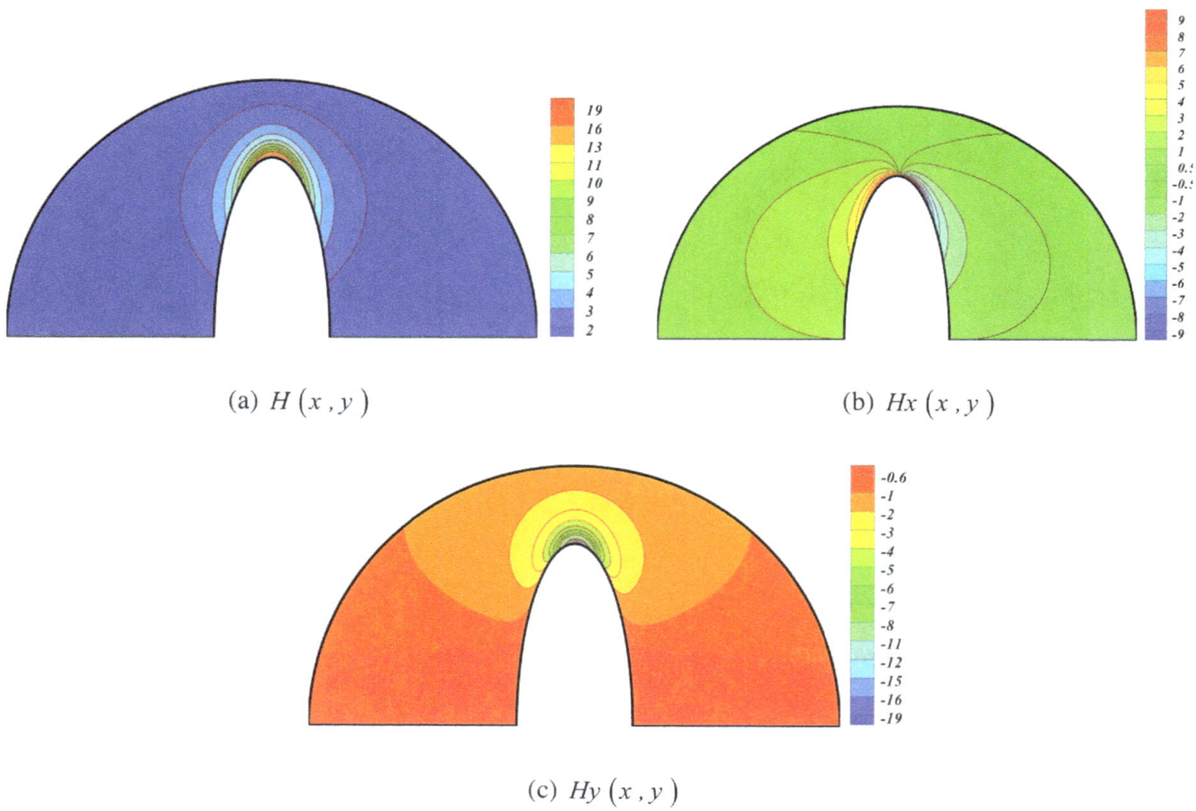

(a) $H(x,y)$

(b) $Hx(x,y)$

(c) $Hy(x,y)$

Figure 16. Impact of nanofluid volume fraction on streamlines (up) and isotherms (bottom) contours (nanofluid ($\phi = 0.04$)(—) and pure fluid ($\phi = 0$) (- - -)) when $Ra = 10^5, Da = 100, Ha = 0$

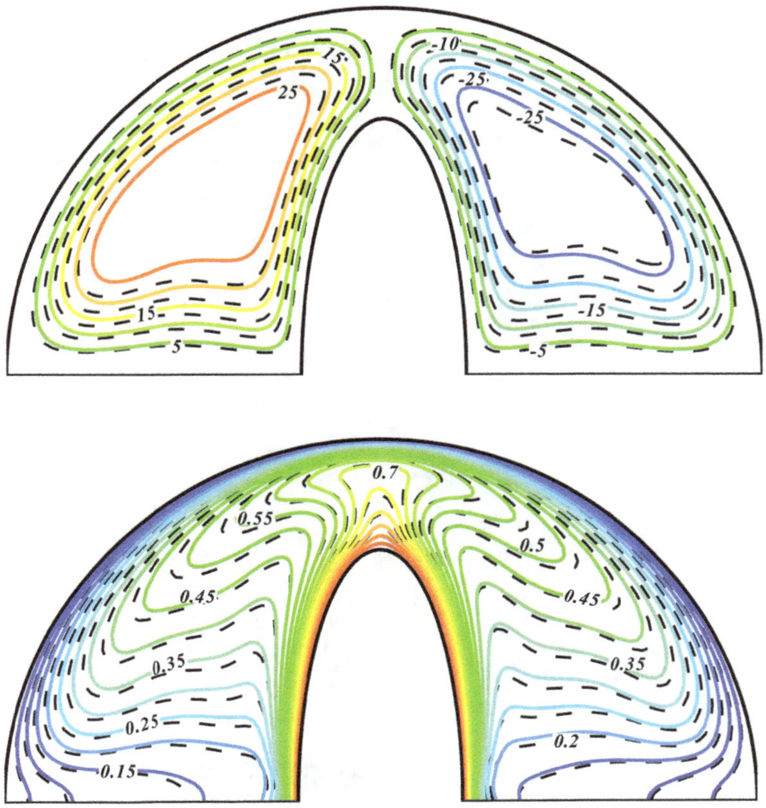

As permeability of porous media enhances, convective heat transfer improves and thermal plume appears. An increase in buoyancy force results in enhance in strength of main eddy and generates thermal plume near the vertical centerline. As Lorentz forces augments, the position of thermal plume become far from vertical centerline. It is fantastic observation in case of Da=100, Ra=10^5, $\left|\Psi_{\max}\right|$ reaches to its maximum value and the main eddy stretch horizontally. Also one powerful thermal plume generates near the $\zeta = 90°$. Applying magnetic field for such case, converts the main eddy to three smaller ones. The middle one rotates counter clock wise. Existence of such eddies generates two thermal plume over the hot elliptic wall.

Rate of heat transfer is depicted in Figure 20. The formula of Nu_{ave} corresponding to important parameters is:

$$\begin{aligned} Nu_{ave} &= 4.37 + 0.15Da^* - 2.77\log\left(Ra\right) + 0.03Ha^* \\ &+0.03Da^*\log\left(Ra\right) - 0.21Da^*Ha^* - 0.39\log\left(Ra\right)Ha^* \\ &+0.07Da^{*2} + 0.51\left(\log\left(Ra\right)\right)^2 + 0.61Ha^{*2} \end{aligned} \quad (64)$$

Figure 17. Influence of Da, Ha *on streamlines (left) and isotherms (right) contours when* $\phi = 0.04, Ra = 10^3$

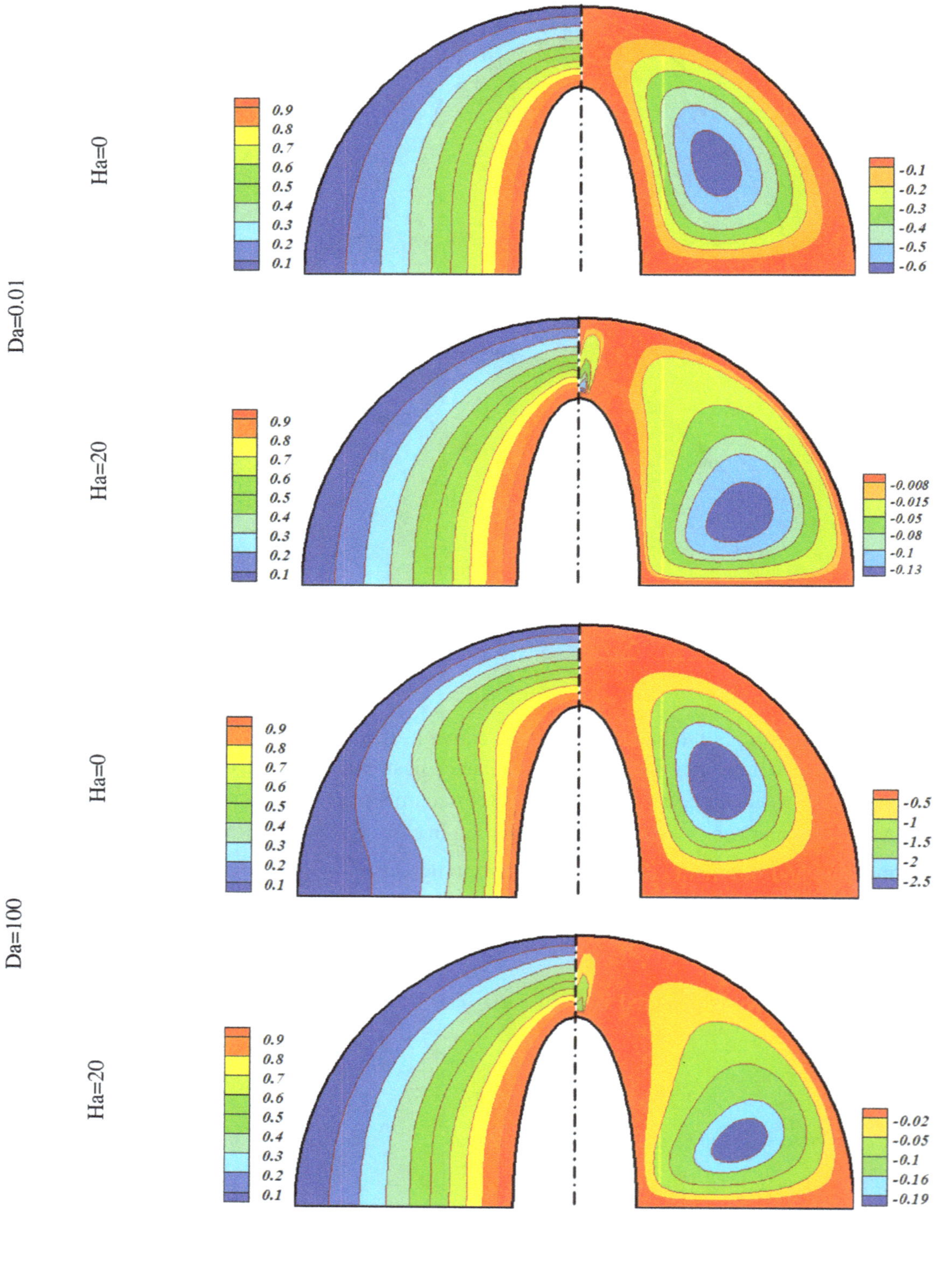

Figure 18. Influence of Da, Ha *on streamlines (left) and isotherms (right) contours when* $\phi = 0.04, Ra = 10^4$

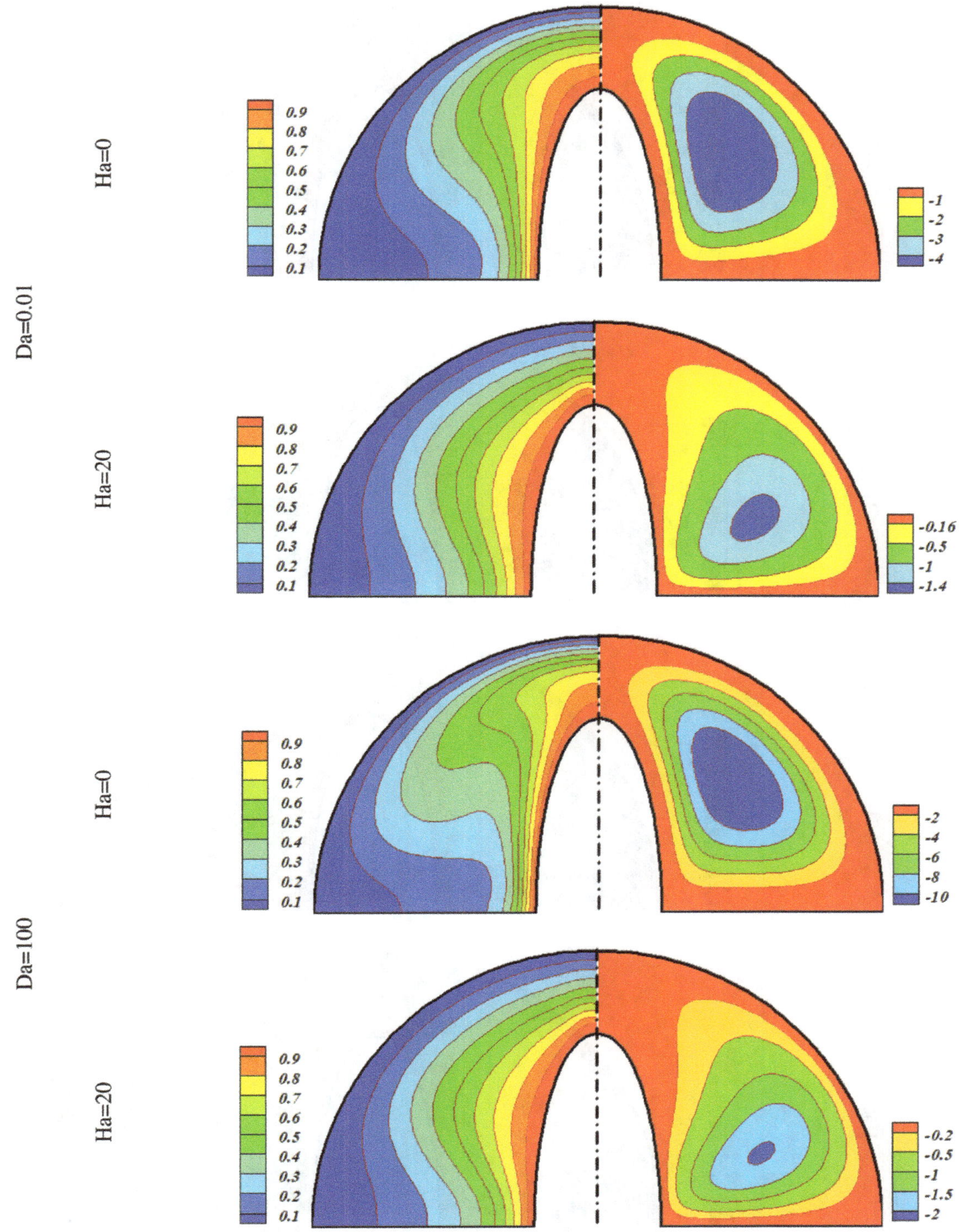

Figure 19. Influence of Da, Ha on streamlines (left) and isotherms (right) contours when $\phi = 0.04, Ra = 10^5$

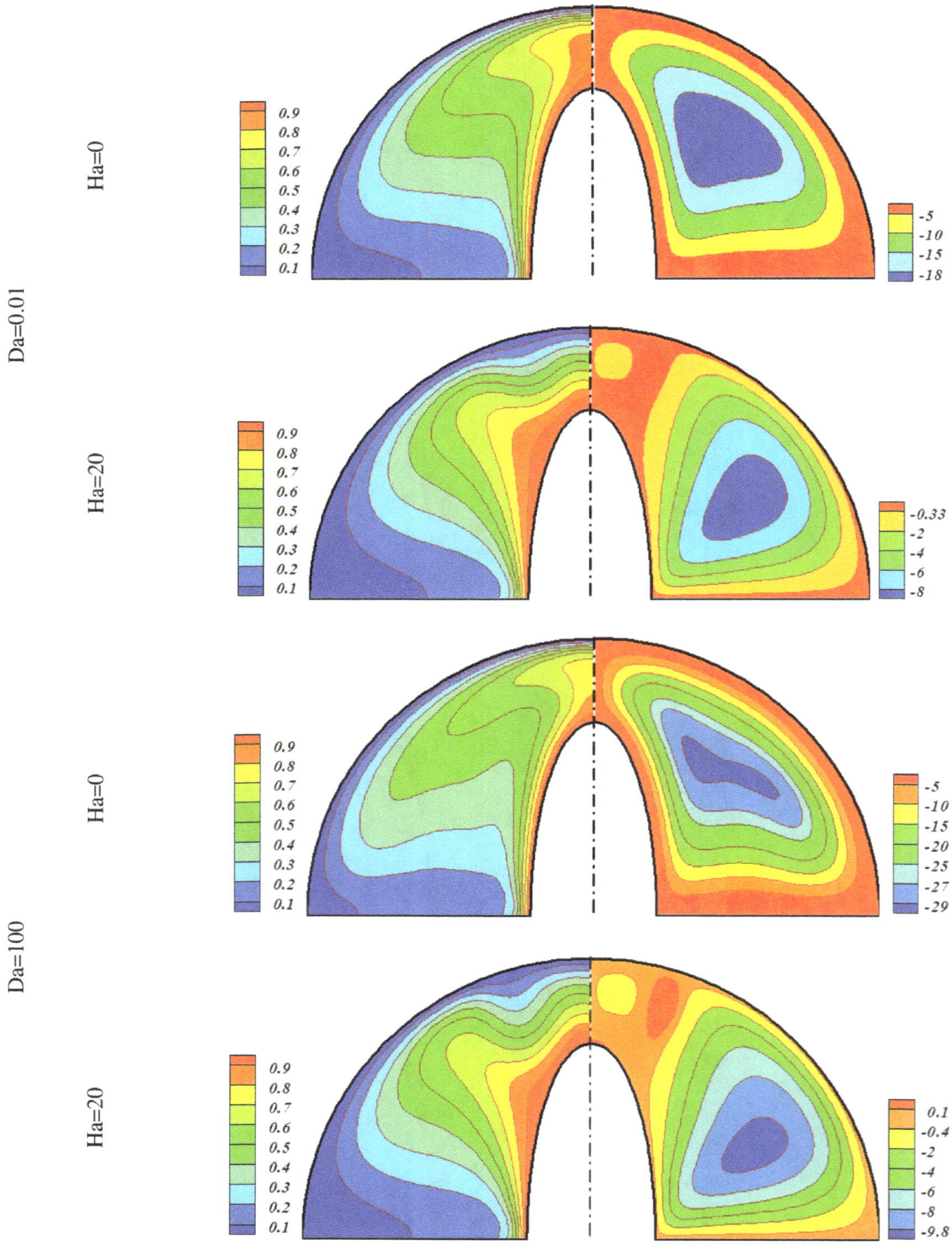

Figure 20. Effects of Da, Ha and Ra on average Nusselt number

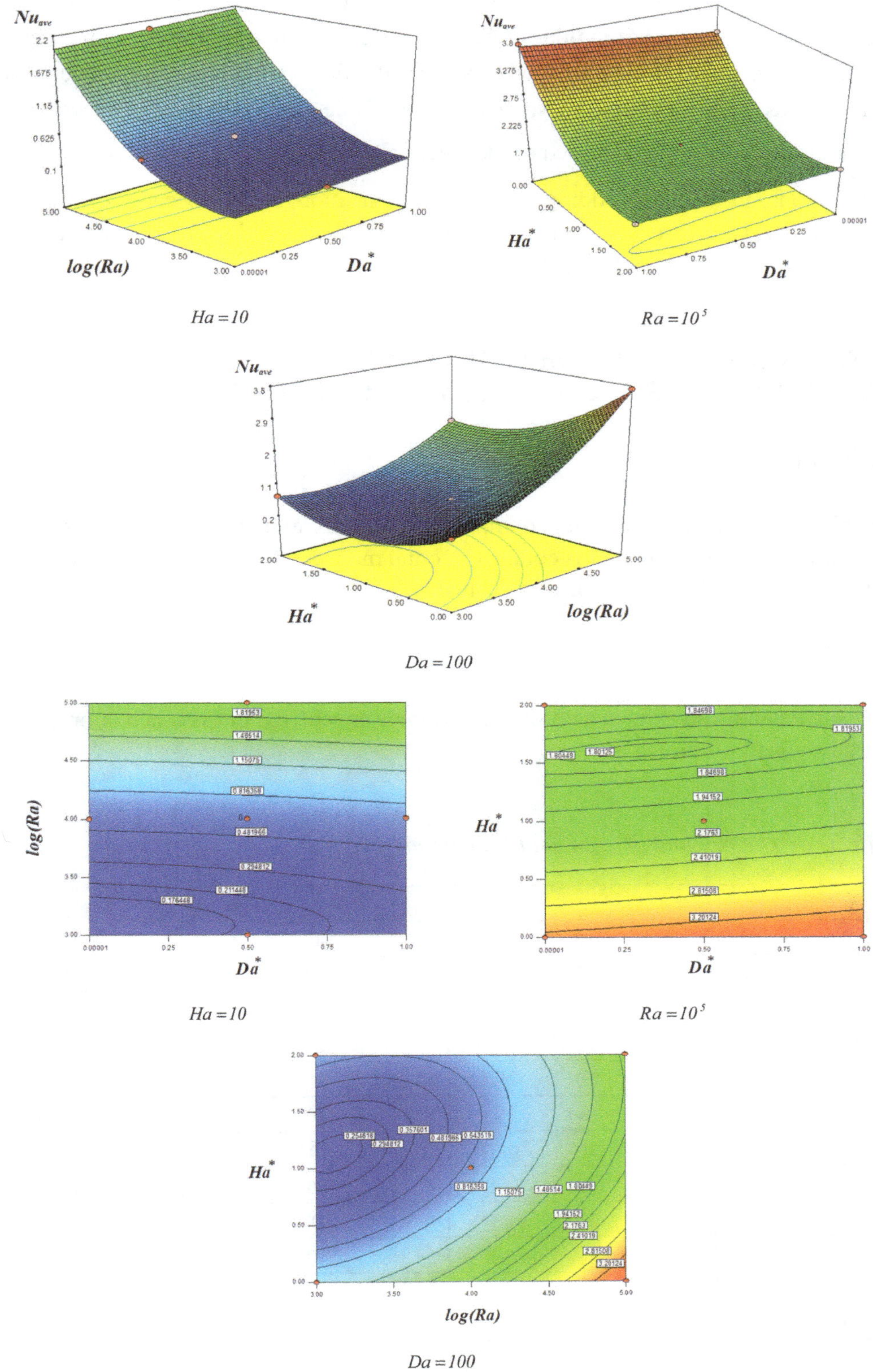

where $Da^* = 0.01Da, Ha^* = 0.1Ha$. Increasing in permeability of porous media results in augments in rate of heat transfer. Enhancing Rayleigh number makes the convective heat transfer to augments. So this non dimension parameter has similar effect on Nusselt number with that of obtained for Darcy number. As Lorentz force increases Nu_{ave} reduces due to domination of conduction mode. Adding Fe_3O_4 nanoparticle into base fluid enhances the Nusselt number. Table 4 demonstrates the influence of Da, Ha and Ra onheattransferimprovement.Thisoutputisdefinedas $E = 100 * \left(Nu_{ave}\big|_{\phi=0.04} - Nu_{ave}\big|_{\phi=0} \right) / Nu_{ave}\big|_{\phi=0}$. In conduction mode, influence of adding nanoparticles has more benefit because of more changes in thermal conductivity. Therefore, heat transfer improvement enhances with enhance of Hartmann number but it detracts with rise of Darcy and Rayleigh numbers.

5. NANOFLUID CONVECTIVE HEAT TRANSFER INTENSIFICATION IN A POROUS CIRCULAR CYLINDER

5.1. Problem Definition

Figure 21 depicts the geometry, boundary condition and sample element. The inner cylinder has constant heat flux condition and outer cylinder is cold. Horizontal magnetic field has been applied in this porous media. Non-Darcy model is utilized for porous media.

5.2. Governing Equation

Steady convective nanofluid flow in a porous enclosure is considered in existence of uniform magnetic field. The PDEs equations are:

Table 4. Effects of Da, Ha and Ra on heat transfer enhancement

Ra	*Da*	*Ha*	*E*
10^3	0.01	0	11.09238
10^3	0.01	20	11.42401
10^4	0.01	0	4.20269
10^4	0.01	20	7.652818
10^5	0.01	0	2.613391
10^5	0.01	20	3.245394
10^3	100	0	5.874633
10^3	100	20	7.615688
10^4	100	0	3.787017
10^4	100	20	6.464888
10^5	100	0	3.757416
10^5	100	20	1.560289

Figure 21. (a, b) Geometry and the boundary conditions with (c) the mesh of Geometry considered in this work; (d) A sample triangular element and its corresponding control volume.

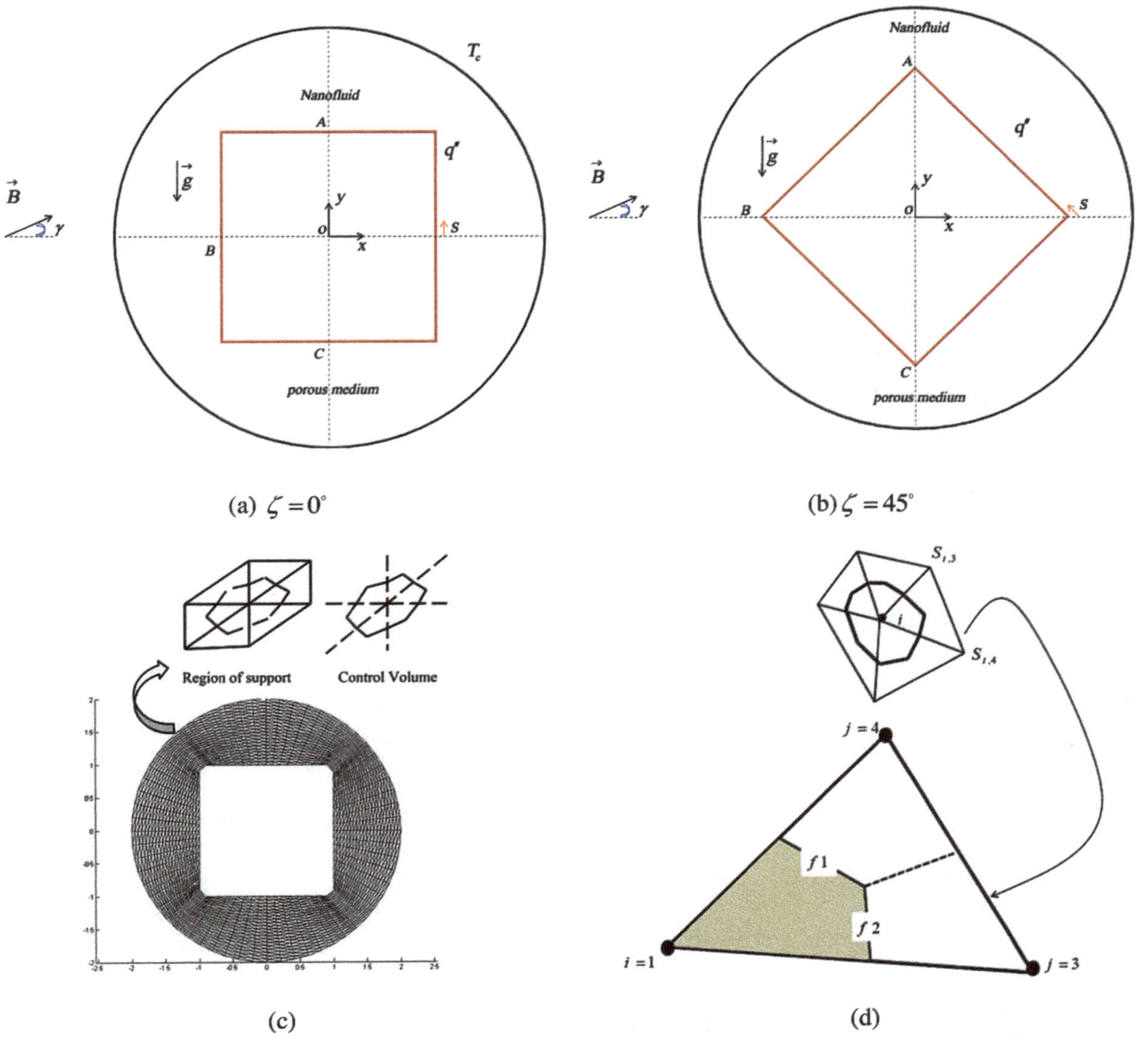

$$\frac{\partial v}{\partial y}+\frac{\partial u}{\partial x}=0 \tag{65}$$

$$\left(v\frac{\partial u}{\partial y}+u\frac{\partial u}{\partial x}\right)=\left[\sigma_{nf}B_xB_yv-\sigma_{nf}B_y^2+\left(\frac{\partial^2 u}{\partial y^2}+\frac{\partial^2 u}{\partial x^2}\right)\mu_{nf}-\frac{\partial P}{\partial x}-\frac{\mu_{nf}}{K}u\right]\left(\rho_{nf}\right)^{-1} \tag{66}$$

$$\rho_{nf}\left(\frac{\partial v}{\partial x}u+\frac{\partial v}{\partial y}v\right)=\mu_{nf}\left(\frac{\partial^2 v}{\partial x^2}+\frac{\partial^2 v}{\partial y^2}\right)-\frac{\partial P}{\partial y}-B_x\sigma_{nf}B_x v+B_y\sigma_{nf}B_x u-\frac{\mu_{nf}}{K}v$$
$$+\left(T-T_c\right)\beta_{nf}g\rho_{nf},\qquad (67)$$
$$B_x=B_o\cos\lambda,\ B_y=B_o\sin\lambda$$

$$\left(\rho C_p\right)_{nf}\left(v\frac{\partial T}{\partial y}+u\frac{\partial T}{\partial x}\right)=k_{nf}\left(\frac{\partial^2 T}{\partial x^2}+\frac{\partial^2 T}{\partial y^2}\right)\qquad (68)$$

$\left(\rho C_p\right)_{nf}, \left(\rho\beta\right)_{nf}$, ρ_{nf}, k_{nf} and σ_{nf} are defined as:

$$\left(\rho C_p\right)_{nf}=\phi\left(\rho C_p\right)_s+(1-\phi)\left(\rho C_p\right)_f\qquad (69)$$

$$\frac{\left(\rho\beta\right)_{nf}}{\left(\rho\beta\right)_f}=\phi\frac{\left(\rho\beta\right)_s}{\left(\rho\beta\right)_f}+(1-\phi)\qquad (70)$$

$$\rho_{nf}=\rho_f(1-\phi)+\rho_s\phi\qquad (71)$$

$$k_{nf}=k_f\left(\frac{k_s+2k_f+2\phi(k_s-k_f)}{k_s-\phi(k_s-k_f)+2k_f}\right)\qquad (72)$$

$$\sigma_{nf}=\left[\left(\frac{\left(2+\sigma_s/\sigma_f\right)-\left(\sigma_s/\sigma_f-1\right)\phi}{3\phi\left(-1+\sigma_s/\sigma_f\right)}\right)^{-1}+1\right]\sigma_f\qquad (73)$$

μ_{nf} is obtained as follows:

$$\mu_{nf}=\left(0.035B^2+3.1B-27886.4807\phi^2+4263.02\phi+316.0629\right)e^{-0.01T}\qquad (74)$$

The properties of nanofluid are provided in table1. Vorticity and stream function should be used to eliminate pressure source terms:

$$\omega + \frac{\partial u}{\partial y} - \frac{\partial v}{\partial x} = 0, \frac{\partial \psi}{\partial x} = -v, \frac{\partial \psi}{\partial y} = u \tag{75}$$

Introducing dimensionless quantities:

$$\left(Y, X\right) = \left(y, x\right) / L, P = \frac{p}{\rho_{nf}\left(\alpha_{nf} / L\right)^2}, U = \frac{uL}{\alpha_{nf}}, V = \frac{vL}{\alpha_{nf}}, \theta = \frac{T - T_c}{\Delta T}, \Delta T = q'' L / k_f \tag{76}$$

The final formulae are:

$$\frac{\partial^2 \Psi}{\partial X^2} + \Omega + \frac{\partial^2 \Psi}{\partial Y^2} = 0, \tag{77}$$

$$\begin{aligned} &\frac{\partial \Omega}{\partial X} U + \frac{\partial \Omega}{\partial Y} V = \Pr \frac{A_5}{A_1} \frac{A_2}{A_4} \left(\frac{\partial^2 \Omega}{\partial Y^2} + \frac{\partial^2 \Omega}{\partial X^2} \right) \\ &+ \Pr Ha^2 \frac{A_6}{A_1} \frac{A_2}{A_4} \left(\frac{\partial U}{\partial X} B_x B_y - \frac{\partial V}{\partial X} B_x^{\ 2} + \frac{\partial U}{\partial Y} B_y^2 - \frac{\partial V}{\partial Y} B_x B_y \right) \\ &+ \Pr Ra \frac{A_3 A_2^2}{A_1 A_4^2} \frac{\partial \theta}{\partial X} - \frac{\Pr}{Da} \left(\frac{A_5 A_2}{A_1 A_4} \right) \Omega, \end{aligned} \tag{78}$$

$$V \frac{\partial \theta}{\partial Y} + \frac{\partial \theta}{\partial X} U = \left(\frac{\partial^2 \theta}{\partial Y^2} + \frac{\partial^2 \theta}{\partial X^2} \right). \tag{79}$$

where dimensionless and constants parameters are illustrated as:

$$\begin{aligned} &\Pr = \upsilon_f / \alpha_f, Ra = g\beta_f q'' L^4 / \left(k_f \upsilon_f \alpha_f\right), Ha = LB_0 \sqrt{\sigma_f / \mu_f}, Da = \frac{K}{L^2}, \\ &A_1 = \frac{\rho_{nf}}{\rho_f}, A_3 = \frac{\left(\rho\beta\right)_{nf}}{\left(\rho\beta\right)_f}, A_5 = \frac{\mu_{nf}}{\mu_f}, A_2 = \frac{\left(\rho C_P\right)_{nf}}{\left(\rho C_P\right)_f}, A_4 = \frac{k_{nf}}{k_f}, A_6 = \frac{\sigma_{nf}}{\sigma_f} \end{aligned} \tag{80}$$

and boundary conditions are:

$$\frac{\partial \theta}{\partial n} = 1.0 \text{ on inner wall}$$

$$\theta = 0.0 \text{ on outer wall}$$

$$\Psi = 0.0 \text{ on all walls} \tag{81}$$

Local and average Nusselt over the inner cylinder can be calculated as:

$$Nu_{loc} = \frac{1}{\theta}\left(\frac{k_{nf}}{k_f}\right) \tag{82}$$

$$Nu_{ave} = \frac{1}{S}\int_0^s Nu_{loc}\, ds \tag{83}$$

5.3. Effects of Active Parameters

Influence of magnetic field on nanofluid transportation in a porous cylinder with inner inclined square obstacle is presented. The working fluid is considered Fe_3O_4-water and its viscosity is a function of ϕ and Ha. Results are demonstrated for several values of volume fraction of Fe_3O_4-water ($\phi = 0$ to 0.04), Darcy number ($Da = 0.001$ to 100), Hartmann number ($Ha = 0$ to 40), Rayleigh number ($Ra = 10^3$ to 10^5) and inclination angle ($\xi = 0^\circ$ and 45°).

Figure 22 demonstrates the impact of ϕ on isotherms and streamlines. Augmenting nanofluid volume fraction leads to augment temperature boundary layer thickness. The nanofluid velocity augments because of enhancing ϕ. Impacts of Ha, ξ, Da, and Ra on hydrothermal behavior are demonstrated in Figures 23, 24 and 25. As nanofluid temperature increases, the nanofluid initiates moving from the inner cylinder to the cold one and dropping along the outer cylinder. Conduction mode is dominant at low Rayleigh and Darcy numbers. So isotherms follow the shape of enclosure. At $\xi = 0^\circ$, one main eddy exists and when the inner cylinder inclined $\left(\xi = 45^\circ\right)$ the main eddy convert to two similar ones. Strength of this main eddy enhances with rise of convective heat transfer. So $\left|\Psi_{max}\right|$ rises with augment of Da, Ra. Also thermal plume appears near the vertical center line when convection mode is dominated. As ξ increases, the distortion of isotherms enhances. As magnetic field augments, $\left|\Psi_{max}\right|$ reduces and the center of main eddy moves to upward. Also Lorentz force makes the thermal plume to diminish.

Figures 26 and 27 illustrate the impact of ξ, Da, Ra and Ha on Nu_{ave}, Nu_{loc}. Respect to active parameters, the following equation can be obtained:

$$\begin{aligned} Nu_{ave} &= 4.58 - 0.66\xi - 2\log\left(Ra\right) - 0.58Da^{*} + 0.18Ha^{*} + 0.19\xi\log\left(Ra\right) \\ &+0.09\xi Da^{*} - 0.04\xi Ha^{*} + 0.18\log\left(Ra\right)Da^{*} - 0.15\log\left(Ra\right)Ha^{*} - 0.41Da^{*}\,Ha^{*} \\ &+0.17\xi^{2} + 0.33\left(\log\left(Ra\right)\right)^{2} + 0.1Rd^{2} + 0.08Ha^{*2} \end{aligned} \tag{84}$$

Figure 22. Effect of volume fraction of nanofluid on Isotherms (left) and streamlines (right) contours ($\phi = 0$ (− −), $\phi = 0.04$ (−)) when $Da = 100, Ra = 10^5$

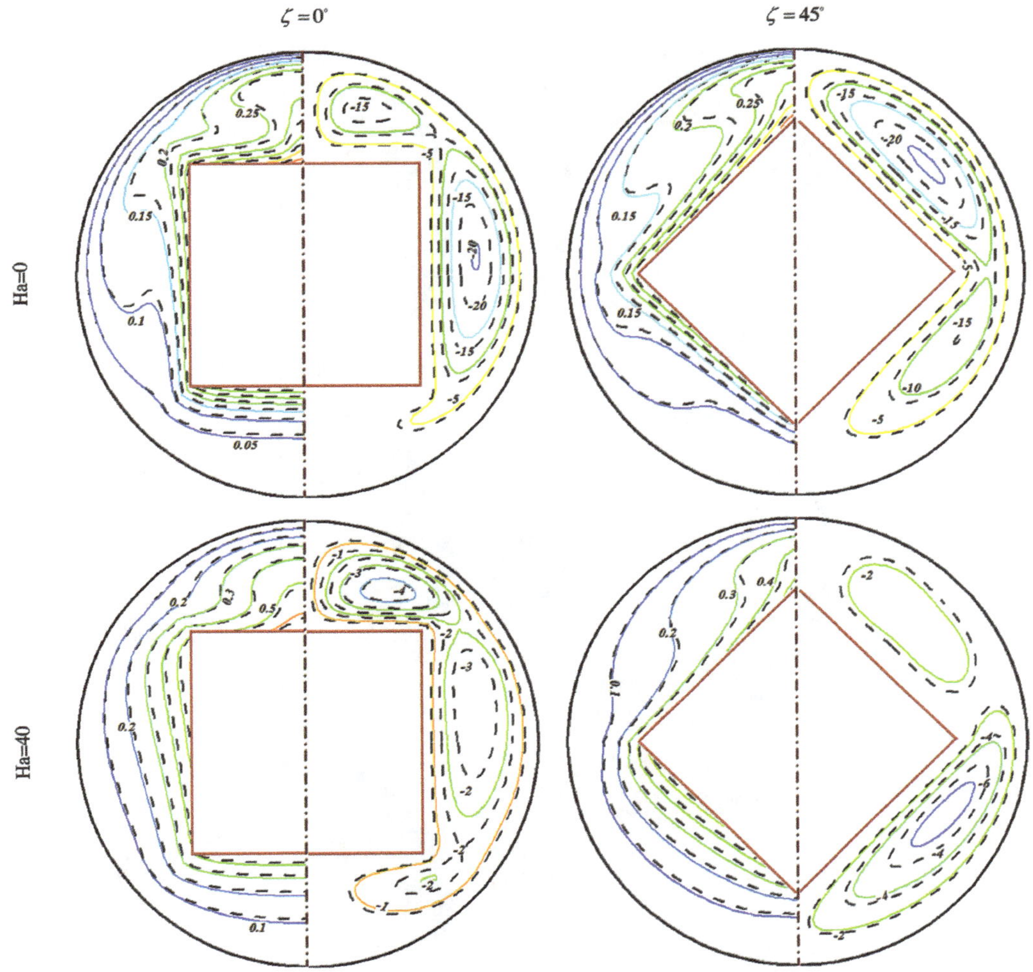

where $Ha^* = 0.1Ha, Da^* = 0.01Da$. The number of extermum in Nu_{loc} is matching to existence of corner of square cylinder and thermal plume. Nu_{ave} increases with augment of Da, Ra, ξ but it decreases with rise Ha .

6. CONVECTIVE FLOW OF NANOFLUID INSIDE A LID-DRIVEN POROUS CAVITY

6.1. Problem Definition

Sample element, boundary condition and geometry are depicted (see Figure 28). The south wall is hot and others are cold. Also the south wall can move horizontally. Porous cavity filled with nanofluid and affected by horizontal magnetic field.

Figure 23. Effect of Darcy and Hartmann numbers on Isotherms (left) and streamlines (right) contours when $\phi = 0.04, Ra = 10^3$

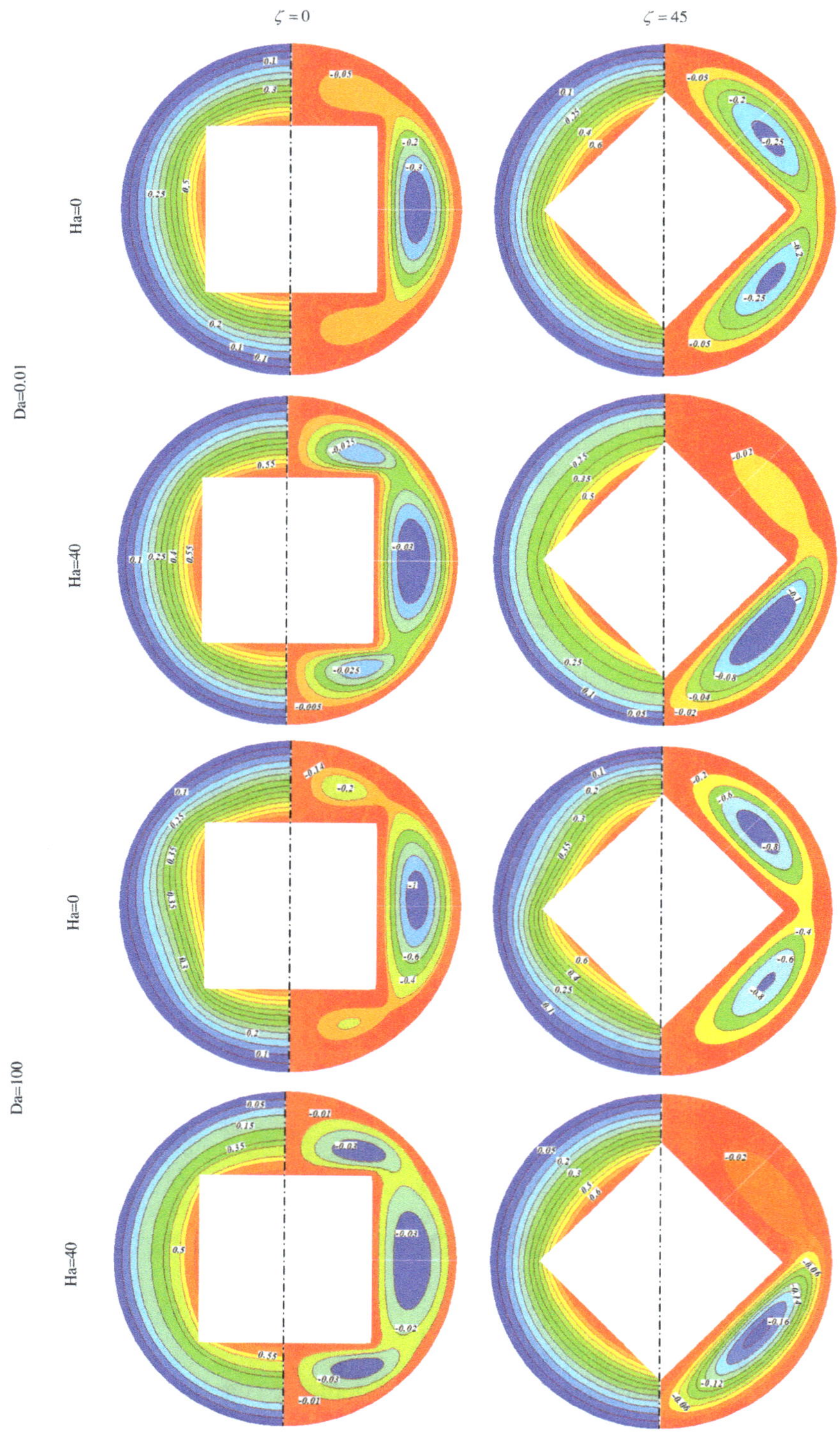

Figure 24. Effect of Darcy and Hartmann numbers on Isotherms (left) and streamlines (right) contours when $\phi = 0.04, Ra = 10^4$

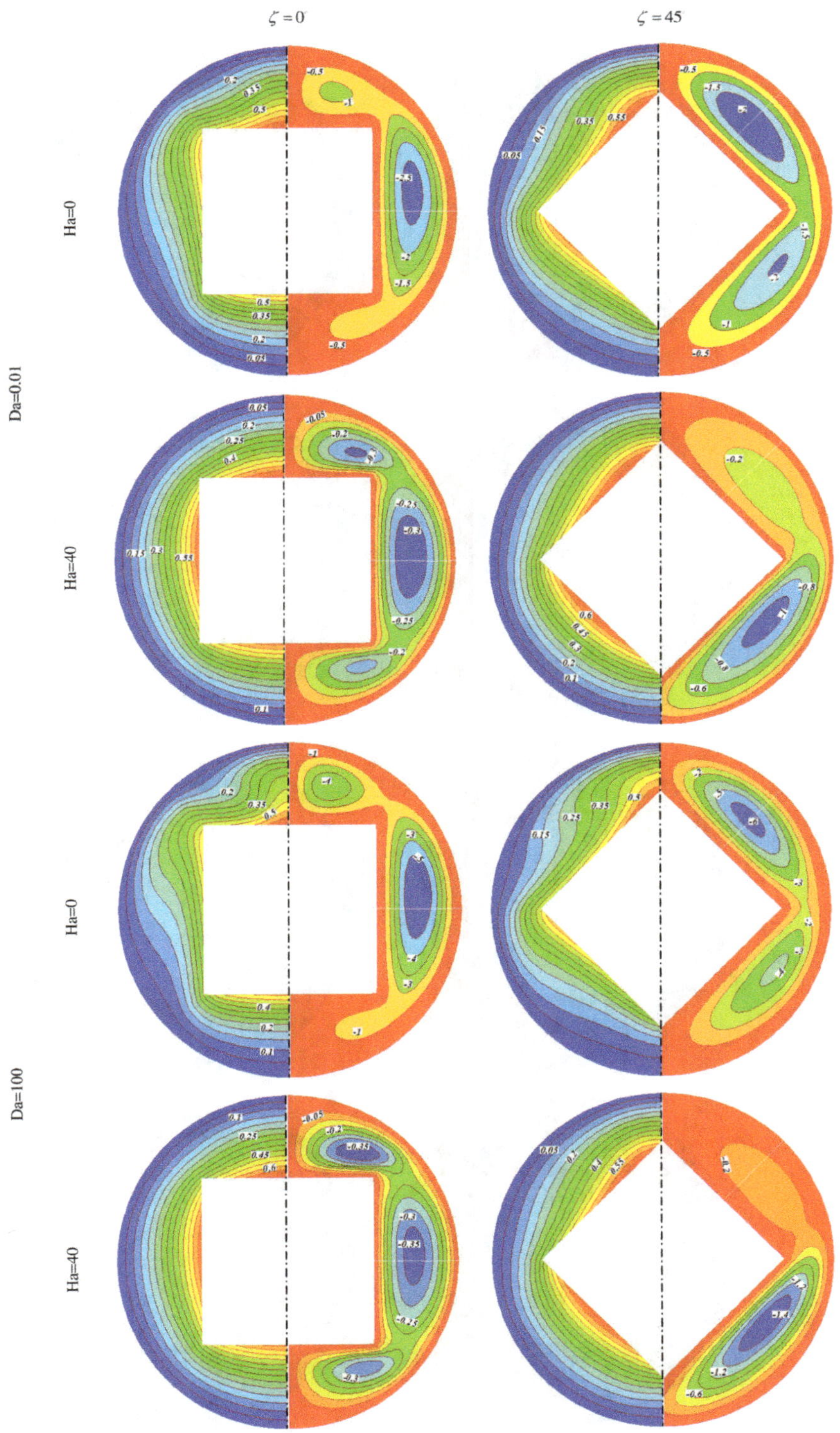

Figure 25. Effect of Darcy and Hartmann numbers on Isotherms (left) and streamlines (right) contours when $\phi = 0.04, Ra = 10^5$

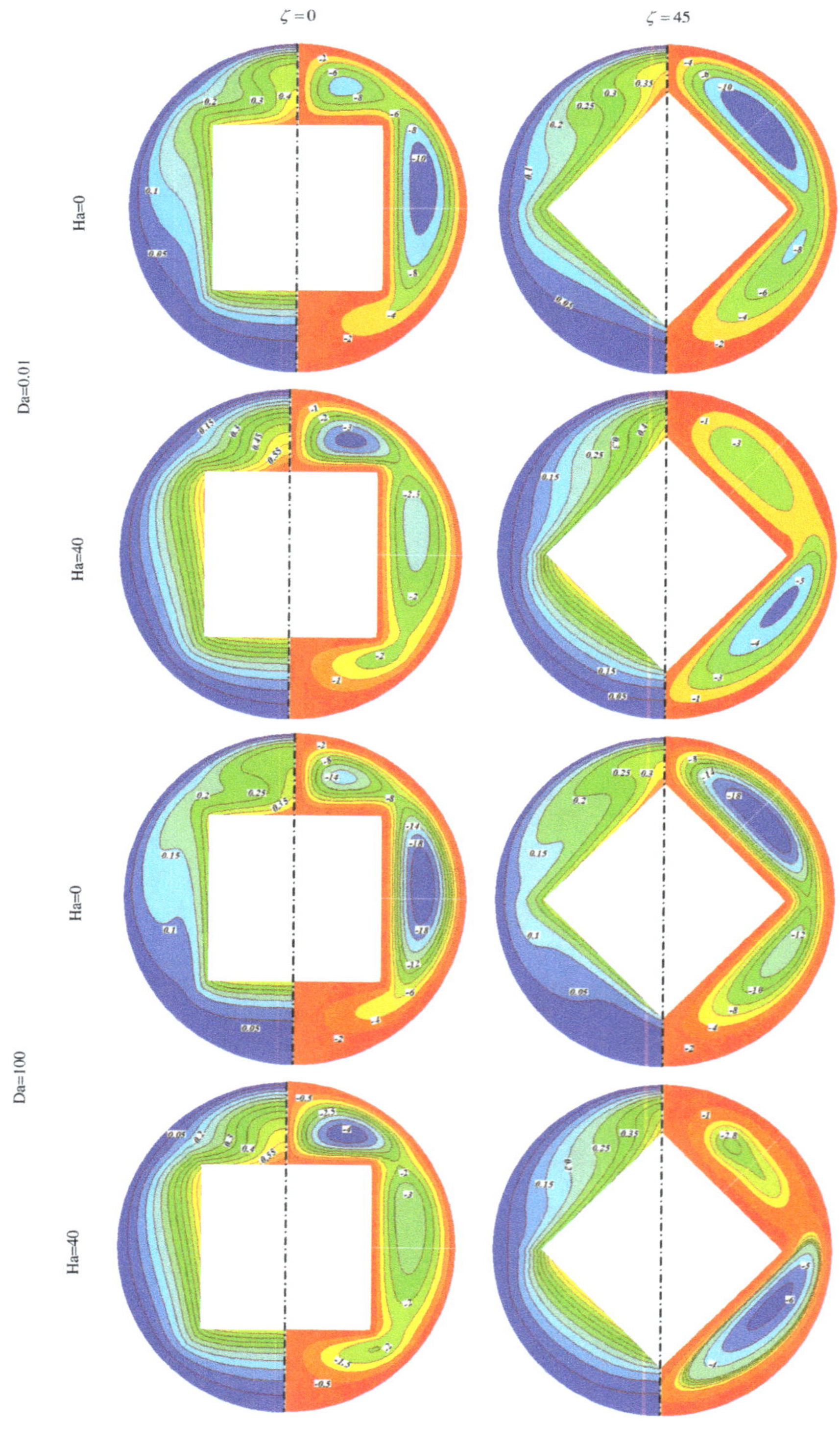

Figure 26a. Effects of inclination angle, Darcy, Rayleigh and Hartmann numbers on local Nusselt number when $\phi = 0.04$

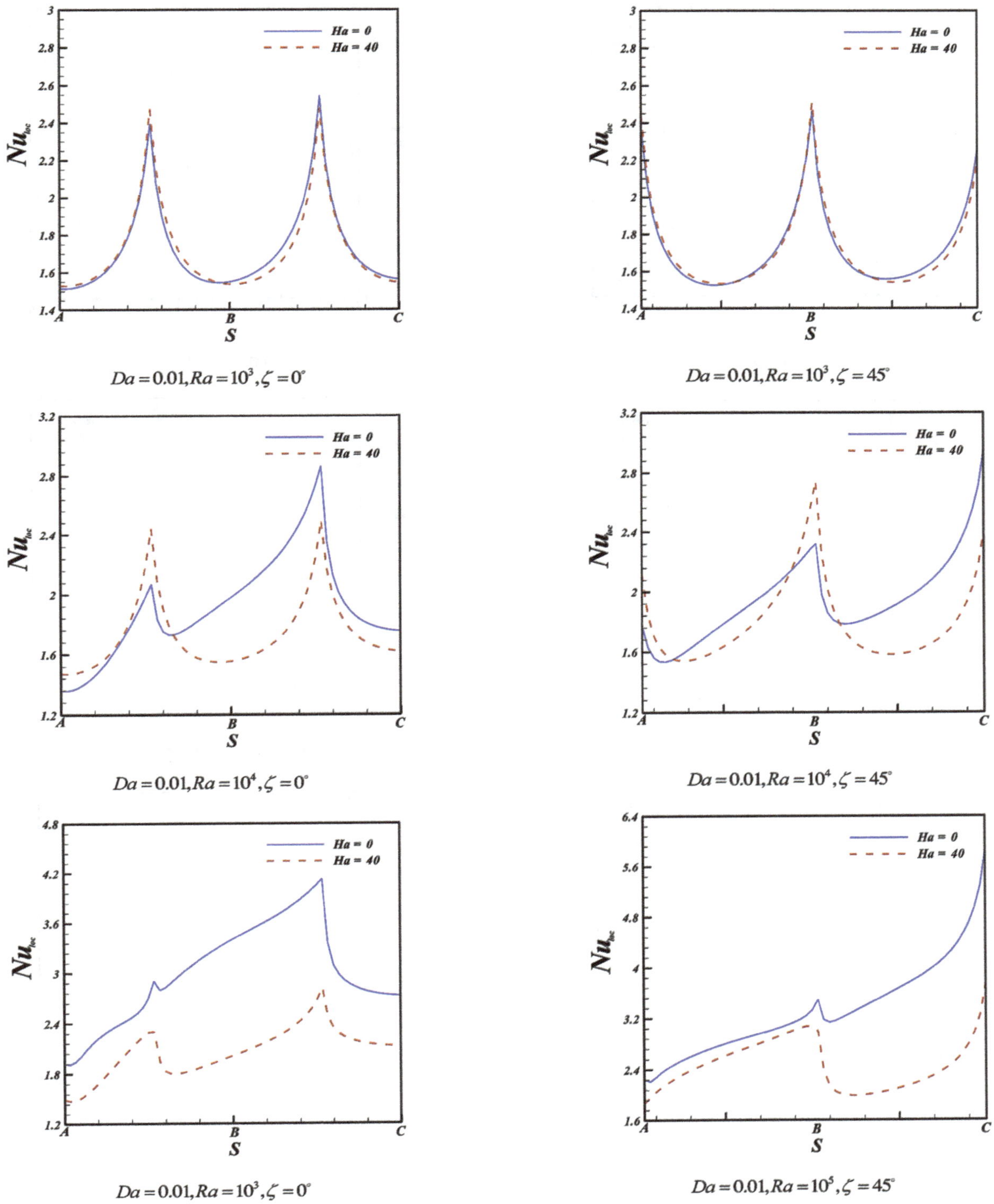

Figure 26b. Effects of inclination angle, Darcy, Rayleigh and Hartmann numbers on local Nusselt number when $\phi = 0.04$

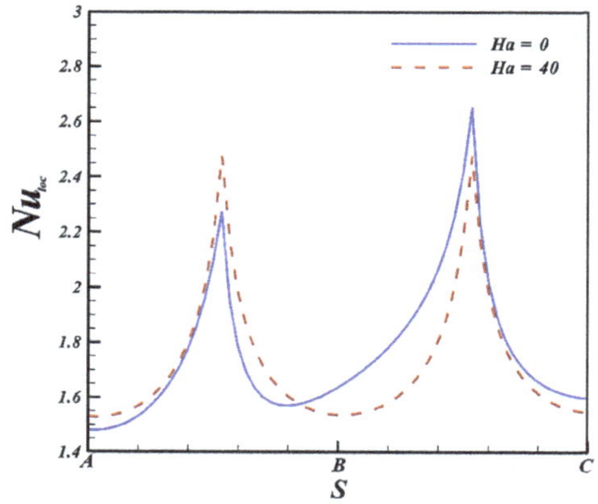

$Da = 100, Ra = 10^3, \zeta = 0^{\circ}$

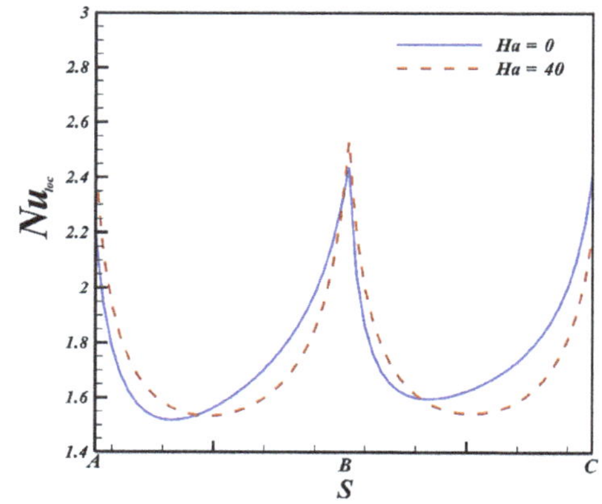

$Da = 100, Ra = 10^3, \zeta = 45^{\circ}$

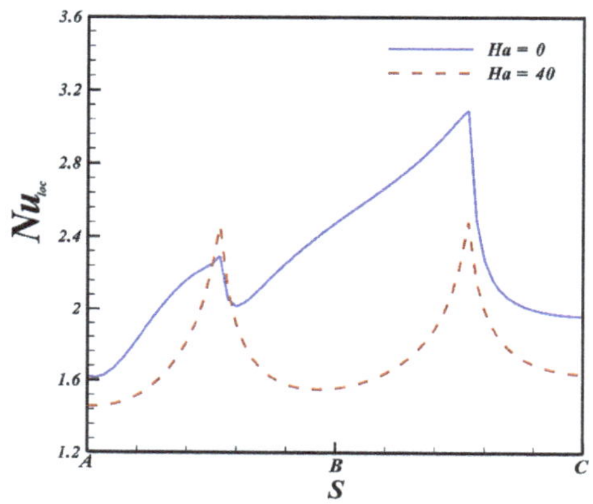

$Da = 100, Ra = 10^4, \zeta = 0^{\circ}$

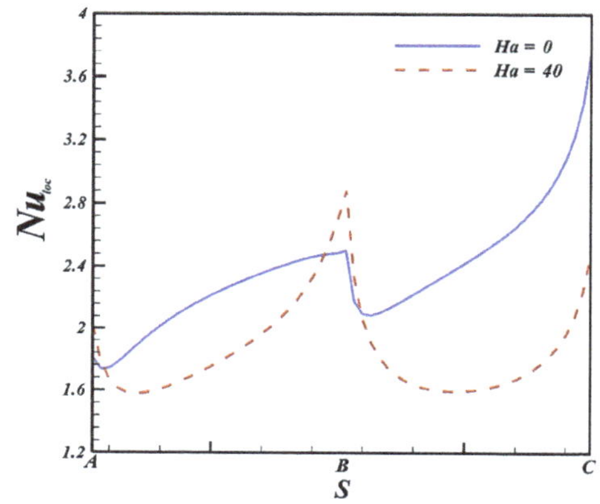

$Da = 100, Ra = 10^4, \zeta = 45^{\circ}$

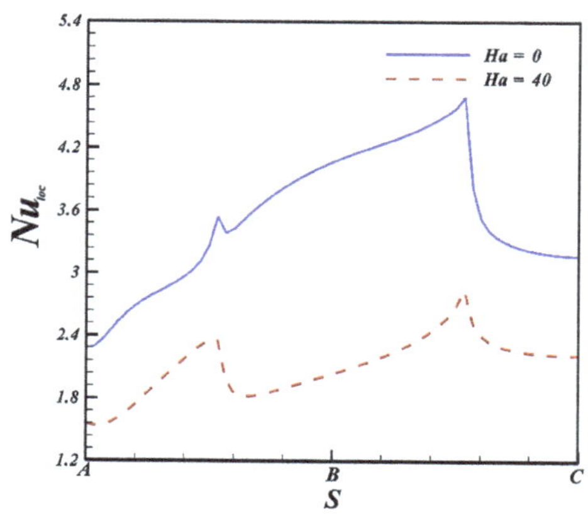

$Da = 100, Ra = 10^5, \zeta = 0^{\circ}$

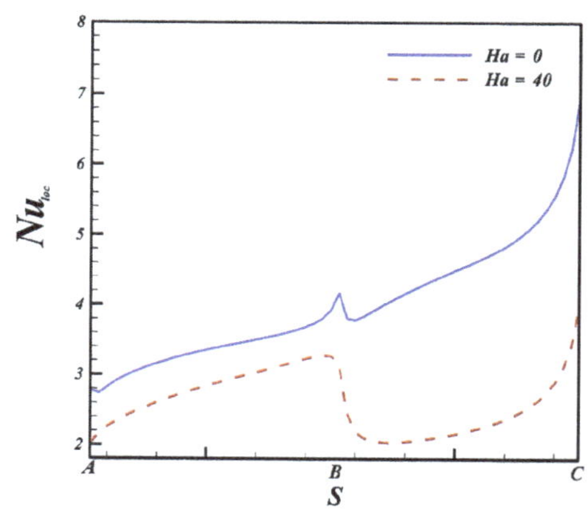

$Da = 100, Ra = 10^5, \zeta = 45^{\circ}$

Figure 27a. Effects of inclination angle, Darcy, Rayleigh and Hartmann numbers on average Nusselt number

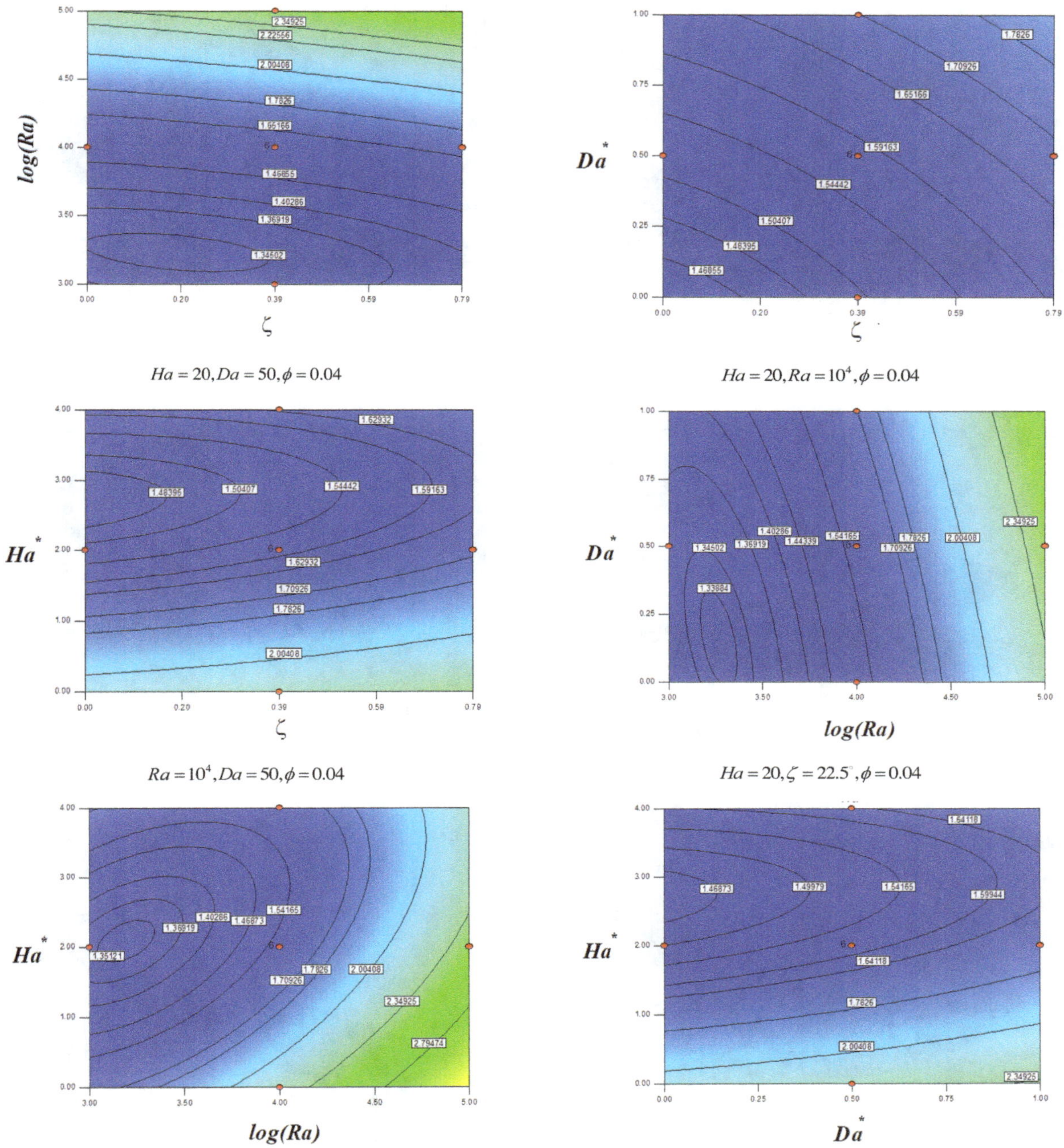

$Ha = 20, Da = 50, \phi = 0.04$

$Ha = 20, Ra = 10^4, \phi = 0.04$

$Ra = 10^4, Da = 50, \phi = 0.04$

$Ha = 20, \zeta = 22.5^{\circ}, \phi = 0.04$

$Da = 50, \zeta = 22.5^{\circ}, \phi = 0.04$

$Ra = 10^4, \zeta = 22.5^{\circ}, \phi = 0.04$

Figure 27b. Effects of inclination angle, Darcy, Rayleigh and Hartmann numbers on average Nusselt number

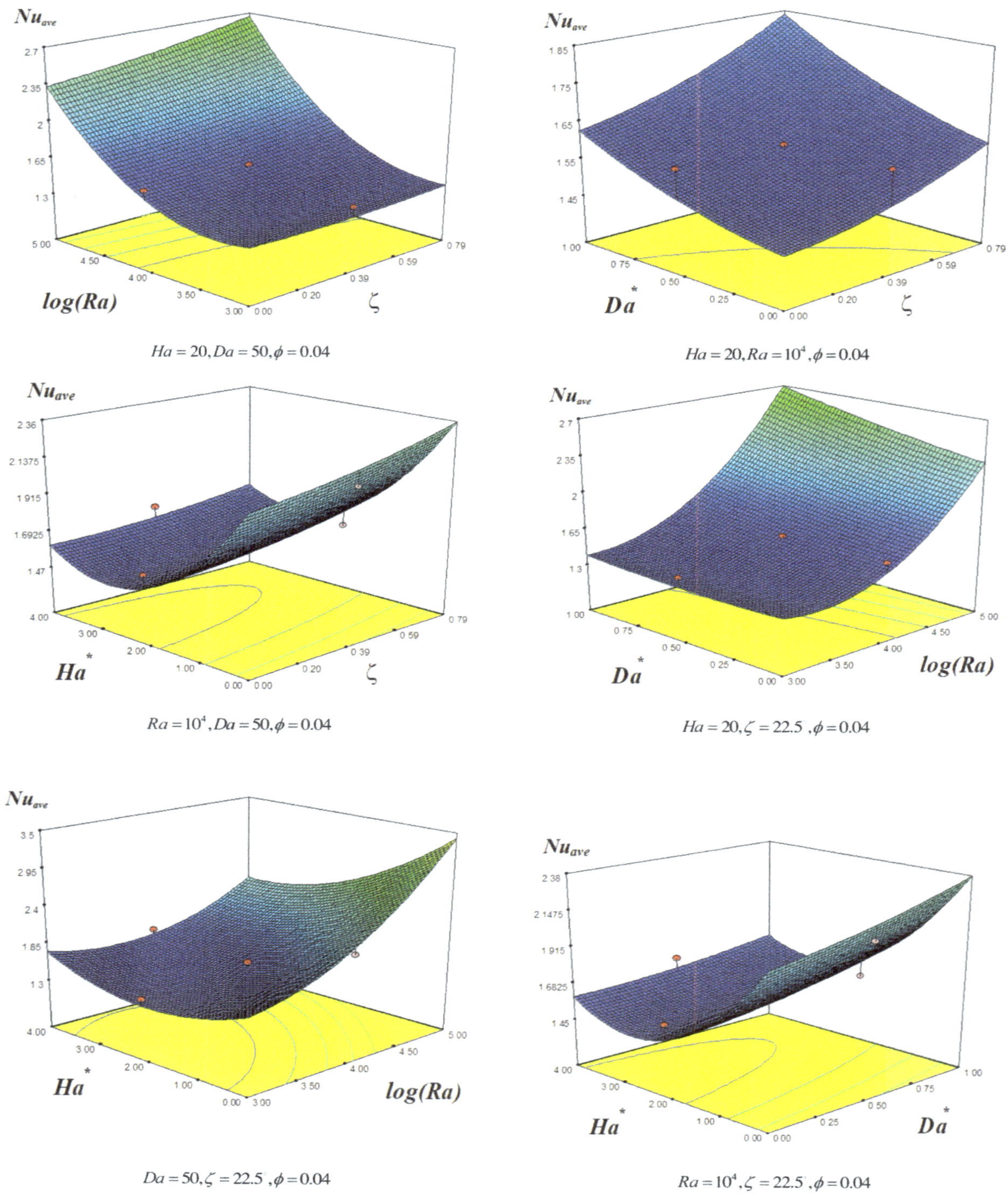

Figure 28. (a) Geometry and the boundary conditions with (b) the mesh of Geometry considered in this work; (c) A sample triangular element and its corresponding control volume.

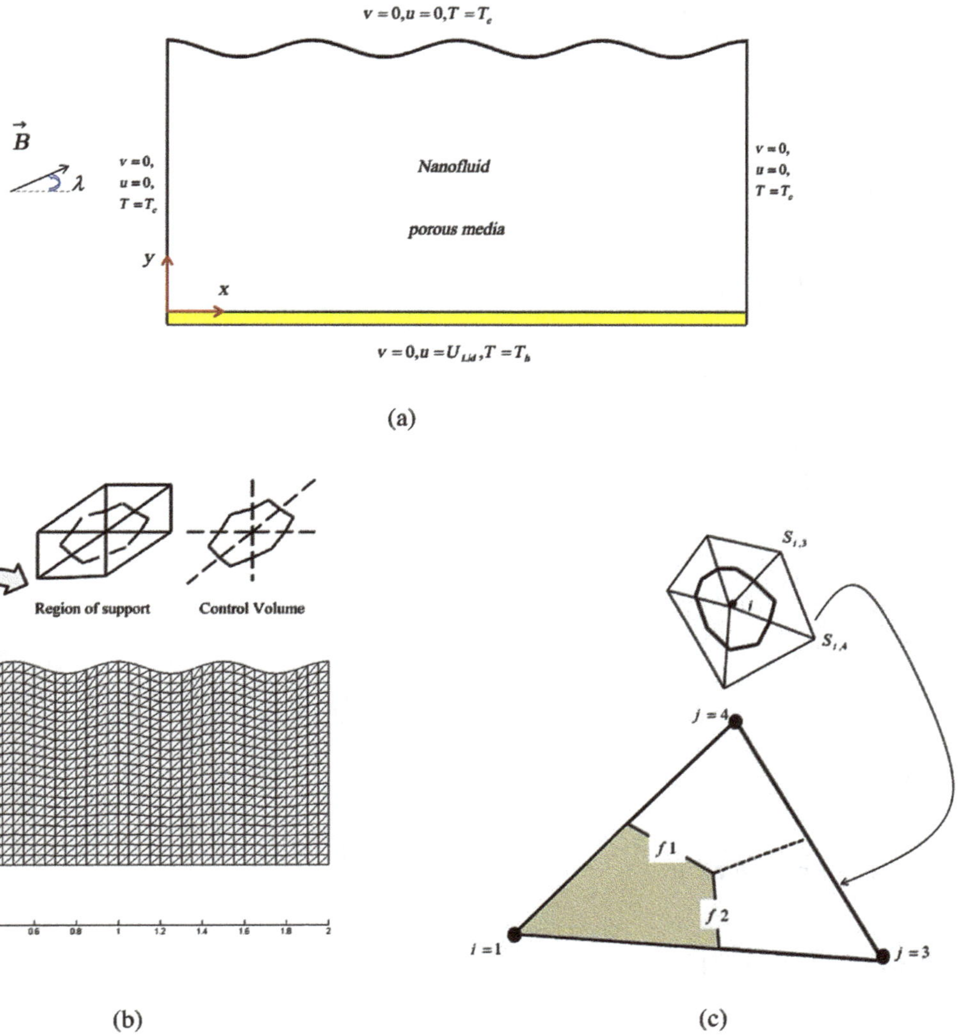

6.2. Governing Equation

Nanofluid forced convective non-Darcy flow is taken into account in presence of uniform magnetic field. The equations are:

$$\frac{\partial v}{\partial y}+\frac{\partial u}{\partial x}=0 \tag{85}$$

$$\left[B_y\sigma_{nf}B_x v - B_y^2\sigma_{nf}u + \left(\frac{\partial^2 u}{\partial y^2}+\frac{\partial^2 u}{\partial x^2}\right)\mu_{nf} - \frac{\partial P}{\partial x}\right] - \frac{\mu_{nf}}{K}u = \rho_{nf}\left(v\frac{\partial u}{\partial y}+\frac{\partial u}{\partial x}u\right) \tag{86}$$

$$\mu_{nf}\left(\frac{\partial^2 v}{\partial y^2}+\frac{\partial^2 v}{\partial x^2}\right)-\frac{\partial P}{\partial y}+B_y\sigma_{nf}uB_x-B_x\sigma_{nf}B_xv-\frac{\mu_{nf}}{K}v=\rho_{nf}\left(\frac{\partial v}{\partial x}u+\frac{\partial v}{\partial y}v\right),\quad B_x=B_o\cos\lambda,\ B_y=B_o\sin\lambda \tag{87}$$

$$k_{nf}\left(\frac{\partial^2 T}{\partial x^2}+\frac{\partial^2 T}{\partial y^2}\right)=\left(\rho C_p\right)_{nf}\left(\frac{\partial T}{\partial x}u+\frac{\partial T}{\partial y}v\right), \tag{88}$$

$\sigma_{nf}, \left(\rho C_p\right)_{nf}$ and ρ_{nf} are:

$$\frac{\sigma_{nf}}{\sigma_f}=1+\frac{3\left(\frac{\sigma_s}{\sigma_f}-1\right)\phi}{\left(\frac{\sigma_s}{\sigma_f}+2\right)-\left(\frac{\sigma_s}{\sigma_f}-1\right)\phi} \tag{89}$$

$$\left(\rho C_p\right)_{nf}=\phi\left(\rho C_p\right)_s+(1-\phi)\left(\rho C_p\right)_f \tag{90}$$

$$\rho_{nf}=\rho_f(1-\phi)+\rho_s\phi, \tag{91}$$

μ_{nf}, k_{nf} can be presented as:

$$\mu_{nf}=\frac{k_{Brownian}}{k_f}\times\frac{\mu_f}{\Pr}+\mu_f\left(1-\phi\right)^{-2.5} \tag{92}$$

$$\begin{gathered}k_{nf}=k_{Brownian}+k_{static}\\ \frac{k_{Brownian}}{k_f}=5\times10^4 g'(d_p,T,\phi)\phi\sqrt{\frac{\kappa_b T}{\rho_p d_p}}\rho_f c_{p,f},\ \frac{k_{static}}{k_f}=1+\frac{3\left(-1+k_p/k_f\right)\phi}{\left(\frac{k_p}{k_f}+2\right)-\left(\frac{k_p}{k_f}-1\right)\phi},\\ g'\left(d_p,T,\phi\right)=Ln\left(T\right)\left(Ln\left(d_p\right)a_2+a_5Ln\left(d_p\right)^2+a_1+a_3Ln\left(\phi\right)+Ln\left(\phi\right)Ln\left(d_p\right)a_4\right)\\ +\left(a_7Ln\left(d_p\right)+a_6+Ln\left(d_p\right)a_9Ln\left(\phi\right)+a_8Ln\left(\phi\right)+a_{10}Ln\left(d_p\right)^2\right),R_f=4\times10^{-8}km^2/W,\\ R_f=\left[-k_p^{-1}+k_{p,eff}^{-1}\right]d_p,\end{gathered} \tag{93}$$

Properties and needed parameters are provided in Tables 1 and 2.

ψ, ω can be defined as:

$$\omega + \frac{\partial u}{\partial y} - \frac{\partial v}{\partial x} = 0, \frac{\partial \psi}{\partial x} = -v, \frac{\partial \psi}{\partial y} = u \tag{94}$$

Introducing dimensionless quantities:

$$\theta = \frac{T - T_c}{\Delta T}, \Delta T = T_h - T_c, \left(X, Y\right) = \frac{\left(x, y\right)}{L}, \Psi = \frac{\psi L}{U_{Lid}},$$
$$V = \frac{v}{U_{Lid}}, \Omega = \frac{\omega}{L U_{Lid}}, U = \frac{u}{U_{Lid}} \tag{95}$$

The final formulae are:

$$\frac{\partial^2 \Psi}{\partial X^2} + \Omega + \frac{\partial^2 \Psi}{\partial Y^2} = 0, \tag{96}$$

$$\frac{\partial \Omega}{\partial Y} V + \frac{\partial \Omega}{\partial X} U = \frac{1}{\mathrm{Re}} \frac{A_5}{A_1} \left(\frac{\partial^2 \Omega}{\partial Y^2} + \frac{\partial^2 \Omega}{\partial X^2} \right)$$
$$+ \frac{Ha^2}{\mathrm{Re}} \frac{A_6}{A_1} \left(\frac{\partial U}{\partial Y} B_y^2 - \frac{\partial V}{\partial X} B_x^{\ 2} + B_y B_x \frac{\partial U}{\partial X} - B_x \frac{\partial V}{\partial Y} B_y \right) \tag{97}$$
$$- \frac{1}{\mathrm{Re}\, Da} \frac{A_5}{A_1} \Omega,$$

$$\frac{A_4}{A_2} \left(\frac{\partial^2 \theta}{\partial X^2} + \frac{\partial^2 \theta}{\partial Y^2} \right) = \Pr \mathrm{Re} \left(V \frac{\partial \theta}{\partial Y} + \frac{\partial \theta}{\partial X} U \right) \tag{98}$$

where dimensionless and constants parameters are:

$$\mathrm{Re} = \frac{\rho_f U_{Lid} L}{\mu_f}, Ha = L B_0 \sqrt{\sigma_f / \mu_f},$$
$$A_1 = \frac{\rho_{nf}}{\rho_f}, A_5 = \frac{\mu_{nf}}{\mu_f}, \tag{99}$$
$$A_2 = \frac{\left(\rho C_p\right)_{nf}}{\left(\rho C_p\right)_f}, A_4 = \frac{k_{nf}}{k_f}, A_6 = \frac{\sigma_{nf}}{\sigma_f}$$

and boundary conditions are:

$\Psi = 0.0$ on all walls

$\theta = 1.0$ on south wall

$\theta = 0.0$ on other walls (100)

Nu_{loc} and Nu_{ave} over the south wall are:

$$Nu_{loc} = \left(\frac{k_{nf}}{k_f}\right)\frac{\partial\theta}{\partial y} \tag{101}$$

$$Nu_{ave} = \frac{1}{L}\int_{r_{in}}^{r_{out}} Nu_{loc}\,dx \tag{102}$$

6.3. Effects of Active Parameters

Nanofluid MHD forced convection in a porous sinusoidal enclosure is examined numerically in this paper. Numerical outputs are presented for various Darcy number ($Da = 0.01$ to 100), Hartmann number ($Ha = 0$ to 20), Reynolds number ($\mathrm{Re} = 100$ to 600) and volume fraction of CuO ($\phi = 0\%$ to 4%).

Figure 29 illustrates the influence of ϕ on isotherms and streamlines. Increasing ϕ leads to increase thermal boundary layer thickness. The nanofluid velocity enhances by adding nanoparticles.

Figures 30 and 31 illustrate the effect of Darcy, Hartmann and Reynolds numbers on hydrothermal behavior. In absence of magnetic field, when Darcy and Reynolds number are low, only one eddy exists in streamline and isotherms are parallel together. As Reynolds number increases, isotherms become denser to lid wall due to increment of convective mode. Also $|\Psi_{\max}|$ augments with rise of Reynolds number. As Darcy number increases, convective mode becomes stronger due to increase in permeability of medium. So temperature gradient over the hot wall increases with augment of Darcy number. Increasing Lorentz forces make the isotherms becomes less dense. Also velocity reduces with rise of Hartmann number.

Influence of important parameters on Nu_{ave} is depicted in Figure 32. The correlation for average Nusselt number is as follows:

$$\begin{aligned} Nu_{ave} &= 7.56 + 4.19\,\mathrm{Re}^* + 0.45Da^* - 0.1Ha^* + 25.6\phi \\ &+0.017\,\mathrm{Re}^*\,Da^* + 0.048\,\mathrm{Re}^*\,Ha^* + 0.45\phi\,\mathrm{Re}^* - 0.0018Da^*Ha^* \\ &+0.6Da^*\phi - 0.048Ha^*Da^* - 0.38\,\mathrm{Re}^{*2} + 0.52Da^{*2} - 0.22Ha^{*2} - 1.35\phi^2 \end{aligned} \tag{103}$$

Figure 29. Influence of nanofluid volume fraction on streamlines (left) and isotherms (right) contours (nanofluid ($\phi = 0.04$)(—) and pure fluid($\phi = 0$) (- - -)) when $Da = 100$

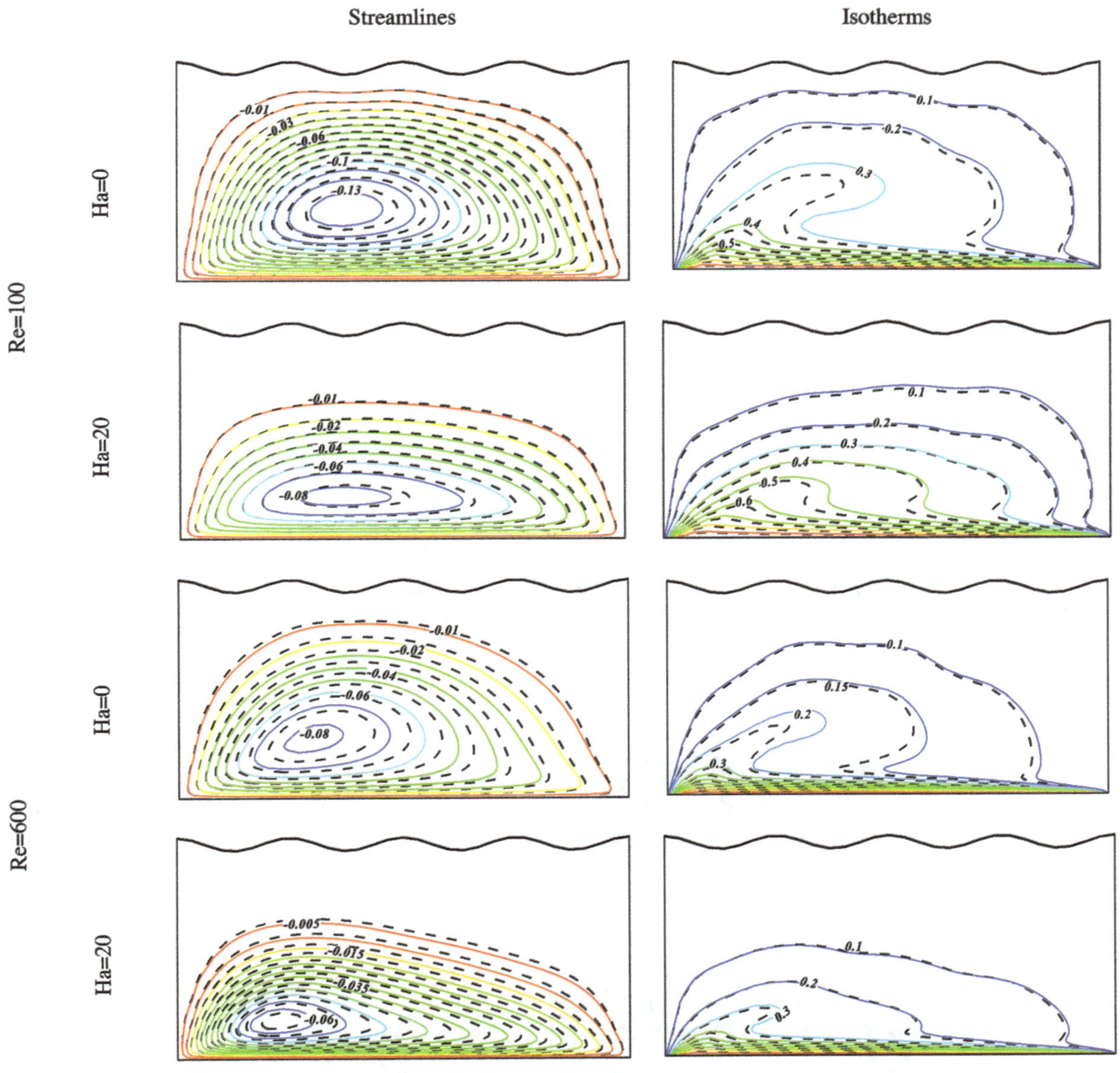

where $\text{Re}^* = 0.01\,\text{Re}, Ha^* = 0.1Ha$. Adding nanoparticles makes Nusselt number to increases due to increase in k_{nf} . Nu_{ave} improves with increase of Darcy and Reynolds numbers because of increment in convective heat transfer. As Ha improves the temperature gradient decreases and in turn Reynolds reduces with augment of Lorentz forces.

Figure 30. Isotherms (left) and streamlines (right) contours for different values of Reynolds and Hartmann numbers when $Da = 0.01, \phi = 0.04$.

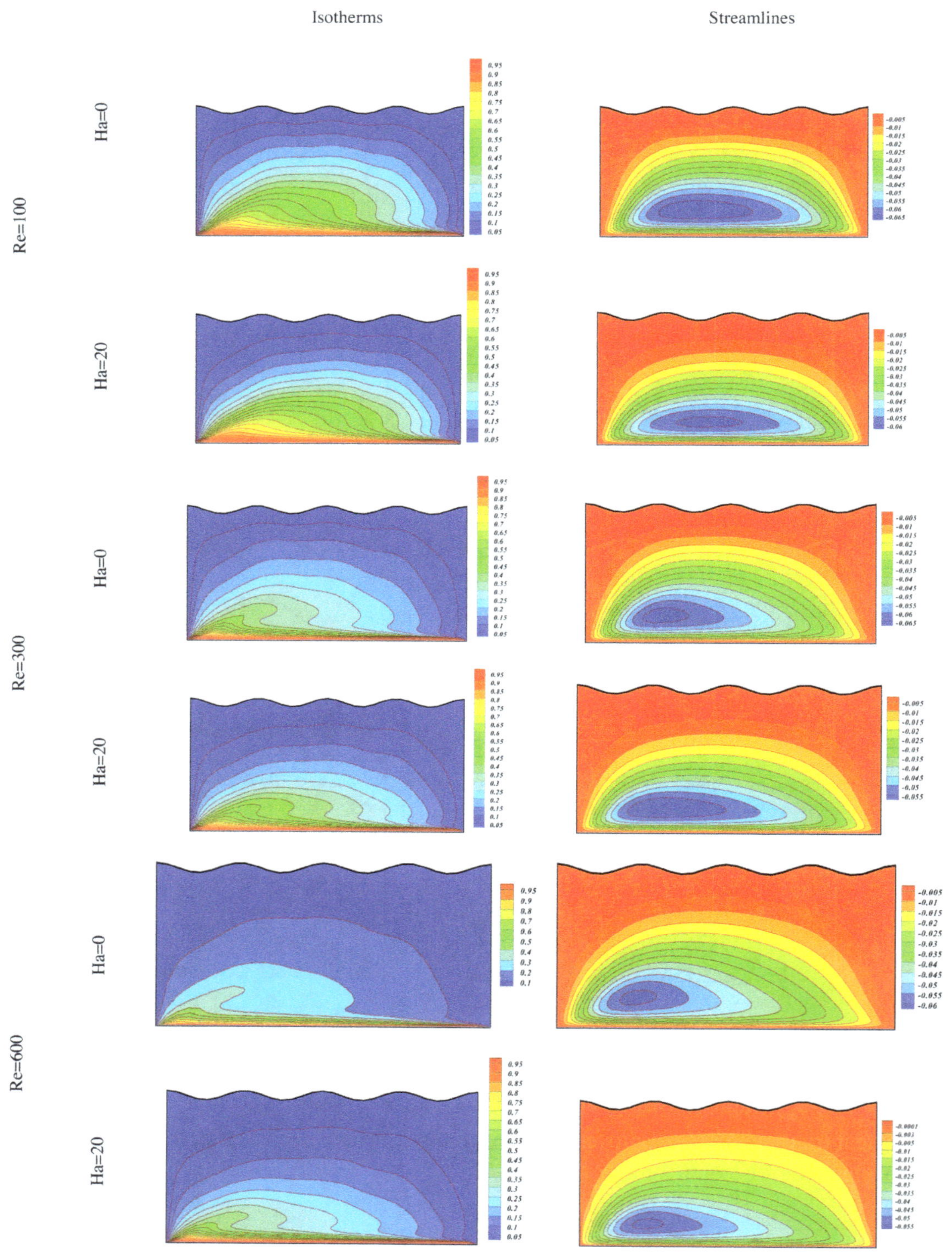

Figure 31. Isotherms (left) and streamlines (right) contours for different values of Reynolds and Hartmann numbers when $Da = 100, \phi = 0.04$.

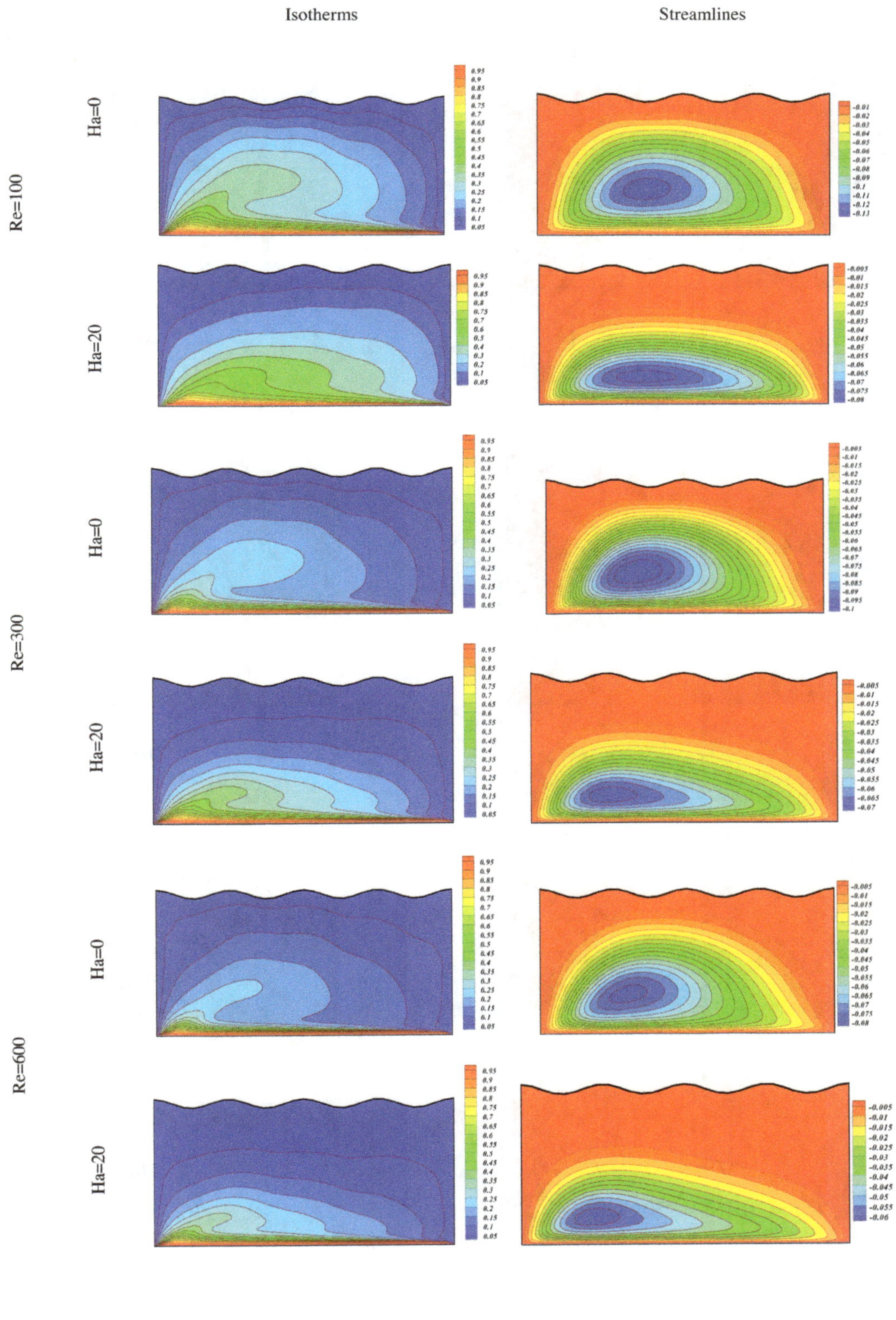

Figure 32a. Influences of the volume fraction of nanofluid, Darcy, Reynolds number and Hartmann numbers on average Nusselt number.

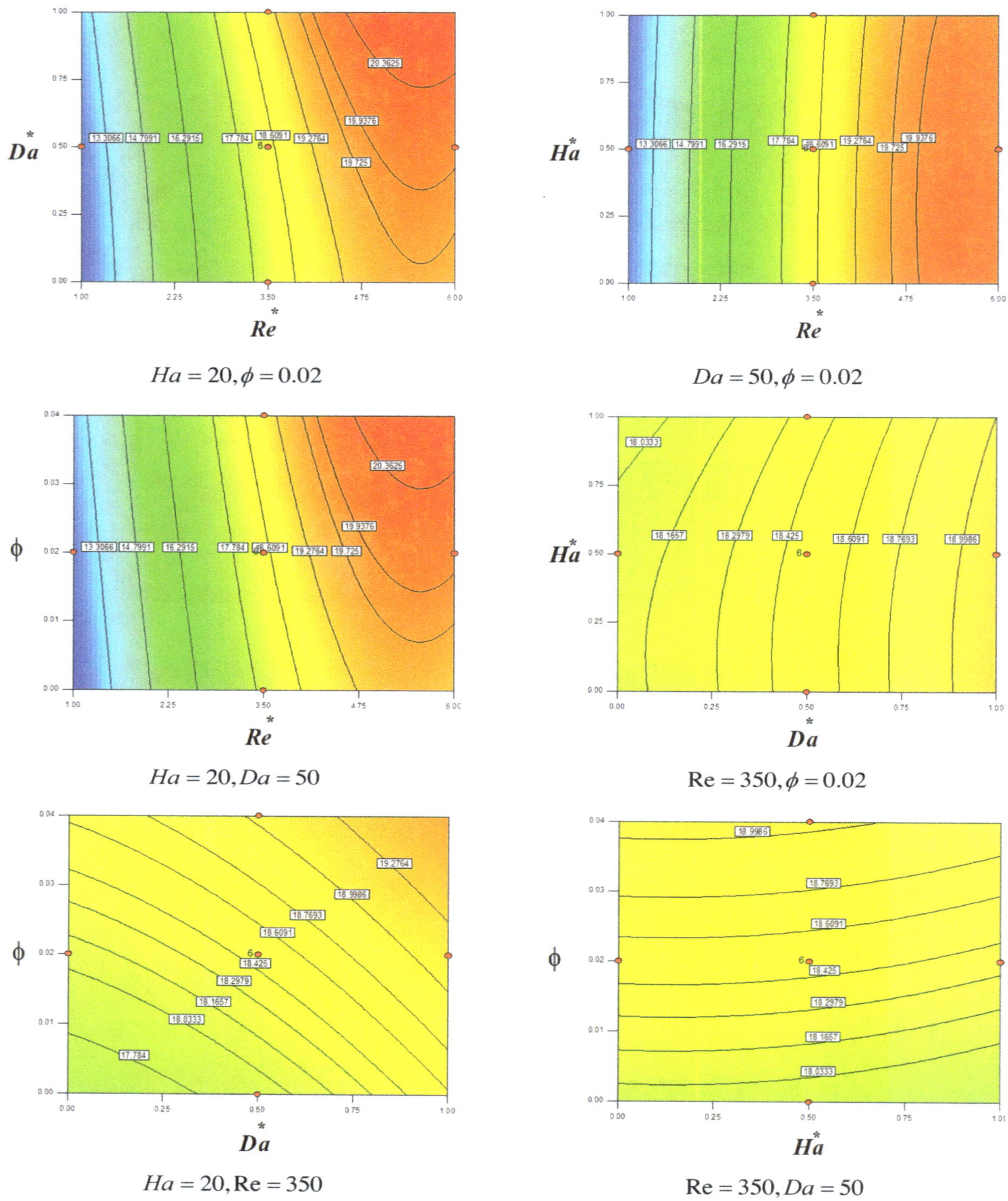

Figure 32b. Influences of the volume fraction of nanofluid, Darcy, Reynolds number and Hartmann numbers on average Nusselt number.

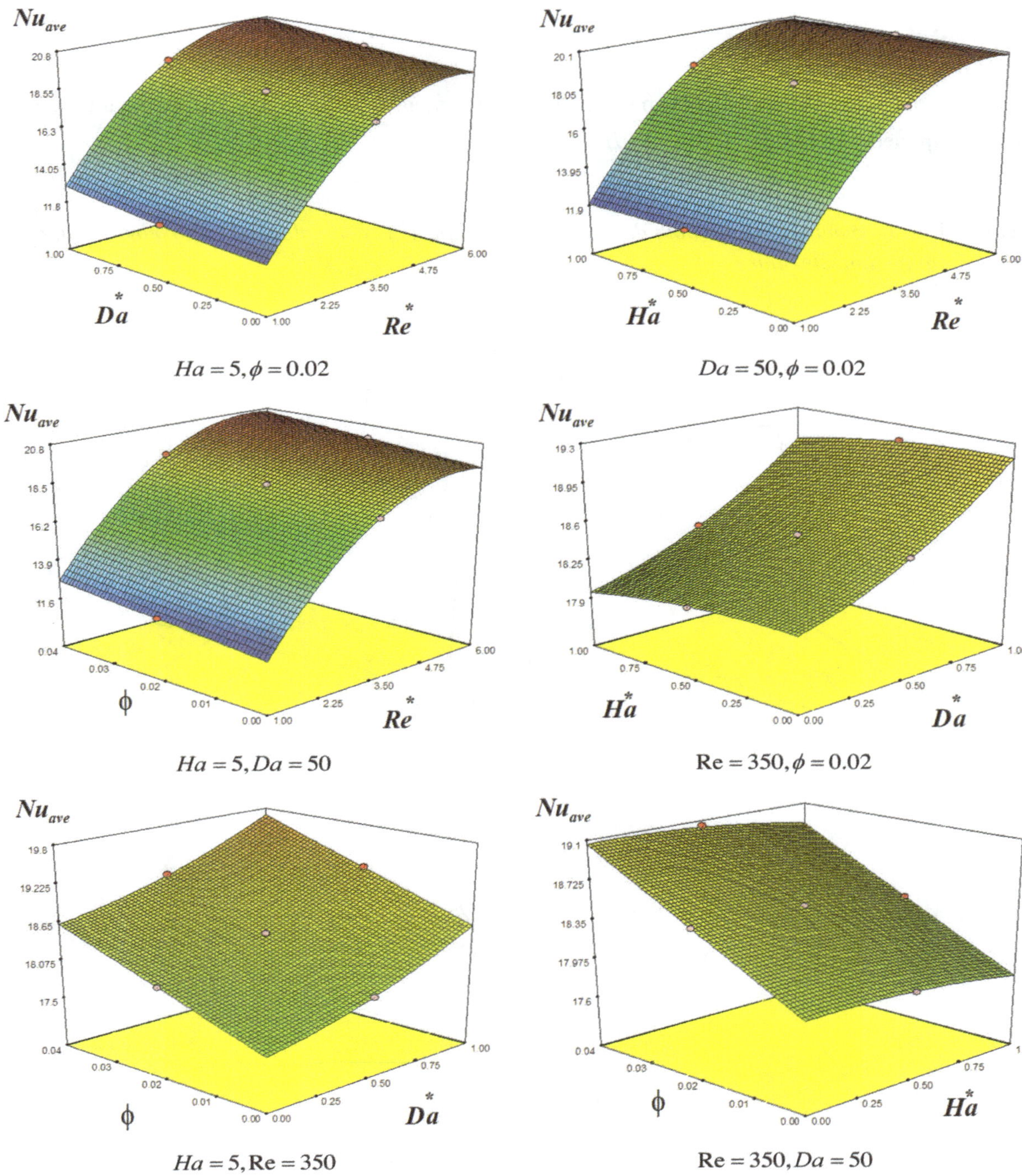

7. NANOFLUID HEAT TRANSFER IN A PERMEABLE ENCLOSURE IN PRESENCE OF VARIABLE MAGNETIC FIELD

7.1. Problem Definition

Geometry, boundary condition and sample element are demonstrated in Figure 33. External magnetic source is applied (see Figure 34). $\overline{H}, \overline{H_x}, \overline{H_y}$ are:

Figure 33. (a)Geometry and the boundary conditions with; (b) A sample triangular element and its corresponding control volume.

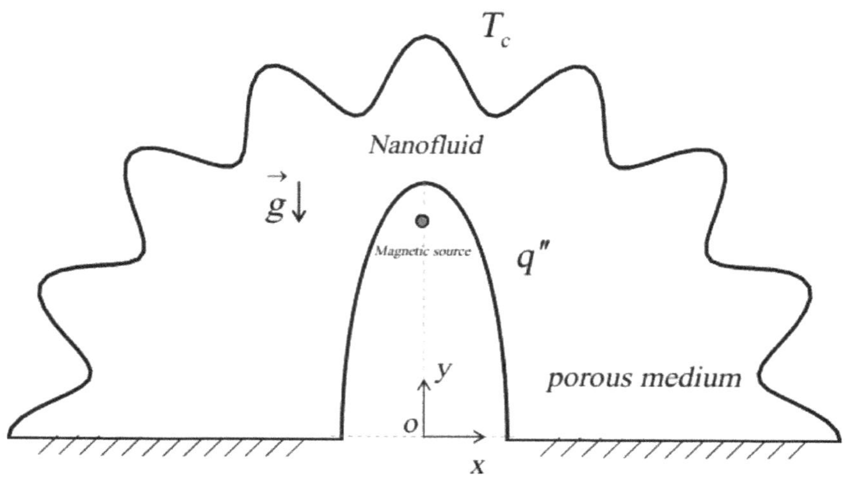

(a)

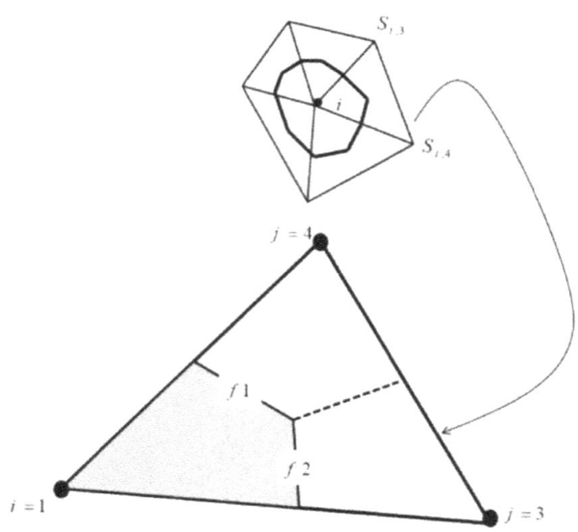

(b)

Figure 34. Contours of the (a) magnetic field strength H; (b) magnetic field intensity component in x direction Hx; (c) magnetic field intensity component in y direction Hy.

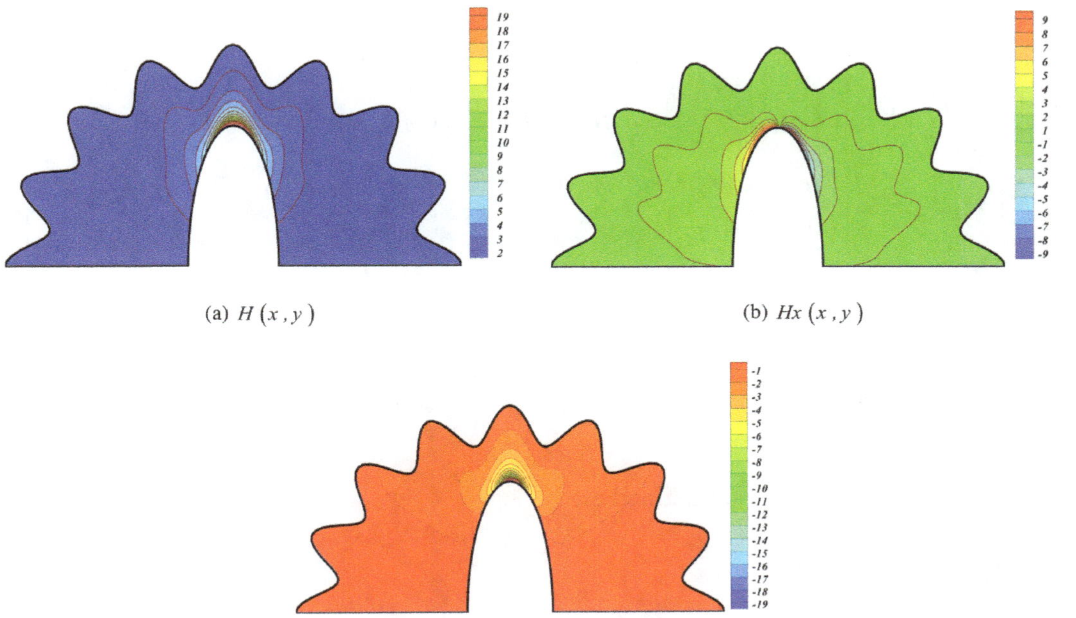

(a) $H(x, y)$ (b) $Hx(x, y)$ (c) $Hy(x, y)$

$$\overline{H}_y = \left(\overline{a} - x\right)\frac{\gamma}{2\pi}\left[\left(\overline{b} - y\right)^2 + \left(\overline{a} - x\right)^2\right]^{-1}, \tag{104}$$

$$\overline{H}_x = \left(y - \overline{b}\right)\frac{\gamma}{2\pi}\left[\left(\overline{b} - y\right)^2 + \left(\overline{a} - x\right)^2\right]^{-1}, \tag{105}$$

$$\overline{H} = \left(\overline{H}_y^2 + \overline{H}_x^2\right)^{0.5}. \tag{106}$$

7.2. Governing Equation

Two dimensional convective non-Darcy flow of nanofluid is considered in existence of external magnetic source. The governing equations are:

$$\frac{\partial u}{\partial x} = -\frac{\partial v}{\partial y}, \tag{107}$$

$$\left(\frac{\partial^2 u}{\partial y^2}+\frac{\partial^2 u}{\partial x^2}\right)\mu_{nf}-\frac{\partial P}{\partial x}-\mu_0^2\sigma_{nf}\bar{H}_y^2 u+\sigma_{nf}\mu_0^2\bar{H}_x\bar{H}_y v-\frac{\mu_{nf}}{K}u=\left(\rho_{nf}\right)\left(\frac{\partial u}{\partial x}u+\frac{\partial u}{\partial y}v\right), \tag{108}$$

$$\begin{aligned}&\mu_{nf}\left(\frac{\partial^2 v}{\partial x^2}+\frac{\partial^2 v}{\partial y^2}\right)-\frac{\partial P}{\partial y}+\mu_0^2\bar{H}_y\sigma_{nf}\bar{H}_x u-\mu_0^2\bar{H}_x\sigma_{nf}\bar{H}_x v-\frac{\mu_{nf}}{K}v\\&+\left(T-T_c\right)\beta_{nf}g\rho_{nf}=\rho_{nf}\left(\frac{\partial v}{\partial x}u+\frac{\partial v}{\partial y}v\right),\end{aligned} \tag{109}$$

$$k_{nf}\left(\frac{\partial^2 T}{\partial y^2}+\frac{\partial^2 T}{\partial x^2}\right)-\frac{\partial q_r}{\partial y}=\left(v\frac{\partial T}{\partial y}+u\frac{\partial T}{\partial x}\right)\left(\rho C_p\right)_{nf},\left[q_r=-\frac{4\sigma_e}{3\beta_R}\frac{\partial T^4}{\partial y},T^4\cong 4T_c^3T-3T_c^4\right]. \tag{110}$$

$\left(\rho C_p\right)_{nf}, \rho_{nf}, \left(\rho\beta\right)_{nf}$ and σ_{nf} are defined as:

$$\left(\rho C_p\right)_{nf}=\left(\rho C_p\right)_f(1-\phi)+\left(\rho C_p\right)_s\phi \tag{111}$$

$$\rho_{nf}=\rho_f(1-\phi)+\rho_s\phi \tag{112}$$

$$\left(\rho\beta\right)_{nf}=\left(\rho\beta\right)_f(1-\phi)+\left(\rho\beta\right)_s\phi, \tag{113}$$

$$\sigma_{nf}=\sigma_f\left[\frac{3\phi\left(\sigma 1-1\right)}{\left(1-\sigma 1\right)\phi+\left(2+\sigma 1\right)}+1\right],\sigma 1=\sigma_s/\sigma_f. \tag{114}$$

μ_{nf} is calculated as follows:

$$\mu_{nf}=\left(0.035\mu_0^2\bar{H}^2+3.1\mu_0\bar{H}-27886.4807\phi^2+4263.02\phi+316.0629\right)e^{-0.01T} \tag{115}$$

k_{nf} can be calculated as:

$$\frac{k_{nf}}{k_f}=\frac{-m\left(k_f-k_p\right)\phi+\left(k_p-k_f\right)\phi+mk_f+k_p+k_f}{mk_f+\left(k_f-k_p\right)\phi+k_f+k_p} \tag{116}$$

Properties of nanofluid are depicted in Table 3.
Different values of shape factors for various shapes of nanoparticles are illustrated in Table 5.
Vorticity and stream function should be used to eliminate pressure source terms:

$$\omega + \frac{\partial u}{\partial y} - \frac{\partial v}{\partial x} = 0, \frac{\partial \psi}{\partial x} = -v, \frac{\partial \psi}{\partial y} = u. \tag{117}$$

Dimensionless parameters are defined as:

$$\begin{aligned} &\left(H_y, H_x, H\right) = \frac{\left(\overline{H_y}, \overline{H_x}, \overline{H}\right)}{\overline{H_0}}, \left(b, a\right) = \frac{\left(\bar{b}, \bar{a}\right)}{L}, \\ &U = \frac{uL}{\alpha_{nf}}, V = \frac{vL}{\alpha_{nf}}, \left(X, Y\right) = \frac{\left(x, y\right)}{L}, \theta = \frac{T - T_c}{\Delta T}, \Delta T = q''L / k_f, \\ &\Psi = \frac{\psi}{\alpha_{nf}}, \Omega = \frac{\omega L^2}{\alpha_{nf}}. \end{aligned} \tag{118}$$

So equations change to:

$$\frac{\partial^2 \Psi}{\partial Y^2} + \Omega + \frac{\partial^2 \Psi}{\partial X^2} = 0, \tag{119}$$

Table 5. The values of shape factor of different shapes of nanoparticles

m	Spherical	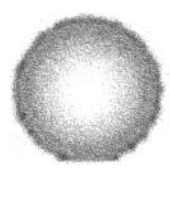	3
	Platelet		5.7
	Cylinder		4.8
	Brick	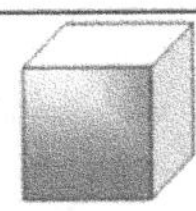	3.7

$$\begin{aligned}&\frac{\partial \Omega}{\partial X}U+V\frac{\partial \Omega}{\partial Y}=\Pr\frac{A_5}{A_1}\frac{A_2}{A_4}\left(\frac{\partial^2\Omega}{\partial Y^2}+\frac{\partial^2\Omega}{\partial X^2}\right)\\&+\Pr Ha^2\frac{A_6}{A_1}\frac{A_2}{A_4}\left(\frac{\partial U}{\partial X}H_yH_x-\frac{\partial V}{\partial X}H_x^{\,2}+\frac{\partial U}{\partial Y}H_y^2-\frac{\partial V}{\partial Y}H_yH_x\right)\\&+\Pr Ra\frac{A_3A_2^2}{A_1A_4^2}\frac{\partial \theta}{\partial X}-\frac{\Pr}{Da}\frac{A_5}{A_1}\frac{A_2}{A_4},\end{aligned}\tag{120}$$

$$\frac{\partial \theta}{\partial X}U+V\frac{\partial \theta}{\partial Y}=\left(\frac{\partial^2\theta}{\partial X^2}+\frac{\partial^2\theta}{\partial Y^2}\right)+\frac{4}{3}\frac{1}{A_4}Rd\frac{\partial^2\theta}{\partial Y^2}.\tag{121}$$

and dimensionless parameters are:

$$\begin{aligned}&Ra_f=g\beta_fL^3\Delta T/\left(\alpha_f\upsilon_f\right),\Pr_f=\upsilon_f/\alpha_f,Ha=L\mu_0H_0\sqrt{\sigma_f/\mu_f},\\&Da=K/L^2,\mathrm{Ec}=\left(\mu_f\alpha_f\right)/\left[\left(\rho C_P\right)_f\Delta T\,L^2\right],\\&A_1=\frac{\rho_{nf}}{\rho_f},A_2=\frac{\left(\rho C_p\right)_{nf}}{\left(\rho C_p\right)_f},A_3=\frac{\left(\rho\beta\right)_{nf}}{\left(\rho\beta\right)_f},\\&A_4=\frac{k_{nf}}{k_f},A_5=\frac{\mu_{nf}}{\mu_f},A_6=\frac{\sigma_{nf}}{\sigma_f}\\&,Rd=4\sigma_eT_c^3/\left(\beta_Rk_f\right)\end{aligned}\tag{122}$$

and boundary conditions are:

on inner wall $\frac{\partial \theta}{\partial n}=1.0$

on outer wall $\theta=0.0$

on other walls $\frac{\partial \theta}{\partial n}=0$

on all walls $\Psi=0.0$ (123)

Nu_{loc}, Nu_{ave} over the hot wall can calculate as:

$$Nu_{loc} = \left(\frac{k_{nf}}{k_f}\right)\left(1 + \left(\frac{k_{nf}}{k_f}\right)^{-1} \frac{4Rd}{3}\right)\frac{1}{\theta}, \tag{124}$$

$$Nu_{ave} = \frac{1}{S}\int_0^S Nu_{loc}\, ds. \tag{125}$$

7.3. Effects of Active Parameters

Impact of non-uniform magnetic field on Fe_3O_4-water flow in a permeable enclosure is simulated. Nanofluid viscosity is estimated according to previous experimental data. Shape effect of nanoparticles on k_{nf} is taken into consideration. CVFEM is utilized to find the effects of radiation parameter $\left(Rd = 0\, to\, 0.8\right)$, Darcy number ($Da = 0.01$ to 100), Rayleigh number ($Ra = 10^3, 10^4, 10^5$), volume fraction of Fe_3O_4-water ($\phi = 0\%$ to 4%), shape of nanoparticle and Hartmann number ($Ha = 0$ to 10).

Impacts of shape of the nanoparticles on Nu_{ave} are presented in Table 6. The maximum Nu_{ave} is obtained for Platelet, followed by Cylinder, Brick and Spherical. So, Platelet nanoparticle has been selected to complete this paper.

Figure 35 shows the impact of adding nanoparticles in to water on hydrothermal treatment. Temperature gradient reduces with increase of ϕ. Velocity augments with adding nanoparticles because of augmentation in the solid movements. Figure 36 demonstrates the effect of radiation parameter on streamline and isotherms contours. Thermal boundary layer thickness rises with augment of Rd. By adding magnetic field, impact of radiation parameter on streamlines becomes no significant. Figures 37, 38 and 39 illustrate the effects of Da, Ra, Ha on isotherms and streamlines. Only one eddy appears in streamline. By increasing Hartmann number, the main eddy moves downward and distortion of isotherms becomes less than before. As buoyancy forces augment, thermal plume generates near the vertical symmetric line. Augmenting Lorentz forces, shifts the thermal plume to left and reduces $\left|\Psi_{\max}\right|$. As permeability of the media augments, convective mode becomes stronger and isotherms shape becomes more complicated.

Effects of significant parameters on Nu_{ave} are depicted in Figure 40. The correlation for Nu_{ave} is:

Table 6. Effect of shape of nanoparticles on Nusselt number when $Da = 100, Ra = 10^5, Rd = 0.8, \phi = 0.04$

Ha		
10	0	
10.1131	12.32892	Spherical
10.15818	12.40613	Brick
10.23001	12.52868	Cylinder
10.28956	12.62989	Platelet

Figure 35. Impact of nanofluid volume fraction on streamlines (up) and isotherms (bottom) contours (nanofluid ($\phi = 0.04$)(—) and pure fluid ($\phi = 0$) ($-\cdot-$)) when $Ra = 10^5, Da = 100, Rd = 0.8$

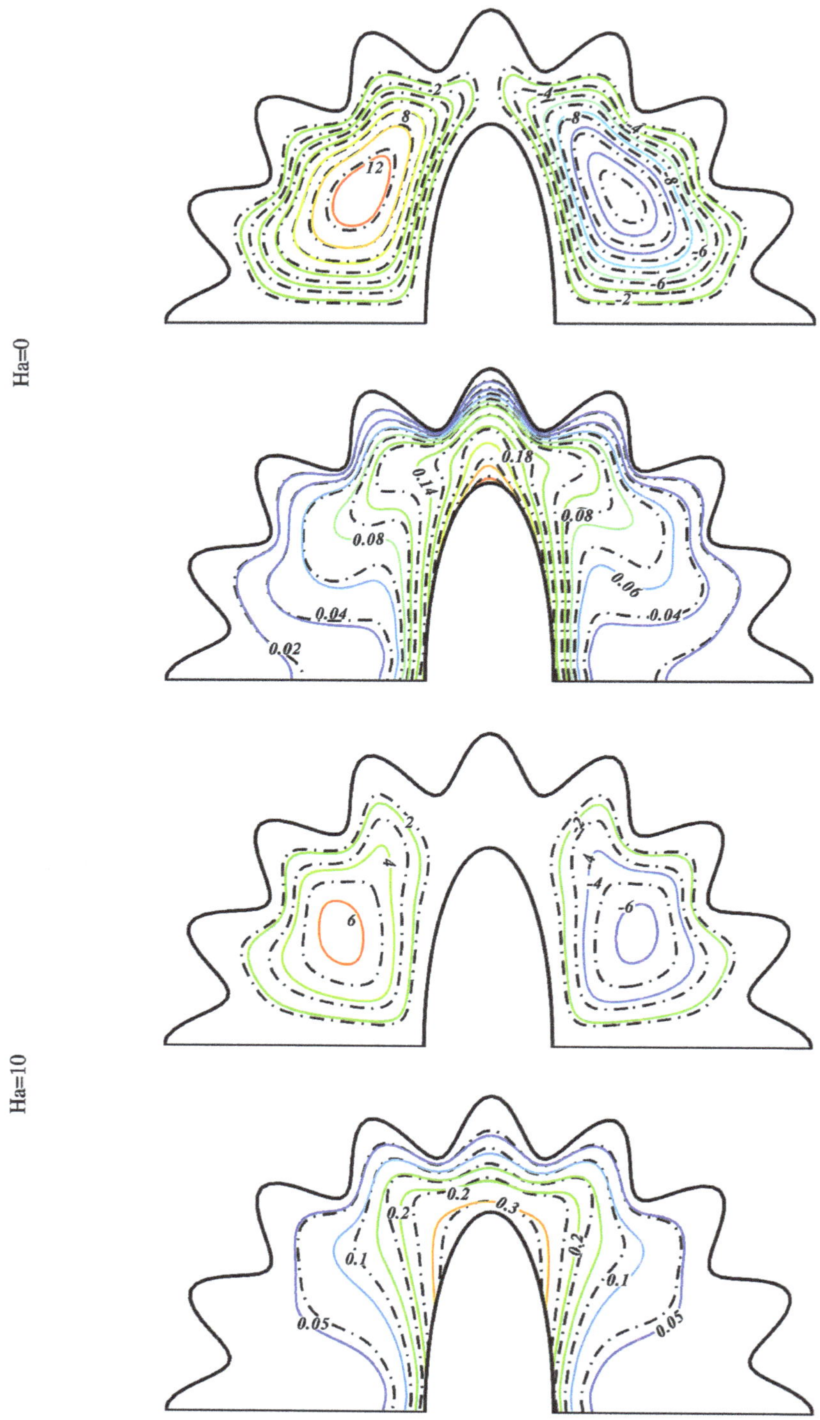

Figure 36. Impact of radiation parameter on streamlines (up) and isotherms (bottom) contours ($Rd = 0.8$ (—), $Rd = 0$ (- - -)) when $Ra = 10^5, Da = 100, \phi = 0.04$

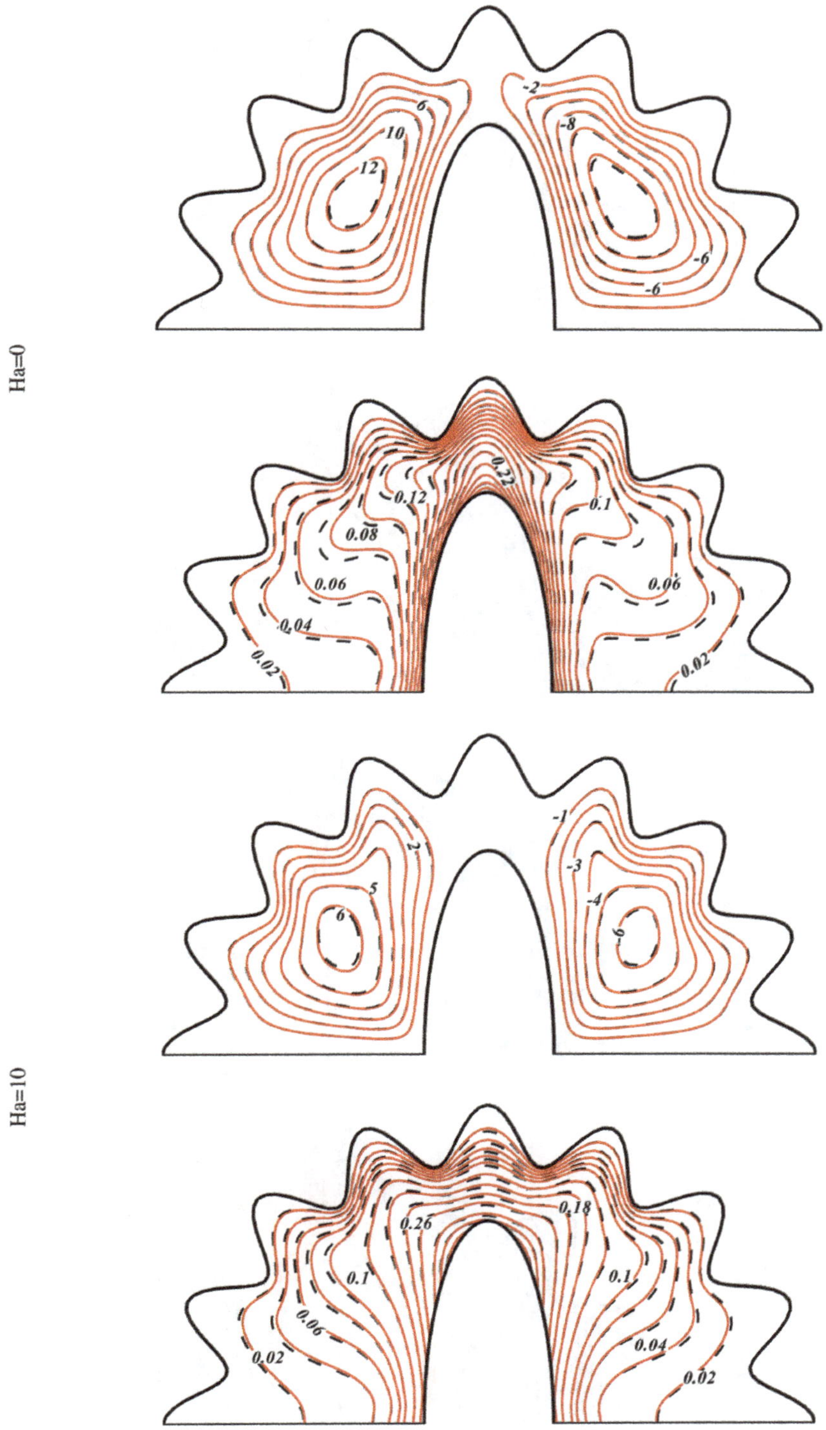

Figure 37. Influence of Da, Ha *on streamlines (right) and isotherms (left) contours when* $\phi = 0.04, Ra = 10^3$

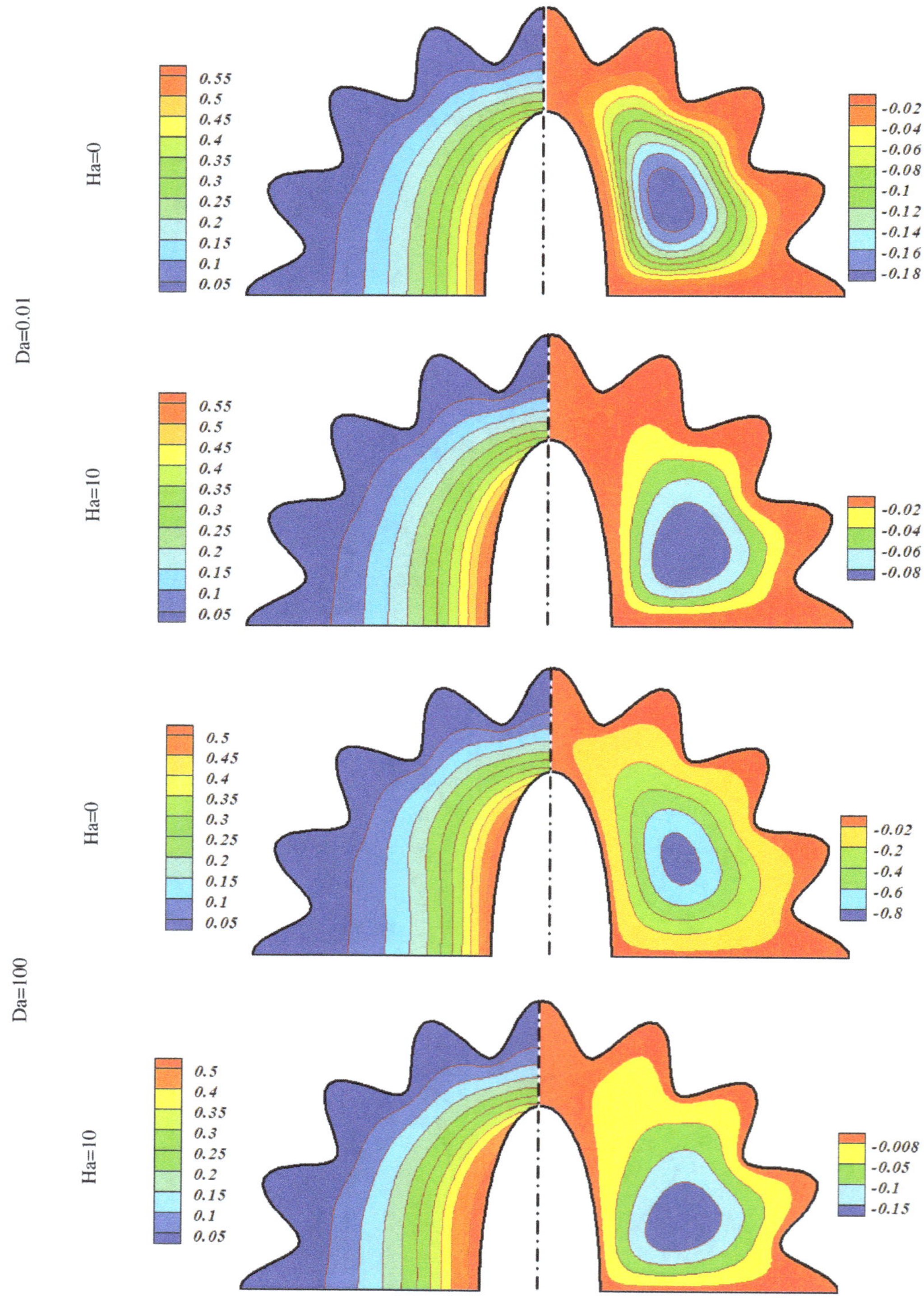

Figure 38. Influence of Da, Ha *on streamlines (right) and isotherms (left) contours when* $\phi = 0.04, Ra = 10^4$

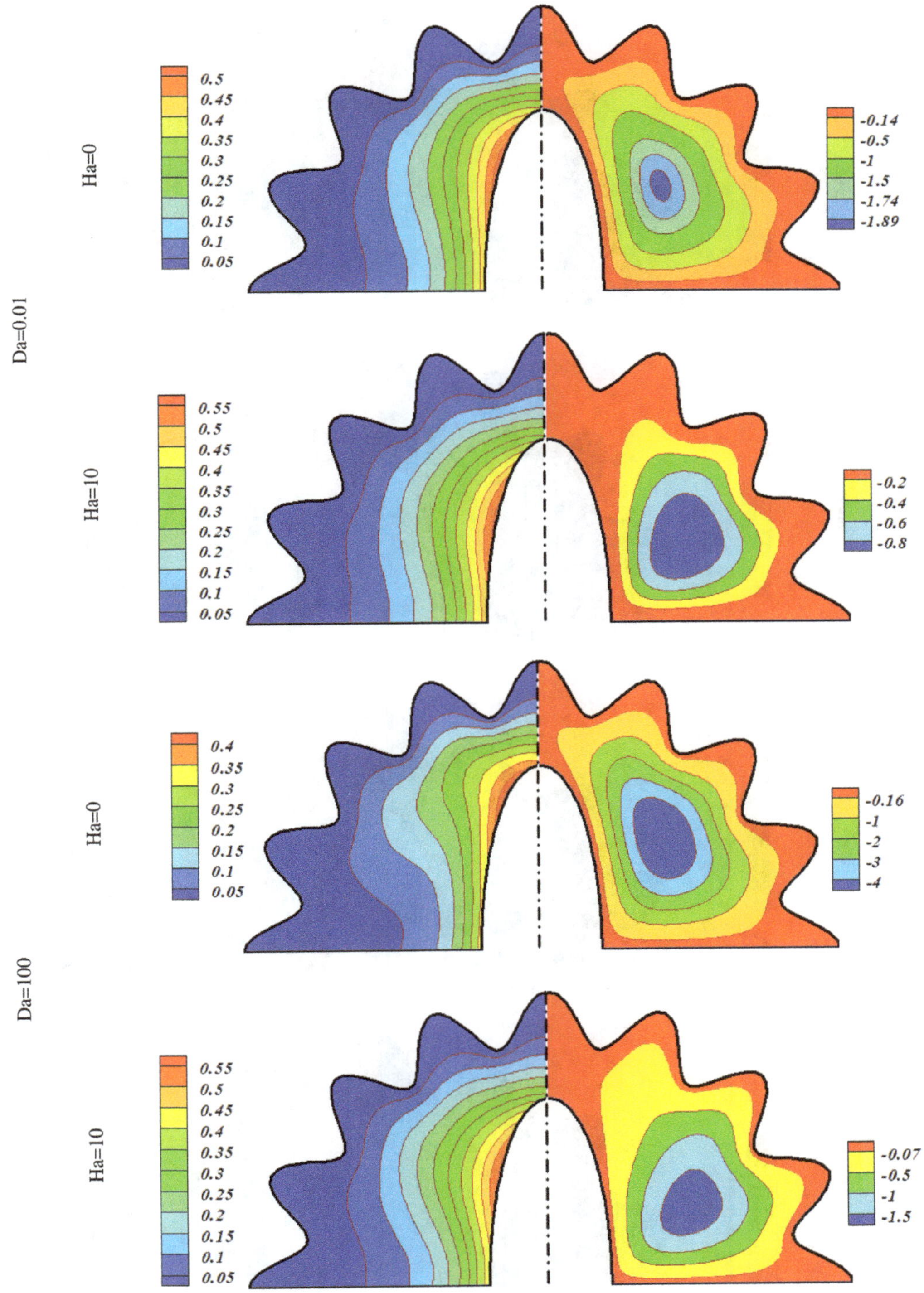

Figure 39. Influence of Da, Ha on streamlines (right) and isotherms (left) contours when $\phi = 0.04, Ra = 10^5$

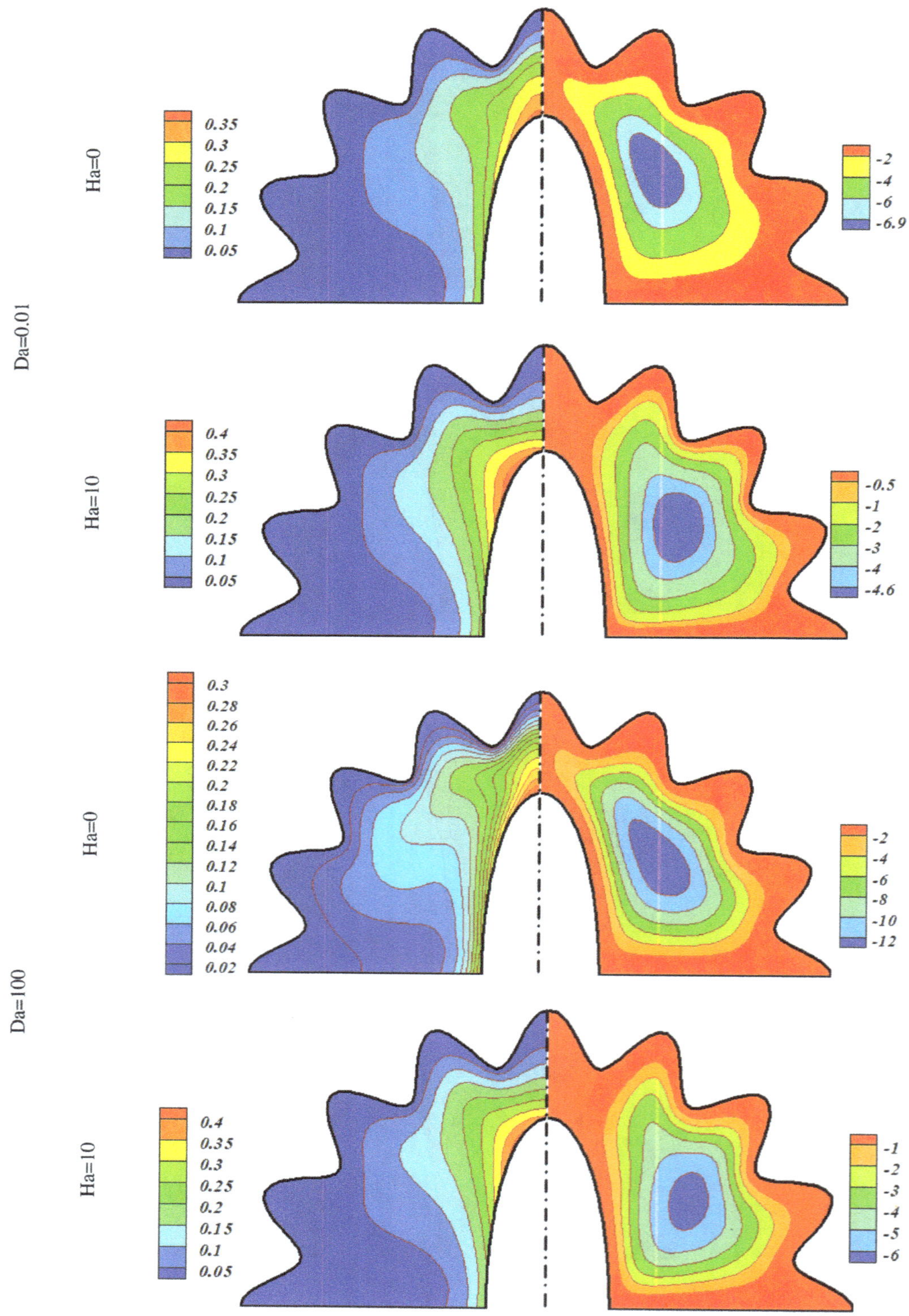

$$\begin{aligned} Nu_{ave} &= 31.9 - 5.2Rd - 15.3\log(Ra) - 0.6Da^{*} + 1.25Ha^{*} \\ &+2.93Rd\log(Ra) + 0.44Rd\,Da^{*} - 0.57Rd\,Ha^{*} \\ &+0.28Da^{*}\log(Ra) - 0.37Ha^{*}\log(Ra) - 0.69Da^{*}Ha^{*} \\ &-4.13Rd^{2} + 2.02(\log(Ra))^{2} + 0.47(Da^{*})^{2} - 0.27(Ha^{*})^{2} \end{aligned} \tag{126}$$

where $Da^{*} = 0.01Da, Ha^{*} = 0.1Ha$. Heat transfer rate enhances with augment of permeability of porous media. Similar treatment is reported for Rayleigh number. Temperature gradient reduces with augment of Hartmann number.

8. EFFECT OF LORENTZ FORCES ON NANOFLUID FLOW IN A POROUS CAVITY BY MEANS OF NON-DARCY MODEL

8.1. Problem Definition

Figure 41 depicts the geometry, boundary condition and sample element. The formula of inner cylinder is:

$$b = \sqrt{1-\varepsilon^{2}}.a \tag{127}$$

where a, b, ε are the major, minor axis of elliptic cylinder and eccentricity for the inner cylinder. The inner cylinder has constant heat flux condition.

8.2. Governing Equation

2D steady convective flow of nanofluid in a porous media is considered in existence of constant magnetic field. The PDEs equations are:

$$\frac{\partial u}{\partial x} + \frac{\partial v}{\partial y} = 0 \tag{128}$$

$$\left(v\frac{\partial u}{\partial y} + \frac{\partial u}{\partial x}u\right) = \left[-\sigma_{nf}B_{y}^{2}u + \sigma_{nf}B_{x}B_{y}v\left(\frac{\partial^{2}u}{\partial y^{2}} + \frac{\partial^{2}u}{\partial x^{2}}\right)\mu_{nf} - \frac{\partial P}{\partial x} - \frac{\mu_{nf}}{K}u\right]\left(\rho_{nf}\right)^{-1} \tag{129}$$

$$\begin{aligned} &\rho_{nf}\left(\frac{\partial v}{\partial x}u + \frac{\partial v}{\partial y}v\right) = +\mu_{nf}\left(\frac{\partial^{2}v}{\partial x^{2}} + \frac{\partial^{2}v}{\partial y^{2}}\right) - \frac{\partial P}{\partial y} + B_{y}\sigma_{nf}B_{x}u - B_{x}\sigma_{nf}B_{x}v - \frac{\mu_{nf}}{K}v \\ &+(T-T_{c})\beta_{nf}g\rho_{nf}, \\ &B_{x} = B_{o}\cos\lambda,\ B_{y} = B_{o}\sin\lambda \end{aligned} \tag{130}$$

Figure 40a. Effects of Da, Ha, Rd and Ra on average Nusselt number

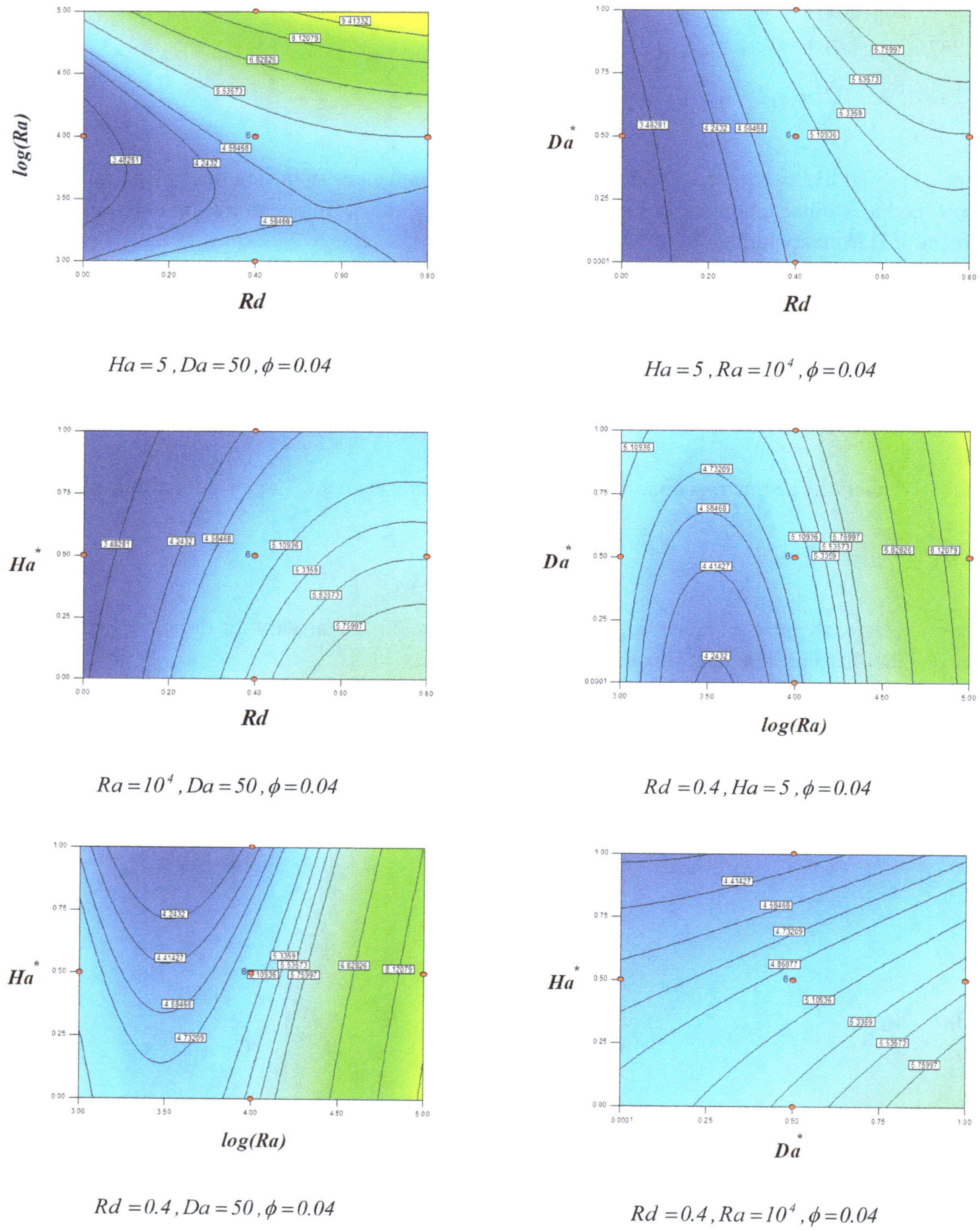

$Ha = 5, Da = 50, \phi = 0.04$

$Ha = 5, Ra = 10^4, \phi = 0.04$

$Ra = 10^4, Da = 50, \phi = 0.04$

$Rd = 0.4, Ha = 5, \phi = 0.04$

$Rd = 0.4, Da = 50, \phi = 0.04$

$Rd = 0.4, Ra = 10^4, \phi = 0.04$

Figure 40b. Effects of Da, Ha, Rd and Ra on average Nusselt number

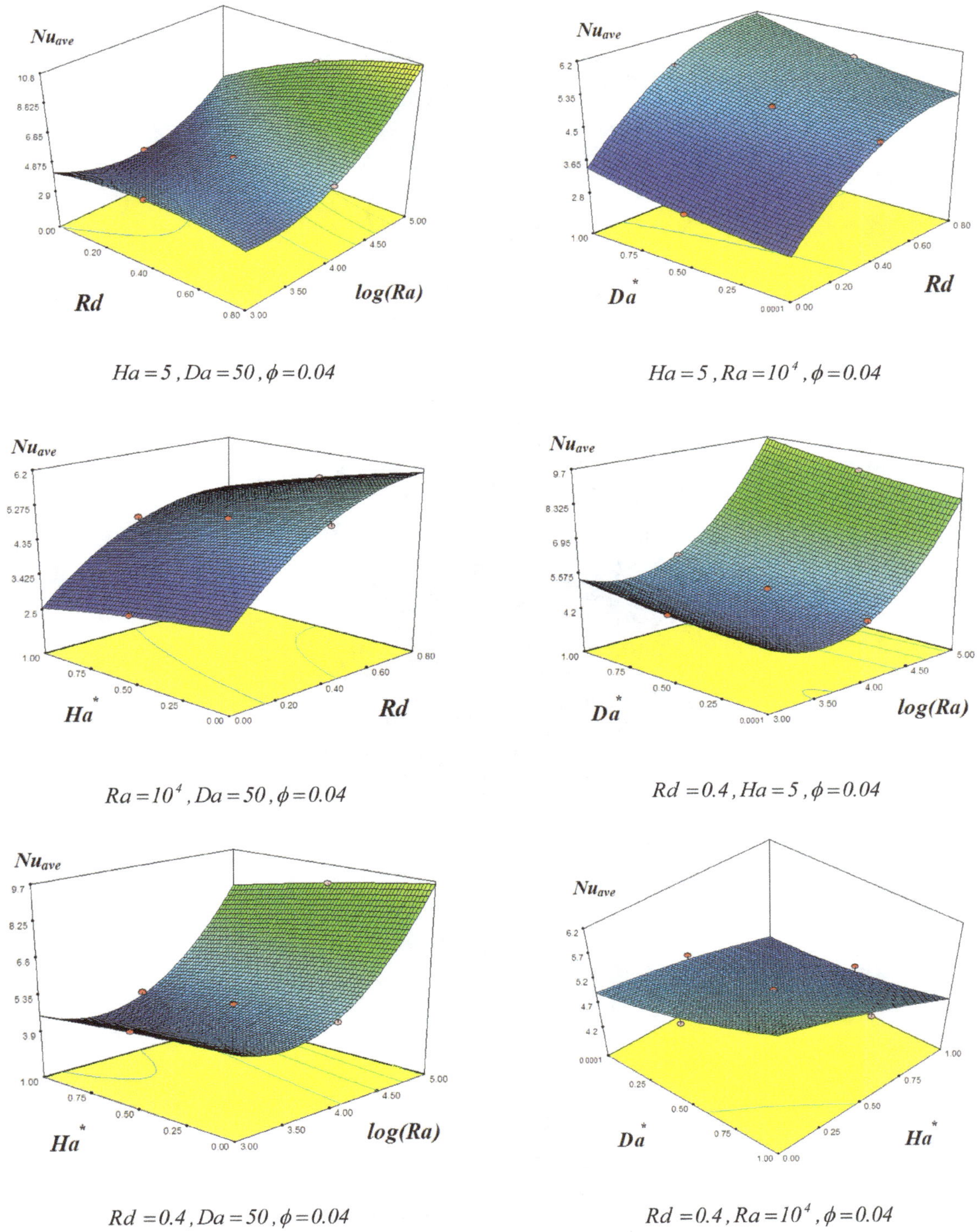

Figure 41. (a)Geometry and the boundary conditions with (b) the mesh of Geometry considered in this work; (c) A sample triangular element and its corresponding control volume.

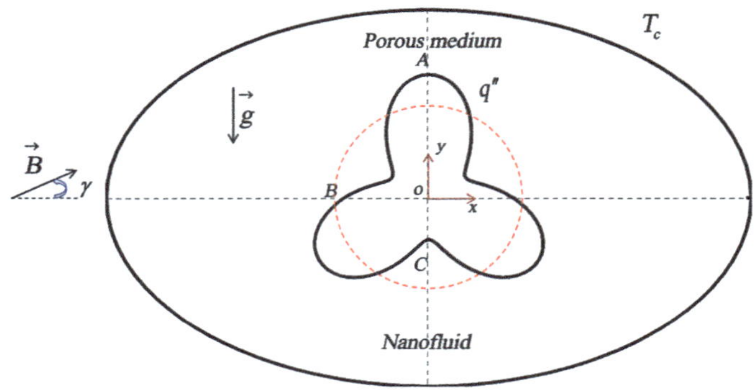

(a)

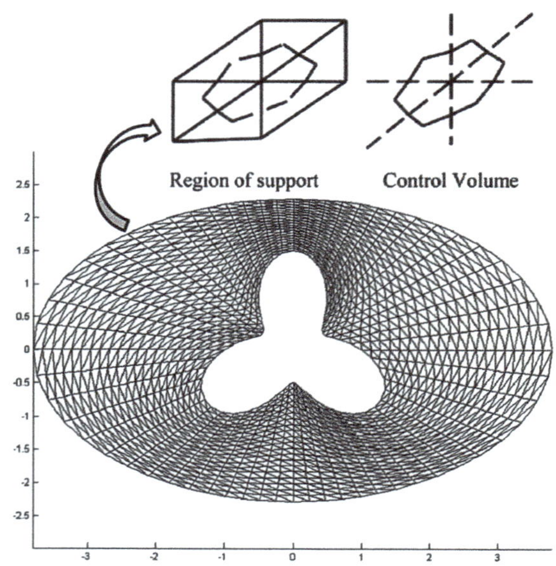

(b)

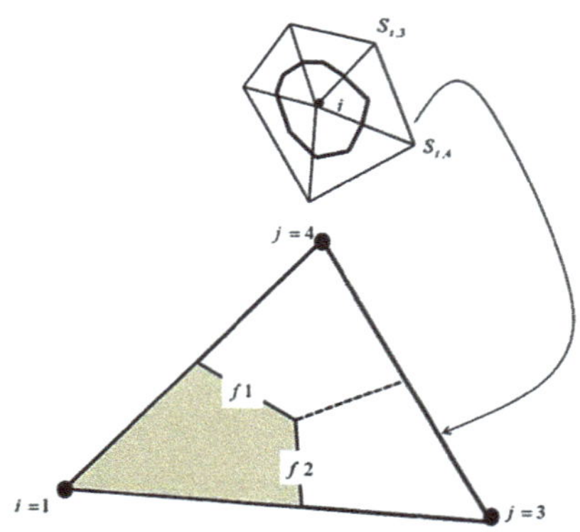

(c)

$$\left(\rho C_p\right)_{nf}\left(v\frac{\partial T}{\partial y}+u\frac{\partial T}{\partial x}\right)=k_{nf}\left(\frac{\partial^2 T}{\partial x^2}+\frac{\partial^2 T}{\partial y^2}\right) \tag{131}$$

$\left(\rho C_p\right)_{nf}$, $\left(\rho\beta\right)_{nf}$, ρ_{nf}, k_{nf} and σ_{nf} are defined as:

$$\left(\rho C_p\right)_{nf}=\left(\rho C_p\right)_f(1-\phi)+\left(\rho C_p\right)_s\phi \tag{132}$$

$$\left(\rho\beta\right)_{nf}=\left(\rho\beta\right)_f(1-\phi)+\left(\rho\beta\right)_s\phi \tag{133}$$

$$\rho_{nf}=\rho_f(1-\phi)+\rho_s\phi \tag{134}$$

$$k_{nf}=k_f\left(\frac{k_s+2k_f+2\phi(k_s-k_f)}{k_s-\phi(k_s-k_f)+2k_f}\right) \tag{135}$$

$$\frac{\sigma_{nf}}{\sigma_f}=1+\frac{3\left(\frac{\sigma_s}{\sigma_f}-1\right)\phi}{\left(\frac{\sigma_s}{\sigma_f}+2\right)-\left(\frac{\sigma_s}{\sigma_f}-1\right)\phi} \tag{136}$$

μ_{nf} is obtained as follows:

$$\mu_{nf}=\left(0.035B^2+3.1B-27886.4807\phi^2+4263.02\phi+316.0629\right)e^{-0.01T} \tag{137}$$

Vorticity and stream function should be used to eliminate pressure source terms:

$$\omega+\frac{\partial u}{\partial y}-\frac{\partial v}{\partial x}=0,\frac{\partial\psi}{\partial x}=-v,\frac{\partial\psi}{\partial y}=u \tag{138}$$

Introducing dimensionless quantities:

$$P = \frac{p}{\rho_{nf}\left(\alpha_{nf} / L\right)^2}, U = \frac{uL}{\alpha_{nf}}, V = \frac{vL}{\alpha_{nf}}, \Theta = \frac{T - T_c}{\Delta T}, \Delta T = q''L / k_f, \left(X, Y\right) = \frac{\left(x, y\right)}{L} \tag{139}$$

The final formulae are:

$$\frac{\partial V}{\partial Y} + \frac{\partial U}{\partial X} = 0, \tag{140}$$

$$\begin{aligned} &U\frac{\partial U}{\partial X} + \frac{\partial U}{\partial Y}V = \Pr\left(\frac{A_5 A_2}{A_1 A_4}\right)\left(\frac{\partial^2 U}{\partial Y^2} + \frac{\partial^2 U}{\partial X^2}\right) \\ &-Ha^2 \Pr\left[\frac{A_6 A_2}{A_1 A_4}\right]\left(B_y^2 U - B_x B_y V\right) - \frac{\partial P}{\partial X} - \frac{\Pr}{Da}\left(\frac{A_5 A_2}{A_1 A_4}\right)\Omega, \end{aligned} \tag{141}$$

$$\begin{aligned} &V\frac{\partial V}{\partial Y} + U\frac{\partial V}{\partial X} = \Pr\left(\frac{\partial^2 V}{\partial Y^2} + \frac{\partial^2 V}{\partial X^2}\right)\left(\frac{A_5 A_2}{A_1 A_4}\right) \\ &-Ha^2 \Pr\left[\frac{A_6 A_2}{A_1 A_4}\right]\left(B_x^2 V - B_x B_y U\right) - \frac{\partial P}{\partial Y} + Ra\Pr\left(\frac{A_3 A_2^2}{A_1 A_4^2}\right)\Theta, \end{aligned} \tag{142}$$

$$\frac{\partial \Theta}{\partial X}U + \frac{\partial \Theta}{\partial Y}V = \left(\frac{\partial^2 \Theta}{\partial X^2}\right) \tag{143}$$

where dimensionless and constants parameters are illustrated as:

$$\begin{aligned} &\Pr = \upsilon_f / \alpha_f, Ra = g\beta_f q'' L^4 / \left(k_f \upsilon_f \alpha_f\right), Ha = LB_0\sqrt{\sigma_f / \mu_f}, Da = \frac{K}{L^2}, \\ &A_1 = \frac{\rho_{nf}}{\rho_f}, A_3 = \frac{\left(\rho\beta\right)_{nf}}{\left(\rho\beta\right)_f}, A_5 = \frac{\mu_{nf}}{\mu_f}, A_2 = \frac{\left(\rho C_P\right)_{nf}}{\left(\rho C_P\right)_f}, A_4 = \frac{k_{nf}}{k_f}, A_6 = \frac{\sigma_{nf}}{\sigma_f} \end{aligned} \tag{144}$$

and boundary conditions are:

$$\frac{\partial \Theta}{\partial n} = 1.0 \text{ on inner wall}$$

$$\Theta = 0.0 \text{ on outer wall}$$

$$\Psi = 0.0 \text{ on all walls} \tag{145}$$

Local and average Nusselt over the hot cylinder can calculate as:

$$Nu_{loc} = \frac{1}{\theta}\left(\frac{k_{nf}}{k_f}\right) \tag{146}$$

$$Nu_{ave} = \frac{1}{S}\int_0^s Nu_{loc}\, ds \tag{147}$$

8.3. Effects of Active Parameters

Natural convection of nanofluid in a porous complex shaped cavity is presented considering Non-Darcy model. Influence of Lorentz forces is taken into account. The working fluid is considered Fe_3O_4-water and its viscosity is a function of nanofluid volume fraction and magnetic field. Results are presented for various values of volume fraction of Fe_3O_4-water ($\phi = 0$ to 0.04), Darcy number ($Da = 0.001$ to 100), Rayleigh number ($Ra = 10^3, 10^4$ and 10^5) and Hartmann number ($Ha = 0$ to 20).

Impacts of Da, Ha and Ra on isotherms and streamlines are demonstrated in Figures 42 and 43. As nanofluid temperature augments, the nanofluid initiates moving from the inner cylinder to the outer one and dropping along the elliptic cylinder. At low Darcy and Rayleigh numbers, conduction is more signification than convection. So isotherms are parallel to each other. One main eddy exists in each side. Strength of this main eddy enhances with rise of convective heat transfer. So $|\Psi_{max}|$ rises with augment of Da, Ra. Also as Rayleigh number augments, the main eddy is stretched vertically and thermal plume appears near the vertical center line. As magnetic field augments $|\Psi_{max}|$ reduces and the center of main eddy moves to upward. Also Lorentz force makes the thermal plume to diminish.

Figures 44 and 45 depict the impact of Da, Ra and Ha on Nu_{loc}, Nu_{ave}. The formula for Nu_{ave} corresponding to active parameters is:

$$\begin{aligned}
Nu_{ave} &= 3.094 - 0.095Da^* - 1.63\log(Ra) + 0.35Ha^* \\
&+0.15Da^*\log(Ra) - 0.12Da^*Ha^* - 0.08\log(Ra)Ha^* \\
&+0.31(\log(Ra))^2 + 0.13Da^{*2} - 0.08Ha^{*2}
\end{aligned} \tag{148}$$

where $Ha^* = 0.1Ha, Da^* = 0.01Da$. The number of extermum in Nu_{loc} profile is matching to existence of thermal plume. Influence of Lorentz forces on Nu_{loc} is more sensible in higher Darcy number. Nusselt number enhances with rise of Da, Ra but it reduces with rise of Hartmann number.

Figure 42. Effect of Rayleigh and Hartmann numbers on Isotherms (left) and streamlines (right) contours when $\phi = 0.04, Da = 0.01$

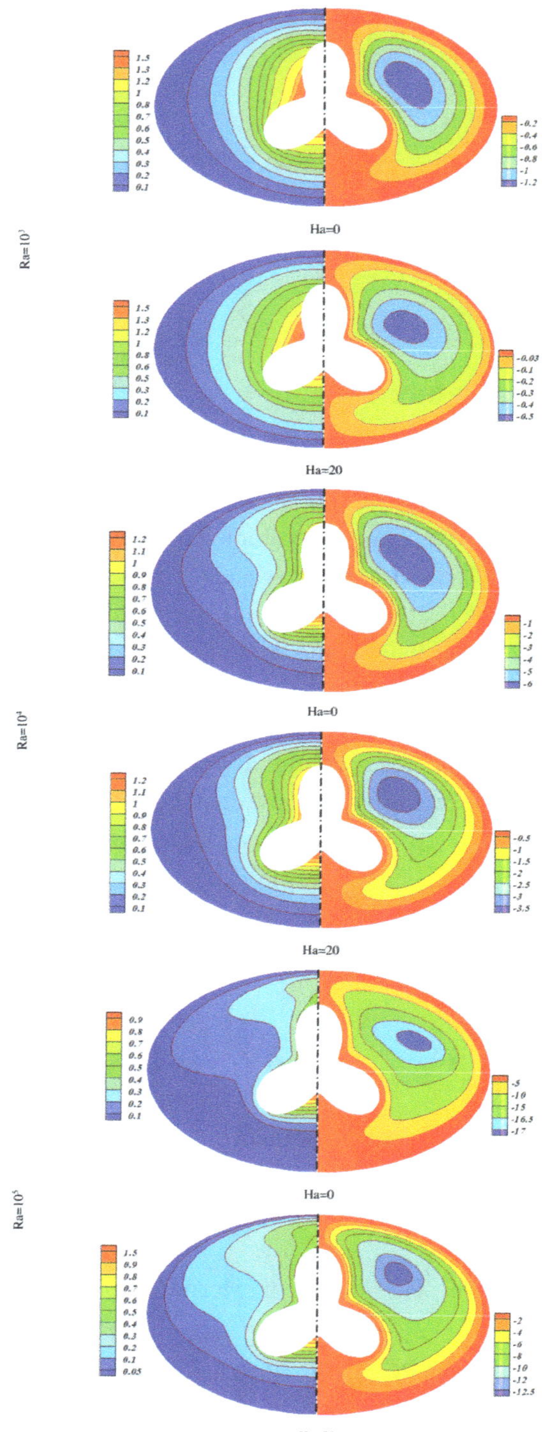

Figure 43. Effect of Rayleigh and Hartmann numbers on Isotherms (left) and streamlines (right) contours when $\phi = 0.04, Da = 100$

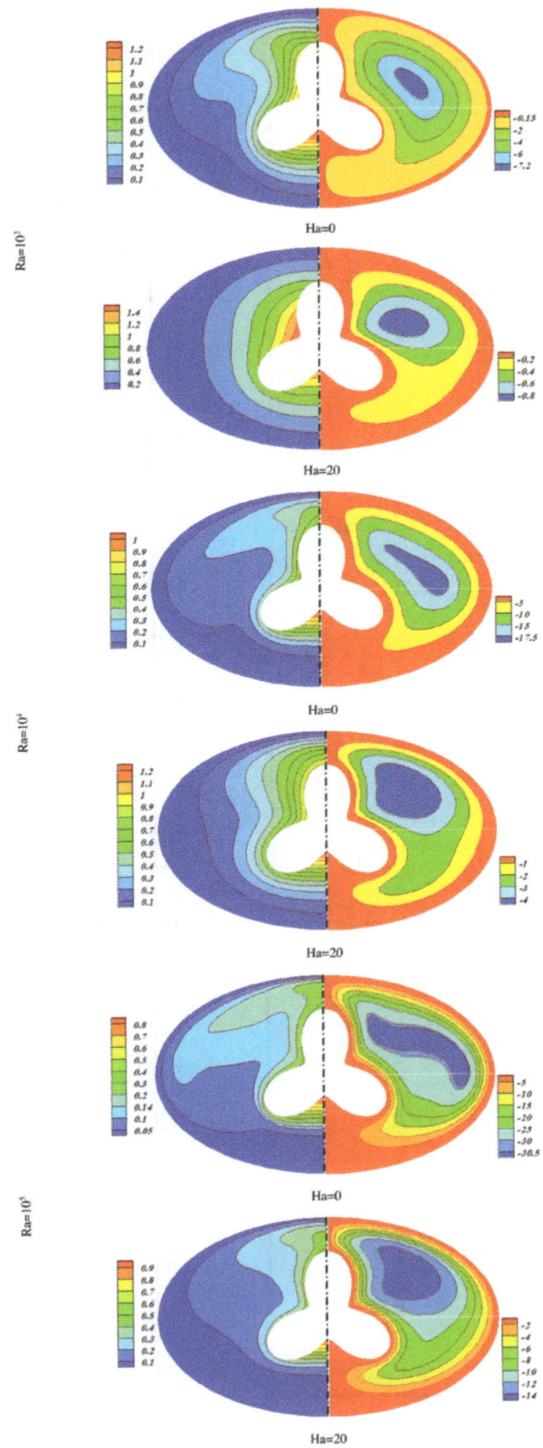

Figure 44. Effects of the Darcy number, Hartmann and Rayleigh numbers on local Nusselt number when $\phi = 0.04$

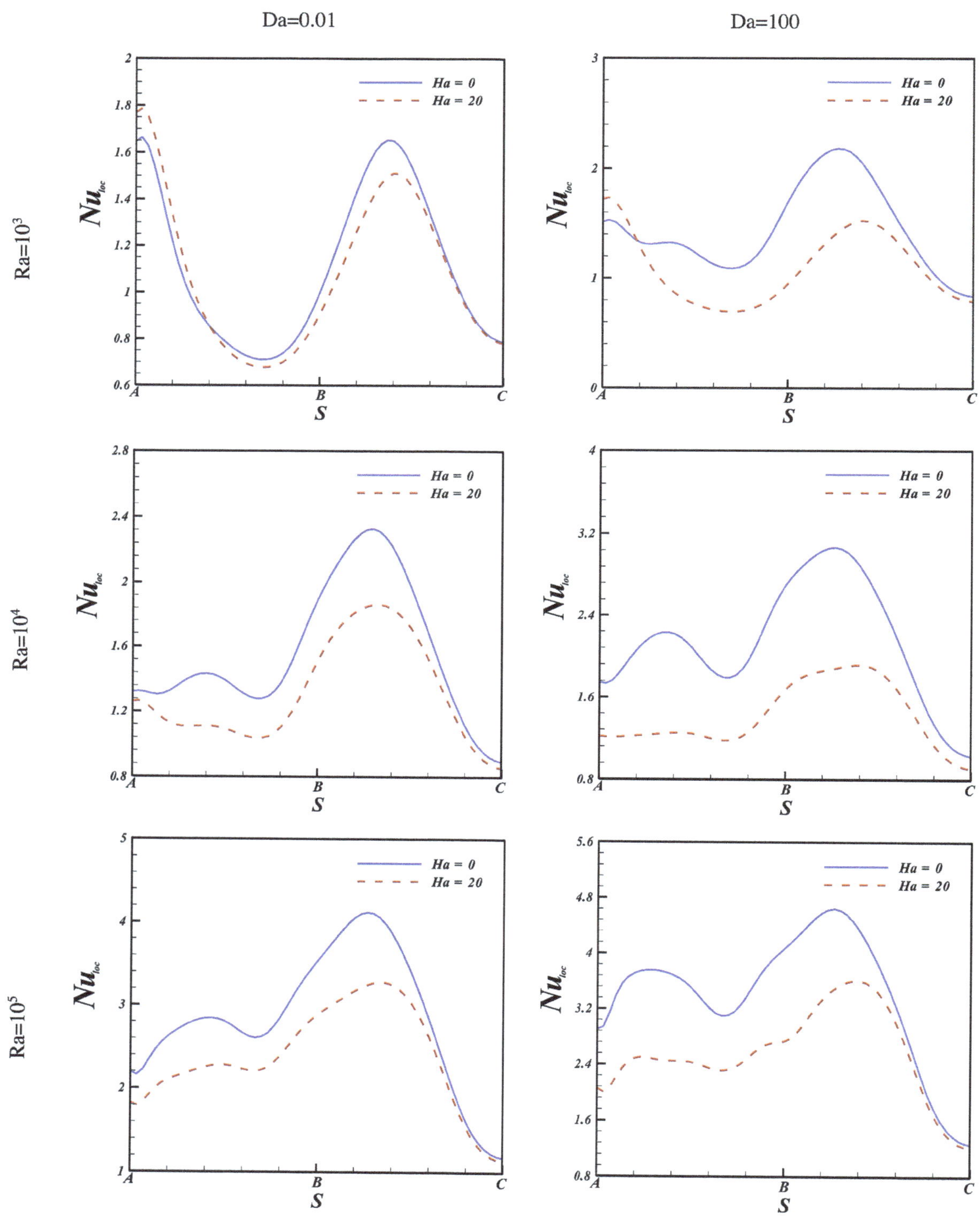

Figure 45. Effects of inclination angle, radiation parameter, Rayleigh and Hartmann numbers on average Nusselt number

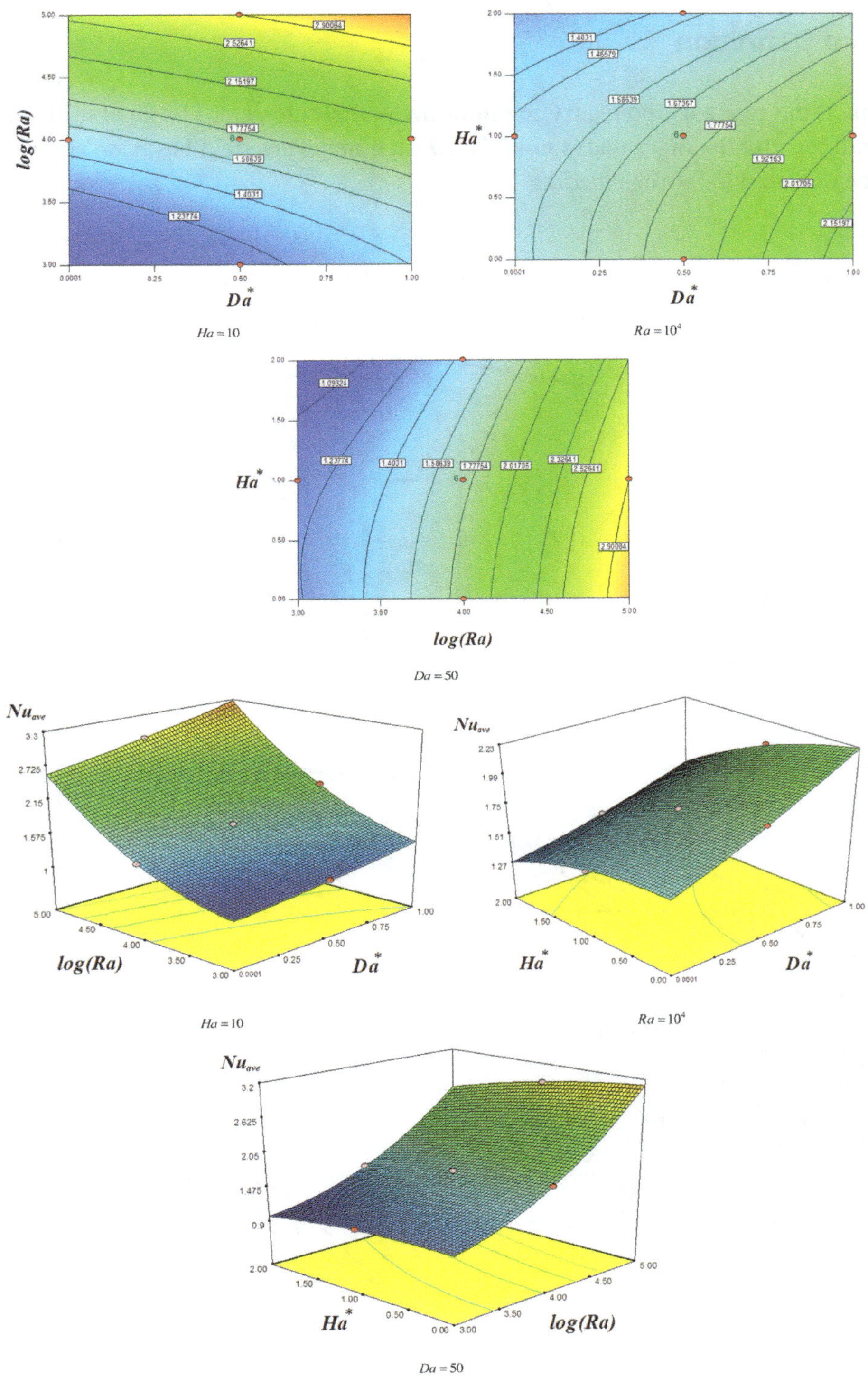

9. NUMERICAL SIMULATION OF Fe_3O_4: WATER NANOFLUID FLOW IN A NON-DARCY POROUS MEDIA

9.1. Problem Definition

Figure 46 depicts the geometry, boundary condition and sample element. The inner cylinder has constant heat flux condition and the outer one is cold. Horizontal magnetic field has been applied. Non-Darcy model has been utilized for porous media.

Figure 46. (a)Geometry and the boundary conditions with (b) the mesh of Geometry considered in this work; (c) A sample triangular element and its corresponding control volume.

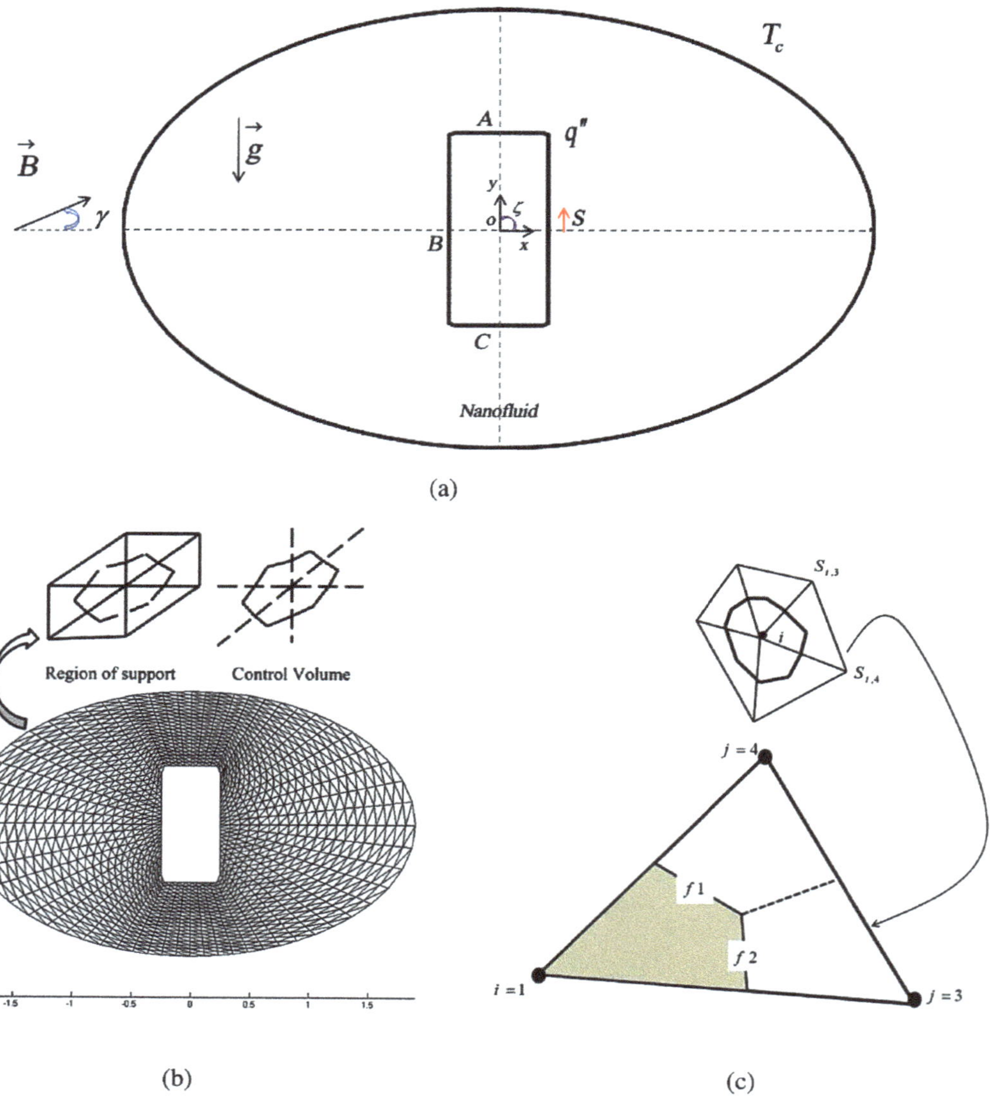

9.2. Governing Equation

2D steady convective flow of nanofluid in a porous media is considered in existence of constant magnetic field. The PDEs equations are:

$$\frac{\partial v}{\partial y}+\frac{\partial u}{\partial x}=0 \tag{149}$$

$$\left(v\frac{\partial u}{\partial y}+u\frac{\partial u}{\partial x}\right)=\left[\sigma_{nf}B_xB_yv-\sigma_{nf}B_y^2+\left(\frac{\partial^2 u}{\partial y^2}+\frac{\partial^2 u}{\partial x^2}\right)\mu_{nf}-\frac{\partial P}{\partial x}-\frac{\mu_{nf}}{K}u\right]\left(\rho_{nf}\right)^{-1} \tag{150}$$

$$\begin{gathered}\rho_{nf}\left(\frac{\partial v}{\partial x}u+\frac{\partial v}{\partial y}v\right)=\mu_{nf}\left(\frac{\partial^2 v}{\partial x^2}+\frac{\partial^2 v}{\partial y^2}\right)-\frac{\partial P}{\partial y}-B_x\sigma_{nf}B_xv+B_y\sigma_{nf}B_xu-\frac{\mu_{nf}}{K}v\\+\left(T-T_c\right)\beta_{nf}g\rho_{nf},\\B_x=B_o\cos\lambda,\ B_y=B_o\sin\lambda\end{gathered} \tag{151}$$

$$\left(\rho C_p\right)_{nf}\left(v\frac{\partial T}{\partial y}+u\frac{\partial T}{\partial x}\right)=k_{nf}\left(\frac{\partial^2 T}{\partial x^2}+\frac{\partial^2 T}{\partial y^2}\right) \tag{152}$$

$\left(\rho C_p\right)_{nf},\left(\rho\beta\right)_{nf},\ \rho_{nf},k_{nf}$ and σ_{nf} are defined as:

$$\left(\rho C_p\right)_{nf}=\phi\left(\rho C_p\right)_s+(1-\phi)\left(\rho C_p\right)_f \tag{153}$$

$$\frac{\left(\rho\beta\right)_{nf}}{\left(\rho\beta\right)_f}=\phi\frac{\left(\rho\beta\right)_s}{\left(\rho\beta\right)_f}+(1-\phi) \tag{154}$$

$$\rho_{nf}=\rho_f(1-\phi)+\rho_s\phi \tag{155}$$

$$k_{nf}=k_f\left(\frac{k_s+2k_f+2\phi(k_s-k_f)}{k_s-\phi(k_s-k_f)+2k_f}\right) \tag{156}$$

$$\sigma_{nf} = \left[\left(\frac{(2+\sigma_s/\sigma_f)-(\sigma_s/\sigma_f-1)\phi}{3\phi(-1+\sigma_s/\sigma_f)}\right)^{-1}+1\right]\sigma_f \tag{157}$$

μ_{nf} is obtained as follows:

$$\mu_{nf} = \left(0.035B^2 + 3.1B - 27886.4807\phi^2 + 4263.02\phi + 316.0629\right)e^{-0.01T} \tag{158}$$

Vorticity and stream function should be used to eliminate pressure source terms:

$$\omega + \frac{\partial u}{\partial y} - \frac{\partial v}{\partial x} = 0, \frac{\partial \psi}{\partial x} = -v, \frac{\partial \psi}{\partial y} = u \tag{159}$$

Introducing dimensionless quantities:

$$P = \frac{p}{\rho_{nf}(\alpha_{nf}/L)^2}, (Y,X) = (y,x)/L, U = \frac{uL}{\alpha_{nf}}, V = \frac{vL}{\alpha_{nf}}, \theta = \frac{T-T_c}{\Delta T}, \Delta T = q''L/k_f \tag{160}$$

The final formulae are:

$$\Omega + \frac{\partial^2 \Psi}{\partial Y^2} = -\frac{\partial^2 \Psi}{\partial X^2}, \tag{161}$$

$$\begin{aligned} &\frac{\partial \Omega}{\partial X}U + \frac{\partial \Omega}{\partial Y}V = \Pr\frac{A_5}{A_1}\frac{A_2}{A_4}\left(\frac{\partial^2\Omega}{\partial Y^2} + \frac{\partial^2\Omega}{\partial X^2}\right) \\ &+\Pr Ha^2\frac{A_6}{A_1}\frac{A_2}{A_4}\left(\frac{\partial U}{\partial X}B_xB_y - \frac{\partial V}{\partial X}B_x^{\,2} + \frac{\partial U}{\partial Y}B_y^2 - \frac{\partial V}{\partial Y}B_xB_y\right) \\ &+\Pr Ra\frac{A_3A_2^2}{A_1A_4^2}\frac{\partial \theta}{\partial X} - \frac{\Pr}{Da}\left(\frac{A_5A_2}{A_1A_4}\right)\Omega, \end{aligned} \tag{162}$$

$$V\frac{\partial \theta}{\partial Y} + \frac{\partial \theta}{\partial X}U = \left(\frac{\partial^2\theta}{\partial Y^2} + \frac{\partial^2\theta}{\partial X^2}\right). \tag{163}$$

where dimensionless and constants parameters are illustrated as:

$$\Pr = \upsilon_f / \alpha_f, Ra = g\beta_f q'' L^4 / \left(k_f \upsilon_f \alpha_f\right), Ha = LB_0\sqrt{\sigma_f / \mu_f}, Da = \frac{K}{L^2},$$
$$A_1 = \frac{\rho_{nf}}{\rho_f}, A_5 = \frac{\mu_{nf}}{\mu_f}, A_3 = \frac{\left(\rho\beta\right)_{nf}}{\left(\rho\beta\right)_f}, A_2 = \frac{\left(\rho C_P\right)_{nf}}{\left(\rho C_P\right)_f}, A_4 = \frac{k_{nf}}{k_f}, A_6 = \frac{\sigma_{nf}}{\sigma_f} \tag{164}$$

and boundary conditions are:

$$\frac{\partial\theta}{\partial n} = 1.0 \text{ on inner wall}$$

$$\theta = 0.0 \text{ on outer wall}$$

$$\Psi = 0.0 \text{ on all walls} \tag{165}$$

where n is normal to surface. Local and average Nusselt over the hot cylinder can calculate as:

$$Nu_{loc} = \frac{1}{\theta}\left(\frac{k_{nf}}{k_f}\right) \tag{166}$$

$$Nu_{ave} = \frac{1}{S}\int_0^s Nu_{loc}\, ds \tag{167}$$

9.3. Effects of Active Parameters

Impact of magnetic field on nanofluid transportation in a porous cylinder with inner inclined square obstacle is investigated. The working fluid is considered Fe_3O_4-water and its viscosity has relationship with ϕ and Ha. Outputs are presented for several values of volume fraction of Fe_3O_4-water ($\phi = 0$ to 0.04), Darcy number ($Da = 0.001$ to 100), Hartmann number ($Ha = 0$ to 20), Rayleigh number ($Ra = 10^3$ to 10^5) and inclination angle ($\xi = 0^\circ$ and 90°).

Impacts of Da, ξ, Ha and Ra on isotherms and streamlines are demonstrated in Figures 47, 48, 49 and 50. As nanofluid temperature increases, the nanofluid initiates moving from the inner cylinder to the cold one and dropping along the outer cylinder. At low Darcy and Rayleigh numbers, conduction is more signification than convection. So isotherms follow the shape of enclosure. One main eddy exists in each side. Strength of this main eddy enhances with rise of convective heat transfer. So $\left|\Psi_{\max}\right|$ rises with augment of Da, Ra. Also thermal plume appears near the vertical center line when convection mode is dominated. As ξ increases, the distortion of isotherms enhances. As magnetic field augments, $\left|\Psi_{\max}\right|$ reduces and the center of main eddy shifts to upward. Also magnetic field makes the thermal plume to diminish.

Figure 47. Effect of Rayleigh and Hartmann numbers on Isotherms (left) and streamlines (right) contours when $\xi = 0, \phi = 0.04, Da = 0.001$

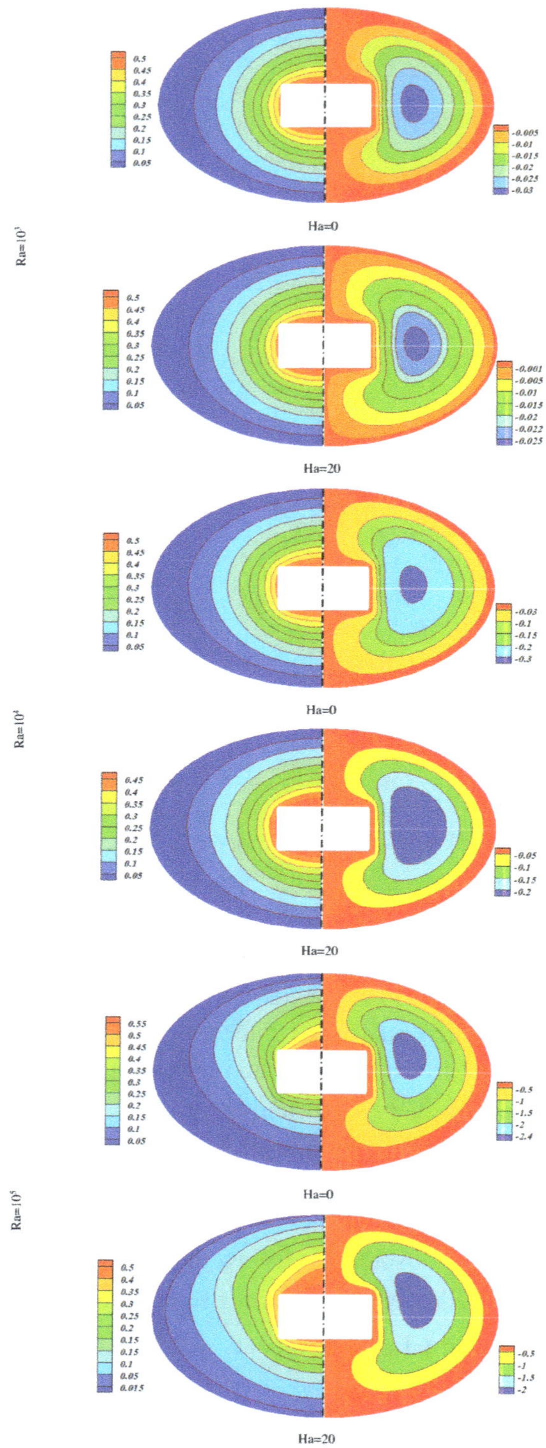

Figure 48. Effect of Rayleigh and Hartmann numbers on Isotherms (left) and streamlines (right) contours when $\xi = 90^{\circ}, \phi = 0.04, Da = 0.001$

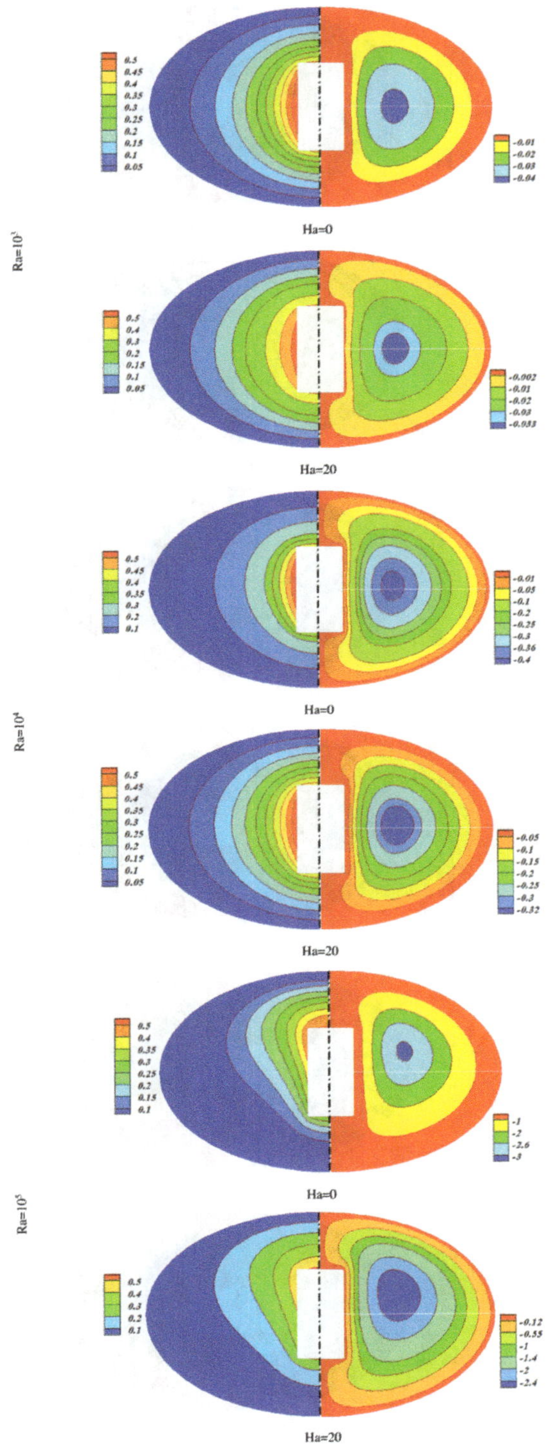

Figure 49. Effect of Rayleigh and Hartmann numbers on Isotherms (left) and streamlines (right) contours when $\xi = 0, \phi = 0.04, Da = 100$

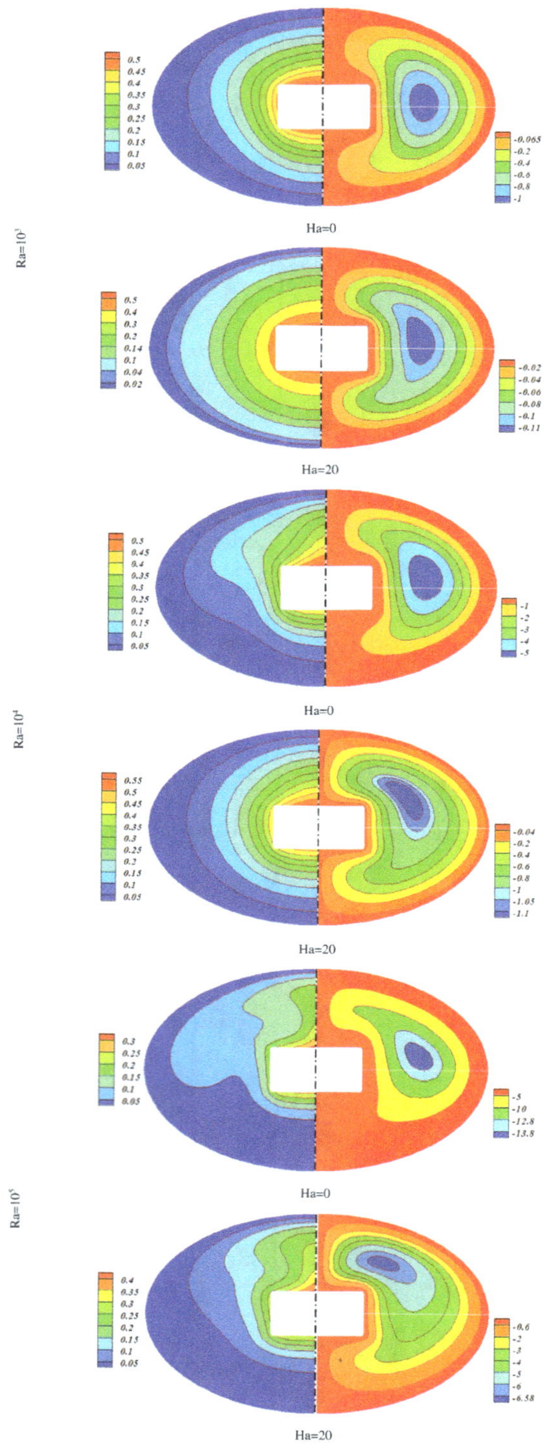

Figure 50. Effect of Rayleigh and Hartmann numbers on Isotherms (left) and streamlines (right) contours when $\xi = 90°, \phi = 0.04, Da = 100$

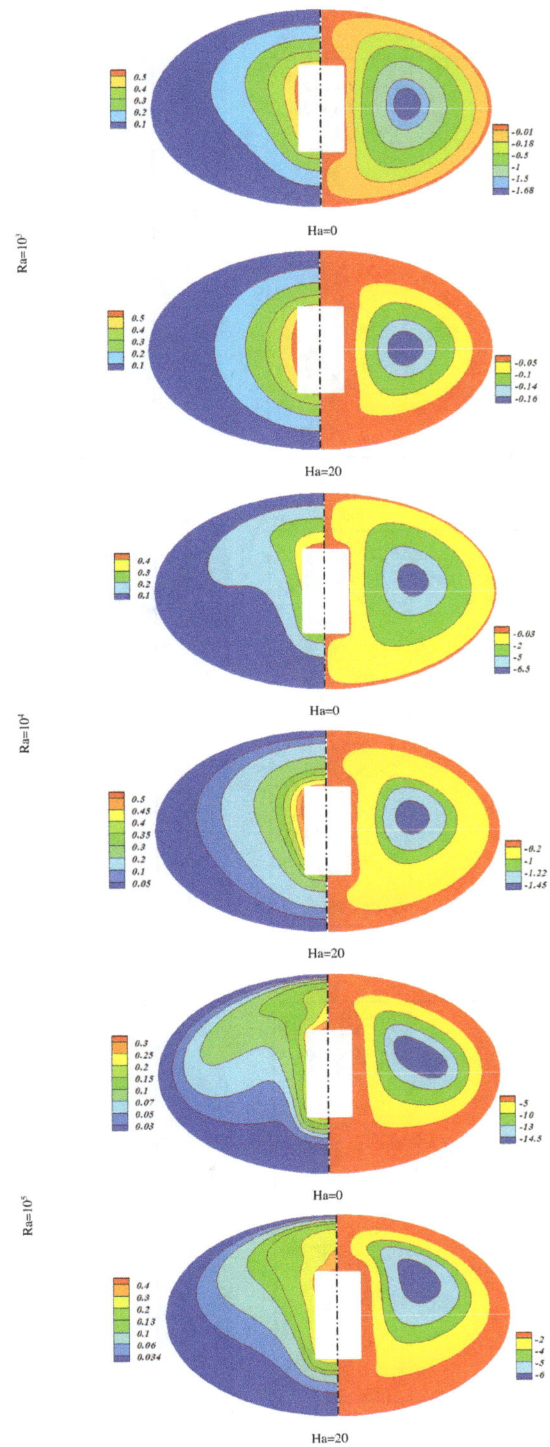

Figure 51. Effects of the Darcy number, Hartmann and Rayleigh numbers on local Nusselt number when $\phi = 0.04, \xi = 0^{\circ}$

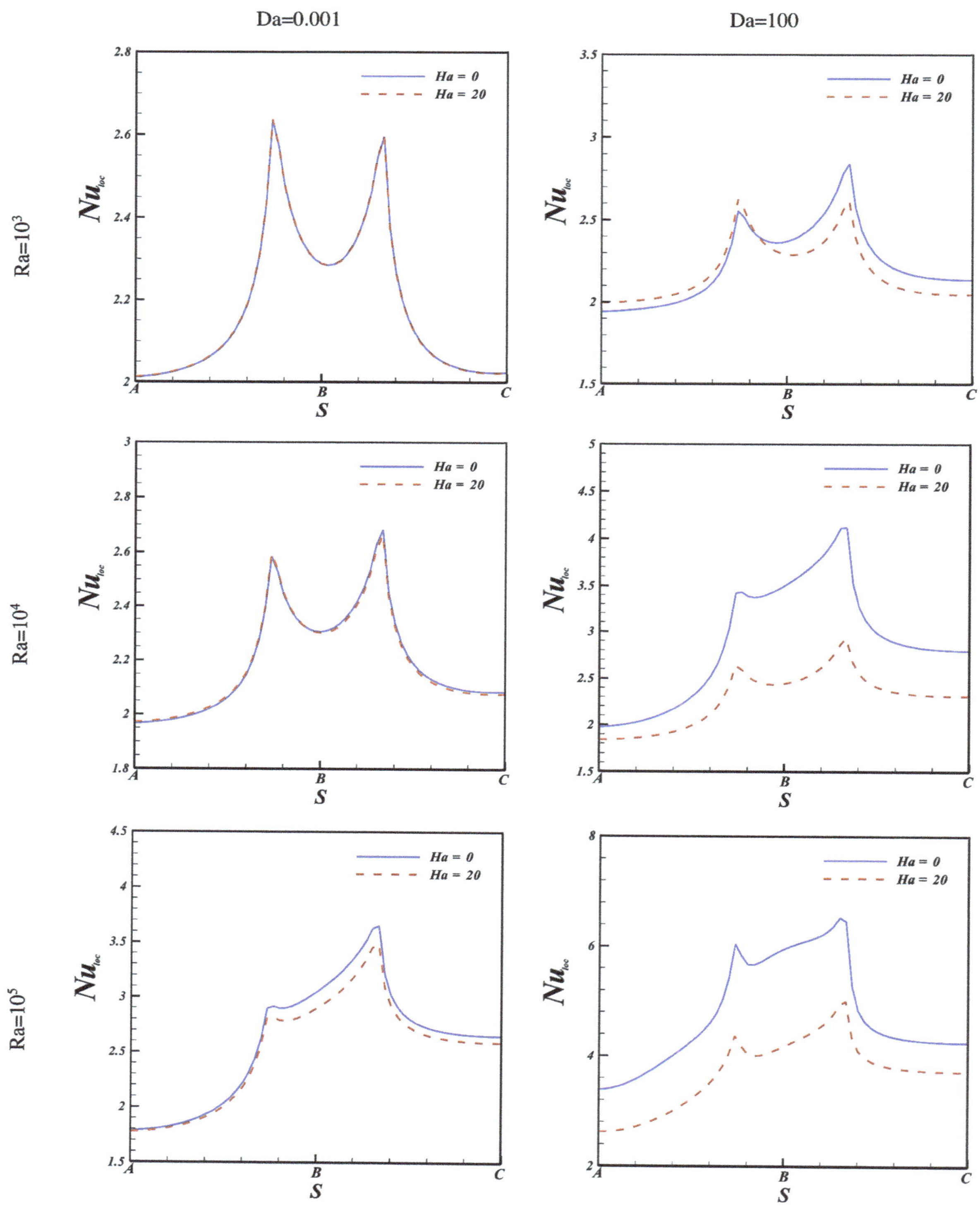

Figure 52. Effects of the Darcy number, Hartmann and Rayleigh numbers on local Nusselt number when $\phi = 0.04, \xi = 90^{\circ}$

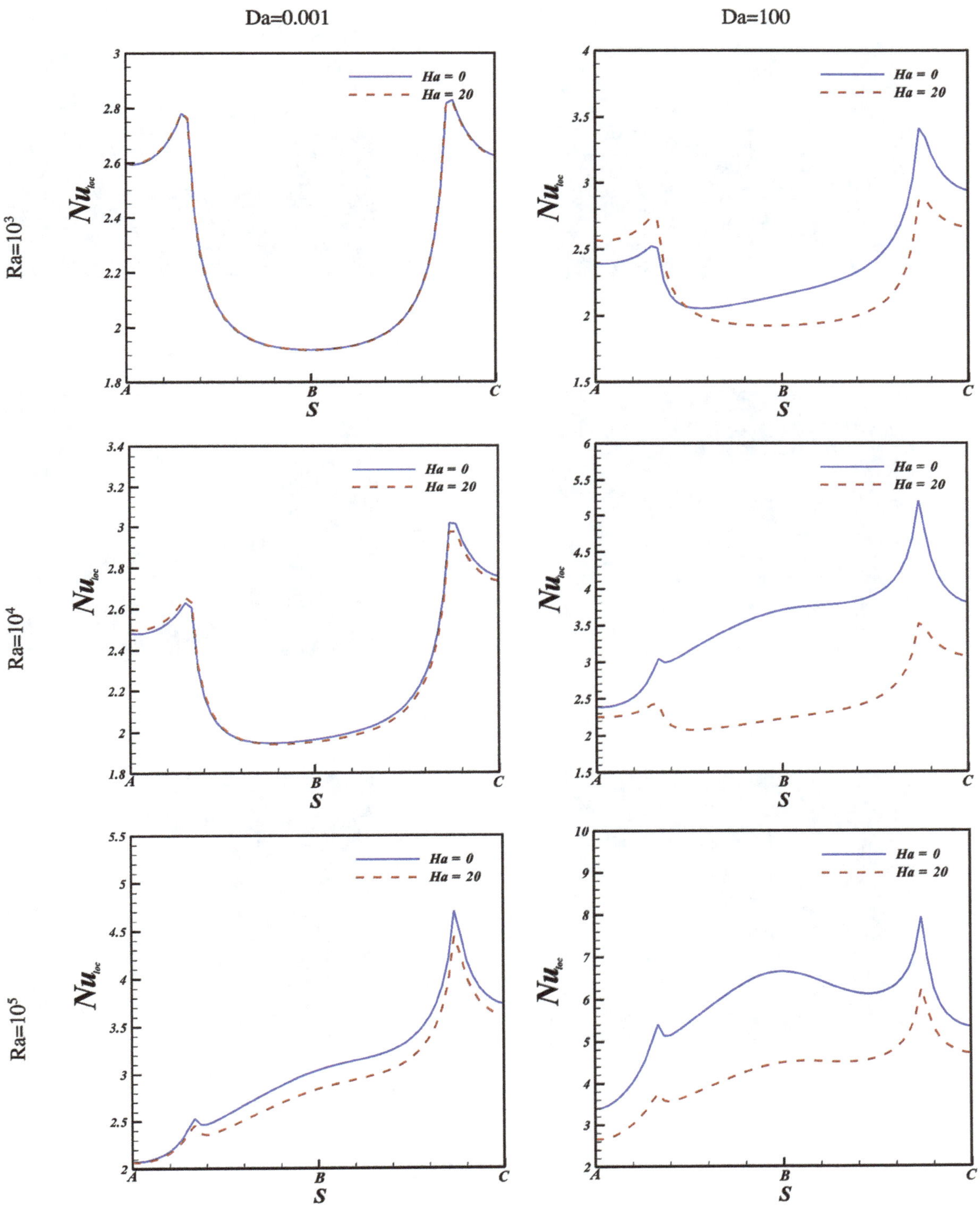

Figure 53a. Effects of inclination angle, radiation parameter, Rayleigh and Hartmann numbers on average Nusselt number

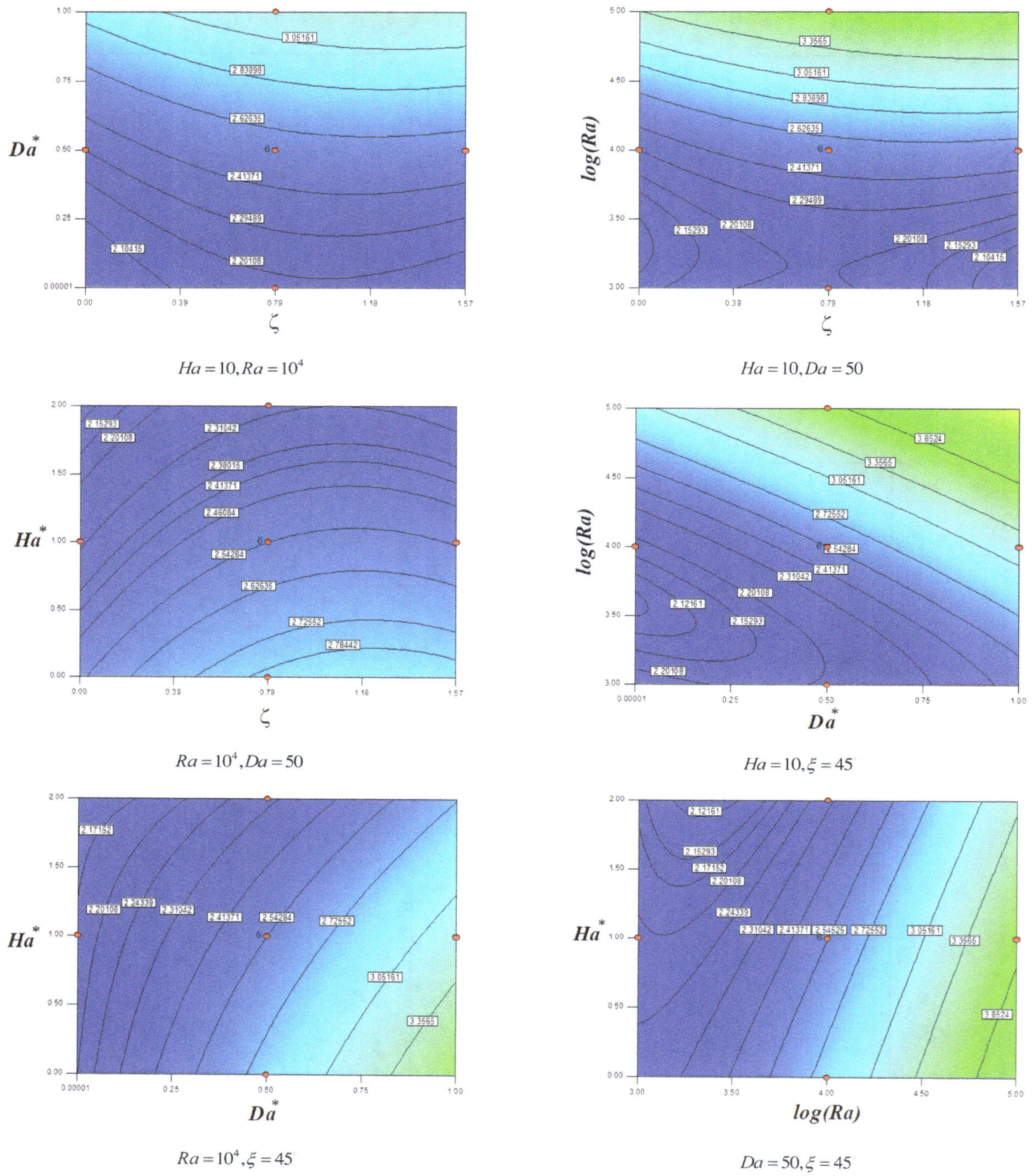

Figure 53b. Effects of inclination angle, radiation parameter, Rayleigh and Hartmann numbers on average Nusselt number

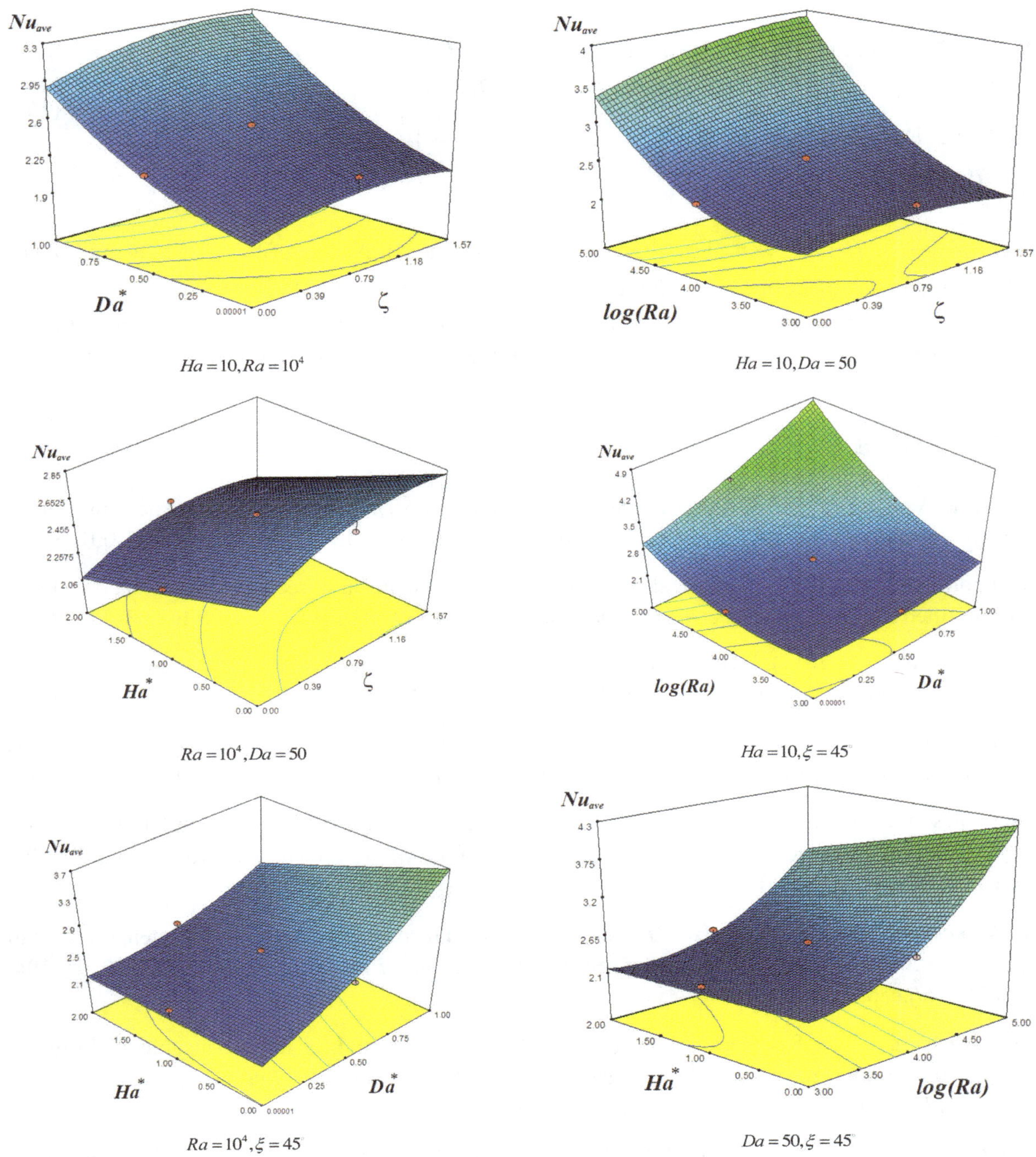

Figures 51, 52 and 53 depicts the impact of ξ, Da, Ra and Ha on Nu_{loc}, Nu_{ave}. The formula for Nu_{ave} corresponding to active parameters is:

$$\begin{aligned} Nu_{ave} &= 7.45 - 0.39\xi - 2.48Da^{*} - 3.14\log\left(Ra\right) + 0.81Ha^{*} \\ &+ 0.14\xi\, Da^{*} + 0.2\xi\log\left(Ra\right) - 0.036\xi Ha^{*} + 0.8\log\left(Ra\right)Da^{*} \\ &- 0.47\log\left(Ra\right)Ha^{*} + 0.82Da^{*}\,Ha^{*} - 0.2\xi^{2} + 0.44\left(\log\left(Ra\right)\right)^{2} + 0.58Da^{*2} \\ &+ 5.079\times10^{-3}Ha^{*2} \end{aligned} \quad (168)$$

where $Ha^{*} = 0.1Ha, Da^{*} = 0.01Da$. The number of extermum in Nu_{loc} profile is matching to existence of thermal plume. Effect of Lorentz forces on Nu_{loc} is more marked in higher permability of porous media. Nusselt number enhances with rise of Da, Ra, ξ but it reduces with rise of Hartmann number. Besides the distance between two maximum points increases with augments of inclination angle.

REFERENCES

Kumar, B. V. R., & Gupta, S. (2003). Free convection in a non-Darcian wavy porous enclosure. *International Journal of Engineering Science*, *41*(16), 1827–1848. doi:10.1016/S0020-7225(03)00113-7

Sheikholeslami, M. (2017). Numerical investigation of MHD nanofluid free convective heat transfer in a porous tilted enclosure. *Engineering Computations*, *34*(6), 1939–1955. doi:10.1108/EC-08-2016-0293

Sheikholeslami, M. (2017). Influence of Lorentz forces on nanofluid flow in a porous cavity by means of Non- Darcy model. *Engineering Computations*, *34*(8), 2651–2667. doi:10.1108/EC-01-2017-0008

Sheikholeslami, M., & Ganji, D. D. (2017). Numerical approach for magnetic nanofluid flow in a porous cavity using CuO nanoparticles. *Materials & Design*, *120*, 382–393. doi:10.1016/j.matdes.2017.02.039

Sheikholeslami, M., & Ganji, D. D. (2017). Numerical analysis of nanofluid transportation in porous media under the influence of external magnetic source. *Journal of Molecular Liquids*, *233*, 499–507. doi:10.1016/j.molliq.2017.03.050

Sheikholeslami, M., & Houman, B. (2017). Rokni, Nanofluid convective heat transfer intensification in a porous circular cylinder, Chemical Engineering & Processing. *Process Intensification*, *120*, 93–104. doi:10.1016/j.cep.2017.07.001

Sheikholeslami, M., & Seyednezhad, M. (2017). Nanofluid heat transfer in a permeable enclosure in presence of variable magnetic field by means of CVFEM. *International Journal of Heat and Mass Transfer*, *114*, 1169–1180. doi:10.1016/j.ijheatmasstransfer.2017.07.018

Sheikholeslami, M., & Shamlooei, M. (2017). Convective flow of nanofluid inside a lid driven porous cavity using CVFEM. *Physica B, Condensed Matter*, *521*, 239–250. doi:10.1016/j.physb.2017.07.005

Sheikholeslami, M., & Zeeshan, A. (2018). Numerical simulation of Fe3O4 -water nanofluid flow in a non-Darcy porous media. *International Journal of Numerical Methods for Heat & Fluid Flow*, *28*(3), 641–660. doi:10.1108/HFF-04-2017-0160

ADDITIONAL READING

Jafaryar, M., Sheikholeslami, M., Li, M., & Moradi, R. (2018). Nanofluid turbulent flow in a pipe under the effect of twisted tape with alternate axis. *Journal of Thermal Analysis and Calorimetry*. doi:10.100710973-018-7093-2

Li, Z., Shehzad, S. A., & Sheikholeslami, M. (2018). An application of CVFEM for nanofluid heat transfer intensification in a porous sinusoidal cavity considering thermal non-equilibrium model. *Computer Methods in Applied Mechanics and Engineering*, *339*, 663–680. doi:10.1016/j.cma.2018.05.015

Li, Z., Sheikholeslami, M., Chamkha, A. J., Raizah, Z. A., & Saleem, S. (2018). Control Volume Finite Element Method for nanofluid MHD natural convective flow inside a sinusoidal annulus under the impact of thermal radiation. *Computer Methods in Applied Mechanics and Engineering*, *338*, 618–633. doi:10.1016/j.cma.2018.04.023

Li, Z., Sheikholeslami, M., Jafaryar, M., Shafee, A., & Chamkha, A. J. (2018). Investigation of nanofluid entropy generation in a heat exchanger with helical twisted tapes. *Journal of Molecular Liquids*, *266*, 797–805. doi:10.1016/j.molliq.2018.07.009

Li, Z., Sheikholeslami, M., Samandari, M., & Shafee, A. (2018). Nanofluid unsteady heat transfer in a porous energy storage enclosure in existence of Lorentz forces. *International Journal of Heat and Mass Transfer*, *127*, 914–926. doi:10.1016/j.ijheatmasstransfer.2018.06.101

Sheikholeslami, M. (2017a). Lattice Boltzmann Method simulation of MHD non-Darcy nanofluid free convection. *Physica B, Condensed Matter*, *516*, 55–71. doi:10.1016/j.physb.2017.04.029

Sheikholeslami, M. (2017b). CuO-water nanofluid free convection in a porous cavity considering Darcy law. *The European Physical Journal Plus*, *132*(1), 55. doi:10.1140/epjp/i2017-11330-3

Sheikholeslami, M. (2017c). Influence of Lorentz forces on nanofluid flow in a porous cavity by means of Non- Darcy model. *Engineering Computations*, *34*(8), 2651–2667. doi:10.1108/EC-01-2017-0008

Sheikholeslami, M. (2017d). Magnetic field influence on nanofluid thermal radiation in a cavity with tilted elliptic inner cylinder. *Journal of Molecular Liquids*, *229*, 137–147. doi:10.1016/j.molliq.2016.12.024

Sheikholeslami, M. (2017e). Numerical simulation of magnetic nanofluid natural convection in porous media. *Physics Letters. [Part A]*, *381*(5), 494–503. doi:10.1016/j.physleta.2016.11.042

Sheikholeslami, M. (2017f). Influence of Lorentz forces on nanofluid flow in a porous cylinder considering Darcy model. *Journal of Molecular Liquids*, *225*, 903–912. doi:10.1016/j.molliq.2016.11.022

Sheikholeslami, M. (2017g). Influence of Coulomb forces on Fe3O4-H2O nanofluid thermal improvement. *International Journal of Hydrogen Energy*, *42*(2), 821–829. doi:10.1016/j.ijhydene.2016.09.185

Sheikholeslami, M. (2017h). Numerical investigation of MHD nanofluid free convective heat transfer in a porous tilted enclosure. *Engineering Computations*, *34*(6), 1939–1955. doi:10.1108/EC-08-2016-0293

Sheikholeslami, M. (2017i). Magnetic field influence on CuO -H2O nanofluid convective flow in a permeable cavity considering various shapes for nanoparticles. *International Journal of Hydrogen Energy*, *42*(31), 19611–19621. doi:10.1016/j.ijhydene.2017.06.121

Sheikholeslami, M. (2018a). Magnetic source impact on nanofluid heat transfer using CVFEM. *Neural Computing & Applications*, *30*(4), 1055–1064. doi:10.100700521-016-2740-7

Sheikholeslami, M. (2018b). Numerical simulation for external magnetic field influence on Fe3O4-water nanofluid forced convection. *Engineering Computations*, *35*(4), 1639–1654. doi:10.1108/EC-06-2017-0200

Sheikholeslami, M. (2018c). Finite element method for PCM solidification in existence of CuO nanoparticles. *Journal of Molecular Liquids*, *265*, 347–355. doi:10.1016/j.molliq.2018.05.132

Sheikholeslami, M. (2018d). Solidification of NEPCM under the effect of magnetic field in a porous thermal energy storage enclosure using CuO nanoparticles. *Journal of Molecular Liquids*, *263*, 303–315. doi:10.1016/j.molliq.2018.04.144

Sheikholeslami, M. (2018e). Influence of magnetic field on Al2O3-H2O nanofluid forced convection heat transfer in a porous lid driven cavity with hot sphere obstacle by means of LBM. *Journal of Molecular Liquids*, *263*, 472–488. doi:10.1016/j.molliq.2018.04.111

Sheikholeslami, M. (2018f). Numerical simulation for solidification in a LHTESS by means of Nano-enhanced PCM. *Journal of the Taiwan Institute of Chemical Engineers*, *86*, 25–41. doi:10.1016/j.jtice.2018.03.013

Sheikholeslami, M. (2018g). Numerical modeling of Nano enhanced PCM solidification in an enclosure with metallic fin. *Journal of Molecular Liquids*, *259*, 424–438. doi:10.1016/j.molliq.2018.03.006

Sheikholeslami, M. (2018h). Numerical investigation of nanofluid free convection under the influence of electric field in a porous enclosure. *Journal of Molecular Liquids*, *249*, 1212–1221. doi:10.1016/j.molliq.2017.11.141

Sheikholeslami, M. (2018i). CuO-water nanofluid flow due to magnetic field inside a porous media considering Brownian motion. *Journal of Molecular Liquids*, *249*, 921–929. doi:10.1016/j.molliq.2017.11.118

Sheikholeslami, M. (2018j). Application of Darcy law for nanofluid flow in a porous cavity under the impact of Lorentz forces. *Journal of Molecular Liquids*, *266*, 495–503. doi:10.1016/j.molliq.2018.06.083

Sheikholeslami, M., Barzegar Gerdroodbary, M., Valiallah Mousavi, S., Ganji, D. D., & Moradi, R. (2018). Heat transfer enhancement of ferrofluid inside an 90o elbow channel by non-uniform magnetic field. *Journal of Magnetism and Magnetic Materials*, *460*, 302–311. doi:10.1016/j.jmmm.2018.03.070

Sheikholeslami, M., & Bhatti, M. M. (2017). Active method for nanofluid heat transfer enhancement by means of EHD. *International Journal of Heat and Mass Transfer*, *109*, 115–122. doi:10.1016/j.ijheatmasstransfer.2017.01.115

Sheikholeslami, M., Darzi, M., & Li, Z. (2018). Experimental investigation for entropy generation and exergy loss of nano-refrigerant condensation process. *International Journal of Heat and Mass Transfer*, *125*, 1087–1095. doi:10.1016/j.ijheatmasstransfer.2018.04.155

Sheikholeslami, M., Darzi, M., & Sadoughi, M. K. (2018). Heat transfer improvement and Pressure Drop during condensation of refrigerant-based Nanofluid; An Experimental Procedure. *International Journal of Heat and Mass Transfer*, *122*, 643–650. doi:10.1016/j.ijheatmasstransfer.2018.02.015

Sheikholeslami, M., & Ghasemi, A. (2018). Solidification heat transfer of nanofluid in existence of thermal radiation by means of FEM. *International Journal of Heat and Mass Transfer*, *123*, 418–431. doi:10.1016/j.ijheatmasstransfer.2018.02.095

Sheikholeslami, M., Ghasemi, A., Li, Z., Shafee, A., & Saleem, S. (2018). Influence of CuO nanoparticles on heat transfer behavior of PCM in solidification process considering radiative source term. *International Journal of Heat and Mass Transfer*, *126*, 1252–1264. doi:10.1016/j.ijheatmasstransfer.2018.05.116

Sheikholeslami, M., Hayat, T., & Alsaedi, A. (2018). Numerical simulation for forced convection flow of MHD CuO-H2O nanofluid inside a cavity by means of LBM. *Journal of Molecular Liquids*, *249*, 941–948. doi:10.1016/j.molliq.2017.10.099

Sheikholeslami, M., Hayat, T., Muhammad, T., & Alsaedi, A. (2018). MHD forced convection flow of nanofluid in a porous cavity with hot elliptic obstacle by means of Lattice Boltzmann method. *International Journal of Mechanical Sciences*, *135*, 532–540. doi:10.1016/j.ijmecsci.2017.12.005

Sheikholeslami, M., Jafaryar, M., Ganji, D. D., & Li, Z. (2018). Exergy loss analysis for nanofluid forced convection heat transfer in a pipe with modified turbulators. *Journal of Molecular Liquids*, *262*, 104–110. doi:10.1016/j.molliq.2018.04.077

Sheikholeslami, M., Jafaryar, M., & Li, Z. (2018a). Second law analysis for nanofluid turbulent flow inside a circular duct in presence of twisted tape turbulators. *Journal of Molecular Liquids*, *263*, 489–500. doi:10.1016/j.molliq.2018.04.147

Sheikholeslami, M., Jafaryar, M., & Li, Z. (2018b). Nanofluid turbulent convective flow in a circular duct with helical turbulators considering CuO nanoparticles. *International Journal of Heat and Mass Transfer*, *124*, 980–989. doi:10.1016/j.ijheatmasstransfer.2018.04.022

Sheikholeslami, M., Jafaryar, M., Saleem, S., Li, Z., Shafee, A., & Jiang, Y. (2018). Nanofluid heat transfer augmentation and exergy loss inside a pipe equipped with innovative turbulators. *International Journal of Heat and Mass Transfer*, *126*, 156–163. doi:10.1016/j.ijheatmasstransfer.2018.05.128

Sheikholeslami, M., Jafaryar, M., Shafee, A., & Li, Z. (2018). Investigation of second law and hydrothermal behavior of nanofluid through a tube using passive methods. *Journal of Molecular Liquids*, *269*, 407–416. doi:10.1016/j.molliq.2018.08.019

Sheikholeslami, M., Li, Z., & Shafee, A. (2018a). Lorentz forces effect on NEPCM heat transfer during solidification in a porous energy storage system. *International Journal of Heat and Mass Transfer*, *127*, 665–674. doi:10.1016/j.ijheatmasstransfer.2018.06.087

Sheikholeslami, M., Li, Z., & Shamlooei, M. (2018). Nanofluid MHD natural convection through a porous complex shaped cavity considering thermal radiation. *Physics Letters. [Part A]*, *382*(24), 1615–1632. doi:10.1016/j.physleta.2018.04.006

Sheikholeslami, M., & Rokni, H. B. (2017). Simulation of nanofluid heat transfer in presence of magnetic field: A review. *International Journal of Heat and Mass Transfer*, *115*, 1203–1233. doi:10.1016/j.ijheatmasstransfer.2017.08.108

Sheikholeslami, M., & Rokni, H. B. (2018a). CVFEM for effect of Lorentz forces on nanofluid flow in a porous complex shaped enclosure by means of Non-equilibrium model. *Journal of Molecular Liquids*, *254*, 446–462. doi:10.1016/j.molliq.2018.01.130

Sheikholeslami, M., Rokni, H.B. (2018b). Magnetic nanofluid flow and convective heat transfer in a porous cavity considering Brownian motion effects. *Physics of Fluids*, *30*(1), doi:10.1063/1.5012517

Sheikholeslami, M., & Sadoughi, M. K. (2017). Mesoscopic method for MHD nanofluid flow inside a porous cavity considering various shapes of nanoparticles. *International Journal of Heat and Mass Transfer*, *113*, 106–114. doi:10.1016/j.ijheatmasstransfer.2017.05.054

Sheikholeslami, M., & Sadoughi, M. K. (2018). Simulation of CuO- water nanofluid heat transfer enhancement in presence of melting surface. *International Journal of Heat and Mass Transfer*, *116*, 909–919. doi:10.1016/j.ijheatmasstransfer.2017.09.086

Sheikholeslami, M., & Seyednezhad, M. (2018). Simulation of nanofluid flow and natural convection in a porous media under the influence of electric field using CVFEM. *International Journal of Heat and Mass Transfer*, *120*, 772–781. doi:10.1016/j.ijheatmasstransfer.2017.12.087

Sheikholeslami, M., Shafee, A., Ramzan, M., & Li, Z. (2018). Investigation of Lorentz forces and radiation impacts on nanofluid treatment in a porous semi annulus via Darcy law. *Journal of Molecular Liquids*, *272*, 8–14. doi:10.1016/j.molliq.2018.09.016

Sheikholeslami, M., Shamlooei, M., & Moradi, R. (2018). Numerical simulation for heat transfer intensification of nanofluid in a porous curved enclosure considering shape effect of Fe3O4 nanoparticles. *Chemical Engineering & Processing: Process Intensification*, *124*, 71–82. doi:10.1016/j.cep.2017.12.005

Sheikholeslami, M., & Shehzad, S. A. (2017a). Thermal radiation of ferrofluid in existence of Lorentz forces considering variable viscosity. *International Journal of Heat and Mass Transfer*, *109*, 82–92. doi:10.1016/j.ijheatmasstransfer.2017.01.096

Sheikholeslami, M., & Shehzad, S. A. (2017b). Magnetohydrodynamic nanofluid convective flow in a porous enclosure by means of LBM. *International Journal of Heat and Mass Transfer*, *113*, 796–805. doi:10.1016/j.ijheatmasstransfer.2017.05.130

Sheikholeslami, M., & Shehzad, S. A. (2018a). Numerical analysis of Fe3O4 –H2O nanofluid flow in permeable media under the effect of external magnetic source. *International Journal of Heat and Mass Transfer*, *118*, 182–192. doi:10.1016/j.ijheatmasstransfer.2017.10.113

Sheikholeslami, M., & Shehzad, S. A. (2018b). CVFEM simulation for nanofluid migration in a porous medium using Darcy model. *International Journal of Heat and Mass Transfer*, *122*, 1264–1271. doi:10.1016/j.ijheatmasstransfer.2018.02.080

Sheikholeslami, M., & Shehzad, S. A. (2018c). Simulation of water based nanofluid convective flow inside a porous enclosure via Non-equilibrium model. *International Journal of Heat and Mass Transfer*, *120*, 1200–1212. doi:10.1016/j.ijheatmasstransfer.2017.12.132

Sheikholeslami, M., & Shehzad, S. A. (2018d). Non-Darcy free convection of Fe3O4-water nanoliquid in a complex shaped enclosure under impact of uniform Lorentz force. *The Chinese Journal of Physiology*, *56*(1), 270–281. doi:10.1016/j.cjph.2017.12.017

Sheikholeslami, M., Shehzad, S. A., Abbasi, F. M., & Li, Z. (2018). Nanofluid flow and forced convection heat transfer due to Lorentz forces in a porous lid driven cubic enclosure with hot obstacle. *Computer Methods in Applied Mechanics and Engineering*, *338*, 491–505. doi:10.1016/j.cma.2018.04.020

Sheikholeslami, M., Shehzad, S. A., & Li, Z. (2018a). Nanofluid heat transfer intensification in a permeable channel due to magnetic field using Lattice Boltzmann method. *Physica B, Condensed Matter*, *542*, 51–58. doi:10.1016/j.physb.2018.03.036

Sheikholeslami, M., Shehzad, S. A., & Li, Z. (2018b). Water based nanofluid free convection heat transfer in a three dimensional porous cavity with hot sphere obstacle in existence of Lorenz forces. *International Journal of Heat and Mass Transfer*, *125*, 375–386. doi:10.1016/j.ijheatmasstransfer.2018.04.076

Sheikholeslami, M., Shehzad, S. A., Li, Z., & Shafee, A. (2018). Numerical modeling for Alumina nanofluid magnetohydrodynamic convective heat transfer in a permeable medium using Darcy law. *International Journal of Heat and Mass Transfer*, *127*, 614–622. doi:10.1016/j.ijheatmasstransfer.2018.07.013

Sheikholeslami, M., & Vajravelu, K. (2017). Forced convection heat transfer in Fe3O4- ethylene glycol nanofluid under the influence of Coulomb force. *Journal of Molecular Liquids*, *233*, 203–210. doi:10.1016/j.molliq.2017.03.026

Sheikholeslami, M., & Zeeshan, A. (2017). Analysis of flow and heat transfer in water based nanofluid due to magnetic field in a porous enclosure with constant heat flux using CVFEM. *Computer Methods in Applied Mechanics and Engineering*, *320*, 68–81. doi:10.1016/j.cma.2017.03.024

Sheikholeslami, M., Zeeshan, A., & Majeed, A. (2018). Control volume based finite element simulation of magnetic nanofluid flow and heat transport in non-Darcy medium. *Journal of Molecular Liquids*, *268*, 354–364. doi:10.1016/j.molliq.2018.07.031

Chapter 9
Magnetic Field Dependent (MFD) Viscosity Effect on Nanofluid Treatment

ABSTRACT

In this chapter, the effect of magnetic field dependent (MFD) viscosity on free convection heat transfer of nanofluid in an enclosure is investigated. A single-phase nanofluid model is utilized considering Brownian motion. The control volume-based finite element method is applied to simulate this problem. The effects of viscosity parameter, Hartmann number, and Rayleigh number on hydrothermal behavior have been examined.

1. INTRODUCTION

Ferrohydrodynamics deals with the study of ferrofluids in the presence of magnetic field and it is a special branch of magnetohydrodynamics. During the last decades, an extensive research work has been done on these fluids since the effect of magnetization has yielded interesting information leading to their diverse fascinating technological applications (Rosensweig, 1985). Ferrofluids consist of colloidal suspensions of single domain magnetic nanoparticles and it has been recognized that they have promising potential for heat transfer applications in electronics, engines, micro and nanoelectromechanical systems (MEMS and NEMS), air-conditioning and ventilation systems (Ganguly, Sen and Puri, 2004). Under the circumstances, the study of thermal convection in ferrofluids is gaining much importance in the recent years. Moreover, many physical properties of these fluids can be tuned by varying the magnetic field. One of the well known phenomena generated by the influence of magnetic fields on ferrofluids is the change of their viscous behavior. Realizing the importance of magnetic field dependent (MFD) viscosity on ferrofluid flows, several studies have been undertaken in the past. The effect of a homogeneous magnetic field on the viscosity of a fluid with solid particles possessing intrinsic magnetic moments has been investigated by Shliomis (1972). The effect of MFD viscosity on the onset of ferroconvection in a rotating ferrofluid layer is discussed by Vaidyanathan et al. (2002). Nanjundappa et al. (2010) have investigated the effect of MFD viscosity on the onset of convection in a ferromagnetic fluid layer in the

DOI: 10.4018/978-1-5225-7595-5.ch009

presence of a vertical magnetic field by considering the bounding surfaces are either rigid-ferromagnetic or stress- free with constant heat flux conditions. Sheikholeslami et al. (2016) considered MFD viscosity effect on natural convection of ferrofluid. Sheikholeslami (2017) studied the magnetic field influence on nanofluid thermal radiation in a cavity with tilted elliptic inner cylinder. Sheikholeslami and Shehzad (2017) studied the thermal radiation of ferrofluid in existence of Lorentz forces considering variable viscosity. Sheikholeslami and Rokni (2017) reported the magnetic nanofluid natural convection in presence of thermal radiation. Sheikholeslami and Abelman (2018) presented the numerical analysis of the effect of magnetic field on Fe3O4-water ferrofluid convection with thermal radiation. Sheikholeslami et al. (2017) investigated the nanofluid radiation and natural convection in an enclosure with elliptical cylinders. Sheikholeslami and Sadoughi (2017) studied the Fe3O4 -water nanofluid flow in porous medium considering MFD viscosity. Magnetic field effect on nanofluid treatment was investigated in recent decade.

2. NATURAL CONVECTION OF MAGNETIC NANOFLUID CONSIDERING MFD VISCOSITY EFFECT

2.1. Problem Definition

The geometry of this problem is shown in Figure 1(a). The heat source is centrally located on the bottom surface and its length L/3. The cooling is achieved by the two vertical walls. The heat source has constant heat flux q'' while the cooling walls have a constant temperature T_c; all the other surfaces are adiabatic. Also, it is also assumed that the uniform magnetic field ($\vec{B} = B_x \vec{e}_x + B_y \vec{e}_y$) of constant magnitude $B = \sqrt{B_x^2 + B_y^2}$ is applied, where $\vec{e}_x$ and $\vec{e}_y$ are unit vectors in the Cartesian coordinate system. The orientation of the magnetic field form an angle θ_M with horizontal axis such that $\theta_M = \cot^{-1}\left(B_x / B_y\right)$. The electric current J and the electromagnetic force F are defined by $J = \sigma\left(\vec{V} \times \vec{B}\right)$ and $F = \sigma\left(\vec{V} \times \vec{B}\right) \times \vec{B}$, respectively.

2.2. Governing Equation

The flow is steady, two-dimensional, laminar and incompressible. The induced electric current and Joule heating are neglected. The magnetic Reynolds number is assumed to be small so that the induced magnetic field can be neglected compared to the applied magnetic field. Neglecting displacement currents, induced magnetic field, and using the Boussinesq approximation, the governing equations of heat transfer and fluid flow for nanofluid can be obtained as follows:

$$\frac{\partial u}{\partial x} + \frac{\partial v}{\partial y} = 0 \tag{1}$$

Figure 1. (a)Geometry and the boundary conditions; (b) A sample triangular element and its corresponding control volume

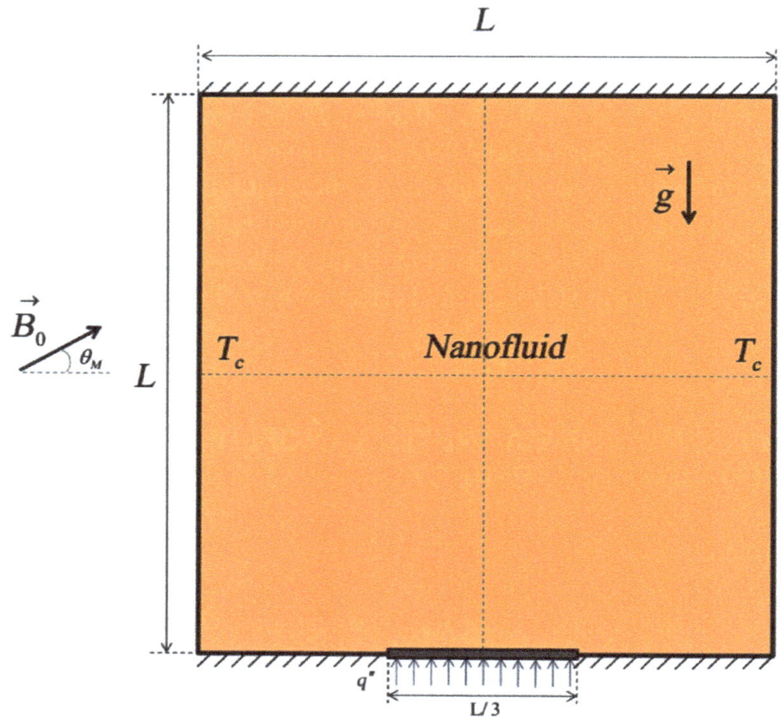

a

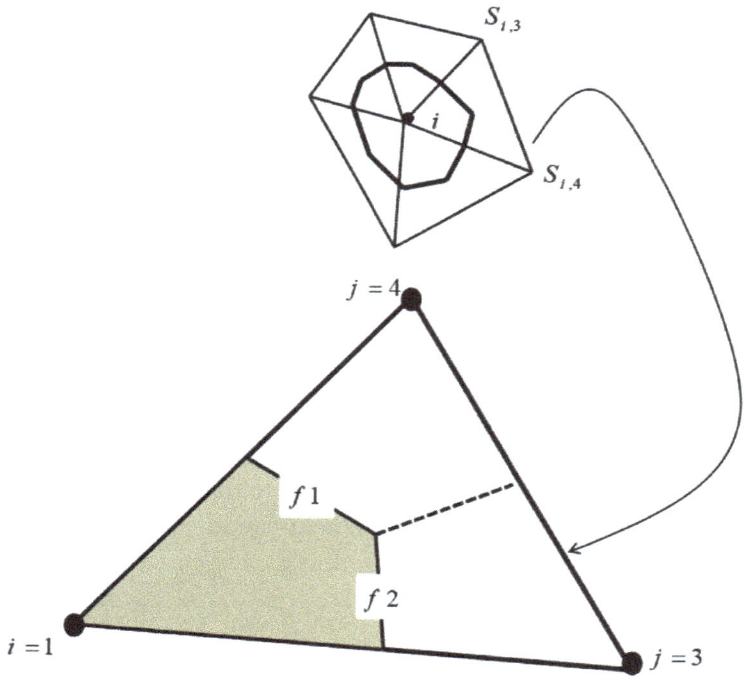

b

$$\rho_{nf}\left(u\frac{\partial u}{\partial x}+v\frac{\partial u}{\partial y}\right)=-\frac{\partial P}{\partial x}+\eta\left(\frac{\partial^2 u}{\partial x^2}+\frac{\partial^2 u}{\partial y^2}\right)$$
$$+\sigma_{nf}B_0^{\,2}\left(v\sin\theta_M\cos\theta_M-u\sin^2\theta_M\right) \tag{2}$$

$$\rho_{nf}\left(u\frac{\partial v}{\partial x}+v\frac{\partial v}{\partial y}\right)=-\frac{\partial P}{\partial y}+\eta\left(\frac{\partial^2 v}{\partial x^2}+\frac{\partial^2 v}{\partial y^2}\right)+\rho_{nf}\beta_{nf}g\left(T-T_c\right)$$
$$+\sigma_{nf}B_0^{\,2}\left(u\sin\theta_M\cos\theta_M-v\cos^2\theta_M\right) \tag{3}$$

$$u\frac{\partial T}{\partial x}+v\frac{\partial T}{\partial y}=\alpha_{nf}\left(\frac{\partial^2 T}{\partial x^2}+\frac{\partial^2 T}{\partial y^2}\right) \tag{4}$$

where $\eta=(1+\vec{\delta}.\vec{B})\mu_{nf}$, the variation of MFD viscosity (δ) has been taken to be isotropic, $\delta_1=\delta_2=\delta_3=\delta$. The effective density (ρ_{nf}), the thermal expansion coefficient (β_{nf}), heat capacitance $(\rho C_p)_{nf}$ and electrical conductivity of nanofluid (σ_{nf}) of the nanofluid are defined as:

$$\rho_{nf}=\rho_f(1-\phi)+\rho_s\phi$$

$$\beta_{nf}=\beta_f(1-\phi)+\beta_s\phi$$

$$(\rho C_p)_{nf}=(\rho C_p)_f(1-\phi)+(\rho C_p)_s\phi \tag{5}$$

$$\frac{\sigma_{nf}}{\sigma_f}=1+\frac{3\left(\sigma_s/\sigma_f-1\right)\phi}{\left(\sigma_s/\sigma_f+2\right)-\left(\sigma_s/\sigma_f-1\right)\phi} \tag{6}$$

Koo-Kleinstreuer-Li (KKL) is used to simulate thermal conductivity of nanofluid:

$$k_{eff}=k_{static}+k_{Brownian} \tag{7}$$

$$\frac{k_{static}}{k_f}=1+\frac{3\left(k_p/k_f-1\right)\phi}{\left(k_p/k_f+2\right)-\left(k_p/k_f-1\right)\phi} \tag{8}$$

$$k_{Brownian} = 5\times10^4 \phi\rho_f c_{p,f}\sqrt{\frac{\kappa_b T}{\rho_p d_p}} g'(T,\phi,d_p) \tag{9}$$

$$R_f + \frac{d_p}{k_p} = \frac{d_p}{k_{p,eff}}, R_f = 4\times10^{-8} km^2 / W \tag{10}$$

$$\begin{aligned} g'\left(T,\phi,d_p\right) &= \left(a_1 + a_2 \ln\left(d_p\right) + a_3 \ln\left(\phi\right) + a_4 \ln\left(\phi\right)\ln\left(d_p\right) + a_5 \ln\left(d_p\right)^2\right)\ln\left(T\right) \\ &+ \left(a_6 + a_7 \ln\left(d_p\right) + a_8 \ln\left(\phi\right) + a_9 \ln\left(\phi\right)\ln\left(d_p\right) + a_{10} \ln\left(d_p\right)^2\right) \end{aligned} \tag{11}$$

with the coefficients a_i (i=0..10) are based on the type of nanopartices, Al_2O_3-water nanofluids has an R^2 of 98%, respectively (Table 1).

The effective viscosity due to micro mixing in suspensions, can be obtained as follows:

Table 1. The coefficient values of Al_2O_3 – Water nanofluid

Coefficient Values	Al_2O_3 – Water
a_1	52.813488759
a_2	6.115637295
a_3	0.6955745084
a_4	4.17455552786E-02
a_5	0.176919300241
a_6	-298.19819084
a_7	-34.532716906
a_8	-3.9225289283
a_9	-0.2354329626
a_{10}	-0.999063481

$$\mu_{eff} = \mu_{static} + \mu_{Brownian} = \mu_{static} + \frac{k_{Brownian}}{k_f} \times \frac{\mu_f}{\Pr_f} \tag{12}$$

where $\mu_{static} = \dfrac{\mu_f}{(1-\phi)^{2.5}}$ is viscosity of the nanofluid, as given originally by Brinkman.

The stream function and vorticity are defined as:

$$u = \frac{\partial \psi}{\partial y},\; v = -\frac{\partial \psi}{\partial x},\; \omega = \frac{\partial v}{\partial x} - \frac{\partial u}{\partial y} \tag{13}$$

The stream function satisfies the continuity Equation (1). The vorticity equation is obtained by eliminating the pressure between the two momentum equations, i.e. by taking y-derivative of Equation (2) and subtracting from it the x-derivative of Equation (3). This gives:

$$\begin{aligned}\frac{\partial \psi}{\partial y}\frac{\partial \omega}{\partial x} - \frac{\partial \psi}{\partial x}\frac{\partial \omega}{\partial y} &= \upsilon_{nf}\left(1+\delta B_0\left(\cos\theta_M + \sin\theta_M\right)\right)\left(\frac{\partial^2 \omega}{\partial x^2} + \frac{\partial^2 \omega}{\partial y^2}\right) \\ &+ \beta_{nf} g\left(\frac{\partial T}{\partial x}\right) + \frac{\sigma_{nf} B_0^{\,2}}{\rho_{nf}}\left(\begin{matrix} -\dfrac{\delta v}{\delta y}\sin\theta_M\cos\theta_M + \dfrac{\delta u}{\delta y}\sin^2\theta_M \\ +\dfrac{\delta u}{\delta x}\sin\theta_M\cos\theta_M - \dfrac{\delta v}{\delta x}\cos^2\theta_M \end{matrix}\right)\end{aligned} \tag{14}$$

$$\frac{\partial \psi}{\partial y}\frac{\partial T}{\partial x} - \frac{\partial \psi}{\partial x}\frac{\partial T}{\partial y} = \alpha_{nf}\left(\frac{\partial^2 T}{\partial x^2} + \frac{\partial^2 T}{\partial y^2}\right) \tag{15}$$

$$\frac{\partial^2 \psi}{\partial x^2} + \frac{\partial^2 \psi}{\partial y^2} = -\omega \tag{16}$$

By introducing the following non-dimensional variables:

$$X = \frac{x}{L}, Y = \frac{y}{L}, \Omega = \frac{\omega L^2}{\alpha_f}, \Psi = \frac{\psi}{\alpha_f}, \Theta = \frac{T - T_c}{\left(q''L / k_f\right)}, U = \frac{uL}{\alpha_f}, V = \frac{vL}{\alpha_f} \tag{17}$$

Using the dimensionless parameters, the equations now become:

$$\frac{\partial \Psi}{\partial Y}\frac{\partial \Omega}{\partial X}-\frac{\partial \Psi}{\partial X}\frac{\partial \Omega}{\partial Y}=\Pr_f\left(\frac{\mu_{nf}}{\mu_f}\frac{\rho_f}{\rho_{nf}}\frac{k_f}{k_{nf}}\frac{(\rho C_p)_{nf}}{(\rho C_p)_f}\right)\left(1+\delta^*\left(\cos\theta_M+\sin\theta_M\right)\right)\left(\frac{\partial^2 \Omega}{\partial X^2}+\frac{\partial^2 \Omega}{\partial Y^2}\right)$$
$$+Ra_f \Pr_f \frac{\beta_{nf}}{\beta_f}\left(\frac{\partial \Theta}{\partial X}\right)+Ha^2 \Pr_f \frac{\sigma_{nf}}{\sigma_f}\frac{\rho_f}{\rho_{nf}}\left(-\frac{\delta V}{\delta Y}\tan\theta_M+\frac{\delta U}{\delta Y}\tan^2\theta_M+\frac{\delta U}{\delta X}\tan\theta_M-\frac{\delta V}{\delta X}\right) \tag{18}$$

$$\frac{\partial \Psi}{\partial Y}\frac{\partial \Theta}{\partial X}-\frac{\partial \Psi}{\partial X}\frac{\partial \Theta}{\partial Y}=\frac{k_{nf}}{k_f}\frac{(\rho C_p)_f}{(\rho C_p)_{nf}}\left(\frac{\partial^2 \Theta}{\partial X^2}\right) \tag{19}$$

$$\frac{\partial^2 \Psi}{\partial X^2}+\frac{\partial^2 \Psi}{\partial Y^2}=-\Omega \tag{20}$$

where $Ra_f = g\beta_f L^4 q'' / \left(k_f \alpha_f \upsilon_f\right)$ is the Rayleigh number for the base fluid, $Ha = LB_x\sqrt{\sigma_f / \mu_f}$ is the Hartmann number and $\Pr_f = \upsilon_f / \alpha_f$ is the Prandtl number for the base fluid. Also $\delta^* = \delta B_0$ is viscosity parameter. The thermo physical properties of the nanofluid are given in Table 2. The boundary conditions as shown in Figure 1 are:

$\partial\Theta/\partial n = -1.0$ on the heat source

$\Theta = 0.0$ on the left and right

$\partial\Theta/\partial n = 0.0$ on all the other adiabatic surfaces

$\Psi = 0.0$ on all solid boundaries (21)

The values of vorticity on the boundary of the enclosure can be obtained using the stream function formulation and the known velocity conditions during the iterative solution procedure. The local Nusselt number of the nanofluid along the heat source can be expressed as:

Table 2. Thermo physical properties of water and nanoparticles

	$\rho(kg / m^3)$	$C_p(j / kgk)$	$k(W / m.k)$	$\beta\times10^5(K^{-1})$	$d_p(nm)$	$\sigma(\Omega\cdot m)^{-1}$
Water	997.1	4179	0.613	21	-	0.05
Al_2O_3	3970	765	25	0.85	47	1×10^{-10}

$$Nu_{local} = \left(\frac{k_{nf}}{k_f}\right)\frac{1}{\theta}\Bigg|_{L/3<X<2L/3,Y=0} \tag{22}$$

The average Nusselt number is evaluated as:

$$Nu_{ave} = \frac{1}{L/3}\int_{L/3}^{2L/3} Nu_{loc}\left(X\right)dX \tag{23}$$

Nusselt ratio is defined as follows:

$$Nu_{ratio} = \frac{Nu_{ave}\big|_{\delta^*=1}}{Nu_{ave}\big|_{\delta^*=0}} \tag{24}$$

2.3. Effects of Active Parameters

In this study magnetohydrodynamic natural convective heat transfer in a cooling system of electronic components is investigated. Magnetic field dependent (MFD) viscosity effect is taken in to account. Brownian motion effect is considered for simulating effective thermal conductivity and viscosity. Influence of active parameters such as: Rayleigh number ($Ra = 10^3, 10^4$ and 10^5), Hartmann number ($Ha = 0, 15, 30$ and 60) and viscosity parameter ($\delta^* = 0, 1$) on flow and heat transfer are examined when $\phi = 0.04, \Pr = 6.2$.

Effects viscosity parameter, Hartmann number and Rayleigh number on isotherms and streamlines are shown in Figures 2, 3 and 4. At low Rayleigh number, the streamlines take the enclosure geometry. By increasing Rayleigh number the prominent heat transfer mechanism is turn from conduction to convection. When the magnetic field is imposed on the enclosure, the velocity field suppressed owing to the retarding effect of the Lorenz force. So intensity of convection weakens significantly. The decelerating effect of the magnetic field is observed from the maximum stream function value. The core vortex is shift downward vertically as the Hartmann number increases. Also imposing magnetic field leads to omit the thermal plume over the bottom wall. At high Hartmann number the conduction heat transfer mechanism is more pronounced. For this reason the isotherms are parallel to each other. Considering MFD viscosity effect increases the thermal boundary layer thickness and in turn Nusselt number decreases with increase of viscosity parameter. Also it can be seen that flow circulation increases with increase of this parameter.

Effects of Hartmann number and Rayleigh number on local and average Nusselt number are shown in Figures 5 and 6. Increasing Rayleigh number leads the Nusselt number to enhance. Increasing Lorentz forces make thermal boundary layer thickness to increase and in turn Nusselt number decreases with enhance of Hartmann number. Figure 7 depicts the effect of Hartmann number and Rayleigh number on Nusselt number ratio. As viscosity parameter increases the rate of heat transfer decreases. This reduction is more pronounced for higher values of Rayleigh number and lower values of Hartmann number. It is interesting observation that increasing Lorentz forces increase the effect of MFD viscosity of Nusselt number. At low Rayleigh number, considering MFD viscosity has no significant effect on Nusselt number.

Figure 2. Effects of viscosity parameter ($\delta^* = 1$ ($---$) and $\delta^* = 0$ (—)) and Hartmann number on isotherms (Right) and streamlines (left) when $\phi = 0.04, Ra = 10^3$

Ha=0

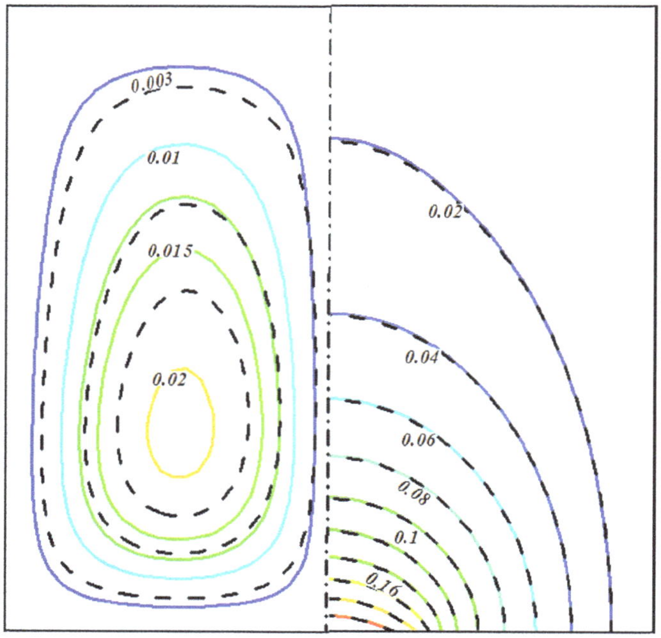

Ha=60

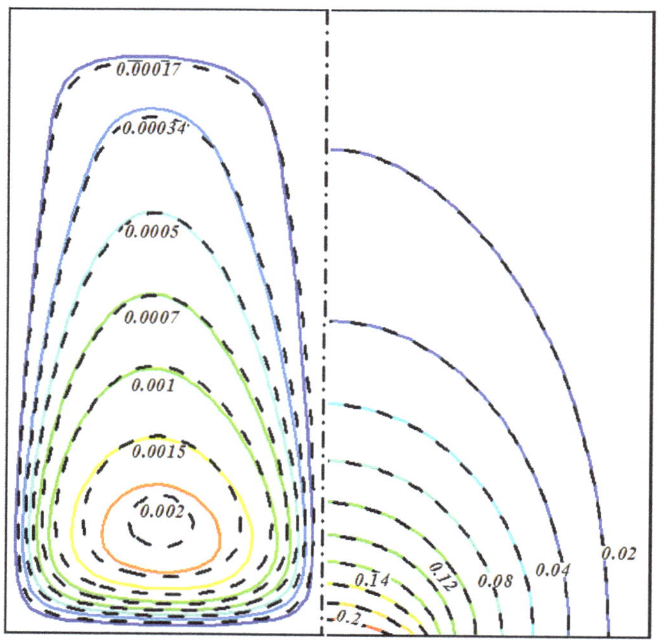

Figure 3. Effects of viscosity parameter ($\delta^* = 1$ (− − −) and $\delta^* = 0$ (—)) and Hartmann number on isotherms (Right) and streamlines (left) when $\phi = 0.04, Ra = 10^4$.

Ha=0

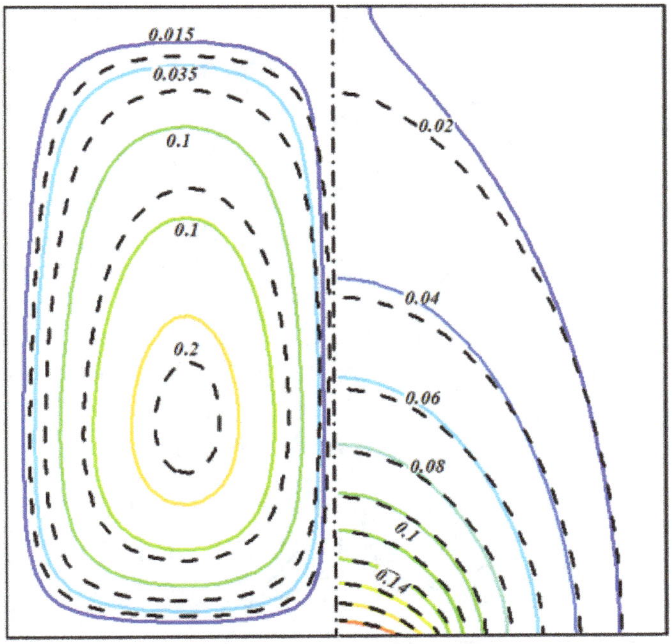

Ha=60

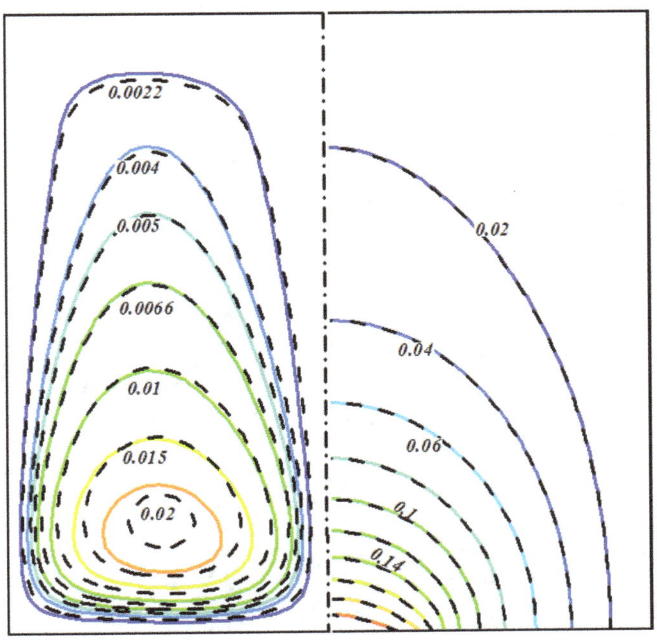

Figure 4. Effects of viscosity parameter ($\delta^* = 1$ (−−−) and $\delta^* = 0$ (—)) and Hartmann number on isotherms (Right) and streamlines (left) when $\phi = 0.04, Ra = 10^5$.

Ha=0

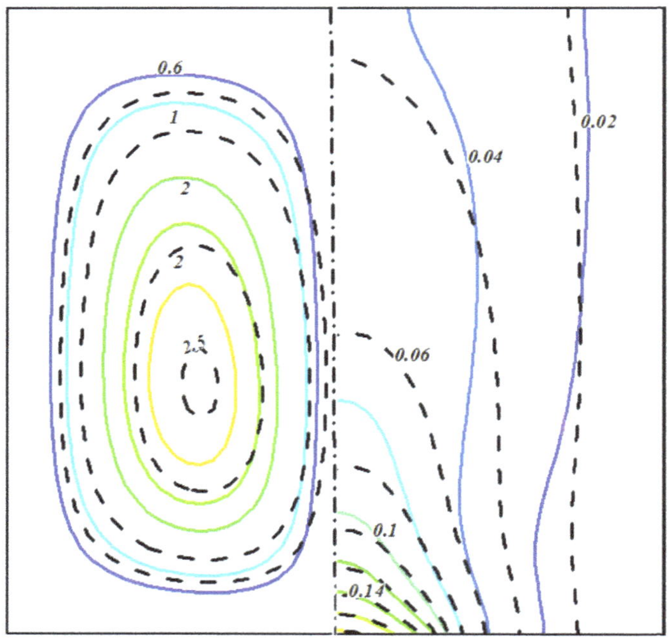

Ha=60

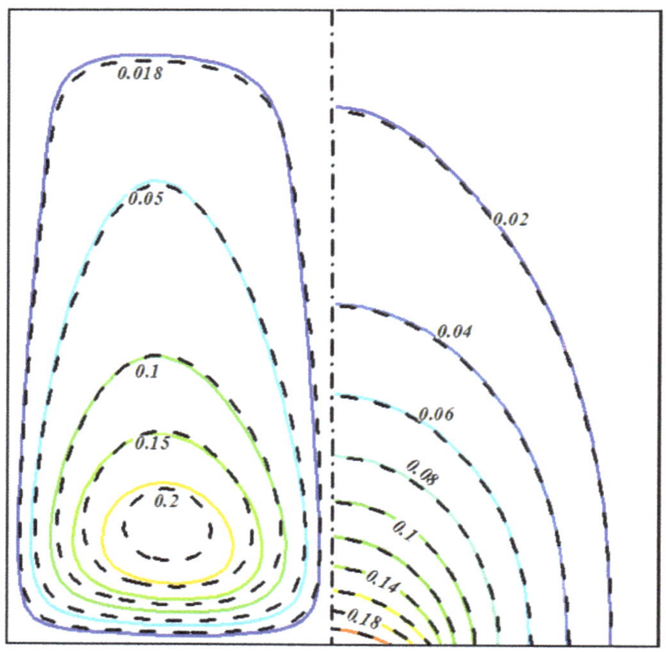

Figure 5. Effects of Hartmann number and Rayleigh number on Local Nusselt number when $\phi = 0.04, \delta^* = 1, \mathrm{Pr} = 6.2$.

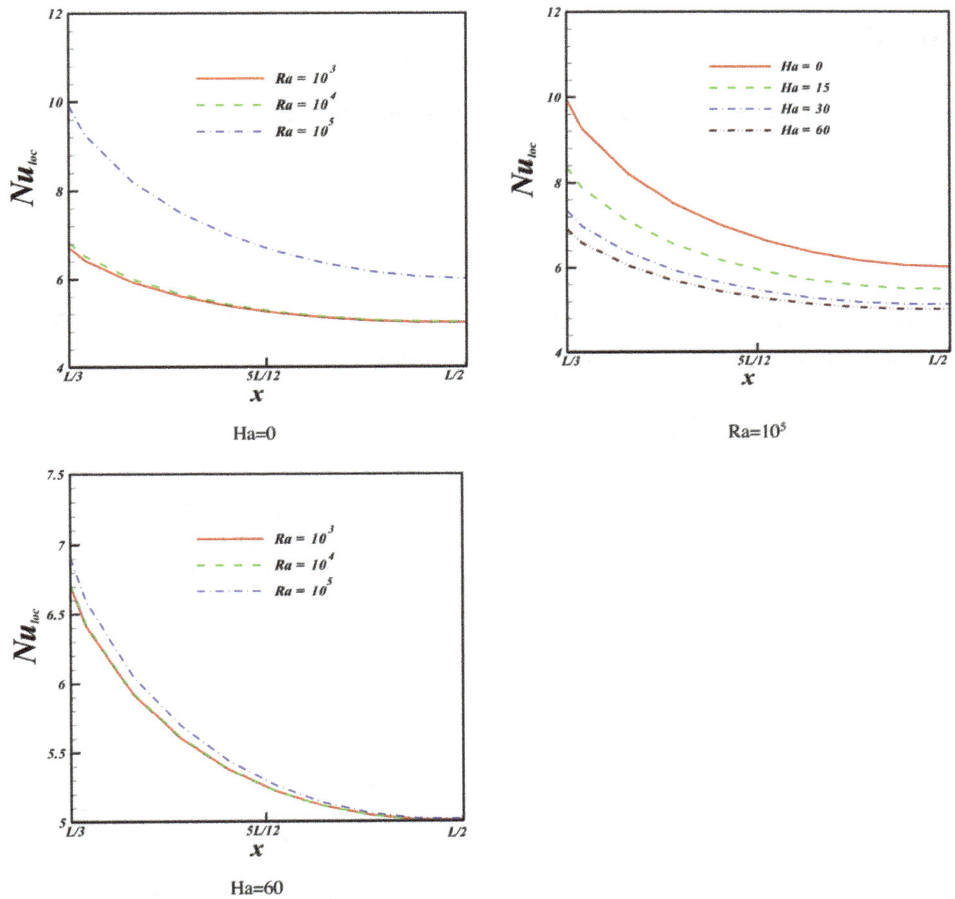

Figure 6. Effects of Hartmann number and Rayleigh number on average Nusselt number when $\phi = 0.04, \delta^* = 1, \mathrm{Pr} = 6.2$.

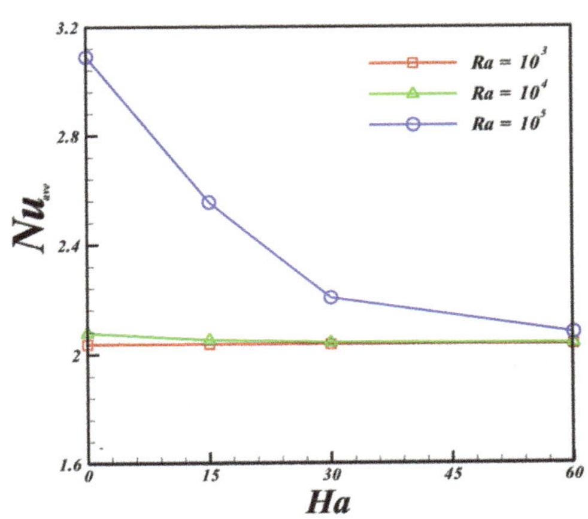

Figure 7. Effects of Hartmann number and Rayleigh number on Nusselt number ratio when $\phi = 0.04, \mathrm{Pr} = 6.2$.

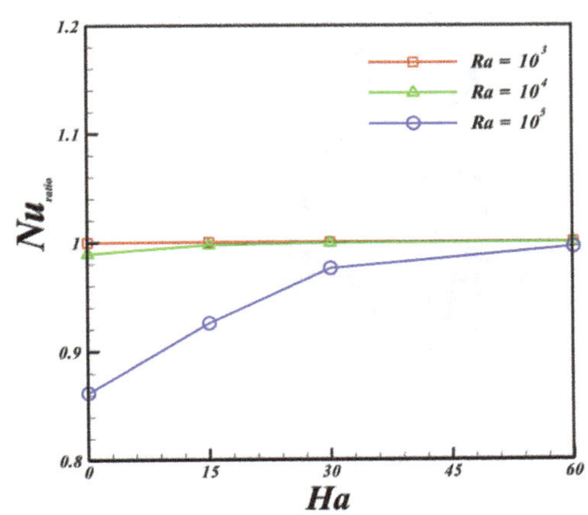

3. MAGNETIC FIELD INFLUENCE ON NANOFLUID THERMAL RADIATION IN A CAVITY WITH TILTED ELLIPTIC INNER CYLINDER

3.1. Problem Definition

Figure 8 depicts the geometry, boundary condition and sample element. The formula of inner cylinder is:

$$b = \sqrt{1 - \varepsilon^2}.a \tag{25}$$

Figure 8. (a)Geometry and the boundary conditions with (b) the mesh of Geometry considered in this work; (c) A sample triangular element and its corresponding control volume.

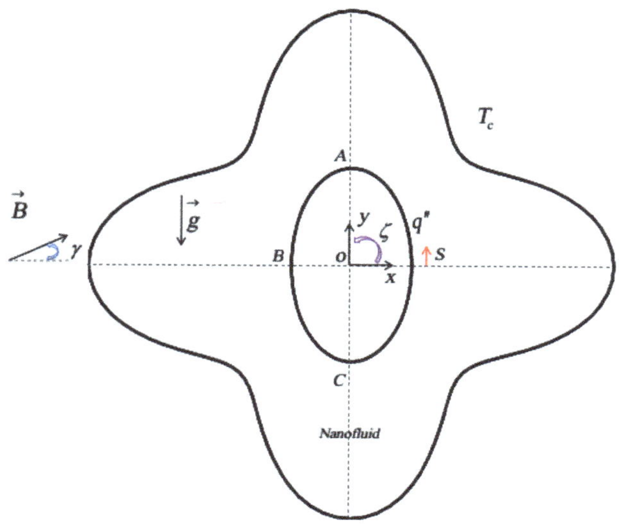

a

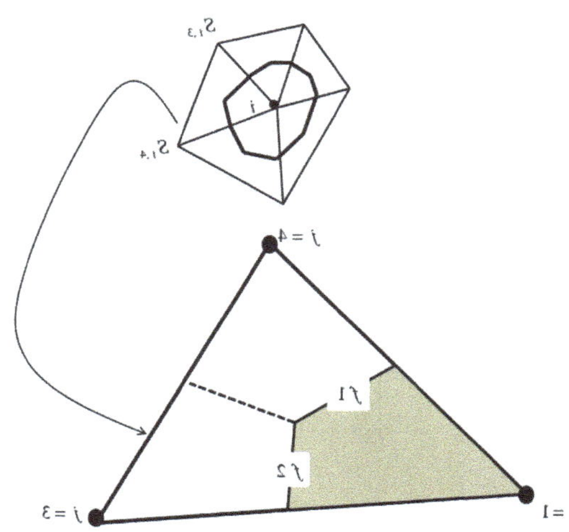

c

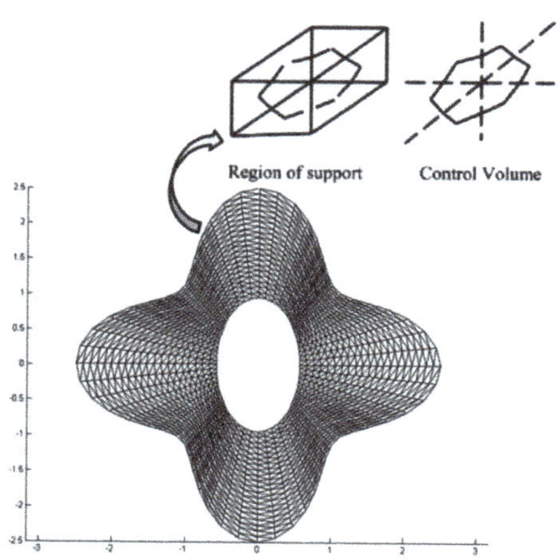

b

where a, b, ε are the major, minor axis of elliptic cylinder and eccentricity for the inner cylinder. The inner cylinder has constant heat flux condition.

3.2. Governing Equation

2D steady Convective flow of nanofluid in a porous media is considered in existence of constant magnetic field. The PDEs equations are:

$$\frac{\partial u}{\partial x} + \frac{\partial v}{\partial y} = 0, \tag{26}$$

$$\left(v\frac{\partial u}{\partial y} + \frac{\partial u}{\partial x}u\right) = \left[-\sigma_{nf}B_y^2 u + \sigma_{nf}B_x B_y v + \left(\frac{\partial^2 u}{\partial y^2} + \frac{\partial^2 u}{\partial x^2}\right)\mu_{nf} - \frac{\partial P}{\partial x}\right]\left(\rho_{nf}\right)^{-1}, \tag{27}$$

$$\begin{gathered}\rho_{nf}\left(\frac{\partial v}{\partial x}u + \frac{\partial v}{\partial y}v\right) = +\mu_{nf}\left(\frac{\partial^2 v}{\partial x^2} + \frac{\partial^2 v}{\partial y^2}\right) - \frac{\partial P}{\partial y} + B_y\sigma_{nf}B_x u - B_x\sigma_{nf}B_x v \\ +\left(T - T_c\right)\beta_{nf}g\rho_{nf}, \\ B_x = B_o\cos\lambda,\ B_y = B_o\sin\lambda,\end{gathered} \tag{28}$$

$$\left(\rho C_p\right)_{nf}\left(v\frac{\partial T}{\partial y} + u\frac{\partial T}{\partial x}\right) = k_{nf}\left(\frac{\partial^2 T}{\partial x^2} + \frac{\partial^2 T}{\partial y^2}\right) - \frac{\partial q_r}{\partial y}, \left[q_r = -\frac{4\sigma_e}{3\beta_R}\frac{\partial T^4}{\partial y}, T^4 \cong 4T_c^3 T - 3T_c^4\right]. \tag{29}$$

$\left(\rho C_p\right)_{nf}, \left(\rho\beta\right)_{nf}, \rho_{nf}, k_{nf}$ and σ_{nf} are defined as:

$$\left(\rho C_p\right)_{nf} = \left(\rho C_p\right)_f (1 - \phi) + \left(\rho C_p\right)_s \phi, \tag{30}$$

$$\left(\rho\beta\right)_{nf} = \left(\rho\beta\right)_f (1 - \phi) + \left(\rho\beta\right)_s \phi, \tag{31}$$

$$\rho_{nf} = \rho_f(1 - \phi) + \rho_s\phi, \tag{32}$$

$$k_{nf} = k_f \left(\frac{k_s + 2k_f + 2\phi(k_s - k_f)}{k_s - \phi(k_s - k_f) + 2k_f} \right), \tag{33}$$

$$\frac{\sigma_{nf}}{\sigma_f} = 1 + \frac{3\left(\frac{\sigma_s}{\sigma_f} - 1\right)\phi}{\left(\frac{\sigma_s}{\sigma_f} + 2\right) - \left(\frac{\sigma_s}{\sigma_f} - 1\right)\phi}. \tag{34}$$

μ_{nf} is obtained as follows:

$$\mu_{nf} = \left(0.035B^2 + 3.1B - 27886.4807\phi^2 + 4263.02\phi + 316.0629\right)e^{-0.01T} \tag{35}$$

Properties of nanofluid exist in Table 3. Vorticity and stream function should be used to eliminate pressure source terms:

$$\omega + \frac{\partial u}{\partial y} - \frac{\partial v}{\partial x} = 0, \frac{\partial \psi}{\partial x} = -v, \frac{\partial \psi}{\partial y} = u. \tag{36}$$

Introducing the dimensionless quantities:

$$\begin{gathered} U = \frac{uL}{\alpha_{nf}}, V = \frac{vL}{\alpha_{nf}}, \theta = \frac{T - T_c}{\Delta T}, \Delta T = q''L / k_f, \\ \left(X, Y\right) = \frac{\left(x, y\right)}{L}, \Psi = \frac{\psi}{\alpha_{nf}}, \Omega = \frac{\omega L^2}{\alpha_{nf}}. \end{gathered} \tag{37}$$

The final formulae are:

$$\frac{\partial^2 \Psi}{\partial Y^2} + \frac{\partial^2 \Psi}{\partial X^2} = -\Omega, \tag{38}$$

Table 3. Thermo physical properties of water and nanoparticles'

	$\rho(kg / m^3)$	$C_p(j / kgk)$	$k(W / m.k)$	$d_p(nm)$	$\sigma\left(\Omega \cdot m\right)^{-1}$
Pure water	997.1	4179	0.613	-	0.05
Fe_3O_4	5200	670	6	47	25000

$$\begin{aligned} U\frac{\partial \Omega}{\partial X}+\frac{\partial \Omega}{\partial Y}V &= \Pr\frac{A_5}{A_1}\frac{A_2}{A_4}\left(\frac{\partial^2\Omega}{\partial Y^2}+\frac{\partial^2\Omega}{\partial X^2}\right) \\ &+\Pr Ha^2\frac{A_6}{A_1}\frac{A_2}{A_4}\left(\frac{\partial U}{\partial X}B_xB_y-\frac{\partial V}{\partial X}B_x^{\,2}+\frac{\partial U}{\partial Y}B_y^2-\frac{\partial V}{\partial Y}B_xB_y\right) \\ &+\Pr Ra\frac{A_3A_2^2}{A_1A_4^2}\frac{\partial \theta}{\partial X}, \end{aligned} \tag{39}$$

$$\frac{\partial \theta}{\partial X}U+\frac{\partial \theta}{\partial Y}V=\left(\frac{\partial^2\theta}{\partial X^2}+\frac{\partial^2\theta}{\partial Y^2}\right)+\frac{4}{3}\frac{1}{A_4}Rd\frac{\partial^2\theta}{\partial Y^2}. \tag{40}$$

where dimensionless and constants parameters are illustrated as:

$$\begin{aligned} &\Pr=\upsilon_f/\alpha_f, Ra=g(\rho\beta)_f\Delta T L^3/(\mu_f\alpha_f), Ha=LB_0\sqrt{\sigma_f/\mu_f}, \\ &A_1=\frac{\rho_{nf}}{\rho_f}, A_2=\frac{(\rho C_p)_{nf}}{(\rho C_p)_f}, A_3=\frac{(\rho\beta)_{nf}}{(\rho\beta)_f}, \\ &A_4=\frac{k_{nf}}{k_f}, A_5=\frac{\mu_{nf}}{\mu_f}, A_6=\frac{\sigma_{nf}}{\sigma_f}, \\ &Rd=4\sigma_e T_c^3/(\beta_R k_f). \end{aligned} \tag{41}$$

and boundary conditions are:

$$\begin{aligned} &\frac{\partial \theta}{\partial n}=1.0 \text{ on inner wall} \\ &\theta=0.0 \text{ on outer wall} \\ &\Psi=0.0 \text{ on all walls} \end{aligned} \tag{42}$$

Local and average Nusselt over the hot cylinder can be calculated as:

$$Nu_{loc}=\left(1+\frac{4}{3}Rd\left(\frac{k_{nf}}{k_f}\right)^{-1}\right)\frac{1}{\theta}\left(\frac{k_{nf}}{k_f}\right), \tag{43}$$

$$Nu_{ave}=\frac{1}{S}\int_0^S Nu_{loc}\,ds. \tag{44}$$

3.3. Effects of Active Parameters

CVFEM is utilized to simulation of magnetic field impact on nanofluid free convection. Radiation source term is taken into account. MFD viscosity is considered. Results are reported for different amount of Radiation parameter ($Rd = 0$ to 0.8), volume fraction of Fe_3O_4-water ($\phi = 0$ to 0.04), Rayleigh number ($Ra = 10^3, 10^4$ and 10^5), inclination angle ($\xi = 0^\circ$ and 90°) and Hartmann number ($Ha = 0$ to 40).

Figure 9 illustrates the impact of radiation parameter on nanofluid hydrothermal treatment. Radiation parameter can enhance the rate of heat transfer. Also augmenting *Rd* makes the nanofluid motion to augment. This impact is more obvious in existence of magnetic field. Influences of ξ, Ha and Ra on hydrothermal characteristics are demonstrated in Figures 10 and 11. As nanofluid temperature augments, the nanofluid begins moving from the warm surface to the outer one and dropping along the outer cylinder. At low Rayleigh number, conduction is more signification than convection. $|\Psi_{max}|$ augments as buoyancy force augments and it reduces as Lorentz force augments. When $\xi = 0^\circ, Ha = 0$, three vortexes appear in streamline and these vortexes convert to two smaller ones in existence of Lorentz forces. As ξ increases, the distortion of isotherms enhances. Thermal plume generates near the vertical centerline. As Lorentz force augments, the two upper vortexes are merge together and thermal plume vanish.

Figures 12, 13 and 14 illustrate the impact of ξ, Rd, Ra and Ha on Nu_{loc}, Nu_{ave}. The formula for Nu_{ave} corresponding to active parameters is:

$$\begin{aligned} Nu_{ave} &= 9.94 - 0.22\xi - 4.44Rd - 4.76\log(Ra) + 0.98Ha^* + 0.4\xi\,Rd \\ &-0.06\xi\log(Ra) - 0.06\xi Ha^* + 1.69\log(Ra)Rd - 0.351\log(Ra)Ha^* \\ &-0.25Rd\,Ha^* - 0.23\xi^2 + 0.72(\log(Ra))^2 + 1.49Rd^2 - 0.06Ha^{*2} \end{aligned} \tag{45}$$

where $Ha^* = 0.1Ha$. The number of extermum in Nu_{loc} profile is matching to existence of thermal plume and shape of outer cylinder. Nu_{ave} enhances with rise of Rd. As buoyancy force augment, temperature decreases and in turn Nu_{ave} augment with rise of Ra. Impact of inclination angle on Nu_{ave} is similar to that of obtained for Ra. Besides, this figure depicts that Nusselt number has reverse relationship with Hartmann number.

4. THERMAL RADIATION OF FERROFLUID IN EXISTENCE OF LORENTZ FORCES CONSIDERING VARIABLE VISCOSITY

4.1. Problem Definition

Figure 15 depicts the geometry, boundary condition and sample element. The formula of inner cylinder is:

$$b = \sqrt{1 - \varepsilon^2 a} \tag{46}$$

Figure 9. Effect of Hartmann number and radiation parameter on Isotherms (left) and streamlines (right) contours ($Rd = 0$ $(-\cdot\cdot-)$, $Rd = 0.8$ $(-)$) *when* $\xi = 90^{\circ}, \phi = 0.04, Ra = 10^{5}$

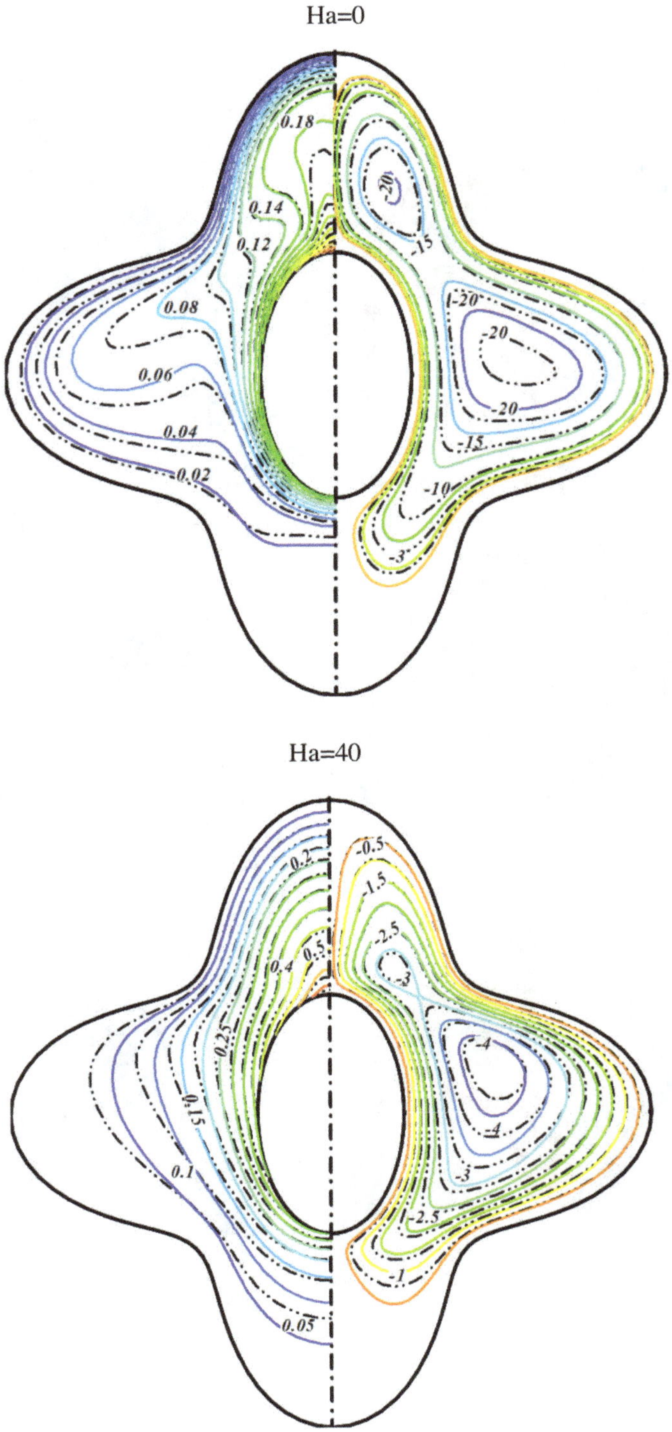

Figure 10. Effect of Rayleigh and Hartmann numbers on Isotherms (left) and streamlines (right) contours when $\xi = 0, \phi = 0.04, Rd = 0.8$

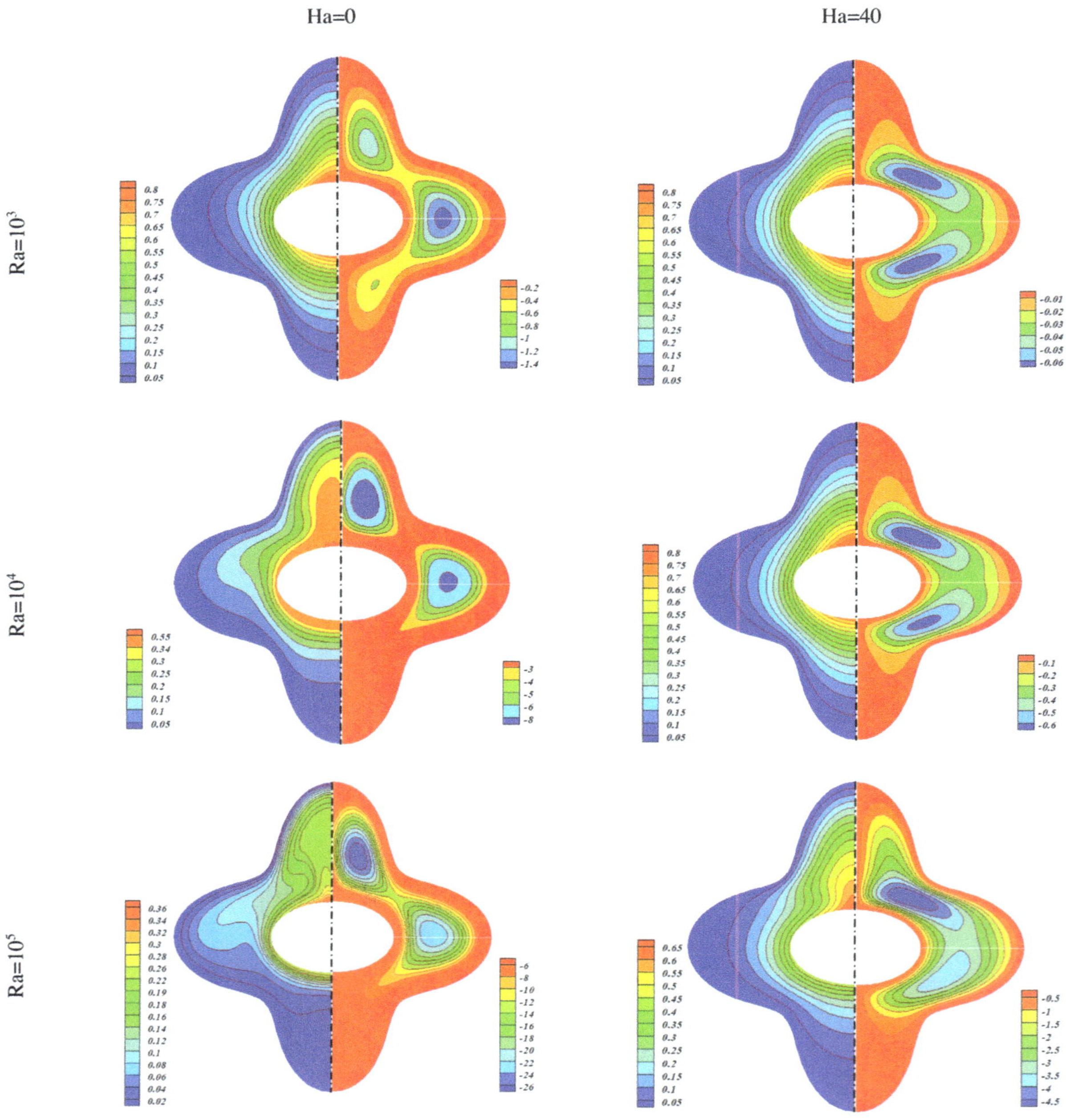

Figure 11. Effect of Rayleigh and Hartmann numbers on Isotherms (left) and streamlines (right) contours when $\xi = 90^{\circ}, \phi = 0.04, Rd = 0.8$

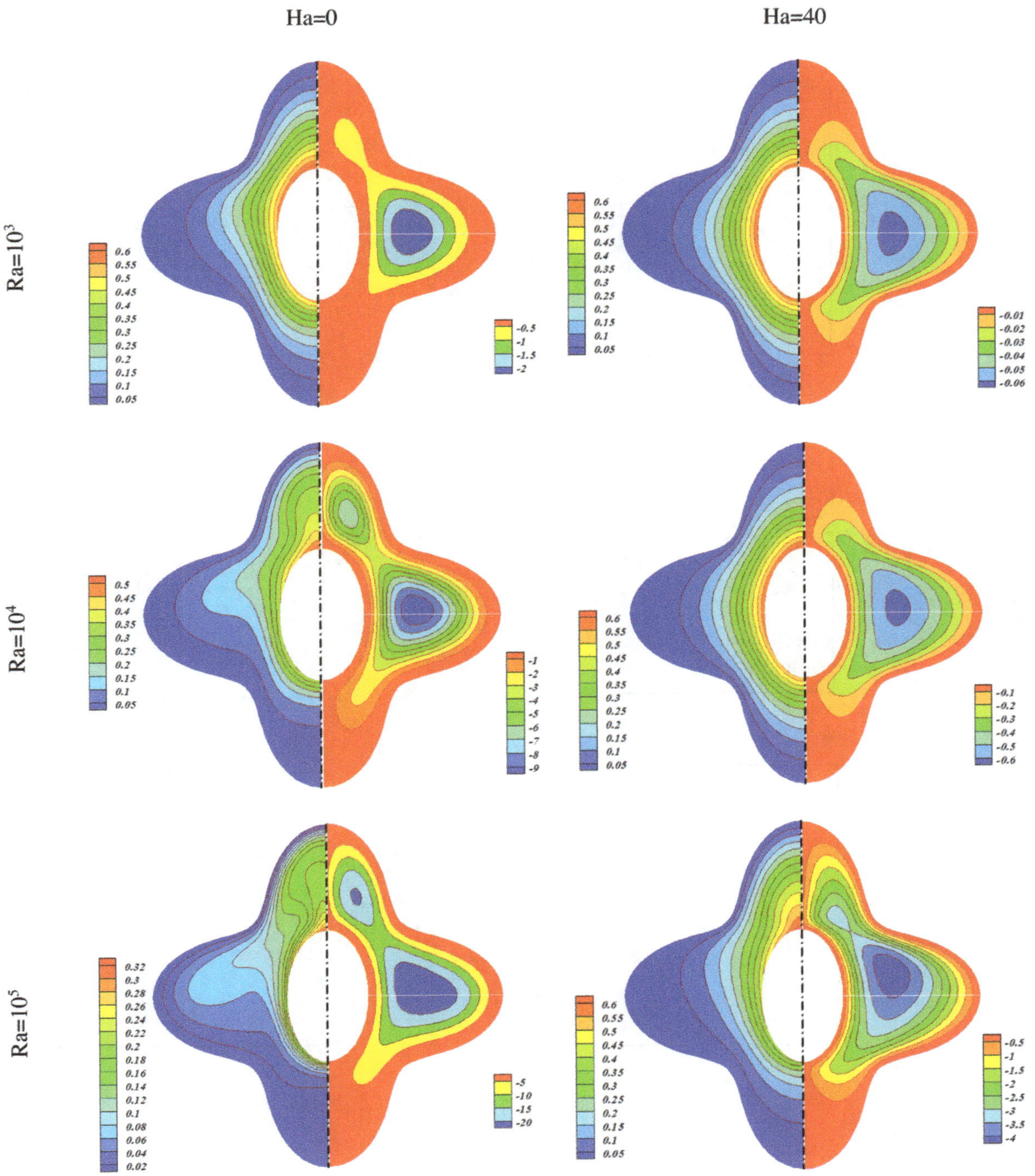

Figure 12a. Effects of the radiation parameter, Hartmann and Rayleigh numbers on local Nusselt number when $\phi = 0.04, \xi = 90^{\circ}$

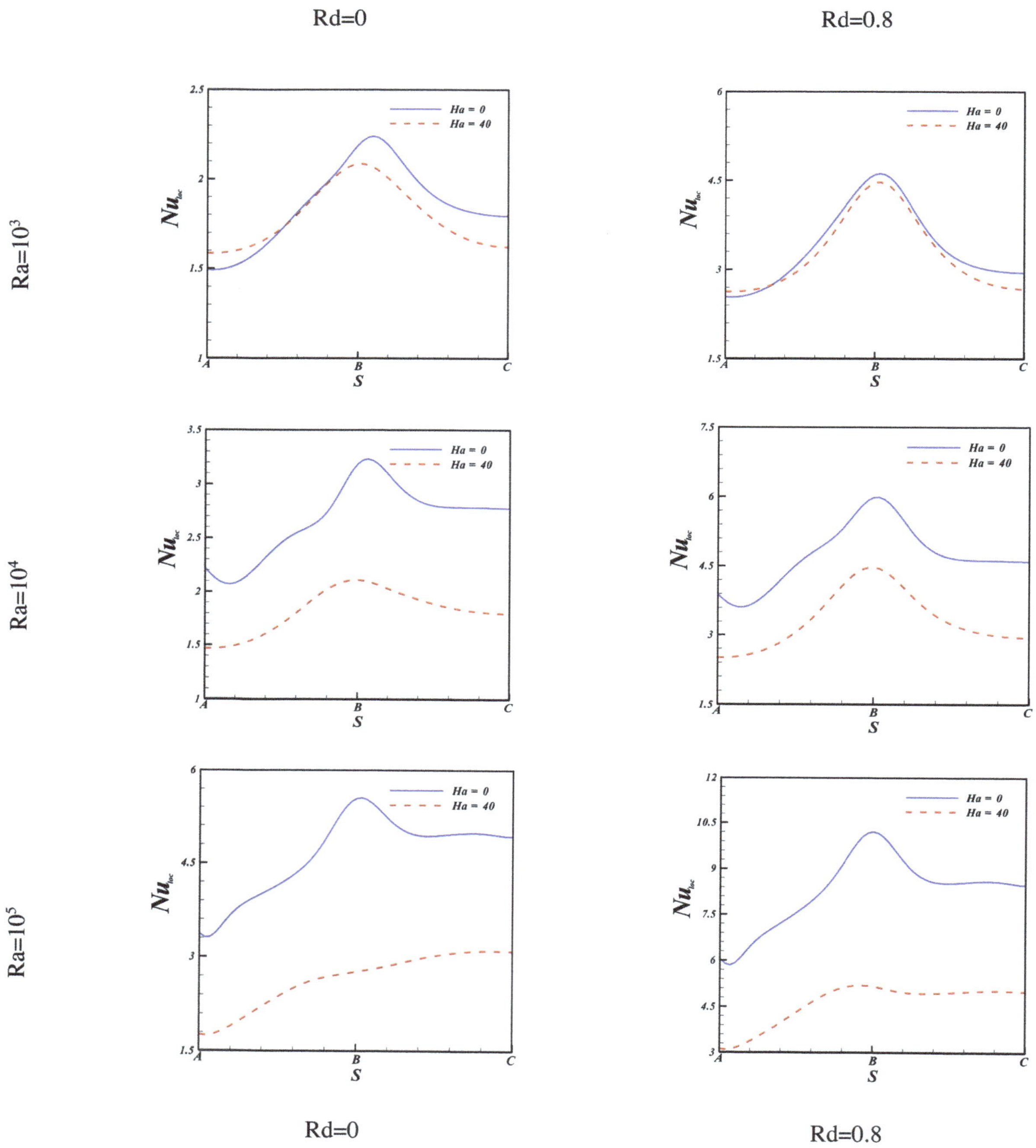

Figure 12b. Effects of the radiation parameter, Hartmann and Rayleigh numbers on local Nusselt number when $\phi = 0.04, \xi = 90^{\circ}$

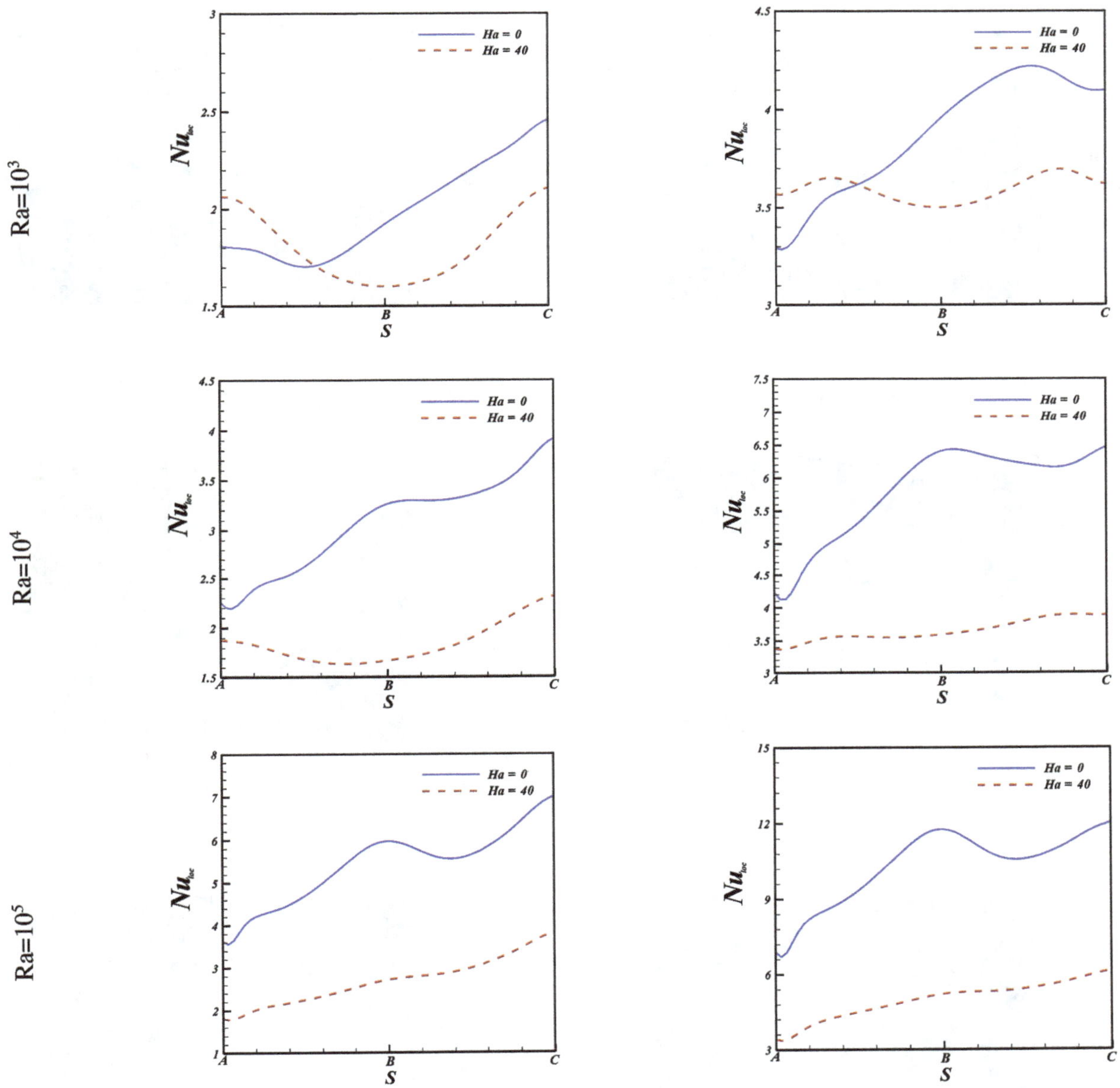

where a, b, ε are the major, minor axis of elliptic cylinder and eccentricity for the inner cylinder. The inner cylinder has constant heat flux condition.

4.2. Governing Equation

2D steady convective flow of nanofluid is considered in existence of constant magnetic field. The PDEs equations are:

Figure 13a. Effects of inclination angle, radiation parameter, Rayleigh and Hartmann numbers on average Nusselt number

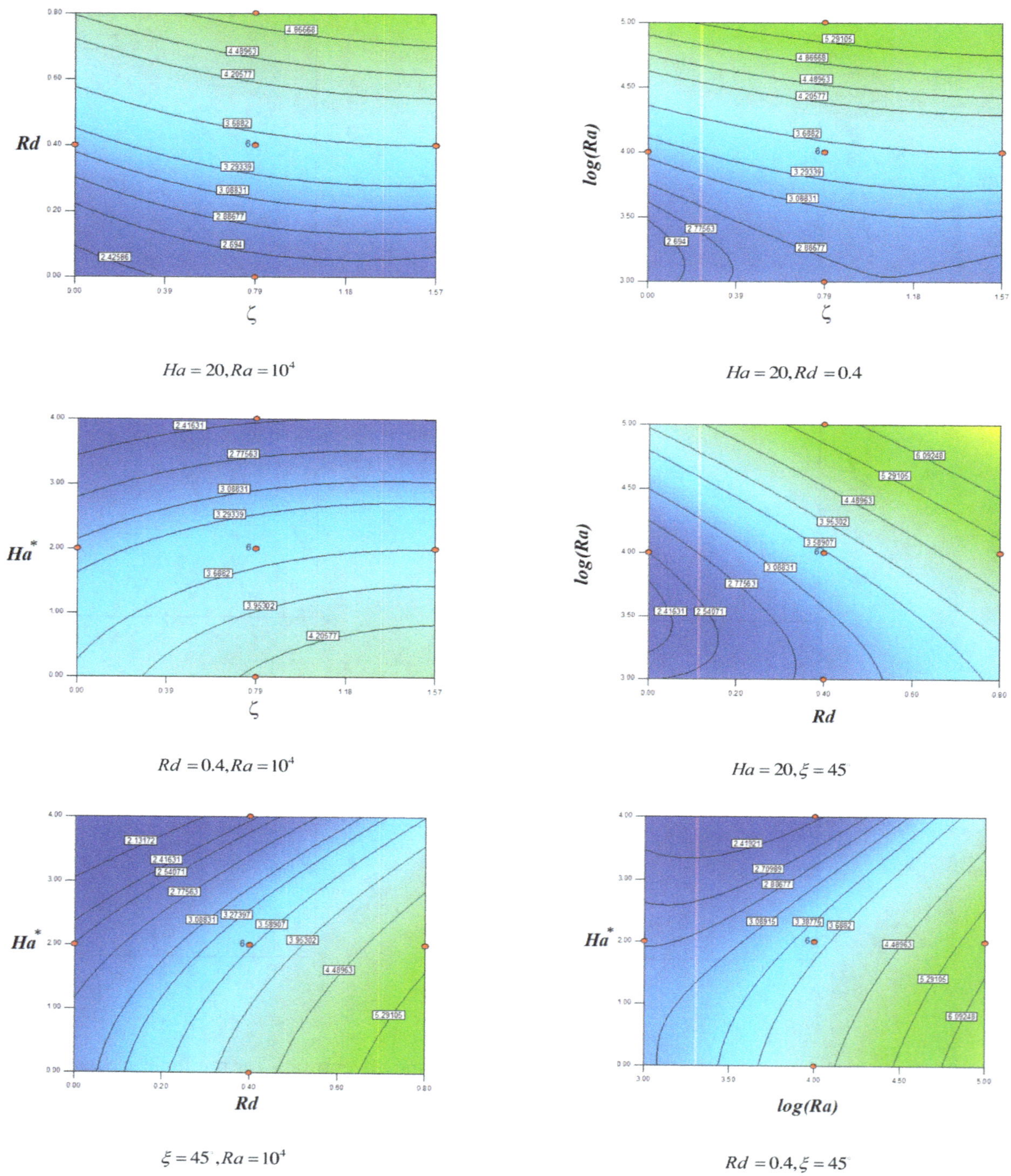

Figure 13b. Effects of inclination angle, radiation parameter, Rayleigh and Hartmann numbers on average Nusselt number

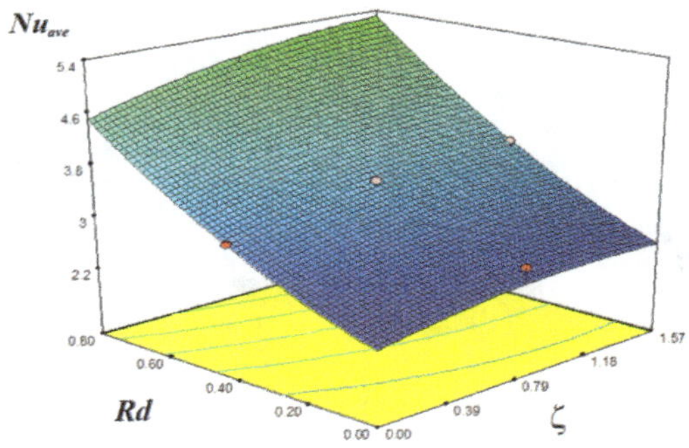

$Ha = 20, Ra = 10^4$

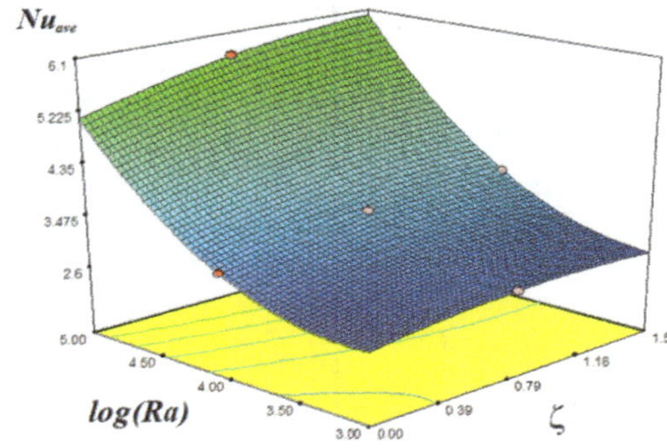

$Ha = 20, Rd = 0.4$

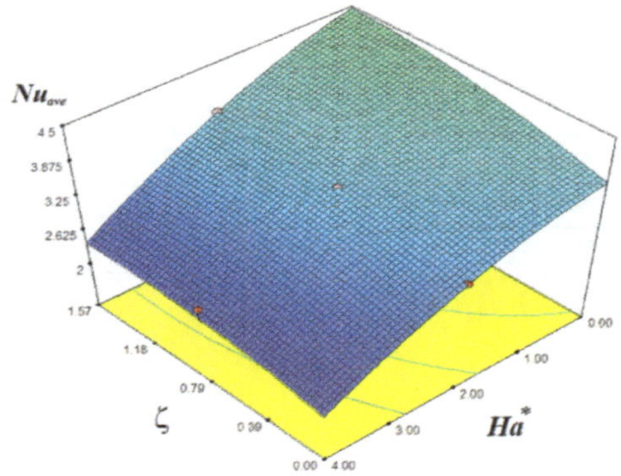

$Rd = 0.4, Ra = 10^4$

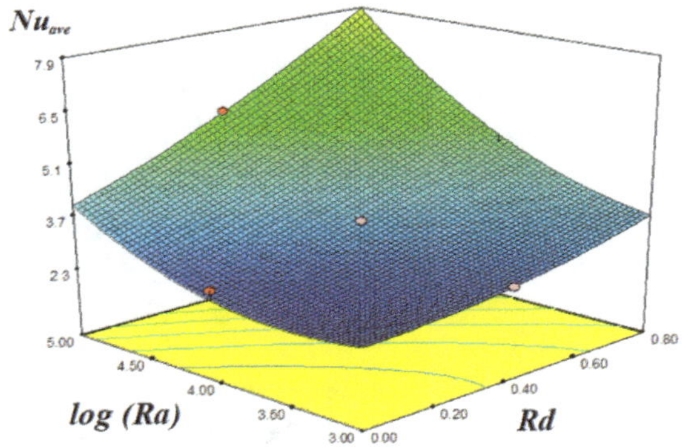

$Ha = 20, \xi = 45^{\circ}$

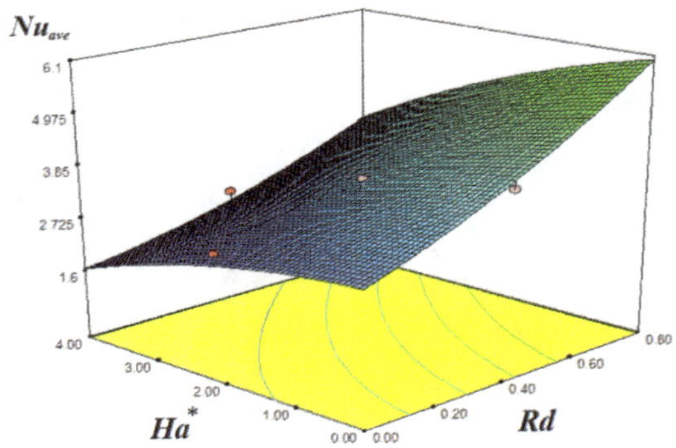

$\xi = 45^{\circ}, Ra = 10^4$

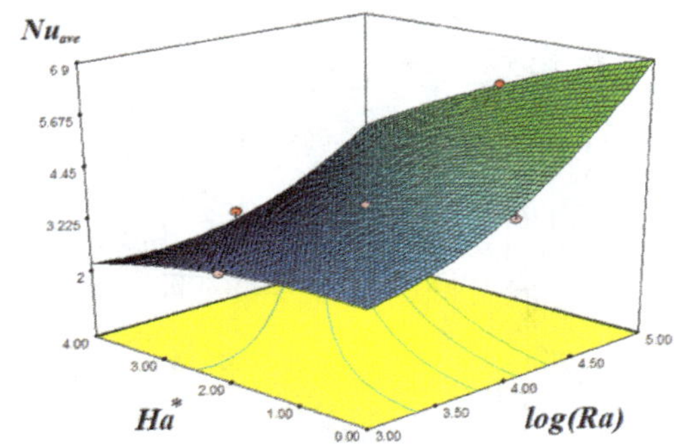

$Rd = 0.4, \xi = 45^{\circ}$

Figure 14. (a)Geometry and the boundary conditions with (b) the mesh of Geometry considered in this work; (c) A sample triangular element and its corresponding control volume.

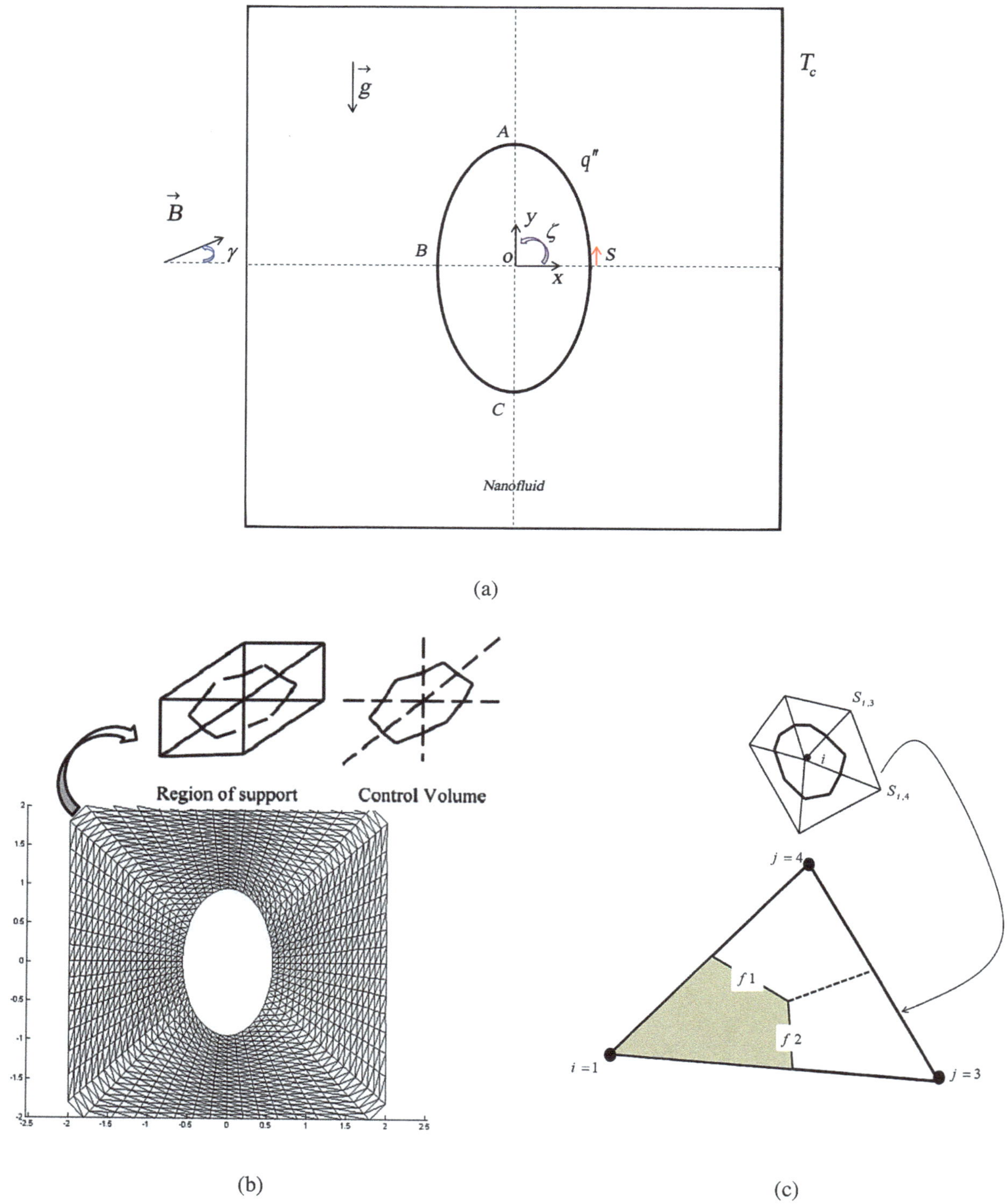

Figure 15. Effect of Hartmann number and radiation parameter on Isotherms (left) and streamlines (right) contours ($Rd = 0$ ($- \cdot\cdot -$), $Rd = 0.8$ ($-$)) when $\xi = 90^{\circ}, \phi = 0.04, Ra = 10^{5}$

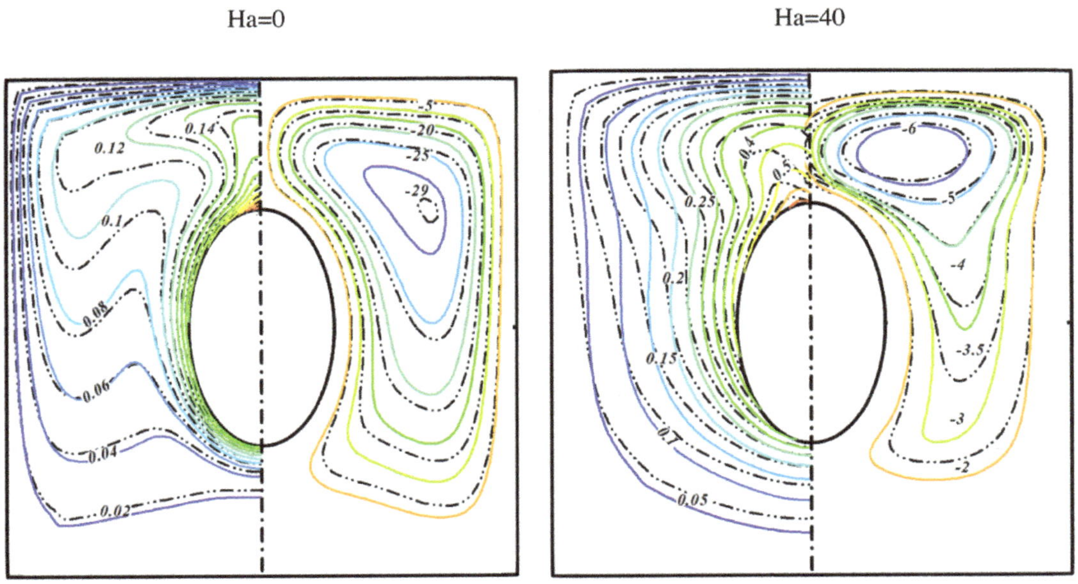

$$\frac{\partial u}{\partial x} + \frac{\partial v}{\partial y} = 0 \tag{47}$$

$$\left(v\frac{\partial u}{\partial y} + \frac{\partial u}{\partial x}u \right) = \left[-\sigma_{nf} B_y^2 u + \sigma_{nf} B_x B_y v \left(\frac{\partial^2 u}{\partial y^2} + \frac{\partial^2 u}{\partial x^2} \right) \mu_{nf} - \frac{\partial P}{\partial x} \right] \left(\rho_{nf} \right)^{-1} \tag{48}$$

$$\begin{gathered} \rho_{nf} \left(\frac{\partial v}{\partial x}u + \frac{\partial v}{\partial y}v \right) = +\mu_{nf} \left(\frac{\partial^2 v}{\partial x^2} + \frac{\partial^2 v}{\partial y^2} \right) - \frac{\partial P}{\partial y} + B_y \sigma_{nf} B_x u - B_x \sigma_{nf} B_x v \\ + \left(T - T_c \right) \beta_{nf} g \rho_{nf}, \\ B_x = B_o \cos\lambda,\ B_y = B_o \sin\lambda \end{gathered} \tag{49}$$

$$\begin{gathered} \left(\rho C_p \right)_{nf} \left(v\frac{\partial T}{\partial y} + u\frac{\partial T}{\partial x} \right) = k_{nf} \left(\frac{\partial^2 T}{\partial x^2} + \frac{\partial^2 T}{\partial y^2} \right) - \frac{\partial q_r}{\partial y}, \\ , \left[q_r = -\frac{4\sigma_e}{3\beta_R}\frac{\partial T^4}{\partial y},\ T^4 \cong 4T_c^3 T - 3T_c^4 \right] \end{gathered} \tag{50}$$

$\left(\rho C_p\right)_{nf}, \left(\beta\right)_{nf}$, ρ_{nf}, k_{nf} and σ_{nf} are defined as:

$$\left(\rho C_p\right)_{nf} = \left(\rho C_p\right)_f (1-\phi) + \left(\rho C_p\right)_s \phi \tag{51}$$

$$\left(\beta\right)_{nf} = \left(\beta\right)_f (1-\phi) + \left(\beta\right)_s \phi \tag{52}$$

$$\rho_{nf} = \rho_f (1-\phi) + \rho_s \phi \tag{53}$$

$$k_{nf} = k_f \left(\frac{k_s + 2k_f + 2\phi(k_s - k_f)}{k_s - \phi(k_s - k_f) + 2k_f} \right) \tag{54}$$

$$\frac{\sigma_{nf}}{\sigma_f} = 1 + \frac{3\left(\frac{\sigma_s}{\sigma_f} - 1\right)\phi}{\left(\frac{\sigma_s}{\sigma_f} + 2\right) - \left(\frac{\sigma_s}{\sigma_f} - 1\right)\phi} \tag{55}$$

μ_{nf} is obtained as follows:

$$\mu_{nf} = \left(0.035B^2 + 3.1B - 27886.4807\phi^2 + 4263.02\phi + 316.0629\right)e^{-0.01T} \tag{56}$$

Vorticity and stream function should be used to eliminate pressure source terms:

$$\omega + \frac{\partial u}{\partial y} - \frac{\partial v}{\partial x} = 0, \frac{\partial \psi}{\partial x} = -v, \frac{\partial \psi}{\partial y} = u \tag{57}$$

Introducing dimensionless quantities:

$$P = \frac{p}{\rho_f \left(\alpha_f / L\right)^2}, U = \frac{uL}{\alpha_f}, V = \frac{vL}{\alpha_f}, \Theta = \frac{T - T_c}{\Delta T}, \Delta T = q'' L / k_f, \left(X, Y\right) = \frac{\left(x, y\right)}{L} \tag{58}$$

The final formulae are:

$$\frac{\partial V}{\partial Y}+\frac{\partial U}{\partial X}=0, \tag{59}$$

$$U\frac{\partial U}{\partial X}+\frac{\partial U}{\partial Y}V=\Pr\left[\frac{\mu_{nf}/\mu_f}{\rho_{nf}/\rho_f}\right]\left(\frac{\partial^2 U}{\partial Y^2}+\frac{\partial^2 U}{\partial X^2}\right)-Ha^2\Pr\left[\frac{\sigma_{nf}/\sigma_f}{\rho_{nf}/\rho_f}\right]\left(B_y^2 U-B_x B_y V\right)-\frac{\partial P}{\partial X}, \tag{60}$$

$$\begin{aligned}V\frac{\partial V}{\partial Y}+U\frac{\partial V}{\partial X}&=\Pr\left(\frac{\partial^2 V}{\partial Y^2}+\frac{\partial^2 V}{\partial X^2}\right)\left[\frac{\mu_{nf}/\mu_f}{\rho_{nf}/\rho_f}\right]\\&-Ha^2\Pr\left[\frac{\sigma_{nf}/\sigma_f}{\rho_{nf}/\rho_f}\right]\left(B_x^2 V-B_x B_y U\right)-\frac{\partial P}{\partial Y}+Ra\Pr\left[\frac{\beta_{nf}}{\beta_f}\right]\Theta,\end{aligned} \tag{61}$$

$$\frac{\partial\Theta}{\partial X}U+\frac{\partial\Theta}{\partial Y}V=\left(1+\frac{4}{3}\left(\frac{k_{nf}}{k_f}\right)^{-1}Rd\right)\left[\frac{(\rho C_P)_f}{(\rho C_P)_{nf}}\frac{k_{nf}}{k_f}\right]\frac{\partial^2\Theta}{\partial Y^2}+\left[\frac{(\rho C_P)_f}{(\rho C_P)_{nf}}\frac{k_{nf}}{k_f}\right]\left(\frac{\partial^2\Theta}{\partial X^2}\right) \tag{62}$$

where dimensionless and constants parameters are illustrated as:

$$\Pr=\upsilon_f/\alpha_f, Ra=g\beta_f q'' L^4/\left(k_f\upsilon_f\alpha_f\right), Ha=LB_0\sqrt{\sigma_f/\mu_f}, Rd=4\sigma_e T_c^3/\left(\beta_R k_f\right), \tag{63}$$

and boundary conditions are:

$$\frac{\partial\Theta}{\partial n}=1.0 \text{ on inner wall}$$

$$\Theta=0.0 \text{ on outer wall}$$

$$\Psi=0.0 \text{ on all walls} \tag{64}$$

Local and average Nusselt over the hot cylinder can calculate as:

$$Nu_{loc}=\left(1+\frac{4}{3}Rd\left(\frac{k_{nf}}{k_f}\right)^{-1}\right)\frac{1}{\theta}\left(\frac{k_{nf}}{k_f}\right) \tag{65}$$

$$Nu_{ave} = \frac{1}{S}\int_{0}^{s} Nu_{loc}\, ds \tag{66}$$

4.3. Effects of Active Parameters

Fe_3O_4-H_2O nanofluid free convection in existence of radiation source term is examined considering magnetic field effect. MFD viscosity is taken into account. Numerical simulation are examined for various amount of Radiation parameter ($Rd = 0$ to 0.8), volume fraction of Fe_3O_4-water, ($\phi = 4$ to 0.04) Rayleigh number ($Ra = 10^3, 10^4$ and 10^5), inclination angle ($\xi = 0^\circ$ and 90°) and Hartmann number ($Ha = 0$ to 40).

Impact of radiation parameter on nanofluid hydrothermal behavior is illustrated in Figure 16. Thermal boundary layer thickness augments with augment of *Rd*. Also augmenting *Rd* makes the nanofluid motion to augment. This influence is more evident in existence of magnetic field.

Influence of ξ, Ha and Ra on streamlines and isotherms are depicted Figures 17 and 18. As nanofluid temperature augments, the nanofluid begins moving from the warm surface to the outer one and dropping along the outer cylinder. At low Rayleigh number, conduction is more signification than convection. $\left|\Psi_{\max}\right|$ augments as buoyancy force augments and it reduces as Lorentz force enhances. When $\xi = 0^\circ, Ha = 0$, as Rayleigh number augments the below vortex diminish and upper one becomes stronger.

As Hartmann number augments, the center of the vortexes move to *x=0*. $\left|\Psi_{\max}\right|$ and temperature gradient enhance with rise of Ra. As ξ increases, the distortion of isotherms augments. Increasing Hartmann number causes the thermal plume to diminish and change the heat transfer mechanism from convection to conduction. As Lorentz force augments, the two upper vortexes are merge together and thermal plume vanish.

Figures19, 20 and 21 illustrate the impact of ξ, Rd, Ra and Ha on Nu_{loc}, Nu_{ave}. The formula for Nu_{ave} corresponding to active parameters is:

$$\begin{aligned} Nu_{ave} &= 6.91 - 0.12\xi - 4.3Rd - 3.5\log\left(Ra\right) + 1.26Ha^{*} + 0.36\xi\, Rd \\ &+ 0.2\xi\log\left(Ra\right) - 0.07\xi Ha^{*} + 1.61\log\left(Ra\right)Rd - 0.31\log\left(Ra\right)Ha^{*} \\ &- 0.45Rd\, Ha^{*} - 0.2\xi^{2} + 0.62\left(\log\left(Ra\right)\right)^{2} + 1.5Rd^{2} - 0.08Ha^{*2} \end{aligned} \tag{67}$$

where $Ha^{*} = 0.1Ha$. The number of extermum in Nu_{loc} profile is matching to existence of thermal plume. Nu_{ave} enhances with rise of Rd. As buoyancy force augment, temperature decreases and in turn Nu_{ave} augment with rise of Ra. Also this figure shows that Nusselt number has direct relationship with inclination angle.

Figure 16. Effect of Rayleigh and Hartmann numbers on Isotherms (left) and streamlines (right) contours when $\xi = 0, \phi = 0.04, Rd = 0.8$

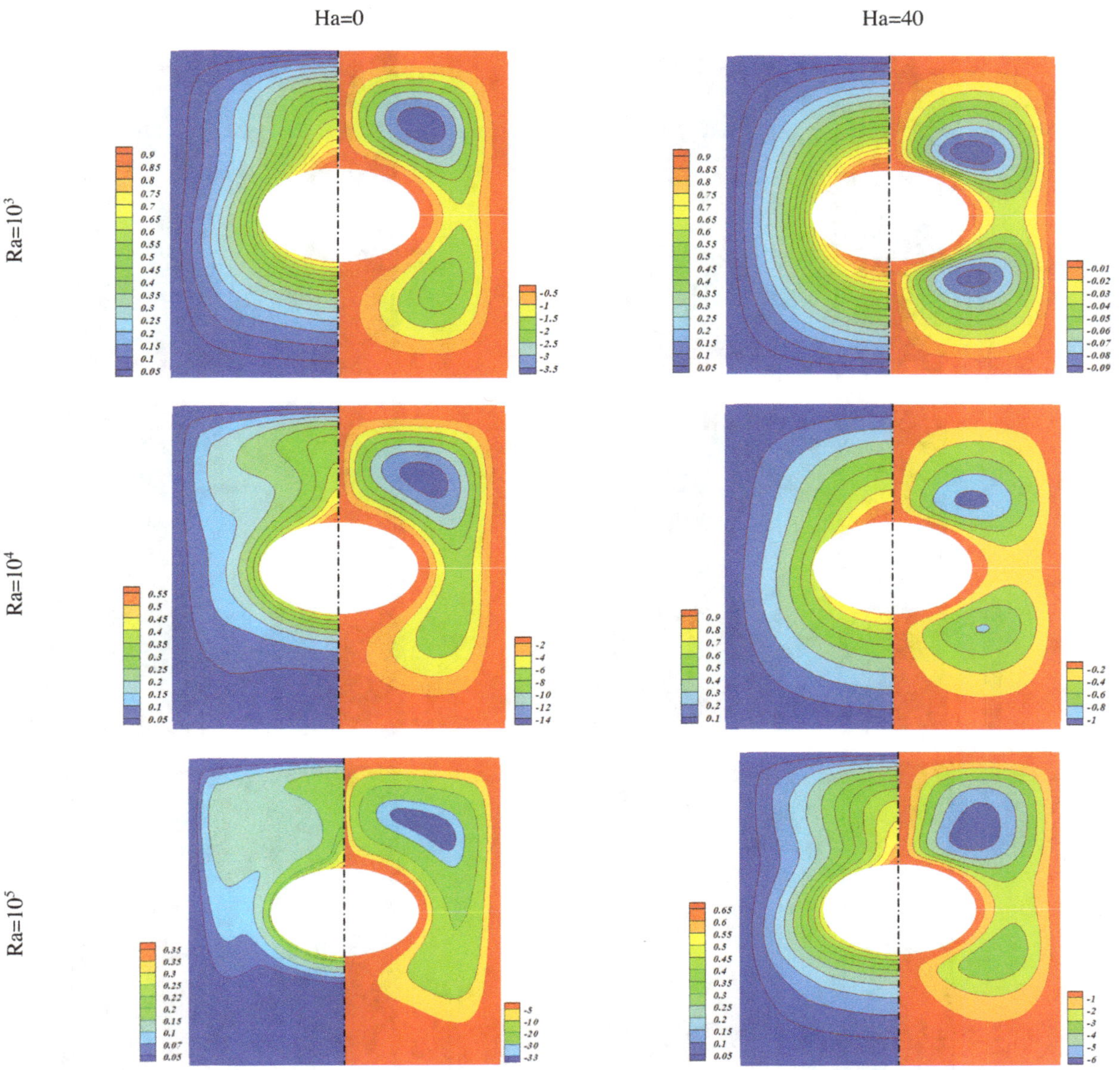

5. MAGNETIC NANOFLUID NATURAL CONVECTION IN PRESENCE OF THERMAL RADIATION CONSIDERING VARIABLE VISCOSITY

5.1. Problem Definition

Figure 22 shows the geometry, boundary condition and sample element. The formula of outer cylinder is:

Figure 17. Effect of Rayleigh and Hartmann numbers on Isotherms (left) and streamlines (right) contours when $\xi = 90^{\circ}, \phi = 0.04, Rd = 0.8$

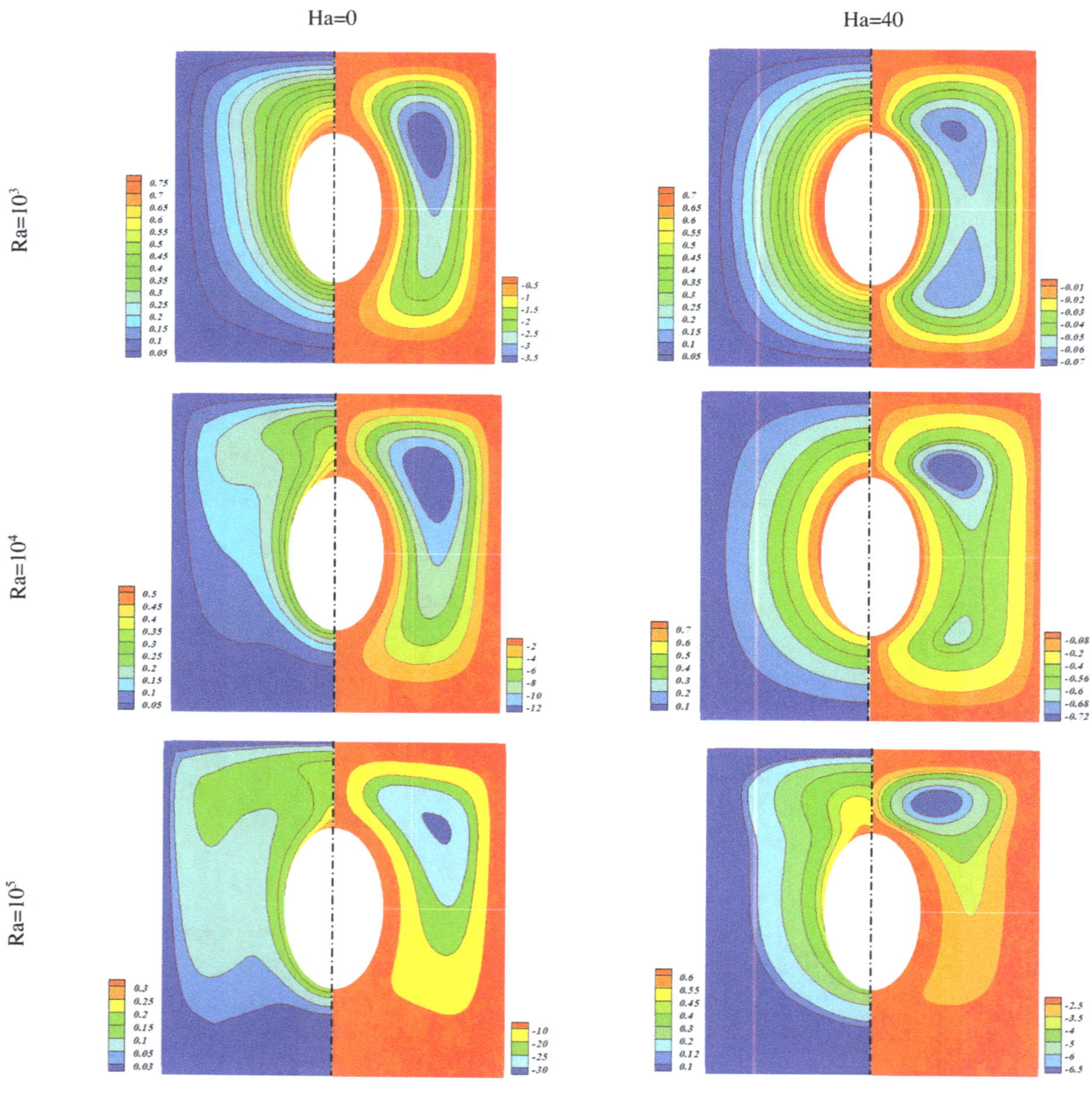

$$r = r_{in} + A\cos\left(N\left(\zeta - \zeta_0\right)\right) \tag{68}$$

in which r_{in}, r_{out} are radius of the base circle and outer cylinder.

Figure 18. Effects of the radiation parameter, Hartmann and Rayleigh numbers on local Nusselt number when $\phi = 0.04, \xi = 0^{\circ}$

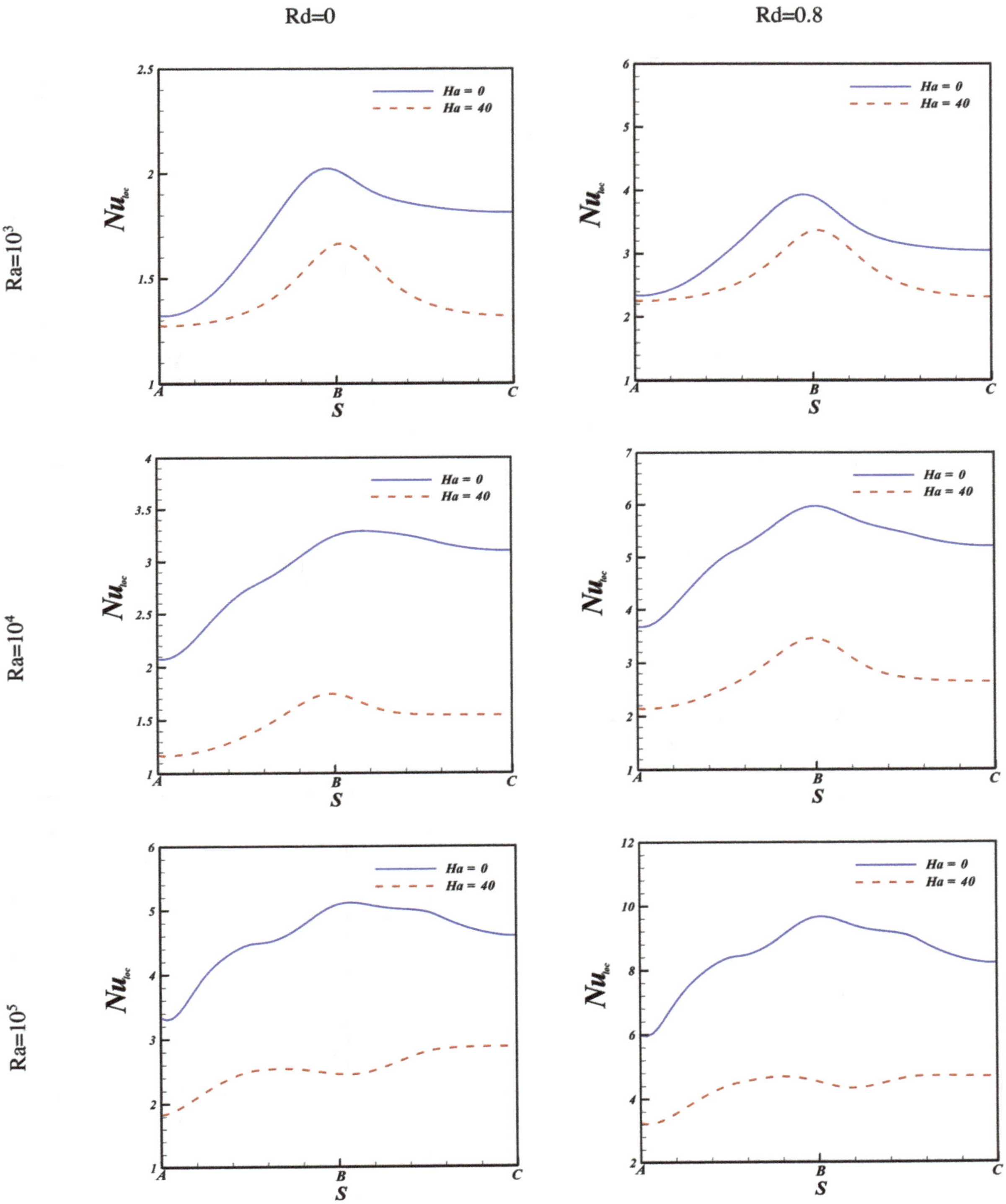

Figure 19. Effects of the radiation parameter, Hartmann and Rayleigh numbers on local Nusselt number when $\phi = 0.04, \xi = 90^{\circ}$

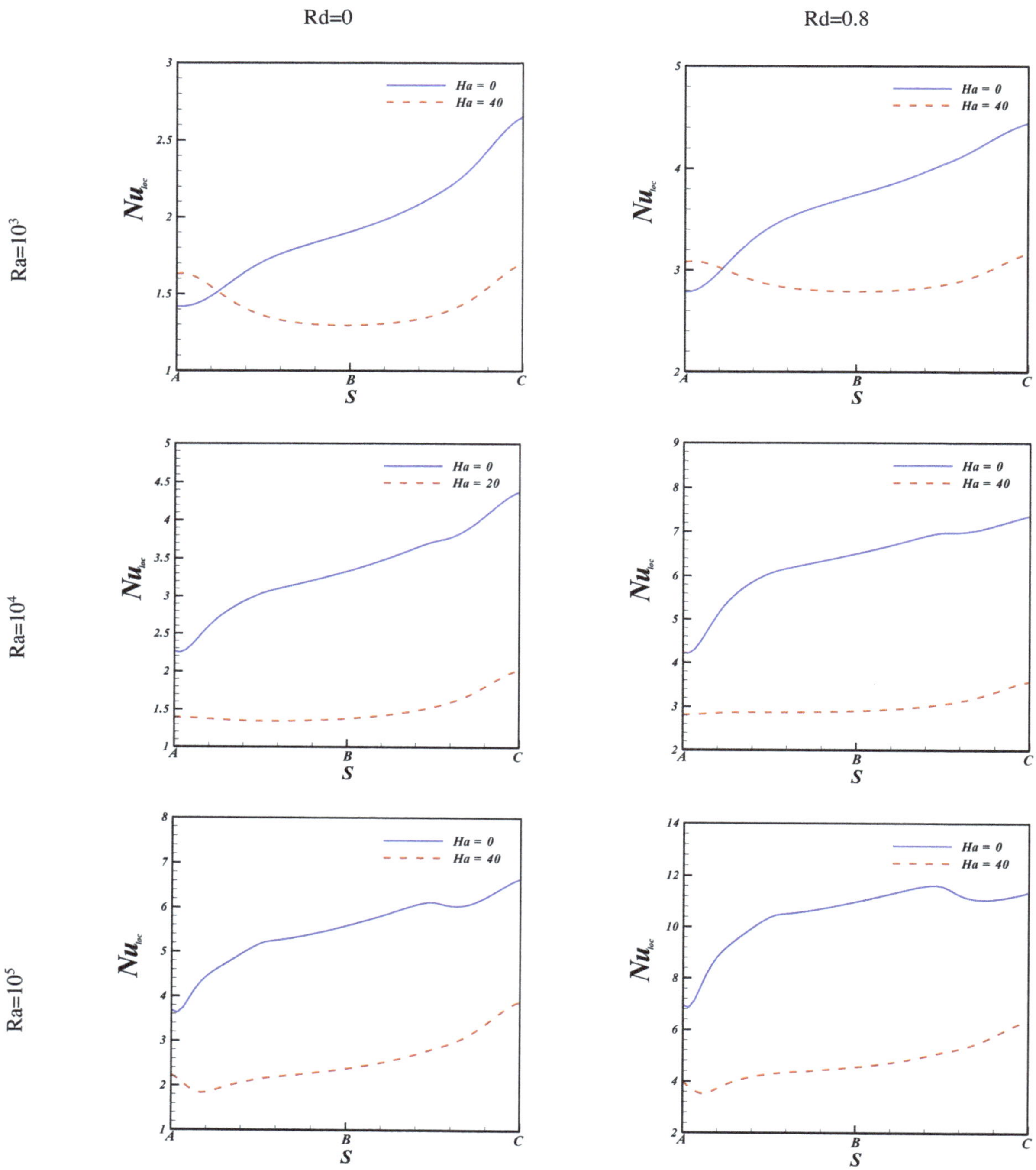

Figure 20a. Effects of inclination angle, radiation parameter, Rayleigh and Hartmann numbers on average Nusselt number

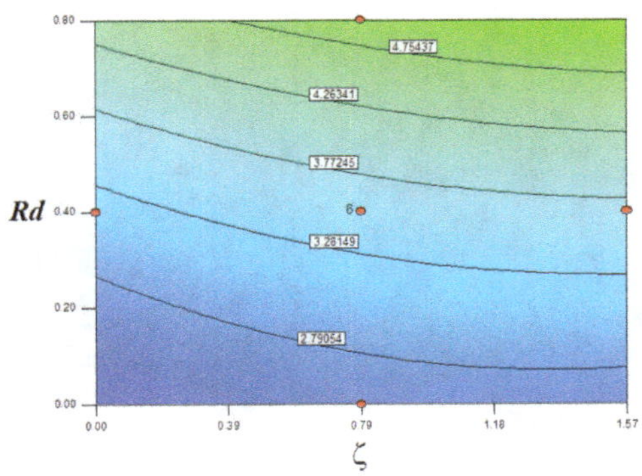

$Ha = 20, Ra = 10^4$

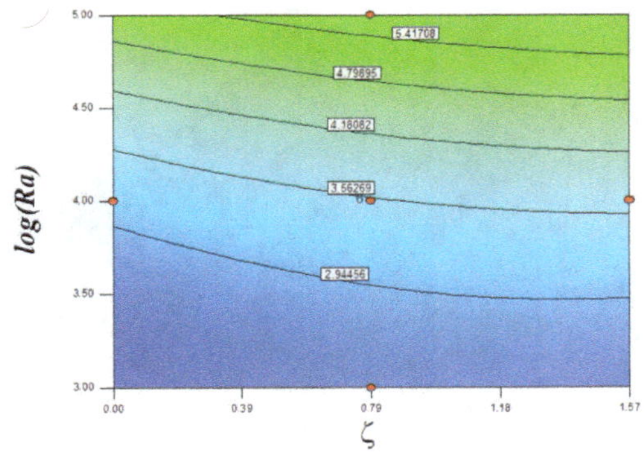

$Ha = 20, Rd = 0.4$

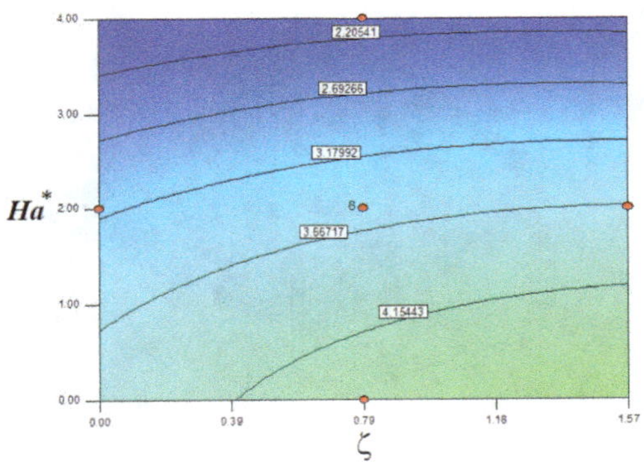

$Rd = 0.4, Ra = 10^4$

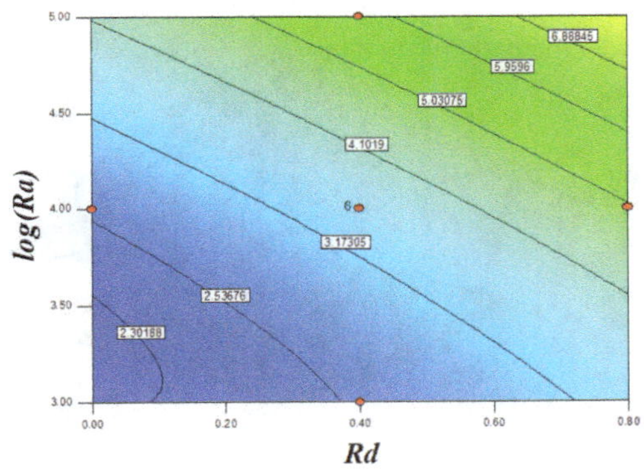

$Ha = 20, \xi = 45^\circ$

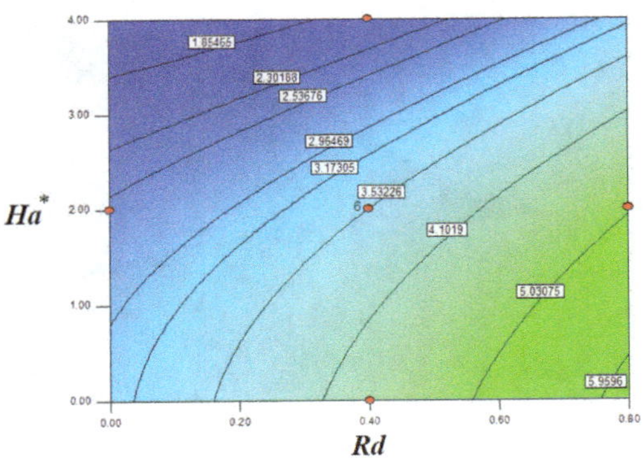

$\xi = 45^\circ, Ra = 10^4$

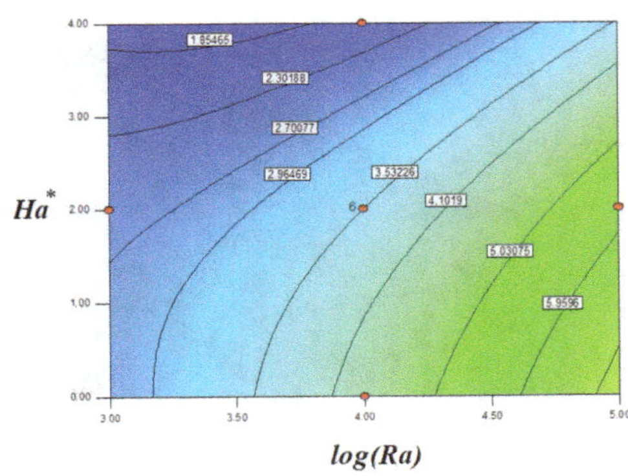

$Rd = 0.4, \xi = 45^\circ$

Figure 20b. Effects of inclination angle, radiation parameter, Rayleigh and Hartmann numbers on average Nusselt number

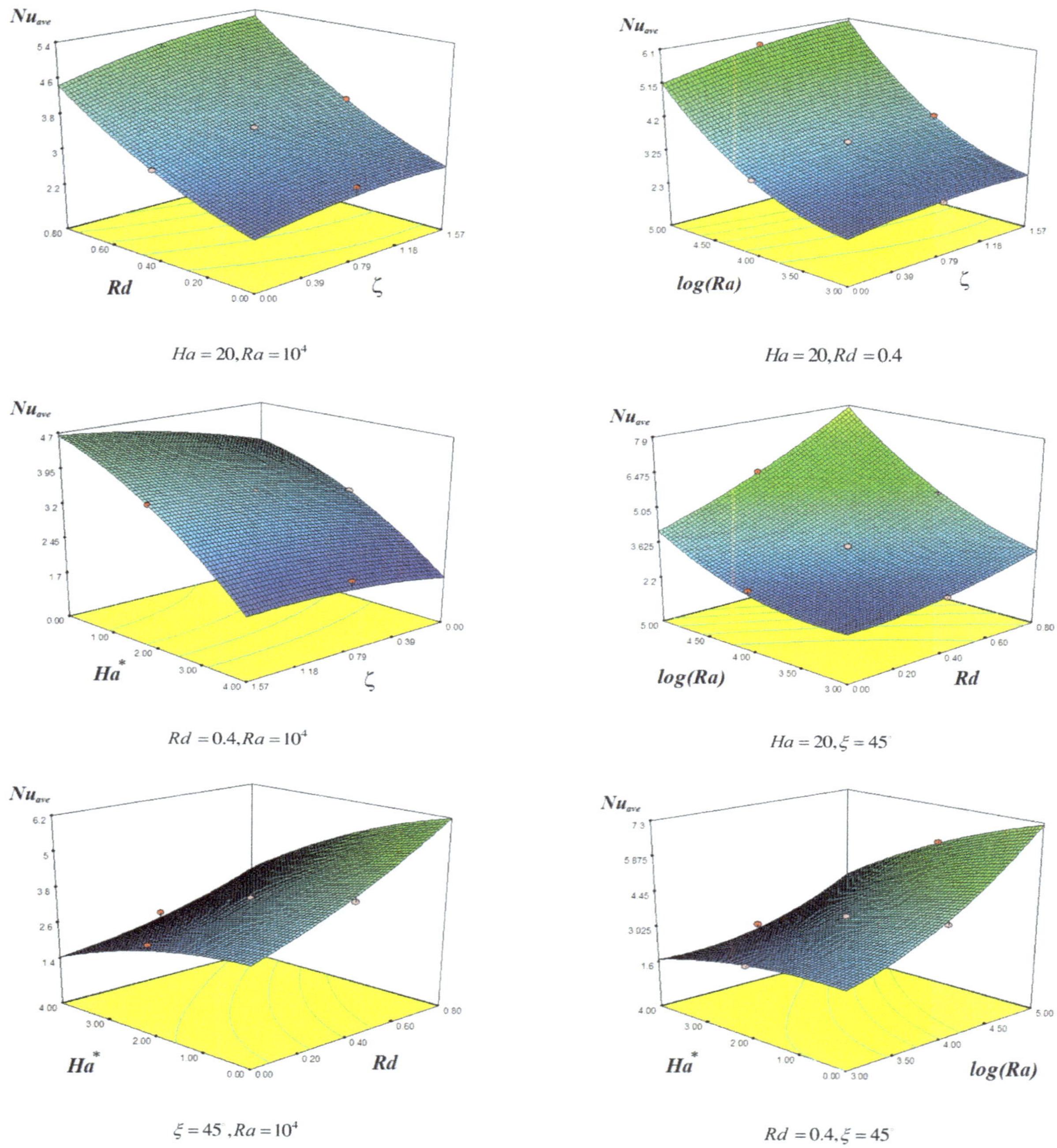

Figure 21. (a)Geometry and the boundary conditions with (b) the mesh of Geometry considered in this work; (c) A sample triangular element and its corresponding control volume.

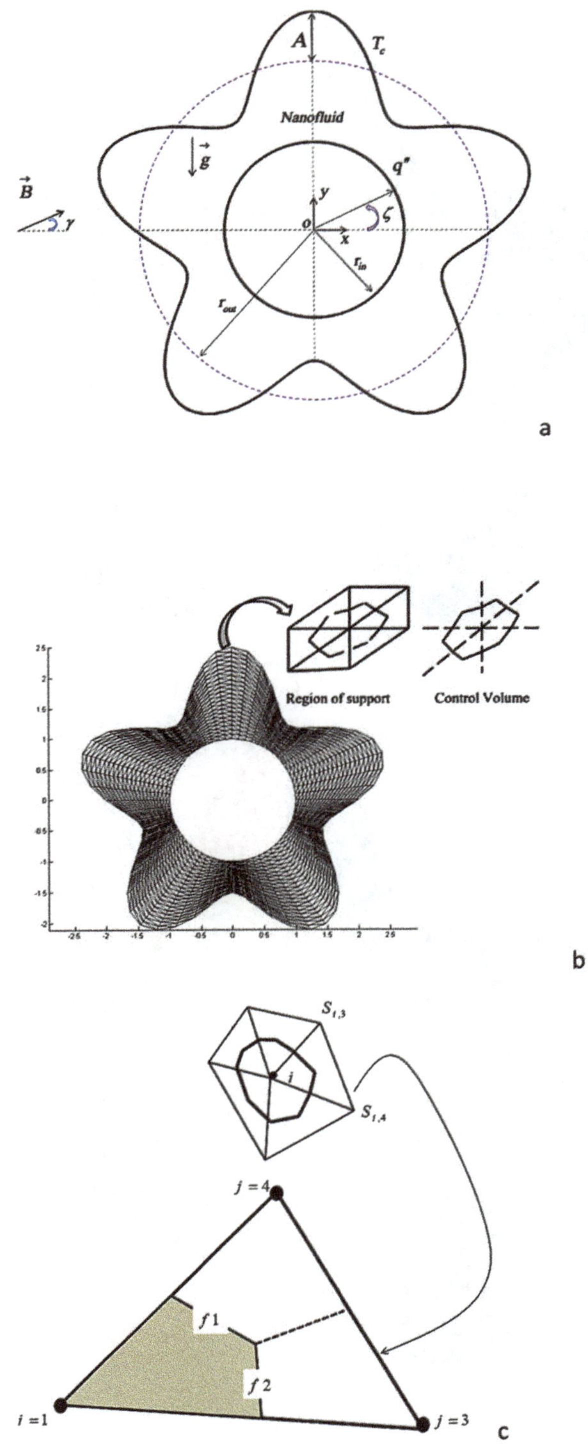

Figure 22. Effect of Hartmann number on isotherms (left) and streamlines (right) contours when (a) Ha=0, (b) Ha=20 and $Ra = 10^3, Rd = 0.8$

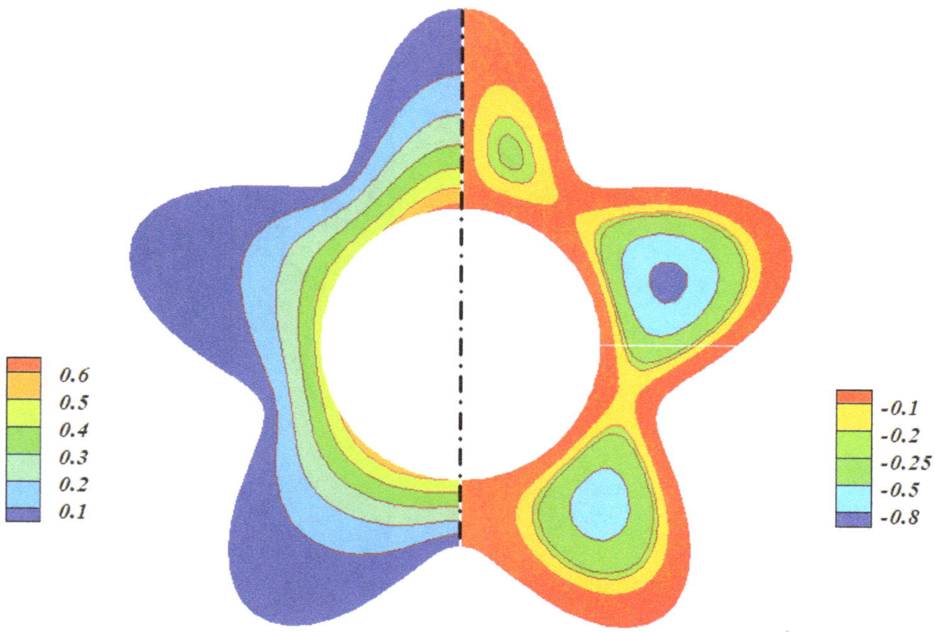

(a)

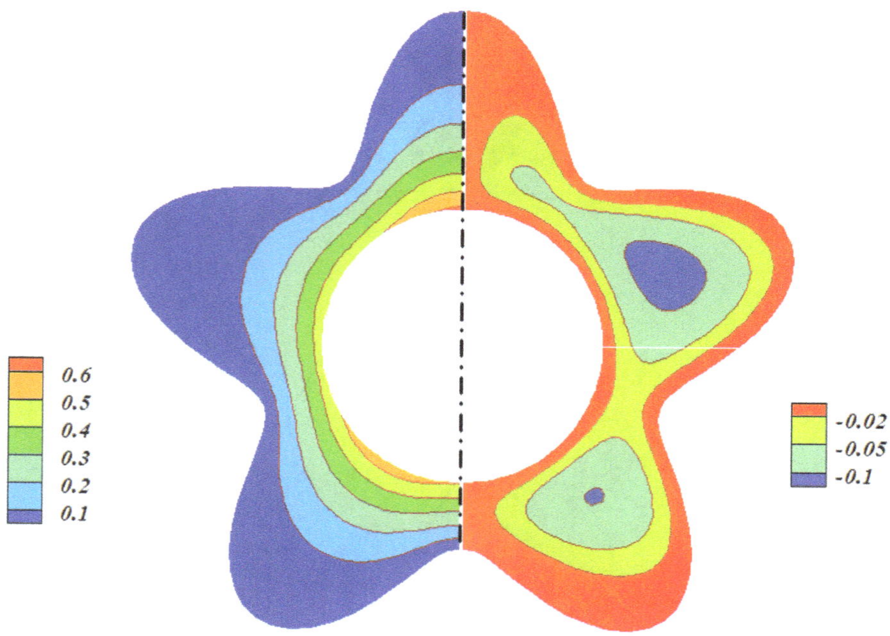

(b)

5.2. Governing Equation

2D steady convective flow of nanofluid is considered in existence of constant magnetic field. The PDEs equations are:

$$\frac{\partial u}{\partial x}+\frac{\partial v}{\partial y}=0 \tag{69}$$

$$\left(v\frac{\partial u}{\partial y}+\frac{\partial u}{\partial x}u\right)=\left[-\sigma_{nf}B_y^2u+\sigma_{nf}B_xB_yv\left(\frac{\partial^2 u}{\partial y^2}+\frac{\partial^2 u}{\partial x^2}\right)\mu_{nf}-\frac{\partial P}{\partial x}\right]\left(\rho_{nf}\right)^{-1} \tag{70}$$

$$\begin{gathered}\rho_{nf}\left(\frac{\partial v}{\partial x}u+\frac{\partial v}{\partial y}v\right)=+\mu_{nf}\left(\frac{\partial^2 v}{\partial x^2}+\frac{\partial^2 v}{\partial y^2}\right)-\frac{\partial P}{\partial y}+B_y\sigma_{nf}B_xu-B_x\sigma_{nf}B_xv\\+\left(T-T_c\right)\beta_{nf}g\rho_{nf},\\B_x=B_o\cos\lambda,\ B_y=B_o\sin\lambda\end{gathered} \tag{71}$$

$$\left(\rho C_p\right)_{nf}\left(v\frac{\partial T}{\partial y}+u\frac{\partial T}{\partial x}\right)=k_{nf}\left(\frac{\partial^2 T}{\partial x^2}+\frac{\partial^2 T}{\partial y^2}\right)-\frac{\partial q_r}{\partial y},\left[q_r=-\frac{4\sigma_e}{3\beta_R}\frac{\partial T^4}{\partial y},T^4\cong 4T_c^3T-3T_c^4\right] \tag{72}$$

$\left(\rho C_p\right)_{nf}$, $\left(\beta\right)_{nf}$, ρ_{nf} and σ_{nf} are defined as:

$$\left(\rho C_p\right)_{nf}=\left(\rho C_p\right)_f(1-\phi)+\left(\rho C_p\right)_s\phi \tag{73}$$

$$\left(\beta\right)_{nf}=\left(\beta\right)_f(1-\phi)+\left(\beta\right)_s\phi \tag{74}$$

$$\rho_{nf}=\rho_f(1-\phi)+\rho_s\phi \tag{75}$$

$$\frac{\sigma_{nf}}{\sigma_f}=1+\frac{3\left(\frac{\sigma_s}{\sigma_f}-1\right)\phi}{\left(\frac{\sigma_s}{\sigma_f}+2\right)-\left(\frac{\sigma_s}{\sigma_f}-1\right)\phi} \tag{76}$$

Properties of Fe_3O_4-water nanofluid is presented in Table 3. Maxwell model and Hamilton–Crosser model for irregular particle geometries by introducing a shape factor can be expressed as

$$\frac{k_{nf}}{k_f} = \frac{k_p + (m+1)k_f - (m+1)\phi(k_f - k_p)}{k_p + (m+1)k_f + \phi(k_f - k_p)} \tag{77}$$

in which k_p and k_f are the conductivities of the particle material and the base fluid. In this equation "m" is shaper factor. Table 4 shows the different values of shape factors for various shapes of nanoparticles.

μ_{nf} is obtained as follows [41]:

$$\mu_{nf} = \left(0.035B^2 + 3.1B - 27886.4807\phi^2 + 4263.02\phi + 316.0629\right)e^{-0.01T} \tag{78}$$

Vorticity and stream function should be used to eliminate pressure source terms:

$$\omega + \frac{\partial u}{\partial y} - \frac{\partial v}{\partial x} = 0, \frac{\partial \psi}{\partial x} = -v, \frac{\partial \psi}{\partial y} = u \tag{79}$$

Introducing dimensionless quantities:

$$P = \frac{p}{\rho_f (\alpha_f / L)^2}, U = \frac{uL}{\alpha_f}, V = \frac{vL}{\alpha_f}, \Theta = \frac{T - T_c}{\Delta T}, \Delta T = q''L / k_f, (X, Y) = \frac{(x, y)}{L} \tag{80}$$

Table 4. The values of shape factor of different shapes of nanoparticles

	Spherical	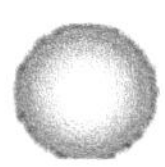	3
	Platelet		5.7
m	Cylinder		4.8
	Brick	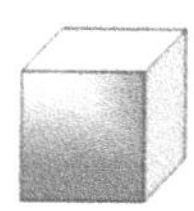	3.7

The final formulae are:

$$\frac{\partial V}{\partial Y}+\frac{\partial U}{\partial X}=0, \tag{81}$$

$$U\frac{\partial U}{\partial X}+\frac{\partial U}{\partial Y}V=\Pr\left[\frac{\mu_{nf}/\mu_f}{\rho_{nf}/\rho_f}\right]\left(\frac{\partial^2 U}{\partial Y^2}+\frac{\partial^2 U}{\partial X^2}\right)-Ha^2\Pr\left[\frac{\sigma_{nf}/\sigma_f}{\rho_{nf}/\rho_f}\right]\left(B_y^2U-B_xB_yV\right)-\frac{\partial P}{\partial X}, \tag{82}$$

$$\begin{aligned}V\frac{\partial V}{\partial Y}+U\frac{\partial V}{\partial X}&=\Pr\left(\frac{\partial^2 V}{\partial Y^2}+\frac{\partial^2 V}{\partial X^2}\right)\left[\frac{\mu_{nf}/\mu_f}{\rho_{nf}/\rho_f}\right]\\&-Ha^2\Pr\left[\frac{\sigma_{nf}/\sigma_f}{\rho_{nf}/\rho_f}\right]\left(B_x^2V-B_xB_yU\right)-\frac{\partial P}{\partial Y}+Ra\Pr\left[\frac{\beta_{nf}}{\beta_f}\right]\Theta,\end{aligned} \tag{83}$$

$$\frac{\partial \Theta}{\partial X}U+\frac{\partial \Theta}{\partial Y}V=\left(1+\frac{4}{3}\left(\frac{k_{nf}}{k_f}\right)^{-1}Rd\right)\left[\frac{(\rho C_P)_f}{(\rho C_P)_{nf}}\frac{k_{nf}}{k_f}\right]\frac{\partial^2\Theta}{\partial Y^2}+\left[\frac{(\rho C_P)_f}{(\rho C_P)_{nf}}\frac{k_{nf}}{k_f}\right]\left(\frac{\partial^2\Theta}{\partial X^2}\right) \tag{84}$$

where dimensionless and constants parameters are illustrated as:

$$\Pr=\upsilon_f/\alpha_f, Ra=g\beta_f q'' L^4/\left(k_f\upsilon_f\alpha_f\right), Ha=LB_0\sqrt{\sigma_f/\mu_f}, Rd=4\sigma_e T_c^3/\left(\beta_R k_f\right), \tag{85}$$

and boundary conditions are:

$$\begin{aligned}&\frac{\partial\Theta}{\partial n}=1.0 \text{ on inner wall}\\&\Theta=0.0 \text{ on outer wall}\\&\Psi=0.0 \text{ on all walls}\end{aligned} \tag{86}$$

Local and average Nusselt over the hot cylinder can calculate as:

$$Nu_{loc}=\left(1+\frac{4}{3}Rd\left(\frac{k_{nf}}{k_f}\right)^{-1}\right)\frac{1}{\theta}\left(\frac{k_{nf}}{k_f}\right) \tag{87}$$

$$Nu_{ave} = \frac{1}{\xi} \int_{0.5\pi}^{1.5\pi} Nu_{loc} \, d\xi \tag{88}$$

5.3. Effects of Active Parameters

Fe_3O_4-water nanofluid magnetohydrodynamic natural convection heat transfer in existence of thermal radiation is studied. MFD viscosity has been considered. Control volume based finite element method is selected to solve the governing equations. Numerical simulation are examined for various values of radiation parameter ($Rd = 0$ to 0.8), Rayleigh number ($\mathrm{R}a = 10^3, 10^4$ and 10^5), volume fraction of nanofluid ($\phi = 0$ to 0.04), shaper of nanoparticles and Hartmann number ($Ha = 0$ to 20). Table 5 shows the shape of nanoparticle effect on Nusselt number. As illustrated in this table, Platelet shape has highest rate of heat transfer. So, Platelet shape selected for further investigation.

Figures 23, 24 and 25 illustrate the effect of Ha and Ra on streamlines and isotherms. As nanofluid temperature augments, the nanofluid begins moving from the warm surface to the outer one and dropping along the outer cylinder, afterwards mounting again at the inner cylinder, generating a revolving vortex inside the cavity. At low Rayleigh number, conduction is more signification than convection. $|\Psi_{max}|$ augments as buoyancy force augments and it reduces as Lorentz force enhances. As buoyancy forces enhances, eddies become stronger and thermal plume generate near the center line. Increasing Hartmann number causes the thermal plume to diminish and change the heat transfer mechanism from convection to conduction.

Figures 26 and 27 demonstrate the influence of Rd, Ra and Ha on Nu_{loc}, Nu_{ave}. The correlation for Nu_{ave} corresponding active parameters is:

$$\begin{aligned} Nu_{ave} &= 18.8 - 5.4Rd - 8.1\log\left(Ra\right) - 1.2Ha^* + 1.68Rd\log\left(Ra\right) \\ &-0.2Rd\,Ha^* - 0.32Ha^*\log\left(Ra\right) + 1.6Rd^2 + 1.07\left(\log\left(Ra\right)\right)^2 - 0.15\left(Ha^*\right)^2 \end{aligned} \tag{89}$$

where $Da^* = 0.01Da, Ha^* = 0.1Ha$. The number of extermum in in Nu_{loc} profile is matching to existence of thermal plume and number of undulation. As buoyancy force and thermal radiation augment, Nu_{ave} enhances. Enhancing Hartmann number causes the nanofluid flow to retard and Nusselt number reduces.

Table 5. Effect of shape of nanoparticles on Nusselt number when $Rd = 0.8, Ra = 10^5, \phi = 0.04$

	Ha	
	0	**20**
Spherical	8.269555	6.119096
Brick	8.336758	6.170139
Cylinder	8.442229	6.250367
Platelet	8.52838	6.316004

Figure 23. Effect of Hartmann number on isotherms (left) and streamlines (right) contours when (a) Ha=0, (b) Ha=20 and $Ra = 10^4, Rd = 0.8$

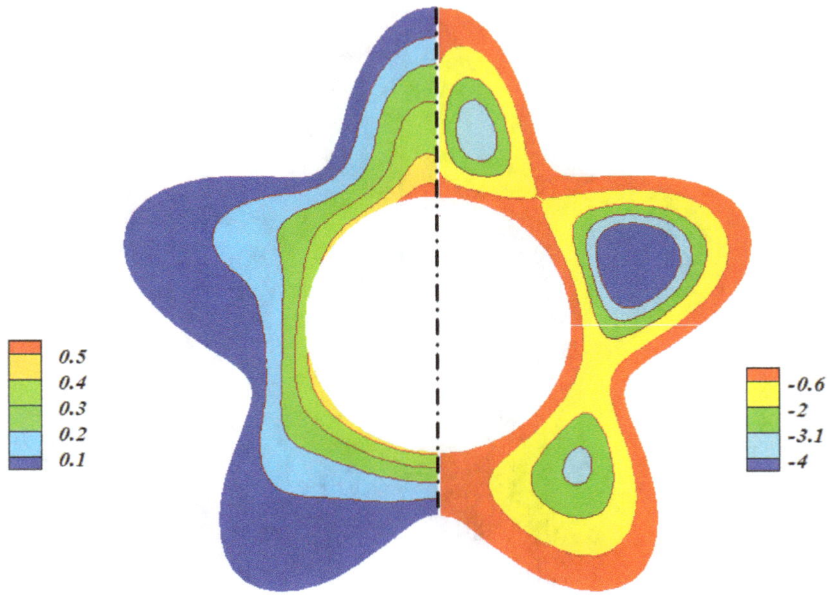

(a)

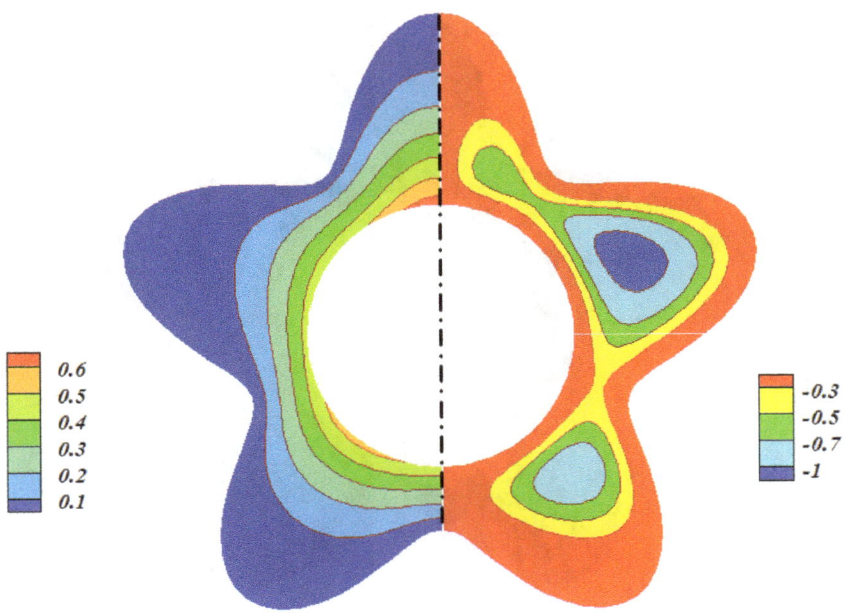

(b)

Figure 24. Effect of Hartmann number on isotherms (left) and streamlines (right) contours when (a) Ha=0, (b) Ha=20 and $Ra = 10^5, Rd = 0.8$

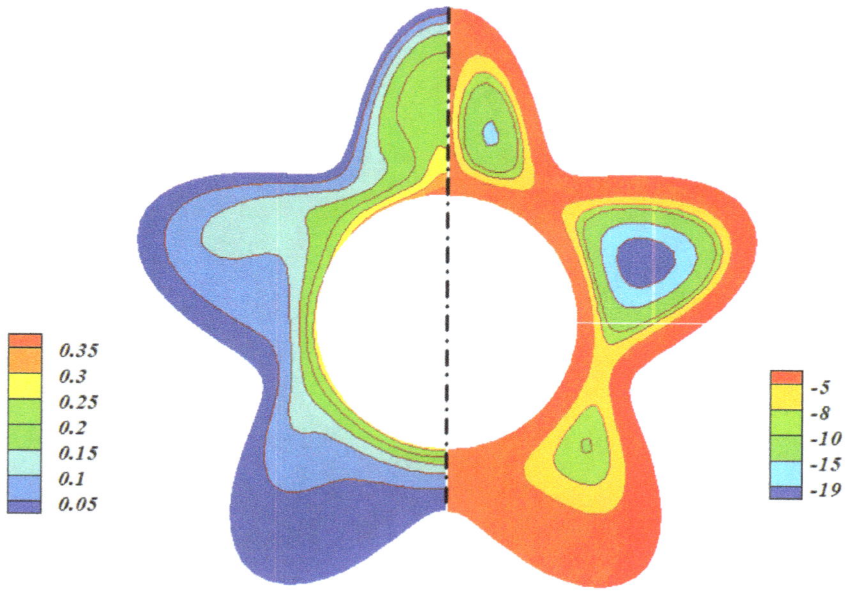

(a)

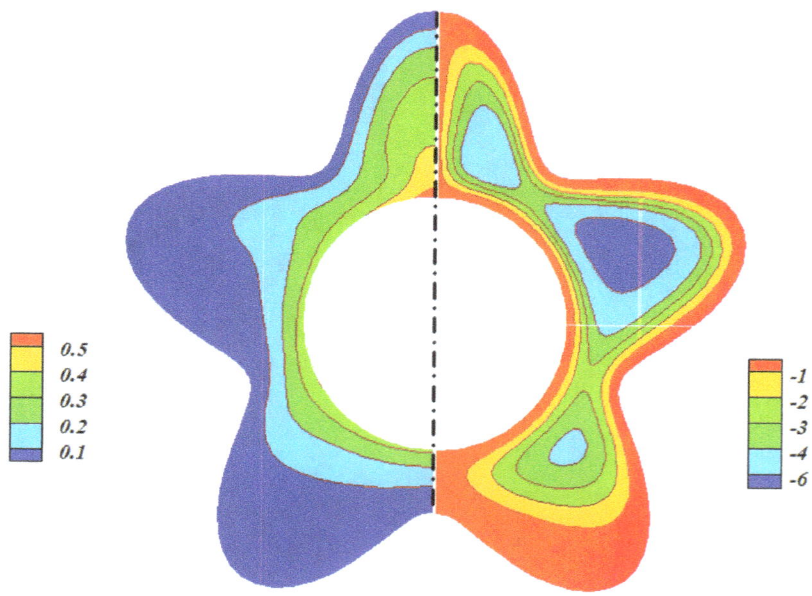

(b)

Figure 25. Effects of radiation parameter, Hartmann number and Rayleigh number on local Nusselt number when (a) $Ra = 10^3, Rd = 0.8$*, (b)* $Ra = 10^4, Rd = 0.8$*, (c)* $Ra = 10^5, Rd = 0.8$*, (d)* $Ra = 10^5, Ha = 0$*, (e)* $Ra = 10^5, Ha = 20$ *and* $\phi = 0.04$

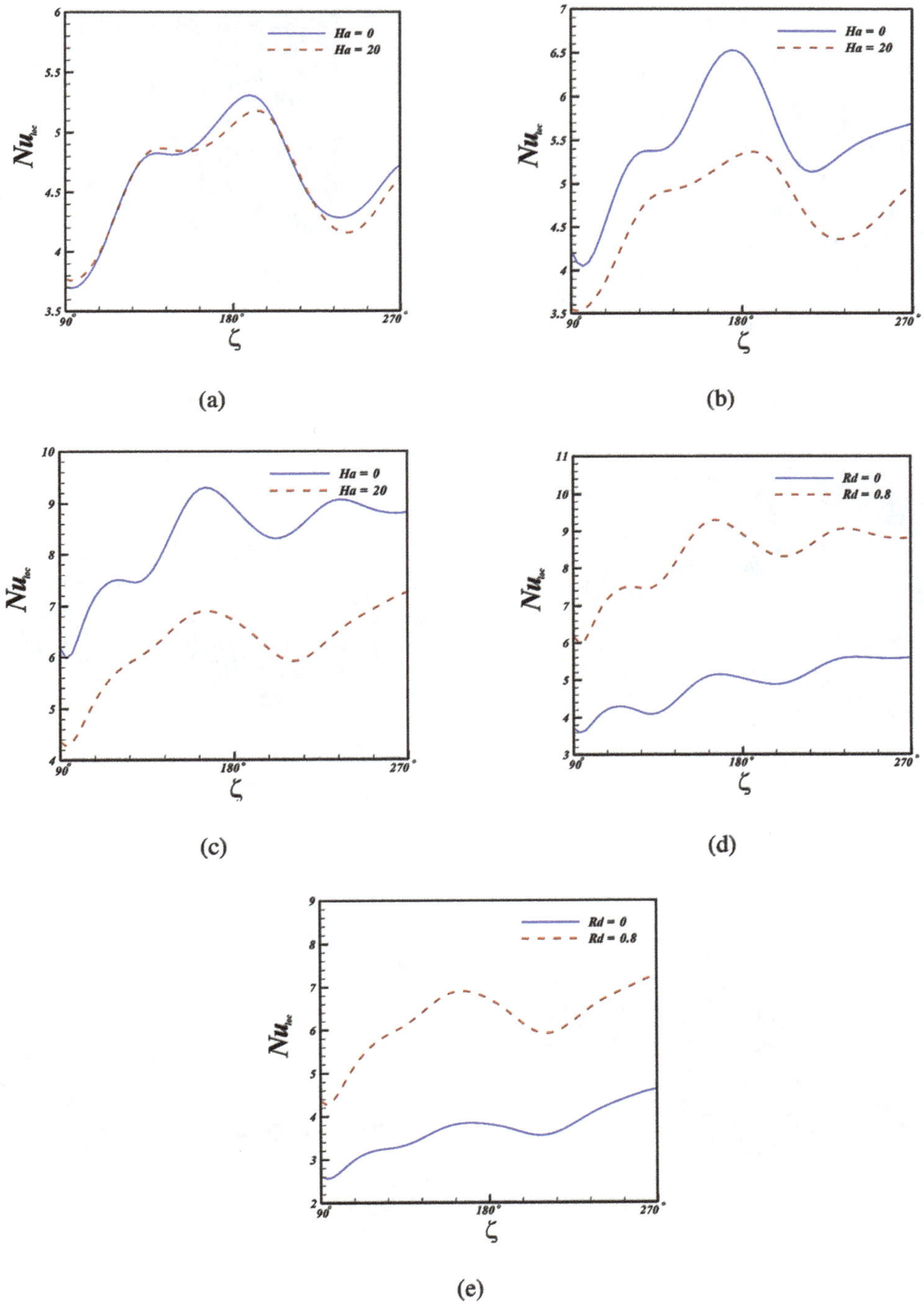

Figure 26. Effects of radiation parameter, Hartmann number and Rayleigh number on average Nusselt number when (a) $Ha = 10, \phi = 0.04$, (b) $Ra = 10^4, \phi = 0.04$, (c) $Rd = 0.4, \phi = 0.04$, (d) $Ha = 10, \phi = 0.04$, (e) $Ra = 10^4, \phi = 0.04$, (f) $Rd = 0.4, \phi = 0.04$

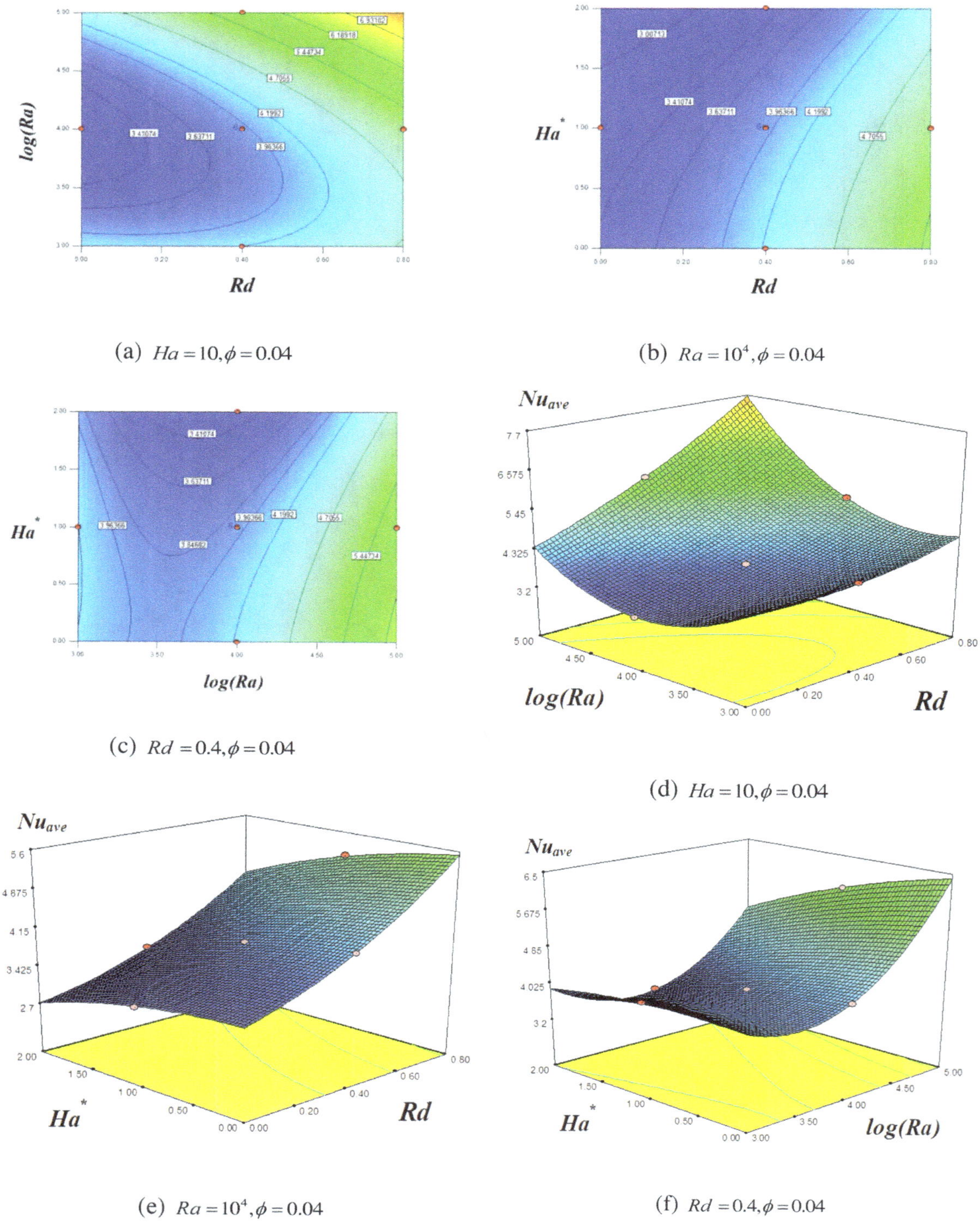

(a) $Ha = 10, \phi = 0.04$

(b) $Ra = 10^4, \phi = 0.04$

(c) $Rd = 0.4, \phi = 0.04$

(d) $Ha = 10, \phi = 0.04$

(e) $Ra = 10^4, \phi = 0.04$

(f) $Rd = 0.4, \phi = 0.04$

Figure 27. (a) Geometry and the boundary conditions with (b) the mesh of Geometry considered in this work; (c) A sample triangular element and its corresponding control volume.

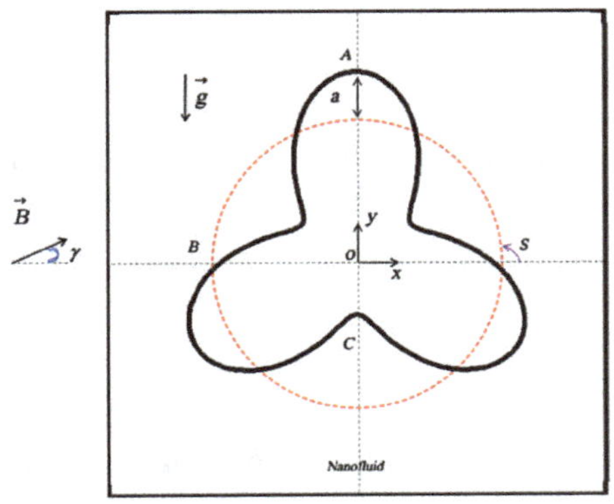

(a)

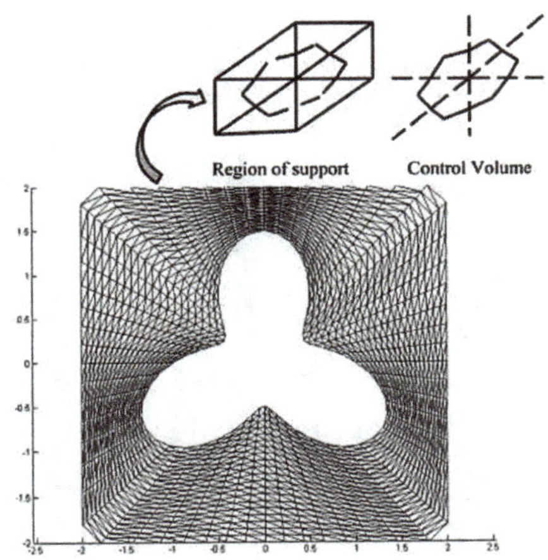

(b)

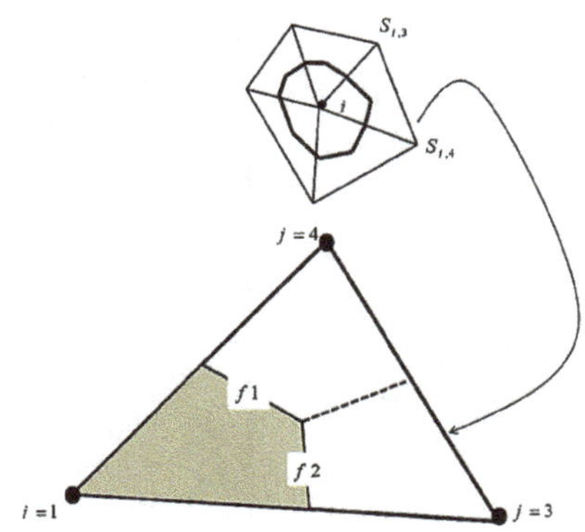

(c)

6. NUMERICAL STUDY OF THE EFFECT OF MAGNETIC FIELD ON FE_3O_4-WATER FERROFLUID CONVECTION WITH THERMAL RADIATION

6.1. Problem Definition

Figure 28 depicts the geometry, boundary condition and sample element. The formula for the inner cylinder is:

$$r = r_{in} + A\cos\left(N\left(\zeta - \zeta_0\right)\right) \tag{90}$$

Figure 28. Effect of Rayleigh number, Hartmann number and radiation parameter on Isotherms (left) and streamlines (right) contours ($Rd = 0\ (-\cdots-)$, $Rd = 0.8\ (-)$) when $N = 3, \phi = 0.04$

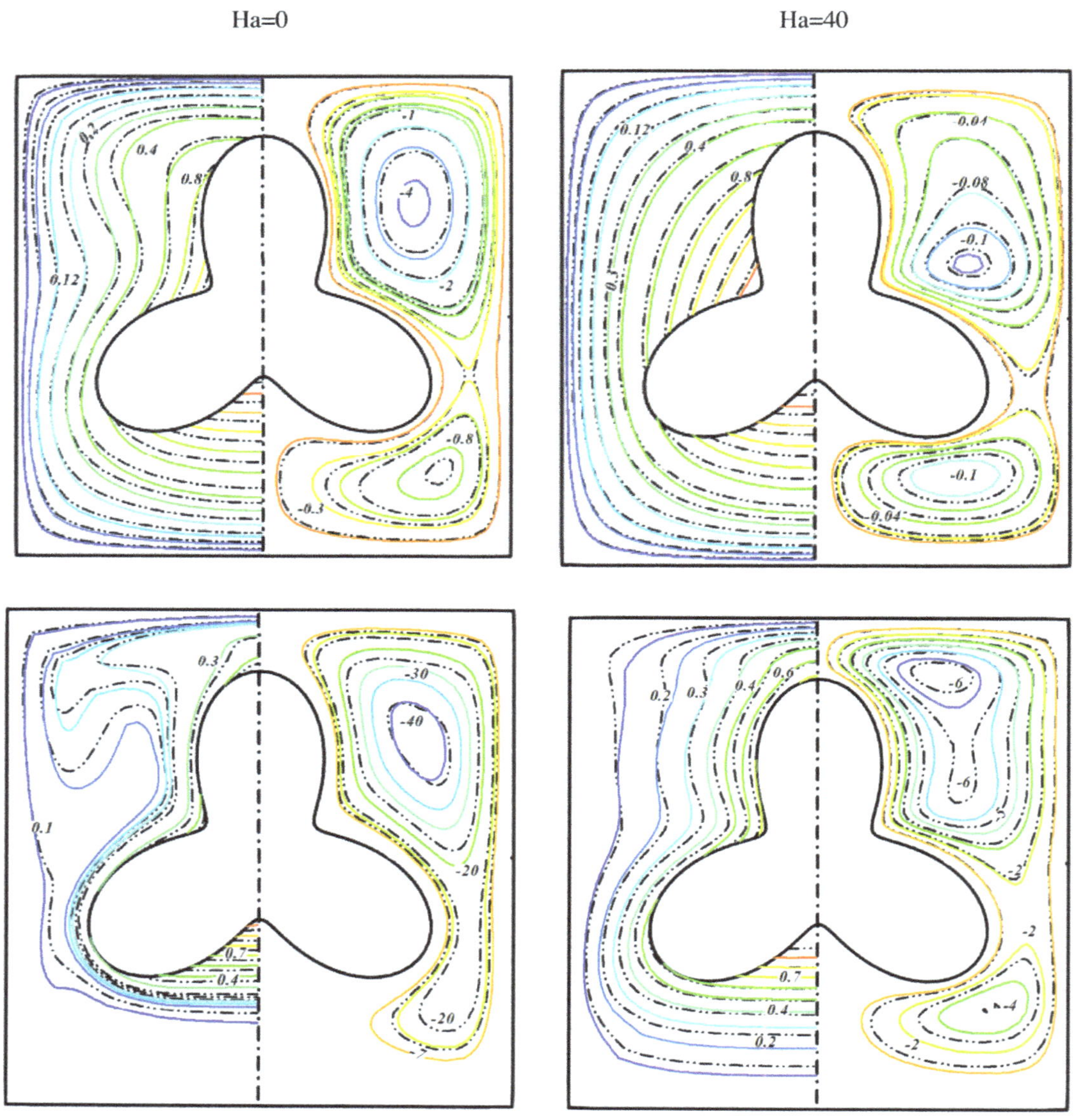

in which r_{in} is radius of the base circle. The inner cylinder has constant heat flux condition.

6.2. Governing Equation

Two-dimensional steady convective flow of nanofluid in a porous medium is considered with constant magnetic field. The PDEs are:

$$\frac{\partial u}{\partial x}+\frac{\partial v}{\partial y}=0 \tag{91}$$

$$\left(v\frac{\partial u}{\partial y}+\frac{\partial u}{\partial x}u\right)=\left[-\sigma_{nf}B_y^2u+\sigma_{nf}B_xB_yv+\left(\frac{\partial^2 u}{\partial y^2}+\frac{\partial^2 u}{\partial x^2}\right)\mu_{nf}-\frac{\partial P}{\partial x}\right]\left(\rho_{nf}\right)^{-1} \tag{92}$$

$$\begin{gathered}\rho_{nf}\left(\frac{\partial v}{\partial x}u+\frac{\partial v}{\partial y}v\right)=+\mu_{nf}\left(\frac{\partial^2 v}{\partial x^2}+\frac{\partial^2 v}{\partial y^2}\right)-\frac{\partial P}{\partial y}+B_y\sigma_{nf}B_xu-B_x\sigma_{nf}B_xv\\ +(T-T_c)\beta_{nf}g\rho_{nf},\\ B_x=B_o\cos\lambda, B_y=B_o\sin\lambda\end{gathered} \tag{93}$$

$$\begin{gathered}\left(\rho C_p\right)_{nf}\left(v\frac{\partial T}{\partial y}+u\frac{\partial T}{\partial x}\right)=k_{nf}\left(\frac{\partial^2 T}{\partial x^2}+\frac{\partial^2 T}{\partial y^2}\right)-\frac{\partial q_r}{\partial y},\\ ,\left[q_r=-\frac{4\sigma_e}{3\beta_R}\frac{\partial T^4}{\partial y},\quad T^4\cong 4T_c^3T-3T_c^4\right]\end{gathered} \tag{94}$$

$(\rho C_p)_{nf},(\rho\beta)_{nf},\rho_{nf},k_{nf}$ and σ_{nf} are defined as:

$$(\rho C_p)_{nf}=(\rho C_p)_f(1-\phi)+(\rho C_p)_s\phi \tag{95}$$

$$\left(\rho\beta\right)_{nf}=\left(\rho\beta\right)_f(1-\phi)+\left(\rho\beta\right)_s\phi \tag{96}$$

$$\rho_{nf}=\rho_f(1-\phi)+\rho_s\phi \tag{97}$$

$$k_{nf} = k_f \left(\frac{k_s + 2k_f + 2\phi(k_s - k_f)}{k_s - \phi(k_s - k_f) + 2k_f} \right) \tag{98}$$

$$\frac{\sigma_{nf}}{\sigma_f} = 1 + \frac{3\left(\frac{\sigma_s}{\sigma_f} - 1\right)\phi}{\left(\frac{\sigma_s}{\sigma_f} + 2\right) - \left(\frac{\sigma_s}{\sigma_f} - 1\right)\phi} \tag{99}$$

μ_{nf} is obtained as follows:

$$\mu_{nf} = (0.035B^2 + 3.1B - 27886.4807\phi^2 + 4263.02\phi + 316.0629)e^{-0.017} \tag{100}$$

Vorticity and stream functions should be used to eliminate pressure source terms:

$$\omega + \frac{\partial u}{\partial y} - \frac{\partial v}{\partial x} = 0, \frac{\partial \psi}{\partial x} = -v, \frac{\partial \psi}{\partial y} = u \tag{101}$$

Introducing dimensionless quantities:

$$P = \frac{p}{\rho_f(\alpha_f / L)^2}, U = \frac{uL}{\alpha_f}, V = \frac{vL}{\alpha_f}, \Theta = \frac{T - T_c}{\Delta T}, \Delta T = q''L / k_f, (X,Y) = \frac{(x,y)}{L} \tag{102}$$

then the final equations are:

$$\frac{\partial V}{\partial Y} + \frac{\partial U}{\partial X} = 0 \tag{103}$$

$$U\frac{\partial U}{\partial X} + \frac{\partial U}{\partial Y}V = \Pr\left[\frac{\mu_{nf} / \mu_f}{\rho_{nf} / \rho_f}\right]\left(\frac{\partial^2 U}{\partial Y^2} + \frac{\partial^2 U}{\partial X^2}\right) - Ha^2 \Pr\left[\frac{\sigma_{nf} / \sigma_f}{\rho_{nf} / \rho_f}\right](B_y^2 U - B_x B_y V) - \frac{\partial P}{\partial X} \tag{104}$$

$$\begin{aligned} &V\frac{\partial V}{\partial Y} + U\frac{\partial V}{\partial X} = \Pr\left(\frac{\partial^2 V}{\partial Y^2} + \frac{\partial^2 V}{\partial X^2}\right)\left[\frac{\mu_{nf} / \mu_f}{\rho_{nf} / \rho_f}\right] \\ &-Ha^2 \Pr\left[\frac{\sigma_{nf} / \sigma_f}{\rho_{nf} / \rho_f}\right](B_x^2 V - B_x B_y U) - \frac{\partial P}{\partial Y} + Ra \Pr\left[\frac{\beta_{nf}}{\beta_f}\right]\Theta \end{aligned} \tag{105}$$

$$\frac{\partial \Theta}{\partial X}U+\frac{\partial \Theta}{\partial Y}V=\left(1+\frac{4}{3}\left(\frac{k_{nf}}{k_f}\right)^{-1}Rd\right)\frac{\partial^2 \Theta}{\partial Y^2}+\left[\frac{\left(\rho C_P\right)_f}{\left(\rho C_P\right)_{nf}}\frac{k_{nf}}{k_f}\right]\left(\frac{\partial^2 \Theta}{\partial X^2}\right) \tag{106}$$

where dimensionless and constants parameters are:

$$\Pr=\upsilon_f/\alpha_f, Ra=g\beta_f q'' L^4/(k_f\upsilon_f\alpha_f), Ha=LB_0\sqrt{\sigma_f/\mu_f}, Rd=4\sigma_\varepsilon T_c^3/(\beta_R k_f) \tag{107}$$

and boundary conditions are:

$$\frac{\partial \Theta}{\partial n}=1.0 \text{ on inner wall}$$

$$\Theta=0.0 \text{ on outer wall}$$

$$\Psi=0.0 \text{ on all walls} \tag{108}$$

Local and average Nusselt over the hot cylinder are determined from:

$$Nu_{loc}=\left(1+\frac{4}{3}Rd\left(\frac{k_{nf}}{k_f}\right)^{-1}\right)\frac{1}{\theta}\left(\frac{k_{nf}}{k_f}\right) \tag{109}$$

$$Nu_{ave}=\frac{1}{S}\int_0^s Nu_{loc}\,ds \tag{110}$$

6.3. Effects of Active Parameters

MHD Fe_3O_4-water nanofluid natural convection with radiation source term is examined. Magnetic field dependent (MFD) viscosity is taken into consideration. Numerical simulations are examined for various values of the radiation parameter ($Rd=0$ to 0.8), volume fraction of Fe_3O_4-water ($\phi=0$ to 0.04), Rayleigh number ($Ra=10^3, 10^4$ and 10^5), number of undulations ($N=3,4$ and 5) and Hartmann number ($Ha=0$ to 40).

Effect of radiation parameter on nanofluid hydrothermal behavior is illustrated in Figure 29. The inner wall temperature increases with increasing *Rd*. Also increasing *Rd* results in increasing nanofluid motion. This influence is more evident when convection flow is dominant.

Influence of N, Ha and Ra on streamlines and isotherms are depicted Figures 30, 31 and 32. As nanofluid temperature increases, the nanofluid begins moving from the warm surface to the outer one

Figure 29. Effect of Rayleigh and Hartmann numbers on Isotherms (left) and streamlines (right) contours when $N = 3, \phi = 0.04, Rd = 0.8$

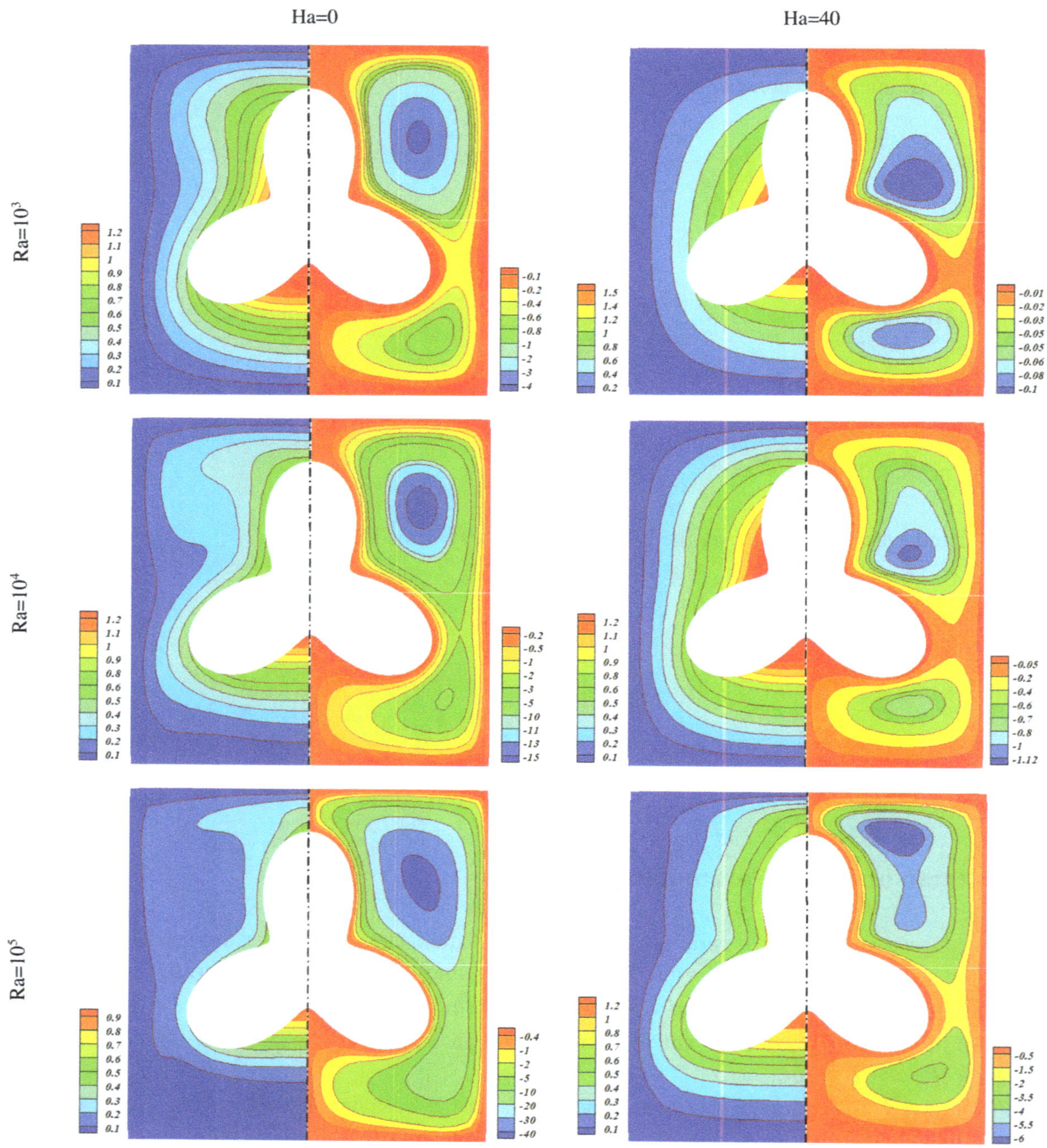

Figure 30. Effect of Rayleigh and Hartmann numbers on Isotherms (left) and streamlines (right) contours when $N = 4, \phi = 0.04, Rd = 0.8$

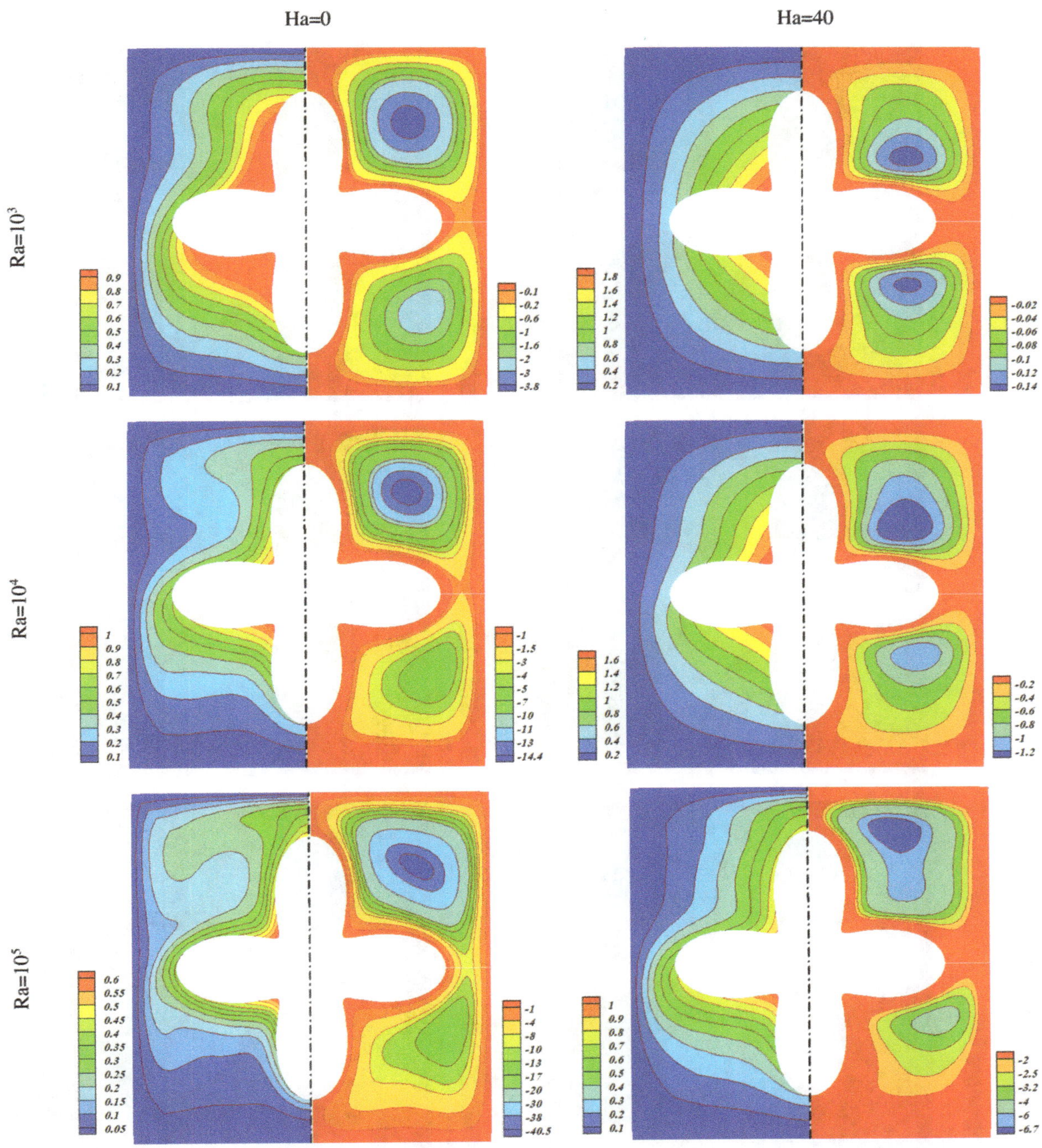

Figure 31. Effect of Rayleigh and Hartmann numbers on Isotherms (left) and streamlines (right) contours when $N = 5, \phi = 0.04, Rd = 0.8$

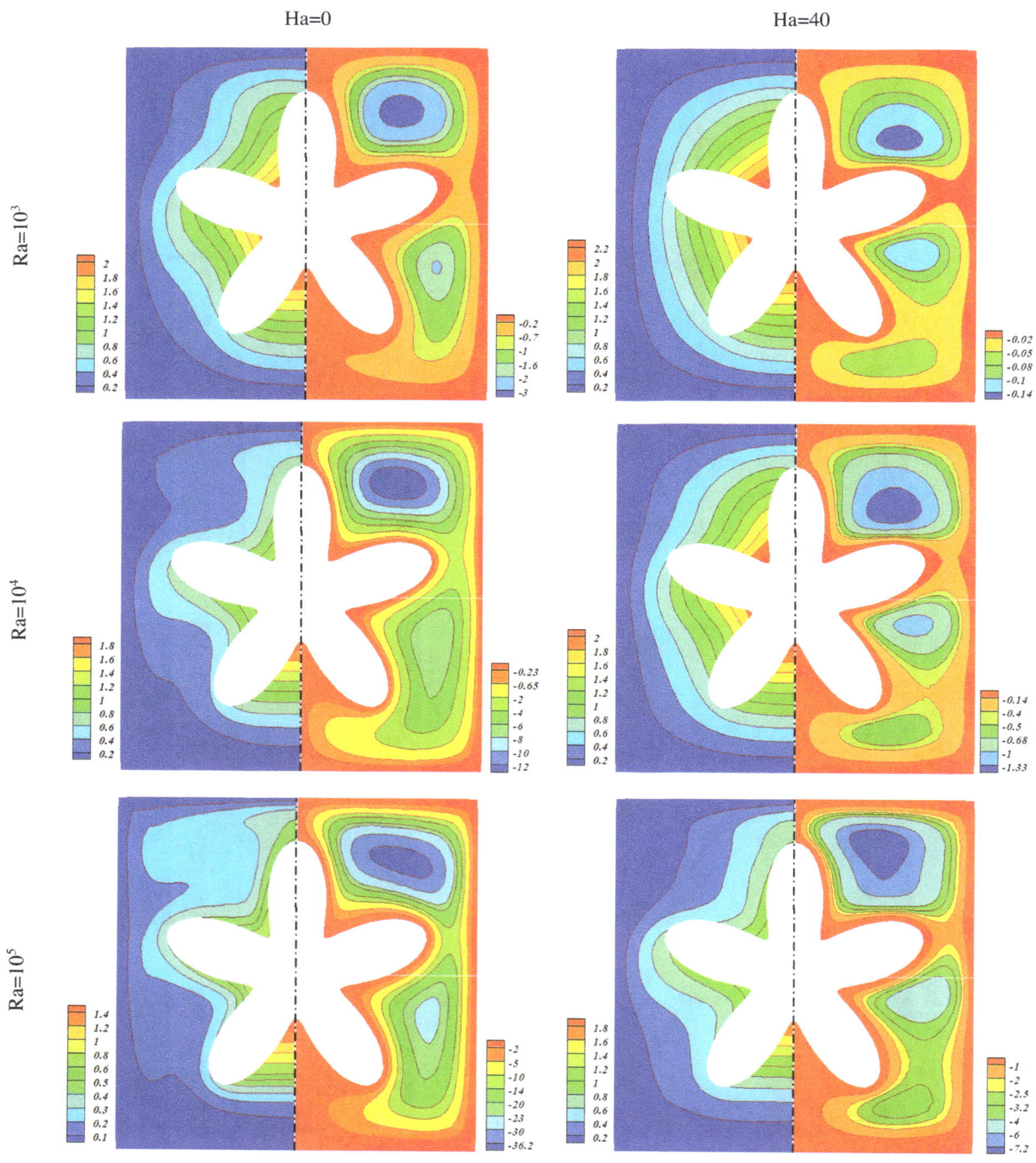

Figure 32. Effects of the number of undulations, Hartmann and Rayleigh numbers on local Nusselt number when $Rd = 0.8, \phi = 0.04$

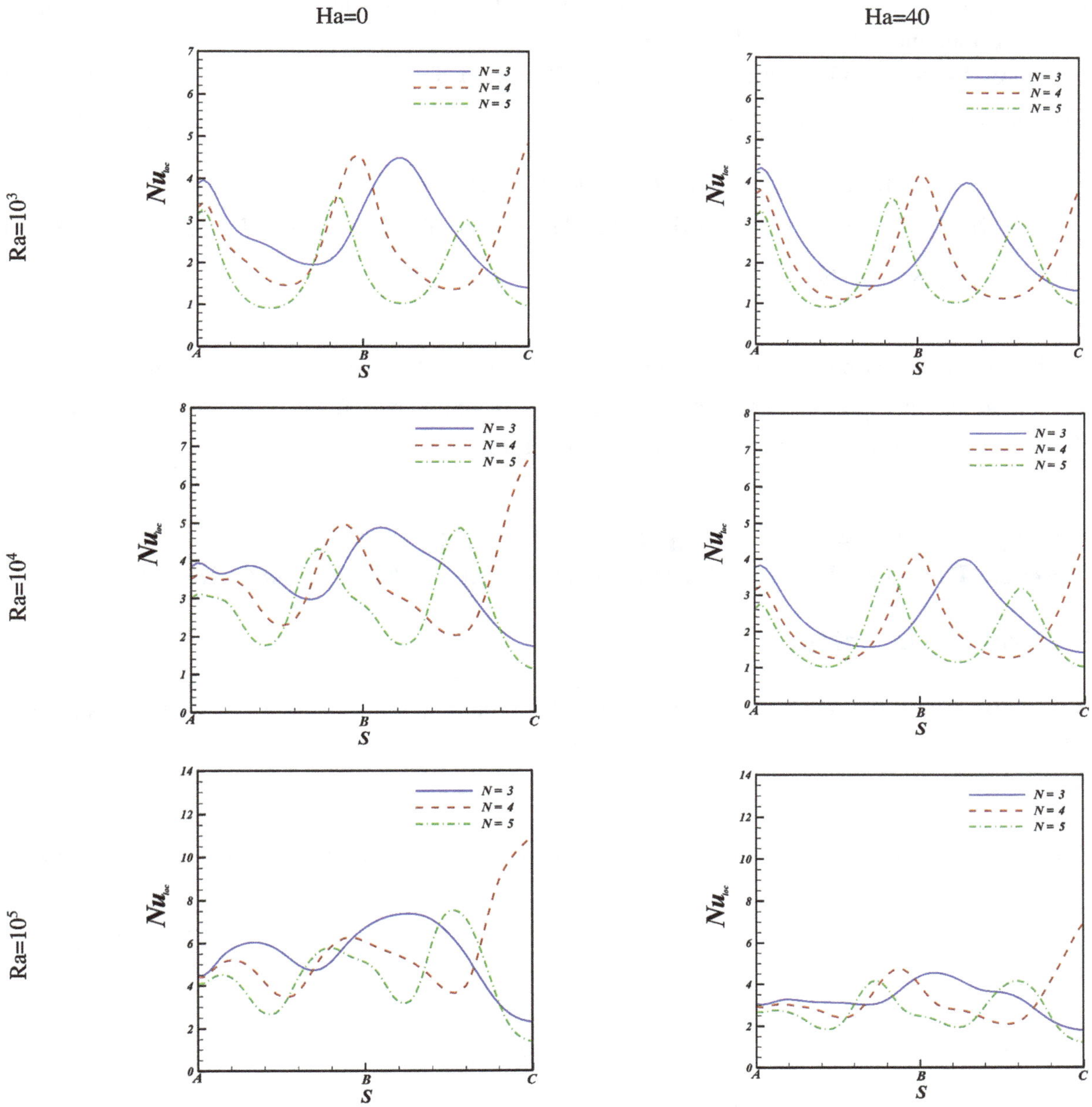

and dropping along the circular cylinder. At low Rayleigh number, conduction is more significant than convection. $|\psi_{max}|$ increase as buoyancy force increases and it decreases as the Lorentz force increases. As Hartmann number increases, the center of the vortices moves to *x=0.* As *Ra* increases, convection becomes stronger. Thus $|\psi_{max}|$ and temperature gradient increase with increasing Ra . As *N* increases, the distortion of isotherms reduces and vortices become weaker. Increasing Hartmann number leads to reduction in the thermal plume and the heat transfer mechanism changes from convection to conduction. So, *Nu* decreases with increasing Lorentz forces. At $N = 5$, increasing the Lorentz force causes the main vortices to convert into three smaller ones. As the Lorentz force increases, the two upper vortices merge together and the thermal plume vanishes.

Figures 33, 34 and 35 illustrate the impact of N, Rd, Ra and Ha on Nu_{loc}, Nu_{ave}. The formula for Nu_{ave} in terms of active parameters is:

$$\begin{aligned} Nu_{ave} &= 3.13 + 0.64N - 1.8\log(Ra) + 0.36Rd - 1.4\times10^{-4}Ha - 0.086N\log(Ra) \\ &-0.25N\,Rd + 2.8\times10^{-3}N\,Ha + 0.46\log(Ra)Rd - 0.016\log(Ra)Ha - 0.015Ra\,Ha \\ &-0.07N^2 + 0.38(\log(Ra))^2 + 1.05Rd^2 + 9.4\times10^{-4}Ha^2 \end{aligned} \tag{111}$$

The number of extrema in the Nu_{loc} profile matches to existence of the thermal plume and number of undulations. Nu_{ave} increases with increasing Rd. As buoyancy forces increase, the temperature decreases and in turn Nu_{ave} increases with increasing Ra.

7. SIMULATION OF NANOFLUID RADIATION AND NATURAL CONVECTION IN AN ENCLOSURE WITH ELLIPTICAL CYLINDERS

7.1. Problem Definition

Figure 36 depicts the geometry, boundary condition and sample element. The formula of inner cylinder is:

Figure 33. Effects of the radiation parameter on local Nusselt number when $Ra = 10^5, \phi = 0.04, Ha = 0$

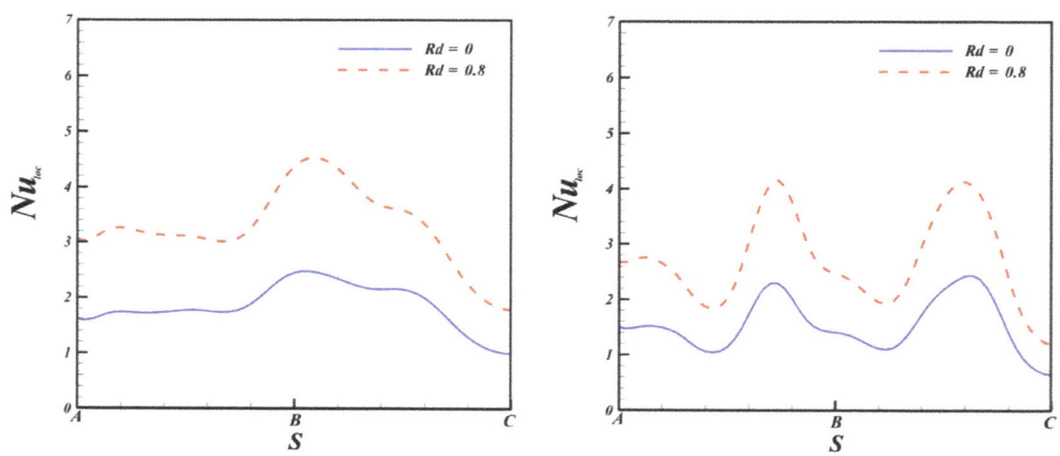

$N = 3$ $N = 5$

Figure 34a. Effects of the number of undulations, radiation parameter, Rayleigh and Hartmann numbers on average Nusselt number when $\phi = 0.04$

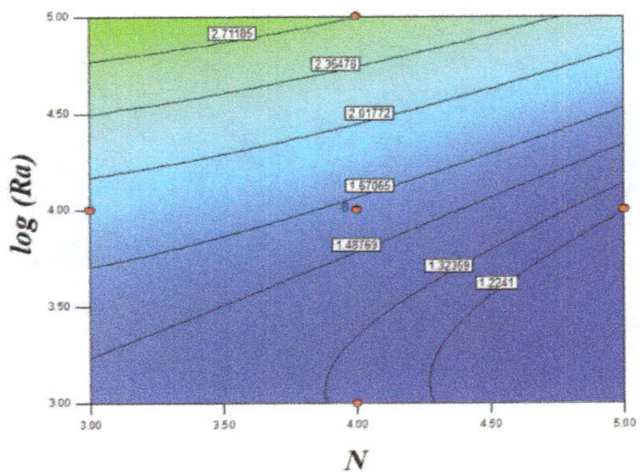

$Ha = 20, Rd = 0.4$

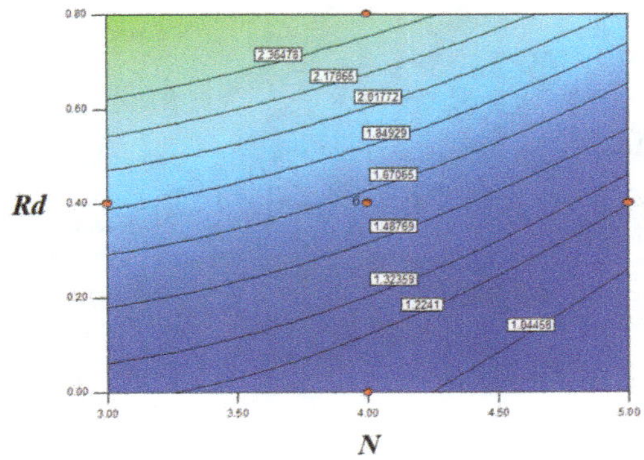

$Ra = 10^4, Ha = 20$

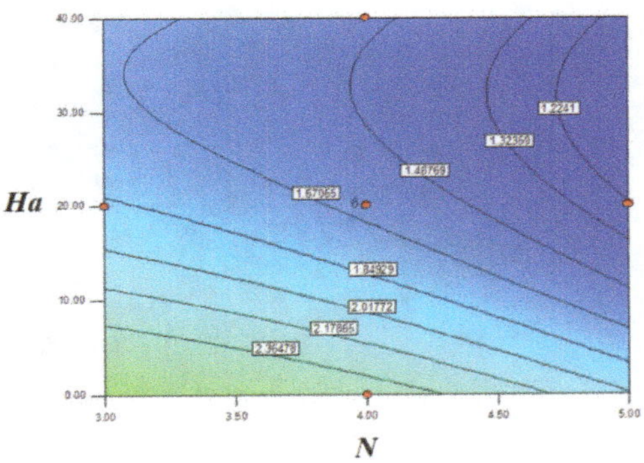

$Ra = 10^4, Rd = 0.4$

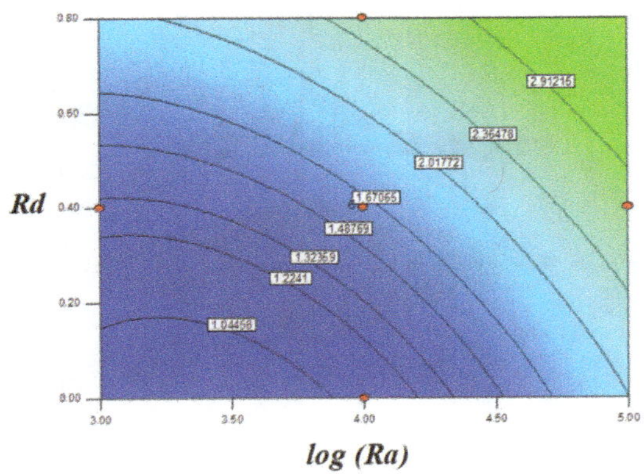

$N = 4, Ha = 20$

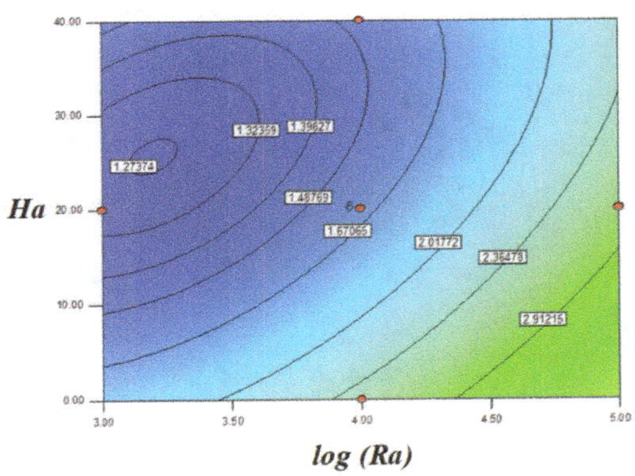

$N = 4, Rd = 0.4$

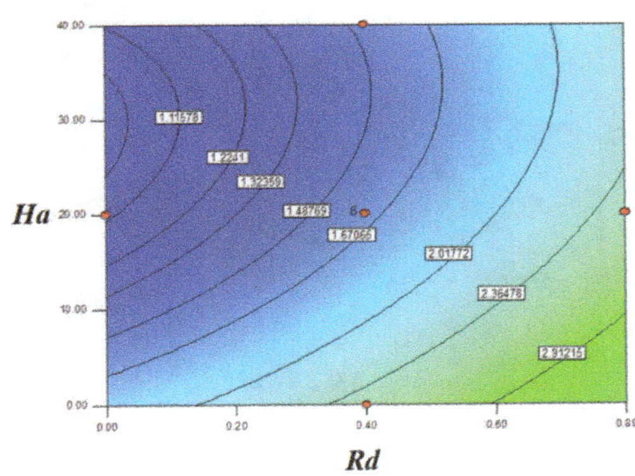

$Ra = 10^4, N = 4$

Figure 34b. Effects of the number of undulations, radiation parameter, Rayleigh and Hartmann numbers on average Nusselt number when $\phi = 0.04$

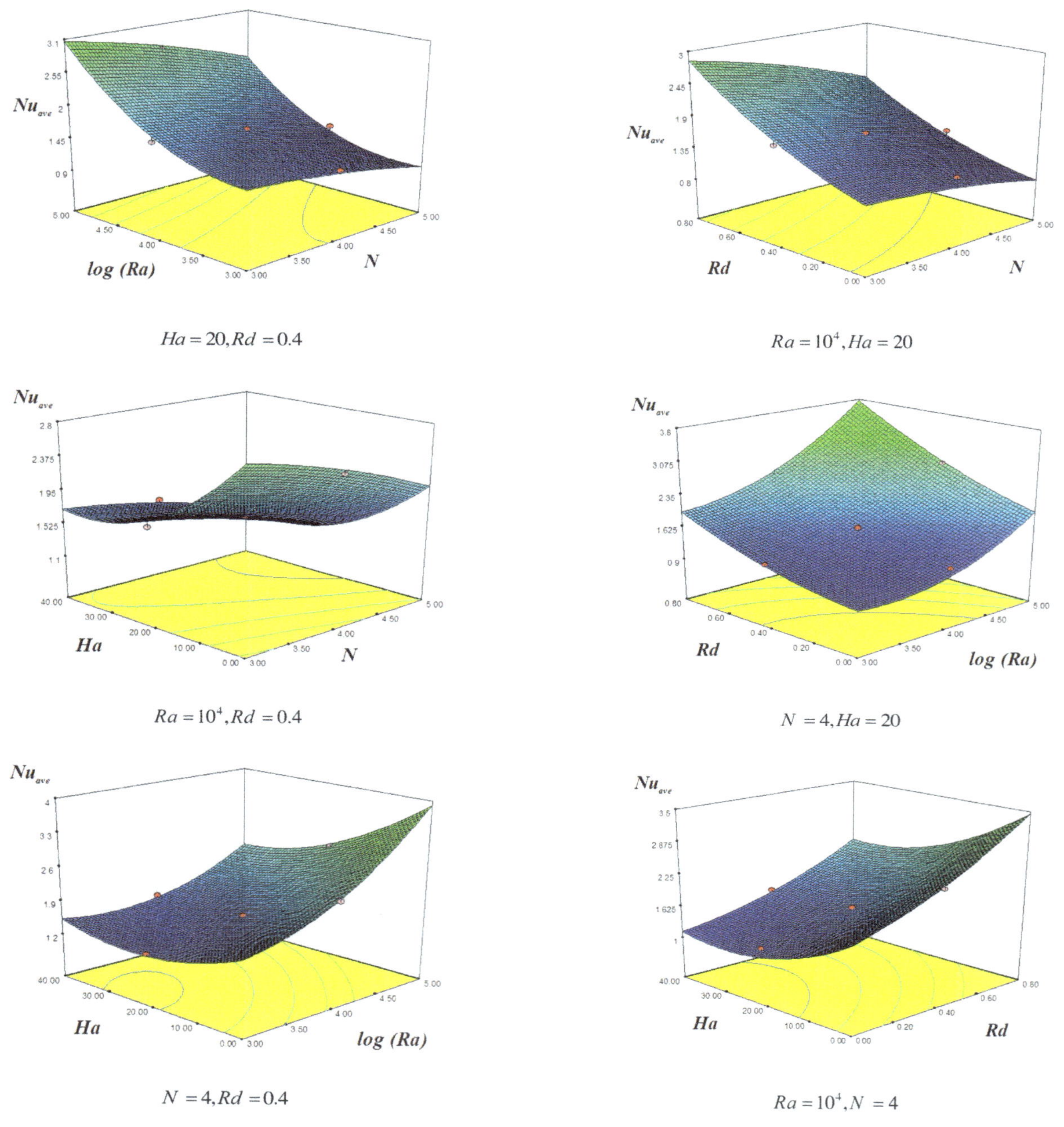

$$b = \sqrt{1-\varepsilon^2}.a \tag{112}$$

where a, b, ε are the major, minor axis of elliptic cylinder and eccentricity for inner cylinder. The inner cylinder has constant heat flux condition.

Figure 35. Effect of Hartmann number and radiation parameter on Isotherms (left) and streamlines (right) contours ($Rd = 0\ (- \cdot\cdot -)$, $Rd = 0.8\,(-)$) when $\xi = 90°, \phi = 0.04, Ra = 10^5$

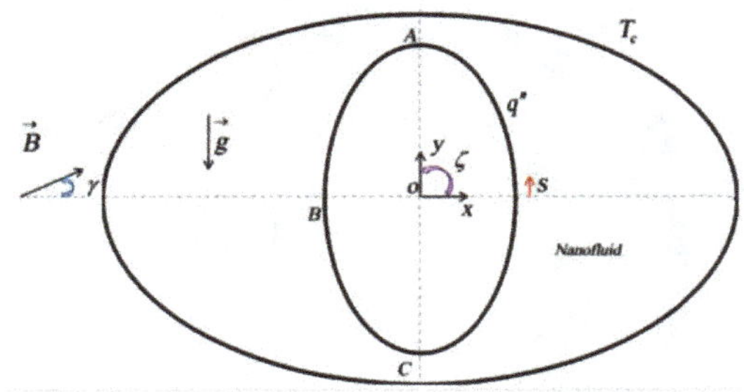

(a)

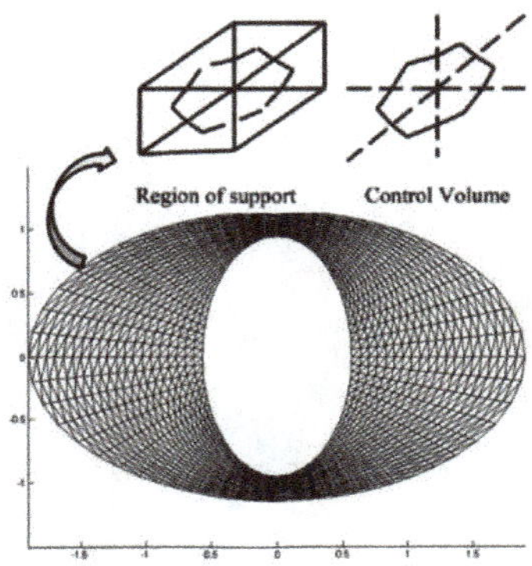

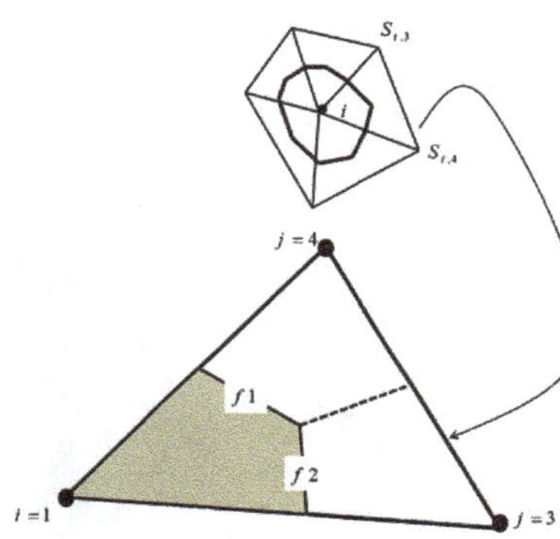

(b) (c)

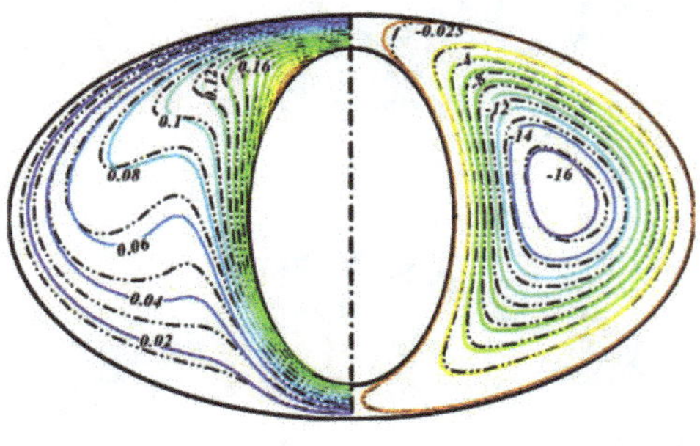

Ha=0

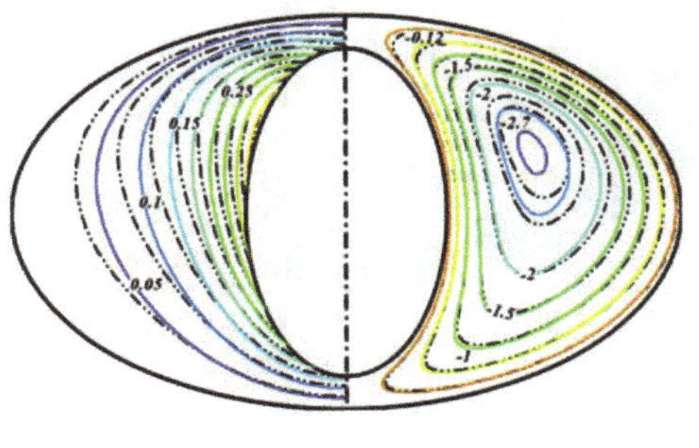

Ha=40

Figure 36. Effect of Rayleigh and Hartmann numbers on Isotherms (left) and streamlines (right) contours when $\xi = 0, \phi = 0.04, Rd = 0.8$

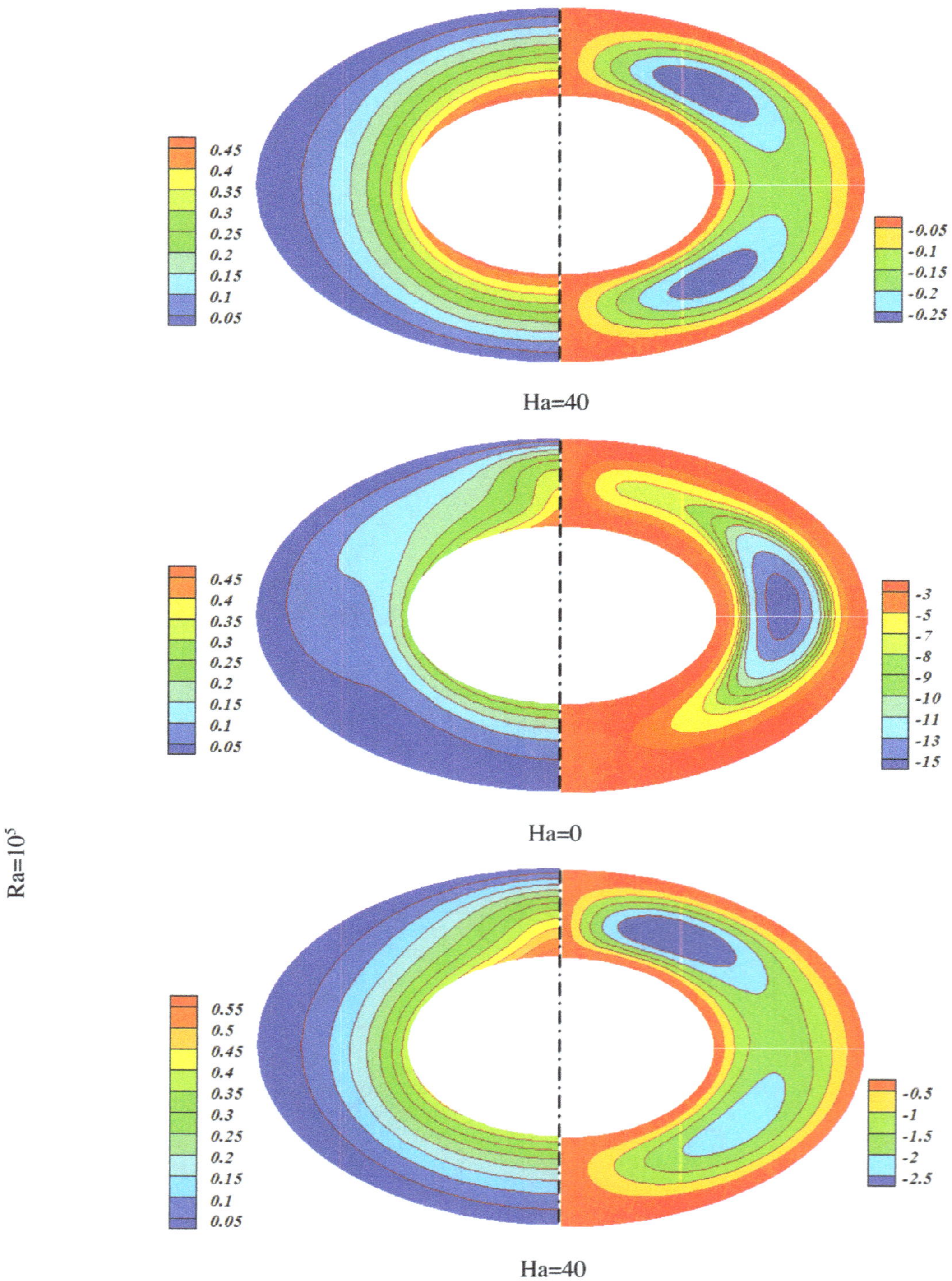

7.2. Governing Equation

2D steady convective flow of nanofluid in a porous media is considered in existence of constant magnetic field. The PDEs can be put into the forms:

$$\frac{\partial u}{\partial x}+\frac{\partial v}{\partial y}=0 \tag{113}$$

$$\left(v\frac{\partial u}{\partial y}+\frac{\partial u}{\partial x}u\right)=\left[-\sigma_{nf}B_y^2u+\sigma_{nf}B_xB_yv\left(\frac{\partial^2 u}{\partial y^2}+\frac{\partial^2 u}{\partial x^2}\right)\mu_{nf}-\frac{\partial P}{\partial x}\right]\left(\rho_{nf}\right)^{-1} \tag{114}$$

$$\begin{gathered}\rho_{nf}\left(\frac{\partial v}{\partial x}u+\frac{\partial v}{\partial y}v\right)=\mu_{nf}\left(\frac{\partial^2 v}{\partial x^2}+\frac{\partial^2 v}{\partial y^2}\right)-\frac{\partial P}{\partial y}+B_y\sigma_{nf}B_xu-B_x\sigma_{nf}B_xv\\ +\left(T-T_c\right)\beta_{nf}g\rho_{nf},\\ B_x=B_o\cos\lambda,\ B_y=B_o\sin\lambda\end{gathered} \tag{115}$$

$$\left(\rho C_p\right)_{nf}\left(v\frac{\partial T}{\partial y}+u\frac{\partial T}{\partial x}\right)=k_{nf}\left(\frac{\partial^2 T}{\partial x^2}+\frac{\partial^2 T}{\partial y^2}\right)-\frac{1}{\left(\rho C_p\right)_{nf}}\frac{\partial q_r}{\partial y},\left[q_r=-\frac{4\sigma_e}{3\beta_R}\frac{\partial T^4}{\partial y},T^4\cong 4T_c^3T-3T_c^4\right] \tag{116}$$

$\left(\rho C_p\right)_{nf},\left(\rho\beta\right)_{nf},\rho_{nf},k_{nf},\sigma_{nf}$ are:

$$\left(\rho C_p\right)_{nf}=\left(\rho C_p\right)_f(1-\phi)+\left(\rho C_p\right)_s\phi \tag{117}$$

$$\left(\rho\beta\right)_{nf}=\left(\rho\beta\right)_f(1-\phi)+\left(\rho\beta\right)_s\phi \tag{118}$$

$$\rho_{nf}=\rho_f(1-\phi)+\rho_s\phi \tag{119}$$

$$k_{nf}=k_f\left(\frac{k_s+2k_f+2\phi(k_s-k_f)}{k_s-\phi(k_s-k_f)+2k_f}\right) \tag{120}$$

$$\frac{\sigma_{nf}}{\sigma_f} = 1 + \frac{3\left(\frac{\sigma_s}{\sigma_f} - 1\right)\phi}{\left(\frac{\sigma_s}{\sigma_f} + 2\right) - \left(\frac{\sigma_s}{\sigma_f} - 1\right)\phi} \tag{121}$$

Here p is the pressure, B_x and B_y the components of magnetic field, q_r the radiative heat flux, T the temperature, σ_e the Stefan--Boltzman constant and β_e mean absorption coefficient, $\left(\rho_{nf}, k_{nf}, \sigma_{nf}\right)$ the (density, thermal conductivity, electrical conductivity) of nanofluid and nanofluid dynamic viscosity. B_x μ_{nf} is obtained as follows:

$$\mu_{nf} = \left(0.035B^2 + 3.1B - 27886.4807\phi^2 + 4263.02\phi + 316.0629\right)e^{-0.01T}, \tag{122}$$

where ϕ is volume fraction and subscript s and f are for solid and fluid.

Vorticity and stream function should be used to eliminate pressure source terms:

$$\omega + \frac{\partial u}{\partial y} - \frac{\partial v}{\partial x} = 0, \frac{\partial \psi}{\partial x} = -v, \frac{\partial \psi}{\partial y} = u \tag{123}$$

Introducing dimensionless quantities:

$$P = \frac{p}{\rho_f\left(\alpha_f / L\right)^2}, U = \frac{uL}{\alpha_f}, V = \frac{vL}{\alpha_f}, \Theta = \frac{T - T_c}{\Delta T}, \Delta T = q''L / k_f, \left(X, Y\right) = \frac{\left(x, y\right)}{L} \tag{124}$$

The final formulae are:

$$\frac{\partial V}{\partial Y} + \frac{\partial U}{\partial X} = 0, \tag{125}$$

$$U\frac{\partial U}{\partial X} + \frac{\partial U}{\partial Y}V = \Pr\left[\frac{\mu_{nf} / \mu_f}{\rho_{nf} / \rho_f}\right]\left(\frac{\partial^2 U}{\partial Y^2} + \frac{\partial^2 U}{\partial X^2}\right) - Ha^2 \Pr\left[\frac{\sigma_{nf} / \sigma_f}{\rho_{nf} / \rho_f}\right]\left(B_y^2 U - B_x B_y V\right) - \frac{\partial P}{\partial X}, \tag{126}$$

$$\begin{aligned} V\frac{\partial V}{\partial Y} + U\frac{\partial V}{\partial X} &= \Pr\left(\frac{\partial^2 V}{\partial Y^2} + \frac{\partial^2 V}{\partial X^2}\right)\left[\frac{\mu_{nf} / \mu_f}{\rho_{nf} / \rho_f}\right] \\ &-Ha^2 \Pr\left[\frac{\sigma_{nf} / \sigma_f}{\rho_{nf} / \rho_f}\right]\left(B_x^2 V - B_x B_y U\right) - \frac{\partial P}{\partial Y} + Ra \Pr\left[\frac{\beta_{nf}}{\beta_f}\right]\Theta, \end{aligned} \tag{127}$$

$$\frac{\partial \Theta}{\partial X}U + \frac{\partial \Theta}{\partial Y}V = \left(1 + \frac{4}{3}\left(\frac{k_{nf}}{k_f}\right)^{-1} Rd\right)\frac{\partial^2 \Theta}{\partial Y^2} + \left[\frac{\left(\rho C_P\right)_f}{\left(\rho C_P\right)_{nf}}\frac{k_{nf}}{k_f}\right]\left(\frac{\partial^2 \Theta}{\partial X^2}\right) \tag{128}$$

where dimensionless and constants parameters are defined as:

$$\Pr = \upsilon_f / \alpha_f, Ra = g\beta_f q'' L^4 / \left(k_f \upsilon_f \alpha_f\right), Ha = LB_0\sqrt{\sigma_f / \mu_f}, Rd = 4\sigma_e T_c^3 / \left(\beta_R k_f\right). \tag{129}$$

and boundary conditions are:

$$\frac{\partial \Theta}{\partial n} = 1.0 \text{ on inner wall}$$

$$\Theta = 0.0 \text{ on outer wall}$$

$$\Psi = 0.0 \text{ on all walls} \tag{130}$$

Local and average Nusselt over the hot cylinder can be calculated as:

$$Nu_{loc} = \left(1 + \frac{4}{3}Rd\left(\frac{k_{nf}}{k_f}\right)^{-1}\right)\frac{1}{\Theta}\left(\frac{k_{nf}}{k_f}\right) \tag{131}$$

$$Nu_{ave} = \frac{1}{S}\int_0^s Nu_{loc}\, ds \tag{132}$$

7.3. Effects of Active Parameters

CVFEM is utilized just to predict hydrodynamic of nanofluid behavior in existence of thermal radiation. The working fluid is Fe_3O_4-water and its viscosity varied with magnetic field. Outputs are presented via various values of volume fraction of Fe_3O_4-water ($\phi = 0$ to 0.04), radiation parameter ($Rd = 0$ to 0.8), inclination angle ($\xi = 0^\circ$ and 90°), Rayleigh number ($Ra = 10^3, 10^4$ and 10^5) and Hartmann number ($Ha = 0$ to 40).

Thermal radiation impact on nanofluid hydrothermal behavior is illustrated in Figure 37. Temperature enhances with rise of *Rd*. Also rising *Rd* causes the nanofluid motion to increase. This effect is more marked at higher Hartmann number. Influences of ξ, Ha and Ra on isotherms and streamlines are demonstrated in Figures 38 and 39. As nanofluid temperature increases, the nanofluid initiates moving from the inner cylinder to the cold one and dropping along the outer cylinder. At low Rayleigh number, conduction is more significant than convection. $\left|\Psi_{\max}\right|$ augments as buoyancy force increases and it reduces when Lorentz force enhances. One eddy appears in streamline and main eddy convert to two

Figure 37a. Effect of Rayleigh and Hartmann numbers on isotherms (left) and streamlines (right) contours when $\xi = 90^{\circ}, \phi = 0.04, Rd = 0.8$

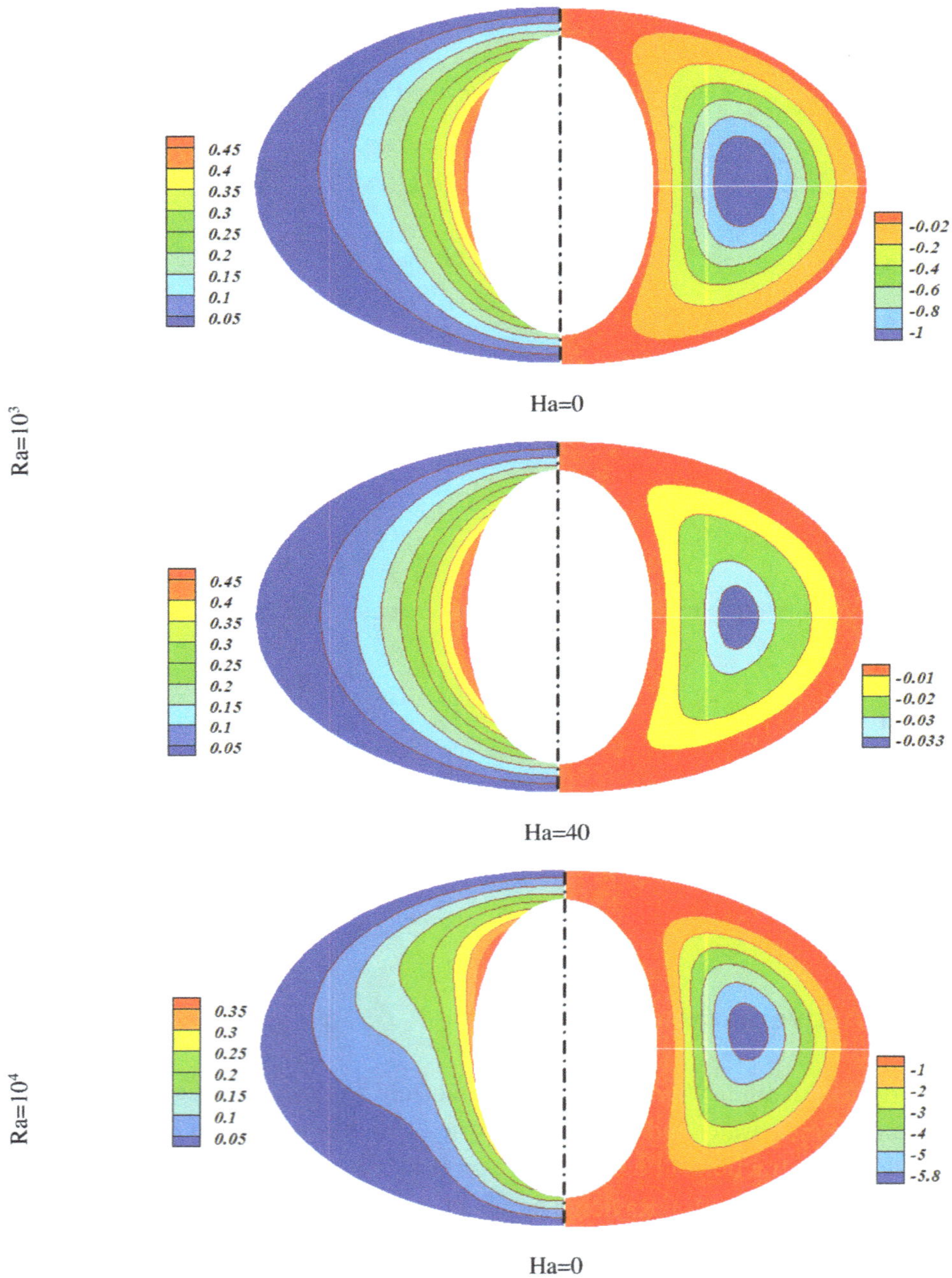

smaller eddies in existence of Lorentz forces. As ξ increases, the distortion of isotherms enhances. Thermal plume generates near the vertical centerline and Lorentz force causes it to diminish.

Figures 40, 41 and 42 show the impact of ξ, Rd, Ra and Ha on Nu_{loc}, Nu_{ave}. The formula for Nu_{ave} corresponding to active parameters is:

Figure 37b. Effect of Rayleigh and Hartmann numbers on isotherms (left) and streamlines (right) contours when $\xi = 90^{\circ}, \phi = 0.04, Rd = 0.8$

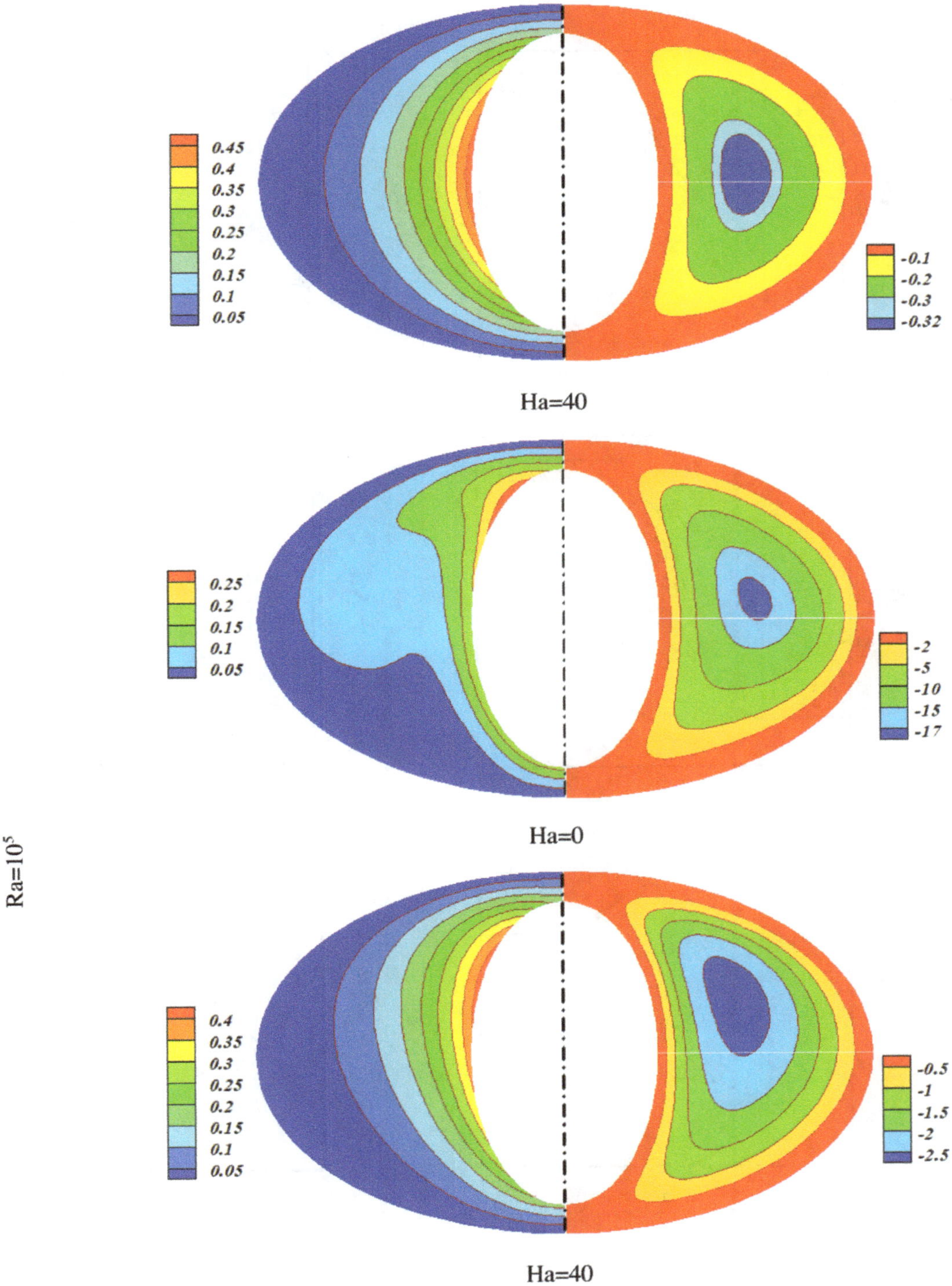

Figure 38. Effects of the radiation parameter, Hartmann and Rayleigh numbers on local Nusselt number when $\phi = 0.04, \xi = 0^{\circ}$

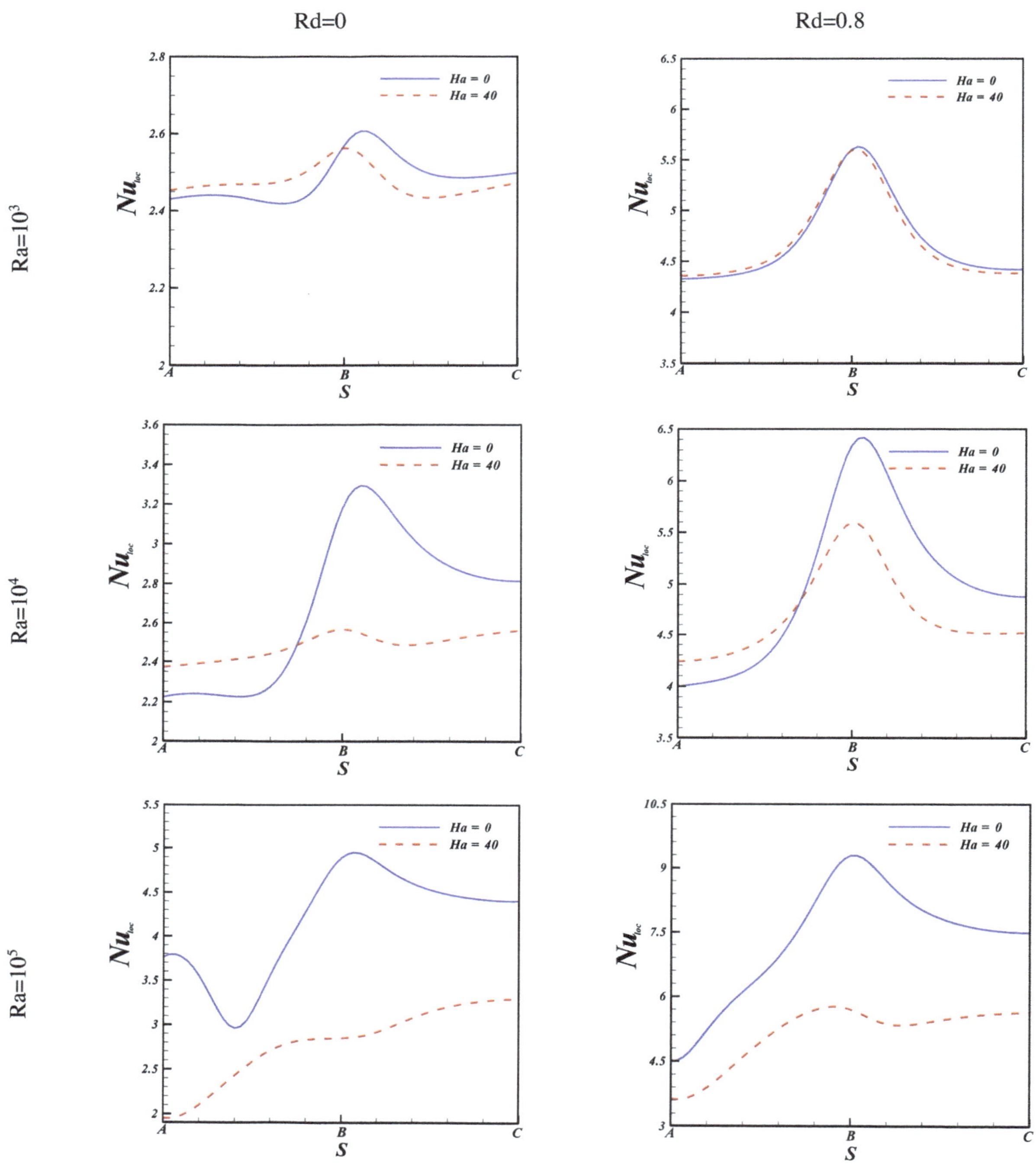

Figure 39. Effects of the radiation parameter, Hartmann and Rayleigh numbers on local Nusselt number when $\phi = 0.04, \xi = 90^{\circ}$

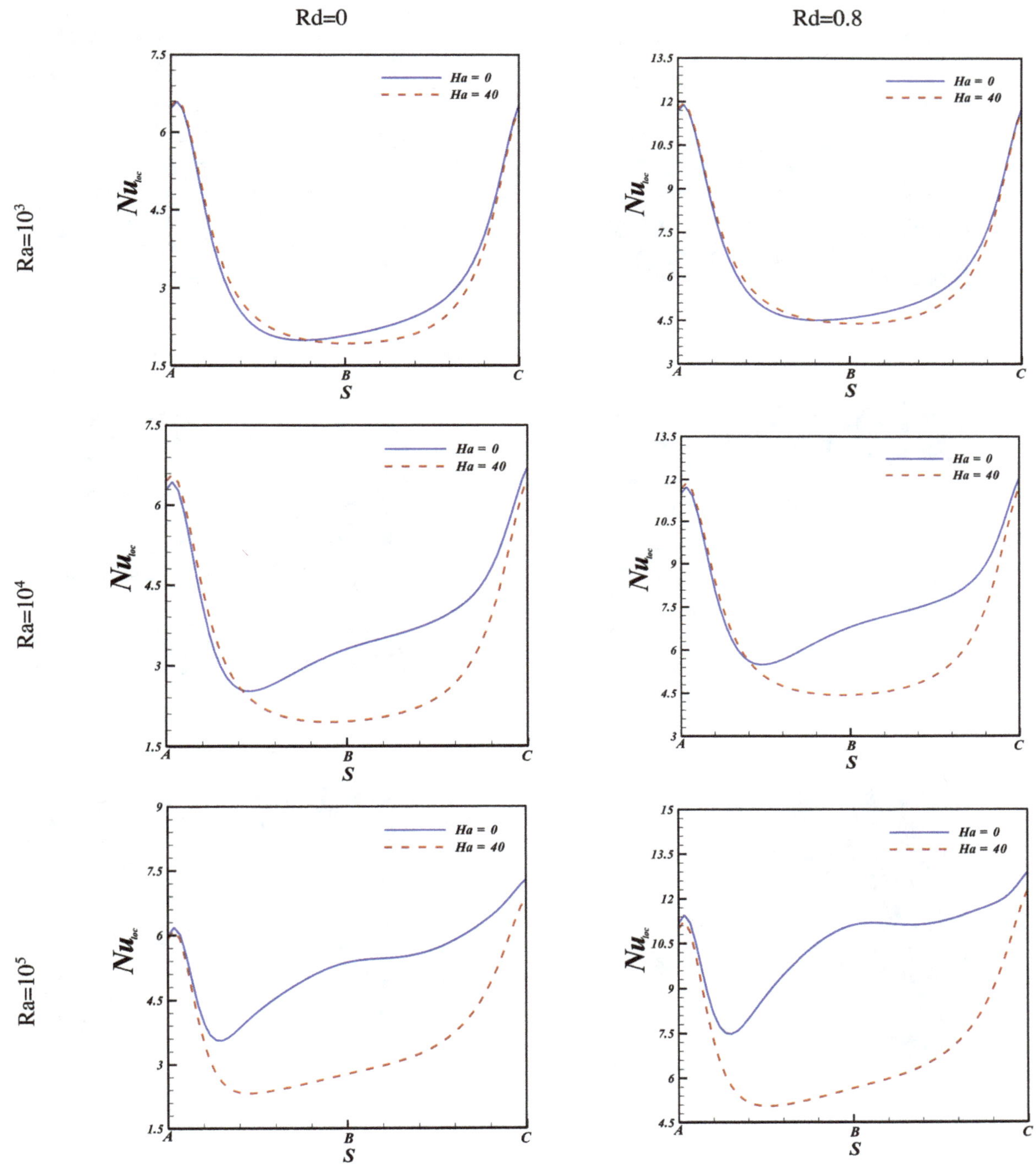

Figure 40a. Effects of inclination angle, radiation parameter, Rayleigh and Hartmann numbers on average Nusselt number

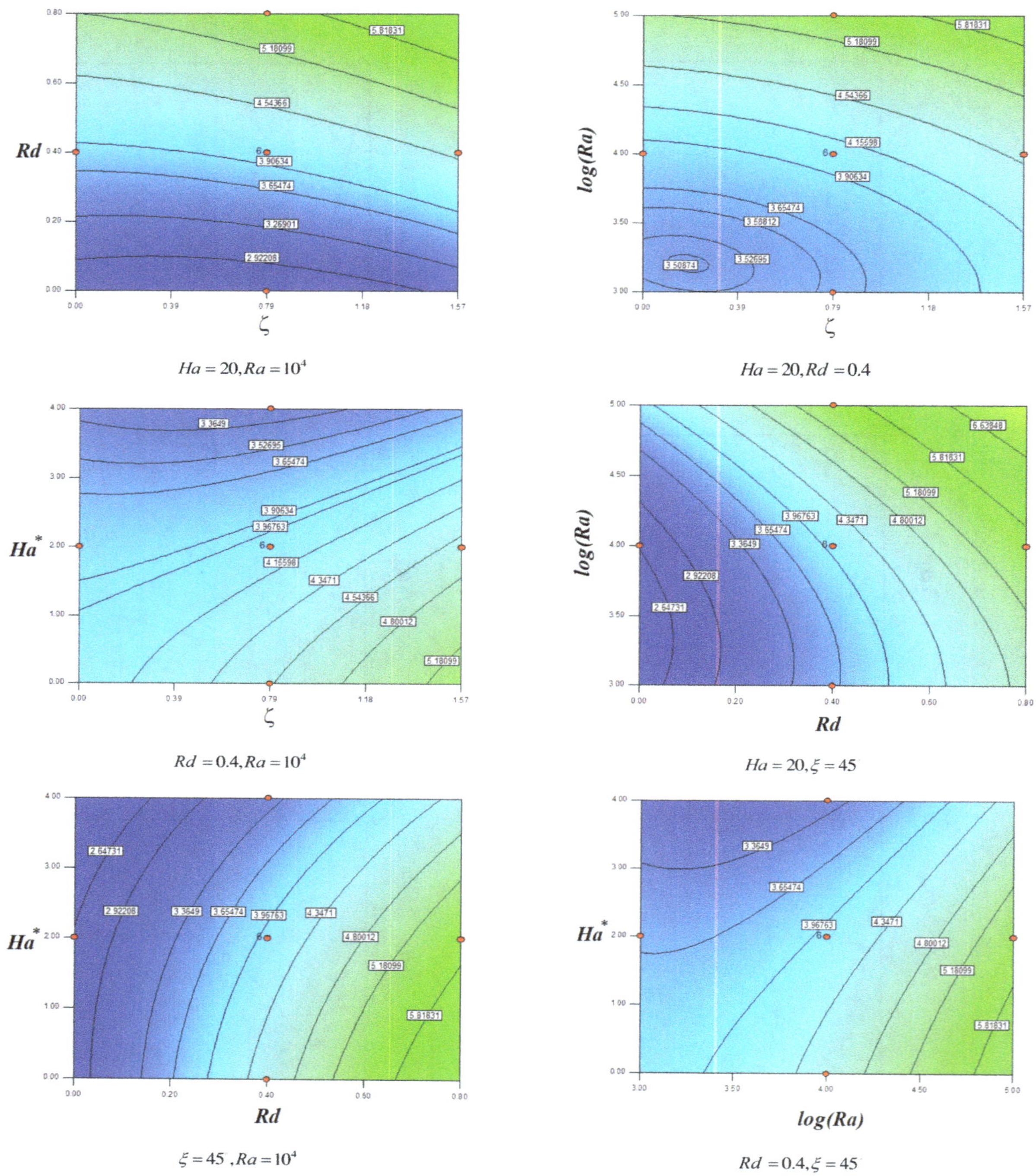

Figure 40b. Effects of inclination angle, radiation parameter, Rayleigh and Hartmann numbers on average Nusselt number

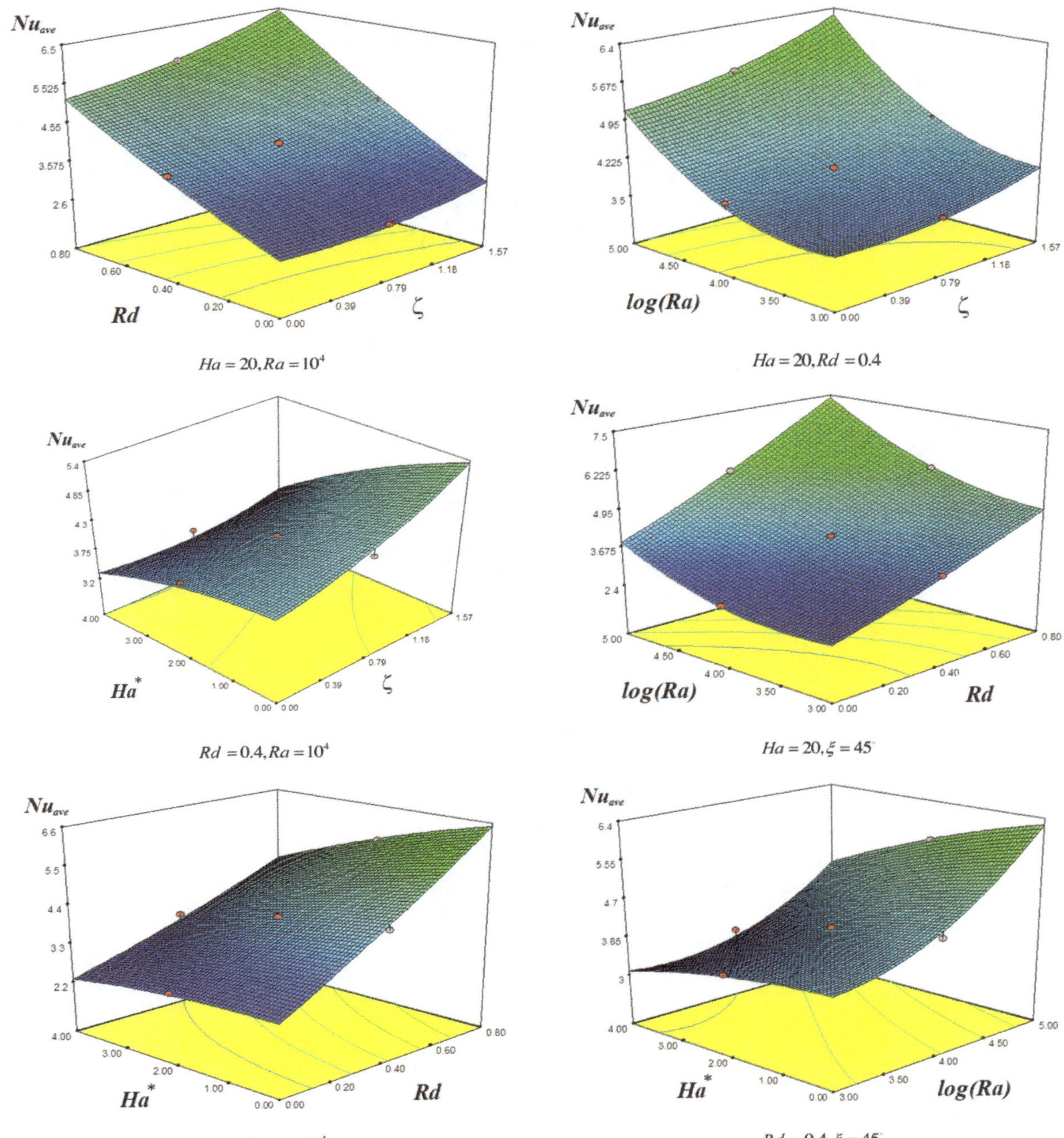

Figure 41. (a)Geometry and the boundary conditions with; (b) A sample triangular element and its corresponding control volume.

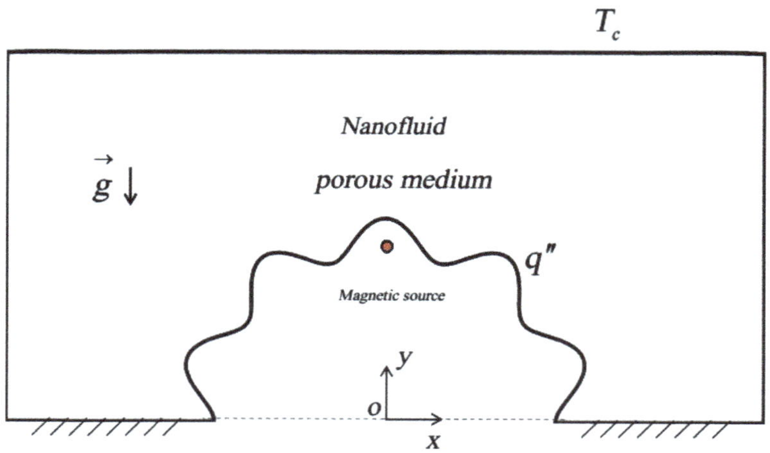

(a)

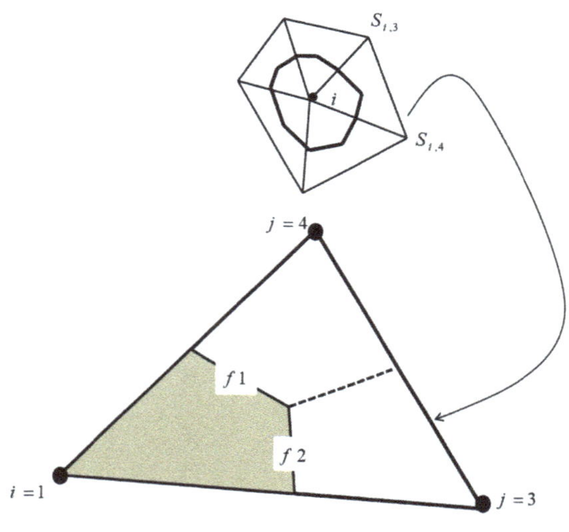

(b)

$$\begin{aligned} Nu_{ave} &= 7.98 - 0.73\xi + 0.62Rd - 3.47\log\left(Ra\right) + 0.62Ha^{*} + 0.77\xi\, Rd \\ &+0.19\xi\log\left(Ra\right) - 0.15\xi Ha^{*} + 0.72\log\left(Ra\right)Rd - 0.12\log\left(Ra\right)Ha^{*} \\ &-0.47Rd\,Ha^{*} + 0.29\xi^{2} + 0.52\left(\log\left(Ra\right)\right)^{2} + 0.69Rd^{2} - 0.034Ha^{*2} \end{aligned} \tag{133}$$

where $Ha^{*} = 0.1Ha$. The number of extermum in in Nu_{loc} profile is matching to existence of thermal plume. Increase in radiation parameter causes Nu_{ave} to enhance. As Ra and ξ augment temperature

Figure 42. Contours of the (a) magnetic field strength H; (b) magnetic field intensity component in x direction Hx; (c) magnetic field intensity component in y direction Hy.

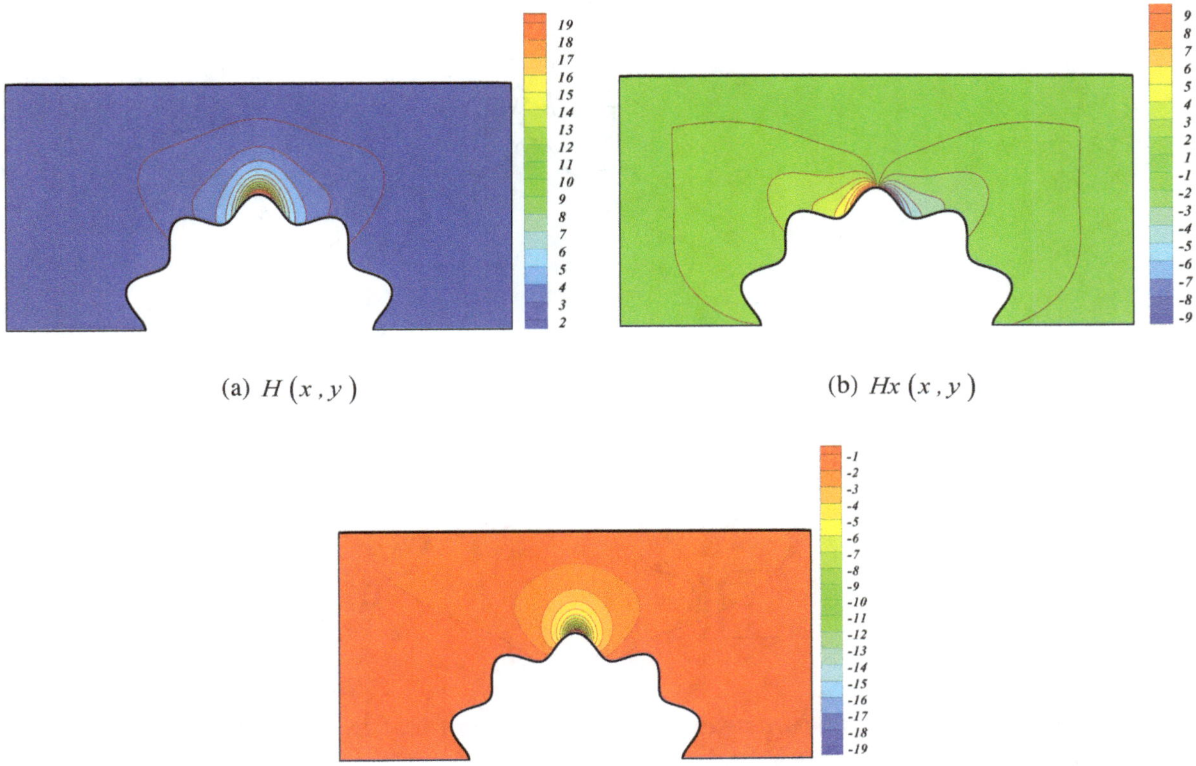

(a) $H(x,y)$ (b) $Hx(x,y)$

(c) $Hy(x,y)$

of inner surface decreases. So Nu_{ave} improves with augment of Ra, ξ. Also, this figure depicts that Nusselt number reduces with rise of Lorentz forces.

8. NUMERICAL MODELING FOR FE3O4: WATER NANOFLUID FLOW IN POROUS MEDIUM CONSIDERING MFD VISCOSITY

8.1. Problem Definition

Geometry, boundary condition and sample element are illustrated (see Figure 43). External magnetic source is applied (see Figure 44). $\overline{H}, \overline{H_x}, \overline{H_y}$ are:

$$\overline{H_y} = \left(\bar{a} - x\right)\frac{\gamma}{2\pi}\left[\left(\bar{b} - y\right)^2 + \left(\bar{a} - x\right)^2\right]^{-1}, \tag{134}$$

Figure 43. Impact of nanofluid volume fraction on streamlines (up) and isotherms (bottom) contours (nanofluid ($\phi = 0.04$)(—) and pure fluid ($\phi = 0$) ($- \cdot -$)) when $Ra = 10^5, Da = 100, Rd = 0.8$

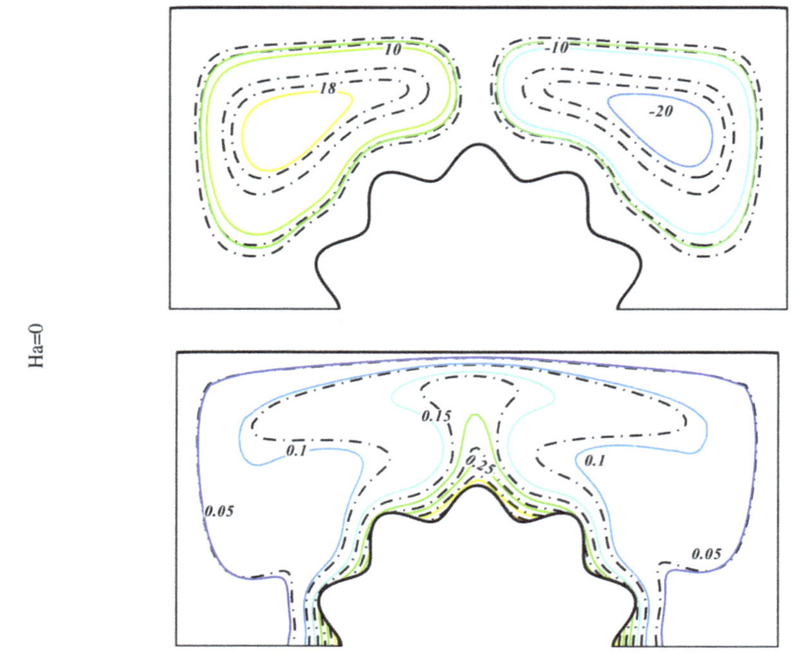

Figure 44. Impact of radiation parameter on streamlines (up) and isotherms (bottom) contours ($Rd = 0.8$ (—), $Rd = 0$ (- - -)) when $Ra = 10^5, Da = 100, \phi = 0.04$

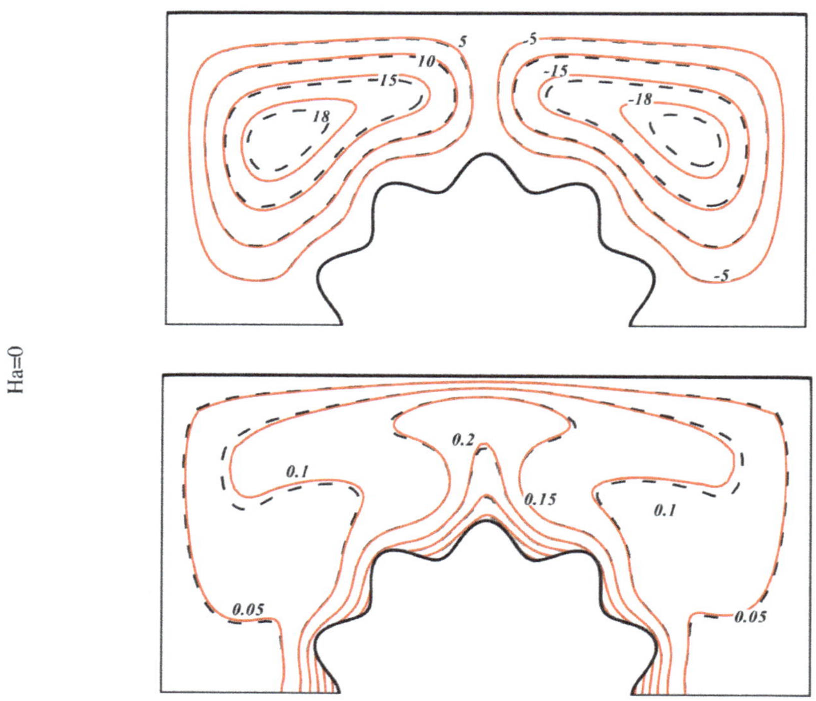

$$\overline{H}_x = \left(y - \bar{b}\right)\frac{\gamma}{2\pi}\left[\left(\bar{b} - y\right)^2 + \left(\bar{a} - x\right)^2\right]^{-1}, \tag{135}$$

$$\overline{H} = \left(\overline{H}_y^2 + \overline{H}_x^2\right)^{0.5}. \tag{136}$$

8.2. Governing Equation

2D steady convective non-Darcy flow of nanofluid is considered in existence of non-uniform magnetic field. The PDEs are:

$$\frac{\partial u}{\partial x} = -\frac{\partial v}{\partial y}, \tag{137}$$

$$\left(\frac{\partial^2 u}{\partial y^2} + \frac{\partial^2 u}{\partial x^2}\right)\mu_{nf} - \frac{\partial P}{\partial x} - \mu_0^2 \sigma_{nf} \overline{H}_y^2 u + \sigma_{nf}\mu_0^2 \overline{H}_x \overline{H}_y v - \frac{\mu_{nf}}{K} u = \left(\rho_{nf}\right)\left(\frac{\partial u}{\partial x} u + \frac{\partial u}{\partial y} v\right), \tag{138}$$

$$\begin{aligned}&\mu_{nf}\left(\frac{\partial^2 v}{\partial x^2} + \frac{\partial^2 v}{\partial y^2}\right) - \frac{\partial P}{\partial y} + \mu_0^2 \overline{H}_y \sigma_{nf} \overline{H}_x u - \mu_0^2 \overline{H}_x \sigma_{nf} \overline{H}_x v - \frac{\mu_{nf}}{K} v \\ &+ \left(T - T_c\right)\beta_{nf} g \rho_{nf} = \rho_{nf}\left(\frac{\partial v}{\partial x} u + \frac{\partial v}{\partial y} v\right),\end{aligned} \tag{139}$$

$$k_{nf}\left(\frac{\partial^2 T}{\partial y^2} + \frac{\partial^2 T}{\partial x^2}\right) - \frac{\partial q_r}{\partial y} = \left(v\frac{\partial T}{\partial y} + u\frac{\partial T}{\partial x}\right)\left(\rho C_p\right)_{nf}, \left[q_r = -\frac{4\sigma_e}{3\beta_R}\frac{\partial T^4}{\partial y}, T^4 \cong 4T_c^3 T - 3T_c^4\right]. \tag{140}$$

$\left(\rho C_p\right)_{nf}, \left(\rho\beta\right)_{nf}, \rho_{nf}$ and σ_{nf} are calculated as:

$$\left(\rho C_p\right)_{nf} = \left(\rho C_p\right)_f (1 - \phi) + \left(\rho C_p\right)_s \phi \tag{141}$$

$$\rho_{nf} = \rho_f (1 - \phi) + \rho_s \phi \tag{142}$$

$$\left(\rho\beta\right)_{nf} = \left(\rho\beta\right)_{f}(1-\phi) + \left(\rho\beta\right)_{s}\phi, \tag{143}$$

$$\sigma_{nf} = \sigma_{f}\left[\frac{3\phi\left(\sigma 1-1\right)}{\left(1-\sigma 1\right)\phi+\left(2+\sigma 1\right)}+1\right], \quad \sigma 1 = \sigma_{s} / \sigma_{f}. \tag{144}$$

μ_{nf} is obtained as follows:

$$\mu_{nf} = \left(0.035\mu_0^2 H^2 + 3.1\mu_0 H - 27886.4807\phi^2 + 4263.02\phi + 316.0629\right)e^{-0.01T} \quad 145$$

k_{nf} can be expressed as:

$$\frac{k_{nf}}{k_f} = \frac{-m\left(k_f - k_p\right)\phi + \left(k_p - k_f\right)\phi + mk_f + k_p + k_f}{mk_f + \left(k_f - k_p\right)\phi + k_f + k_p} \tag{146}$$

Properties of nanofluid are depicted in Table 3.

Different values of shape factors for various shapes of nanoparticles are illustrated in Table 4. Vorticity and stream function should be used to eliminate pressure source terms:

$$\omega + \frac{\partial u}{\partial y} - \frac{\partial v}{\partial x} = 0, \frac{\partial \psi}{\partial x} = -v, \frac{\partial \psi}{\partial y} = u. \tag{147}$$

Dimensionless parameters are defined as:

$$\begin{gathered}\left(H_y, H_x, H\right) = \frac{\left(\overline{H_y}, \overline{H_x}, \overline{H}\right)}{\overline{H_0}}, \left(b, a\right) = \frac{\left(\bar{b}, \bar{a}\right)}{L}, \Psi = \frac{\psi}{\alpha_{nf}}, \Omega = \frac{\omega L^2}{\alpha_{nf}}, \\ U = \frac{uL}{\alpha_{nf}}, V = \frac{vL}{\alpha_{nf}}, \left(X, Y\right) = \frac{\left(x, y\right)}{L}, \theta = \frac{T - T_c}{\Delta T}, \Delta T = q'' L / k_f.\end{gathered} \tag{148}$$

So equations change to:

$$\frac{\partial^2 \Psi}{\partial Y^2} + \Omega + \frac{\partial^2 \Psi}{\partial X^2} = 0, \tag{149}$$

$$\frac{\partial \Omega}{\partial X}U + V\frac{\partial \Omega}{\partial Y} = \Pr \frac{A_5}{A_1}\frac{A_2}{A_4}\left(\frac{\partial^2 \Omega}{\partial Y^2} + \frac{\partial^2 \Omega}{\partial X^2}\right)$$
$$+\Pr Ha^2 \frac{A_6}{A_1}\frac{A_2}{A_4}\left(\frac{\partial U}{\partial X}H_y H_x - \frac{\partial V}{\partial X}H_x^{\ 2} + \frac{\partial U}{\partial Y}H_y^2 - \frac{\partial V}{\partial Y}H_y H_x\right) \quad (150)$$
$$+\Pr Ra \frac{A_3 A_2^2}{A_1 A_4^2}\frac{\partial \theta}{\partial X} - \frac{\Pr}{Da}\frac{A_5}{A_1}\frac{A_2}{A_4},$$

$$\frac{\partial \theta}{\partial X}U + V\frac{\partial \theta}{\partial Y} = \left(\frac{\partial^2 \theta}{\partial X^2} + \frac{\partial^2 \theta}{\partial Y^2}\right) + \frac{4}{3}\frac{1}{A_4}Rd\frac{\partial^2 \theta}{\partial Y^2}. \quad (151)$$

and dimensionless parameters are

$$Ra_f = g\beta_f L^3 \Delta T / \left(\alpha_f \upsilon_f\right), \Pr_f = \upsilon_f / \alpha_f, Ha = L\mu_0 H_0 \sqrt{\sigma_f / \mu_f},$$
$$Da = K / L^2, Rd = 4\sigma_e T_c^3 / \left(\beta_R k_f\right), A_1 = \frac{\rho_{nf}}{\rho_f}, A_2 = \frac{\left(\rho C_p\right)_{nf}}{\left(\rho C_p\right)_f}, A_3 = \frac{\left(\rho\beta\right)_{nf}}{\left(\rho\beta\right)_f}, \quad (152)$$
$$A_4 = \frac{k_{nf}}{k_f}, A_5 = \frac{\mu_{nf}}{\mu_f}, A_6 = \frac{\sigma_{nf}}{\sigma_f}$$

and boundary conditions are:

on inner wall $\frac{\partial \theta}{\partial n} = 1.0$

on outer wall $\theta = 0.0$

on other walls $\frac{\partial \theta}{\partial n} = 0$

on all walls $\Psi = 0.0$ (153)

Nu_{loc}, Nu_{ave} over the hot wall can calculate as:

$$Nu_{loc} = \left(\frac{k_{nf}}{k_f}\right)\left(1 + \left(\frac{k_{nf}}{k_f}\right)^{-1}\frac{4Rd}{3}\right)\frac{1}{\theta}, \quad (154)$$

$$Nu_{ave} = \frac{1}{S}\int_{0}^{S} Nu_{loc}\, ds. \tag{155}$$

8.3. Effects of Active Parameters

Non uniform magnetic field influence on Fe_3O_4-water flow in a porous enclosure is simulated. Shape factor effect on k_{nf} is estimated according to previous experimental data. is taken into consideration. CVFEM is utilized to find the effects of radiation parameter $\left(Rd = 0\, to\, 0.8\right)$, Darcy number ($Da = 0.01$ to 100), Rayleigh number ($Ra = 10^3, 10^4, 10^5$), volume fraction of Fe_3O_4 ($\phi = 0\%$ to 4%), shape of nanoparticle and Hartmann number ($Ha = 0$ to 10).

Effect of shape factor on Nu_{ave} is illustrated in Table 6. According to this table, selecting Platelet shaped nanoparticles leads to maximum Nusselt number. So, Platelet nanoparticle has been selected for further investigation. Figure 43 depicts the influence of adding nanoparticles in to water on velocity and temperature fields. Temperature gradient decreases with augment of ϕ. Velocity of the fluid increases with increase of volume fraction of nanofluid. Figure 44 reveals the influence of radiation parameter on velocity and temperature fields. Thermal boundary layer thickness increases with increase of Rd. This parameter is more pronounced in absence of magnetic field. Figures 45, 46, and 47 illustrate the impact of Da, Ra, Ha on streamlines and isotherms. When conduction is prominent, just one vortex occurs in streamline. As buoyancy force enhances, thermal plume generates near the vertical symmetric line. As Darcy number increases, convective mode becomes stronger and the primary vortex converts to two vortexes. As Lorentz forces increases, the velocity of nanofluid decreases and the main vortex moves downward. Also convective heat transfer becomes weak by adding magnetic field. Increasing Ha, shifts the thermal plume to left side.

Effect of significant parameters on Nu_{ave} is demonstrated in Figure 48. The correlation for Nu_{ave} is as follows:

$$\begin{aligned} Nu_{ave} &= 21.6 - 3.79Rd - 10.44\, og\left(Ra\right) - 0.41Da^{*} + 1.02Ha^{*} + 2.14Rd\log\left(Ra\right) \\ &+0.33Rd\, Da^{*} - 0.44Rd\, Ha^{*} + 0.22Da^{*}\log\left(Ra\right) - 0.29Ha^{*}\log\left(Ra\right) \\ &-0.72Da^{*}Ha^{*} - 2.98Rd^{2} + 1.39\left(\log\left(Ra\right)\right)^{2} + 0.45\left(Da^{*}\right)^{2} - 0.23\left(Ha^{*}\right)^{2} \end{aligned} \tag{156}$$

Table 6. Effect of shape of nanoparticles on Nusselt number when $Da = 100, Ra = 10^5, Rd = 0.8, \phi = 0.04$

	Ha	
	0	**10**
Spherical	9.40168	7.463479
Brick	9.469653	7.504079
Cylinder	9.575843	7.568335
Platelet	9.662061	7.621197

Figure 45. Influence of Da, Ha on streamlines (right) and isotherms (left) contours when $\phi = 0.04, Ra = 10^3$

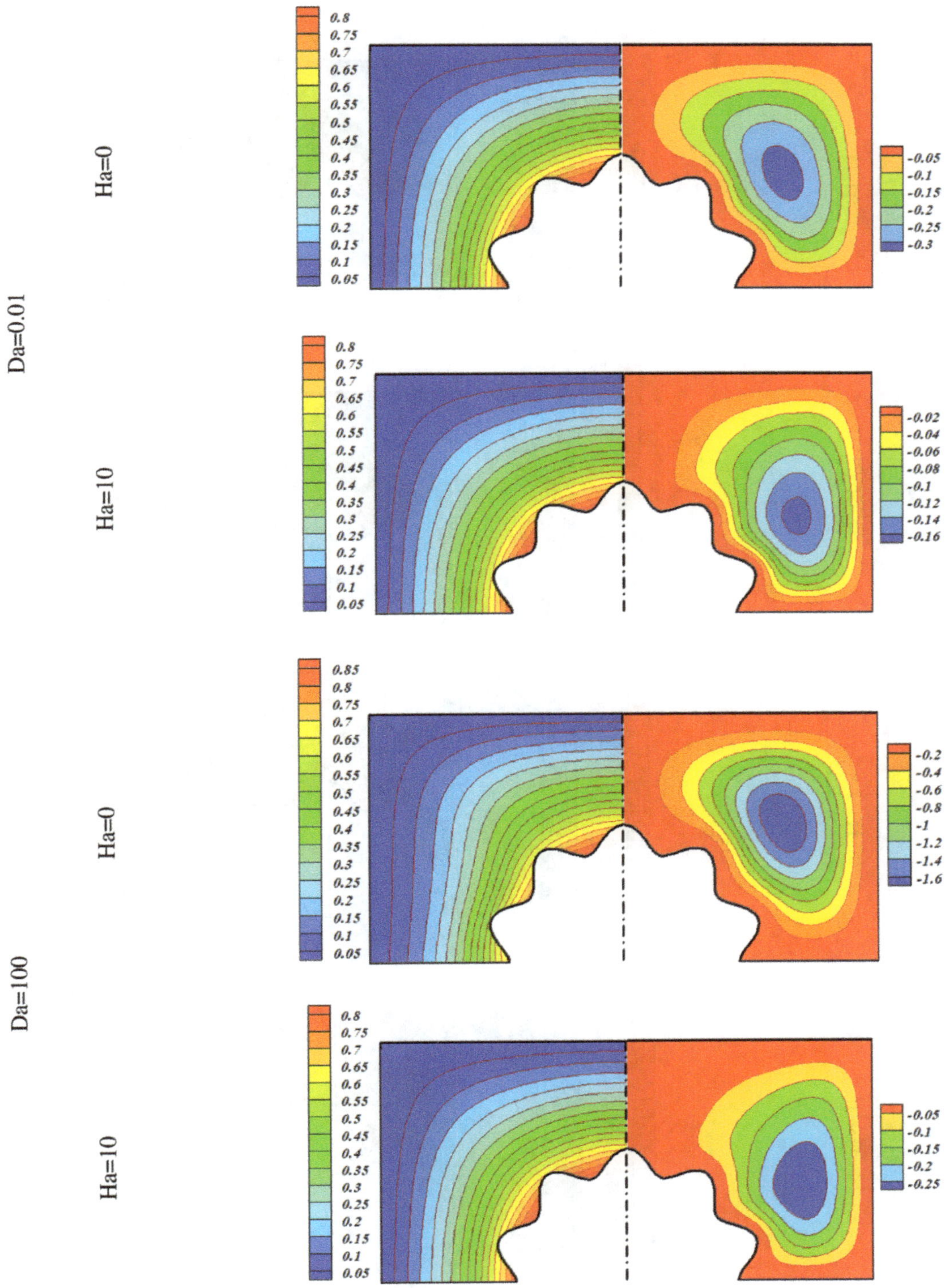

Figure 46. Influence of Da, Ha on streamlines (right) and isotherms (left) contours when $\phi = 0.04, Ra = 10^4$

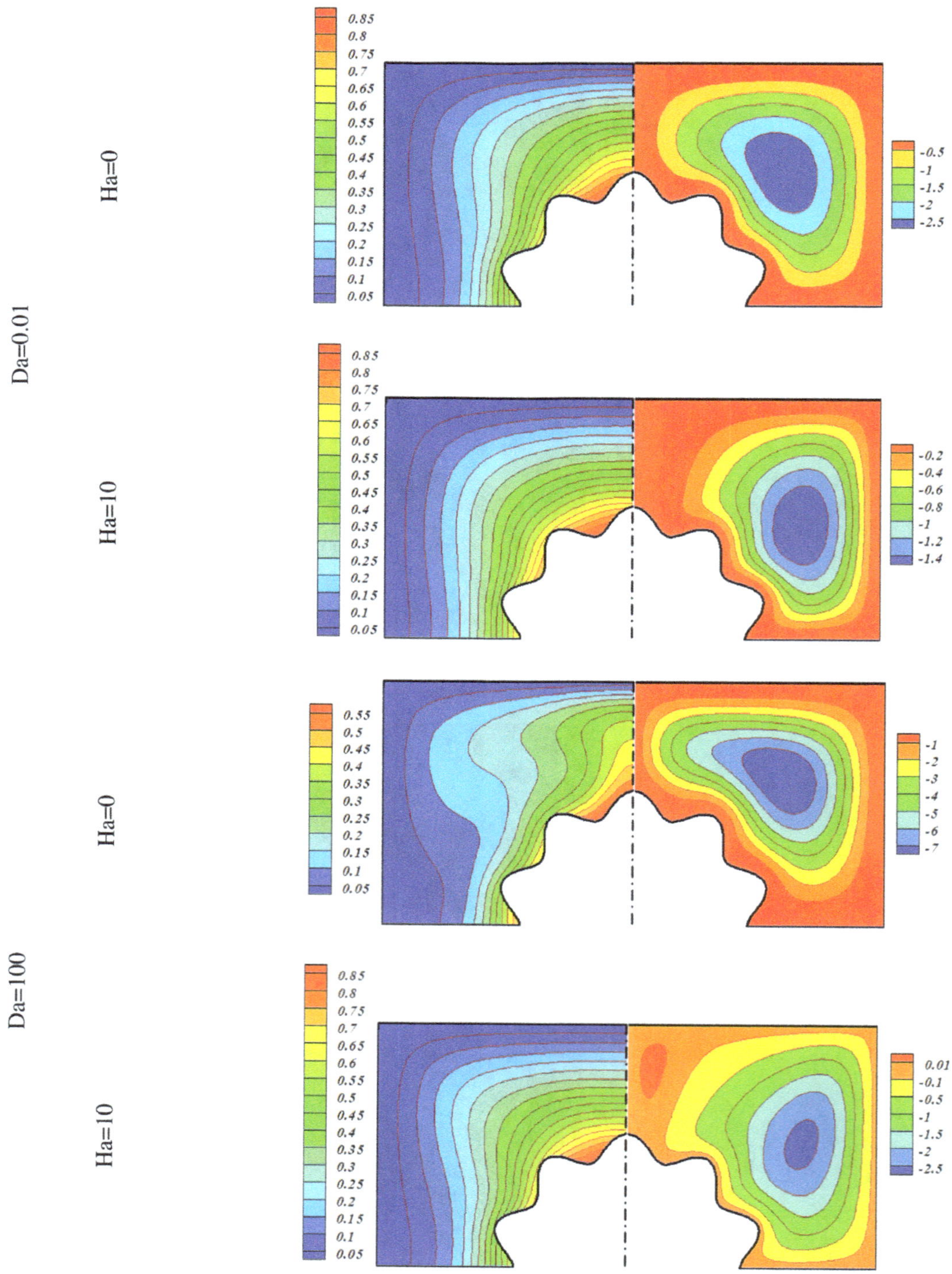

Figure 47. Influence of Da, Ha *on streamlines (right) and isotherms (left) contours when* $\phi = 0.04, Ra = 10^4$

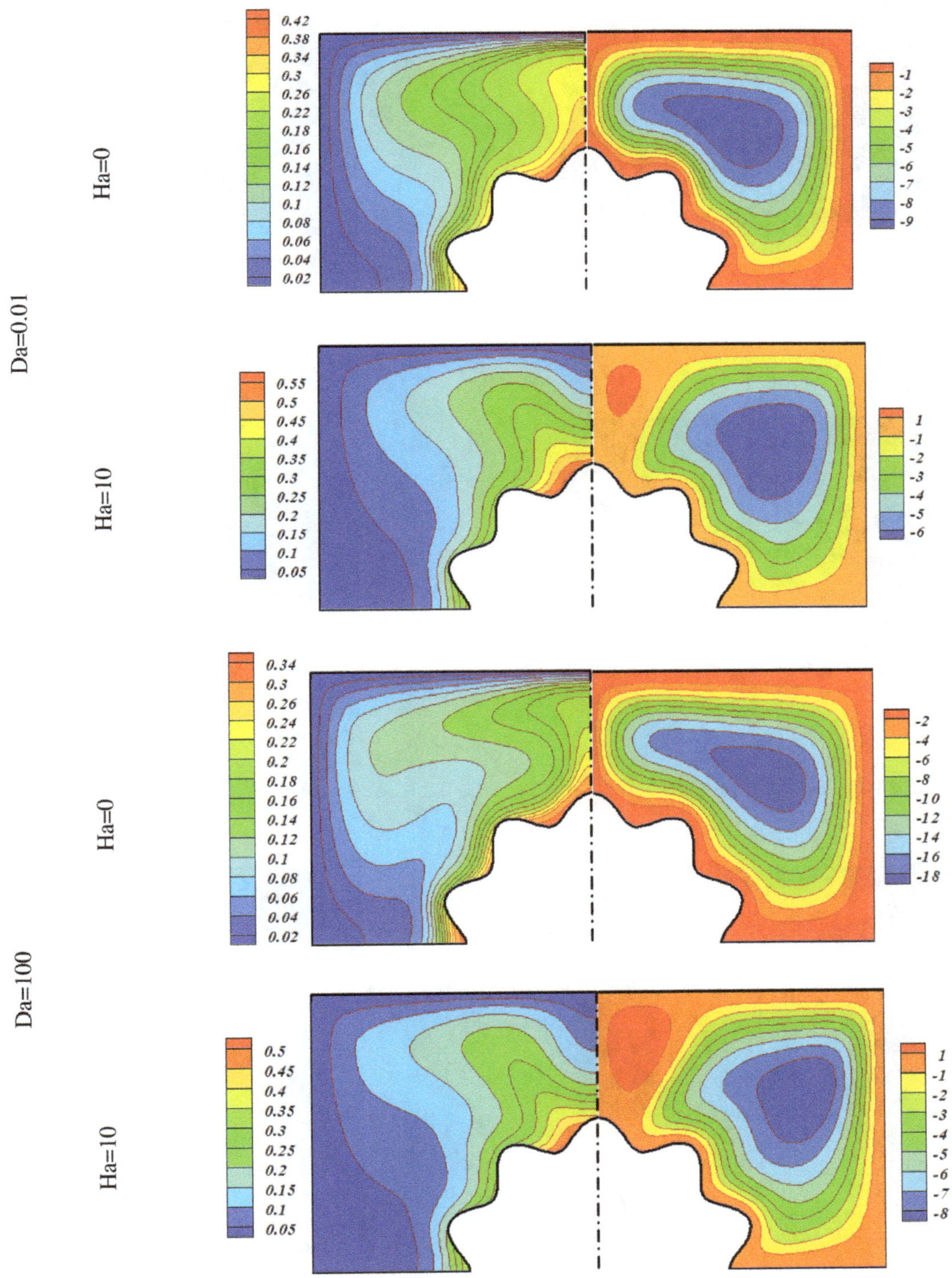

Figure 48a. Effects of Da, Ha, Rd and Ra on average Nusselt number

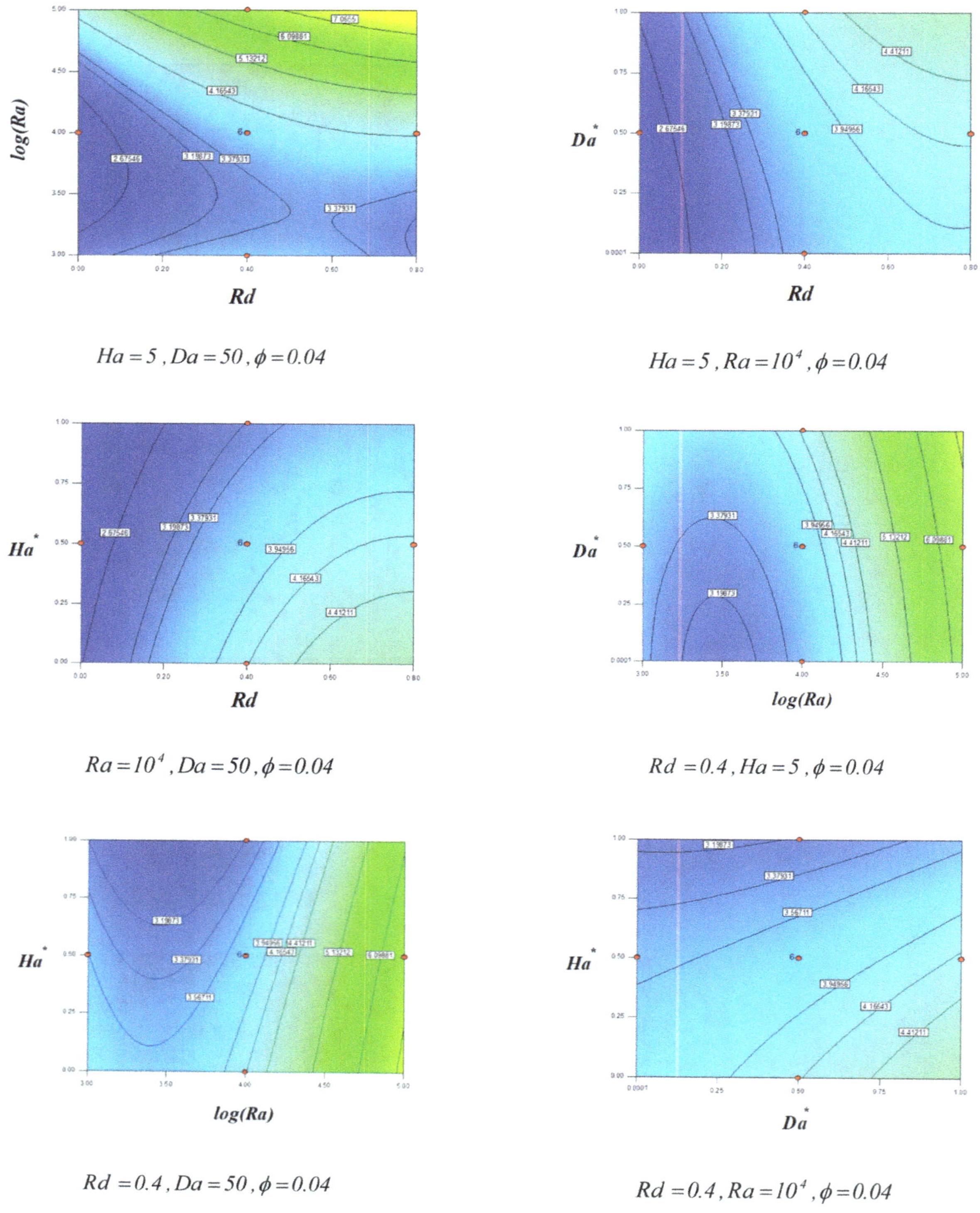

Figure 48b. Effects of Da, Ha, Rd and Ra on average Nusselt number

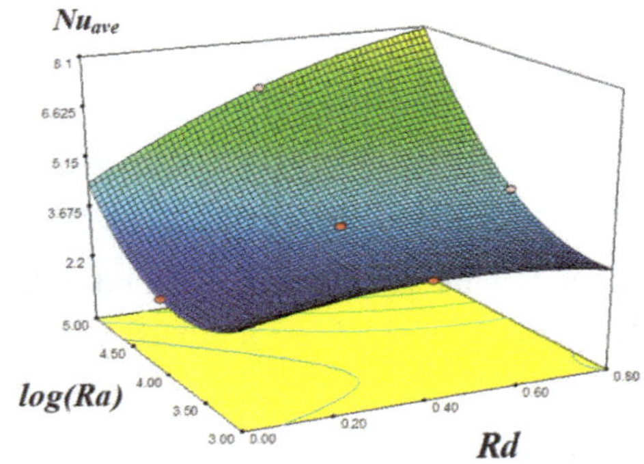

$Ha=5\,, Da=50\,, \phi=0.04$

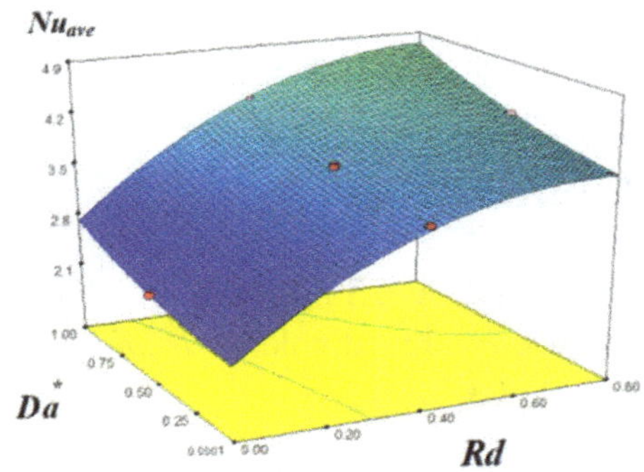

$Ha=5\,, Ra=10^4\,, \phi=0.04$

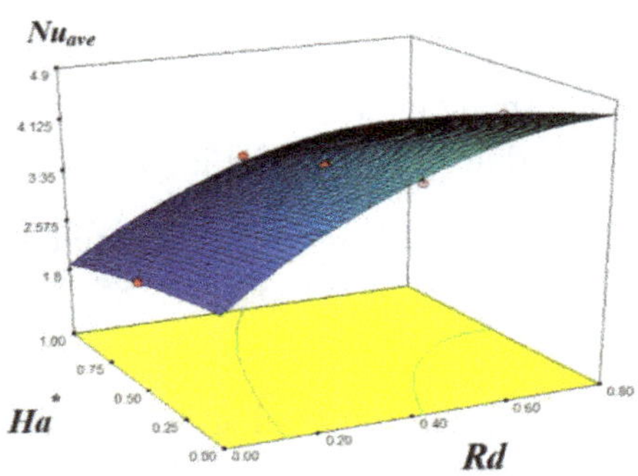

$Ra=10^4\,, Da=50\,, \phi=0.04$

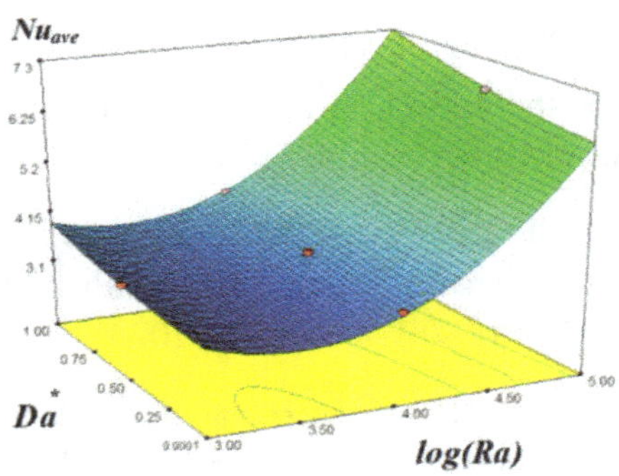

$Rd=0.4\,, Ha=5\,, \phi=0.04$

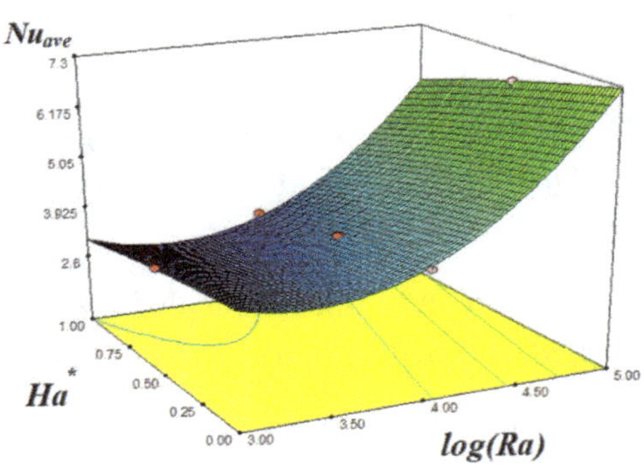

$Rd=0.4\,, Da=50\,, \phi=0.04$

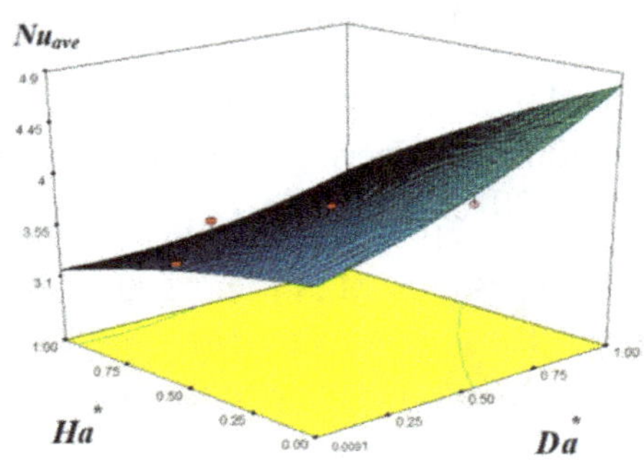

$Rd=0.4\,, Ra=10^4\,, \phi=0.04$

where $Da^{*} = 0.01Da, Ha^{*} = 0.1Ha$. Heat transfer rate enhances with rise of permeability of porous media and Rayleigh number due to increase in convective mode. Nusselt number reduces with augment of Lorentz forces.

REFERENCES

Ganguly, R., Sen, S., & Puri, I. K. (2004). Heat transfer augmentation using a magnetic fluid under the influence of a line dipole. *Journal of Magnetism and Magnetic Materials*, *271*(1), 63–73. doi:10.1016/j.jmmm.2003.09.015

Nanjundappa, C. E., Shivakumara, I. S., & Ravisha, M. (2010). The onset of ferroconvection in a horizontal saturated porous layer heated from below and cooled from above with constant heat flux subject to MFD viscosity. *International Communications in Heat and Mass Transfer*, *37*(9), 1246–1250. doi:10.1016/j.icheatmasstransfer.2010.08.011

Rosensweig, R. E. (1985). *Ferrohydrodynamics*. Cambridge, UK: Cambridge University Press.

Sheikholeslami, M. (2017). Magnetic field influence on nanofluid thermal radiation in a cavity with tilted elliptic inner cylinder. *Journal of Molecular Liquids*, *229*, 137–147. doi:10.1016/j.molliq.2016.12.024

Sheikholeslami, M., & Abelman, S. (2018). Numerical study of the effect of magnetic field on Fe3O4-water ferrofluid convection with thermal radiation. *Engineering Computations*, *35*(5), 1855–1872. doi:10.1108/EC-06-2017-0202

Sheikholeslami, M., Hayat, T., & Alsaedi, A. (2017). On simulation of nanofluid radiation and natural convection in an enclosure with elliptical cylinders. *International Journal of Heat and Mass Transfer*, *115*, 981–991. doi:10.1016/j.ijheatmasstransfer.2017.07.119

Sheikholeslami, M., & Houman, B. (2017). Rokni, Magnetic nanofluid natural convection in presence of thermal radiation considering variable viscosity. *The European Physical Journal Plus*, *132*(5), 238. doi:10.1140/epjp/i2017-11498-4

Sheikholeslami, M., Rashidi, M. M., Hayat, T., & Ganji, D. D. (2016). Free convection of magnetic nanofluid considering MFD viscosity effect. *Journal of Molecular Liquids*, *218*, 393–399. doi:10.1016/j.molliq.2016.02.093

Sheikholeslami, M., & Shehzad, S. A. (2017). Thermal radiation of ferrofluid in existence of Lorentz forces considering variable viscosity. *International Journal of Heat and Mass Transfer*, *109*, 82–92. doi:10.1016/j.ijheatmasstransfer.2017.01.096

Sheikholeslami, M. K. (2017). Sadoughi, Numerical modeling for Fe3O4 -water nanofluid flow in porous medium considering MFD viscosity. *Journal of Molecular Liquids*, *242*, 255–264. doi:10.1016/j.molliq.2017.07.004

Shliomis, M. I. (1972). Effect of viscosity of magnetic suspensions. *Soviet Phys. FETP*, *34*, 1291–1294.

Vaidyanathan, G., Sekar, R., & Ramanathan, A. (2002). Effect of magnetic field dependent viscosity on ferroconvection in rotating medium. *Indian Journal of Pure and Applied Physics*, *40*, 159–165.

ADDITIONAL READING

Jafaryar, M., Sheikholeslami, M., Li, M., & Moradi, R. (2018). Nanofluid turbulent flow in a pipe under the effect of twisted tape with alternate axis. *Journal of Thermal Analysis and Calorimetry*. doi:10.100710973-018-7093-2

Li, Z., Shehzad, S. A., & Sheikholeslami, M. (2018). An application of CVFEM for nanofluid heat transfer intensification in a porous sinusoidal cavity considering thermal non-equilibrium model. *Computer Methods in Applied Mechanics and Engineering*, *339*, 663–680. doi:10.1016/j.cma.2018.05.015

Li, Z., Sheikholeslami, M., Chamkha, A. J., Raizah, Z. A., & Saleem, S. (2018). Control Volume Finite Element Method for nanofluid MHD natural convective flow inside a sinusoidal annulus under the impact of thermal radiation. *Computer Methods in Applied Mechanics and Engineering*, *338*, 618–633. doi:10.1016/j.cma.2018.04.023

Li, Z., Sheikholeslami, M., Jafaryar, M., Shafee, A., & Chamkha, A. J. (2018). Investigation of nanofluid entropy generation in a heat exchanger with helical twisted tapes. *Journal of Molecular Liquids*, *266*, 797–805. doi:10.1016/j.molliq.2018.07.009

Li, Z., Sheikholeslami, M., Samandari, M., & Shafee, A. (2018). Nanofluid unsteady heat transfer in a porous energy storage enclosure in existence of Lorentz forces. *International Journal of Heat and Mass Transfer*, *127*, 914–926. doi:10.1016/j.ijheatmasstransfer.2018.06.101

Sheikholeslami, M. (2017a). Lattice Boltzmann Method simulation of MHD non-Darcy nanofluid free convection. *Physica B, Condensed Matter*, *516*, 55–71. doi:10.1016/j.physb.2017.04.029

Sheikholeslami, M. (2017b). CuO-water nanofluid free convection in a porous cavity considering Darcy law. *The European Physical Journal Plus*, *132*(1), 55. doi:10.1140/epjp/i2017-11330-3

Sheikholeslami, M. (2017c). Influence of Lorentz forces on nanofluid flow in a porous cavity by means of Non- Darcy model. *Engineering Computations*, *34*(8), 2651–2667. doi:10.1108/EC-01-2017-0008

Sheikholeslami, M. (2017d). Magnetic field influence on nanofluid thermal radiation in a cavity with tilted elliptic inner cylinder. *Journal of Molecular Liquids*, *229*, 137–147. doi:10.1016/j.molliq.2016.12.024

Sheikholeslami, M. (2017e). Numerical simulation of magnetic nanofluid natural convection in porous media. *Physics Letters. [Part A]*, *381*(5), 494–503. doi:10.1016/j.physleta.2016.11.042

Sheikholeslami, M. (2017f). Influence of Lorentz forces on nanofluid flow in a porous cylinder considering Darcy model. *Journal of Molecular Liquids*, *225*, 903–912. doi:10.1016/j.molliq.2016.11.022

Sheikholeslami, M. (2017g). Influence of Coulomb forces on Fe3O4-H2O nanofluid thermal improvement. *International Journal of Hydrogen Energy*, *42*(2), 821–829. doi:10.1016/j.ijhydene.2016.09.185

Sheikholeslami, M. (2017h). Numerical investigation of MHD nanofluid free convective heat transfer in a porous tilted enclosure. *Engineering Computations*, *34*(6), 1939–1955. doi:10.1108/EC-08-2016-0293

Sheikholeslami, M. (2017i). Magnetic field influence on CuO -H2O nanofluid convective flow in a permeable cavity considering various shapes for nanoparticles. *International Journal of Hydrogen Energy*, *42*(31), 19611–19621. doi:10.1016/j.ijhydene.2017.06.121

Sheikholeslami, M. (2018a). Magnetic source impact on nanofluid heat transfer using CVFEM. *Neural Computing & Applications*, *30*(4), 1055–1064. doi:10.100700521-016-2740-7

Sheikholeslami, M. (2018b). Numerical simulation for external magnetic field influence on Fe3O4-water nanofluid forced convection. *Engineering Computations*, *35*(4), 1639–1654. doi:10.1108/EC-06-2017-0200

Sheikholeslami, M. (2018c). Finite element method for PCM solidification in existence of CuO nanoparticles. *Journal of Molecular Liquids*, *265*, 347–355. doi:10.1016/j.molliq.2018.05.132

Sheikholeslami, M. (2018d). Solidification of NEPCM under the effect of magnetic field in a porous thermal energy storage enclosure using CuO nanoparticles. *Journal of Molecular Liquids*, *263*, 303–315. doi:10.1016/j.molliq.2018.04.144

Sheikholeslami, M. (2018e). Influence of magnetic field on Al2O3-H2O nanofluid forced convection heat transfer in a porous lid driven cavity with hot sphere obstacle by means of LBM. *Journal of Molecular Liquids*, *263*, 472–488. doi:10.1016/j.molliq.2018.04.111

Sheikholeslami, M. (2018f). Numerical simulation for solidification in a LHTESS by means of Nano-enhanced PCM. *Journal of the Taiwan Institute of Chemical Engineers*, *86*, 25–41. doi:10.1016/j.jtice.2018.03.013

Sheikholeslami, M. (2018g). Numerical modeling of Nano enhanced PCM solidification in an enclosure with metallic fin. *Journal of Molecular Liquids*, *259*, 424–438. doi:10.1016/j.molliq.2018.03.006

Sheikholeslami, M. (2018h). Numerical investigation of nanofluid free convection under the influence of electric field in a porous enclosure. *Journal of Molecular Liquids*, *249*, 1212–1221. doi:10.1016/j.molliq.2017.11.141

Sheikholeslami, M. (2018i). CuO-water nanofluid flow due to magnetic field inside a porous media considering Brownian motion. *Journal of Molecular Liquids*, *249*, 921–929. doi:10.1016/j.molliq.2017.11.118

Sheikholeslami, M. (2018j). Application of Darcy law for nanofluid flow in a porous cavity under the impact of Lorentz forces. *Journal of Molecular Liquids*, *266*, 495–503. doi:10.1016/j.molliq.2018.06.083

Sheikholeslami, M., Barzegar Gerdroodbary, M., Valiallah Mousavi, S., Ganji, D. D., & Moradi, R. (2018). Heat transfer enhancement of ferrofluid inside an 90o elbow channel by non-uniform magnetic field. *Journal of Magnetism and Magnetic Materials*, *460*, 302–311. doi:10.1016/j.jmmm.2018.03.070

Sheikholeslami, M., & Bhatti, M. M. (2017). Active method for nanofluid heat transfer enhancement by means of EHD. *International Journal of Heat and Mass Transfer*, *109*, 115–122. doi:10.1016/j.ijheatmasstransfer.2017.01.115

Sheikholeslami, M., Darzi, M., & Li, Z. (2018). Experimental investigation for entropy generation and exergy loss of nano-refrigerant condensation process. *International Journal of Heat and Mass Transfer*, *125*, 1087–1095. doi:10.1016/j.ijheatmasstransfer.2018.04.155

Sheikholeslami, M., Darzi, M., & Sadoughi, M. K. (2018). Heat transfer improvement and Pressure Drop during condensation of refrigerant-based Nanofluid; An Experimental Procedure. *International Journal of Heat and Mass Transfer*, *122*, 643–650. doi:10.1016/j.ijheatmasstransfer.2018.02.015

Sheikholeslami, M., & Ghasemi, A. (2018). Solidification heat transfer of nanofluid in existence of thermal radiation by means of FEM. *International Journal of Heat and Mass Transfer*, *123*, 418–431. doi:10.1016/j.ijheatmasstransfer.2018.02.095

Sheikholeslami, M., Ghasemi, A., Li, Z., Shafee, A., & Saleem, S. (2018). Influence of CuO nanoparticles on heat transfer behavior of PCM in solidification process considering radiative source term. *International Journal of Heat and Mass Transfer*, *126*, 1252–1264. doi:10.1016/j.ijheatmasstransfer.2018.05.116

Sheikholeslami, M., Hayat, T., & Alsaedi, A. (2018). Numerical simulation for forced convection flow of MHD CuO-H2O nanofluid inside a cavity by means of LBM. *Journal of Molecular Liquids*, *249*, 941–948. doi:10.1016/j.molliq.2017.10.099

Sheikholeslami, M., Hayat, T., Muhammad, T., & Alsaedi, A. (2018). MHD forced convection flow of nanofluid in a porous cavity with hot elliptic obstacle by means of Lattice Boltzmann method. *International Journal of Mechanical Sciences*, *135*, 532–540. doi:10.1016/j.ijmecsci.2017.12.005

Sheikholeslami, M., Jafaryar, M., Ganji, D. D., & Li, Z. (2018). Exergy loss analysis for nanofluid forced convection heat transfer in a pipe with modified turbulators. *Journal of Molecular Liquids*, *262*, 104–110. doi:10.1016/j.molliq.2018.04.077

Sheikholeslami, M., Jafaryar, M., & Li, Z. (2018a). Second law analysis for nanofluid turbulent flow inside a circular duct in presence of twisted tape turbulators. *Journal of Molecular Liquids*, *263*, 489–500. doi:10.1016/j.molliq.2018.04.147

Sheikholeslami, M., Jafaryar, M., & Li, Z. (2018b). Nanofluid turbulent convective flow in a circular duct with helical turbulators considering CuO nanoparticles. *International Journal of Heat and Mass Transfer*, *124*, 980–989. doi:10.1016/j.ijheatmasstransfer.2018.04.022

Sheikholeslami, M., Jafaryar, M., Saleem, S., Li, Z., Shafee, A., & Jiang, Y. (2018). Nanofluid heat transfer augmentation and exergy loss inside a pipe equipped with innovative turbulators. *International Journal of Heat and Mass Transfer*, *126*, 156–163. doi:10.1016/j.ijheatmasstransfer.2018.05.128

Sheikholeslami, M., Jafaryar, M., Shafee, A., & Li, Z. (2018). Investigation of second law and hydrothermal behavior of nanofluid through a tube using passive methods. *Journal of Molecular Liquids*, *269*, 407–416. doi:10.1016/j.molliq.2018.08.019

Sheikholeslami, M., Li, Z., & Shafee, A. (2018a). Lorentz forces effect on NEPCM heat transfer during solidification in a porous energy storage system. *International Journal of Heat and Mass Transfer*, *127*, 665–674. doi:10.1016/j.ijheatmasstransfer.2018.06.087

Sheikholeslami, M., Li, Z., & Shamlooei, M. (2018). Nanofluid MHD natural convection through a porous complex shaped cavity considering thermal radiation. *Physics Letters. [Part A]*, *382*(24), 1615–1632. doi:10.1016/j.physleta.2018.04.006

Sheikholeslami, M., & Rokni, H. B. (2017). Simulation of nanofluid heat transfer in presence of magnetic field: A review. *International Journal of Heat and Mass Transfer*, *115*, 1203–1233. doi:10.1016/j.ijheatmasstransfer.2017.08.108

Sheikholeslami, M., & Rokni, H. B. (2018a). CVFEM for effect of Lorentz forces on nanofluid flow in a porous complex shaped enclosure by means of Non-equilibrium model. *Journal of Molecular Liquids*, *254*, 446–462. doi:10.1016/j.molliq.2018.01.130

Sheikholeslami, M., Rokni, H.B. (2018b). Magnetic nanofluid flow and convective heat transfer in a porous cavity considering Brownian motion effects. *Physics of Fluids*, *30*(1), doi:10.1063/1.5012517

Sheikholeslami, M., & Sadoughi, M. K. (2017). Mesoscopic method for MHD nanofluid flow inside a porous cavity considering various shapes of nanoparticles. *International Journal of Heat and Mass Transfer*, *113*, 106–114. doi:10.1016/j.ijheatmasstransfer.2017.05.054

Sheikholeslami, M., & Sadoughi, M. K. (2018). Simulation of CuO- water nanofluid heat transfer enhancement in presence of melting surface. *International Journal of Heat and Mass Transfer*, *116*, 909–919. doi:10.1016/j.ijheatmasstransfer.2017.09.086

Sheikholeslami, M., & Seyednezhad, M. (2018). Simulation of nanofluid flow and natural convection in a porous media under the influence of electric field using CVFEM. *International Journal of Heat and Mass Transfer*, *120*, 772–781. doi:10.1016/j.ijheatmasstransfer.2017.12.087

Sheikholeslami, M., Shafee, A., Ramzan, M., & Li, Z. (2018). Investigation of Lorentz forces and radiation impacts on nanofluid treatment in a porous semi annulus via Darcy law. *Journal of Molecular Liquids*, *272*, 8–14. doi:10.1016/j.molliq.2018.09.016

Sheikholeslami, M., Shamlooei, M., & Moradi, R. (2018). Numerical simulation for heat transfer intensification of nanofluid in a porous curved enclosure considering shape effect of Fe3O4 nanoparticles. *Chemical Engineering & Processing: Process Intensification*, *124*, 71–82. doi:10.1016/j.cep.2017.12.005

Sheikholeslami, M., & Shehzad, S. A. (2017a). Thermal radiation of ferrofluid in existence of Lorentz forces considering variable viscosity. *International Journal of Heat and Mass Transfer*, *109*, 82–92. doi:10.1016/j.ijheatmasstransfer.2017.01.096

Sheikholeslami, M., & Shehzad, S. A. (2017b). Magnetohydrodynamic nanofluid convective flow in a porous enclosure by means of LBM. *International Journal of Heat and Mass Transfer*, *113*, 796–805. doi:10.1016/j.ijheatmasstransfer.2017.05.130

Sheikholeslami, M., & Shehzad, S. A. (2018a). Numerical analysis of Fe3O4 –H2O nanofluid flow in permeable media under the effect of external magnetic source. *International Journal of Heat and Mass Transfer*, *118*, 182–192. doi:10.1016/j.ijheatmasstransfer.2017.10.113

Sheikholeslami, M., & Shehzad, S. A. (2018b). CVFEM simulation for nanofluid migration in a porous medium using Darcy model. *International Journal of Heat and Mass Transfer*, *122*, 1264–1271. doi:10.1016/j.ijheatmasstransfer.2018.02.080

Sheikholeslami, M., & Shehzad, S. A. (2018c). Simulation of water based nanofluid convective flow inside a porous enclosure via Non-equilibrium model. *International Journal of Heat and Mass Transfer*, *120*, 1200–1212. doi:10.1016/j.ijheatmasstransfer.2017.12.132

Sheikholeslami, M., & Shehzad, S. A. (2018d). Non-Darcy free convection of Fe3O4-water nanoliquid in a complex shaped enclosure under impact of uniform Lorentz force. *The Chinese Journal of Physiology*, *56*(1), 270–281. doi:10.1016/j.cjph.2017.12.017

Sheikholeslami, M., Shehzad, S. A., Abbasi, F. M., & Li, Z. (2018). Nanofluid flow and forced convection heat transfer due to Lorentz forces in a porous lid driven cubic enclosure with hot obstacle. *Computer Methods in Applied Mechanics and Engineering*, *338*, 491–505. doi:10.1016/j.cma.2018.04.020

Sheikholeslami, M., Shehzad, S. A., & Li, Z. (2018a). Nanofluid heat transfer intensification in a permeable channel due to magnetic field using Lattice Boltzmann method. *Physica B, Condensed Matter*, *542*, 51–58. doi:10.1016/j.physb.2018.03.036

Sheikholeslami, M., Shehzad, S. A., & Li, Z. (2018b). Water based nanofluid free convection heat transfer in a three dimensional porous cavity with hot sphere obstacle in existence of Lorenz forces. *International Journal of Heat and Mass Transfer*, *125*, 375–386. doi:10.1016/j.ijheatmasstransfer.2018.04.076

Sheikholeslami, M., Shehzad, S. A., Li, Z., & Shafee, A. (2018). Numerical modeling for Alumina nanofluid magnetohydrodynamic convective heat transfer in a permeable medium using Darcy law. *International Journal of Heat and Mass Transfer*, *127*, 614–622. doi:10.1016/j.ijheatmasstransfer.2018.07.013

Sheikholeslami, M., & Vajravelu, K. (2017). Forced convection heat transfer in Fe3O4- ethylene glycol nanofluid under the influence of Coulomb force. *Journal of Molecular Liquids*, *233*, 203–210. doi:10.1016/j.molliq.2017.03.026

Sheikholeslami, M., & Zeeshan, A. (2017). Analysis of flow and heat transfer in water based nanofluid due to magnetic field in a porous enclosure with constant heat flux using CVFEM. *Computer Methods in Applied Mechanics and Engineering*, *320*, 68–81. doi:10.1016/j.cma.2017.03.024

Sheikholeslami, M., Zeeshan, A., & Majeed, A. (2018). Control volume based finite element simulation of magnetic nanofluid flow and heat transport in non-Darcy medium. *Journal of Molecular Liquids*, *268*, 354–364. doi:10.1016/j.molliq.2018.07.031

Chapter 10
Influence of Melting Surface on Nanofluid Convective Heat Transfer

ABSTRACT

In this chapter, the effect of melting surface heat transfer on magnetohydrodynamic nanofluid free convection is analyzed by means of control volume-based finite element method (CVFEM). KKL model is taken into account to obtain viscosity and thermal conductivity of CuO-water nanofluid. The roles of melting parameter, nanofluid volume fraction, Hartmann and Rayleigh numbers are illustrated.

1. INTRODUCTION

Heat transfer generated during the melting process of a stretching sheet has various industrial applications which involve preparation of semi-conductor materials, magma solidification, melting of permafrost and thawing of frozen ground, etc. Epstein and Cho (1976) examined the characteristics of melting heat transfer in the laminar flow over a flat plate. The work of Epstein and Cho (1976) has been extended by the various researchers. Cheng and Lin (2007) investigated the melting effect in mixed convective heat transfer with aiding and opposing external flows from the vertical plate in a liquid-saturated porous medium. Ishak et al. (2010) studied melting heat transfer in steady laminar flow over a moving surface. Melting heat transfer in boundary layer stagnation-point flow towards a stretching/shrinking sheet in a micropolar fluid has been analyzed by Yacob et al. (2011). Recently, Ahmad and Pop (2014) have studied the effect of melting parameter on the dual solutions for the problem of mixed convection boundary-layer flow past a vertical flat surface embedded in a fluid-saturated porous medium. They considered only opposing flow which shows that the effect of surface melting reduces the heat transfer rate and expedites the boundary-layer separation at the solid–liquid interface. Hayat et al. (2013) analyzed the stagnation-point flow of couple stress fluid with melting heat transfer. Sheikholeslami and Rokni (2017) studied the melting heat transfer influence on nanofluid flow inside a cavity in existence of magnetic field. Sheikholeslami and Sadoughi (2018) simulated the CuO- water nanofluid heat transfer enhancement in presence of melting surface. Sheikholeslami et al. (2017) investigated CuO-water

DOI: 10.4018/978-1-5225-7595-5.ch010

nanofluid magnetohydrodynamic natural convection inside a sinusoidal annulus in presence of melting heat transfer. Also, nanotechnology utilized for other applications

2. MELTING HEAT TRANSFER INFLUENCE ON NANOFLUID FLOW INSIDE A CAVITY IN EXISTENCE OF MAGNETIC FIELD

2.1. Problem Definition

Figure 1 depicts the geometry, boundary condition and sample element. The bottom wall is hot wall $\left(T = T_h\right)$ and the top one is melting surface $\left(T = T_m\right)$. Other walls are adiabatic. Horizontal magnetic field has been applied. The enclosure is field with nanofluid.

2.2. Governing Equation

2D steady convective nanofluid flow is considered in existence of constant magnetic field. The PDEs are:

$$\frac{\partial v}{\partial y} + \frac{\partial u}{\partial x} = 0 \tag{1}$$

$$\left(\rho_{nf}\right)\left(v\frac{\partial u}{\partial y} + u\frac{\partial u}{\partial x}\right) = \left[B_y \sigma_{nf} v B_x - B_y^2 \sigma_{nf} u + \left(\frac{\partial^2 u}{\partial y^2} + \frac{\partial^2 u}{\partial x^2}\right)\mu_{nf} - \frac{\partial P}{\partial x}\right] \tag{2}$$

$$\begin{gathered}\rho_{nf}\left(v\frac{\partial v}{\partial y} + \frac{\partial v}{\partial x}u\right) = uB_y \sigma_{nf} B_x + \left(T - T_c\right)\beta_{nf} g \rho_{nf} - vB_x \sigma_{nf} B_x \\ -\frac{\partial P}{\partial y} + \mu_{nf}\left(\frac{\partial^2 v}{\partial y^2} + \frac{\partial^2 v}{\partial x^2}\right), \\ B_x = B_o \cos\lambda,\ B_y = B_o \sin\lambda\end{gathered} \tag{3}$$

$$\left(\rho C_p\right)_{nf}\left(v\frac{\partial T}{\partial y} + u\frac{\partial T}{\partial x}\right) = k_{nf}\left(\frac{\partial^2 T}{\partial x^2} + \frac{\partial^2 T}{\partial y^2}\right) \tag{4}$$

$\left(\rho C_p\right)_{nf}$, ρ_{nf}, $\left(\rho\beta\right)_{nf}$ and σ_{nf} are calculated as:

$$\left(\rho C_p\right)_{nf} = \phi\left(\rho C_p\right)_s + (1-\phi)\left(\rho C_p\right)_f \tag{5}$$

Figure 1. (a)Geometry and the boundary conditions with (b) the mesh of geometry considered in this work; (c) A sample triangular element and its corresponding control volume.

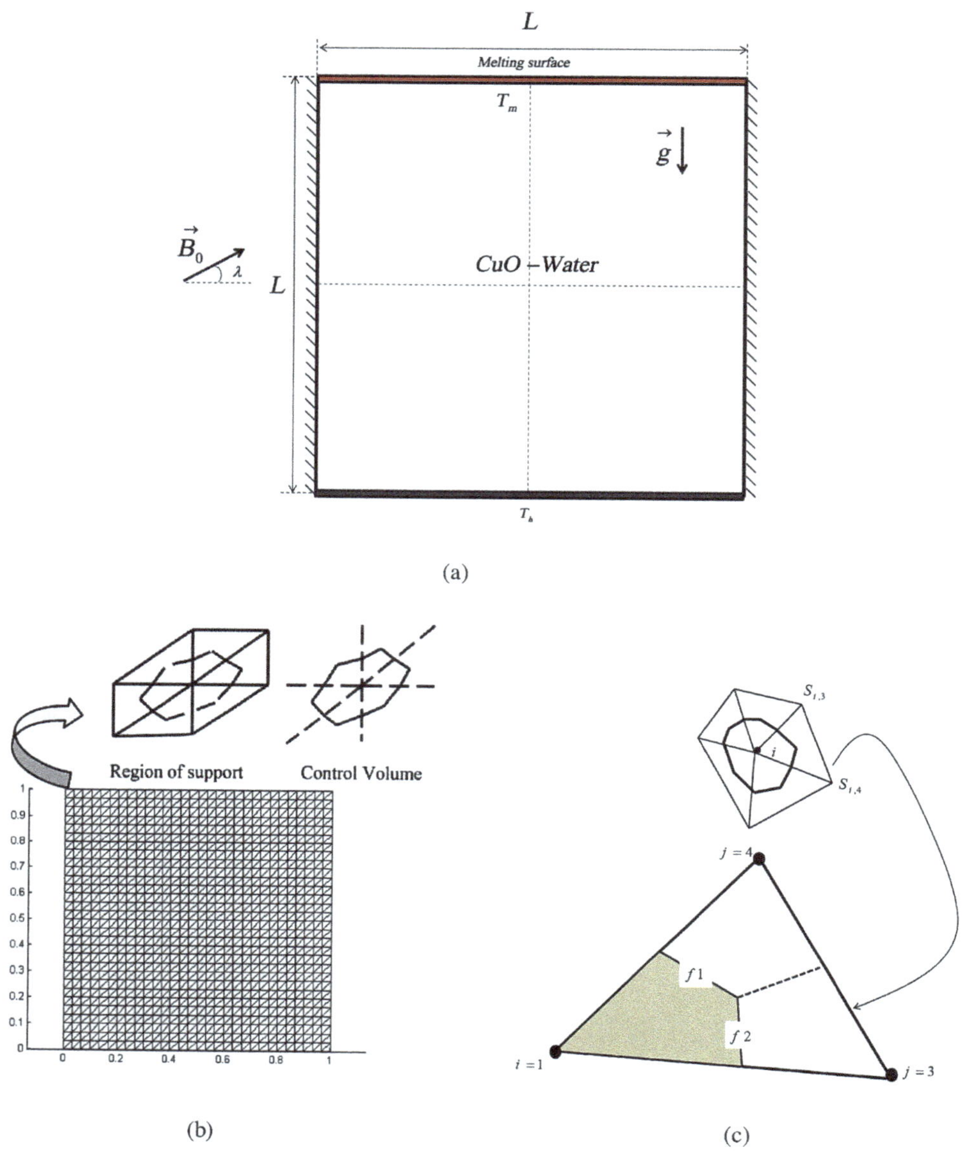

$$\rho_{nf} = \phi\rho_s + (1-\phi)\rho_f \tag{6}$$

$$\left(\rho\beta\right)_{nf} = \phi\left(\rho\beta\right)_s + (1-\phi)\left(\rho\beta\right)_f \tag{7}$$

$$\frac{\sigma_{nf}}{\sigma_f} = \left(\frac{+\left(2+\sigma_s / \sigma_f\right) - \phi\left(-1+\sigma_s / \sigma_f\right)}{3\phi\left(-1+\sigma_s / \sigma_f\right)} \right)^{-1} + 1 \tag{8}$$

k_{nf}, μ_{nf} are calculated via KKL model:

$$\begin{aligned}
&\frac{k_{nf}}{k_f} = 1 + \frac{3\left(\frac{k_p}{k_f} - 1\right)\phi}{\left(\frac{k_p}{k_f} + 2\right) - \left(\frac{k_p}{k_f} - 1\right)\phi} + 5\times 10^4 g'(\phi, T, d_p)\phi\rho_f c_{p,f}\sqrt{\frac{\kappa_b T}{d_p \rho_p}} \\
&, R_f = 4\times 10^{-8} km^2 / W \\
&, R_f = -d_p(1/k_p - 1/k_{p,eff}), \\
&g'\left(\phi, d_p, T\right) = Ln\left(T\right)\left(a_1 + a_3 Ln\left(\phi\right) + a_2 Ln\left(d_p\right) + a_5 Ln\left(d_p\right)^2 + a_4 \ln\left(d_p\right) Ln\left(\phi\right)\right) \\
&+ \left(a_6 + a_8 Ln\left(\phi\right) + a_7 Ln\left(d_p\right) + a_{10} Ln\left(d_p\right)^2 + a_9 Ln\left(\phi\right) \ln\left(d_p\right)\right)
\end{aligned} \tag{9}$$

$$\mu_{nf} = \frac{\mu_f}{\left(1-\phi\right)^{2.5}} + \frac{k_{Brownian}}{k_f} \times \frac{\mu_f}{\Pr} \tag{10}$$

Properties and needed parameters are provided Tables 1 and 2.
Vorticity and stream function should be used to eliminate pressure source terms:

$$\omega + \frac{\partial u}{\partial y} - \frac{\partial v}{\partial x} = 0, \frac{\partial \psi}{\partial x} = -v, \frac{\partial \psi}{\partial y} = u \tag{11}$$

Introducing dimensionless quantities:

$$\begin{aligned}
&P = \frac{p}{\rho_{nf}\left(\alpha_{nf} / L\right)^2}, U = \frac{uL}{\alpha_{nf}}, V = \frac{vL}{\alpha_{nf}}, \\
&\theta = \frac{T - T_m}{T_h - T_m}, \left(X, Y\right) = \frac{\left(x, y\right)}{L}, \Psi = \frac{\psi}{\alpha_{nf}}, \Omega = \frac{\omega L^2}{\alpha_{nf}}
\end{aligned} \tag{12}$$

The final formulae are:

$$\Omega + \frac{\partial^2 \Psi}{\partial Y^2} + \frac{\partial^2 \Psi}{\partial X^2} = 0, \tag{13}$$

Table 1. The coefficient values of CuO – Water nanofluid

Coefficient Values	CuO – Water
a_1	-26.5933108
a_2	-0.403818333
a_3	-33.3516805
a_4	-1.915825591
a_5	6.421858E-02
a_6	48.40336955
a_7	-9.787756683
a_8	190.245610009
a_9	10.9285386565
a_{10}	-0.72009983664

Table 2. Thermo physical properties of water and nanoparticles

	$\rho(kg/m^3)$	$C_p(j/kgk)$	$k(W/m.k)$	$\beta\times10^5(K^{-1})$	$d_p(nm)$	$\sigma(\Omega\cdot m)^{-1}$
Water	997.1	4179	0.613	21	-	0.05
CuO	6500	540	18	29	45	10^{-10}

$$\frac{\partial\Omega}{\partial X}U+V\frac{\partial\Omega}{\partial Y}=\Pr\frac{A_5}{A_1}\frac{A_2}{A_4}\left(\frac{\partial^2\Omega}{\partial Y^2}+\frac{\partial^2\Omega}{\partial X^2}\right)$$
$$+\Pr Ha^2\frac{A_6}{A_1}\frac{A_2}{A_4}\left(-\frac{\partial V}{\partial X}B_xB_x+\frac{\partial U}{\partial X}B_xB_y-\frac{\partial V}{\partial Y}B_yB_x+\frac{\partial U}{\partial Y}B_yB_y\right) \tag{14}$$
$$+\Pr Ra\frac{A_3A_2^2}{A_1A_4^2}\frac{\partial\theta}{\partial X},$$

$$\frac{\partial \theta}{\partial X}U+\frac{\partial \theta}{\partial Y}V=\left(\frac{\partial^2 \theta}{\partial X^2}+\frac{\partial^2 \theta}{\partial Y^2}\right) \tag{15}$$

Boundary conditions are:

$\theta = 0$ on top wall

$\theta = 1.0$ on bottom wall

$\frac{\partial \theta}{\partial n} = 0.0$ on other walls

$\Psi = 0.0$ Right, left and bottom walls (16)

and in melting surface, we have:

$$\left(\frac{\partial \theta}{\partial n}\right)\Bigg|_{Y=1}=\frac{A_2}{A_1}\frac{1}{\delta}\left(L+c_s\left(T_m-T_0\right)\right)\frac{\partial \Psi}{\partial n}\Bigg|_{Y=1} \tag{17}$$

where dimensionless and constants parameters are illustrated as:

$$\begin{aligned}&\Pr = \upsilon_f/\alpha_f, Ra = g\beta_f \Delta T L^3/\left(\upsilon_f \alpha_f\right), Ha = LB_0\sqrt{\sigma_f/\mu_f}, \delta = \frac{\left(\rho C_p\right)_f}{\rho_f}\frac{\left(T_h - T_m\right)}{1+c_s\left(T_m - T_0\right)}\\ &A_1 = \frac{\rho_{nf}}{\rho_f}, A_5 = \frac{\mu_{nf}}{\mu_f}, A_2 = \frac{\left(\rho C_P\right)_{nf}}{\left(\rho C_P\right)_f}, A_4 = \frac{k_{nf}}{k_f}, A_6 = \frac{\sigma_{nf}}{\sigma_f}, A_3 = \frac{\left(\rho\beta\right)_{nf}}{\left(\rho\beta\right)_f}\end{aligned} \tag{18}$$

It should be mentioned that δ is related to Stefan numbers.

Local and average Nusselt over the hot wall can calculate as:

$$Nu_{loc}=\left(\frac{k_{nf}}{k_f}\right)\frac{\partial \theta}{\partial y} \tag{19}$$

$$Nu_{ave}=\frac{1}{L}\int_0^L Nu_{loc}\,dx \tag{20}$$

2.3. Effects of Active Parameters

Impacts of melting heat transfer on nanofluid magnetohydrodynamic flow in a cavity is examined. μ_{nf}, k_{nf} of CuO-water nanofluid are estimated by means of KKL model. Graphs and tables are depicted for different amounts of CuO-water volume fraction ($\phi = 0$ to $\phi = 0.04$), Melting parameter ($\delta = 0$ to 0.8), Hartmann number ($Ha = 0$ to 40) and Rayleigh number ($Ra = 10^3$ to 10^4).

Impact of adding nanoparticles in the base fluid on velocity and temperature contours is depicted in Figure 2. Temperature gradient decreases with augment of ϕ. $|\Psi_{\max}|$ augments with adding nanoparticles because of increment in the solid movements. In presence of melting heat transfer and magnetic field, effect of adding nanoparticles on isotherms becomes negligible.

Figure 3 depicts the impact of Rayleigh and Hartmann numbers in absence of melting heat transfer. There is only one eddy in streamline. In low Rayleigh number, conduction mechanism is dominant. As Ra increases the distortion of isotherms enhances close to the hot wall. Adding magnetic field makes isotherms to become parallel. Figures 4 and 5 demonstrate the influence of Ra and Ha on streamlines and isotherms in existence of melting process. In presence of melting heat transfer the primary eddy diminishes and increasing Lorentz forces generates three layers for streamline. Increasing melting parameter augments the strength of bottom eddy.

Figure 6 illustrates the impact of δ, Ra and Ha on Nu_{ave}. The formula for Nu_{ave} corresponding to active parameters is:

$$\begin{aligned} Nu_{ave} &= 0.425 + 0.122\delta - 1.63\log(Ra) + 0.35Ha^* + 1.5\log(Ra)\delta \\ &-0.58\delta Ha^* - 0.66\log(Ra)Ha^* + 0.45Ha^{*2} + 0.53(\log(Ra))^2 - 4.04\delta^2 \end{aligned} \tag{21}$$

where $Ha^* = 0.1Ha$. As melting paramtere augments, temperature gradient enhances and in turn Nusselt number enhances. Increasing bouyancy forces leads the thermal boudary layer thicknesss to reduce. So Nusselt number increases with enhance of Ra. As Hatmann number augments isotherms become parallel to each others. Therefore Nusselt number has opposite relationship with Ha.

3. SIMULATION OF CUO- WATER NANOFLUID HEAT TRANSFER ENHANCEMENT IN PRESENCE OF MELTING SURFACE

3.1. Problem Definition

Figure 7 depicts the geometry, boundary condition and sample element. The inner wall is hot wall $(T = T_h)$ and the outer one is melting surface $(T = T_m)$. Other walls are adiabatic. Horizontal magnetic field has been applied. The enclosure is field with nanofluid.

Figure 2. Influence of nanofluid volume fraction on streamlines (left) and isotherms (right) contours (nanofluid ($\phi = 0.04$)(—) and pure fluid ($\phi = 0$) (- - -)) when $Ra = 10^4$

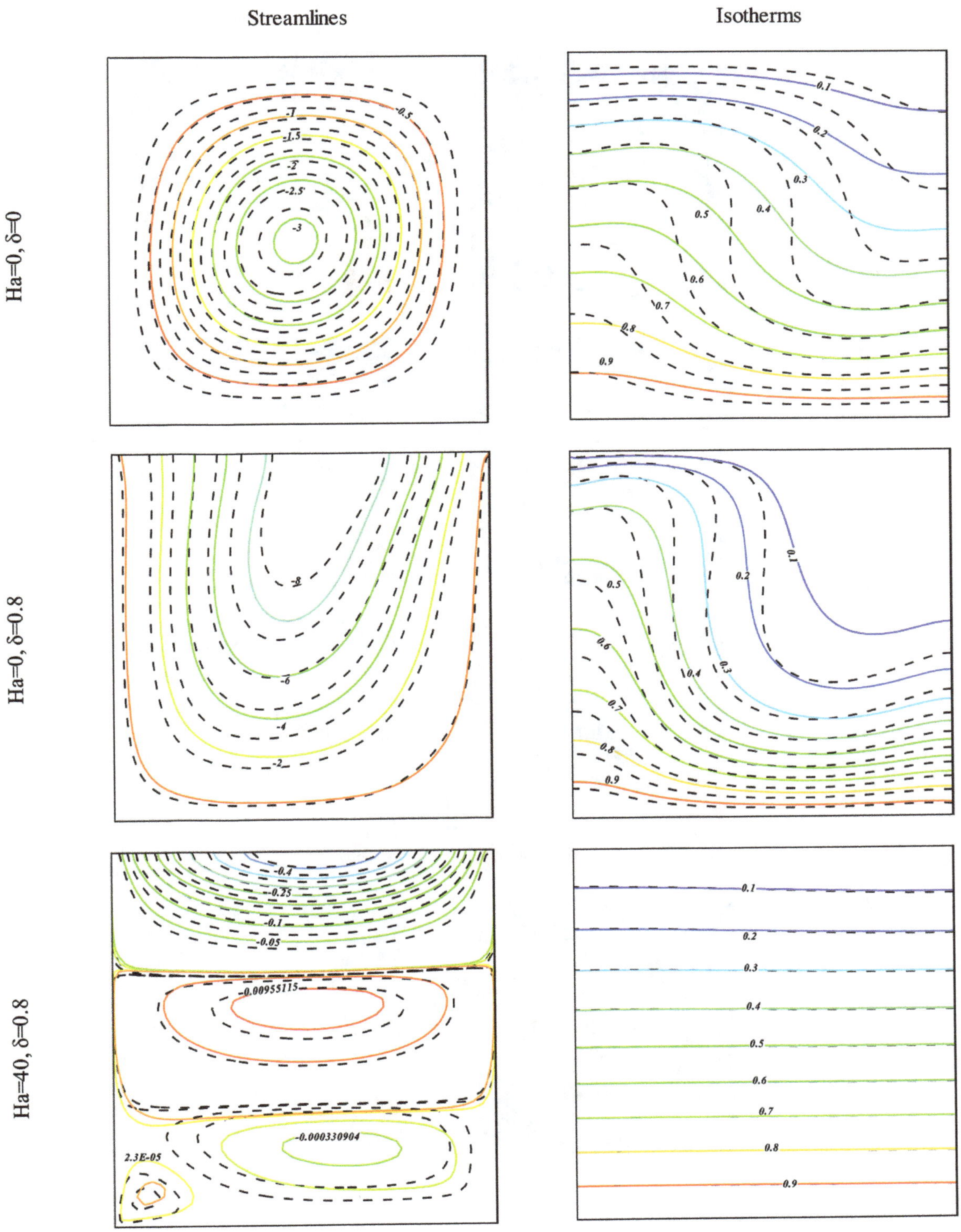

Figure 3. Streamlines (left) and isotherms (right) contours for different values of Rayleigh and Hartmann numbers when $\delta = 0, \phi = 0.04$.

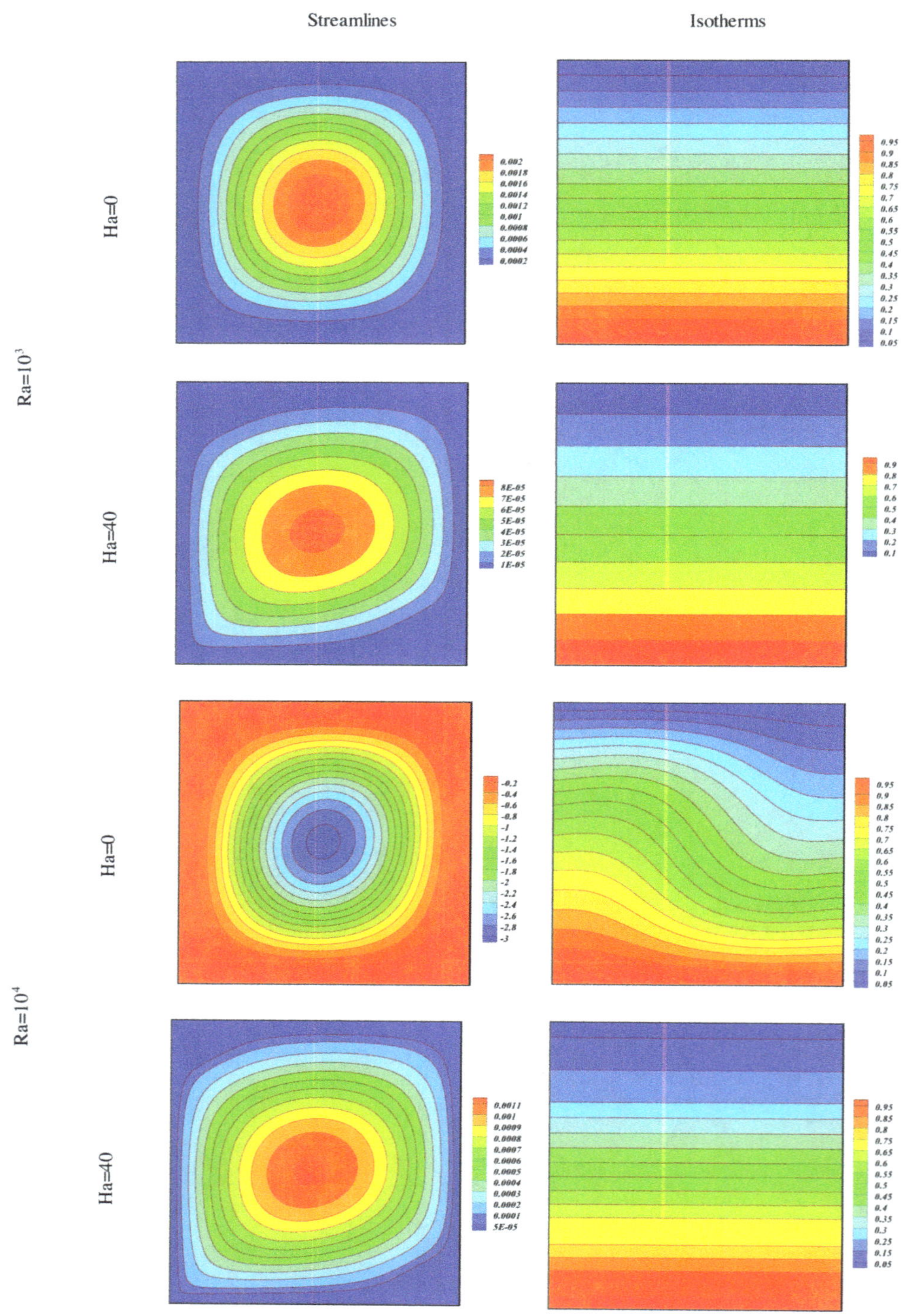

Figure 4. Streamlines (left) and isotherms (right) contours for different values of Rayleigh and Hartmann numbers when $\delta = 0.2, \phi = 0.04$.

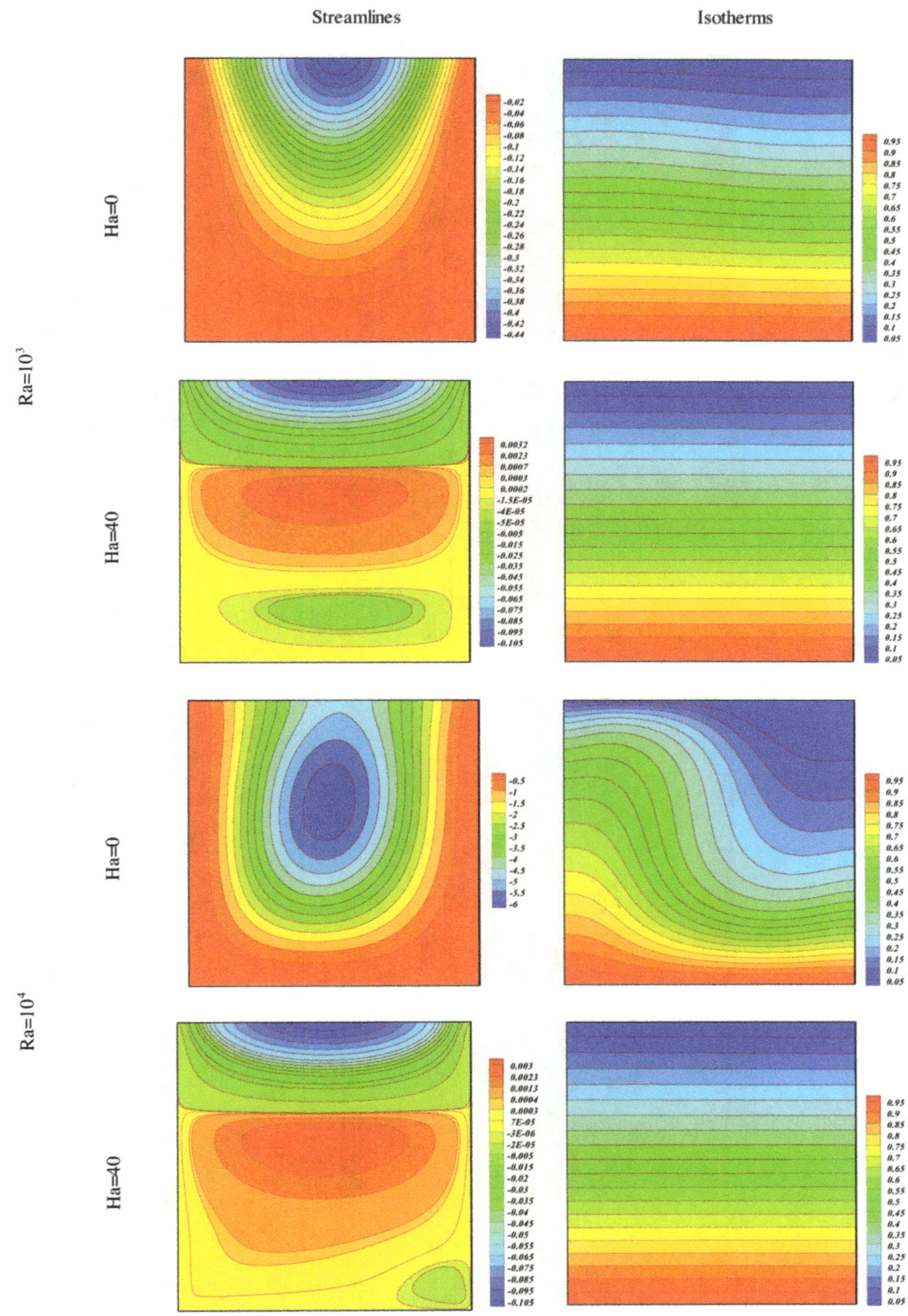

Figure 5. Streamlines (left) and isotherms (right) contours for different values of Rayleigh and Hartmann numbers when $\delta = 0.8, \phi = 0.04$.

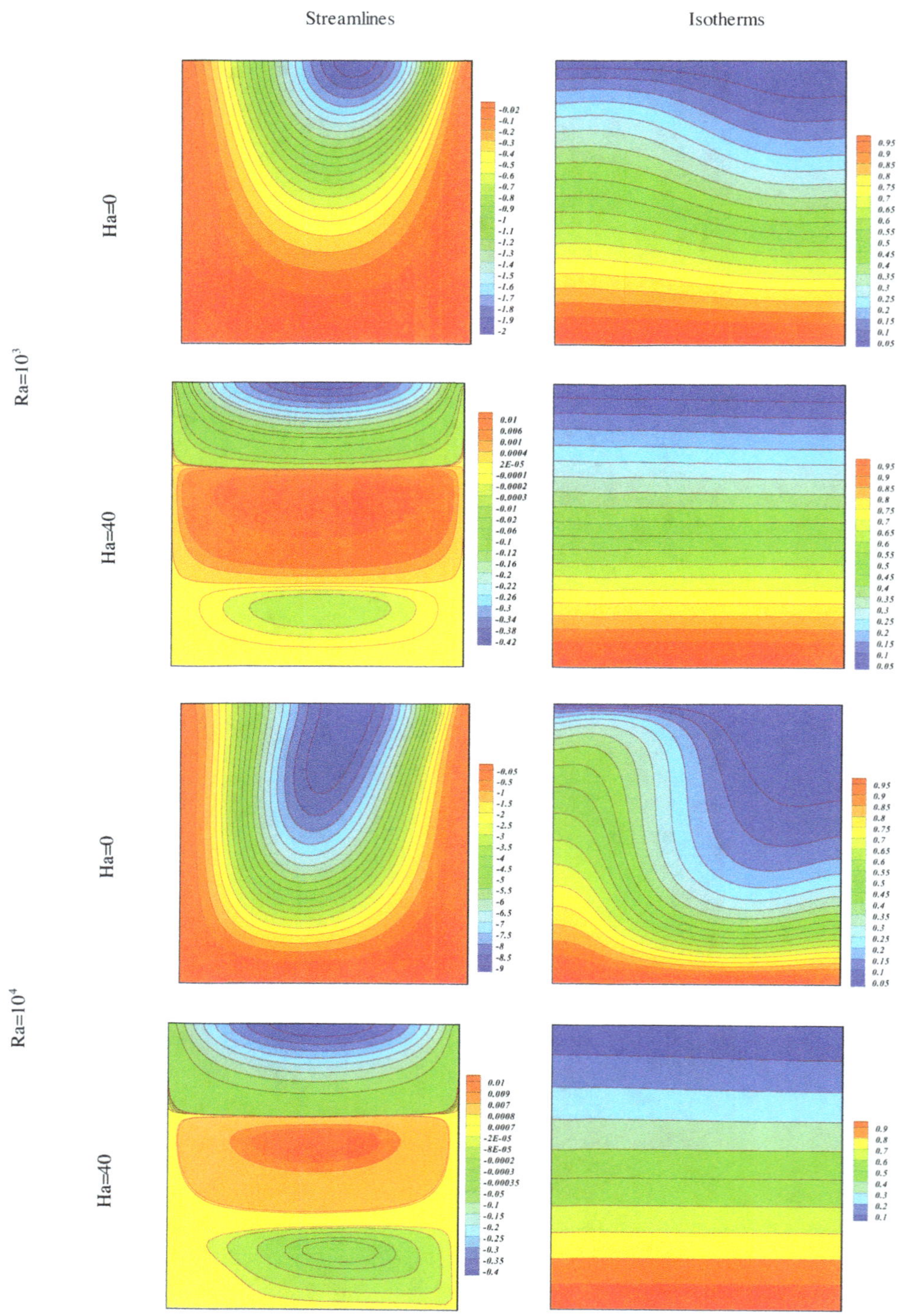

Figure 6. Influences of the melting parameter, Reynolds number and Hartmann numbers on average Nusselt number.

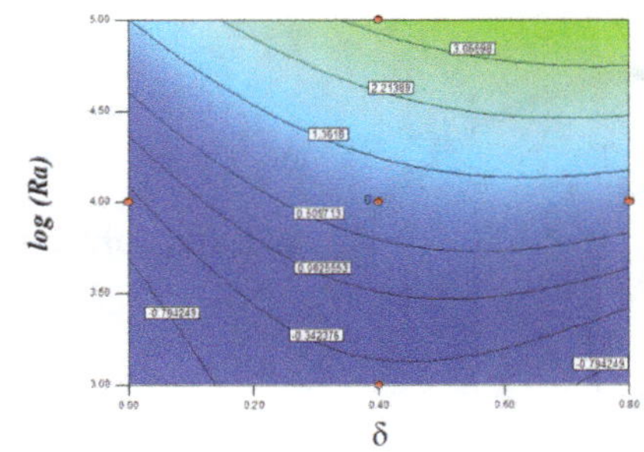

$Ha = 20, \phi = 0.04$

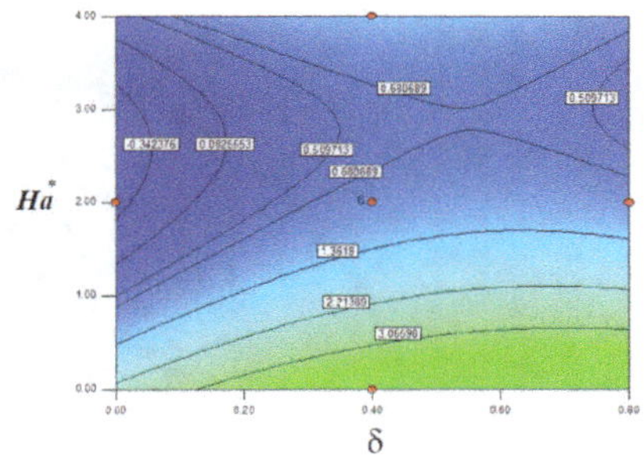

$Ra = 5500, \phi = 0.04$

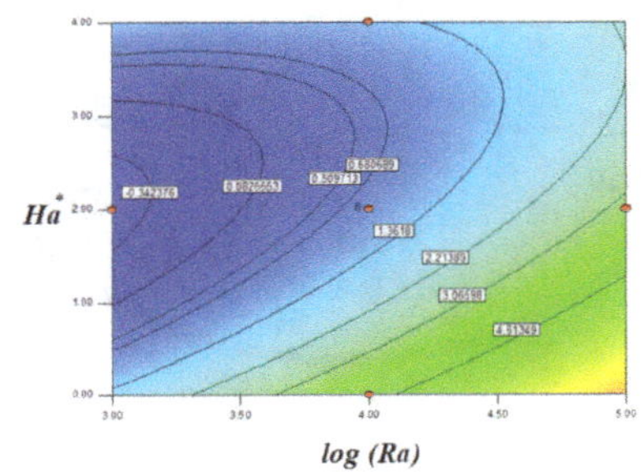

$\delta = 0.4, \phi = 0.04$

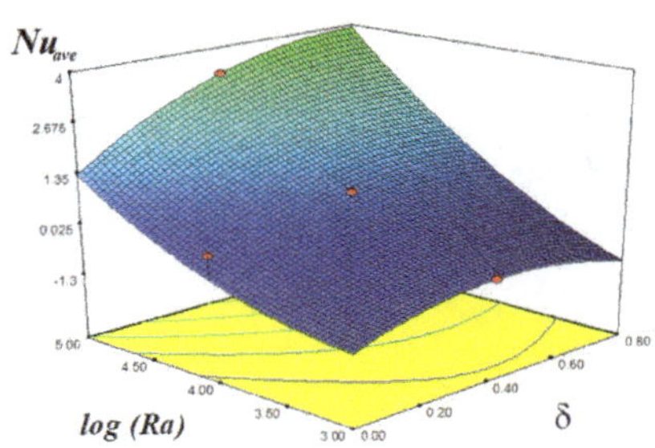

$Ha = 20, \phi = 0.04$

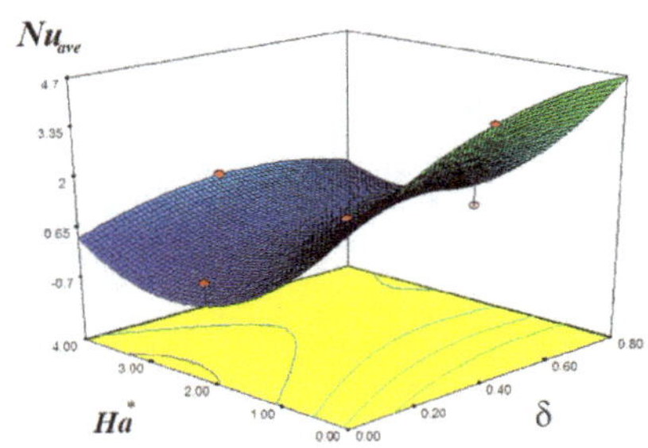

$Ra = 5500, \phi = 0.04$

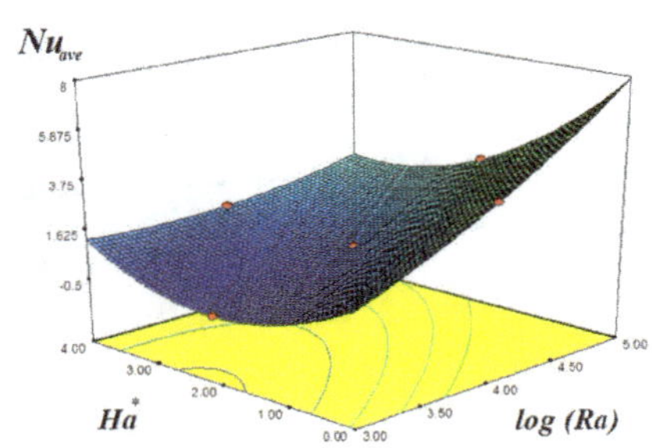

$\delta = 0.4, \phi = 0.04$

Figure 7. (a)Geometry and the boundary conditions (b) A sample triangular element and its corresponding control volume.

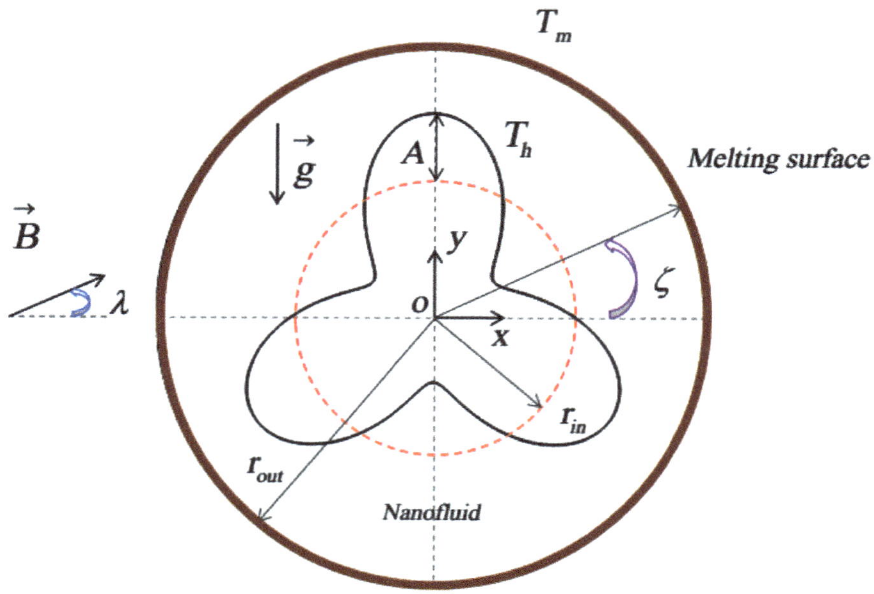

(a)

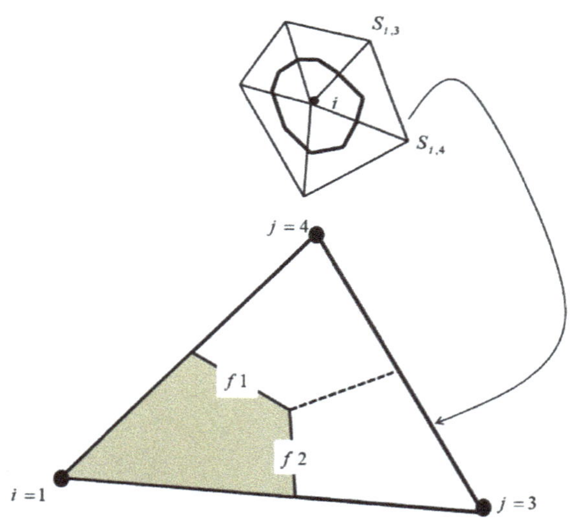

(b)

3.2. Governing Equation

Nanofluid steady convective flow is considered in presence of constant magnetic field. The PDEs are:

$$\frac{\partial v}{\partial y}+\frac{\partial u}{\partial x}=0 \tag{22}$$

$$\left(\rho_{nf}\right)\left(v\frac{\partial u}{\partial y}+u\frac{\partial u}{\partial x}\right)=\left[B_y\sigma_{nf}vB_x-B_y^2\sigma_{nf}u+\mu_{nf}\left(\frac{\partial^2 u}{\partial y^2}+\frac{\partial^2 u}{\partial x^2}\right)-\frac{\partial P}{\partial x}\right] \tag{23}$$

$$\rho_{nf}\left(\frac{\partial v}{\partial y}v+\frac{\partial v}{\partial x}u\right)=uB_y\sigma_{nf}B_x+\left(T-T_c\right)\beta_{nf}g\rho_{nf}-vB_x\sigma_{nf}B_x+\left(\frac{\partial^2 v}{\partial x^2}+\frac{\partial^2 v}{\partial y^2}\right)\mu_{nf}-\frac{\partial P}{\partial y}, \tag{24}$$
$$B_x=B_o\cos\lambda, B_y=B_o\sin\lambda$$

$$\left(\rho C_p\right)_{nf}\left(v\frac{\partial T}{\partial y}+u\frac{\partial T}{\partial x}\right)=k_{nf}\left(\frac{\partial^2 T}{\partial x^2}+\frac{\partial^2 T}{\partial y^2}\right) \tag{25}$$

$\left(\rho C_p\right)_{nf}$, ρ_{nf}, $\left(\rho\beta\right)_{nf}$ and σ_{nf} are calculated as:

$$\left(\rho C_p\right)_{nf}=\phi\left(\rho C_p\right)_s+(1-\phi)\left(\rho C_p\right)_f \tag{26}$$

$$\rho_{nf}=\rho_f(1-\phi)+\phi\rho_s \tag{27}$$

$$\frac{\left(\rho\beta\right)_{nf}}{\left(\rho\beta\right)_f}=\phi\frac{\left(\rho\beta\right)_s}{\left(\rho\beta\right)_f}+(1-\phi) \tag{28}$$

$$\sigma_{nf}=\left[\left(\frac{\left(2+\sigma_s/\sigma_f\right)-\left(\sigma_s/\sigma_f-1\right)\phi}{3\phi\left(-1+\sigma_s/\sigma_f\right)}\right)^{-1}+1\right]\sigma_f \tag{29}$$

k_{nf}, μ_{nf} are calculated via KKL model:

$$\begin{aligned}&\frac{k_{nf}}{k_f}=1+\frac{3\left(\frac{k_p}{k_f}-1\right)\phi}{\left(\frac{k_p}{k_f}+2\right)-\left(\frac{k_p}{k_f}-1\right)\phi}+5\times10^4 g'(\phi,T,d_p)\phi\rho_f c_{p,f}\sqrt{\frac{\kappa_b T}{d_p\rho_p}}\\&,R_f=4\times10^{-8}km^2/W\\&,R_f=-d_p(1/k_p-1/k_{p,eff}),\\&g'\left(\phi,d_p,T\right)=Ln\left(T\right)\left(a_1+a_3Ln\left(\phi\right)+a_2Ln\left(d_p\right)+a_5Ln\left(d_p\right)^2+a_4\ln\left(d_p\right)Ln\left(\phi\right)\right)\\&+\left(a_6+a_8Ln\left(\phi\right)+a_7Ln\left(d_p\right)+a_{10}Ln\left(d_p\right)^2+a_9Ln\left(\phi\right)\ln\left(d_p\right)\right)\end{aligned}\tag{30}$$

$$\mu_{nf}=\frac{\mu_f}{\left(1-\phi\right)^{2.5}}+\frac{k_{Brownian}}{k_f}\times\frac{\mu_f}{\Pr}\tag{31}$$

Vorticity and stream function should be used to eliminate pressure source terms:

$$\omega+\frac{\partial u}{\partial y}-\frac{\partial v}{\partial x}=0,\frac{\partial\psi}{\partial x}=-v,\frac{\partial\psi}{\partial y}=u\tag{32}$$

Introducing dimensionless quantities:

$$\begin{aligned}&P=\frac{p}{\rho_{nf}\left(\alpha_{nf}/L\right)^2},U=\frac{uL}{\alpha_{nf}},V=\frac{vL}{\alpha_{nf}},\theta=\frac{T-T_m}{T_h-T_m},\\&\left(X,Y\right)=\frac{\left(x,y\right)}{L},\Psi=\frac{\psi}{\alpha_{nf}},\Omega=\frac{\omega L^2}{\alpha_{nf}}\end{aligned}\tag{33}$$

The final formulae are:

$$\frac{\partial^2\Psi}{\partial Y^2}+\Omega+\frac{\partial^2\Psi}{\partial X^2}=0,\tag{34}$$

$$\begin{aligned}&\frac{\partial\Omega}{\partial X}U+V\frac{\partial\Omega}{\partial Y}=\Pr\frac{A_5}{A_1}\frac{A_2}{A_4}\left(\frac{\partial^2\Omega}{\partial Y^2}+\frac{\partial^2\Omega}{\partial X^2}\right)\\&+\Pr Ha^2\frac{A_6}{A_1}\frac{A_2}{A_4}\left(-B_x\frac{\partial V}{\partial X}B_x+B_y\frac{\partial U}{\partial X}B_x-\frac{\partial V}{\partial Y}B_yB_x+B_y\frac{\partial U}{\partial Y}B_y\right)\\&+\Pr Ra\frac{A_3A_2^2}{A_1A_4^2}\frac{\partial\theta}{\partial X},\end{aligned}\tag{35}$$

$$\frac{\partial \theta}{\partial X}U+\frac{\partial \theta}{\partial Y}V=\left(\frac{\partial^2 \theta}{\partial X^2}+\frac{\partial^2 \theta}{\partial Y^2}\right) \tag{36}$$

Boundary conditions are:

$\frac{\partial \theta}{\partial n}=0.0$ on other walls

$\theta=0$ on outer wall

$\theta=1.0$ on inner wall

$\Psi=0.0$ on all walls except melting surface (37)

and in melting surface, we have:

$$\frac{\partial \theta}{\partial n}=\frac{A_2}{A_1}\frac{1}{\delta}\left(L+c_s\left(T_m-T_0\right)\right)\frac{\partial \Psi}{\partial n} \tag{38}$$

where dimensionless and constants parameters are illustrated as:

$$\begin{aligned}&\Pr=\upsilon_f/\alpha_f, Ra=g\beta_f\Delta TL^3/\left(\upsilon_f\alpha_f\right), Ha=LB_0\sqrt{\sigma_f/\mu_f}, \delta=\frac{\left(\rho C_p\right)_f}{\rho_f}\frac{\left(T_h-T_m\right)}{1+c_s\left(T_m-T_0\right)}\\&A_1=\frac{\rho_{nf}}{\rho_f}, A_5=\frac{\mu_{nf}}{\mu_f}, A_2=\frac{\left(\rho C_P\right)_{nf}}{\left(\rho C_P\right)_f}, A_6=\frac{\sigma_{nf}}{\sigma_f}, A_4=\frac{k_{nf}}{k_f}, A_3=\frac{\left(\rho\beta\right)_{nf}}{\left(\rho\beta\right)_f}\end{aligned} \tag{39}$$

It should be mentioned that δ is related to Stefan numbers.

Nu_{loc} and Nu_{ave} over the cold wall can calculate as:

$$Nu_{loc}=\left(\frac{k_{nf}}{k_f}\right)\frac{\partial \theta}{\partial r} \tag{40}$$

$$Nu_{ave}=\frac{1}{0.5\pi}\int_0^{0.5\pi}Nu_{loc}\,d\xi \tag{41}$$

3.3. Effects of Active Parameters

Nanofluid in existence of melting heat transfer is presented under the effect of Lorentz forces. KKL model is selected to calculate properties of nanofluid. Outputs are demonstrated for various number of undulation ($N = 3$ to 5), volume fraction of nanofluid ($\phi = 0$ to 0.04), Hartmann number ($Ha = 0$ to 40), Melting parameter ($\delta = 0$ to 0.2) and Rayleigh number ($Ra = 500$ to 5000).

Figure 8 presents the influence of using nanofluid instead of base fluid. Velocity increases with using nanofluid due to increment in the solid movements. In existence of magnetic field and melting heat transfer, impact of adding nanoparticles on isotherms becomes insignificant. Temperature gradient reduces by using nanofluid.

Influences of Hartmann and Rayleigh numbers without melting heat transfer are illustrated in Figures 9, 10 and 11. Streamline has just one rotating eddy. Conduction mode is prominent in Rayleigh number. As buoyancy force enhances the distortion of isotherms increase close to the hot wall. Increasing Hartmann number causes isotherms to become parallel. In existence of melting heat transfer the primary eddy diminishes and enhancing Lorentz forces generates three layers for streamline. The bottom eddy becomes stronger with augment of melting parameter. As number of undulation increases, temperature gradient near the hot wall enhances.

Effects of δ, Ra, N and Ha on average Nusselt number are depicted in Figure 12. The following correlation can be offered:

$$\begin{aligned} Nu_{ave} &= 0.068 + 0.505N - 0.15\delta + 0.08Ra^{*} + 0.11Ha^{*}0.13N\delta \\ &+0.002N\,Ra^{*} - 0.021NHa^{*} + 0.056Ra^{*}\delta - 0.14\delta Ha^{*} - 0.008Ra^{*}Ha^{*} \\ &-0.0485N^{2} + 0.49\delta^{2} - 0.007Ra^{*2} - 0.011Ha^{*2} \end{aligned} \tag{42}$$

where $Ha^{*} = 0.1Ha, Ra^{*} = 0.001Ra$. Temperature gradient increases with increase of melting paramtere. So, Nusselt number has direct relationship with δ . Augmenting Raylight number leads reduce in thermal boudary layer thicknesss. Therefore, Nusselt number enhances with augment of Ra . As Lorentz forces augments, conduction mechanisim become stronger. Thus, Nu_{ave} reduces with increase of hartmann number. Besides, rate of heat transfer augments with rise of N .

4. CUO-WATER NANOFLUID MAGNETOHYDRODYNAMIC NATURAL CONVECTION INSIDE A SINUSOIDAL ANNULUS IN PRESENCE OF MELTING HEAT TRANSFER

4.1. Problem Definition

Figure 13 depicts the geometry, boundary condition and sample element. The inner wall is hot wall $\left(T = T_{h}\right)$ and the outer one is melting surface $\left(T = T_{m}\right)$. Other walls are adiabatic. Horizontal magnetic field has been applied. The enclosure is field with nanofluid.

Figure 8a. Isotherms (left) and streamlines (right) contours for different values of Rayleigh and Hartmann numbers when $\delta = 0, \phi = 0.04, N = 3$.

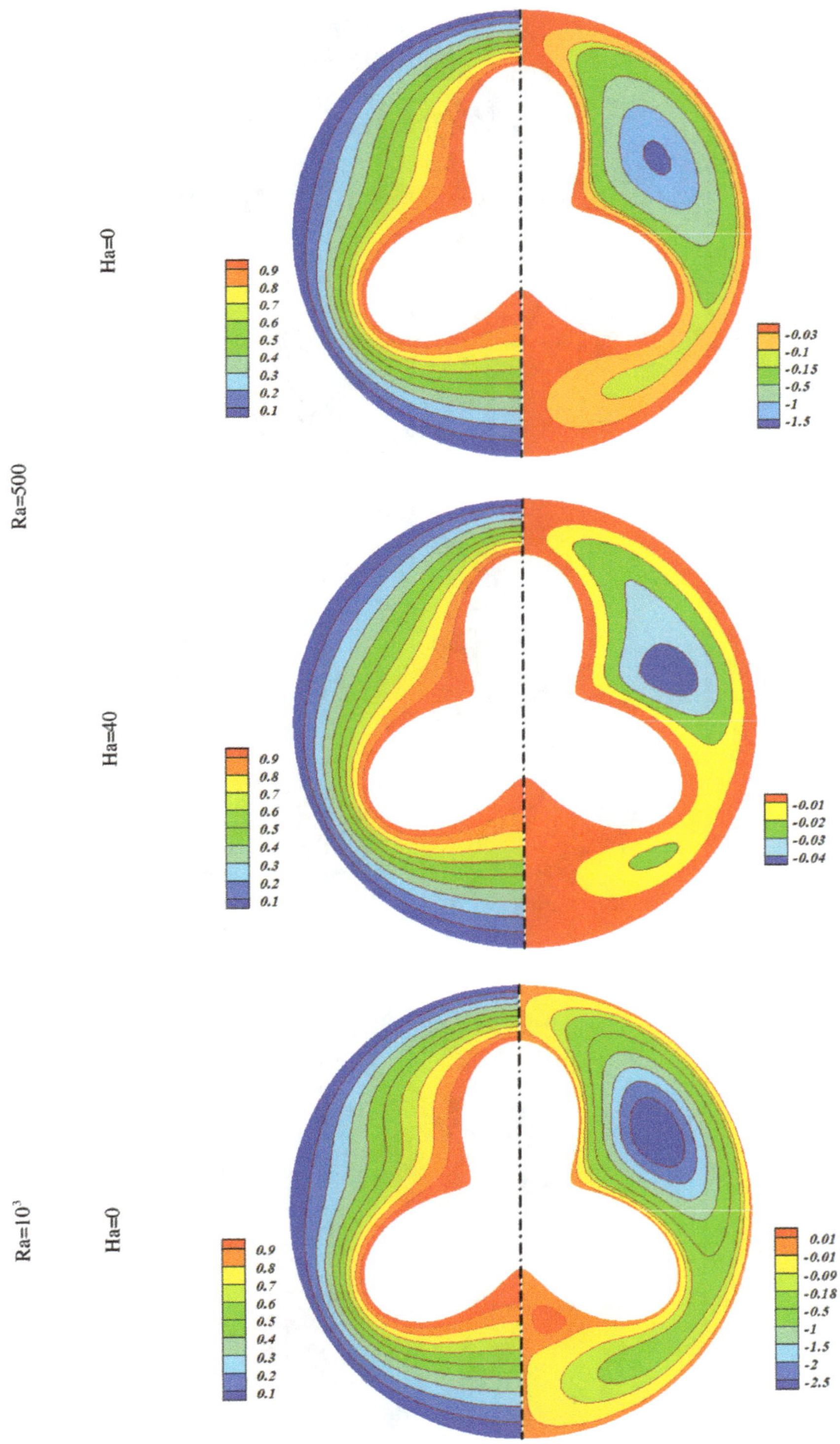

Figure 8b. Isotherms (left) and streamlines (right) contours for different values of Rayleigh and Hartmann numbers when $\delta = 0, \phi = 0.04, N = 3$.

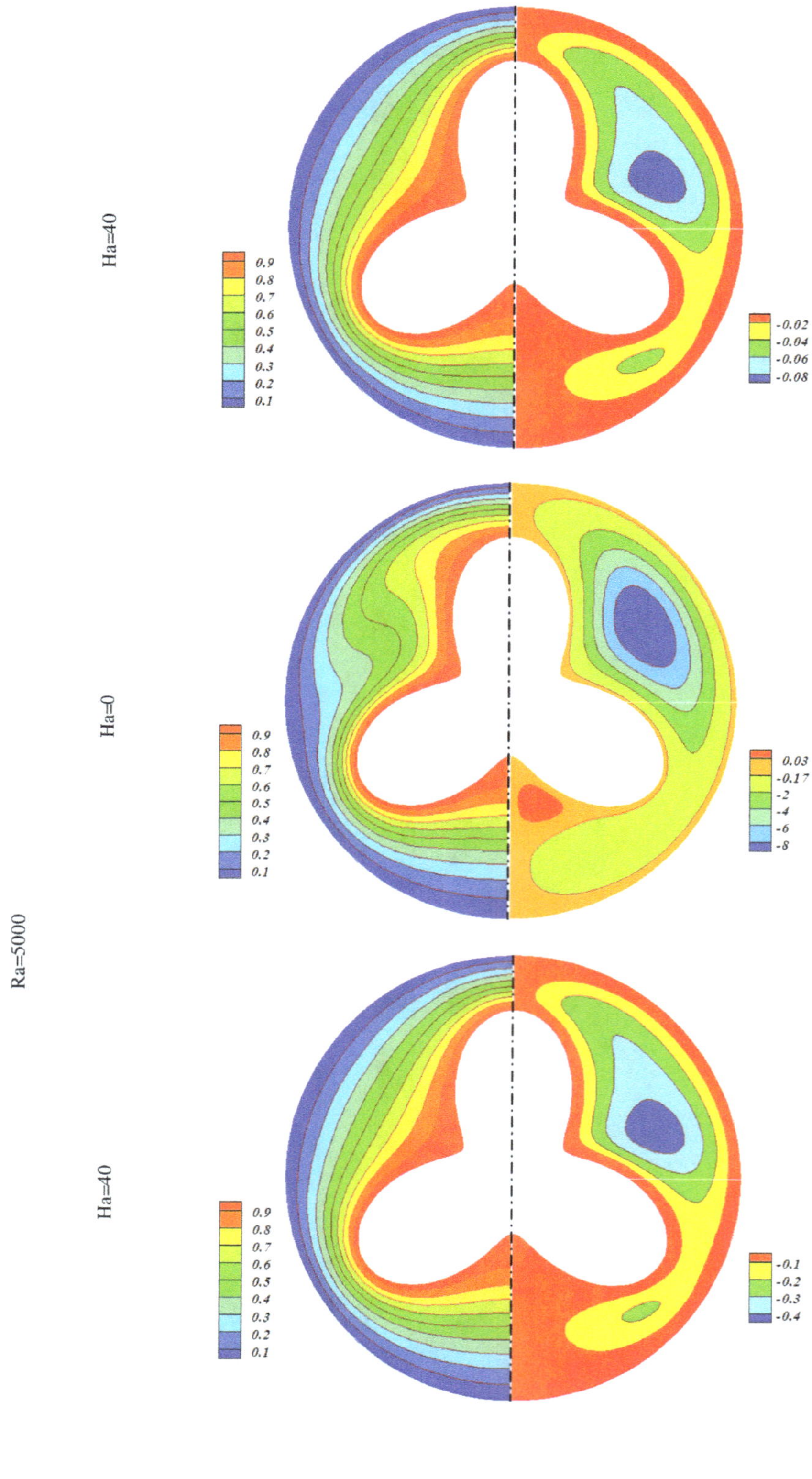

Figure 9a. Isotherms (left) and streamlines (right) contours for different values of Rayleigh and Hartmann numbers when $\delta = 0.2, \phi = 0.04, N = 3$.

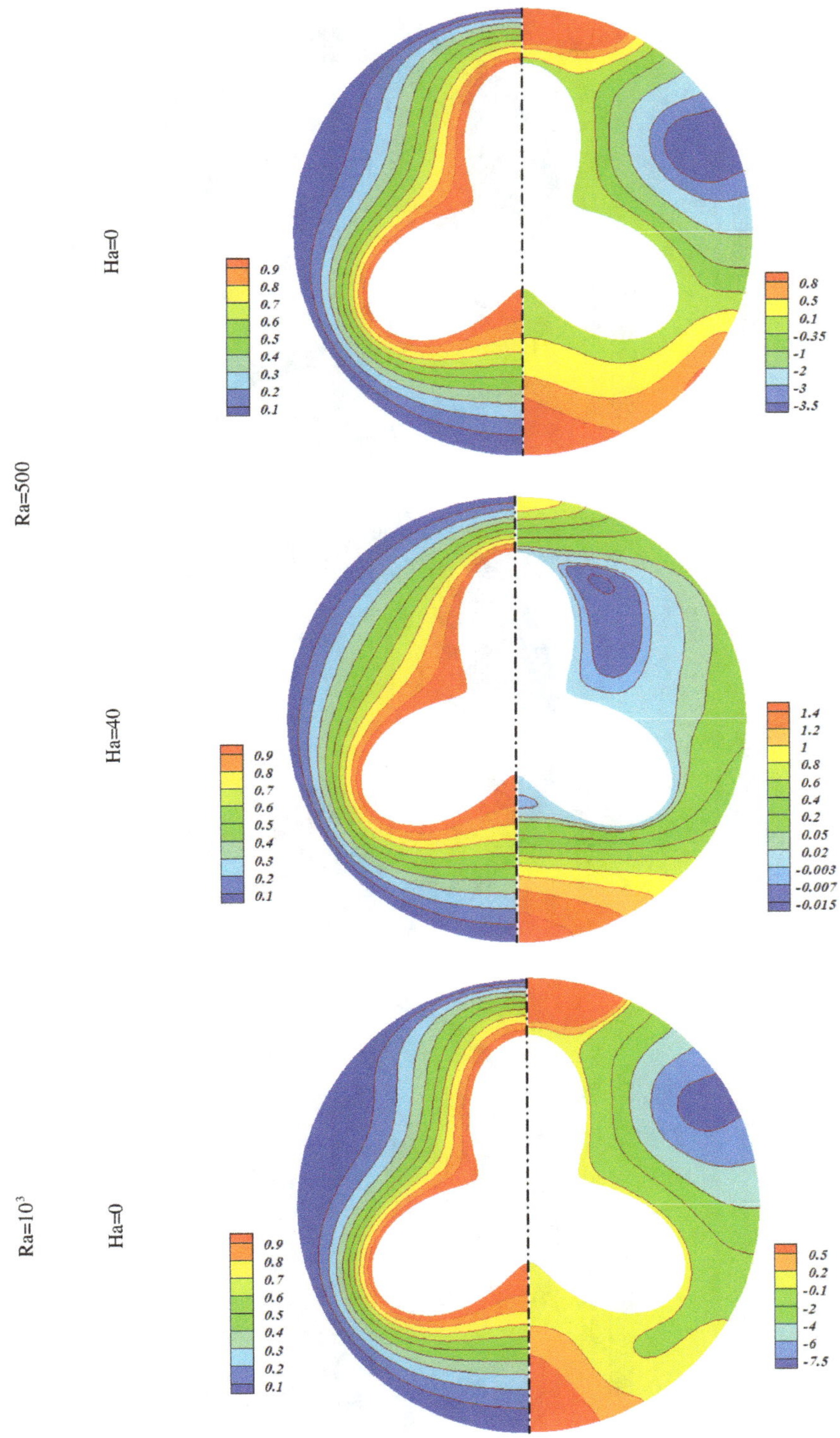

Figure 9b. Isotherms (left) and streamlines (right) contours for different values of Rayleigh and Hartmann numbers when $\delta = 0.2, \phi = 0.04, N = 3$.

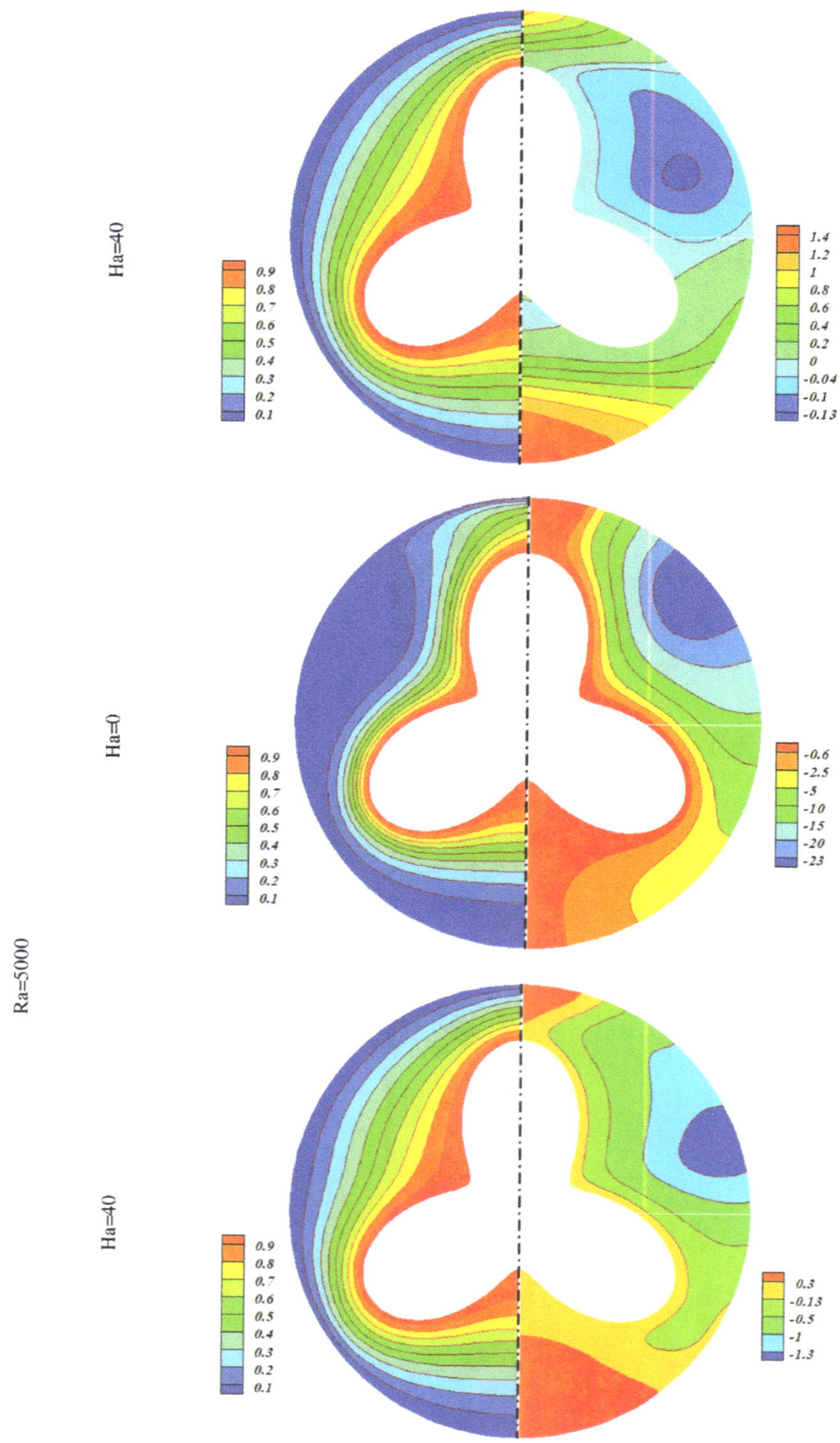

Figure 10a. Isotherms (left) and streamlines (right) contours for different values of Rayleigh and Hartmann numbers when $\delta = 0, \phi = 0.04, N = 5$.

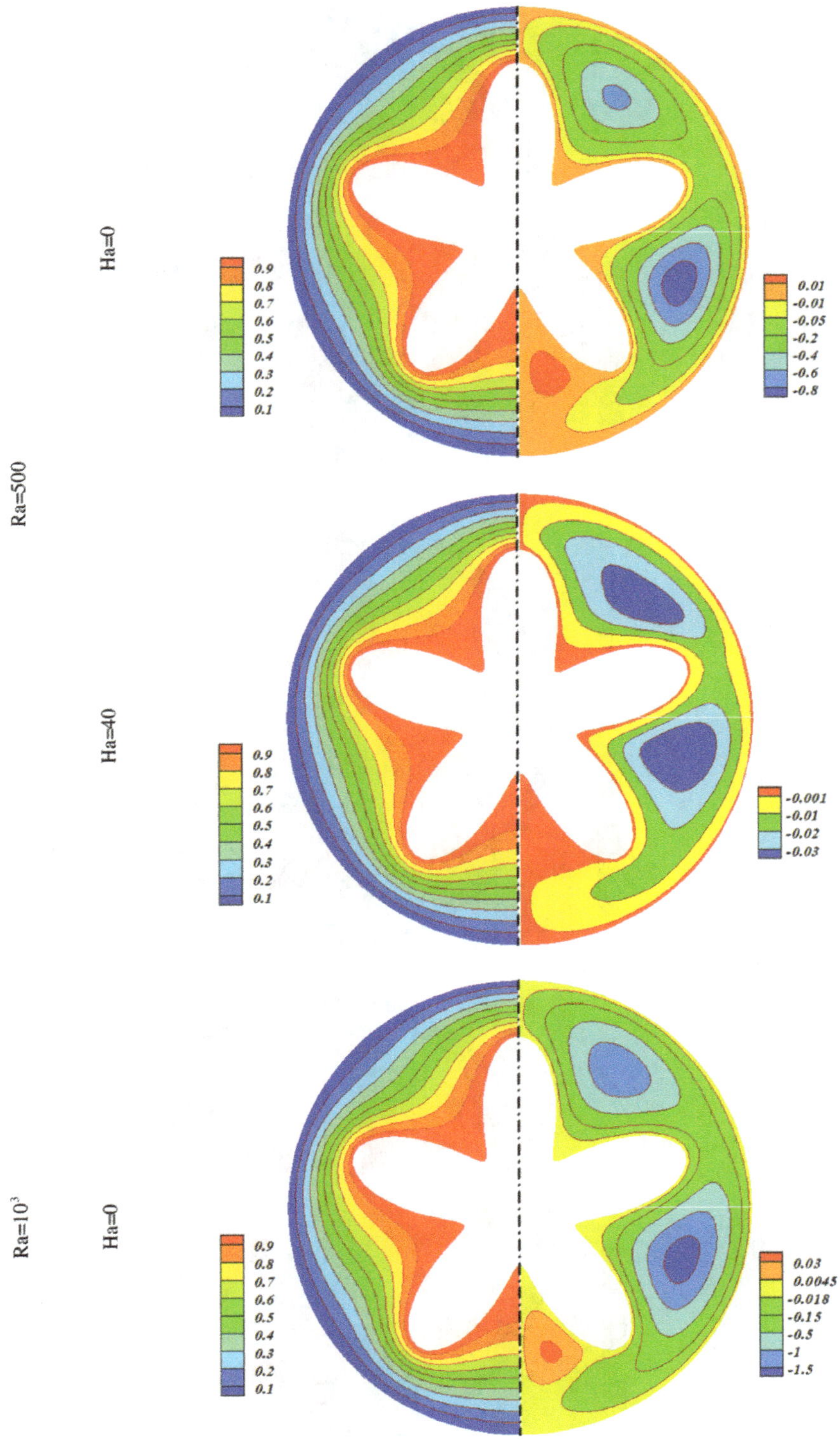

Figure 10b. Isotherms (left) and streamlines (right) contours for different values of Rayleigh and Hartmann numbers when $\delta = 0, \phi = 0.04, N = 5$.

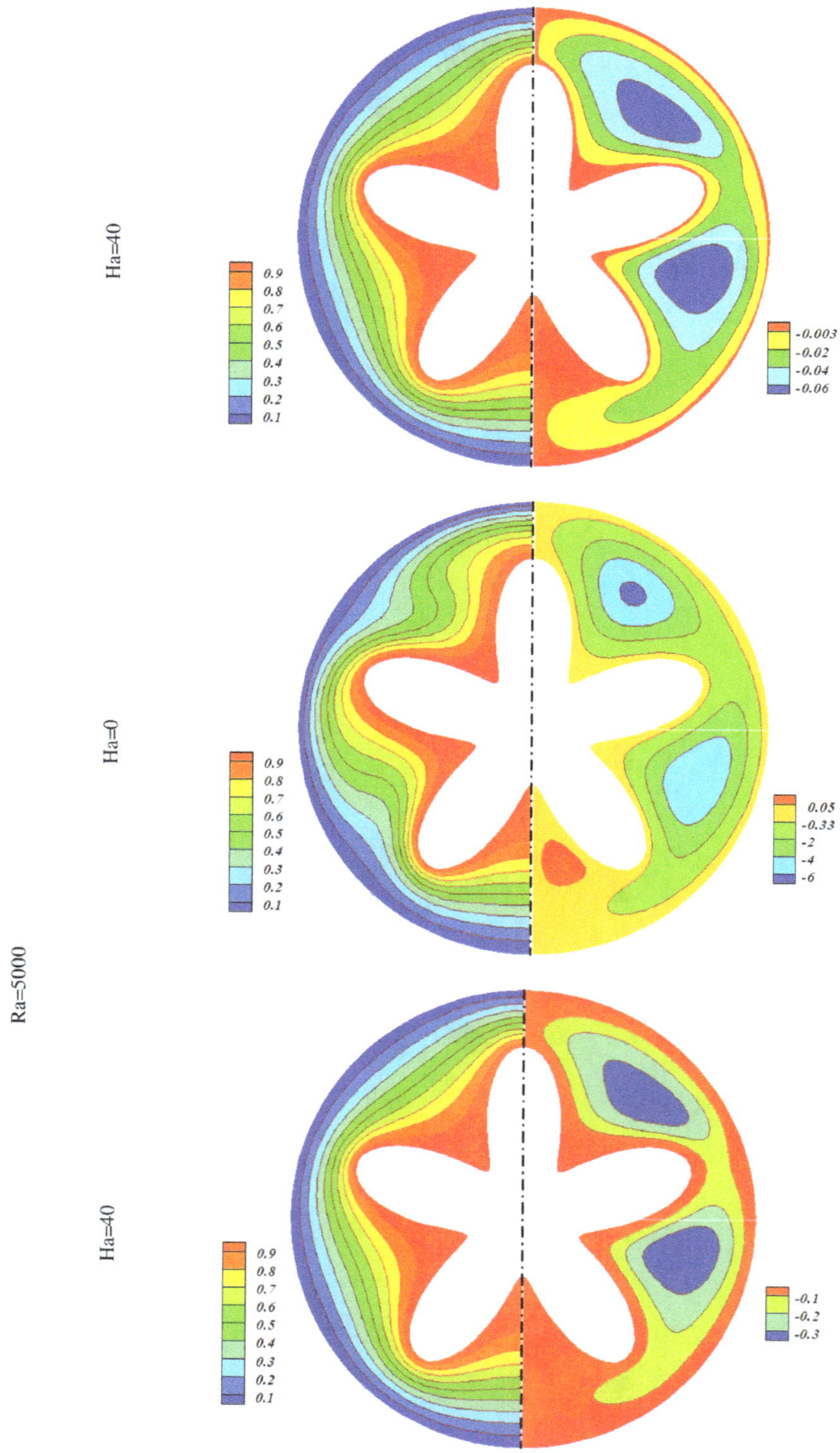

Figure 11a. Isotherms (left) and streamlines (right) contours for different values of Rayleigh and Hartmann numbers when $\delta = 0.2, \phi = 0.04, N = 5$.

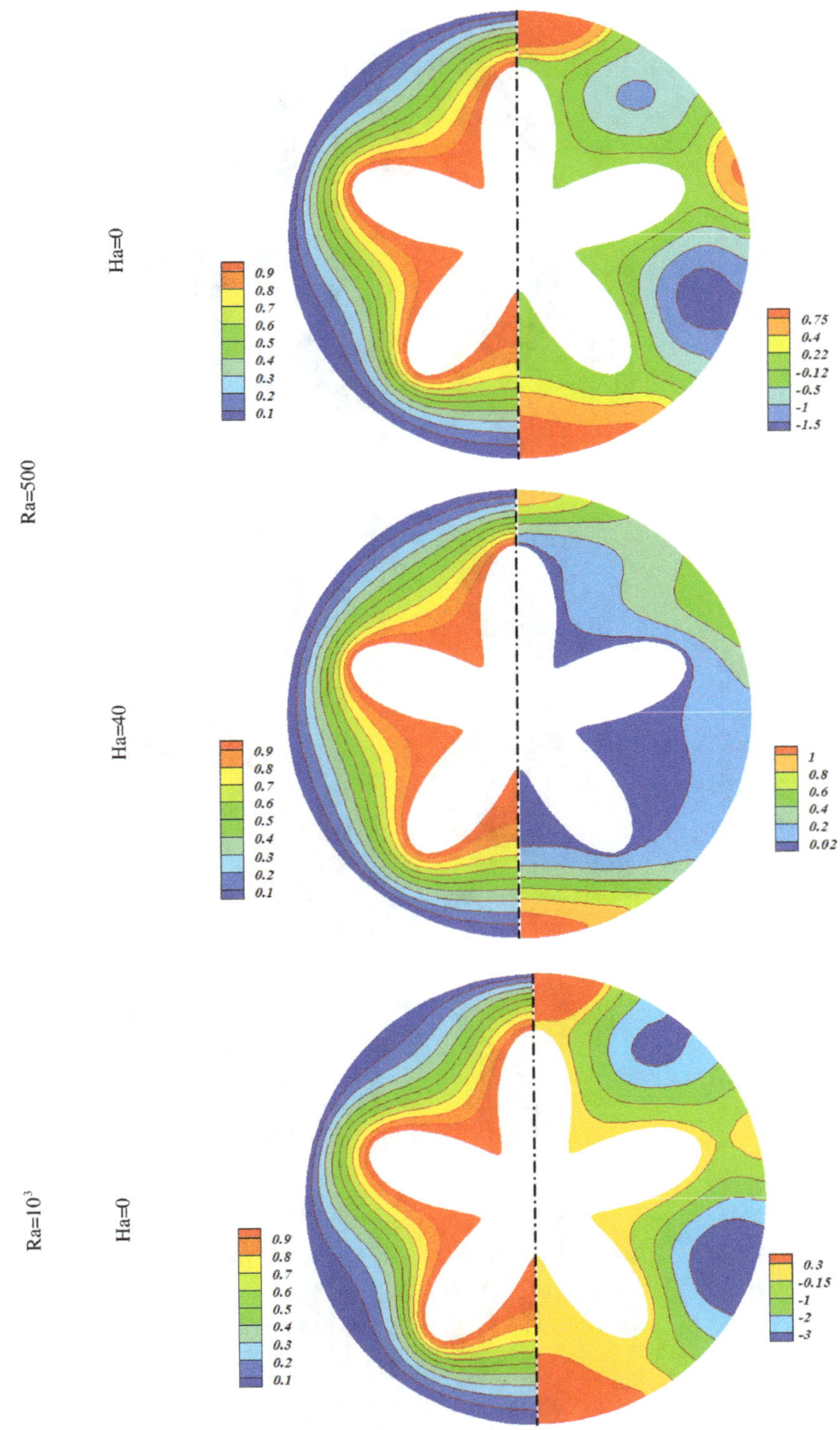

Figure 11b. Isotherms (left) and streamlines (right) contours for different values of Rayleigh and Hartmann numbers when $\delta = 0.2, \phi = 0.04, N = 5$.

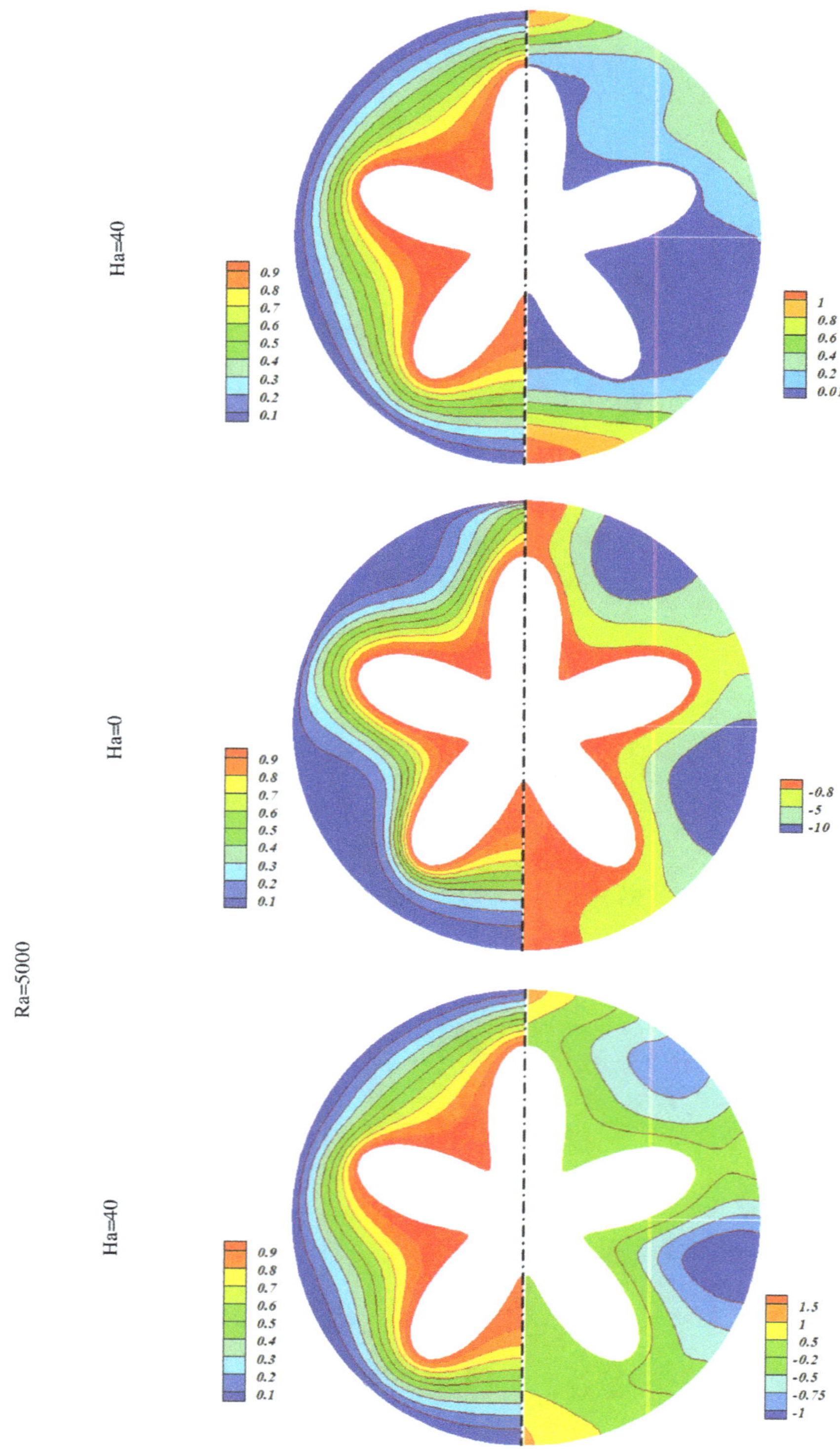

Figure 12. Influences of the number of undulation, Darcy, Reynolds number and Hartmann numbers on average Nusselt number.

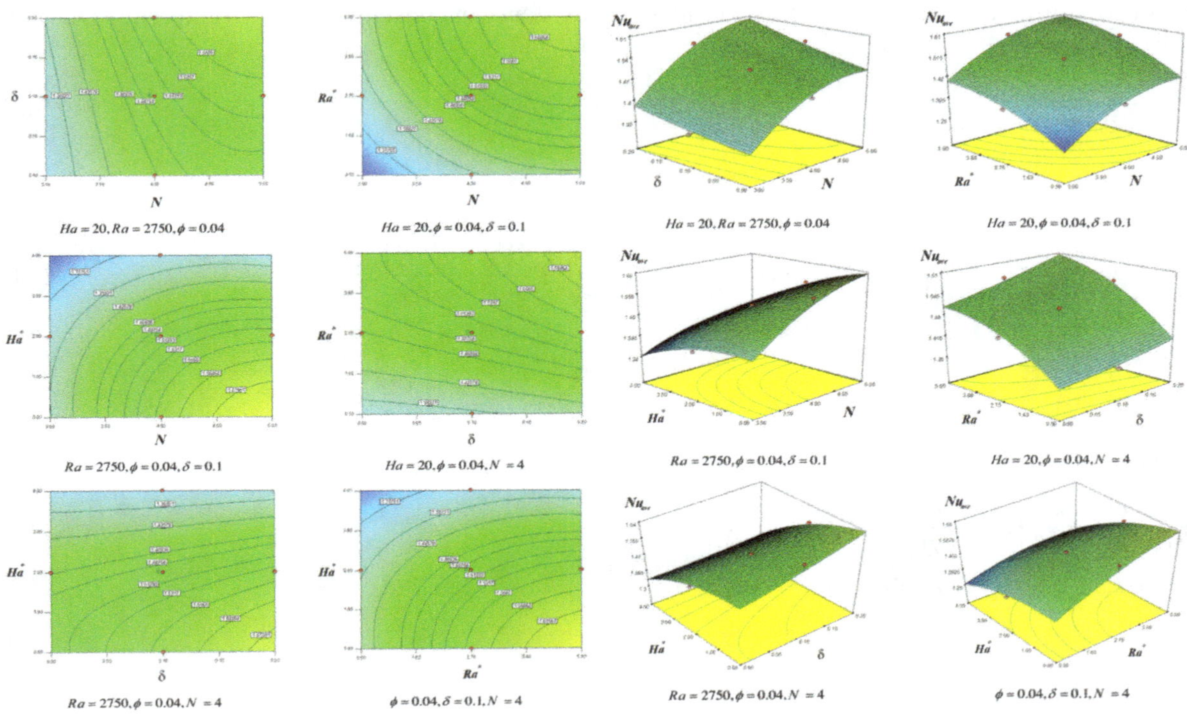

Figure 13. (a)Geometry (b) Sample element.

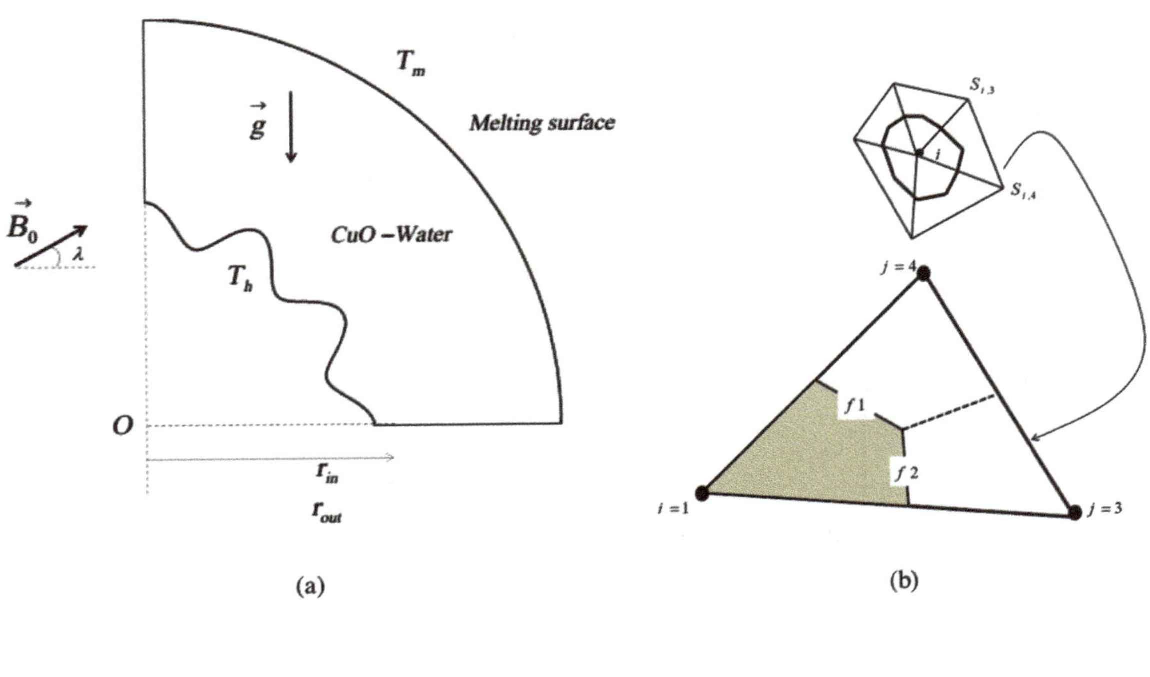

4.2. Governing Equation

2D steady convective nanofluid flow is considered in existence of constant magnetic field. The PDEs are:

$$\frac{\partial v}{\partial y}+\frac{\partial u}{\partial x}=0 \tag{43}$$

$$\left(\rho_{nf}\right)\left(v\frac{\partial u}{\partial y}+u\frac{\partial u}{\partial x}\right)=\left[B_{y}\sigma_{nf}vB_{x}-B_{y}^{2}\sigma_{nf}u+\mu_{nf}\left(\frac{\partial^{2}u}{\partial y^{2}}+\frac{\partial^{2}u}{\partial x^{2}}\right)-\frac{\partial P}{\partial x}\right] \tag{44}$$

$$\begin{aligned}&\rho_{nf}\left(\frac{\partial v}{\partial y}v+\frac{\partial v}{\partial x}u\right)=uB_{y}\sigma_{nf}B_{x}+\left(T-T_{c}\right)\beta_{nf}g\rho_{nf}-vB_{x}\sigma_{nf}B_{x}\\&+\left(\frac{\partial^{2}v}{\partial x^{2}}+\frac{\partial^{2}v}{\partial y^{2}}\right)\mu_{nf}-\frac{\partial P}{\partial y},\\&B_{x}=B_{o}\cos\lambda,\ B_{y}=B_{o}\sin\lambda\end{aligned} \tag{45}$$

$$\left(\rho C_{p}\right)_{nf}\left(v\frac{\partial T}{\partial y}+u\frac{\partial T}{\partial x}\right)=k_{nf}\left(\frac{\partial^{2}T}{\partial x^{2}}+\frac{\partial^{2}T}{\partial y^{2}}\right) \tag{46}$$

$\left(\rho C_{p}\right)_{nf}$, ρ_{nf}, $\left(\rho\beta\right)_{nf}$ and σ_{nf} are calculated as:

$$\left(\rho C_{p}\right)_{nf}=\phi\left(\rho C_{p}\right)_{s}+(1-\phi)\left(\rho C_{p}\right)_{f} \tag{47}$$

$$\rho_{nf}=\rho_{f}(1-\phi)+\phi\rho_{s} \tag{48}$$

$$\left(\rho\beta\right)_{nf}=\phi\left(\rho\beta\right)_{s}+(1-\phi)\left(\rho\beta\right)_{f} \tag{49}$$

$$\sigma_{nf}=\sigma_{f}\left[\left(\frac{+\left(2+\sigma_{s}/\sigma_{f}\right)-\phi\left(-1+\sigma_{s}/\sigma_{f}\right)}{3\phi\left(-1+\sigma_{s}/\sigma_{f}\right)}\right)^{-1}+1\right] \tag{50}$$

k_{nf}, μ_{nf} are calculated via KKL model:

$$\frac{k_{nf}}{k_f} = 1 + \frac{3\left(\frac{k_p}{k_f} - 1\right)\phi}{\left(\frac{k_p}{k_f} + 2\right) - \left(\frac{k_p}{k_f} - 1\right)\phi} + 5 \times 10^4 g'(\phi, T, d_p)\phi\rho_f c_{p,f}\sqrt{\frac{\kappa_b T}{d_p \rho_p}}$$
$$, R_f = 4 \times 10^{-8} km^2 / W$$
$$, R_f = -d_p(1/k_p - 1/k_{p,eff}), \tag{51}$$
$$g'(\phi, d_p, T) = Ln(T)\left(a_1 + a_3 Ln(\phi) + a_2 Ln(d_p) + a_5 Ln(d_p)^2 + a_4 \ln(d_p) Ln(\phi)\right)$$
$$+\left(a_6 + a_8 Ln(\phi) + a_7 Ln(d_p) + a_{10} Ln(d_p)^2 + a_9 Ln(\phi)\ln(d_p)\right)$$

$$\mu_{nf} = \frac{\mu_f}{(1-\phi)^{2.5}} + \frac{k_{Brownian}}{k_f} \times \frac{\mu_f}{\Pr} \tag{52}$$

Vorticity and stream function should be used to eliminate pressure source terms:

$$\omega + \frac{\partial u}{\partial y} - \frac{\partial v}{\partial x} = 0, \frac{\partial \psi}{\partial x} = -v, \frac{\partial \psi}{\partial y} = u \tag{53}$$

Introducing dimensionless quantities:

$$P = \frac{p}{\rho_{nf}\left(\alpha_{nf} / L\right)^2}, U = \frac{uL}{\alpha_{nf}}, V = \frac{vL}{\alpha_{nf}}, \theta = \frac{T - T_m}{T_h - T_m},$$
$$(X, Y) = \frac{(x, y)}{L}, \Psi = \frac{\psi}{\alpha_{nf}}, \Omega = \frac{\omega L^2}{\alpha_{nf}} \tag{54}$$

The final formulae are:

$$\Omega + \frac{\partial^2 \Psi}{\partial Y^2} + \frac{\partial^2 \Psi}{\partial X^2} = 0, \tag{55}$$

$$\frac{\partial \Omega}{\partial X}U + V\frac{\partial \Omega}{\partial Y} = \Pr\frac{A_5}{A_1}\frac{A_2}{A_4}\left(\frac{\partial^2 \Omega}{\partial Y^2} + \frac{\partial^2 \Omega}{\partial X^2}\right)$$
$$+\Pr Ha^2\frac{A_6}{A_1}\frac{A_2}{A_4}\left(-\frac{\partial V}{\partial X}B_xB_x + B_y\frac{\partial U}{\partial X}B_x - \frac{\partial V}{\partial Y}B_yB_x + B_y\frac{\partial U}{\partial Y}B_y\right) \tag{56}$$
$$+\Pr Ra\frac{A_3A_2^2}{A_1A_4^2}\frac{\partial \theta}{\partial X},$$

$$\frac{\partial \theta}{\partial X}U + \frac{\partial \theta}{\partial Y}V = \left(\frac{\partial^2 \theta}{\partial X^2} + \frac{\partial^2 \theta}{\partial Y^2}\right) \tag{57}$$

Boundary conditions are:

$$\frac{\partial \theta}{\partial n} = 0.0 \text{ on other walls}$$

$$\theta = 0 \text{ on outer wall}$$

$$\theta = 1.0 \text{ on inner wall}$$

$$\Psi = 0.0 \text{ on all walls except melting surface} \tag{58}$$

and in melting surface, we have:

$$\frac{\partial \theta}{\partial n} = \frac{A_2}{A_1}\frac{1}{\delta}\left(L + c_s\left(T_m - T_0\right)\right)\frac{\partial \Psi}{\partial n} \tag{59}$$

where dimensionless and constants parameters are illustrated as:

$$\Pr = \upsilon_f / \alpha_f, Ra = g\beta_f \Delta T L^3 / \left(\upsilon_f \alpha_f\right), Ha = LB_0\sqrt{\sigma_f / \mu_f}, \delta = \frac{\left(\rho C_p\right)_f}{\rho_f}\frac{\left(T_h - T_m\right)}{1 + c_s\left(T_m - T_0\right)}$$
$$A_1 = \frac{\rho_{nf}}{\rho_f}, A_5 = \frac{\mu_{nf}}{\mu_f}, A_2 = \frac{\left(\rho C_P\right)_{nf}}{\left(\rho C_P\right)_f}, A_6 = \frac{\sigma_{nf}}{\sigma_f}, A_4 = \frac{k_{nf}}{k_f}, A_3 = \frac{\left(\rho\beta\right)_{nf}}{\left(\rho\beta\right)_f} \tag{60}$$

It should be mentioned that δ is related to Stefan numbers.

Local and average Nusselt over the cold wall can calculate as:

$$Nu_{loc} = \left(\frac{k_{nf}}{k_f}\right)\frac{\partial \theta}{\partial r} \tag{61}$$

$$Nu_{ave} = \frac{1}{0.5\pi}\int_0^{0.5\pi} Nu_{loc}\, d\xi \tag{62}$$

4.3. Effects of Active Parameters

Nanofluid flow in a half sinusoidal annulus due to magnetic field in presence of melting surface is examined. μ_{nf}, k_{nf} of CuO-water nanofluid are estimated by means of KKL model. Graphs and tables are depicted for different amounts of CuO-H_2O volume fraction ($\phi = 0$ to 0.04), Melting parameter ($\delta = 0$ to 0.2), Rayleigh number ($Ra = 500$ to 5000) and Hartmann number ($Ha = 0$ to 40).

Impact of adding CuO nanoparticles in water on velocity and temperature contours is depicted in Figure 14. Temperature gradient decreases with augment of ϕ. $|\Psi_{max}|$ augments with adding nanoparticles because of increment in the solid movements. In presence of melting heat transfer and magnetic field, effect of adding nanoparticles on isotherms becomes negligible.

Figure 14. Impact of adding CuO in water on streamlines and isotherms (nanofluid ($\phi = 0.04$)(—) and pure fluid ($\phi = 0$) (- - -)) when $Ra = 5000$, $\delta = 0.2$

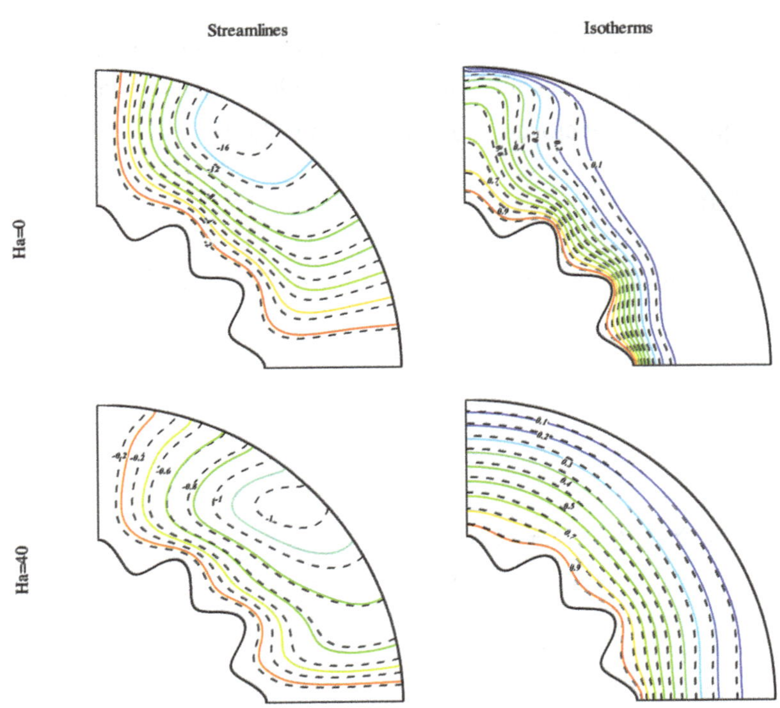

Figures 15 and 16 depict the impact of Rayleigh and Hartmann numbers in absence of melting heat transfer. There is only one eddy in streamline. In low Rayleigh number, conduction mechanism is dominant. As Ra increases the distortion of isotherms enhances close to the hot wall. Adding magnetic field makes isotherms to become parallel. In presence of melting heat transfer the primary eddy diminishes and increasing Lorentz forces generates three layers for streamline. Increasing melting parameter augments the bottom eddy.

Figure 17 illustrates the impact of δ, Ra and Ha on Nu_{ave}. The formula for Nu_{ave} corresponding to active parameters is:

$$\begin{aligned} Nu_{ave} &= 0.799 + 0.305\delta + 0.19Ra^{*} - 0.16Ha^{*} + 0.003Ra^{*}\delta - 0.0039\delta Ha^{*} \\ &-0.028Ra^{*}Ha^{*} + 0.058\delta^{2} - 0.012Ra^{*2} + 0.039Ha^{*2} \end{aligned} \tag{63}$$

where $Ha^{*} = 0.1Ha, Ra^{*} = 0.001Ra$. As melting paramtere augments, temperature gradient enhances and in turn Nusselt number enhances. Increasing bouyancy forces leads the thermal boudary layer thicknesss to reduce. So Nusselt number increases with enhance of Ra. As Hatmann number augments isotherms become parallel. Therefore Nusselt number has opposite relationship with *Ha*.

MHD NANOFLUID NATURAL CONVECTION INSIDE A HALF ANNULUS WITH MELTING SURFACE

Problem Definition

Figure 18 depicts the geometry, boundary condition and sample element. The inner wall is hot wall $\left(T = T_{h}\right)$ and the outer one is melting surface $\left(T = T_{m}\right)$. Other walls are adiabatic. Horizontal magnetic field has been applied. The enclosure is field with nanofluid.

Governing Equation

2D steady convective nanofluid flow is considered in existence of constant magnetic field. The PDEs are:

$$\frac{\partial v}{\partial y} + \frac{\partial u}{\partial x} = 0 \tag{64}$$

$$\left(\rho_{nf}\right)\left(v\frac{\partial u}{\partial y} + u\frac{\partial u}{\partial x}\right) = \left[B_{y}\sigma_{nf}vB_{x} - B_{y}^{2}\sigma_{nf}u + \left(\frac{\partial^{2}u}{\partial y^{2}} + \frac{\partial^{2}u}{\partial x^{2}}\right)\mu_{nf} - \frac{\partial P}{\partial x}\right] \tag{65}$$

Figure 15. Isotherms and streamlines for various Ha, Ra *when* $\delta = 0, \phi = 0.04$.

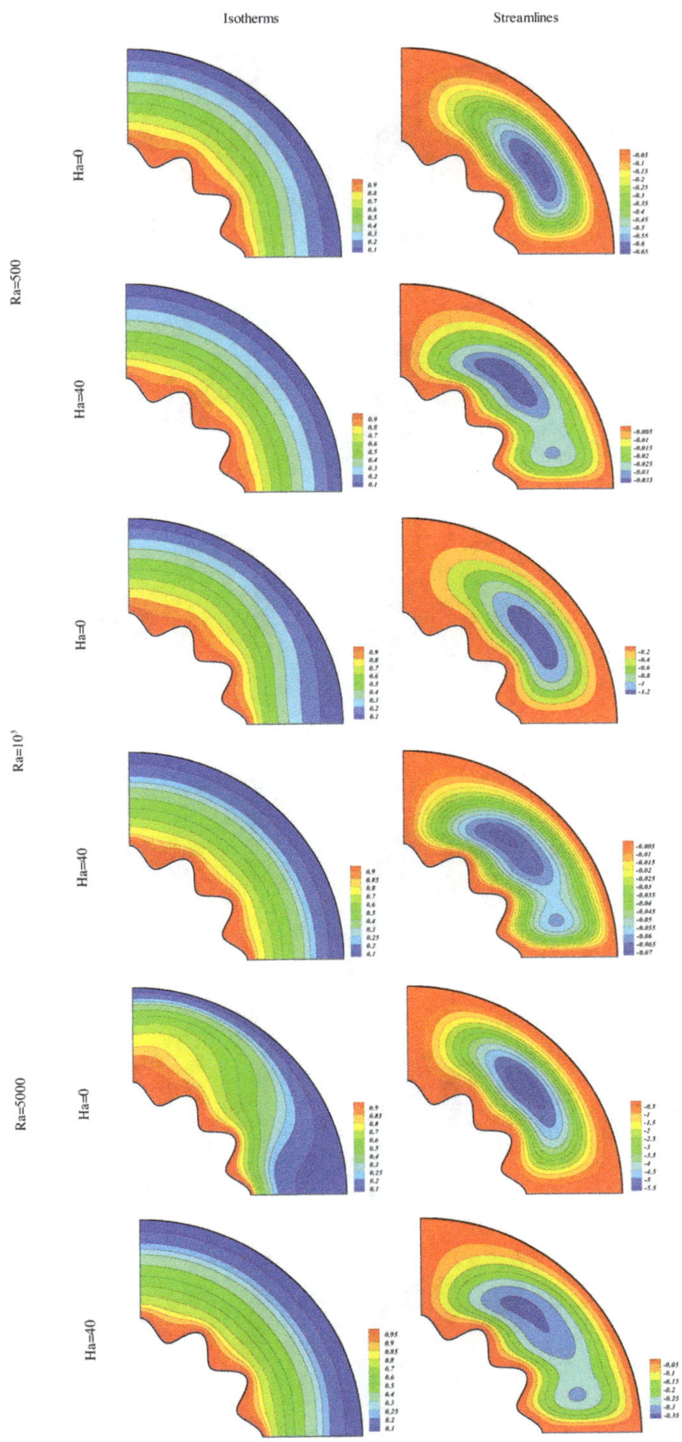

Figure 16. Isotherms and streamlines for various Ha, Ra *when* $\delta = 0.2, \phi = 0.04$.

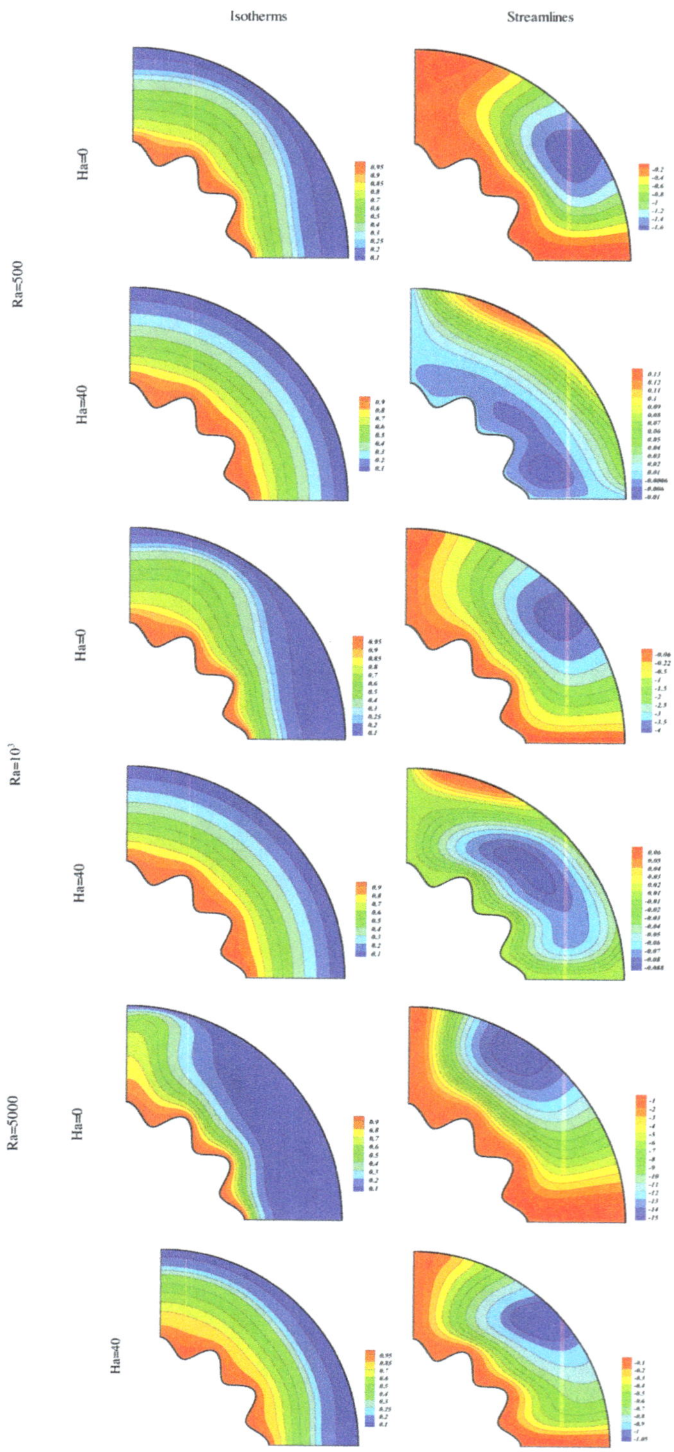

Figure 17. Impacts of the δ, Ha, Ra *on* Nu_{ave}.

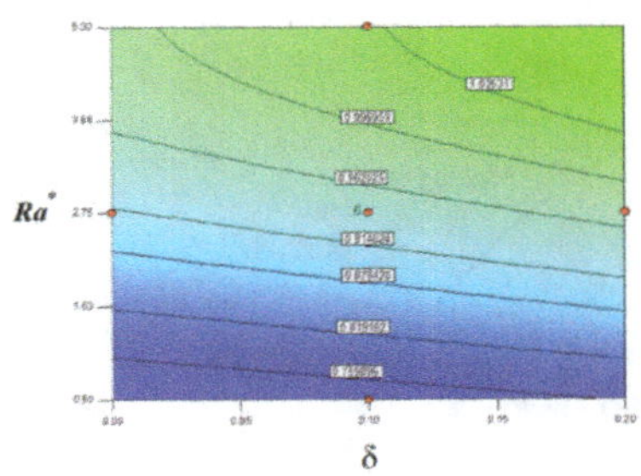

$Ha = 20, \phi = 0.04$

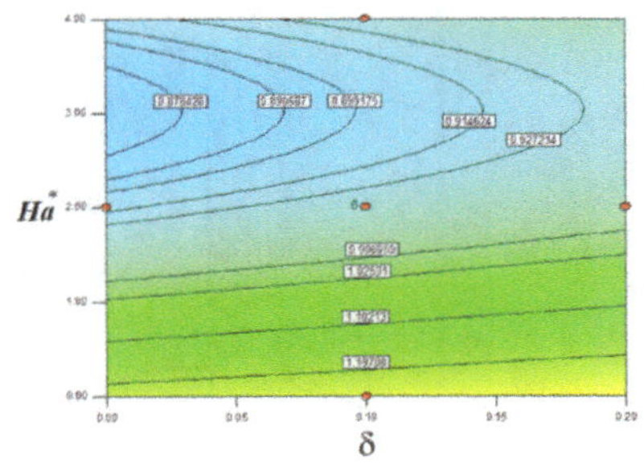

$Ra = 2750, \phi = 0.04$

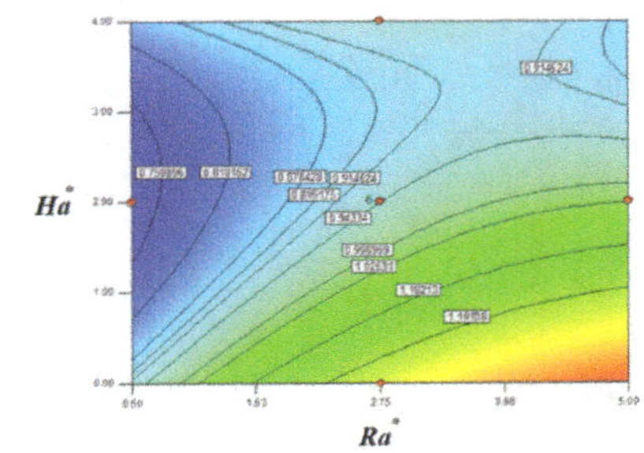

$\delta = 0.1, \phi = 0.04$

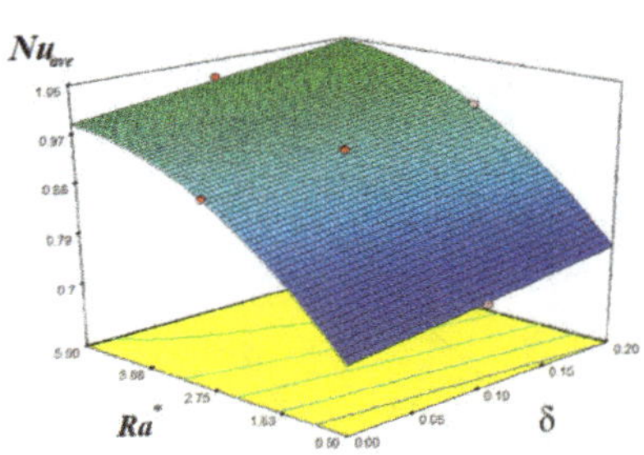

$Ha = 20, \phi = 0.04$

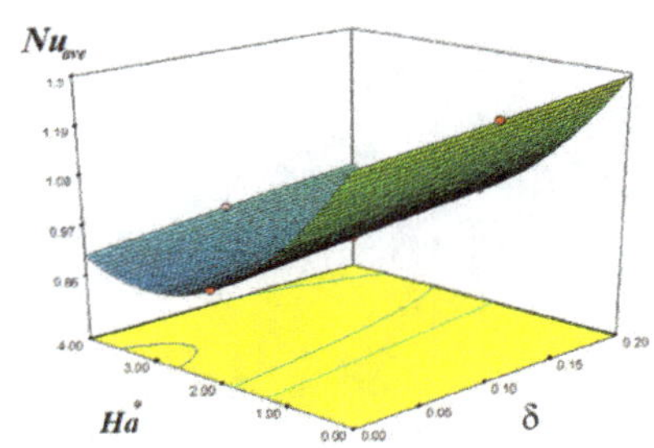

$Ra = 2750, \phi = 0.04$

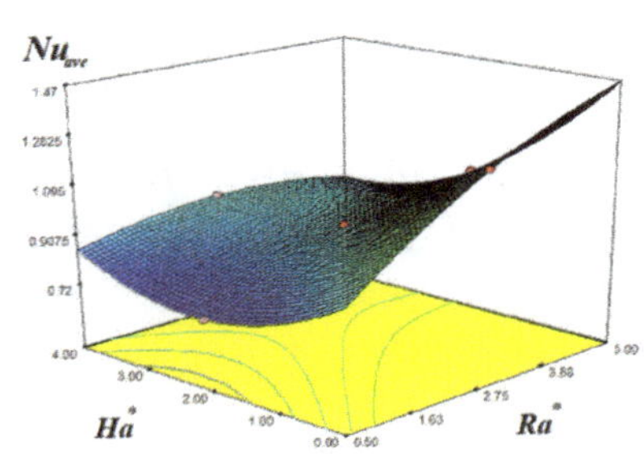

$\delta = 0.1, \phi = 0.04$

Figure 18. (a)Geometry and the boundary conditions with (b) the mesh of Geometry considered in this work; (c) A sample triangular element and its corresponding control volume.

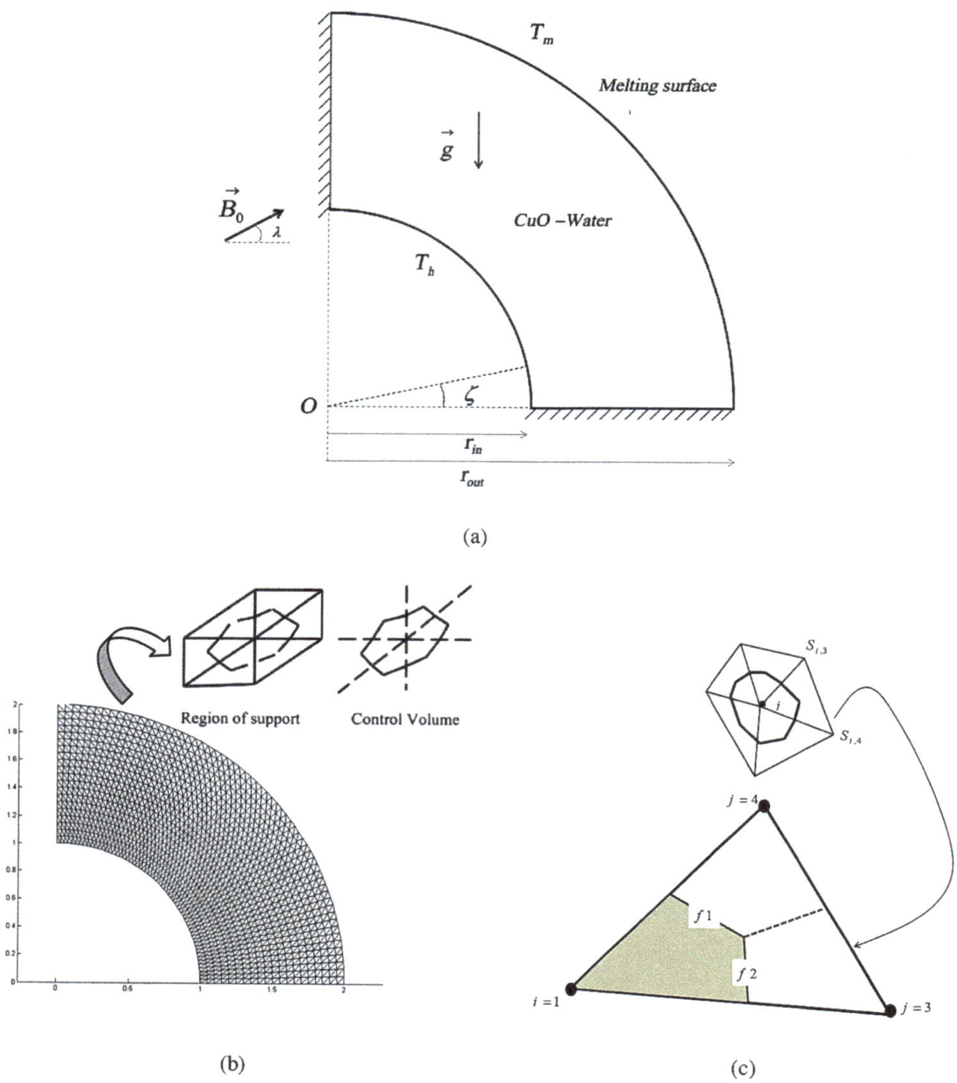

$$\rho_{nf}\left(v\frac{\partial v}{\partial y}+\frac{\partial v}{\partial x}u\right)=uB_y\sigma_{nf}B_x+\left(T-T_c\right)\beta_{nf}g\rho_{nf}-vB_x\sigma_{nf}B_x-\frac{\partial P}{\partial y}+\mu_{nf}\left(\frac{\partial^2 v}{\partial y^2}+\frac{\partial^2 v}{\partial x^2}\right),\quad B_x=B_o\cos\lambda,\ B_y=B_o\sin\lambda \tag{66}$$

$$\left(\rho C_p\right)_{nf}\left(v\frac{\partial T}{\partial y}+u\frac{\partial T}{\partial x}\right)=k_{nf}\left(\frac{\partial^2 T}{\partial x^2}+\frac{\partial^2 T}{\partial y^2}\right) \tag{67}$$

$\left(\rho C_p\right)_{nf}$, ρ_{nf}, $\left(\rho\beta\right)_{nf}$ and σ_{nf} are calculated as:

$$\left(\rho C_p\right)_{nf} = \phi\left(\rho C_p\right)_s + (1-\phi)\left(\rho C_p\right)_f \tag{68}$$

$$\rho_{nf} = \phi\rho_s + (1-\phi)\rho_f \tag{69}$$

$$\left(\rho\beta\right)_{nf} = \phi\left(\rho\beta\right)_s + (1-\phi)\left(\rho\beta\right)_f \tag{70}$$

$$\frac{\sigma_{nf}}{\sigma_f} = \left(\frac{+\left(2+\sigma_s/\sigma_f\right) - \phi\left(-1+\sigma_s/\sigma_f\right)}{3\phi\left(-1+\sigma_s/\sigma_f\right)}\right)^{-1} + 1 \tag{71}$$

k_{nf}, μ_{nf} are calculated via KKL model:

$$\begin{aligned}
&\frac{k_{nf}}{k_f} = 1 + \frac{3\left(\frac{k_p}{k_f}-1\right)\phi}{\left(\frac{k_p}{k_f}+2\right)-\left(\frac{k_p}{k_f}-1\right)\phi} + 5\times10^4 g'(\phi,T,d_p)\phi\rho_f c_{p,f}\sqrt{\frac{\kappa_b T}{d_p\rho_p}}\\
&, R_f = 4\times10^{-8} km^2/W\\
&, R_f = -d_p(1/k_p - 1/k_{p,eff}),\\
&g'\left(\phi,d_p,T\right) = Ln\left(T\right)\left(a_1 + a_3 Ln\left(\phi\right) + a_2 Ln\left(d_p\right) + a_5 Ln\left(d_p\right)^2 + a_4 \ln\left(d_p\right) Ln\left(\phi\right)\right)\\
&+\left(a_6 + a_8 Ln\left(\phi\right) + a_7 Ln\left(d_p\right) + a_{10} Ln\left(d_p\right)^2 + a_9 Ln\left(\phi\right)\ln\left(d_p\right)\right)
\end{aligned} \tag{72}$$

$$\mu_{nf} = \frac{\mu_f}{\left(1-\phi\right)^{2.5}} + \frac{k_{Brownian}}{k_f}\times\frac{\mu_f}{\Pr} \tag{73}$$

Vorticity and stream function should be used to eliminate pressure source terms:

$$\omega + \frac{\partial u}{\partial y} - \frac{\partial v}{\partial x} = 0, \frac{\partial\psi}{\partial x} = -v, \frac{\partial\psi}{\partial y} = u \tag{74}$$

Introducing dimensionless quantities:

$$P = \frac{p}{\rho_{nf}\left(\alpha_{nf} / L\right)^2}, U = \frac{uL}{\alpha_{nf}}, V = \frac{vL}{\alpha_{nf}}, \theta = \frac{T - T_m}{T_h - T_m},$$
$$\left(X, Y\right) = \frac{\left(x, y\right)}{L}, \Psi = \frac{\psi}{\alpha_{nf}}, \Omega = \frac{\omega L^2}{\alpha_{nf}} \tag{75}$$

The final formulae are:

$$\Omega + \frac{\partial^2 \Psi}{\partial Y^2} + \frac{\partial^2 \Psi}{\partial X^2} = 0, \tag{76}$$

$$\begin{aligned} \frac{\partial \Omega}{\partial X} U + V \frac{\partial \Omega}{\partial Y} &= \Pr \frac{A_5}{A_1} \frac{A_2}{A_4} \left(\frac{\partial^2 \Omega}{\partial Y^2} + \frac{\partial^2 \Omega}{\partial X^2} \right) \\ &+ \Pr Ha^2 \frac{A_6}{A_1} \frac{A_2}{A_4} \left(-\frac{\partial V}{\partial X} B_x B_x + \frac{\partial U}{\partial X} B_x B_y - \frac{\partial V}{\partial Y} B_y B_x + \frac{\partial U}{\partial Y} B_y B_y \right) \\ &+ \Pr Ra \frac{A_3 A_2^2}{A_1 A_4^2} \frac{\partial \theta}{\partial X}, \end{aligned} \tag{77}$$

$$\frac{\partial \theta}{\partial X} U + \frac{\partial \theta}{\partial Y} V = \left(\frac{\partial^2 \theta}{\partial X^2} + \frac{\partial^2 \theta}{\partial Y^2} \right) \tag{78}$$

Boundary conditions are:

$\frac{\partial \theta}{\partial n} = 0.0$ on other walls

$\theta = 0$ on outer wall

$\theta = 1.0$ on inner wall

$\Psi = 0.0$ on all walls except melting surface (79)

and in melting surface, we have:

$$\frac{\partial \theta}{\partial n}=\frac{A_2}{A_1}\frac{1}{\delta}\left(L+c_s\left(T_m-T_0\right)\right)\frac{\partial \Psi}{\partial n} \tag{80}$$

where dimensionless and constants parameters are illustrated as:

$$\begin{gathered}\Pr=\upsilon_f/\alpha_f, Ra=g\beta_f\Delta TL^3/\left(\upsilon_f\alpha_f\right), Ha=LB_0\sqrt{\sigma_f/\mu_f}, \delta=\frac{\left(\rho C_p\right)_f}{\rho_f}\frac{\left(T_h-T_m\right)}{1+c_s\left(T_m-T_0\right)}\\ A_1=\frac{\rho_{nf}}{\rho_f}, A_5=\frac{\mu_{nf}}{\mu_f}, A_2=\frac{\left(\rho C_P\right)_{nf}}{\left(\rho C_P\right)_f}, A_4=\frac{k_{nf}}{k_f}, A_6=\frac{\sigma_{nf}}{\sigma_f}, A_3=\frac{\left(\rho\beta\right)_{nf}}{\left(\rho\beta\right)_f}\end{gathered} \tag{81}$$

It should be mentioned that δ is related to Stefan numbers.

Local and average Nusselt over the hot wall can calculate as:

$$Nu_{loc}=\left(\frac{k_{nf}}{k_f}\right)\frac{\partial \theta}{\partial r} \tag{82}$$

$$Nu_{ave}=\frac{1}{0.5\pi}\int_0^{0.5\pi} Nu_{loc}\, d\xi \tag{83}$$

5.3. Effects of Active Parameters

Effect of melting heat transfer on nanofluid MHD flow in a half annulus is examined. μ_{nf}, k_{nf} of CuO-water nanofluid are estimated by means of KKL model. Graphs and tables are depicted for different amounts of CuO-H_2O volume fraction ($\phi=0$ to $\phi=0.04$), Melting parameter ($\delta=0$ to 0.8), Hartmann number ($Ha=0$ to 40) and Rayleigh number ($Ra=10^3$ to 10^4).

Impact of adding nanoparticles in the base fluid on velocity and temperature contours is depicted in Figure 19. Temperature gradient decreases with augment of ϕ. $\left|\Psi_{\max}\right|$ augments with adding nanoparticles because of increment in the solid movements. In presence of melting heat transfer and magnetic field, effect of adding nanoparticles on isotherms becomes negligible.

Figures 20 and 21 depicts the impact of Rayleigh and Hartmann numbers in absence of melting heat transfer. There is only one eddy in streamline. In low Rayleigh number, conduction mechanism is dominant. As Ra increases the distortion of isotherms enhances close to the hot wall. Adding magnetic field makes isotherms to become parallel. In presence of melting heat transfer the primary eddy diminishes and increasing Lorentz forces generates three layers for streamline. Increasing melting parameter augments the bottom eddy.

Figure 19. Influence of nanofluid volume fraction on streamlines (left) and isotherms (right) contours (nanofluid ($\phi = 0.04$)(—) and pure fluid($\phi = 0$) (- - -)) when $Ra = 5000$

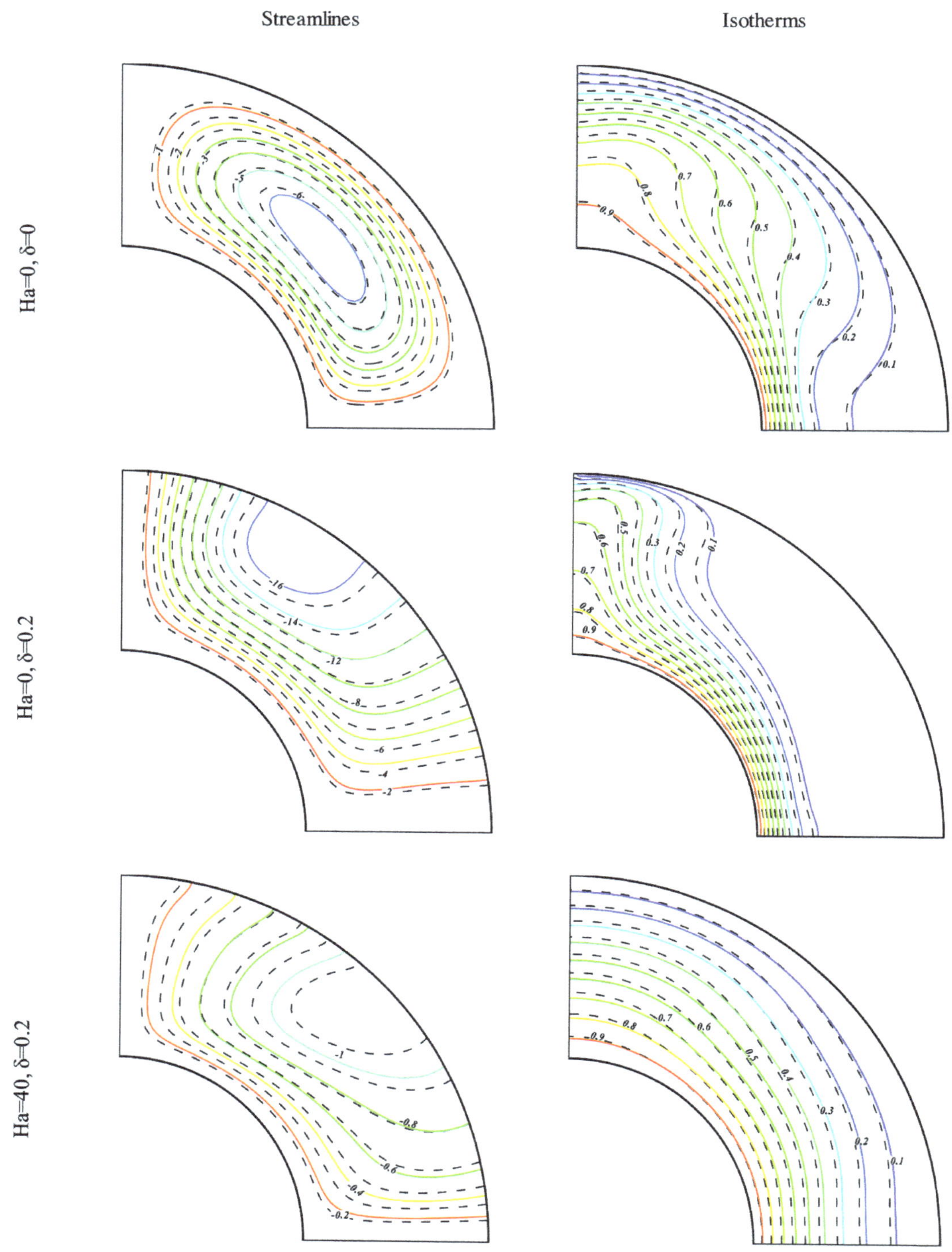

Figure 20. Isotherms (left) and streamlines (right) contours for different values of Rayleigh and Hartmann numbers when $\delta = 0, \phi = 0.04$.

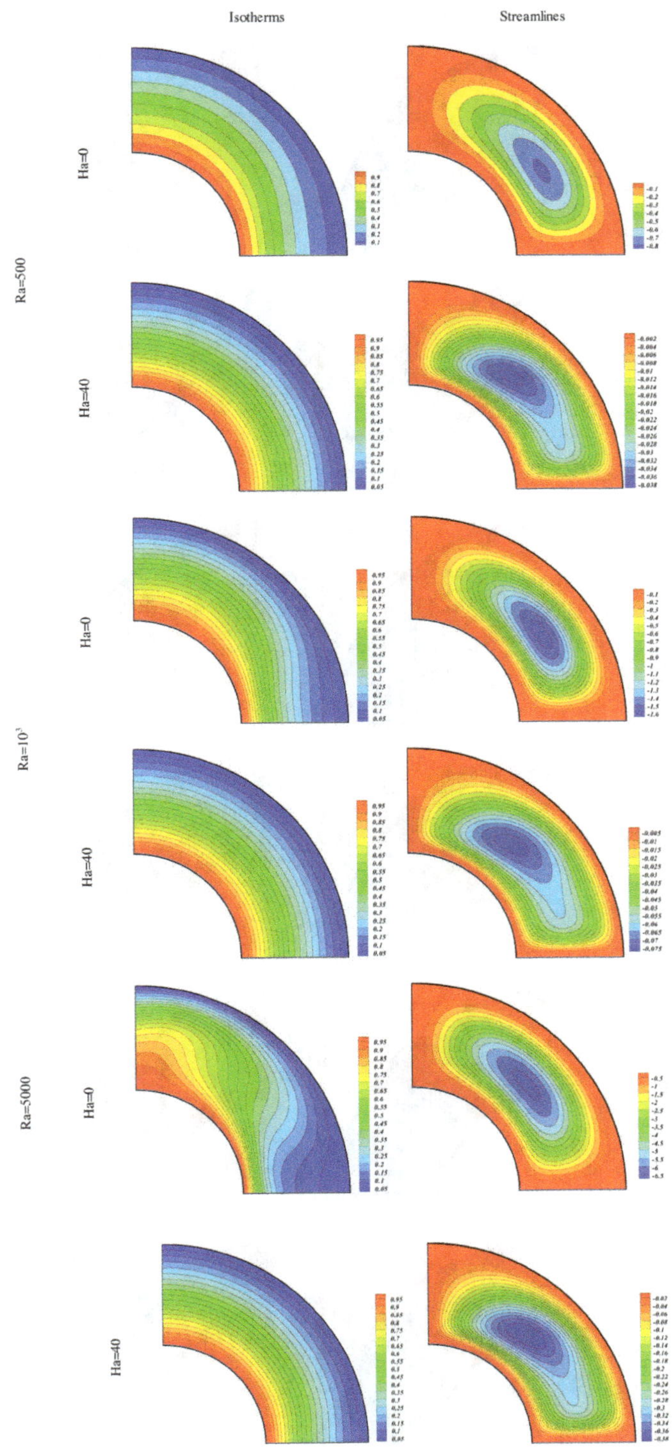

Figure 21. Isotherms (left) and streamlines (right) contours for different values of Rayleigh and Hartmann numbers when $\delta = 0.2, \phi = 0.04$.

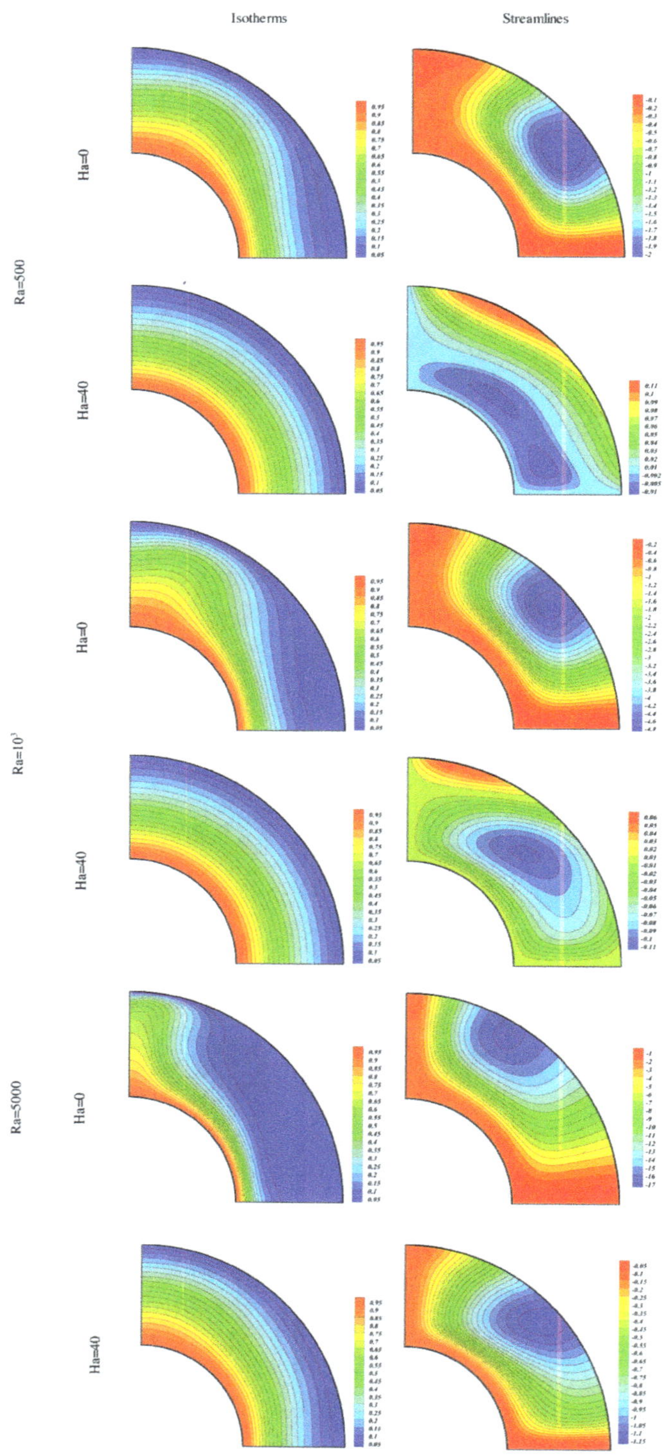

Figure 22. Influences of the Darcy, Reynolds number and Hartmann numbers on average Nusselt number.

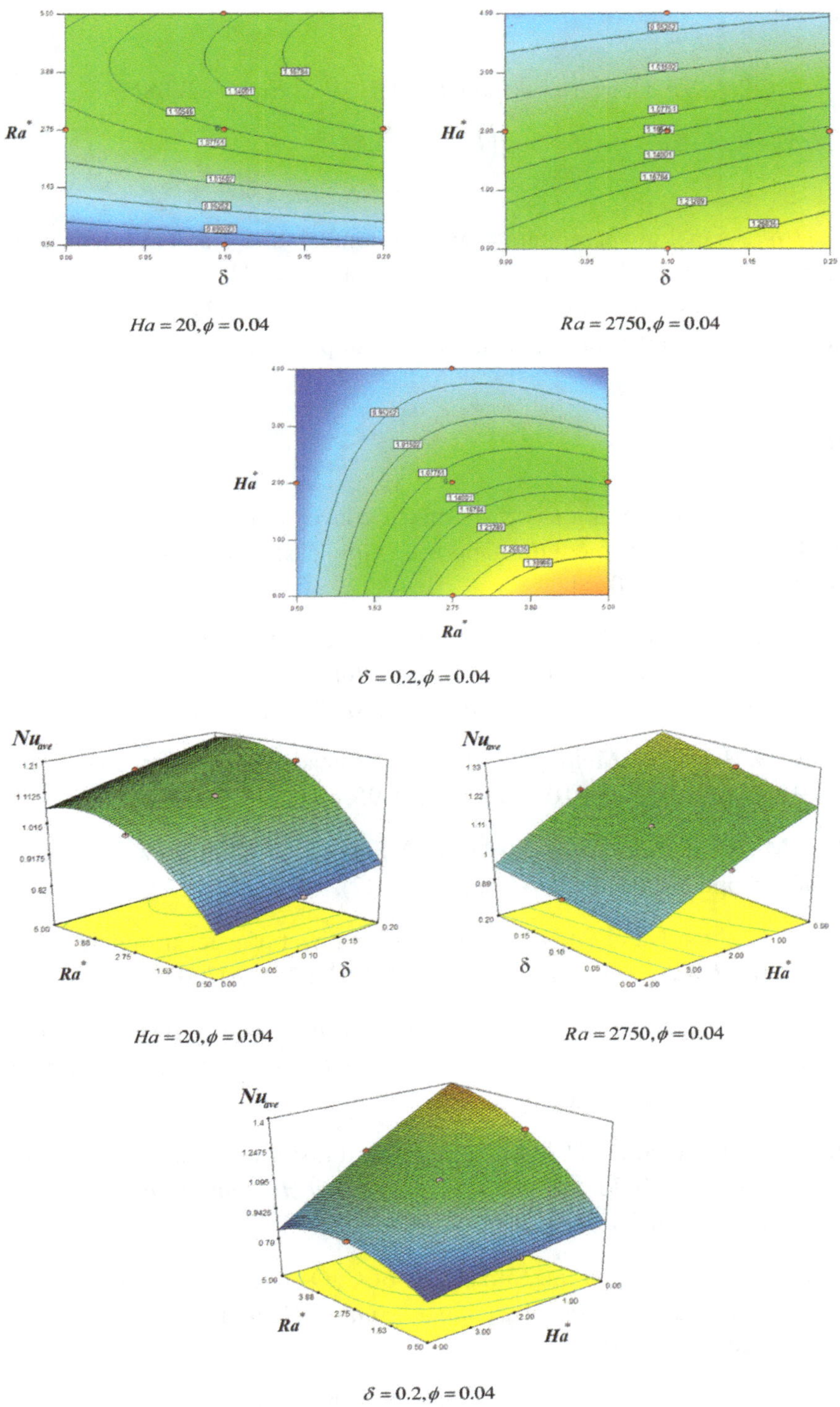

Figure 22 illustrates the impact of δ, Ra and Ha on Nu_{ave}. The formula for Nu_{ave} corresponding to active parameters is:

$$\begin{aligned}Nu_{ave} &= 0.425 + 0.122\delta - 1.63\log\left(Ra\right) + 0.35Ha^{*} + 1.5\log\left(Ra\right)\delta - 0.58\delta Ha^{*} \\ &-0.66\log\left(Ra\right)Ha^{*} + 0.45Ha^{*2} + 0.53\left(\log\left(Ra\right)\right)^{2} - 4.04\delta^{2}\end{aligned} \tag{84}$$

where $Ha^{*} = 0.1Ha$. As melting paramtere augments, temperature gradient enhances and in turn Nusselt number enhances. Increasing bouyancy forces leads the thermal boudary layer thicknesss to reduce. So Nusselt number increases with enhance of Ra. As Hatmann number augments isotherms become parallel. Therefore Nusselt number has opposite relationship with *Ha*.

REFERENCES

Ahmad, S., & Pop, I. (2014). Melting effect on mixed convection boundary layer flow about a vertical surface embedded in a porous medium: Opposing flow case. *Transport in Porous Media, 102*(3), 317–323. doi:10.100711242-014-0291-x

Cheng, W. T., & Lin, C. H. (2007). Melting effect on mixed convective heat transfer with aiding and opposing external flows from the vertical plate in a liquid-saturated porous medium. *International Journal of Heat and Mass Transfer*, *50*(15-16), 3026–3034. doi:10.1016/j.ijheatmasstransfer.2006.12.018

Epstein, M., & Cho, D. H. (1976). Melting heat transfer in steady laminar flow over a flat plate. *Journal of Heat Transfer*, *98*(3), 531–533. doi:10.1115/1.3450595

Hayat, T., Mustafa, M., Iqbal, Z., & Alsaedi, A. (2013). Stagnation-point flow of couple stress fluid with melting heat transfer. Appl. Math. Mech. – Engl.Ed., 34(2), 167–176.

Ishak, A., Nazar, R., Bachok, N., & Pop, I. (2010). Melting heat transfer in steady laminar flow over a moving surface. *Heat and Mass Transfer*, *46*(4), 463–468. doi:10.100700231-010-0592-8

Sheikholeslami, Ellahi, & Fetecau. (2017). *CuO-water nanofluid magnetohydrodynamic natural convection inside a sinusoidal annulus in presence of melting heat transfer*. Hindawi, Mathematical Problems in Engineering. doi:10.1155/2017/5830279

Sheikholeslami, M., & Rokni, H. B. (2017). Melting heat transfer influence on nanofluid flow inside a cavity in existence of magnetic field. *International Journal of Heat and Mass Transfer*, *114*, 517–526. doi:10.1016/j.ijheatmasstransfer.2017.06.092

Sheikholeslami, M. K. (2018). Sadoughi, Simulation of CuO- water nanofluid heat transfer enhancement in presence of melting surface. *International Journal of Heat and Mass Transfer*, *116*, 909–919. doi:10.1016/j.ijheatmasstransfer.2017.09.086

Yacob, N. A., Ishak, A., & Pop, I. (2011). Melting heat transfer in boundary layer stagnationpoint flow towards a stretching/shrinking sheet in a micropolar fluid. *Computers & Fluids*, *47*(1), 16–21. doi:10.1016/j.compfluid.2011.01.040

ADDITIONAL READING

Jafaryar, M., Sheikholeslami, M., Li, M., & Moradi, R. (2018). Nanofluid turbulent flow in a pipe under the effect of twisted tape with alternate axis. *Journal of Thermal Analysis and Calorimetry*. doi:10.100710973-018-7093-2

Li, Z., Shehzad, S. A., & Sheikholeslami, M. (2018). An application of CVFEM for nanofluid heat transfer intensification in a porous sinusoidal cavity considering thermal non-equilibrium model. *Computer Methods in Applied Mechanics and Engineering*, *339*, 663–680. doi:10.1016/j.cma.2018.05.015

Li, Z., Sheikholeslami, M., Chamkha, A. J., Raizah, Z. A., & Saleem, S. (2018). Control Volume Finite Element Method for nanofluid MHD natural convective flow inside a sinusoidal annulus under the impact of thermal radiation. *Computer Methods in Applied Mechanics and Engineering*, *338*, 618–633. doi:10.1016/j.cma.2018.04.023

Li, Z., Sheikholeslami, M., Jafaryar, M., Shafee, A., & Chamkha, A. J. (2018). Investigation of nanofluid entropy generation in a heat exchanger with helical twisted tapes. *Journal of Molecular Liquids*, *266*, 797–805. doi:10.1016/j.molliq.2018.07.009

Li, Z., Sheikholeslami, M., Samandari, M., & Shafee, A. (2018). Nanofluid unsteady heat transfer in a porous energy storage enclosure in existence of Lorentz forces. *International Journal of Heat and Mass Transfer*, *127*, 914–926. doi:10.1016/j.ijheatmasstransfer.2018.06.101

Sheikholeslami, M. (2017a). Lattice Boltzmann Method simulation of MHD non-Darcy nanofluid free convection. *Physica B, Condensed Matter*, *516*, 55–71. doi:10.1016/j.physb.2017.04.029

Sheikholeslami, M. (2017b). CuO-water nanofluid free convection in a porous cavity considering Darcy law. *The European Physical Journal Plus*, *132*(1), 55. doi:10.1140/epjp/i2017-11330-3

Sheikholeslami, M. (2017c). Influence of Lorentz forces on nanofluid flow in a porous cavity by means of Non- Darcy model. *Engineering Computations*, *34*(8), 2651–2667. doi:10.1108/EC-01-2017-0008

Sheikholeslami, M. (2017d). Magnetic field influence on nanofluid thermal radiation in a cavity with tilted elliptic inner cylinder. *Journal of Molecular Liquids*, *229*, 137–147. doi:10.1016/j.molliq.2016.12.024

Sheikholeslami, M. (2017e). Numerical simulation of magnetic nanofluid natural convection in porous media. *Physics Letters. [Part A]*, *381*(5), 494–503. doi:10.1016/j.physleta.2016.11.042

Sheikholeslami, M. (2017f). Influence of Lorentz forces on nanofluid flow in a porous cylinder considering Darcy model. *Journal of Molecular Liquids*, *225*, 903–912. doi:10.1016/j.molliq.2016.11.022

Sheikholeslami, M. (2017g). Influence of Coulomb forces on Fe3O4-H2O nanofluid thermal improvement. *International Journal of Hydrogen Energy*, *42*(2), 821–829. doi:10.1016/j.ijhydene.2016.09.185

Sheikholeslami, M. (2017h). Numerical investigation of MHD nanofluid free convective heat transfer in a porous tilted enclosure. *Engineering Computations*, *34*(6), 1939–1955. doi:10.1108/EC-08-2016-0293

Sheikholeslami, M. (2017i). Magnetic field influence on CuO -H2O nanofluid convective flow in a permeable cavity considering various shapes for nanoparticles. *International Journal of Hydrogen Energy*, *42*(31), 19611–19621. doi:10.1016/j.ijhydene.2017.06.121

Sheikholeslami, M. (2018a). Magnetic source impact on nanofluid heat transfer using CVFEM. *Neural Computing & Applications*, *30*(4), 1055–1064. doi:10.100700521-016-2740-7

Sheikholeslami, M. (2018b). Numerical simulation for external magnetic field influence on Fe3O4-water nanofluid forced convection. *Engineering Computations*, *35*(4), 1639–1654. doi:10.1108/EC-06-2017-0200

Sheikholeslami, M. (2018c). Finite element method for PCM solidification in existence of CuO nanoparticles. *Journal of Molecular Liquids*, *265*, 347–355. doi:10.1016/j.molliq.2018.05.132

Sheikholeslami, M. (2018d). Solidification of NEPCM under the effect of magnetic field in a porous thermal energy storage enclosure using CuO nanoparticles. *Journal of Molecular Liquids*, *263*, 303–315. doi:10.1016/j.molliq.2018.04.144

Sheikholeslami, M. (2018e). Influence of magnetic field on Al2O3-H2O nanofluid forced convection heat transfer in a porous lid driven cavity with hot sphere obstacle by means of LBM. *Journal of Molecular Liquids*, *263*, 472–488. doi:10.1016/j.molliq.2018.04.111

Sheikholeslami, M. (2018f). Numerical simulation for solidification in a LHTESS by means of Nano-enhanced PCM. *Journal of the Taiwan Institute of Chemical Engineers*, *86*, 25–41. doi:10.1016/j.jtice.2018.03.013

Sheikholeslami, M. (2018g). Numerical modeling of Nano enhanced PCM solidification in an enclosure with metallic fin. *Journal of Molecular Liquids*, *259*, 424–438. doi:10.1016/j.molliq.2018.03.006

Sheikholeslami, M. (2018h). Numerical investigation of nanofluid free convection under the influence of electric field in a porous enclosure. *Journal of Molecular Liquids*, *249*, 1212–1221. doi:10.1016/j.molliq.2017.11.141

Sheikholeslami, M. (2018i). CuO-water nanofluid flow due to magnetic field inside a porous media considering Brownian motion. *Journal of Molecular Liquids*, *249*, 921–929. doi:10.1016/j.molliq.2017.11.118

Sheikholeslami, M. (2018j). Application of Darcy law for nanofluid flow in a porous cavity under the impact of Lorentz forces. *Journal of Molecular Liquids*, *266*, 495–503. doi:10.1016/j.molliq.2018.06.083

Sheikholeslami, M., Barzegar Gerdroodbary, M., Valiallah Mousavi, S., Ganji, D. D., & Moradi, R. (2018). Heat transfer enhancement of ferrofluid inside an 90o elbow channel by non-uniform magnetic field. *Journal of Magnetism and Magnetic Materials*, *460*, 302–311. doi:10.1016/j.jmmm.2018.03.070

Sheikholeslami, M., & Bhatti, M. M. (2017). Active method for nanofluid heat transfer enhancement by means of EHD. *International Journal of Heat and Mass Transfer*, *109*, 115–122. doi:10.1016/j.ijheatmasstransfer.2017.01.115

Sheikholeslami, M., Darzi, M., & Li, Z. (2018). Experimental investigation for entropy generation and exergy loss of nano-refrigerant condensation process. *International Journal of Heat and Mass Transfer*, *125*, 1087–1095. doi:10.1016/j.ijheatmasstransfer.2018.04.155

Sheikholeslami, M., Darzi, M., & Sadoughi, M. K. (2018). Heat transfer improvement and Pressure Drop during condensation of refrigerant-based Nanofluid; An Experimental Procedure. *International Journal of Heat and Mass Transfer*, *122*, 643–650. doi:10.1016/j.ijheatmasstransfer.2018.02.015

Sheikholeslami, M., & Ghasemi, A. (2018). Solidification heat transfer of nanofluid in existence of thermal radiation by means of FEM. *International Journal of Heat and Mass Transfer*, *123*, 418–431. doi:10.1016/j.ijheatmasstransfer.2018.02.095

Sheikholeslami, M., Ghasemi, A., Li, Z., Shafee, A., & Saleem, S. (2018). Influence of CuO nanoparticles on heat transfer behavior of PCM in solidification process considering radiative source term. *International Journal of Heat and Mass Transfer*, *126*, 1252–1264. doi:10.1016/j.ijheatmasstransfer.2018.05.116

Sheikholeslami, M., Hayat, T., & Alsaedi, A. (2018). Numerical simulation for forced convection flow of MHD CuO-H2O nanofluid inside a cavity by means of LBM. *Journal of Molecular Liquids*, *249*, 941–948. doi:10.1016/j.molliq.2017.10.099

Sheikholeslami, M., Hayat, T., Muhammad, T., & Alsaedi, A. (2018). MHD forced convection flow of nanofluid in a porous cavity with hot elliptic obstacle by means of Lattice Boltzmann method. *International Journal of Mechanical Sciences*, *135*, 532–540. doi:10.1016/j.ijmecsci.2017.12.005

Sheikholeslami, M., Jafaryar, M., Ganji, D. D., & Li, Z. (2018). Exergy loss analysis for nanofluid forced convection heat transfer in a pipe with modified turbulators. *Journal of Molecular Liquids*, *262*, 104–110. doi:10.1016/j.molliq.2018.04.077

Sheikholeslami, M., Jafaryar, M., & Li, Z. (2018a). Second law analysis for nanofluid turbulent flow inside a circular duct in presence of twisted tape turbulators. *Journal of Molecular Liquids*, *263*, 489–500. doi:10.1016/j.molliq.2018.04.147

Sheikholeslami, M., Jafaryar, M., & Li, Z. (2018b). Nanofluid turbulent convective flow in a circular duct with helical turbulators considering CuO nanoparticles. *International Journal of Heat and Mass Transfer*, *124*, 980–989. doi:10.1016/j.ijheatmasstransfer.2018.04.022

Sheikholeslami, M., Jafaryar, M., Saleem, S., Li, Z., Shafee, A., & Jiang, Y. (2018). Nanofluid heat transfer augmentation and exergy loss inside a pipe equipped with innovative turbulators. *International Journal of Heat and Mass Transfer*, *126*, 156–163. doi:10.1016/j.ijheatmasstransfer.2018.05.128

Sheikholeslami, M., Jafaryar, M., Shafee, A., & Li, Z. (2018). Investigation of second law and hydrothermal behavior of nanofluid through a tube using passive methods. *Journal of Molecular Liquids*, *269*, 407–416. doi:10.1016/j.molliq.2018.08.019

Sheikholeslami, M., Li, Z., & Shafee, A. (2018a). Lorentz forces effect on NEPCM heat transfer during solidification in a porous energy storage system. *International Journal of Heat and Mass Transfer*, *127*, 665–674. doi:10.1016/j.ijheatmasstransfer.2018.06.087

Sheikholeslami, M., Li, Z., & Shamlooei, M. (2018). Nanofluid MHD natural convection through a porous complex shaped cavity considering thermal radiation. *Physics Letters. [Part A]*, *382*(24), 1615–1632. doi:10.1016/j.physleta.2018.04.006

Sheikholeslami, M., & Rokni, H. B. (2017). Simulation of nanofluid heat transfer in presence of magnetic field: A review. *International Journal of Heat and Mass Transfer*, *115*, 1203–1233. doi:10.1016/j.ijheatmasstransfer.2017.08.108

Sheikholeslami, M., & Rokni, H. B. (2018a). CVFEM for effect of Lorentz forces on nanofluid flow in a porous complex shaped enclosure by means of Non-equilibrium model. *Journal of Molecular Liquids*, *254*, 446–462. doi:10.1016/j.molliq.2018.01.130

Sheikholeslami, M., Rokni, H.B. (2018b). Magnetic nanofluid flow and convective heat transfer in a porous cavity considering Brownian motion effects. *Physics of Fluids*, *30*(1), doi:10.1063/1.5012517

Sheikholeslami, M., & Sadoughi, M. K. (2017). Mesoscopic method for MHD nanofluid flow inside a porous cavity considering various shapes of nanoparticles. *International Journal of Heat and Mass Transfer*, *113*, 106–114. doi:10.1016/j.ijheatmasstransfer.2017.05.054

Sheikholeslami, M., & Sadoughi, M. K. (2018). Simulation of CuO- water nanofluid heat transfer enhancement in presence of melting surface. *International Journal of Heat and Mass Transfer*, *116*, 909–919. doi:10.1016/j.ijheatmasstransfer.2017.09.086

Sheikholeslami, M., & Seyednezhad, M. (2018). Simulation of nanofluid flow and natural convection in a porous media under the influence of electric field using CVFEM. *International Journal of Heat and Mass Transfer*, *120*, 772–781. doi:10.1016/j.ijheatmasstransfer.2017.12.087

Sheikholeslami, M., Shafee, A., Ramzan, M., & Li, Z. (2018). Investigation of Lorentz forces and radiation impacts on nanofluid treatment in a porous semi annulus via Darcy law. *Journal of Molecular Liquids*, *272*, 8–14. doi:10.1016/j.molliq.2018.09.016

Sheikholeslami, M., Shamlooei, M., & Moradi, R. (2018). Numerical simulation for heat transfer intensification of nanofluid in a porous curved enclosure considering shape effect of Fe3O4 nanoparticles. *Chemical Engineering & Processing: Process Intensification*, *124*, 71–82. doi:10.1016/j.cep.2017.12.005

Sheikholeslami, M., & Shehzad, S. A. (2017a). Thermal radiation of ferrofluid in existence of Lorentz forces considering variable viscosity. *International Journal of Heat and Mass Transfer*, *109*, 82–92. doi:10.1016/j.ijheatmasstransfer.2017.01.096

Sheikholeslami, M., & Shehzad, S. A. (2017b). Magnetohydrodynamic nanofluid convective flow in a porous enclosure by means of LBM. *International Journal of Heat and Mass Transfer*, *113*, 796–805. doi:10.1016/j.ijheatmasstransfer.2017.05.130

Sheikholeslami, M., & Shehzad, S. A. (2018a). Numerical analysis of Fe3O4 –H2O nanofluid flow in permeable media under the effect of external magnetic source. *International Journal of Heat and Mass Transfer*, *118*, 182–192. doi:10.1016/j.ijheatmasstransfer.2017.10.113

Sheikholeslami, M., & Shehzad, S. A. (2018b). CVFEM simulation for nanofluid migration in a porous medium using Darcy model. *International Journal of Heat and Mass Transfer*, *122*, 1264–1271. doi:10.1016/j.ijheatmasstransfer.2018.02.080

Sheikholeslami, M., & Shehzad, S. A. (2018c). Simulation of water based nanofluid convective flow inside a porous enclosure via Non-equilibrium model. *International Journal of Heat and Mass Transfer*, *120*, 1200–1212. doi:10.1016/j.ijheatmasstransfer.2017.12.132

Sheikholeslami, M., & Shehzad, S. A. (2018d). Non-Darcy free convection of Fe3O4-water nanoliquid in a complex shaped enclosure under impact of uniform Lorentz force. *The Chinese Journal of Physiology*, *56*(1), 270–281. doi:10.1016/j.cjph.2017.12.017

Sheikholeslami, M., Shehzad, S. A., Abbasi, F. M., & Li, Z. (2018). Nanofluid flow and forced convection heat transfer due to Lorentz forces in a porous lid driven cubic enclosure with hot obstacle. *Computer Methods in Applied Mechanics and Engineering*, *338*, 491–505. doi:10.1016/j.cma.2018.04.020

Sheikholeslami, M., Shehzad, S. A., & Li, Z. (2018a). Nanofluid heat transfer intensification in a permeable channel due to magnetic field using Lattice Boltzmann method. *Physica B, Condensed Matter*, *542*, 51–58. doi:10.1016/j.physb.2018.03.036

Sheikholeslami, M., Shehzad, S. A., & Li, Z. (2018b). Water based nanofluid free convection heat transfer in a three dimensional porous cavity with hot sphere obstacle in existence of Lorenz forces. *International Journal of Heat and Mass Transfer*, *125*, 375–386. doi:10.1016/j.ijheatmasstransfer.2018.04.076

Sheikholeslami, M., Shehzad, S. A., Li, Z., & Shafee, A. (2018). Numerical modeling for Alumina nanofluid magnetohydrodynamic convective heat transfer in a permeable medium using Darcy law. *International Journal of Heat and Mass Transfer*, *127*, 614–622. doi:10.1016/j.ijheatmasstransfer.2018.07.013

Sheikholeslami, M., & Vajravelu, K. (2017). Forced convection heat transfer in Fe3O4- ethylene glycol nanofluid under the influence of Coulomb force. *Journal of Molecular Liquids*, *233*, 203–210. doi:10.1016/j.molliq.2017.03.026

Sheikholeslami, M., & Zeeshan, A. (2017). Analysis of flow and heat transfer in water based nanofluid due to magnetic field in a porous enclosure with constant heat flux using CVFEM. *Computer Methods in Applied Mechanics and Engineering*, *320*, 68–81. doi:10.1016/j.cma.2017.03.024

Sheikholeslami, M., Zeeshan, A., & Majeed, A. (2018). Control volume based finite element simulation of magnetic nanofluid flow and heat transport in non-Darcy medium. *Journal of Molecular Liquids*, *268*, 354–364. doi:10.1016/j.molliq.2018.07.031

About the Author

Mohsen Sheikholeslami works at the Babol Noshirvani University of Technology's Department of Mechanical Engineering in Iran. He is head of Renewable energy systems and nanofluid applications in heat transfer Laboratory at Babol Noshirvani University of Technology. His research interests are Nanofluid, CFD, simulation, Mesoscopic modeling, Nonlinear Science, Magnetohydrodynamic, Ferrohydrodynamic, Electrohydrodynamic and Heat exchangers. He has written several papers and books in various fields of mechanical engineering. He is also the first author of the books "Applications of Nanofluid for Heat Transfer Enhancement", "Application of semi analytical methods for nanofluid flow and heat transfer", "Hydrothermal Analysis in Engineering Using Control Volume Finite Element Method" and "External Magnetic Field Effects on Hydrothermal Treatment of Nanofluid" which are published in ELSEVIER. According to the reports of Thomson Reuters (Clarivate Analytics), he has been selected as a Web of Science Highly Cited Researcher (Top 0.01%) in 2016 and 2017.

Index

N

P

S

T

www.ingramcontent.com/pod-product-compliance
Lightning Source LLC
LaVergne TN
LVHW060250011026
849161LV00011B/2687
* 9 7 8 1 5 2 2 5 7 5 9 5 5 *